2021 IEEE Workshop on Wide Bandgap Power Devices and Applications in Asia (WiPDA Asia 2021)

Wuhan, China
25-27 August 2021

IEEE Catalog Number: CFP21O09-POD
ISBN: 978-1-6654-4817-8

**Copyright © 2021 by the Institute of Electrical and Electronics Engineers, Inc.
All Rights Reserved**

Copyright and Reprint Permissions: Abstracting is permitted with credit to the source. Libraries are permitted to photocopy beyond the limit of U.S. copyright law for private use of patrons those articles in this volume that carry a code at the bottom of the first page, provided the per-copy fee indicated in the code is paid through Copyright Clearance Center, 222 Rosewood Drive, Danvers, MA 01923.

For other copying, reprint or republication permission, write to IEEE Copyrights Manager, IEEE Service Center, 445 Hoes Lane, Piscataway, NJ 08854. All rights reserved.

*** This is a print representation of what appears in the IEEE Digital Library. Some format issues inherent in the e-media version may also appear in this print version.

IEEE Catalog Number:	CFP21O09-POD
ISBN (Print-On-Demand):	978-1-6654-4817-8
ISBN (Online):	978-1-6654-1851-5

Additional Copies of This Publication Are Available From:

Curran Associates, Inc
57 Morehouse Lane
Red Hook, NY 12571 USA
Phone: (845) 758-0400
Fax: (845) 758-2633
E-mail: curran@proceedings.com
Web: www.proceedings.com

Table of Contents

893025   **Modeling and Comparison of Switching Loss between SiC MOSFETs with Current Source and Voltage Source Gate Driver**   *1*
Quan Zheng, Yong Kang, Cai Chen

893399   **Power Cycling Capabilities of Bond Buffer Technologies for Wide Bandgap Power Devices**   *6*
Nan Jiang, Haitao Zhang, Jianing Wang, Chengguo Li, Jinhao Cai

893410   **Modeling and Suppression of Crosstalk of SiC MOSFET in Bidirectional Buck/Boost Converter**   *11*
Hao Zhang, Runquan Meng, Dingbang Zhang, Yingying Ding, Ziniu Wu

893424   **An Accurate Crosstalk Evaluation and Prediction Method for SiC MOSFET Considering Nonlinear Capacitance and Stray Parameters**   *17*
Huaqing Li, Chengzi Yang, Longyang Yu, Haoyuan Jin, Xingshuo Liu, Laili Wang

893426   **Analysis of the Influence of Vibration and Thermal Vibration Coupling on the Power Module**   *23*
Jiajia Guan, Chi Zhang, Cai Chen, Yong Kang

893432   **Improved Breakdown Characteristics for AlN/GaN/InGaN Coupling Channel HEMTs with SiNx Removal and Backfill Technique**   *27*
Hao Lu, Xiaohua Ma, Bin Hou, Ling Yang, Yue Hao

893433   **Analysis of GaN HEMT Degradation under RF Overdrive Stress**   *31*
YuHan Xie,   Yan Ren, Chang Liu, YiQiang Chen, RongSheng Chen

893434   **Comparison of Two Types of Single Gate Drivers for SiC MOSFET Stacks in Flyback Converters**   *36*
Rui Wang, Hongbo Zhao, Stig Munk-Nielsen

893439   **Over-Voltage and Oscillation Suppression Circuit with Switching Losses Optimization and Clamping Energy Feedback for SiC MOSFET**   *41*
ChengZi Yang, Huaqing Li, Haoyuan Jin, Longyang Yu, Laili Wang, Yunqing Pei

893443   **Degradation behavior and mechanism of SiC power MOSFET by total dose irradiation under different gate voltages**   *46*
Kexin Gao, Yiqiang Chen*, huaizhi Zheng*, Xinbing Xu, Min Liao, Meng Lu

893444   **Investigation on Parameter Extraction for An Improved Fourier-Series-Based NPT IGBT Model**   *51*
Yifei Ding, Xin Yang, Jun Wang, Chunming Tu, Guoyou Liu

893450   **Soft-Switching Resonant Active Clamp Flyback Converter based-on GaN HEMTs for MHz High Step-Up Applications**   *57*
Wuji MENG, Lin LI, Fanghua ZHANG, Jianjun SHU

893453 **A Lossless and Passive Voltage Spikes Clamping Circuit for SiC HERIC Inverter** 63
Yong Li, Shanxu Duan, Qiqi Li

893454 **Design of Aging Test System for SiC MOSFET Modules** 68
Chaoyue Shen, Fei Wang, Zhenye Wang, Zhong Ye

893455 **Effects of p-type Islands Configuration on the Electrical Characteristics of the 4H-SiC Trench MOSFETs with Integrated Schottky Barrier Diode** 74
Fei Yang, Lixin Tian, hanwei Shen*, Guoguo Yan, Xingfang Liu, Wanshun Zhao, Lei Wang, Guosheng Sun, Junmin Wu,Feng Zhang, Yiping Zeng

893457 **Characteristics of SiC MOSFET in a Wide Temperature Range** 79
Mengyu Zhu, Laili Wang, Huaqing Li, Chengzi Yang, Dingkun Ma, Fengtao Yang

893466 **Fractional-Order Model Predictive Control of SiC PFC Converter** 83
Qihui Fu, Zishun Peng, Zipeng Ke, Huimin Quan, Zhenxing Zhao, Zeng Liu, Yuxing Dai, Jun Wang

893468 **GaN HEMT with Current-Driven Gate and its Driving Circuit Design** 89
Owen Song, Rafael Garcia

893469 **Comparison Study of Parasitic Inductance, Capacitance and Thermal Resistance for Various SiC Packaging Structures** 94
Yue Xie, Yifan Zhang, Cai Chen, Yong Kang

893479 **Off-State Negative Differential Capacitance AlGaN/GaN Heterostructures in cryogenic temperature** 99
Siyu Liu, Jiejie Zhu, Jingshu Guo, Minhan Mi,Xiaohua Ma, Yue Hao, Jielong Liu, Yilin Chen

893480 **Deep energy levels investigation on high resistivity bulk monocrystalline diamond** 103
Yutian Wang, Fangzhou Zhao, Hui Guo*, Qian Sun

893507 **Power Semiconductor IGBT Packaging Technology and Reliability** 113
Yameng Sun, Shizhao Wang, Lianghao Xue, Zheng Feng, Rui Li, Sheng Liu

893521 **Research on A Novel Parallel Resonant DC Link Soft switching Inverter Based on SiC MOSFET** 117
Si Li, Ming Yang, Yu Ma, Dianguo Xu

893529 **Comparison of the Influence of Reverse Conduction on EMI of WBG and Si Devices** 123
Ru Zhang, Wenjie Chen, YuXuan Chen, Yue Cao, Ruitao Yan, Xu Yang

893539 **A Novel AC/DC Single-Phase Bridgeless SEPIC PFC Converter With Reduced Conduction Losses and Simple Structure** 127
Xiang Lin, Shumin Ding, Deliang Wu, Jian Luo

893545 A Single-Stage Modular DCX with High Voltage Conversion Ratio Based on 132
High Frequency LLC Resonant Converter
Yueshi Guan, Zhaoliang Wen, Yijie Wang, Dianguo Xu

893554 Ultra-thin Coupled Inductor for a GaN-Based CRM Buck Converter 138
Ming hua, Junyu Chen, Guolin Xu, Hongfei Wu

893557 Comparison Study on Short Circuit Capability of 1.2 kV Split-Gate 143
MOSFET and Split-Source MOSFET with Integrated JBS Diode
Hongyi Xu, Chaobiao Lin, Na Ren, Xinhui Gan, Liping Liu, Zhengyun Zhu, Li Liu, Qing
Guo, Jianxin Ji, Kuang Sheng*

893575 Analysis of Crosstalk and Suppression Methods for Enhancement-Mode 147
GaN HEMTs in A Phase-Leg Topology
Haihong Qin, Wenlu Wang, Feifei Bu, Zihe Peng, Ao Liu, Song Bai

893590 Comparing Hexagonal and Circular Cell Designs for SiC MPS Diode: The 153
Curvature Effect on Avalanche Capability
Li Liu, Na Ren, Jiupeng Wu, Zhengyun Zhu, Hongyi Xu, Qing Guo, Kuang Sheng*

893593 A Real-Time Self-Learning High Performance Control for Megahertz 157
GaN-based DC-DC Converter
Jing Chen, Jing Chen, Yong Kang

893595 15kV Press Pack SiC IGBT 162
*DU Yujie, TANG Xinling, WANG Liang, ZHAO Zhibin, YANG Xiaolei，YANG Fei, WU
Junmin*

893601 Degradation mechanism of D-mode GaN HEMT based on high temperature 167
reverse bias stress
Meng Lu, Yiqiang Chen, Min Liao*, Chang Liu, Shuaizhi Zheng, Kexin Gao*

893603 Comprehensive Investigations on Paralleling Operation of SiC MOSFETs 171
based on Subcircuit Model in MATLAB/SIMULINK
Yuqi Wei, Dereje Woldegiorgis, Xia Du, Venkata Samhitha Machireddy, Alan Mantooth

893607 Mode Switchover Strategy for Multi-port Energy Router Based on State 179
Flow Diagram
Jingwen Zheng, Zhiguo Wei, Zaixun Ling, Yu Guo, Ping Xiong, Yiqun Kang

893609 A Dynamic Current Sharing Method in Multi-chip SiC Power Module Using 184
Stacked DBC Bridges and Decoupling Capacitors Based on the Original
Simple Module Layout
Jianwei Lv, Chi Zhang, Cai Chen, Yong Kang

893623 Power Loop Inductance Extraction with High Order Polynomial Fitting 189
Algorithm for SiC MOSFET Power Module Characterization
Zhikun Wang, Saijun Mao, Shuhao Yang, Wenyu Li, Yujie Ding, Keqiu Zeng

893624 **Automated SiC MOSFET Power Module Switching Characterization Test Platform** 195
Shuhao Yang, Saijun Mao, Zhikun Wang, Xi Lu, Hansen Chen, Keqiu Zeng

893629 **LLC Resonant Converter Based on Trench Gate SiC MOSFET** 202
YumingZhou, JinkunChu, JiahuiZhou

893630 **A Review of the Crosstalk Suppression Methods for SiC MOSFETs in the Phase-leg Circuit Configuration** 206
Yujie Ding, Saijun Mao, Zhikun Wang, Shuhao Yang, Wenyu Li, Keqiu Zeng

893635 **Single Pulse Short-Circuit Failure Mechanism of 1200V Asymmetric Trench SiC MOSFETs** 213
Zhaoxiang Wei, Jiaxing Wei, Xiaowen Yan, Hua Zhou, Hao Fu, Siyang Liu*, Weifeng Sun

893638 **Modeling and Analysis of the Switching Characteristics Difference for Paralleling SiC MOSFETs in Multichip Power Modules** 217
Wenyu Li, Saijun Mao, Zhikun Wang, Shuhao Yang, Yujie Ding, Keqiu Zeng

893639 **Multiple UIS Ruggedness of 1200V Asymmetric Trench SiC MOSFETs** 224
Jiayue Liu, Xiaochuan Deng*, Xu   Li, Xuan Li, Zhiqiang Li, Hongling Lu

893641 **Design, Fabrication and Characterization of 6.5kV/100A 4H-SiC PiN power rectifier** 228
Mengling Ta, Xiaochuan Deng*, Rui Hu, Xuan Li, Zhiqiang Li, Hongling Lu

893642 **Dynamic gate leakage current of p-GaN Gate AlGaN/GaN HEMT under positive bias Conditions** 232
Yu Sun, Maojun Wang, Wen Lei, Chun Han

893643 **An Integrated GaN-Based Converter Based on Cooling-System-Inductor Structure for point-of-load converters** 236
Longyang Yu , Wei Mu, Huaqing Li, Chengzi Yang, Chenya Wang, Laili Wang

893644 **A Survey on Modeling of SiC IGBT** 241
Yuwei Wu, Laili Wang, Jianpeng Wang, Feng Zhang

893645 **A Layout Optimization Method to Reduce Commutation Inductance of Multi-Chip Power Module Based on Genetic Algorithm** 247
Yu Zhou, Yu Chen, Hongyi Gao, Chengmin Li, Haoze Luo, Wuhua Li, Xianging He

893646 **A GaN-based High Power Density Power Optimizer for Solar-powered Aircraft Applications** 253
Peng Chen, Tao Liu, Yujie Cheng, Hongfei Wu, Jianxin Zhu

893649 **Influence of Al/CucorAl wire bonding on reliability of SiC devices** 259
ChaoFang, XiangTang ,GuangyuanQin, HaotaoKe, YiboWu, JingZhang, GuiqinChang, HaihuiLuo

893650   **An Efficient Voltage Step-up/down Partial Power Converter(SUD-PPC) using Wide Bandgap devices**   264
Chao Liu, Zhe Zhang, Michael A. E. Andersen

893653   **The Method for Decoupling the Parasitic Inductance of the Laminated Busbar with SiC MOSFETs in Parallel**   270
Shaolin Yu, Jianing Wang, Xing Zhang, Yuanjian Liu, Zhaoyang Wei

893658   **EMI Noise Reduction in GaN-based Full-bridge LLC Converter**   276
Yue Cao, Yuxuan Chen, Xingwei Huang, Pengyuan Ren, Wenjie Chen, Xu Yang

893661   **The influence of hydrogen annealing on minority carrier lifetimes in 4H-SiC**   281
Ruijun Zhang, Rongdun Hong, Jiafa Cai, Xiaping Chen, Dingqu Ling, Mingkun Zhang, Shaoxiong Wu, Yuning Zhang, Jingrui Han, Zhengyun Wu, Feng Zhang

893662   **Comparative Study of Thermal Performance of a SiC MOSFET Power Module Integrated with Vapor Chamber for Traction Inverter Applications**   285
Wei Mu, Binyu Wang, Shenghe Wang, Haoyuan Jin, Huaqing Li, Laili Wang

893663   **Optimized Parameter Selection Method of Driving Circuit for SiC MOSFET**   290
Haihong Qin, Sixuan Xie, Feifei Bu, Shishan Wang, Wenming Chen, Dafeng Fu

893666   **Improved One Cycle Control for Three-Phase Three-Wire VIENNA Rectifier**   296
Junnan Gu, Xikun Chen, Ruiying Li, Borui Liu, Ni Zheng

893669   **Research on the strategy of parallel wide range bidirectional DC-DC converter**   302
Zehui Peng, Xikun Chen, Borui Liu, Yongjian Chen, Junnan Gu, Ruiying Li

893671   **Analysis of an Output Series High Voltage Gain Impedance Source Circuit Based on SiC Switch**   307
Qing Cheng, Wei Wang, Yueshi Guan, Tingting Yao, Dianguo Xu

893676   **A Compact Model for Si/SiC IGBT Implemented in LTspice**   313
Md Maksudul Hossain, Arman Ur Rashid, Yuqi Wei, H. Alan Mantooth

893678   **Temperature-Dependent Current Collapse and Gate Leakage in AlGa/NGaN HEMTs With Si-rich SiN Interlayer**   318
Jielong Liu, Yuwei Zhou, Minhan Mi, Jiejie Zhu, Siyu Liu, Qing zhu, Pengfei Wang, Hong Wang, Xiaohua Ma, Yue Hao

893681   **Design and Verification of Gate Driver for 6.5 kV SiC MOSFET Module**   322
Yijian Wang, Lin Liang, Hai Shang, Lubin Han

893683   **Soft Precharging Method for Four-Level Hybrid-Clamped Converter**   327
Yihui Zhao, Jianyu Pan, Yao Luo, Jian Li

893689   **Design and Research on Package Insulation of Highvoltage Silicon Carbide Module**   333
Yang Zhou, Ling Sang, Xinling Tang, Hao Shi

893690   **Adaptive Digital Technique Assisted Hard Switching Fault Detection for SiC MOSFETs**   338
Saravanan Dhanasekaran, Vamshi Krishna Miryala, Kamalesh Hatua

893695   **Short-circuit Protection Circuit of SiC MOSFET Based on Drain-source Voltage Integral**   344
Hong Li, Yuting Wang, Zhidong Qiu, Zuoxing Wang, Xiaofei Hu, Jia Zhao

893696   **Failure Analysis of 200V p-GaN HEMT under Unclamped Inductive Switching Conditions**   350
Junjie Ye, Li Xuan*, Yangyang Wu, Xiaochuan Deng, Zhiqiang Li, Bo Zhang

893697   **Low Roughness SiC Trench Formed by ICP Etching with Sacrificial Oxidation and Ar Annealing Treatment**   354
Changwei Zheng, Zhicheng Wang, Shasha Jiao, Qijun Liu, Yehui Luo, Jieqin Ding, Chengzhan Li

893698   **Research on Threshold Voltage Hysteresis of D-mode and Fully Recessed E-mode AlGaN/GaN MIS-HEMTs with HfO$_2$ Dielectric**   358
Zicheng Yu, Chi Sun, Xiaoyu Ding, Xing Wei, Weining Liu, Li Zhang, Zhang Chen, Guohao Yu, Baoshun Zhang

893700   **Investigation of the Insulation Failure of Power Modules by Observation of Electrical Trees**   363
Kaixuan Li, Boya Zhang, Xingwen Li, Haotao Ke

893704   **A Novel GaN MIS-HEMT with a Source-connected Clamp Electrode for Suppressing Short-channel effect**   367
Yijun Shi, Shan Wu, Hongyue Wang, Zhiwei Fu, Si Chen, Bin Zhou

893920   **Resonant Gate Driver with Wide Range Adjustment of Driving Speed**   371
Hao Peng, Han Peng, Qiaozhi Yue

893938   **A Predictive Method for Switching Time of Nanosecond Pulsed Power System of Ohmic Loads Using SiC MOSFETs**   376
Yifei Luo, Xin Li*, Fei Xiao , Zenan Shi, Ruitian Wang, Feng Xie

893950   **Active Magnetic Bearing Amplifier Design based on SiC Devices**   382
Gang Cao, Hongbo Sun, Gao Yang, Dong Jiang

893958   **An Integrated Buck-Boost Converter with SRC for Wide Input Voltage**   387
Yanqing Wang, Yutao Lou, Xiang Guo, Changle Xu, Xudong Zou, Yong Kang

894026   **Design of a High Power Density Bidirectional AC/DC Converter Based on GaN**   393
Jiajia Guan, Zhiwei Wang, Ziyan Tang, Jianwei Lv, Cai Chen, Yong Kang

894147   **Influence of the Interface Traps Distribution on I-V and C-V Characteristics of SiC MOSFET Evaluated by TCAD Simulations**   *398*
Yumeng Cai, Hao Xu, Peng Sun, Zhibin Zhao, Zhong Chen

894196   **High Breakdown Voltage AlGaN/GaN HEMT with Graded Fluorine Ion Implantation Terminal in Thick Passivation Layer**   *403*
Siyu Deng, Xiaorong Luo*, Jie Wei*, Yanjiang Jia, Tao Sun, Lufan Xi, Zhuolin Jiang, Kemeng Yang, Qinfeng Jiang, Bo Zhang

894238   **Homogeneous-Flux Transmitter Coil Design with Improved Position Tolerance**   *407*
Yunfeng Liu, Yi Dou, Ziwei Ouyang, Michael A. E. Andersen

894245   **Review of soft-switching high-frequency GaN-based single-phase Bridgeless Rectifier**   *411*
Yunfeng Liu, Ziwei Ouyang, Michael A. E. Andersen

894288   **A Low Winding Loss Magnetic Circuit Structure Design of Planar Inductor for GaN-based Totem-Pole PFC**   *417*
Pengyuan Ren, Wenjie Chen, Xingwei Huang, Yue Cao, Yuxuan Chen, Xu Yang

894296   **Design Methodology of SiC MOSFET Based Bidirectional CLLC Resonant Converter for Wide Battery Voltage Range**   *423*
Mingjie Liu, Xuehua Wang, Jiangtao Xu

894320   **Design of a 10kW, High-Frequency Dual Active Bridge Converter Using SiC Devices**   *428*
Haoyuan Jin, Huaqing Li, Junduo Wen, Chengzi Yang, Hang Kong, Laili Wang

894325   **Evaluating Switching Performance of GaN HEMT Using Analytical Modeling**   *434*
Yingzhe Wu, Shan Yin, Hui Li, Minghai Dong, Xi Liu, Yuhua Cheng

894330   **A Novel SiC Trench MOSFET Structure with Enhanced Short Circuit Robustness**   *440*
Chongyu Jiang, Hongyi Xu, Na Ren*, Qing Guo, Kuang Sheng

894331   **A High Power Density Chip-on-Chip Gan-based Module with Ultra-Low Parasitic Inductance**   *444*
Yi Zhang, Zongheng Wu, Cai Chen, Yong Kang, Han Peng

894347   **Modeling and Experimental Verification of Common Mode Crosstalk with Shielded Cables in Power Converter System**   *449*
Ruizhou Xue, Xuejun Pei, Chunyu Yang, Yi Yu

894353   **Analytical Averaged Loss Model of a Three-level NPC-type Converter With SiC Devices**   *454*
Xinyue Guo, Yue Xie, Cai Chen, Yong Kang

894357 **Dual-Side Three-stage Asymmetric Phase Shift Strategy for Bidirectional** *459*
Inductive Power Transfer System with SiC Power Module
Haowen Chen, Changsong Chen, Mengjie Jiang, Shuran Jia, Xuezheng Huang

894384 **A Synchronous Boot-strapping Technique with Increased On-time and** *462*
Improved Efficiency for High-side Gate-drive Power Delivery
Nathan M.Ellis, Rahul Iyer, Robert C.N.Pilawa-Podgurski

894391 **Single-Pulse Avalanche Failure Characterization of Single and Paralleled** *467*
SiC MOSFETs
Hua Mao, Huaping Jiang*, Guanqun Qiu, Yifu Zhang, Xiaohan Zhong, Hao Feng, Li Ran

894392 **DC Transform Circuit Design Based on Multiplier Rectification** *472*
Penghui Yin, Xuehua Wang, Xinbo Ruan

894397 **Impact of Gate Resistances on Switching on Behaviors of Si/SiC Hybrid** *478*
Switch
Xiaofeng Jiang, Huaping Jiang*, Hongyu Yu, Jinhong Jiang, Hao Feng, Hua Mao, Lei Tang,
Xiaohan Zhong, Li Ran

894398 **The Influence of Dynamic Threshold Voltage Drift on Third Quadrant** *483*
Characteristics of SiC MOSFET
Lei Tang, Huaping Jiang*, Hua Mao, Zebing Wu, Xiaohan Zhong, Xiaowei Qi, Li Ran

894410 **An Optimal Design Scheme of Intermediate Bus Voltage for two-stage LLC** *488*
Resonant Converter Based on SiC MOSFET
Feng Wang, Xuehua Wang, Xinbo Ruan

894430 **Modeling and Design of A 10MHz Class $\Phi 2$ Inverter** *493*
Yongzhi Liu, Yiyang Yan, JiaJia Guan, Cai Chen, Yu Chen, Yong Kang

894558 **An Automated Electro-Thermal-Mechanical Co-Simulation Methodology** *499*
Based on PSpice-MATLAB-COMSOL for SiC Power Module Design
Yayong Yang, Yuxin Ge, Zhiqiang (Jack) Wang, Yong Kang

894585 **An Accurate Analytical Model for Motor Terminal Overvoltage Prediction** *504*
and Mitigation in SiC Motor Drives
Neng Wang, Cheng Qian, Zhiqiang (Jack) Wang, Yong Kang

894625 **An Improved Desaturation Protection Method with Self-Adaptive** *510*
Blanking-Time for Silicon Carbide (SiC) Power MOSFETs
Jiawei Li, Cheng Qian, Zhiqiang (Jack) Wang, Yong Kang

894627 **Analysis of Dynamic Current Balancing in Multichip SiC Power Modules** *516*
Based on Coupled Parasitic Network Model
Yuxin Ge, Yayong Yang, Cheng Qian, Zhiqiang (Jack) Wang, Yong Kang

894634 **A Compact 175 ℃ High Temperature Gate Driver with Isolated Power** *522*
Supply and Advanced Protection for HybridPACK Drive SiC Module
Cheng Qian, Neng Wang, Yayong Yang, Zhiqiang (Jack) Wang, Yong Kang

894635 **An Optimized Parameter Design Method for Desaturation Protection Circuit** 527
towards Fast Response Speed and Strong Noise Immunity
Cheng Qian, Zhiqiang (Jack) Wang, Yong Kang

894636 **Datasheet Driven Turn Off Overvoltage Prediction for Silicon Carbide** 532
Power MOSFETs Based on Theoretical Analysis
Cheng Qian, Yuxin Ge, Zhiqiang (Jack) Wang, Yong Kang

894975 **650V 4H-SiC VDMOSFET with Additional n Region: A Simulation Study** 537
Xiuxiu Gao, Chengzhan Li, Xiaoping Dai

Author Index

IEEE Workshop on Wide Bandgap Power Devices and Applications in Asia (WiPDA Asia 2021)

Aug 25-27, 2021 Wuhan · China

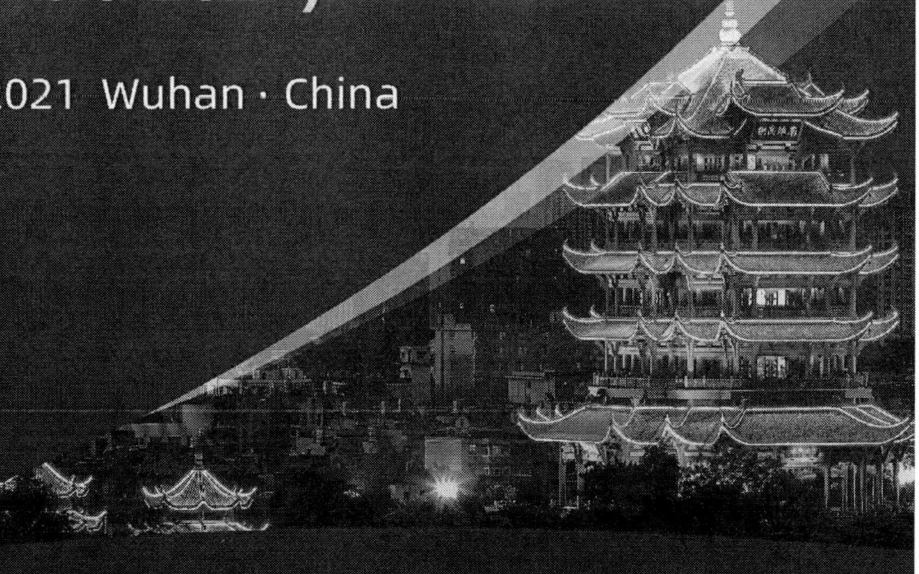

PROGRAM

Proceedings of

2021 IEEE Workshop on Wide Bandgap Power Devices and Applications in Asia

(WiPDA Asia)

Publisher: Institute of Electrical and Electronics Engineers, Inc.

Keynote Speakers

✧ Fred C. Lee

University Distinguished Professor Emeritus, Virginia Tech, USA

Member of National Academy of Engineering

IEEE Fellow

Topic: Next Generation of Power Supplies: EV On-Board Charger

Dr. Lee is a University Distinguished Professor Emeritus at Virginia Tech. He is a member of the *U.S. National Academy of Engineering*, an academician of Taiwan's *Academia Sinica*, and a foreign member of the *Chinese Academy of Engineering*, China. Dr. Lee founded the Center for power electronics and led a program that encompasses research, technology development, educational outreach, industry collaboration, and technology transfer. To date, more than 230 companies worldwide have benefited from this industry partnership program.

Dr. Lee has supervised to completion 89 Ph.D. and 93 M.S. students. He holds over 100 US patents, and has published over 330 journal articles and more than 760 refereed technical papers. His research interests include high-frequency power conversion, magnetics and EMI, distributed power systems, renewable energy, power quality, high-density electronics packaging and integration, and modeling and control.

Dr. Lee is a fellow of the US National Academy of Inventor, and the recipient of the 2015 IEEE Medal in Power Engineering "for contributions to power electronics, especially high-frequency power conversion."

✧ Alan Mantooth

University Distinguished Professor, University of Arkansas, USA

Past-President of IEEE Power Electronics Society

IEEE Fellow

Topic: Designing Wide Bandgap Power Electronic Systems

H. Alan Mantooth received the B.S.E.E. and M.S.E.E. degrees from the University of Arkansas in 1985 and 1986, and the Ph.D. degree from Georgia Tech in 1990. He then joined Analogy, a startup company in Oregon, where he focused on semiconductor device modeling and the research and development of modeling tools and techniques. In 1998, he joined the faculty of the Department of Electrical Engineering at the University of Arkansas, Fayetteville, where he currently holds the rank of Distinguished Professor. His research interests now include analog and mixed-signal IC design & CAD, semiconductor device modeling, power electronics, power electronics packaging, and cybersecurity. Dr. Mantooth helped establish the National Center for Reliable Electric Power Transmission (NCREPT) at the UA in 2005.

Professor Mantooth serves as the Executive Director for NCREPT as well as two of its centers of excellence: the NSF Industry/University Cooperative Research Center on GRid-connected Advanced Power Electronic Systems (GRAPES) and the Cybersecurity Center on Secure, Evolvable Energy Delivery Systems (SEEDS) funded by the U.S. Department of Energy. In 2015, he also helped to establish the UA's first NSF Engineering Research Center entitled Power Optimization for Electro-Thermal Systems (POETS) that focuses on high power density systems for electrified transportation applications. Dr. Mantooth has co-founded three companies in design automation (Lynguent), IC design (Ozark Integrated Circuits), and cybersecurity (Bastazo) as well as advising a fourth in power electronics packaging (Arkansas Power Electronics International) to maturity and acquisition as a board member. Dr. Mantooth holds the 21st Century Research Leadership Chair in Engineering.

✧ Yong Kang

Professor,

Huazhong University of Science and Technology, China

Topic: The Research of High Frequency, High Efficiency and High Power Density (3H) Application for GaN Devices

Professor Yong Kang received the B.E., M.E., and Ph.D. degrees from the Huazhong University of Science and Technology (HUST), Wuhan, China, in 1988, 1991, and 1994, respectively. He joined the School of Electrical and Electronic Engineering, HUST in 1994, where he became a professor in 1998. He is the Vice Chairman of the Power Electronics Society of China Electrotechnical Society, the Vice Chairman with the China UPS Standard Committee, and the Associate Editor of the Journal of Power Electronics.

Professor Yong Kang has authored or coauthored more than 200 technical articles published in journals and conferences and holds more than 30 Chinese patents. His research interests include power electronic converter, wide bandgap semiconductor device packaging, integration and its application, renewable energy generation systems, ac and dc drivers and electromagnetic compatibility. Yong Kang has presided over or participated in 8 National Natural Science Foundation of China projects, and won national, provincial and ministerial and international awards 8 times. In 2001, Yong Kang was awarded the special government allowance of the State Council, and was awarded the honorary title of "Zhongda Scholar" in 2005.

WiPDA Asia 2021 Aug 25-27, 2021 Wuhan, China

✧ Mingxiang Chen

Professor,

Huazhong University of Science and Technology, China

Topic: Development and Applications of Packaging Materials for Power Semiconductor Devices

Mingxiang Chen, Professor of Huazhong University of Science and Technology (HUST). He received his B. S. and M. S. in Material Engineering from Wuhan University of Science and Technology, and his Ph. D. in Physical Electronics from HUST. After his doctoral study, he was awarded the postdoctoral fellowship with Professor C. P. Wong at Georgia Institute of Technology, where he conducted studies on nano packaging. Now his research focuses on electronic packaging materials & micro/nano fabrication. He has published over 50 peer-reviewed papers and holds over 10 patents (some have been licenced to the industry).

✧ Harufusa Kondo

Senior Technical Advisor

Mitsubishi Electric, Power Device Works, Japan

Topic: Technology Trends of SiC Chips and Modules

Harufusa Kondo received the B.S., M.S., and Ph.D. degrees from Osaka University, JAPAN. In 1985, he joined the LSI R&D Laboratory, Mitsubishi Electric Corporation, where he had been engaged in the design of system VLSI's for digital communication. In 2003, he moved to the Optical and High-frequency Device Works as a manager of Optical Transceiver. Since 2009, he has been working at Power Device Works for the development of DIPIPM, Industrial IGBT modules, and high-voltage modules including SiC. He is currently the senior technical advisor at Power Device Works, Mitsubishi Electric, Japan.

⬦ Song Bai

Director

State Key Laboratory of Wide Bandgap Semiconductor

Power Electronic Devices, China

Topic: Development of High Voltage SiC Power MOSFETs

Song Bai received the B. S. degree in physics from Peking University, in 1997 and the Ph. D degree in physics from University of Pittsburgh, in 2003. In 2004, he joined Nanjing Electronic Devices Institute where he currently heads research on SiC power device development at State Key Laboratory of Wide-Bandgap Semiconductor Power Electronic Devices. He is the author or coauthor of over 30 publications and holds more than 20 patents. His present research interests are in developing high-voltage power devices of wide-bandgap semiconductors.

✧ Stig Munk-Nielsen

Professor,

Aalborg University, Denmark

Topic: WBG Power Devices and Digital Design Framework: Challenges, Possibilities, Opportunities

Stig Munk-Nielsen is currently Professor at the Department of Energy Technology, Aalborg University, Denmark. Since 2008 Stig worked with circuits for monitoring of high power IGBTs voltage drop for failure analysis purpose and the team managed to install monitoring systems in off-shore wind turbine in 2018. Since 2013, Stig has secured funding for a die packaging team and laboratory facilities for 10 kV SiC devices and later on the team did numerous application designs with GaN, Si and other SiC devices. Since 2013, where the team initially simulated and included the R,L,C parasitics of power module layouts we are building a second version of the packaging facilities. We expect the new laboratory facilities is a key enabler to the goal of extending the experience with digital design framework. In the newly funded project CoDE we hire 5 PhDs and want to use the digital framework in more applications but also to include mechanical wear out in simulations. The application examples include megawatt PtX converters and pump drive systems technology. A number of industrial related projects is conducted in parallel.

Modeling and Comparison of Switching Loss between SiC MOSFETs with Current Source and Voltage Source Gate Driver

Quan Zheng
School of Electrical and Electronic Engineering
Huazhong University of Science and Technology
Wuhan, China
zhengquan@hust.edu.cn

Cai Chen
School of Electrical and Electronic Engineering
Huazhong University of Science and Technology
Wuhan, China
caichen@hust.edu.cn

Yong Kang
School of Electrical and Electronic Engineering
Huazhong University of Science and Technology
Wuhan, China
ykang@hust.edu.cn

Abstract—**Previous research showed that the switching loss of SiC MOSFET (SM) with current source driver (CSD) is lower than that with voltage source driver (CSD). But sometimes the switching loss of SM with CSD is higher than that with VSD. This paper constructs an evaluation index *K* through theoretical derivation to compare the switching loss between SiC MOSFETs(SMs) with CSD and VSD under different conditions. *K* is the ratio of current injected into the gate of the SM during the Miller platform stage between CSD and VSD. When the load is determined, the charging current of CSD and the gate driving resistance of VSD greatly affect the value of K, thereby affecting the switching loss between SiC MOSFETs with CSD and VSD, which is verified by double pulse test (DPT).**

Keywords—SiC MOSFET (SM), gate driver, switching loss, modeling

I. Introduction

The gate driver can improve the efficiency of SiC MOSFET (SM) converter. After traditional voltage source driver (VSD), resonant gate driver appeared [1]. Inductor is used to replace the gate driving resistance to eliminate gate driver loss. However, the maximum gate voltage will be higher than gate driver power voltage, which is uncontrollable. To solve this problem, voltage closed-loop resonant gate driver is proposed, and its final gate voltage is independent of resonant inductor and capacitor. However, the limitation of resonant gate driver is that it cannot provide a large initial charging current, so current source driver (CSD) came into being, using four mosfets and a charging inductor to form basic circuit structure [2]. CSD charges the inductor firstly, and then charges the gate of SM using characteristic that current of the inductor remains constant in a short time. CSD in this paper is based on this scheme. Some studies are conducted for comparison of gate driver methods. Literature [3] compared continuous and discontinuous CSD, but did not compare with VSD. Infineon had developed a CSD IC and compared it with VSD, but this IC is not suitable for SM power devices with faster switching speed.

Previous research showed that switching loss of SM with CSD is lower than that with VSD. But when the charging current of CSD is small or the Miller platform voltage is low, the switching loss of SM with CSD is higher than that with VSD. Therefore, this paper constructs an evaluation index *K* to compare the switching loss between SMs with CSD and VSD under different conditions. *K* is the ratio of current injected into the gate of SM during the Miller platform stage between CSD and VSD. When the load is determined, the

charging current of CSD and the gate driving resistance of VSD greatly affect the value of K. In actual situations, when the gate driving resistance of VSD is determined, the model can be used to calculate the critical CSD charging current value that makes the switching loss of SM with CSD lower than that with VSD.

II. Current Source Gate Driver

The CSD used is shown in Fig. 1, which is composed of controllable mosfets S_1-S_4 (including its body diode) and inductor L_H. There are three working stages for CSD during the turn on and turn off of SM. Three working states of CSD are explained in the following content.

The first stage of SM turn on process is shown in Fig. 2(a). S_1, S_4 are turned on. S_2, S_3 are turned off. the inductor L_H is charged, and the inductor current I_H rises and reaches the expected charging current value at the end of this stage. The second stage is shown in Fig2. (b). S_1 is turned on. S_2, S_3, and S_4 are turned off. The inductor L_H starts to charge the gate of SM until the gate voltage of SM reaches V_{CC} and is clamped. The third stage is shown in Fig2. (c). S_2 is turned on. S_1, S_3, and S_4 are turned off. The gate voltage of SM remains constant at V_{CC}, and the inductor current I_H begins to decrease until it reaches zero.

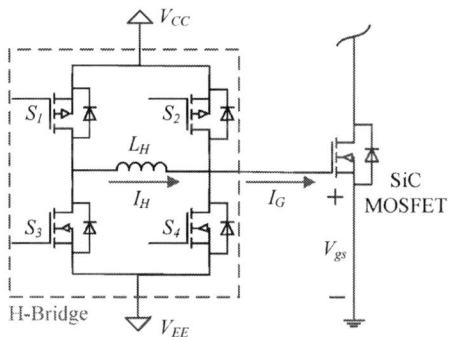

Fig. 1.   The schematic diagram of CSD

978-1-6654-4817-8/21 $31.00 © 2021 IEEE

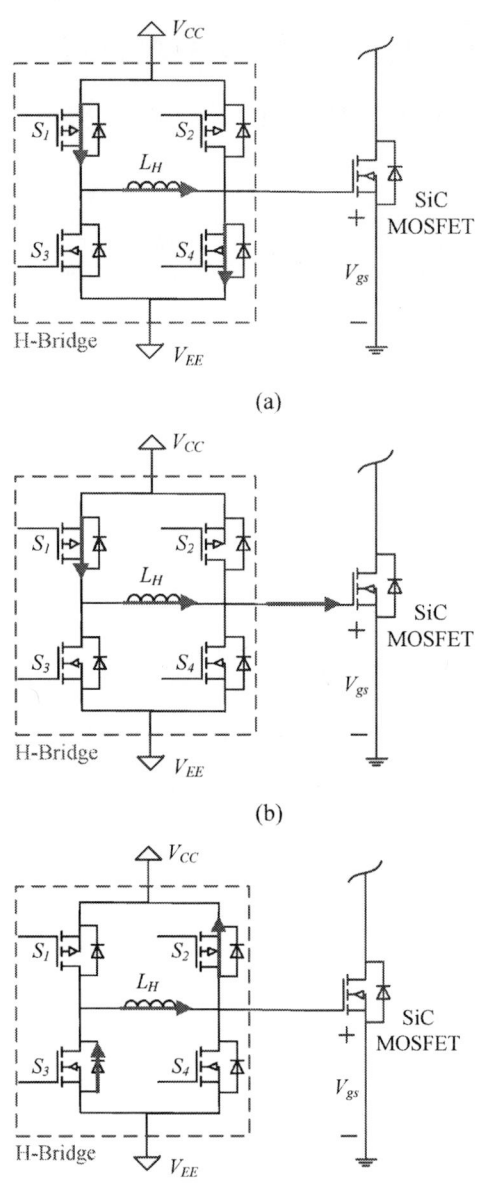

Fig. 2. Turn on process of CSD. (a) First stage. (b) Second stage. (c) Third stage

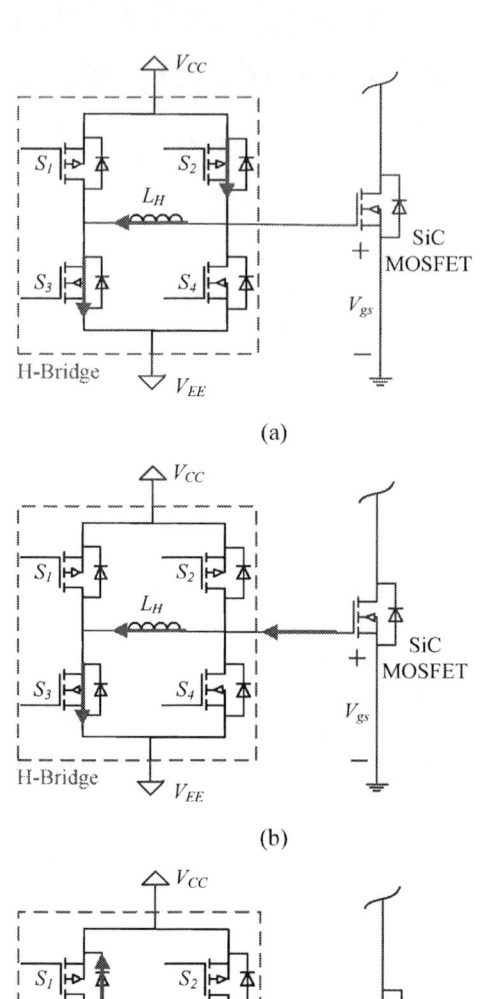

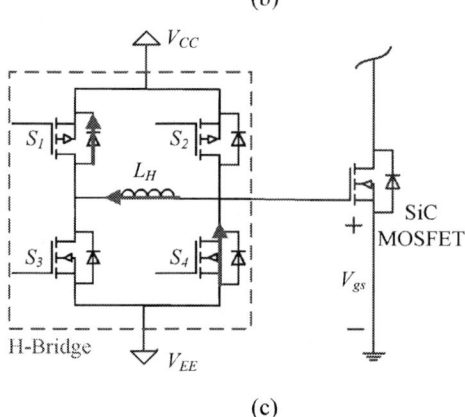

Fig. 3. Turn off process of CSD. (a) First stage. (b) Second stage. (c) Third stage

The first stage of SM turn off process is shown in Fig. 3(a). S_2, S_3 are turned on. S_1, S_4 are turned off, the inductor L_H is charged, and the inductor current I_H rises and reaches the expected charging current value at the end of this stage. The second stage is shown in Fig. 3(b). S_3 is turned on. S_1, S_2, and S_4 are turned off. The inductor L_H starts to charge the gate of SM until the gate voltage of SM reaches V_{EE} and is clamped. The third stage is shown in Fig3. (c). S_4 is turned on. S_1, S_2, and S_3 are turned off. The gate voltage of the SM remains constant at V_{EE}, and the inductor current I_H begins to decrease until it reaches zero.

III. MODELING OF SWITCHING LOSS

The switching time of SM depends on the charging time of the junction capacitance. The comparison of switching loss between SMs with CSD and VSD can be transformed into comparison of switching time, which depends on the gate charging current of SM.

Taking turn on process of SM as example, the switching loss evaluation index K of SM with CSD and VSD is derived in the following, and the turn off process is the same. The ideal turn on waveform of SM is shown in Fig.4.

978-1-6654-4817-8/21 $31.00 © 2021 IEEE

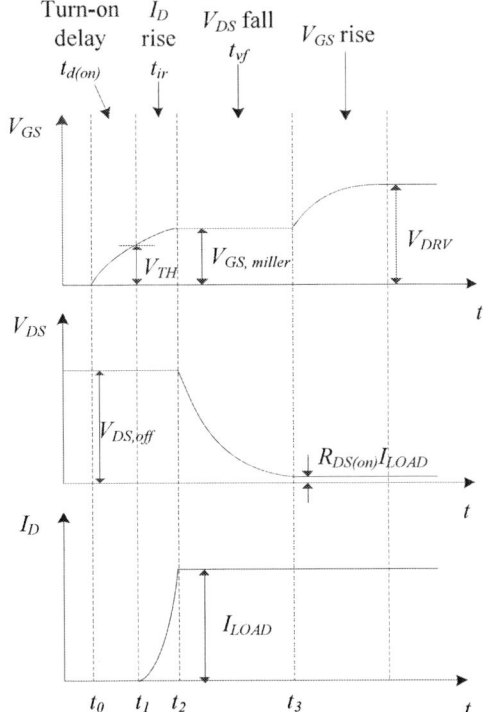

Fig. 4. The ideal turn-on waveform of SiC MOSFET (SM)

The area where voltage V_{DS} and current I_D overlap is mainly in the two stages of I_D rise and V_{DS} fall. If the change process of I_D and V_{DS} is approximately linearized, it can be concluded that the turn on switching loss of SM at switching frequency f_s is:

$$P_{on} = 0.5 \cdot f_s \cdot V_{DS} \cdot I_{LOAD} \cdot (t_{2-1} + t_{3-2})$$

In practical applications, it is found that the Miller platform t_{3-2} has a longer time than t_{2-1}, which results in most of the switching loss exist in the Miller platform stage. Therefore, the switching loss of the Miller platform stage is used to replace the switching loss of entire turn on process, so as to avoid the complicated analysis of the entire switching process.

At the Miller platform stage, the charging current of VSD is I_{G-VDRV} and charging time of SM with VSD is T_{VDRV}. $R_{Gate,tot}$ ($=R_{Gate,ext}+R_{Gate,int}$) is the total gate driving resistance of the VSD. $R_{Gate,ext}$ is the external driving resistance of VSD, which can be changed as required. $R_{Gate,int}$ is the internal driving resistance of SM.

$$I_{G-VDRV} = \frac{V_{DRV} - V_{GS,Miller}}{R_{Gate,ext} + R_{Gate,int}} \tag{1}$$

$$T_{VDRV} = C_{RSS} \cdot \frac{V_{DS,off}}{I_{G-VDRV}} \tag{2}$$

At the Miller platform stage, the charging current of CSD is I_H which is preset, and charging time of SM with CSD is T_{CDRV}.

$$T_{CDRV} = C_{RSS} \cdot \frac{V_{DS,off}}{I_H} \tag{3}$$

The ratio K of the charging time of SM with CSD and VSD in the Miller platform stage is:

$$K = \frac{T_{CDRV}}{T_{VDRV}} = \frac{I_{G-VDRV}}{I_H} = \frac{V_{DRV} - V_{GS,Miller}}{I_H \left(R_{Gate,ext} + R_{Gate,int} \right)} \tag{4}$$

V_{DRV} is the driving voltage of VSD. $V_{GS,Miller}$ is the Miller platform voltage which is related to the load and can be expressed as:

$$V_{GS,Miller} = V_{TH} + \sqrt{\frac{I_{LOAD}}{N}} \tag{5}$$

The parameter N is related to the characteristics of SM products. V_{TH} is the gate threshold voltage of SM.

Through above analysis, it is found that the value of K depends on the charging current of CSD I_H, the gate driving resistance of VSD $R_{Gate,tot}$ ($=R_{Gate,ext} + R_{Gate,int}$) and the Miller platform voltage $V_{GS,Miller}$.

The Miller platform voltage is mainly affected by the load current. The specific calculation is as follows:

As shown in Fig.5, taking two points (I_{D1}-V_{GS1}, I_{D2}-V_{GS2}) on the transfer characteristic curve of SM and approximating the transfer characteristic curve as a parabola, there are:

$$I_{D1} = K \cdot (V_{GS1} - V_{TH})^2 \tag{6}$$

$$I_{D2} = K \cdot (V_{GS2} - V_{TH})^2 \tag{7}$$

It can be deduced:

$$V_{TH} = \frac{V_{GS1} \cdot \sqrt{I_{D2}} - V_{GS2} \cdot \sqrt{I_{D1}}}{\sqrt{I_{D2}} - \sqrt{I_{D1}}} \tag{8}$$

$$N = \frac{I_{D1}}{\left(V_{GS1} - V_{TH} \right)^2} \tag{9}$$

$$V_{GS,Miller} = V_{TH} + \sqrt{\frac{I_{LOAD}}{N}} \tag{10}$$

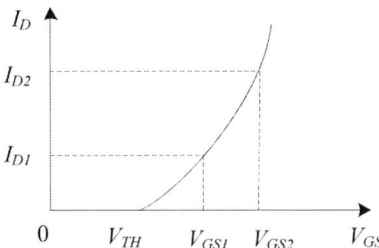

Fig. 5. the transfer characteristic curve of SM

When V_{TH} and N are determined, for any load current, the corresponding Miller platform voltage $V_{GS,Miller}$ can be calculated. At this time, the main factors that affect the value of K are I_H and $R_{Gate,tot}$ ($=R_{Gate,ext} + R_{Gate,int}$).

When $K>1$, $T_{CDRV}>T_{VDRV}$, the switching loss of SM with CSD is higher than that with VSD. When $K=1$, $T_{CDRV} = T_{VDRV}$, the switching loss of SM with CSD and with VSD are about the same. When $K <1$, $T_{CDRV} < T_{VDRV}$, the switching loss of SM with CSD is lower than that with VSD.

978-1-6654-4817-8/21 $31.00 © 2021 IEEE

IV. EXPERIMENTAL DESIGN

In order to verify the model and conclusions, different I_H and $R_{Gate,tot}$ is selected, and three cases of $K>1$, $K<1$, $K=1$ are constructed for experimental verification.

The 1200V/60A SM of CREE company is selected for 600V/40A DPT. According to the above, when the working condition of the selected SM is 600V/40A and the driving voltage is 18V/0V, the gate threshold voltage V_{TH} is 5.27V, and the Miller platform voltage $V_{GS,\,Miller}$ is 10.73V. In order to construct the three cases of $K>1$, $K<1$ and $K=1$, different charging current of CSD I_H and different gate driving resistance of VSD $R_{Gate,ext}$ are selected as shown in Table I.

The picture of DPT platform, VSD board, CSD board are shown in Figure.6, Figure.7 and Figure.8 respectively.

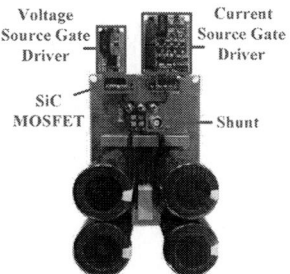

Fig. 6.   DPT platform

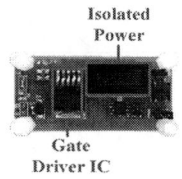

Fig. 7.   VSD board

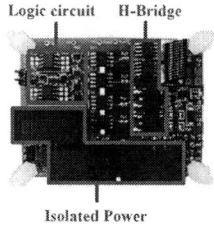

Fig. 8.   CSD board

V. EXPERIMENTAL RESULTS

The key to experimental verification is the switching loss between SMs with CSD and VSD at $K=1$ respectively. When $K=1$, the charging current of CSD I_H is 2A, and the turn on and turn off waveforms of SM with CSD are shown in Fig.9 and Fig.10. The driving resistance of VSD $R_{Gate,tot}$ (=$R_{Gate,ext}$ + $R_{Gate,int}$) is 3.6 Ω, $R_{Gate,int}$ is 1.8 Ω, $R_{Gate,ext}$ is 1.8 Ω, and the turn on and turn off waveforms of SM with VSD are shown in Fig.11 and Fig.12.

After processing the data, when $K=1$, the switching loss of SM with CSD is 1362μJ , the switching loss of SM with VSD is 1346μJ. Because the switching loss of Miller platform stage is used to replace the switching loss of entire switching

process, the error range is tolerable. When $K>1$ and $K<1$, the switching loss of SM with the two gate drivers are shown in Table I. The experimental results of the three cases verify the correctness of the model.

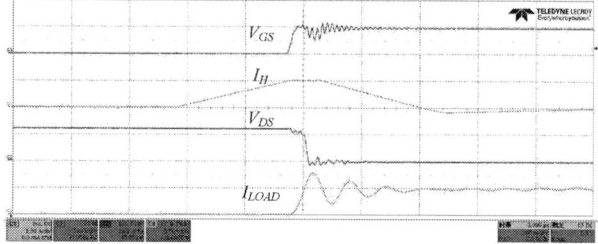

Fig. 9.   Turn on waveform of SM with CSD at I_H=2A

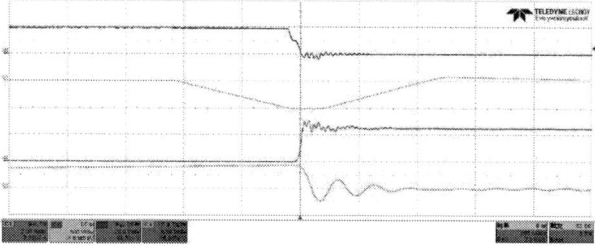

Fig. 10. Turn off waveform of SM with CSD at I_H=2A

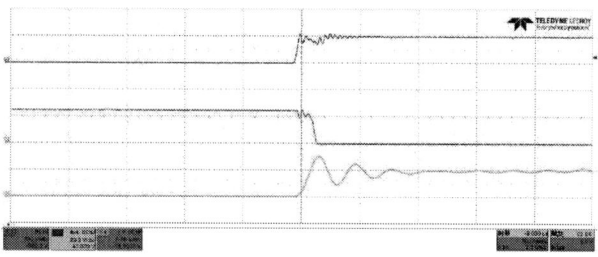

Fig. 11. Turn on waveform of SM with VSD at $R_{Gate,ext}$ =1.8Ω

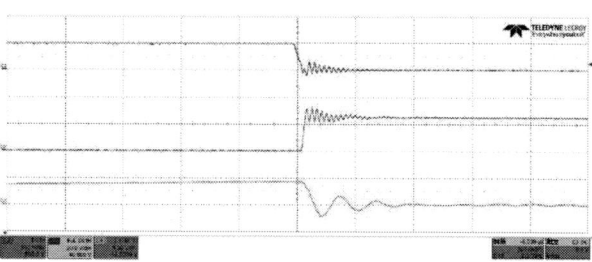

Fig. 12. Turn off waveform of SM with VSD at $R_{Gate,\,ext}$ =1.8Ω

TABLE I.   THREE CASES OF K>1, K<1, AND K=1

600V 40A	CSD		VSD			
	I_H	Total switching loss	$R_{Gate,ext}$	$R_{Gate,int}$	$R_{Gate,tot}$	Total switching loss
$K>1$			1Ω	1.8Ω	2.8Ω	1153μJ
$K=1$	2A	1360μJ	1.8Ω	1.8Ω	3.6Ω	1346μJ
$K<1$			7.2Ω	1.8Ω	9Ω	1811μJ

VI. CONCLUSIONS

This paper discusses the two most widely used gate drivers for SM, namely current source driver and voltage source driver, and constructs an evaluation index K to judge

978-1-6654-4817-8/21 $31.00 © 2021 IEEE

switching loss between SMs with CSD and VSD, which is affected by V_{TH}, $V_{GS,Miller}$, I_{LOAD}, I_H and $R_{Gate,tot}$ ($=R_{Gate,ext} + R_{Gate,int}$). The designed 600V/40A DPT well verifies the correctness of the model and the error range is tolerable. In actual situations, when the gate driving resistance of VSD is determined, the model can be used to calculate the critical CSD charging current value that makes the switching loss of SM with CSD lower than that with VSD.

REFERENCES

[1] Maksimovic D. " A MOS gate driver with resonant transitions," IEEE Power Specialist Conference (PESC'9 I), San Antonio, 1991.

[2] W. Eberle, Z. Zhang, Y. Liu and P. C. Sen, "A Current Source Gate Driver Achieving Switching Loss Savings and Gate Energy Recovery at 1-MHz," in IEEE Transactions on Power Electronics, vol. 23, no. 2, pp. 678-691, March 2008.

[3] Z. Zhang, J. Fu, Y. Liu and P. C. Sen, "Comparison of continuous and discontinuous Current Source Drivers for high frequency applications," 2010 IEEE Energy Conversion Congress and Exposition, 2010, pp. 2434-2440.

Power Cycling Capabilities of Bond Buffer Technologies for Wide Bandgap Power Devices

Nan Jiang
Institute of Semiconductors
Guangdong Academy of Sciences
Guangzhou, China
jiangnan@gdisit.com

Haitao Zhang
High-Frequency High-Voltage Device
and Integrated Circuits Center
Institute of Microelectronics of Chinese
Academy of Sciences
Beijing, China
zhanghaitao19@mails.ucas.edu.cn

Jianing Wang
College of Electrical Engineering and
Automation
Hefei University of Technology
Hefei, China
jianingwang@hfut.edu.cn

Chengguo Li
Institute of Semiconductors
Guangdong Academy of Sciences
Guangzhou, China
line 5: email address or ORCID

Jinhao Cai
College of Electrical Engineering and
Automation
Hefei University of Technology
Hefei, China
caijinhao@mail.hfut.edu.cn

Abstract—Wide bandgap power devices with excellent performance over traditional silicon power devices have been introduced as the prime candidate for power electronics applications. However, interconnections on the chip topside in the traditional packaging are now limiting the lifetime of wide bandgap power devices. It is necessary to replace Al bond wires with Cu bond wires, ribbons, and lead-frames with the help of bond buffer technologies to fulfill the requirements of wide bandgap power devices under high temperature operation conditions. The reliability performances of different bond buffer technologies and bonding materials are reviewed. The Cu bonding material shows the most robust power cycling capability. The weak point of the packaging of wide bandgap power devices has been changed from the bond material to the substrate or the metallization layer.

Keywords—bond buffer, reliability, SiC device, GaN device, packaging.

I. Introduction

In the power electronic applications, Si power devices which dominate for decades have reached the material limitation of silicon. Thus, wide bandgap (WBG) material with lower intrinsic carrier concentration, higher thermal conductivity, and larger saturated electron drift velocity has been introduced in this field [1]. With the outstand material parameter, WBG power chips with high switching speed, low power losses and high operation temperature were developed. However, the excellent performance of WBG power chips, especially the high operation temperature is limited by the state-of-the-art packaging technology.

Sinter technologies were applied to enhance the reliability of die-attaches for power devices operate under high temperature conditions [2][3]. The lifetime of chip topside interconnections can be extended by replacing the bond wire material from Al to Cu [4]. However, the Cu wire bonding technology requires a more robust metallization layer for the chip to avoid the damage of the chip during the bonding process [4]. Unfortunately, chips metallized with a thick Cu layer are not well available in the power semiconductor industry. Therefore, bond buffer technologies were developed to enable the ultrasonic welding process for

Cu bonded wires/ribbons for IGBT and SiC devices [5][6][7][8]. In this paper, several bond buffer technologies were reviewed, power cycling test results and failure mechanism were analyzed to figure out the suitable bonding technology for wide bandgap power devices under high temperature operation conditions.

Power Cycling (PC) tests are applied to estimate the lifetime of power devices, especially the reliability of the interconnections. The power device is mounted on a water/air cooled heatsink and a power source is applied to load current on the device. Then the chip loaded with current would heat up the device under test (DUT), which leads to the uneven temperature distribution inside the device. The temperature distribution combines with the difference of the coefficient of thermal expansion (CTE) of material would generate strain/stress concentration inside the DUT. By switching the power source on and off periodically, a junction temperature swing ΔTj is induced. During PC test, junction temperature, voltage, thermal resistance and power loss are monitored. The End of Life (EoL) criterion are 5% increase of forward voltage and 20% increase of thermal resistance. The cycle number to EoL (Nf) is used for reliability comparison and lifetime estimation.

In this paper, several bond buffer technologies were reviewed, PC test results and failure mechanisms were analyzed to figure out the suitable bonding technology for WBG power devices under high temperature operation conditions.

II. Bond Buffer Technology

A. ABB Bond Buffer

In the standard module, several thick aluminum wires are bonded on the thin aluminum metallization of the silicon chips with an ultrasonic bonding process. The mismatch of the thermal expansion coefficient (CTE) between bond wires and silicon chips combined with the thermal impact applied would lead to bond fatigues.

To improve the power cycling performance of the bond wires, a reliable bond buffer technology was introduced by A. Hamidi et al., in late 1990s, which consisted a thin Mo

---

Project supported by GDAS' Project of Science and Technology Development (Grant No. 2020GDASYL-20200102023）.

plate soldered on the chip as a strain buffer, and Al wires were bonded on the Mo plate to archive electrically contact [9].

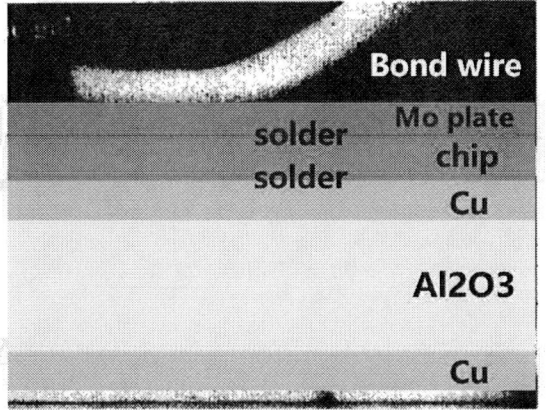

Fig.1. The bond buffer structure introduced by A. Hamidi [9].

In the power cycling test with test conditions of ton=0.9s, toff=1.3s and ΔTj=60K, an three times improvement of Nf performance was archived.

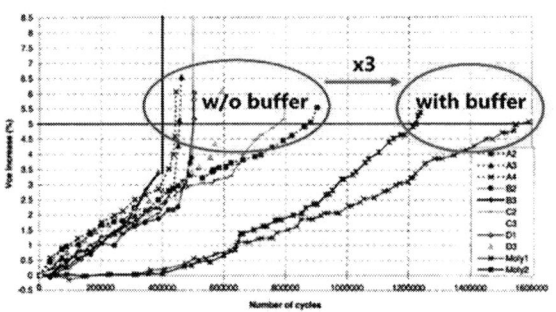

Fig.2. The improved power cycling performance of the bond buffer technology introduced by A. Hamidi [9].

The failure mode in this bond buffer technology was Rth increase and Vce increase. The SAM imaging revealed the solder delamination between the Mo plate and the chip surface. Thus, solder fatigue was confirmed as the main failure mechanism in the bond buffer technology.

Fig.3. The reinforced bond buffer bonded with Cu bond wires introduced by E. Özkol [7].

Cu wire bonding contains the advantage of the process flexibility of the wire bonding method. While due to high stiffness of Cu and large ultrasonic energy demands in the bonding process, the Cu wire bonding requires a much more robust top metallization in order to protect the chip against the damage of the fine structures under the bond pad. Thus in 2015, E. Özkol et al. introduced an improved bond buffer method by replacing the buffer layer from Mo to Cu [7]. The Mo plate applied in the previous method with the quite close CTE value to that of Si and SiC was able to reduce the stress at the interface. However, the poor thermal conductivity and the relatively high price of the Mo material restricted its applications. Thus, a 1 mm thick Cu block as the buffer layer was sintered on the chip to enable Cu wire bonding process, while the chip was soldered to the ceramic substrate [7]. The Cu block would reduce the temperature of the wires bonded on top of the plate due to the large heat capacity of the block. Thus would extend the power cycling lifetime of the bond wires. On the other hand, the Cu block would buffer the mechanical impacts of the wire bonding process to reduce the defect on the chip.

The power cycling tests showed an enhanced performance by a factor of 20 compared to the cycle number of standard power modules with conditions of ton=0.36s, toff=0.36s and ΔTj=60K. The degradation of the die attach during the power cycling test caused the increase of ΔTj, which led to the failure of the module.

B. Danfoss Bond Buffer

The DBB method which mainly consists of a thin Cu foil which is sintered onto the upper surface of the semiconductor. Furthermore the sintering technology was also used to replace the soldering of the connection between the die and the DBC substrate [5]. The thickness of the Cu foil was minimized compared to the previous mentioned ABB method in order to reduce thermal-mechanical stresses during power cycles.

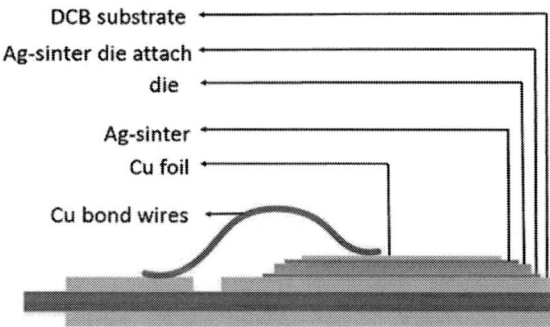

Fig.4 The bond buffer structure introduced by J. Rudzki [5].

By integrating the Ag-sintering and the Cu foil, the failure mechanism of the standard module such as bond wire lift-offs and solder delamination were eliminated.

Further research of the reliability performance of the DBB method was published in 2018 [10]. Test conditions of ton= 1s, ΔTj= 130 K were applied for the power cycling test. The modules bonded with Cu wires with the assist of DBB method show a significant increase of Nf by a factor of more than 60 compared to the Nf of the standard modules bonded with Al bond wires [10].

The Vce, Rth and ΔTj measurements of one DUT applied with DBB method during power cycling test are demonstrated in Fig.5. Vce with 5% of increase is observed

978-1-6654-4817-8/21 $31.00 © 2021 IEEE

in the figure, which indicates the failure on the topside of the chip. On the other hand, the Rth increase was caused by the degradation of Ag sintered die attaches. Both failures led to the increase of junction temperature swing during the test.

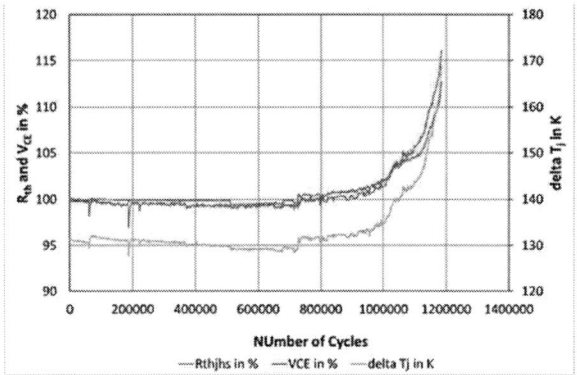

Fig.5. Vce, Rth and ΔTj trends of one DUT from Set 6 during power cycling test [10].

Failure analysis is demonstrated in Fig.6, an initial failure is related to the DCB substrate, which is similar as the reported test results of baseplate-free DUTs directly bonded with Cu bond wires [11]. The ceramic cracks combined with the delamination of the Cu layer of the substrate resulted in the crack formation within the Ag-sinter layer under the chip, which is the main reason of the Rth increase during the power cycling test.

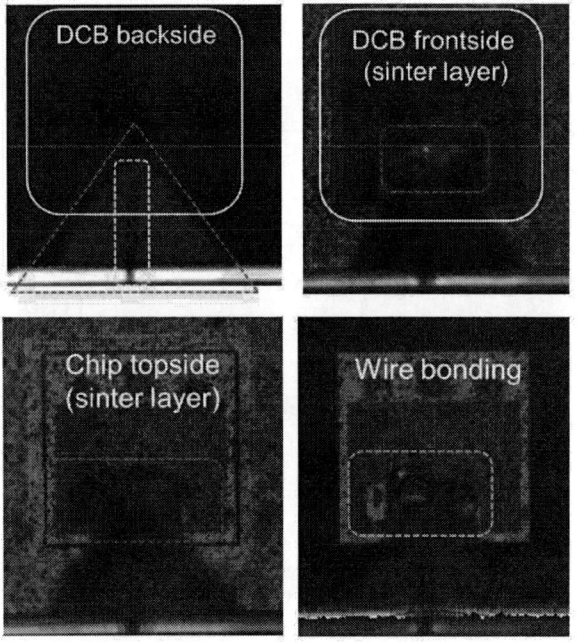

Fig.6. SAM images of the DUT with DBB method after power cycling test [10].

On the topside of the chip, the sintered Ag layer under the Cu foil degraded from the upper edge to the bottom side, which can be explained by the temperature distribution of the chip topside. The Cu wire bonding can only be observed on the remaining Cu foil on the chip. The degradation under the Cu foil is the main reason of the Vce increase during power

cycles, which is similar to the phenomenon of solder degradation of the standard module.

Cu ribbon was introduced to improve the reliability performance of DBB method [12]. 16 times increase of Nf compared to Al-bonded DUT was reached. The failure mode of Cu ribbon DUT is similar to what has been described from Cu wires bonded DUT [10]. The results confirm that the module reliability is not limited by the top interconnection with DBB method, but instead by the ceramic substrate and the metallization structure of the chip.

C. CIC Bond Buffer

Cu is an excellent buffer material for chip topside interconnection with high thermal/electrical conductivity. But the CTE of Cu is much higher compared to the value of Si. So a Cu-Invar-Cu composite material with the matched CTE with Si was presented. The Invar-core (FeNi36) has a both-sided Cu-plating with a total thickness of 150 μm (Fig.7). The layer-ration is 12.5/75/12.5 vol. % resulting in a 112.5 μm core with 2 x 18.75 μm Cu [13].

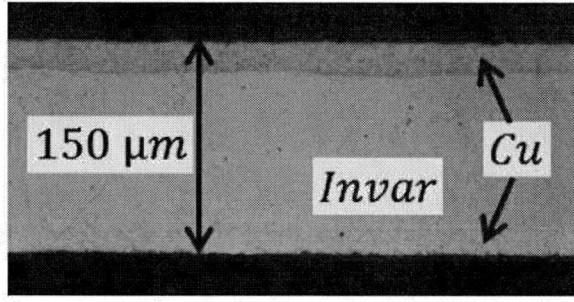

Fig.7. Cross section of the CIC composite material [13].

For the CTE value of the CIC composite material is constant between 75 °C and 225 °C (Fig.8). For higher values the Invar effect is not valid anymore and the CTE rises.

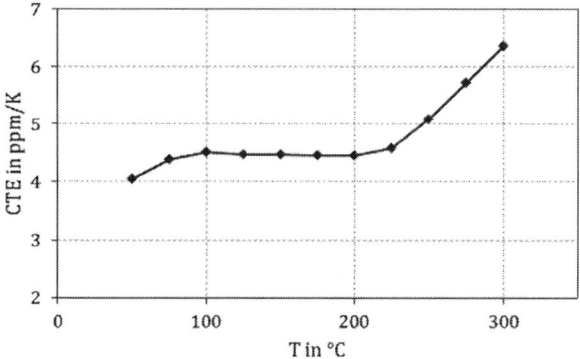

Fig.8. CTE of CIC-composite material over temperature [13].

The deformation in z-direction for all combinations with CIC and Cu ribbons were compared. The warpage is depending on the difference in CTE between the Si Chip and the ribbon material. Therefore the displacement for the Cu samples is larger than CIC samples. Higher stress during thermal cycling for the Cu ribbons is the consequence, which would lead to the shorter lifetime.

978-1-6654-4817-8/21 $31.00 © 2021 IEEE   8

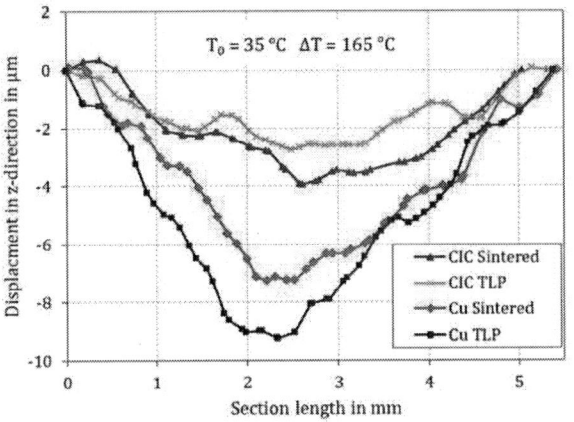

Fig.9. Deformation in z-direction along the measuring path compared for all materials for $\Delta T = 165\,°C$ [13].

For high temperature operations with high reliability, three major concepts such as CIC buffer, Cu-sintering and Cu wire bonding were applied for SiC module design, which showed an excellent CTE matching behavior in a wide temperature range [14].

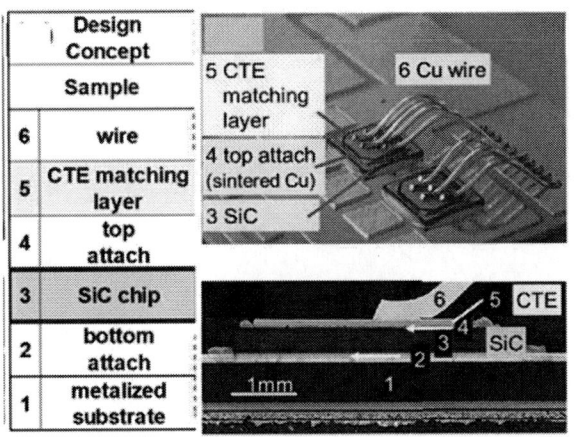

Fig.10. Module structure with Cu wire bonding, CIC buffer layer, SiC chip, Cu sintering and ceramic substrate [14].

Power cycling test with conditions of ton=1s, ΔTj=135 K, Tjmax=200°C was performed. The test focused on the lifetime of the layers close to the SiC chip, wire bonding, and the sintered joint materials. The power cycling lifetime of the SiC devices applied with CIC buffer materials under high temperature test conditions was improved by a factor of 100 compared to the soldered and Al wire bonded SiC devices [14].

Cross-sectional observations through a scanning electron microscope (SEM) were performed before and after the PCT. The analysis was conducted by focusing on bonded stitch of the Cu wire.

Although no wire lift-offs were observed in both cases, cracks were detected at edges of the wire. The clear differences of the crack propagation path are confirmed. In Sample C (Cu layer), the cracks pass through the two layers containing the Cu matching layer and sintered Cu, and they reach the chip surface. In Sample D (CIC layer), the cracks

stop at the upper Cu/Invar interface and propagate horizontally, which is due to the stiffness of the Invar.

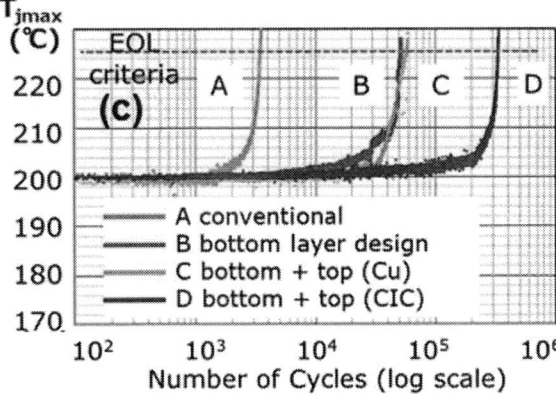

Fig.11. Power cycling results of DUTs with different bonding technology.

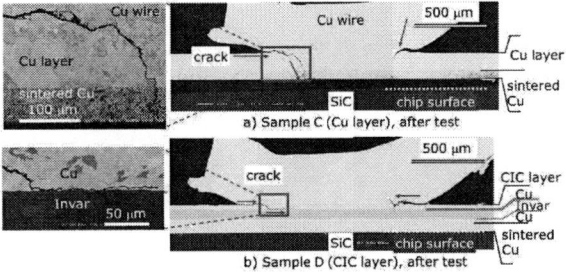

Fig.12. Longitudinal cross-sectional SEM images of DUTs applied with Cu buffer and CIC buffer after power cycling test.

III. Conclusion

Bond buffer technologies with a Mo plate and solder layers, a Cu foil and Ag-sintered layers, a Cu-Invar-Cu plate and Cu-sintered layers are evaluated in this paper. From the reported results, the bond buffer system consisted of a Cu-Invar-Cu plate and Cu-sintered layers showed the power cycling improvement of a factor of 100 compared to the conventional power packaging. Which is due to the matched CTE with SiC chip.

Different failure mechanisms were observed. The DBC substrate and the sinter layer became the weak point in the power device, which should be improved for the future development of the power device.

References

[1] M. N. Yoder, "Wide bandgap semiconductor materials and devices," IEEE Transactions on Electron Devices, vol. 43, no. 10, pp. 1633-1636, 1996

[2] U. Scheuermann et al., "The road to the next generation power module - 100% solder free design," In: Proc. CIPS2008, Nuremberg, Germany.

[3] K. Guth et al., "New assembly and interconnects beyond sintering methods," In: Proc. CIPS2010, Nuremberg, Germany.

[4] D. Siepe et al., "The future of wire bonding is? Wire bonding!," In: Proc. CIPS2010, Nuremberg, Germany.

[5] J. Rudzki et al., "Novel Cu-bond contacts on sintered metal buffer for power module with extended capabilities," In: Proc. CIPS2012, Nuremberg, Germany.

[6] A. Streibel et al., "Reliability of SiC MOSFET with Danfoss Bond Buffer Technology in Automotive Traction Power Modules," In: Proc. PCIM2019, Nuremberg, Germany.

[7] E. Özkol et al., "Enhanced power cycling performance of IGBT modules with a reinforced emitter contact," Microelectron. Rel., vol. 55, pp. 912–918, 2015.

[8] N. Hiroshi et al., "SiC module operational at 200 °C with high power-cycling capability using fatigue-free chip surface packaging technologies," In: Proc. PCIM2020, Nuremberg, Germany.

[9] A. Hamidi et al., "Reliability and lifetime evaluation of different wire bonding technologies for high power IGBT modules," Microelectron. Rel., vol. 39, pp. 1153–1158, 1999.

[10] N. Jiang et al., "Investigation of power cycling capability of a novel Cu wire bonded interconnection system," In: Proc. PCIMAisa2018, Shanghai, China.

[11] N. Heuck at al., "Lifetime Analysis of Power Modules with New Packaging Technologies," In: Proc. ISPSD2015, Hong Kong, China.

[12] S. Behrendt at al., "Feasibility of copper-based ribbon bonding as an assembly method for advanced power modules," In: Proc. PCIM2018, Nuremberg, Germany.

[13] M. Feißt at al., "Power Chip Interconnections Based on TLP, Sintering and CTE-Matched Conductors," In: Proc. CIPS2018, Stutgart, Germany.

[14] N. Notsu at al., "SiC module operational at 200 °C with high power-cycling capability using fatigue-free chip surface packaging technologies," In: Proc. PCIM2020, Nuremberg, Germany.

Modeling and Suppression of Crosstalk of SiC MOSFET in Bidirectional Buck/Boost Converter

Hao Zhang, Runquan Meng, Dingbang Zhang, Yingying Ding, Ziniu Wu
School College of Electrical and Power Engineering
Taiyuan University of Technology
Shanxi, China
1432441246@qq.com; mengrunquan@126.com; 592300037@qq.com; 49474619@qq.com; 723991319@qq.com

Abstract—**Due to the influence of parasitic parameters, there will be crosstalk when SiC MOSFET is used in bidirectional Buck/Boost converter, which affects the safe and reliable operation of the converter. Aiming at the crosstalk voltage of bidirectional Buck/Boost converter in operation, a simulation model with parasitic parameters is established under LTspice, and then the corresponding experimental test device is built to verify the correctness of the model. Finally, the influence of parasitic inductance parameters on crosstalk voltage is explored in LTspice model, and how to optimize the device and wiring layout to suppress crosstalk in bidirectional Buck/Boost converter is analyzed.**

Keywords—*SiC MOSFET，parasitic parameters，crosstalk suppression，Buck/Boost*

I. INTRODUCTION

With the development of intelligent distribution network, bi-directional DC-DC converter as an important device of power conversion, its reliability, efficiency, power density and other performance requirements have reached a new height. The traditional Si semiconductor devices can no longer meet the requirements of the future power grid. The third generation wide band gap semiconductor devices represented by SiC MOSFET have attracted the attention of scholars at home and abroad for their advantages of high voltage withstand, good thermal stability and low on impedance[1-3]. Due to its simple and efficient topology, bidirectional Buck/Boost converter has a good prospect in electric vehicles, energy storage systems and other fields[4]. However, with the introduction of SiC MOSFET devices and the improvement of switching speed and frequency, there will be crosstalk caused by parasitic parameters between the main Buck mode power transistor and the main Boost mode power transistor in bidirectional Buck/Boost converter. The crosstalk phenomenon here refers to that in the process of switching, a switch tube will cause the gate-source of other switches in the off state to produce a damped and oscillatory disturbance voltage, which will affect the normal operation of the converter[5]. Due to the higher switching speed of SiC devices, and the smaller difference between the gate turn-on threshold voltage of SiC MOSFET and the minimum allowable turn-off negative voltage [6], so the crosstalk problem is more serious in SiC devices. Its impact on the converter is mainly manifested in the positive and negative crosstalk voltage spikes. If the positive voltage spike is too high, the device will be misled to turn on, which will increase the switching loss and reduce the efficiency of the converter, or cause the bridge arm short circuit and harm the DC bus power supply. If the negative voltage spike is too large, the gate of the device may be damaged and the converter will not work normally[7].

At present, the influence of parasitic inductance, capacitance and drive circuit impedance on crosstalk needs to be further studied. The existing crosstalk suppression methods can be divided into three kinds[1]: The first is to reduce the switching speed of the power transistor. Although this method suppresses the crosstalk from the root, it is not conducive to give full play to the advantages of high switching speed and low switching loss of SiC devices, which will reduce the efficiency of the converter. The second method is to apply a suitable negative voltage to the gate and source of turn off power transistor, and ensure that both the positive and negative voltage spikes are within the safe range. However, the negative safe voltage that the gate and source of SiC MOSFET can withstand is generally within -10V, so it is difficult to meet the driving negative pressure that can be selected in this method. The third basic idea is to improve the drive circuit, change the gate drive circuit impedance value[8] and PCB routing, or provide a low impedance branch between the gate and source of SiC MOSFET to reduce the crosstalk voltage caused by displacement current[9]. Although this method is more complex, it is more feasible and practical than the first two methods.

In order to optimize the design of SiC MOSFET driver circuit, it is necessary to accurately grasp the influence of parasitic parameters on crosstalk. This paper studies the simulation model of SiC MOSFET with parasitic parameters from modeling, verifies the accuracy of the model through experiments, and then goes back to the model, changes the value of each parameter, and grasps its influence on crosstalk voltage, so as to provide the basis for the design of driving circuit.

II. CROSSTALK IN BIDIRECTIONAL CONVERTERS

Fig. 1(a) and (b) are the topological schematic diagrams of Buck and Boost circuits respectively. It can be seen that Buck circuit and Boost circuit are different in structure only in the position of inductor, switch tube and diode, but in function, one can step-down and the other can step-up. In the application of energy storage system, the battery usually has both charging and discharging functions, so the converter is required to be able to step-down and step-up. The original method is to use a step-down DC-DC converter and a step-up DC-DC transformer together. Although this method can meet the requirements, it takes into account the cost, reliability and efficiency, It is gradually replaced by Bidirectional DC-DC converter.

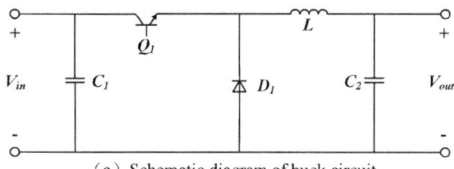

（a）Schematic diagram of buck circuit

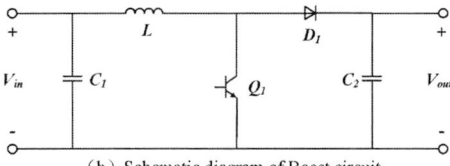

(b) Schematic diagram of Boost circuit

Fig. 1. Schematic diagram of Buck circuit and Boost circuit

As shown in Fig. 2, the bidirectional Buck/Boost converter is formed by replacing the power transistor and diode in Buck or Boost circuit with the combination of switch transistor anti parallel diode. When Q_1 turns on periodically and Q_2 turns off all the time, it can be equivalent to a Buck circuit, which can convert the high voltage input at $DC1$ side into the low voltage output at $DC2$ side, which is the step-down mode; when Q_1 turns off all the time and Q_2 turns on periodically, it can be equivalent to a Boost circuit, which can convert the low voltage input at $DC2$ side into the high voltage output at $DC1$ side, which is the step-up mode.

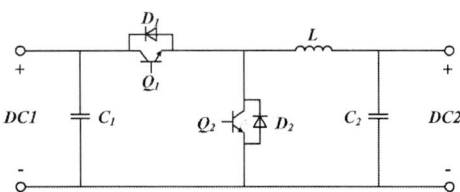

Fig. 2. Circuit diagram of bidirectional Buck/Boost converter

The bidirectional Buck/Boost converter can meet the demand of bi-directional power conversion with few power electronic devices, and has good reliability, efficiency and practicability. It has a broad application prospect in low-voltage applications. With the development of semiconductor devices, SiC devices gradually take the place of Si devices by virtue of its advantages of high voltage, low loss and good thermal stability. Generally speaking, the switching speed of SiC devices is more than ten times that of Si devices. Therefore, the influence of parasitic parameters caused by device packaging, pins and PCB routing can not be ignored. Because the switches Q_1 and Q_2 are a kind of bridge structure, when using SiC MOSFET, parasitic parameters will bring some problems at high switching speed, among which crosstalk is the most serious. For example, during the switching process, the periodically conducting switch Q_1 will produce crosstalk voltage to the gate and source of Q_2 which is always in the off state. In serious cases, Q_2 will be misled to turn on, resulting in short circuit of $DC1$ side power supply.

Fig. 3 shows the circuit structure of bidirectional Buck/Boost converter with parasitic parameters [1]. Among them, both Q_1 and Q_2 are SiC MOSFET, which are packaged as T0-247-3. The parasitic parameters of Q_1 can be divided into two parts. The first part is about the gate-drain parasitic capacitance C_{gd1}, the gate-source parasitic capacitance C_{gs1}, the gate-drain parasitic capacitance C_{ds1} and the gate parasitic resistance R_{g1_in}, drain parasitic inductance L_{d1_in}, gate parasitic inductance L_{g1_in} and source parasitic inductance L_{s1_in} inherent in the device package and pins. They cannot be changed from the moment the device is made. The second part is the parasitic inductance L_{d1_ex}, L_{g1_ex}, L_{s1_ex} caused by external PCB wiring, and the additional gate drive circuit impedance R_{g1_ex}, which can be changed in the use of the device. The reliability of the converter can be greatly improved by optimizing the design of PCB wiring. Similarly,

the parasitic parameters of Q_2 include the inter electrode parasitic capacitance C_{gd2}, C_{gs1}, C_{ds1} and gate parasitic resistance R_{g2_in} caused by packaging, parasitic inductance L_{d2_in}, L_{g2_in}, L_{s2_in}, and the parasitic inductors L_{d2_ex}, L_{g2_ex}, L_{s2_ex} in the external circuit, and the external gate drive circuit impedance R_{g2_ex}.

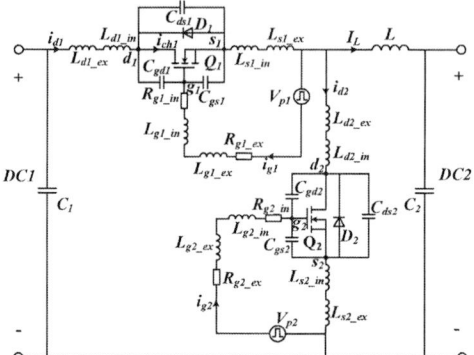

Fig. 3. Bidirectional Buck/Boost converter with parasitic parameters

III. PRINCIPLE ANALYSIS OF SWITCHING PROCESS

When the bidirectional Buck/Boost converter works in the Boost mode, the switch Q_2 turns on periodically, the diode D_1 turns on all the time, and the drain and source of the switch Q_1 bear the back voltage. Therefore, the influence of the positive crosstalk voltage can be ignored, and the negative crosstalk is the main adverse effect of the crosstalk voltage.

Taking the Buck mode of the converter as an example, the principle of crosstalk voltage in the switching process is explained. Fig.4 shows the piecewise linearization waveform of main parameters in the switching process of bidirectional Buck/Boost converter. For the convenience of explanation, let the total parasitic inductance of Q_1 drain $L_{d1} = L_{d1\_in} + L_{d1\_ex}$. Similarly, there are Q_1 gate and source total parasitic inductances L_{g1}, L_{s1} and gate total driving resistance R_{g1}, Q_2 drain, gate and source total parasitic inductances L_{d2}, L_{g2}, L_{s2} and gate total driving resistance R_{g2}.

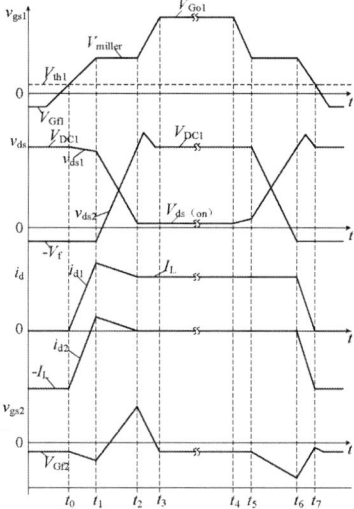

Fig. 4. Main parameters piecewise linearized waveform

978-1-6654-4817-8/21 $31.00 © 2021 IEEE

A. Turn-on process of Q_1

Stage 1 [t_0, t_1]: at t_0, the gate-source voltage v_{gs1} of power transistor Q_1 reaches its threshold voltage, and the conduction begins. According to Kirchhoff's law, equations (1) to (5) can be established. The drain current i_{d1} of Q_1 increases from zero, while the drain current i_{d2} of Q_2 decreases linearly from -I_L. There are voltages on the parasitic inductors L_{d1}, L_{s1}, L_{d2} and L_{s2}, and the drain-source voltage v_{ds1} of Q_1 decreases. Due to the existence of the parasitic inductor L_{s2}, the gate-source voltage v_{gs2} of Q_2 increases negatively. When i_{d1} increases to the inductance current I_L, the parasitic diode D_2 enters the reverse recovery phase, i_{d2} begins to increase in the positive direction, and i_{d1} continues to rise. When the current flowing through D_2 reaches the maximum value of its reverse recovery current, the first phase ends.

$$V_{DC1} = v_{ds1} + v_{ds2} + (L_{d1} + L_{s1})\frac{di_{d1}}{dt} + (L_{d2} + L_{s2})\frac{di_{d2}}{dt} \quad (1)$$

$$V_{Gf2} = v_{gs2} + L_{s2}\frac{di_{d2}}{dt} \quad (2)$$

$$i_{d1} = i_{ch1} + C_{ds1}\frac{dv_{ds1}}{dt} - C_{gd1}\frac{dv_{gd1}}{dt} \quad (3)$$

$$i_{d2} = i_{d1} - I_L \quad (4)$$

$$V_{Go1} = v_{gs1} + R_{g1}i_{g1} + (L_{g1} + L_{s1})\frac{di_{g1}}{dt} + L_{s1}\frac{di_{d1}}{dt} \quad (5)$$

In the formula, V_{DC1} is the input voltage at the high voltage side, V_{Gf2} is the turn-off voltage of Q_2, I_{ch1} is the channel current of Q_1, and V_{Go1} is the turn-on voltage of Q_1.

Stage 2 [t_1, t_2]: at t_1, the gate-source voltage v_{gs1} of Q_1 rises to the Miller voltage v_{miller}, and the reverse recovery current i_f of diode D_2 reaches its peak. After that, i_f gradually decreases, v_{ds1} gradually decreases, and the drain-source voltage v_{ds2} of Q_2 gradually increases. In this process, C_{gd2} charges, and the displacement current flowing through C_{gd2} is about $C_{gd2}dv_{ds2}/dt$. This current flows through the parasitic capacitor C_{gs2}, causing the gate potential of Q_2 to rise, and v_{gs2} begins to increase in a positive direction. With the decrease of v_{ds1}, v_{gs2} continues to increase, and equations (6) and (7) can be established. Until vds1 drops to the on state voltage $V_{ds(on)}$ of Q_1, the phase ends.

$$i_{g2} = C_{gs2}\frac{dv_{gs2}}{dt} + C_{gd2}\frac{dv_{gd2}}{dt} \quad (6)$$

$$V_{Gf2} = v_{gs2} + R_{g2}i_{g2} + (L_{g2} + L_{s2})\frac{di_{g2}}{dt} + L_{s2}\frac{di_{g2}}{dt} \quad (7)$$

In the formula, i_{g2} is the gate current of Q_2.

Stage 3 [t_2, t_3]: at t_2, v_{gs1} continues to rise from Miller voltage, and v_{ds2} rises to bus voltage V_{DC1}. Due to the effect of parasitic inductance and capacitance, v_{ds2} oscillates and overshoot, and finally stabilizes at bus voltage V_{DC1}. At the same time, the displacement current through C_{gd2} changes accordingly, C_{gs2} discharges, and Q_2 gate-source voltage decreases rapidly. At t_3, when v_{gs1} rises to Q_1 turn-on voltage V_{Go1}, Q_1 turns on completely, and this phase ends. At this time,

v_{gs2} returns to Q_2 turn off voltage V_{Gf2}, and Q_1 turn on process ends.

B. Turn-off process of Q_1

Stage 1 [t_4, t_5]: at t_4, when the off signal of Q_1 comes, the gate-source voltage v_{gs1} begins to decrease. According to Kirchhoff's law, there is equation (8). In this process, because v_{gs1} is always greater than the Miller voltage v_{miller}, Q_1 is still in the saturated conduction state.

$$V_{Gf1} = v_{gs1} + R_{g1}i_{g1} + (L_{g1} + L_{s1})\frac{di_{g1}}{dt} + L_{s1}\frac{di_{d1}}{dt} \quad (8)$$

In the formula, V_{Gf1} is turn-off voltage of Q_1.

Stage 2 [t_5, t_6]: at t_5, v_{gs1} decreases to v_{miller}, and the drain-source voltage v_{ds1} of Q_1 begins to increase from $V_{ds(on)}$, while the drain-source voltage v_{ds2} of Q_2 begins to decrease from V_{DC1}, with a change rate of dv_{ds2}/dt. When the parasitic capacitor C_{gd2} is discharged, the displacement current flowing through C_{gd2} is about $C_{gd2}dv_{ds2}/dt$, which flows through the parasitic capacitor C_{gs2}, resulting in the negative increase of v_{gs2}. With the decrease of v_{ds2}, C_{gs2} continues to charge and v_{gs2} continues to increase negatively. When v_{ds2} drops to -V_f, diode D_2 turns on and v_{gs2} reaches the peak value, and this phase ends.

Stage 3 [t_6, t_7]: at t_6, v_{ds2} drops to -V_f and maintains, diode D_2 starts to turn on, meanwhile i_{d1} decreases in positive direction, i_{d2} increases in negative direction, v_{gs2} decreases in negative direction. At the same time, v_{ds1} oscillates and overshoot under the influence of parasitic inductance and capacitance. When i_{d2} increases to - I_L, the phase ends. At t_7, v_{gs1} drops to the Q_1 threshold voltage. Then v_{gs1} gradually drops to the turn off voltage V_{Gf1}, v_{gs2} returns to V_{Gf2}, and the Q_1 turn off process ends completely.

IV. ESTABLISHMENT OF SIMULATION MODEL

In order to explore the influence of parasitic parameters on bidirectional Buck/Boost converter based on SiC MOSFET, the simulation model of converter with parasitic parameters is established under LTspice, as shown in Fig. 5, and its schematic diagram is shown in Fig. 3. Select Infineon's SiC MOSFET Power Transistor (IMW120R045M1) to extract the parasitic parameters. According to the datasheet, the main parameters are shown in Table I. The parasitic capacitances C_{gs}, C_{ds} and C_{gd} can be determined by equation (9).

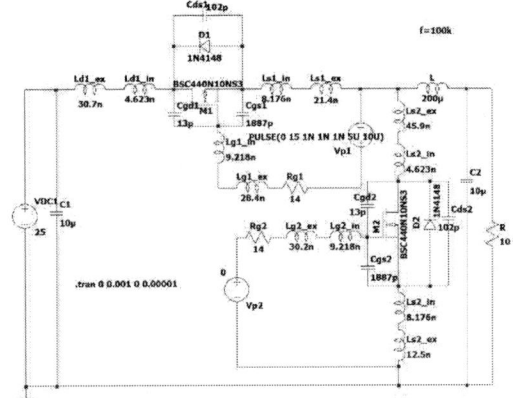

Fig. 5. Buck/Boost simulation model with parasitic parameters

$$\begin{cases} C_{gs} = C_{iss} - C_{rss} \\ C_{ds} = C_{oss} - C_{rss} \\ C_{gd} = C_{rss} \end{cases} \tag{9}$$

TABLE I.   MAIN PARAMETERS OF IMW120R045M1

Parameter	Value	Parameter	Value
$V_{GS(th)}$	4.5V	C_{iss}	1900pF
V_{Go}	15V	C_{oss}	115pF
V_{Gf}	0V	C_{rss}	13pF

The experimental device of bi-directional converter is built as shown in Fig. 6. The crosstalk voltage varies with the voltage and switching frequency. When the input voltage V_{DC1} increases in Buck mode, the crosstalk voltage spike on Q_2 also increases during Q_1 switching. In order to ensure the safety of the experiment, the input voltage is controlled at the appropriate value to verify the correctness of the model. The converter parameters are shown in Table II.

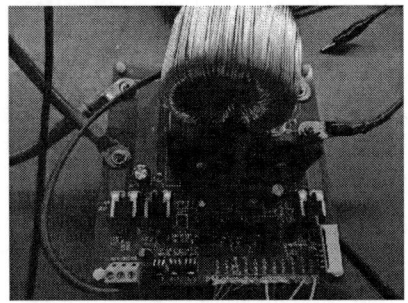

Fig. 6. Bidirectional Buck/Boost converter based on SiC

TABLE II.   PARAMETERS OF BIDIRECTIONAL BUCK / BOOST CONVERTER

Parameter	Value	Parameter	Value
V_{DC1}	25V	L	200μH
f	100kHz	C_1	10μF
R	10Ω	C_2	10μF

Fig. 7 and Fig. 8 show the simulation and experimental waveforms of Q_1 gate-source voltage v_{gs1}, drain-source voltage v_{ds1}, Q_2 gate-source voltage v_{gs2} and drain-source voltage v_{ds2} in the turn-on and turn-off process of Q_1 respectively.

It can be seen that the simulation results of the switch model can accurately reflect the actual operation of the converter. When the circuit works in the Buck mode, when Q_1 is turned on and off, due to the existence of parasitic parameters, there will be a damped and scillatory crosstalk voltage between the gate and source of Q_2, and with the increase of switching frequency and voltage level, the influence of crosstalk voltage will be intensified.

Among them, the reason why the simulation and experimental waveforms are not completely consistent is that there are inevitable errors between the parasitic parameters extracted by SiC MOSFET and the actual values, and the coupling inductance between the wires is not taken into account, and the device itself also has some manufacturing errors.

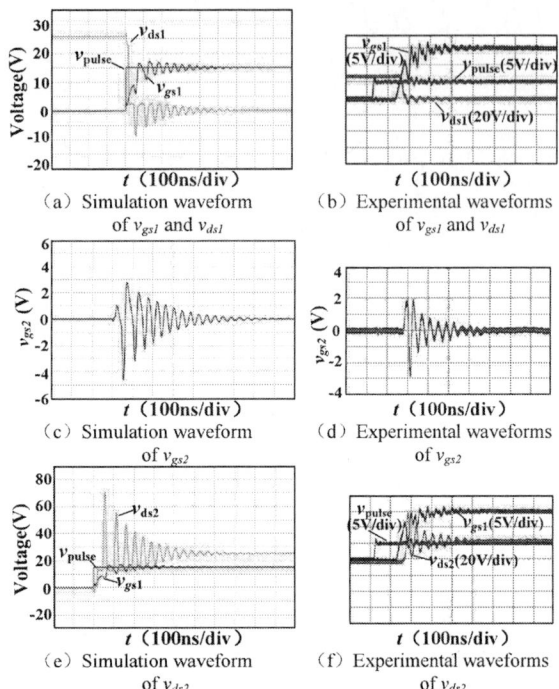

Fig. 7. Waveform comparison between simulation and experiment in turn on process

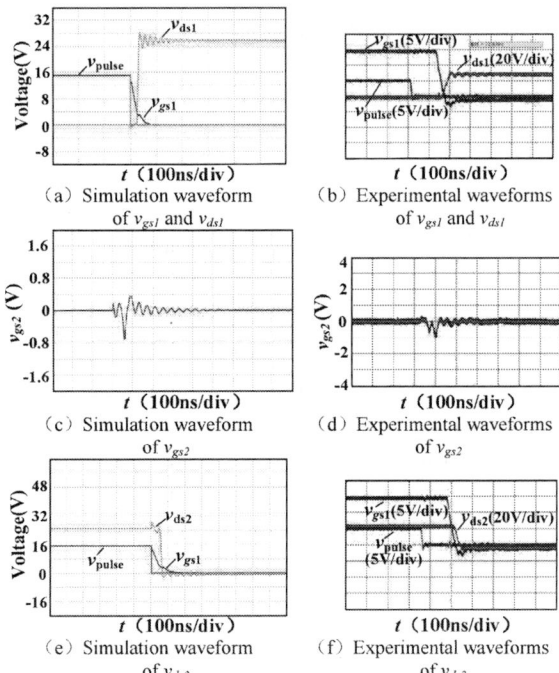

Fig. 8. Comparison of simulation and experimental waveforms in turn off process

V.   INFLUENCE AND SUPPRESSION OF PARASITIC PARAMETERS

Crosstalk voltage exists in the form of attenuation oscillation, with positive and negative voltage spikes. The values of L_{s1_ex}, L_{g1_ex}, L_{d1_ex}, L_{s2_ex}, L_{g2_ex} and L_{d2_ex} in LTspice model of bidirectional Buck/Boost converter with parasitic parameters are changed to observe the positive and negative voltage spikes of crosstalk voltage in Q_2 during the turn-on and turn-off of Q_1.

Fig. 9 and Fig. 10 show the positive crosstalk voltage spike v_{gs2_max+} between the gate and source of Q_2 in the process of Q_1 turning on and negative crosstalk voltage spike v_{gs2_max-} varies with parasitic inductance. Among them, Fig. 9(a) shows the influence of the changes of L_{s1_ex}, L_{g1_ex} and L_{d1_ex} on v_{gs2_max+}. It can be seen that L_{s1_ex} and L_{d1_ex} are negatively correlated with v_{gs2_max+}, while L_{g1_ex} is positively correlated with v_{gs2_max+}. However, the influence of these three parameters is very slight and can be ignored. Fig. 9(b) reflects the influence of the changes of L_{s2_ex}, L_{g2_ex}, and L_{d2_ex}, on v_{gs2_max+}. It can be seen that L_{s2_ex}, is positively correlated with v_{gs2_max+} and has a greater influence. A smaller change in L_{s2_ex}, can cause a significant change in v_{gs2_max+}. There was a negative correlation between L_{g2_ex} and v_{gs2_max+}. Increasing the value of L_{g2_ex} within a reasonable range could slightly inhibit v_{gs2_max+}. There was no correlation between L_{d2_ex} and v_{gs2_max+}.

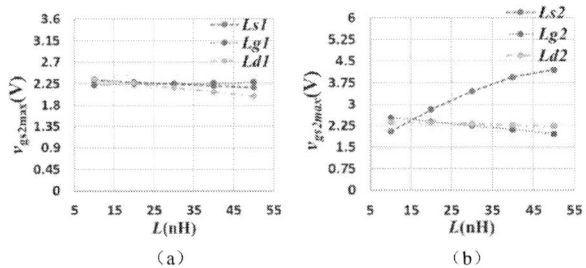

Fig. 9. Comparison of v_{gs2_max+} simulation results in Q_1 opening process

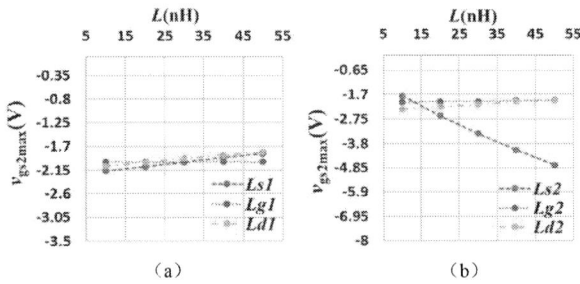

Fig. 10. Comparison of v_{gs2_max-} simulation results in Q_1 opening process

Fig. 10(a) shows the influence of the changes of L_{s1_ex}, L_{g1_ex} and L_{d1_ex} on v_{gs2_max-}. It can be seen that L_{s1_ex} and L_{d1_ex} are negatively correlated with v_{gs2_max-}. Increasing the values of L_{s1_ex} and L_{d1_ex} within a reasonable range can inhibit v_{gs2_max-} to a certain extent. There was no correlation between L_{g1_ex} and v_{gs2_max-}. Fig. 10(b) shows the influence of the changes of L_{s2_ex}, L_{g2_ex} and L_{d2_ex} on v_{gs2_max-}. It can be seen that L_{s2_ex} is positively correlated with v_{gs2_max-}, which is the main parasitic inductance parameter affecting v_{gs2_max-}. There was no correlation between L_{g2_ex} and v_{gs2_max-}. There is a negative correlation between L_{d2_ex} and v_{gs2_max-}. Increasing the value of L_{d2_ex} within a reasonable range can inhibit v_{gs2_max-}.

Fig. 11 and Fig. 12 show the variation of the positive and negative crosstalk voltage spikes between the gate and source of Q_2 with the parasitic inductance during Q_1 turn off. Similar

to the turn-on process, the crosstalk voltage spike is mainly affected by L_{s2_ex}, which has a positive correlation trend.

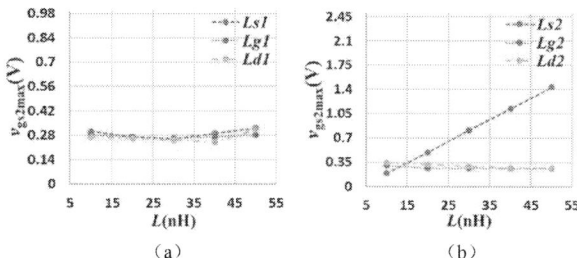

Fig. 11. Comparison of v_{gs2_max+} simulation results in Q_1 shutdown process

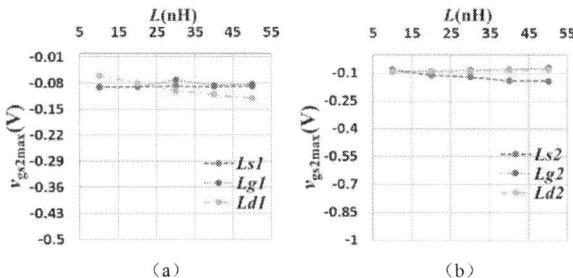

Fig. 12. Comparison of v_{gs2_max-} simulation results in Q_1 shutdown process

It can be found that the crosstalk voltage during the turn-on and turn off of Q_1 is mainly affected by the source parasitic inductance L_{s2_ex} of Q_2. The larger the L_{s2_ex} is, the greater the crosstalk voltage spike is and the greater the harm is. The influence of gate parasitic inductors L_{g1_ex} and L_{g2_ex} on the crosstalk voltage is very small, even can produce a small beneficial effect. And because the gate PCB wiring is to connect the low-voltage side control circuit and the high-voltage side main circuit, in order to prevent the creepage from the high-voltage side to the low-voltage side, the safety distance of the wiring can be given priority, and its impact on the crosstalk voltage can be ignored. In the process of device placement and PCB routing, the main consideration is to reduce L_{s2_ex} as much as possible, so as to prevent the crosstalk voltage spike from exceeding Q_2 turn-on threshold and misleading, resulting in short circuit on the power side and damaging the power supply and converter. Secondly, consider L_{s1_ex}, because in the process of Q_2 on and off, although the drain and source of Q_1 bear negative pressure and can not conduct, the negative crosstalk voltage may still break down the device gate, causing damage to SiC MOSFET and affecting the normal operation of the converter. Then consider L_{d1_ex} and L_{d2_ex}, and finally consider L_{g1_ex} and L_{g2_ex}.

VI. CONCLUSION

Aiming at the crosstalk phenomenon in the design of bidirectional Buck/Boost converter, this paper establishes a simulation model with parasitic parameters based on LTspice, and verifies the effectiveness of the model through the experimental test. Aiming at the influence of parasitic parameters caused by PCB routing on the positive and negative crosstalk voltage spikes, the corresponding parameters are changed under the model to master the influence law. The main conclusions are as follows:

978-1-6654-4817-8/21 $31.00 © 2021 IEEE

1）Bidirectional Buck/Boost converter mainly has two working modes. The crosstalk voltage in Boost mode is mainly affected by the negative crosstalk voltage spike, which may damage the SiC MOSFET.

2）In Buck mode, excessive positive crosstalk voltage spike may lead to device misoperation, and the influence of parasitic parameters caused by PCB routing on crosstalk voltage mainly comes from L_{s2_ex}. In the process of device placement and PCB routing, it is mainly considered to reduce L_{s2_ex} as much as possible, so as to prevent crosstalk voltage spike from exceeding Q_2 turn-on threshold and misoperation. Otherwise, the power supply side may be short circuited and the power supply and converter may be damaged.

ACKNOWLEDGMENT

This paper and its related research work are supported by Shanxi Province Science and Technology Major Special Project (20181102028);National Natural Science Foundation of China under Project (U1610121).

REFERENCES

[1] Ba Tengfei, Li Yan, Liang Mei. The Effect of Parasitic Parameters on Gate-Source Voltage of SiC MOSFET[J]. Transactions of China Electrotechnical Society, 2016,31(13):64-73.

[2] Ke Junji, Zhao Zhibin, Xie Zongkui, et al. Analytical Switching Transient Model for Silicon Carbide MOSFET under the Influence of Parasitic Parameters[J]. Transactions of China Electrotechnical Society, 2018,33(08):1762-1774.

[3] Li H , Zhong Y , Yu R , et al. Assist Gate Driver Circuit on Crosstalk Suppression for SiC MOSFET Bridge Configuration[J]. IEEE Journal of Emerging and Selected Topics in Power Electronics, 2020, 8(2):1611-1621.

[4] Wu Shaolei, Xiao Jianhong, Feng Yu, et al. Multi-mode control based bidirectional Buck-Boost DC/DC converter for energy storage[J]. Electronic Measurement Technology, 2019, 042(010):22-27.

[5] Liang Mei, Li Yan, Zhao Hongyan, et al. Analysis for Crosstalk of SiC MOSFET with Different Packages in a Phase-Leg Configuration and a Low Gate Turn-Off Impedance Driver[J]. Transactions of China Electrotechnical Society, 2017,32(18):162-174.

[6] Peftitsis D , Rabkowski J . Gate and Base Drivers for Silicon Carbide Power Transistors: An Overview[J]. IEEE Transactions on Power Electronics, 2016, 31(10): 7194-7213.

[7] Chen Ying, Li Chenmin, Lu Zhebie, et al. Modeling of SiC MOSFET Crosstalk Voltage in Half Bridge Circuit[J]. Journal of Chinese Electrical Engineering Science, 2020,40(06):1775-1787.

[8] Wu Lei, Liang Jian. SiC MOSFET Gate Drive Circuit Design Based on Dynamic Gate Resistance[J]. Electric Drive, 2020,50(11):117-121.

[9] Li Hui, Huang Zhangjian, Liao Xinglin, et al. An Improved Si C MOSFET Gate Driver Design for Crosstalk Suppression in a Phase-Leg Configuration[J]. Transactions of China Electrotechnical Society, 2019,34(02):275-285.

An Accurate Crosstalk Evaluation and Prediction Method for SiC MOSFET Considering Nonlinear Capacitance and Stray Parameters

Huaqing Li
Power Electronics and Renewable Energy Research Center
Xi'an Jiaotong University
Xi'an, P.R. China
lihuaqing@stu.xjtu.edu.cn

Chengzi Yang
Power Electronics and Renewable Energy Research Center
Xi'an Jiaotong University
Xi'an, P.R. China
lemonyang@stu.xjtu.edu.cn

Longyang Yu
Power Electronics and Renewable Energy Research Center
Xi'an Jiaotong University
Xi'an, P.R. China
382856693@qq.com

Haoyuan Jin
Power Electronics and Renewable Energy Research Center
Xi'an Jiaotong University
Xi'an, P.R. China
jinhaoyuan@stu.xjtu.edu.cn

Xingshuo Liu
Power Electronics and Renewable Energy Research Center
Xi'an Jiaotong University
Xi'an, P.R. China
liuxingshuo@stu.xjtu.edu.cn

Laili Wang
Power Electronics and Renewable Energy Research Center
Xi'an Jiaotong University
Xi'an, P.R. China
llwang@mail.xjtu.edu.cn

Abstract—In high-frequency applications of SiC MOSFET, crosstalk restricts the switching speed, increases additional switching losses and reduces the system stability. This paper proposes an accurate crosstalk evaluation and prediction method for SiC MOSFET, which considers nonlinear capacitance and stray parameters. The proposed method is a programmed prediction and evaluation process, including analysis of stray parameters, nonlinear capacitance analysis and measurement of nonlinear transfer capacitance. First, the influence of drain-to-gate stray capacitance $C_{dg'}$ caused by PCB layout and probes on crosstalk peak voltage is analyzed in details for the first time. Second, detailed analysis and explanation are made to reveal the disadvantages and inaccuracy of the traditional method which regards the reverse transfer capacitance as a constant. The simulation results show that the error of crosstalk peak voltage under different transfer capacitance values is very large. In addition, the influence of drain-source capacitance is also studied deeply. Third, this paper considers another challenging problem, the accurate measurement of nonlinear reverse transfer capacitance and propose a practical transfer capacitance big-signal measurement method for SiC MOSFETs based on dynamic transient, which can overcome the shortcomings of the frequency measurement method of small signal and shows high-performance and easy-operation. Finally, the performance of this proposed crosstalk evaluation and prediction method is verified and compared experimentally by a double-pulse test (DPT) platform.

Keywords—crosstalk evaluation, SiC MOSFET, stray parameters, nonlinear capacitance, big-signal measurement

I. INTRODUCTION

Compared to Si MOSFET, the inter-electrode capacitance is much smaller than SiC MOSFET, whose switching speed is faster and the change rate of drain-source voltage is relatively higher [1]. Therefore, in a phase-leg configuration, the upper and lower devices will interact with each other during the switching transient and generate voltage spikes between gate and source terminals, resulting in the device misleading or damage. The limiting gate voltage of SiC MOSFET is relatively low [2], and the gate voltage spike of switch transistor is easy to damage the device, which makes the high-frequency applications of SiC MOSFET face unprecedented

challenges [3]. Therefore, the study of crosstalk has been becoming a key point in the design of high frequency and high efficiency converters.

There are many scholars focus their attention on the suppression technology of crosstalk, which can be divided into two categories. One is passive suppression methods [4]; another is active suppression methodology. For the former, the traditional method is to directly paralleling the buffer capacitance with several nanofarads level between the gate and the source terminals, or use negative pressure shutdown. Furthermore, the authors of [5] propose an improved active crosstalk suppression capability, which could effectively suppress the crosstalk by increasing the gate source capacitance and reducing the gate resistance. Meanwhile, this method shows high performance for that additional capacitor is activated only when crosstalk occures, which ensures the high-speed performance of SiC MOSFET. In addition, the authors of [6] propose a new gate driver based on magnetic coupling to suppress the crosstalk voltage, which keeps the gate-source voltage in a safe range, whether it is positive or negative spurious pulse voltages.

It is worth noting that people pay too much attention on the suppression of crosstalk, which often increases the burden of driver design [7]. Furthermore, the new drive crosstalk structure always occupies a large space of the whole drive circuit. There are few people focus their attention on the evaluation and prediction of crosstalk spike voltage. Several key factors need to be considered in accurate crosstalk voltage prediction. First and foremost, the nonlinear transfer capacitance C_{dg} is an important source of crosstalk voltage. Accurate characterization and modeling of nonlinear transfer capacitance C_{dg} are the key factors to precise prediction [8]. In previous studies, the nonlinear C_{dg} capacitance was always regarded as a constant linear value, thus ignoring the influence of nonlinearity. In addition, conventional nonlinear capacitance modeling is based on the C-v_{ds} curve in off-state of the SiC MOSFET datasheet, which is inaccurate and unscientific. Therefore, practical and high-performance measurement methods should be studied deeply. Second, the stray parameters should to be discussed in details, which also does a great influence on the peak voltage of crosstalk.

The work was supported by 2018 steadily supports research projects from Key Laboratory of National Defense Science and Technology. Project number:614221720180401.

The corresponding author is Laili Wang (llwang@mail.xjtu.edu.cn).

978-1-6654-4817-8/21 $31.00 © 2021 IEEE

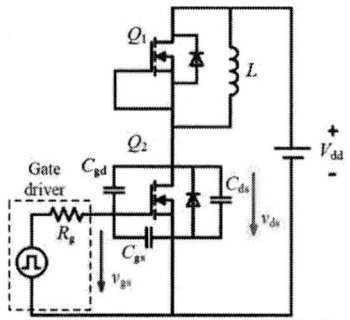

Fig. 1. Conventional double pulse test (DPT) circuit with a clamped inductive load.

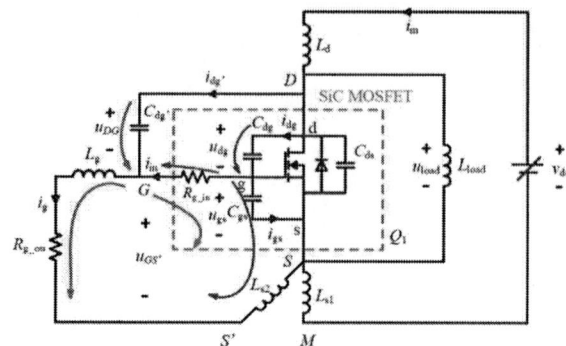

Fig. 2. The equivalent circuit of upper device Q_1 when the lower device Q_2 is turning on.

In this paper, an accurate crosstalk evaluation and prediction method for SiC MOSFET considering nonlinear capacitance and stray parameters is proposed. This method is a programmed prediction and evaluation process, including the discussion of stray parameters, nonlinear capacitance analysis and measurement of nonlinear transfer capacitance. It's worth noting the influence of parasitic parameters especially drain-to-gate stray capacitance $C_{dg'}$ caused by PCB layout and probes on crosstalk peak voltage is analyzed in details for the first time, which shows a great impact on the peak voltage of the crosstalk. In addition, detailed analysis and explanation are made to reveal the disadvantages and inaccuracy of the traditional method which regards the reverse transfer capacitance as a constant. The simulation results show that the error of crosstalk peak voltage under different transfer capacitance values is very large. Meanwhile, the influence of drain-source capacitance is also studied deeply. Lastly, we consider another challenging problem, the accurate measurement of nonlinear reverse transfer capacitance. We propose a practical transfer capacitance big-signal measurement method for SiC MOSFETs based on dynamic transient, which can overcome the shortcomings of the frequency measurement method of small signal and shows high performance and easy to operation. Finally, the performance of this proposed crosstalk evaluation and prediction method is verified and compared experimentally by a double-pulse test (DPT) platform, which shows well performance.

The rest of this paper is organized as follows: the impact of stray parameters and nonlinear capacitance on crosstalk is described and discussed in Section II. Section III presents the proposed practical transfer capacitance big-signal method. The performance of this proposed method on the crosstalk prediction and evaluation is verified and compared experimentally by a double-pulse test (DPT) platform in Section IV, and Section V shows the conclusions.

II. IMPACT OF STRAY PARAMETERS AND NONLINEAR CAPACITANCE ON CROSSTALK

A. Impact of stray parameters

Specifically, during the turn-on switching process of the lower device, the drain-to-source voltage of the upper device will increase sharply. A large transient current will be generated under the action of C_{dg}, $C_{dg'}$ and voltage change rate, which will inject into the gate driver circuit and give rise to voltage overshoot. Fig. 1 is a conventional half-bridge circuit topology, in which Q_1 and Q_2 are two SiC MOSFETs in the phase-leg configuration, while Q_1 is the synchronous device

and Q_2 is the device under test (DUT). Fig. 2 illustrates the equivalent circuit of upper device Q_1 when the lower device is turning on. A clamped inductance is connected in parallel with the source and drain electrodes of Q_1. L_d and L_{s1} are the bus parasitic inductance in main power loop, which can cause voltage oscillations and spikes. While $C_{dg'}$ is the stray capacitance between the gate driver loop and main power loop, which is resulting from layout and have a great impact on the overshoot voltage in gate driver loop. L_g and L_{s2} are the parasitic inductance in gate driver loop caused by PCB layout and component wiring. R_{g_in} is the inside gate resistance of the SiC MOSFET chip and the R_{g_ou} is the total resistance of the drive circuit, including the extra and internal resistance of the driver chip. v_{ds} is the drain-to-source voltage caused by the rapidly voltage change of the lower device Q_2. The energy stored in C_{dg}, and $C_{dg'}$ are released to the driver loop and gate-to-source capacitance C_{gs}, whose power flow direction is shown in the purple and red curves in Fig. 2. According to Kirchhoff current law, the equations of the node G and g can be expressed as:

$$\begin{cases} C_{dg'} \dfrac{du_{DG}}{dt} - i_g + i_{in} = 0 \\ C_{dg} \dfrac{du_{dg}}{dt} - C_{gs} \dfrac{du_{gs}}{dt} - i_{in} = 0 \end{cases} \quad (1)$$

Where i_{in} is the current flow through the internal resistance R_{g_in} of the SiC MOSFET chip and i_g is the current flow through L_g. u_{gs}, u_{dg} and u_{DG} are the voltage of C_{gs}, C_{dg} and $C_{dg'}$ respectively. According to Kirchhoff voltage law, the equations of mesh voltage can be expressed as:

$$\begin{cases} u_{DG} = u_{dg} + i_{in} R_{g\_in} \\ (L_g + L_{s2}) \dfrac{di_g}{dt} + i_g R_{g\_ou} + i_{in} R_{g\_in} - u_{gs} \dfrac{dC_{gs}}{dt} = 0 \\ L_d \dfrac{di_m}{dt} + L_{s1} \dfrac{di_m}{dt} + u_{load} - v_{ds} = 0 \end{cases} \quad (2)$$

Where i_m is the current flow through the internal L_g in the main power circuit and u_{load} is the voltage of L_{load}. $u_{GS'}$ is the overshoot voltage test point of crosstalk, which can be described as:

$$u_{GS'} = L_g \frac{di_g}{dt} + i_g R_{g\_ou} \quad (3)$$

978-1-6654-4817-8/21 $31.00 © 2021 IEEE

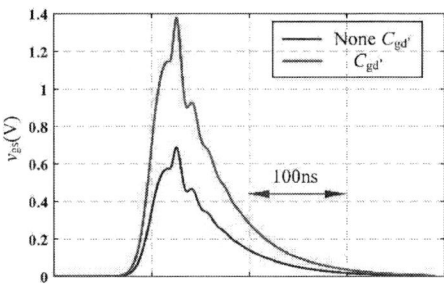

Fig. 3. The influence on crosstalk voltage spike with stray capacitance $C_{dg'}$ caused by PCB layout and probes.

In order to verify the effectiveness of the crosstalk voltage spike with stray capacitance $C_{dg'}$ caused by PCB layout and probes, a simulation related to crosstalk voltage of a phase-lag converter is carried out by using the software LT-Spice. In this analysis, 1.7kV 40A SiC MOSFETs from CREE/*Wolfspeed* with type C2M0080170P and a clamped inductance with 150μH are applied. The bus voltage V_{dd} is 600V, the gate driver voltage is 19V and the turn-off voltage is 0V. The reverse transfer capacitance is regarded as a constant as in previous study and the stray capacitance caused by PCB layout can be obtained from ANSYS Q3d. The stray capacitance caused by probes can be achieved by the analysis and modeling of probes and testing circuits [9]. Fig. 3 shows the corresponding results of influence on crosstalk voltage spike with stray capacitance $C_{dg'}$ caused by PCB layout and probes. It is evident to find that the crosstalk voltage spike difference is close to 1V compared with no parasitic parameters. So, in extreme conditions, the crosstalk voltage spike caused by stray parameters will damage the driver and converter.

From the above analysis, it can be concluded that the crosstalk mechanism is strongly coupled and complex due to the influence of parasitic parameters. Therefore, the parasitic parameters need to be fully captured and considered when predicting the peak crosstalk voltage.

B. Impact of Nonlinear Capacitance C_{dg}

The three inter-electrode capacitances C_{dg}, C_{ds}, C_{gs} of SiC MOSFET have a great influence on the switching transient process. In particular, the drain-to-source nonlinear capacitance C_{dg} plays an important role on the generation of crosstalk, which may directly affect the peak current and voltage.

It should be pointed out that in most of the previous studies on predicting crosstalk voltage, the nonlinear reverse transfer capacitor C_{dg} is always linearized, that is, C_{dg} is regarded as a constant C_{dg_con}, whose value can be obtained from datasheet. In this case, the maximum crosstalk peak voltage v_{gs_peak} can be roughly derived, that is [10]

$$v_{gs\_peak} = (R_{g\_in} + R_{g\_ou})C_{dg\_con} \cdot \frac{dv_{ds}}{dt} \cdot$$
$$(1 - e^{-\frac{V_{dd}}{\frac{dv_{ds}}{dt}C_{iss}(R_{g\_in}+R_{g\_ou})}}) \qquad (4)$$

Where R_{g_in} is the internal gate resistance of the SiC MOSFET chip and the R_{g_ou} is the total resistance of the drive circuit, including extra and internal resistance of the driver chip. V_{dd} is the bus voltage and C_{iss} is the input capacitance of the upper device, which can be described as

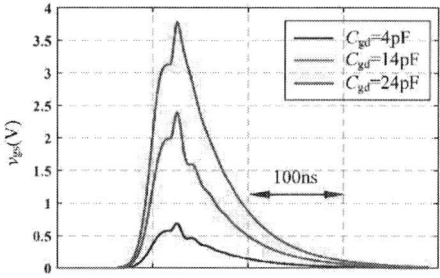

Fig. 4. The simulation of peak crosstalk voltage under different value of nonlinear C_{dg}

$$C_{iss} = C_{rss} + C_{gs} \qquad (5)$$

Actually, the nonlinearity of reverse transfer capacitance C_{dg} has a great influence on the peak crosstalk voltage. For 1.7kV 40A SiC MOSFET from CREE/*Wolfspeed* with type C2M0080170P, the typical capacitance is 4 pF. However, due to the fact that the value of C_{dg} is changing rapidly ranging from 0V to 900V with drain-source voltage v_{ds} from 700pF to 4pF. Therefore, the prediction error of crosstalk spike caused by the numerical error of nonlinear capacitance can't be ignored. The simulation of peak crosstalk voltage under different value of nonlinear C_{dg} is shown in Fig. 4. The bus voltage V_{dd} is 600V, the drive voltage is 19V and the turn-off voltage is 0V. The drive resistance is 20Ω. The comparison results show that the positive crosstalk peak caused by the constant transfer capacitance C_{gd} 14pF is 4 times of that by datasheet 4pF. Meanwhile, the positive crosstalk peak caused by the constant transfer capacitance C_{gd} 24pF is 1.6 times of that by 14pF. Therefore, the value of transfer capacitance C_{gd} is sensitive to the positive crosstalk voltage, which determines the crosstalk peak voltage and restricts the design and size of the gate driver. In order to accurately evaluate and predict crosstalk of the phase-leg configuration, accurate and simple capacitance characterization is need.

C. Impact of Nonlinear Capacitance C_{ds}

In order to further explore the essential influence factors of crosstalk voltage peak, drain-source capacitance C_{ds} is also considered. The value of drain-source capacitance C_{ds} changes rapidly with the increase of drain-source voltage v_{ds}, which is one of the important factors that affect the dynamic electrical characteristic curve. Meanwhile, the impact of nonlinear capacitance C_{ds} on the spike voltage of crosstalk needs to be analyzed in details. Fig. 5 shows the simulation results on the peak voltage of crosstalk under different values of C_{ds}. The results show that the value of C_{ds} have little real bearing on crosstalk voltage.

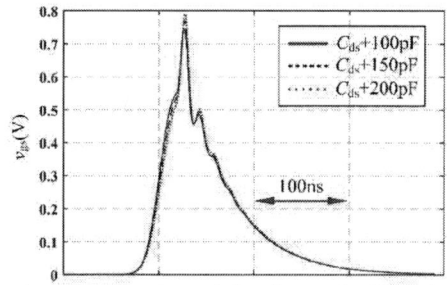

Fig. 5. The simulation of peak crosstalk voltage under different value of nonlinear drain-source capacitance C_{ds}.

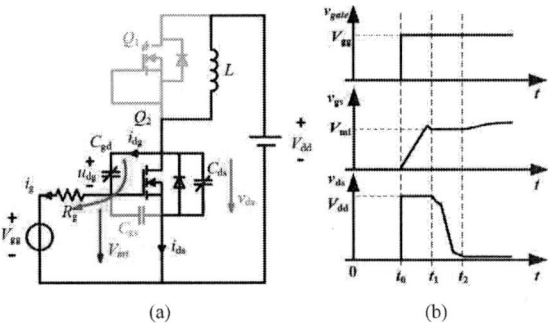

(a)   (b)

Fig. 6. (a) illustrates the turn-on process of DUT with a large drive resistance R_g and (b) displays the waveforms of the drain-source voltage v_{ds} and gate-source voltage v_{gs} in the turn-on process of the DUT.

III. PROPOSED PRACTICAL TRANSFER CAPACITANCE BIG - SIGNAL MEASUREMENT METHOD

A. Basic Principle of the Proposed Method

The proposed measurement method is based on a regular double pulse test (DPT) circuit with a clamped inductive load L, which is easy to design and operate, as is shown in Fig.1. Q_1 and Q_2 are two SiC MOSFETs in the phase-leg configuration, while Q_1 is the synchronous device and Q_2 is the device under test (DUT). Compared to previous static capacitance measurement, the proposed methodology focuses on the nonlinear capacitance during dynamic transient in an actual switching process, so the most realistic C_{dg} values can be captured directly. Fig. 6 (a) illustrates the turn-on process of DUT with a large drive resistance R_g Fig. 6 (b) displays the waveforms of the drain-source voltage v_{ds} and gate-source voltage v_{gs} in the turn-on process of the DUT. It is obviously that the gate-source voltage v_{gs} remains constant V_{mi} for a period of time during the switching process, which is called plateau area. C_{dg} can be obtained by only measuring the drain-source voltage v_{ds} and V_{mi} in plateau period that is convenient to operate. The nonlinear electric characteristics of C_{dg} element is

$$Q = C_{dg}(u_{dg})u_{dg} \qquad (6)$$

Where Q is the stored charge of C_{dg} and u_{dg} is the drain-gate voltage. The current i_{dg} flows through C_{dg} can be described as

$$i_{dg} = \frac{\partial Q}{\partial t} = \frac{\partial C_{dg}(u_{dg})u_{dg}}{\partial t} \qquad (7)$$

Due to the fact that v_{gs} remains constant during plateau period, all of the gate current i_g flows into C_{dg}, as is shown in Fig. 6 (a). In addition, i_g can be calculated in the gate drive loop

$$i_{dg} = i_g = \frac{V_{mi} - V_{gg}}{R_g} \qquad (8)$$

By putting transforming (3) into (2), that is

$$\frac{\partial C_{dg}(u_{dg})u_{dg}}{\partial t} = \frac{V_{mi} - V_{gg}}{R_g} \qquad (9)$$

In all of the previous research, the nonlinear capacitance C_{dg} was separated out directly from formula (7), that is $i_{dg} = C_{dg}du_{dg}/dt$. Actually, C_{dg} is a function of u_{dg} and changes rapidly that shows a strong nonlinear feature in the large signal model. As a result, this paper points out for the first time that C_{dg} can't be separated out independently in the

process of characterizing transfer capacitance and a new solution is presented as follows.

V_{ds} and time t present a monotonic functional relationship during plateau period. To narrow the initial time effect of plateau period and minimize the misalignment caused by probes propagation delay, t is normalized and standardized, described as a function of u_{dg}

$$t = \varphi(u_{dg}) \qquad (10)$$

Thus, put (10) into (9), that is

$$\frac{\partial C_{dg}(u_{dg})u_{dg}}{\partial \varphi(u_{dg})} = \frac{V_{mi} - V_{gg}}{R_g} \qquad (11)$$

The differential equation of formula (11) is transformed into

$$\frac{\partial C_{dg}(u_{dg})u_{dg}}{\partial u_{dg}} \cdot \frac{\partial u_{dg}}{\partial \varphi(u_{dg})} = \frac{V_{mi} - V_{gg}}{R_g} \qquad (12)$$

and

$$\frac{\partial C_{dg}(u_{dg})}{\partial u_{dg}} + \frac{1}{u_{dg}} C_{dg}(u_{dg}) = \frac{V_{mi} - V_{gg}}{R_g} \cdot \frac{1}{u_{dg}} \cdot \frac{\partial \varphi(u_{dg})}{\partial u_{dg}} \qquad (13)$$

Particularly, formula (13) is a general first order linear differential equation and it is easy to solve as

$$C_{dg}(u_{dg}) = \frac{C_0 + \dfrac{V_{mi} - V_{gg}}{R_g} \cdot \varphi(u_{dg})}{u_{dg}} \qquad (14)$$

Where C_0 is the initial constant introduced in the process of solving. Moreover, it should be noted that C_0 can be determined in the beginning of plateau area when drain-source voltage v_{ds} is about to drop. In addition, compared with drain-source capacitance C_{ds}, the value of C_{dg} is much smaller.

Obviously, the final expression (14) is very simple and practical, only by acquiring V_{mi} and data points of v_{ds}-t curve during plateau period, the value of C_{dg} varying with v_{ds} can be obtained continuously.

B. High-performance of the Proposed Method

The proposed practical transfer capacitance big-signal measurement method for SiC MOSFETs based on dynamic transient is easy to operation, which shows well performance that can achieve the real value of capacitance in dynamic switching process. First and foremost, the measurement circuit is simple and easy to design, only using the traditional double pulse test topology. Second, the curve of reverse transfer capacitance C_{dg} versus drain-source voltage v_{ds} in the datasheet of SiC MOSFET is a small signal method, which is measured under the signal of specific frequency, and the capacitance value will vary with the change of external signal frequency. Furthermore, three are too many passive components in the measurement circuit of the small signal method used in datasheet description, which is sensitive to frequency variation and small signal method applying specific high frequency signal is complicated in principle. Third, only v_{gs} and v_{ds} need to be measured by probes, other electrical quantities do not need to be achieved, which avoids complex voltage and current phase alignment. In addition, the

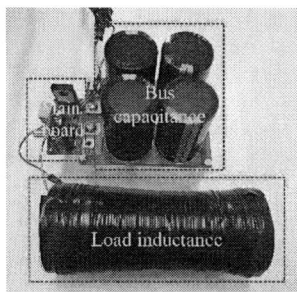

Fig. 7. The double pulse test workbench with a clamped inductive load.

 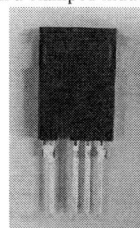

(a)   (b)

Fig. 8. (a) is the designed small-size gate driver and (b) is 1.7kV 40A SiC MOSFETs from CREE/*Wolfspeed* with type C2M0080170P.

final calculation formula of transfer nonlinear capacitance is simple, and the capacitance value changing continuously with voltage can be obtained only through the data points on v_{ds}-t curve in the process of switching transient.

IV. CROSSTALK PREDICTION AND EXPERIMENTAL VERIFICATION

In order to demonstrate the applicability of the proposed method, 1.7kV 40A SiC MOSFETs from CREE/*Wolfspeed* with type C2M0080170P are implied, as is shown in Fig. 8 (b). Moreover, the double pulse test workbench is established as is shown in Fig. 7. A clamped inductive load with 150 μH is connected in parallel with the upper device. The oscilloscope is MSO646-BW-1000A with a bandwidth of 1GHz and the high-voltage differential probes THDP0100 from Tektronix are used. In addition, the designed small-size gate driver is shown in Fig. 8 (a) with a short power loop, which is aimed at realizing the miniaturization of the converter. It's worth noting that to better observe the crosstalk phenomenon, a resistance of 20Ω are connected in parallel at both ends of the gate and source terminals of the upper device in the phase-leg configuration, which is different from the common double pulse test.

To verify the performance of the proposed method on the

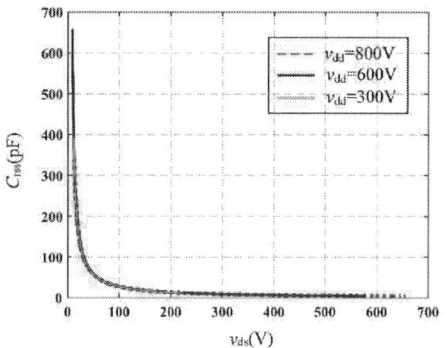

Fig. 10. The relationship of reverse transfer capacitance C_{rss} decreases and drain–source voltage v_{ds}.

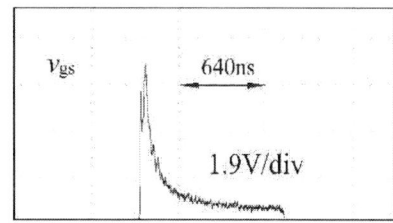

Fig. 11. The experimental result of the crosstalk waveform under the DC bias voltage 800V.

nonlinear reverse transfer capacitance measurement for different bus voltages, 800V and other test-voltage conditions are applied to the device, respectively. It should be noted that the driving resistance employed in each group of experiments is the same, that is R_g=200Ω. Fig. 9 (a)-(c) display the experimental waveforms of the gate-to-source voltage v_{gs} and the drain-to-source voltage v_{ds} under different types of DC bias voltage conditions. In Fig. 9 (c), during a certain period, the gate-to-source voltage v_{gs} remains unchanged and the Miller plateau keeps for about 80 ns during the turn on transient. As is shown in Fig. 10, from the experimental results, it can be found that reverse transfer capacitance C_{rss} decreases with an increase in the drain–source voltage v_{ds}. The experimental results indicate that the nonlinear reverse transfer capacitance measurement for different bus voltages corresponded with each other, which shows high consistency. The value of reverse transfer capacitance C_{rss} is a few larger than the value in datasheet about several picofarad on average.

In order to verify the correctness of the proposed crosstalk peaks evaluation and prediction, a simulation related to crosstalk voltage of a phase-lag converter is carried out by

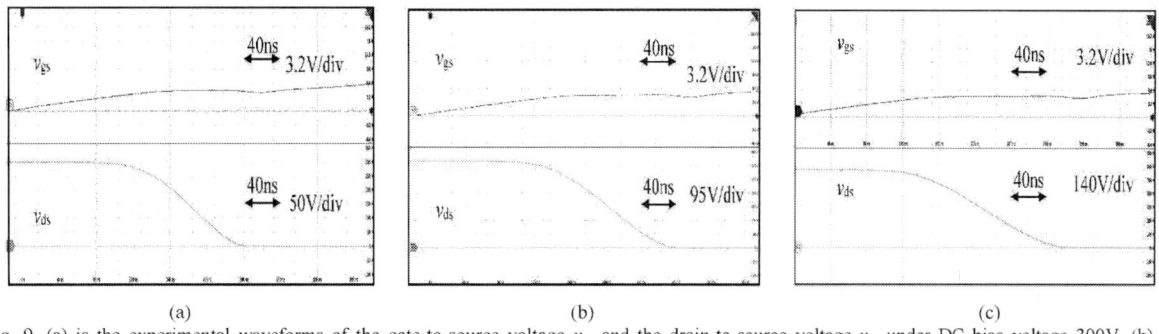

(a)   (b)   (c)

Fig. 9. (a) is the experimental waveforms of the gate-to-source voltage v_{gs} and the drain-to-source voltage v_{ds} under DC bias voltage 300V, (b) is the experimental waveforms of the gate-to-source voltage v_{gs} and the drain-to-source voltage v_{ds} under DC bias voltage 600V and (c) is the experimental waveforms of the gate-to-source voltage v_{gs} and the drain-to-source voltage v_{ds} under DC bias voltage 800V.

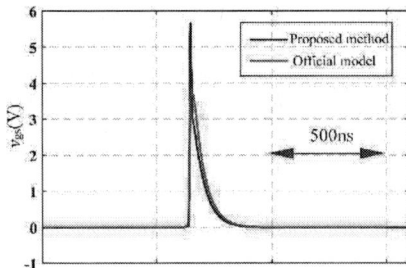

Fig. 12. The crosstalk spike prediction and evaluation results by the proposed method and official model, respectively.

using the software LT-Spice. The bus voltage V_{dd} is 800V, the gate driver voltage is 19V and the turn-off voltage is 0V, for which the first pulse will keep 2.875μs with the high level voltage 1.875μs and the second pulse will also keep 2.875μs with the low level voltage 1μs. It should be noted that the driving resistance R_g employed in this experiment is 200Ω. The stray capacitance caused by PCB layout can be obtained from ANSYS Q3d. The stray capacitance caused by probes can be achieved by the analysis and modeling of probes and testing circuits.

The experimental result of the crosstalk waveform is shown in Fig. 11, which shows a 6V crosstalk voltage spike and it damages the stability of the gate driver and converter. Fig. 12 shows the crosstalk spike prediction and evaluation results, in which the magenta line represents the simulation waveform by the official model and the blue line represents the simulation waveform by the proposed method. In addition, the simulation crosstalk peak voltage by the proposed method is 5.7V, having an error of 0.3V. Meanwhile the simulation crosstalk peak voltage by the official model is 3.8V, having an error of 2.2V. It is obvious that the simulation crosstalk waveform by the proposed method shows a good agreement with the experimental and testifies high performance of the improved method.

V. CONCLUSIONS

This paper proposes an accurate crosstalk evaluation and prediction method for SiC MOSFET considering nonlinear capacitance and stray parameters, which is a programmed prediction and evaluation process, including analysis of stray parameters, nonlinear capacitance analysis and measurement of nonlinear transfer capacitance. First, the influence of drain-to-gate stray capacitance $C_{dg'}$ caused by PCB layout and probes on crosstalk peak voltage is analyzed in details for the first time. Second, detailed analysis and explanation are made to reveal the disadvantages and inaccuracy of the traditional method which regards the reverse transfer capacitance as a constant. The simulation results show that the error of crosstalk peak voltage under different transfer capacitance values is very large. In addition, the influence of drain-source capacitance is also studied deeply. Third, we consider another challenging problem, the accurate measurement of nonlinear reverse transfer capacitance. This paper proposes a practical transfer capacitance big-signal measurement method for SiC MOSFETs based on dynamic transient, which can overcome the shortcomings of the frequency measurement method of small signal and shows high performance and easy to operation. Finally, the performance of this proposed crosstalk evaluation and prediction method is verified and compared experimentally by a double-pulse test (DPT) platform, which shows well performance.

REFERENCES

[1] W. Zhang, Z. Zhang, F. Wang, E. V. Brush and N. Forcier, "High-Bandwidth Low-Inductance Current Shunt for Wide-Bandgap Devices Dynamic Characterization," in *IEEE Trans. Power Electron.*, vol. 36, no. 4, pp. 4522-4531, April 2021.

[2] L. Zhang, X. Yuan, X. Wu, C. Shi, J. Zhang and Y. Zhang, "Performance Evaluation of High-Power SiC MOSFET Modules in Comparison to Si IGBT Modules," in *IEEE Trans. Power Electron.*, vol. 34, no. 2, pp. 1181-1196, Feb. 2019.

[3] J. Orti Gonzalez and O. Alatise, "Impact of BTI-Induced Threshold Voltage Shifts in Shoot-Through Currents From Crosstalk in SiC MOSFETs," in *IEEE Trans. Power Electron.*, vol. 36, no. 3, pp. 3279-3291, March 2021.

[4] H. Tang, H. Shu-Hung Chung, J. Wing-To Fan "Passive Resonant Level Shifter for Suppression of Crosstalk Effect and Reduction of Body Diode Loss of SiC MOSFETs in Bridge Legs," in *IEEE Trans. Power Electron.*, vol. 35, no. 7, pp. 7204-7225, July 2020.

[5] P. Wang, L. Zhang, X. Lu, H. Sun, W. Wang and D. Xu, "An Improved Active Crosstalk Suppression Method for High-Speed SiC MOSFETs," in *IEEE Trans. Ind. Appli.*, vol. 55, no. 6, pp. 7736-7744, Nov.-Dec. 2019.

[6] B. Zhang, S. Xie, J. Xu, Q. Qian, Z. Zhang and K. Xu, "A Magnetic Coupling Based Gate Driver for Crosstalk Suppression of SiC MOSFETs," in *IEEE Trans. Ind Electron.*, vol. 64, no. 11, pp. 9052-9063, Nov. 2017.

[7] C. Li et al., "High Off-State Impedance Gate Driver of SiC MOSFETs for Crosstalk Voltage Elimination Considering Common-Source Inductance," in *IEEE Trans. Power Electron.*, vol. 35, no. 3, pp. 2999-3011, March 2020.

[8] H. Li, Y. Jiang, Z. Qiu, Y. Wang and Y. Ding, "A Predictive Algorithm for Crosstalk Peaks of SiC MOSFET by Considering the Nonlinearity of Gate-Drain Capacitance," in *IEEE Trans. Power Electron.*, vol. 36, no. 3, pp. 2823-2834, March 2021.

[9] Z. Zeng, X. Zhang, F. Blaabjerg and L. Miao, "Impedance-Oriented Transient Instability Modeling of SiC mosfet Intruded by Measurement Probes," in *IEEE Trans. Power Electron.*, vol. 35, no. 2, pp. 1866-1881, Feb. 2020.

[10] Z. Zhang, F. Wang, L. M. Tolbert and B. J. Blalock, "Active Gate Driver for Crosstalk Suppression of SiC Devices in a Phase-Leg Configuration," in *IEEE Trans. Power Electron.*, vol. 29, no. 4, pp. 1986-1997, April 2014.

978-1-6654-4817-8/21 $31.00 © 2021 IEEE

Analysis of the Influence of Vibration and Thermal Vibration Coupling on the Power Module

Jiajia Guan
School of Electrical and Electronic Engineering
Huazhong University of Science and Technology
Wuhan,China
jiajiaguan@hust.edu.cn

Chi Zhang
School of Electrical and Electronic Engineering
Huazhong University of Science and Technology
Wuhan,China
chizhang_zach@hust.edu.cn

Cai Chen
School of Electrical and Electronic Engineering
Huazhong University of Science and Technology
Wuhan,China
caichen@hust.edu.cn

Yong Kang
School of Electrical and Electronic Engineering
Huazhong University of Science and Technology
Wuhan,China
ykang@hust.edu.cn

Abstract—**This paper uses finite element simulation to study the influence of the area, thickness and material of the substrate on the solder layer of the power module under a random vibration environment. The results show that when the area of the substrate increases, the thickness decreases, and the density increases, the stress of the solder layer increases and it is more prone to failure. In addition, the influence of random vibration on the power module is studied under the premise of thermal load. It is found that the thermal load will make the influence of random vibration on the solder layer smaller.**

Keywords—random vibration, thermo-vibration coupling, power module

I. INTRODUCTION

Automobiles, spacecraft, etc. are affected by vibration all the time during operation. The intensity of vibration is also an important criterion for judging whether the working environment is bad. According to the statistics of the US Air Force, about 55% of the faults related to the electronic equipment hardware and the working environment are related to structural vibration and shock [1]. However, there are still few studies on the impact of vibration on circuit boards or power modules. This leads to a lack of consideration of vibration when designing the circuit board or power module structure. Therefore, it is very important to do a research of the impact of vibration and shock on electronic equipment and power modules, and to optimize the structure of the electronic equipment and power modules based on the research results.

At present, there have been many researches on the influence of temperature and temperature cycle on power modules, and some papers study the effect of vibration on PCB circuit boards[2, 3,4]. D. Liang, V. Samavatian et al. [5,6] analyzed the effect of vibration direction of the stress of PCB circuit board, and the results showed that the stress is the smallest when the load direction is orthogonal to the circuit board; L. Tao [7]analyzed the random vibration stress on different assembly methods of multiple power modules and optimized the distributed assembly; Rajaguru P et al. [8] conducted a finite element simulation analysis on the influence of random vibration on IGBT power modules, and optimized the modeling and simulation methods, which significantly improved the simulation speed.

However, there are still few studies on the influence of structural vibration and shock on power modules, and there are very few studies on the influence of thermal field and vibration field coupling on power modules. Therefore, this article hopes to start with the impact of the vibration field on the power module, and expand it to do a preliminary analysis of the impact of the thermal field and the vibration field coupling on the power module based on this. Finally, a conclusion is given to guide the design of the power module.

This paper carried out the following works. Section II introduces the establishment and parameters of the model in this paper, and then analyzes its harmonic response. Section III first analyzes the solder layer stress changes with the substrate parameters under random vibration excitation, and then analyzes the solder layer stress changes corresponding to different substrate parameters under the prestress of the temperature field and random vibration excitation. Finally, a conclusion is reached through the comparative analysis of the results of multiple groups of experiments. Section IV is the conclusion and future work.

II. FINITE ELEMENT ANALYSIS FOR MODAL

A. Modaling and Set-ups

In order to simplify the analysis, this article establishes a two-chip power module model. The modal used in this study is as in Fig 1. The model includes seven layers, which are chip, chip solder layer, DBC top copper, ceramic, DBC bottom copper, substrate solder layer, and substrate. In addition, there are four positioning holes for imposing constraints. The materials and parameters of each layer are shown in Table I. The model is established in ANSYS EM, and then imported into ANSYS Workbench software for subsequent analysis.

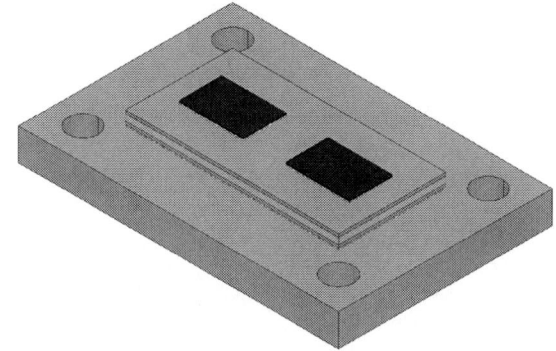

Fig 1 Power module model.

978-1-6654-4817-8/21 $31.00 © 2021 IEEE

TABLE I. MOUDLE COMPONENT MATERIALS AND PROPERTIES

Layers	Materials	ρ (kg/m3)	E(GPa)	V
Chip	SiC	3099.6	410	0.14
Chip Solder	SnAg	7400	65	0.3
DBC	Cu	8960	110	0.34
Ceramic	AlN	3250	300	0.24
Substrate Solder	SnAg	7400	65	0.3
Substrate	Cu	8960	110	0.34

When the power module is externally excited, it will make a forced movement. The frequency of the excitation must avoid the power module's natural frequency, otherwise it will cause resonance and cause huge damage to the power module. Before structural analysis, we need to know the first few natural frequencies of the module, which is the basis for subsequent analysis. Therefore, it is necessary to perform modal simulation first to know the natural frequency of the module. Considering the actual application, fixed supports are applied to the four positioning holes during the simulation, and Table II shows the obtained first six-order natural frequencies.

TABLE II. NATURAL FREQUENCIES OF VARIOUS MODES

Mode	Frequency(Hz)
1	26197
2	36201
3	53157
4	61137
5	63086
6	71616

According to the motion equation of the object:

$$[M]\{\ddot{X}\} + [C]\{\dot{X}\} + [K]\{X\} = \{F_{(t)}\} \quad (1)$$

Where [M] is quality matrix; {X} is displacement matrix; [C] is the matrix of damping; [K] is the matrix of Stiffness and {F(t)} is force matrix.

For a simple system, a single degree freedom system and without damping, the natural frequency $fn = \frac{1}{2\pi}\sqrt{\frac{K}{M}}$, for a damped systems $fd = fn\sqrt{1 - \varepsilon^2}$.

Where ε is the damping coefficient.

Therefore, stiffness, mass of the object, and damping coefficient all affect the natural frequency of the object. Before designing a power module, it is necessary to fully consider the possible working environment of the module and make adjustments to these three quantities so that the natural frequency of the module avoids the frequency of external excitation that may be suffered.

B. Harmonic Response Analysis of the Specifications

Harmonic response analysis is to apply a series of periodic sinusoidal excitations of different frequencies to a linear system, and analyze its periodic response (steady-state response) under periodic excitation, that is, it does not consider the transient response when the excitation is first added to the system. Through the harmonic response analysis, the natural frequencies and weak parts excited by the system under a specific load can be obtained, and the structural response in the whole process can also be obtained. In this paper, a sine frequency sweep experiment is performed on the power module in the three directions of +X, +Y, and +Z to observe the stress peaks at different frequencies.

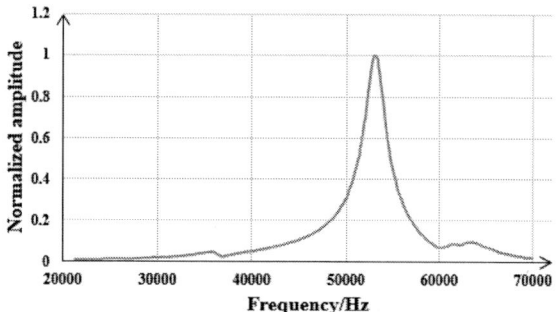

Fig 2 +X amplitude response.

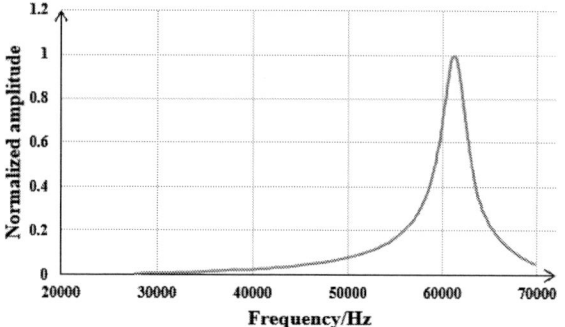

Fig 3 +Y amplitude response.

Since the first-order natural frequency and the sixth-order natural frequency of the power module are 26197 Hz and 71616 Hz, respectively, the frequency range of the sine sweep experiment is 20kHz-70kHz. The simulation results are shown in Figure 2-4. Through Figure 2, it can be seen that when excitation is applied in the +X direction, the stress peaks only at the third-order natural frequency, and the module bears the maximum stress. In Figure 3, when excitation is applied in the +Y direction, the stress peaks only at the fifth-order natural frequency, and the module bears the maximum stress. In Figure 4, when excitation is applied in the +Z direction, the stress at the first natural frequency and the fifth natural frequency appears maximum, and the stress peak appears at the first natural frequency. At this time, the module bears the maximum stress . To sum up, not modules bear the greatest stress at all natural frequencies, which is related to the direction of excitation. This can lead to the concept of participation coefficient. At the peak of stress, the participation coefficient is the largest. At other natural frequencies, the participation coefficient is small, so the stress is also small. Take the frequency sweep experiment in the +X direction as an example. At the third natural frequency, the

978-1-6654-4817-8/21 $31.00 © 2021 IEEE

power module participation coefficient in the +X direction is large, so the stress peak appears at this time. At other natural frequencies, the participation coefficient is small, so the stress is also small.

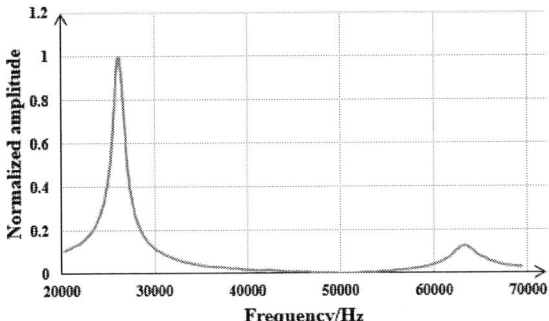

Fig 4 +Z amplitude response.

III. VIBRATION AND THERMAL VIBRATION COUPLING

A. Random Vibration Impact Analysis

In the power module, the most vulnerable layer is the solder layer. Therefore, this section mainly studies the changes in the solder layer stress when other factors change, hoping to provide guidance for the actual production process. Among the many layers of the power module, the easier to change is the substrate, including the thickness, area, and density of the substrate. Therefore, this section uses the controlled variable method to study the changes in the solder layer stress when these three key factors change, and give corresponding explanations according to the formula.

As a common test method, random vibration test is commonly used in the test standards of various components. According to the ECPE Guideline AGQ324 standard, this paper carries out random vibration test on the established model, and Table III is the tested power spectral density. The four positioning holes of the power module still impose fixed support constraints.

TABLE III. RANDOM VIBRATION POWER SPECTRAL DENSITY

Frequency(Hz)	PSD level ($(m/s^2)^2/Hz$)
10	10
100	10
300	0.51
500	20
2000	20

Firstly, when the substrate thickness changes, the chip solder stress and the substrate solder stress change trend are studied, and the results are shown in Figure 5. When the substrate's thickness increases, the chip solder stress and the substrate solder stress both show a downward trend. When the substrate thickness is small, the substrate solder stress is greater than the chip solder stress. When the substrate thickness is large, the chip solder stress is greater than the substrate solder stress, and an intersection occurs when the substrate thickness is equal to 2.8mm. Then, when the substrate density changes, the chip solder stress and the substrate solder stress change trend are studied, and the results

are shown in Figure 6. With the substrate's density increases, the chip solder stress and substrate solder stress both show an upward trend, and the chip solder stress has always been greater than the substrate solder stress. When the substrate thickness is small, the substrate solder stress and chip solder stress change little. When the substrate area changes, the comparison of chip solder stress and substrate solder stress is shown in Figure 7. It shows that once the substrate area increases, the chip solder stress and the substrate solder stress both increase, especially the substrate solder stress increases significantly. Therefore, in the actual design of the power module, the area of the substrate cannot be increased blindly to enhance the heat dissipation capacity, which may cause excessive solder stress on the substrate and failure.

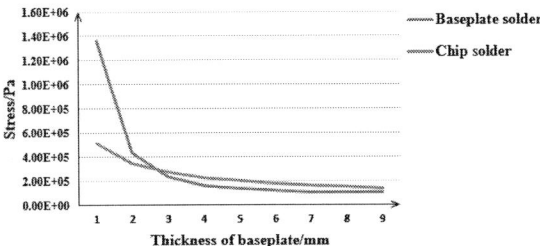

Fig 5 Solder stress vs Thickness of baseplate.

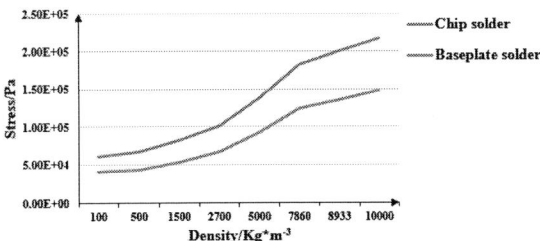

Fig 6 Solder stress vs Density of baseplate.

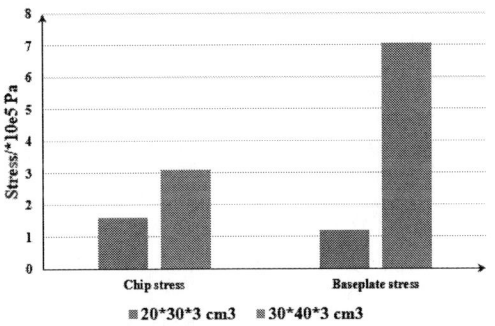

Fig 7 Solder stress vs Area of baseplate.

When a random vibration load is applied, the sharp corners of the solder layer will cause stress concentration, so the corners will have greater stress; Increasing the substrate's thickness, M is increases, and K increases, but $\sqrt{\frac{K}{M}}$ increases, the low-order natural frequency becomes larger, so the solder layer stress decreases and the power module's life is increases; Increasing the substrate's area, M is increases, and K decreases, $\sqrt{\frac{K}{M}}$ decreases, and the low-order natural frequency becomes smaller, so stress of the solder layer is increased.

978-1-6654-4817-8/21 $31.00 © 2021 IEEE   25

Then the power module's life is decreased; When only the baseplate material is changed, the M and $\sqrt{\frac{K}{M}}$ of the power module will change. The substrate should be made of material with a lower density to keep the low-order natural frequency away from the applied load frequency.

B. Analysis of the Influence of Thermal Field on Vibrations

The previous part studied the influence of random vibration on the power module. This is a study under a single physical field, which is not completely in line with the actual situation. Because in the actual working environment, the power module is not only affected by vibration, but also by the temperature field. Therefore, this section mainly studies how the influence of random vibration on the power module changes after the temperature field is added, that is, consider the interaction of the two physical fields. First, perform thermal simulation according to the loss of the chips and the dissipation conditions to obtain the heat generation of the power module and the stress of each part in the temperature field, and then use this as the prestress to conduct random vibration research.

In traditional thinking, temperature does not have any effect on vibration. However, according to the simulation results (Figure 8), the higher the chip power consumption, the higher the temperature of the power module, the lower the solder stress caused by vibration. That is, there is an interaction between the temperature field and the vibration field, and the two cannot be completely separated for analysis, and then the results are simply added as the final result. Especially when carrying out life prediction, it is not possible to add the individual stresses under the two physics fields as the final stress, and then estimate the life, which will make the result too small.

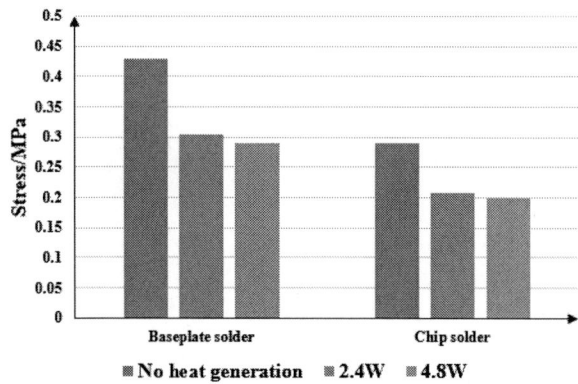

Fig 8 Effect of thermal field on stress of vibration field

The influence of the thermal field on the stress of the vibration field can be explained by the following two aspects: 1) Temperature will affect the elastic modulus of the material, thereby affecting the natural frequency and the stress under vibration. 2) Thermal stress will affect the structural rigidity of the power module. At this time, thermal stress is used as the

internal force of the structure and has an effect on the rigidity of the structure, thus affecting the stress under vibration.

IV. CONCLUSIONS AND FUTURE WORK

As there are few studies on the impact of vibration on power modules, this paper considers the structural design of power modules from a mechanical point of view, and mainly studies the impact of vibration on the stress of the solder layer of power modules. The results show that the substrate change has a greater impact on the solder layer of the power module. Specifically, as the substrate thickness increases, the solder layer stress decreases. As the substrate density increases, the stress of the solder layer is increased. When the substrate area increases, the solder layer stress increases.

This paper preliminarily studies the influence of temperature field on the stress of the solder layer under random vibration through simulation. The results show that when the heating power of the chip increased, the solder layer's stress gradually decreases under the same random vibration power spectrum density. Therefore, when predicting the life of the power module, the total stress should not be a simple superposition of the stress in the respective physical fields, and the interaction between temperature and vibration field should be considered. This is also the future research content.

REFERENCES

[1] Dave S. Steinberg. Vibration Analysis for electronic Equipment.Wiley-Interscience, 1973, pp. 134-145.

[2] Noor Muhammad, Zhigeng Fang, Muhammad Shoaib,Remaining useful life (RUL) estimation of electronic solder joints in rugged environment under random vibration,Microelectronics Reliability,Volume 107,2020,113614,ISSN 0026-2714,https://doi.org/10.1016/j.microrel.2020.113614.

[3] B. Wang et al., "Response prediction and verification for PCB with package due to thermal and random vibration coupling effects," 2009 4th International Microsystems, Packaging, Assembly and Circuits Technology Conference, Taipei, 2009, pp. 401-404, doi: 10.1109/IMPACT.2009.5382202.

[4] J. S. Karppinen, J. Li and M. Paulasto-Krockel, "The Effects of Concurrent Power and Vibration Loads on the Reliability of Board-Level Interconnections in Power Electronic Assemblies," in IEEE Transactions on Device and Materials Reliability, vol. 13, no. 1, pp. 167-176, March 2013, doi: 10.1109/TDMR.2012.2226462.

[5] D. Liang, Q. Wu, D. Ghaderi and J. M. Guerrero, "Analysis of Multilayered Power Module Packaging Behavior Under Random Vibrations," in IEEE Transactions on Components, Packaging and Manufacturing Technology, doi: 10.1109/TCPMT.2020.2995735.

[6] V. Samavatian, A. Masoumian, M. Mafi, M. Lakzaei and D. Ghaderi, "Influence of Directional Random Vibration on the Fatigue Life of Solder Joints in a Power Module," in IEEE Transactions on Components, Packaging and Manufacturing Technology, vol. 9, no. 2, pp. 262-268, Feb. 2019, doi: 10.1109/TCPMT.2018.2838148.

[7] L. Tao, X. Hui, Z. Junhua and W. Hongqin, "Random vibration simulation and structural optimization for DC/DC converter modules assembly," 2016 17th International Conference on Electronic Packaging Technology (ICEPT), Wuhan, 2016, pp. 616-619, doi: 10.1109/ICEPT.2016.7583209.

[8] Rajaguru P , Lu H , Bailey C , et al. Modelling and analysis of vibration on power electronic module structure and application of model order reduction[J]. Microelectronics Reliability, 2020, 110:113697.

978-1-6654-4817-8/21 $31.00 © 2021 IEEE

Improved Breakdown Characteristics for AlN/GaN/InGaN Coupling Channel HEMTs with SiN$_x$ Removal and Backfill Technique

Hao Lu
School of Microelectronics
Xidian University
Xi'an 710071, China
luhao@stu.xidian.edu.cn

Xiaohua Ma
School of Microelectronics
Xidian University
Xi'an 710071, China
xhma@xidian.edu.cn

Bin Hou
School of Microelectronics
Xidian University
Xi'an 710071, China
houbinme@163.com

Ling Yang
School of Microelectronics
Xidian University
Xi'an 710071, China
yangling@xidian.edu.cn

Yue Hao
School of Microelectronics
Xidian University
Xi'an 710071, China
yhao@xidian.edu.cn

Abstract—In this work, we report high breakdown characteristics of AlN/GaN/InGaN coupling channel HEMTs (CC-HEMTs) using by silicon nitride (SiN$_x$) passivation layer removal and backfill technique (PRB). Owing to the modulated barrier stress and blocked interface leakage path, the fabricated HEMTs exhibit excellent electrical characteristics, including a high breakdown voltage of 340 V. The small-signal measurements revealed that the current gain cutoff frequency (f_T) and maximum oscillation frequency (f_{max}) are 12.3 and 35.9 GHz, respectively. A high Johnson's figure of merit (JFoM) of 4.2 THz·V is achieved, benchmark against state-of-the-art (SOA) report of existing GaN HEMTs platform. These results have proved that the great potential of AlN/GaN/InGaN CC-HEMTs with reasonable device design applied for the next generation radio frequency front-end modules (FEMs) and power conversion integrated circuits.

Keywords—*AlN barrier; high breakdown field; coupling channel; Gallium nitride; HEMT*

I. INTRODUCTION

High breakdown field, high electron velocity, and excellent thermal management performance have enabled the GaN-based HEMT advantageous for high output power and high frequency (millimeter and sub-millimeter wave band) applications [1]. When the operating frequency is increased to millimeter-wave band (mmWave-beyond 30 GHz), the short channel effect, sensitive to the aspect ratio (L_g/t_{bar}) of transistors, will be more severe as the gate length (L_g) decreased. In order to maintain the aspect ratio, the thickness of the barrier should be decreased. However, the electron density (n_s) will be degraded with the t_{bar} decreased for the conventional barrier AlGaN barrier. Therefore, the selection of barrier material is preferring a strong polarization effect to alleviate the degradation of carrier transport with the barrier dimension variation. In view of this, AlN manifests the greatest advantage as a vertical scaling-down barrier material for mmWave applications due to its strong polarization effects and ultra-wide bandgap (UWBG) among group III-Nitrides barrier material [2]-[4]. However, the breakdown voltage of the AlN-based HEMTs is undesirable, due to the undesirable growth quality and high sheet charge density in high Al-content systems [5]. Recently, we proposed the AlN/GaN/InGaN strong coupling channel HEMT for enhanced transconductance and RF linearity [6]. However, the

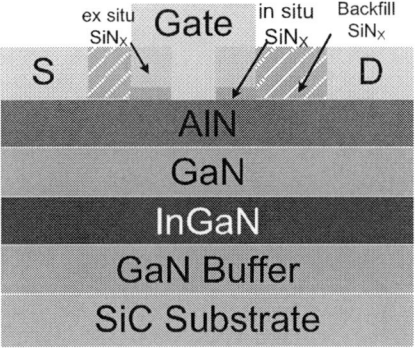

Fig. 1. (a) Cross-sectional schematic of the AlN/GaN/InGaN CC-HEMTs with the SiN$_x$ PRB process.

breakdown voltage of the CC-HEMTs still need to improve for their practical application.

In this work, we have reported high breakdown characteristics of AlN/GaN/InGaN coupling channel HEMTs (CC-HEMTs) using by *in-situ* silicon nitride (SiN$_x$) passivation layer removal and backfill technique (PRB). Benefitting from the modulated barrier stress and suppressed interface leakage path, the fabricated three-terminal HEMTs show excellent electrical characteristics, including a high breakdown voltage of 340 V. The RF measurements revealed that the current gain cutoff frequency (f_T) of 12.3 GHz and maximum oscillation frequency (f_{max}) of 35.9 GHz. A high Johnson's figure of merit (JFoM) of 4.2 THz·V was achieved, benchmark against state-of-the-art (SOA) report of existing GaN HEMTs platform. These results have proved that the AlN/GaN/InGaN CC-HEMTs with reasonable device design can be applied for the next generation radio frequency front-end modules (FEMs) and power conversion integrated circuits.

II. DESIGN CONCEPT OF THE CC-HEMT

As shown in Fig. 1, the epitaxial stacks are consisted of Fe-doped GaN buffer layer, unintentionally doped (UID) GaN layer, high-speed InGaN channel layer, GaN channel layer, UWBG AlN barrier, and *in situ* SiN$_x$ cap layer to enhance AlN barrier robustness and alleviate the relaxation [6]. The channel coupling degree and interface barrier height are modulated by the spatial distance between the two coupling channels and the Indium mole fraction, respectively. Hence, the design of the

978-1-6654-4817-8/21 $31.00 © 2021 IEEE

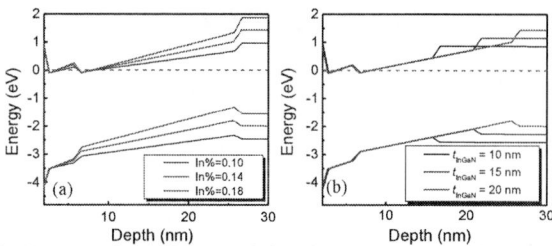

Fig. 2. Energy band diagrams of the AlN/GaN/InGaN coupling-channel heterostructures with (a) the In content and (b) the thickness of InGaN variation.

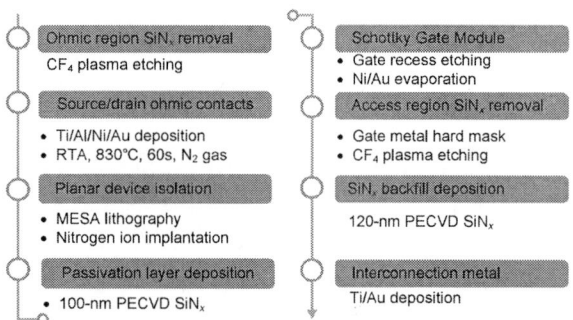

Fig. 3. Process flow of the CC-HEMTs with the PRB process. The Ni/Au gate metal stack was utilized as etching hard mask to remove in situ SiNx/ex situ SiNx passivation layer.

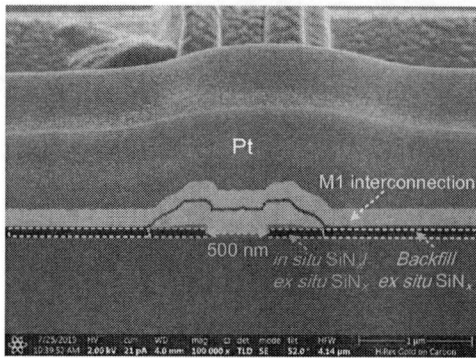

Fig. 4. Cross-sectional FIB/SEM micrograph of the fabricated 0.5-μm AlN/GaN/InGaN CC-HEMTs with the SiN$_x$ PRB process.

structural parameters is important for the coupling of the two channels.

Fig. 2 depicts the impact of InGaN channel thickness and Indium mole fraction on the energy band profile. As shown in Fig. 2 (a), When the In mole fraction is 18%, the barrier height between the two channels is increased, then the electron transfer between the two channels becomes difficult, resulting in a weak channel coupling. However, when the In mole fraction is 10%, the ΔE_C between the InGaN channel and UID GaN buffer is small. In this regime, the electron will spill over into the GaN buffer under a high electric field, resulting in source-drain punch-through and insufficient gate modulation efficiency. As shown in Fig. 2(b), when the thickness of the InGaN channel (t_{InGaN}) is thinner than 20 nm, the ΔE_C between the InGaN channel and UID GaN buffer is narrow, which will degrade the carrier confinement. Therefore, the proper thick InGaN channel should be considered. However, it should be noted that the actual growth should be also considered. The over-high In composition and over-thick InGaN channel will degrade the channel conductivity. The high In composition will degrade the mobility due to the increased carrier alloyed disorder scattering [1]. For the InGaN channel with a given In composition, the thicker of the InGaN channel is, the lower the mobility will be due to the strain relaxation. In summary, considering the carrier confinement, strong coupling and actual growth conditions, the t_{InGaN} of 20 nm and Indium content of 14% are selected. The structural factors of the coupling channel play a big role in linearity improvement. Therefore, the relatively low barrier height (ϕ_B) and the short the spatial distance between the channels within the proper range will increase the electron transfer, resulting in an improvement in linearity.

III. DEVICE STRUCTURE AND FABRICATION

Fig. 3 depicts the overall process flow for the AlN/GaN/InGaN CC-HEMT employed the PRB technique, which began with the in situ SiN$_x$ layer removal in the S/D region using CF$_4$ plasma etching for ohmic contact. Ti/Al/Ni/Au (20 nm/160 nm/55 nm/45 nm) metal stack without pre-metallization etch was then deposited by electron-beam evaporation and rapid thermally annealed at 830 °C for 50 s in N$_2$ ambient. Multi-energy nitrogen ion implantation was performed to accomplish planar electrical isolation, and 100 nm plasma-enhanced chemical vapor deposition (PECVD) SiN$_x$ layer was deposited to passivate the surface states. The T-gate foot was defined by NBL electron-beam lithography and etched using CF$_4$-based plasma to remove in situ SiN$_x$ and PECVD SiN$_x$, followed by Ni/Au (45 nm/255 nm) gate metal for the Schottky contact was deposited using physical vapor deposition (PVD). Then the *in situ* SiN$_x$/PECVD SiN$_x$ layers in the access region were removed using the gate metal stack as a hard mask [7]. It should be noted that the composite SiN$_x$ layers are reserved to ensure high channel conductivity under the gate. 120-nm PECVD SiN$_x$ as compressive stressor were backfilled onto the access region to modulate the barrier stress and block the leakage pathway causing from the *in situ* SiN$_x$. The devices have a gate-drain spacing (L_{gd}) of 3.4 μm, a gate length (L_g) of 500 nm, and a T-gate field plate with the extension above the SiN$_x$ layer of 1.3 μm. The focus-ion-beam scanning electron micrograph (FIB-SEM) imaging of the fabricated HEMT is revealed in Fig. 4. It can be observed that the deposition of the gate metal stack is uniform, the device dimension is consistent with the design value, and there is no obvious demarcation between the backfilled SiN$_x$ and the original SiN$_x$ under the gate.

IV. RESULTS AND DISCUSSION

The DC breakdown I-V characteristics were measured using by Agilent B1505 semiconductor parameter analyzer. As shown in Fig. 5(a), three terminal OFF-state breakdown measurement of AlN/GaN/InGaN CC-HEMT with the gate-drain spacing (L_{gd}) of 3.4 μm was indicated. The gate voltage for the breakdown test is set as $V_{GS} = V_{th} - 3$ V. The breakdown voltage defined at 1 mA/mm is 340 V. This indicates that the ultra-high breakdown field of 1 MV/cm is gained.

S-parameters of the AlN/GaN/InGaN CC-HEMTs were estimated using Agilent 8363B network analyzer, which the measured frequency sweeps ranging from 100 MHz to 40

978-1-6654-4817-8/21 $31.00 © 2021 IEEE

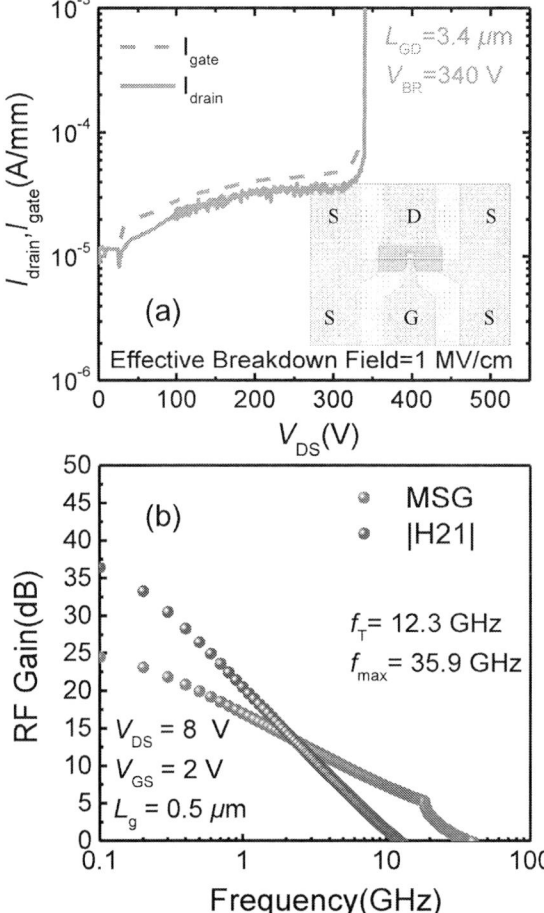

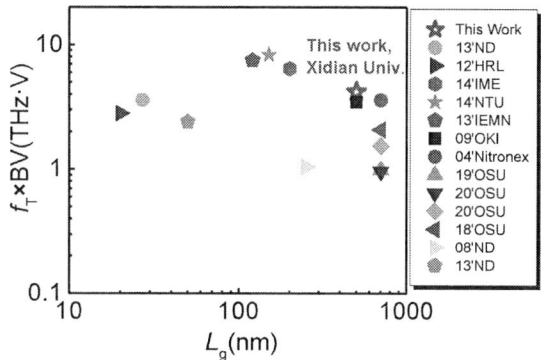

Fig. 6. Benchmark of $f_T \times$ BV versus gate length with the results reported for GaN-based HEMTs.

Fig. 5. (a) Three terminal off-state breakdown measurement of AlN/GaN/InGaN CC-HEMT with gate-drain spacing (L_{gd}) of 3.4 microns, indicates that the high breakdown field of 1 MV/cm. The inset is the device layout structure. (b) RF small signal gain as a function of measured frequency of the proposed devices at $V_{DS} = 8$ V.

GHz using an S-O-L-T on-wafer calibration. Fig. 5(b) demonstrated the small signal characteristics of the fabricated CC-HEMTs using the SiN$_x$ PRB process at $V_{DS} = 8$ V. The peak f_T and f_{max} of the 0.5-μm gate length coupling-channel HEMTs are 12.3 and 35.9 GHz, respectively, resulting in a high JFoM of 4.2 THz·V.

Fig. 6 lists the benchmark of $f_T \times$ BV versus gate length with the results reported for GaN-based HEMTs [8]-[18]. These results highlight a significant improvement of the J-FOM for the gate length larger than 500 nm in the present study.

V. CONCLUSION

In this article, we report high breakdown voltage of AlN/GaN/InGaN coupling channel HEMTs (CC-HEMTs) using by silicon nitride (SiN$_x$) passivation layer removal and backfill technique (PRB). Due to the modulated barrier stress and blocked interface leakage path causing by the *in situ* SiN$_x$, the fabricated HEMTs exhibit excellent electrical characteristics, including a high breakdown voltage of 340 V. The small-signal RF characteristics measurements revealed that the peak f_T and f_{max} are 12.3 and 35.9 GHz, respectively. A high Johnson's figure of merit of 4.2 THz·V is achieved,

benchmark against state-of-the-art (SOA) report of existing GaN HEMTs platform for the gate length larger than 500 nm. These results have proved that the great potential of AlN/GaN/InGaN CC-HEMTs with reasonable device design applied for the next generation radio frequency front-end modules (FEMs) and power conversion integrated circuits.

ACKNOWLEDGMENT

This work was supported by the National Key R&D Program of China under Grant 2018YFB1802100; in part by the National Natural Science Foundation of China under Grant 62104184; in part by the Nature Science Foundation of Shaanxi Province under Grants 2020JM-191 and 2018HJCG-20. The authors would like to thank Chupeng Yi, Ziyue Zhao, and Wenliang Liu for their help with the microwave test. The authors would also like to appreciate Gang Chen, Jiawei Zhang, Ning Liang, Juan Dou, and Xiufen Huang for their support for the device fabrication.

REFERENCES

[1] U. K. Mishra, L. Shen, T. E. Kazior and Y. Wu, "GaN-Based RF Power Devices and Amplifiers," in Proceedings of the IEEE, vol. 96, no. 2, pp. 287-305, Feb. 2008.

[2] K. Shinohara et al., "220GHz fT and 400GHz fmax in 40-nm GaN DH-HEMTs with re-grown ohmic," 2010 International Electron Devices Meeting, 2010, pp. 30.1.1-30.1.4..

[3] Y. Tang et al., "Ultrahigh-Speed GaN High-Electron-Mobility Transistors With fT/fmax of 454/444 GHz," in IEEE Electron Device Letters, vol. 36, no. 6, pp. 549-551, June 2015.

[4] K. Harrouche, R. Kabouche, E. Okada and F. Medjdoub, "High Power AlN/GaN HEMTs with record power-added-efficiency >70% at 40 GHz," 2020 IEEE/MTT-S International Microwave Symposium (IMS), 2020, pp. 285-288.

[5] H. Lee, D. Piedra, M. Sun, X. Gao, S. Guo and T. Palacios, "3000-V 4.3-mΩcm² InAlN/GaN MOSHEMTs With AlGaN Back Barrier," in IEEE Electron Device Letters, vol. 33, no. 7, pp. 982-984, July 2012.

[6] H. Lu et al., "AlN/GaN/InGaN Coupling-Channel HEMTs for Improved g$_m$ and Gain Linearity," in IEEE Transactions on Electron Devices, Early Access.

[7] T. Deguchi, T. Kikuchi, M. Arai, K. Yamasaki and T. Egawa, "High On/Off Current Ratio p-InGaN/AlGaN/GaN HEMTs," in IEEE Electron Device Letters, vol. 33, no. 9, pp. 1249-1251, Sept. 2012.

[8] M. L. Schuette, A. Ketterson, B. Song, E. Beam, T. M. Chou, M. Pilla, H. Q. Tserng, X. Gao, S. P. Guo, P. J. Fay, H. G. Xing, and P. Saunier, "Gate-recessed integrated E/D GaN HEMT technology with fT/fmax >300 GHz," IEEE Electron Device Lett., vol. 34, no. 6, pp. 741–743, Jun. 2013.

[9] K. Shinohara, D. Regan, A. Corrion, D. Brown, V. Lee, P. M. Asbeck, I. Alvarado-Rodriguez, M. Cunningham, C. Butler, A. Schmitz, S. Kim, B. Holden, D. Chang, A. Margomenos, and M. Micovic, "Deeply-

scaled E/D-mode GaN-HEMTs for sub-mm-wave amplifiers and mixed-signal applications," in *Proc. CSICS*, Oct. 2012, pp. 1–4.

[10] S. Huang et al., "High-f_{MAX} High Johnson's Figure-of-Merit 0.2-um Gate AlGaN/GaN HEMTs on Silicon Substrate With AlN/SiNx Passivation," in IEEE Electron Device Letters, vol. 35, no. 3, pp. 315-317, March 2014.

[11] K. Ranjan, S. Arulkumaran, G. I. Ng, and S. Vicknesh, "High Johnson's figure of merit (8.32 THz·V) in 0.15 μm conventional T-gate AlGaN/GaN HEMTs on silicon," *Appl. Phys. Express*, vol. 7, pp. 044102-1–044102-4, Mar. 2014.

[12] F. Medjdoub, B. Grimbert, D. Ducatteau, and N. Rolland, "Record combination of power-gain cut-off frequency and three-terminal breakdown voltage for GaN-on-Silicon devices," *Appl. Phys. Express*, vol. 6, pp. 044001-1–044001-4, Apr. 2013.

[13] S. H. Sohel et al., "Polarization Engineering of AlGaN/GaN HEMT With Graded InGaN Sub-Channel for High-Linearity X-Band Applications," in IEEE Electron Device Letters, vol. 40, no. 4, pp. 522-525, April 2019.

[14] S. H. Sohel et al., "Linearity Improvement With AlGaN Polarization-Graded Field Effect Transistors With Low Pressure Chemical Vapor Deposition Grown SiNx Passivation," in IEEE Electron Device Letters, vol. 41, no. 1, pp. 19-22, Jan. 2020.

[15] S. H. Sohel et al., "X-Band Power and Linearity Performance of Compositionally Graded AlGaN Channel Transistors," in IEEE Electron Device Letters, vol. 39, no. 12, pp. 1884-1887, Dec. 2018.

[16] J. W. Johnson, E. L. Piner, A. Vescan, R. Therrien, P. Rajagopal, J. C. Roberts, J. D. Brown, S. Singhal, K. J. Linthicum, "12 W/mm AlGaN-GaN HFETs on silicon substrates," in IEEE Electron Device Letters, vol. 25, no. 7, pp. 459-461, July 2004.

[17] S. Hoshi, M. Itoh, T. Marui, H. Okita, Y. Morino, I. Tamai, F. Toda, S. Seki, and T. Egawa, "12.88 W/mm GaN High Electron Mobility Transistor on Silicon Substrate for High Voltage Operation," Applied Physics Express, vol. 2, no. 6, pp. 061001, June. 2009.

[18] T. Zimmermann, D. Deen, T. Cao, J. Simon, P. Fay, D. Jena, and H. G. Xing, "AlN/GaN Insulated-Gate HEMTs With 2.3 A/mm Output Current and 480 mS/mm Transconductance, " in *IEEE Electron Device Letters*, vol. 29, no. 7, pp. 661-664, July 2008.

Analysis of GaN HEMT Degradation under RF Overdrive Stress

YuHan Xie
The School of Electronic and Information Engineering
South China University of Technology
Guangzhou, China
201921012524@mail.scut.edu.cn

RongSheng Chen
The School of Electronic and Information Engineering
South China University of Technology
Guangzhou, China
chenrs@scut.edu.cn

Chang Liu
The Science and Technology on Reliability Physics and Application of Electronic Component Laboratory
The No.5 Electronics Research Institute of the Ministry of Industry and Information Technology
Guangzhou, China
xd_liuchang@163.com

YiQiang Chen
The Science and Technology on Reliability Physics and Application of Electronic Component Laboratory
The No.5 Electronics Research Institute of the Ministry of Industry and Information Technology
Guangzhou, China
yiqiang-chen@hotmail.com

Yan Ren
The Science and Technology on Reliability Physics and Application of Electronic Component Laboratory
The No.5 Electronics Research Institute of the Ministry of Industry and Information Technology
Guangzhou, China
reny@ceprei.com

Abstract—This work concentrated at the degradation characteristics of GaN HEMT devices when exposed to RF overdrive stress. According to results of the experiment, the drain current of GaN HEMT decreases by 35.6% from 877 mA to 564mA after RF overstress. Moreover, after the experiment, the threshold voltage has a 0.26-V positive shift, and the maximum transconductance decreases. The gate leakage current increases by three times more than that of original ones. The gate-lag characteristics also deteriorates with the increase of experimental time. After measuring the high frequency performance, the return loss of the device degrades. Based on the measurement of leakage current, the height of Schottky barrier before and after the experiment was calculated and the captured charge density was extracted as 1.11×10^{14} cm^{-2} by combining with the threshold voltage drift.

Index Terms—RF overdrive stress, GaN HEMT, Schottky barrier, reliability

I. INTRODUCTION

There are many excellent advantages in Gallium Nitride (GaN) semiconductor material. It is an excellent material for radio frequency, sensors, optoelectronics, power electronics due to its characteristics such as high carrier concentration, high electron drift velocity, large bandgap, good heat resistance and excellent radiation resistance, and it shows a huge application prospect in the information communication, electronic products and other aspects[1, 2]. The reliability problem of GaN devices has been restricting the development of this promising technology. A lot of previous research has been done to investigate the mechanism of GaN HEMT degradation[3-7].However, only a few reports have mentioned the RF stress reliability analysis of devices.

In order to enable GaN devices to achieve higher RF power levels, device designs are constantly being optimized, it becomes increasingly important to evaluate the long-term reliability of devices in practice. In general, the degradation of RF stress tends to be more serious than that of DC stress. When the device works under continuous RF conditions, it is more vulnerable to high channel temperature, high current density, high electric field and other factors[8]. RF overdrive stress test is usually used to infer the reliability of equipment components. The RF overdrive stress can help us to explore the failure behaviors of microwave devices and the RF power limit level that the devices can withstand[9]. In addition, by calculating and characterizing the device defects before and after overdrive stress, such as the generation of traps or charge capture, we can have a deeper understanding of the mechanism of device degradation and failure[10].

After evaluating the influence of RF overdrive stress on GaN HEMTs, we discovered that the electrical characteristics, reflectance coefficient, and gate-lag all degraded.Furthermore, the Schottky barrier changes and the captured charge density before and after the experiment are obtained by derivation. The analysis results can provide reference value for the RF applications of GaN devices.

II. EXPERIMENT DETAILS

A. Device Specifications

GaN MMIC internal matched power transistor is used in the experiment, which is suitable for standard communication frequency band, and the operating frequency band ranges are from 1.8 GHz to 2.3 GHz.

Fig. 1 and 2 are surface morphology and schematic diagram of the device, respectively. The Silicon Carbide (SiC) serves as the substrate, with Gallium Nitride (GaN) serving as the buffer layer, Aluminum Gallium Nitrogen (AlGaN) serving as the barrier layer, and a passivation layer covering the surface. This device has a gate length of 0.5um and a gate width of 4mm.

This work was supported in part by the Key Realm R&D Program of Guangdong Province under Grant 2020B010173001, in part by the Key Realm R&D Program of Guangdong Province under Grant 2020B010171002, and in part by the Project of Science and Technology on Reliability Physics and Application Technology of Electronic Component Laboratory under Grant 6142806190402.

978-1-6654-4817-8/21 $31.00 © 2021 IEEE

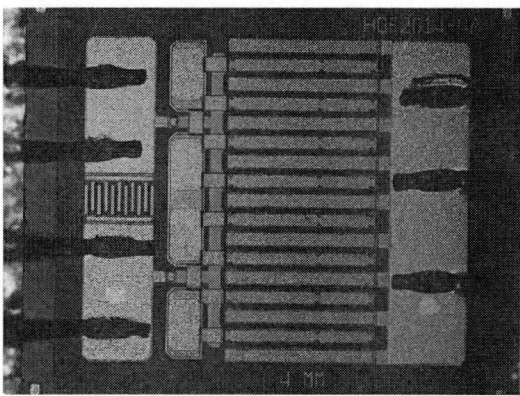

Fig. 1. surface morphology of GaN HEMT

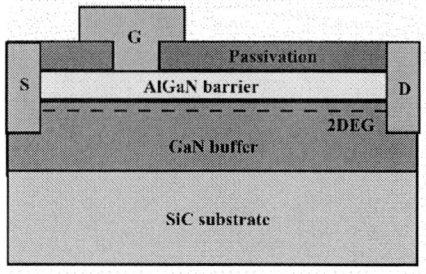

Fig. 2. The cross section structure of device

B. Experimental Setup

Fig. 3 is the block diagram of this test system. The test system includes a signal source, a DUT, a power amplifier, a DC power supply, a directional coupler, two attenuators and two load type power meters.

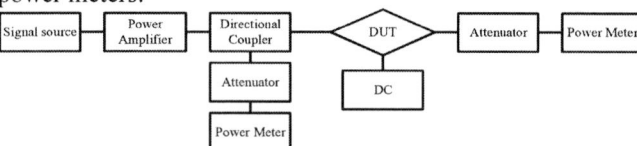

Fig. 3. Block diagram of the test system.

During the RF overdrive stress duration, the GaN device was biased at V_{ds}=28 V and V_{gs}=-1.5 V, and it was kept at room temperature. The microwave signal source generates a continuous wave (CW) of 2 GHz, after the input signal passes through the power amplifier, the directional coupler divides it into two signals. One signal reaches the power meter through the attenuator, which is used to monitor the input power, and the other signal is transmitted to the input end of the tested fixture. Another power meter monitors and records the output power. The purpose of using attenuators is to prevent the power meter from being damaged by excessive output power.

To determine the saturation point of the output power, the device's power gain was measured before the experiment(Fig. 4). The input power was set to 15 dB more than the saturated input power, then began monitoring the RF input and output power to ensure that GaN device is operating under normal conditions.

In this experiment, a semiconductor analyzer (model: Agilent B1500A) was used to measure electrical characteristics, and a network analyzer (model: Agilent N524A) was used to measure RF performance.

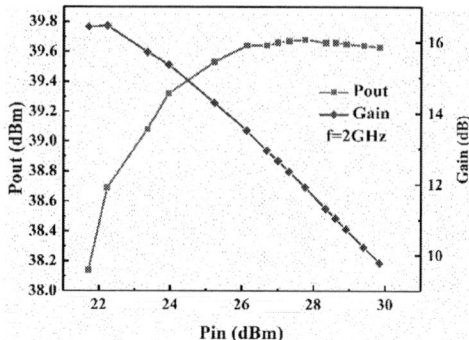

Fig. 4. RF power measurement at 2 GHz.

III. RESULTS AND DISCUSSION

A. DC characteristic

Fig. 5 depicts the device's output characteristics (I_{ds}–V_{ds}). With a 0.25 V step, the gate-to-source voltage (V_{gs}) ranges from -2.5 to -1 V. The output current is larger than that after RF overdrive stress, which decreases from 877 mA to 573 mA under the circumstances of V_{gs} = -1 V and V_{ds} = 10 V. This phenomenon is similar to earlier published research[11]. This degradation may be caused by the effective flat-band voltage shift of the device, which means the increase of the surface state density of the device[12].

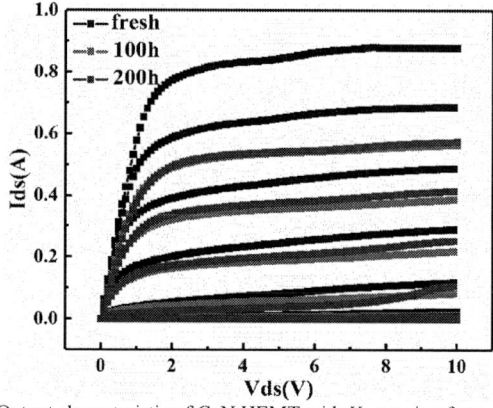

Fig. 5. Output characteristic of GaN HEMT with V_{gs} ranging from -2.5V to -1.0 V.

According to the device's transfer curve, it can be seen that the maximum transconductance of the device decreases from 0.65 S to 0.44 S, and the threshold voltage(V_{th}) moves forward from -1.88 V to -1.62 V(Fig. 6).

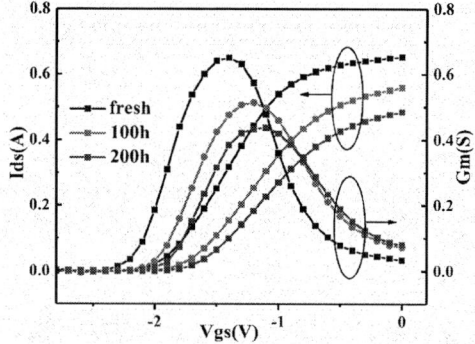

Fig. 6. Transfer and transconductance characteristics of GaN HEMT with V_{ds} = 1V.

The gate leakage current (I_{gd}-V_{gd} & I_{gs}-V_{gs}) shows that after 100 h stress, compared with the fresh gate leakage current of the device, it has increased significantly[13], but the leakage current after 200 h RF stress is little different from that after 100 h stress.

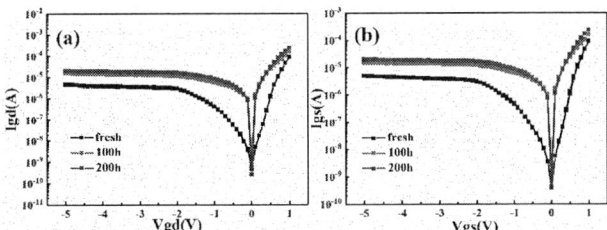

Fig. 7. Gate leakage characteristics of GaN HEMT with V_g ranging from -5V to 1 V (a) I_{gd}-V_{gd} (b) I_{gs}-V_{gs}.

B. Gate-Lag Performance

The gate-lag characteristic measurement is frequently used in device trap analysis because the gate-lag is mostly related to the surface capture effect[14, 15]. This could be due to trapping in the device's AlGaN barrier layer and on the AlGaN surface[16].

Fig. 8 shows the gate-lag measurement result before and after experiment. Convert the gate voltage from -8V to 0 V to switch GaN HEMT from the off-state to the on-state after 100 seconds in the off-state. A large number of electrons will flow into the region near the drain below the gate of the device. There are a few traps on the barrier layer's surface and within the barrier layer. During this process, the electrons will be captured by these traps. The device switches on instantaneously, and the electrons trapped in the trap do not have enough time to release. After a short period of time, the drain current can be restored[17].With the increase of stress time, the time taken for the drain current to reach saturation becomes longer and the gate-lag degradation is more significant.

The degradation of gate-lag characteristics may be caused by the increase of traps, which is consistent with the positive drift of threshold voltage of the device.

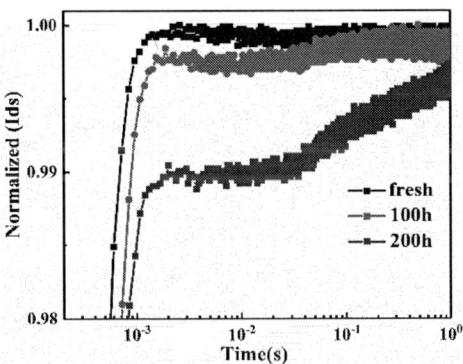

Fig. 8. Gate-lag characteristics of GaN HEMT with Vg = -8V change to 0V.

C. RF performance

The RF measurements were performed and registering S parameter values corresponding to the frequency of 1.8 GHz to 2.3GHz as shown in Fig. 9. Results of high frequency measurements show a slight degradation in S_{11} and S_{22} parameters after RF overdrive stress (measurement conditions V_{ds}=28 V, V_{gs}=-1.5 V, power level -5 dBm, frequency 1.8-2.3 GHz). Thus it can be seen that overdrive stress has certain influence on the matching level of device in the 2 GHz ~2.3 GHz frequency band.

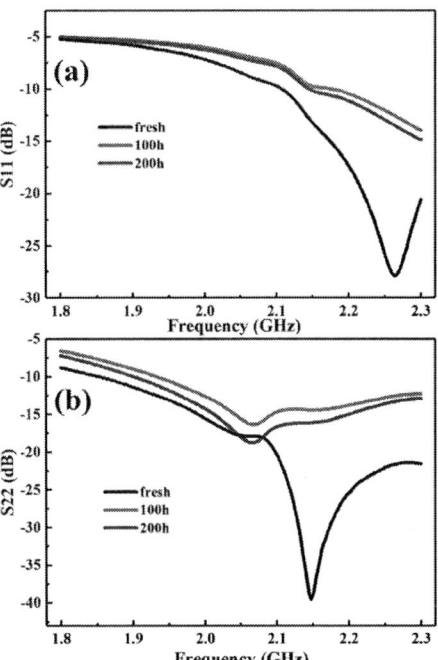

Fig. 9. Measurement of return loss in the frequency range 1.8 GHz to 2.3 GHz. (a)S_{11} (b)S_{22}.

D. Extraction of capture charge density

The Schottky characteristic is the most important criterion for evaluating the performance of GaN HEMT devices. Fig. 7 depicts the device's gate current characteristics before and after the experiment. The Schottky barrier height φ and the ideal factor n can be calculated using the positive I-V characteristics.

The Schottky barrier current is represented as the following equation based on the thermionic emission theory and the current transfer formula[18]:

$$I = AA^*T^2 \exp\left(-\frac{q\phi}{kT}\right)\left[\exp\left(\frac{q(V-IR_s)}{kT}\right)-1\right] \quad (1)$$

The effective Richardson constant is A^*, A is the area of contact with Schottky, and it is 33.74 A/cm²/K² at room temperature. R_S is the equivalent series resistance, ϕ is the Schottky barrier height. When $V > 3kT/q$, Eq.1 can be simplified as:

$$I = AA^*T^2 \exp\left(-\frac{q\phi}{kT}\right)\exp\left(\frac{qV}{nkT}\right) \quad (2)$$

where n is the ideal factor. Take the logarithm of both sides of this equation:

$$\ln I = \ln\left(AA^*T^2\right) - \frac{q\phi}{kT} + \frac{qV}{nkT} \quad (3)$$

The linear fitting of In I and V can be used to obtain n and ϕ according to Eq. 3.

$$\phi = \frac{kT}{q}\left[\ln(AA^*T^2) - Intercept\right] \quad (4)$$

$$n = \frac{q}{kT} \cdot \frac{1}{Slope} \quad (5)$$

Fig. 10 shows the linear fitting relationship between leakage current and forward voltage in semi-logarithmic coordinates,

978-1-6654-4817-8/21 $31.00 © 2021 IEEE

the calculated values of n and ϕ are provided in Table 1. The Schottky barrier height of the device reduces dramatically after the experiment, as seen in Table 1. The ideal factor n is larger than the initial state. The results show that the experiment changes the interface state of the metal-semi contact and weakens the Schottky contact. Moreover, as the density of traps below the gate increases, the channels for auxiliary electrons to tunnel through the Schottky barrier increase. As the height of the barrier decreases, the virtual gate effect is enhanced because electrons in the channel are more likely to leak from the gate and be caught by surface states between the gate and drain. So the RF overdrive stress leads to a reduction of Schottky barrier height and further increases gate leakage current.

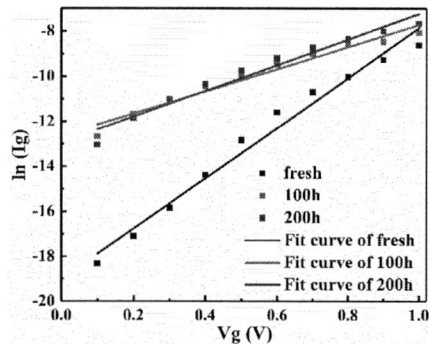

Fig. 10. ln(Ig) versus forward Vg

TABLE I. Schottky barrier height and ideal factor

The experimental time	Barrier height (V)	n
Fresh	0.51	3.47
100h	0.35	7.87
200h	0.36	6.78

The threshold voltage of GaN HEMT can be expressed as [19]:

$$V_{th}(n) = \phi - \Delta E_C(n) - \frac{qN_d d_d^2}{2\varepsilon(n)} - \frac{q\sigma(n)}{\varepsilon(n)}d \qquad (4)$$

Where ΔE_c is the conduction band discontinuous, N_d is the barrier layer doping concentration, σ is the charge density, and d_d is the barrier layer thickness. It can be considered that the drift of threshold voltage before and after the experiment is due to charge capture effect. Therefore, the captured charge density in the AlGaN barrier or GaN buffer can be expressed as:

$$-\Delta\sigma = \frac{\varepsilon(n)}{q(d_d + d_i)}\left(\Delta V_{th} - \Delta\phi\right) \qquad (5)$$

where ΔV_{th} is the threshold voltage shift, and d_i is the undoped spacer layer thickness. Substituting the data in Table 1, the captured charge density can be calculated as 9.35×10^{13} cm^{-2} for overdrive stress time of 100 hours and 1.11×10^{14} cm^{-2} for 200 hours.

E. Mechanism of RF Stress degradation

An energy band diagram before and after the experiment is shown to describe the degradation mechanism of device's internal defects under RF overdrive stress (Fig. 11).

Compared with the untested device, the electrons captured by the tested device increased significantly. This is related to the production of a high intensity electric field, induction of high energy carriers, and an increase in the number of traps in the barrier layer and on the barrier layer's surface under the RF overdrive stress[20][21].

The resulting traps capture electrons, leading to a virtual gate effect[22]. The additional traps also lead to transconductance degradation and positive threshold voltage drift of the device. Due to the increase of captured electrons, the height of the Schottky barrier of the device decreases, and the device degrades.

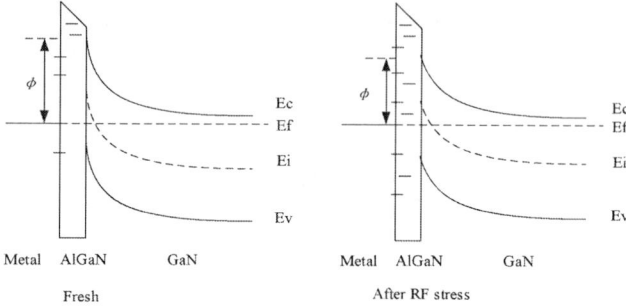

Fig. 11. GaN HEMT energy band diagram before and after experiment

IV. Conclusion

In conclusion, 15 dB RF overdrive stress leads a degradation of the device, including a drop in saturation output current and transconductance, a positive shifts of threshold voltage and a considerable increase in leakage current. The RF measurement results show that the return loss of the device is larger than the original. Moreover, the height of the Schottky barrier drops from 0.51 V to 0.35 V after 100 h experiment, but almost no change between 100 h and 200 h. Based on the drift of threshold voltage, the captured charge densities at 100 hours and 200 hours are calculated as 9.35×10^{13} cm^{-2} and 1.11×10^{14} cm^{-2}, respectively.

REFERENCES

[1]. Zeng, F., et al., A Comprehensive Review of Recent Progress on GaN High Electron Mobility Transistors: Devices, Fabrication and Reliability. Electronics, 2018. 7(12).

[2]. Roccaforte, et al., An Overview of Normally-Off GaN-Based High Electron Mobility Transistors. Materials, 2019. 12(10): p. 1599.

[3]. Kikkawa, T., et al. An over 200-W output power GaN HEMT push-pull amplifier with high reliability. in Microwave Symposium Digest, 2004 IEEE MTT-S International. 2004.

[4]. Lee, C., et al. Reliability evaluation of AlGaN/GaN HEMTs grown on SiC substrate. in High Performance Devices, 2002. Proceedings. IEEE Lester Eastman Conference on. 2002.

[5]. Nguyen, C., N.X. Nguyen and D.E. Grider, Drain current compression in GaN MODFETs under large-signal modulation at microwave frequencies. Electronics Letters, 1999. 35(16): p. 1380-1382.

[6]. Koley, G., et al., Electrical bias stress related degradation of AlGaN/GaN HEMTs. Electronics Letters, 2003. 39(16): p. 1217-1218.

[7]. Mizutani, et al., A study on current collapse in AlGaN/GaN HEMTs induced by bias stress. Electron Devices, IEEE Transactions on, 2003.

[8]. Hot-Electron Degradation of AlGaN/GaN High-Electron Mobility Transistors During RF Operation: Correlation With GaNBuffer Design. IEEE Electron Device Letters, 2015.

[9]. Borgarino, M., et al., Reliability physics of compound semiconductor transistors for microwave applications. Microelectronics Reliability, 2001. 41(1): p. 21-30.

[10]. Caesar, M., et al. Generation of traps in AlGaN/GaN HEMTs during RF- and DC-stress test. in Reliability Physics Symposium. 2012.

[11]. Saugnon, D., et al., Fully automated RF-thermal stress workbench with S-parameters tracking for GaN reliability analysis, in European Microwave Integrated Circuits Conference - Proceedings. 2018, IEEE: NEW YORK. p. 17-20.

[12]. Hasegawa, H., et al., Mechanisms of current collapse and gate leakage currents in AlGaN/GaN heterostructure field effect transistors. Journal of Vacuum Science & Technology B, 2003. 21.

[13]. Rao, H.P. and G. Bosman, Study of RF Reliability of GaN HEMTs Using Low-Frequency Noise Spectroscopy. IEEE Transactions on Device & Materials Reliability, 2012. 12(1): p. 31-36.

[14]. Faqir, et al., Investigation of High-Electric-Field Degradation Effects in AlGaN/GaN HEMTs. Electron Devices, IEEE Transactions on, 2008.

[15]. Tirado, J.M., J.L. Sanchez-Rojas and J.I. Izpura, Trapping Effects in the Transient Response of AlGaN/GaN HEMT Devices. IEEE Transactions on Electron Devices, 2007. 54: p. p.410-417.

[16]. Mitrofanov, O., Mechanisms of gate lag in GaN/AlGaN/GaN high electron mobility transistors. Superlattices and Microstructures, 2003. 34(1-2): p. 33-53.

[17]. Meneghesso, G., et al., Current Collapse and High-Electric-Field Reliability of Unpassivated GaN/AlGaN/GaN HEMTs. IEEE Transactions on Electron Devices, 2006. 53(12): p. 2932-2941.

[18]. Turuvekere, S., et al., Gate Leakage Mechanisms in AlGaN/GaN and AlInN/GaN HEMTs: Comparison and Modeling. IEEE Transactions on Electron Devices, 2013. 60(10): p. 3157-3165.

[19]. Kranti, A., S. Haldar and R.S. Gupta, An accurate charge control model for spontaneous and piezoelectric polarization dependent two-dimensional electron gas sheet charge density of lattice-mismatched AlGaN/GaN HEMTs. Solid State Electronics, 2002. 46(5): p. 621-630.

[20]. Vetury, R., et al., The impact of surface states on the DC and RF characteristics of AlGaN/GaN HFETs. Electron Devices IEEE Transactions on, 2001. 48(3): p. 560-566.

[21]. Joh, J. and J. Alamo, RF Power Degradation of GaN High Electron Mobility Transistors. Electron Devices Meeting .iedm.technical Digest.international, 2011.

[22]. Murao, Y., et al. A Study of GaN HEMTs Current Collapse Impacts on Doherty Multistage PA Linearity. in Compound Semiconductor Integrated Circuit Symposium. 2015.

Comparison of Two Types of Single Gate Drivers for SiC MOSFET Stacks in Flyback Converters

Rui Wang[a], Hongbo Zhao[a] and Stig Munk-Nielsen[a]
[a] Department of Energy Technology
Aalborg University
Aalborg, Denmark
{rwa, hzh, smn}@et.aau.dk

Abstract—In some high-voltage low-power applications, silicon carbide metal-oxide-semiconductor field-effect transistor (SiC MOSFET) stack is required to withstand the high voltage. Instead of using separated gate driver for each SiC MOSFET, the single gate driver only requires one standard gate driver to drive all the series connected SiC MOSFET in the stack with the help of some passive components. Despite of its advantage of cost-efficiency, however, it is vulnerable to the voltage oscillation of power loop due to its working principal. Therefore, in this paper, the single gate driver is analyzed in a flyback converter application since an abrupt power loop voltage dropping would occur when it is operating in discontinuous conduction mode (DCM), and two types of single gate drivers are compared. Besides, in order to obtain better switching performance, suitable design guidelines are provided for both two types. Finally, the experimental results verify the correctness of the analysis.

Keywords—SiC MOSFET stack; single gate driver; flyback converter

I. INTRODUCTION

Silicon carbide (SiC) devices are well developed as the substitutes of low voltage Silicon (Si) counterparts nowadays because of the higher switching frequency while less power loss. In the commercial market, SiC metal-oxide-semiconductor field-effect transistors (MOSFETs) with 1.7kV and below blocking voltages are easily available. However, when it comes to higher voltage level, for example, 10kV SiC MOSFETs, are costly to obtain due to the high price, which makes SiC MOSFET stack consisting of series connected low voltage devices necessary in some high-voltage low-power applications.

Generally, in the gate driver design of SiC MOSFET stack, each SiC MOSFET requires one separated power supply and one corresponding gate driver. To further reduce the cost and greatly simplify the design, the concept of single gate driver has been proposed, which only requires one standard gate driver to drive all the devices in the stack with the help of some passive components. There exist many relevant applications. In [1, 2], two series connected SiC MOSFETs as a solid-state circuit breaker show good voltage balancing performance using a single gate driver, and the stability is well analyzed. In [3], based on it, the gate side closed-loop compensation circuit is further added, which makes it adapt to the variation of the dc bus voltage. In [4], the voltage balancing of two SiC MOSFETs under varied dc-bus voltages is also achieved with the aid of limiting snubber circuits. In [5, 6], the single gate driver is combined with passive snubber circuit, which realizes the driving and voltage balancing of more series

connected SiC MOSFETs. In [7], the performances of different types of single gate drivers without voltage balancing schemes are shown in the application of high voltage nanosecond pulse generator.

Although the single gate driver has the advantage of cost-efficiency, it is vulnerable to the voltage oscillation of power loop due to its working principal. Specifically, when a flyback converter is operating in discontinuous conduction mode (DCM), the power loop voltage of main switch will suddenly drop some degree during off state, which may cause the false turn-on if the single gate driver is applied to drive the SiC MOSFET stack as the main switch. Therefore, in this paper, the single gate driver for SiC MOSFET stack is analyzed in a flyback converter application, and the comparison of two types of single gate drivers is made. Besides, in order to obtain better switching performance, suitable design guidelines are provided for both types.

In Section II, the operation of flyback converter is firstly introduced, followed by the topologies of two types of single gate drivers. In Section III, based on theoretical analysis, the comparison and design guidelines of two types are given. In Section IV, the experimental results are given. Finally, the conclusion is given in Section V.

II. TWO TYPES OF SINGLE GATE DRIVERS

A. Operation of flyback converter

Here, the basic flyback converter is defined as shown in Fig. 1, which consists of input voltage power source V_{in}, leakage inductance L_σ, a transformer with primary inductance L_p and turns ratio of 1:1, main switch T, diode D, output capacitor C_{out} and load R. When the transferred power is low, the flyback converter could be operating in DCM mode, and the corresponding simplified switching waveforms are shown in Fig. 2.

When the driving signal is turned from "off" to "on", T is turned on and the drain-source voltage v_{ds} drops to zero from V_{in}. The primary current i_p starts to increase from zero, while D is not conducted and the secondary current remains zero. After a period, the driving signal is changed to "off" and v_{ds} rises to $V_{in}+V_{out}$ (V_{out} is the output voltage) since D is conducted, where some oscillation may occur due to the energy dissipation of L_σ, as shown in the dash line. i_p drops rapidly from the maximum value I_p to zero and i_s rises correspondingly to its maximum value I_s. Then the energy stored in the transformer starts to transfer to the output terminal, and i_s starts to decrease to zero. Once i_s reaches zero, v_{ds} abruptly drops to V_{in} from $V_{in}+V_{out}$ since D is not conducted again, and some oscillation could also occur as shown in the

978-1-6654-4817-8/21 $31.00 © 2021 IEEE

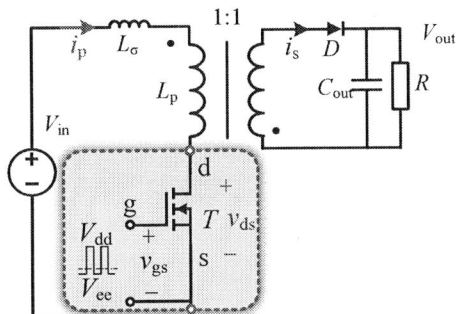

Fig. 1. Basic topology of a flyback converter.

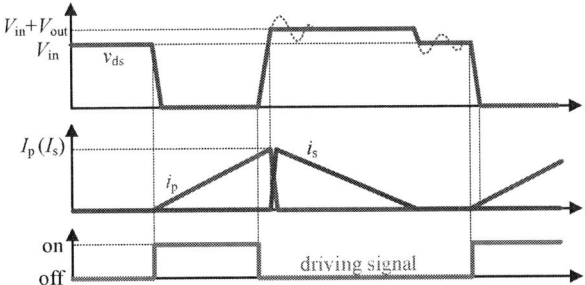

Fig. 2. Switching waveforms.

dash line. After that, the driving signal is turned to "on" again and it means the beginning of a new switching period.

B. Two types of single gate driven SiC MOSFET stack

In some high-voltage low-power applications, the switch T in Fig.1 is required to withstand the high voltage of $V_{in}+V_{out}$, which means a transistor with much higher blocking voltage should be chosen. However, it is costly to obtain due to the high price. Therefore, since SiC MOSFETs with 1.7kV blocking voltage and below are easily available, using SiC MOSFET stack consisting of series connected low voltage devices is a good option.

In order to further simplify the gate driver design and save the cost, the single gate driver is applied here to drive the SiC MOSFET stack, as shown in Fig. 3, and three SiC MOSFETs in series $T_1 \sim T_3$ are taken as an example. In addition, a passive voltage balancing circuit is applied as well. Such a circuit takes the advantage of RCD snubber circuit to balance the voltages, and additional diodes are used for the automatic balancing purpose of capacitors.

In general, there are two categories of single gate drivers according to their configurations for driving upper devices. One category still requires additional power supplies to assist the driving of upper devices. The other category just adopts coupling capacitors to drive upper devices, which is simpler and preferred. Here in our case, two types of capacitor-driving single gate drivers suitable for such an application are compared, named Type I and Type II respectively. In both two types, T_3 is driven by one standard gate driver, and T_i ($i=1, 2$) is driven by driving capacitor C_{ai} ($i=1, 2$). R_{gi} is the gate resistor of T_i, and R_{si} is the static voltage balancing resistor. Diodes D_{ai}, D_{bi} are used for gate protection. The difference is, Type I uses the driving capacitor which is connected between

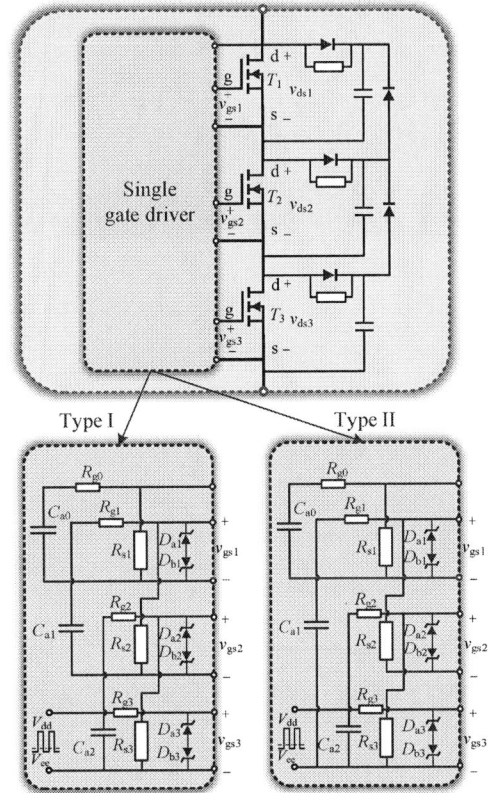

Fig. 3. Two types of single gate drivers

the gate terminal of corresponding SiC MOSFET to the source terminal of lower adjacent one, while Type II uses the driving capacitor which is connected between the gate terminal of corresponding SiC MOSFET to the ground.

III. ANALYSIS AND COMPARISON

A. Analysis of Type I

The switching of the SiC MOSFET stack starts from the switching of the bottom SiC MOSFET T_3. With Type I, when T_3 is turned on, v_{ds3} will drop to zero rapidly, resulting in a current i_{g2} flowing from C_{a2} to the gate side of T_2 since such a formula exists:

$$R_{gi}i_{gi} + v_{gsi} + v_{ds(i+1)} = v_{Cai}, \quad i=1, 2 \tag{1}$$

i_{g2} will contribute to the turn-on of T_2, and its value could be obtained by using the following relationship:

$$i_{gi} = -\frac{C_{ai}dv_{Cai}}{dt} = \frac{C_{gsi}dv_{gsi}}{dt}, \quad i=1, 2 \tag{2}$$

When T_2 is turned on, the rapid dropping of v_{ds2} will cause the current i_{g1} flowing from C_{a1} to the gate side of T_1 as well. Therefore, T_3, T_2 and T_1 will be turned on successively.

Combining (1) and (2), it is solved that:

$$\frac{dv_{gsi}}{dt} \approx -\frac{dv_{ds(i+1)}/dt}{C_{gsi}/C_{ai}+1}, \quad i=1, 2 \qquad (3)$$

The total delay t_I with Type I from the initial moment of v_{ds3} dropping to that of v_{ds1} dropping is calculated as:

$$t_I \approx (\frac{C_{gs1}}{C_{a1}}+1)\frac{V_{ee}-V_{dd}}{dv_{ds2}/dt} + (\frac{C_{gs2}}{C_{a2}}+1)\frac{V_{ee}-V_{dd}}{dv_{ds3}/dt} \qquad (4)$$

Since it is a flyback application operating in DCM, an abrupt dropping of v_{dsi} ($i=1\sim3$) will occur when i_s reaches zero during the off state of SiC MOSFET stack. By defining the voltage drop as ΔV, the induced voltage $\Delta v_{gsi}(I)$ of v_{gsi} with Type I is obtained as:

$$\Delta v_{gsi}(I) \approx \frac{\Delta V}{C_{gsi}/C_{ai}+1}, \quad i=1, 2 \qquad (5)$$

The parameter t_I and $\Delta v_{gsi}(I)$ are the major concerns of this paper, since t_I determines the switching speed of the SiC MOSFET stack, and $\Delta v_{gsi}(I)$ decides whether the false turn-on of the SiC MOSFET stack occurs.

B. Analysis of Type II

Different with Type I, the gate side relationships of T_1 and T_2 with Type II are as follows:

$$\begin{cases} R_{g1}i_{g1} + v_{gs1} + v_{ds2} + v_{ds3} = v_{Ca1} \\ R_{g2}i_{g2} + v_{gs2} + v_{ds3} = v_{Ca2} \end{cases} \qquad (6)$$

Since (2) is still valid here, by combining (2) and (6), it is solved that:

$$\begin{cases} \dfrac{dv_{gs2}}{dt} \approx -\dfrac{dv_{ds3}/dt}{C_{gs2}/C_{a2}+1} \\[3mm] \dfrac{dv_{gs1}}{dt} \approx -\dfrac{dv_{ds3}/dt+dv_{ds2}/dt}{C_{gs1}/C_{a1}+1} \end{cases} \qquad (7)$$

Therefore, the total delay t_{II} with Type II from the initial moment of v_{ds3} dropping to that of v_{ds1} dropping can be calculated as:

$$t_{II} \approx (\frac{C_{gs2}}{C_{a2}}+1)\frac{V_{ee}-V_{dd}}{dv_{ds3}/dt} + (\frac{C_{gs1}}{C_{a1}}-\frac{C_{gs2}}{C_{a2}})\frac{V_{ee}-V_{dd}}{dv_{ds2}/dt+dv_{ds3}/dt} \qquad (8)$$

When the voltage drop ΔV occurs, the induced voltage $\Delta v_{gsi}(II)$ with Type II is obtained as:

$$\begin{cases} \Delta v_{gs2}(II) \approx \dfrac{\Delta V}{C_{gs2}/C_{a2}+1} \\[3mm] \Delta v_{gs1}(II) \approx \dfrac{2\Delta V}{C_{gs1}/C_{a1}+1} \end{cases} \qquad (9)$$

Similarly, t_{II} and $\Delta v_{gsi}(II)$ are also concerned here. Based on above, the comparison of Type I and Type II will be made in the following part.

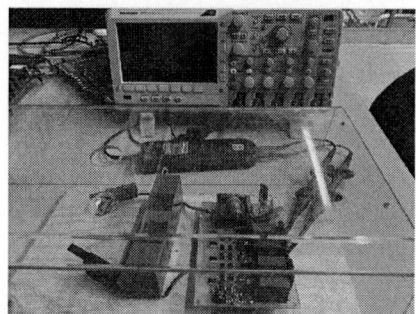

Fig. 4. Photograph of experimental platform

TABLE I. NAMES AND PARAMETERS OF KEY DEVICES

Name	Parameter
T_i ($i=1\sim3$)	C2M1000170J (1700V/5.3A)
D_{ai} ($i=1\sim3$)	PDZVTR18B
D_{bi} ($i=1\sim3$)	PDZVTR6.2B
R_{g0}, R_{g1}, R_{g2}	50Ω
R_{g3}	10Ω
C_{a0}, C_{a2}	94pF
C_{a1} in Type I, C_{a1} in Type II	94pF, 47pF
R_{si} ($i=1\sim3$)	500kΩ

C. Comparison

In this design with 1:1 turns ratio, ΔV is equal to V_{out} if the oscillation is neglected, and the value of $\Delta v_{gsi}+V_{ee}$ should be below the value of threshold voltage of SiC MOSFET. Once ΔV and Δv_{gsi} are confirmed, C_{ai} ($i=1, 2$) of Type I and Type II could be chosen according to (5) and (9), respectively. It is concluded that, if Δv_{gsi} is required to be identical in both types, C_{a1} in Type I should be twice as much as that in type II, while C_{a2} is the same in both types.

After C_{ai} is confirmed, comparing (4) and (8), it is concluded that t_{II} is much less than t_I. Therefore, if a higher switching speed is required in such a SiC MOSFET stack, Type II would be a better choice. However, SiC MOSFETs with Type I are turned on and off sequentially and one lags another, which facilitates the voltage balancing of passive voltage balancing circuit. In contrast, the switching of upper SiC MOSFETs with Type II depends, which burdens the parameter selection if a good voltage balancing performance is expected. Therefore, the parameter selection of the whole circuit with Type I is easier than that with Type II.

IV. EXPERIMENTAL VERIFICATION

In order to verify the correctness of above theoretical analysis, an experimental platform is established based on the schematics in Fig. 1 and Fig. 3, and its photograph is shown in Fig. 4. The voltages are measured with the high-voltage differential probes (Tektronix P5200A, 50MHz) and captured with the oscilloscope (Tektronix DPO 2014, 100MHz). Besides, the names of key devices and their corresponding parameters are listed in Table I.

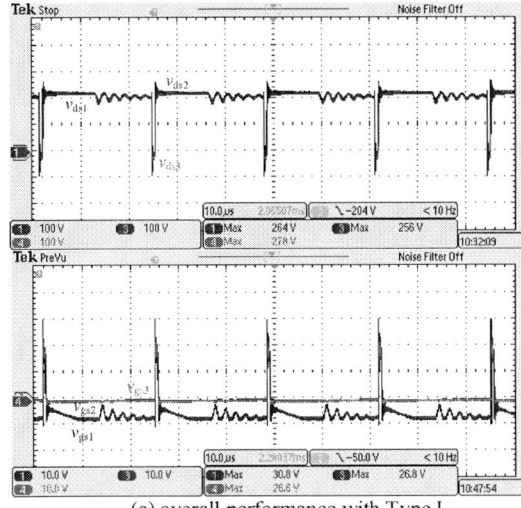

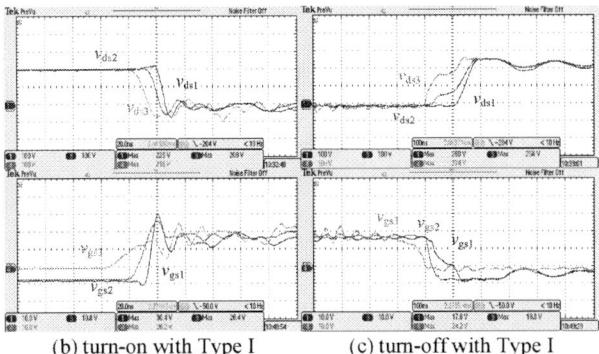

(a) overall performance with Type I

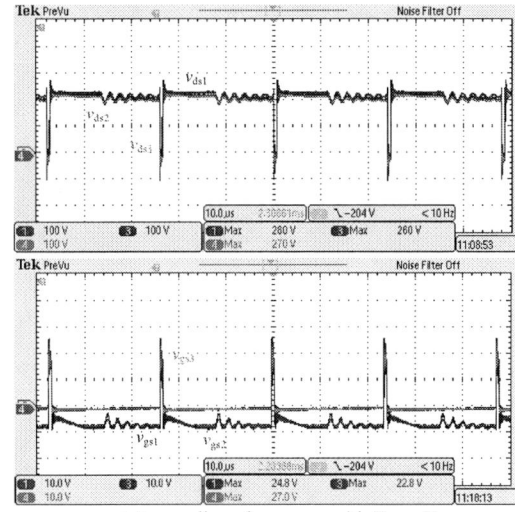

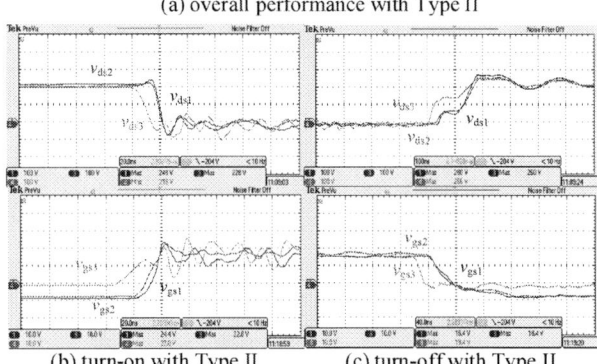

(a) overall performance with Type II

(b) turn-on with Type I   (c) turn-off with Type I

Fig. 5. Switching performance with Type I
(v_{ds1}, v_{ds2}, v_{ds3}: 100V/div; v_{gs1}, v_{gs2}, v_{gs3}: 10V/div)

(b) turn-on with Type II   (c) turn-off with Type II

Fig. 6. Switching performance with Type II
(v_{ds1}, v_{ds2}, v_{ds3}: 100V/div; v_{gs1}, v_{gs2}, v_{gs3}: 10V/div)

With Type I single gate driver, the switching performance of three SiC MOSFETs in the stack under V_{dc}=600V is shown in Fig. 5. It is seen from Fig. 5(a) that good voltage balancing of v_{dsi} (i=1~3) is achieved despite of the existence of some oscillations. In addition, the DCM operation of flyback converter induces a voltage spike of $\underline{v}_{gsi}$ (i=1, 2), which is well limited below 0V to avoid false turn-on phenomenon. In order to facilitate the observation, the turn-on and turn-off waveforms are enlarged, respectively shown in Fig. 5(b) and Fig. 5(c). It is found that t_{I} is 20ns, while the delay from the initial moment of v_{ds3} rising to that of v_{ds1} rising is 100ns.

With Type II single gate driver, the switching performance of three SiC MOSFETs under V_{dc}=600V is shown in Fig. 6. From Fig. 6(a), good voltage balancing of v_{dsi} (i=1~3) is performed as well, however, not as good as that with Type I. Apart from that, induced voltage spike of $\underline{v}_{gsi}$ (i=1, 2) is nearly the same as that with Type I since the value of C_{a1} is chosen half as that in Type I according to the analysis in Section III. The enlarged turn-on and turn-off waveforms are shown in Fig. 6(b) and Fig. 6(c), respectively. It is seen that t_{II} is 14ns, while the delay from the moment of v_{ds3} rising to that of v_{ds1} rising is 40ns. Apparently, both values are much less than those with Type I. However, in this case, v_{ds1} reaches a little bit earlier to the clamping value than v_{ds2}, which is the factor that slightly influences the voltage balancing. Therefore, the parameters should be carefully reselected when it comes to other conditions.

Based on above illustrations, it is concluded that the experimental results are consistent with the theoretical analysis, which well verifies the correctness of the analysis to Type I and Type II single gate drivers in this paper.

V. CONCLUSION

In this paper, the single gate driver is analyzed in a flyback converter application considering it is vulnerable to the voltage oscillation of power loop. Two types of single gate drivers named Type I and Type II are theoretically compared and suitable design guidelines are provided, which are further verified by the experimental results. It is concluded that, both two types show good voltage balancing performance, while some capacitors in Type II for driving purpose should be smaller than those in Type I considering the induced gate voltage spike. Besides, if a higher switching speed is required, Type II would be a better choice; if not, Type I would be preferred since the circuit parameter selection with Type I could be easier.

REFERENCES

[1] Ren, Y., et al. (2017). "A Compact Gate Control and Voltage-Balancing Circuit for Series-Connected SiC MOSFETs and Its Application in a DC Breaker." IEEE Transactions on Industrial Electronics 64(10): 8299-8309.

[2] Ren, Y., et al. (2020). "Stability Analysis and Improvement for SSCB with Single-gate Controlled Series-connected SiC MOSFETs." IEEE Transactions on Industrial Electronics: 1-1.

[3] Yang, C., et al. (2020). "A Gate Drive Circuit and Dynamic Voltage Balancing Control Method Suitable for Series-Connected SiC mosfets." IEEE Transactions on Power Electronics 35(6): 6625-6635.

[4] R. Wang, L. Liang, Y. Chen and Y. Kang (2020). "A Single Voltage-Balancing Gate Driver Combined With Limiting Snubber Circuits for Series-Connected SiC MOSFETs," IEEE Journal of Emerging and Selected Topics in Power Electronics 8(1): 465-474.

[5] Wang, Z., et al. (2019). Series SiC MOSFETs with Single Gate Driver Based on Capacitance Coupling and Passive Snubber Circuits. 2019

22nd International Conference on Electrical Machines and Systems (ICEMS).

[6] Jørgensen, A. B., et al. (2020). "Analysis of cascaded silicon carbide MOSFETs using a single gate driver for medium voltage applications." IET Power Electronics 13(3): 413-419.

[7] Pang, L., et al. (2019). "A Compact Series-Connected SiC MOSFETs Module and Its Application in High Voltage Nanosecond Pulse Generator." IEEE Transactions on Industrial Electronics 66(12): 9238-9247.

Over-Voltage and Oscillation Suppression Circuit with Switching Losses Optimization and Clamping Energy Feedback for SiC MOSFET

Chengzi Yang, Huaqing Li, Haoyuan Jin, Longyang Yu, Laili Wang, Yunqing Pei
School of Electrical Engineering
Xi'an Jiaotong University
Xi'an, P.R. China
Email: lemonyang@stu.xjtu.edu.cn

Abstract — **The normal snubber circuit decoupling the power loop inductance is a cost-effective method to solve the severe turn-off oscillation problem of SiC MOSFETs. However, this will also significantly increase the turn-on loss and decrease the performance of SiC MOSFET. This paper proposes a SiC MOSFET turn-off oscillation reduction circuit and the proposed circuit shows good switching losses reduction characteristics. By using the clamping capacitors, the oscillation is greatly reduced and the oscillation energy is stored in these capacitors. At the same time, the clamping capacitors will only work until the drain-source voltage is larger than the DC-bus voltage, and this feature is the benefit to the switching losses reduction during the turn-on transient. Once the SiC MOSFET circuit turned into the static state, the over-voltage and oscillation energy storage in the clamping capacitors can be feedback which helps further increase the efficiency of the whole system. Experimental and comparison studies were made to verify the effectiveness of the proposed turn-off oscillation reduction circuit. Experimental and comparison results show that turn-off oscillation reduction circuit has excellent turn-off over-voltage and oscillation suppression performance, and can greatly reduce total switching losses.**

Index Terms—**SiC MOSFETs, turn-off over-voltage and oscillation, snubber circuits, turn-on and off switching loss**

I. INTRODUCTION

SiC MOSFET is a third-generation wide-bandgap power semiconductor device, which shows better performance compared with Si power devices [1]. These are all conducive to further improving the performance of converters. However, the high di/dt of SiC MOSFETs can cause oscillation problems during the turn-off process, thereby reducing the voltage utility coefficient of SiC MOSFETs, exacerbating EMC problem, and reducing the reliability of the whole system.

The turn-off oscillation problem is often solved in three ways which are using the active gate driver (AGD) decreases the turn-off di/dt, reducing the power loop parasitic inductance by using advanced layout, decoupling the power loop inductance by using passive snubber.

The active gate driver normally changes the gate driver resistors, driver voltages, or gate driver currents [2]-[4] during the turn-on and off process in real-time to improve the switching losses and reduce the significant turn-off over-voltage and oscillation of SiC MOSFET. Using the active gate drivers is a good way to reduce turn-off oscillation and

fully optimize the low switching loss characteristics of SiC MOSFETs [5]. However, due to the complexity of the gate drive circuit structure and corresponding control, the active gate driver mostly stays in scientific research and has not been widely used in practical applications.

Overvoltage and oscillation are caused by the voltage induced by the parasitic inductance of power loop and oscillate together with the SiC MOSFET output capacitor. Therefore, reducing this parasitic inductance is the most direct way to solve the turn-off over-voltage and oscillation problem. The previous work was devoted to optimizing the inductance of the bus-bar through different layout methods [6]-[8]. However, like the snubber circuit, although the oscillation is effectively reduced the turn-on switching losses also greatly increase.

The snubber circuits such as the RC snubber circuits are the normal ways to handle the severe turn-off over-voltage and oscillation of SiC MOSFET [9]-[13]. For the half-bridge circuit, the decoupling capacitor which only uses the high-frequency capacitors is widely used [10] [11]. The decoupling capacitors can effectively reduce SiC MOSFET's turn-off over-voltage and oscillation and is very easy to use. But this method will also cause a lower frequency oscillation problem. The higher order snubber circuit is the solution to reduce the lower frequency oscillation at the same time. An analytical method using the detailed mathematical analysis is used for optimal parameter selection in [9]. In order to further reduce the turn-off overshoot and oscillation of SiC MOSFET, the third-order snubber circuits are discussed and the parameters selection method are studied in [12]. Furthermore, the multi-order snubber circuit for SiC MOSFET oscillation reduction has the better performances, but at the same time has a complex parameters design process. No matter what kind of snubber circuits are used in phase-leg configurations, due to the loop parasitic inductance which causes the severe oscillation problem is decoupled, the switching loss of SiC MOSFET during turn-on process will greatly increase. The detailed comparison result in [12] shows that the snubber circuit is used in the half bridge configurations, the turn-on losses of SiC MOSFETs can increase to 1.65 times compared with that without snubber. At the same time, the total switching losses including the turn-on and off switching losses will also increase [13].

According to the discussion above, to solve the severe turn-off over-voltage and oscillation problem, at the same time, optimize the turn-on and turn-off switching losses. In this paper, the turn-off over-voltage and oscillation reduction snubber circuit is proposed. The proposed circuit

---

The work was supported by 2018 steadily supports research projects from Key Laboratory of National Defense Science and Technology. Project number:614221720180401.

has the switching losses optimization function as the clamping capacitors will only work until the drain-source voltage larger than the DC-bus voltage. The proposed snubber circuit can effectively reduce the turn-off oscillation and over-voltage problems, and reduce the total losses

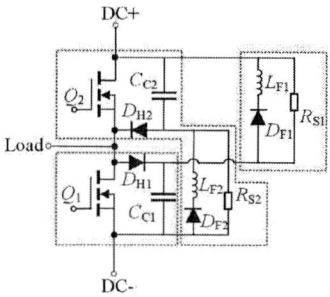

Fig.1. Proposed oscillation reduction circuit.

including the total switching losses and the snubber losses compared with RC snubber circuits.

II. THE PROPOSED SNUBBER CIRCUIT TOPOLOGY AND THE BASIC OPERATION PRICIPLE

A. The Proposed Snubber Circuit Topology

The proposed turn-off over-voltage and oscillation snubber circuit topology are shown in Fig.1. The circuit has a symmetrical structure, half of which is composed of a start-up resistor R_{S2} which has a large value and only works during start-up period to charge the clamping capacitor C_{C2}, a diode and capacitor branch is composed of a clamping capacitor C_{C2} and a small power diode D_{H2} to reduce the turn-off oscillation, an energy feedback branch composed of the feedback inductor L_{F2} and diode D_{F2} and work after the turn-off transient ends.

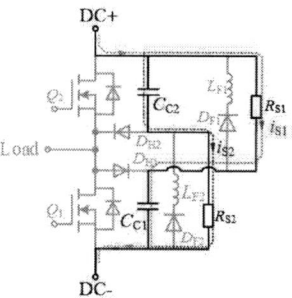

Fig.2. Start-up transient of proposed circuit.

B. The Basic Operation Principle

1) Start-up process

When the circuit is powered on, the voltage between DC+ and DC- rises, and clamping capacitors C_{C1} and C_{C2} voltage rises through the start-up resistors R_{S1} and R_{S2} which have a large value. As depicted in Fig.2, the snubber capacitor C_{C1} is charged by i_{s1} and C_{C2} is charged by i_{s2} until the voltage between them reaches V_{DC}, and the start-up process of the proposed snubber circuit ends. The value of R_{S1} and R_{S2} is large, at least several kΩ, to ensure R_{S1} and R_{S2} can be treated as an open circuit and not participate in the circuit during transient process.

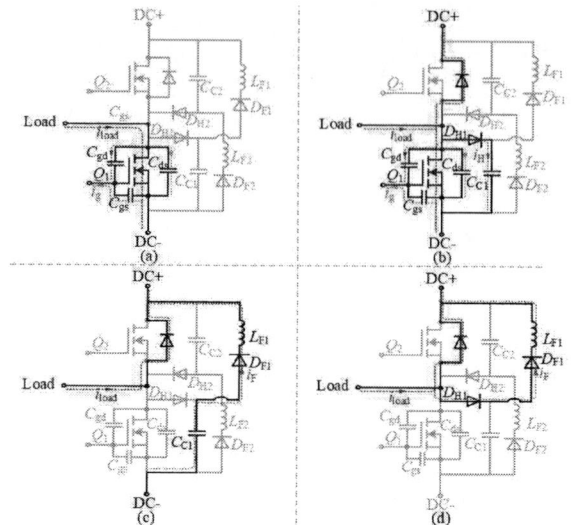

Fig.3. Turn-off transient of the proposed oscillation reduction circuit

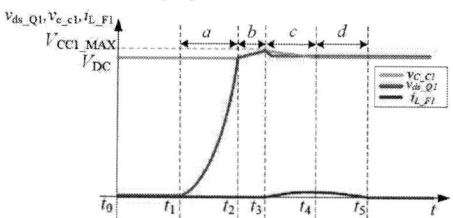

Fig.4. Typical waveforms of the proposed oscillation reduction circuit during turn-off.

2) Turn-on transient

Taking the low-side devices as an example to analyze the turn-on transient. When v_{ds_Q1} is lower than the DC bus voltage, the energy harvest diode D_{H1} keeps off during the whole process. Thus the proposed circuit has almost no impact when the parasitic parameters are ignored during the turn-on transient.

3) Turn-off transient

Taking the low-side devices as an example to analyze the turn-off transient. The operation stages during the turn-off process can be divided into four parts, are depicted in Fig.3. The typical waveforms of the proposed circuit during turn-off are shown in Fig.4.

a) Voltage rise period (t_1-t_2): Once the drive circuit works, the drain source voltage of lower side SiC MOSFET Q_1 rises and is lower than the DC-bus voltage V_{DC} in this period. It should be noticed that the voltage between C_{C1} is V_{DC} after the start-up process. Therefore, the diode D_{H1} is on the off state. During voltage rise period, the proposed turn-off snubber circuit has almost no effect on the low side SiC MOSFET Q_1.

b) Over-voltage clamping period (t_2-t_3): As soon as, v_{ds_Q1} equals to V_{DC}, D_{H1} turns on and can be treated as a wire by which connects C_{C1} to the drain of the low side SiC MOSFET Q_1. At this time, C_{C1} clamps v_{ds_Q1} to V_{DC} and the over-voltage and oscillation during turn-off is well suppressed.

c) Energy feedback to source side period (t_3-t_4)： When it comes to t_3, Q_1 is turned off. v_{ds_Q1} goes back to the DC-bus voltage V_{DC} and energy harvest diode D_{H1} is turned off. It

978-1-6654-4817-8/21 $31.00 © 2021 IEEE

should be noticed that, C_{C1} almost absorbs all the parasitic inductance energy during turn-off, V_{cc1} increases to V_{CC1_MAX} which is larger than V_{DC}. After the over-voltage clamping period ends, the energy stored in C_{C1} begins feedback to DC side through an energy feedback branch which is composed of D_{F1} and L_{F1}. During this period, V_{CC1} drops back to V_{DC} and this period end at t_4.

d) Energy feedback to load side period (t_4-t_5): In this period, when the v_{CC1} falls to V_{DC}, v_{CC1} is clamped at V_{DC} by the energy storage diode D_{H1}. At the same time, the energy in the feedback inductance L_{F1} is released to the load side by a new path. This period ends until i_F drops to zero.

III. EXPERIMENTAL RESULTS

To achieve the proposed oscillation reduction circuit performance verification using in SiC MOSFETs, the test platform and results analysis are established in this part. Fig.5 (a) shows the circuit of test platform, which will be used in the whole oscillation reduction circuit performance

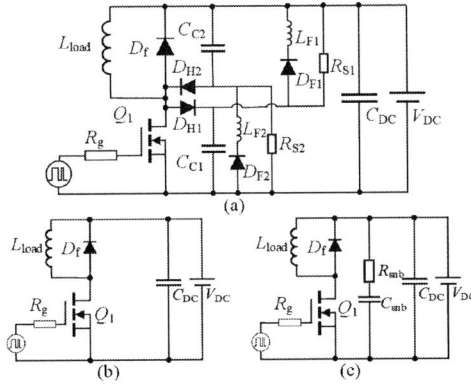

Fig.5. DPT test platform, (a) with proposed oscillation reduction circuit circuit, (b) normal circuit, (c) with RC snubber circuit.

TABLE I
COMPONENTS SELECTION IN THIS PAPER

Components	Values	Detailed
Q_1	1200V, 52A	IMW120R045M1 from Infineon
D_f	1200V, 50A	UJ3D1250K from UnitedSiC
R_g	5Ω	0805 resistor from Yageo
R_{S1}, R_{S2}	5.2kΩ	0805 resistors
L_{load}	150uH	air core inductor
C_{CC1}, C_{CC2}, and C_{snb}	1000V,66nF	MLCCs from KEMET
R_{snb}	1Ω	1206 resistor from Yageo
D_{H1}, D_{H2}, D_{F1}, D_{F2}	600V, 2A	CSD01060A from CREE/Wolfspeed

verification. Table I shows the relevant specific double pulse test experimental parameters, in which both of the starting resistances are made up of four series 1.3kΩ resistances. Meanwhile, two pieces of 33nF multilayer ceramic chip capacitance (MLCCs) are paralleled to make up the clamped capacitance C_{C1} and C_{C2}.

A. The double-pulse test

Fig.6 shows the final experiment results of the proposed oscillation reduction circuit, including drain-source voltage of Q_1(v_{ds_Q1}), the voltage between the clamped capacitance C_{C1}(v_{CC1}), the gate to source voltage of Q_1(v_{gs}), and the drain current of Q_1(i_d).

Affected by the parasitic inductance L_{loop} of the power circuit, when the drain current i_d increases, v_{ds_Q1} will drop from 238V, as is shown in Fig.6 (b). This phenomenon will enormously decrease the turn on switching loss of Q_1. Fig.6 (c) and (d) depict an enlarged view of the shutdown process at different time scales. The maximum transient voltage of v_{ds_Q1} during shutdown process is 538V, and the overshoot voltage is clamped with advantage, which is only 88V. Meanwhile, the ringing containing gate to source voltage ringing, drain to source voltage ringing and leakage current ringing are all integrally repressed. Fig.6 (d) also displays that v_{cc1} remains at the DC bus voltage V_{DC} until v_{ds_Q1} rises to V_{DC}. When the voltage v_{ds_Q1} goes higher than V_{DC}, C_{c1} acts, v_{cc1} increases, thus the overshoot voltage of Q_1 is clamped. With the passages of time, the turn off process ends, oscillation reduction circuit enters the feedback process and v_{cc1} reduces to V_{DC} ceaselessly.

B. Comparison studies

In this part, we make the analysis and comparison of the proposed oscillation reduction circuit adding the normal circuit without additional snubber (Fig.5 (b)) and DPT with RC snubber circuit (Fig.5 (c)) in detail.

Fig.7 shows the typical waveforms comparison under 450V 30A conditions, including the drain to source voltage of Q_1 (v_{ds_Q1}), the drain current of Q_1 (i_d), and the power losses (p). Fig.7 (a) shows that the normal circuit generates major overshoot voltage, and the ringing that will decrease the voltage utilization of SiC MOSFETs, even lead to unreliable turn-off process, and cause additional turn-off loss. Both of the RC snubber circuit and proposed circuit can solve this problem effectively. The overshoot voltage of the proposed circuit is 88V and the overshoot voltage is smaller than the RC snubber circuit. However, both the turn off switching loss of the proposed circuit and RC snubber circuit are close to 200 which are 190.4uJ and 194uJ, respectively. Fig.7 (b) shows that during the turn on process,

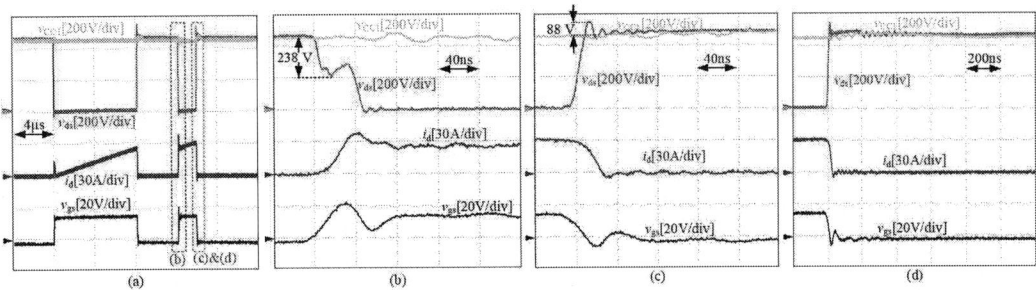

Fig.6. Turn-off transient of the proposed oscillation reduction circuit. (a) DPT experimental results, (b) zoom-in view of turn-on transient, (c)&(d) zoom in view of turn-off transient.

978-1-6654-4817-8/21 $31.00 © 2021 IEEE

affected by the RC snubber circuit, v_{ds_Q1} remains unchanged basically in the value of bus voltage and when the fall period of the voltage starts. Also, the turn on loss of upper bridge arm power device Q_1 using RC snubber circuit is 483.3uJ and the value is greater than 290.7uJ in normal circuit and 281.2uJ in oscillation reduction circuit circuit. This result fully displays the disadvantage that RC snubber circuit will produce huge conduction loss of main power device. Like ordinary circuits, oscillation reduction circuit also makes full use of L_{loop} to reduce the turn on switching loss.

Fig.8 shows the comparisons of the overshoot voltage during turn off process. The overshoot voltage of the normal circuit can be as high as 338V. Under different bus voltages, the maximum buffer voltage of oscillation reduction circuit and RC buffers are the same. From this point of view, these two methods can well control voltage spikes and ringing. And through comparison, it can be found that the effect of this method is better.

Fig.9 shows the loss analysis and comparison under different power bus voltages. Fig. 9 shows that the E_{on} of oscillation reduction circuit is similar to that of normal circuit, and its value is obviously smaller than E_{on} of RC buffer circuit. On the other hand, the turn off switching loss E_{off} of oscillation reduction circuit and RC buffer circuit is obviously less than that of normal circuit. The total switching loss E_{total} of main power devices using oscillation reduction circuit is the smallest. With the increasing of load current I_D and the decreasing of supply bus voltage, the advantage of low switching loss will become more important. When the DC bus voltage is 450V, different from RC buffer circuit, oscillation reduction circuit can reduce the total switching loss by 30.3% under 30A load current. When DC bus voltage is 300V, there will be a 42% decrease. And when DC bus voltage is 150V, there will be a 54% decrease.

IV. CONCLUSIONS

This paper proposes a spike voltage and oscillation restraint circuit using in SiC MOSFET. By using this circuit, the spike voltage and oscillation during turn off process can be effectively clamped, and the switching losses of the main power devices can also be greatly reduced. A SiC MOSFET based DPT prototype was established, experimental results showed that oscillation reduction circuit has potential clamped ability as good as the RC snubber circuit. However, compared to RC snubber circuit, it can reduce 30.3% total switching losses when the supply bus voltage is 450V and 54% total switching when the supply bus voltage is 150V.

REFERENCES

[1] Zhao, T. , et al. "Comparisons of SiC MOSFET and Si IGBT Based Motor Drive Systems." Industry Applications Conference IEEE, 2007.

[2] Marzoughi, et al. "Active Gate-Driver Withdv/dtController for Dynamic Voltage Balancing in Series-Connected SiC MOSFETs." IEEE Transactions on Industrial Electronics 66.4(2019):2488-2498.

[3] McHale, and L. Alexander . "An Integrated IGBT Active Gate Driver with Fast Feed-Forward Variable Current." (2016).

[4] [1] Wang, Z. , et al. "A di/dt Feedback-Based Active Gate Driver for Smart Switching and Fast Overcurrent Protection of IGBT Modules." IEEE Transactions on Power Electronics 29.7(2014):3720-3732.

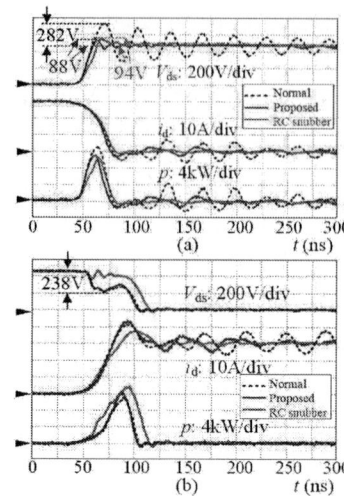

Fig.7. Waveforms comparison under 450V 30A, (a) turn-off process, (b) turn-on process.

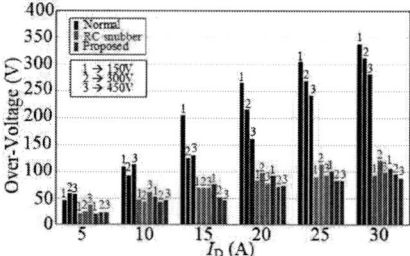

Fig.8. Turn-off over-voltage.

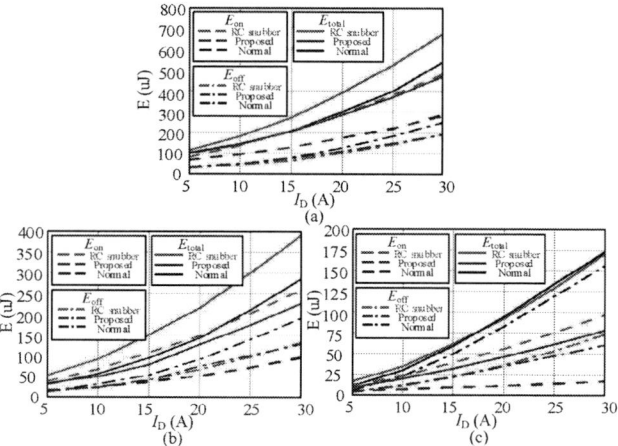

Fig.9. Switching Losses comparison in different DC Voltage (a) 450V, (b) 300V, (c) 150V.

[5] Zhang F , Yang X , Ren Y , et al. "A hybrid active gate drive for switching loss reduction and boltage balancing of series-connected IGBTs," *IEEE Trans. Power Electron.*. vol. 32, no. 10, pp. 7469-7481, Oct., 2017.

[6] Kangping, W. , et al. "An optimized layout with low parasitic inductances for GaN HEMTs based DC-DC converter." Conference Proceedings - IEEE Applied Power Electronics Conference and Exposition - APEC.

[7] Aberg, B. , et al. "Estimation and minimization of power loop inductance in 135 kW SiC traction inverter." 2018 IEEE Applied Power Electronics Conference and Exposition (APEC) IEEE, 2018.

[8] Moorthy, Rsk , et al. "Estimation, Minimization, and Validation of Commutation Loop Inductance for a 135-kW SiC EV Traction Inverter." IEEE Journal of Emerging and Selected Topics in Power Electronics 8.1(2020):286-297.

[9] Chen, Z. , et al. "Optimal DC-Link RC Snubber Design for SiC MOSFET Applications." 2019 IEEE Energy Conversion Congress and Exposition (ECCE) IEEE, 2019.

[10] Wu, Y. , et al. "Impact of RC Snubber on Switching Oscillation Damping of SiC MOSFET with Analytical Model." IEEE Journal of Emerging and Selected Topics in Power Electronics PP.99(2019):1-1.

[11] Liu, S. , H. Lin , and T. Wang . "Comparative Study of Three Different Passive Snubber Circuits for SiC Power MOSFETs." 2019 IEEE Applied Power Electronics Conference and Exposition (APEC) IEEE,2019.

Degradation behavior and mechanism of SiC power MOSFETs by total ionizing dose irradiation under different gate voltages

Kexin Gao
Department of Materials Science and Engineering
Xiangtan University
Hunan, China
201921001356@smail.xtu.edu.cn

Yiqiang Chen*
Science and Technology on Reliability Physics and Application of Electronic Component Laboratory
The No.5 Electronics Research Institute of the Ministry of Industry and Information Technology Guangzhou, Guangdong, China
yiqiang-chen@hotmail.com

Shuaizhi Zheng*
Department of Materials Science and Engineering
Xiangtan University
Hunan, China
shuaizhi@xtu.edu.cn

Min Liao
Department of Materials Science and Engineering
Xiangtan University
Hunan, China
mliao@xtu.edu.cn

Xinbing Xu
Science and Technology on Reliability Physics and Application of Electronic Component Laboratory
The No.5 Electronics Research Institute of the Ministry of Industry and Information Technology Guangzhou, Guangdong, China
xuxb11@163.com

Meng Lu
Department of Materials Science and Engineering
Xiangtan University
Hunan, China
201921001457@smail.xtu.edu.cn

Abstract—In this work, the degradation behavior of the SiC power MOSFETs under total ionizing dose (TID) irradiation at different gate voltages was investigated. To simulate the radiation environment, ^{60}CO was used as the γ-ray source, the dose rate was 50rad/s, and the cumulative absorbed dose was 1M Rad. The experimental results show that when a gate voltage is impressed to the test device, the threshold voltage is obviously negative shift, the output characteristic and capacitance curve also have negative shift phenomenon. Meanwhile, the device with positive gate voltage has more obvious shift than negative bias voltage. Furthermore, the blocking characteristics of the device had a noticeable change after irradiation, it means that the TID effect has an effect on the body diode. Changes in gate capacitance were also analyzed. The reason for this degradation could be that the high-energy particle radiation exciting plenty of electron-hole pairs in the oxide layer of MOSFET, the electrons flow out of the metal electrode after radiation, and the holes are captured by the oxide layer to form a new interface trap charge. This paper may have some reference value for the application of SiC power MOSFET in aerospace.

Keywords—*total ionizing dose irradiation (TID), traps, different gate voltages, silicon carbide (SiC) power MOSFETs*

I. INTRODUCTION

SILICON carbide (SiC) material as a wide band gap semiconductor material, has the advantages of higher electron drift saturation rate, higher breakdown electric field and lower dielectric constant compared with Si [1].In recent years, SiC has gradually replaced Si material as the preferred material for high temperature, high pressure, low power consumption and ultra-

This work was supported by the National Key Research and Development Plan (2020YFF0218500) and the Key Realm R&D Program of Guangdong Province (2020B010171002,2020B010173001, 2018B010142001 and 2019B010143002).

high speed switching devices for its excellent material properties [2][3].In such a trend that SiC power devices are gradually replacing Si devices, their applications are more and more extensive, which means that they need to work stably in a wide variety of environments, so the reliability study in the severe environment is essential.

With the rapid development of the aviation industry, more and more MOSFETs are used in space equipment [16]. However, due to the incidence of many high-energy rays and particles in the space environment. If the equipment is operated for a long time in the irradiation environment, the incidence of high-energy rays and particles will generate a large amount of charge accumulation in the semiconductor material, resulting in a total ionizing dose effect that leads to a decline in device performance or even failure [4][5].Therefore, it is essential to study the reliability of semiconductor devices in an irradiated environment.

Recently, some pieces of literature have focused on the influence of irradiation on silicon carbide power MOSFET. A. Akturk et al. examined the radiation effects of different radiation doses on SiC MOSFETs and found that gamma-ray radiation will cause changes in the electrical characteristics of SiC MOSFETs, and the interface state will change with the increase in the cumulative radiation dose [6]. High temperature annealing can restore the electrical parameters of the SiC MOSFETs degradation caused by the radiation effect, which was discovered by T. Yokoseki et al. By comparison of the temperature and the time began to recover electrical parameters, found as compared with Si MOSFET, the interface traps and oxide trapped charge of SiC MOSFET are more stable. In order to find the optimal structure of the radiation resistant SiC MOSFETs, S. Mitomo et al. studied the influence of the thickness of the gate oxide layer and the nitridation process on the radiation resistance of the SiC MOSFETs [7].

In this paper, the total ionizing dose irradiation test is used to simulate the space environment to study the anti-irradiation ability of SiC MOSFETs devices. The degradation behavior of the silicon carbide power devices was investigated under total ionizing dose irradiation with different gate voltages. Our experiment data demonstrated that the degradation of the silicon carbide power MOSFETs under different gate voltage conditions is different. Therefore, we finally compare and analyze the causes of electrical performance degradation. This research may be helpful to the design and application of anti-irradiation devices.

II. EXPERIMENTAL

The commercial devices used in this experiment were n-type SiC power MOSFETs manufactured by Cree company. And the device is 1.7 kV(@25°C) standard TO-247-3 package device. Fig. 1(a) is the cross-section schematic draw of this experimental device. The Circuit symbol diagram of the device with terminal definitions and the actual photograph of the experimental device are given in Fig. 1(b), and Fig. 1(c).

In order to simulate the radiation environment, ^{60}Co was used as the γ ray radiation source , the dose rate was 50 rad/s, and the cumulative total dose was 1M rad. Moreover, to explore the influence of the TID effect on the electrical performance of the SiC MOSFETs under different working conditions, we set three gate voltage conditions for the device during the irradiation process: +10V, no external voltage, and -11V. In addition, in order to determine whether the continuous application of the gate voltage has an effect on the experiment, we continue to apply the positive gate voltage (+10V) for the same duration to a fresh device.

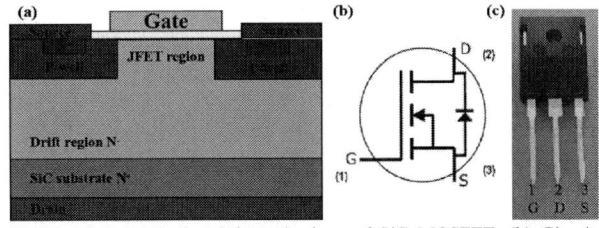

Fig. 1 (a) Cross-section Schematic draw of SiC MOSFET, (b) Circuit symbol diagram of the device, and (c) an actual photograph of the experimental device

III. RESULTS AND DISCUSSION

A. Transformation of Electrical Characteristics under TID effect

The output and transfer characteristics are the basic static electrical properties of silicon carbide power MOSFETs [14]. Therefore, in this experiment, the output and transfer curves of SiC power MOSFET are used as a tool to measure the variation of static electrical parameters of the device. Then these two curves are seen from Fig. 2-3.

During the irradiation process, for the device with positive gate voltage applied, it can be seen clearly in Fig. 2-3(a), both the output and transfer characteristic curves showed a significant negative shift.

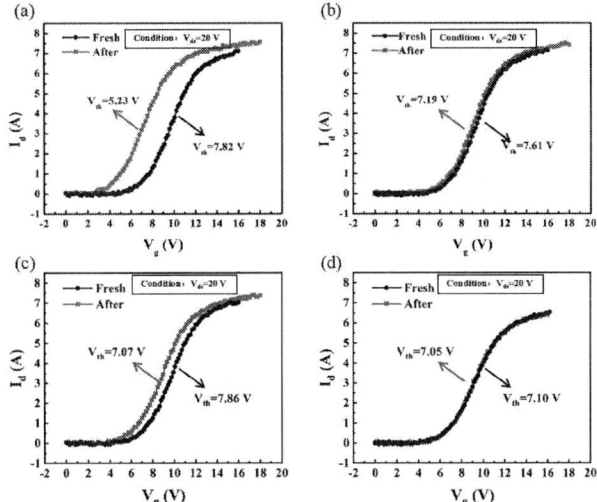

Fig. 2 Transfer characteristics $(I_d\text{-}V_g)$ of 1.7 kV 5 A SiC power MOSFETs: (a) V_{gs}=+10 V with irradiation; (b) no external voltage with irradiation; (c) V_{gs}=-11 V with irradiation; (d) V_{gs}=+10 V with non-irradiation.

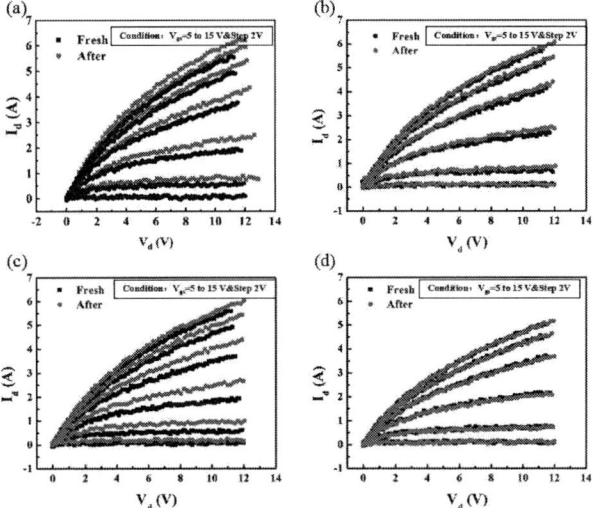

Fig. 3 Output characteristics $(I_d\text{-}V_d)$ of 1.7 kV 5 A SiC power MOSFETs: (a) V_{gs}=+10 V with irradiation; (b) no external voltage with irradiation; (c) V_{gs}=-11 V with irradiation; (d) V_{gs}=+10 V with non-irradiation.

The two characteristic curves of the device with negative gate voltage can also observe the phenomenon of negative drift as shown in Fig. 2-3(c), but it is not as obvious as Fig. 2-3(a). The phenomenon of Fig. 2-3(b) without applied voltage and Fig. 2-3(d) with only continuous positive gate voltage applied but not irradiated is even less obvious. The threshold voltage (V_{th}) is marked on Fig. 2. In this experiment, we used the linear extrapolation method to extract the threshold voltage [11] , which takes an inflection point as the tangent line, and the intersection point between the tangent line and the X axis is the threshold voltage value. Where, this inflection point is the point on the transition curve corresponding to the V_g with the maximum transconductance (g_{mmax}) [12].

It serves to show from Fig. 2 that the V_{th} also exhibits negative drift. This indicates that the device is damaged during the irradiation process, and the increase of positive

978-1-6654-4817-8/21 $31.00 © 2021 IEEE

charge defects in the gate oxygen layer leads to the decrease of threshold voltage.

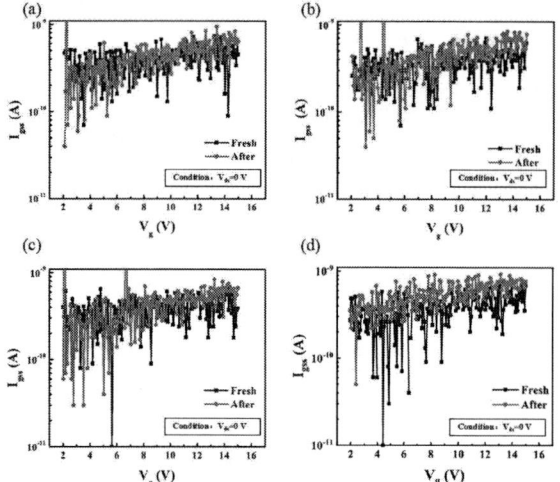

Fig. 4 Gate leakage current characteristics $(I_{gss}\text{-}V_{gs})$ of 1.7 kV 5 A SiC power MOSFETs: (a) V_{gs}=+10 V with irradiation; (b) no external voltage with irradiation; (c) V_{gs}=-11 V with irradiation; (d) V_{gs}=+10 V with non-irradiation.

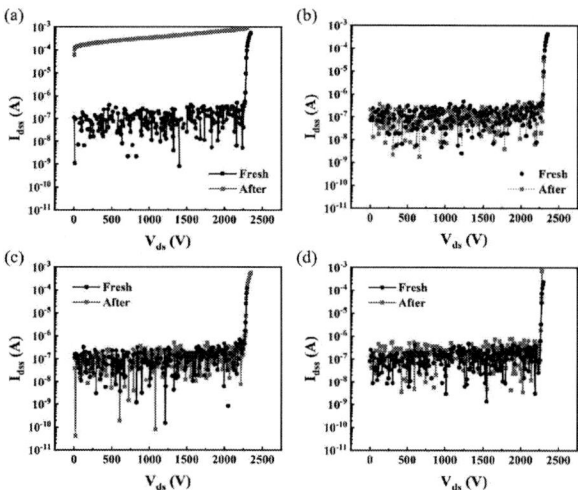

Fig. 5 Blocking characteristics $(I_{dss}\text{-}V_d)$ of 1.7 kV 5 A SiC power MOSFETs :(a) V_{gs}=+10 V with irradiation; (b) no external voltage with irradiation; (c) V_{gs}=-11 V with irradiation; (d) V_{gs}=+10 V with non-irradiation.

In addition, in order to study the influence of the total ionizing dose radiation effect on the gate voltage layer and body diode, the gate leakage current curve and the blocking characteristic curve are shown in Fig. 4 and Fig. 5. It serves to show from Fig. 4 that the gate leakage current only has a little increase, which confirmed that the gate pressure layer of the device has good radiation resistance. However, as show in Fig. 5(a) that the blocking characteristic curve of the device after the positive gate voltage irradiation is different from that of other devices. It can be inferred that the body diode of the device represented by Fig. 5(a) was broken down after the experiment. And the comparative experiments of Fig. 5(a) and Fig. 5(d) can prove that the breakdown phenomenon is not caused by the continuous application of gate voltage, but the total dose effect under the positive gate voltage.

B. Variation of C-V Characteristics under TID effect

MOS capacitor structure is the core of MOSFET. From the capacitance-voltage relationship of the device, that is, the C-V characteristic curve, we can get a great deal of information about the MOS devices and gate oxide. A large amount of information at the interface between the MOS device and the gate oxide layer can be obtained from the capacitance-voltage relationship of the device, which is the C-V characteristic curve. The measured C_g–V_g characteristic curve can be used to determine the location of device damage [9]. During the measurement, to measure the gate capacitance (C_g), V_s=V_d=0 V, an AC small signal (V_{ac}) was given to the gate. The C_g-V_g characteristic of silicon carbide power MOSFET can be roughly separated into five parts, each of which corresponds to a different state of the device channel and Junction Field-Effect Transistor (JFET) region [10], The part I, II, III and IV of gate capacitance (C_g) can be expressed as:

$$C_g = C_{oc} + C_{oj} \tag{1}$$

$$C_g = \cfrac{1}{\cfrac{1}{C_{oc}} + \cfrac{1}{C_{dc}}} + C_{oj} \tag{2}$$

$$C_g = \cfrac{1}{\cfrac{1}{C_{oc}} + \cfrac{1}{C_{dc}}} + \cfrac{1}{\cfrac{1}{C_{oj}} + \cfrac{1}{C_{dj}}} \tag{3}$$

$$C_g = C_{oc} + \cfrac{1}{\cfrac{1}{C_{oj}} + \cfrac{1}{C_{dj}}} \tag{4}$$

Where the oxide capacitance and the depletion capacitance of the channel and the JFET region can be indicated as C_{oc}, C_{dc}, C_{oj}, and C_{dj}, respectively [9].

It can be seen from the above formula that the II region of the C_g-V_g characteristic curve corresponds to the degradation of the device channel region. In addition, the part of III and IV correspond to the degradation of the JFET region [13].Based on this characteristic, the damage location of gate oxygen layer can be determined by segmental C_g–V_g curve.

As shown in Fig. 6, for the irradiated device, the part I, II and III shifted negatively, which explains that the JFET and channel region were damaged during the irradiation. Among them, the device with positive gate voltage has the most obvious shift, followed by the device with negative gate voltage, and the device without voltage has only a slight shift. This indicates that during total ionizing dose irradiation, the JFET region and the channel of the device is significantly damaged. Moreover, it can be seen from the drift direction of capacitance curve that the reason for degradation is that the defects with positive charge in the gate oxygen layer increase [8].

This is consistent with the previous conclusions drawn through electrical performance.

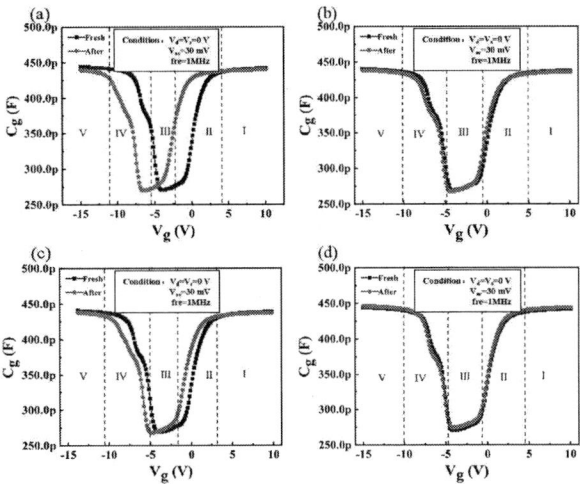

Fig. 6 Variations of the C_g-V_g characteristic of 1.7 kV 5 A SiC power MOSFETs: (a) V_{gs}=+10 V with irradiation; (b) no external voltage with irradiation; (c) V_{gs}=-11 V with irradiation; (d) V_{gs}=+10 V with non-irradiation.

C. Degradation Mechanism of SiC MOSFETs

To sum up, both the transfer and output characteristic curves of the irradiated device degrade to varying degrees, and the reason for this degradation phenomenon is that when the γ-ray incidents on device, there will be ionization damage inside the device. Some of the bound electrons in the semiconductor material break free after absorbing the energy of the incident particle, create electron hole pairs [16]. Since the mobility of electrons is higher than that of holes in semiconductor materials, the generated electrons in the electron-hole pairs will quickly flow out of the metal electrode, and the holes with low mobility are captured by the oxidation layer traps during the process of moving towards to the SiO$_2$/SiC interface, and the fixed positive charge of the oxide layer is formed [8]. This process is shown in the Fig.7. The oxide layer traps charges far away from the interface. The electric field generated by the positive charge can be considered to have the same effect as the positive gate electric field. It will induce negative charges in the device channel, invert the channel surface, and cause a negative drift of the threshold voltage.

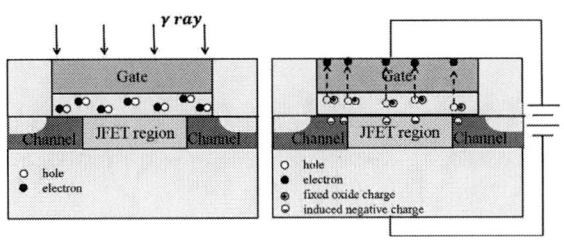

Fig. 7 Degradation process of SiC power MOSFET under total ionizing dose irradiation

Whether gate voltage is positive or negative, the breakdown of electron-hole pair will be accelerated.

However, the experimental results obtained by applying positive pressure and negative pressure are not the same. This is because the negative voltage causes the electrons of the hole pair to move towards the SiO$_2$/SiC interface, while the

holes move towards the gate electrode. SiO2/ oxide traps are less, so the number of holes captured is relatively small.

In addition, in the vertical MOSFET structure, the thicker the drift region, the lower the doping concentration, the larger the expansion space of the depletion region, and the higher the drain-source breakdown voltage of the device. As shown in Fig. 5, after irradiation the breakdown voltage of the device is significantly reduced. This is because the oxide trapped charge induced by irradiation will reduce the width of the depletion region, which causes a decrease in the breakdown voltage, which is consistent with the conclusions drawn from Fig. 1 and Fig. 2.

IV. CONCLUSIONS

The changes of the SiC MOSFETs' electrical characteristics before and after the total ionizing dose irradiation experiment under different gate voltage conditions are studied. These electrical characteristics are significantly affected by the total ionizing dose irradiation effect. TID effect will reduce the device performance, for example, the negative shift of threshold voltage and C_g-V_g curve. The degree of degradation caused by irradiation under different gate voltage conditions is different. In addition, the blocking characteristics of the device subjected to the irradiation experiment under the condition of positive gate voltage deteriorated. Moreover, through the comparison of the gate capacitance curve before and after, the damage area of the device is determined. The study maybe offers some important insights into the design of radiation-resistant devices and the application of SiC MOSFETs in aviation, aerospace, and other fields.

REFERENCES

[1] Fayyaz A, Romano G, Urresti J, et al. A comprehensive study on the avalanche breakdown robustness of silicon carbide power MOSFETs[J]. Energies, 2017, 10(4): 452.

[2] Fayyaz A, Castellazzi A, Romano G, et al. UIS failure mechanism of SiC power MOSFETs[C]//2016 IEEE 4th Workshop on Wide Bandgap Power Devices and Applications (WiPDA). IEEE, 2016: 118-122.

[3] Abbate C, Busatto G, Iannuzzo F. Unclamped repetitive stress on 1200 V normally-off SiC JFETs[J]. Microelectronics Reliability, 2012, 52(9-10): 2420-2425.

[4] Hughes H L, Giroux R R, Space radiation affects MOSFET's[J]. Electronics, 1964, 37(32): 58-60.

[5] Barmakov Y N, Butin V I, Butina A V. MOSFETs application for TID measurements[C]//2017 International Multi-Conference on Engineering, Computer, and Information Sciences (SIBIRCON). IEEE, 2017: 254-256.

[6] Akturk A, Mcgarrity J M, Potbhare S, et al. Radiation Effects in Commercial 1200 V 24 A Silicon Carbide Power MOSFETs[J]. IEEE Transactions on Nuclear Science, 2012, 59(6): 3258-3264.

[7] Mitomo S, Matsuda T, Murata K, et al. Optimum structures for gamma-ray radiation resistant SiC-MOSFETs[J]. Physica Status Solidi, 2017, 214(4).

[8] Ausman G A, McLean F B. Electron-hole Pairs Creation Energy in Si O$_2$[J]. Appl Phys, 1975, 26:173-175.

[9] Wei J, Liu S, Ye R, et al. Interfacial damage extraction method for SiC power MOSFETs based on CV characteristics[C]//2017 29th International Symposium on Power Semiconductor Devices and IC's (ISPSD). IEEE, 2017: 359-362.

[10] Zhou X, Su H, Yue R, et al. A deep insight into the degradation of 1.2-kV 4H-SiC MOSFETs under repetitive unclamped inductive switching stresses[J]. IEEE Transactions on Power Electronics, 2017, 33(6): 5251-5261.

[11] Dobrescu L, Petrov M, Dobrescu D, Ravariu C. Threshold Voltage Extraction Methods for MOS Transistors[J]. International Semiconductor Conference, 2000, 1:371-374.

978-1-6654-4817-8/21 $31.00 © 2021 IEEE

[12] Tsividis Y. Operation and Modeling of the MOS Transistor[M]. McGraw-Hill, Inc., 1987.

[13] Wang J, Chen Y, Feng J, et al., Trap Analysis Based on Low-Frequency Noise for SiC Power MOSFETs Under Repetitive Short-Circuit Stress[J], IEEE Journal of the Electron Devices Society, 2020,8, 145-151.

[14] Yang X, Chen Y Q, Hou B, et al. Degradation Behavior and Defect Analysis for SiC Power MOSFETs Based on Low-Frequency Noise Under Repetitive Power-Cycling Stress[J]. IEEE Transactions on Electron Devices, 2020, 68(2): 666-671.

[15] Xie T, Ge H, Lv Y, et al., The impact of total ionizing dose on RF performance of 130 nm PD SOI I/O nMOSFETs[J]. Microelectronics Reliability, 2021, 116: 114001.

[16] Ali K B, Gammon P M, Chan C W, et al. Single event effects and total ionising dose in 600V Si-on-SiC LDMOS transistors for rad-hard space applications[C]//2017 47th European Solid-State Device Research Conference (ESSDERC). IEEE, 2017: 236-239.

Investigation on Parameter Extraction for An Improved Fourier-Series-Based NPT IGBT Model

Yifei Ding[a], Xin Yang[a], Jun Wang[a], Chunming Tu[a], Guoyou Liu[b]

[a] College of Electrical and Information Engineering, Hunan University, Changsha, China, 410082
[b] CRRC Zhuzhou Electric Locomotive Institute Co., Ltd., Zhuzhou, China, 412000
Email: xyang@hnu.edu.cn

Abstract—With the wide application of IGBTs in recent years, the accuracy of IGBT models has been improved constantly. Physics-based IGBT models have shown great advantages on the simulation accuracy. However, the unavailability of device parameters is one of the main factors that hinder the application of those physics-based models. In this paper, a NPT IGBT chip is designed with the assistance of two-dimensional (2-D) finite-element method (FEM) model implemented in Sentaurus TCAD, which enables access to the intrinsic device parameters. This way, the IGBT parameters of the FEM model can provide a benchmark for accuracy evaluations of device parameters. An optimization-based parameter extraction procedure is proposed. This procedure, using an improved Fourier-series-based IGBT model, enables an efficient extraction of 10 parameters of the NPT IGBT requiring only IGBT turn-off waveforms in a simple clamped inductive load test. Validation with simulated switching characteristic results under different operating conditions demonstrates the usefulness and robustness of the extracted parameters. By comparing the extracted parameters with the intrinsic parameters set in TCAD, the stability of the extracted parameters is discussed in detail.

Keywords—Insulated gate bipolar transistor (IGBT), model, parameter extraction, particle swarm optimization, switching transient.

I. INTRODUCTION

IGBT is one kind of widely-used power semiconductor devices. IGBT model can help circuit designers to predict circuit behavior, understand device fundamentals and design chips [1]. Therefore, accurate modeling of IGBT devices is indispensable. The physics-based IGBT model can reflect internal changes of devices during the switching process such as the carrier concentration distribution and electric field intensity [2], and can have a good performance under different working conditions. The physics-based IGBT model is not only has a fast simulation speed, but also has a high precision, which can be used in closed-loop gate drive design simulation, device losses estimation, and junction-temperature-sensitive physical parameters verification [3].

Early physics-based IGBT models, such as the Hefner model [4] and Kraus model [5], can produce good switching results but cannot capture accurate transient behavior of carrier distribution due to the assumptions on the carrier storage region (CSR). The lumped-charge model has fast simulation speed [6], but it cannot provide carrier distributions close to actual chips in the base region because of its piecewise simplification in the base region. Palmer et al developed the Fourier-series-based (FSB) IGBT model [7-9] which can not only accurately simulate the switching waveform of IGBTs, but also provide accurately carrier dynamics information in

the base region. The FSB IGBT model has been proven to be very efficient in terms of simulation speed and it can be realized in the circuit simulation platform [10]. However, device manufacturers generally seldom provide complete chip design parameters and the parameters used in physics-based models are difficult to be extracted from the corresponding datasheets, which discourage power electronic engineers from using those physical IGBT models.

A precise physics-based model requires precise parameters. Thus, parameter extraction has received great attention in previous studies. Hefner proposed a parameter extraction method consists of 5 programs in [11], and an improved parameter extraction method through Hefner model contains of 6 test circuits and 6 algorithms is proposed in [12]. However, these methods require precise and complex experiments, which are inconvenient for electronic engineers. In [13-15], the device parameters are extracted by extrapolating datasheet information and estimating correlation equations. However, the results of the extraction method lack accuracy because datasheet information only represents mean values and some parameters obtained by empirical estimation have appreciable errors [16].

Using optimization algorithm to extract parameters has been realized in references [8, 16], and only simple device waveform measurements are needed to obtain device parameters. The method is more convenient, but they only provides the nicely-fitted switching waveforms and the stability of the extracted parameters is not discussed. Additionally, it is very likely to obtain good electrical characteristics by using different parameter sets, but accurate physical characteristics can only be obtained by using parameter sets that agree with the intrinsic parameter set. Therefore, our goal here is to stably and accurately obtain device intrinsic parameters of the IGBT. Although experimental verification can well prove the practicability of the extracted parameters, there is no target device parameter set that can be used as a reference. Therefore, a NPT IGBT chip is designed with the assistance of two-dimensional (2-D) finite-element method (FEM) model implemented in Sentaurus TCAD and the parameters set in TCAD simulation are used as a reference. Further, an improved IGBT model in reference [3], which is proved to be in good agreement with the FEM model, is chosen.

This paper propose a parameter extraction procedure using the improved FSB NPT IGBT model. The proposed parameter extraction procedure based on PSO algorithm is convenient and only IGBT device turn-off waveforms in a simple clamped inductive load test is required. By comparing the switching characteristics of the FSB IGBT model with those of the finite element result under different operating

978-1-6654-4817-8/21 $31.00 © 2021 IEEE

conditions, the robustness of the extracted parameters is verified. Furthermore, the extracted parameters are compared with the parameters set in TCAD, and the stability of the extracted parameters is discussed in detail by analyzing the relative errors of the extracted parameters and the physical characteristics of IGBT device.

II. THE IMPROVED FOURIER-SERIES-BASED IGBT MODEL

A. The Solution of the Ambipolar Diffusion Equation

Under the high-level injection condition, ambipolar diffusion equation (ADE) can described carrier dynamics:

$$D \frac{\partial^2 p(x,t)}{\partial x^2} = \frac{p(x,t)}{\tau_{HL}} + \frac{\partial p(x,t)}{\partial t} \qquad (1)$$

where D is the ambipolar diffusion coefficient, τ_{HL} is the high-level carrier lifetime in the drift region, $p(x,t)$ is the hole concentration at position x and time t.

The method of solving the ADE by Fourier series is proposed by Leturcq [17]. The form of ADE is a partial differential equation (PDE) which is difficult to solve. By decomposing $p(x,t)$ using Fourier series, ADE can be converted into ordinary differential equations (ODEs), which can be solved by numerical simulation software (MATLAB/Simulink [18]), as well as by circuit simulation software (PSpice [10]).

In order to solve the set of ODEs, the carrier density gradients $\partial p/\partial x|_{1,2}$, boundary positions $x_{1,2}$ and their derivatives with respect to time $dx_{1,2}/dt$ are required as the boundary conditions. Where x_1 is the boundary position at anode end of CSR and x_2 is the boundary position at cathode end of CSR. Carrier density gradients can be calculated by the boundary currents [18]. x_1 is equal to 0 at the collector side of IGBT. The Boundary position x_2 and its derivative with respect to time can be obtained by depletion layer voltage V_{d2}.

B. Mitigating Gibbs Phenomenon and Considering 2-D effects

The carrier distribution of 90% volume in the base region can be approximately considered as 1-D distribution [7]. Because Fourier series only retain a finite number of terms, Gibbs phenomenon occurs, which affects the accuracy of the model the 1-D FSB IGBT model at turn-on as shown in Figs. 1 and 2. The carrier concentration of a IGBT in the base region is very low at the beginning of turn-on process. Due to the rapid decline of V_{CE}, the depletion layer contracted rapidly. At this time, the carrier near the anode in the CSR will accumulate rapidly, but the carrier concentration near the cathode in CSR remains still at a low level as shown in Fig. 3(a), which will be influenced by the Gibbs phenomenon. Gibbs phenomenon will affect the electric field intensity in the low carrier concentration region [3]. Therefore, the FSB IGBT model will be greatly affected by Gibbs phenomenon at turn-on.

On the contrary, this 1-D FSB model is almost unaffected by Gibbs phenomenon in the turn-off process. When IGBT is in the turn-off process, due to the rapid increase of V_{CE}, the depletion layer expanded rapidly, resulting in the rapid sweep out of the carriers near the cathode in the CSR. At this time, the carrier concentration at the end of CSR (close to the emitter side) is still at a high level as shown in Fig. 3(b). Therefore, the model will not be affected by Gibbs phenomenon basically at turn-off.

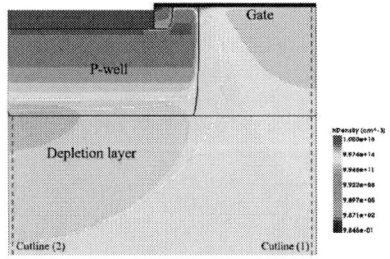

Fig. 1. Hole density distributions under the gate and the P-well in a 2-D IGBT FEM model implemented in Sentaurus TCAD.

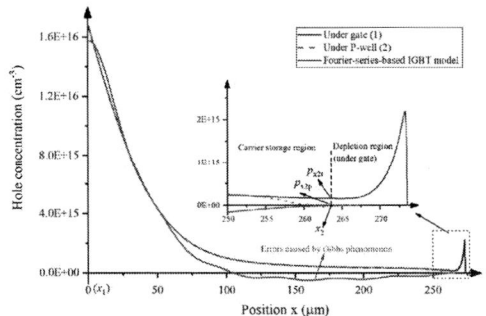

Fig. 2. Hole concentrations in the turn-on process under the gate (cutlines (1)) and the Pwell (cutlines (2)) of the FEM model in Fig. 1. With the hole concentration of the FSB IGBT Model.

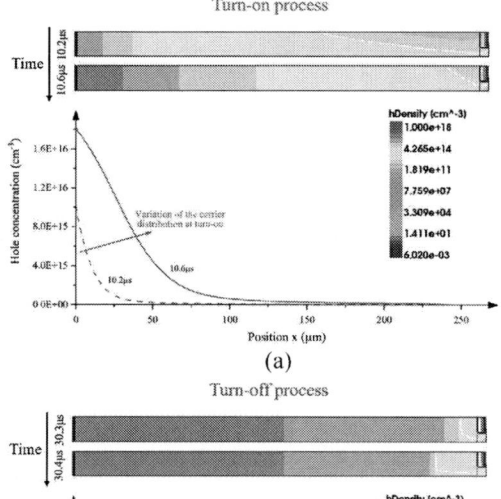

(a)

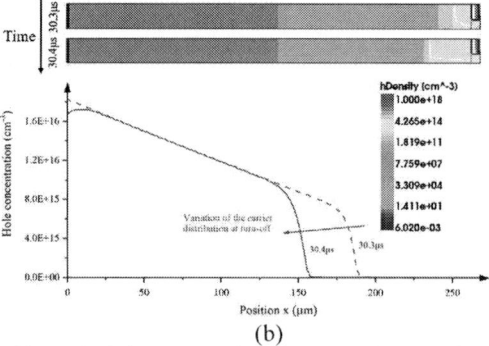

(b)

Fig. 3. Change in hole concentration of the TCAD simulation in the switching process. (a) Turn-on. (b) Turn-off. IGBT starts to turn on at 10 μs and starts to turn off at 30 μs.

978-1-6654-4817-8/21 $31.00 © 2021 IEEE

TABLE I
IGBT MODEL PARAMETER LIST

Symbol	Units	Description
K_{psat}	AV^{-2}	MOS transconductance coefficient
λ	V^{-1}	Short-channel parameter
C_{OX}	$nFcm^{-2}$	Oxide capacitance
l_m	μm	Inter-cell half-width
A	cm^2	Device area
a_i	-	Ratio of intercell to total die area
N_B	cm^{-3}	Doping concentration of N-drift region
W_B	μm	Width of N- drift region
τ	μS	Lifetime
H_P	$cm4/s$	Hole recombination coefficient in emitter

The carrier distribution under the cathode in the base region is affected by the MOS 2-D structure and presents a 2-D distribution. The carrier concentration under the P-well is very low ($p_{x2p} \approx 0$) due to the existence of the depletion layer, and, there is a relatively high carrier concentration under the gate ($p_{x2t} > 0$) due to the existence of the MOS channel as shown in Fig. 2. Due to the lifting of P_{x2} by 2-D effects, the effect of the Gibbs phenomenon can be effectively mitigated, which has been realized in references [3, 19 and 20]. The model in reference [19] is a 2-D model and the calculation cost of iteration will be very high, the improved model in the reference [3] is selected as the model used for parameter extraction. The boundary current in the improved model is calculated as follows:

$$I_{MOS} = \frac{b}{b+1} I_C \qquad (2)$$

$$b = \frac{\mu_n}{\mu_p} \qquad (3)$$

where μ_n and μ_p are electron mobility and hole mobility. MATLAB/Simulink is selected as the software platform to implement the FSB IGBT model.

A 2-D FEM model corresponding to the FSB IGBT model is implemented in Sentaurus TCAD. It has been verified that the improved FSB IGBT model can fit the FEM model well in both switching and physical characteristics under the same structure [3]. Among the IGBT model parameters, the MOS threshold voltage V_{TH} and the gate-emitter capacitance C_{GE} are obtained directly from FEM. The method of obtaining the remaining IGBT model parameters, as shown in Table I, will be discussed in detail in Section III.

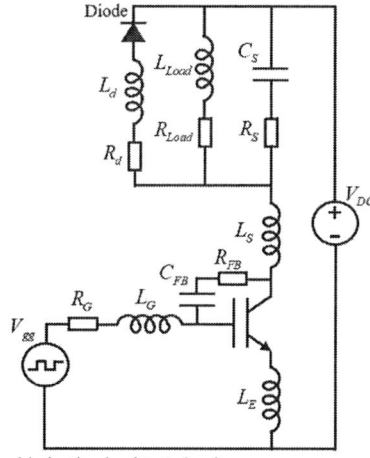

Fig. 4. Clamped inductive load test circuit.

III. PARAMETER EXTRACTION PROCEDURE

A. Selection of the Target Function

The freewheel diode basically does not affect the turn-off waveform of IGBT, thus this paper extracts NPT IGBT parameters by fitting V_{CE}, I_C and V_G at turn-off and extracts freewheel diode parameters by fitting V_{CE}, I_C and V_G in the turn-on process. In this way, IGBT parameters can be extracted more stably. The test circuit is shown in Fig. 4. The advantage of this approach is that the accuracy of voltage and current can be ensured, and the instantaneous power dissipation can be calculated through the product of V_{CE} and I_C, so as to ensure the accuracy of power loss estimation.

In order to fit the V_{CE}, I_C and V_G simultaneously, normalization processing is required. This is because the size range of these traces is different, and direct identification without normalization will lead to different weights for different traces as it is difficult to fit the traces with a small weight ratio. The normalization process is as follows:

$$V'_{CE} = \frac{V_{CE}}{V_{DC}}$$

$$I'_C = \frac{I_C}{I_F} \qquad (4)$$

$$V'_G = \frac{V_G}{V_{DC}}$$

Therefore, the sum of the squared errors for the normalized current and voltage at turn-on/turn-off is selected as the target as:

$$f_e = SSE(V'_{CE}) + SSE(I'_C) + SSE(V'_G) \qquad (5)$$

TABLE II
PARAMETERS OBTAINED THROUGH THE PARAMETER EXTRACTION (PE) PROCEDURE

	K_{psat}	λ	C_{OX}	l_m	A	a_i	N_B	W_B	τ	H_P
TCAD (standard)				6.5	1.00	0.62	5.00	268	39.0	
PE1	6.4	1.0	13.1	6.0	0.96	0.58	5.30	284	71.0	10.0
PE2	6.0	2.8	26.2	8.1	0.70	0.47	6.98	247	39.9	12.5
PE3	6.5	2.7	24.1	6.5	1.04	0.35	5.16	290	47.4	8.9
PE4	6.4	1.5	13.5	5.3	1.30	0.43	3.95	300	25.8	7.2
PE5	5.4	6.7	15.0	1.0	0.73	0.66	4.91	285	54.7	13.9
Units	AV^{-2}	$10^{-3}V^{-1}$	$nFcm^{-2}$	μm	cm^2	-	$10^{13}cm^{-3}$	μm	μs	$10^{-13}cm^4/s$
Mean relative error				22.4%	19.2%	22.5%	14.3%	8.2%	36.0%	

978-1-6654-4817-8/21 $31.00 © 2021 IEEE

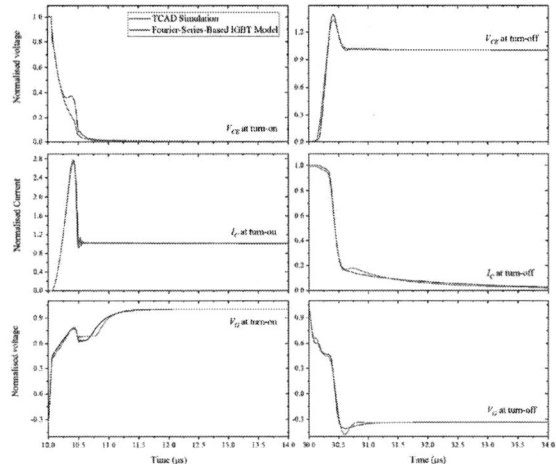

Fig. 5. Comparison of V_{CE}, I_C and V_G of the FSB IGBT model with the TCAD simulation at 750V/75A/300K.

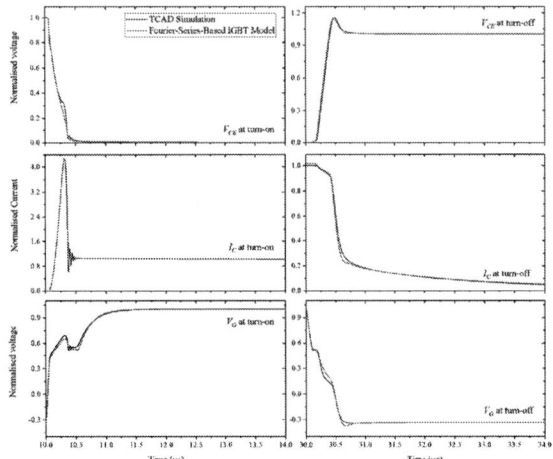

Fig. 6. Comparison of V_{CE}, I_C and V_G of the FSB IGBT model with the TCAD simulation at 700V/30A/300K.

$$SSE(x) = \sum_{k=1}^{M} (x_{sim}(k) - x_{TCAD}(k))^2 \qquad (6)$$

where M is the number of sampling points, x_{sim} is the data extracted from the IGBT model, x_{TCAD} is the data extracted from the TCAD simulation. In this paper, particle swarm optimization (PSO) algorithm is used for automatic optimization.

B. Acquisition of Target Waveform Data

Starting from the point where the supply voltage signal on the gate begins to change, waveforms within 4 µs are collected at a sampling frequency of 200Mhz with 801 sampling points within each sampling interval. Only turn-off waveforms are sampled when extracting IGBT parameters, and only turn-on waveforms are sampled when extracting diode parameters.

IV. RESULTS AND DISCUSSION

The parameter extraction procedure selects V_{CE}, I_C and V_G of the TCAD simulation at 750V/75A/300K as the goal. PSO algorithm is realized in MATLAB, through the 'sim' function to achieve the interaction with Simulink. The FSB IGBT model use M=13 Fourier terms. The estimation of the range of the parameters is obtained by the method in reference [8].

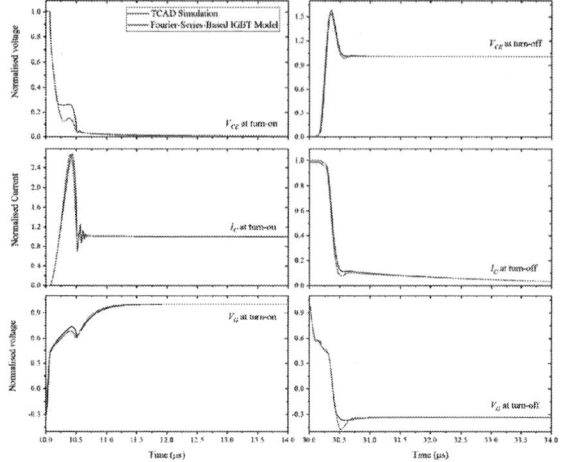

Fig. 7. Comparison of V_{CE}, I_C and V_G of the FSB IGBT model with the TCAD simulation at 450V/60A/300K.

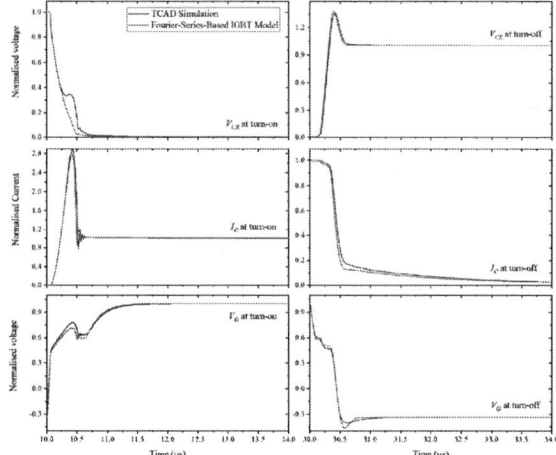

Fig. 8. Comparison of V_{CE}, I_C and V_G of the FSB IGBT model with the TCAD simulation at 750V/75A/270K.

The temperature correlation of the parameters is shown in reference [7].

The parameter extraction procedure is intentionally repeated for 5 times to investigate the stability of the extracted parameters, and their results are shown in Table II.

W_B and N_B are key parameters of the FSB IGBT model, which determine the width and the doping concentration in drift region respectively. The mean relative errors (MREs) of the extracted N_B and W_B are less than 15%. As the key parameters of the FSB IGBT model, the extracted N_B and W_B are very close to the intrinsic parameters. The extracted l_m, A and a_i are also close to the parameters set in the FEM model, and their MREs are less than 23%. The MRE of τ is relatively large, which is the same as the result in literature [8]. The reason may be that the carrier lifetime has no obvious influence on the waveform in the switching process of the FSB IGBT model compared with other parameters. To sum up, combined with the improved FSB IGBT model, the proposed parameter extraction process can accurately and steadily extract most IGBT parameters.

As can be seen from Figs. 5-7, the waveform of I_C at turn-on, V_{CE} at turn-off and V_G is in good agreement with TCAD simulation. At turn-on, V_{CE} has obvious bumps at 750V/75A and 450V/60A, but the bump is less obvious at 700V/30A. This is resulted from the diode reverse recovery. During this

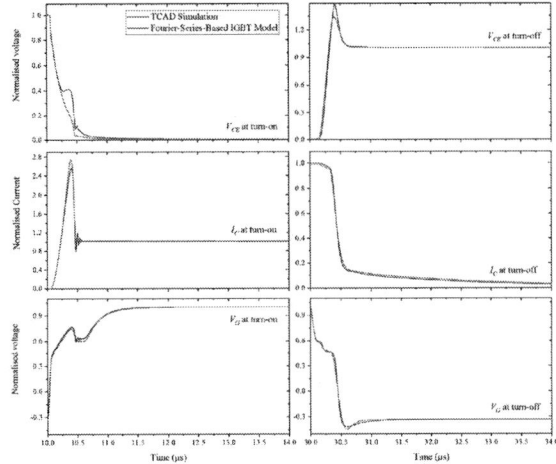

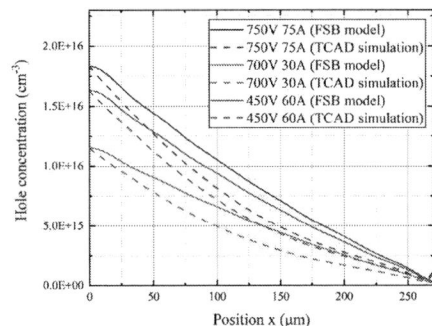

Fig. 9. Comparison of V_{CE}, I_C and V_G of the FSB IGBT model with the TCAD simulation at 750V/75A/360K.

Fig. 10. Comparison of carrier concentration distribution of the FSB IGBT model with the TCAD simulation at on-state using parameters extracted through PE3.

process, I_C will increase rapidly, resulting in an increase of voltage drop in the CSR, forming an obvious bump. The FSB IGBT model cannot simulate this voltage drop in the CSR well. Thus, V_{CE} of the FSB IGBT model is smaller than that of the TCAD simulation during the diode reverse recovery phase, which is discussed in detail in reference [3]. The voltage drop in the CSR is positively correlated with I_C. That is why this bump phenomenon is not obvious at 700V/30A. The waveform results of the FSB IGBT model at different temperatures are basically consistent with the TCAD simulation as shown in Figs 5, 8 and 9. All the parameter sets in table II can provides a nicely-fitted switching waveforms, and the FSB IGBT model used in Figs. 5-9 utilizes the parameters obtained through PE3.

Fig. 10 shows the on-state carrier distribution of the FSB IGBT model under different operating conditions. The results show that the carrier distribution of the FSB IGBT model can basically fit that of the FEM simulation.

To sum up, the extracted parameters can be applied to the model at different operating voltages, currents and temperatures, which proves the robustness of the FSB IGBT model using the extracted parameters. The proposed parameter extraction procedure can make the model accurate in terms of electrical and physical characteristics simultaneously.

V. CONCLUSION

A parameter extraction procedure using the improved NPT FSB IGBT model is proposed in this paper. By comparing the FSB IGBT model with the TCAD simulation under different voltages, different currents and different temperatures, the robustness of the FSB IGBT model using extracted parameters and the availability of the extracted parameters are verified. In addition, the stability of the extracted parameters is discussed.

ACKNOWLEDGMENT

This work was supported in part by National Key R&D Program of China (2018YFB1201802-3) and the National Natural Science Foundation of China 52077071.

REFERENCES

[1] K. Sheng, B. W. Williams, and S. J. Finney, "A review of IGBT models," *IEEE Trans. Power Electron.*, vol. 15, no. 6, pp. 1250–1266, Nov. 2000.

[2] S. Ji, Z. Zhao, T. Lu, L. Yuan, and H. Yu, "HVIGBT physical model analysis during transient," *IEEE Trans. Power Electron.*, vol. 28, no. 5, pp. 2616–2624, May 2013.

[3] X. Yang, Y. Ding, J. Wang, G. Liu and P. R. Palmer, "An Improved Fourier-Series-Based IGBT Model by Mitigating the Effect of Gibbs Phenomenon at Turn on", *IEEE Transactions on Electron Devices*, vol., no., pp.1-6, 2022.

[4] A. R. Hefner, Jr., "Modeling buffer layer IGBTs for circuit simulation," *IEEE Trans. Power Electron.*, vol. 10, no. 2, pp. 111–123, Mar. 1995.

[5] R. Kraus, P. Turkes, and J. Sigg, "Physics-based models of power semiconductor devices for the circuit simulator SPICE," in *Proc. IEEE PESC*, Fukuoka, Japan, 1998, vol. 2, pp. 1726–1731.

[6] Y. Duan, F. Iannuzzo and F. Blaabjerg, "A New Lumped-Charge Modeling Method for Power Semiconductor Devices," *IEEE Trans. Power Electron.*, vol. 35, no. 4, pp. 3989-3996, April 2020.

[7] P. R. Palmer, E. Santi, J. L. Hudgins, X. Kang, J. C. Joyce, and P. Y. Eng, "Circuit simulator models for the diode and IGBT with full temperature dependent features," *IEEE Trans. Power Electron.*, vol. 18, no. 5, pp. 1220–1229, Sep. 2003.

[8] A. T. Bryant, X. Kang, E. Santi, P. R. Palmer, and J. L. Hudgins, "Two-step parameter extraction procedure with formal optimization for physics-based circuit simulator IGBT and PIN diode models," *IEEE Trans. Power Electron.*, vol. 21, no. 2, pp. 295–309, Mar. 2006.

[9] X. Kang, A. Caiafa, E. Santi, J. L. Hudgins, and P. R. Palmer, "Characterization and modeling of high-voltage field-stop IGBTs," *IEEE Trans. Ind. Appl.*, vol. 39, no. 4, pp. 922–928, Jul./Aug. 2003.

[10] X. Kang, E. Santi, J. Hudgins, P. Palmer, and J. F. Donlon, "Parameter extraction for a physics-based circuit simulator IGBT model," in *Proc. IEEE Applied Power Electronics Conf. (APEC'03)*, Feb. 2003, pp. 946–952.

[11] A. R. Hefner and S. Bouché, "Automated parameter extraction software for advanced IGBT modeling," in *Proc. 7th Workshop Comput. Power Electron. (COMPEL)*, Blacksburg, VA, Jul. 2000, pp. 10–18.

[12] A. Claudio, M. Cotorogea, and M. A. Rodriguez, "Parameter extraction for physics-based IGBT models by electrical measurement," in *Proc. IEEE 33rd Annu. Power Electronics Specialists Conf. (PESC'02)*, Jun. 2002, pp. 1295–1300.

[13] X. Kang, E. Santi, J. Hudgins, P. Palmer, and J. F. Donlon, "Parameter extraction for a physics-based circuit simulator IGBT model," in *Proc. IEEE Applied Power Electronics Conf. (APEC'03)*, Feb. 2003, pp. 946–952.

[14] S.C. Yuan and C.C. Zhu: "IGBT SPICE model with nondestructive parameters extraction and measured verification", *IEE Proc. - Electr. Power App.*, vol. 150, No. 5, pp. 575-579, Sept. 2003

[15] J. Sigg, P. Türkes, and R. Kraus, "Parameter extraction methodology and validation for an electro-thermal physics-based NPT IGBT model," in *Proc. IEEE Ind. Applicat. Society Annual Meeting*, New Orleans, LA, Oct. 5–9, 1997.

[16] R. Chibante, A. Araujo, and A. Carvalho, "Finite-element modeling and optimization-based parameter extraction algorithm for NPT-

IGBTs," *IEEE Trans. Power Electron.*, vol. 24, no. 5, pp. 1417–1427, May 2009.

[17] P. Leturcq, "A study of distributed switching processes in IGBTs and other power bipolar devices," in *Proc. Power Electron. Specialists Conf.*, vol. 1. Jun. 1997, pp. 139–147.

[18] X. Yang, M. Otsuki, and P. R. Palmer, "Physics-based insulated-gate bipolar transistor model with input capacitance correction," *IET Power Electron.*, vol. 8, no. 3, pp. 417–427, 2015.

[19] A. T. Bryant, "Simulation and optimization of diode and IGBT interaction in a chopper cell," Ph.D. dissertation, Queens' College, Univ. Cambridge, Cambridge, U.K., Jan. 2005.

[20] L. Lu, A. Bryant, J. L. Hudgins, P. R. Palmer, and E. Santi, "Physics based model of planar-gate IGBT including MOS side two-dimensional effects," *IEEE Trans. Ind. Appl.*, vol. 46, no. 6, pp. 2556–2567, Nov./Dec. 2010.

[21] R. C. Eberhart and J. Kennedy, "A new optimizer using particle swarm theory," in *Proc. 6th Int. Symp. Micromach. Human Sci.*, 1995, pp. 39–43.

Soft-switching Resonant Active Clamp Flyback Converter based-on GaN HEMTs for MHz High Step-up Applications

Wuji MENG
Department of Electrical Engineering
Nanjing University of Aeronautics and Astronautics
Nanjing, China
mengwuji1122@nuaa.edu.cn

Lin LI
Department of Electrical Engineering
Nanjing University of Aeronautics and Astronautics
Nanjing, China
nuaa_lilin@126.com

Fanghua ZHANG
Department of Electrical Engineering
Nanjing University of Aeronautics and Astronautics
Nanjing, China
zhangfh@nuaa.edu.cn

Jianjun SHU
Department of Electrical Engineering
Nanjing University of Aeronautics and Astronautics
Nanjing, China
shujianjun@nuaa.edu.cn

Abstract—For MHz GaN-based power converters, zero-voltage switching (ZVS) is one of the critical strategies to avoid excessive switching losses of hard-switching. Active clamp flyback (ACF) is one of the most commonly used isolated topologies with the capability of zero-voltage turn on when the switching sequence of the clamping switch is appropriately controlled. However, sharp current drops of leakage inductor current occur after the turn-off actions of the clamping switch, especially in high frequency high step-up applications. This leads to insufficient energy to discharge the junction capacitance of the main switch to achieve zero-voltage turn on. In this paper, the reasons for the current drops and ZVS loss are explained in detail. The related parameters are then analyzed with quantitative derivations. Then, a resonant voltage-doubling solution is proposed with the capability of ZVS in full-load range, optimized transformer designs, and low ringing. The circuit modes and the conditions of ZVS for both switches are discussed, along with the design considerations of the key parameters. All the proposed analyses are verified on 1 MHz 300W active clamp flyback converter prototypes with 28 V input and 180V output.

Keywords— gallium nitride, active clamp flyback, soft switching, ZVS, high frequency, resonant

I. INTRODUCTION

Gallium Nitride (GaN) high electron mobility transistors (HEMTs) have been widely used for the superior conduction and switching performance [1-2]. The converters based on GaN HEMTs could be operated at higher switching frequencies due to the ultra-low junction capacitance. The switching frequencies of GaN HEMTs range from hundreds of kHz to several MHz, even up to 5 MHz in [3]. High switching frequency benefits the reduction of the converter volume to achieve higher power density.

To avoid the high hard-switching losses caused by high switching frequency, soft switching strategies are preferred in the designs of high frequency converters. Active clamp flyback (ACF) is one of the most commonly-used isolated soft-switching topologies. The clamping capacitor suppresses the high voltage spikes caused by leakage inductor energy during turn-off transitions of the devices. Meanwhile, ZVS turn-on of the main switches can be achieved with appropriate control strategies for the active clamp switches. However, the conventional analysis and design methods based on Si devices

may be no longer applicable for ACF converters with high switching frequency up to 1 MHz. On one hand, the switching transients and commutation time are relatively large comparing to the short switching periods. On the other hand, the assumptions that some of the key voltage and current variables are nearly constant in the transients are no longer valid due to the small inductance and capacitance. Moreover, the influence of parasitic parameters cannot be ignored.

It is pointed out in [4] that the waveforms of magnetizing and leakage inductor current in ACF at switching frequencies of MHz are different from those at low frequencies, which leads to different ZVS conditions. Then, the improved design methods of magnetizing inductance, turn ratio, and clamping capacitance are proposed. It is also explained in [5] that the magnetizing current changes significantly in the dead time, which leads to a large dead time deviation according to the conventional design methods based on the assumptions of constant magnetizing current during the switching transients. Although improved dead time design methods are proposed, the impacts of parasitic parameters were not fully considered in [5].

The leakage inductor current drops were observed in high switching frequency ACF converters for adapter applications in [6]. The current drops, which are caused by the discharging of the secondary side diode junction capacitance, occur after the turn-off action of the main switch. This leads to a different initial condition of the leakage inductor and clamping capacitor resonance, resulting in the ZVS loss. Unfortunately, the current drops will become more severe in high step-up applications since the equivalent capacitance of the secondary side diode will be much higher when converted to the primary side with the coefficient of the square of the turn ratio [7].

It is demonstrated in [8] that the leakage inductor current drop leads to the reduction of resonant energy and deteriorates the realization of ZVS. The resonant parameter design method with the consideration of leakage inductor current drop is proposed. A similar phenomenon is found in [9], where the influence is reduced by the optimization of the secondary side diodes. However, the switching frequencies are both below

978-1-6654-4817-8/21 $31.00 © 2021 IEEE

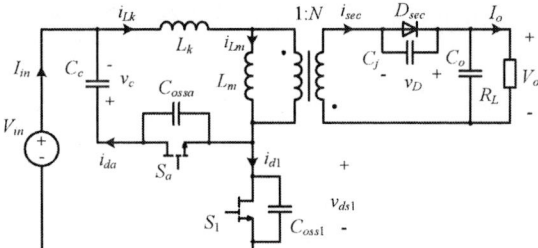

Fig. 1. Topology of the conventional active clamp flybak converter.

200 kHz in [8] and [9]. The proposed methods are no longer effective when the switching frequency increases to 1 MHz or higher.

This paper focuses on the ZVS problem of ACF converters in high frequency high step-up applications. The reason of leakage inductor current drops and ZVS loss is explained in detail. Then, a solution with the resonant voltage-doubling rectifier, which is suitable for high frequency high step-up applications, is proposed, along with the analysis of operation modes and ZVS design methods. All the theoretical analyses are verified on 300 W ACF prototypes.

II. ACTIVE CLAMP FLYBACK CONVERTERS AT HIGH SWITCHING FREQUENCY

A. Conventional ACF Converters and ZVS Conditions

Fig. 1 shows a conventional ACF topology based on GaN HEMTs. S_1 and S_a are the main switch and active clamp switch, respectively. L_m is the magnetizing inductor, while L_k is the leakage inductor of the transformer. The turn ratio of the transformer is 1:N. C_c is the clamping capacitor and D_{sec} is the secondary side diode. C_{oss1}, C_{ossa}, and C_j are the junction capacitors of S_1, S_a, and D_{sec}, respectively. V_{in} and V_o are the input and output voltage, respectively. The schematic key waveforms are shown in Fig. 2, in which v_{gs1} and v_{gsa} are the driving signals of the main switch and active clamp switch, respectively. v_{ds1} is the drain-source voltage of the main switch and v_{Cc} is the voltage across the clamping capacitor. i_{Lm} and i_{Lk} are the magnetizing current and leakage inductor current of the primary side of the transformer, respectively.

After S_a is turned off at t_6, C_{ossa} charges and C_{oss1} discharges until the voltage on C_{oss1} reaches 0 at t_7. Then, S_1 starts to reverse conduct the current and the leakage current i_{Lk} increases linearly. i_{Lk} is in the reverse direction during the time interval of t_7 to t_9. Therefore, the zero-voltage turn-on of S_1 can be achieved if S_1 is turned on between t_7 and t_9.

As can be seen from Fig. 2, the absolute value of i_{Lk} at t_6 should be large enough so that the energy stored in L_k could be able to complete the charging and discharging of C_{ossa} and C_{oss1}. Therefore, the ZVS condition is obtained as

$$\left| i_{Lk}\left(t_6 \right) \right| > \sqrt{\frac{2C_{oss}\left(N \cdot V_{in} + V_o \right)^2}{L_k}} \tag{1}$$

B. Leakage Inductor Current Drop and ZVS Loss

For the conventional ACF shown in Fig. 1 with high switching frequencies, sharp leakage inductor current drops may occur during t_1 and t_3 after the turn-off action of S_a at t_1 as shown in Fig. 2. There are two commutation stages after t_1 as shown in Fig. 3 (a) and (b), where C_c is simplified as a

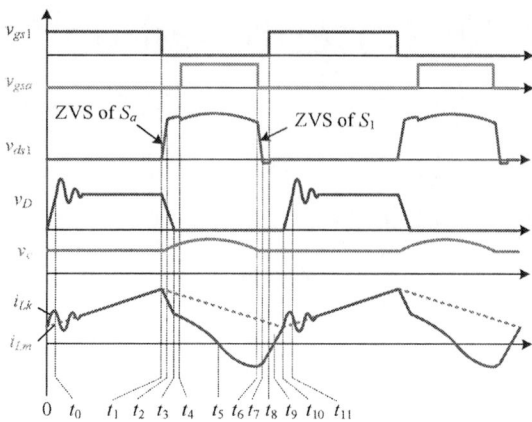

Fig. 2. Key waveforms of ACF converter in high switching frequency.

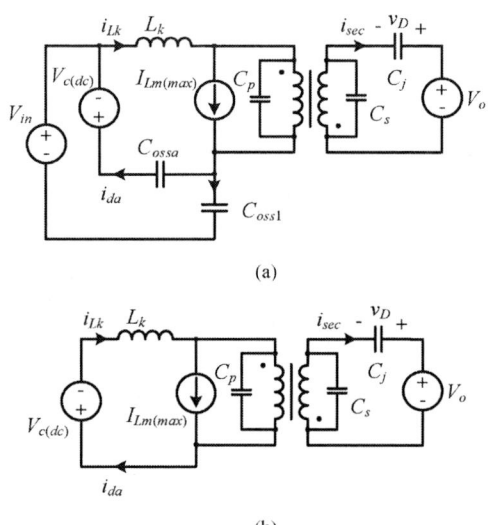

Fig. 3. Equivalent circuits of two commutation stages. (a) $t_1 \sim t_2$. (b) $t_2 \sim t_3$.

constant voltage source $V_{c(min)}$ and L_m is simplified as a constant current source $I_{Lm(max)}$. In addition, C_p and C_s are the winding parasitic capacitance of the primary side and secondary side of the transformer, respectively.

According to Fig. 2, i_{Lk} charges C_{oss1} and discharges C_{ossa} during the time interval of t_1 to t_2, while i_{Lm} - i_{Lk} discharges C_p and the equivalent capacitance of C_j and C_s. On one hand, i_{Lk} will be remarkably larger than i_{Lm} - i_{Lk} since i_{Lk} is close to i_{Lm}. On the other hand, the equivalent capacitance of C_j and C_s will be N^2 times their original values when converted to the primary side. Therefore, the commutation time of Stage I will be much shorter than that of Stage II and only Stage II is analyzed hereinafter.

According to Fig. 3 (b), the time required for v_D, which is the voltage across the secondary side diode, to drop from V_{in} + V_o' to 0 is

$$\Delta t = \frac{\pi}{2} \cdot \sqrt{L_k C_{sec}} . \tag{2}$$

Furthermore, the variation of i_{Lk} during this period is derived as

978-1-6654-4817-8/21 $31.00 © 2021 IEEE

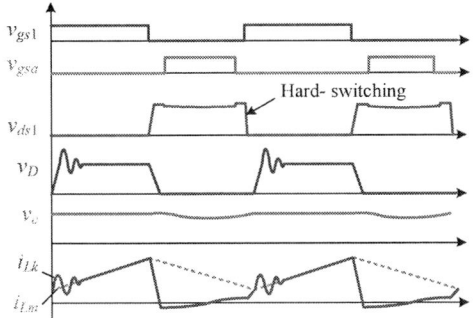

Fig. 4. Waveforms of ZVS loss.

$$\Delta i_{Lk} = \left(V_{in} + V_o{'}\right) \cdot \sqrt{\frac{C_{sec}}{L_k}}, \tag{3}$$

where $V_o{'}$ and C_{sec} satisfies

$$\begin{cases} V_o{'} = \dfrac{V_o}{N} \\ C_{sec} = C_p + N^2 \cdot \left(C_s + C_j\right) \end{cases} . \tag{4}$$

The turn ratio N is usually large in high step-up applications, leading to much higher equivalent capacitance of C_j and C_s when converted to the primary side than C_p. Thus, equation (4) could be simplified as

$$\begin{aligned} \Delta i_{Lk} &= \left(V_{in} + \frac{V_o}{N}\right) \cdot \sqrt{\frac{C_p + N^2 \cdot \left(C_s + C_j\right)}{L_k}} \\ &\approx \left(N \cdot V_{in} + V_o\right) \cdot \sqrt{\frac{C_s + C_j}{L_k}} \end{aligned} . \tag{5}$$

According to (5), the leakage inductor current drops are determined by the input voltage V_{in}, the output voltage V_o, the turn ratio N, the leakage inductance L_k and the sum of the secondary side capacitance C_j and C_s. With the increase of the switching frequency, L_m and L_k are significantly reduced. Moreover, C_s increases after the adoption of PCB winding and interleaved winding arrangements. These will lead to a very large Δi_{Lk}, especially in high step-up applications where the turn ratio N tends to be higher.

The excessive Δi_{Lk} in the commutation time shown in Fig. 3 (b) leads to ZVS loss of the main switch S_1, which is depicted in Fig. 4. After the turn-off action of S_1, i_{Lk} decreases rapidly and soon below zero before S_a is turned on, which leads to an unexpected resonant status during the on-time of S_a. As a result, S_1 remains in forward conduction state when the driving signal of S_1 turns to the high level and S_1 is hard turned on.

III. RESONANT VOLTAGE-DOUBLING ACTIVE CLAMP FLYBACK CONVERTERS

A. Resonant Voltage-doubling ACF and Operating Principles

It can be seen from (5) that the leakage inductor current drops of the conventional ACF converters increases with the increase of N, C_j, C_s, and the decrease of L_k for certain input

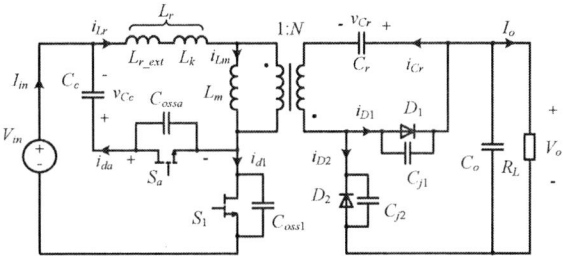

Fig. 5. Topology of resonant voltage-doubling ACF converter.

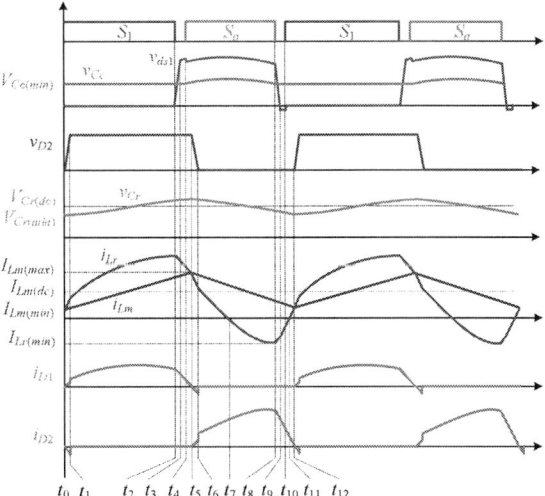

Fig. 6. Key waveforms of RVD ACF converter.

and output voltage. Therefore, the current drops can be suppressed by approaches such as reducing N, C_j, C_s or increasing L_k. Among all the approaches, the reduction of C_j could be achieved by the selection of the secondary side diodes with lower junction capacitance. C_s could be reduced with optimized winding arrangements. However, each approach alone has limited effects. Thus, the combination of different approaches is preferred and the resonant voltage-doubling (RVD) ACF converter is proposed as a solution in high frequency high step-up applications.

Fig. 5 shows the topology of RVD ACF converter, in which S_1 is the main switch and S_a is the active clamp switch. L_r is the primary side resonant inductor, which consists of the primary side leakage inductor L_k and the external resonant inductor L_{r_ext}. C_r is the secondary side resonant voltage-doubling capacitor, while D_1 and D_2 are the secondary side rectifier diodes. C_{j1} and C_{j2} are the junction capacitors of D_1 and D_2, respectively. Compared with the conventional ACF topology, RVD ACF converter provides high voltage step-up capability while the turn ratio of the transformer is reduced by half. Meanwhile, the leakage inductor current drops caused by commutation could be reduced as the equivalent leakage inductor of the primary side is increased by adding an additional inductor in series with primary side winding.

The schematic diagram of key waveforms of RVD ACF converter is presented in Fig. 6, where v_{gs1} and v_{gsa} are the driving signals of S_1 and S_a, respectively. v_{Cc}, v_{Cr}, and v_{D1} are the voltage across C_c, C_r, and D_1, respectively. I_{Lr} is the resonant inductor current, while i_{D1} and i_{D2} are the current flows through D_1 and D_2, respectively. For RVD ACF converter, one switching cycle could be divided into 7 different modes as shown in Fig. 6.

Mode I: $t_0 \sim t_1$.

Secondary side commutation stage. The junction capacitance of D_1 (C_{j1}) discharges and the junction capacitance of D_2 (C_{j2}) charges until v_{D1} falls to 0.

Mode II: $t_1 \sim t_2$.

L_r and C_r resonance stage. L_r resonates with C_r until S_1 is turned off at t_2.

Mode III: $t_2 \sim t_5$.

Primary side commutation and i_{Lr} falling stage. After the turn-off action of S_1, i_{Lr} starts to fall until it is equal to i_{Lm} at t_5.

Mode IV: $t_5 \sim t_6$.

Secondary side commutation stage. The junction capacitance of D_1 (C_{j1}) charges and the junction capacitance of D_2 (C_{j2}) discharges until v_{D2} falls to 0.

Mode V: $t_6 \sim t_8$.

L_r, C_c and C_r resonance stage. L_r resonates with C_c and C_r until S_a is turned off at t_2.

Mode VI: $t_8 \sim t_9$.

Primary side commutation stage. C_{oss1} discharges and C_{ossa} charges until v_{ds1} falls to 0.

Mode VII: $t_9 \sim t_{12}$.

i_{Lr} rising stage. i_{Lr} rises linearly until it is equal to i_{Lm} at t_{12}.

B. ZVS Conditions of Resonant Voltage-doubling ACF

According to Mode III, the ZVS of S_a could be easily achieved as long as the primary side commutation time after the turn-off action of S_1 is less than the dead time t_{dead}. In fact, i_{Lr} at the turn-off time of S_1 is usually relatively large, so the primary side commutation time of $t_3 \sim t_4$ is much shorter than that of $t_8 \sim t_9$, which makes it easier for S_a to achieve zero-voltage turn-on than S_1. Therefore, the zero-voltage turn-on of S_1 is the critical design for RDV ACF converters.

Two ZVS conditions for S_1 could be obtained by combining Mode VI and VII in Fig. 6. Firstly, the resonant energy of L_r at t_8 should be large enough to complete the charging and discharging of C_{ossa} and C_{oss1}. In other words, the absolute value of the minimum of i_{Lr} ($|I_{Lr(min)}|$) should be large enough to ensure v_{ds1} falls to 0 at t_9. Secondly, S_1 should be still in reverse conduction state at the end of the dead time, which is to say, i_{Lr} should be below zero at t_{10}.

The waveforms of the resonant inductor current (i_{Lr}) and drain-source voltage of S_1 (v_{ds1}) after the turn-off action of S_a are depicted in Fig. 7 (a) and (b). It is known from Fig. 7 that i_{Lr} falls to zero at $t = t_{min}$, while v_{ds1} reaches its minimum value $V_{ds1(min)}$ if the transitions of modes are not considered. Therefore, t_{min} and $V_{ds1(min)}$ is derived as (6) and (7) according to Fig. 5 and Fig. 6.

$$t_{min} = t_8 + \arctan \frac{\omega_p L_r \left| I_{Lr(min)} \right|}{\left| V_o{}' - V_{Cc(dc)} - V_{Cr(dc)}{}' \right|}. \quad (6)$$

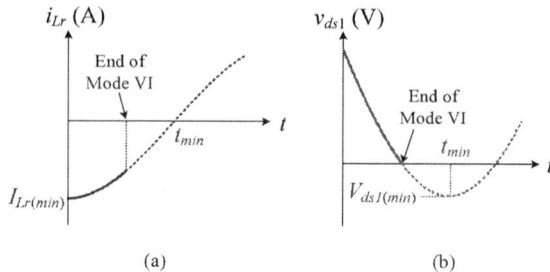

Fig. 7. Waveforms of the resonant inductor current and the drain-source voltage of the main switch. (a) Resonant inductor current i_{Lr}. (b) drain-source voltage of the main switch v_{ds1}.

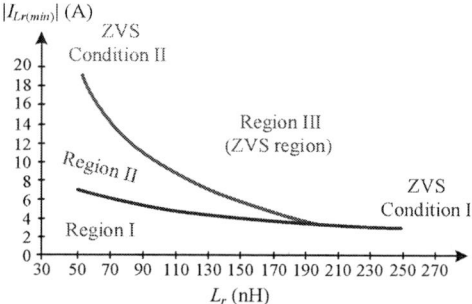

Fig. 8. ZVS conditions of RVD ACF converter.

$$\begin{aligned}
V_{ds1(min)} &= v_{ds1(t8,t9)}\left(t_{min}\right) \\
&= -\sqrt{\frac{L_r}{2C_{oss}} I_{Lr(min)}{}^2 + \left(V_o{}' - V_{Cc(dc)} - V_{Cr(dc)}{}'\right)^2} \\
&\quad + V_o{}' + V_{in} - V_{Cr(dc)}{}'.
\end{aligned} \quad (7)$$

As can be seen from Fig. 7 (a) and (b), t_{min} decreases and $V_{ds1(min)}$ increases with the decrease of $|I_{Lr(min)}|$. Thus, the absolute value of $I_{Lr(min)}$ should be large enough in order to ensure a negative $V_{ds1(min)}$. Accordingly, one of the ZVS conditions is derived as

$$\left| I_{Lr(min)} \right| > \sqrt{\frac{2C_{oss}}{L_r}} \cdot \frac{1}{\sqrt{\left(V_o{}' + V_{in} - V_{Cr(dc)}{}'\right)^2 - \left(V_o{}' - V_{Cc(dc)} - V_{Cr(dc)}{}'\right)^2}}. \quad (8)$$

To obtain a negative resonant inductor current when S_1 is turned on at the end of the dead time, i_{Lr} at t_{10} should be lower than 0, which leads to the other ZVS condition expressed as (9).

$$\left| I_{Lr(min)} \right| > \Delta i_{Lr(t8,t9)} + \frac{V_{in} + V_o{}' - V_{Cr(dc)}{}'}{L_r}\left(t_{dead} - \Delta t_{(t8,t9)}\right) \quad (9)$$

Where $t_{dead} = t_{10} - t_8$ is the dead time.

Two curves of the relations between $|I_{Lr(min)}|$ and L_r are then obtained according to (8) and (9), which divide the plane into three regions as shown in Fig. 8. For Region I and Region II, it is obvious that the values of L_r and $I_{Lr(min)}$ cannot satisfy both of the two conditions in (8) and (9). Therefore, only in Region III can the main switch S_1 realize zero-voltage turn-on.

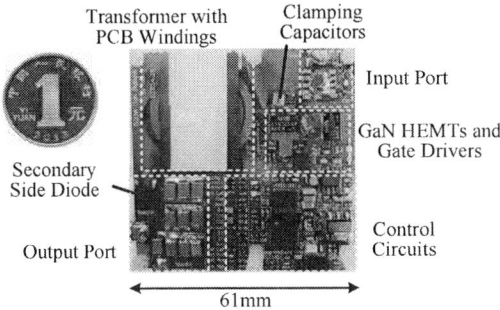

Fig. 9. Photo of the conventional ACF prototype.

TABLE I.   KEY PARAMETERS OF THE PROTOTYPE

Key Parameters	Values
Input Voltage V_{in}	28 V
Output Voltage V_o	180 V
Output Power P_o	300 W
Switching Frequency f_s	1 MHz
Magnetizing Inductance L_m	1 μH
Turn Ratio 1: N	1:6
GaN HEMT	GS61008P
Output Capacitance C_{oss} of GaN HEMT	288 pF
Secondary Side Diode	LSIC2SD065C08A
Junction Capacitance C_j of Diode	84 pF

Additionally, it is illustrated in Fig. 9 that larger $|I_{Lr(min)}|$ is required for the ZVS of S_1 with smaller L_r, which leads to higher turn-off loss of S_a and reverse conduction loss of S_1. On the contrary, although the ZVS of S_1 is easy to be achieved when L_r is large, the converter may suffer from large inductor volume and thus low power density. Therefore, it is necessary to consider the trade-off in the design of L_r and $|I_{Lr(min)}|$.

IV. EXPERIMENTAL VARIFICATION

A. Convertional ACF

A conventional ACF converter prototype is built for the verification of the aforementioned analyses, which is shown in Fig. 9. The key parameters of the prototype are listed in TABLE I. Fig. 10 (a) and (b) show the experimental waveforms with the leakage inductance of 15 nH and 40 nH, respectively. It can be seen from Fig. 10 (a) that remarkable leakage inductor current drops of nearly 30 A are observed on the waveform of i_{Lk} after the turn-off action of S_1. This results in insufficient resonant energy during the on-time of S_a and then to the ZVS loss.

When the leakage inductance is increased to 40 nH, the current drops are significantly reduced and the ZVS of S_1 is achieved. However, the converter suffers from severe oscillation after S_1 is turned off as shown in Fig. 10 (b). The oscillation is caused by the resonance of L_k and C_j and will increase the voltage and current stresses and the conduction loss of the devices as well as the winding loss of the transformer, leading to low efficiency and low reliability of the converters.

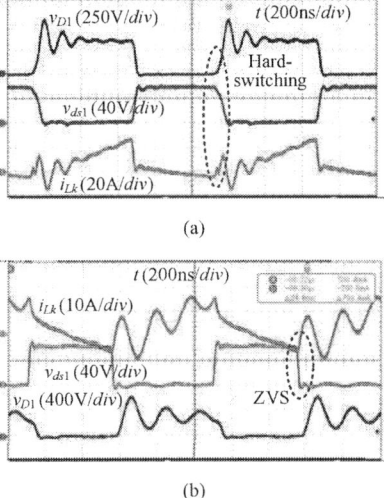

(a)

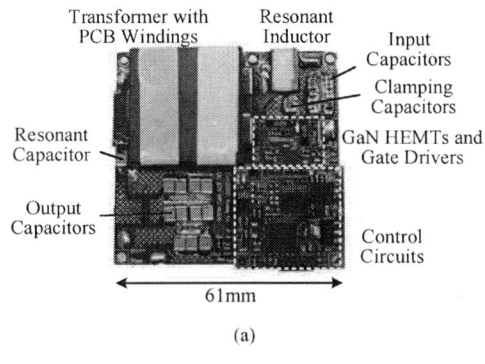

(b)

Fig. 10. Experimental waveforms of conventional ACF converter. (a) L_k = 15 nH and C_c = 1.6 μF. (b) L_k = 40 nH and C_c = 630 nF.

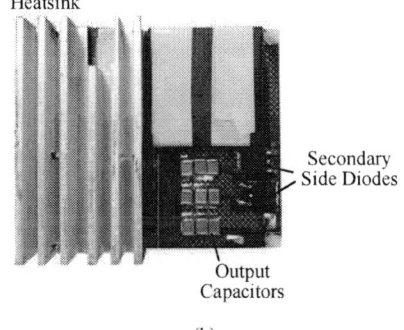

(a)

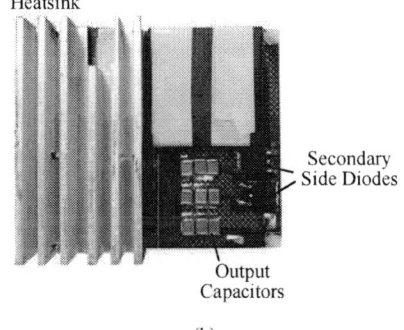

(b)

Fig. 11. Photo of RVD ACF prototype. (a) Top view. (b) Bottom view.

B. RVD-ACF Converter

An RVD ACF converter prototype based on GaN HEMTs, which is shown in Fig. 11, is also built in this paper for comparison. Since the voltage-doubling rectifier is adopted in this prototype, the turn ratio of the transformer is reduced to 1:3. The primary side leakage inductance of the transformer is 20 nH, while the external resonant inductance is 100 nH. This prototype is designed in the size of half-brick and with the peak efficiency of 93%.

Fig. 12 shows the experimental waveforms of RVD ACF converter with different output power. It is observed on the waveforms of v_{ds1} and v_{gs1} that the drain-source voltage of S_1

978-1-6654-4817-8/21 $31.00 © 2021 IEEE

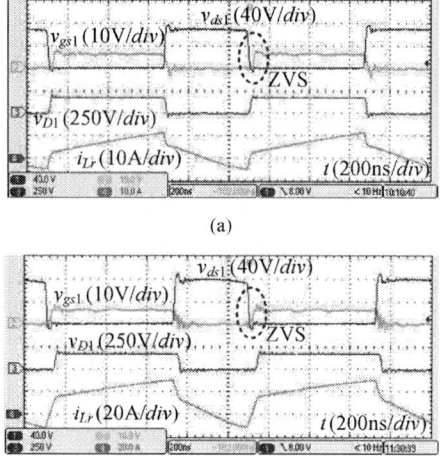

(a)

(b)

Fig. 12. Experimental waveforms of conventional ACF converter. (a) Output power is 100 W. (b) Output power is 300 W.

falls to 0 before it is turned on, which indicates the ZVS of S_1 in the full load range. Also, L_r resonates with C_r during the conduction time of both S_1 and S_a, avoiding the unexpected oscillation between L_r and the junction capacitance of the secondary side diodes.

V. Conclusions

In this paper, the soft switching problem of GaN-based active clamp flyback converter in high switching frequency and high step-up applications is studied. The cause of ZVS loss in conventional ACF is explained. A resonant voltage-doubling rectifier solution is then proposed. The operation principle and the ZVS conditions are presented. All the analyses are verified by experimental results.

The conclusions can be drawn as follows: 1) The discharging of the junction capacitance of the secondary diodes in the conventional ACF converters leads to the drops of leakage inductor current; 2) The current drops is determined by the values of the leakage inductance, the total capacitance

of the secondary side and the turn ratio of the transformer; 3) The current drops become much more severe with high frequency and high step-up ratio, which will lead to the ZVS loss of the main switches; 4) The resonant voltage-doubling ACF converter could effectively solve this problem and achieve zero-voltage turn-on of the primary side switches in the full-load range.

Acknowledgment

This work is supported by National Natural Science Foundation of China (No.51777094).

References

[1] E. A. Jones, F. Wang, and B. Ozpineci, "Application-based review of GaN HFETs," in *Proc. IEEE Workshop Wide Bandgap Power Devices Appl.*, Oct. 2014, pp. 24-29.

[2] R. Quay, R. Reiner, B. Weiss, S. Mueller, F. Benkhelifa and P. Waltereit, "Overview and Recent Progress on the Development of Compact GaN-based Power Converters," in *Proc. Power Electron. Compon. Appl.; 7. ETG-Symposium*, April 2017, pp. 1-6.

[3] Z. Zhang, K. D. T. Ngo and J. L. Nilles, "A 30-W flyback converter operating at 5 MHz," in *Proc. IEEE Appl. Power Electron. Conf. Expo.*, March 2014, pp. 1415-1421.

[4] X. Liu, R. Burgos, B. Sun and D. Boroyevich, "Wide-input-voltage-range dual-output GaN-based isolated DC-DC converter for aerospace applications," in *Proc. IEEE Appl. Power Electron. Conf. Expo.*, March 2017, pp. 279-286.

[5] H. Wen, D. Jiao and J. Lai, "Optimal Design Methodology for High Frequency GaN Based Step-up LLC Resonant Converter," in *Proc. IEEE Int. Future Energy Electron. Conf.*, Nov. 2019, pp. 1-5.

[6] L. Xue and J. Zhang, "Design considerations of highly-efficient active clamp flyback converter using GaN power ICs," in *Proc. IEEE Appl. Power Electron. Conf. Expo.*, March 2018, pp. 777-782.

[7] X. Zhao, C. Chen, J. Lai and O. Yu, "Circuit Design Considerations for Reducing Parasitic Effects on GaN-Based 1-MHz High-Power-Density High-Step-Up/Down Isolated Resonant Converters," *IEEE J. Emerg. Sel. Topics Power Electron.*, vol. 7, no. 2, pp. 695-705, June 2019.

[8] Y. Chen, Y. Sun, M. Tian, L. Wang and H. Jin, "Analysis and Design of a Bidirectional High Step-up Active Clamp Flyback Converter for Dielectric Elastomer Actuator," in *Proc. Int. Conf. Power Electron.*, Nov. 2019, pp. 1-5.

[9] P. Liu, "Design consideration of active clamp flyback converter with highly nonlinear junction capacitance," in *Proc. IEEE Appl. Power Electron. Conf. Expo. (APEC)*, March 2018, pp. 783-790.

A Lossless and Passive Voltage Spikes Clamping Circuit for SiC HERIC Inverter

Yong Li
State key laboratory of strong electromagnetic engineering and new technology
Huazhong University of Science and Technology
Wuhan, China
yong_li@hust.edu.cn

Shanxu Duan
State key laboratory of strong electromagnetic engineering and new technology
Huazhong University of Science and Technology
Wuhan, China
duanshanxu@hust.edu.cn

Qiqi Li
State key laboratory of strong electromagnetic engineering and new technology
Huazhong University of Science and Technology
Wuhan, China
lqq@hust.edu.cn

Abstract—**When HERIC(High Efficient and Reliable Inverter Concept) inverter operates at the non-unit power factor, there are larger turn-off voltage spikes on freewheeling switches than operating at the unit power factor, which increases switching loss, devices voltage stress, and electromagnetic interference. Moreover, the turn-off voltage spikes are much larger in some high-frequency applications, owing to the larger di/dt. This paper analyzes the turn-off voltage spikes of freewheeling switches when the HERIC inverter operates at the non-unit power factor. On this basis, a passive clamping circuit is proposed, which is simple, cheap, and lossless, and its principle is analyzed. Finally, a 10kW prototype was used to verify the proposed circuit.**

Keywords—*HERIC, Voltage spike, Clamping circuit, SiC*

I. INTRODUCTION

Nowadays, the photovoltaic(PV) system has been the most popular renewable distributed energy source for residential applications. However, the high proportion of distributed energy sources will cause disturbance to the voltage and frequency of the power grid. Therefore, PV inverters are required to have reactive power injection and absorption capability [1]. HERIC inverter is a single-phase transformerless grid-connected inverter topology, which is widely used in household single-phase distributed PV systems. Considering restrictions of layout, increased switching frequency, and increased solar system power, the influence of PCB parasitic inductance and switch junction capacitance on turn-off voltage spikes cannot be ignored [2]. Due to the zero reverse recovery characteristic of SiC Schottky Barrier Diodes (SiC SBD) [3], the turn-off voltage spikes of freewheeling switches are eliminated when HERIC inverter operates at unity power factor. However, when HERIC inverter injects and absorbs reactive power, reverse recovery is not the dominant reason for freewheeling switches voltage spikes. Therefore, SiC MOSFET or Hybrid IGBT (The device co-packages a silicon-based IGBT with a SiC Schottky barrier diode) is not enough. Therefore, it is desirable to exploit efficient, passive, and lossless voltage spikes clamping method for freewheeling switches in HERIC inverter.

There are currently two main types of solutions: the first is to design an active clamping circuit, the second is to design a passive snubber circuit or clamping circuit. The first type of method is the active neutral-point clamped HERIC [4, 5]. In those topologies, the freewheeling switches are clamped to the neutral point of the DC bus, so that the voltage spikes of the freewheeling switches are suppressed. But due to the active devices are used, the auxiliary circuits are more complex and

expensive than the passive circuits. To absorb the voltage spikes and keep down costs, a passive overvoltage absorption circuit is proposed [2]. The overvoltage absorption circuit is only composed of two diodes and a snubber RC network. However, due to the static voltages on snubber capacitors and resistors are the same as the DC bus voltage, the losses of the overvoltage absorption circuit are large, and there has to be a trade-off between the losses and voltage spikes. Moreover, high voltage on snubber capacitors and snubber resistors make the circuit large. H6.5 is a good topology to avoid freewheeling switches voltage spikes, by using a diode clamp to the bus [6]. However, the topology needs a special IGBT module packaging six IGBTs and five diodes which is more expensive than the discrete device. Moreover, there is a price to pay for reducing the number of diodes from six to five, namely one additional junction at freewheeling during reactive power [6]. In this paper, a lossless and passive freewheeling switches voltage spikes clamping absorption circuit for HERIC inverter is proposed. The power losses in the proposed circuit can be neglectable, the voltages on capacitors and resistors in the proposed circuit are lower than state-of-the-art snubber circuits, and the voltage spikes suppression is remarkable.

The rest of this paper is organized as follows. The reason for the freewheeling switches voltage spikes in HERIC inverter is discussed in Section II. In Section III, the principle of the proposed circuit is presented. Section IV presents experimental results on a 10kW PV single-phase inverter prototype to validate the proposed clamping circuit. Finally, conclusions are drawn in Section V.

II. REASON FOR FREEWHEELING SWITCHES VOLTAGE SPIKES

HERIC inverter with parasitic inductances is shown in Fig. 1. L_A is the parasitic inductance from point A to point C, L_B is the parasitic inductance from point B to point D, L_{busP} is the parasitic inductance of the positive bus, L_{busN} is the parasitic inductance of the negative bus, and L_f is the filter inductor.

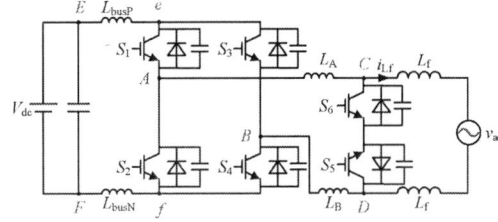

Fig. 1 Simplified circuit of HERIC considering parasitic inductances and junction capacitances

In the actual circuit, the parasitic inductances L_A, L_B, L_{busP}, and L_{busN} are larger than other parasitic inductances. To simplify the analysis, only the other parasitic inductances are neglected when analyzing the turn-off voltage spike of Freewheeler $S_5 \sim S_6$. The key waveforms of HERIC inverter are shown in Fig. 2. v_m is the modulation wave, $V_{GS1} \sim V_{GS4}$ are the main driving signals of the switching devices $S_1 \sim S_4$, and V_{GS5}, V_{GS6} are the driving signals of the freewheeling devices S_5 and S_6. V_{GS1} and V_{GS4} are synchronization signals, and V_{GS2} and V_{GS3} are synchronization signals.

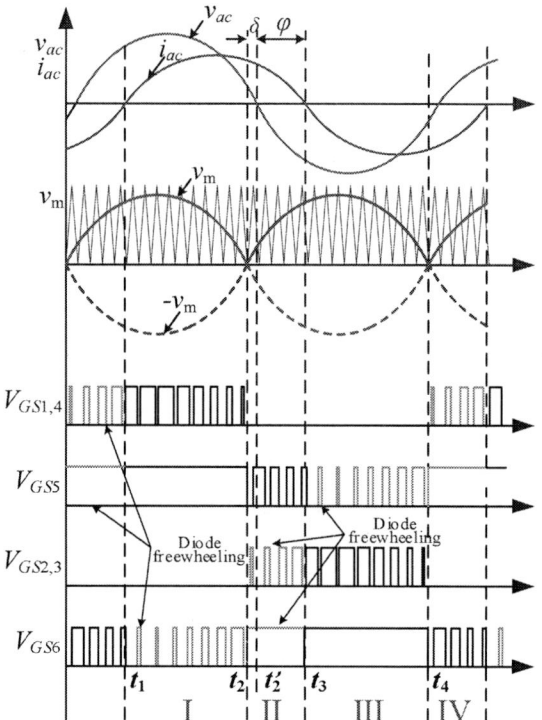

Fig. 2   Key waveforms of modulation strategy for HERIC operating at non-unit power factor.

In Fig. 2, v_{ac} and i_{ac} are grid voltage and grid current, respectively, φ is the power factor angle output by the converter, and δ is the angle between the HERIC bridge voltage v_{AB} and the grid voltage v_{ac}, which can generally be ignored. A grid frequency cycle is divided into four sectors I-IV. Due to the symmetry of the circuit, only the turn-off voltage spikes of S_5 in sector I and sector II are analyzed. In sector I, S_5 remains on and S_6 remains off while the anti-parallel diode of S_6 freewheels. Therefore, there are no turn-off voltage spikes in sector I if the anti-parallel diode of S_6 reverse recovery is neglectable, such as SiC SBD. The equivalent circuits of different modes in sector II are shown in respectively. Because the grid voltage and the grid current in sector II has different signs, the filter inductor freewheels through the switches S_2 and S_3, and is charged through the freewheeling switch S_5.

Charging mode in sector II. v_{ac} is negative and i_{ac} is positive at this stage. The main switches $S_1 \sim S_4$ are off, the freewheeling switch S_5 is on, and the freewheeling switch S_6 is off while the anti-parallel diode of S_6 is on. The filter inductor is charged by v_{ac} through the anti-parallel diodes of the freewheeling switch S_6 and the freewheeling switch S_5. The circuit of the mode is shown in Fig. 3(a).

Commutation mode in sector II. S_5 is turned off at t_2, the current of L_f is approximately unchanged, the current of S_5 decreases, and the current of the junction capacitor of S_5 increases. At the same time, the current flowing through L_A L_B, L_{busP}, and L_{busN} rise rapidly. L_A L_B, L_{busP}, and L_{busN} and the junction capacitances of the freewheeling switch S_5 and main switches S_2 and S_3 resonate. When the voltage of the junction capacitance of S_2 and S_3 resonates to zero, and after the anti-parallel diodes of the switches S_2 and S_3 are turned on, the currents of L_A L_B, L_{busP}, and L_{busN} still rise. The voltages on L_A L_B, L_{busP}, and L_{busN} are shown in Fig. 3(b). Therefore, the voltage on the freewheeling switch S_5 will further rise to produce a turn-off voltage spike. The resulting turn-off voltage spike can be represented by equation (1).

$$
\begin{aligned}
v_{ce5} &= \left(v_E + \Delta v_{LB} + \Delta v_{LbusP} \right) - \left(v_F - \Delta v_{LA} - \Delta v_{LbusN} \right) \\
&= v_E - v_F + \Delta v_{LA} + \Delta v_{LB} + \Delta v_{LP} + \Delta v_{LbusN} \\
&= V_{dc} + L_A \frac{di_{LA}}{dt} + L_B \frac{di_{LB}}{dt} \\
&\quad + L_{busP} \frac{di_{LbusP}}{dt} + L_{busN} \frac{di_{LbusN}}{dt}
\end{aligned}
\tag{1}
$$

Freewheeling mode in sector II. The commutation between the freewheeling switch S_5 and the anti-parallel diodes of the main switches S_2 and S_3 is completed. The freewheeling switches S_5 and S_6 are both off, and the anti-parallel diodes of main switches S_2 and S_3 are conducting. Filter inductor L_f freewheels through the anti-parallel diode of S_2, L_A, and L_B and the anti-parallel diode of S_3, as shown in Fig. 3(c).

The analysis of sector IV is consistent with that of sector II. Due to the large di/dt on L_A L_B, L_{busP}, and L_{busN}, the turn-off voltage spikes of the freewheeling switches can be large without any clamping circuit. And turn-off voltage spikes of the freewheeling switches mainly exist in sector II and sector IV. Fig. 4 Shows the real voltage spikes of S_5 in sector II. In particular, when the power factor angle φ increases to 180°, sector II and sector IV each occupy an angle of 180°, the HERIC inverter works as a single-phase rectifier. Through analysis, it is concluded that a clamp absorption circuit needs to be designed to suppress the turn-off voltage spikes of the freewheeling switches.

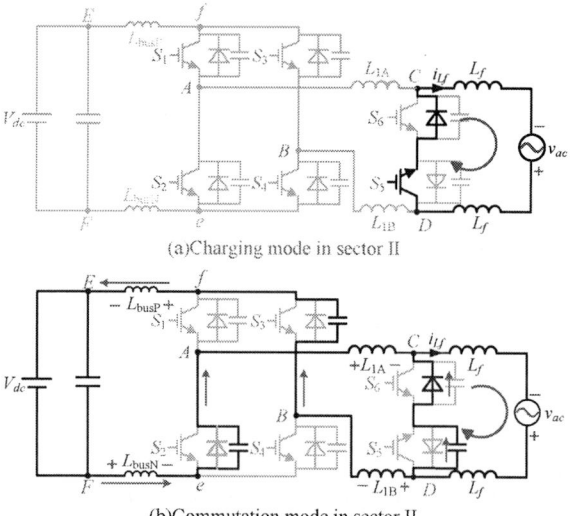

(a)Charging mode in sector II

(b)Commutation mode in sector II

978-1-6654-4817-8/21 $31.00 © 2021 IEEE   64

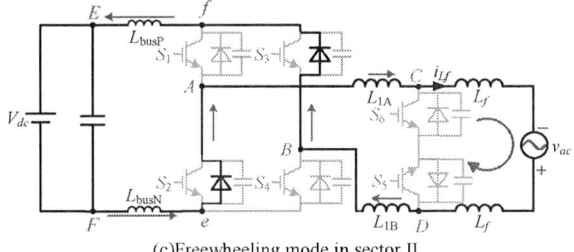

(c)Freewheeling mode in sector II.

Fig. 3 The equivalent circuit of each mode in sector II

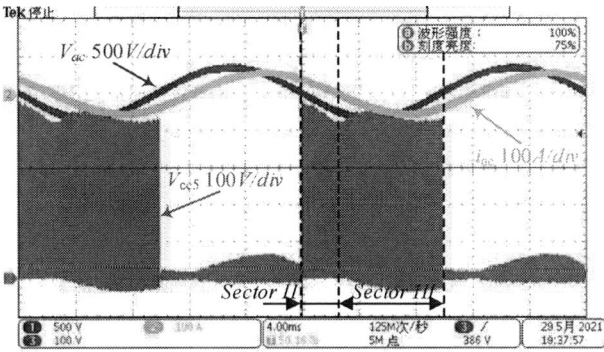

Fig. 4 Voltage spikes waveforms of a 10 kW HERIC inverter prototype when injecting 6 kvar reactive power

III. PRINCIPLES OF PROPOSED CIRCUIT

A passive and lossless clamping absorption circuit proposed in this paper is shown in Fig. 5. The passive and lossless clamping absorption circuit is composed of a positive bus RCD clamping absorption circuit and a negative bus RCD clamping absorption circuit.

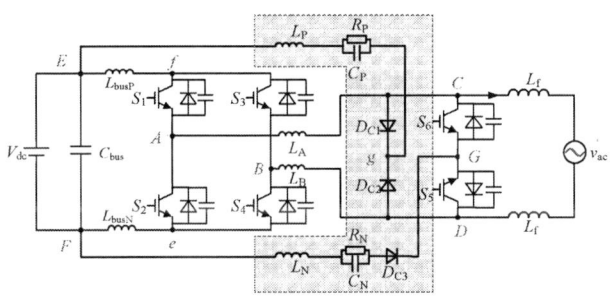

Fig. 5 Passive and lossless clamping absorption circuit

The positive bus clamping absorption circuit includes clamping diodes D_{C1}, D_{C2}, RC networks R_N and C_N, and parasitic inductance L_p. The negative bus clamping absorption circuit includes clamping diode D_{C3}, RC networks R_P and C_P, and parasitic inductance L_N. The following takes the turn-off commutation process of S_5 when the grid current i_{Lf} is positive as an example to theoretically analyze the working principle and clamping effect of the proposed circuit. Fig. 6 shows the equivalent circuit with the proposed clamping circuit of each mode in sector II.

Charging mode in Sector II. v_{ac} is negative and i_{ac} is positive at this stage. The main switches $S_1\sim S_4$ are off, the freewheeling switch S_5 is on, and the freewheeling switch S_6

is off while the anti-parallel diode of S_6 is on. The filter inductor is charged by v_{ac} through the anti-parallel diodes of the freewheeling switch S_6 and the freewheeling switch S_5. v_{AB} is zero. The circuit of the mode is shown in Fig. 6 (a). The proposed clamping circuit has no power loss in this mode.

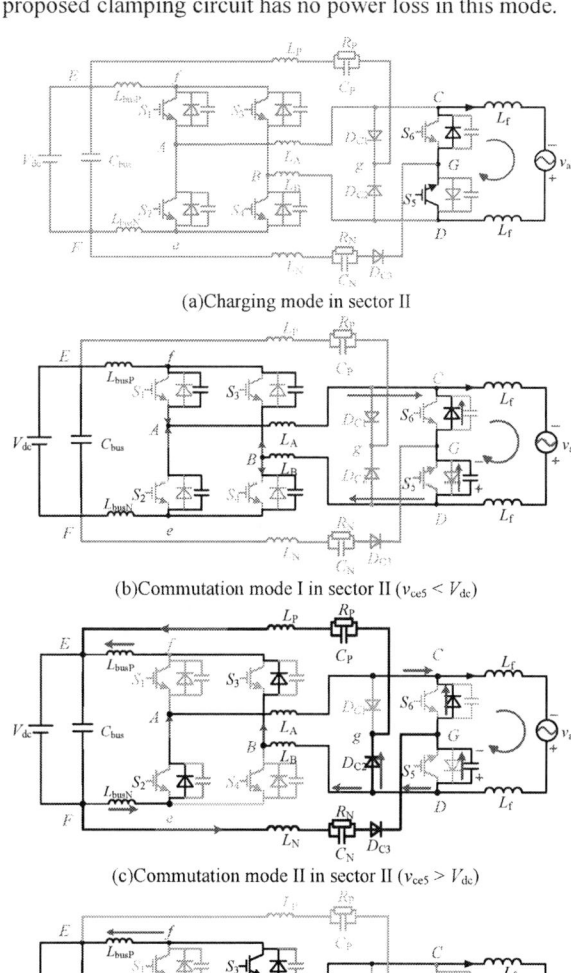

(a)Charging mode in sector II

(b)Commutation mode I in sector II ($v_{ce5} < V_{dc}$)

(c)Commutation mode II in sector II ($v_{ce5} > V_{dc}$)

(d)Freewheeling mode in sector II.

Fig. 6 The circuit with proposed clamping circuit of each mode in sector II

Commutation mode I in sector II. The current of L_f is approximately unchanged, but the current of S_5 decreases, so that the current of junction capacitance of S_5 and $S_1\sim S_3$ increase, and the potential at point A decreases, and the potential at point B increases until the voltage v_{AB} is equal to V_{dc}. The circuit of the mode is shown in Fig. 6 (b).

Commutation mode II in sector II. When v_{AB} larger than V_{dc}, clamping diodes D_{C2} and D_{C3} are turned on. The positive clamping circuit clamps the potential at point D to the positive bus through R_P, C_P, and D_{C2}, limiting the voltage rise at point D and shunting with L_B. The negative bus clamping circuit clamps the potential at point G to the negative bus through R_N, C_N, and D_{C3}, limiting the voltage drop at point G and shunting

978-1-6654-4817-8/21 $31.00 © 2021 IEEE

with L_A at the same time. But the parasitic inductance L_P and L_N should be considered. The smaller the parasitic inductance, the more significant the spikes suppression, because the parasitic inductance makes the high-frequency impedance of the clamping circuit larger, which means voltage drop on clamping circuit larger. The circuit of the mode is shown in Fig. 6 (c).

Freewheeling mode in sector II. The commutation between the freewheeling switch S_5 and the anti-parallel diodes of the main switches S_2 and S_3 is completed. The freewheeling switches S_5 and S_6 are both off, and the anti-parallel diodes of main switches S_2 and S_3 are on. Filter inductor L_f freewheels through the anti-parallel diode of S_2, L_A, L_B, and the anti-parallel diode of S_3, as shown in Fig. 6 (d). In this mode, the potentials of point D and point E are equal, and potentials of point G and Point F are equal, therefore the voltages of capacitors C_P and C_N are zero, and there is no power loss in the proposed circuit.

On the one hand, both the positive and negative bus clamping circuits have no static loss when the freewheeling switches are completely turned off, because the positive and negative bus clamping circuits are cut off by snubber capacitors, on the other hand, the clamping circuits have no static loss when the freewheeling switches are on, because the clamping diodes D_{C1}, D_{C2}, and D_{C3} are blocked. Therefore, the clamping circuit only works briefly in the dynamic process of commutation. In commutation mode II, part of the voltage spikes energy is fed back to the DC bus through the proposed clamping circuit, and part is consumed in the form of heat on the absorption resistors R_p and R_N. A reasonable selection of the absorption resistance can make the loss of the clamping circuits is almost zero. Moreover, voltages on capacitors C_P and C_N are V_{DE} and V_{FG}, which are much smaller than V_{dc}. That means cheap and small capacitors, resistors and diodes can be used in the proposed circuit. Because of the small power loss and low voltage, the proposed circuit can be designed to be very compact, which means the parasitic inductances of the proposed circuit can be small.

IV. EXPERIMENTAL VERIFICATIONS

For experimental verification of the analysis of the effect of the proposed circuit on turn-off overvoltage suppression, a 10 kW HERIC inverter prototype, as shown in Fig. 7, has been built and tested. TABLE I shows the parameters of the prototype and proposed clamping circuit.

TABLE I. PARAMETERS OF THE PROTOTYPE AND PROPOSED CLAMPING CIRCUIT

Parameters	Values
DC bus voltage Vdc	200V to 520V
Rated Power	10kW
Snubber Capacitors C_P and C_N	4.7 nF (0805)
Snubber Resistors R_P and R_N	22 Ω (0805)
Clamping Diodes D_{C1}, D_{C2} and D_{C3}	UF1M (SMA)

Fig. 8 shows the experimental waveforms of HERIC inverter without any snubber or clamping circuit when the output power is 7kW, reactive power is 6.3kvar and DC bus voltage V_{dc} is 400V. The voltage spike of freewheeling switch S_5 is 92V, and the overshoot is 23.9%. Fig. 9 shows the

experimental waveforms of HERIC inverter with the overvoltage absorption circuit [2]. The voltage spike of freewheeling switch S_5 is 66V, and the overshoot is 16.49%. Fig. 10 shows the experimental waveforms of HERIC inverter with the proposed clamping circuit. The voltage spike of freewheeling switch S_5 is 53V, and the overshoot is 13.25%. Fig. 11, Fig. 12, and Fig. 13 show IR images of HERIC inverter with the three methods respectively. It is obvious that the power loss on the overvoltage absorption circuit is large, and the temperatures of snubber resistors and capacitors are very high. That means lower efficiency, large volume, and more cost. Due to the neglectable power loss, the Temperature of the proposed clamping circuit is just 40.6 degrees, much lower than others.

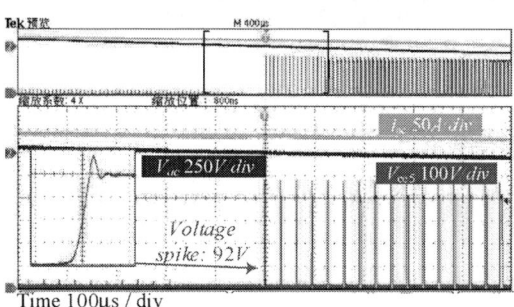

Fig. 7 10 kW HERIC Inverter with proposed clamping circuit prototype.

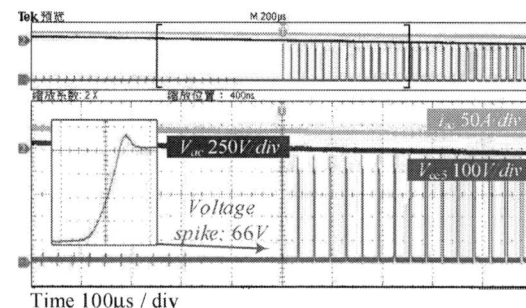

Fig. 8 Experimental waveforms of HERIC inverter without any snubber or clamping circuit when the output power is 7kW, reactive power is 6.3kvar

Fig. 9 Experimental waveforms of HERIC inverter with the overvoltage absorption circuit [2] when the output power is 7kW, reactive power is 6.3kvar

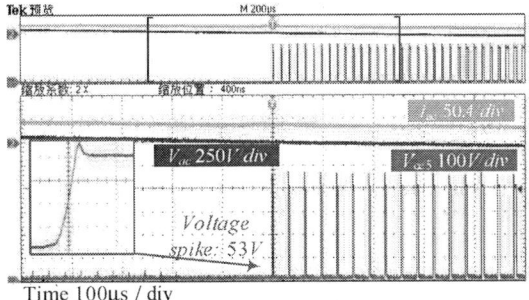

Time 100μs / div

Fig. 10 Experimental waveforms of HERIC inverter with proposed clamping circuit when the output power is 7kW, reactive power is 6.3kvar

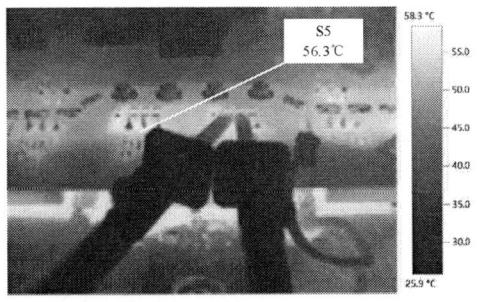

Fig. 11 IR image of HERIC inverter without any snubber or clamping circuit when the output power is 7kW, reactive power is 6.3kvar (ambient temperature is 30 degrees)

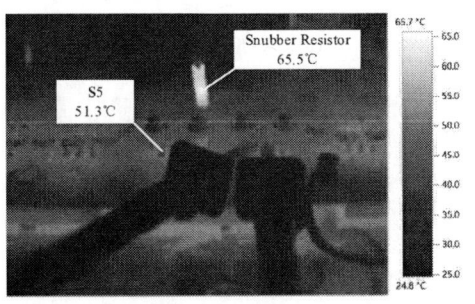

Fig. 12 IR image of HERIC inverter with the overvoltage absorption circuit [2] when the output power is 7kW, reactive power is 6.3kvar (ambient temperature is 30 degrees)

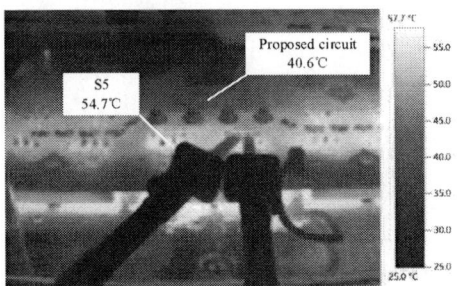

Fig. 13 IR image of HERIC inverter with proposed clamping circuit when the output power is 7kW, reactive power is 6.3kvar (ambient temperature is 30 degrees)

Therefore, the voltage spike of freewheeling switch S5 is decreased dramatically by using the proposed clamping circuit, but the power loss and the cost are almost not increased. The effect of the proposed circuit on turn-off overvoltage suppression. is be verified. The comparison of different methods is shown in TABLE II.

TABLE II.   COMPARISON OF DIFFERENT METHODS

Voltage Spikes suppression method	Spikes suppression effect	Power loss	cost
the overvoltage absorption circuit [2]	Good	Poor	Fair
H6.5 [6]	Excellent	Good	Poor
Proposed clamping circuit	Good	Excellent	Good

V. CONCLUSIONS

This paper points out that the reason for the larger turn-off voltage spikes of the freewheeling switches is the high di/dt of the parasitic inductances during the commutation. This phenomenon is more obvious in some high-frequency applications using SiC MOSFET or Hybrid IGBT, owing to the higher di/dt. On this basis, a lossless and passive clamping circuit is proposed to suppress the voltage spikes. And its principle is introduced in detail. At last, an experimental prototype is built to verify the effect of the proposed circuit on turn-off overvoltage suppression. The experimental results show that when the proposed clamping circuit is used, the voltage spikes are reduced by 40V almost without additional loss and cost. The proposed clamping circuit can also be extended to any topology containing bidirectional switches. Further research is ongoing.

REFERENCES

[1] IEEE Std 1547-2018 IEEE Standard for Interconnection and Interoperability of Distributed Energy Resources with Associated Electric Power Systems Interfaces. New York, USA: IEEE, 2018

[2] Zhang Peng. "Overvoltage absorption circuit and single-phase Heric inverter topology". Chinese Patent CN210608913U, May 22, 2020.

[3] Camuso Gianluca, L. Giorgia, U. Florin, L. EvelynZero, C. Terry, C. Max. "reverse recovery in SiC and GaN Schottky diodes: A comparison," *2016 28th International Symposium on Power Semiconductor Devices and ICs (ISPSD)*, Chicago, 2018, pp. 71-74.

[4] K. K. Sateesh, K. Annamalai, Subrahmanyam. N. "Bidirectional Clamping-Based H5, HERIC, and H6 Transformerless Inverter Topologies With Reactive Power Capability," *in IEEE Transactions on Industry Applications*, vol. 56, no. 5, pp. 5119-5128, Sep. 2020.

[5] W. Hui, W. Zhenxi, T. Zhongting, H. Hua, Y. Yongheng, B. Frede. "An Improved Modulation Strategy for the Active Voltage Clamping HERIC Inverter," *2019 IEEE Applied Power Electronics Conference and Exposition (APEC)*, Baltimore, 2019.

[6] Michael. Frisch, Temesi Ernö. "New 3-Level Topology for Efficient Solar Applications," in *Power Electronics Europe*, No. 7, pp. 29-31, Oct. 2014.

Design of Aging Test System for SiC MOSFET Modules

Chaoyue Shen
SMEA
Shanghai University
Shanghai, China
SCY9897@163.com

Fei Wang
SMEA
Shanghai University
Shanghai, China
f.wang@shu.edu.cn

Zhenye Wang
SMEA
Shanghai University
Shanghai, China
zywang219@foxmail.com

Zhong Ye
InventChip Technology
Co.,Ltd
Shanghai, China
zhong.ye@inventchip.com.cn

Abstract—**Power devices have been widely used in power electronic converters in various applications, such as new energy power generation, electric vehicle charging stations and motor drives. Among power devices, SiC MOSFETs are gaining significant attention and gradually entering the market due to their high speed, high temperature and high power characteristics, but their long-term reliability still remains a top concern. The power cycling test is the most effective and important aging test for device reliability assessment. This paper presents a versatile and energy-saving dual three-phase bridge power cycling aging test system for SiC Modules. It includes the circuit topology, control strategy, heat dissipation design. High-voltage R_{DS_ON} measurement circuit is described in detail, which monitors all power devices' turn-on resistance in real time for health assessment. A 300kW system is designed and tests are conducted to validate the feasibility and effectiveness of the proposed system under different conditions.**

Keywords—*SiC module, reliability, power cycling, aging test system*

I. INTRODUCTION

Silicon carbide (SiC) materials are well known for their high speed, high temperature and high power characteristics. They perform significantly better than Si materials in power device designs. Converters using SiC devices can greatly reduce the volume of magnetic components and heatsinks. As such, the converters' system cost be reduced. SiC technologies have now been widely used in applications, new energy power generation, electric vehicle charging stations and motor drives. [1]. While the SiC technology is getting more mature, the reliability of SiC power devices still remain a top concern in current stage, which is mainly caused by the problem of device's aging [2],[3]. The aging problem may result in the damage of devices and converters prematurely. Device aging research plays an important role in new technology development. An aging test station, which can emulate real working conditions and device stress and establish the relationship between the working conditions and the reliability indexes, will be a significant tool for the reliability research.

To design an aging test system for power devices, it is necessary to have an accelerated method, which can generate thermal cycling impacts on the devices and accurately monitor the power device aging characterization [4],[5]. So far there are two types of power device aging test systems. One is simply real converters designed with the aging parameter monitoring function [6]. This method generates high power losses since loads are required, so it is not suitable for long time aging test in a large scale. The other is a low-frequency pulse test system, which implements the power cycle aging test scheme based on current source

injection or electronic loads with voltage source [7]-[9]. This method can reduce lower losses, but it is not based on real switching operation conditions. Furthermore, adjusting the current direction and amplitude is not easy. In general, neither the above schemes can meet the requirements of emulating real working conditions and low power loss, simultaneously.

To meet the demands of both emulating real working conditions and low power loss, a new aging test system is proposed[10]. The system is constructed with two bi-directional converters connected in cascade with energy recycling capability. Power can flow between the two converters circularly with proper current control strategy, which realizes the two requirements mentioned above. However, this type of aging test systems is mainly designed for IGBT modules so far, and not directly applicable for SiC modules [11]. This paper proposes a new power cycling aging test system for SiC modules with real-time Ron monitoring capability to facilitate the investigation on the aging process.

II. SYSTEM OVERVIEW

A. Circuit Topology

The purpose of the power cycling test system proposed in this paper is to provide real working conditions for targeted SiC modules, which should be very similar to real converters, rather than to impose simple pulse currents for aging tests.

As shown in Fig. 1, the proposed power cycling aging test system is composed of two three-phase bridges, in which each terminals of the three-phase bridge Unit A and Unit B are connected in parallel through inductors. In view of the rated voltage of mainstream commercial SiC modules, e.g., 1200 V, the maximum operating voltage in practical applications is generally designed at 800 V. Similarly, for single bridge modules, the maximum rated current and hard switching frequency are set at 400 A and 65 kHz, respectively, for practical applications. Therefore, the

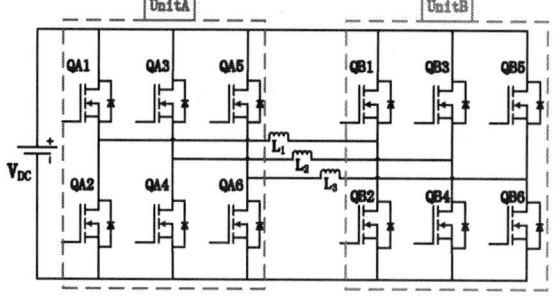

Fig. 1. Circuit topology of the aging test system.

This work is supported by National Natural Science Foundation of China (51977126).

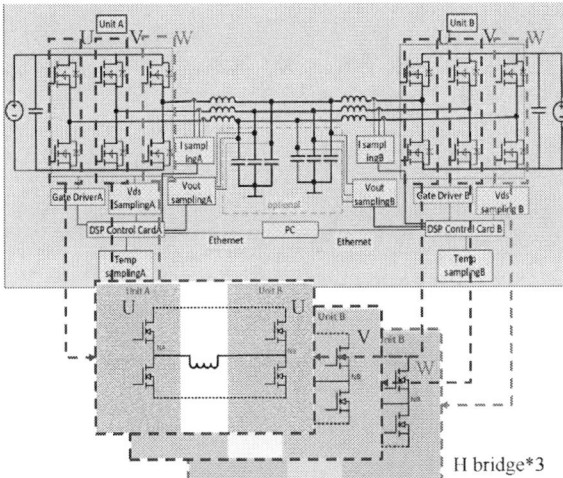

Fig. 2. Control Strategy Diagram.

working parameters of the proposed testing system, including maximum input voltage, cycling current and switching frequency, are determined correspondingly.

B. Control Strategy

As shown in Fig. 2, the system can be treated as three independent H-bridge circuits. In order to emulate the real working conditions of three-phase converters, the three H-bridge circuits can be controlled in 120° phase apart each other. Each H-bridge is controlled independently to emulate one phase of a three-phase converter. Beneficially, if one of the H-bridge is damaged during the aging test, the other two H-bridges will not be affected. In order to emulate the working scenario of three-phase applications, only the frequency and amplitude of the inductor current should be controlled.

To address the proposed control strategy of inductor currents, the U-phase is taken as an example, as the other two phases are the same. In Fig. 3, the duty cycles of the upper switches in phase U are defined as D_{A1} and D_{B1}, respectively. Assuming that D_{A1} is greater than D_{B1}, current will flow through the inductor from left to right. The current will increase on the general trend, as shown in Fig. 4. Therefore, the desired inductor current can be obtained by controlling the duty cycle difference between D_{A1} and D_{B1}. Specifically, the duty cycles of all the power switches in unit B can be fixed at 0.5 simply, while whose in unit A are adjusted around 0.5 through closed-loop control in real time. Consequently, by giving expected reference currents for the three-phase inductors, practical working conditions can be emulated by the proposed aging test system.

III. HEAT DISSIPATION DESIGN

Power device losses are the main losses in power electronic converters, which consist of condition losses and switching losses. The losses are converted into heat, which is dissipated through device package, heatsinks and fans. In view of the setup designed for targeted voltage, current and frequency, the preliminary thermal design is even more important.

Based on the mechanisms aforementioned, a reverse method for thermal design is implemented to ensure DUTs

work within a safe operation area. Specific steps are as follows: First, thermal models are established based on Finite Element Simulation, including design of heatsink and fans. Then, in view of the limitation that maximum temperature resistance of SiC modules is 175 ℃, the maximum allowable loss of every chip in modules can be obtained. The finite element thermal simulation for a SiC module at approximately 175℃ is shown in Fig. 5. Finally, based on the maximum loss and loss model, maximum cycling current and frequency can be obtained reversely for each DUT.

Therefore, it is essential to build an accurate loss model for the DUT. In this paper, based on the loss model for a single switching cycle derived in advance, the average loss model for this system can then be derived.

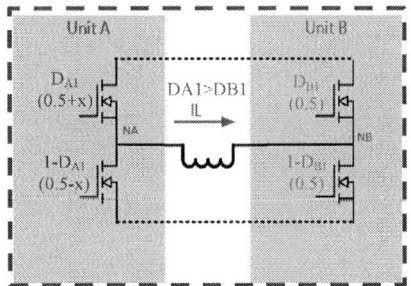

Fig. 3. U-phase state when D_{A1} is greater than D_{B1}

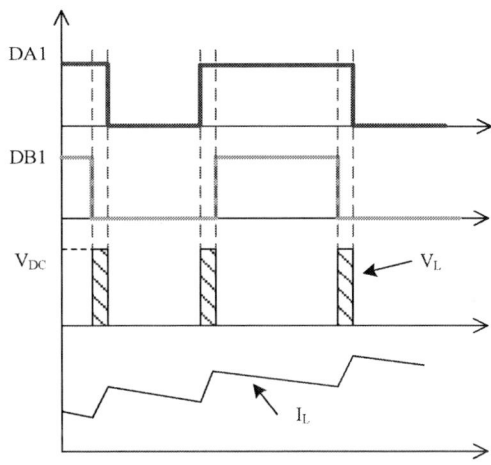

Fig. 4. Corresponding inductor voltage and current When D_{A1} is greater than D_{B1}.

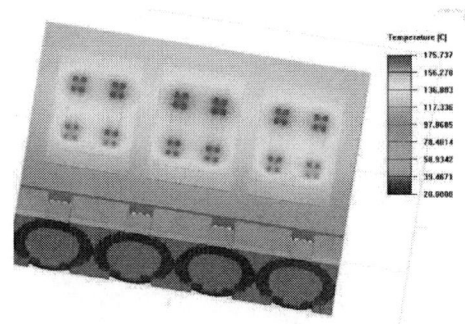

Fig. 5. Finite element thermal simulation of chip junction temperature 175℃.

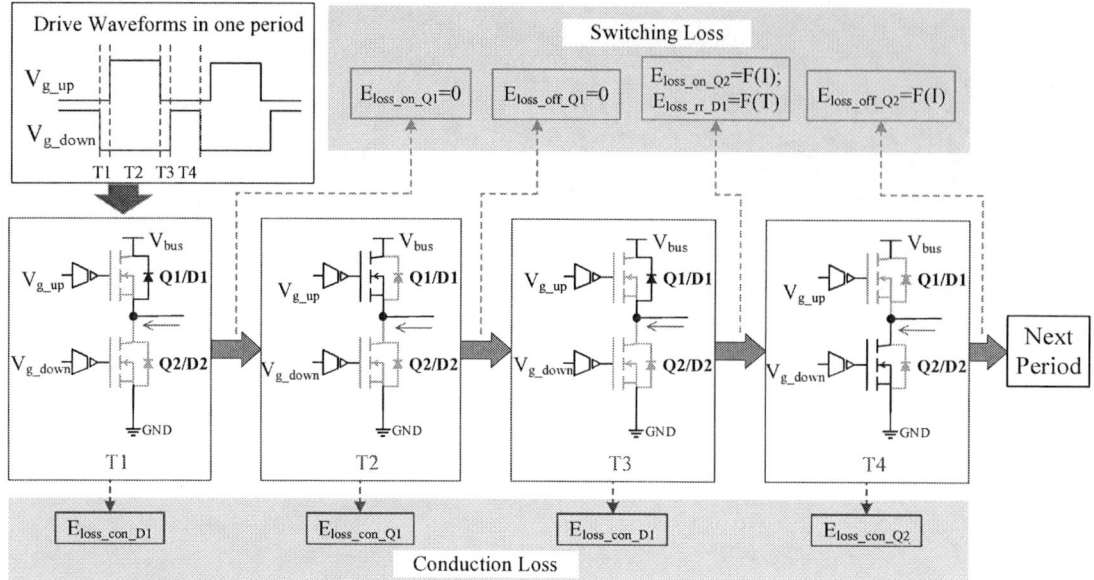

Fig. 6.   Schematic diagram of loss analysis in one period of the DUT.

To prevent bridge switches from shooting through, a deadtime is always inserted. This will increase the complexity of the loss model establishment. As shown in Fig. 6, the DUT goes through four states during one switching cycle, expressed as T1, T2, T3 and T4 respectively, in which T1 and T3 are the deadtime. D1 is on during T1 and T3 states, Q1 and Q2 are on during T2 and T4 state respectively. Therefore, the on-state loss of D1, Q1 and Q2 can be expressed as

$$E_{loss\_con\_D1} = IV_f T_{DeadTime} \tag{1}$$

$$E_{loss\_con\_Q1} = I^2 R_{DS\_ON} (D / F_S - T_{DeadTime}) \tag{2}$$

$$E_{loss\_con\_Q2} = I^2 R_{DS\_ON} ((1-D) / F_S - T_{DeadTime}) \tag{3}$$

where I is the cycling current, V_f is the forward voltage of diode, R_{DS_ON} is the on-state resistance of the MOSFET, D is the duty cycle of the upper switch Q1, F_S is the switching frequency, $T_{DeadTime}$ is the deadtime.

In addition, during the transition of the four states mentioned above, some of the devices will also generate losses at the turn-on or turn-off edges. During the transition between T3 and T4, D1 and Q2 generate reverse recovery loss and turn-on loss respectively, which is expressed as $E_{loss\_rr\_D1}$ and $E_{loss\_on\_Q2}$. During the transition between T4 and T1, Q2 generate turn-off loss, which is expressed as $E_{loss\_off\_Q2}$. All the switching losses can be expressed as

$$E_{loss\_on}(I) = F(I) \cdot (\frac{V_{DS}}{V_{DS\_TEST}})^M \cdot$$
$$[1 + K(T - 25°C)] \cdot \frac{E_{on}(R_{Gon})}{E_{on}(R_{Gon\_TEST})} \tag{4}$$

$$E_{loss\_off}(I) = F(I) \cdot (\frac{V_{DS}}{V_{DS\_TEST}})^M \cdot$$
$$[1 + K(T - 25°C)] \cdot \frac{E_{off}(R_{Goff})}{Eoff(R_{Goff\_TEST})} \tag{5}$$

$$E_{loss\_rr}(T_J) = F(T_J) \cdot (\frac{V_{DS}}{V_{DS\_TEST}})^M \cdot$$
$$[1 + K(I_D - 400A)] \cdot \frac{E_{rr}(R_{Grr})}{E_{rr}(R_{Grr\_TEST})} \tag{6}$$

where I is the cycling current, V_{DS} is the on-state voltage, $E_{on}(R_{Gon})$ is the turn-on and $E_{off}(R_{Gon})$ is the turn-off loss of MOSFET, $E_{rr}(R_{Grr})$ is the reverse loss of diode gained from official manual, respectively. F(I) and F(T_J) are the functions between corresponding switching losses and current I (or junction temperature T_J) created by linear interpolation method.

It is worth to note that no switching loss is generated by Q1 at the specific current direction due to ZVS.

Therefore, based on the analysis above, the total losses for the DUT in one switching cycle can be expressed as

$$E_{loss\_total} = 2E_{loss\_con\_D1} + E_{loss\_con\_Q1} +$$
$$E_{loss\_con\_Q1} + E_{loss\_on\_Q2} + E_{loss\_off\_Q2} \tag{7}$$

Eventually, the average loss model for the DUT is expressed as

$$P_{loss\_AVE} = 2F_0 \sum_{N=0}^{\frac{F_S}{2F_0}} E_{loss\_total}(I(N)) \tag{8}$$

where F_0 is the carrier frequency, $I(N)$ is the current at different sampling moments.

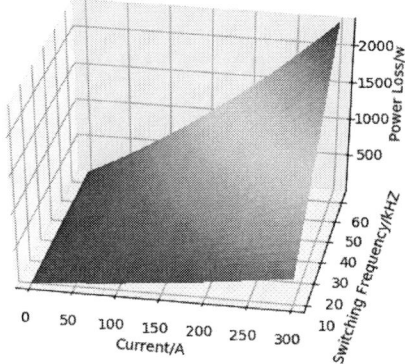

Fig. 7. 3D model between power loss, switching frequency and cycling current.

Based on the derived average loss model, a 3D model is established by Python to obtain the maximum frequency and cycling current, as shown in Fig.1. As a result, a safe operation range of the power cycling test system can be defined.

IV. R_{DS_ON} SAMPLING CIRCUIT

The main purpose of the power cycling setup is to accelerate devices' aging and obtain their reliability data in a relatively short time. Therefore, reasonable aging characteristic parameters are used to determine the aging situation of devices. Several researches has shown that many parameters of SiC power MOSFETs can be treated as degradation indicators, such as gate leakage current (I_{GSlk}), drain leakage current (I_{DSlk}), threshold voltage (V_{GSth}), channel resistance (R_{DS_ON}), body diode forward voltage drop (V_{fwd}), and thermal impedance (Z_{THjc}) [12]. In view of the factors such as difficulty of monitoring, equipment costs, and the samples installation, drain-source on-state resistance R_{DS_ON} is considered as the best aging characteristic parameter for on-line monitoring [13].

For the proposed setup aforementioned, the voltage and current are sampled at high frequencies. Therefore, sampling bandwidth and accuracy are the main challenges of the sampling circuit design. The R_{DS_ON} is measured indirectly by using the ratio between the on-state voltage V_{DS} and load current I. The continued load current is measured with a hall current sensor. However, V_{DS} is a switching voltage. It is challenging to extract on-stage V_{DS} from high voltage switching pulses. Therefore, the following discussion primarily focuses on the real-time on-state voltage measurement circuit.

To explain the principle of V_{DS} measurement circuit, the upper switch Q1 is taken as an example and the circuit is shown in Fig. 8 and Fig. 9. The measurement circuit consists of two parts: one is constant current sources and the other is differential circuit. The constant current source is implemented with Zener diode D3, PNP transistor Q3 and resistance R1. The circuit generates a 0.625 mA DC current,

$$I_{eq} = \frac{V_{D3} + V_{BE\_Q3}}{R_1} = \frac{13 - 0.5}{20k} = 0.625mA \tag{9}$$

where V_{D3} is the regulated voltage value of the Zener diode D3, about 13 V, V_{BE_Q3} is the base-emitter voltage of the transistor Q3, about -0.5 V, and R1 is the resistance value of 20kΩ. Similarly, D3, Q4 and R2 form another constant current source $I2$, whose value matches closely with $I1$. Since $I1$ and $I2$ values are the same or almost the same, the forward voltage drops of D1 and D2 closely match as well, which excludes the impact of diode D1 and D2's voltage variation on V_{DS} measurement due to temperature.

When the lower switch Q2 is turned off and the upper switch Q1 is turned on, the on-state voltage of *Q1* can be measured. As shown in Fig. 8, in this time interval, $I1$ flows through Q1 to the midpoint of the bridge arm while $I2$ flows directly to the midpoint of the bridge arm. At this moment, the positive and negative inputs of the differential circuit are the drain and source voltage of MOSFET respectively, then the output is the needed real-time on-state voltage.

When Q1 is turned off and Q2 is turned on, Q1 is in a non-measurement state. As shown in Fig. 9, the drain and source voltage of Q1 are denoted as V_H and V_L respectively, where the value of V_H is V_{DC} and V_L approximate zero voltage. In this time interval, diode D1 blocks the high voltage V_H. The current source $I1$ will pass through resistor R3 and capacitor C1. To prevent damage to the differential amplifier from excessive or reverse voltages at the positive input, a Schottky diode *D4* is necessary to clamp the positive input voltage within a safe range of 0-5 V. As a result, the measurement circuit can not only measure the on-state voltage in real time, but also can isolated the circuit from high voltage in the off-state of the MOSFET.

The V_{DS} measurement of the lower switches is similar as the upper switches. With the help of such measurement circuits, the V_{DS} of each DUT can be monitored in real time. R_{DS_ON} can be gained subsequently.

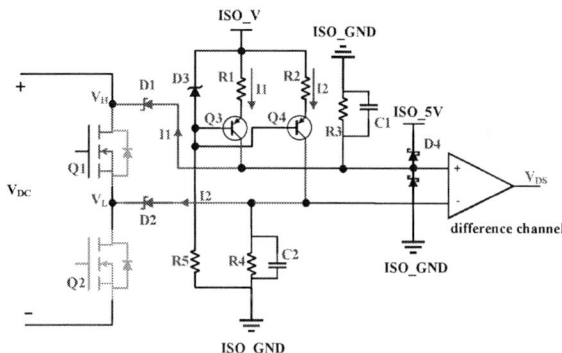

Fig. 8. Sampling circuit state of Q1 when Q1 is on and Q2 is off.

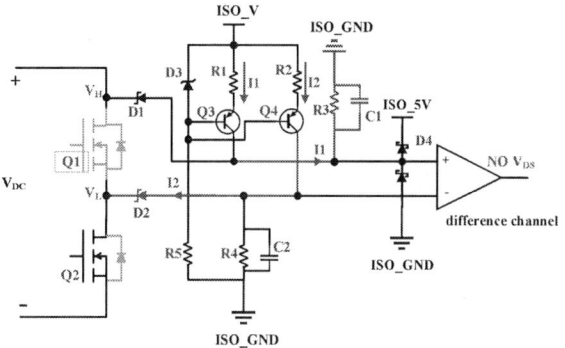

Fig. 9. Sampling circuit state of Q1 when Q1 is off and Q2 is on.

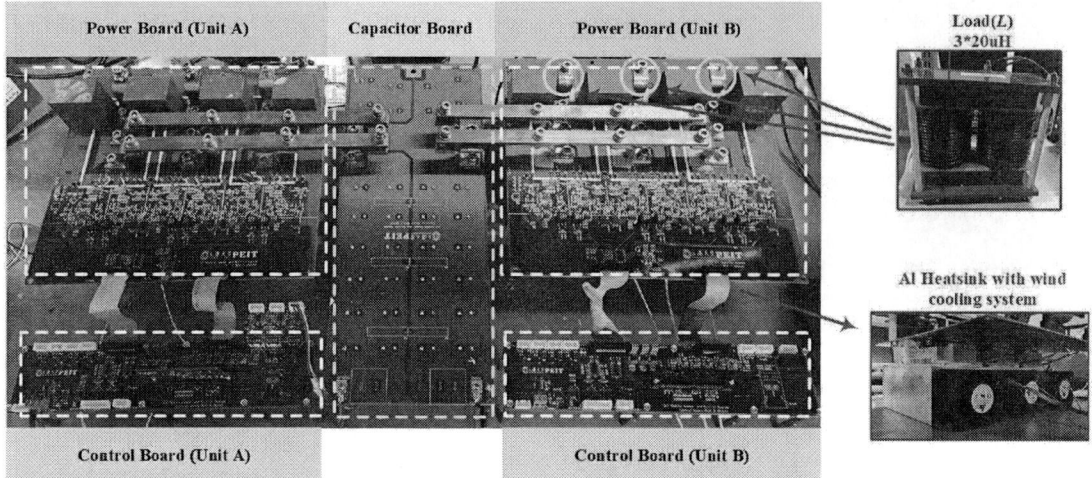

Fig. 10. Prototype of the power cycling aging test system.

V. EXPERIMENT RESULTS

A. Experiment Platform

Fig. 10 shows a prototype of the power cycling aging test system. A power board mainly consists of three parts: main circuit, V_{DS} sampling circuit and driver circuit. A control board is mainly used to regulate such as sampling and driver signals, and communicate with host computer. In this system, the selected SiC MOSFET module for aging test is a commercial SiC MOSFET with the rated voltage of 1200 V, conduction resistance of 6.5 mΩ, and 62mm package.

Fig. 11 shows the gate signals of the DUTs in U-phase with the help of simulation. It can be seen that the duty cycles of the power switches in unit B are fixed at 0.5, and the duty cycles in unit A are adjusted around 0.5 through closed-loop control in real time. In addition, the waveform of the inductor current matches the analysis results shown in Fig. 4. Note since the current ripple could be significant, V_{DS} and I_L should be sampled at the same time to minimize Ron measurement error.

As mentioned in section II, based on the proposed control strategy, real working conditions, such as soft-starting and different current waveform, can be emulated. It is perfect for module power cycling. However, to emulate acceleration and deceleration of motors or grid connected inverters/converters, back-EMF or AC line voltage should be taken into account, which requires duty cycle to be adjusted between 0 and 1, but in either way, the operation can be emulated by tracking the predefined reference current. As shown in Fig. 12 and Fig. 13, the three-phase inductor currents under working conditions of soft-starting and load change are emulated, respectively.

B. V_{DS_ON} Sampling Result

As mentioned in section IV, R_{DS_ON} is measured indirectly by calculating the ratio between the on-state voltage V_{DS} and load current I in real time. Fig. 14 shows the V_{DS_ON} sampling result example in U-phase of the proposed power cycling aging test under the operating condition.

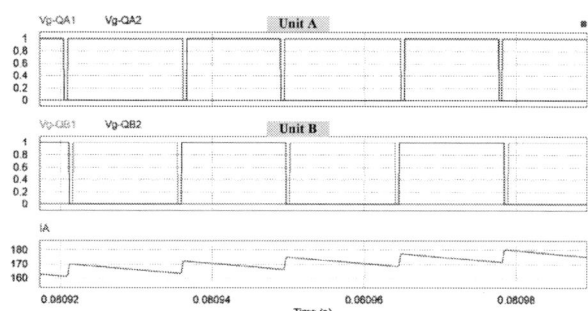

Fig. 11. Switching signals of the DUTs in U-phase.

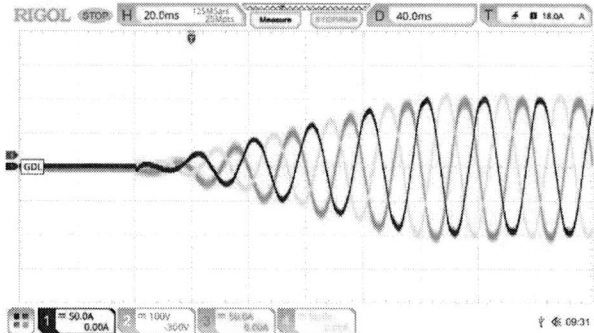

Fig. 12. Current waveform under the condition of soft-starting.

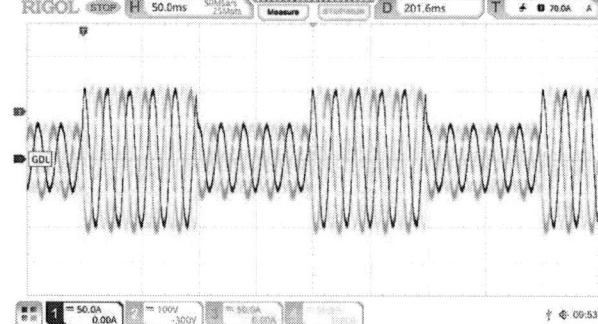

Fig. 13. Current waveform under the condition of load change.

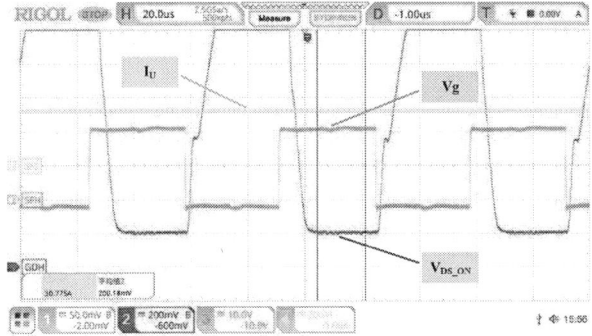

Fig. 14. Result of the power cycling aging test under the specific operating Condition.

VI. CONCLUSION

In this paper, the new power cycling test system for SiC modules has been proposed. The main circuit topology and control strategy has been presented. To ensure the 450A power cycling capability, heat dissipation design has been analyzed in detail. The real-time V_{DS_ON} measurement circuit has been proposed and validated, by which R_{DS_ON} can be measured indirectly. With the proposed real-time measurement circuit, wear-out condition of the SiC modules can be monitored in real time.

Furthermore, it is worth mentioning that the obtained R_{DS_ON} is one of the TSEPs of SiC MOSFET. In the following research, the proposed system can also be used to predict junction temperature of power devices, which is not discussed in this paper so far.

The proposed system can provide real working conditions for target SiC modules to operate at high voltage and high current switching mode, but consumes much less power. As such, the wear-out process of the SiC module under various industrial applications can be monitored. In addition, by means of the proposed control strategy, the testing system can be treated as three independent H-bridge circuits, which can increase test flexibility with different configuration. Different operating conditions and operation modes have been presented.

The feasibility and effectiveness of the proposed system under the conditions of high frequency and current has been demonstrated. The system is a versatile and energy-saving tool for SiC MOSFET module aging test and it would help manufactures to save resource and accelerate module development.

REFERENCES

[1] Q. Peng, Q. Jiang, Y. Yang, T. Liu, H. Wang, and F. Blaabjerg, "On the stability of power electronics-dominated systems: Challenges and potential solutions," IEEE Trans. Ind. Appl., vol. 55, no. 6, pp. 7657–7670, Nov./Dec. 2019.

[2] J. W. Palmour, "Silicon carbide power device development for industrial markets", *Proc. IEEE Int. IEDM*, pp. 1.1.1-1.1.8, Dec. 2014

[3] J. Millán, P. Godignon, X. Perpiñà, A. Pérez-Tomás and J. Rebollo, "A survey of wide bandgap power semiconductor devices", *IEEE Trans. Power Electron.*, vol. 29, pp. 2155-2163, May 2014.

[4] U. Choi, F. Blaabjerg and S. Jørgensen, "Study on Effect of Junction Temperature Swing Duration on Lifetime of Transfer Molded Power IGBT Modules," in *IEEE Transactions on Power Electronics*, vol. 32, no. 8, pp. 6434-6443, Aug. 2017, doi: 10.1109/TPEL.2016.2618917.

[5] U. Choi, F. Blaabjerg and S. Jørgensen, "Power Cycling Test Methods for Reliability Assessment of Power Device Modules in Respect to Temperature Stress," in *IEEE Transactions on Power Electronics*, vol. 33, no. 3, pp. 2531-2551, March 2018, doi: 10.1109/TPEL.2017.2690500.

[6] Baker N, Munk-Nielsen S, Bęczkowski S. Test setup for long term reliability investigation of Silicon Carbide MOSFET[C]. 15th European Conference on Power Electronics and Applications (EPE), Lille, 2013.

[7] A. Hanif and F. Khan, "Degradation Detection of Thermally Aged SiC and Si Power MOSFETs using Spread Spectrum Time Domain Reflectometry (SSTDR)," 2018 IEEE 6th Workshop on Wide Bandgap Power Devices and Applications (WiPDA), Atlanta, GA, 2018, pp. 18-23, doi: 10.1109/WiPDA.2018.8569099.

[8] J. Chen, X. Jiang, Z. Li, H. Yu, J. Wang and Z. J. Shen, "Investigation on Effects of Thermal Stress on SiC MOSFET Degradation through Power Cycling Tests," 2020 IEEE Applied Power Electronics Conference and Exposition (APEC), New Orleans, LA, USA, 2020, pp. 1106-1110, doi: 10.1109/APEC39645.2020.9124249.

[9] Baba S, Gieraltowski A, Jasinski M T, et al. Active Power Cycling Test Bench for SiC Power MOSFETs - Principles, Design and Implementation[J]. IEEE Transactions on Power Electronics, 2021, 36(3) : 2661-2675.

[10] Ma K, Song Y. Power-Electronic-Based Electric Machine Emulator Using Direct Impedance Regulation[J]. IEEE Transactions on Power Electronics, 2020, 35(10) : 10673-10680.

[11] Choi, U. M. , S. Jrgensen , and F. Blaabjerg . "Advanced Accelerated Power Cycling Test for Reliability Investigation of Power Device Modules." IEEE Transactions on Power Electronics, 31.12(2016):8371-8386.

[12] J. P. Kozak, K. D. T. Ngo, D. J. DeVoto, and J. J. Major, "Impact of accelerated stress-tests on SiC MOSFET precursor parameters," in Proc. 2nd Int. Symp. 3D Power Electron. Integr. Manuf., Jun. 2018, pp. 1–5.

[13] E. Ugur, F. Yang, S. Pu, S. Zhao, and B. Akin, "Degradation assessment and precursor identification for SiC MOSFETs under high temp cycling," IEEE Trans. Ind. Appl., vol. 55, no. 3, pp. 2858–2867, May 2019.

Effects of p-type Islands Configuration on the Electrical Characteristics of the 4H-SiC Trench MOSFETs with Integrated Schottky Barrier Diode

Fei Yang[1], Lixin Tian[1]
1. State Key Laboratory of Advanced Power Transmission Technology Global Energy Interconnection Research Institute Co., Ltd. Beijing, China

Zhanwei Shen[2]*, Guoguo Yan[2], Xingfang Liu[2,3], Wanshun Zhao[2], Lei Wang[2], Guosheng Sun[2,3]
2. Key Laboratory of Semiconductor Material Sciences Institute of Semiconductors, Chinese Academy of Sciences Beijing, China
3. College of Materials Science and Opto-Electronic Technology University of Chinese Academy of Science Beijing, China
*Email: zwshen@semi.ac.cn

Junmin Wu[1]
1. State Key Laboratory of Advanced Power Transmission Technology Global Energy Interconnection Research Institute Co., Ltd. Beijing, China

Feng Zhang[4], and Yiping Zeng[2,3]
2. Key Laboratory of Semiconductor Material Sciences Institute of Semiconductors, Chinese Academy of Sciences Beijing, China
3. College of Materials Science and Opto-Electronic Technology University of Chinese Academy of Science Beijing, China
4. Department of Physics, Xiamen University Xiamen, China

Abstract—**In this work, two types of trench gate MOSFETs embedded by Schottky Barrier Diode (SBD) are demonstrated by numerical simulations. The presented structures feature a p-type buried layer (BL) inside the drift region, leading to excellent balance between MOSFET and body diode characteristics of the device. Also, effects of p-buried layers position with respect to the gate and source trenches on the device's performances are revealed. The p-type islands embedded below the bottom of the trench regions (BLSI-MOS2) are found effective in shielding the contacts and reducing the leakage current even at high temperature of 225 °C. Furthermore, the BL is short to source contacts or floating so as to justify the connection's effects on the dynamic characteristics. The results demonstrate that the reverse recovery charges (Q_{rr}) of the BLSI-MOS2 can be reduced to approximately zero when BL is grounded, and the critical short-circuit energy can be increased compared to that of the structure with floating BL. These results can provide guidance for design and modeling of 4H-SiC SBD-integrated trench MOSFET.**

Keywords—*Silicon Carbide; trench MOSFET; integrated Schottky Barrier Diode; short-circuit capability*

II. INTRODUCTION

Metal-oxide-semiconductor field effect transistor (MOSFET) in 4H-SiC has been the most attractive power device because of the excellent material attribute such as wide band gap and high critical field [1-6]. However, the development of 4H-SiC MOSFET is hindered by high chip cost as compared with the rated power device in silicon. Another fatal flaw is that the reliability degradation developed by stacking faults evolution when body pn diode is activated in SiC MOSFET [7]. In consideration of these factors, it is beneficial to combine Schottky barrier diode (SBD) with MOSFET in one die so as to reduce the chip manufacturing cost of a module and suppress the bipolar conduction effect [8-11]. Nonetheless, high electric field and high temperature cause the Schottky contacts deteriorating and hence leads to a leakier

MOSFET [12]. The p-type islands embedded into the drift layer were found effective in shielding the contacts and reducing the leakage current even at high temperature [13]. However, there are still few investigations into the effects of p-type islands configuration on both MOSFET and SBDs' electrical characteristics.

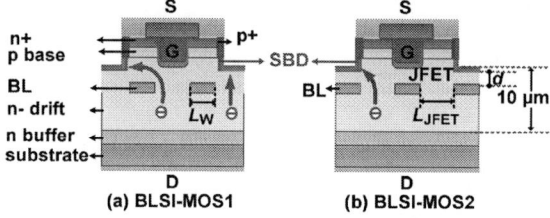

Fig. 1. Device structures of (a) the BLSI-MOS1 and (b) the BLSI-MOS2 with buried p-doped layers.

In this paper, we propose two types of SBD-integrated trench MOSFETs accompanied by p-type buried-layer (BLSI-MOS, see Fig. 1) of medium voltage class via simulations. Even though the BLSI-MOS has embedded SBD, the active area of MOSFET is not consumed due to the effect of buried grid. Thus, they demonstrate the advantage of reduced specific on-resistance ($R_{on,sp}$) compared to the developed double-trench MOSFET (DT-MOS) without SBD [14]. However, the position of the buried-layer (BL) with respect to the contacts has an influence on the electrical properties of MOSFET and SBD. Meanwhile, the BL is short to source contacts or floating so as to justify the connection's effects on the dynamic characteristics.

III. RESULTS AND DISCUSSIONS

A. Device Structure and Working Principle

The p-type islands are embedded below the bottom of the p-base regions in BLSI-MOS1, while they are embedded below the bottom of the trench regions in BLSI-MOS2. The two devices were reproduced utilizing Sentaurus software [15]. Via the shielding effect of the BL regions, both the trench oxide interface and the Schottky contacts surface get protected from the high blocking voltage, accordingly gaining decreased electric field strength in them. The breakdown voltage mainly depends on the pn junction of BL indicated in Fig. 1. In the third-quadrant conduction mode, the electron flows into the n-drift layers and is collected at the Schottky contacts. Thus the conduction of the body pn diodes is effectively suppressed. All the devices have a 10 μm thick n- drift layer (from the n buffer region to the trench bottom) with the donor dopant concentration of 7×10^{15} cm^{-3}. The widths of the buried layers and the JFET regions are L_W and L_{JFET} in BLSI-MOS1 and BLSI-MOS2, respectively. d is the gap between the trench bottom and the BL, showing a value of 0.5 μm. Furthermore, the simulator coverd the mobility degradation, Auger and Schockley-Read-Hall models [16-19]. Meanwhile, the simulated devices included positive charge of 1×10^{12} cm^{-2}, which were confined at the SiC/SiO$_2$ interface. The inversion channel electron mobility is evaluated to be 30 cm^2/V·s for all devices. The doping concentration of BL is 5×10^{17} cm^{-3}. In consideration of tunneling and Schottky barrier lowering models [1], Schottky barrier height shows the value of 1.17 eV for the comparisons of the static characteristics.

B.   Comparison of Static Characteristics

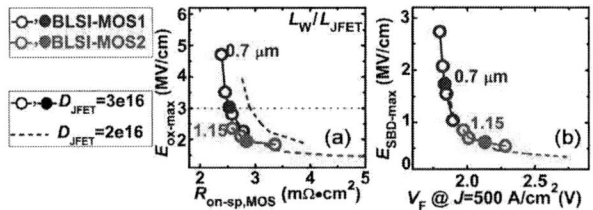

Fig. 2. (a) The compromise curve between $E_{ox\text{-}max}$ (V_{DS} = 1200 V) and $R_{on,sp}$ (V_{GS} = 15 V). (b) The compromise curve between $E_{SBD\text{-}max}$ (V_{DS} = 1200 V) and V_F (J = 500 A/cm^2) for the two devices. Full symbols represent the optimized structures with L_W = 1.0 μm and L_{JFET} = 0.95 μm, for the BLSI-MOS1 and BLSI-MOS2, respectively.

Figures 2(a) and (b) separately plot the tradeoffs between the oxide maximal electric field ($E_{ox\text{-}max}$) and $R_{on,sp}$, and the tradeoffs between the maximal electric field ($E_{SBD\text{-}max}$) in Schottky contacts and the SBD forward voltage drop (V_F). $R_{on,sp}$ and $E_{ox\text{-}max}$ of MOSFET are determined by applying the gate source voltage of V_{GS} = 15 V and the drain source voltage of V_{DS} = 1200 V, respectively. Likewise, when V_{DS} = 1200 V or the current density of J = 500 A/cm^2, $E_{SBD\text{-}max}$ and V_F of SBD are attained, respectively. L_W is 1.3 μm, 1.1 μm, 1.0 μm, 0.9 μm and 0.7 μm for the BLSI-MOS1. L_{JFET} is varied from 0.85 μm to 1.15 μm with a step of 0.1 μm for the BLSI-MOS2. The two devices also involve different JFET doping concentrations (D_{JFET}) of 2×10^{16} cm^{-3} and 3×10^{16} cm^{-3}. $R_{on,sp}$ of the BLSI-MOS1 is comparable to that of the DT-MOS, which is revealed in our previous work [13]. Therefore, no electrical degradation of first-quadrant characteristics can be achieved in the proposed BLSI-MOS1. However, with decreasing L_W, $E_{ox\text{-}max}$ of BLSI-MOS1 is readily increased beyond 3 MV/cm which is not beneficial for long-term stability [20]. It should also be noted that the BL configuration has effects on minimizing $E_{ox\text{-}max}$. Because of the well-shielded effect under

the trench bottom, high electric field can be further mitigated in the BLSI-MOS2. However, the BLSI-MOS2 presents relatively flat tradeoffs between $R_{on,sp}$ and $E_{ox\text{-}max}$ with the D_{JFET} of 2×10^{16} cm^{-3}. Owing to the two narrow path of the JFET current flow, $R_{on,sp}$ is increased with decreasing L_{JFET}. From Fig. 2(a), an excellent compromise between $E_{ox\text{-}max}$ and $R_{on,sp}$ is obtained in BLSI-MOS2 with the D_{JFET} of 3×10^{16} cm^{-3}. Consequently, superior charge compensation effect is justified between the JFET region and the BL region.

Furthermore, the BLSI-MOS1 exhibits lower V_F than the BLSI-MOS2. This is attributed to the reason that the BLSI-MOS1 features wider carrier conduction path in the JFET region, and the carrier transport length is kept short in it. Thus, high threshold forward voltage can be obtained for the diode of BLSI-MOS1 without the turning-on of the intrinsic pin diode, because of the fact that large unipolar current density can be attained if the spreading resistance under the SBD contacts is decreased in a way [8, 10]. Nonetheless, $E_{SBD\text{-}max}$ of BLSI-MOS1 exhibits an approximate value of 1.7 MV/cm at L_W = 1.0 μm and D_{JFET} of 3×10^{16} cm^{-3}, and it increases steadily with decreasing L_W for the BLSI-MOS1. From this point of view, small L_W is not beneficial for improvement of the blocking reliability of the integrated devices. Anyway, the configuration of p-type islands generate advantage due to its simpleness of layout pattern, and thus result in more desirable characteristics of tradeoffs and the suppressed conduction of the bipolar diode.

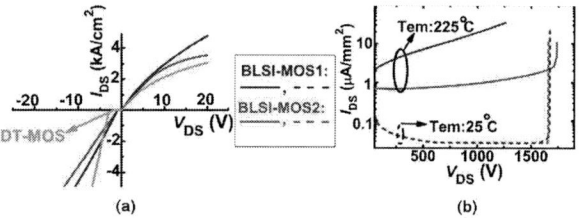

Fig. 3. (1) I-V curves in MOSFET and diode-mode operation of the optimized devices. (b) The off-state current density of the optimized MOSFETs at a different operation temperature.

The I-V characteristics for MOSFET and diode-mode operation of the optimized devices of the optimized MOSFETs are compared in Fig. 3(a). For the BLSI-MOS1, it shows superior on-state performance of the MOSFET than that of the BLSI-MOS2 and the MOSFET without SBD, which may be attributed to the highest current density distribution in the JFET and drift regions. The calculated value of $R_{on,sp}$ is 2.51 mΩ·cm^2 in BLSI-MOS1, which yields a 19% decrease in $R_{on,sp}$ for that of DT-MOS (3.1 mΩ·cm^2). With regard to the low $E_{ox\text{-}max}$ device of BLSI-MOS2, $R_{on,sp}$ is 2.84 mΩ·cm^2 and is 8% lower than that of DT-MOS. There results demonstrate that even though a decrement of the chip area is performed in the BLSI-MOS, the device can give identical on-state current density compared to the DT-MOS. More importantly, the total die size can be saved in comparison to the approach of utilizing the discrete SiC-SBDs and individual MOSFET for the conduction suppression of the intrinsic pin diode. Additionally, in body-diode-mode operation of the optimized devices, the BLSI-MOS1 exhibits higher unipolar current density and low forward voltage drop than the BLSI-MOS2. Especially, it is worth noting that the unipolar current flow is higher than 4 kA/cm^2 and bipolar conduction is successfully inhibited in the

978-1-6654-4817-8/21 $31.00 © 2021 IEEE   75

BLSI-MOS structures. However, as indicated in Fig. 3(b), the BLSI-MOS1 clearly shows higher leak current at 225 °C compared with the BLSI-MOS2. It is attributed to the reason that deterioration phenomenon occurs regionally between SiC and Schottky contacts, in consideration of both the tunneling mode and barrier lowering mechanisms. Also, more SBD areas are effectively shared out in the source trenches of the BLSI-MOS1 in comparison to the BLSI-MOS2.

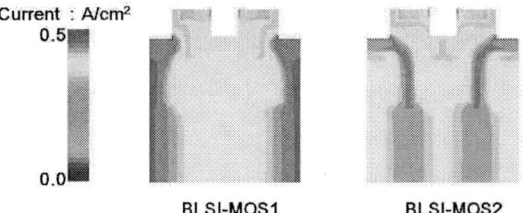

Fig. 4. Leak current in one-unit-cell of the devices at V_{DS} = 1200 V and 225 °C.

Figure 4 shows the separate off-state current distribution of the BLSI-MOSs at V_{DS} = 1200 V and 225 °C. It is noticeable that the BLSI-MOS1 presents quite high leak current at high temperature due to the weak shielding effect of the BL in the off-state. The BLSI-MOS2 allows narrower electron conduction path and longer transport distance compared with the BLSI-MOS1. Therefore, the p-type islands configuration must be optimized in order to reduce the leakage current at the Schottky contacts for the SBD-integrated MOSFET in high temperature applications.

C. Comparison of Dynamic Characteristics

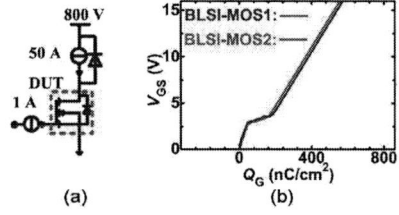

Fig. 5. (a) A simulated circuit for Q_G, and (b) comparisons of the Q_G.

The mixed-mode features in the Sentaurus were used in order to conduct the dynamic switching performances [21]. Figs. 5(a) and (b) separately show the simulated circuit for Q_G and comparisons of the Q_G characteristic at different V_{GS}. A 800 V DC voltage is applied to the drain terminals connected with a 50 A current-source. It can be observed that the BLSI-MOS1 and BLSI-MOS2 have gate-drain charge (Q_{GD}) of 128.5 nC/cm² and 92.3 nC/cm², respectively. Additionally, when V_{GS} = 15 V, the BLSI-MOS2 has slightly lower gate charges of 525.7 nC/cm² as compared to that of 542.7 nC/cm² for the BLSI-MOS1. As a matter of fact, the BLSI-MOS2 shows lower degree of interlacing portion between the drift regions and the gate contacts.

Then, the reverse recovery characteristics of the MOFETs are conducted by employing a double-pulse testing circuit as illustrated in Fig. 6(a). The maximum gate voltage is 15 V. An output current of 30 A is also applied. Two identical MOSFETs (Q1 and Q2) are series-connected, and in the meantime, gate is shorted to the source in Q2. Because of the

insignificant effect of minority-carrier injection into the drift regions, both of the two devices show significantly low reverse recovery charges (Q_{rr}) with the values of 2.2 nC/cm² and 3.2 nC/cm², for the BLSI-MOS1 and BLSI-MOS2, respectively. In order to justify the effect of the BL connection on the switching transients, the BL is short to source (grounded) or floating in the devices. As shown in Fig. 6(b), the connection has negligible effect on the reverse recovery characteristics of the BLSI-MOS1. However, the BLSI-MOS2 with grounded BL exhibits soft reverse recovery transient, leading to approximately zero for the Q_{rr}. This may be attributed two reasons described as following. First, the area ratio between the BL and the electrodes of the MOSFETs is larger in BLSI-MOS2 than that in BLSI-MOS1. Secondly, the gate-to-source and the drain-to-source capacitances are increased in the two BLSI-MOS devices owing to the shorted BL-source terminals. Then, the current rising time is increased, and a high degree of displacement current can be integrated into the lower arm device of Q1. Accordingly, the current decreases in a soft mode for the diode of the upper arm, indicating the reduced reverse recovery charges in it.

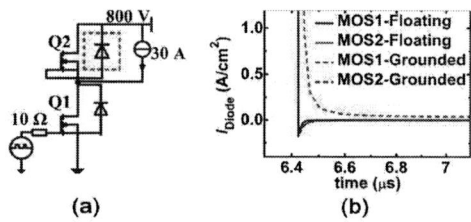

Fig. 6. (a) The double pulse testing circuit, and (b) comparisons of the reverse recovery transients for the BLSI-MOS1 and the BLSI-MOS2.

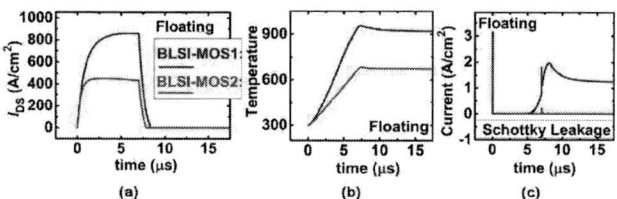

Fig. 7. (a) Short-circuit waveforms, (b) maximum lattice temperature and (c) Schottky leakage curves in BLSI-MOS1 and (b) BLSI-MOS2 with floating BL at a DC voltage of 600 V and V_{GS} = 15/0 V.

Comparisons of short-circuit (SC) characteristics between the two MOSFETs are carried out via electrothermal simulations [22]. Fig. 7 shows the SC waveforms, maximum lattice temperature and Schottky leakage curves in BLSI-MOS with floating BL at a DC voltage of 600 V and V_{GS} = 15/0 V. The BLSI-MOS1 exhibits larger saturation current density than the BLSI-MOS2, which is consistent with the result attained by the aforementioned comparisons. Therefore, on the basis of self-heating effect of the SC device, the peak value of the device temperature in BLSI-MOS1 is higher than that in BLSI-MOS2. From Fig. 7(b), any further increase in SC time intervals in BLSI-MOS1 will cause the temperature to be increased beyond the melting point of approximately 933 K for the overlay metals of aluminum [23]. For the same reason, the SBD leakage current of the BLSI-MOS1 is higher than that of BLSI-MOS2 when the devices are switched off, demonstrating similar leakage characteristics revealed by Fig. 3(b).

Figure 8 shows the electric field and lattice temperature (T) distributions in the two devices with floating BL at a DC voltage of 600 V and short-circuit time of 7 μs. As indicated by Fig. 8(a), the peak electric field of gate oxide and SBD contacts in BLSI-MOS1 is above 0.5 MV/cm and is higher than that in BLSI-MOS2. The device suffers from significant power dissipation during the SC transient. With increasing SC time intervals, the BLSI-MOS1 exhibits a higher T than the BLSI-MOS2. Thus, the coexistence of large power heat and high-intensity electric field in the former causes a severe leakage current at the SBD contact surfaces, indicating a degraded SC reliability in the device.

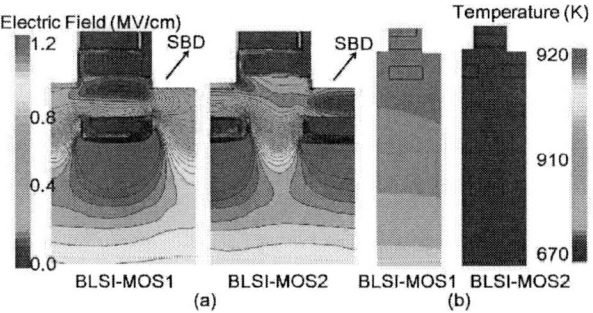

Fig. 8. (a) (a) Electric field and (b) lattice temperature distributions in the two devices with floating BL at a DC voltage of 600 V and short-circuit time of 7 μs.

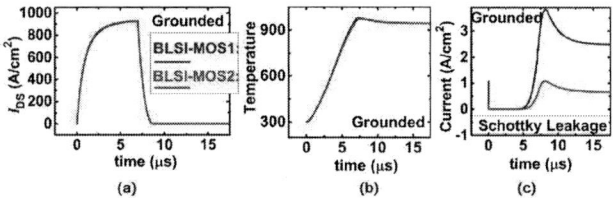

Fig. 9. (a) Short-circuit waveforms, (b) maximum lattice temperature and (c) Schottky leakage curves in BLSI-MOS1 and (b) BLSI-MOS2 with grounded BL at a DC voltage of 600 V and V_{GS} = 15/0 V.

The effect of the BL connection on the SC characteristics is also revealed. Fig. 9 plots the SC waveforms in BLSI-MOS with grounded BL at a DC voltage of 600 V and V_{GS} = 15/0 V. The saturation current density of BLSI-MOS1 is slightly increased as compared to that of the floating devices (Fig. 7(a)), resulting in larger lattice temperature and Schottky leakage current in it. However, it is noticeable that the saturation current density of BLSI-MOS2 is dramatically improved and is comparable to that of BLSI-MOS1. Then, the maximum lattice temperature reaches beyond 933 K in comparison to that of the floating BL device. According to findings from previously confirmed works, the MOSFETs with floating p-shield regions boast charge storage effect, and the negative charge extends the depletion region around the p-shield, leading to the degradation of on-state performances of the device [24-25]. Herein, it is speculated that the depletion region around the BL may not be influenced by the connection of the BL due to the large gap between the p-base layers and the BL, and the gap between the two neighboring BL. However, effect of the widened depletion region of the floating BL in BLSI-MOS2 may be more severe as compared to that of the grounded BL. Meanwhile, the potential in the BL below the SBD trenches is pinned to zero, which provide suppression effect on the extension of the depletion layer near the SBD contacts. Accordingly, the increased saturation current density of BLSI-MOS2 occurs and causes amounts of heat in the chip.

IV. CONCLUSION

In conclusion, 4H-SiC SBD-integrated trench MOSFET with buried p-type regions in the drift layer is characterized by simulation, for the purpose of revealing the effects of p-type islands configuration on the devices' characteristics. It is shown that the BLSI-MOS1, with BL located below the p-base regions, exhibits a 10% decrement in V_F and a 12% decrease in $R_{on,sp}$ compared to the BLSI-MOS2. However, owing to the well-shielded trench corner, quite low leak current is demonstrated in blocking BLSI-MOS2 at 225 °C. In the switching mode, the BLSI-MOS1 and BLSI-MOS2 have gate-drain charge (Q_{GD}) of 128.5 nC/cm² and 92.3 nC/cm², respectively. Meanwhile, the BLSI-MOS2 with grounded BL exhibits soft reverse recovery transient, causing approximately zero for the Q_{rr}. Also, the BLSI-MOS2 with grounded BL shows improved critical short-circuit energy compared to the device with floating BL. However, the maximum lattice temperature reaches beyond the melting point for the overlay metals of aluminum. These results indicate that the configuration of the buried p-type layers should be carefully considered in terms of improving device performances over a broad temperature range.

ACKNOWLEDGMENT

This work was supported in part by the Science & Technology Program of the State Grid Corporation of China Co., Ltd., under Grant No. 5500-202058402A-0-0-00.

REFERENCES

[1] T. Kimoto and J. A. Cooper, Fundamentals of silicon carbide technology: growth, characterization, devices, and applications. Singapore: Wiley, Nov. 2014.

[2] B. J. Baliga, *Silicon Carbide Power Devices.* Singapore: World Scientific, Jan. 2006.

[3] J. A. Cooper, Jr., and A. Agarwal, "SiC power-switching devices—The second electronics revolution?" Proc. IEEE, vol. 90, no. 6, pp. 956–968, Jun 2002.

[4] Z. Shen, F. Zhang, S.Dimitrijev, J. Han, L. Tian, G. Yan, Z. Wen, W. Zhao, L. Wang, X. Liu, G. Sun and Y.Zeng, "Prediction of High-Density and High-Mobility Two-Dimensional Electron Gas at AlxGa1-xN/4H-SiC Interface," Mat. Sci. Forum, vols. 897, pp. 719–722, May 2017.

[5] Q. Zhang, M. Gomez, C. Bui, and E. Hanna, "1600V 4H-SiC UMOSFETs with dual buffer layers," in *Proc. ISPSD*, Santa Barbara, CA, USA, pp. 211-214, May 2005.

[6] Z. Shen, F. Zhang, S.Dimitrijev, J. Han, G. Yan, Z. Wen, W. Zhao, L. Wang, X. Liu, G. Sun and Y.Zeng, "Comparative study of electrical characteristics for n-type 4H–SiC planar and trench MOS capacitors annealed in ambient NO," Chin. Phys. B, vol. 26, no. 10, p. 107101, Aug 2017.

[7] A. Agarwal, H. Fatima, S. Haney, and S. H. Ryu, "A New Degradation Mechanism in High-Voltage SiC Power MOSFETs," *IEEE Electron Device Lett.*, vol. 28, no. 7, pp. 587–589, Jun. 2015.

[8] S. Hino, H. Hatta, K. Sadamatsu, Y. Nagahisa, S. Yamamoto, T. Iwamatsu, Y. Yamamoto, M. Imaizumi, S. Nakata, and S. Yamakawa, "Demonstration of SiC-MOSFET Embedding Schottky Barrier Diode for Inactivation of Parasitic Body Diode," *Mat. Sci. Forum*, vols. 897, pp. 477–482, May. 2017.

[9] F. J. Hsu, C. T. Yen, C. C. Hung, H. T. Hung, C. Y. Lee, L. S. Lee, Y. F. Huang, T. L. Chen, and P. J. Chuang, "High Efficiency High Reliability SiC MOSFET with Monolithically Integrated Schottky Rectifier," in *Proc. ISPSD*, Sapporo, Japan, Jun. 2017, pp. 45-48.

[10] Y. Kobayashi, N. Ohse, T. Morimoto, M. Kato, T. Kojima, M. Miyazato, M. Takei, H. Kimura, and S. Harada, "Body PiN diode inactivation with low on-resistance achieved by a 1.2 kV-class 4H-SiC SWITCH-MOS," in *IEDM Tech. Dig.*, San Francisco, CA, USA, Dec. 2017, pp. 9–1.

[11] F. J. Hsu, C. T. Yen, C. C. Hung, H. T. Hung, C. Y. Lee, L. S. Lee, Y. F. Huang, T. L. Chen, and P. J. Chuang, "High Efficiency High Reliability SiC MOSFET with Monolithically Integrated Schottky Rectifier," in *Proc. ISPSD*, Sapporo, Japan, Jun. 2017, pp. 45-48.

[12] W. Sung, and B. J. Baliga, "On Developing One-Chip Integration of 1.2 kV SiC MOSFET and JBS Diode (JBSFET)," *IEEE Trans. Ind. Electron.*, vol. 64, no. 10, pp. 8206-8212, Apr. 2017.

[13] L. Tian, F. Yang, Z. Shen, F. Zhang, X. Liu, G. Yan, W. Zhao, L. Wang, G. Sun, J. Wu, and Y. Zeng, "Low leakage current and high unipolar current density in a 4H-SiC trench gate MOSFET with integrated Schottky barrier diode," in *IEEE SSLChina: IFWS*, Shenzhen, China, Nov 2020, pp. 58-62.

[14] T. Nakamura, Y. Nakano, M. Aketa, R. Nakamura, S. Mitani, H. Sakairi, and Y. Yokotsuji, "High performance SiC trench devices with ultra-low Ron," in *IEDM Tech. Dig.*, Washington, DC, USA, Dec. 2011, pp. 599–601.

[15] *TCAD Sentaurus Device Manual*, Synopsys, Inc., Mountain View, CA, USA, 2013.

[16] M. Ruff, H. Mitlehner, and R. Helbig, " SiC devices: physics and numerical simulation," *IEEE Trans. Electron. Dev.*, vol. 41, no. 6, pp. 1040–1054, Jun. 1994.

[17] A. Galeckas, J. Linnros, V. Grivickas, U. Lindefelt, and C. Hallin, "Auger recombination in 4H-SiC: Unusual temperature behavior," *Appl. Phys. Lett.*, vol. 71, no. 22, pp. 3269–3271, Dec. 1997.

[18] J. Pernot, S. Contreras, and J. Camassel, "Electrical transport properties of aluminum-implanted 4H–SiC," *J. Appl. Phys.*, vol. 98, no. 2, 2005, Art. no. 023706.

[19] S. Kagamihara, H. Matsuura, T. Hatakeyama, T. Watanabe, M. Kushibe, T. Shinohe, and K. Arai, "Parameters required to simulate electric characteristics of SiC devices for n-type 4H–SiC," *J. Appl. Phys.*, vol. 96, no. 10, pp. 5601-5606, Nov. 2004.

[20] Z. Shen, F. Zhang, G. Yan, Z. Wen, W. Zhao, L. Wang, X. Liu, G. Sun and Y.Zeng, "High-frequency Switching Properties and Low Oxide Electric Field and Energy Loss in a Reverse-Channel 4H-SiC UMOSFET," *IEEE Trans. Electron. Dev.*, vol. 67, no. 10 (2020), pp. 4046–4053.

[21] H. Jiang, J. Wei, X. Dai, C. Zheng, M, Ke, X. Deng, Y. Sharma, I. Deviny, and P. Mawby, "SiC MOSFET with built-in SBD for reduction of reverse recovery charge and switching loss in 10-kV applications," in *Proc. ISPSD*, Sapporo, Japan, Jun. 2017, pp. 49-52.

[22] M. Okawa, R. Aiba, T. Kanamori, H. Yano, N. Iwamuro, and S. Harada, "Experimental and Numerical Investigations of Short-Circuit Failure Mechanisms for State-of-the-Art 1.2 kV SiC Trench MOSFETs," in *Proc. ISPSD*, Shanghai, China, May. 2019, pp. 167–170.

[23] J. Liu, G. Zhang, B. Wang, W. Li, and J. Wang, "Gate Failure Physics of SiC MOSFETs Under Short-Circuit Stress," *IEEE Electron Device Lett.*, vol. 41, no. 1, pp. 103-106, Jan. 2020.

[24] J. Wei, M. Zhang, H. Jiang, H. Wang, and K. J. Chen, "Dynamic degradation in SiC trench MOSFET with a floating p-shield revealed with numerical simulations," *IEEE Trans. Electron Devices*, vol. 64, no. 6 (2017), pp. 2592–2598.

[25] Y. Kagawa, N. Fujiwara, K. Sugawara, R. Tanaka, Y. Fukui, Y. Yamamoto, N. Miura, M. Imaizumi, S. Nakata, and S. Yamakawa, "4H-SiC trench MOSFET with bottom oxide protection," *Mater. Sci. Forum*, vols. 778–780, pp. 919–922.

Characteristics of SiC MOSFET in a Wide Temperature Range

Mengyu Zhu
College of Artificial Intelligence
Xi'an Jiaotong University Xi'an, China
zmy3118104006@stu.xjtu.edu.cn

Laili Wang
College of Artificial Intelligence
Xi'an Jiaotong University Xi'an, China
llwang@mail.xjtu.edu.cn

Huaqing Li
College of Artificial Intelligence
Xi'an Jiaotong University Xi'an, China
lihuaqing@stu.xjtu.edu.cn

Chengzi Yang
College of Artificial Intelligence
Xi'an Jiaotong University Xi'an, China
lemonyang@stu.xjtu.edu.cn

Dingkun Ma
College of Artificial Intelligence
Xi'an Jiaotong University Xi'an, China
mdk1014@stu.xjtu.edu.cn

Fengtao Yang
College of Artificial Intelligence
Xi'an Jiaotong University Xi'an, China
yangfengtao@stu.xjtu.edu.cn

Abstract—In order to expand the application of SiC MOSFET in a wide temperature range, the performance of SiC MOSFET from 25 °C to 425 °C is studied in this paper by taking the 1.2 kV QPM3-1200-0013D as an example. Different from the device characteristics given in the data sheet or previous articles at only three discrete temperatures or temperatures below 250 °C, the temperature-dependent static characteristics of SiC MOSFET are remeasured and nonlinearly characterized in a wide temperature range in this paper. In addition, considering the improvement effect of SiC Schottky barrier diode on reliability of SiC MOSFET's body diode, the temperature-dependent static characteristics of SiC Schottky barrier diode are also analyzed in this paper, which provides guidance for the high-temperature pre-design and application of SiC MOSFET.

Index Terms—SiC MOSFET, SiC SBD, high temperature

I. INTRODUCTION

SiC device have superior advantages in high temperature applications such as energy exploration, aerospace and military fields. There have been many in-depth studies on temperature-dependent characteristics of SiC MOSFET in a wide temperature rang have been carried out [1]–[3].

The device characteristics of SiC MOSFETs are temperature-dependent, key parameters such as threshold voltage and on-resistance will affect the switching performance with the vary of temperature [4]. Studies have shown that the theoretical operating temperature of SiC MOSFETs can be up to 500 °C [2], [5]. However, the variation of device characteristics at high temperatures cannot be obtained directly from the datasheet, which brings difficulties to the pre-design of the system at high temperatures. Therefore, it is necessary to remeasure the device characteristics of SiC MOSFET over a wide temperature range for the accurate characteristic-analyzing and model-establishing.

In addition, the body diode of SiC MOSFET (SiC MBD) may experience bipolar degradation after long-term forward voltage bias. Therefore, it is necessary to equip both SiC MBD and SiC Schottky barrier diode (SiC SBD) to ensure

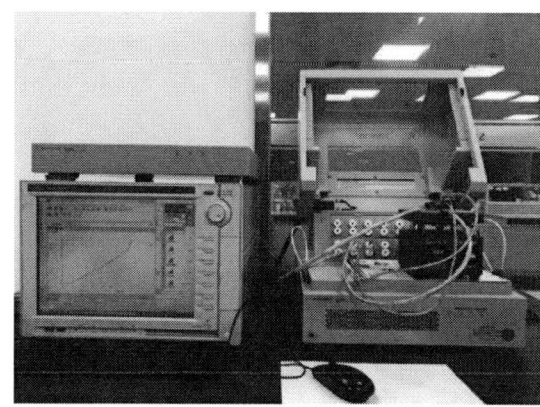

Fig. 1. Agilent B1505A static characteristic test system

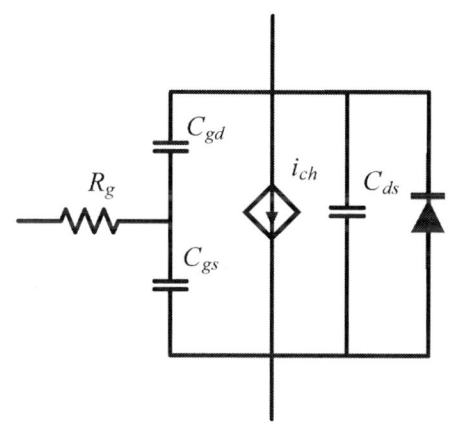

Fig. 2. The equivalent device model of SiC MOSFET

the reliability of the system, especially in high temperature applications.

II. TEMPERATURE-DEPENDENT STATIC CHARACTERIZATION OF SiC MOSFET

Static characteristics of SiC MOSFET will be measured and characterized in this section. Fig. 2 shows the device equivalent model of SiC MOSFET. In this model, SiC MOSFET is composed with three voltage-dependent capacitances (C_{gs}, C_{ds} and C_{gd}), a controlled current source i_{ch}, and an anti-paralleled MOSFET's body diode. R_g is the lumped gate resistance. SiC MOSFET QPM3-1200-0013D is selected as an example for study.

The temperature dependent threshold voltage V_{th} of SiC MOSFET from 25 °C to 425 °C under the test condition of $I_d = 26$ mA is shown in Fig. 3(a), and the relationship with temperature can be characterized by

$$V_{th}(T) = V_{th,25} + k_1(T - 25)^{k_2} \qquad (1)$$

The negative temperature coefficient of V_{th} causes $V_{th,425} \approx 0.5 \sim 0.6 V_{th,25}$, which makes converter prone to crosstalk at high temperatures. Therefore, it is necessary to further increase the negative turn-off voltage or use crosstalk suppression circuits to prevent device from being damaged during high-temperature operation.

Compared to the exponentially increased on-state resistance R_{ds} of Si counterpart, the R_{ds} of SiC MOSFET rises slowly with the increase of drain-source breakdown voltage, hence, SiC MOSFET can achieve a higher breakdown voltage while remaining a low R_{ds}. Since R_{ds} decreases with the increase of gate-source voltage V_{gs}, a larger V_{gs} should be selected as much as possible without exceeding the limit. This paper will use the recommended voltage $V_{gs} = 15$ V for experiments. Under the test conditions of $V_{gs} = 15$ V, $I_d = 70$ A, R_{ds} is measured at different temperatures, the experimental results and the fitting results by (2) are shown in Fig. 3(b).

$R_{ds,425} \approx 4 \sim 5 R_{ds,25}$, this will cause the conduction loss of the device to increase drastically at high temperatures, and it cannot be improved by increasing V_{gs}.

$$R_{ds}(T) = R_{ds,25} \cdot \left(\frac{T}{25}\right)^k \qquad (2)$$

The transfer characteristic of SiC MOSFET at different temperatures within 25 °C $\sim$ 425 °C are measured and shown in Fig. 4(a). Channel current I_d as a function of V_{gs} and junction temperature T can be expressed by (3). It should be noted that (3) is only applicable to the saturation region ($V_{ds} \geq V_{ds\_sat}$).

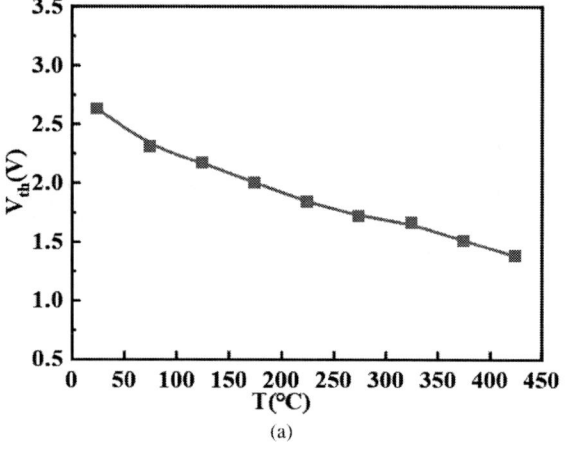 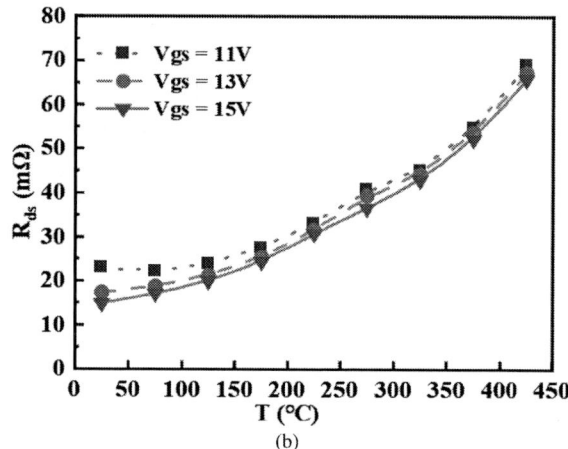

Fig. 3. The temperature-dependent (a) threshold voltage (b) on-state resistance of 1.2 kV SiC MOSFET.

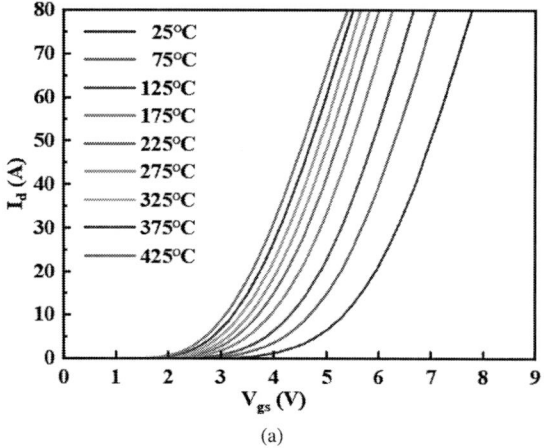 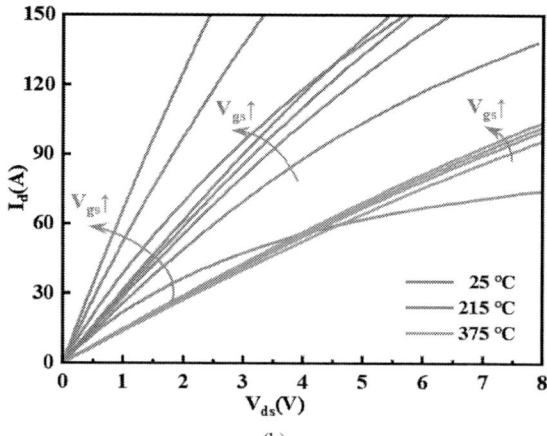

Fig. 4. The temperature-dependent (a) transfer characteristic (b) output characteristic of 1.2 kV SiC MOSFET.

$$I_d(V_{gs}, Vds, T) = \begin{cases} k_1(V_{gs} - V_{th}(T) - k_2 V_{ds}) \cdot V_{ds} \cdot (1 + \lambda V_{gs})^2 & 25\,°C \leq T < 215\,°C \\ k_1(V_{gs} - V_{th}(T) - k_2 V_{ds}) \cdot V_{ds} \cdot (1 + \lambda V_{gs}) & 215\,°C \leq T < 325\,°C \\ k_1(V_{gs} - V_{th}(T) - k_2 V_{ds}) \cdot V_{ds} - \lambda V_{gs} V_{ds} & T \geq 325\,°C \end{cases} \quad (4)$$

$$I_d(V_{gs}, T) = k_1(V_{gs} - V_{th}(T))^2 \quad (3)$$

When $V_{gs} > V_{th}$, $V_{ds} < V_{ds\_sat}$, SiC MOSET is working in the linear region, the temperature-dependent output characteristic curves of SiC MOSFET are shown in Fig. 4(b) (Only the images when $V_{gs} = 8\,V$, $10\,V$, $12\,V$, $15\,V$ at three separate temperature points of 25 °C, 215 °C and 375 °C are listed). Since the transconductance g_m has negative temperature characteristics [4], the control effect of V_{gs} on I_d will become smaller as the temperature rises. Therefore, at high temperatures, the output characteristic curves under different V_{gs} almost overlap. The channel current in linear region at different temperatures can be formulated by (4).

III. TEMPERATURE-DEPENDENT STATIC CHARACTERIZATION OF POWER DIODES

The body diode of SiC MOSFET may experience bipolar degradation after long-term forward voltage bias. This not only destroys the thermal design, but may also cause the degradation of the SiC MOSFET's blocking voltage, which will seriously affect the reliability of the system. Therefore, it is necessary to equip SiC Schottky barrier diode as a freewheeling diode in applications with high-reliability requirements.

A. Temperature-dependent forward characteristics of SiC MBD and SiC SBD

The parasitic body diode of SiC MOSFET is a typical kind of p-i-n junction diode, while SiC Schottky barrier diode is a metal-semiconductor hetero-junction diode formed by rectifying contacts.

For SiC MBD, the difference in the concentration of free electrons between N region and P region decreases with the increasing temperature, resulting in a decrease in built-in voltage. On the other hand, the equivalent forward conduction resistance R_D of SiC MBD hardly changes with temperature. So the forward conduction voltage V_{F_on} required to handle the same current will decrease with temperature. The temperature-dependent forward characteristics of SiC MBD are demonstrated in Fig. 5(a).

Taking 1.2 kV CPW5-1200-Z050B as an example to study the temperature-dependent static characteristics of SiC SBD. The trend of the forward characteristics of SiC SBD with increasing temperature is shown in Fig. 5(b). The main current transport mechanism of SiC SBD is the current formed by thermionic emission process. The forward conduction voltage drop of SiC SBD can be expressed as

$$V_F = V_{FS} + V_{R_D} = \frac{kT}{q}\ln\frac{I_F}{I_S} + R_D I_F \quad (5)$$

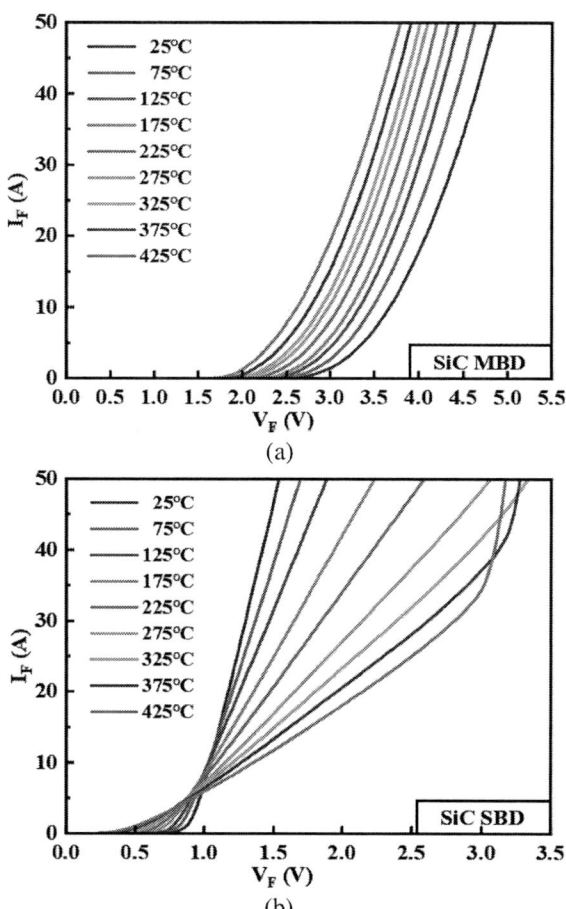

Fig. 5. The temperature-dependent forward characteristics of (a) 1.2 kV SiC MOSFET's body diode (b) 1.2 kV SiC Schottky barrier diode.

As temperature rises, the Schottky contact forward voltage of SiC SBD decreases, while the equivalent series resistance R_D of SiC SBD has positive temperature coefficient. Therefore, in small current conduction mode ($< 5\,A$), the forward conduction voltage of SiC SBD decreases. Conversely, the V_{F_on} increases with temperature in high current conduction mode, since the voltage across the R_D become higher.

I_F of SiC MBD and SiC SBD can be expressed by

$$I_{F\_M}(T, V_F) = I_S e^{\frac{qV_F^\lambda}{kT}} = A_1 T^{\gamma_1} e^{-\frac{b_1}{T}} e^{\frac{k_1 V_F^{\lambda_1}}{T}} \quad (6)$$

$$I_{F\_S}(T, V_F) = I_S e^{\frac{qV_F^\lambda}{kT}} = A_2 T^2 e^{-\frac{b_2}{T}} e^{\frac{k_2 V_F^{\lambda_2}}{T}} \quad (7)$$

The equivalent forward conduction resistance R_D and the forward conduction voltage V_{F_on} of diodes under different temperatures are shown in Fig. 6(a) and (b). Because the

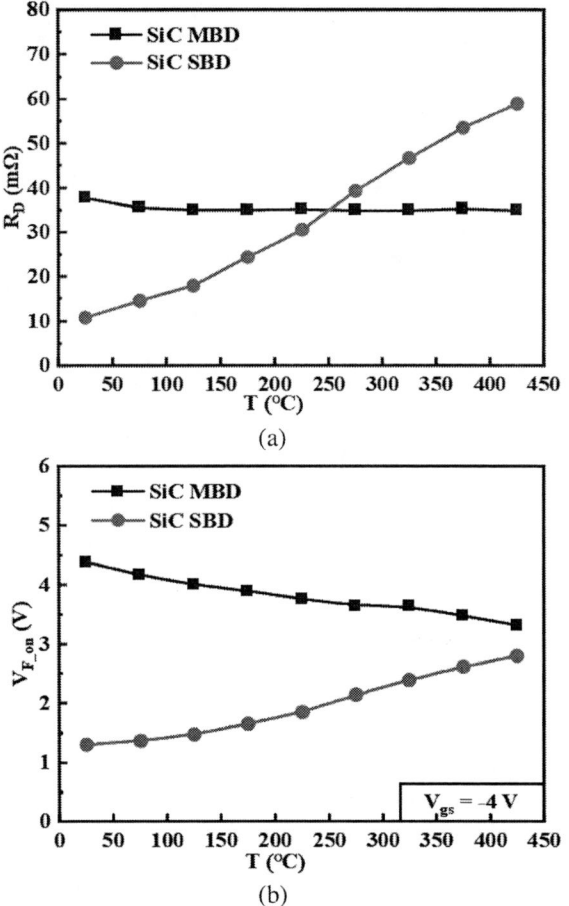

Fig. 6. The temperature-dependent (a) equivalent forward conduction resistance (b) forward conduction voltage of 1.2 kV SiC MOSFET's body diode and 1.2 kV SiC Schottky barrier diode.

barrier height of SiC SBD is lower than that of SiC MBD, the turn-on threshold voltage and forward voltage drop of SBD at room temperature are lower than that of SiC MBD. Therefore, equipped with SiC SBD can also reduce the conduction loss of the diode.

B. Junction capacitors of SiC MBD and SiC SBD

The junction capacitor of SiC device is an important parameter related to switching characteristics. Under high voltage conditions, the temperature-dependent built-in voltage is negligible, the junction capacitor of SiC power diodes can be expressed as

$$C = \frac{C_0}{\sqrt{1 + \frac{v^x}{V_{B1}}}} + C_H \qquad (8)$$

IV. CONCLUSION

SiC devices can work at high temperatures. This paper selects 1.2 kV QPM3-1200-0013D and 1.2 kV CPW5-1200-Z050B as examples, analyze the static characteristics of SiC MOSFET in a wide temperature range of 25 °C ∼ 425 °C, as

well as the temperature-dependent forward characteristics of SiC MOSFET's body diode and SiC Schottky barrier diode. The experiment results provides guidance for the pre-design and application of SiC MOSFETs at high temperatures.

REFERENCES

[1] M. Chen, H. Wang, D. Pan, X. Wang, and F. Blaabjerg, "Thermal Characterization of Silicon Carbide MOSFET Module Suitable for High-Temperature Computationally-Efficient Thermal-Profile Prediction," IEEE Journal of Emerging and Selected Topics in Power Electronics, pp. 1-1, 2020.

[2] T. Funaki et al., "Power Conversion With SiC Devices at Extremely High Ambient Temperatures," IEEE Transactions on Power Electronics, vol. 22, no. 4, pp. 1321-1329, 2007.

[3] A. Tsibizov, I. Kovacevic-Badstubner, B. Kakarla, and U. Grossner, "Accurate Temperature Estimation of SiC Power mosfets Under Extreme Operating Conditions," IEEE Transactions on Power Electronics, vol. 35, no. 2, pp. 1855-1865, 2020.

[4] S. Ji, S. Zheng, F. Wang, and L. M. Tolbert, "Temperature-Dependent Characterization, Modeling, and Switching Speed-Limitation Analysis of Third-Generation 10-kV SiC MOSFET," IEEE Transactions on Power Electronics, vol. 33, no. 5, pp. 4317-4327, 2018.

[5] D. J. Spry, P. G. Neudeck, L. Chen, D. Lukco, C. W. Chang, and G. M. Beheim, "Prolonged 500 °C Demonstration of 4H-SiC JFET ICs With Two-Level Interconnect," IEEE Electron Device Letters, vol. 37, no. 5, pp. 625-628, 2016.

[6] R. Kraus and A. Castellazzi, "A Physics-Based Compact Model of SiC Power MOSFETs," IEEE Transactions on Power Electronics, vol. 31, no. 8, pp. 5863-5870, 2016.

[7] H. Li, X. Zhao, K. Sun, Z. Zhao, G. Cao, and T. Q. Zheng, "A Non-Segmented PSpice Model of SiC mosfet With Temperature-Dependent Parameters," IEEE Transactions on Power Electronics, vol. 34, no. 5, pp. 4603-4612, 2019.

[8] J. Sun, L. Yuan, R. Duan, Z. Lu, and Z. Zhao, "A semi-physical semi-behavioral analytical model for switching transient process of SiC MOSFET module," IEEE Journal of Emerging and Selected Topics in Power Electronics, pp. 1-1, 2020.

[9] T. Ishigaki, S. Hayakawa, T. Murata, T. Masuda, T. Oda, and Y. Takayanagi, "Diode-less SiC power module with Countermeasures sgainst bipolar degradation to achieve ultrahigh power density,"IEEE Transactions on Electron Devices, vol. 67, no. 5, pp. 2035–2043,May. 2020.

[10] R. Fujita, K. Tani, K. Konishi, and A. Shima, "Failure of switchingoperation of SiC-MOSFETs and effects of stacking faults on safeoperation area,"IEEE Transactions on Electron Devices, vol. 65, no. 10,pp. 4448–4454, Dec. 2018

Fractional-Order Model Predictive Control of SiC PFC Converter

Qihui Fu
College of Electrical and Information Engineering
Hunan University
Changsha, China
fuqihui@hnu.edu.cn

Zhenxing Zhao, Zeng Liu
College of Electrical and Information Engineering
Hunan Institute of Engineering
Xiangtan, China
22006@hnie.edu.cn

Zishun Peng, Zipeng Ke
National-Local Joint Engineering Laboratory for Digitalize Electrical Design Technology
Wenzhou University
Wenzhou, China
pzshnu@hnu.edu.cn

Yuxing Dai
National-Local Joint Engineering Laboratory for Digitalize Electrical Design Technology
Wenzhou University
Wenzhou, China
daiyx@hnu.edu.cn

Huimin Quan
College of Electrical and Information Engineering
Hunan University
Changsha, China
hmquan@hnu.edu.cn

Jun Wang
College of Electrical and Information Engineering
Hunan University
Changsha, China
junwang@hnu.edu.cn

Abstract—Due to the fractional-order characteristic of the PFC circuit, the traditional integer-order equation cannot describe the circuit model accurately. To increase the model description accuracy, the fractional-order theory is employed to describe the SiC PFC circuit model in this research, and then the model predictive control approach is applied to improve the PFC circuit's performance. The experimental results verify the feasibility of the proposed method.

Keywords—fractional-order, model predictive control, power factor correction, SiC MOSFET

I. INTRODUCTION

The creation of new energy is improving as the global energy shift becomes more green and low-carbon. Due to the enormous number of power electronic devices linked to the power grid, major power quality issues such as harmonics have arisen. One of the most effective ways to fix the problem of power quality is to use Power Factor Correction (PFC) technology. As a result, the study of high-precision PFC converter control technology plays a vital role in promoting global energy transformation.

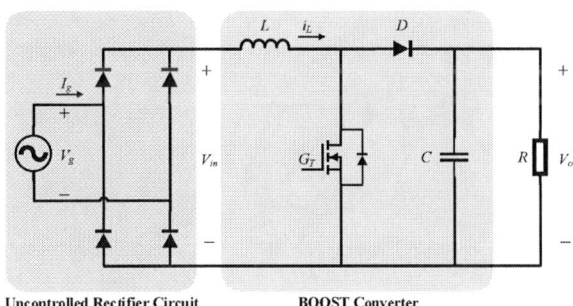

Fig. 1. Topology of power factor correction converter.

Model predictive control (MPC) is widely used in power electronics because of its good dynamic performance and strong robustness. MPC predicts and optimizes the future behavior of the system through the model describing the dynamic behavior of the system, and achieves a good control effect combined with error feedback correction. At present, the predictive model is still based on the traditional integer-

order differential equation modeling, which cannot accurately describe the time-varying nonlinear control object of the PFC converter in the actual physical system. In fact, the inductance and capacitance of the PFC converter show fractional-order characteristics [1]. As a result, integrating fractional order theory with PFC circuit modeling can increase the description accuracy of the predictive model, and hence the performance of the model predictive control algorithm.

In recent decades, fractional calculus theory is more and more widely used in industry. Researchers in the field of power electronics have gradually introduced fractional-order into circuit topology modeling. Several researchers have explored the fractional-order modeling and analytic approaches for boost circuits in continuous conduction mode (CCM)[2] and discontinuous conduction mode (DCM)[3]. Some researchers have proposed predictive control based on a process model for PFC circuits to improve power factor and stability[4]. The fractional-order model may more precisely and flexibly suit the true features of the control object, improving the control accuracy and effect of MPC that is dependent on the process model. At present, some researchers have combined fractional order with the MPC algorithm and applied it to a 2-D organization crane system[5], isolated microgrid[6], and so on.

The fractional calculus equation model can more correctly and flexibly reflect the properties of the actual system than the classic integer-order approach. However, there is no common definition of fractional order, and there is no clear analytical solution, which prevents fractional-order from being widely used in engineering. The idea of fractional calculus almost appears at the same time as integer calculus. Due to the lack of specific application background, the research progress of fractional calculus is slow. At present, the research of fractional calculus theory is not mature, there are still some theoretical problems to be solved, and there is no unified definition form of fractional calculus. So far, the three most widely used definitions in the research of fractional calculus are as follows: Grünwald-Letnikov(GL) definition, Riemann-Liouville(RL) definition, and Caputo definition[7].

The fractional calculus equation has a long storage characteristic, so it has no simple analytical solution and is difficult to discretize. As a result, numerical techniques are commonly utilized in fractional-order system analysis,

978-1-6654-4817-8/21 $31.00 © 2021 IEEE

potentially resulting in a significant increase in computing effort. The three different definitions of fractional order also lead to different numerical calculation methods, which are suitable for different application scenarios, such as the scheme of discretization of fractional calculus operator based on second-order infinite impulse response (IIR) filter[8], or the steady-state periodic solutions and transient analytical solutions of the state variables of the fractional-order buck converter are obtained by using the equivalent small parameters[9].

In this paper, a novel fractional-order MPC for SiC PFC converter is proposed based on the fractional equivalent circuit model of different switching states and the Oustaloup filtering algorithm [10]. The control strategy can improve the power quality and power utilization of the PFC converter.

II. FRACTIONAL-ORDER MODEL PREDICTIVE CONTROL

MPC establishes a system predictive model to forecast the future changes of control variables in the system under each switch state and determines the optimal switch state of the system according to the defined optimization criteria.

A. Establishment of Predictive Model

In the paper, as shown in Fig.1, the PFC converter includes an uncontrolled rectifier circuit and a BOOST converter. It works in CCM. The equivalent circuit diagram of two switch states of the BOOST circuit is shown in Fig.2.

Since the actual capacitance and inductance have fractional characteristics, the relationship between current and voltage at both ends of the inductor is shown below:

$$V = L\frac{d^\alpha i}{dt^\alpha} \tag{1}$$

where α is fractional order and $0 < \alpha \le 1$ (when $\alpha = 1$, it is integer order).

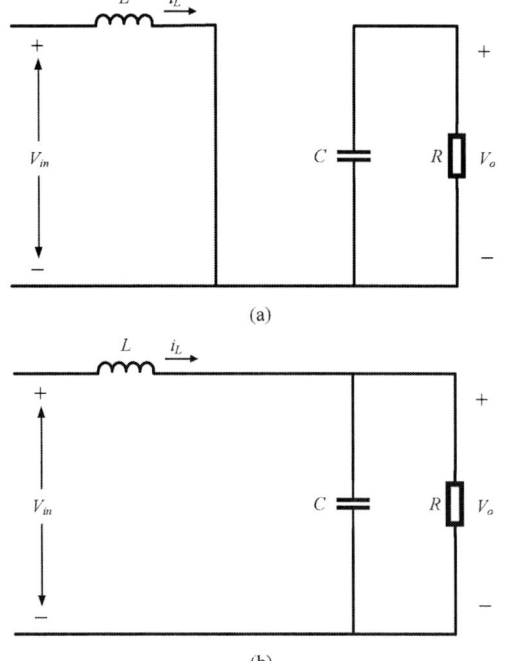

Fig. 2. The inductive current in the equivalent circuit of the PFC converter. (a) When the switch is on. (b) When the switch is off.

According to Fig.2(a), when the switch is on, the fractional differential equation of inductance current is as follows:

$$\frac{d^\alpha i_L}{dt^\alpha} = \frac{V_{in}}{L} \tag{2}$$

According to Fig.2(b), when the switch is off, the fractional differential equation of inductance current is as follows:

$$\frac{d^\alpha i_L}{dt^\alpha} = \frac{V_{in} - V_o}{L} \tag{3}$$

where V_{in} is the input voltage, V_o is the output voltage, L is the inductance value, i_L is the inductance current.

From the commutative law of fractional calculus operator:

$$\frac{d^\alpha i_L}{dt^\alpha} = \frac{d^{\alpha-1}(\frac{di_L}{dt})}{dt^{\alpha-1}} \tag{4}$$

According to Euler's algorithm, at the $(n+1)$-th cycle:

$$\frac{di_L}{dt} = \frac{\Delta i_L}{\Delta t} = \frac{i_L(n+1) - i_L(n)}{T_S} \tag{5}$$

where T_s is the switching period.

By introducing equation 4 and equation 5 into equation 2, the following results can be obtained:

$$i_{L1}(n+1) = \frac{d^{\alpha-1}(\frac{V_{in}(n)}{L})}{dt^{\alpha-1}} \cdot T_S + i_L(n) \tag{6}$$

1) Oustaloup filtering algorithm

Oustaloup filtering algorithm is a kind of numerical differential processing algorithm for unknown signals by constructing continuous filters. The transfer function of the continuous filter is shown below::

$$G_f(s) = K\prod_{k=-N}^{N}\frac{s+\omega_k'}{s+\omega_k} \tag{7}$$

According to the selected fitting frequency band (w_b,w_h), the zeros, poles, and gain of the filter can be obtained:

$$\begin{cases} K = \omega_h^\alpha \\ \omega_k' = \omega_b(\frac{\omega_h}{\omega_b})^{\frac{k+N+\frac{1}{2}(1-\alpha)}{2N+1}} \\ \omega_k = \omega_b(\frac{\omega_h}{\omega_b})^{\frac{k+N+\frac{1}{2}(1+\alpha)}{2N+1}} \end{cases} \tag{8}$$

The transfer function model of the continuous filter can be determined by formula 7 and formula 8. The signal $y(t)$ is filtered by the Oustaloup filter, and the output signal can be considered as the approximation of $\dfrac{d^{\alpha-1}y}{dt^{\alpha-1}}$.

2) Prediction model processed by Oustaloup filtering algorithm

p^{α} is used to represent the α-order differential operator d^{u}/dt^{α} after being processed by the Oustaloup filtering algorithm.

Therefore, when the switch is on:

$$i_{L1}(n+1) = p^{1-\alpha}\left(\frac{V_{in}(n)}{L}\right)\cdot T_S + i_L(n) \tag{9}$$

In the same way, when the switch is off:

$$i_{L2}(n+1) = p^{1-\alpha}\left(\frac{V_{in}(n)-V_0(n)}{L}\right)\cdot T_S + i_L(n) \tag{10}$$

where T_s is the reciprocal of MOSFET operating frequency.

B. Design of Cost Function

Based on the above two operation modes, the cost function of fractional order predictive control is as follows:

$$C_1 = \left| i_{L1}(n+1) - i_{ref}(n+1) \right| \tag{11}$$

$$C_2 = \left| i_{L2}(n+1) - i_{ref}(n+1) \right| \tag{12}$$

where $I_{ref}(n+1)$ represents the reference current signal.

According to the principle of current equivalent, the switch turns on when $C_1 < C_2$, and turns off when $C_1 > C_2$.

When the PFC converter is in a steady state, the actual inductor current can accurately track the given reference current signal after the nth switching cycle ends. Because the sampling frequency of the system is much greater than the frequency of the input terminal change, it is assumed that this reference current will not change much within a sampling period, that is $i_{ref}(n+1) = i_{ref}(n)$. The expression of reference current $i_{ref}(n+1)$ is as follows:

$$i_{ref}(n+1) = K_{PI}\left|\sin\left[\omega_{line}(n+1)\right]\right| \tag{13}$$

where K_{PI} is the output of the PI regulator, which is determined by the voltage outer loop. $\left|\sin\left[\omega_{line}(n+1)\right]\right|$ is the unidirectional pulsating DC voltage after rectification.

C. Overall control design of PFC converter

The overall control strategy includes two parts: the outer voltage loop with PI control and the inner current loop with fractional-order MPC, as shown in Fig.2. The signal is discretized by using the Oustaloup filtering algorithm to construct a continuous filter. Then the signal is output to the drive port of the switch through the cost function.

First, the voltage outer loop needs to adjust the DC output voltage of the PFC through PI regulation, and the current inner loop uses fractional model predictive control. Establish a fractional-order prediction model based on the Oustaloup filtering algorithm to predict the current value of the next sampling period, set the cost function of the predictive control as the absolute value of the difference between the reference current and the actual inductor current, and select the switch with the optimal output. The state controls the turn-on or turn-off of the switch tube of the PFC converter and realizes the rolling optimization of predictive control through continuous sampling and calculation.

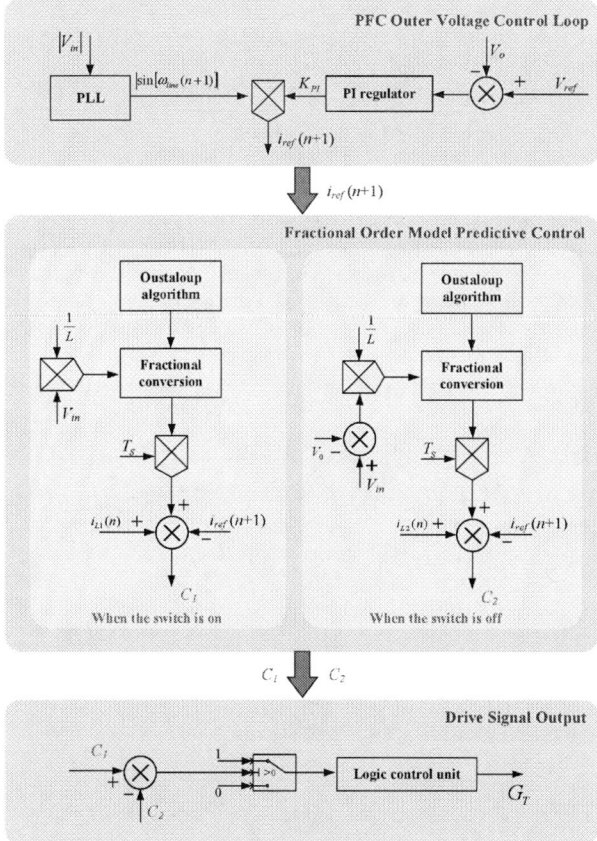

Fig. 3. Control block diagram of fractional-order MPC

III. SIMULATION VERIFICATION OF FRACTIONAL-ORDER MPC

This section uses a visual simulation tool called Simulink in MATLAB software to verify the efficiency of the fractional-order model predictive control algorithm developed in this work. The DC load power is 1kW, the input AC voltage amplitude is 220V and the frequency is 50Hz, the output DC voltage is 400V, and the output load is 160Ω. Table 1 lists the other circuit parameters.

For the fractional-order predictive control, when the switching frequency of the system is low, the predictive control will have a large error. Therefore, when using predictive control, it is necessary to choose a higher switching frequency. In order to reduce the switching loss of the switch, SiC MOSFET is selected in this paper. When the switching frequency increases, the system will produce large electromagnetic interference due to the influence of parasitic capacitance and parasitic inductance. In this paper, the switching frequency of 100kHz is taken as an example to study the proposed control algorithm.

TABLE I.   PARAMETERS OF THE EXPERIMENTAL PLATFORM

Parameter	Value
Rated power	1kW
Input ac voltage	220 V
Output dc voltage	400V
Output inductance	1000μH
Output capacitance	1000μF
Output load	160Ω
Input voltage frequency	50Hz
Switching frequency	100kHz

When $\alpha=0.7$, the simulation results of the input current, input voltage, and output voltage of the system are shown in Fig.4. The peak value of input voltage is 311V, the peak value of input current is 9A, the overshoot of output voltage is 11.8%, the ripple of the output voltage is 1.1%, and the regulation time is 0.5s. The output voltage can quickly track the reference voltage, and the input current THD is 5.82%. The input active power is 1081W, the input reactive power is 10.18W, that is, the power factor is 0.999556.

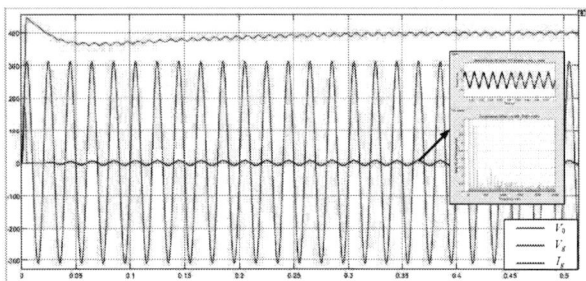

Fig. 4.   Block diagram of fractional-order MPC

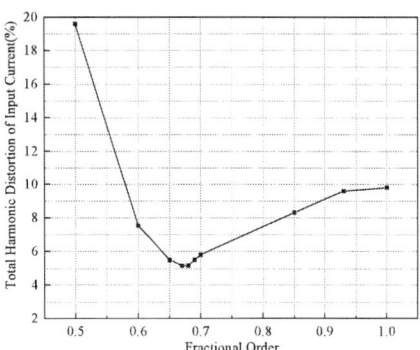

Fig. 5.   Block diagram of fractional-order MPC

When the fractional order is changed, the predictive model characteristics of the fractional-order circuit will also change. The closer the model characteristics are to the actual circuit characteristics, the better the control effect of fractional order MPC will be. Therefore, different fractional orders can be adjusted in simulation, and the experimental results under different fractional orders can be obtained. In order to get the best value range of fractional order, this part selects 10 data between the step distance (0.5,1) for simulation, and according to these experimental data, draws the relationship curve between fractional-order and input current THD, as shown in Fig.5.

According to Fig.5, when the fractional order is between 0.65 and 0.7, the control effect is the best.

IV. EXPERIMENTAL VERIFICATION

Comparative experiments on a power factor corrector converter experimental platform based on SiC MOSFETs, as shown in Fig. 6, validate the fractional-order model predictive control strategy. The load power is 1kW, the input AC voltage amplitude is 220V and the frequency is 50Hz, the output DC voltage is 400V, and the switching frequency is 100kHz. Table 1 lists all of the other key parameters in the experiment.

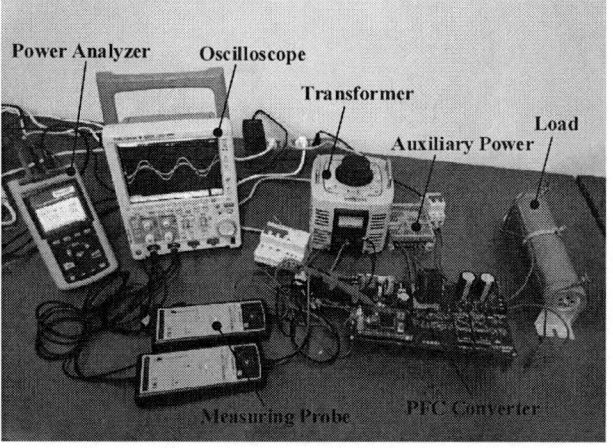

Fig. 6.   PFC experiment platform.

In this section of the system, six fractional orders are selected according to the way of an arithmetic sequence to describe the changing trend of the system when the fractional order changes. Under different fractional orders α, voltage and current waveforms and their corresponding harmonic content and power factor are shown in Fig.7.

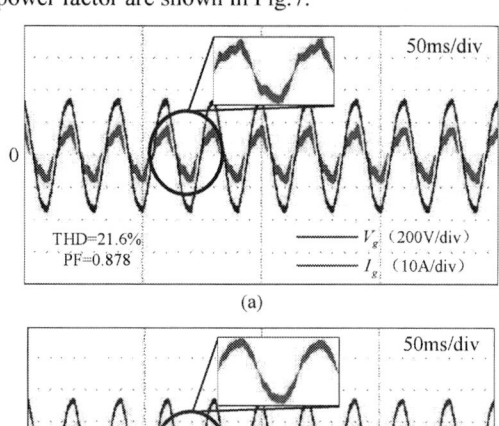

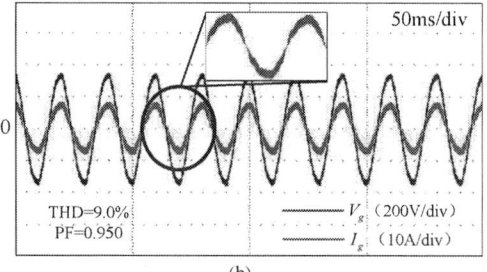

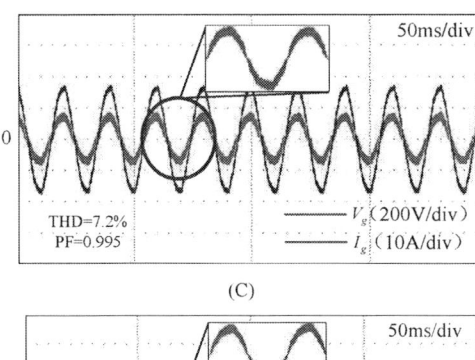

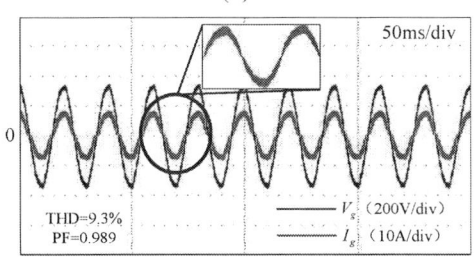

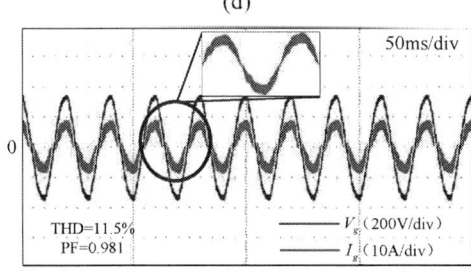

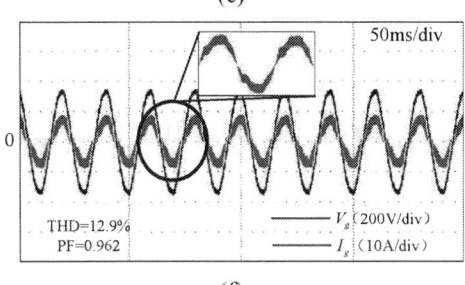

Fig. 7. Experimental results under diffirent Fractional order α. (a) α= 0.5.(b) α= 0.6. (c) α= 0.7. (d) α= 0.8. (e) α= 0.9. (f) Integer order α=1.

In order to make the experimental results more intuitive, the data in Fig.7 is sorted out to get Table 2. The relationship between power factor and total harmonic distortion of PFC converter and fractional order is shown in Fig.8.

TABLE II. EXPERIMENTAL RESULTS

Order	THD	Power factor
Fractional order (α=0.5)	21.6%	0.878
Fractional order (α=0.6)	9.0%	0.950
Fractional order (α=0.7)	7.2%	0.995
Fractional order (α=0.8)	9.3%	0.989
Fractional order (α=0.9)	11.5%	0.981
Integer order (α=1)	12.9%	0.962

According to the chart analysis of the experimental results, we can see that:

- The simulation results are verified by experiments, and the general trend of Fig.8 is consistent with Fig.5.

- The analysis shows that when the order of fractional order MPC is in the range of (0.65,1), the control effect is better than that of integer-order MPC.

- when the order is about 0.7, the THD of the input current harmonic is the lowest and the power factor is the highest.

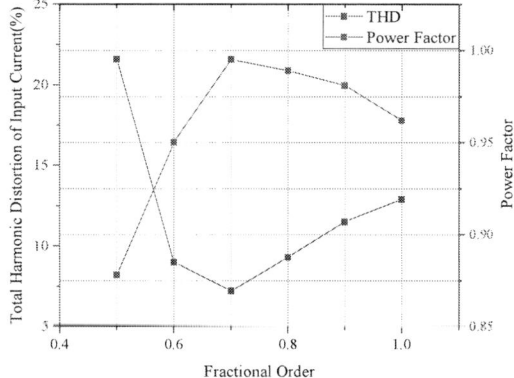

Fig. 8. THD of input current and power factor under different orders.

V. CONCLUSION

In this paper, a novel fractional-order predictive control for SiC PFC converter is proposed, which combines fractional-order theory, Oustaloup filtering algorithm, and integer-order MPC. The experimental results show that the SiC PFC converter with fractional-order MPC achieves higher power quality and power utilization. The proposed fractional-order MPC can be extended to other power electronic topologies.

ACKNOWLEDGMENT

The authors would like to gratefully acknowledge the Project (SKLD21KM05) supported by the State Key Laboratory of Power System and Generation Equipment.

REFERENCES

[1] M. C. Tripathy, D. Mondal, K. Biswas, and S. Sen, "Experimental studies on realization of fractional inductors and fractional-order bandpass filters,"Int. J. Circuit Theory Appl., vol. 43, no. 9, pp. 1183–1196, Sep. 2015.

[2] X. Chen, Y. Chen, B. Zhang and D. Qiu, "A Modeling and Analysis Method for Fractional-Order DC–DC Converters," in IEEE Transactions on Power Electronics, vol. 32, no. 9, pp. 7034-7044, Sept. 2017, doi: 10.1109/TPEL.2016.2628783.

[3] Y. Chen, X. Chen, J. Hu, B. Zhang, and D. Qiu, "A symbolic analysis method for fractional-order boost converter in discontinuous conduction mode," IECON 2017 - 43rd Annual Conference of the IEEE Industrial Electronics Society, Beijing, 2017, pp. 8738-8743, doi: 10.1109/IECON.2017.8217536.

[4] Zou Q, Jin Q, Zhang R. Design of fractional order predictive functional control for fractional industrial processes[J]. Chemometrics and Intelligent Laboratory Systems, 2016, 152: 34-41.

[5] Singh A, Agrawal H. A fractional model predictive control design for 2-d gantry crane system[J]. Journal of Engineering Science and Technology, 2018, 13(7): 2224-2235.

[6] Chen M R, Zeng G Q, Dai Y X, et al. Fractional-Order model predictive frequency control of an islanded microgrid[J]. Energies, 2019, 12(1): 84.

[7] Wan Y. Design and optimzation of fractional controller in manufacturing system with application of multilink robot and time delay chatter [J]. Dissertations &Theses-Gradworks, 2011.

978-1-6654-4817-8/21 $31.00 © 2021 IEEE

[8] Yangquan Chen and Blas M. Vinagre. A new IIR-type digital fractional order differentiator[J]. Signal Processing, 2003, 83, pp. 2359-2365.

[9] X. Li, Y. Chen, X. Chen, B. Zhang, and D. Qiu, "An analytical approach for obtaining the transient solution of the fractional-order buck converter in CCM," IECON 2017 - 43rd Annual Conference of the IEEE Industrial Electronics Society, 2017, pp. 6859-6863, doi: 10.1109/IECON.2017.8217199.

[10] Oustaloup A, Levron F, Mathieu B, et al. Frequency-band complex noninteger differentiator: characterization and synthesis[J]. IEEE Transactions on Circuits and Systems I: Fundamental Theory and Applications, 2000, 47(1): 25-39.

GaN HEMT with Current-Driven Gate and its Driving Circuit Design

Owen Song
Power & Sensor Systems
Infineon Semiconductors (Shenzhen) Company Limited
Shenzhen, China
Owen.Song@infineon.com

Rafael Garcia
Power & Sensor Systems
Infineon Semiconductors Austria AG
Villach, Austria
Rafael. Garcia@infineon.com

Abstract—there are two gate configurations, i.e. voltage-driven and current-driven, of GaN HEMT in power electronics industry. The characteristics of the two configurations are discussed in the paper, current-driven configuration is more robust and reliable compared to its counterpart based on the analysis. The gate current requirements of the current-driven GaN transistor are depicted and simple but reliable gate driving circuits are recommended in the paper. Since the fast turn on/off speed of GaN transistors and unsymmetrical PCB layouts the oscillations between GaN transistors in parallel are unavoidable, the design of gate driving circuit when GaN transistors working in parallel is critical for the reliability of the whole system, the paper provides the guidelines and solutions for gate driving circuits of GaN transistors working in parallel.

Keywords—GaN HEMT, power electronics, driving circuit, parallel

I. INTRODUCTION

With the advancement of power electronics the efficiency and power density of SMPS (Switched Mode Power Supply) are getting higher and higher, however two factors hinder its further improvements: the power loss of magnetic elements and power loss of power semiconductors. Regarding to the former many core materials, structures of magnetic core and winding approaches were researched [1]-[6] to reduce magnetic and copper loss, respectively. For the latter on the one hand more novel topologies, e.g. LLC, totem-pole PFC are researched [7]-[10] to decrease the loss of power switches, on the other hand state-of-the-art silicon power MOSFET, e.g. CoolMOS™ with super junction technology developed by Infineon Technology , nearly touches the theoretical limitation of A*Ron of Silicon provided in[11] and [12], which means it's more and more difficult to reduce the loss of silicon power switch further, therefore huge attentions have been paid to wide bandgap devices including GaN (Gallium Nitride) HEMT (high electron mobility transistor) and SiC (Silicon Carbide) MOSFETs.

GaN is a very promising material for power semiconductors because of its higher bandgap and electron mobility. Efforts worldwide have led to significant progress in the realization of reliable GaN transistors on cheap silicon substrates and this has been key to their economic success. Technically, the wide bandgap of GaN allows higher electric field strengths and thus results in more compact high voltage switches with faster transition speed, lower $R_{DS(on)}$ and smaller parasitic capacitors compared to silicon switches, thus far surpassing state-of-the-art silicon transistors in terms of all relevant Figures of Merit (FOM).

There are two types of gate configuration of GaN HEMT in the market, i.e. voltage-driven and current-driven. And it was well know that it should be more careful to drive a GaN HEMT compared to Silicon MOSFET. The paper will address these topics thus the rest of the paper is organized as follows: section II presents the two gate configurations of GaN

transistor and its comparison, section III shows the gate driving requirements and recommended gate driving circuits of current-driven GaN transistor, the design guidelines of gate driving circuit of GaN transistors working in parallel are provided in section IV and the paper is summarized in section V.

II. TWO GATE CONFIGURATIONS OF GaN TRANSISTORS

A. Voltage-driven and current-driven

Because of two kinds of mental on the gate of high voltage enhancement-mode GaN used in power electronics, two gate configurations are formed, as shown in figure 1.

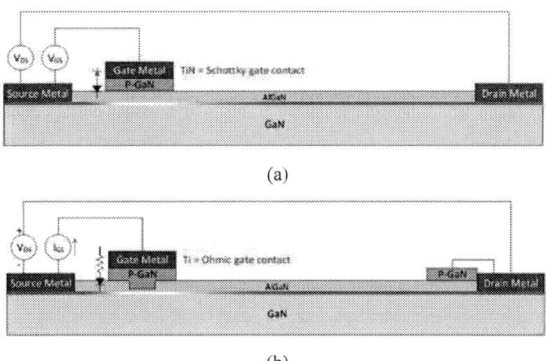

(a)

(b)

Fig. 1. Two gate configurations including (a) voltage-driven and (b) current driven

In figure 1 (a) TiN (Titanium Nitride) is used as gate metal as a result a schottkey gate (a Zener diode in reverse series with a diode) is formed, which results in a voltage-driven configuration similar to silicon MOSFET: the transistor is turned on or off only by gate voltage not considering charge and discharge of parasitic capacitors. On the contrary, Ti is employed as gate mental in figure 1 (b), thus a ohmic gate (a resistor in series with a diode) is shaped, which results in current-driven gate structure: the transistor is turned on or off by both gate current and voltage.

In order to understand the differences of the two gate configurations, the equivalent electrical circuits of them are illustrated in figure 2.

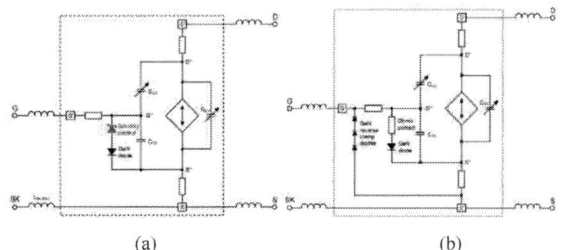

(a)   (b)

978-1-6654-4817-8/21 $31.00 © 2021 IEEE

Fig. 2. Equivalent electrical cicuit of (a) voltage-driven and (b) current driven gate configuration

Based on the equivalent electrical circuits some comparisons will be made to show the pros and cons of the two gate configurations.

B. Comparisons of the two gate configurations

In terms of voltage-driven GaN HEMT shown in figure 2 (a), the GaN transistor is turned on or off through the high or low level of gate voltage, which is same to that of silicon MOSFET, and in theory gate current is not needed. However due to the existence of parasitic capacitors the gate current is needed to charge or discharge them at transition time. The gate voltage should be much higher than threshold voltage (generally 1.2V at room temperature) to achieve full enhancement, which is about 5~6V. However the breakdown voltage of the Zener diode between gate and source is around 3~6V, therefore the gate voltage should be less than 7~10V (the sum of breakdown voltage of Zener diode and forward voltage of GaN diode which is around 3.5V), or the gate will break down resulting in permanent failure of the GaN transistor. The gate voltage margin for users to ensuring GaN transistor full enhancement while keeping it within the range the gate performs reliably is quite narrow, as a result more efforts should be put into the gate driving circuit design and relevant PCB layout, even additional gate driving circuit used to clamp the max. gate voltage should be employed, which is not beneficial to cost effective and fast go-to-market strategy for products using GaN transistor.

On the contrary, current-driven GaN HEMT gate structure showed in figure 2 (b) is composed of a 3.5 V GaN diode and a resistor with a few ohms in series, thus the gate voltage is clamped to be around 3.5V when a high level gate voltage is exerted between gate and source, as a result the gate is self-protected provided the gate current is lower than the max. value listed in the datasheet which is around 2A. The gate should be driven by a current with average 3~10mA to keep "on" condition and peak 100~500mA with tens of nanosecond for fast turn on/off of the transistor, since gate voltage is clamped to be around 3.5V at room temperature, the power loss of gate driving is just tens of mW in average so its impact on power loss of GaN transistor is negligible. Based on figure 2 (b) additional benefit of this current-driven gate configuration is when the GaN transistor is turned off with a negative gate to source voltage due to the low gate threshold, the gate voltage will be clamped to be around -10.5V (three GaN didoes in series) to protect the gate from further low voltage, the reliability of the gate driving is further enhanced.

In a conclusion the current-driven GaN HEMT gate configuration is more reliable and robust compared to its counterpart, especially for those GaN transistors working at high dv/dt and/or di/dt conditions and with high voltage noise/spike at gate.

III. Design of Driving Circuit of Current-driven GaN HEMT

As mentioned previously a constant gate current is needed to keep the GaN transistor with current-driven gate configuration "on" , and transient pulse currents are required to turn the GaN transistor on and off quickly, as depicted in figure 3. Generally t1 in figure 3 is tens of nanosecond to make sure the parasitic capacitors of GaN transistor are fully

charged and Iss a few mA to ensure GaN transistor is in full enhancement.

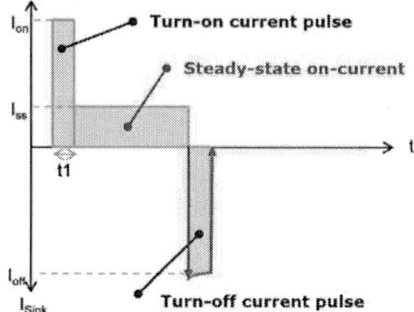

Fig. 3. Gate current to turn on and off the GaN transistor with current-driven gate configuration

An approach with low cost and less element quantity to achieve aforementioned gate current waveform is illustrated in figure 4.

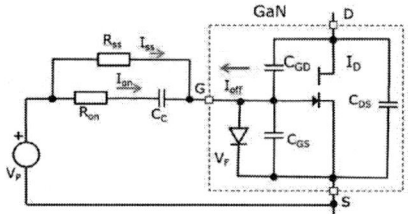

Fig. 4. Gate driving circuit recommended

In figure 4 Vp, the output voltage of a normal gate driver, is recommended to be from 5V to 12V, though it can be higher the driving loss will be higher accordingly. Then R_{on} and R_{ss} can be calculated approximately as:

$$R_{on} = \frac{V_p}{I_{on}} \qquad (1)$$

$$R_{ss} = \frac{V_p - V_F}{I_{ss}} \qquad (2)$$

In which V_F is around 3.5V at room temperature, I_{on} and I_{ss} can obtained in the datasheet of GaN transistor. Another benefit of this gate driving circuit is during turn off phase the gate voltage is clamped to be negative, as shown in figure 5. Since the threshold voltage of GaN transistor is only about 1.2V and will be lower at high junction temperature the negative gate voltage is essential for GaN transistor to keep 'off' steadily.

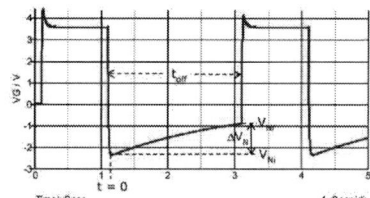

Fig. 5. Gate voltage waveform when Vp is 12V

The initial negative voltage V_{Ni} can be calculated:

$$V_G(0) = V_{Ni} = V_F - V_p * \frac{c_c}{c_g} \qquad (3)$$

In which Cg is the equivalent parasitic capacitance between gate and source.

978-1-6654-4817-8/21 $31.00 © 2021 IEEE

The initial negative voltage V_{Ni} will decrease with the time constant:

$$\tau = R_{ss} * (C_c + C_g) \qquad (4)$$

And at the end of the off-phase the final V_{Nf} is:

$$V_G(t_{off}) = V_{Nf} = V_{Ni} * e^{-t_{off}/\tau} \qquad (5)$$

Regarding to those GaN HEMTs employed in half-bridge topology, i.e. totem-pole bridgeless PFC or LLC, normally they works at high frequency with limited dead time between high and low side switch to achieve high power density at the same time high efficiency. The stability of bias voltage and tight channel to channel delay time of the gate driver ICs that driving GaN transistors are key factors for robust and reliable performance of GaN HEMTs. In figure 6 2PCS single channel gate driver (Driver1 & Driver 2) with tight channel delay time or 1 PCS double channel gate driver (Driver 3) with tight channel to channel delay time is recommended to drive GaN transistors in half-bridge topology.

An excellent bias voltages generation circuit is recommended as: the bias voltages of driver 1 and driver 2 or driver 3 are generated by an auxiliary power circuit whose input voltage coming from another gate driver, and the input of this gate drive can be from MCU or RC oscillation circuit, as shown in figure 7. The benefits of the aux. power include: the positive (Vcsx to GND_refx, x is 1 or 2) and negative (GNDx to GND_refx) bias voltage are generated simultaneously, and the positive and negative bias voltages can be adjusted smartly based on the output load of the system if employing MCU to generate driving signal: for example at light load the value of Vcs1 and Vcs2 will be reduced, as a result gate driving currents are reduced according to equation (1) and (2) to decrease the power loss caused by gate driving, the efficiency can be increased especially at high frequency and light load condition.

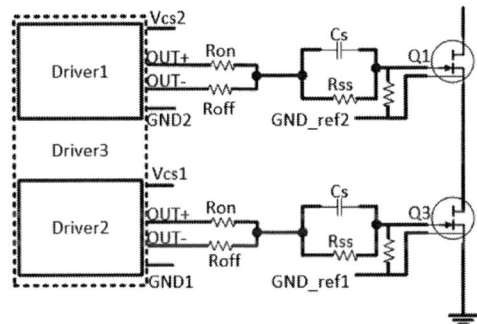

Fig. 6. GaN transistors in half-bridge configuration and its dirving circuits

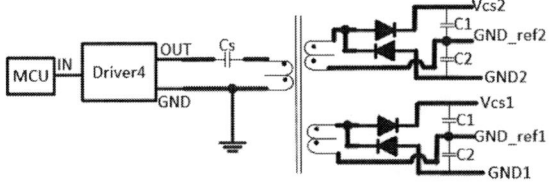

Fig. 7. Bias voltages generation circuit

IV. Gate Driving Circuit Design When GaN Transistors Working in Parallel

When two or more GaN HEMTs have to be working in parallel in half-bridge configuration in high power SMPS applications, designers should pay more attention to the gate driving circuit because, on the one hand high voltage spikes at the gate is easily generated by the high transition speed of GaN transistor through inductive and/or capacitive coupling while gate threshold voltage is quite low compared to silicon MOSFET, as a result the possibility of shoot through between high and low side GaN transistors is high. On the other hand the high frequency oscillations at gates due to even slight imbalance in gate threshold and small circuit parasitic parameters will reduce the reliability of GaN transistors or even damage the parts eventually.

In order to understand the gate oscillations thoroughly a half bridge configuration with two GaN transistors in parallel in high and low side, respectively, was set up, as shown in figure 8.

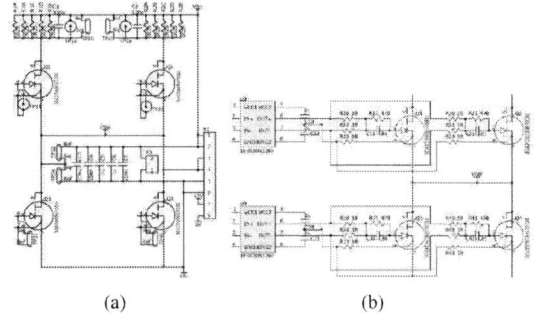

(a)   (b)

Fig. 8. Two GaN transistors in parallel in half bridge configuration (a) and driving circuits of GaN transistors (b)

Taking it as an example when half bridge works at hard switching mode. Figure 9 depicts the drain currents of high side GaN transistors (Ch1 & Ch2), switch node voltage referenced to 400V (Ch3, which is also the V_{SD} of high side GaN transistors) and total switch node current (Ch4), the scaling of each channel was explained in the figure.

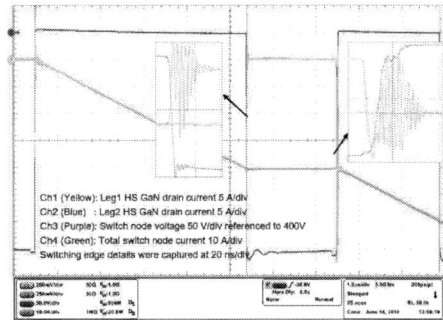

Fig. 9. Waveforms of GaN transistors in parallel in half bridge configuration working in hard swithing mode

The strong oscillations appear when GaN transistors are turned off and turned on at the condition switch mode current is 40A in figure 9, even though the current sharing between the two GaN transistors is perfect before they are turned off or turned on again. To confirm the oscillations occur between the two GaN transistors in parallel the waveforms around turning on of GaN transistors are zoomed in and depicted in figure 10.

In figure 10, the turning on edge at 40 A starts in phase, then goes to 180° phase shift and finishes in phase again. During 180° phase shift section of the oscillations, the currents through the devices reach very high values but the current across the switch node of the half bridge configuration (green

978-1-6654-4817-8/21 $31.00 © 2021 IEEE

curve in figure 10) doesn't show an apparent change, which means the oscillation only occurs between two GaN transistors in parallel.

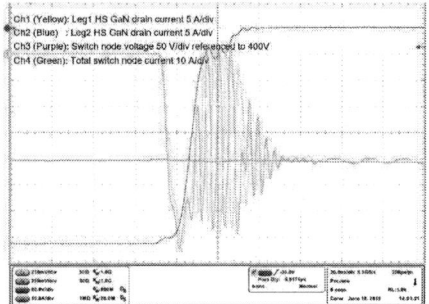

Fig. 10. Waveforms zoomed in when GaN transitors in parallel are turned on at total 40A

And during 180° phase shift section of the oscillations, the waveform of the drain voltage curve (purple curve in figure 10) shows that the devices have difficulty to fully turn on. As a result the oscillations will cause EMI issue, high power loss issue and even damage the GaN transistors if the amplitudes of the oscillations are large enough. Although the tests are implemented with current-driven gate configuration GaN transistors, the phenomena exist in voltage-driven gate configuration GaN transistors in parallel as well, the paper will not show the similar results.

To address the oscillations issue, small surface mount CM (common mode) inductors with high CM path impedance in the range of hundreds of ohm in MHz frequency range but very low DCR (direct current resistor) are suggested in the G/KS circuit of each device, as shown in figure 11.

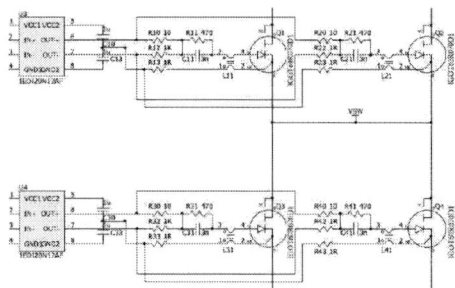

Fig. 11. Gate driving circuit with CM inductors

Experiments with Bourns SRF2012-361YA CM inductors showed positive results: no sustained oscillations were observed up to 80 A total switch node current, as shown in figure 12. The current sharing, even during transients, looks very good.

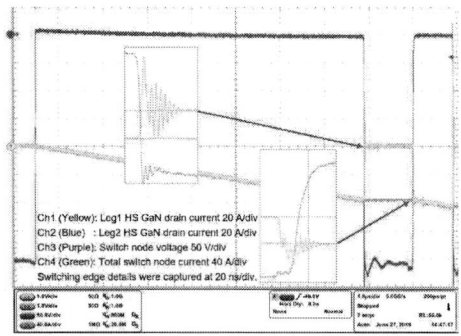

Fig. 12. Waveforms of GaN transistors in parallel with CM inductors at gate in half bridge configuration working in hard switching mode

Compared to those in fig. 9 and 10 the oscillations in figure 12 are in phase, amplitudes of which are smaller and rapidly mitigate in less than 30nS even in 80A total current condition. The soft switching operation from 500 kHz to 1 MHz is checked as well and the results are nearly same to those in hard switching operation, which verifies that the CM inductors solution is effective in both hard switching and soft switching applications.

Besides the CM inductors are recommended to be employed, other approaches to achieve robust and reliable performance of GaN transistors in parallel includes :

(1) perfectly symmetrical driving and power circuit layout;

(2) GaN transistors with matched gate thresholds;

(3) power loops and gate driving loops must be as small as possible especially for voltage-driven GaN transistors and

(4) optimized turn on/off speed of GaN transistors.

V. CONCLUSIONS

Two gate configurations, i.e. voltage-driven and current-driven, of GaN HEMT employed in power electronics are discussed, pros and cons of each gate configuration are analyzed, current-driven gate configuration is more robust and reliable compared to its counterpart. The design of driving circuit of the current-driven gate configuration GaN HEMT was elaborated especially for those working in topologies with half bridge configuration including totem-pole PFC, LLC, etc. a novel auxiliary power supply was introduced to provide stable bias voltages to gate drivers, at the same time it has the capability to optimize efficiency at light load condition.

GaN transistors have to work in parallel in high power applications, however, because of high turn on/off speed and unavoidable unsymmetrical PCB layouts, the oscillations appear at transition time with high current. The severe oscillations will reduce the reliability of GaN transistors and the whole power electronics system, or even damage the GaN transistors in a short time. CM inductors with hundreds of ohm at MHz frequency but a few ohm at working frequency are recommended to be used in driving circuit, the test results show that the oscillations can mitigate effectively, and other suggestions are provided as well to suppress the oscillations.

REFERENCES

[1] Teng Sun, Tadashi Takahashi, Hobie Yun, et al., "Magnetic materials and design trade-offs for high inductance density, high-Q and low-cost power and EMI filter inductors," 2016 IEEE 66th Electronic Components and Technology Conference.

[2] Mahbubur Rahman Kiran, Omar Farrok, Md. Rabiul Islam, Jianguo Zhu, Abbas Z. Kouzani and M. A. Parvez Mahmud, "Characterization of a high frequency transformer with HTS winding and amorphous magnetic core for power converter application," Proceedings of 2020 IEEE International Conference on ID20281 Applied Superconductivity and Electromagnetic Devices, Tianjin, China, 2020.

[3] Ismagilov F., Vavilov V., Gusakov D., "Hybrid magnetic core for perspective supply system of the aircraft," 2018 International Conference on Industrial Engineering, Applications and Manufacturing (ICIEAM).

[4] Dianbo Fu, Shuo Wang, "Novel concepts for high frequency high efficiency transformer design,", 2011 IEEE Energy Conversion Congress and Exposition, 2011.

[5] Matthias J. Kasper , Luca Peluso , Gerald Deboy , Gustavo Knabben , Thomas Guillod , and Johann W. Kolar , "Ultra-high power density server supplies employing gan power semiconductors and pcb-integrated magnetics,", CIPS 2020, 11th International Conference on Integrated Power Electronics Systems

[6] G. C. Knabben, J. Scḧafer, L. Peluso, J. W. Kolar, M. J. Kasper, G. and G. Deboy, "New PCB winding "snake-core"matrix transformer for ultra-compact wide dc input voltage range hybrid b+dcm resonant server power supply," 2018 IEEE International Power Electronics and Application Conference and Exposition (PEAC).

[7] Dongcheng Huang, Dianbo Fu and Frad. C. Lee, "High switching frequency, high efficiency CLL resonant converter with synchronous rectifier" 2009 IEEE Energy Conversion Congress and Exposition.

[8] Ke Zhu, Matt O'Grady, Jonathan Dodge, John Bendel and John Hostetler, "1.5 kW Single phase ccm totem-pole pfc using 650v SiC cascodes", 2016 IEEE 4th Workshop on Wide Bandgap Power Devices and Applications (WiPDA).

[9] Qingyun Huang, Qingxuan ma, Ruiyng Yu,Tianxiang Chen, Alex Q. Huang and Zhuoran Liu, "Improved analysis, design and control for interleaved dual-phase ZVS GaN-based totem-pole PFC rectifier with coupled inductor", 2018 IEEE Applied Power Electronics Conference and Exposition (APEC).

[10] Qingyun Huang, Qingxuan Ma, Pengkun Liu, lex Q. Huang and Michael de Rooij, "99% Efficient 2.5 kW four-level flying capacitor multilevel gan totem-pole PFC", IEEE journal of emerging and selected topics in power electronics, 12 Jan. 2021.

[11] Wataru Saito, "Theoretical limits of superjunction considering with charge imbalance margin", Proceedings of the 27th International Symposium on Power Semiconductor Devices & IC's, May 10-14, 2015, Kowloon Shangri-La, Hong Kong.

[12] Don Disney and Gary Dolny, "JFET depletion in superjunction devices", 20th International Symposium on Power Semiconductor Devices and IC's, 2008.

Comparison Study of Parasitic Inductance, Capacitance and Thermal Resistance for Various SiC Packaging Structures

Yue Xie
School of Electrical and Electronic Engineering
Huazhong University of Science and Technology
Wuhan, China
xie_yue@hust.edu.cn

Yifan Zhang
School of Electrical and Electronic Engineering
Huazhong University of Science and Technology
Wuhan, China
z_yifan@hust.edu.cn

Cai Chen
School of Electrical and Electronic Engineering
Huazhong University of Science and Technology
Wuhan, China
caichen@hust.edu.cn

Yong Kang
School of Electrical and Electronic Engineering
Huazhong University of Science and Technology
Wuhan, China
ykang@hust.edu.cn

Abstract—**Parasitic parameters of packaging structure can affect the performance of silicon carbide (SiC) devices. Previous researches on SiC power modules lack a comprehensive comparison of different kinds of packaging structures. In this paper, six kinds of power modules are designed under the same standard, and their parasitic inductance, parasitic capacitance, and thermal resistance are extracted by finite element simulation for comparison. According to the result, the double-side-cooling structure has the smallest thermal resistance. The chip-on-chip structure shows extremely small parasitic inductance and capacitance, but this structure is complex. The full-shielding structure has extremely small parasitic capacitance and medial parasitic inductance, while its thermal resistance is larger than other structures. The rest three structures show mediocre performance.**

Keywords—*SiC power module, parasitic inductance, parasitic capacitance, thermal resistance, Packaging.*

I. Introduction

Silicon carbide (SiC) power devices have many excellent characteristics, but the parasitic parameters will limit their performance. For example, the parasitic inductance will affect the switching speed and loss [1], the parasitic capacitance will affect the conducted EMI [2], and thermal resistance will affect the junction temperature [3]. Therefore, the packaging structure makes significant sense for SiC devices. Many packaging structures are proposed by researchers to deal with the parasitic issues.

Currently, single DBC layer structures are most widely used in commercial SiC power modules. To achieve low parasitic inductance, a hybrid packaging structure is proposed in [4], which can reduce the parasitic inductance by mutual inductance cancellation. Literature [5] optimized a hybrid structure to reduce parasitic capacitance by partial shielding, and in [6] the middle layer of the half-bridge power module was full-shielded to reduce CM EMI further. For high-temperature conditions, a double-side-cooling structure was proposed [7]. And for smaller sizes, a chip-on-chip module was studied in [8]. However, different types of structures lack comparison study in a unified standard. This paper aims to compare the parasitic inductance, capacitance, thermal resistance of different packaging structures in a unified standard, to clarify their advantages and disadvantages, and guide power module design.

This paper is organized as follows: in section II, the design principles and structures of different SiC power modules are presented. In section III, the parasitic inductance and parasitic capacitance of various designed SiC power modules are studied. In section IV, the thermal resistance of designed structures is studied. In section V, the normalized comparison of parasitic inductance, capacitance, and thermal resistance is made, and the conclusion is given.

II. Structure Design of SiC Power Modules

In his paper, 6 kinds of half-bridge SiC power module are designed under the same principle:

- Using the same bear die SiC MOSFETs and the same bear die SiC diode. In this paper CPM212000080B SiC MOSFET and CPW41200S010B SiC diode are used in power module design.

- Using the same kind of DBC, solder, and the other packaging materials try to be consistent.

- The insulation distances of each copper layer are the same, and the distance of each element has the same principle.

- The terminals of each power module have the same length.

- The rest detailed designs try to be consistent.

Under these design principles, 6 SiC power modules are designed, Fig. 1 shows the 3D structures of each power module. Fig. 1 (a) is the single-DBC structure (SD structure), which is most widely used in commercial SiC power modules. Fig. 1 (b) is the multi-DBC structure (MD structure), which is corresponding to the packaging structure presented in [4]. This structure can achieve low parasitic inductance by multi-inductance canceling. Fig. 1 (c) is called as partial-shielded structure (PS structure) in this paper, which is corresponding to the structure presented in [5], aimed to reduce CM EMI by

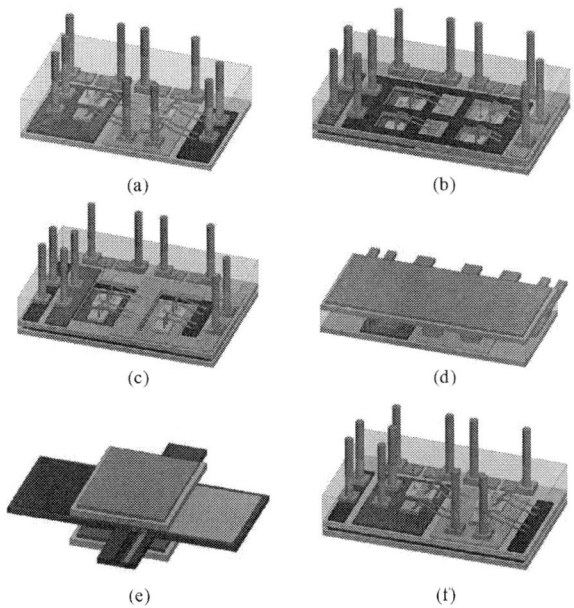

(a)

(b)

(c)

(d)

(e)

(f)

Figure 1. 3D model of designed SiC power module for comparison: (a) single-DBC structure (SD Structure), (b) multi-DBC structure (MD Structure), (c) partial-shielded structure (PS Structure), (d) double-side-cooling structure (DSC Structure), (e) chip-on-chip structure (COC Structure), (f) full-shielded structure (FS Structure).

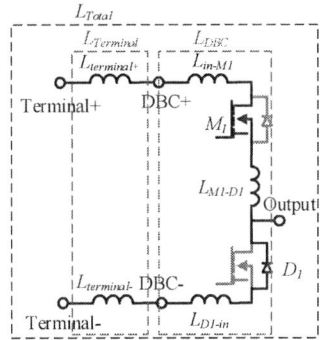

Figure 2. The equivalent circuit of a SiC power module.

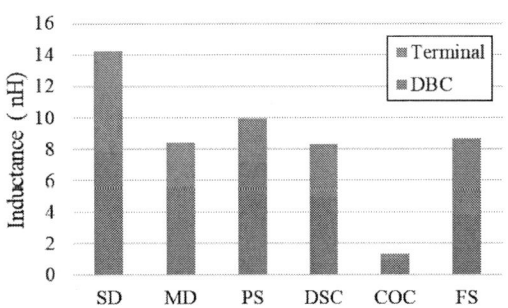

Figure 3. Parasitic inductance of SiC power modules.

layer optimizing. Fig. 1 (d) is double-side-cooling structure (DSC structure) referred the power module in [7]. Fig. 1 (e) is chip-on-chip structure (COC structure) referred the power module in [8]. Fig. 1 (f) is called full-shielded structure (FS structure) in this paper, which refers to the structure in [6].

The 6 SiC power modules in Fig. 1 can cover most of the existing packaging structure for SiC power modules. A comprehensive comparison is made between these designed SiC power modules, in order to investigate the advantages and disadvantages of existing packaging structures. The compared characteristics include parasitic inductance, which affects the switching performance, parasitic capacitance, which affects the CM EMI, and thermal resistance, which affects heat sinking.

III. COMPARISON STUDY OF PARASITIC INDUCTANCE AND CAPACITANCE

A. Comparison of Parasitic Inductance

The parasitic inductance of DBC and bonding wire is the major inductance of a SiC power module, and the other inductance main contributed by terminals. Therefore, the inductance caused by terminals are also included in this paper. The parasitic inductance of 6 SiC power modules are extracted by ANSYS Q3D, and the parasitic inductances of DBC and terminal are counted separately.

The equivalent circuit of a SiC power module, considering parasitic inductance, is shown in Fig. 2. Inductance L_{DBC} is the total inductance of DBC, which is the summary of the self-inductance and the mutual-inductance of each copper layer on the DBC. Inductance $L_{Terminal}$ is the parasitic inductance caused by terminals, which is calculated by L_{total} minus L_{DBC}, where L_{total} is the total inductance of the whole power module.

The simulation results of each SiC power module are shown in Fig. 3. When considering only the parasitic

inductance of DBC, L_{DBC} of SD-structure is the largest, because of the lack of mutual inductance cancellation design. On the contrary, the COC structure has the lowest L_{DBC}, thanks to its small commutation loop so that it can make full use of mutual inductance cancellation. L_{DBC} of other structures are between that of SD structure and COC structure, since they can achieve mutual inductance cancellation to a certain extent, but not as good as COC structure. When $L_{Terminal}$ is considered, the results are similar. In general, the SD structure has the largest L_{total} and the COC structure has the lowest L_{total}, L_{total} of the other structures are medial.

B. Comparison of Parasitic Capacitance

In a SiC power module, there are parasitic capacitances between each copper layer. Among them, C_{DC+}, C_{out}, C_{DC-} are the main capacitance, as shown in Fig. 4. Many studies show C_{out} is the dominant factor that can affect CM EMI [2]. Therefore, C_{out} of 6 designed SiC power modules are selected for comparison in this paper. Each power module used 0.3mm/0.6mm/0.3mm, Cu/AlN/Cu DBC. The parasitic capacitances are calculated by ANSYS Maxwell, and only DBC is considered in the simulation.

The simulation results are shown in Fig.5. It can be seen that the SD structure has a medial C_{out} 15pF. The C_{out} of MD structure and the DSC structure are about 80% larger than that of SD structure because their areas of the middle layer are larger. The C_{out} of PS structure decreases over 50% compared to the SD structure, thanks to the optimization design of the middle layer in this structure, so that a part of out copper layer can be shielded by DC- copper layer. The COC structure and the FS structure can utilize the shielding effect sufficiently, thus extremely small parasitic capacitance can be achieved.

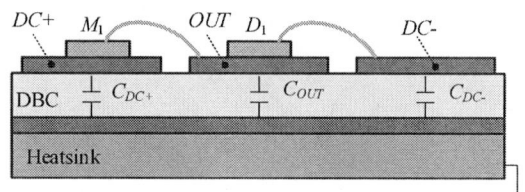

Figure 4. Parasitic capacitance schematic of SiC power modules.

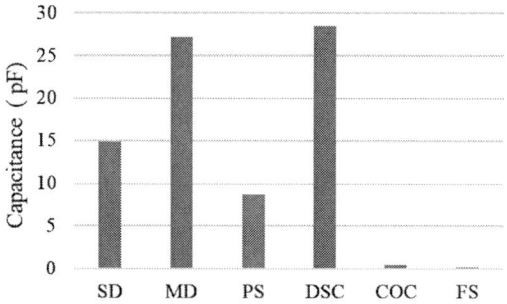

Figure 5. Comparison of parasitic inductance.

IV. COMPARISON STUDY OF THERMAL RESISTANCE

A. Thermal Resistance Calculation Method

SiC power modules need additional parts to achieve better heat dissipation, including baseplate, TIM, heatsink, etc. In this paper, thermal simulation is carried out by assembling the designed power module with similar cooling parts. The setup of thermal simulation is given in Fig. 6, the setup for single-side-cooling structure and double-side-cooling structure is a little different. SD structure, MD structure, PS structure, and FS structure can only achieve single side cooling, the setup for them is shown in Fig. 6 (a). Where the power module is soldered on a baseplate, and the baseplate is contacted to the heatsink through a layer of TIM. In practice, the heat caused

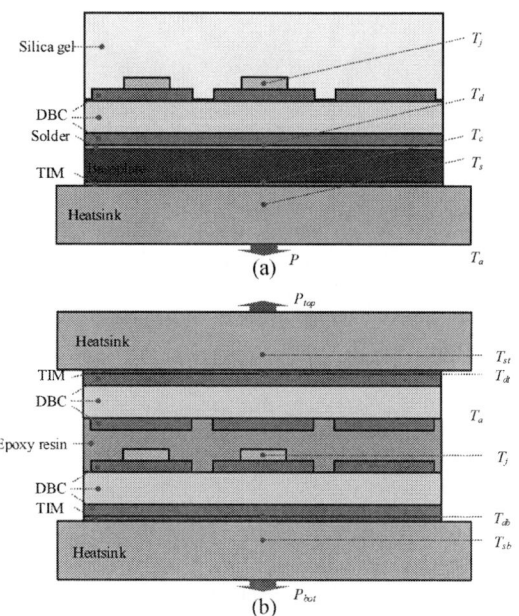

Figure 6. Thermal simulation setup: (a) for single-side-cooling structure, (b) for double-side-cooling structure.

by SiC chips is mainly dissipated through the heatsink. Thus in thermal simulation, the bottom surface of the heatsink is set as heat exchange surface, while the other surfaces are set as thermal insulation. The equivalent heat transfer coefficient is 500 W/(m²·K), which is corresponding to forced convection cooling. DSC structure and COC structure used the setup shown in Fig. 6 (b), because they can achieve double-side cooling. In practice, double-side-cooling power modules are usually without the baseplate, thus the baseplate is omitted in thermal simulation. The top surface and bottom surface of heatsinks are set as heat exchange surface, and the equivalent heat transfer coefficient is also 500 W/(m²·K). The ambient temperature is 20°C, and the parameters of materials are listed in Tab. I.

TABLE I. PARAMETERS OF SiC POWER MODULE IN THERMAL SIMULATION

Parts	Parameters		
	Material	*Thickness (mm)*	*Thermal conductivity (W/m·K)*
DBC	Cu/AlN/Cu	0.3/0.6/0.3	400/170/400
Solder	$Pn_{63}\ Sn_{37}$	0.1	50
Baseplate	Cu	2	400
TIM	Silicone grease	0.1	3
Heatsink	Cu	30×30×3	400

In the thermal simulation, several points are measured and the thermal resistance between them are calculated, the measured points are shown in Fig. 6. In Fig. 6 (a), T_j is the average junction temperature of two SiC MOSFETs, T_d is the average DBC temperature (the average temperature of two measured points, each point is on the bottom surface of DBC, right below each SiC chip), T_c is the average case temperature, T_s is the average heatsink temperature (the measured point is 1mm below the top surface of heatsink), P is the total loss power. In Fig. 6 (b), the measured points are similar to that in Fig. 6 (a), P_{bot} is the heat flux through the bottom surface and P_{top} is the heat flux through the top surface. The thermal resistances are calculated by (1)-(7), (1)-(4) are for single side cooling structures, and (5)-(7) are for double side cooling structures.

$$R_{th(j-d)} = \frac{(T_j - T_d)}{P} \tag{1}$$

$$R_{th(d-c)} = \frac{(T_d - T_c)}{P} \tag{2}$$

$$R_{th(c-s)} = \frac{(T_c - T_s)}{P} \tag{3}$$

$$R_{th(s-a)} = \frac{(T_s - T_a)}{P} \tag{4}$$

$$R_{th(j-d)} = \frac{1}{{P_{bot}}\big/{(T_{jb} - T_{db})} + {P_{top}}\big/{(T_{jt} - T_{dt})}} \tag{5}$$

$$R_{th(d-s)} = \cfrac{1}{\cfrac{P_{bot}}{(T_{db} - T_{sb})} + \cfrac{P_{top}}{(T_{dt} - T_{st})}} \qquad (6)$$

$$R_{th(s-a)} = \cfrac{1}{\cfrac{P_{bot}}{(T_{sb} - T_{a})} + \cfrac{P_{top}}{(T_{st} - T_{a})}} \qquad (7)$$

B. Comparison of Thermal Resistance

The results of the thermal simulation are shown in Fig. 7. Fig. 7 (a) shows the thermal resistance from junction to case $R_{th(j-c)}$, which represents the heat dissipation capability of each power module only. It needs to point out that single-side cooling structures are simulated with baseplate, while double-side cooling structures are simulated without baseplate, thus DSC-structure and COC-structure do not have $R_{th(d-c)}$. It can be seen that $R_{th(j-c)}$ of SD-structure, MD-structure, and PS-structure are similar. The DSC structure has the smallest $R_{th(j-c)}$ because each SiC chip can achieve double-side cooling, thus this structure has the largest heat dissipation area between SiC chips and DBC. On the contrary, in COC structure, $R_{th(d-c)}$ is the largest. It is because each SiC chip can only achieve single-side cooling, although the whole power module can achieve double-side cooling. Besides, in this structure, half of the SiC chips use the source pad for heat sinking, which has smaller area compared to the drain pad. Thus the $R_{th(d-c)}$ of COC-structure is even larger than that of single-side cooling structures. FS structure has two layers of DBC, and the SiC chips are soldered on the top DBC, thus its $R_{th(d-c)}$ is larger than SD structure.

Fig. 7 (b) shows the total thermal resistance from junction to the environment $R_{th(j-a)}$, which reflects the heat dissipation capability when other heat dissipation parts are considered. Compared to $R_{th(j-c)}$, $R_{th(c-a)}$ makes the main part of $R_{th(j-a)}$, which is much larger. Therefore, $R_{th(j-a)}$ of SD structure, MD

structure, PS structure, and FS structure are closed, because their heatsinks are the same. DSC structure has the lowest $R_{th(j-a)}$, because it can extend the surface of heatsink to two times within the same area by double-side cooling, thus $R_{th(s-a)}$ can decrease a lot. Thanks to achieving double-side cooling, COC structure has the second-lowest $R_{th(j-a)}$, although it has the largest $R_{th(j-c)}$.

The thermal simulation results of designed SiC power modules are shown in Fig. 8. The loss power of each SiC MOSFET is 20W, the other parameters are the same as mentioned above. It can be seen that the highest temperatures of single-side cooling structures are very close, including SD structure, MD structure, PS structure, and FS structure. DSC structure has the lowest temperature and COC structure has the second-lowest temperature, due to double-side cooling.

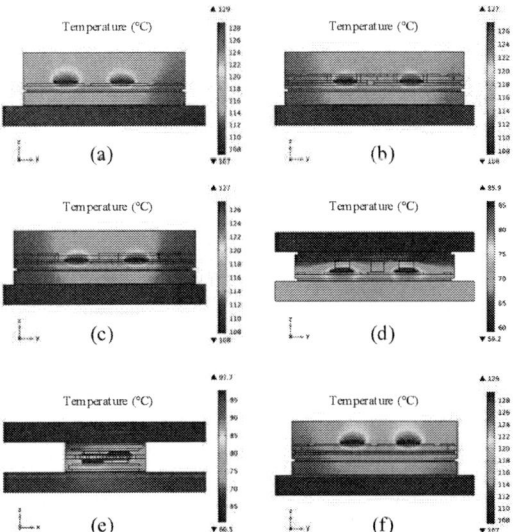

Figure 8. Thermal simulation results of designed SiC power modules: (a) SD structure, (b) MD structure, (c) PS structure, (d) DSC structure, (e) COC structure, (f) FS structure.

V. CONCLUSION

In this paper, 6 kinds of SiC power modules are designed under the same design principles for comparison study. Based on the designed power modules, the parasitic inductance, parasitic capacitance, and thermal resistance of each structure are extracted by finite element simulations. As the normalized comparison chart shows in Fig.9, The SD structure is mediocre, but its structure is most simple. Compared to SD

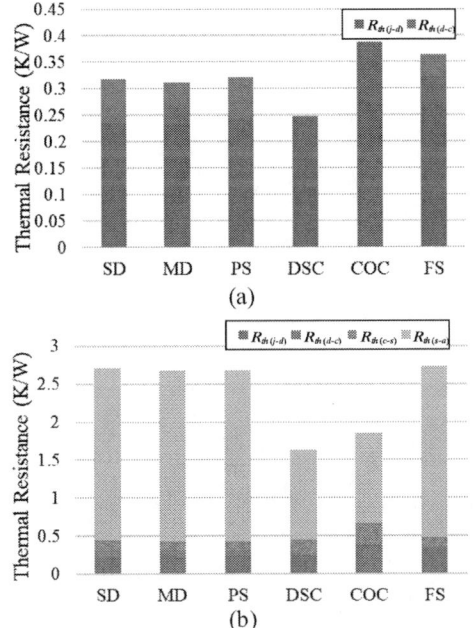

Figure 7. Thermal resistance of designed SiC power modules: (a) junction to case, (b) junction to environment.

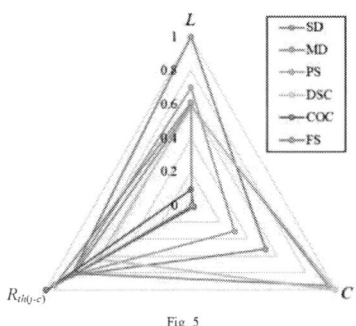

Fig. 5

Figure 9. Normalization comparison of parasitic inductance, parasitic capacitance and thermal resistance.

structure, MD structure has smaller parasitic inductance but larger parasitic capacitance. PS structure goes one step further than MD structure by optimizing the parasitic capacitance. The DSC Structure has the smallest $R_{th(j-c)}$, thus it is suitable for the condition with temperature requirement. The COC structure shows excellent performance, but its structure is complex thus the manufacturing technique and reliability should be considered. The FS structure has extremely small parasitic capacitance and medial parasitic inductance, while its $R_{th(j-c)}$ is larger than other structures.

REFERENCES

[1] Y. Xie, C. Chen, Y. Yan, Z. Huang, and Y. Kang, "Investigation on Ultra-low Turn-off Losses Phenomenon for SiC MOSFETs with Improved Switching Model," IEEE Transactions on Power Electronics, pp. 1-1, 2021.

[2] Y. Xie, C. Chen, Z. Huang, T. Liu, Y. Kang, and F. Luo, "High Frequency Conducted EMI Investigation on Packaging and Modulation for a SiC-Based High Frequency Converter," IEEE Journal of Emerging and Selected Topics in Power Electronics, vol. 7, no. 3, pp. 1789-1804, 2019.

[3] S. Tanimoto and K. Matsui, "High Junction Temperature and Low Parasitic Inductance Power Module Technology for Compact Power

Conversion Systems," IEEE Transactions on Electron Devices, vol. 62, no. 2, pp. 258-269, 2015.

[4] C. Chen, Y. Chen, Y. Li, Z. Huang, T. Liu, and Y. Kang, "An SiC-Based Half-Bridge Module With an Improved Hybrid Packaging Method for High Power Density Applications," IEEE Transactions on Industrial Electronics, vol. 64, no. 11, pp. 8980-8991, 2017.

[5] Y. Xie, Z. Huang, C. Chen, and Y. Kang, "An EMI performance improved stacked substrate packaging structure with ultra-low parasitics for SiC half-bridge power module," in PCIM Europe 2019; International Exhibition and Conference for Power Electronics, Intelligent Motion, Renewable Energy and Energy Management, 2019, pp. 1-6.

[6] T. Huber, A. Kleimaier, and R. Kennel, "Ultra-low inductive power module design with integrated common mode noise shielding," in 2017 19th European Conference on Power Electronics and Applications (EPE'17 ECCE Europe), 2017, pp. P.1-P.9.

[7] Z. Liang, "Integrated double sided cooling packaging of planar SiC power modules," in 2015 IEEE Energy Conversion Congress and Exposition (ECCE), 2015, pp. 4907-4912.

[8] G. Regnat et al., "Silicon carbide power chip on chip module based on embedded die technology with paralleled dies," in 2015 IEEE Energy Conversion Congress and Exposition (ECCE), 2015, pp. 4913-4919.

Off-State Negative Differential Capacitance AlGaN/GaN Heterostructures in cryogenic temperature

Siyu Liu, Jiejie Zhu, Jingshu Guo, Minhan Mi,Xiaohua Ma, Yue Hao
School of Microelectronics, Xidian University
Xi'an 710071, People's Republic of China
xhma@xidian.edu.cn

Jielong Liu, Yilin Chen
School of Advanced Materials and Nanotechnology, Xidian University
Xi'an 710071, People's Republic of China

Abstract—There are few reports on the degradation phenomenon and mechanism of AlGaN/GaN heterojunction field-effect transistors (HFETs) operating at cryogenic temperature. In this paper, the off-state capacitance of AlGaN / GaN HFETs at cryogenic temperature and the room temperature was investigated. The negative differential capacitance at 77 K and 300 K is 11.26 F and 0.69 F, respectively. The capacitance was tested at different temperatures, which showed that the negative differential capacitance came from low-temperature conditions. At room temperature, the resistance was increased sharply under the gate of the device off-state. At this time, the current leakage channel was turned on, the phenomenon of negative differential capacitance was caused by an accumulation of electrons. Through the *C-V* test at different frequencies at 77 K, it reflects the relationship between the trap level, the off-state leakage current, and the negative differential capacitance. This article provides suggestions for the application of GaN-based radio frequency devices in cryogenic temperature environments.

Keywords—cryogenic temperature; negative differential capacitance; AlGaN/GaN; HEMT

I. INTRODUCTION

GaN-based heterostructure field-effect transistors (HFETs) have the advantages of high-power density, high efficiency, and compact size, and are therefore candidates for high-speed electronic power devices.[1-3] The reliability of device is critical to commercialization. Therefore, the temperature-dependent study is an essential part of the comprehensive characterization of the AlGaN/GaN HFETs. With the continuous deepening of research, AlGaN/GaN HFETs devices are gradually moving from the laboratory to practical applications, especially in the fields of national defense construction, aviation, aerospace, and space exploration. However, due to the wide range of temperature changes in the space environment, and the large range of changes, when the spacecraft flies in space, it will be in a constant alternation of extremely low temperature and extremely high temperature. For example, in the absence of light, the temperature of outer space will be lower than minus 200°C, and in the case of light, the temperature of the outer space environment will be higher than 100°C. For the spacecraft to maintain long-term stable operation, the devices used need to have very high-temperature reliability[4, 5]. Not only can they meet the working requirements under high-temperature conditions, but the device performance should also maintain a high degree of consistency under low and extremely low-temperature conditions. Therefore, the research on the temperature reliability of GaN HFETs devices, in addition to the current mature high-temperature field, also needs to pay great attention to the changes in device performance at low or even extremely low temperatures[6-8].

The papers published in the past shows that GaN-based devices working at cryogenic temperature mainly focused on the following aspects. Better performance at cryogenic temperature than room temperature, for example, due to the higher drain current and trans-conductance[9], the lower leakage current[10]. Kink effect at a cryogenic temperature [11] and current collapse [12, 13]. Transmission characteristics are also used to evaluate interface quality at cryogenic temperature [14, 15]. Use temperature-dependent gate current characteristics to study the gate leakage mechanism [16]. However, there are little attention has been paid the negative differential capacitance.

In this paper, the capacitance characteristics of the device at 77K and 300K were compared, and it is found that the device had an abnormal negative differential capacitance in the off state. A variable frequency C-V test was performed on the device, and the source of the negative differential capacitance was analyzed by establishing a test model. In addition, the characteristics of trapping and emitting electrons at the device interface at low temperature are also analyzed. The equipment under low temperature is studied, which provides Reference for the application and reliability analysis of GaN-based HFET at low temperature.

II. DEVICES AND EXPERIMENTS

A. Devices fabrication

The epitaxial layers of AlGaN/GaN are grown on sapphire substrates by metal organic chemical vapor deposition(MOCVD). The material structure from bottom to top consists of a 1.3-μm unintentionally doped (UID) GaN layer, a 1-nm AlN intermediate layer and a 20-nm thick unintentionally doped $Al_{0.3}Ga_{0.7}N$ barrier layer. Relying on the technology of Cl_2-based reactive ion etching (RIE) to perform mesa isolation. Hall measurement at 300 K indicates 7.2×10^{12} cm^{-2} for a sheet carrier density and 2105 $cm^2/V \cdot s$ for a mobility. Through the technology of e-beam evaporation of Ti/Al/Ni/Au, lift-off process, and rapid thermal annealing at 840 °C in N_2 for 30 s twice, the source and drain ohmic contacts were formed. The ohmic contact resistance of approximately 0.33Ω • mm is obtained by using transmission line passivation and plasma enhanced chemical vapor deposition (PECVD). The area of the gate foot was defined by removing Si_3N_4 layer with ICP etch in CF_4 plasma. And 445-nm Ni/Au metal was evaporated to serve as a T-gate.

In this paper, the tested devices are with 20 μm of L_g, 100 μm of W_g, 3 μm of L_{gs} and 3 μm of L_{gd}. Fig. 1 shows the cross-section of a scaled recess-gate AlGaN/GaN HFETs.

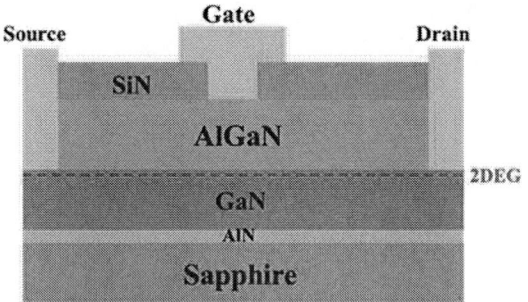

Fig. 1. Cross-section view of the device structure utilized in the low-temperature experiments.

B. DC characteristic

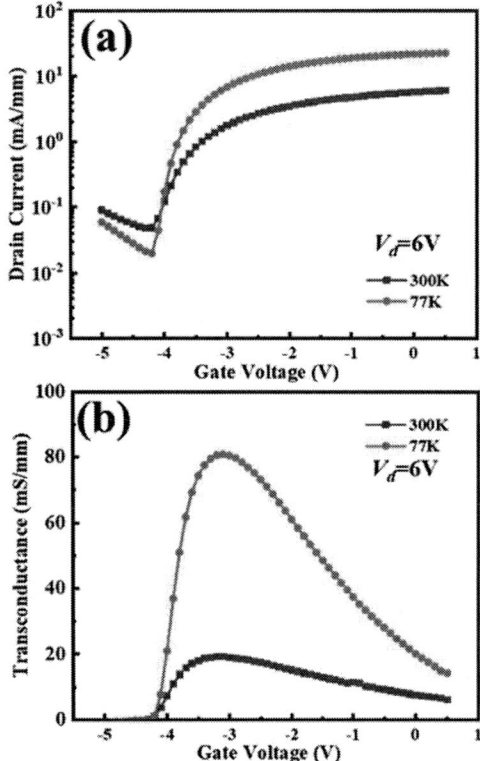

Fig. 2. L_g=20 μm, L_{gs}=L_{gd}=3 μm: (a) The relationship between gate voltage and drain current at 300 K and 77 K. (b)The transfer characteristics of the devices at 300 K and 77 K.

As shown in Fig. 2, the transfer characteristics of devices with a gate length of 20 μm were tested at 300 K and 77 K. The maximum saturation current of device at 300 K was 6.1 mA/mm, the current of off-state was 9.1×10^{-2} mA/mm and the peak trans-conductance is 19.1 mS/mm at V_d=0.1 V. The maximum saturation current of device at 77 K was 22.6 mA/mm, the current of off-state was 5.9×10^{-2} mA/mm and the peak trans-conductance is 80.8 mS/mm at V_d=0.1 V. Compared with 300 K, the device had larger saturation current, transconductance peak value, and smaller off-state leakage at 77 K. This is because the field-effect mobility increases in low-temperature. As Fig. 2(a) indicates, the increased rate of

the off-state leakage with the change of the gate voltage is higher at 77 K, so the capacitance test at different temperatures was performed on this device.

III. RESULTS AND DISCUSSION

The results and discussion of the devices capacitance at 300 K and 77 K are analyzed in this section.

A. Off-state capacitance

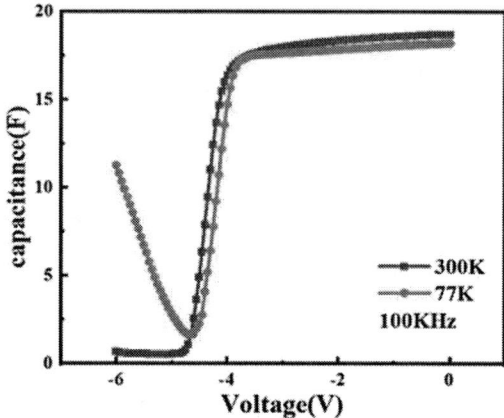

Fig. 3. C-V characteristics at 100 kHz: Abnormal negative differential capacitance in low-temperature conditions.

Fig. 3 shows the C-V curves measured at 100 kHz in different temperature with 300 K and 77 K. Compared with the results at 300 K, the C-V curves showed a decrease in accumulation capacitance and a positive shift of 0.15 V at 77 K. Furthermore, the C-V curve at 77 K showed an abnormal negative differential capacitance when the AlGaN/GaN heterostructure biased at off-state. It has never been observed from the reports about GaN-based HFETs recently. When the gate voltage is -6 V, the measured capacitance was 11.26 F at 77 K and 0.69 F at 300 K, respectively. The channel is not turned on under the off-state, there is no accumulation of electrons in the area under the gate, and the negative differential capacitance cannot be observed. The hypothesis has been proposed that the existence of other conductive channels will cause the phenomenon of negative differential capacitance.

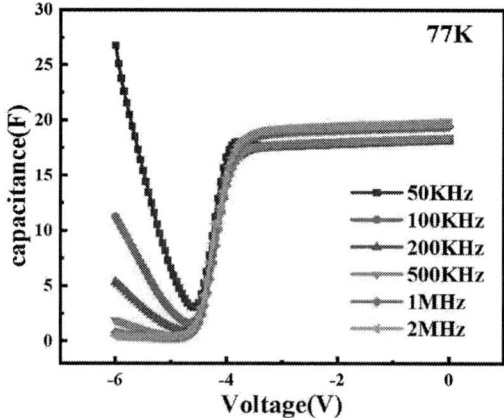

Fig. 4. C-V characteristics at 77 K: As the C-V test frequency decreases, the negative differential capacitance gradually increases

The C-V frequency conversion test was performed on the device under 77 K, and it was found that the phenomenon of

978-1-6654-4817-8/21 $31.00 © 2021 IEEE

negative differential capacitance appeared in the cut-off state. As the test frequency increases, the phenomenon of negative differential capacitance is gradually suppressed. At 2 MHz, even the negative differential capacitance can no longer be monitored. This is because the process of capturing and releasing electrons in some interface states is easier to observe when the test frequency is lower. As the test frequency increases, some electrons are too late to fill the trap, so the phenomenon of negative differential capacitance cannot be observed. The gate and source of the device in the off state are not turned on, so the negative differential capacitance may be caused by surface leakage.

The relationship between trap emission time constant and temperature is as follows:

$$\tau_{it} = (\sigma_t N_c v_{th})^{-1} exp(\frac{E_C - E_T}{kT}) \qquad (1)$$

Where τ_{it} is trap emission time constant, σ_t is trap capture cross-section, N_c is GaN conduction band effective density of states, hot-carrier rate v_{th}=2.6×10^7 cm/s and E_T is energy levels of charge traps.

Compared with room temperature, the trap emission time constant is larger at low temperature. This leads to a longer period for electrons to be released after traps capture electrons at 77 K. In the long-term alternating current signal test, the trap captures electrons at a higher speed than the trap emits electrons, and the electrons continue to accumulate in the trap to form a negative differential capacitance.

B. Equivalent circuit model of device

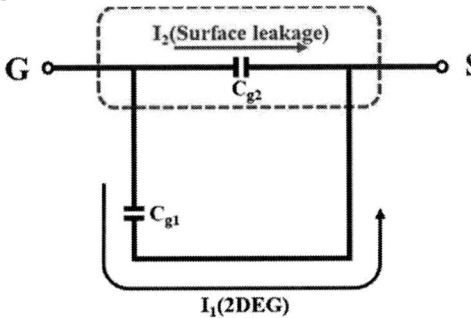

Fig. 5. Equivalent circuit model of AlGaN/GaN HFETs.

Figure 5 indicates the equivalent circuit model of AlGaN/GaN HFETs. When the device is in the on state, the gate and source are turned on through the two-dimensional electron gas (2DEG) channel. At this time, the measured capacitance is the under-gate capacitance C_{g1}. In the off state, the 2DEG channel is not turned on. At this time, only the leakage channel can cause the gate of the source to be turned on. Therefore, the measured capacitance at this time is the capacitance of the surface leakage channel. C_{g2}. The *C-V* test shows the linear relationship between the capacitance and the gate voltage. The test model is:

$$C_{total} = \frac{I_{max}}{2\pi f V_{ac}} = \frac{I_{max1} + I_{max2}}{2\pi f V_{ac}} \qquad (2)$$

Where C_{total} is measured capacitance, I_{max} is the maximum current in a single period, f is the frequency of test, V_{ac} is AC voltage. In the actual test, Imax is composed of the device on-current Imax1 and the surface leakage current I_{max2}.

When the device is in the off state, the channel is not turned on, so only the surface leakage current I_{max2} will affect the final test result. As the gate voltage decreases, I_{max2} gradually increases, so the negative differential capacitance will gradually increase. In addition, the model also explains the phenomenon in Figure 4: as the test frequency increases, the phenomenon of negative differential capacitance becomes less and less obvious. When the frequency increases to a certain amount, under the condition that the other parameters remain unchanged, the influence of Imax on the capacitance C_{total} becomes smaller and smaller, which causes the negative differential capacitance to be unobserved in the high-frequency test.

IV. CONCLUSION

In this article, the influence of surface leakage on negative differential capacitance has been investigated by low-temperature measurements. It was confirmed that surface leakage plays a significant effect on degrading the low-temperature performance of AlGaN/GaN HFETs. The appearance of negative differential capacitance is due to the fact that the constant of trap emission at cryogenic temperature is higher than room temperature. The test frequency will also affect the observation of negative differential capacitance. This work provides important suggestions for the application of GaN-based radio frequency devices in cryogenic temperature environments.

ACKNOWLEDGMENT

This work was supported by National Key Research and Development Project, Grant 2020YFB1807403, National Natural Science Foundation of China, Grant 61904135, 11690042, Fundamental Research Funds for the Central Universities under grant NO. QTZX2172, and supported by the Innovation Fund of Xidian University.

REFERENCES

[1] Y. F. Wu et al., "30-W/mm GaN HEMTs by Field Plate Optimization," IEEE Electron Device Letters, vol. 25, no. 3, pp. 117-119, 2004.

[2] J. Millan, P. Godignon, X. Perpina, A. Perez-Tomas, and J. Rebollo, "A Survey of Wide Bandgap Power Semiconductor Devices," IEEE Transactions on Power Electronics, vol. 29, no. 5, pp. 2155-2163, 2014.

[3] E. A. Jones, F. F. Wang, and D. Costinett, "Review of Commercial GaN Power Devices and GaN-Based Converter Design Challenges," IEEE Journal of Emerging and Selected Topics in Power Electronics, vol. 4, no. 3, pp. 707-719, 2016.

[4] J. Yang, X. Zhang, and C. J. I. Lv, "Study on the reliability of AlGaN/GaN HEMTs at high temperature," 2011.

[5] Y. Zhang, S. Feng, H. Zhu, X. Gong, L. Shi, and C. Guo, "Determining Drain Current Characteristics and Channel Temperature Rise in GaN HEMTs," IEEE Transactions on Device and Materials Reliability, vol. 14, no. 4, pp. 978-982, 2014.

[6] M. W. Pospieszalski and S. Weinreb, "FET's and HEMT's at Cryogenic Temperatures - Their Properties and Use in Low-Noise Amplifiers," 1987 IEEE MTT-S International Microwave Symposium Digest, 2010.

[7] E. Cha, A. Pourkabirian, J. Schleeh, N. Wa De Falk, and J. Grahn, "Cryogenic low-noise InP HEMTs: A source-drain distance study," in Compound Semiconductor Week, 2016.

[8] Nidhi, T. Palacios, A. Chakraborty, S. Keller, and U. K. Mishra, "Study of Impact of Access Resistance on High-Frequency Performance of AlGaN/GaN HEMTs by Measurements at Low Temperatures," IEEE Electron Device Letters, vol. 27, no. 11, pp. 877-880, 2006.

[9] M. A. Alim, A. A. Rezazadeh, and C. Gaquiere, "Temperature Effect on DC and Equivalent Circuit Parameters of 0.15- μm Gate Length GaN/SiC HEMT for Microwave Applications," IEEE

Transactions on Microwave Theory and Techniques, vol. 64, no. 11, pp. 3483-3491, 2016.

[10] L. Xia, A. Hanson, T. Boles, and D. Jin, "On reverse gate leakage current of GaN high electron mobility transistors on silicon substrate," Applied Physics Letters, vol. 102, no. 11, 2013.

[11] R. Cuerdo et al., "The Kink Effect at Cryogenic Temperatures in Deep Submicron AlGaN/GaN HEMTs," vol. 30, no. 3, pp. p.209-212, 2009.

[12] L. Ching-Hui, W. Wen-Kai, L. Po-Chen, L. Cheng-Kuo, C. Yu-Jung, and C. Yi-Jen, "Transient pulsed analysis on GaN HEMTs at cryogenic temperatures," IEEE Electron Device Letters, vol. 26, no. 10, pp. 710-712, 2005.

[13] H. F. Sun and C. R. Bolognesi, "Anomalous behavior of AlGaN/GaN heterostructure field-effect transistors at cryogenic temperatures: From current collapse to current enhancement with cooling," Applied Physics Letters, vol. 90, no. 12, 2007.

[14] F. Meng et al., "Transport characteristics of AlGaN/GaN/AlGaN double heterostructures with high electron mobility," Journal of Applied Physics, vol. 112, no. 2, 2012.

[15] J. Zhang et al., "Enhanced transport properties in InAlGaN/AlN/GaN heterostructures on Si (111) substrates: The role of interface quality," Applied Physics Letters, vol. 110, no. 17, 2017.

[16] Y. D. Du, W. H. Han, W. Yan, and F. H. J. C. P. L. Yang, "Impact of CHF3 Plasma Treatment on AlGaN/GaN HEMTs Identified by Low-Temperature Measurement," vol. 31, no. 4, p. 048501, 2014.

Deep energy levels investigation on high resistivity monocrystalline diamond

Yutian Wang
school of microelecronics
Xidian University
Xi'an, China
ytwang@xidian.edu.cn

Fangzhou Zhao
school of microelecronics
Xidian University
Xi'an, China
fzzhao_1@stu.xidian.edu.cn

Hui Guo*
school of microelecronics
Xidian University
Xi'an, China
guohui@mail.xidian.edu.cn

Qian Sun
school of microelecronics
Xidian University
Xi'an, China
qsun_2020@stu.xidian.edu.cn

Abstract—Three independent methods, isothermal transient spectroscopy (ITS), Current-DLTS (C-DLTS) and optical-excited DLTS (O-DLTS), were preformed to search defect information of a transparent undoped bulk monocrystalline diamond. Activation energies from 0.407 eV to 1.184 eV and capture cross section in the range from 10^{-12} cm^{-2} to 10^{-18} cm^{-2} have been obtained from ITS, C-DLTS and O-DLTS. We found that more activation energies can be fitted from ITS than current DLTS and O-DLTS. The fitted activation energies from C-DLTS and O-DLTS are supported by ITS, indicating ITS as a suitable technology for the high resistivity diamond.

Keywords—Diamond; Deep level transient spectroscopy, Energy level, Optical deep level transient spectroscopy, Isothermal transient spectroscopy

I. INTRODUCTION

Diamond is a new wide-bandgap semiconductor. The large band gap and high rigidity make diamond as a promising material for manufacturing special detector working in rough environment [1,27] and device for quantum technology [2,3,26]. In the past decade many reports have been focused on energy levels in doped and undoped diamond film but few investigations about bulk diamond because of immature fabrication technology [4,5,6,12,13]. Recently, accompanying improvements in technology of growing mono-crystalline diamond, the bulk undoped mono-crystalline diamond began to have the applications of new type photoelectric devices, such as UV photodetector [7,8,9].Therefore, the basic features in mono-crystalline undoped bulk diamond, such as energy levels, concentration of traps and cross section, are important for those devices' designing and fabrication.

Thus, to investigate energy levels information of undoped bulk diamond, we use current deep level transient spectroscopy (C-DLTS), optical-excited DLTS (O-DLTS) and isothermal transient capacitance spectroscopy (ITS) to measure defect information [5,15,16]. C-DLTS and O-DLTS will measure thermally stimulated current in a temperature scan range. The ITS will test transient capacitance varying time windows (Tw) to collect signal at a fixed temperature[14,25]. The difference of the test will be explained in next section.

II. EXPERIMENT

The homo-epitaxial bulk diamond was fabricated by a high-power microwave plasma-assisted chemical vapor deposition system (MPCVD). One HPHT-Ib diamond wafer was used to grow diamond in the direction (100). CH$_4$ and H$_2$ were recourse gas and carrier gas, respectively. The ambient temperature was 950 $^{\circ}$C. A bulk diamond was cut by laser from the wafer and the size of diamond is shown in Fig. 1. The Raman spectrum was recorded by a HORIBA-HR-Evolution Raman spectrometer to test crystallization quality. As shown in Fig. 1, only one sharp single peak appears at 1332.5 cm^{-1} in the range of 1000 cm^{-1} to 2500 cm^{-1}. This peak position shows crystallization quality of sample is similar to natural unstressed diamond monocrystalline (Grade: IIa) [17]. Small FWHM stands for a few defect concentrations and a lattice distortion.

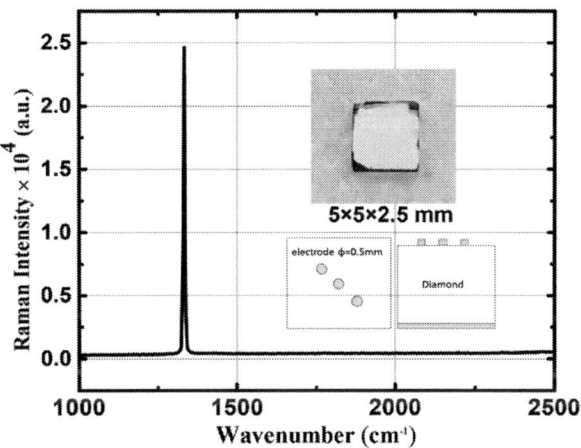

Fig. 1.   Raman spectrum of undoped diamond.Inset; Image of the diamond sample and the size is 5 mm × 5 mm × 2.5 mm.

Activation energies are investigated by a Deep Level Transient Spectroscopy (FT 1030, Phys Tech, GmbH). The measurements were organized as follow. First, current versus voltage (I-V) characteristic measurements under dark and light were tested. C-DLTS scanned transient from 180 K to 760 K and O-DLTS was performed from 150 K to 550 K was utilized to detect defect energy levels. Then, the ITS tests scanning electron emission time under variation temperature, respectively [18] to verify ITS test results of C-DLTS and O-DLTS.

978-1-6654-4817-8/21 $31.00 © 2021 IEEE

III. Results and Discussion

A. Static test

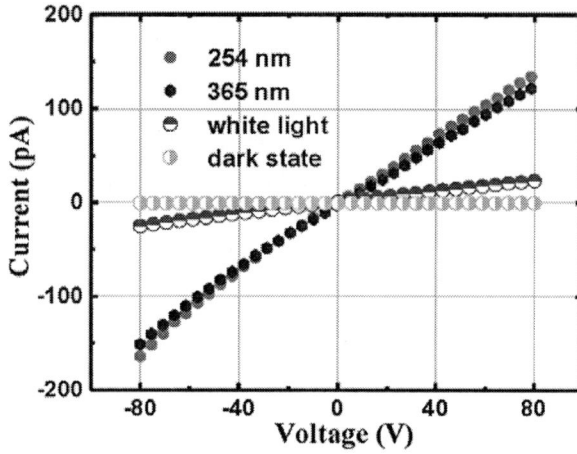

Fig. 2. I-V characteristics of diamond under dark, white light, 254 nm and 365 nm LED illumination.

Before C-DLTS, O-DLTS and ITS, we test IV characteristics under dark state, white light, 365 nm and 254 nm LED array excited light conditions by a Tektronix MPS150 probe station which are shown in Fig. 2. UV light can strongly increase current density than white light and dark state conditions, but the light excited currents under the light of 254 nm and 365 nm are the almost same. Thus, the light current comes from the traps in the band gap and the light motivation can increase signal response effectively. The measurable I-V signals proved current recording can be effective methods. That is reason for C-DLTS and O-DLTS used the current mode. Generally, recording capacitance is more accurate than current mode. The capacitance mode of DLTS also be attempted for this research but ratio of signal to noise is bad. One possible reason is thickness of the sample. Although diamond has bigger thermal conductivity than other semiconductors, the 2.5 mm thickness will lead a temperature gradient in the process of heating. Thus, it is hard to accurately record a small capacitance variation in each short temperature scanning process. Another possible reason is hard to prepare a good ohmic contact for the high electrical resistivity diamond. To solve problems, we adopt two techniques. The first is double Schottky junctions on front side and backside with a great area ratio, such as two back-to-back Schottky diodes shown in Fig. 1. Thus, the small area region will have main contribution for the capacitance which is used for the materials of hard to preparing ohmic contact. More details can be seen in reference [24]. The second is ITS scanning with variation of the period width (Tw) of correcting capacitance at fixed temperature which avoid problem of temperature gradient and has longer recording time than DLTS in capacitance mode. In addition, with a continuously temperature scan of ITS, the ITS curves can also provide accurate defect information.

Thus, the DLTS section organized as follow. The current-DLTS and optical-excited DLTS were firstly preformed to search DLTS signals. Then, ITS works in capacitance mode were recorded as a means of verification.

B. Temperature scanning for C-DLTS and O-DLTS

C-DLTS experiments were performed with the 30 V bias voltage (UR), -20 V pulse voltage (UP) and 100 µs pulse time. The temperature scanning is from 180 K to 750 K. As shown in the Fig. 3. there are two peaks at 540 K and 726 K respectively. Then, illuminated sample with a single 405 nm LED, temperature scanning was repeated from 150 K to 550 K with 15 V UR and -6V UP. For light case of 405 nm (as shown in the Fig. 4, light was kept on during test to increase signal to noise ratio and filling pulse time is also 100 µs. Two peaks were found at 200 K and 540 K respectively. The peak at 540 K was observed by C-DLTS and O-DLTS to confirm the validity of the tests. The intensity of peak at 200 K is stronger than the peak at 540 K in the O-DLTS.

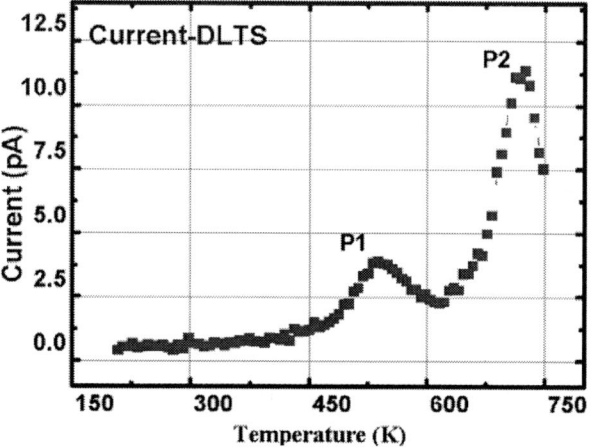

Fig. 3. C-DLTS spectrum obtain dark illumination.

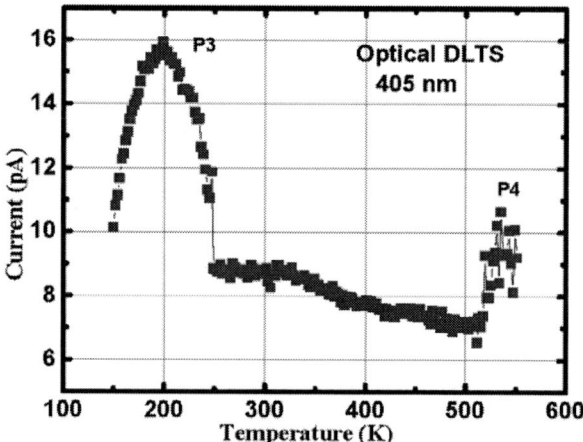

Fig. 4. O-DLTS spectrum illuminated with 405 nm light.

The Arrhenius curves fitting are show in the Fig. 5 and Fig. 6. The peak at 540 K is broader than the peak at 726 K, so two levels can be fitted at 540 K and only one peak at 726 K. The peak at 540 K which was also reported by O-DLTS in boron doped diamond homo-epitaxial layer which originates from threading dislocation [19]. The energy levels are Ec-0.774 eV, Ec-0.822 eV and Ec-1.101eV respectively, corresponding to the cross section areas are 8.96×10^{-17} cm^{-2}, 1.28×10^{-16} cm^{-2} and 2.98×10^{-18} cm^{-2}. For O-DLTS spectrum, only one level can be fitted at the 200 K peak since the points composed of the 540 K peak are too discrete to be fitted.

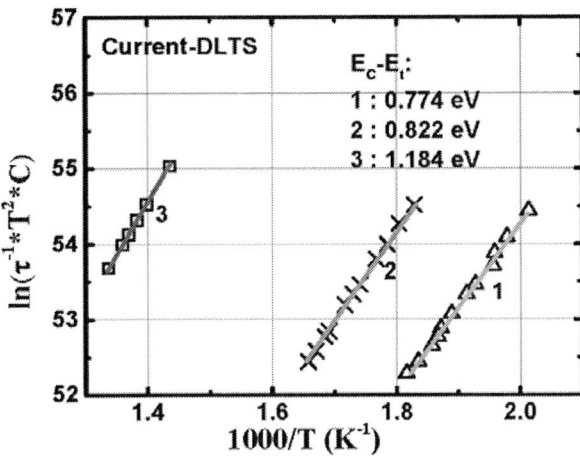

Fig. 5. Arrhenius plot of current-DLTS (C-DLTS), from which activation energies can be obtained.

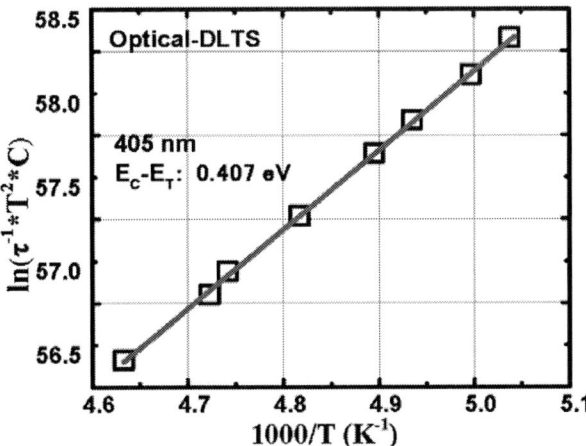

Fig. 6. Arrhenius plot of optical-excited-DLTS (O-DLTS), from which activation energies can be obtained.

C. Isothermal Transient Spectroscopy

ITS scans are recorded from 298 K to 450 K by 10 K as a step, which is the most used working temperature range of electronic instruments. The UR and UP is 20 V and -10 V respectively. The scanning period is also from 0.001 s to 20 s. As shown in Fig. 7, the whole ITS temperature scan results can be apparently divided into two parts by 0.1 s. They are main emission plateau before 0.1 s and the weak emission plain after 0.1 s. All the ITS curves have done Arrhenius fitting which can be seen in the supplement materials. We summarized activation energy, capture cross section and defect concentration in table 1. Several activation energies were considered from the same defect energy level when their energy difference is within 0.05 eV. Thus, the capture cross section and defect concentration float to some extent. However, in contract to the big difference of capture cross section from fitted energy levels, defect concentration is relatively stable at 10^{15} cm^{-3}.

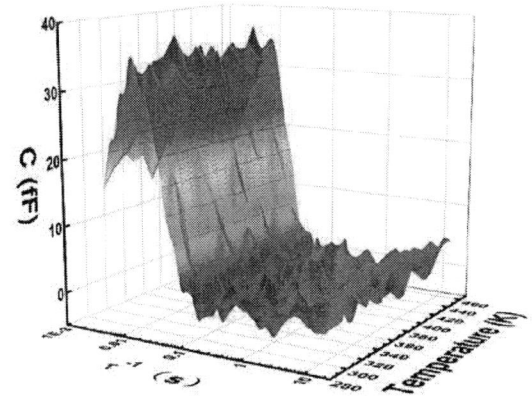

Fig. 7. ITS spectra scans in temperature from 290 K to 400 K.

IV. DISCUSSION

All the fitted activation energies and capture cross section by C-DLTS, O-DLTS and ITS are summarized in the table 1. All calculated defect concentration shows the low concentration of defect that proves good crystallization quality of the sample. Activation energies fitted 0.407 eV and 0.774 eV and 0.882 eV, which are from C-DLTS and O-DLTS, can be consider as same one of 0.42 eV, 0.77 eV and 0.82 eV in ITS with the lager cross section and concentration. The reason for this is over this paper, but we attribute it to ITS's better ratio of signal to noise and thermostability in recording process.

TABLE I. SUMMARY OF DEEP LEVELS OBSERVED BY CURRENT-DLTS, O-DLTS AND ITS OF UNDOPED DIAMOND

Test Mode	E_C-E_T(eV)	Capture cross section(cm^{-2})	Defect concentration(cm^{-3})
Temperature scan	1.184	2.98×10^{-18}	1.79×10^{11}
	0.822	1.28×10^{-16}	1.64×10^{12}
	0.774	8.96×10^{-17}	8.23×10^{11}
Optical temperature scan (405nm LED)	0.407	8.05×10^{-16}	4.77×10^{13}
ITS with temperature	0.42	4.03-5.50×10^{-17}	5.76-5.84×10^{15}
	0.44	0.732-17.8×10^{-17}	3.89-5.84×10^{15}
	0.50	1.16-19.4×10^{-16}	4.23-5.29×10^{15}
	0.55	0.103-16.5×10^{-16}	2.24-4.90×10^{15}
	0.58	6.17-34.0×10^{-16}	3.42-6.32×10^{15}
	0.615	0.756-197×10^{-16}	2.28-6.32×10^{15}
	0.65	0.622-19.1×10^{-15}	2.49-3.91×10^{15}
	0.68	0.129-22.7×10^{-14}	2.18-6.03×10^{15}
	0.71	0.432-12.4×10^{-14}	1.73-3.88×10^{15}
	0.73	0.514-36.6×10^{-14}	0.372-3.79×10^{15}
	0.77	0.309-16.5×10^{-13}	2.32-3.18×10^{15}
	0.83	3.53-18.3×10^{-12}	2.13-2.59×10^{15}

In table 1, the energy level at 1.1 eV was reported by Borchi et. al. In their report, this energy level was found after β and γ radiation. So, the energy level is probably from single defects [6]. In our case, this level was observed at 726 K with quite small capture cross section and low concentration. Energy at 0.82 (0.83) eV was assigned to vacancy-type defects [20]. Beside the known energies, ITS provides some new energies from 0.4 eV to 0.83 eV. In those energies, energy at 0.7 eV has been reported in sample metal-diamond-like atomic-scale composite films, which possibly from the point defect [21]. 0.45 eV, 0.55 eV and

0.65eV have also reported in the undoped hydrogenated polycrystalline CVD diamond film [6]. Some activation energies are close to each other, such as 0.55 eV, 0.587 eV, 0.65 eV, 0.68 eV, 0.774 eV and 0.78 eV. Although the origin of the energy difference has not been identified, it is temporarily assigned to different electron states of the same defect. To further identify origin of activation energies, other experimental methods, such as electron paramagnetic resonance (EPR) and positron annihilation spectrum (PAS) are needed [22,23].

It is worthy of saying that fitted cross section of the same energy level from different ITS curves have several magnitude difference. In general, the extrapolated the capture cross section from Arrhenius curves exists several magnitudes uncertainties which can explain the discrepancy of capture cross section for the similar energy level with different capture cross section [17]. It is supposed that the discrepancy is due to a small overlapping effect with a neighboring signal. The discrepancy of capture cross section will be further investigated by varying filling voltage pulses and larger temperature range with a higher sensitivity of current or capacitance meters.

V. CONCLUSIONS

In this letter, defect in undoped monocrystalline diamond were investigated by C-DLTS, O-DLTS and ITS. The Arrhenius plot shows that there are four energy levels which are under conduct band bottom 0.407 eV, 0.774 eV, 0.822 eV and 1.184 eV obtained by C-DLTS and O-DLTS which are also observed in ITS. Meanwhile more energy levels can be disentangled in the range of 0.4 eV to 0.7 eV from ITS indicating ITS as a better method for the high resistivity diamond. Those information is useful for further investigating the defects in undoped bulk monocrystalline diamond material and developing optical electric device.

REFERENCES

[1] Schmid, G.J & Koch, Jeffrey & Lerche, R.A & Moran, M.J. (2004). A neutron sensor based on single crystal CVD diamond. Nuclear Instruments and Methods in Physics Research Section A Accelerators Spectrometers Detectors and Associated Equipment. 527. 554. 10.1016/j.nima.2004.03.199.

[2] Jelezko, Fedor. (2006). Processing quantum information in diamond. Journal of Physics: Condensed Matter. 18. S807. 10.1088/0953-8984/18/21/S08.

[3] Hausmann, Birgit & Khan, Mughees & Zhang, Yinan & Babinec, Tom & Martinick, Katie & McCutcheon, Murray & Hemmer, Phil & Loncar, Marko. (2009). Fabrication of Diamond Nanowires for Quantum Information Processing Applications. Diamond and Related Materials. 19. 621-629. 10.1016/j.diamond.2010.01.011.

[4] Polyakov, V.I. & Rukovishnikov, A.I. & Avdeeva, L.A. & Kun'kova, Z.E. & Varnin, V.P. & Teremetskaya, Irina & Ralchenko, V.G.. (2006). UV Schottky photodiode on boron-doped CVD diamond films. Diamond and Related Materials. 15. 1972-1975. 10.1016/j.diamond.2006.08.008.

[5] Bruzzi, Mara & Menichelli, David & Pirollo, S. & Sciortino, Silvio. (2000). Photo-induced deep level analysis in undoped CVD diamond films. Diamond and Related Materials. 9. 1081-1085. 10.1016/S0925-9635(00)00197-7.

[6] Garin, Boris & Polyakov, V.I. & Rukovishnikov, A.I. & Avdeeva, L.A. & Derkach, Vadim & Parshin, V.V. & Ralchenko, V.G.. (2006). Dielectric loss and energy distribution of the shallow levels in CVD diamonds. Diamond and Related Materials. 15. 1917-1920. 10.1016/j.diamond.2006.08.023.

[7] Chaonan, Lin & Lu, Ying-Jie & Yang, Xun & Tian, Yong - Zhi & Gao, Chao - Jun & Sun, Jun - Lu & Dong, Lin & Zhong, Fang & Shan, Chong-Xin. (2018). Diamond - Based All - Carbon

Photodetectors for Solar - Blind Imaging. Advanced Optical Materials. 6. 10.1002/adom.201800068.

[8] Chen, Yan-Cheng & Lu, Ying-Jie & Chaonan, Lin & Tian, Yong-Zhi & Gao, Chaojun & Dong, Lin & Shan, Chong-Xin. (2018). Self-Powered Diamond/β-Ga 2 O 3 Photodetectors for Solar-Blind Imaging. Journal of Materials Chemistry C. 6. 10.1039/C8TC01122B.

[9] Chaonan, Lin & Tian, YongZhi & Gao, ChaoJun & Fan, MingMing & Yang, Xun & Dong, Lin & Shan, ChongXin. (2019). Diamond based photodetectors for solar-blind communication. Optics Express. 27. 29962. 10.1364/OE.27.029962.

[10] Davanloo, F. & Iosif, M.C. & Camase, D.T. & Collins, C.B. & Agee, F.J.. (2000). Development of a diamond treated photoconductive semiconductor switch for use in Blumlein pulsers. 73 - 77. 10.1109/MODSYM.2000.896168.

[11] Davanloo, F. & Iosif, M.C. & Camase, D.T. & Collins, C.B.. (2001). Development of high power photoconductive semiconductor switches treated with amorphous diamond coatings. Fuel Cells Bulletin. 337-340 vol.1. 10.1109/PPPS.2001.1002061.

[12] Gaudin, Olivier & Whitfield, Michael & Foord, John & Jackman, Richard. (2001). Deep level transient spectroscopy of CVD diamond: The observation of defect states in hydrogenated films. Diamond and Related Materials - DIAM RELAT MATER. 10. 610-614. 10.1016/S0925-9635(00)00418-0.

[13] Mitromara, N. & Evans-Freeman, J. & Gädtke, C. & May, P.. (2008). High resolution Laplace deep level transient spectroscopy of p-type polycrystalline diamond. Physica Status Solidi (a). 205. 2184-2189. 10.1002/pssa.200879710.

[14] Okushi, Hideyo & Tokumaru, Yozo. (1980). Isothermal Capacitance Transient Spectroscopy for Determination of Deep Level Parameters. Japanese Journal of Applied Physics. 19. L335-L338. 10.1143/JJAP.19.L335.

[15] Polyakov, V.I & Rukovishnikov, A.I. & Rossukanyi, N.M & Pereverzev, V.G & Pimenov, S.M & Carlisle, John & Gruen, Dieter & Loubnin, E.N. (2003). Charge-based deep level transient spectroscopy of undoped and nitrogen-doped ultrananocrystalline diamond films. Diamond and Related Materials. 12. 1776-1782. 10.1016/S0925-9635(03)00203-6.

[16] Giri, Pravat. (2002). Metastablility of Interstitial Clusters in Ion-Damaged Silicon Studied by Isothermal Capacitance Transient Spectroscopy. Defects and Diffusion Forum. 210-212. 1-14. 10.4028/www.scientific.net/DDF.210-212.1.

[17] Elsherif, Osama & Vernon-Parry, Karen & Evans-Freeman, J. & May, P. (2012). Effect of doping on electronic states in B-doped polycrystalline CVD diamond films. Semiconductor Science and Technology. 27. 065019.

[18] Tokuda, Yutaka & Nakamura, Wakana & Terashima, Hiroshi. (2006). Isothermal deep-level transient spectroscopy study of metastable defects in hydrogen-implanted n-type silicon. Materials Science in Semiconductor Processing. 9. 288-291. 10.1016/j.mssp.2006.01.053.

[19] Whitfield, Michael & Lansley, Stuart & Gaudin, Olivier & McKeag, Robert & Rizvi, Nadeem & Jackman, Richard. (2001). High-speed diamond photoconductors: A solution for high rep-rate deep-UV laser applications. Diamond and Related Materials. 10. 650-656. 10.1016/S0925-9635(00)00532-X.

[20] Nebel, Christoph & Zeisel, R. & Stutzmann, Martin. (2001). Space charge spectroscopy of diamond. Diamond and Related Materials - DIAM RELAT MATER. 10. 639-644. 10.1016/S0925-9635(00)00489-1.

[21] Pimenov, S. & Beloglazov, A. & Konov, V. & Karabutov, A.V. & Nassisi, V. & Polyakov, V.. (1997). Excimer laser-induced electron emission from diamond films. Diamond and Related Materials - DIAM RELAT MATER. 6. 1650-1657. 10.1016/S0925-9635(97)00043-5.

[22] Watt, G.A. & Newton, Mark & Baker, John. (2001). EPR and optical imaging of the growth-sector dependence of radiation-damage defect production in synthetic diamond. Diamond and Related Materials. 10. 1681-1683. 10.1016/S0925-9635(01)00395-8.

[23] Hu, Xiaojun & Ye, J.S. & Liu, H.J. & Mariazzi, Sebastiano & Brusa, R.s. (2008). A positron annihilation study on the defect properties of doped diamond films. Thin Solid Films. 516. 1699-1702. 10.1016/j.tsf.2007.05.016.

[24] Mallik, Kanad & Falster, R. & Wilshaw, Peter. (2004). Schottky diode back contacts for high frequency capacitance studies on

semiconductors. Solid-State Electronics. 48. 231-238. 10.1016/S0038-1101(03)00315-0.

[25] Okushi, Hideyo & Tokumaru, Yozo. (1981). Isothermal Capacitance Transient Spectroscopy. Japanese Journal of Applied Physics. 20. 261. 10.7567/JJAPS.20S1.261.

[26] Smallwood, Christopher & Ulbricht, Ronald & Day, Matthew & Schröder, Tim & Bates, Kelsey & Autry, Travis & Diederich, Geoffrey & Bielejec, Edward & Siemens, Mark & Cundiff, Steven. (2021). Hidden Silicon-Vacancy Centers in Diamond. Physical Review Letters. 126. 10.1103/PhysRevLett.126.213601.

[27] Imanishi, Shoichiro and Kudara, Ken and Ishiwata, Hitoshi and Horikawa, Kiyotaka and Amano, Shotaro and Iwataki, Masayuki and Morishita, Aoi and Hiraiwa, Atsushi and Kawarada, Hiroshi. Drain Current Density Over 1.1 A/mm in 2D Hole Gas Diamond MOSFETs With Regrown p++-Diamond Ohmic Contacts. (2021) Doi 10.1109/LED.2020.3047522.

Supplement Material

Transient capacity test.

For the bulk diamond sample, in the current mode, the transient curve with or without light have obvious difference. As the figure shown at 300 K , the UV light irradiated curve is smooth otherwise too noisy.Bias voltage UR is 20 V and pluse Voltage is -20V， the filling time is 100 μs.

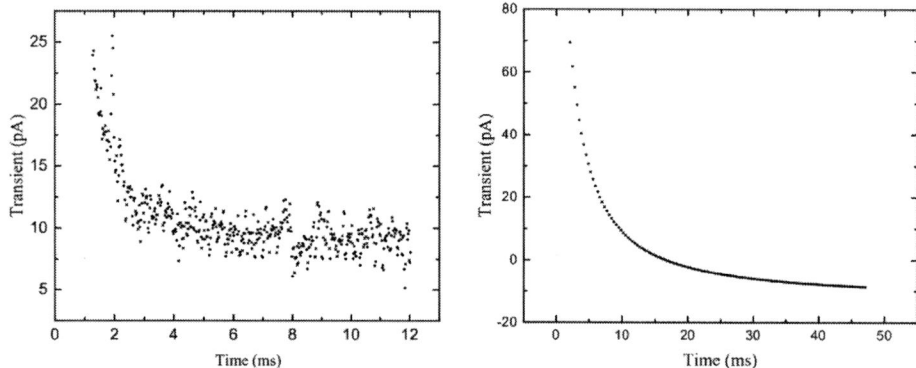

Fig. transient curve in dark (left)and with 405 nm light irradiation (right).

In generally speaking, the current mode less sensitivity than the traditional capacitance mode. Thus, for the high resistant sampel, the capacitance mode could not be optimized choose. So we use traditional current mode for investigate defect information at higher temperature region and optical DLTS at low temperature region.

Isothermal Transient Spectroscopy

To further investigate energy levels in diamond, we did continuous time isothermal transient spectroscopy from 298 K to 450 K with a 10 K step. The bias voltage is 20 V and the plus voltage is -10 V. The scan time range from 0.1ms to 20 s.

So the measurement of each temperature scan and corresponding arrhenius fitting are shown as follow:

298K

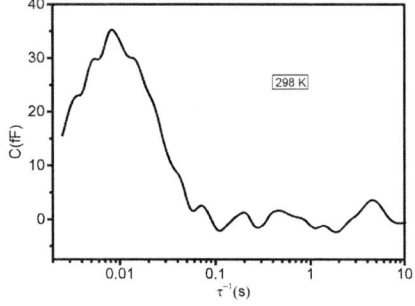
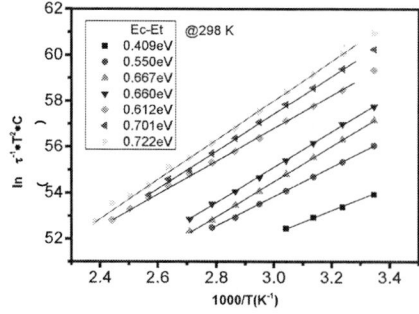

309 K

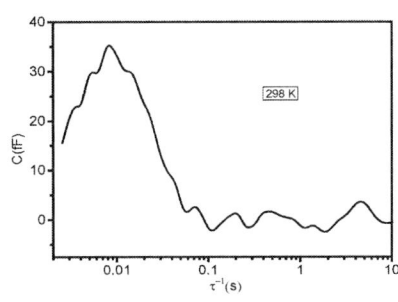 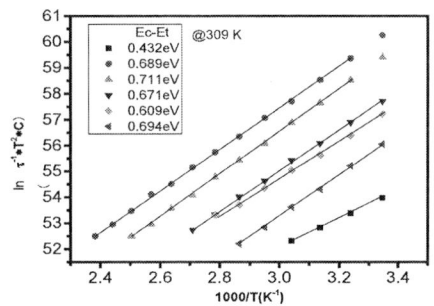

318 K

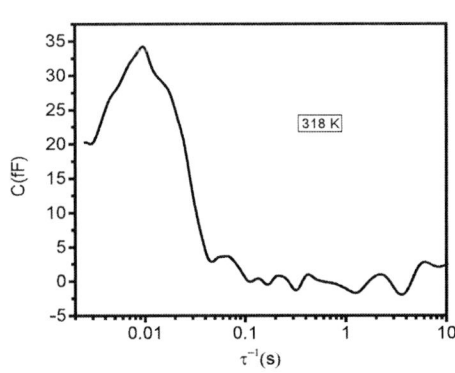 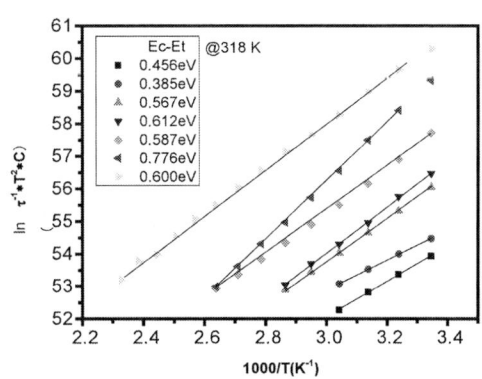

328 K

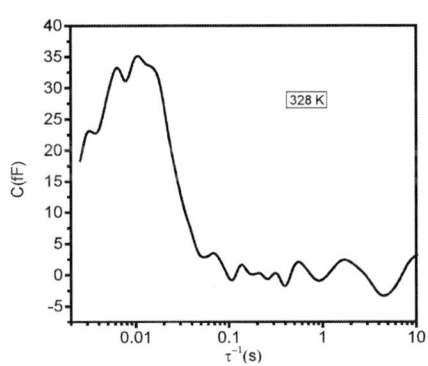 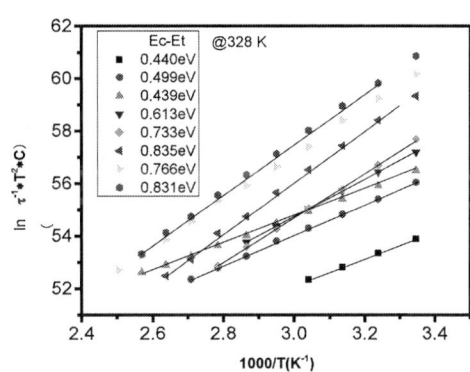

339 K

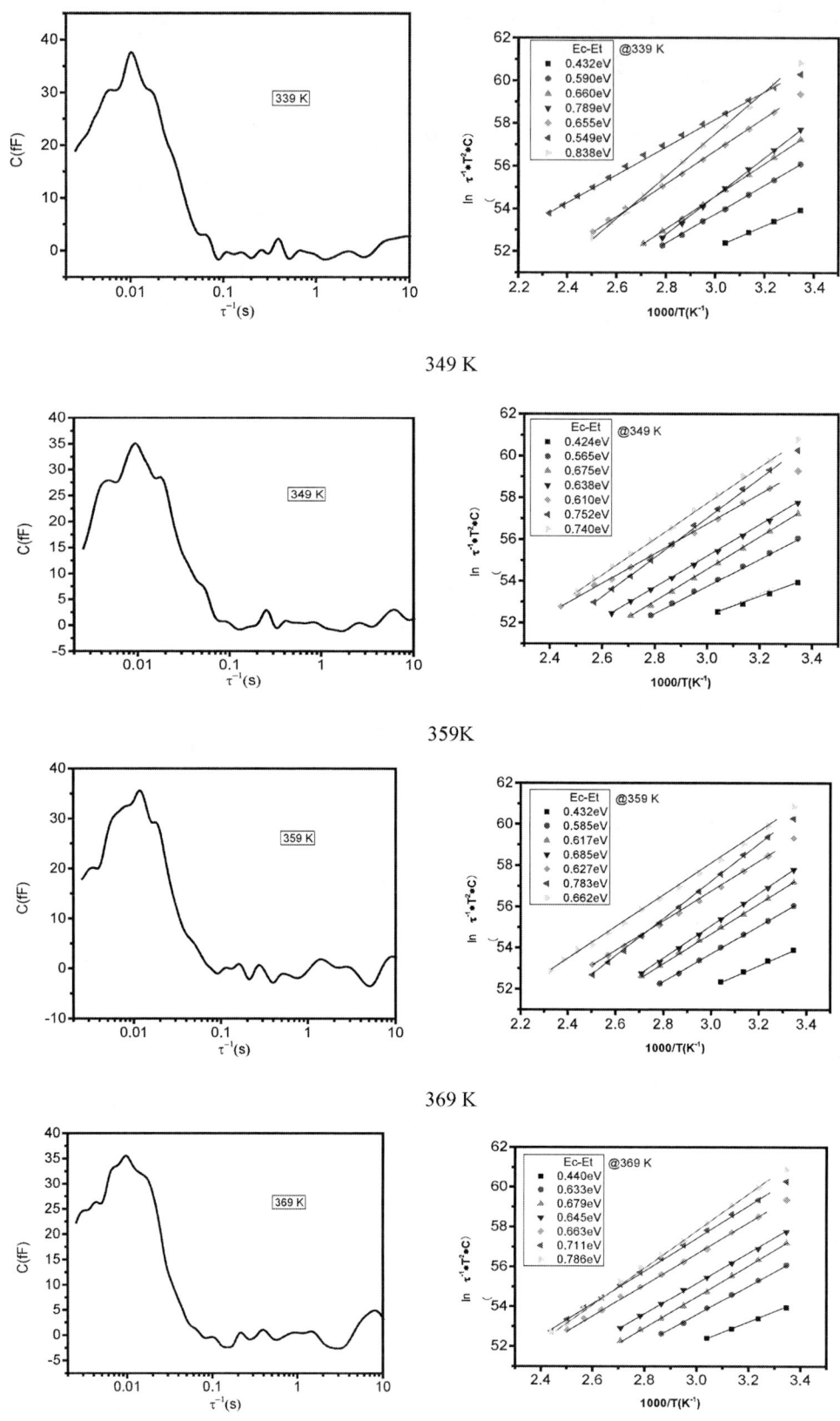

349 K

359K

369 K

379 K

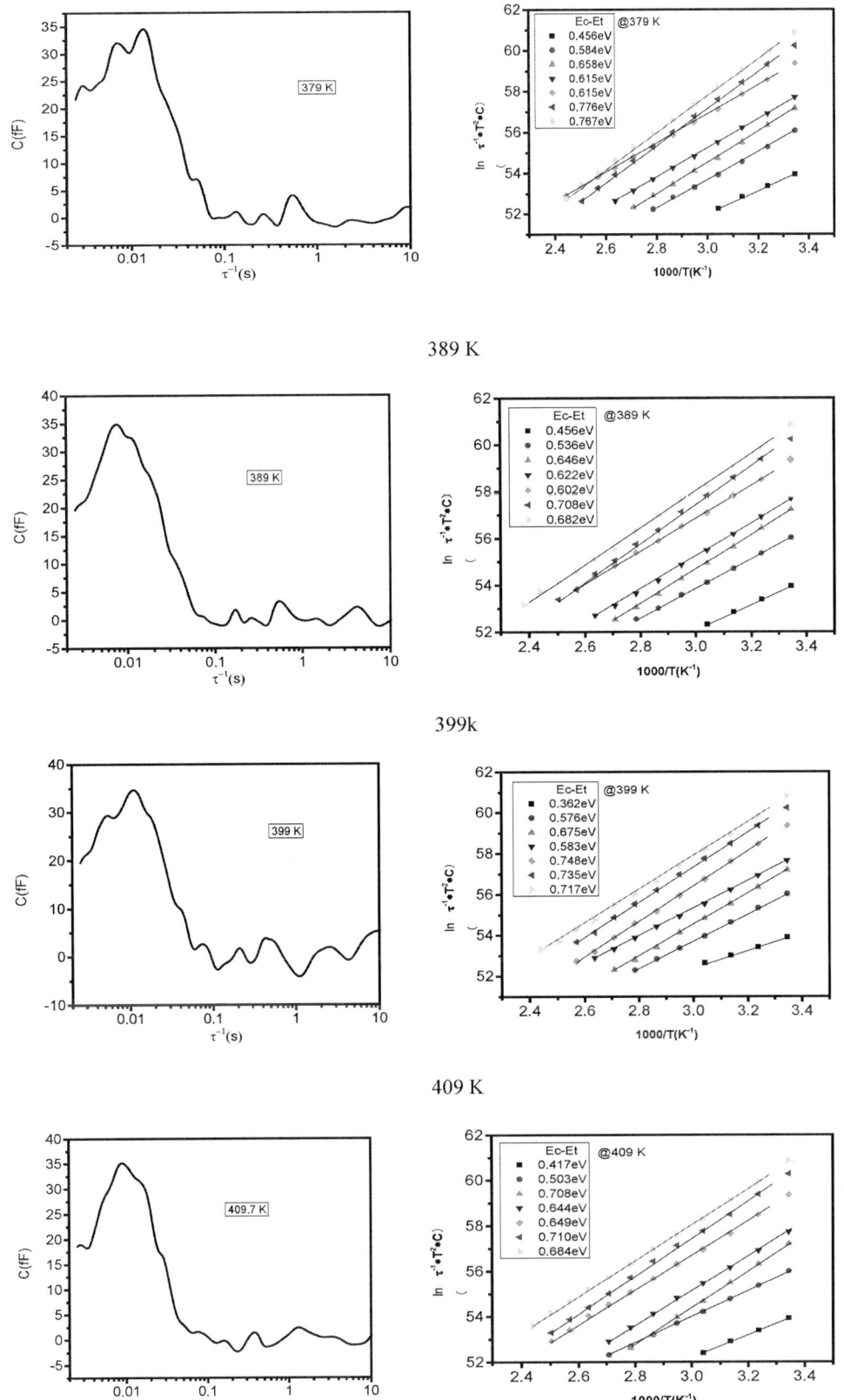

389 K

399k

409 K

978-1-6654-4817-8/21 $31.00 © 2021 IEEE

419 K

429 K

439 K

450 K

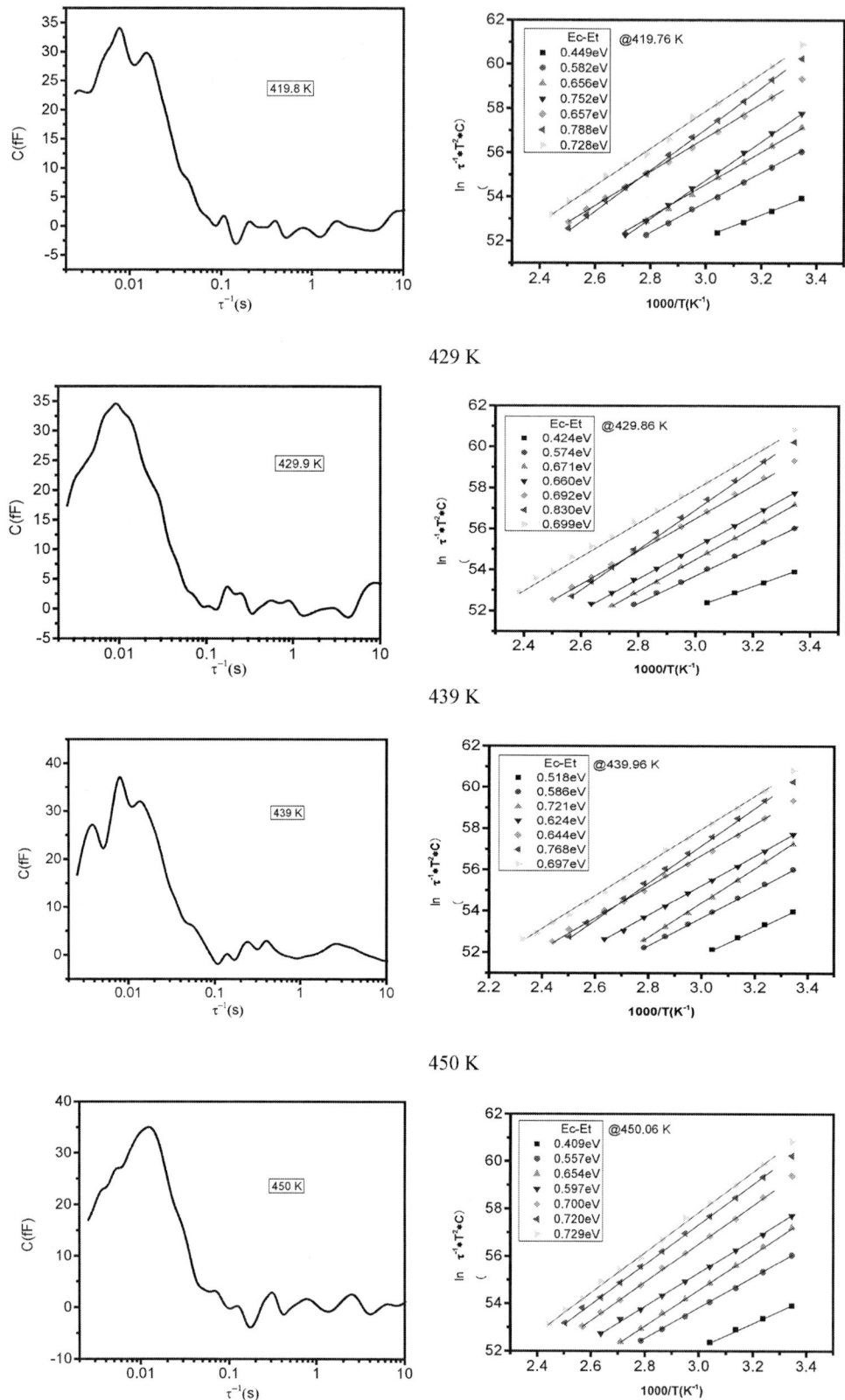

Power Semiconductor IGBT Packaging Technology and Reliability

Yameng Sun
The Institute of Technological Sciences
Wuhan University
Wuhan 430072, China
sunyameng @whu.edu.cn

Shizhao Wang
School of Power and Mechanical Engineering
Wuhan University
Wuhan 430072, China
wangshizhao@whu.edu.cn

Lianghao Xue
School of Power and Mechanical Engineering
Wuhan University
Wuhan 430072, China
lance_xue@whu.edu.cn

Zheng Feng
The Institute of Technological Sciences
Wuhan University
Wuhan 430072, China
fzheng_kite@163.com

Rui Li
The Institute of Technological Sciences
Wuhan University
Wuhan 430072, China
2019106520010@whu.edu.cn

Sheng Liu
School of Power and Mechanical Engineering
Wuhan University
Wuhan 430072, China
shengliu @whu.edu.cn

Abstract—Recently, IGBT has found an increasingly wide utilization, resulting in higher power density and thermal flux difficulty, which promotes the researches of new materials and structures to face up to the challenges of thermal reliability.

As the crucial step in the packaging process, the reflow process is also an important part of the quality control, while typical IGBT module will result in large residual stress and warpage due to various temperature characteristics and the thermal mismatch of packaging materials. Existing researches mainly improve the quality of welding by changing the control mode, however, the cost is high and the manufacture is difficult.

In this work, the thermal conduction of double-sided cooling IGBT modules in the reflow process and the reliability in the presence of voids were analyzed. Furthermore, transient temperature gradient was explored by a thermal simulation model, meanwhile, the reflow temperature curve of solder layer, the temperature gradient curve, and the influence of the void rate on the junction temperature of the chip was optimized and evaluated. The result indicates that the temperature gradient of the solder layer reduced by 50%, and the junction temperature of chip increased by approximately 3°C when the void rate changed from 0% to 3%.

Keywords—Packaging, IGBT, Reflow Process, Void Rate

I. INTRODUCTION

Although DSC-IGBT module can be used in the field of high current and voltage because of its excellent heat dissipation performance, the poor welding quality in the process of module manufacturing leads to solder voids which increase the junction temperature of the chip[1-3]. In order to improve the welding quality and reduce the junction temperature, some optimized packaging processes are gradually adopted, but reliability problems such as large temperature gradient and solder void began to appear with a high probability, seriously affecting the yield and reliability[4]. Due to complex components, thermal resistance of components in reflux process is large and the heat transfer is difficult[5]. Especially for the top layer solder, it can't reach the setting reflow temperature, the welding quality is poor, resulting in voids, and the internal temperature of the module is uneven with the accumulation of power cycle[6].

The packaging process is very important for IGBT module, which can maintain good electrical performance of

the module, and support and protect the module. As the name suggests, process mechanics is to explore the thermodynamic behavior of materials in the process. The concept of process mechanics was proposed by Sheng Liu and Yuhai Mei in 1993. In 1996, Liu and Wang[7] introduced the concept of process mechanics into microelectronic packaging analysis for the first time. In fact, process mechanics is to establish a multi-physical coupling model and optimize the process parameters and module structure according to the methods of mechanics, fluid, heat transfer and material science.

In this work, the thermal conduction of double-sided cooling IGBT modules in reflow process and the reliability in the presence of voids were analyzed. Furthermore, transient temperature gradient was explored by a thermal simulation model, meanwhile, the reflow temperature curve of solder layer, the temperature gradient curve, and the influence of the void rate on the junction temperature of chip was optimized and evaluated.

II. FINITE ELEMENT ANALYSIS

A. Study on temperature curve of reflow process

The packaging structure of DSC-IGBT module plays a decisive role in setting the reflow temperature curve. The thermal resistance makes the actual temperature of each component can't reach the setting reflow temperature curve, especially for the solder layer, the temperature difference will affect the quality of reflow soldering. It is necessary to analyze finite element analysis of reflow process, including steady-state thermal analysis and transient thermal analysis. The numerical analysis of transient temperature field in packaging process is necessary to optimize process parameters, which can analyze the influence factors such as structure sizes, material properties and boundary conditions.

In this paper, the three-dimensional model of the package structure with IGBT chip is established and the line AB is drawn, as shown in Fig.1. The size of IGBT chip is 3.88mm × 3.88mm × 07mm, the thickness of solder layer is 0.15mm. The reflow process is operated in the vacuum reflow furnace and carried out under the protection of nitrogen. Except for the bottom surface of the copper substrate, the rest of the outside are heat dissipated with nitrogen, and the nitrogen is under the condition of natural convection.

978-1-6654-4817-8/21 $31.00 © 2021 IEEE

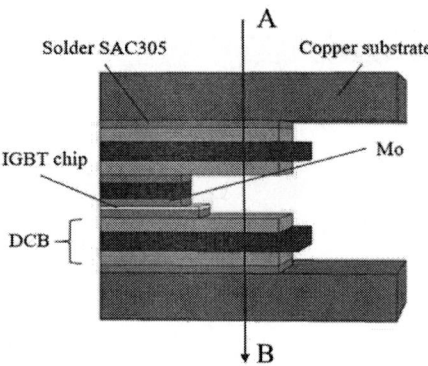

Fig.1. Finite element structure diagram of DSC-IGBT model and definition of line AB

The reflow curve shown in the Fig.2 is used as the load curve to heat load the substrate. The reflow time is 280s and the maximum temperature is 237 °C. In order to control the research variables, it is assumed that all solder layers are reflowed together without considering the welding sequence. At the same time, lead-free solder SAC305 is used, and solders are numbered from top to bottom, as shown in the Fig.3. The material thermophysical parameters of each component are shown in Table 1.

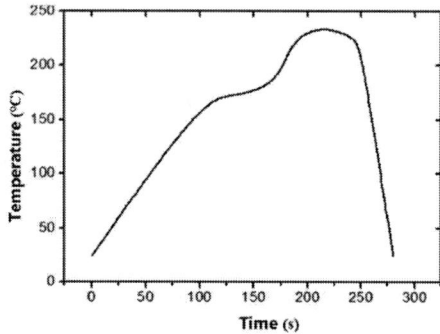

Fig.2. Reflow process temperature setting curve

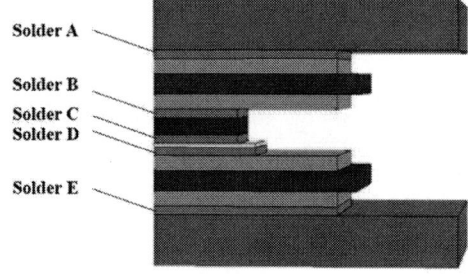

Fig.3. Solder layer number definition

B. Void failure analysis of solder layer

The reliability of the module after packaging can be predicted by the junction temperature of the chip. The effect of the void rate(θ) on the heat dissipation of the module is studied. It is assumed that the void of solder layer appears in the form of a single cavity.

III. RESULTS AND ANALYSIS

A. Thermal analysis results of reflow process

The simulation result of dsc-igbt module transient temperature is shown in Fig.4. In the preheating, holding and reflowing stages, the temperature gradients are within 5 °C. During the heating, the heat is transferred from bottom to top, and the temperature of the bottom copper substrate is the highest. Due to the thermal resistance of each component, the temperature of the uppermost copper substrate is the lowest. At the same time, compared with other components, the corner temperature of DBC is lower, which is due to the lower thermal conductivity of Al_2O_3.

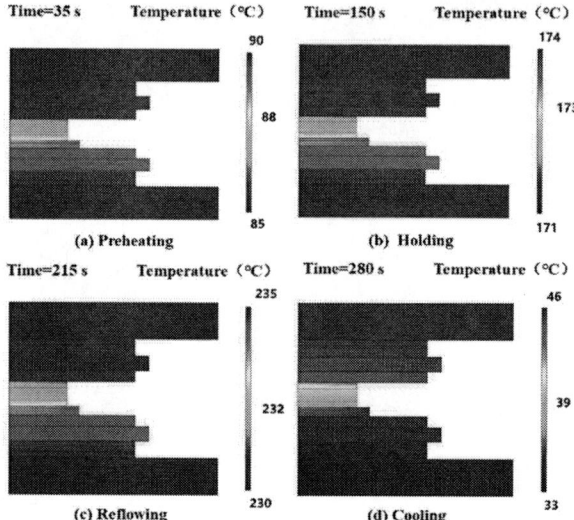

Fig.4. Transient temperature distribution in reflow stage

Fig.5 shows the temperature distribution of each solder layer when the reflow temperature reaches the peak. In each solder layer, the temperature uniformity of single solder is good, and the temperature gradient is not more than 2 °C. Moreover, the solder temperature gradient is the smallest at the position far away from the chip.

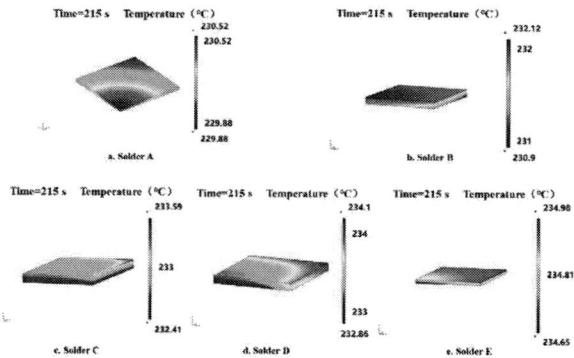

Fig.5. Transient temperature distribution of solder layer at peak temperature

In reflow process, it should be noted that the whole module has obvious temperature gradient. When the reflow process reaches the peak temperature, the temperature gradient between the components of dsc-igbt is shown in Fig.6. The gradient is the largest at the position of two segments of DBC ceramic layer. The existence of DBC ceramic layer will limit the heat conduction, which will lead to the solder layer can't reach the standard reflow temperature. At the same time, it is found that the temperature gradient of the first section is much larger than that of the second section, because it is close to the heat source.

It can be seen from Fig.6 that when the reflow temperature reaches the peak, the difference between the maximum temperature and the minimum temperature of the component is 6 °C, especially for the solder layer near point A, it is difficult to reach the standard reflow soldering temperature. Fig.7 shows the reflow curves of solder layers A, B, C, D and E in the reflow process. Because the thickness of solder layer in the module is thin and the temperature gradient of single layer is small, the temperature curve experienced by the center point of solder layer is selected as the reflow curve of solder layer. During the reflow process, the temperature of solder layer E is similar to that of the bottom of copper substrate, and the maximum temperature difference is 0.84 °C. With the increase of the linear distance from the copper substrate, the peak temperature of solder layer will gradually increase. The maximum temperature difference is solder layer A, and the temperature difference reaches 4.61 °C. The temperature difference between the upper solder layer and the copper substrate is too large, the solder layer can't reach the setting temperature, which has a bad impact on the welding quality.

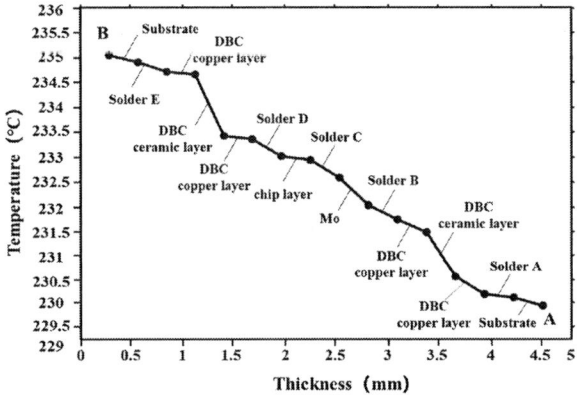

Fig.6. Transient temperature gradient between components at peak temperature

In order to improve the welding quality and make the solder layer experience the ideal temperature curve as much as possible, it is necessary to optimize the temperature curve in the reflow process. The temperature curve in the reflow furnace is optimized and compared with the standard temperature curve. As shown in Fig.8, the temperature curve of optimized solder layer a is close to the standard reflow temperature curve, the maximum temperature difference is only 2 °C, and the temperature difference during cooling is less than 5 °C than the original reflow process.

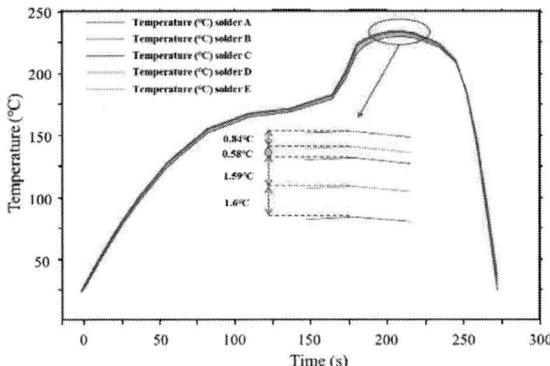

Fig.7. Temperature curve of solder layers during reflow process

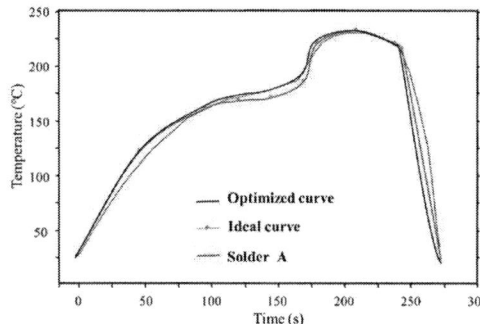

Fig.8. Comparison between optimized solder layer A temperature curve and ideal curve

B. Effect of void ratio on thermal performance of modules

Fig.9 shows that when the void ratio θ_C and θ_D of solder increase from 0% to 3%, the junction temperature of chip without void is 85.27 °C. When the void ratio of solder layer C and D is between 0% and 0.5%, the junction temperature increases very slowly. When the void ratio of solder layer C is more than 0.5%, the junction temperature of chip will change abruptly. On the contrary, the solder layer D will change abruptly when it is more than 1.5%. After the mutation, the junction temperature of the chip increases rapidly in direct proportion to the void ratio, but the effect of solder layer D is smaller. When the void rate of the solder layer reaches 3%, the junction temperature is 88.05 °C and 86.41 °C, and the chip is not overheating and fails.

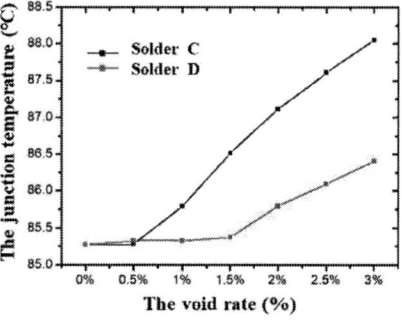

Fig.9. Influence of void ratio and solder layer position on chip junction temperature

IV. CONCLUSIONS

The thermal simulation of the reflow process of dsc-igbt module was established. During the heating process, the temperature gradient between the modules is within 5 °C. The results showed that there are a temperature gradient in the multilayer solder layer of the module, and it was proved that the ceramic layer in the two parts of DBC substrate had a relatively obvious temperature gradient during the reflow process.

The temperature setting curve of reflow process was optimized, so that the temperature of solder layer could reach the set temperature of reflow process, and the maximum temperature difference was no more than 1 °C, which effectively improved the welding quality.

The void ratio of solder layer increased slowly at the beginning. However, when the void ratio exceeded a certain value, the junction temperature of the chip would change abruptly, which helped to improve the junction temperature of the chip when the void ratio was within 0% ~ 3%.

TABLE I.   THERMOPHYSICAL PARAMETERS OF MATERIALS

Component	Material Properties	
	Thermal conductivity [W/(m·K)]	Specific heat capacity [J/(kg·K)]
Chip	124	713
Solder	57	217
DBC (96%Al$_2$O$_3$)	25	880
Mo	138	250

ACKNOWLEDGMENT

This work was supported by the National Natural Science Foundation of China (Grant Nos. 51727901 and U1501241), the National Key R&D Program of China (No. 2017YFB1103904), and the Hubei Provincial Natural Science Foundation of China under Grant No. 2020CFA032.

REFERENCES

[1] H. Guojun and A. Tay, "On the relative contribution of temperature, moisture and vapor pressure to delamination in a plastic IC package during lead-free solder reflow[C]," IEEE Electronic Components and Technology Conference, 2005, pp.172-178.

[2] Y. E. Huang, D. Hagen and G. Dody, "Effect of solder reflow temperature profile on plastic package delamination[C]," IEEE Electronics Manufacturing Technology Symposium, 1998, pp.105-111.

[3] D. C. Katsis and J. D. V. Wyk, "Void-induced thermal impedance in power semiconductor modules: Some transient temperature effects[J]," IEEE Transactions on Industry Applications, 2003, 39 (5): 1239-1246.

[4] D. S. Kim, Y. Qiang, T. Shibutani, N. Sadakata, T. Inoue, "Effect of void formation on thermal fatigue reliability of lead-free solder joints[C]," IEEE Thermal and Thermomechanical Phenomena in Electronic Systems, 2004, pp.325-329.

[5] R. Shook and V. Sastry, "Influence of preheat and maximum temperature of the solder-reflow profile on moisture sensitive IC's[C]," IEEE Electronic Components and Technology Conference, 1997, pp.1041-1048.

[6] P. P. Paret, D. J. Devoto and S. Narumanchi, "Reliability of emerging bonded interface materials for large-area attachments[J]," IEEE Transactions on Components Packaging Manufacturing Technology, 2015, 6(1): 40-49.

[7] W. Jianjun and L. Sheng, "Sequential processing mechanics modeling for a model IC package[J]," IEEE Transactions on Components Packaging and Manufacturing Technology, 1997, 20 (4): 335-342.

978-1-6654-4817-8/21 $31.00 © 2021 IEEE

Research on A Novel Parallel Resonant DC Link Soft Switching Inverter Based on SiC MOSFET

1st Si Li
School of Electrical Engineering and Automation
Harbin Institute of Technology (HIT)
Harbin, China
13080819608@163.com

2nd Ming Yang
School of Electrical Engineering and Automation
Harbin Institute of Technology (HIT)
Harbin, China
yangming@hit.edu.cn

3rd Yu Ma
School of Electrical Engineering and Automation
Harbin Institute of Technology (HIT)
Harbin, China
hiteemy@163.com

4th Dianguo Xu
School of Electrical Engineering and Automation
Harbin Institute of Technology (HIT)
Harbin, China
xudiang@hit.edu.cn

Abstract—In order to realize a silicon carbide (SiC) MOSFET soft switching inverter with low loss, low stress and easy to design, a parallel resonant DC link (PRDCL) soft switching inverter topology is adopted in this paper. In this topology, the switches in the main inverter circuit realize zero voltage (ZV) and zero current (ZC) turn-on and -off, which is lossless; The switches in the auxiliary circuit realize quasi-ZC turn-on and quasi-ZV turn-off, which helps to significantly reduce switching losses. The topology is convenient to control without setting inductor current threshold. And it avoids using coupled inductor and split capacitor which reduce power density, and the latter will also bring neutral point potential drift. And its control is simple and the parameter design only depends on the DC power supply voltage and load/bus current peak value. In this paper, according to the commutation law of bus current, the working mode is distinguished and resonance parameters and delay time selection are introduced. At last, the 5kW/16kHz inverter prototype is made by adopting SiC MOSFET as switching device, and the soft switching characteristics are verified by experimental results.

Keywords—SiC MOSFET, soft switching, zero current (ZC), zero voltage (ZV), parallel resonant DC link inverter (PRDCLI)

I. INTRODUCTION

Historically, the innovation and development in the field of power electronics are inseparable from power electronic devices. In recent years, with the continuous maturity of wide bandgap devices represented by silicon carbide (SiC) devices, wide bandgap inverters applied in switching power supply, new energy grid connection, motor drive and other occasions have gradually become the research focus[1]-[3]. However, when the switching frequency of the wide bandgap inverter is hundreds of kilohertz or even tens of kilohertz, its switching loss also increases rapidly with the increase of switching frequency[4]-[5]. In order to reduce the loss and electromagnetic interference of the wide bandgap inverter under high switching frequency, soft-switching technology is a way worth exploring.

According to the position of the auxiliary circuit, the soft switching inverter is distinguished as auxiliary resonant commutated pole inverter (ARCPI)[6]-[7] and resonant DC link inverter (RDCLI)[8]-[13]. Because the RDCLI own the merits of succinct circuit topology and easy control mode, experts in

this field attach great importance to it, which is a key research direction to soft switching inverter at present.

The research of RDCLI starts from [8]. In response to the challenge of switch device voltage stress and discrete pulse modulation strategy, active clamped RDCLI (ACRDCLI)[9]-[10] appears, but its resonant inductor is located on DC bus, which leads to increased conduction loss. Therefore, parallel RDCLI (PRDCLI)[11]-[13] appears. However, there are still some defects in the existing PRDCLIs, such as setting inductor current threshold to increase control complexity; The split capacitor with neutral point potential drift is used as auxiliary power supply; Using coupled magnetic elements increases design difficulty and reduces power density.

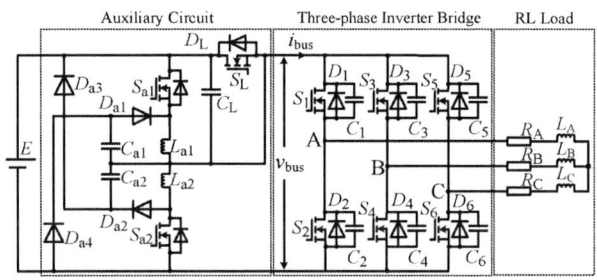

Fig. 1. Novel PRDCLI circuit topology

In view of some shortcomings of the current PRDCLI circuit topology, this paper adopts the novel PRDCLI circuit topology[14]-[15] (shown as the Fig.1), which has the following characteristics: 1) There is no series capacitor on DC bus, which avoids the problem of neutral point potential change; 2) The auxiliary resonant circuit has no resonant threshold limit; 3) Coupling magnetic elements are not needed, so there is no need to consider the turns ratio and magnetizing current reset problem; 4) The main switch realize zero voltage (ZV) and zero current (ZC) turn-on and -off, the bus switch realizes ZVZC turn-on and quasi-ZV turn-off and the auxiliary switch realizes quasi-ZC turn-on and quasi-ZV turn-off; 5) According to the requirements of PWM modulation and soft-switching operation of inverter, the DC bus voltage can cross zero at any time for the required time, and various modulation strategies can be adopted flexibly.

This work is sponsored by Project 51991385 supported by National Natural Science Foundation of China.

978-1-6654-4817-8/21 $31.00 © 2021 IEEE

II. COMMUTATION LAW OF BUS CURRENT

Fig. 1 shows that bus current i_{bus} actually participates in the commutation process of auxiliary circuit, and the magnitude and direction of bus current i_{bus} depend on single-phase load current and switching state of inverter.For explaining the commutation process clearly, the commutation law of bus current i_{bus} is analyzed below.

The inverter is connected with resistive inductive load, and under unit power factor, the waveforms of three-phase modulation wave, three-phase load current and bus current i_{bus} in power frequency cycle are shown in Fig. 2, where colored solid lines respectively represent three-phase modulation waves and colored dotted lines respectively represent three-phase load currents, and the solid black line represents the envelope of bus current i_{bus}. The abscissa is the electrical angle corresponding to time.

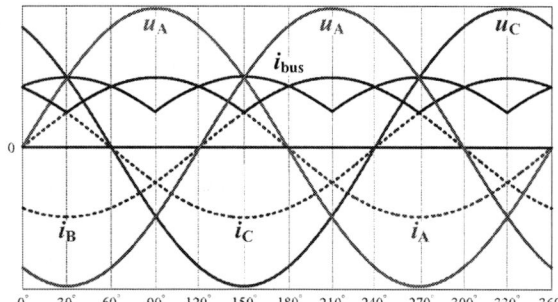

Fig. 2. Waveform of three-phase modulation wave, three-phase load current and bus current

In the [0°, 30°] interval as shown in Fig. 2, in a switching cycle T_s, the three-phase modulation wave is compared with the triangular carrier to generate the trigger pulse of the main switch of the upper bridge leg of each phase, and the corresponding switching states are shown in Fig. 3.

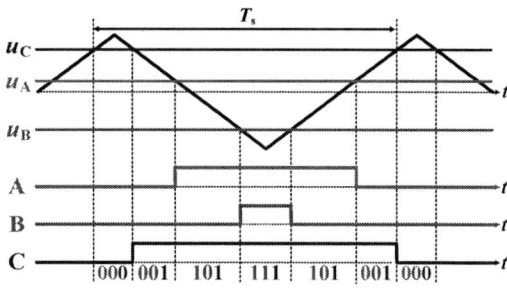

Fig. 3. Trigger pulse and switching state of main switch of upper leg on each phase in [0°, 30°] interval

According to the single-phase load current direction in the [0°, 30°] interval in Fig. 2 and the switching state in the [0°, 30°] interval in Fig. 3, the circuit working mode in each switching state in the [0°, 30°] interval can be drawn as shown in Fig. 4, and the bus current i_{bus} in each switching state can be further obtained. From Fig. 4, we can know that the variation law of i_{bus} in the [0°, 30°] interval is as follows:

$$0_{000} \rightarrow +i_C \rightarrow -i_B \rightarrow 0_{111} \rightarrow -i_B \rightarrow +i_C \rightarrow 0_{000} \quad (1)$$

Where $+i_C$ represents C phase load current and $-i_B$ represents B phase load current inversion; 0_{000} means bus

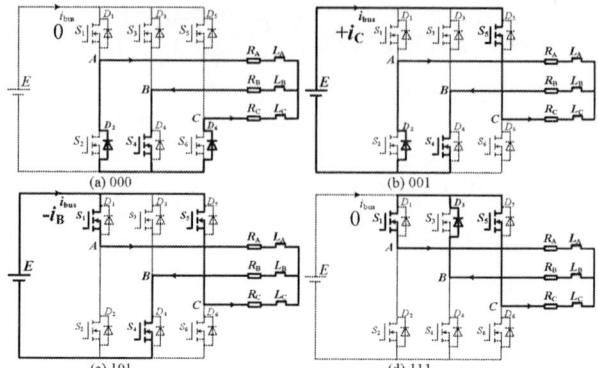

Fig. 4. Circuit working mode and bus current in each switching state in [0°, 30°] interval

current is zero in switching state 000, and 0_{111} means bus current is zero in switching state 111.

The variation law of bus current i_{bus} in other intervals in Fig. 2 can be analogized. The variation law of i_{bus} in each interval in the whole power frequency cycle is summarized in Table I. Combining Fig.2 and Table I, it can be known that: 1) In the whole power frequency cycle, there is always a case where the i_{bus} is zero; 2) In each 60° interval, commutation occurs between two phases with larger modulation amplitude (absolute value); 3) In the switching state located on the left and right sides of the switching state 000, the i_{bus} is the single-phase load current flowing through the phase with the largest modulation amplitude; In the switching state located on the left and right sides of the switching state 111, the i_{bus} is the inverse of the single-phase load current flowing through the phase with the second largest modulation amplitude.

TABLE I.   COMMUTATION LAW OF BUS CURRENT

Interval	Variation law of bus current i_{bus}
$\left[0°, 30°\right] \cup \left[330°, 360°\right]$	$0_{000} \rightarrow +i_C \rightarrow -i_B \rightarrow 0_{111} \rightarrow -i_B \rightarrow +i_C \rightarrow 0_{000}$
$\left[30°, 90°\right]$	$0_{000} \rightarrow +i_A \rightarrow -i_B \rightarrow 0_{111} \rightarrow -i_B \rightarrow +i_A \rightarrow 0_{000}$
$\left[90°, 150°\right]$	$0_{000} \rightarrow +i_A \rightarrow -i_C \rightarrow 0_{111} \rightarrow -i_C \rightarrow +i_A \rightarrow 0_{000}$
$\left[150°, 210°\right]$	$0_{000} \rightarrow +i_B \rightarrow -i_C \rightarrow 0_{111} \rightarrow -i_C \rightarrow +i_B \rightarrow 0_{000}$
$\left[210°, 270°\right]$	$0_{000} \rightarrow +i_B \rightarrow -i_A \rightarrow 0_{111} \rightarrow -i_A \rightarrow +i_B \rightarrow 0_{000}$
$\left[270°, 330°\right]$	$0_{000} \rightarrow +i_C \rightarrow -i_A \rightarrow 0_{111} \rightarrow -i_A \rightarrow +i_C \rightarrow 0_{000}$

III. WORKING PRINCIPLE

The working principle of auxiliary circuit is expounded by taking the commutation of bus current i_{bus} from $+i_C$ to $-i_B$ in [0°, 30°] interval as an example, and the operation principle of auxiliary circuit in other intervals is similar. The hypothetical conditions are as follows to simplify the analysis: 1) The devices used in the circuit topology are ideal devices that do not consider the effects of parasitic parameters; 2) switching frequency f_s >> output AC frequency f_o, so bus current i_{bus} remains unchanged in a switching state; 3) $C_L = C_a$, $C_{a1} = C_{a2} = C_b$, $L_{a1} = L_{a2} = L$, $C_1 - C_6$ is equal to C_s and $C_L = 3C_s = C_a$. Fig. 5 shows the sequence of the trigger pulses of each switch. Fig. 6 and Fig. 7 respectively show the key working waveforms and equivalent circuits of the adopted soft switching topology under the switching sequence shown in Fig. 5. Working modes are briefly described as follows:

978-1-6654-4817-8/21 $31.00 © 2021 IEEE

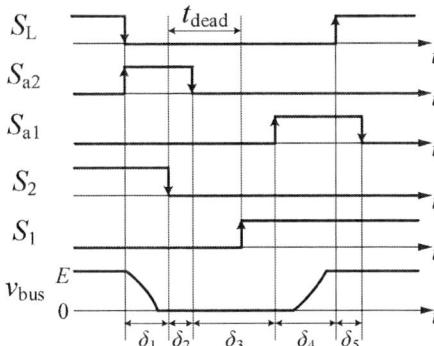

Fig. 5. Modulation strategy

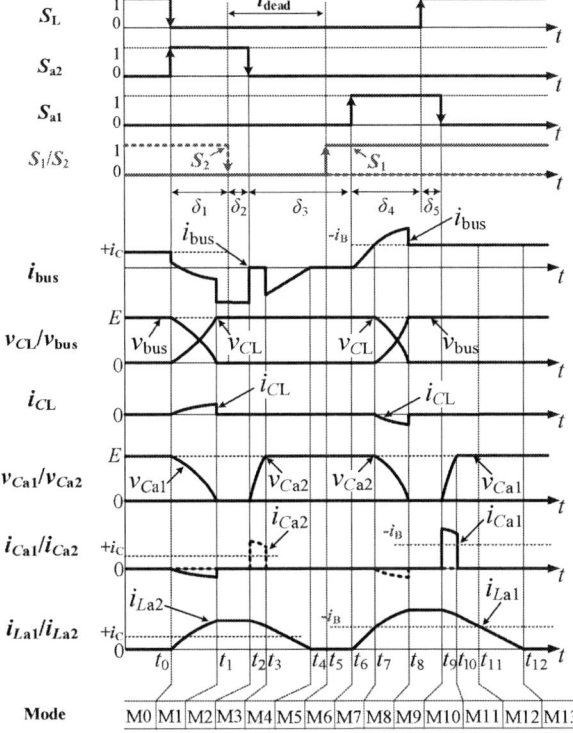

Fig. 6. Key working waveform

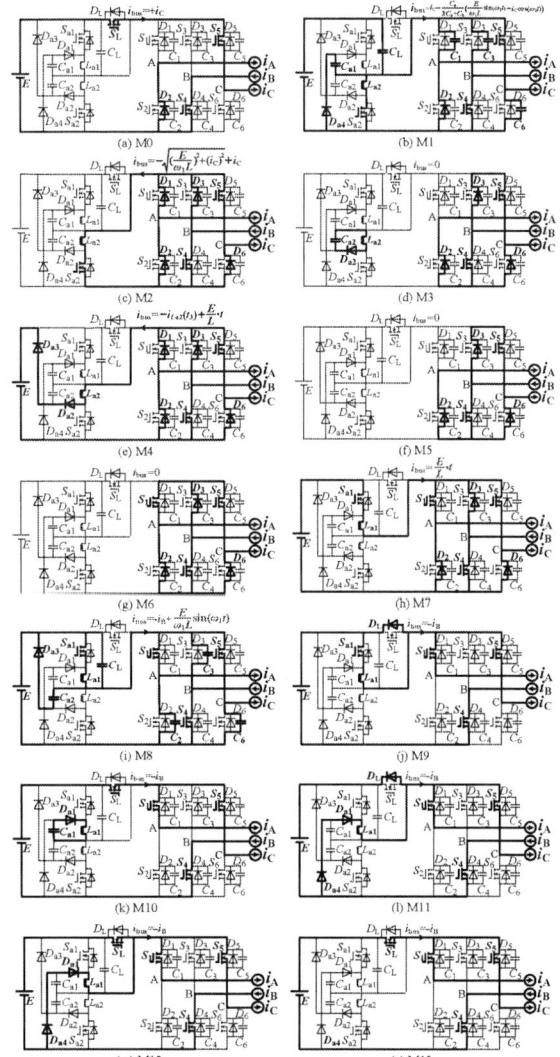

Fig. 7. Working modes

1) Mode M0 [~ t_0]: before t_0, S_L conducts, S_{a1} and S_{a2} are turned off, and the switching state of three-phase inverter bridge at this time is 001.

2) Mode M1 [t_0, t_1]: when t_0 comes, S_L is turned off and S_{a2} is turned on and D_{a4} conducts. C_L, C_{a1}, C_1, C_3 and C_6 resonate with L_{a2}. Under the restrictions of C_L, C_{a1}, C_1, C_3 and C_6, v_{SL} cannot be mutated, and S_L realizes quasi-ZV turn-off. Under the restriction of L_{a2}, i_{Sa2} slowly rises resonally from zero, and S_{a2} realizes quasi-ZC turn-on.

3) Mode M2 [t_1, t_2]: At t_1, D_{a4} stops conducting, D_1, D_3 and D_6 starts to conduct, and i_{La2} reaches the maximum i_{La2max}. Inverter circuit circulates through D_3 and D_6, and i_{La2} circulates in the loop formed by L_{a2} and S_{a2} and D_1, D_3, D_6 with constant i_{La2max}. In this period, S_2 is turned off, and the voltage at both ends of S_2 is kept at zero, and S_2 realizes ZVZC turn-off.

4) Mode M3 [t_2, t_3]: At t_2, S_{a2} is turned off, D_{a2} starts to conduct, L_{a2} resonates with C_{a2}, under the restriction of C_{a2}, v_{Sa2} cannot mutate, so S_{a2} realizes quasi-ZV turn-off.

5) Mode M4 [t_3, t_4]: At t_3, D_{a3} starts to conduct. L_{a2} feeds back energy to the DC power supply by D_{a2}, D_{a3} and D_1, D_3, D_6.

6) Mode M5 [t_4, t_5]: At t_4, D_{a2} and D_{a3} stop conducting, and the inverter circuit circulates through D_3 and D_6.

7) Mode M6 [t_5, t_6]: At t_5, S_1 is turned on, and the voltage at both ends of S_1 is kept at zero, so S_1 realizes ZVZC turn-on. Inverter circuit circulates through D_2, D_3 and D_6. From t_1 to t_6, the DC bus voltage keeps zero, which provides conditions for the main switch to realize ZVZC switching, and the time for the DC bus voltage to keep zero can be arbitrarily set by freely selecting the turn-on time of S_{a1} as required, that is, adjusting the δ_3 time.

8) Mode M7 [t_6, t_7]: At t_6, S_{a1} is turned on, and bus current i_{bus} is commutated from D_2, D_3 and D_6 to L_{a1} in turn. Under the restriction of L_{a1}, i_{Sa1} cannot be mutated, then S_{a1} realizes quasi-ZC turn-on.

978-1-6654-4817-8/21 $31.00 © 2021 IEEE   119

9) Mode M8 [t_7, t_8]: At t_7, D_2, D_3 and D_6 stop conducting, D_{a3} starts to conduct, and L_{a1} resonates with C_L, C_{a2} and C_2, C_3 and C_6. C_2, C_3 and C_6 are charged, while C_L and C_{a2} are discharged.

10) Mode M9 [t_8, t_9]: at t_8, D_{a3} stops conducting then D_L starts to conduct. i_{La1} reaches the maximum value i_{La1max}. At this time, i_{La1max} is divided into two parts, one part is responsible for maintaining bus current i_{bus}, and the other part circulates in the loop formed by L_{a1}, D_L and S_{a1} in the form of resonance current. The turn-on of D_L creates ZVZC turn-on condition for S_L.

11) Mode M10 [t_9, t_{10}]: at t_9, S_{a1} is turned off, D_L stops conducting, bus current i_{bus} flows through S_L, L_{a1} and C_{a1} starts resonate. Under the restriction of C_{a1}, v_{Sa1} cannot mutate, and S_{a1} realizes quasi-ZV turn-off.

12) Mode M11 [t_{10}, t_{11}]: at t_{10}, D_{a4} and D_L start to conduct. L_{a1} feeds back energy to DC power supply through D_{a1}, D_{a4} and D_L.

13) Mode M12 [t_{11}, t_{12}]: at t_{11}, D_L stops conducting. Bus current i_{bus} starts commutation from L_{a1} to S_L.

14) Mode M13 [t_{12}, ~]: at t_{12}, i_{La1} drops to zero, bus current i_{bus} commutation ends, and M12 ends. After this mode ends, the inverter circuit enters the switching state 101.

IV. PARAMETER DESIGN

When designing parameters, from the perspective of reducing switching losses, the difference between the current change rate when each switch is turned on and the voltage change rate when it is turned off should be fully considered.

1) The current change rates di/dt_{Sa1} and di/dt_{Sa2} during the turn-on progress of S_{a1} and S_{a2} are:

$$di/dt_{S_{a1}} = di/dt_{S_{a2}} = \frac{E}{L} \qquad (2)$$

2) The voltage change rate dv/dt_{Sa2} during the turn-off progress of S_{a2} is:

$$dv/dt_{S_{a2}} = \frac{1}{C_b} \cdot \left(E \cdot \sqrt{\frac{2C_a + C_b}{L}} \right) \qquad (3)$$

3) The voltage change rate dv/dt_{Sa1} during the turn-off progress of S_{a1} is:

$$dv/dt_{S_{a1}} = \frac{1}{C_b} \cdot \left(E \cdot \sqrt{\frac{2C_a + C_b}{L}} + I_{busmax} \right) \qquad (4)$$

4) The voltage change rate dv/dt_{SL} during the turn-off progress of S_L is:

$$dv/dt_{S_L} = \frac{E}{\sqrt{(2C_a + C_b)L}} \cdot \sin \omega t + \frac{I_{busmax}}{2C_a + C_b} \cdot \cos \omega t \quad (5)$$

E=200V and I_{busmax}=40A are assumed. To reduce the switching loss during the turn-on process, $(di/dt)_{set}$=50A/μs is assumed, and (2) must be less than or equal to $(di/dt)_{set}$, that is, $L \geq 4$μH, so taking L=4μH. To reduce the switching loss during the turn-off process, $(dv/dt)_{set}$=800V/μs is assumed, and (3)-(5) must be less than or equal to $(dv/dt)_{set}$. Through mathematical comparison, it can be found that (4) is larger than (3) and (5), so $(4) \leq (dv/dt)_{set}$ is ok. According to (4), $C_a \leq 3.2 \times 10^7 C_b^2 - 3.7C_b + 8 \times 10^{-8}$, and taking C_b=10nF, C_a=45nF.

When the resonance parameters are determined, the key delay time is determined by (6)-(8):

$$\delta_1 \geq t_{0-1} = \frac{\pi}{2} \cdot \sqrt{(2C_a + C_b)L} \qquad (6)$$

$$\delta_3 \geq t_{2-3} + t_{3-4} = \frac{1}{\omega_2} \cdot \arcsin\left(\frac{E}{\omega_2 L i_{La2max}}\right) + \frac{L}{E} \cdot i_{La2}(t_3) \quad (7)$$

$$\delta_4 \geq t_{6-7} + t_{7-8} = \frac{L}{E} \cdot I_{busmax} + \frac{\pi}{2} \cdot \sqrt{(2C_a + C_b)L} \qquad (8)$$

The key delay time can be easily obtained by substituting resonance parameters into the above equations: $\delta_1 \geq 0.99$μs, $\delta_4 \geq 1.79$μs, $\delta_3 \geq 1.05$μs, so taking δ_1=1.00μs, δ_4=1.80μs, and δ_3=1.10μs. As long as the rest of the delay time exists, taking $\delta_2 = \delta_5$=0.10μs.

V. EXPERIMENTAL VERIFICATION

To verify the correctness of the working principle and parameter design, the 5kW/16kHz inverter prototype of the novel PRDCLI based on SiC MOSFET is made. Fig. 8 shows the experimental platform, the control board and the front and back of the power board.

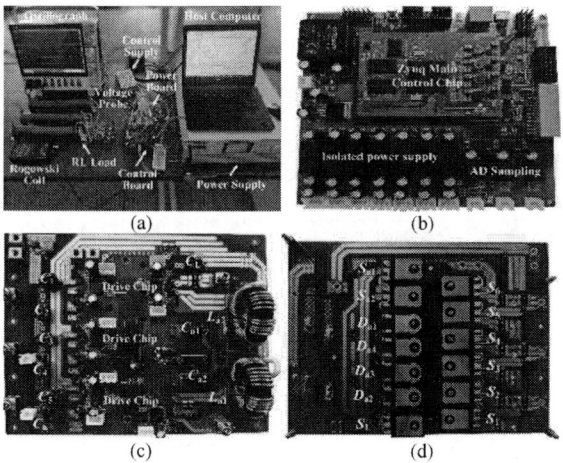

Fig. 8. The photos of experimental prototype

Fig. 9(a) and Fig. 9(b) respectively show the voltage and current waveforms when the S_1 is turned on and off. Fig. 9(c) and Fig. 9(d) respectively show the voltage and current waveforms when the S_L is turned on and off. Fig. 9(e) and Fig. 9(f) respectively show the voltage and current waveforms when the S_{a1} is turned on and off. It can be seen that the S_1 is turned on/off by ZVZC. And the S_L is turned on by ZVZC and turned off by quasi-ZV. And the S_{a1} is turned on by quasi-ZC and turned off by quasi-ZV.

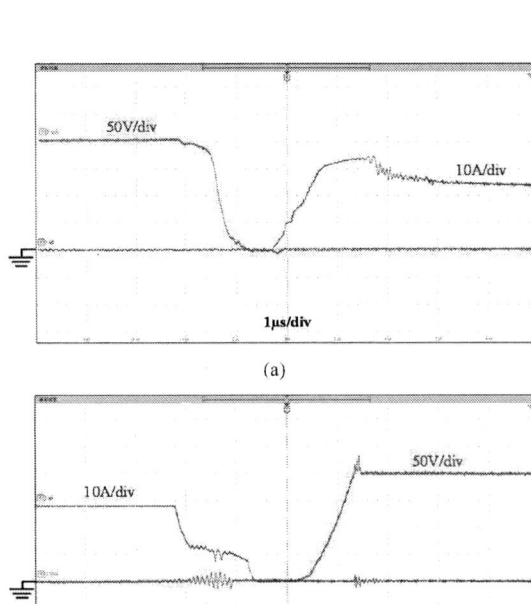

(a)

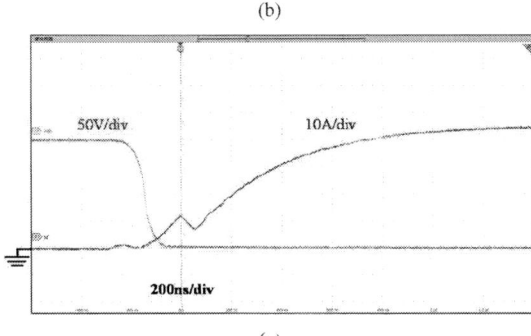

(b)

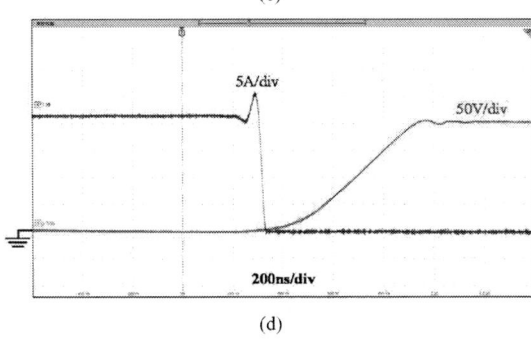

(c)

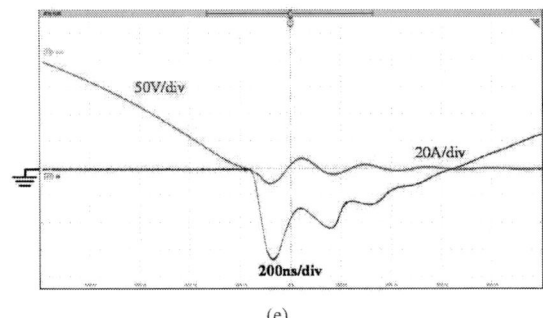

(d)

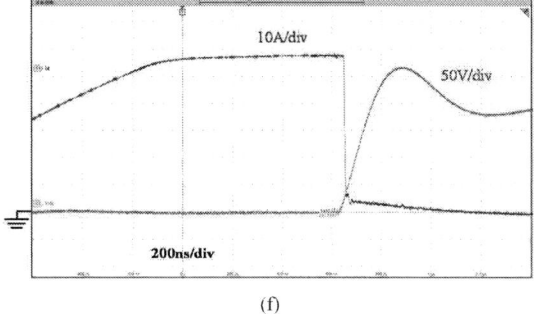

(e)

(f)

Fig. 9. Experimental waveforms

VI. CONCLUSION

A novel PRDCLI based on SiC MOSFET is studied in this paper. The commutation law of bus current, the working principle and the parameter design are given. The novel PRDCLI based on SiC MOSFET has the following advantages: 1) Two huge capacitors located in DC link are omitted, so there is no neutral point potential imbalance phenomenon; 2) The auxiliary circuit has no resonant threshold limit; 3) There is no magnetic element such as coupled inductor or transformer; 4) All switching devices have achieved soft turn-on and turn-off, which is conducive to improving power density.

REFERENCES

[1] M. Yang, Z. Lyu, D. Xu, J. Long, S. Shang, P. Wang and D. Xu, "Resonance Suppression and EMI Reduction of GaN-Based Motor Drive With Sine Wave Filter," *IEEE Transactions on Industry Applications*, vol. 56, no. 3, pp. 2741-2751, 2020.

[2] N. He, M. Chen, J. Wu, N. Zhu and D. Xu, "20-kW Zero-Voltage-Switching SiC-mosfet Grid Inverter With 300 kHz Switching Frequency," *IEEE Transactions on Power Electronics*, vol. 34, no. 6, pp. 5175-5190, 2019.

[3] C. Chen, Y. Chen, Y. Tan, J. Fang, F. Luo and Y. Kang, "On the Practical Design of a High Power Density SiC Single-Phase Uninterrupted Power Supply System," *IEEE Transactions on Industrial Informatics*, vol. 13, no. 5, pp. 2704-2716, 2017.

[4] F. Xu, B. Guo, Z. Xu, L. M. Tolbert, F. Wang and B. J. Blalock, "Paralleled Three-Phase Current-Source Rectifiers for High-Efficiency Power Supply Applications," *IEEE Transactions on Industry Applications*, vol. 51, no. 3, pp. 2388-2397, 2015.

[5] S. Safari, A. Castellazzi and P. Wheeler, "Experimental and Analytical Performance Evaluation of SiC Power Devices in the Matrix Converter," *IEEE Transactions on Power Electronics*, vol. 29, no. 5, pp. 2584-2596, 2014.

[6] A. Charalambous, X. Yuan and N. McNeill, "High-Frequency EMI Attenuation at Source With the Auxiliary Commutated Pole Inverter," *IEEE Transactions on Power Electronics*, vol. 33, no. 7, pp. 5660-5676, 2018.

[7] D. Ma, P. Wang, R. Wang, S. Li and Q. Sun, "Hybrid SVPWM Modulation Strategy for Auxiliary Resonant Commutated Pole Inverter," *IEEE Journal of Emerging and Selected Topics in Power Electronics*, doi: 10.1109/JESTPE.2020.3031643.

[8] D. M. Divan, "The resonant DC link converter-a new concept in static power conversion," *IEEE Transactions on Industry Applications*, vol. 25, no. 2, pp. 317-325, 1989.

[9] R. Li, Z. Ma and D. Xu, "A ZVS Grid-Connected Three-Phase Inverter," *IEEE Transactions on Power Electronics*, vol. 27, no. 8, pp. 3595-3604, 2012.

[10] X. Qi and X. Ruan, "A novel two-amplitude control strategy for the active clamped resonant DC link inverter," *in 2006 37th IEEE Power Electronics Specialists Conference*, 2006, pp. 1-5.

[11] M. Stunda and L. Ribickis, "Evaluation of Quasi-resonant DC Link Topologies for Soft Switching of Multiple DC-Inputs Three Phase Inverter," *in 2018 20th European Conference on Power Electronics and Applications (EPE'18 ECCE Europe)*, 2018, pp. 1-10.

[12] M. Turzynski, P. J. Chrzan, M. Kolincio and S. Burkiewicz, "Quasi-resonant DC-link voltage inverter with enhanced zero-voltage switching control," *in 2017 19th European Conference on Power Electronics and Applications (EPE'17 ECCE Europe),* 2017, pp. 1-8.

[13] Q. Wang, G. Guo, Y. Wang and J. Chen, "An Efficient Three-Phase Resonant DC-Link Inverter With Low Energy Consumption," *IEEE Transactions on Power Electronics,* vol. 36, no. 1, pp. 702-715, 2021.

[14] E. Chu, H. Xie, J. Bao, Z. Chen and Y. Kang, "Resonant Inductance Design and Loss Analysis of a Novel Resonant DC Link Inverter," *IEEE Transactions on Power Electronics,* vol. 35, no. 2, pp. 1392-1405, 2020.

[15] E. Chu, H. Xie, Z. Chen, J. Bao, Y. Zhou and H. Zhang, "Parallel Resonant DC Link Inverter Topology and Analysis of Its Operation Principle," *IEEE Journal of Emerging and Selected Topics in Power Electronics,* vol. 8, no. 3, pp. 3124-3138, 2020.

Comparison of the Influence of Reverse Conduction on EMI of WBG And Si Devices

Ru Zhang
State Key Laboratory of Electrical Insulation and Power Equipment
Xi'an Jiaotong University
Xi'an China
me854009399@stu.xjtu.edu.cn

Wenjie Chen
State Key Laboratory of Electrical Insulation and Power Equipment
Xi'an Jiaotong University
Xi'an China
cwj@mail.xjtu.edu.cn

YuXuan Chen
State Key Laboratory of Electrical Insulation and Power Equipment
Xi'an Jiaotong University
Xi'an China
cyx1998@stu.xjtu.edu.cn

Yue Cao
State Key Laboratory of Electrical Insulation and Power Equipment
Xi'an Jiaotong University
Xi'an China
Yue1403300584@stu.xjtu.edu.cn

Ruitao Yan
State Key Laboratory of Electrical Insulation and Power Equipment
Xi'an Jiaotong University
Xi'an China
yanruitao@stu.xjtu.edu.cn

Xu Yang
State Key Laboratory of Electrical Insulation and Power Equipment
Xi'an Jiaotong University
Xi'an China
yangxu@mail.xjtu.edu.cn

Abstract—**Wide bandgap (WBG) devices are widely used in power electronics. However, it brings electromagnetic interference (EMI) problems. To compare the EMI generated by wide bandgap (WBG) devices and Si counterparts, a datasheet-based method predicting the EMI of different devices is proposed here. What's more, a parameter *n* is defined to analyze and compare theEMI of Si, SiC and GaN on high-frequency. The result reveals that the larger *n* is, the more serious effect of it on EMI. The DPT experiments were carried out, which verified the accuracy of the above analysis.**

Keywords—Electromagnetic interference (EMI), gallium nitride (GaN), reverse conduction, silicon carbide (SiC).

I. INTRODUCTION

High frequency, integration, modularization and intelligence are the main development directions of power electronic devices in the future. WBG devices due to their higher band gap, breakdown field strength, electron saturation rate and electron mobility[1], can well adapt to and meet the requirements of power electronics development. They are widely used in synchronous motors, LLC resonant converters, communication power supplies, wireless charging[1] and other fields. As we all know, because of higher switching frequency, the EMI of WBG devices are more serious. A lot of work have been done about the influence of WBG devices on EMI.

Not only that, reverse conduction is an important characteristic of WBG devices, which will also affect EMI. A lot of scholars have studied the influence of reverse conduction on switching losses[2]-[3] and switching processes[4]-[6]. For example, Xiucheng Huang and Di Han respectively compared the efficiency of cascade GaN and MOSFET[2] and the losses of diodes and GaN devices due to reverse conduction[3]. Paper [5] analyzes the surge current causing by reverse conduction of GaN HEMT. But there are several paper study and compare its effect on the EMI.

Some scholars [7]-[8] analyzed the effect of diode's reverse conduction. However, mechanism of switching devices is different from it. A method to predict EMI through

simulation has been proposed[9], which has a certain degree of accuracy and convenience. However, because the models of simulation software such as pspice and saber are not accurate enough, the reverse conduction voltage cannot be seen in the simulation results, so it is not suitable for this paper. Di han[10] proposed a method of EMI prediction based on the switching waveform, which can accurately calculate the EMI of the noise source. However, it does not consider the reverse conduction effect, so modeling of the switching waveform is not accurate, which affects the calculated accuracy of EMI.

Therefore, the following work has been done in this paper. It considers the reverse conduction of switching waveform, and accurately models the switching process. Based on this, the analytical expression of the EMI spectrum is derived. And a parameter n is defined to compare it. It has the following advantages: First, the EMI modeling of the system can help us to understand the generation mechanism and conduction path of the interference in the system. Secondly, It can also provide a basis for filter design, thereby reduce the cost of the product, and finally realize the EMC.

This paper is mainly composed of the following four parts. Section II analyzes the switching waveform and the EMI mathematical model. The effect of voltage, time, and the defined parameter n on the EMI spectrum envelope are analyzed in Section III. What more, the EMI of WBG devices are also compared. To verify the above analysis, DPT experiments were built in Section IV. The results of tests and analysis are consistent. Section V summarizes the proposed mathematical analysis method of EMI.

II. CONCEPT AND THEORETICAL ANALYSIS

MOSFETs are commonly used power electronics as power switches, like DC/DC converters. But it has a problem of reverse conduction. Although the conduction mechanism of GaN HEMT is different from Si, it also has problem of reverse conduction. Enhanced GaN HEMT is a symmetrical device, and the driving voltage can be applied to the gate source or the gate drain to turn on the it. When the GaN transistor is turned off, if a reverse current flows through it, since the gate-source is short-circuited at this time, the reverse current will charge the gate-drain junction capacitance C_{gd}. When V_{gd} reaches V_{th} of the switch under the load condition, the GaN will turn on. After that, V_{gd} remains V_{th}, and GaN devices maintains reverse

---

This work is supported by the National Nature Science Foundation of China under project 51977175.

conduction. Since the cascaded GaN HEMT is a series connection of an enhancement mode GaN device and a Si MOSFET, V_{sd} is the sum of parasitic body diode and two-dimensional electron gas. Therefore, different devices have different reverse conduction voltages and time due to different materials and conduction mechanisms.

Due to the existence of the reverse conduction characteristic of the device, the loss of the circuit during the dead time increases. It will not only affect the efficiency, but also have a certain impact on the switching process. Due to the difference in control, the voltage at which the device achieves soft switching is also different, and reverse conduction has different effects on the switching waveform under different control conditions. Therefore, this paper studys the switching voltage waveform under hard switching conditions based on the device manual.

In this paper, to analyze EMI more simply and vividly, it can be obtained by FFT of the time-domain waveform and equivalent infinitesimal substitution, and the EMI spectrum envelope expression shown in formula (1) can be obtained.

$$S(f)=\begin{cases} \dfrac{f_s}{\pi^2 f^2}\left|\dfrac{V_1}{t_r}+\dfrac{V_1}{t_f}+\dfrac{V_{sd}}{t_{n1}}+\dfrac{V_{sd}}{t_{n2}}\right| & \dfrac{1}{2\pi}\left(\dfrac{1}{t_{n1}}+\dfrac{1}{t_{n2}}\right)<f \\[2ex] \dfrac{f_s}{\pi^2 f^2}\left|\dfrac{V_1}{t_r}+\dfrac{V_1}{t_f}+2\pi f V_{sd}\right| & \dfrac{1}{\pi t_f}<f\le\dfrac{1}{2\pi}\left(\dfrac{1}{t_{n1}}+\dfrac{1}{t_{n2}}\right) \\[2ex] \dfrac{f_s}{\pi^2 f^2}\left|\pi f V_1+\dfrac{V_1}{t_f}+2\pi f V_{sd}\right| & \dfrac{1}{\pi t_f}<f\le\dfrac{1}{\pi t_r} \\[2ex] \dfrac{f_s}{\pi f}\left|V_1+V_{sd}\right| & \dfrac{2f_s(V_1+V_{sd})}{\left|V_1(e^{-j2\pi f_sB}-1)-V_{sd}(e^{-j2\pi f_sB}-e^{-j2\pi f_sA})\right|}<f\le\dfrac{1}{\pi t_f} \\[2ex] \dfrac{2f_s(V_1+V_{sd})}{\left|V_1(e^{-j2\pi f_sB}-1)-V_{sd}(e^{-j2\pi f_sB}-e^{-j2\pi f_sA})\right|} & f\le\dfrac{2f_s(V_1+V_{sd})}{\left|V_1(e^{-j2\pi f_sB}-1)-V_{sd}(e^{-j2\pi f_sB}-e^{-j2\pi f_sA})\right|} \end{cases}$$

(1)

$$A=DT+\frac{t_r}{2}+\frac{t_f}{2}+\frac{2D'T'+t_{m1}+t_{m2}}{4} \qquad (2)$$

$$B=DT+\frac{t_r}{2}+\frac{t_f}{2}+\frac{2D'T'+t_{m1}+t_{m2}}{4} \qquad (3)$$

$$C=DT+\frac{t_r}{2}+\frac{t_f}{2}+\frac{2D'T'+t_{m1}+t_{m2}}{4} \qquad (4)$$

Where, V_1 is the bus voltage. D is the duty cycle. T is the period, and V_{sd} is the reverse conduction voltage. t_r and t_f are switching time. t_{m1} and t_{m2} are the voltage rise and fall times of reverse conduction, and $D'T$ is the duration of reverse conduction, and its value is related to the dead time.

To prove the method, the simulated spectrum is compared with calculated spectrum envelope. And the result match well. What's more, the calculated spectral envelope is polylines with different slopes. The influence of it on EMI is mainly reflected in the fifth part of the spectrum envelope.

III. THE THEOREICAL ANLYSIS OF EMI

For theoretical analysis, the reverse conduction parameters of the switching waveform: V_{sd}, t_{m1}, t_{m2}, should be compared. Since t_{m1} and t_{m2} are related to the dead time and device characteristics, it is difficult to compare their sizes. In the analysis of this article, for simplicity, let $t_{m1}<t_{m2}$.

A. Theoretical analysis of reverse conduction voltage

The impact of different voltage on EMI shown in Fig. 2. With the V_{sd} increasing, the amplitude of the EMI increases. It can be seen from formula (1) that this is because as the f_s increases, the proportion of EMI generated by the V_{sd}

increases. Therefore, the EMI generated by a device with a high V_{sd} is high in the entire frequency range.

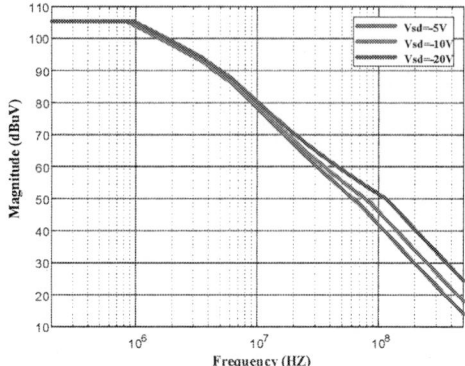

Fig2. The influence V_{sd} on EMI.

B. Theoretical analysis of reverse conduction time

It can be seen that the EMI spectrum envelope changes with time t_{n1} as shown in Fig. 3. As time t_{n1} decreases, the amplitude of the envelope point of the EMI spectrum increases. What's more, the influence of t_n on EMI is only reflected in the highest frequency. Observing of Fig. 2 and Fig. 3, influence of t_{n1} on EMI is far inferior to that of V_{sd}. This is because V_{sd} not only affects EMI in the full frequency range, but also the effect of V_{sd} in the ultra-high frequency band on EMI is equivalent to the sum of the effects of t_{n1} and t_{n2} on EMI.

Through the observation of formula (1), t_{n1} and t_{n2} have the same position in the formula and the same effect. Therefore, t_{n2} has the same effect on EMI as t_{n1}. Although t_{n2} and t_{n1} have different effects on EMI. This paper will only analyze t_{n1}. A conclusion can also be drawn, when other conditions are the same, the shorter the reverse conduction time t_{n1} and t_{n2}, the smaller the EMI caused by the reverse conduction.

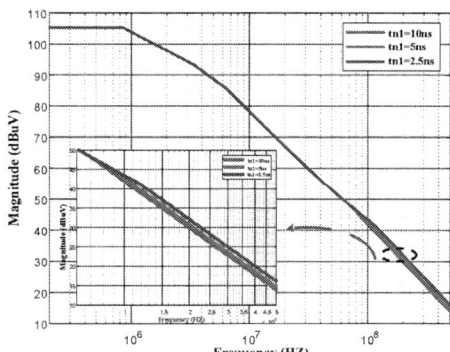

Fig3. The influence of t_{n1} on EMI.

C. Theoretical analysis of n

Fig. 2 and Fig. 3 respectively show the EMI spectrum envelope with V_{sd} and t_{n1}, and have obtained the conclusion that EMI varies with t_{n1}, t_{n2} and V_{sd}. According to the above conclusions, when other conditions are the same, the EMI of the device under different reverse conduction voltage or time can be compared. However, due to the intrinsic characteristics and materials of different devices, V_{sd} and t_{n1} are also different. Therefore, the parameter n of the reverse conduction ratio is defined for comparison as shown in formula (5).

$$n = \frac{V_{sd}}{t_{rr}} \tag{5}$$

The spectral envelopes for varying n is shown in Fig. 4. It can be seen, with n increasing, the EMI increases. And the higher the frequency, the more its amplitude increases. It can also be seen, n is the largest parameter for fourth and fifth frequency bands. Among them, the change of EMI in the fourth band of is caused by t_n, and EMI in the fifth frequency band is caused by the change of V_{sd}.

It can also be seen that when n changes by the same multiple, the EMI amplitude changes disproportionately. When the value of n decreases, the EMI decreases little or even unchanged when changing the same multiple of n. And when n is small enough, the EMI spectrum envelope is no longer divided into five segments but becomes four segments. It shows that when V_{sd} is small and t_n is long, the reverse conduction effect no longer affects the EMI.

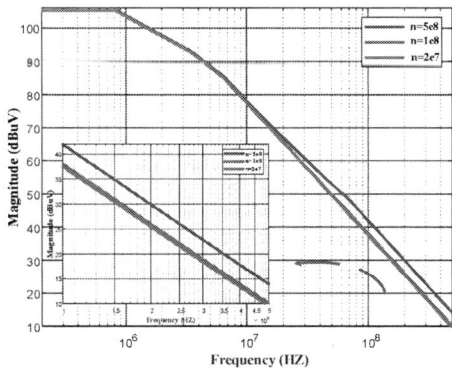

Fig4. The influence of n on EMI.

D. Theoretical analysis of different devices

WBG devices are widely used equipment due to faster switching speeds and higher switching frequencies. In order to better solve the EMI problem caused by WBG devices, the first step is to analyze the difference between the noise source of WBG devices and Si counterparts. The reverse conduction phenomenon is a problem caused by freewheeling and dead time in the switching process of the device. It will bring unnecessary losses. Many articles have analyzed this problem. However, at the same time, it will also have a serious impact on the EMI of the device. Therefore, this section focuses on comparing the EMI problems of WBG devices and Si counterparts.

It can be seen that V_{sd} and t_n of different devices are all different. Based on the above analysis, the magnitude of EMI of different devices under the same conditions can be compared by comparing the value of n. Therefore, the reverse conduction ratio n should be calculated. The larger the value of n, the more serious EMI.

TABLE I
PARASITIC PARAMETERS OF DIFFERENT DEVICES

Transistor	(VA)	(V_{sd})	(t_{rr})	(n)
Si MOSFET (IPB65R190CFDA)	650V 17.5A	0.9V	120ns	0.0075
SiC MOSFET (C3M0060065D)	600V 37A	5V	33ns	0.1516
Cascade GaN HEMT	650V 16A	2.4V	17ns	0.1412

(TPH3206)				
Emode GaN HEMT (GS66504B)	650V 15A	1.7V	12ns	0.1417

Further comparing SiC devices and GaN counterparts, the V_{sd} of SiC mosfet is the largest, so the EMI in the low frequency and highest frequency are both greater. However, due to its long reverse recovery time, the increase in EMI in the fourth frequency band is relatively small. Since cascade GaN is a cascade of emode GaN and Si MOSFET, its reverse conduction time and voltage are relatively long compared with emode GaN. Since n is roughly equal, the influence on EMI mainly depends on V_{sd}. Therefore, compared with emode GaN, cascade GaN has greater low-frequency EMI.

IV. EXPERIMENTAL ANALYSIS

To verify the above analysis, four DPT circuits are tested at 10kHz. And the parameters are shown in Table II. The experiment is tested at 300V input. The output voltages are measured with oscilloscope. The EMI can be measured by test receiver through a pair of LISN.

TABLE II
PARAMETER OF PCB BOARD

V_{in}	f_s	V_{out}	C_{in}	C_{out}
300V	10KHz	300V	5uF	220pF

The EMI of the four devices is shown in Fig. 6. It can be seen the EMI of WBG devices is high due to their faster switching speeds. Among them, the spikes on the frequency spectrum are caused by switching oscillations. Since the parasitic capacitance is small, its oscillation will be more serious.

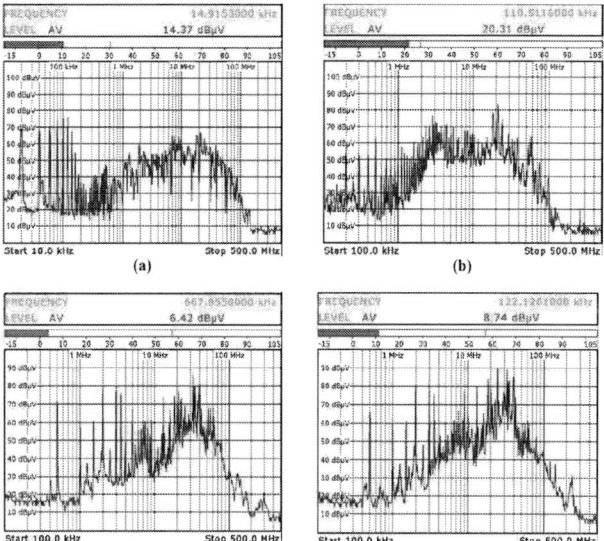

Fig.6 Comparison of common mode EMI of different devices: (a) Si (b) SiC (c) Emode GaN (d) Cascade GaN.

Comparing the high-frequency EMI of four different devices, the Si device has higher high-frequency EMI and slower decline due to the reverse conduction ratio n. As for the SiC MOSFET, due to its higher reverse conduction voltage, its EMI at high frequencies is also more serious. Under the

978-1-6654-4817-8/21 $31.00 © 2021 IEEE

same other conditions, the parameters n of cascade GaN and emode GaN are not much different, so their EMI is not much different. The error of the above experiment is mainly caused by the switching oscillation caused by parasitic parameters. Therefore, it is necessary to reduce the parasitic inductance of the PCB wiring process as much as possible.

V. CONCLUSION

Based on the predicting method, this article analyzes the influence of V_{sd} and t_n on EMI. To compare the high frequency EMI, a parameter n is defined. And got the conclusion that as the parameter n increases, the high frequency EMI gradually increases. As the n of different devices increase gradually, their EMI is becoming more and more serious. Experiments and simulations have verified it.

REFERENCES

[1] H. Wen, J. Gong, X. Zhao, C. Yeh and J. Lai, "Analysis of Diode Reverse Recovery Effect on ZVS Condition for GaN-Based LLC Resonant Converter," in IEEE Transactions on Power Electronics, vol. 34, no. 12, pp. 11952-11963, Dec. 2019, doi: 10.1109/TPEL.2019.2909426.

[2] W. Lee, D. Han, W. Choi and B. Sarlioglu, "Reducing reverse conduction and switching losses in GaN HEMT-based high-speed permanent magnet brushless dc motor drive," 2017 IEEE Energy Conversion Congress and Exposition (ECCE), 2017, pp. 3522-3528, doi: 10.1109/ECCE.2017.8096628.

[3] D. Zhang et al., "Linear Equivalent Model for VHF Class Φ2 Inverter Based on Spectrum Quantification Method to Reduce GaN Reverse Conduction Loss," in IEEE Access, vol. 9, pp. 61635-61645, 2021, doi: 10.1109/ACCESS.2021.3074637.

[4] S. Li et al., "Numerical Study of Novel GaN HEMTs With Integrated SBDs for Ultrahigh Reverse Conduction Capability," in IEEE Transactions on Electron Devices, vol. 68, no. 2, pp. 931-933, Feb. 2021, doi: 10.1109/TED.2020.3046174.

[5] Y. Liu, S. Han, S. Yang and K. Sheng, "Surge Current Capability of GaN E-HEMTs in Reverse Conduction Mode," 2019 31st International Symposium on Power Semiconductor Devices and ICs (ISPSD), 2019, pp. 439-442, doi: 10.1109/ISPSD.2019.8757634.

[6] J. S. Glaser and D. Reusch, "Comparison of deadtime effects on the performance of DC-DC converters with GaN FETs and silicon MOSFETs," 2016 IEEE Energy Conversion Congress and Exposition (ECCE), 2016, pp. 1-8, doi: 10.1109/ECCE.2016.7854939.

[7] X. Yuan, S. Walder, and N. Oswald, "EMI generation characteristics of SiC and Si diodes: Influence of reverse-recovery characteristics," IEEE Trans. Power Electron., vol. 30, no. 3, pp. 1131–1136, Mar. 2015.

[8] J. Wang, H. S. H. Chung, and R. T. H. Li, "Characterization and experimental assessment of the effects of parasitic elements on the MOSFET switching performance," IEEE Trans. Power Electron., vol. 28, no. 1, pp. 573–590, Jan. 2013.

[9] I. Stevanović, B. Wunsch, G. L. Madonna and S. Skibin, "High-Frequency Behavioral Multiconductor Cable Modeling for EMI Simulations in Power Electronics," in IEEE Transactions on Industrial Informatics, vol. 10, no. 2, pp. 1392-1400, May 2014, doi: 10.1109/TII.2014.2307198.

[10] D. Han, S. Li, Y. Wu, W. Choi and B. Sarlioglu, "Comparative Analysis on Conducted CM EMI Emission of Motor Drives: WBG Versus Si Devices," in IEEE Transactions on Industrial Electronics, vol. 64, no. 10, pp. 8353-8363, Oct. 2017, doi: 10.1109/TIE.2017.2681968.

A Novel AC/DC Single-Phase Bridgeless SEPIC PFC Converter With Reduced Conduction Losses and Simple Structure

Xiang Lin
Department of Electrical Engineering
Shanghai University
Shanghai, China
Xianglin@shu.edu.cn

Shumin Ding
State Grid Shanghai Municipal Electric Power Company
Shanghai, China
smding09@163.com

Deliang Wu
Department of Electrical Engineering
Shanghai University
Shanghai, China
wudeliang@shu.edu.cn

Jian Luo
Department of Electrical Engineering
Shanghai University
Shanghai, China
Luojian@shu.edu.cn

Abstract — **Bridgeless rectifier as front-end power factor correction (PFC) converter has been obtained abundant popularity due to its praiseworthy merits of optimized power losses produced by power semiconductors. In this technical paper, novel bridgeless SEPIC rectifier with improved losses and concise circuit structure is proposed by using bidirectional power switch. Compared to conventional SEPIC rectifier, proposed SEPIC rectifier features an optimized efficiency via lessening the count of semiconductors in power-flowing path. Compared to state-of-art bridgeless SEPIC rectifiers, proposed bridgeless SEPIC rectifier possesses concise topology structure and less component. The discontinuous current mode (DCM) is presented, which has the merits of naturally PFC and the zero-current switching are analyzed. Finally, a 100W simulated model are constructed aiming to certify superiority of this new rectifier.**

Keywords—Bridgeless SEPIC rectifier, discontinuous current mode (DCM), power factor (PF), simple structure

I. INTRODUCTION

Single-phase front-end rectifier with power factor correction (PFC) has been widespread for switching-mode power supplies for purpose of suppressing input current harmonics limited by IEC 61000-3-2 criteria, etc. Traditional full-bridge PFC converter typically composed of diode bridge and DC-DC topology is the most commonly option. Whereas, the traditional PFC converters will produce unsatisfied conduction losses in semiconductors, due to the overmuch conducting semiconductors in power flowing path [1].

Aiming to lessen count of conducting semiconductors, the bridgeless PFC converters have become an effective technological method [2]. In the past decade, numerous bridgeless PFC converters, for example, boost, buck, buck-boost. etc. have been proposed and investigated depended on different demands [2-7]. Amongst multiple bridgeless PFC converters, bridgeless SEPIC PFC converter has gotten lots of popularity for its merits of continuous input current, low inrush current, easy implementation of galvanic isolation and high power factor (PF) [5].

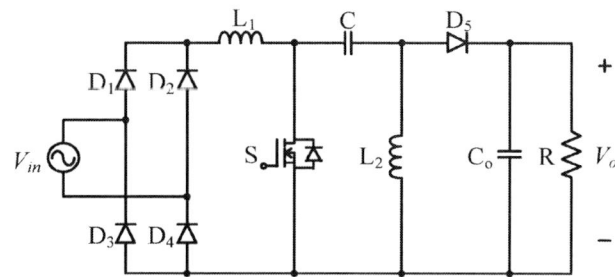

Fig. 1. Traditional SEPIC rectifier.

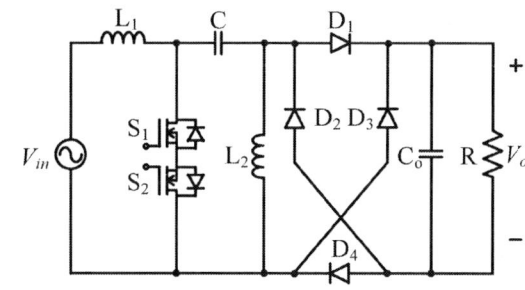

Fig. 2. Proposed new bridgeless SEPIC rectifier.

As a consequence, lots of bridgeless SEPIC rectifiers have been invented in last decade for improving efficiency of traditional SEPIC rectifier which has been presented in Fig. 1 [5,8-17]. All the state-of-art bridgeless SEPIC PFC converter are outstanding innovations with unique superiority in different demands. Nevertheless, most of them uses whole or partly two SEPIC topologies, being active in the positive and negative line period, separately [5,8-16]. These characteristic will inevitably increase the count of components, particularly the large-volume magnetic inductor and bulk electrolytic capacitor, which will degenerate power density performance. A common topology used in positive and negative input line period will contribute to concise topology structure as with the bridgeless SEPIC PFC rectifier [17]. However, one power switch of forementioned bridgeless SEPIC PFC rectifier in [17] is floating which makes for complex drive circuit.

978-1-6654-4817-8/21 $31.00 © 2021 IEEE

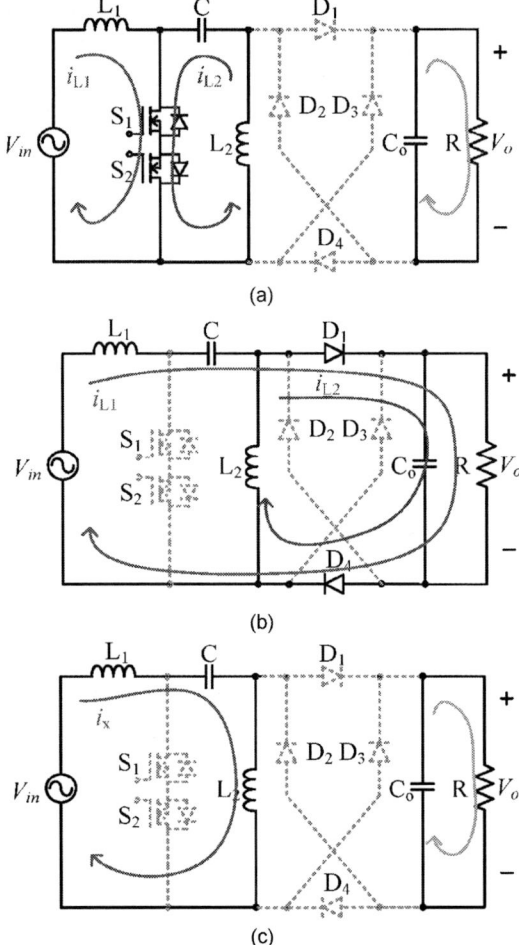

Fig. 3. Switching operation stages in proposed new SEPIC rectifier for positive half input line (a) Switches are ON, (b) Switches are OFF, (c) DCM stage.

A new bridgeless SEPIC rectifier by using bidirectional switch exhibited in Fig. 2 is invented and investigated. Compared with traditional SEPIC PFC rectifier, this proposed new SEPIC rectifier owns fewer count of conducting semiconductors in power-flow path, availing improved efficiency. Compared to state-of-art bridgeless SEPIC PFC converters in [5, 8-16], the proposed converters features simple circuit structure and utilizes less magnetic inductors and bulk electrolytic capacitor. Compared with bridgeless SEPIC PFC rectifier [17], drive circuit in proposed rectifier should be simpler, since the bidirectional power switch owns common source terminal. Notably, proposed new SEPIC rectifier could be able to work in the condition of step-up voltage as with the converter in [5].

II. ANALYSIS OF THE NOVEL BRIDGELESS SEPIC RECTIFIER

Detailed working principle is exhibited for understanding. Comprehensive comparison between proposed new SEPIC rectifier and other state-of-art bridgeless SEPIC rectifiers

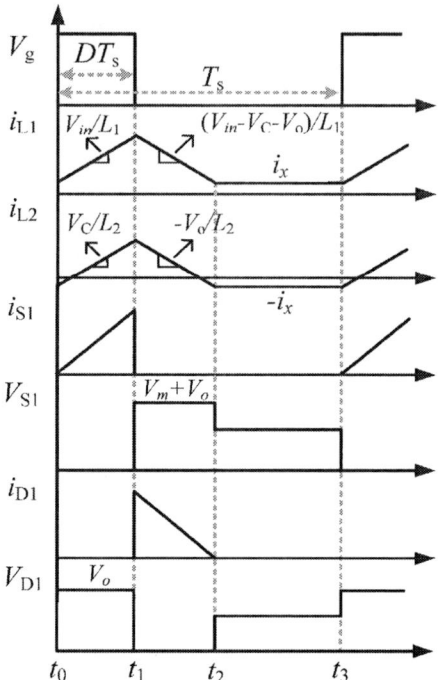

Fig. 4. Key time-domain waveforms in proposed bridgeless SEPIC converter in positive half input line.

are investigated for the purpose of expressing the advantages of this proposed new bridgeless SEPIC rectifier.

A. Working Stages

Since new bridgeless SEPIC PFC rectifier operates in discontinuous current mode (DCM), there are three stages in a switching cycle as illustrated in Fig. 3. Its corresponding pivotal waveforms are given by Fig. 4. Notably, V_g refers to control voltage waveform, i_x is considered as ideal constant current.

Stage I presented in Fig. 3 (a): When two switches S_1, S_2 are active, input source V_{in} charges inductor L_1. Inductor L_2 is charged by the capacitor C through two switches S_1 and S_2. The output capacitor C_o release power energy for output load. During this sub-interval, the currents i_{L1} and i_{L2} could be derived as

$$V_{in} = L_1 \frac{di_{L1}}{dt} \tag{1}$$

$$V_C = L_2 \frac{di_{L2}}{dt} \tag{2}$$

Stage II shown in Fig. 3 (b): In this interval, two switches S_1, S_2 both are turned OFF. Inductor L_1 together with input voltage source V_{in} charge capacitor C, as well as load R, simultaneously. Simultaneously, inductor L_2 is discharged by load R. During this interval, the inductor currents can be derived by

978-1-6654-4817-8/21 $31.00 © 2021 IEEE

TABLE I
NUMBER OF COMPONENTS AMONG THE BRIDGELESS SEPIC PFC
RECTIFIERS

The SEPIC PFC rectifier	Diode	Switch	Inductor	Capacitor	Total
Proposed	4	2F	2L	2	10
Converter in [5]	2	2F	3L	4	11
Converter in [8]	2	2N	3L	3	10
Converter in [9]	5	2N	3L	3	13
Converter in [10]	4	2N	4L	4	14
Converter in [11]	4	2N	4L	3	13
Converter I in [12]	4	2F	3L	4	13
Converter II in [12]	2	4F	3L	4	13
Converter in [13]	5	1N	3L	3	12
Converter in [14]	5	2N	3L	4	14
Converter in [15]	12	2F	3L	8	25
Converter in [16]	2	2N	1L2T	3	10
Converter in [17]	3	1N1F	2L	2	9

L-Inductor, T-Transformer, F-Floating, N-Non-floating.

TABLE II
COMPARISON AMONG VOLTAGE RATIO, VOLTAGE STRESSES AND DRIVER

SEPIC rectifier	Voltage stresses		Isolated Driver	Normalized voltage ratio
	Switch	Output diode		
Traditional	V_m+V_o	V_m+V_o	No	M
Proposed	V_m+V_o	V_o	Yes	$M>1$
Rectifier in [5]	$V_m+V_o/2$	$V_m+V_o/2$	Yes	$M>2$
Rectifier in [8]	V_m+V_o	V_m+V_o	No	M
Rectifier in [9]	V_m+V_o	V_m+V_o	No	M
Rectifier in [10]	V_m+V_o	V_m+V_o	No	M
Rectifier in [11]	V_m+V_o	V_m+V_o	No	M
Rectifier I in [12]	$V_m+V_o/2$	$V_m+V_o/2$	Yes	M
Rectifier II in [12]	$V_m+V_o/2$	$V_m+V_o/2$	Yes	M
Rectifier in [13]	V_m+V_o	V_m+V_o	No	M
Rectifier in [17]	V_m+V_o	V_m+V_o	Yes	M

$$V_{in} - V_C - V_o = L_1 \frac{di_{L1}}{dt} \quad (3)$$

$$-V_o = L_2 \frac{di_{L2}}{dt} \quad (4)$$

Stage III exhibited in Fig. 3 (c): Two switches S_1, S_2 are still maintained OFF. Current through inductor L_1, L_2 are both constant, while the output bulk capacitor C_o maintains output voltage invariable.

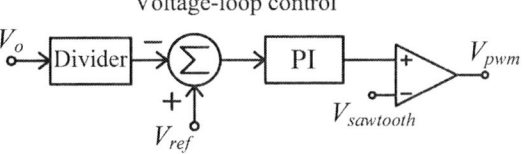

Fig. 5. The diagram of single voltage-loop control method.

B. Comparison Between the Proposed and State-of-art Bridgeless SEPIC PFC Rectifiers

The comprehensive comparison between proposed new SEPIC rectifier and other state-of-art bridgeless SEPIC rectifiers has been given by this subsection.

Comparison about count of components has been exhibited by the Table I. Clearly, it could be found that the proposed new SEPIC rectifier uses only ten components including two inductors and one output electrolytic capacitor, in Table I. Hence, proposed bridgeless SEPIC PFC rectifier possesses a competitive circuit topology in terms of volume and count of component.

In the Table II, comparison in terms of voltage stresses, drive and voltage ratio are provided. It can be found that the voltage stresses across switches in this proposed bridgeless rectifier are same with converters in [8-11,13,17], while its voltage stresses across output fast-recovery diodes are lower. It could be also found that proposed converter must use isolated drive because of the floating power switches. In addition, the normalized voltage ratio of this proposed bridgeless PFC rectifier is keeping with other traditional bridgeless SEPIC PFC converters [5,8-13,17], while the voltage ratio of this proposed rectifier must be larger than 1.

III. SIMULATED VALIDATION

Simulation has been presented for purpose of verifying superiority of the new bridgeless SEPIC rectifier. Its key operation parameters are V_{in}=110V/50Hz, V_o=200V, f_s=40kHz, C=1uF, L_1=300uH, L_2=100uH, C_o=470uF and P_o=100W. In simulated model, the D_1 and D_2 are selected as fast-recovery diodes with voltage drop 1.5V. D_3 and D_4 are slow-recovery diode with voltage drop 1V. The conduction resistance of the power switches are 0.03Ω. In addition, since the proposed new SEPIC rectifier is compelled in DCM, which owns inherently power factor correction, one voltage loop presented by Fig. 5 are used for this new bridgeless SEPIC PFC rectifier.

Fig. 6 provides the simulated waveforms of input current and voltage in this proposed new SEPIC converter. Obviously, input current in this new rectifier is sinusoidal, which keeps same phase as its input voltage in Fig. 6. At same time, its simulated power factor could reach up to 0.997, which is very high among state-of-art bridgeless SEPIC PFC rectifiers. Therefore, this proposed bridgeless SEPIC PFC rectifier owns extremely high PFC capability.

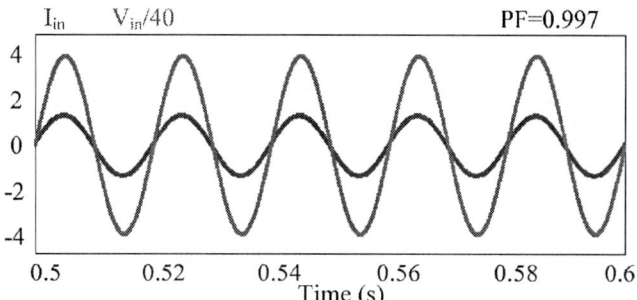

Fig. 6. Simulation waveforms of input current and voltage.

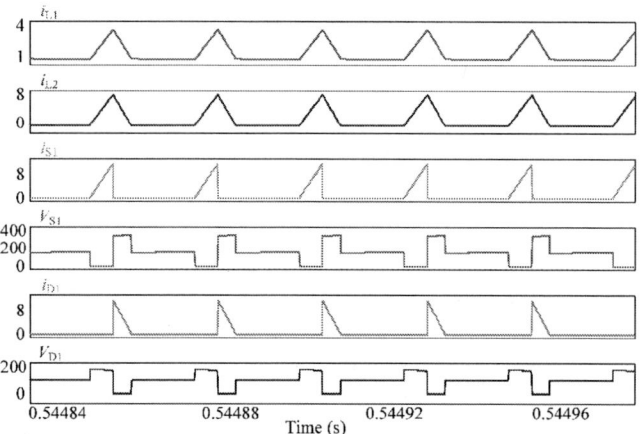

Fig. 7. Simulated pivotal waveforms of this new bridgeless SEPIC rectifier.

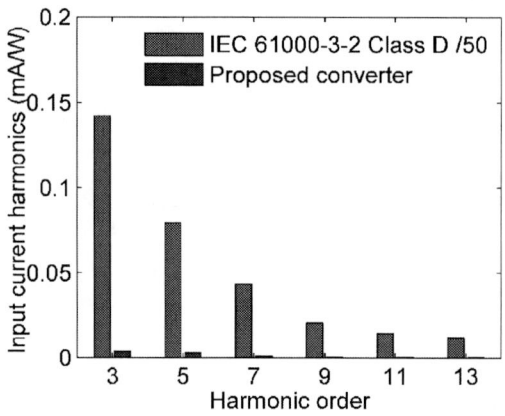

Fig. 8. Simulation of harmonics versus the IEC 61000-3-2 international criteria.

Fig. 7 exhibits simulated pivotal waveforms in this new SEPIC rectifier. Clearly, simulated results are keeping with forementioned design. Also, it validates this proposed new SEPIC rectifier could work in DCM, availably.

Fig. 8 provides simulation input current harmonics of the new SEPIC PFC rectifier versus IEC 61000-3-2 international criteria. Obviously, simulated input current harmonics in this proposed rectifier can meet IEC 61000-3-2 international criteria, completely.

Additionally, simulated efficiency in this proposed bridgeless SEPIC PFC rectifier at 100W output is 98.3%. At same operation parameters condition, simulated efficiency in traditional SEPIC PFC rectifier is 97%. Thereby, the proposed bridgeless SEPIC PFC rectifier could improve efficiency effectively. Notably, the efficiency and power density could be further improved by using GaN HEMT/ SiC MOSFET in the future experiment.

IV. CONCLUSION

In the technical paper, new bridgeless SEPIC converter has been invented, investigated and validated. Operation principle, comparison and simulated results are provided. It validates that compared with traditional SEPIC rectifier, proposed new SEPIC rectifier could reduce conduction losses in semiconductors. Compared with state-of-art bridgeless SEPIC rectifiers, proposed new SEPIC rectifier possesses concise circuit topology. Additionally, this proposed new bridgeless SEPIC rectifier can generate extremely high power factor, and could meet IEC 61000-3-2 international criteria readily.

REFERENCES

[1] L. Huber, Y. Jang and M. M. Jovanović, "Performance evaluation of bridgeless PFC boost rectifiers," *IEEE Trans. Power. Electron*, vol. 23, no. 3, pp.1381-1390, May. 2008.

[2] Y. Jang and M. M. Jovanović, "A bridgeless boost rectifier with optimized magnetic utilization," *IEEE Trans. Power. Electron*, vol. 24, no. 1, pp.85-93, Jan. 2009.

[3] X. Lin and F. Wang, "New bridgeless buck PFC converter with improved input current and power factor", *IEEE Trans. Ind. Electron*, vol. 65, no. 10, pp. 7730-7740, Oct. 2018.

[4] W. Wang, H. Liu, S. Jiang and D. Xu, "A novel bridgeless buck-boost PFC converter," in *Proc. IEEE Power Electron. Spec. Conf.*, 2008, pp. 1304–1308.

[5] E. H. Ismail, "Bridgeless SEPIC rectifier with unity power factor and reduced conduction losses," *IEEE Trans. Ind. Electron*, vol. 56, no. 4, pp. 1147-1157, Apr. 2009.

[6] V. Bist and B. Singh, "A unity power factor bridgeless isolated Cuk converter-fed brushless DC motor drive," *IEEE Trans. Ind. Electron.*, vol. 62, no. 7, pp. 4118-4129, Jul. 2015

[7] J. W. Shin, S. J. Choi and B. H. Cho, "High-efficiency bridgeless flyback rectifier with bidirectional switch and dual output windings," *IEEE Trans. Power. Electron*, vol. 29, no. 9, pp.4752-4762, Sep. 2014.

[8] M. R. Sahid, A. H. M. Yatim and T. Taufik, "A new AC-DC converter using bridgeless SEPIC," in *Proc. Annual. Conf. On IEEE Ind. Electron. Society*, 2010, pp. 286–290.

[9] A. J. Sabzali, E. H. Ismail, M. A. Al-Saffar and A. A. Fardoun, "New bridgeless DCM sepic and Cuk PFC rectifiers with low conduction and switching losses," *IEEE Trans. Ind. Appl*, vol. 47, no. 2, pp. 873-881, Mar/Apr. 2011.

[10] S. Singh, B. Singh, G. Bhuvaneswari and V. Bist, "A power quality improved bridgeless converter-based computer power supply," *IEEE Trans. Ind. Appl*, vol. 52, no. 5, pp. 4385-4394, Sep/Oct. 2016.

[11] H. Ma, Y. Li, J. Lai, C. Zheng, and J. Xu, "An improved bridgeless SEPIC converter without circulating losses and input-voltage sensing," *IEEE J. Emerg. Sel. Topics Power Electron.*, vol. 6, no. 3, pp. 1447–1455, Sep. 2018.

[12] P. J. S. Costa, C. H. I. Font and T. B. Lazzarin, "A family of single-phase voltage-doubler high-power-factor SEPIC rectifiers operating in DCM," *IEEE Trans. Power. Electron*, vol. 32, no. 6, pp.4279-4290, Jun. 2017.

[13] M. K. R Noor, A. Ponniran, M. A. Z. A. Rashid, J. N. Jumadril, M.H. Yatim, M. A. N. Kasiran, A. A. Bakar, S. M. Shah, K. S. Muhammad and J. Itoh, "Modified single-switch bridgeless PFC SEPIC structure by eliminating circulating current and power quality improvement," *IET Power. Electron*, vol. 12, no. 14, pp.3792-3801, Nov. 2019.

[14] A. M. A. Gabri, A. A. Fardoun and E. H. Ismail, "Bridgeless PFC-Modified SEPIC rectifier with extended gain for universal input voltage applications," *IEEE Trans. Power. Electron*, vol. 30, no. 8, pp.4272-4282, Aug. 2015.

[15] P. J. S. Costa, C. H. I. Font and T. B. Lazzarin, "Single-phase hybrid switched-capacitor voltage-doubler SEPIC PFC rectifiers," *IEEE Trans. Power. Electron*, vol. 33, no. 6, pp.5118-5130, Jun. 2018.

[16] B. Singh and R. Kushwaha, "A PFC based EV battery charger using a bridgeless isolated SEPIC converter," *IEEE Trans. Ind. Appl*, vol. 56, no. 1, pp. 477-487, Jan/Feb. 2020.

[17] M. Mahdavi and H. Farzanehfard, "Bridgeless SEPIC PFC rectifier with reduced components and conduction losses," *IEEE Trans. Ind. Electron*, vol. 58, no. 9, pp. 4153-4160, Sep. 2011.

A Single-Stage Modular DCX with High Voltage Conversion Ratio Based on High Frequency LLC Resonant Converter

Yueshi Guan
School of Electrical Engineer and Automation
Harbin Institute of Technology
Harbin, China
guanyueshi@hit.edu.cn

Zhaoliang Wen
School of Electrical Engineer and Automation
Harbin Institute of Technology
Harbin, China
wenzhaoliang98@outlook.com

Yijie Wang
School of Electrical Engineer and Automation
Harbin Institute of Technology
Harbin, China
wangyijie@hit.edu.cn

Dianguo Xu
School of Electrical Engineer and Automation
Harbin Institute of Technology
Harbin, China
xudiang@hit.edu.cn

Abstract—Data centers are increasingly requiring high-efficiency power supplies with large current output capability. And the high frequency resonant converter is more attractive in industrial application for its high power density and high efficiency. This paper presents a modular and linear extendable single-stage DCX with high voltage conversion ratio which based on high frequency LLC resonant converter. The converter using a planar matrix transformer has successfully achieved ZVS for half bridge switches as well as ZCS for the secondary rectifier circuit. Meanwhile, GaN devices are implemented since compared to conventional high voltage Si MOS, they show much lower parasitic capacitance. A 500 kHz, 400-6 V, 600 W isolated prototype with high conversion ratio has been built and tested as well, which achieved 94.1% peak efficiency.

Keywords—Isolated modular converter, High frequency, LLC resonant converter, Planar matrix transformer.

I. INTRODUCTION (HEADING 1)

The CPUs, GPUs, memories, and other chipsets in present server racks are all powered from VRMs，which input is a 12V bus. It's in the early 1990s, when data centers power consumption was far less than today's usage, that this kind of 12V bus architecture was proposed [1]. Apparently, due to the increasing current requirement of nowadays data centers power supply, the conduction loss for 12V bus is excessive and it's necessary to mitigate the conduction loss in power distribution. One of the methods is to implement a higher voltage distribution bus such as 400V so that the power dissipated on the wire resistance becomes smaller [2]-[6]. In order to improve the distribution efficiency, a 400–6V unregulated single-stage modular DC-DC converter (DCX) is proposed to take place of the multiple-stage DC-DC converters. Directly reducing the voltage from 400V to 6V not only reduces the transmission current of the bus which reduces the transmission loss, but also makes the structure of the converter more simple. Furthermore, modular design allows the converters to accommodate wider input voltages.

The single-stage modular DCX with high conversion ratio presented in this paper is based on LLC resonant converter. LLC converters are favored in telecommunication and data center applications because of their smaller switching losses,

easier operation at high frequencies which leads to smaller magnetic components [7]-[9]. The advantages of LLC converter can be summarized as the following:

1) when working in the inductive area, it can ensure that the switch of the inverter circuit achieves ZVS;

2) synchronous rectifiers in secondary side can achieve ZCS;

3) can be designed to operate in high frequency which leads to smaller magnetic components.

Therefore, LLC circuit is very suitable as an intermediate power conversion topology in the next-generation data center power supply system. [10]. At the same time, the development of GaN devices offers the opportunity to further decrease the switch loss at high frequency and increase the converter's power density [11]-[13].

In this paper, the design process of a 500 kHz, 400-6 V, 600 W isolated LLC resonant converter is deduced in section II. Besides, the design and analysis of the planar matrix transformer is presented in section III. In section IV, a prototype has been built to verify the performance of the 500 kHz, 400-6 V, 600 W isolated single-stage modular DC-DC converter..

II. THEORETICAL ANALYSIS OF THE PROPOSED CONVERTER

A. The Architecture of the Proposed Converter

The architecture of the proposed converter is shown in Fig. 1. And Fig. 2 shows the circuit schematic of one modular LLC converter. The whole structure includes two LLC circuits with primary side in series and secondary side in parallel. The converter gets its input from a 400 V bus, and its output voltage is 6 V. Half-bridge circuit is adopted in primary side of the topology, while secondary side rectification implements a full-wave rectifier circuit with center tap. Both LLC resonant circuits are modularly designed and each of them contains a matrix transformer which makes the circuit divided into 8 parallel outputs with a rated current of 100 A in total. It can significantly reduce the input voltage by applying such kind of architecture so that the step-down ratio of each LLC resonant converter is relatively lower as well, which, at the same time, simplifies the design difficulty of the converter's

978-1-6654-4817-8/21 $31.00 © 2021 IEEE

parameters. Furthermore, lower conversion ratio provides the opportunity to design a matrix transformer with fewer primary and secondary turns, as a result that the copper loss of the transformer can be lessened. In addition, the structure of 8 parallel outputs effectively shunts the bus current, thus greatly reducing the current stress of each synchronous rectifier, which means cheaper rectifier switches, more flexible printed circuit board design as well as heat dissipation design. The converter runs at a fixed operating frequency of 500 kHz, which maximizes the efficiency benefits [14]. Compared with traditional power supplies, higher operating frequency means faster electromagnetic conversion speed so that the energy stored in the transformer per unit time can be significantly reduced. Consequently, the volume of the transformer can be greatly reduced. Given that the operating frequency is high, GaN devices are implemented since compared to conventional high voltage Si MOS, they show much lower parasitic capacitance, which guarantees a higher switching speed but smaller loss. Each LLC converter is designed to be modular and liner extendable, helping the device adapt to a wider input range. Moreover, the zero voltage turn on of the primary side half bridge and the zero current turn off of the secondary side synchronous rectifier are realized under the full range of load. Such soft switching feature not only reduces the switching loss during the power conversion process, but also reduces the electromagnetic interference of the device [15].

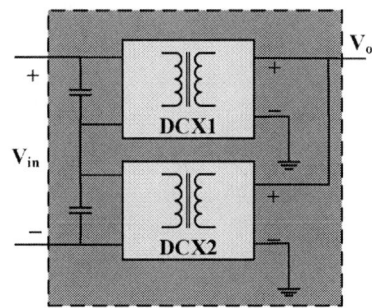

Fig. 1 Structure of the proposed converter

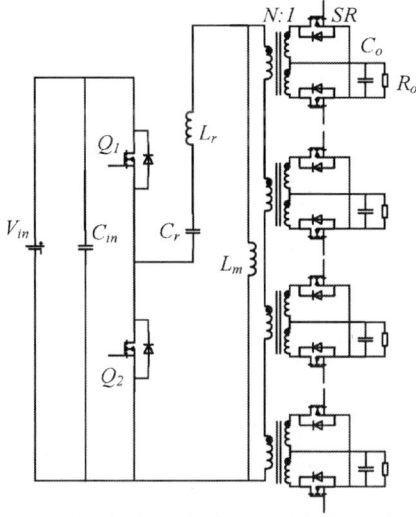

Fig. 2 Circuit schematic of one modular LLC DCX

B. Design Process Considering the Leakage Inductance

Since the transformer turns of the LLC converter with high step-down ratio is large, even a small secondary side leakage

inductance will have a considerable impact after being equivalent to the primary side. Therefore, it is quite necessary to take in the influence of secondary side leakage inductance when designing parameters. The improved LLC AC circuit model is presented in Fig. 3. Some variables are defined in Fig. 3, in which V_{in_ac} represents the effective value of the fundamental wave of the output by the half-bridge, L_p^e is the inductance measured from the primary side when the secondary side is open, L_r^e is the inductance measured from the primary side when the secondary side is shorted, M_v is the virtual gain introduced by considering the leakage inductance of the secondary side, V_{l_ac} and R_l are the output voltage and equivalent load of the secondary side respectively. The voltage gain expression of the LLC DCX can be obtained as shown in equation (1), while parameters such as ω_p^e, ω_r^e and so on are defined in equation (2).

$$M_1^e = \frac{M_v V_{l\_ac}}{V_{in\_ac}} = \left| \frac{k_l^e \dfrac{\omega^2}{\omega_r^{e2}} \sqrt{(k_l^e+1)/k_l^e}}{(\dfrac{\omega^2}{\omega_p^{e2}}-1) + jk_l^e \dfrac{\omega}{\omega_r^e}(\dfrac{\omega^2}{\omega_r^{e2}}-1)Q_l^r} \right| \quad (1)$$

$$\begin{cases} \omega_p^e = 1/\sqrt{L_p^e C_r} \, , L_p^e = L_m + L_{lkp} \\ \omega_r^e = 1/\omega\sqrt{L_r^e C_r} \, , L_r^e = L_{lkp} + L_m\,//\,n^2 L_{lks} \\ k_l^e = \left(L_p^e - L_r^e\right)/L_r^e \\ M_v = \sqrt{\left(L_p^e - L_r^e\right)/L_r^e} \\ R_l^e = \dfrac{8n^2}{\pi^2}\dfrac{R_o}{M_v^2} \\ Q_l^e = \sqrt{\dfrac{L_r^e}{C_r}}\dfrac{1}{R_l^e} \end{cases} \quad (2)$$

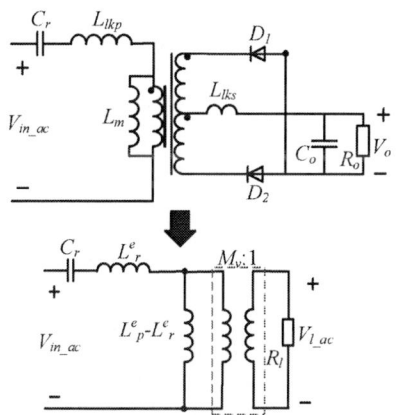

Fig. 3 LLC AC circuit model

There are two important variables in equation (2), where k_l^e is the inductance coefficient and Q_l^e represents the quality factor. The larger k_l^e guarantees not only a wider frequency range for adjusting the voltage gain, but also lower conduction loss and switching loss of the GaN devices near the resonance frequency. Q_l^e should be as large as possible because it leads to a smaller primary current. The overall design process is as follows:

1) Determine the transformation ratio n and voltage gain M_1^e;

2) Select the appropriate inductance coefficient k_l^e according to the working frequency range;

3) The maximum quality factor Q_l^e that determines the maximum value of the LLC gain curve is indicated in the following.

$$Q_l^e = \frac{0.95}{k_l^e M_{1\max}^e} \cdot \sqrt{k_l^e + \frac{M_{1\max}^{e\,2}}{M_{1\max}^{e\,2} - 1}} \tag{3}$$

4) Derive the value of C_r and L_r^e from the expression of quality factor Q_l^e and resonant frequency.

5) Calculate L_p from k_l^e and L_r^e.

The transformation ratio n of proposed LLC DCX is determined by equation (4), where V_{in} and V_o are rated input voltage and rated output voltage respectively. And the value of k_l^e is 3. As a result, Q_l^e can be determined as 1.09, then C_r is 10.7 nF and L_r^e is 9.4 uH.

$$n = \frac{V_{in}}{2V_o} \tag{4}$$

III. Designn Considerations of the Matrix Transformer

Matrix transformers have proven to be one of the best options for LLC DCX to avoid conduction and terminal losses associated with high frequencies [16]. Through the combination of multiple transformer units, it reduces the transformation ratio of each unit and simplifies the overall winding structure. Such simplification is important for PCB winding-based planar transformer design, since a complex winding structure will increase the difficulty of PCB design and increase the loss at the connection of series windings. In addition, the matrix transformer also has features of high step-down ratio and high current shunt capability. [10], [17], [18].

A. Achievement of the Magnetic Inductance

Given that the input voltage of the proposed module is 200 V, and the output voltage is 6 V, the overall transformer turn ratio of a module is supposed to be 16:1. The proposed modular LLC DCX has a two-core matrix transformer, and each core contains two units connected in series with a turns ratio of 4:1 respectively. Hence, a 4-layer PCB is utilized for the primary winding, where the 4:1 turn ratio can be achieved by directly connecting windings of different layers in series and avoid windings in the same layer composing parallel paths and leading to current sharing issues. Another 2-layer PCB is implemented for each secondary winding to realize a full-wave rectifier circuit with center tap. At the same time, the secondary windings on top layer and bottom layer are connected in parallel. Besides, in this design, ferrite core material 3F36 (from FERROXCUBE, E38/8/25) is chosen because of its low loss in the designed operating frequency. By equation (5), where U_1 is the amplitude of the primary sinusoidal voltage, f is the operating frequency, N_1 is the

turns of the primary side and A_e is the effective area, the peak flux density B_m can be derived as 28 mT. Then the core loss can be calculated by equation (6), where P_v is the power loss density that can be obtained from datasheet and V_{core} is volume of the core.

$$B_m = \frac{U_1}{4.44 f N_1 A_e} \tag{5}$$

$$P_{core} = P_v V_{core} \tag{6}$$

Fig. 4 shows the structure of one unit of the matrix transformer. By establishing a reluctance model, the magnetizing inductance of the transformer can be better calculated. As is illustrated in Fig. 5 the reluctance of the sectional cores R_{c1}, R_{c2}, R_{cp} and the reluctance of the air gap R_g are specified by the following:

$$\begin{cases} R_g = \dfrac{g_p}{\mu_0 \left(b_c + g_p \right)\left(l_w + g_p \right)} \\[2mm] R_{c1} = \dfrac{l_{c1}}{\mu_r \mu_0 b_c l_w} \\[2mm] R_{c2} = \dfrac{l_{c2}}{\mu_r \mu_0 b_c l_w} \\[2mm] R_{cp} = R_{c1} \end{cases} \tag{7}$$

where μ_0 is the permeability of the air, μ_r is the relative permeability of the core, g_p is the air gap between the core and the plate. The definitions of l_{c1}, l_{c2}, l_w, b_w and b_c are illustrated in Fig. 4, which can be obtained from the datasheet. Considering the fringing effect, the literature [19] points out an effective cross section of air gap with dimensions a by b would become $(a+g)(b+g)$, where g is the length of the air gap. This consideration is included in the expression of . Therefore, the expression for the magnetic inductance L_m calculation based on the equivalent reluctance model is as follows:

$$L_m = \frac{N_1^2}{R_{cp} + R_{c1} + 2R_{c2} + 2R_g} \tag{8}$$

Considering the value of L_m is already designed in section II, the air gap length of the core can be deduced from the equation (8). Define S as follows:

$$S = \frac{N_1^2}{2L_m} - \frac{R_{cp}}{2} - \frac{R_{c1}}{2} - R_{c2} \tag{9}$$

Combining (7) and (9), the air gap length g_p is calculated by (10).

$$g_p = \frac{1 - \mu_0 S\left(b_c + l_w\right) + \sqrt{\mu_0^2 S^2 \left(b_c - l_w\right)^2 - 2\mu_0 S\left(b_c + l_w\right) + 1}}{2\mu_0 S} \tag{10}$$

B. Analysis of the Leakage Inductance

The energy stored in the transformer leakage inductance includes the energy between the turns of the primary and secondary windings and the energy in the window air. Within the winding area, the direction of leakage flux is almost parallel to the winding's cross section. Therefore, the leakage inductance referred to the primary side can be calculated by (11).

$$E = \frac{1}{2}\int_V B \cdot H dV = \frac{1}{2}L_k I_p^2 \qquad (11)$$

where V is the effective volume and I_p is the current in the primary winding. In order to obtain the leakage inductance of the presented planar transformer, the total energy stored in the primary windings E_p, the secondary windings E_s, and the window air E_{air} needs to be analyzed and deduced.

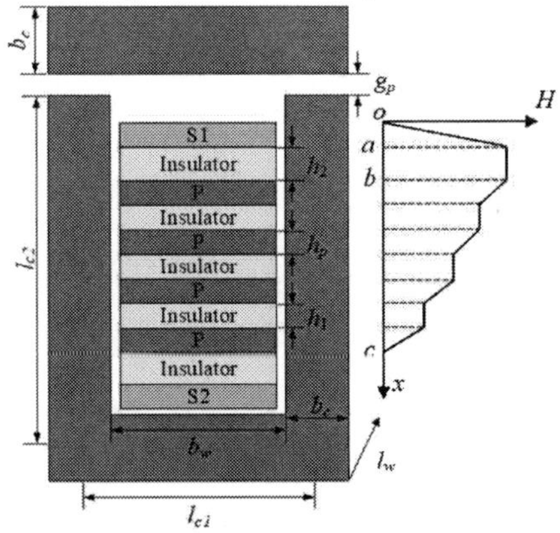

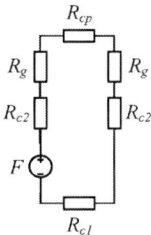

Fig. 4 Coil structure of one unit

Fig. 5 The magnetic reluctance model of one unit

Based on Fig. 4, h_1 is the separation distance between primary layers. h_2 is the separation distance between the secondary winding and the primary winding. And h_p is the thickness of the copper foil. Since the magnetic pressure drop in the magnetic core is approximately zero, the entire magnetomotive force falls on the air path when magnetic core with high permeability is used. Therefore, based on the Ampere Loop Law, the magnetic field strength generated by the secondary winding can be written as

$$H_s = \frac{N_p I_p x}{h_p b_w} \qquad (12)$$

where N_p represents the number of primary turns. Then the energy stored in the secondary windings E_s can be obtained by (13)

$$E_s = \frac{1}{2}\mu_0 l_w b_w \int_0^{h_p} H_s^2 dx = \frac{\mu_0 l_w h_p}{6b_w}\left(N_p I_p\right)^2 \qquad (13)$$

Similarly, the energy stored in the window air between the secondary winding and the primary winding can be shown as follows:

$$E_{air} = \frac{\mu_0 l_w h_2}{2b_w}\left(N_p I_p\right)^2 \qquad (14)$$

As for the energy stored in the primary windings E_p, literature [20] derives an energy calculation method for the step curve in the b-c segment shown in Fig. 4. Based on the equation given in the literature, the energy stored in the primary windings E_p of the planar transformer presented in this paper can be calculated by (15).

$$E_p = \frac{1}{6}\mu_0 \frac{l_w}{b_w}\left[2h_p N_p^3 + h_1\left(2N_p^3 - 3N_p^2 + N_p\right)\right]I_p^2 \quad (15)$$

Combining equation (11), (13), (14), and (15), the leakage inductance is as follows:

$$L_k = \frac{2\left(E_p + E_s + E_{air}\right)}{I_p^2} \qquad (16)$$

Parameters of the designed transformer are presented in Table I, where the required air gap length for magnetizing inductance g_p and the leakage inductance of the designed matrix transformer L_k are shown.

TABLE I   PARAMETERS OF DESIGNED TRANSFORMER

Symbol	Parameter	Value
N_p	Primary turns	4
h_1	Separation distance 1	0.5mm
h_2	Separation distance 2	0.8mm
h_p	Thickness of the copper foil	0.07mm
g_p	The air gap	0.16mm
L_m	Magnetic inductance	28.3uH
L_k	Leakage inductance	0.68uH

C. Optimization of the Windings

The alternating current flowing in the conductor will generate an alternating magnetic field in the conductor which induces eddy currents. At the same time, the eddy current will generate an opposite magnetic field to cancel the ac current. As a result, the current presents an uneven distribution inside the conductor, which is manifested in that the current density near the surface of the conductor is higher than that in the center of the conductor. This is called skin effect. And for two adjacent conductors with reverse current flowing, the proximity effect will cause the current density distribution to tend to the adjacent surfaces of the two conductors. Under the combined action of the skin effect and the proximity effect, the current distribution in the transformer windings shows an uneven trend, which leads to the effective conductive cross-sectional area of the conductor decreases and the AC impedance increases [21].

Based on Dowell's hypothesis and the general field solution of the current density distribution of an infinite foil conductor in a single layer, the expression for m layer AC resistance is derived as [22], [23]

$$\frac{R_{ac}}{R_{dc}} = \frac{Q}{2}\left[\frac{\sin hQ + \sin Q}{\cos hQ - \cos Q} + (2m-1)^2 \cdot \frac{\sin hQ - \sin Q}{\cos hQ + \cos Q}\right] \quad (17)$$

978-1-6654-4817-8/21 $31.00 © 2021 IEEE

$$Q = \frac{h\sqrt{F_1}}{\Delta} \qquad (18)$$

especially, m refers to the number of winding layers between the plane with zero magnetomotive force and the plane with maximum magnetomotive force. For the infinitely foil conductor, $F_1 = 1$. h is the conductor height and Δ is the skin depth that can be described as follows:

$$\Delta = \sqrt{\frac{k\rho}{\pi f \mu}} \qquad (19)$$

where k is temperature coefficient, ρ is material's resistivity, f is the operating frequency and μ is material's permeability. According to the equation (17), (18) and (19), the ac resistance then can be calculated.

Since the amplitude of the ac current on the secondary side is very large and the circuit is working on a high frequency, it's important to reduce the ac resistance of the secondary side. Literature [24] indicates an approach so that, in this paper, a 2-layer PCB is implemented for each secondary winding in the center-tapped rectifier structure, while the secondary windings on top layer and bottom layer are connected in parallel. As a result, the ac resistance on the secondary side is reduced and the efficiency of the entire device has been improved.

IV. EXPERIMENTAL RESULTS OF THE PROTOTYPE

Based on the above analysis, a 500 kHz single-stage unregulated DCX prototype with 2 modular LLC resonant converters with primary side in series and secondary side in parallel is built to verify the performance. Fig. 6 is the picture of a modular converter entity. The input voltage is 400 V, while the output voltage is 6 V. The rated power of the entire device is 600 W. The magnetic inductance L_m and resonant inductance L_r are 28.3 uH and 9.4 uH respectively. C_r is designed to be 10.7 nF and NP0 capacitors are adopted due to its better performance under high operating frequency. Also, GaN devices GS66508B from GaN-Systems is implemented in the primary side half bridge. And the overall conversion ratio of the device is 64:1 while the turn ratio of each element in the matrix transformer is 4:1. Besides, ferrite core material 3F36 (from FERROXCUBE, E38/8/25) is chosen because of its low loss in the designed operating frequency and suitable size for winding design. TI's UCC24624 synchronous rectifier controller is implemented to drive the SRs on the output side. Fig. 7 presents the current waveform of the resonant cavity on the input side. Furthermore, the zero voltage turn on of the primary side half bridge and the zero current turn off of the secondary side synchronous rectifier are realized under the full range of load, which are shown in Fig. 8 and Fig. 9 respectively. The overall efficiency of the designed converter is depicted in Fig. 10, where the peak efficiency of the converter is 94.1%.

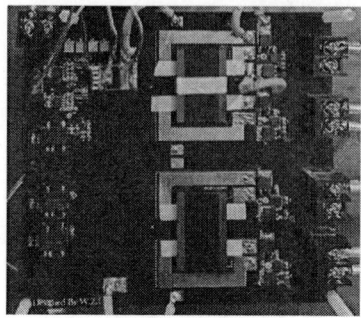

Fig. 6 one modular LLC DCX

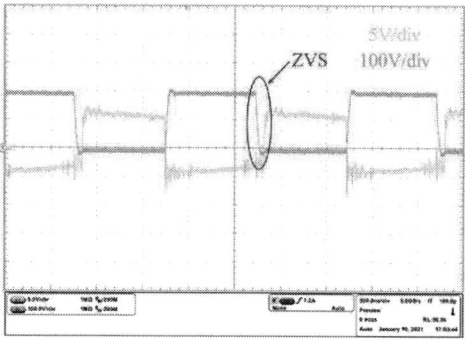

Fig. 7 Current waveform of the resonant cavity

Fig. 8 ZVS for the primary switches

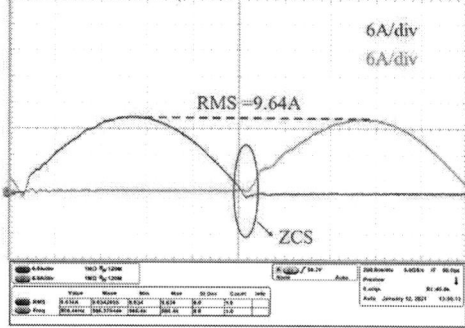

Fig. 9 ZCS for the synchronous rectifier

978-1-6654-4817-8/21 $31.00 © 2021 IEEE

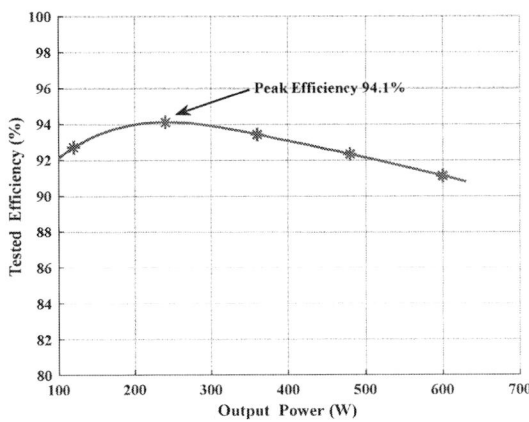

Fig. 10 Tested efficiency

REFERENCES

[1] M. Andres, A. Bieswanger, F. Bosco, G. F. Goth, H. Hering, W. Kostenko, T. B. Mathias, T. Pohl, H. Wen "IBM zEnterprise energy management" 2012.Jan/March.

[2] T. Babasaki, T. Tanaka, and Y. Nozaki, "Developing of higher voltage direct-current power-feeding prototype system," in Proc. IEEE 31st Int. Telecommun. Energy Conf., 2009, pp. 1–5.

[3] A. Matsumoto, A. Fukui, T. Takeda, and M. Yamasaki, "Development of 400-V dc output rectifier for 400-V dc power distribution system in telecom sites and data centers," in Proc. IEEE 32nd Int. Telecommun. Energy Conf., 2010, pp. 1–6.

[4] A. Pratt, P. Kumar, and T. V. Aldridge, "Evaluation of 400 V DC distribution in telco and data centers to improve energy efficiency," in Proc. IEEE Int. Telecommun. Energy Conf., 2007, pp. 32–39.

[5] G. AlLee and W. Tschudi, "Edison redux: 380 Vdc brings reliability and efficiency to sustainable data centers." IEEE Power Energy Mag., vol. 10, no. 6, pp. 50–59, Oct. 2012.

[6] M. Noritake, K. Hirose, M. Yamasaki, T. Oosawa, and H. Mikami, "Evaluation results of power supply to ICT equipment using HVdc distribution system," in Proc. IEEE 32nd Int. Telecommun. Energy Conf., 2010, pp. 1–8.

[7] B. Lu, W. Liu, Y. Liang, F. C. Lee, and J. D. Van Wyk, "Optimal design methodology for LLC resonant converter," in Proc. 21st Annu. IEEE Appl. Power Electron. Conf. Expo. (APEC), Mar. 2006, p. 6.

[8] D. Fu, "Topology investigation and system optimization of resonant converters," Ph.D. dissertation, Dept. Elect. Comput. Eng., Virginia Polytechn. Inst. State Univ., Blacksburg, VA, USA, 2010.

[9] D. Reusch and F. C. Lee, "High frequency bus converter with low loss integrated matrix transformer," in Proc. 27th Annu. IEEE Appl. Power Electron. Conf. Expo. (APEC), Feb. 2012, pp. 1392–1397.

[10] M. Mu and F. C. Lee, "Design and Optimization of a 380-12 v High-Frequency, High-Current LLC Converter with GaN Devices and Planar Matrix Transformers," IEEE J. Emerg. Sel. Top. Power Electron., vol. 4, no. 3, pp. 854–862, 2016.

[11] Z. Liu, X. Huang, M. Mu, Y. Yang, F. C. Lee, and Q. Li, "Design and evaluation of GaN-based dual-phase interleaved MHz critical mode PFC converter," in Proc. IEEE Energy Convers. Congr. Expo. (ECCE), Pittsburgh, PA, USA, Sep. 2014, pp. 611–616.

[12] W. Zhang et al., "Evaluation and comparison of silicon and gallium nitride power transistors in LLC resonant converter," in Proc. IEEE Energy Convers. Congr. Expo. (ECCE), Sep. 2012, pp. 1362–1366.

[13] X. Huang, Z. Liu, Q. Li, and F. C. Lee, "Evaluation and application of 600 V GaN HEMT in cascode structure," in Proc. 28th Annu. IEEE Appl. Power Electron. Conf. Expo. (APEC), Mar. 2013, pp. 1279–1286.

[14] H. Shi, X. Wu and M. Xia, "Analysis of MHz 380V-12V DCX with Low FoM Device," 2019 10th International Conference on Power Electronics and ECCE Asia (ICPE 2019 - ECCE Asia), Busan, Korea (South), 2019, pp. 1821-1829.

[15] F. C. Lee, S. Wang, P. Kong, C. Wang, and D. Fu, "Power architecture design with improved system efficiency, EMI and power density," in Proc. IEEE PESC, 2008, pp. 4131–4137.

[16] D. Reusch and F. C. Lee, "High frequency bus converter with low loss integrated matrix transformer," Conf. Proc. - IEEE Appl. Power Electron. Conf. Expo. - APEC, pp. 1392–1397, 2012.

[17] C. Yan, F. Li, J. Zeng, T. Liu, and J. Ying, "A novel transformer structure for high power, high frequency converter," in Proc. IEEE Power Electron. Specialists Conf., Orlando, FL, USA, Jun. 2007, pp. 214–218.

[18] D. Huang, S. Ji, and F. C. Lee, "LLC resonant converter with matrix transformer," IEEE Trans. Power Electron., vol. 29, no. 8, pp. 4339–4347, Aug. 2014.

[19] W. G. Hurley and W. H. Wolfle, Transformers and Inductors for Power Electronics. Hoboken, NJ, USA: Wiley, Aug. 2013.

[20] J. Zhang, Z. Ouyang, M. C. Duffy, M. A. E. Andersen and W. G. Hurley, "Leakage Inductance Calculation for Planar Transformers With a Magnetic Shunt," in IEEE Transactions on Industry Applications, vol. 50, no. 6, pp. 4107-4112, Nov.-Dec. 2014, doi: 10.1109/TIA.2014.2322140.

[21] Z. Ouyang, O. C. Thomsen and M. A. E. Andersen, "Optimal Design and Tradeoff Analysis of Planar Transformer in High-Power DC–DC Converters," in IEEE Transactions on Industrial Electronics, vol. 59, no. 7, pp. 2800-2810, July 2012, doi: 10.1109/TIE.2010.2046005.

[22] J. Ferreira, "Improved analytical modeling of conductive losses in magnetic components," IEEE Trans. Power Electron., vol. 9, no. 1, pp. 127–131, Jan. 1994.

[23] X. Nan and C. R. Sullivan, "An improved calculation of proximity-effect loss in high-frequency windings of round conductors," in Proc. IEEE Power Electron. Spec. Conf., 2003, pp. 853–860.

[24] M. H. Ahmed, F. C. Lee, Q. Li and M. d. Rooij, "Design Optimization of Unregulated LLC Converter with Integrated Magnetics for Two-Stage 48V VRM," 2019 IEEE Energy Conversion Congress and Exposition (ECCE), Baltimore, MD, USA, 2019, pp. 521-528, doi: 10.1109/ECCE.2019.8912785.

978-1-6654-4817-8/21 $31.00 © 2021 IEEE

Ultra-thin Coupled Inductor for a GaN-Based CRM Buck Converter

Ming Hua[1], Junyu Chen[2], Guolin Xu[1], Hongfei Wu[2]
[1]Nanjing Research Institute of Electronics Technology, Nanjing, China
[2]College of Automation Engineering, Nanjing University of Aeronautics and Astronautics, Nanjing, China
Email: huamingnuaa@163.com

Abstract—A interleaved Buck converter with wide input voltage range and critical conduction mode based on coupled inductor is studied. Due to the limitation of high-frequency filter inductor on the size of power module, the flux distribution characteristics of coupled inductor and its influence on the volume and loss of magnetic components are deeply studied. With ultra-thin and low loss as objective, a four-pole planar coupled inductor structure based on PCB winding is proposed. Meanwhile, the design method of its structure parameters is given as well as a design example. Finally, an experimental prototype with height of only 7mm, power density of 790W/in³ and maximum efficiency of 99% is developed. The experimental results show the effectiveness of the proposed scheme.

Keywords—*coupled inductor, wide-input voltage, magnetic integration, planar inductor*

I. INTRODUCTION

With the rapid development of renewable energy generation, electric vehicles, aerospace, data center and other industries, DC power module with high efficiency, high power density, light weight and miniaturization has increasingly become a common demand [1]. High-efficiency ultra-miniaturization power module can not only reduce the volume and weight of power system, but also develope energy efficiency. In addition, input voltage of the power module usually changes in a wide range in order to meet the requirements of different working conditions, so that wide voltage adaptability is also a challenge for DC module.

Buck converter is the simplest DC/DC converter, which is widely used in all kinds of non-isolated step-down applications. Increasing switching frequency is an effective way to reduce the size of passive components and improve the power density of Buck converter. However, it will increase switching loss and driving loss so that the efficiency will be greatly decreased, which makes it difficult for Buck converter based on silicon devices to make a major breakthrough in efficiency and power density [2]. GaN devices have better switching performance, switching and driving losses greatly reduced [3]. Based on GaN devices and operating the Buck converter in critical conduction mode (CRM), zero voltage switching (ZVS) of all switches can be realized naturally, so that the switching frequency can be increased by orders of magnitude, reaching MHz or even higher. This makes active switching devices no longer the main factor that restricts the power density and efficiency. However, the high switching frequency and CRM operation bring a new challenge to the design of high-frequency filter inductor. Skin effect and proximity effect make AC resistance and loss of traditional wire-wound winding increase greatly.

PCB winding and planar magnetic technology has been proved to be an effective solution for high-frequency magnetic components [4]. PCB winding can realize complex winding structures that traditional wire-wound winding can't achieve, which is conducive to optimization design of winding structure, and planar magnetic core can reduce the height and size of the magnetic component. However, even if the planar magnetic core is used, the height of the core is still much higher than the thickness of PCB and the height of surface mounted components, which makes the core still the key factor limiting the power density and volume of the converter. For interleaved CRM Buck converter, a solution based on negative coupled inductor is proposed in [5]. Studies have shown that the negative coupled inductor can improve dynamic response speed, reduce volume of magnetic components and improve power density [6]. In CRM situation, it has the advantages of expanding ZVS range, reducing cycle energy and decreasing switching frequency, which reduces conduction and turn-off losses [7]. However, the coupled inductors use mostly EI core [7]. The design method of planar coupled inductor with EI structure core based on PCB winding has been studied, but its height is still much higher than that of PCB and surface mounted components.

In this paper, an interleaved CRM Buck converter with GaN devices adopted is studied for wide-input high-density DC module. Starting from the study of flux analysis based on common mode component and differential mode component of inductance current, a four-pole planar coupled inductor structure is proposed. Its advantages of magnetic flux distribution characteristics to the ultra-thin design are deeply studied, and size parameters design method of the planar inductor with PCB winding is given based on the finite element simulation, which realizes the ultra-thin, high efficiency and high power density of the power module.

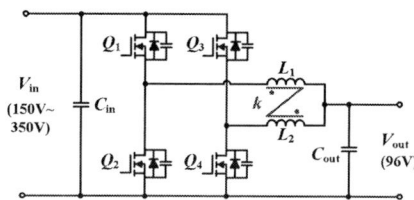

Fig. 1. Topology of interleaved Buck converter.

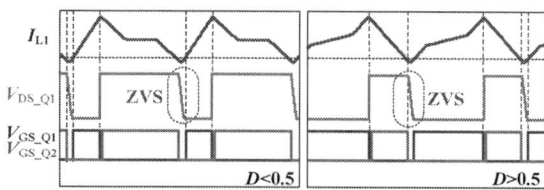

Fig. 2. Key waveforms.

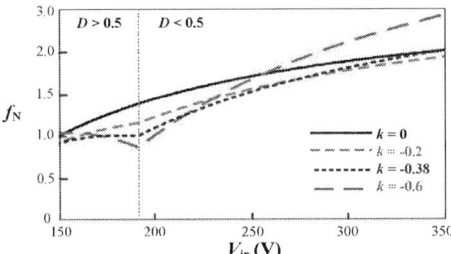

Fig. 3. Relationship between f_N and k.

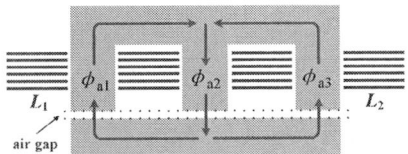

Fig. 4. Coupled inductor with EI structure.

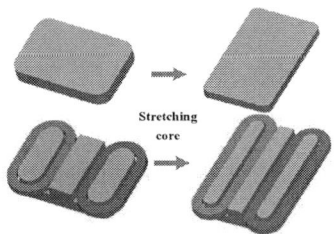

Fig. 5. Stretching core with EI structure.

II. CRM BUCK WITH NEGATIVE COULED INDUCTOR

A. Topology and Working Principle

The topology of interleaved Buck converter is shown in Fig. 1. Current ripple is reduced by interleaving two phases. Both two phases operation under CRM and ZVS is achieved by the resonance between the inductor and the switch junction capacitors.

Fig. 2 shows the waveforms of CRM Buck converter with negative coupled inductor, where V_{DS_Q1} is the drain-source voltage of Q_1 and I_{L1} is the current of L_1. V_{GS_Q1} and V_{GS_Q2} are driving signals of Q_1 and Q_2, respectively.

B. Frequency Reduction Effect of Negative coupling

Coupling the two filter inductors can reduce the switching frequency [7]. Fig. 3 shows the relationship between switching frequency and coupling coefficient under different input voltages. The switching frequency under 150V input and coupling coefficient $k = 0$ is taken as the reference to conduct per unit transformation so that a series of curves of per-unit switching frequency f_N with respect to V_{in} is obtained. Obviously, proper coupling coefficient can reduce the switching frequency over the full voltage range.

III. ULTRA-THIN STRUCTURE OF PLANAR COUPLED INDUCTORS

A. Realization of Ultra-thin Planar Inductor

Inductor is not only the main source of loss for CRM Buck converter, but its height also determines the height of the converter. Fig. 4 shows the structure diagram of two-phase negative coupled inductor based on EI core which is widely used at present [8]. The height of magnetic columns is determined by PCB thickness, air gap and the avoidance

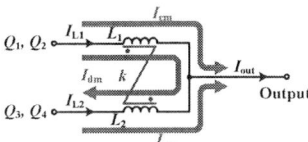

Fig. 6. Inductor current decomposition.

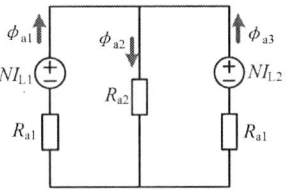

Fig. 7. Magnetic circuit model of EI structure.

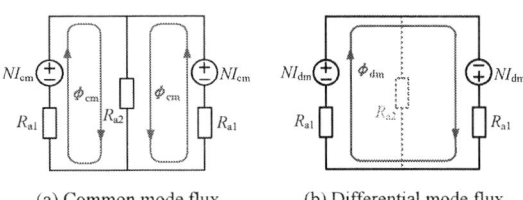

(a) Common mode flux   (b) Differential mode flux

Fig. 8. Flux decomposition of EI structure.

distance between windings and air gap, which is not easy to adjust. Only by minimizing the thickness of upper and lower magnetic plates, the height of the core can be reduced. This can be achieved in two ways: (1) adjusting the length of the core to reduce the thickness of the magnetic board without changing its cross-sectional area; (2) reducing magnetic flux in the magnetic board.

As shown in Fig. 5, the thickness of the magnetic board can be reduced with its cross-sectional area unchanged by stretching the length of magnetic core to increase the length-width ratio of magnetic columns. The magnetic flux density of the magnetic boards and columns will not be changed after stretching the core, but it will increase the length of the winding, significantly increasing winding loss, and the footprint of the inductor will also be larger, affecting the improvement of power density. At the same time, too large length-width ratio is not conducive to the overall design of the power module.

B. Flux Analysis Method

Fig. 6 shows the diagram of inductor current decomposition. Actually, the current of the coupled inductor can be decomposed into two parts, i.e., common mode component I_{cm} and differential mode component I_{dm}. The direction of I_{cm} is flowing to the output side, while the direction of I_{dm} is circulating from L_1 to L_2. I_{cm} and I_{dm} can be expressed as

$$\begin{cases} I_{cm} = \dfrac{I_{L1} + I_{L2}}{2} \\ I_{dm} = \dfrac{I_{L1} - I_{L2}}{2} \end{cases} \quad (1)$$

Similar to the decomposition of current, the magnetic flux of inductor can also be decomposed into the differential mode flux ϕ_{dm} generated by I_{dm} and the common mode flux ϕ_{cm} generated by I_{cm}.

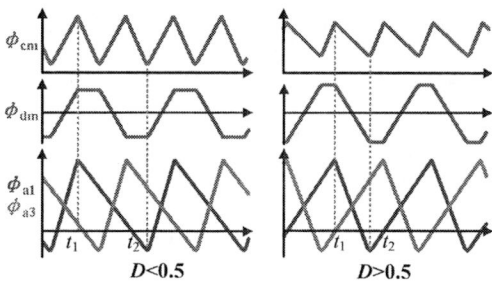

Fig. 9. Flux waveforms.

Fig. 7 shows the magnetic circuit model of coupled inductor with EI structure, ignoring reluctance of magnetic core, magnetic leakage, etc. The magnetomotive force can be decomposed to obtain the path of ϕ_{cm} and that of ϕ_{dm}, as shown in Fig. 8. NI_{cm} is the magnetomotive force generated by I_{cm}, and NI_{dm} is the magnetomotive force generated by I_{dm}. The two magnetomotive forces are of equal magnitude and in the same direction, as shown in Fig. 8(a). Due to symmetry, ϕ_{cm} flows from the two side columns to the middle column respectively. In Fig. 8(b), the two magnetomotive forces are also of equal magnitude, but in the opposite direction. Thus, ϕ_{dm} circulates between the two side columns and does not flow through the middle column.

According to the magnetic circuit model, the expressions of ϕ_{dm} and ϕ_{cm} can be deduced as

$$\begin{cases} \phi_{cm} = \dfrac{(1+k)LI_{cm}}{N} \\ \phi_{dm} = \dfrac{(1-k)LI_{dm}}{N} \end{cases} \qquad (2)$$

Fig. 9 gives flux waveforms. It can be seen that the maximum values of ϕ_{cm} and ϕ_{dm} are obtained at t_1 and their minimum values appears at t_2. In addition, it is noted that ϕ_{cm} contains DC bias flux while ϕ_{dm} is AC flux.

ϕ_{a1} and ϕ_{a3} are the magnetic flux on each side of the magnetic board. We have

$$\begin{cases} \phi_{a1} = \phi_{cm} + \phi_{dm} \\ \phi_{a3} = \phi_{cm} - \phi_{dm} \end{cases} \qquad (3)$$

Therefore, both ϕ_{cm} and ϕ_{dm} exist in the magnetic board of EI structure, which leads to the superposition effect of magnetic flux. The inherent defects of the structure limit the thickness reduction of the magnetic board.

C. Four-pole Planar Coupled Inductor

The key to reduce the height of magnetic core is to reduce the thickness of magnetic board, and reducing the flux in magnetic board is one of the effective measures. Based on the analysis of the distribution paths and synthesis characteristics of common mode flux and differential mode flux, it is found that the magnetic flux in magnetic board can be effectively reduced if there is only common mode flux or differential mode flux in the magnetic board, which can be realized by optimizing the core structure.

Fig. 10 shows the optimization process of core structure. First of all, the magnetic core with EI structure can be abstracted into two parts. One part provides common mode flux path and the other part provides differential mode flux

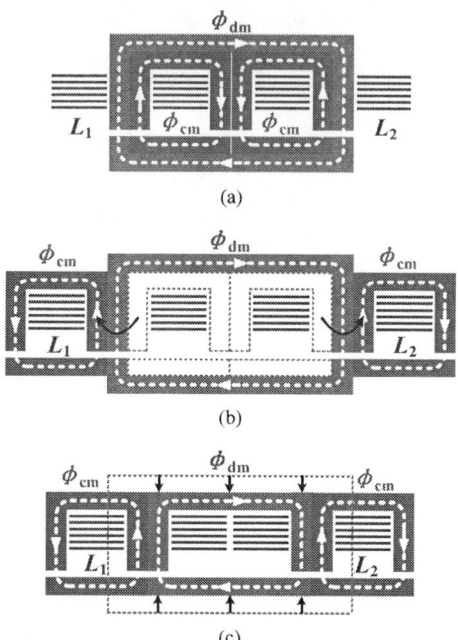

Fig. 10. Optimization process of core structure. (a) Separating flux paths. (b) Flipping common mode flux paths. (c) Reducing height.

path, as shown in Fig. 10(a). Then, keeping the differential mode flux path unchanged, the two common mode flux paths are flipped 180 degrees to the outside of the differential mode flux path, as shown in Fig. 10(b). Finally, by optimizing the volume and height of the differential mode flux path, the magnetic flux path diagram of the novel four-pole structure can be obtained, as Fig. 10(c) shows. Comparing Fig. 10(a) with Fig. 10(c), it can be seen that the four-pole structure realizes the separation of ϕ_{dm} and ϕ_{cm} in the magnetic board, reducing the height and volume of the magnetic core without changing the length and width of the inductor.

In addition, since ϕ_{dm} is AC flux, another advantage of the four-pole structure is that the DC bias flux is eliminated in the middle of the magnetic board, which is conducive to reducing the core loss.

IV. OPTIMIZATION DESIGN AND COMPARISON OF PLANAR COUPLED INDUCTORS

A. Dimension Parameter Design

In this paper, the proposed four-pole structure is used to design the size parameters of the planar coupled inductor. Fig. 11 gives the diagram of its dimension parameters. The self-inductance and k are designed as 10.2uH and -0.38, respectively. The overall height of the magnetic core is 7mm and 6-layer PCB windings are used. The core material adopts DMR51W. In order to make full use of the occupied area, the length of the magnetic board is extended to be flush with the winding to increase the cross-sectional area of the magnetic board.

Considering the thickness of PCB, the length of air gap and the avoidance distance between winding and air gap, the thickness of magnetic board is set as 2mm. Other key parameters to be determined are r, l and c, which satisfy the following equations:

978-1-6654-4817-8/21 $31.00 © 2021 IEEE

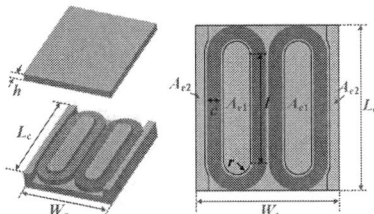

Fig. 11. Dimension parameters diagram.

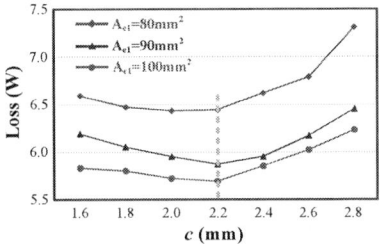

Fig. 12. Inductor loss under different c and A_{e1}.

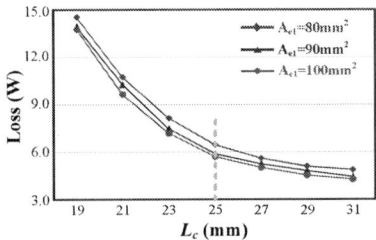

Fig. 13. Inductor loss under different L_c and A_{e1}.

$$\begin{cases} L_c = 2r + 2c + l \\ A_{e1} = 2rl + \pi r^2 \end{cases} \quad (4)$$

Where L_c is the length of the core and A_{e1} is cross-sectional area of the inner column. Since the ratio of A_{e2} to A_{e1} is related to k, when the above parameters are determined, the side pillar width can be obtained according to k, thus the core width W_c is also determined.

First of all, since (4) contains multiple parameters, it is necessary to fix one variable first and then scan the other parameters. Considering the overall layout of the power module and its restrictions on the size of the magnetic core, L_c is preliminarily selected as 25mm. L_c and r can be solved according to (4). Then Maxwell 3D is used to establish the coupled inductor model, and its loss can be obtained through simulation. The results is shown in Fig. 12. It can be seen that, the inductor loss decreases at the beginning and then increases with the enlargement of winding width. When c is larger than 2.2mm, further increase of c can no longer reduce the inductor loss. Therefore, winding width c is chose as 2.2mm.

Then, taking L_c as the independent variable, the curve of inductor loss related to L_c is obtained through simulation, as shown in Fig. 13. It is easy to see that in the range of 19mm ~ 31mm, the inductor loss gradually decreases with the increase of L_c. However, the increase of L_c will make the inductor occupy a larger area, so it is necessary to compromise the size and loss. With comprehensive consideration, L_c and A_{e1} are selected as 25mm and 90mm² respectively.

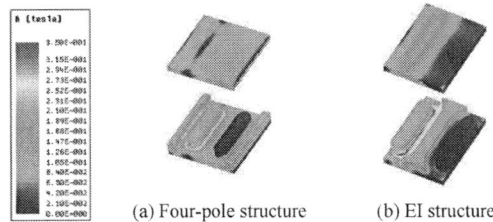

(a) Four-pole structure   (b) EI structure

Fig. 14. Flux density distribution of two structures.

TABLE II. LOSS COMPARISON

	Four-pole structure	EI structure
Size (mm³)	25.0x24.4x7.0	
Core loss (W)	3.47	12.21
Winding loss (W)	2.40	2.11
Total loss (W)	5.87	14.32

B. Comparison

In order to verify the advantages of the proposed four-pole structure, finite element simulation was conducted on the coupled inductors of the four-pole structure and the EI structure respectively under the same size parameters. Fig. 14 gives the magnetic flux density distribution when the current of L_1 reaches peak value (i.e., t1). Compared with EI structure, the magnetic flux density of four-pole structure is much smaller and better distributed.

Table I shows the loss comparison of two structures under the same size. It can be seen that core loss of four-pole structure is much smaller than that of EI structure. Due to the position difference of the no-winding column, the influence of air gap on the AC resistance of the winding is different, so that winding loss of four-pole structure is slightly larger. However, as far as the total loss is concerned, the four-pole structure has considerable advantages.

V. EXPERIMENTAL RESULTS AND ANALYSIS

In order to verify the performance of the ultra-thin coupled inductor with four-pole structure and the rationality of its design method, the magnetic core is fabricated. A prototype of interleaved Buck converter with wide input range (i.e., 150V ~ 350V) and constant output voltage (i.e., 96V) is built as shown in Fig. 15, and its height is only 7mm.

The test is carried out and the experimental waveforms at full load (i.e., 700W) is shown in Fig.16. It is fairly to see that the converter works under CRM, and the switch Q_1 realizes ZVS.

Fig.17 shows the efficiency curves measured under different input voltages and load conditions. Because all switches use GaN devices and ZVS is realized, as well as the advantages of four-pole-structure coupled inductor, the prototype achieves the goal of high efficiency. As Fig.16 shows, the efficiency achieves 97.8% at 270V under full load. Since the converter works under CRM, the lower the input voltage, the slower the switching frequency is. Therefore, the switching loss and core loss are smaller and the efficiency is higher when input voltage is low. The peak efficiency under light load reaches 99% at 150V input.

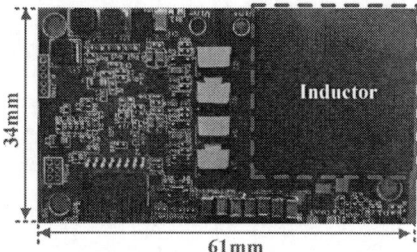

Fig. 15. Prototype.

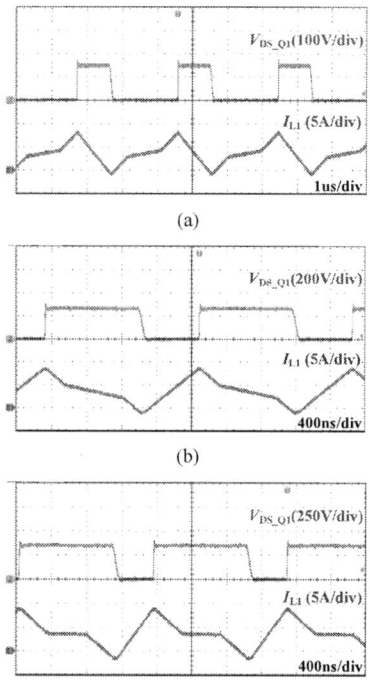

(a)

(b)

(c)

Fig. 16. Experimental waveforms. (a) Vin=150V. (b) Vin=270V. (c) Vin=350V.

VI. CONCLUSION

In order to meet the requirements of high efficiency, high power density and light weight of DC power module, the ultra-thin planar coupled inductor for interleaved CRM Buck converter is studied. A magnetic flux analysis method based on the common mode and differential mode components of two-phase inductor current is presented. The magnetic flux characteristics of traditional EI structure and its structural

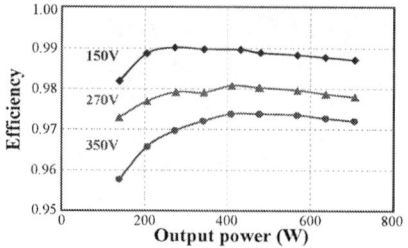

Fig. 17. Efficiency curves.

defects in ultra-thin design are analyzed in detail. Based on the concept of separating common mode flux and differential mode flux, an ultra-thin four-pole structure for planar coupled inductor is proposed, which has the advantages of low magnetic flux in magnetic board, low height and low loss. A dimension parameter design method based on finite element simulation is given. Finally, a prototype of CRM Buck converter is built. The power density reaches 790 W/in^3 with the peak efficiency of 99%. The experimental results show the good performance of the researched planar coupled inductor with four-pole structure.

REFERENCES

[1] Q. Liu, Q. Qian, B. Ren, S. Xu, W. Sun and L. Yang, "A two-stage Buck-Boost integrated LLC converter with extended ZVS range and reduced conduction loss for high-frequency and high-efficiency applications," IEEE Journal of Emerging and Selected Topics in Power Electronics, vol. 9, no. 1, pp. 727-743, Feb. 2021.

[2] F. C. Lee and Q. Li, "High-frequency integrated point-of-load converters: overview," IEEE Transactions on Power Electronics, vol. 28, no. 9, pp. 4127-4136, Sept. 2013.

[3] X. Huang, Z. Liu, Q. Li and F. C. Lee, "Evaluation and application of 600 V GaN HEMT in cascode structure," IEEE Transactions on Power Electronics, vol. 29, no. 5, pp. 2453-2461, May 2014.

[4] J. Zou, H. Wu, Y. Liu, L. Yang and X. Xu, "Optimal design of integrated planar inductor for a hybrid Totem-Pole PFC converter," in IEEE Energy Conversion Congress and Exposition, Detroit, MI, USA, 2020, pp. 1560-1564.

[5] X. Huang, F. C. Lee, Q. Li and W. Du, "High-frequency high-efficiency GaN-based interleaved CRM bidirectional Buck/Boost converter with inverse coupled inductor," IEEE Transactions on Power Electronics, vol. 31, no. 6, pp. 4343-4352, June 2016.

[6] S. Kimura, Y. Itoh, W. Martinez, M. Yamamoto and J. Imaoka, "Downsizing effects of integrated magnetic components in high power density DC-DC converters for EV and HEV applications," IEEE Transactions on Industry Applications, vol. 52, no. 4, pp. 3294-3305, July-Aug. 2016.

[7] M. Fu, C. Fei, Y. Yang, Q. Li and F. C. Lee, "Optimal design of planar magnetic components for a two-stage GaN-based DC-DC Converter," IEEE Transactions on Power Electronics, vol. 34, no. 4, pp. 3329-3338, April 2019.

Comparison Study on Short Circuit Capability of 1.2 kV Split-Gate MOSFET and Split-Source MOSFET with Integrated JBS Diode

Hongyi Xu[1], Chaobiao Lin[1], Na Ren[*1 3], Xinhui Gan[2], Liping Liu[2], Zhengyun Zhu[1], Li Liu[1], Qing Guo[1], Jianxin Ji[2] and Kuang Sheng[1]

1 College of Electrical Engineering, Zhejiang University, Hangzhou, China
2 China Resources Microelectronics Limited, Wuxi 214061, China
3 Hangzhou Innovation Center, Zhejiang University, Hangzhou, China
* ren_na@zju.edu.cn

Abstract—The 1.2 kV split-gate and split-source MOSFET with integrated JBS diode is fabricated. The two types of devices with different Schottky widths (1 μm and 2 μm) are designed and static characteristics are compared. Additionally, the short circuit capability of two types of devices is compared. The failure mechanism of devices is analyzed. The integrated Schottky diode leakage current at high temperature leads to the destruction of devices.

Keywords—SiC MOSFET, integrated JBS diode, short circuit

I. INTRODUCTION

The SiC MOSFET, operating at high switching frequency, has been proved to reduce the volume and weight of power equipment. In application, the body diode of SiC MOSFET is always used as a freewheeling diode which could improve the power density in the system [1]. However, the internal body diode in SiC MOSFET is a PN diode which not only has high turn-on voltage but raises the risk of bipolar degradation. To eliminate these effects caused by the PN body diode of a 4H-SiC MOSFET[2], a monolithic Junction-Barrier-Schottky (JBS) diode integrated SiC MOSFET has been proposed [3]. Before being accepted in the application, the new technology must be fully assessed regarding performance and all reliability aspects.

Usually, the JBS cell is integrated into the source region of the MOSFET cell, which has been studied by a couple of groups [3] [4] [5]. In this work, the JBS integrated MOSFETs with two kinds of integration, i.e. JBS cell integrated into source region (named Split-Source MOSFET) and JBS cell integrated into gate region (named Split-Gate MOSFET) are investigated. The most important reliability aspect, i.e., short circuit robustness is also compared via experiment and simulation. The device failure mechanisms are studied and the method of improving short circuit ruggedness of the JBS integrated MOSFET is proposed. In addition, the short circuit tests are carried out to investigate the short circuit capability of the two types of devices. The detailed short circuit current waveforms after turning off are examined.

II. FABRICATION AND STATIC CHARACTERISTICS

The devices are fabricated on 4H-SiC and N-doping concentration is 8×10^{15} cm^{-3} and thickness is 12 μm. The Schottky contact is formed after the interlayer opening and before forming the ohmic contact between gate polysilicon and metal. Junction barrier MOSFET is fabricated is compatible without additional steps. The Schottky barrier could be integrated into the JFET region and the source region. The Schottky barrier in the JFET region is named as

Split-Gate MOSFET (SG-MOSFET). The Schottky barrier in the source region is named as Split-Source MOSFET (SS-MOSFET). As shown in Fig.1, 1.2 kV SiC Split-Source MOSFET (SS-MOSFET) and Split-Gate MOSFET (SG-MOSFET) are designed and fabricated in this work.

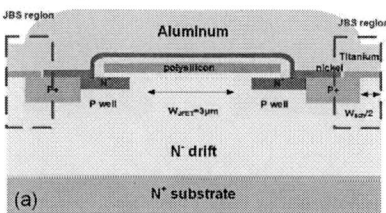

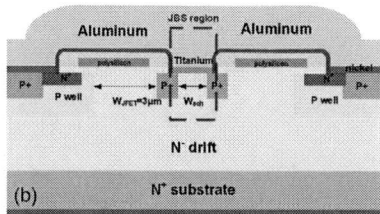

Fig. 1. Schematic of SiC Split-Source MOSFET (SS-MOSFET) and Split-Gate MOSFET (SG-MOSFET) with monolithic JBS diode integration.

To evaluate two different structures, the widths of the P-plus in Split-Source MOSFET are widened and the cell pitch of Split-Source MOSFET and Split-Gate MOSFET (SG-MOSFET) is comparable. The output characteristics are compared for the two devices and the impacts of Schottky width design on the output and blocking characteristics are also studied.

The output characteristics and body diode characteristics for the SS-MOSFET and SG-MOSFET with 2 μm Schottky width design are compared in Fig. 2(a). In the 3rd quadrant, the integrated JBS diode starts to conduct current at 1 V voltage. Lower turn-on voltage in the 3rd quadrant reduces the conduction loss during the freewheeling state. Furthermore, to check the current did not flow through the channel, I-V curves under different V_{gs} bias (0V and -5V) in the 3rd quadrant are tested. The IV curve coincides with different V_{gs} bias (0V and -5V) and the current did not flow through the channel. Due to the shallower junction depth of P+ than P-well junction, the Schottky current could spread between two P-well regions. And the resistance of JBS in SG-MOSFET is lower than JBS in SS-MOSFET.

This research was funded by grants from the Guangdong Key Research and Development Program of China (2019B010143001) and National Natural Science Foundation of China (52007165).

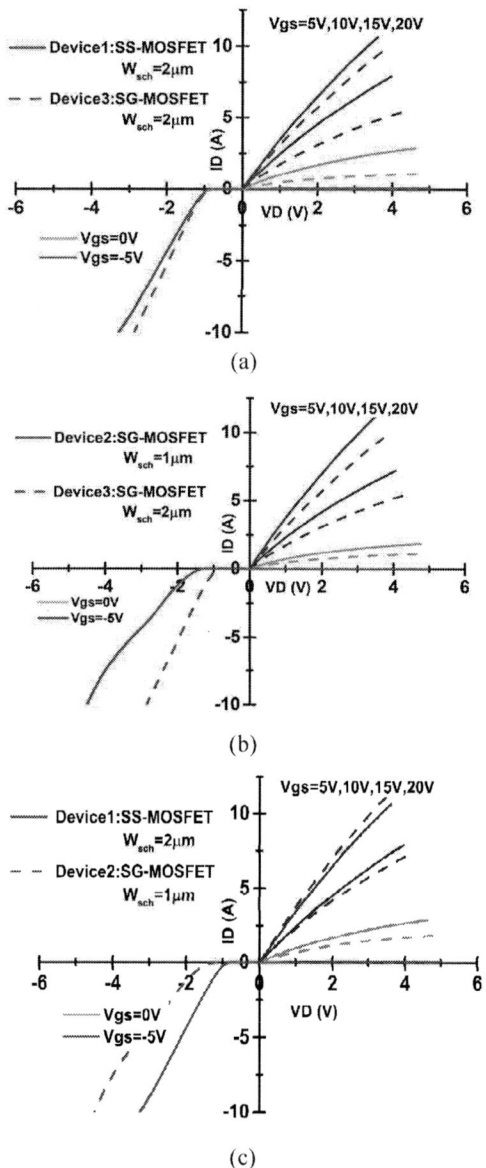

Fig. 2. (a) Comparison of output characteristics for the SS-MOSFET and SG-MSOFET with 2 μm Schottky width design. (b) Comparison of output characteristics for the SG-MOSFET devices with different Schottky width designs (1 μm and 2 μm). (c) Comparison of output characteristics for the SG-MOSFET devices with different Schottky width designs (1 μm and 2 μm).

The output characteristics for the SG-MOSFET devices with different Schottky width W_{sch} designs (1 μm and 2 μm) are compared in Fig. 2(b). Ascribe to the narrower pitch size in the device of W_{sch}=1 μm, the output characteristic is better than the device with W_{sch}=2 μm. For the 1 μm Schottky width design, the diode turn-on voltage is increased from 1 V to 1.4 V due to the pinch-off effect provided by the adjacent P+ regions. However, the body diode current at 5A in a device with W_{sch}=1 μm reaches 3V and is not good for JBS diodes.

The blocking characteristics of the three devices (SS-MOSFET with W_{sch} = 2 μm, SG-MOSFET with W_{sch}= 1 μm

and 2 μm) are shown in Fig. 3. SG-MOSFET with W_{sch} = 2 μm has a higher leakage current through the Schottky contact as the protection from the P well and the P+ region is weak.

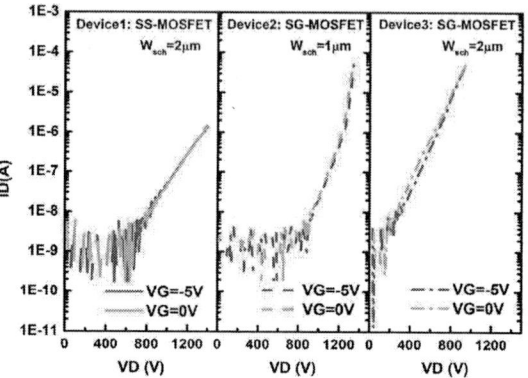

Fig. 3. The blocking characteristics of the three devices (SS-MOSFET with W_{sch} = 2 μm, SG-MOSFET with W_{sch} = 1 μm and 2 μm). SG-MOSFET with W_{sch} = 2μm has higher leakage current through the Schottky contact. The leakage current is affected by the $V_{gs\text{-}off}$ in SG-MOSFET illustrating the channel potential is influenced by the high drain voltage.

III. SHORT CIRCUIT TEST

Short circuit tests are carried out and the current waveforms of SS-MOSFET are shown in Fig. 4. The short circuit duration time increases from 3 μs to 6.3 μs. Device failure occurs after the short circuit time reaches 6.3 μs. When the short circuit time reaches 4 μs, the current tail after gate turn-off appears. This current tail with a small peak, which is distinct from the traditional MOSFET. The maximum short circuit withstand time of SS-MOSFET at 600V bus voltage is 6.1 μs and a unique current tail with a small peak after gate turn-off is observed. The drain current cannot be successfully suppressed after turn-off. And the increasing drain current leading to the destruction of the devices.

While SG-MOSFET with W_{sch} = 1 μm has a comparable short circuit withstand time (5.8 μs) with SS-MOSFET from Fig. 5. However, in Fig. 6, SG-MOSFET with W_{sch} = 2 μm has a shorter short circuit withstand time (SCWT=4 μs). Compare with SG-MOSFET with W_{sch} = 1 μm, SG-MOSFET with W_{sch} = 2 μm has serious Schottky leakage as shown in Fig.2(b). The Schottky barrier height with W_{sch} = 2 μm is lower than that with W_{sch} = 1 μm. Thus, the electron could get over at lower temperature in SG-MOSFET with W_{sch} = 2 μm. The mechanism of the current tail in SG-MOSFET and SS-MOSFET during the short circuit is discussed in the next section which could also illustrate the different failure waveforms in SG-MOSFET with W_{sch} = 2 μm. The last non-destruction waveforms of three devices are shown in Fig. 7.

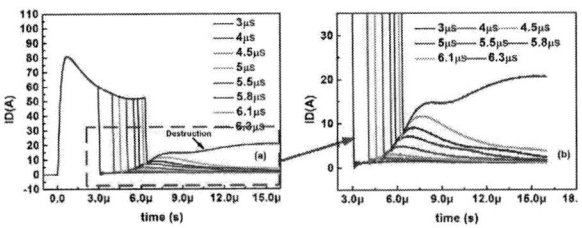

Fig. 4. (a) Short circuit current waveforms of Device 1. (b) Regional enlarged view of the current tail after gate voltage turns off.

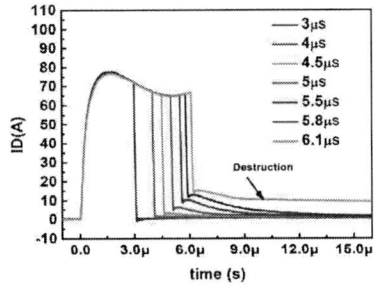

Fig.5.Short circuit current waveforms of SG-MOSFET with $W_{sch} = 1$ μm. The failure occurs at 6.1 μs, which is comparable to Device 1. The tail current is due to the thermionic emission and tunneling current through the Schottky barrier.

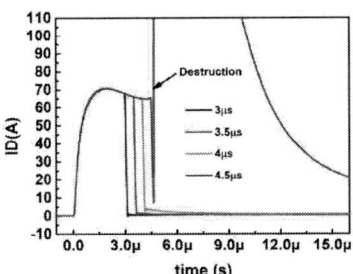

Fig. 6. Short circuit current waveforms of SG-MOSFET with $W_{sch} = 2$ μm. Failure occurs at 4.5 μs. The shorter SC withstand time is due to weaker protection for the SBD from the P well and P+ region, and much higher Schottky leakage.

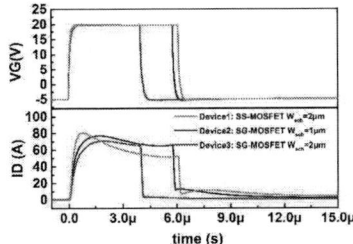

Fig. 7. Last non-destruction waveforms of three devices. The short circuit capability of the Device 1 and Device 2 are close. Thanks to the strong pinch-off effect provided by the adjacent P+ region, Device 2 has higher junction barrier in JBS cell, which results in an improved SC capability than Device 3.

IV. DISCUSSION

The simulation of SG-MOSFET and SS-MOSFET short circuit procedure has been taken to illustrate the mechanism for two different types of current tail.

The short circuit simulated drain current waveforms of SG-MOSFET are shown in Fig.9(a). The device reaches the maximum temperature at point A. At point B, the gate signal turns off (0V). The temperature of the JFET region decreases while the temperature of the Schottky region is increasing and the current mainly leaks through the Schottky contact due to the thermionic emission and tunneling at such high temperature.

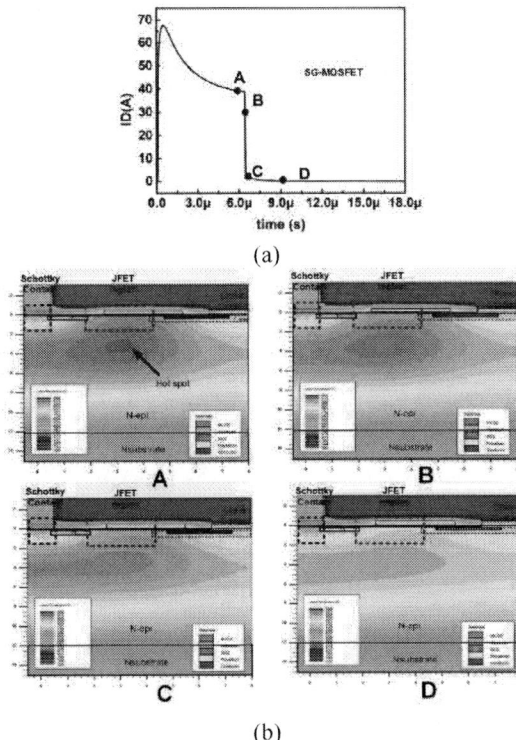

Fig.9 (a)Simulated short circuit waveform. (b)Simulation result of temperature distribution of SS-MOSFET at a different time (A-D points from the current waveform).

The short circuit simulated drain current waveforms of SS-MOSFET are shown in Fig.10(a). The JFET region is close to the SBD region, the hot spot is always located under the JFET region, the current tail is still observed from the waveform but not that significant as that in SS-MOSFET. At point C, the temperature at SBD still increases and surpass that in the JFET region causing the high current tail. When the heat is dissipated within the device, the temperature starts to decrease at point D. In SS-MOSFET, the temperature at the Schottky barrier is close to the hot spot. The temperature difference between the two sites is increased with the maximum temperature increment. Due to the electron could get over at lower temperature in SG-MOSFET with $W_{sch} = 2$ μm，the current tail would disappear in the failure at low temperature. And the short circuit withstanding time could be improved by using higher work-function metal to replace Titanium as Schottky metal.

978-1-6654-4817-8/21 $31.00 © 2021 IEEE

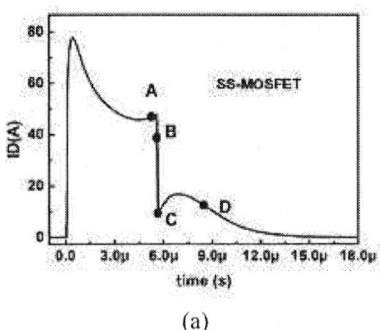

(a)

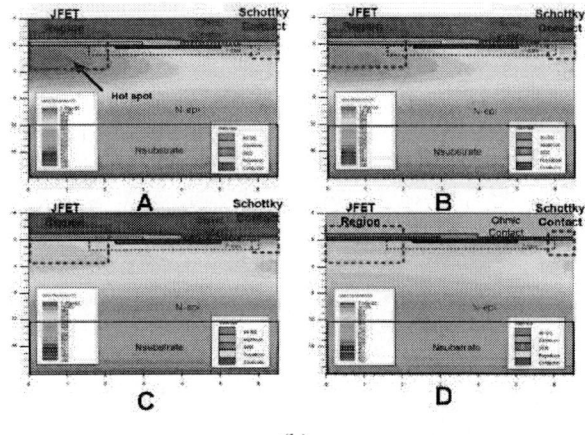

(b)

Fig.10 (a)Simulated short circuit waveform. (b) Simulation result of temperature distribution of SG-MOSFET at different time (A-D points from the current waveform).

V. CONCLUSION

In this work, 1.2kV SS MOSFET and SG MOSFET are fabricated. The SG-MOSFET has better body diode characteristics than SS MOSFET under the same Schottky width (W_{sch} = 2 μm). However, the wider W_{sch} in SG-MOSFET reduces blocking voltage from 1400V to 1100V.

The short circuit robustness is compared via experiment and simulation. The SCWT is comparable for Device 1 (SG-MOSFET, W_{sch} = 2 μm) and Device 2 (SG-MOSFET, W_{sch} = 1 μm) under the drain voltage of 600V. Schottky barrier height has a significant effect on the short circuit robustness of SiC MOSFET with integrated Schottky diode through the comparison of devices with different Schottky widths. The mechanisms of drain current tail after gate signal turning off are studied. And the method of improving short circuit ruggedness of the JBS integrated MOSFET is proposed.

ACKNOWLEDGMENT

The authors thank the Technology Innovation and Training Center, Polytechnic Institute, Zhejiang University, Hangzhou, Zhejiang Province, China for providing the Keysight B1505A power devices analyzer equipment.

REFERENCES

[1] T Kimoto, "Material science and device physics in SiC technology for high-voltage power devices." Japanese Journal of Applied Physics, vol.54.no.4, pp.040103. Apr, 2015.doi: 10.7567/jjap.54.040103

[2] Z. Zhu, et.at, "Investigation on Surge Current Capability of 4H-SiC Trench-Gate MOSFETs in Third Quadrant Under Various VGS Biases," in IEEE Journal of Emerging and Selected Topics in Power Electronics, doi: 10.1109/JESTPE.2020.3028094

[3] C. Yen et al., "1700V/30A 4H-SiC MOSFET with low cut-in voltage embedded diode and room temperature boron implanted termination," 2015 IEEE 27th International Symposium on Power Semiconductor Devices & IC's (ISPSD), Hong Kong, 2015, pp. 265-268.

[4] F. Hsu, C. Yen, C. Hung, K. Chu, L. Lee and C. Lee, "Short-Circuit Ruggedness Analysis of SiC JMOS and DMOS," 2019 31st International Symposium on Power Semiconductor Devices and ICs (ISPSD), 2019, pp. 255-258, doi: 10.1109/ISPSD.2019.8757630.

[5] X. Jiang et al., "Comparative evaluation of surge current capability of the body diode of SiC JMOS, SiC DMOS, and SiC Schottky barrier diode," 2020 IEEE Applied Power Electronics Conference and Exposition (APEC), 2020, pp. 1111-1115, doi: 10.1109/APEC39645.2020.9124530.

978-1-6654-4817-8/21 $31.00 © 2021 IEEE

Analysis of Crosstalk and Suppression Methods for Enhancement-Mode GaN HEMTs in A Phase-Leg Topology

Haihong Qin
College of Automation Engineering
Nanjing University of Aeronautics and
Astronautics
Nanjing, China
qinhaihong@nuaa.edu.cn

Wenlu Wang
College of Automation Engineering
Nanjing University of Aeronautics and
Astronautics
Nanjing, China
906747352@qq.com

Feifei Bu
College of Automation Engineering
Nanjing University of Aeronautics and
Astronautics
Nanjing, China
pufeifei@nuaa.edu.cn

Zihe Peng
College of Automation Engineering
Nanjing University of Aeronautics and
Astronautics
Nanjing, China
821489301@qq.comm

Ao Liu
State Key Laboratory of Wide
Bandgap Semiconductor Power
Electronic Devices
Nanjing Electronic Devices Institute
Nanjing, China
15851831604@163.com

Song Bai
State Key Laboratory of Wide
Bandgap Semiconductor Power
Electronic Devices
Nanjing Electronic Devices Institute
Nanjing, China
13809020747@163.com

Abstract—**When using traditional driver circuits, enhanced gallium nitride power device (eGaN HEMT) suffers from serious crosstalk problems in a phase-leg topology. An improved active Miller clamp driver circuit is designed in this paper to suppress the influence of crosstalk. First, the crosstalk mechanism is analyzed, and the influence of various factors on crosstalk were evaluated experimentally. Then, we describe the the improved active Miller clamp driver circuit's operating principle and give the optimized parameter design method. Finally, the effect of the improved active Miller clamp method to suppress crosstalk is verified.**

Keywords—eGaN HEMT, crosstalk suppression, gate driver, high-speed switching

I. INTRODUCTION

Enhanced gallium nitride power device (eGaN HEMT) is the representative of the third generation wide band-gap semiconductor devices. Compared with Si device, it has smaller junction capacitor, lower on-resistor, better high temperature resistor and faster switching speed. GaN based power electronic converter has the advantages of increasing switching frequency and reducing volume of power converter. Therefore, eGaN HEMT has very broad application prospects in aerospace, new energy power generation and other fields [1]-[3].

When power device is applied to a half-bridge topology, high dv/dt of turning-on switch will be coupled with parasitic capacitor of another switch in the half-bridge, resulting in positive crosstalk voltage at gate and source. This positive crosstalk voltage spike probably bring about the complementary switch that should turn off turns on by mistake, resulting in the danger of bridge-arm through. In particular, the gate-source threshold voltage of GaN device is about 1.2V, which is more prone to misdirect conduction. Similarly, high dv/dt of turning-off switch will be coupled with parasitic capacitor of another switch in the half-bridge, resulting in negative crosstalk voltage at gate-source of the complementary switch. This negative crosstalk voltage is likely to exceed withstand range of gate-source voltage,

resulting in device damage or accelerated performance degradation [4]-[5]. Thus, to ensure high switching speed characteristics of GaN and guarantee reliability of phase-leg configuration, crosstalk suppression methods need to be studied.

There have been plenty of work on bridge-arm crosstalk suppression currently, which can be mainly divided into two kinds.

1) Adopt negative gate turn-off voltage [6]-[8]. During switching-off process, positive crosstalk voltage will be superimposed with the negative bias voltage, thus reducing positive crosstalk voltage amplitude. However, it aggravates the impact of negative crosstalk and is very likely to damage power devices. And for eGaN HEMT, this method will increase reverse conduction loss of its dead time.

2) Control gate drive impedance [9]-[11]. Miller current can flow through a low impedance branch by parallel connection of external capacitor between gate and source. It can effectively suppress crosstalk phenomenon. However, switching speed of power device will decrease and switching loss will increase simultaneously , so it is not recommended. The active Miller clamp suppression method reduces the impedance of the driving turn-off circuit by paralleling the auxiliary switch at gate and source. It controls the auxiliary switch to turn on only when crosstalk occurs to reduce gate drive impedance. Thus, crosstalk voltage fluctuation will be suppressed under the premise of not affecting the switching speed of power device.

In summary, for eGaN HEMT, crosstalk is a serious problem that limits its high-frequency advantages, and proposing a reliable suppression method of crosstalk is very important to the design of half-bridge circuit. In this paper, crosstalk mechanism is analyzed first. Then, through the establishment of a crosstalk voltage evaluation platform, the impact factors of the crosstalk voltage are evaluated experimentally. Finally, the principle analysis and key parameter design guidelines of the improved crosstalk

---

Supported by State Key Laboratory of Wide-Bandgap Semiconductor Power Electronic Devices (2019KF001).

suppression method are given, and its effectiveness is verified by experimental results.

II. PRINCIPLE AND MODELING OF CROSSTALK

A. Principle of Crosstalk

The schematic diagram of crosstalk voltage generation process when Q_1 turns on is shown in Fig. 1. When Q_1 turns on, the mid point potential quickly drops to zero, and V_{DS2} quickly rises to V_{DC}. The high dV_{DS2}/dt causes Miller capacitor C_{GD2} to charge rapidly. Part of Miller current flows to the mid point through gate-source parasitic capacitor C_{GS2} and part through turn-off driving resistor R_{G2} flows to the midpoint, resulting in the generation of the positive crosstalk voltage V_{GS_on} between gate and source of Q_2. V_{GS_on} may cause Q_2 to turn on untimely. And it is likely to bring about the bridge-arm passes through. Fig. 2 illustrates the waveform of gate-source voltage behaviour when Q_1 turns on.

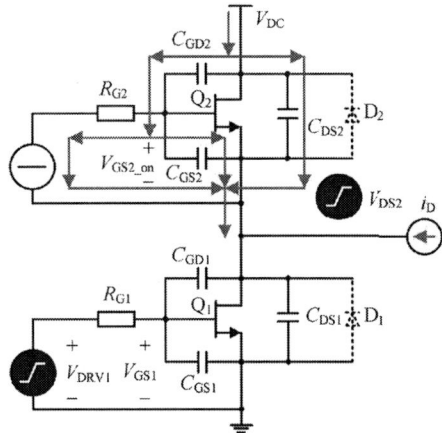

Fig. 1. The schematic diagram of crosstalk voltage generation process when Q_1 turns on.

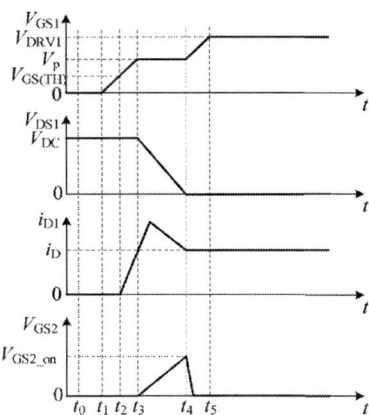

Fig. 2. Gate-source voltage behaviour when Q_1 turns on.

Similarly, Fig. 3 shows the schematic diagram of crosstalk voltage generation process when Q_1 turns off. The high dV_{DS2}/dt causes fast reverse charging of C_{GD2}. Part of the charging current flows to the drain of Q_2 through C_{GS2} and part through R_{G2} flows to the drain of Q_2, resulting in a negative crosstalk voltage V_{GS2_off}. If V_{GS2_off} exceeds withstand range of gate-source voltage, the switch will suffer from damage.

Fig. 4 illustrates the waveform of gate-source voltage behaviour during switching-off process of Q_1.

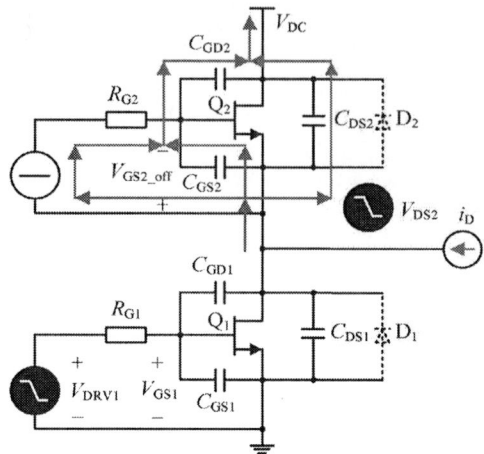

Fig. 3. The schematic diagram of crosstalk voltage generation process when Q_1 turns off.

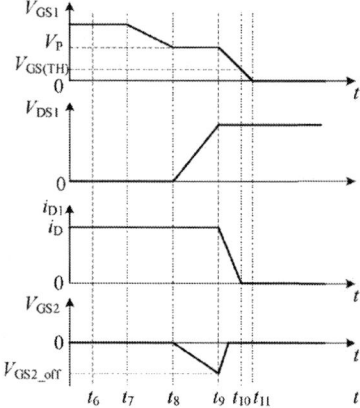

Fig. 4. Gate-source voltage behaviour when Q_1 turns off.

B. Modeling of Crosstalk

The equivalent circuit representing positive crosstalk behavior during Q_1 turning on transient is shown in Fig. 5.

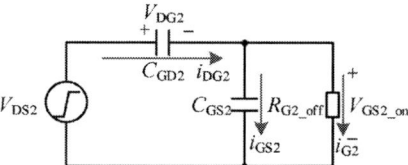

Fig. 5. The equivalent circuit representing positive crosstalk behavior during Q_1 turning on transient.

According to KVL and KCL discipline, the following formula can be obtained from Fig. 5:

$$-C_{GD2}\frac{dV_{DS1}}{dt} = \left(C_{GS2} + C_{GD2}\right)\frac{dV_{GS2\_on}}{dt} + \frac{V_{GS2\_on}}{R_{G2\_off}} \qquad (1)$$

where C_{GD2} is Miller capacitor of Q_2, V_{DS1} is the drain-source voltage of Q_1, C_{GS2} is the gate-source capacitor of Q_2, R_{G2_off} is the turn-off driving resistor of Q_2 and V_{GS2_on} is the positive crosstalk voltage.

V_{GS2_on} can be expressed as

$$V_{GS2\_on} = -R_{G2\_off} \cdot C_{GD2} \cdot \frac{dV_{DS1}}{dt} \cdot (1 - e^{\frac{-t}{R_{G2\_off}(C_{GD2}+C_{GS2})}}) \quad (2)$$

$V_{DC} = V_{DS1} + V_{DS2}$, where V_{DS2} is the drain-source voltage of Q_2. When V_{DS1} drops to zero, V_{GS2_on} reaches the maximum $V_{GS2\_on(max)}$:

$$V_{GS2\_on(max)} = R_{G2\_off} \cdot C_{GD2} \cdot \frac{V_{DC}}{T_{DS1}} \cdot (1 - e^{\frac{-T_{DS1}}{R_{G2\_off}(C_{GD2}+C_{GS2})}}) \quad (3)$$

where T_{DS1} is the time when V_{DS1} drops from V_{DC} to zero.

It can be seen from (3) that the influencing factors of $V_{GS2\_on(max)}$ include R_{G2_off}, C_{GS2} and T_{DS1}, which also means switching on speed of Q_1.

Similarly, When V_{DS1} rises to V_{DC}, negative crosstalk voltage spike $V_{GS2\_off(max)}$ is:

$$\left| V_{GS2\_off(max)} \right| = R_{G2\_off} \cdot C_{GD2} \cdot \frac{V_{DC}}{T_{DS2}} \cdot (1 - e^{\frac{-T_{DS2}}{R_{G2\_off} \cdot C_{GS2}}}) \quad (4)$$

It can be seen from (4) that the influencing factors of $V_{GS2\_off(max)}$ include R_{G2_off}, C_{GS2} and T_{DS2}, which also means switching off speed of Q_1.

III. EVALUATION RESULTS OF IMPACT FACTORS

As shown in Fig. 6, evaluation of factors affecting crosstalk is conducted by a double pulse experiment platform. The power device adopts GS66506T (650V/22.5A) of GaN Systems. During the experiment, by short-circuiting gate-source of top switch, the crosstalk effect on gate-source of top switch caused by switching action of bottom switch was investigated.

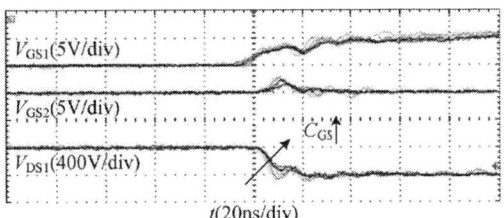

Fig. 6. Experimental setup of DPT.

A. Gate-source capacitor

When load current is 6A and bus voltage is 400V, crosstalk voltage is evaluated when gate-source external capacitor is 0pF, 200pF, 400pF, 600pF, 800pF and 1nF. Fig. 7 is the typical switching waveform and Fig. 8 presents the relationship curve between crosstalk voltage of top switch and gate-source external capacitor.

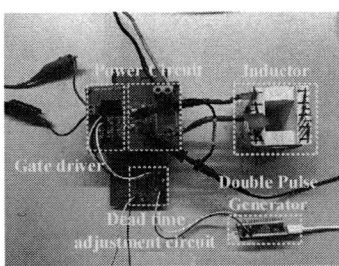

(a) Q_1 turn-on waveform

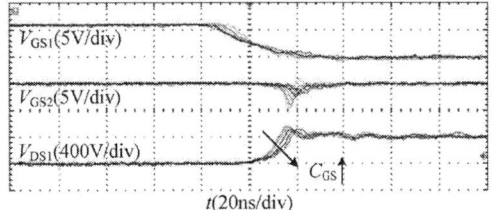

(b) Q_1 turn-off waveform

Fig. 7. Crosstalk voltage waveform of top switch under different gate-source external capacitor.

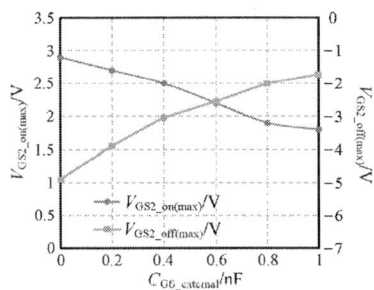

Fig. 8. The relationship curve between the crosstalk voltage of top switch and the gate-source external capacitor.

During turn-on process of Q_1, as external gate-source capacitor increases, gate-source voltage rises slower and the oscillation slows down. Meanwhile, the drain-source voltage drop rate decreases, positive crosstalk voltage of Q_2 decreases correspondingly. Under the dual influence of drain-source voltage change rate and gate-source capacitor itself, when gate-source external capacitor increases from 0nF to 1nF, positive crosstalk voltage decreases from 2.9V to 1.8V. In the same way, during turn-off process of Q_1, when external capacitor increases, negative crosstalk voltage increases significantly from −4.9V to −1.8V.

Although increasing gate-source external capacitor has a significant suppression effect on crosstalk voltage, it is not suitable to be adopted because it does so at the expense of significantly increasing switching loss and sacrificing the performance advantages of eGaN HEMT's high switching speed. Fig. 9 illustrates the relationship curve between turn-on loss E_{on}, turn-off loss E_{off} and total loss E_{total} of bottom switch and gate-source external capacitor.

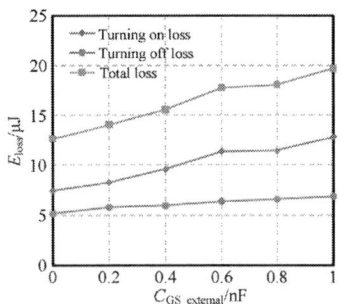

Fig. 9. The relationship curve between E_{on}, E_{off} and E_{total} of bottom switch and gate-source external capacitor.

B. Turn-on driving resistor

Fig. 10 shows the crosstalk voltage waveform under different turn-on driving resistor when load current is 6A and bus voltage is 400V. The relationship curve of crosstalk voltage of top switch with the change of turn-on drive resistor is shown in Fig. 11.

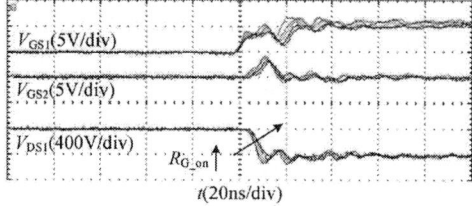

(a) Q_1 turn-on waveform

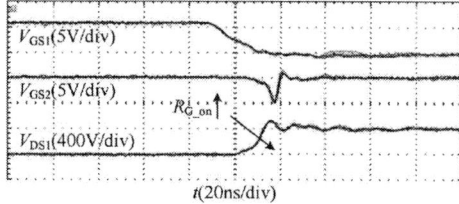

(b) Q_1 turn-off waveform

Fig. 10. Crosstalk voltage waveform of top switch under different turn-on driving resistor.

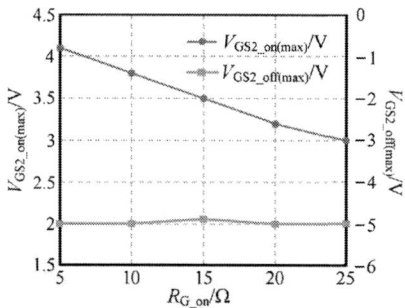

Fig. 11. The relationship curve between the crosstalk voltage of top switch and the the turn-on driving resistor.

It can be seen that during turn-on process of Q_1, as turn-on drive resistor increases, the rate of gate-source voltage slows down and the oscillation decreases. Due to the increase of turn-on drive resistor, the discharge speed of gate-drain capacitor slows down, which results in a slower rate of drain-source voltage rise, and further reduces positive crosstalk voltage of Q_2. Turn-on driving resistor itself has no effect on crosstalk voltage, and it mainly affects crosstalk voltage indirectly by changing the rate of drain-source voltage. The negative crosstalk voltage remains unchanged because turn-on driving resistor has no effect on switching off speed.

From above analysis, although increasing turn-on driving resistor has a significant suppression effect on crosstalk voltage, it is not suitable to be adopted because it will reduce switching speed, resulting in increasing loss. Fig. 12 is the relationship curve between E_{on}, E_{off} and E_{total} of bottom switch and turn-on driving resistor. When turn-on drive resistor increases from 5Ω to 25Ω, the total switching loss of bottom switch increases from 7.798μJ to 13.331μJ, increased by 71%.

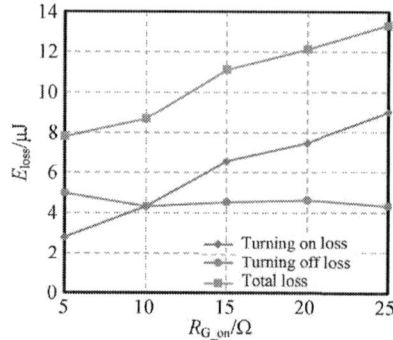

Fig. 12. The relationship curve between E_{on}, E_{off} and E_{total} of bottom switch and turn-on driving resistor.

C. Turn-off driving resistor

On the one hand, turn-off driving resistor indirectly affects crosstalk voltage by affecting dV_{DS}/dt during turn-off process. On the other hand, it will directly affect crosstalk voltage to a certain extent because turn-off drive resistor is in the driving turn-off loop.

Fig. 13 and Fig. 14 illustrate the influence of turn-off driving resistor. During turn-on process of Q_1, as turn-off driving resistor increases, positive crosstalk voltage increases from 2.9V to 3.7V. Since turn-off drive resistor basically has no influence on switching-on speed, the effect of turn-off driving resistor on positive crosstalk voltage is mainly due to the impact of drive turn-off loop impedance. During turn-off process of Q_1, under the combined action of two influencing factors, the influence of turn-off driving resistor itself is more significant, so the final result is that the absolute value of negative crosstalk voltage of Q_2 increases when turn-off driving resistor increases. With turn-off drive resistor increasing from 5Ω to 25Ω, the negative crosstalk voltage changes from −3.7V to −6.3V.

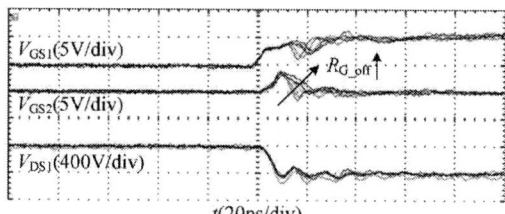

(a) Q_1 turn-on waveform

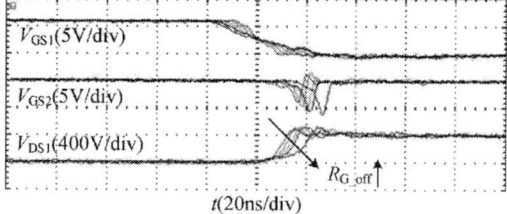

(b) Q_1 turn-off waveform

Fig. 13. Crosstalk voltage waveform of top switch under different turn-off driving resistor.

978-1-6654-4817-8/21 $31.00 © 2021 IEEE

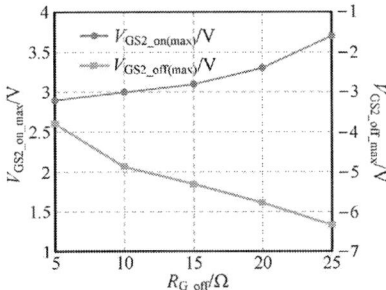

Fig. 14. The relationship curve between the crosstalk voltage of top switch and the turn-off gate resist.

The relationship curve between E_{on}, E_{off} and E_{total} of bottom switch and turn-off driving resistor is illustrated in Fig. 15. Reducing turn-off driving resistor has obvious effects on suppressing the crosstalk voltage and reducing the loss, but simultaneously it will cause a increased drain-source voltage spike when power device turns off.

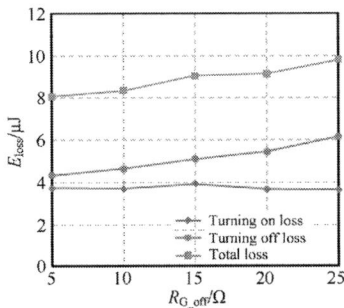

Fig. 15. The relationship curve between E_{on}, E_{off} and E_{total} of bottom switch and turn-off driving resistor.

Through the evaluation of various impact factors, it shows that only the adjustment of some circuit parameters can not effectively suppress crosstalk voltage, because it will bring about the problem of increased switching loss or increased spikes. Thus, it is necessary to seek effective suppression method.

IV. ACTIVE SUPPRESSION METHOD

The schematic diagram of crosstalk suppression circuit used in this paper is shown in Fig. 16. Auxiliary switches Q_{aux1}, Q_{aux2} are connected in series with auxiliary capacitor C_{aux1} and C_{aux2} respectively, and then connected in parallel to gate and source of the corresponding power device.

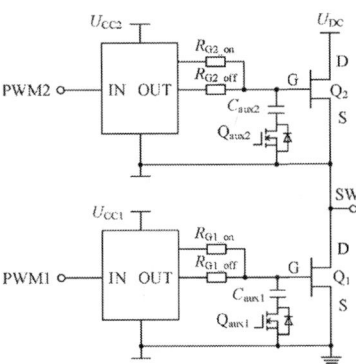

Fig. 16. Active Miller clamp circuit.

According to the drvinging sequence diagram shown in Fig. 17, after the auxiliary switch turns on, the auxiliary capacitor is equivalent to increasing gate-source capacitor, so as to suppress occurrence of crosstalk. Since the auxiliary capacitor only works when crosstalk occurs, the switching characteristics of power device are not affected.

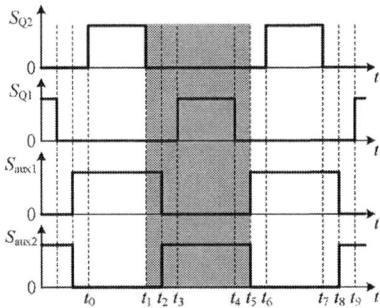

Fig. 17. Driving sequence diagram.

Fig. 18 shows the relationship curve between the crosstalk voltage and the auxiliary capacitor. With the increase of the auxiliary capacitor, the induced crosstalk voltage first decreases relatively in a slower rate, then decreases quickly, and at last decreases slowly too. Through trade-off, we finally choose the auxiliary capacitor value of $1\mu F$.

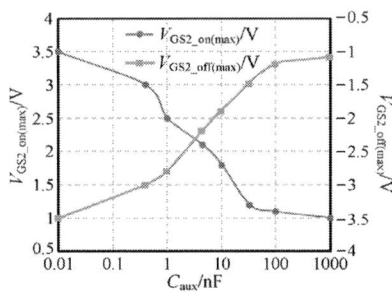

Fig. 18. The relationship curve between the crosstalk voltage and the auxiliary capacitor.

Table I shows the total loss of auxiliary switch Q_{aux2} and power switch Q_2 in a single switching cycle when $C_{aux2}=1\mu F$. As can be seen from the table, the loss of crosstalk suppression circuit is very small, and the impact on the system is negligible.

TABLE I. Q_{aux2} TOTAL LOSS AND Q_2 TOTAL LOSS

$C_{aux2}/\mu F$	Total Loss/μJ		Q_{aux2}/Q_2
	Q_{aux2}	Q_2	
1	0.025	97.975	0.00026

Fig. 19 and Fig. 20 compare the crosstalk voltage waveform without or with active Miller clamp circuit under the operating condition of 400V/8A. After the active suppression method is applied, crosstalk voltages are reduced from 3.5V and −3.5V to 1V and −1.3V respectively. The active suppression method realizes the effective suppression of crosstalk fluctuation.

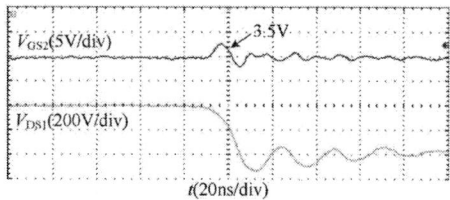

(a) Positive crosstalk voltage

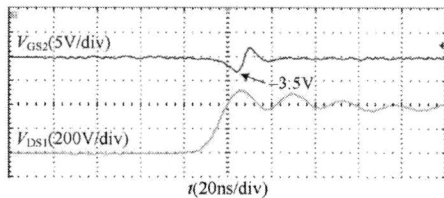

(b) Negative crosstalk voltage

Fig. 19. Crosstalk behavior without active Miller clamp circuit.

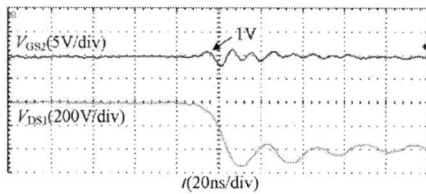

(a) Positive crosstalk voltage

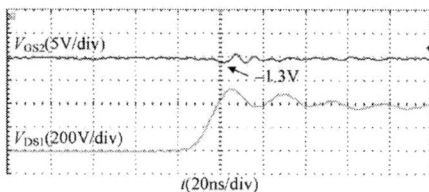

(b) Negative crosstalk voltage

Fig. 20. Crosstalk behavior when the auxiliary capacitor is 1μF.

V. CONCLUSION

According to the theoretical analysis and experimental research on crosstalk suppression in this paper, we can draw the following conclusions:

First, although increasing gate-source external capacitor or turn-on driving resistor has a significant suppression effect on crosstalk voltage, it is not suitable to be adopted because it will reduce switching speed and increase switching loss. Reducing turn-off driving resistor has obvious effects on suppressing crosstalk voltage and reducing loss, but

simultaneously it will cause a increase in the drain-source voltage spike when the power device turns off.

Second, combining turn-off loop impedance control and active Miller clamp to form an improved active Miller clamp crosstalk suppression method, which reduces gate-source equivalent impedance only during crosstalk voltage generation stage, does not affect normal operation of the power switch, and its loss can be ignored.

Third, by optimizing the value of the auxiliary capacitor in active suppression method, the crosstalk suppression effect can be enhanced.

REFERENCES

[1] E. A. Jones, F. Wang and B. Ozpineci, "Application-based review of GaN HFETs," 2014 IEEE Workshop on Wide Bandgap Power Devices and Applications, 2014, pp. 24-29.

[2] H. Qin, Z. Zhu, W. Dai, K. Xu and D. Fu, et al, "Influence of parasitic inductance on switching characteristics of SiC MOSFET," in Journal of Nanjing University of Aeronautics & Astronautics, vol. 49, no. 4, pp. 531-539, Aug. 2017.

[3] H. A. Mantooth, K. Peng, E. Santi and J. L. Hudgins, "Modeling of Wide Bandgap Power Semiconductor Devices — Part I," in IEEE Transactions on Electron Devices, vol. 62, no. 2, pp. 423-433, Feb. 2015.

[4] E. A. Jones, F. Wang, D. Costinett, Z. Zhang and B. Guo, "Cross conduction analysis for enhancement-mode 650-V GaN HFETs in a phase-leg topology," 2015 IEEE 3rd Workshop on Wide Bandgap Power Devices and Applications (WiPDA), Blacksburg, VA, USA, 2015, pp. 98-103.

[5] R. Xie, H. Wang, G. Tang, X. Yang and K. J. Chen, "An analytical model for false turn-on evaluation of high-voltage enhancement-mode GaN transistor in bridge-leg configuration," in IEEE Transactions on Power Electronics, vol. 32, no. 8, pp. 6416-6433, Aug. 2017.

[6] Z. Zhang, F. Wang, L. M. Tolbert and B. J. Blalock, "Active gate driver for crosstalk suppression of SiC devices in a phase-leg configuration," in IEEE Transactions on Power Electronics, vol. 29, no. 4, pp. 1986-1997, April 2014..

[7] R. Kelley, A. Ritenour, D. Sheridan and J. Casady, "Improved two-stage DC-coupled gate driver for enhancement-mode SiC JFET," 2010 Twenty-Fifth Annual IEEE Applied Power Electronics Conference and Exposition (APEC), Palm Springs, CA, USA, 2010, pp. 1838-1841.

[8] C. Liu, Z. Zhang, Y. Liu, Y. Si and Q. Lei, "Smart self-driving multilevel gate driver for fast switching and crosstalk suppression of SiC MOSFETs," in IEEE Journal of Emerging and Selected Topics in Power Electronics, vol. 8, no. 1, pp. 442-453, March 2020.

[9] Y. Li, M. Liang, J. Chen, T. Q. Zheng and H. Guo, "A low gate turn-off impedance driver for suppressing crosstalk of SiC MOSFET based on different discrete packages," in IEEE Journal of Emerging and Selected Topics in Power Electronics, vol. 7, no. 1, pp. 353-365, March 2019.

[10] E. Aeloiza, A. Kadavelugu and R. Rodrigues, "Novel bipolar active miller clamp for parallel SiC MOSFET power modules," 2018 IEEE Energy Conversion Congress and Exposition (ECCE), Portland, OR, USA, 2018, pp. 401-407.

[11] H. Qin, Z. Zhu, D. Wang, H. Xie and H. Xu, "Method of crosstalk suppression applied in silicon carbide based converters," in Journal of Nanjing University of Aeronautics & Astronautics, vol. 49, no. 6, pp. 872-882, Dec. 2017.

Comparing Hexagonal and Circular Cell designs for SiC MPS Diode: The Curvature Effect on Avalanche Capability

Li Liu[1], Na Ren[1,2]*, Jiupeng Wu[1,2], Zhengyun Zhu[1],
Hongyi Xu[1,2], Qing Guo[1] and Kuang Sheng[1,2]
[1] *College of Electrical Engineering, Zhejiang Univerisity*, Hangzhou, China
[2] *Hangzhou Innovation Center, Zhejiang Universit*, Hangzhou, China
Email: ren_na@zju.edu.cn

Abstract—In this work, the avalanche ruggedness of 1200V/2A SiC MPS diodes with hexagonal and circular cells are studied through experiment, modeling analysis, and 3D-TCAD simulation. The experimental result demonstrated that the avalanche energy/current capability of the circular cell-designed MPS diode (MPS-B) exhibit higher than the hexagonal cell-designed one (MPS-A). Further analysis and simulation results both reveal the MPS-A diode with hexagonal cell design will suffer higher junction temperature rise (ΔT_j) due to the more serious curvature effect and subsequent current/heating crowding during avalanche duration. This is adverse to the avalanche ruggedness. Therefore, it is recommend that a smoother design (such as the circular cell design in this case) is employed to replace the corner area and to alleviate the curvature effect, which helps to enhance the avalanche reliability of the SiC device.

Keywords—*SiC, MPS diode, Avalanche capability, Hexagonal cell, Circular cell, Curvature effect.*

I. INTRODUCTION

SiC MPS diode has achieved tremendous progress since commercialization in 2005[1]. With negligible reverse recovery, and a good trade-off between forward voltage drop and reverse leakage current, the MPS diode exhibits superior performance[2, 3] and has been widely employed in high voltage, high-frequency, and high-temperature applications.[4] To further enhance the device competitiveness, Many efforts[5-9] are made to improve the performance and reliability of the diode, including structure optimization design, etc. Hexagonal cell design is adopted in Infineon`s products[9]. However, the electric field crowding at the corners of the hexagonal P+ regions due to the curvature effect is seldom studied, which may have great impacts on the breakdown performance and avalanche robustness.

In this work, the avalanche capability of 1200V/2A SiC MPS diodes with two cell designs, i.e., hexagonal and circular cells are compared, analyzed, and investigated via experiment and TCAD simulation. Unclamped Inductive Switching (UIS) tests are carried out to characterize the avalanche energy and current capability of investigated MPS diode, and then a detail analysis of experimental results are follow. The electric field distributions of two cell structure under avalanche breakdown condition are studied through 3D-TCAD simulation, the influence of curvature effect on avalanche capability is discussed. Finally, suggestion for optimizing and improving avalanche reliability of SiC device is also given out.

II. DEVICE STRUCTURE AND UIS TEST SETUP

A. Device Structure

The cross-sectional schematic view of the hexagonal cell-designed MPS diode (named MPS-A) and circular cell-designed MPS diode (named MPS-B) are shown in Fig. 1 (a) and (b), respectively. The structural parameters are set the same. The doping concentration and thickness of the drift layer are $8 \times 10^{15} \text{cm}^{-3}$ and 12μm, and the P+ regions of the cell are composed of outer P+ rings (width: P=1.73μm, spacing: N=3μm) and inner P+ islands (width: W=7μm).

The two cell-designed 1200V/2A SiC MPS diodes are fabricated on a 4-inch 4H-SiC wafer, and then packaged into TO-220. The active region is 1mm^2.

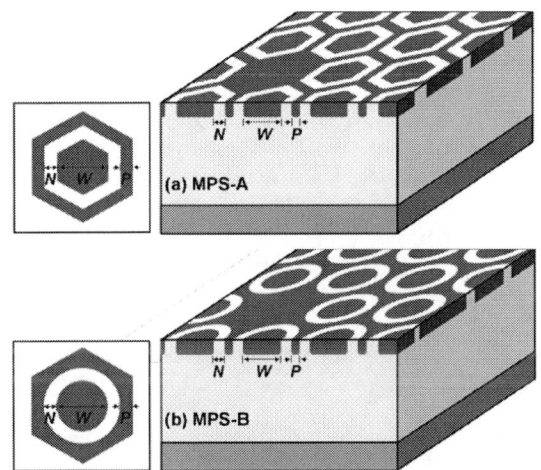

Fig. 1. Cross-sectional views of SiC MPS diode with two cell designs (a) MPS-A: Hexagonal cell, and (b) MPS-B: Circular cell

B. UIS Setup

The standard UIS tests are conducted to get the avalanche capability of investigated MPS diodes. Fig. 2 shows the UIS test bench and circuit schematic. In this circuit, the device under test (DUT) is connected in parallel with a high-voltage SiC MOSFET. The DC bus voltage is set at 100 V, and the load inductors(L) are selected to be 1mH and 5mH. When the MOSFET switch turns on, the inductor is charged to a peaked current. When the MOSFET switch turns off, the inductor current flows through the DUT diode from cathode to anode. The device enters the avalanche breakdown mode and

978-1-6654-4817-8/21 $31.00 © 2021 IEEE

consumes the energy stored in the inductive load. During the UIS test, the on-state time (t_{on}) of the MOSFET switch is increased gradually by adjusting the gate pulse signal, thereby increasing the inductor peak current step by step. DUT withstands higher and higher avalanche energy until the device failure occurs.

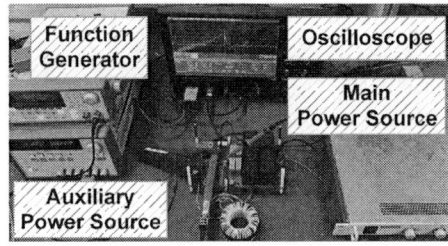

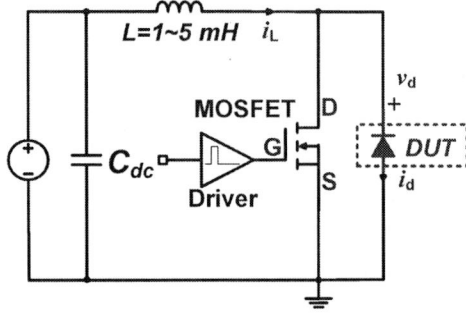

Fig. 2. UIS test bench and circuit diagram

III. EXPERIMENTAL RESULT OF UIS TEST

The UIS tests are performed for the two devices to investigate the avalanche current/energy capabilities. Fig. 3(a) and (b) show the experimental results of avalanche current and voltage waveforms of MPS-A diode when tested under conditions of L=1mH and 5mH, respectively. The on-state time (t_{on}) of the MOSFET switch is increased gradually to increase the peak inductor current (i.e. avalanche current I_{ave}) as well as the dissipated avalanche energy (E_{ave}). The maximum E_{ave} and I_{ave} that the device can withstand are calculated or extracted from the last test waveforms before device failure.

It can be seen from Fig. 3, the maximum I_{ave} of MPS-A diode under the two load inductors are 10.29A (t_{on}=113.8μs, 1mH) and 5.26A (t_{on}=280μs, 5mH), repectively. And the maximum E_{ave} is defined as the avalanche energy capability, which can be calculated by equation (1).

$$E_{ave} = \int_0^{t_{ave}} v_d(t) \cdot i_d(t)\, dt \qquad (1)$$

where, v_d is the diode avalanche voltage, i_d is the diode avalanche current and t_{ave} is the avalanche duration.

The avalanche energy capability and avalanche current capability for MPS-A diode and MPS-B diode are summarized and compared in Fig. 4(a)~(d). Obviously, the MPS-B diode has better avalanche capability both in terms of avalanche energy (140~220mJ/cm^{-2} higher) and avalanche current (6~8A/cm^{-2} higher). The experimental results illustrate that the circular cell design is helpful to improve the avalanche capability of the device.

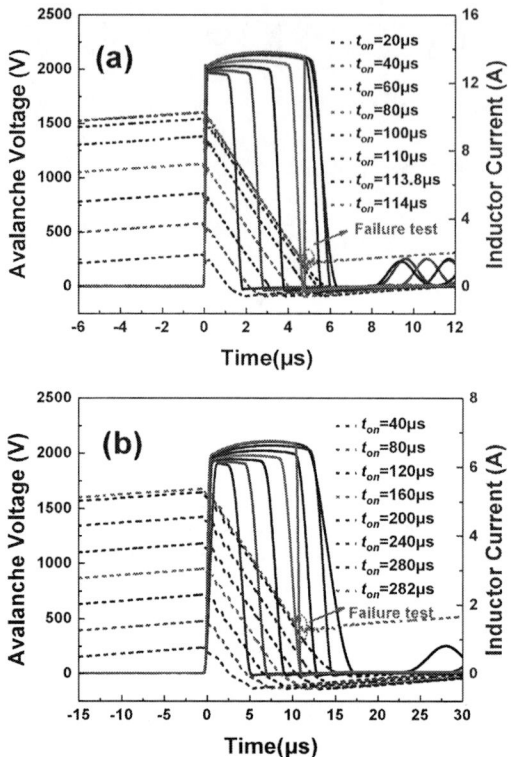

Fig. 3. UIS test results of the investigated SiC MPS-A diode: avalanche voltage and inductor current waveforms, tested at (a) 1mH inductance and (b) 5mH inductance, respectively.

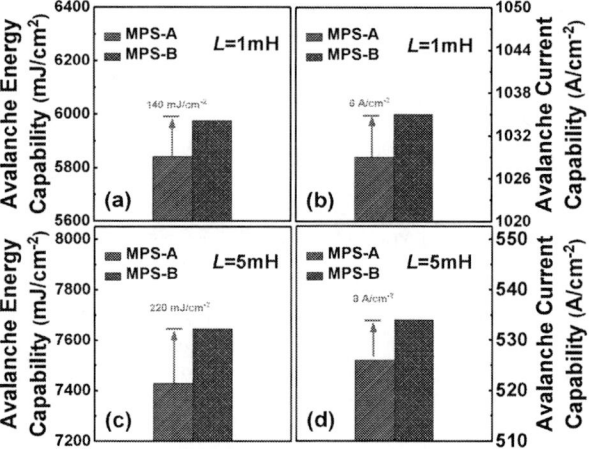

Fig. 4. The comparisons of avalanche energy/current capabilities of MPS-A and MPS-B diodes.

IV. ANALYSIS AND SIMULATION

In this section, detailed analysis and 3D-TCAD simulation are carried to further study the impact of cell structure on avalanche capability. Fig. 5(a) shows the avalanche waveforms of MPS-A and MPS-B diodes (t_{on}=110us, 1mH). The avalanche current waveforms of the two devices almost overlap, while their avalanche voltage waveforms show different conditions that MPS-A diode has almost the same initial avalanche voltage v_{d_intial} but higher maximum voltage v_{d_max} than MPS-B diode. Comparing the other avalanche

978-1-6654-4817-8/21 $31.00 © 2021 IEEE   154

voltage waveforms tested at various t_{on}, shown in Fig. 5(b) and (c), this phenomenon still holds. The initial avalanche voltage of two MPS diode are the same, which means the avalanche breakdown voltage is similar to the two devices at room temperature T_{j0}. After the initial, the junction temperature T_j of the device starts to rise due to the energy dissipation. A V-T model[10, 11] to estimate the T_j during avalanche duration can be expressed by

$$v_d = v_{d\_intial} + I_d \cdot \rho_{sc}$$
$$\approx v_{d\_intial0}\{1 + \beta \cdot (T_j - T_{j0})\} + I_d \cdot \rho_{sc0} \quad (2)$$

where, v_{d_intial} and $v_{d\_intial0}$ are the breakdown voltage at T_j and T_{j0}. β is the temperature coefficient of the avalanche voltage. ρ_{sc} and ρ_{sc0} are the space charge resistance at T_j and T_{j0}. From equation (2), the junction temperature rise (ΔT_j) can be calculated by using (3)

$$\Delta T_J = T_J + T_{J0} \cdot \rho_{sc}$$
$$= (v_d - v_{d\_intial0} - I_d \cdot \rho_{sc0})/\beta v_{intial0} \quad (3)$$

Simultaneously, the avalanche voltage rises along with the junction temperature due to its positive temperature coefficient[11]. However, the MPS-A diode shows a higher increase rate and higher maximum voltage (v_{d_max}) during the avalanche pulse. This means the MPS-A diode has a higher junction temperature rise (ΔT_j) when suffering the same energy (energy from the inductor is the same for the two devices). Thus, the premature failure due to critical temperature will cause a reduced avalanche capability for the MPS-A diode.

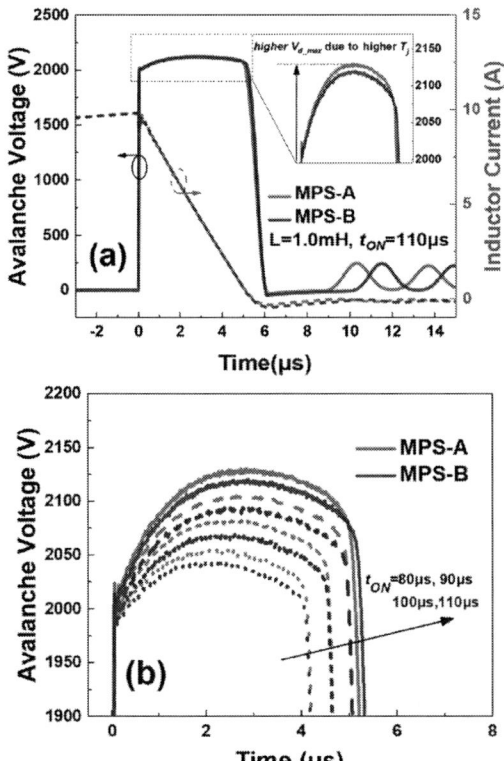

Fig. 5. The comparisons of (a) avalanche voltage/current waveforms (t_{on}=110µs), (b) the avalanche voltage increase (t_{on}=80-110µs) of MPS-A and MPS-B diodes.

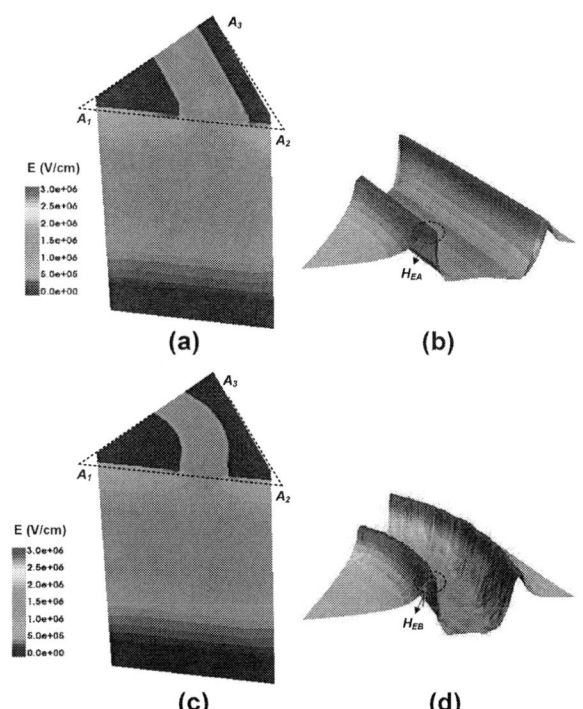

Fig. 7. The Simulation results of 3D electric field distribution under reverse-blocking breakdown and 2D planar electric field distribution on a 0.55µm plane from the anode surface (face $A_1A_2A_3$) for the (a)~(b) the SiC MPS-A diode with hexagonal cell design, and (c)~(d) the SiC MPS-B diode with circular cell design.

The 3D simulations are performed to study the device behavior in avalanche conditions. Fig. 7(a)~(d) show the Simulation results of 3D electric field distribution under reverse-blocking breakdown and 2D planar electric field distribution on a 0.55µm plane from the anode surface (face $A_1A_2A_3$). The corners of hexagonal P+ islands (H_{EA} marked in Fig. 7(b)) are higher than the edges of circular P+ islands (H_{EB} marked in Fig. 7(d)) due to the curvature effect. This more unbalanced electric field distribution can lead to more serious localized current crowding and unbalanced current distribution, which means a reduced effective power dissipation area and heat crowding happened[8, 12]. Therefore, the junction temperature for the MPS-A diode will rise higher during the avalanche duration. Since the thermal runaway is the common failure mechanism of avalanche for SiC diode, the MPS-A diode with the hexagonal cell gets failed at a lower current/energy than that of the MPS-B diode with reduced curvature effect.

This phenomenon illustrates a smoother design (such as the circular cell design) should be used to replace the corner area and to alleviate the electric field crowding issue, which helps to enhance the avalanche reliability for the SiC device.

V. CONCLUSION

In this work, the avalanche ruggedness of 1200V/2A SiC MPS diodes with hexagonal and circular cells are investigated through single unclamped inductive switching (UIS) test. The experiment results demonstrate that a circular cell-designed MPS diode (MPS-B) exhibits higher avalanche energy/current capability than the hexagonal cell-designed one (MPS-A). A

V-T model is applied to analyze the junction temperature rise (ΔT_j) during the avalanche condition. It is found that the hexagonal cell-designed diode (MPS-A) has a higher junction temperature rise when suffering the same energy, thereby leading to premature failure. Simulation results show that more unbalanced electric field distribution for MPS-A diode due to curvature effect can lead to more serious localized current crowding and unbalanced current distribution under avalanche stress events. This means a reduced effective power dissipation area and heat crowding. comparing two cell-designed MPS diodes, this work proves that the curvature effect has an impact on the avalanche robustness. It is also suggested that a smoother design (such as the circular cell design) should be used to replace the corner area and to alleviate the curvature effect, which helps to enhance the avalanche robustness of the SiC device.

ACKNOWLEDGMENT

This research was funded by grants from the Guangdong Key Research and Development Program of China (2019B010143001) and National Natural Science Foundation of China (52007165) & Power Electronics Science and Education Development Program of Delta Group. The authors also would like to thank the Technology Innovation and Training Center, Polytechnic Institute, Zhejiang University, Hangzhou, Zhejiang Province, China, for providing the measurement equipment Agilent B1505A.

REFERENCES

[1] R. Rupp, M. Treu, S. Voss, F. Bjork, and T. Reimann, ""2nd Generation" SiC Schottky diodes: A new benchmark in SiC device ruggedness," in *2006 IEEE International Symposium on Power Semiconductor Devices and IC's*, 4-8 June 2006 2006, pp. 1-4, doi: 10.1109/ISPSD.2006.1666123.

[2] O. Harmon, T. Basler, and F. Björk, "Advantages of the 1200 V SiC Schottky Diode with MPS Design," pp. 34-37, 2015, doi: 10.13140/RG.2.1.4951.0486.

[3] J. Wu, N. Ren, H. Wang, and K. Sheng, "1.2kV 4H-SiC Merged PiN Schottky Diode with Improved Surge Current Capability," *IEEE Journal of Emerging and Selected Topics in Power Electronics,* pp. 1-1, 2019, doi: 10.1109/JESTPE. 2019.2921970.

[4] M. Treu *et al.*, "A surge current stable and avalanche rugged SiC merged pn Schottky diode blocking 600V especially suited for PFC applications," in *Materials science forum*, 2006, vol. 527: Trans Tech Publ, pp. 1155-1158.

[5] S. Palanisamy, S. Fichtner, J. Lutz, T. Basler, and R. Rupp, "Various structures of 1200V SiC MPS diode models and their simulated surge current behavior in comparison to measurement," in *2016 28th International Symposium on Power Semiconductor Devices and ICs (ISPSD)*, 12-16 June 2016 2016, pp. 235-238, doi: 10.1109/ISPSD.2016.7520821.

[6] N. Ren and K. Sheng, "An Analytical Model With 2-D Effects for 4H-SiC Trenched Junction Barrier Schottky Diodes," *IEEE Transactions on Electron Devices,* vol. 61, no. 12, pp. 4158-4165, 2014, doi: 10.1109/TED.2014.2365519.

[7] N. Ren, J. Wang, and K. Sheng, "Design and Experimental Study of 4H-SiC Trenched Junction Barrier Schottky Diodes," *IEEE Transactions on Electron Devices,* vol. 61, no. 7, pp. 2459-2465, 2014, doi: 10.1109/TED.2014.2320979.

[8] L. Liu, J. Wu, N. Ren, Q. Guo, and K. Sheng, "1200-V 4H-SiC Merged p-i-n Schottky Diodes With High Avalanche Capability," *IEEE Transactions on Electron Devices*, pp. 1-6, 2020, doi: 10.1109/TED.2020.3007136.

[9] T. Basler, R. Rupp, R. Gerlach, B. Zippelius, and M. Draghici, "Avalanche Robustness of SiC MPS Diodes," in *PCIM Europe 2016; International Exhibition and Conference for Power Electronics, Intelligent Motion, Renewable Energy and Energy Management*, 2016: VDE, pp. 1-8.

[10] Y. Okuto, "Junction Temperatures under Breakdown Condition," *Japanese Journal of Applied Physics*, vol. 8, no. 7, pp. 917-922, 1969/07 1969, doi: 10.1143/jjap.8.917.

[11] N. Ren *et al.*, "Investigation on single pulse avalanche failure of SiC MOSFET and Si IGBT," vol. 152, pp. 33-40, 2019.

[12] L. Liu *et al.*, "Single Pulse Avalanche Robustness and Analysis for 1200-V SiC Junction Barrier Schottky Diode," in *2020 17th China International Forum on Solid State Lighting & 2020 International Forum on Wide Bandgap Semiconductors China (SSLChina: IFWS)*, 23-25 Nov. 2020 2020, pp. 90-93, doi: 10.1109/SSLChinaIFWS51786. 2020.9308741.

A Real-Time Self-Learning Control for Megahertz GaN-based DC-DC Converter

Jing Chen
The State Key Laboratory of Advanced Electromagnetic Engineering and Technology
Huazhong University of Science and Technology (HUST)
Hubei, China
jingchen@hust.edu.cn

Yu Chen
The State Key Laboratory of Advanced Electromagnetic Engineering and Technology
Huazhong University of Science and Technology (HUST)
Hubei, China
ayu03@163.com

Yong Kang
The State Key Laboratory of Advanced Electromagnetic Engineering and Technology
Huazhong University of Science and Technology (HUST)
Hubei, China
ykang@mail.hust.edu.cn

Abstract—With the vigorous development of the wide bandgap power devices, the switching frequency of DC-DC converter is improved to MHz. But it leads to the problem that the high-performance but complicated adaptive control methods cannot be completed in such a short switching cycle. To solve it, this paper proposes a real-time self-learning control methodology, which can be implemented at MHz switching frequency by FPGA and its dynamic control performance can be real-time enhanced. A GaN-based buck converter prototype with 1 MHz switching frequency is built to demonstrate the feasibility of the proposed real-time self-learning control methodology.

Keywords—wide bandgap power device, GaN-based DC-DC converter, MHz switching frequency, real-time, self-learning control

I. INTRODUCTION

With the vigorous development of wide bandgap devices, increasing the switching frequency of DC-DC converters is a feasible way to improve power density. For example, the operating frequency of GaN-based DC-DC converters has reached megahertz (MHz)[1-4]. However, increasing switching frequency not only reduces the volume of passive devices, but also shortens the switching cycle, making high-performance but complicated adaptive control methodologies hard to be completed[5-6].

To achieve high-performance control at MHz, a neural network (NN)-based control algorithm was proposed in [7], where the NN was trained to fit the high performance control law by offline training. Due to the parallel structure of the designed NN controller, the Field Programmable Gate Array (FPGA) that has parallel computing ability, was used for fast implementation and 1MHz was successfully realized in [7].

However, when there is parameter deviation between the design based model and the actual model, the control performance will inevitably decrease[7]. For this reason, real-time optimization is required to ensure the high control performance. Online training methods for NN controller were proposed in [8-9], however, since the NN control calculation and the NN parameter training involve considerate amount of calculation, they cannot be implemented at MHz switching frequency. In this regard, this paper proposes a real-time self-learning control algorithm to enhance the control performance under parameter deviation at MHz switching frequency. To adapt the MHz switching frequency, the training rules of NN controller are properly integrated and simplified to reduce the

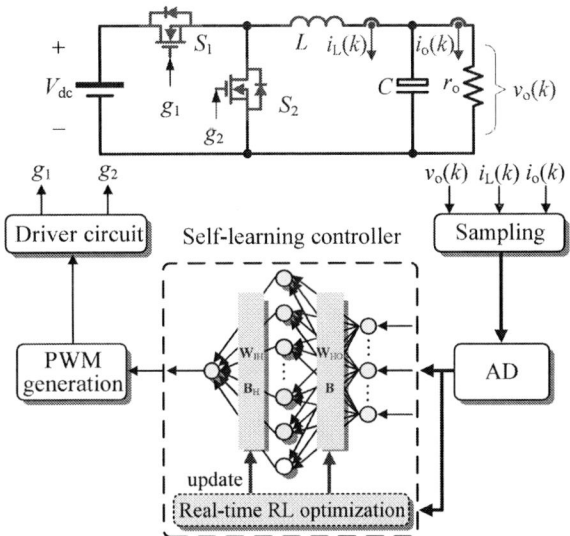

Fig. 1. Flowchart of buck converter with the proposed real-time self-learning control algorithm

real-time calculation. FPGA timing optimization design is also utilized to parallel the current round of the NN control calculation and the previous round of the NN parameter training, so as to avoid extra execution time. In addition, parameter sharing is also realized between the NN control calculation and the NN parameter training to further reduce the calculation burden. With such a design, the proposed self-training control method is able to be implemented within 4 FPGA clock cycles and 1MHz switching frequency is successfully realized by a low-cost FPGA.

The remainder of this paper is organized as follows: in Section II, the NN controller is briefly reviewed and the proposed real-time self-learning control method is proposed; in Section III, the fast FPGA implementation scheme and the experimental results under 1MHz switching frequency are addressed; finally, the conclusion is given in Section IV.

II. THE PROPOSED REAL-TIME SELF-LEARNING CONTROL

A. Modeling and control of buck converter

To develop the proposed control method, the discrete buck model is built as[7]:

978-1-6654-4817-8/21 $31.00 © 2021 IEEE

$$\begin{bmatrix} i_L(k+1) \\ v_o(k+1) \end{bmatrix} = \underbrace{\begin{bmatrix} a_{11} & a_{12} \\ a_{21} & a_{22} \end{bmatrix}}_{\mathbf{A}_d} \begin{bmatrix} i_L(k) \\ v_o(k) \end{bmatrix} + \underbrace{\begin{bmatrix} b_1 \\ b_2 \end{bmatrix}}_{\mathbf{B}_d} d(k) \quad (1)$$

where $\mathbf{A}_d = e^{\mathbf{A}T_s}$, $\mathbf{A} = \begin{bmatrix} 0 & -1/L \\ 1/C & -1/r_oC \end{bmatrix}$ and $\mathbf{B}_d = \int_0^{T_s} e^{\mathbf{A}(T_s-\tau)} d\tau \begin{bmatrix} v_{dc} \\ L \\ 0 \end{bmatrix}$;

$d(k)$, $v_o(k)$ and $i_L(k)$ are the values of i_L, v_o and d of the k-th horizon.

Based on the model (1), a variety of high-performance nonlinear controllers such as sliding mode controller[3] or model predictive controller[4] can be designed. But no matter which method is adopted, the essence is to send the k-th state variables $\mathbf{X}(k)$ to the controller, and calculates the $(k+1)$-th duty ratio $d(k+1)$ according to a certain algorithm $f(\mathbf{X}(k))$, as:

$$d(k+1) = f(\mathbf{X}(k)) \quad (2)$$

To achieve high performance control under MHz switching frequency, this paper utilizes the control method in [7], where a single-hidden–layer neutral network (NN) is utilized to imitate the input-output mapping relationship of the high performance explicit model predictive control law generated at different operating points. Such a method converts the serial online operation process into the fast parallel form. With the parallel implementation of FPGA, the NN controller is able to be implemented in 3 FPGA clock cycles, 1MHz switching frequency was achieved. Based on this method, a high-speed NN controller is built as Fig. 1. The expression of the NN controller is[7]:

$$\mathbf{H} = \text{ReLU}\left[\mathbf{W}_{IH}\mathbf{X}(k) + \mathbf{B}_H\right] \quad (3)$$

$$d_{RL}(k+1) = \text{ReLU}\left[\mathbf{W}_{HO}\mathbf{H} + \mathbf{B}\right]$$
$$\triangleq f_\varphi(\mathbf{X}(k)) \quad (4)$$

$$\text{ReLU}(x) = \begin{cases} 0 & x \le 0 \\ x & x > 0 \end{cases} \quad (5)$$

where $\mathbf{X}(k) = [v_{dc}(k), i_L(k), v_o(k), i_o(k)]$, ReLU denotes the activation function. $\mathbf{W}_{IH}$, $\mathbf{B}_H$, $\mathbf{W}_{HO}$, $\mathbf{B}$ denote the weight matrices of the NN controller. $\mathbf{H}$ denotes the output matrix of the hidden layer. N_H is the number of hidden layer neurons. φ denotes $\mathbf{B}$, $\mathbf{W}_{HO}$, $\mathbf{B}_H$ and $\mathbf{W}_{IH}$.

B. Real-time self-learning algorithm design

As mentioned above, since the NN controller is trained offline on the basis of the rated converter parameters, the control performance will inevitably be deteriorated when the actual parameters mismatch the rated ones. Therefore real-time self-learning algorithm should be implemented to adjust the NN parameter φ to handle the parameter mismatch, achieving "self-learning". In addition, "self-learning" can continuously enhance the control strategy real-time automatically to realize better control performance.

As a result of the inertia of the power electronic converter, the duty cycle $d(k+1)$ applied at the $(k+1)$-th horizon has an impact on subsequent horizons, so the instantaneous and long-term control effects of $d(k+1)$ should be evaluated. For this consideration, state variables of the subsequent N horizons are recorded to evaluate the control performance of $d(k+1)$, and the reward function is thus designed as:

$$R_{RL}(k+1) = \begin{cases} -P \cdot e_v(k+i+1)^2 & e_v(k+i+1) \cdot e_v(k+i) < 0 \\ -\sum_{i=1}^{N} \gamma_1^{i-1} e_v(k+i+1)^2 & \text{others} \end{cases} \quad (6)$$

where $e_v(k+i+1) = V_{ref} - v_o(k+i+1)$ denotes the output voltage error of the $(k+i+1)$-th horizon. (6) can quantitatively evaluate two typical dynamic characteristics:

- Overshoot: $e_v(k+i+1) \cdot e_v(k+i) < 0$ indicates there is overshoot, which means the control effect of $d(k+1)$ on the $(k+i+1)$-th horizon is extremely poor. Therefore, a large penalty coefficient P ($P > 0$) is applied to increase $R_{RL}(k+1)$.

- No overshoot: $e_v(k+i+1) \cdot e_v(k+i) \ge 0$ indicates there is no overshoot in the $(k+i+1)$-th horizon. Accumulating the output voltage errors of N horizons to get $R_{RL}(k+1)$, and the discount factor γ_1 is utilized to weight the time factor.

$R_{RL}(k+1)$ can be optimized by adjusting φ according to the gradient of $R_{RL}(k+1)$ with respect to φ. According to the chain rule:

$$\frac{\partial R_{RL}(k+1)}{\partial \varphi} = \frac{\partial R_{RL}(k+1)}{\partial d(k+1)} \frac{\partial d(k+1)}{\partial f_\varphi(\mathbf{X}(k))} \frac{\partial f_\varphi(\mathbf{X}(k))}{\partial \varphi} = \frac{\partial R_{RL}(k+1)}{\partial d(k+1)} \frac{\partial f_\varphi(\mathbf{X}(k))}{\partial \varphi} \quad (7)$$

where $\partial R_{RL}(k+1)/\partial d(k+1)$ can be expressed as:

$$\frac{\partial R_{RL}(k+1)}{\partial d(k+1)} = \begin{cases} 2P \cdot e_v(k+i+1) \cdot \dfrac{\partial v_o(k+i+1)}{\partial d(k+1)} & e(k+i+1) \cdot e(k+i) < 0 \\ 2\sum_{i=1}^{N} \gamma_1^{i-1} \cdot e_v(k+i+1) \cdot \dfrac{\partial v_o(k+i+1)}{\partial d(k+1)} & \text{else} \end{cases} \quad (8)$$

Since $v_o(k+i+1)$ and $i_o(k+i+1)$ have been recorded, (8) can be simplified as:

$$\frac{\partial R_{RL}(k+1)}{\partial d(k+1)} = \begin{cases} 2P \cdot e_v(k+i+1) \cdot b_2 \cdot a_{22}^{i-1} & e_v(k+i+1) \cdot e_v(k+i) < 0 \\ 2\sum_{i=1}^{N} \gamma_1^{i-1} \cdot e_v(k+i+1) \cdot b_2 \cdot a_{22}^{i-1} & \text{else} \end{cases} \quad (9)$$

Subscribing (9) into (7), yielding:

$$\frac{\partial R_{RL}(k+1)}{\partial \varphi} = \begin{cases} 2P \cdot e_v(k+i+1) \cdot b_2 \cdot a_{22}^{i-1} \cdot \dfrac{\partial g_\varphi(\mathbf{X}(k))}{\partial \varphi} & e_v(k+i+1) \cdot e_v(k+i) < 0 \\ 2\sum_{i=1}^{N} \gamma_1^{i-1} \cdot e_v(k+i+1) \cdot b_2 \cdot a_{22}^{i-1} \cdot \dfrac{\partial g_\varphi(\mathbf{X}(k))}{\partial \varphi} & \text{else} \end{cases} \quad (10)$$

$\partial f_\varphi(\mathbf{X}(k))/\partial \varphi$ in (10) can be directly calculated by the gradient descent algorithm [7-9]. According to (4), $\partial f_\varphi(\mathbf{X}(k))/\partial \mathbf{W}_{HO}$ can be expressed as:

$$\frac{\partial f_\varphi(\mathbf{X}(k))}{\partial \mathbf{W}_{HO}} = \text{sgn}[f_\varphi(\mathbf{X}(k))] \quad (11)$$

where $\text{sgn}[\cdot]$ denotes the gradient of ReLU.

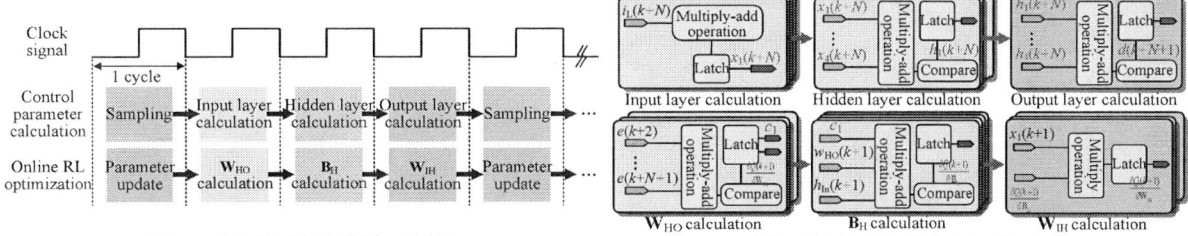

(a) Execution Flowchart of submodules   (b) Parallel implementation of submodules

Fig. 2. Fast FPGA RTL implementation.

Subscribing (11) into (10), and defining $\zeta_1=2P\cdot b_2$ and $\gamma_2=\gamma_1\cdot a_{22}$, yielding:

$$\frac{\partial R_{RL}(k+1)}{\partial \mathbf{W}_{HO}}=\begin{cases} \underbrace{\zeta_1\cdot a_{22}{}^{i-1}\cdot e_v(k+i+1)\cdot \text{sgn}[f_\varphi(\mathbf{X}(k))]}_{c_1}\cdot\mathbf{H} & e(k+i+1)\cdot e(k+i)<0 \\ \underbrace{\sum_{i=1}^{N}\zeta_2\cdot \gamma_2{}^{i-1}\cdot e_v(k+i+1)\cdot \text{sgn}[f_\varphi(\mathbf{X}(k))]}_{c_2}\cdot\mathbf{H} & \text{else} \end{cases}$$

(12)

$\partial f_\varphi(\mathbf{X}(k))\big/\partial\mathbf{B}_H$ and $\partial f_\varphi(\mathbf{X}(k))\big/\partial\mathbf{W}_{IH}$ can be successively deduced as:

$$\frac{\partial R_{RL}(k+1)}{\partial\mathbf{B}_H}=\begin{cases} c_1\cdot\mathbf{W}_{HO}\cdot\text{sgn}[\mathbf{H}] & e_v(k+i+1)\cdot e_v(k+i)<0 \\ c_2\cdot\mathbf{W}_{HO}\cdot\text{sgn}[\mathbf{H}] & \text{others}\end{cases}$$ (14)

$$\frac{\partial R_{RL}(k+1)}{\partial\mathbf{W}_{IH}}=\frac{\partial R_{RL}(k+1)}{\partial\mathbf{B}_H}\cdot\mathbf{X}(k)$$ (15)

It can be seen that (14)-(15) can be obtained by performing extra simple calculations on the basis of (13)-(14), respectively, which greatly reduces the online calculation burden. Then $\mathbf{W}_{IH}$, $\mathbf{W}_{HO}$ and $\mathbf{B}_H$ update as:

$$\mathbf{W}_{IH}\leftarrow\mathbf{W}_{IH}-\eta_{on}\frac{\partial R(k+1)}{\partial\mathbf{W}_{IH}}$$ (16)

$$\mathbf{W}_{HO}\leftarrow\mathbf{W}_{HO}-\eta_{on}\frac{\partial R(k+1)}{\partial\mathbf{W}_{HO}}$$

$$\mathbf{B}_H\leftarrow\mathbf{B}_H-\eta_{on}\frac{\partial R(k+1)}{\partial\mathbf{B}_H}$$

where η_{on} is the update rate.

III. Fast FPGA Implementation and Experimental Verification

A. Fast FPGA Implementation

The online execution of the proposed real-time self-learning controller includes the forward control calculation and backward self-learning calculation. The forward calculation is to obtain the control duty cycle for real-time control. It can be subdivided into four submodules: state variable sampling submodule, input layer calculation submodule, hidden layer calculation submodule and output layer calculation submodule. Each submodule takes 1 FPGA clock cycle, so the total consuming time for the forward calculation is 4 FPGA clock cycles[7]. The backward self-learning can also be divided into 4 parts, namely $\mathbf{W}_{IH}$, $\mathbf{W}_{HO}$ and $\mathbf{B}_H$ partial derivative calculations in (13-15) and parameter update calculations in (16).

To minimize the online execution time and achieve MHz switching frequency, this paper designs a highly parallel implementation scheme as shown in Fig. 2:

- Since the forward calculation involves parameters $\mathbf{W}_{IH}$, $\mathbf{W}_{HO}$ and $\mathbf{B}_H$, the update of $\mathbf{W}_{IH}$, $\mathbf{W}_{HO}$ and $\mathbf{B}_H$ should be paralleled with the sampling submodule; as shown in Fig. 2(a), the previous round of the partial derivative calculation of $\mathbf{W}_{IH}$, $\mathbf{W}_{HO}$ and $\mathbf{B}_H$ in (13)-(15) should be paralleled with the current round forward calculation of the three layers, respectively. Such a design can avoid the conflict between control calculation and self-learning and minimize the online execution time. Thus the proposed real-time self-learning control system can be implemented in 4 FPGA clock cycles.

- As shown in Fig. 2(b), the intermediate calculation results in the partial derivative calculation are fully multiplexed to simplify the calculation. Calculation in the same layer is parallel, for example $\mathbf{W}_{HO}$'s partial derivative calculation can be divided into N parallel execution calculation units, and the implementation of each unit only includes simple zero-crossing comparison, multiplication and addition operations; the partial derivative of $\mathbf{W}_{IH}$ directly uses the $\mathbf{W}_{HO}$'s intermediate calculation results c_1 or c_2; the partial derivative calculation of $\mathbf{B}_H$ also directly uses the partial derivative result of $\mathbf{W}_{IH}$.

With the above design, when N_H and N are set as 8 and 4, the proposed control only needs to perform 25 zero-crossing comparisons, 109 multiplications and 192 addition-subtraction, which not only reduces the resource requirement of FPGA, but also can be completed within 4 FPGA clock cycles. As a result, the proposed self-learning control algorithm is particularly suitable for MHz switching frequency occasions.

B. Prototype Design

To demonstrate the effectiveness of the proposed real-time self-learning control algorithm, a 1kW, 1MHz GaN-based buck prototype is built in Fig. 3(a). Intel's low-cost FPGA EP4C15F23C8 is used for digital implementation. The circuit parameters of the prototype are given in Table I.

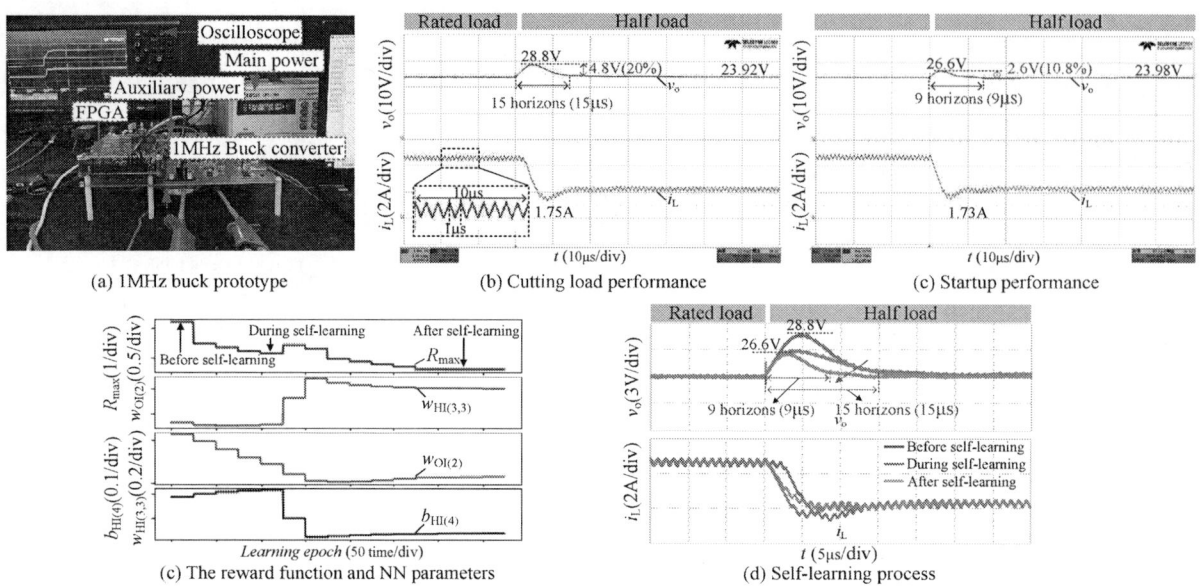

(a) 1MHz buck prototype

(b) Cutting load performance

(c) Startup performance

(c) The reward function and NN parameters

(d) Self-learning process

Fig. 3. 1 MHz buck prototype and the experimental verification.

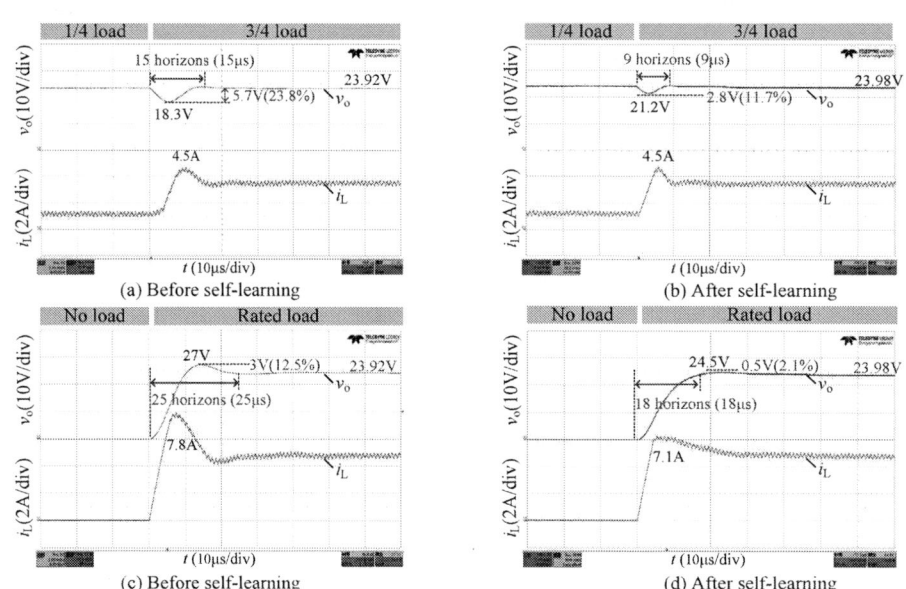

(a) Before self-learning

(b) After self-learning

(c) Before self-learning

(d) After self-learning

Fig. 4. Experimental comparisons under rated parameters.

TABLE I.   BUCK PROTOTYPE PARAMETER

Variable	Parameter	Value
L	Inductor	29 μH
C	Capacitor	1.1 μF
r_o	Rated load resistor	5 Ω
f_s	Switching frequency	1MHz
V_{in}	Rated input voltage	48 V
V_{ref}	Reference output voltage	24 V

C. Experimental Verification

The experimental results under the rated parameters (namely, L=29μH and C=1.1μF) and under parameter deviation are performed to demonstrate the proposed control can real-time enhance the control performance under both rated parameters and deviated parameters.

1) Experimental results under rated converter parameters

Fig. 3 and Fig. 4 show the waveforms under rated converter parameters. Fig. 3(b) gives the steady-state waveform, the waveform of i_L indicates that the switching cycle is 1μs, that is, the switching frequency is 1MHz. Fig. 3(b) and 3(c) give the cutting load waveforms before and after self-

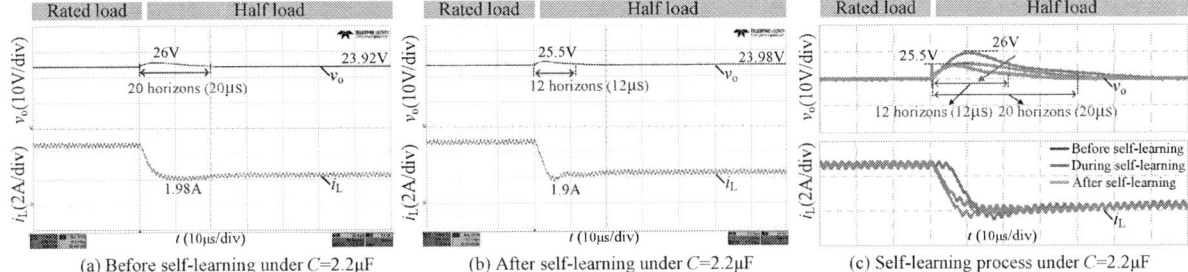

(a) Before self-learning under $C=2.2\mu F$

(b) After self-learning under $C=2.2\mu F$

(c) Self-learning process under $C=2.2\mu F$

Fig. 5. Experimental results under parameter deviation.

learning, it is seen that the self-learning reduced the settling time from 15 horizons (15μs) to 9 horizons (9μs), the voltage overshoot is reduced from 4.8V to 2.6V and the steady state error is reduced by 0.06V. Fig. 3(d) and Fig. 3(e) further gives the self-learning process, Fig. 3(d) verifies that the settling time and voltage overshoot are gradually reduced. Such results coincide with the optimized maximal reward R_{max} and NN parameters in Fig. 3(e).

Fig. 4 further shows the comparisons during adding load and startup, both proving that the control performance is enhanced after self-learning. The above results all demonstrate that the proposed self-learning control algorithm can real-time enhance the control performance under the rated parameters.

2) Experimental results under the deviated parameters

To prove the proposed self-learning method can enhance the control performance when the circuit parameters deviate greatly. Fig. 5 gives the waveforms under $C=2.2\mu F$ (with +100% deviation). It is seen from Fig. 5(a) that the doubled capacitance prolonged the dynamic regulation process to 20 horizons (20μs). After self-learning, Fig. 5(b) demonstrates the dynamic settling time is optimized to 12μs and voltage overshoot is reduced by 0.5V. Fig. 5(c) gives the self-learning process, which intuitively proves that the self-learning can gradually enhance the control performance with parameter deviation.

IV. CONCLUSION

To achieve adaptive high-performance control at MHz switching frequency, this paper proposes a real-time self-learning control algorithm. To adapt the MHz switching frequency, the self-learning rules are properly integrated and simplified; FPGA timing optimization design is also implemented to parallel the control calculation and the control optimization; parameter sharing is also realized between the control calculation and the control optimization. With such a design, the proposed self-training control can be completed within 4 FPGA clock cycles. The 1MHz experimental results under rated parameter and deviated parameter demonstrate that the proposed real-time self-learning control algorithm can real-time enhance the control performance under both rated parameters and deviated parameters, achieving small steady-state error, fast dynamic response with no overshoot and adaptive to parameter deviation merits.

REFERENCES

[1] P. He, A. Mallik, A. Sankar and A. Khaligh, "Design of a 1-MHz High-Efficiency High-Power-Density Bidirectional GaN-Based CLLC Converter for Electric Vehicles," *IEEE Trans. Veh. Technol.*, vol. 68, no. 1, pp. 213-223, Jan. 2019.

[2] S. Ji, D. Reusch, and F. C. Lee, "High-frequency high power density 3-D integrated gallium-nitride-based point of load module design," *IEEE Trans. Power Electron.*, vol. 28, no. 9, pp. 4216–4226, Sep. 2013.

[3] A. Hariya et al., "Circuit Design Techniques for Reducing the Effects of Magnetic Flux on GaN-HEMTs in 5-MHz 100-W High Power-Density LLC Resonant DC–DC Converters," *IEEE Trans. Power Electron.*, vol. 32, no. 8, pp. 5953-5963, Aug. 2017.

[4] K. Kruse, M. Elbo and Z. Zhang, "GaN-based high efficiency bidirectional DC-DC converter with 10 MHz switching frequency," in *Proc. IEEE Appl. Power Electron. Conf. Expo* (APEC), Tampa, FL, USA, 2017, pp. 273-278.

[5] R. Wai, M. Chen and Y. Liu, "Design of Adaptive Control and Fuzzy Neural Network Control for Single-Stage Boost Inverter," *IEEE Trans. Ind. Electron.*, vol. 62, no. 9, pp. 5434-5445, Sept. 2015.

[6] F. Lin, C. Chang and P. Huang, "FPGA-Based Adaptive Backstepping Sliding-Mode Control for Linear Induction Motor Drive," *IEEE Trans. Power Electron.*, vol. 22, no. 4, pp. 1222-1231, July 2007.

[7] J. Chen, Y. Chen, L. Tong, L. Peng and Y. Kang, "A Backpropagation Neural Network-Based Explicit Model Predictive Control for DC–DC Converters with High Switching Frequency," *IEEE Jour. of Emer. and Sel. Top. in Pow. Electron.*, vol. 8, no. 3, pp. 2124-2142, Sept. 2020.

[8] S. Cong and Y. Liang, "PID-Like Neural Network Nonlinear Adaptive Control for Uncertain Multivariable Motion Control Systems," *IEEE Trans. Ind. Electron.*, vol. 56, no. 10, pp. 3872-3879, Oct. 2009.

[9] D. Xu, J. Liu, X. Yan and W. Yan, "A Novel Adaptive Neural Network Constrained Control for a Multi-Area Interconnected Power System With Hybrid Energy Storage," *IEEE Trans. Ind. Electron.*, vol. 65, no. 8, pp. 6625-6634, Aug. 2018.

[10] M. Schenke, W. Kirchgässner and O. Wallscheid, "Controller Design for Electrical Drives by Deep Reinforcement Learning: A Proof of Concept," *IEEE Trans. Ind. Inform.*, vol. 16, no. 7, pp. 4650-4658, July 2020.

[11] M. Gheisarnejad and M. H. Khooban, "IoT-Based DC/DC Deep Learning Power Converter Control: Real-Time Implementation," *IEEE Trans. Power Electron.*, vol. 35, no. 12, pp. 13621-13630, Dec. 2020.

[12] R. Mao, R. Cui and C. L. P. Chen, "Broad Learning With Reinforcement Learning Signal Feedback: Theory and Applications," *IEEE Trans. Neural Netw. Learn. Syst.*, early access.

15kV Press Pack SiC IGBT

DU Yujie[a], TANG Xinling[a], WANG Liang[a], ZHAO Zhibin[b], YANG Xiaolei[c], YANG Fei[a], WU Junmin[a],

[a] State Key Laboratory of Advanced Power Transmission Technology, Global Energy Interconnection Research Institute Co.,Ltd, Beijing 102209, China, hsmscdyj@163.com

[b] North China Electric Power University, Beijing, China

[c] Nanjing Electronic Devices Institute, Nanjing, China

Abstract

Press pack is easier to use in series and has a short circuit failure mode, which makes it especially suitable for the application requirements of high voltage and high power devices in the power system. The SiC IGBT device itself has the characteristics of high voltage, high temperature and high power density. The application of press pack SiC IGBT devices will greatly promote the technological innovation of power electronics technology. This article gives a solution for press pack SiC IGBTs, and the 15kV SiC IGBT press pack module sample is prepared based on 15kV SiC IGBT chips independently developed by this research group. The static and switching characteristics of the module are tested, and the test results show that the module has good conduction characteristics and a low leakage current of 24.7μA at 15kV.

1. Introduction

The press pack module has been widely used in power systems for the characteristics of easy large-scale parallel grouping of power chips, short circuit failure and double-sided heat dissipation. With the development of intelligent, safe and flexible power systems, Si IGBT devices cannot meet the stringent characteristics requirements of power semiconductor devices in power systems, due to the material limitations. the silicon carbide (SiC) devices, as the third-generation semiconductors, which have the characteristics of higher voltage, larger capacity, higher efficiency and higher junction temperature have become the most potential power semiconductor devices in the power system[1][2]. Combining the advantages of press pack and SiC devices to develop high-voltage and large-capacity press pack SiC modules will greatly promote the application of SiC devices process.

As a bipolar device, SiC IGBT devices will better take advantage of the high voltage of SiC materials, and the voltage of SiC IGBT devices currently reported have reached to be 27.5kV[3]. In terms of SiC IGBT module packaging, North Carolina State University and Cree combined 15kV SiC IGBT and 10kV SiC JBS into modules through series and parallel packages, and tried to use the modules in medium-voltage three-phase converters[4][5]. There are also attempts for press-pack SiC modules, Zhejiang University and the University of Arkansas analyzed challenges in the realization of press pack SiC MOSFET, and proposed a press pack packaging structure based on 1200V SiC MOSFET[6]. When it comes to press pack of SiC IGBT, high voltage level requires a complete insulation coordination design, and the evolution of the characteristics of SiC devices under different pressures needs further study, to sum up that the press pack of SiC IGBT devices is facing great challenges.

In this paper, a solution to press pack module of 15kV SiC IGBT was proposed，and the structure of the module is shown in Fig. 1.1. The package structure fully considers the insulation coordination requirements under 15kV high voltage, and a highly reliable pressure equalization structure is necessary due to the small area and high electric field strength of SiC chip. Based on the self-developed chip of the research group, a 15kV SiC IGBT module sample was prepared, at the same time, a high-voltage SiC test platform was built, and the static and switching characteristics of the 15kV SiC IGBT module were obtained finally. The results verify the overall insulation coordination of the module and provide important data for the preparation and application of high voltage and high power SiC IGBT devices.

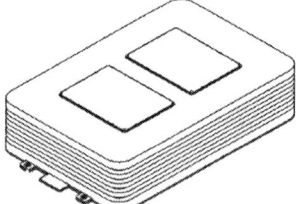

Fig. 1.1 15kV SiC IGBT press pack module

2. Challenges in Press Pack SiC IGBT

According to the realization method, the press pack structures are mainly divided into elastic press pack and the rigid press pack. The disc spring structure in elastic press pack can effectively balance the pressure between the multiple sub-units, which can reduce dimensional tolerance requirements for packaged components compared with rigid press pack. In short, the elastic press pack structure is easier to achieve the overall pressure balance of the semiconductor power device, and especially suitable for SiC devices with high power density and small chip area. Nevertheless, there are many challenges in the design and implementation of high-voltage press pack SiC devices：

The voltage level of the 15kV SiC IGBT device and the electric field strength of the chip terminal far exceed those of the existing devices, which poses challenges for the external and internal insulation design of the device. According to the requirements of IEC60664-1, 15kV voltage level leads to a significant increase in device clearance and creepage distance, and external insulation requires a more complex umbrella group structure. In addition, 15kV SiC IGBT has high voltage level and small terminal distance. It is very necessary to study the electric field distribution of the chip terminal in different insulating media, combine the insulation characteristics of the material, and reasonably design the terminal insulation protection structure.

The active area of SiC chip is much smaller than that of silicon chip, which puts forward higher requirements for the design of crimping structure. SiC materials are different from Si materials in terms of hardness and elastic modulus. The pressure range that SiC chips can withstand and the changes in the characteristics of SiC chips under different pressures need to be further studied。

High-reliability packaging material selection and packaging process are essential to give full play to the advantages of high-voltage and high-temperature operation of SiC devices. At present, the highest voltage level of Si IGBT devices on the market is 6.5kV, and the voltage level of thyristors can reach 8.5kV. The existing packaging Material selection and technology cannot meet the packaging requirements of 15kV SiC IGBT. Therefore, it is necessary to carry out research on the selection of packaging materials and the development of supporting packaging processes;

Similar to Si devices, SiC devices also have electrical balance problems when multiple chips are used in parallel, and due to the faster switching speed, SiC devices will be more sensitive to chip parameter dispersion and package parasitic parameters under high di/dt and du/dt. In order to realize multi-chip parallel current sharing of SiC devices, on the one hand, it is necessary to study the influence of chip parameter dispersion on device characteristics and perform chip parameter screening; on the other hand, package parasitic parameters should be reduced as much as possible, and solutions such as symmetrical layout should be adopted.

Reduce the parasitic parameter difference between the branches in the device. It is worth noting that in the crimping device, the requirement of higher withstands voltage level for higher clearance and creepage distance will lead to an increase in parasitic inductance, which is in contradiction with the low-inductance compact package structure design.

3. A Solution to Press Pack SiC IGBT

3.1 Package design

A 15kV SiC IGBT module is proposed based on the elastic press pack structure, and the cross-sectional view of the module is shown in Fig. 3.1. Considering the different operation stresses between SiC IGBT and diode, this might result in different failure modes. The module uses two sub-modules, one of the sub-modules is SiC IGBT, and the other is SiC diode, for the purpose of reducing the mutual influence of different failure modes of SiC IGBT and SiC diode when used in parallel.

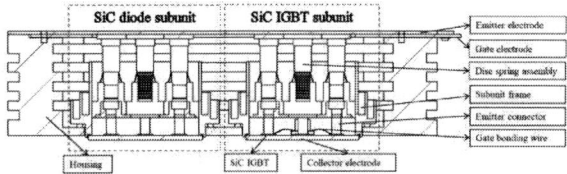

Fig. 3.1 Section view of 15kV SiC IGBT press-pack module

The structure of 15kV press pack SiC IGBT includes housing, gate PCB, emitter metal plate, SiC IGBT subunit and the SiC diode subunit. The housing provides external insulation and mechanical support for the entire module, and the requirement of housing material's CTI is ≥600. The section view of housing is shown in Fig. 3.2, the height H, the outer surface creep age distance L1 and the inner surface cree page distance L2 shall meet the insulation coordination requirements of the 15kV voltage level.

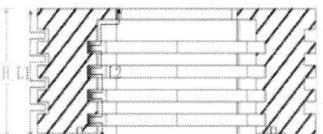

Fig. 3.2 Section view of housing

Taking the SiC IGBT subunit as an example, the subunit includes subunit frame, collector metal plate, SiC chips, gate bonding wire, emitter metal connector and disc spring assembly. The subunit frame is made of nylon, and a certain proportion of flame retardant and stabilizer is added. The requirement of subunit frame material's CTI is ≥600, and the section view of housing is shown in Fig. 3.3, the height D and surface creepage distance L shall meet the requirements of insulation coordination. The subunit frame is formed by high-precision open-mold injection molding, with a size deviation of not more than 0.1mm, and provides support and positioning functions for the disc spring assembly and the emitter metal connector. The disc spring assembly adopts the structure of opposing laminated coil springs, which has the advantages of large load and small deformation.

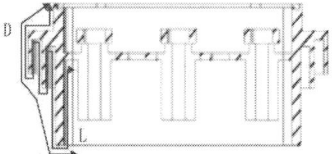

Fig. 3.3 Section view of subunit frame

In order to achieve press pack of SiC IGBT, the SiC chip is connected with the collector metal plate and the emitter metal connector by double-sided welding or sintering. As a result, the module power circuit realizes a short-circuit failure mode, which is more suitable for need of module series connection for high-voltage and high-power applications. At the same time, the double-sided welding or sintering structure avoids micro gaps between the surface of SiC chip and the contact surface of the insulating medium during the impact or vibration process, thereby improving the long-term reliability of the device.

3.2 Module Preparation

The 15kV SiC IGBT module was prepared based on press pack structure mentioned above. Considering that the current of the SiC IGBT chip used is small, the metal opening area of the emitter is only 0.37mm×0.24mm (×2), it is difficult to carry out the emitter welding or sintering process. In the actual manufacturing process of the module, the emitter of SiC IGBT chip is connected to the direct bond copper (DBC) by aluminum bonding wire, and then connected to the disc spring assembly through the emitter metal connector, and finally connected to the external electrode through the emitter metal plate. As a result, the module does not have a short-circuit failure mode, and the actual SiC IGBT module prepared is shown in Fig. 3.4.

Fig. 3.4 15kV SiC IGBT module

The current capacity and chip size of SiC IGBT will increase under the iteration of chip technology. Combined with the double-sided welding or sintering process, the SiC IGBT press pack module with short-circuit failure mode will be prepared subsequently.

4. Module characteristics

A test fixture was designed to evaluate the static and switching characteristics of the 15kV SiC IGBT module，and the structure is shown in Fig. 4.1. The four corner bolts are designed to apply pressure to the module, and the housing can effectively clamp the pressure, to avoid excessive pressure on the module and make the pressure evenly distributed. In addition, a high-voltage SiC test platform was built to meet the test requirements of 15kV SiC IGBT devices.

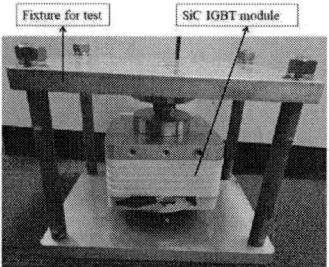

Fig. 4.1 Fixture used to evaluate the characteristics of 15kV SiC IGBT module

4.1 Static characteristics

The forward characteristic of 15kV SiC IGBT module is obtained by pulse method test, and the output characteristic curve at room temperature is shown in Fig.4.2. Under the conditions of V_{GE}=20V and I_C=10A, $V_{CE(sat)}$ is about 5.7V, which indicate good conduction characteristics.

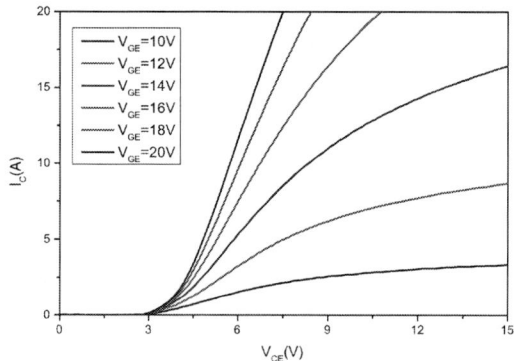

Fig. 4.2 Output characteristic curve

The blocking characteristic curve of the SiC IGBT module at room temperature is shown in Fig. 4.3, and the leakage current of the module at 15kV can be obtained as 24.7μA. It should be noted that the leakage current of DBC, silicone gel and sub-unit frame in the package structure is not obvious under low voltage, but cannot be ignored at 15kV. In addition, due to the impact of the accuracy of the test equipment, there are certain errors in the test results.

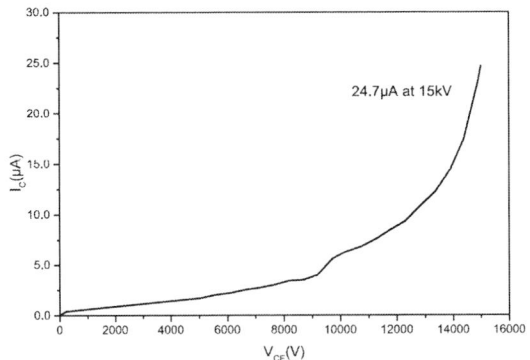

Fig. 4.3 blocking characteristic curve

4.2 Switching characteristics

The switching waveform of the 15kV SiC IGBT module at room temperature is obtained through the high-voltage SiC dynamic test platform test. The test conditions are V_{CC}=7000V, V_{GE}=-5V/+20V, R_G=90Ω, and the turn on waveform of the module is shown in Fig.4.4. In the turn on transient, the rise time t_r is 128ns, the di/dt is 55.9A/μs, the du/dt is 6387V/μs, and the turn on energy E_{on} is 58.2mJ.

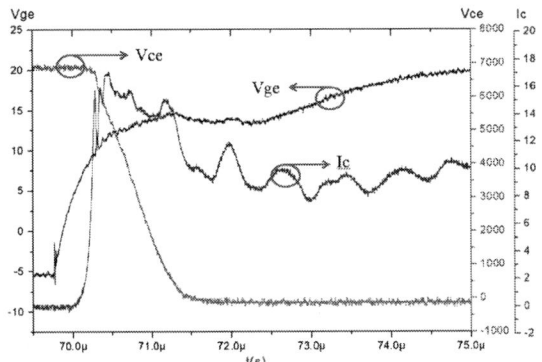

Fig. 4.4 The turn on waveform of 15kV SiC IGBT

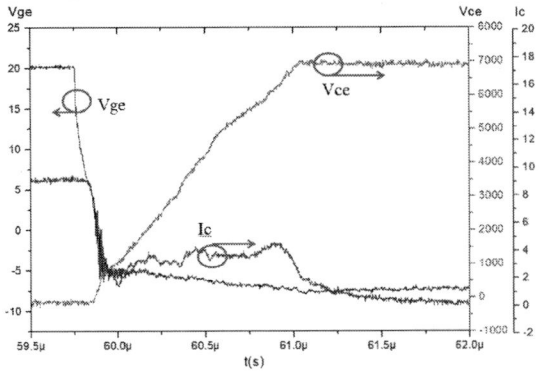

Fig. 4.5 The turn off waveform of 15kV SiC IGBT

The turn off waveform is shown in Fig. 4.5, it can be seen that there is distortion in the current waveform in the turn-off transient. The di/dt of the first segment is 34.3A/μs, the di/dt of the second segment is 25.6A/μs, the du/dt is 5578V/μs, and the turn off energy E_{off} is 9.2mJ.

The similar current distortion also exists during the switching test of Si IGBT [7]. The main reason is that the distributed capacitance current $\Delta i_{C\_C\sigma}$ generated by the parasitic capacitance $C_{\sigma1}$ of the test circuit and the junction capacitance $C_{\sigma2}$ of the test device under the action of du/dt during the switching process is superimposed on the collector current i_C. The correlation between the measured current i_{C_test}, i_C, $\Delta i_{C\_C\sigma}$ and the total distributed capacitance C_σ and the IGBT collector-emitter voltage change rate du_{ce}/dt is:

$$\Delta i_{\_C_\sigma} = i_{c\_test} - i_c = -C_\sigma \frac{du_{ce}}{dt} \tag{4-1}$$

For the 15kV SiC IGBT module, a higher voltage level means a higher du/dt, and the test current is relatively small, which highlights the role of distributed capacitance current and ultimately leads to the distortion of the turn off current in Fig. 4.5. It can be obtained from the formula (4-1) that the influence of distributed capacitance exists in both the turn on and turn off process. Through calculation, it can be obtained that in the turn on transient, the distributed capacitance introduces additional current, which causes the turn on current and turn on energy test results to be larger, and in the turn off transient, the distributed capacitance causes the turn off current and turn off energy test results to be smaller. Therefore, reducing the distributed capacitance of the test loop is necessary to accurately evaluate the switching characteristics of high voltage SiC devices.

5. Conclusions

This paper presents a high insulation, high reliability 15kV press pack solution, and a 15kV SiC IGBT module is prepared to study the static and switching characteristics. Static characteristics test results show that when V_{GE}=20V, I_C=10A, the on-state voltage drop $V_{CE(sat)}$ is about 5.7V, and the leakage current under 15kV is 24.7μA. The switching waveforms of the module show that reducing parasitic parameters is essential for the accurate evaluation of the switching characteristics of high-voltage SiC devices, and the distributed capacitance in the test platform needs to be further optimized.

Acknowledgments

This work is supported by the National Key Research and Development Program (2016YFB0400500). Meanwhile, thanks for the simulation support of State Key Laboratory of Advanced Power Transmission Technology.

978-1-6654-4817-8/21 $31.00 © 2021 IEEE

References

[1]Lee H , Smet V , Tummala R . A Review of SiC Power Module Packaging Technologies: Challenges, Advances, and Emerging Issues[J]. IEEE Journal of Emerging and Selected Topics in Power Electronics, 2019, PP(99):1-1.

[2] Fengze, Hou, Wenbo, et al. Review of Packaging Schemes for Power Module[J]. IEEE Journal of Emerging and Selected Topics in Power Electronics, 2019, 8(1):223-238.

[3] Michael, J, O'Loughlin, et al. 27 kV, 20 A 4H-SiC n-IGBTs[J]. Materials Science Forum, 2015, 821/823:847-850.

[4] Kadavelugu A , Bhattacharya S , Ryu S H , et al. Characterization of 15 kV SiC n-IGBT and its application considerations for high power converters[C]// Energy Conversion Congress & Exposition. IEEE, 2013.

[5] Madhusoodhanan S , Mainali K , Tripathi A K , et al. Power Loss Analysis of Medium-Voltage Three-Phase Converters Using 15-kV/40-A SiC N-IGBT[J]. IEEE Journal of Emerging & Selected Topics in Power Electronics, 2016, 4(3):902-917.

[6] Zhu N , Mantooth H A , Xu D , et al. A Solution to Press-Pack Packaging of SiC MOSFETs[J]. IEEE Transactions on Industrial Electronics, 2017, PP(10):1-1.

[7] Tang X , Zhang Y , Sai Z , et al. Influence of Stray Capacitance on Dynamic Test Results of Power Semiconductor devices[C]// 2019 4th IEEE Workshop on the Electronic Grid (eGRID). IEEE, 2019.

[8] Yujie D, Cui L, Sun S, et al. Influence of Distributed Capacitance on Switching Characteristics of 6500V SiC MOSFET[C]// SSLChina: IFWS. 2020.

Degradation mechanism of D-mode GaN HEMT based on high temperature reverse bias stress

Meng Lu
Department of Materials Science and
Engineering
Xiangtan University
Xiangtan,China
201921001457@smail.xtu.edu.cn

Yiqiang Chen*
Science and Technology on Reliability
Physics and Application of Electronic
Component Laboratory
The No.5 Electronics Research Institute
of the Ministry of Industry and
Information Technology Guangzhou
Guangzhou,China
yiqiang-chen@hotmail.com

Min Liao*
Department of Materials Science and
Engineering
Xiangtan University
Xiangtan,China
mliao@xtu.edu.cn

Chang liu
Science and Technology on Reliability
Physics and Application of Electronic
Component Laboratory
The No.5 Electronics Research Institute
of the Ministry of Industry and
Information Technology
Guangzhou
Guangzhou,China
xd_liuchang@163.com

Shuaizhi Zheng
Department of Materials Science and
Engineering
Xiangtan University
Xiangtan,China
shuaizhi@xtu.edu.cn

Kexin Gao
Department of Materials Science and
Engineering
Xiangtan University
Xiangtan,China
201921001356@smail.xtu.edu.cn

Abstract—This paper systematically discusses the degradation of reliability of depleted gallium nitride devices under short to long-term high temperature reverse bias (HTRB) stress. The electrical parameters of the device during the whole process were obtained by the stress-measure-stress alternation experimental scheme. We believe that the Schottky barrier decreases and the channel conduction resistance increases in the process of high temperature and high field stress. This conclusion can be verified by the gate-lag characteristics. It is a new phenomenon that capacitor degradation can be divided into three stages in the experiment. Further analysis shows that the device has three processes: *dielectric layer consumption — AlGaN/GaN layer interface donor state increase—AlGaN barrier layer degradation FN emission and trap-assisted tunneling.* There is also a positive correlation between capacitance and gate leakage. When the channel is rapidly depleted, the capacitance decreases gradually and the leakage reaches saturation. Current saturation is mainly because the external voltage has little effect on the longitudinal electric field between Schottky electrode and Ohm electrode, but has a great effect on the transverse electric field. The conclusion of this paper will be of reference significance to the manufacturing process and the life of the device under normal operation.

Keywords—high temperature reverse bias stress (HTRB), capacitance-voltage characteristics (C-V), traps, D-mode GaN HEMTs.

I. INTRODUCTION

GaN is a wide band gap semiconductor with large polarity, and AlGaN/GaN heterojunction can obtain excellent electrical performance, which makes AlGaN/GaN HEMTs have the advantages of high frequency and high concentration of two-dimensional electron gas (2DEG) [1][2]. However, under high frequencies or high temperature reverse bias (HTRB), the device which has mature technological progresses is still subject to trap induced limitations. HTRB stress is an effective way to investigate the reliability of the gate and drain-side region under the gate of the device.

Many groups are investigating the degradation under high temperature and negative bias [9]. Among them，Shanjie Li *et al.* [9] used E-mode MIS HEMTs for HTRB and NBTI experiments at different temperatures. He regarded p-GaN layer hole-emission as the results of threshold voltage (V_{th}) positive shift. Matteo Meneghini *et al.* [3] used D-mode MIS HEMTs for negative gate voltage under high temperature. He emphasized that trap state depletion occurs in the region SiN/AlGaN and in the gate insulation layer after applying negative gate voltage and high temperature, and threshold voltage drift occurs in the negative direction. J. Robertson *et al.* [9]believe that as the Group III nitrides material has a large band gap, most interface states exist below the conduction band, so the emission time constant is long, which ultimately leads to the threshold voltage instability in Si high-k MOSFETs.

Some recent works have reported that significant recoverable threshold voltage negative drift when devices are applied to negative gate voltage at high temperatures[3]. The fixed charge in the insulating layer [5-6] and the border trap [7] and interface state of the insulating layer /AlGaN interface are related to the threshold voltage shift mechanism. It is worth noting that it has not been clear which interface trap state change causes the threshold voltage drift, and how interface states affect device performance.

This paper presents aging test of AlGaN/GaN HEMTs with time in HTRB experiments at the same bias. The degradation law of the device can be effectively captured by applying high temperature reverse deviational stress from short to long time. The V_{th} of the device has a large positive drift in a relatively short time, while the high temperature bias has little effect on the threshold voltage of the device for several hundred hours. However, as the conduction resistance increasing, the output current has further current collapse effect. At the same time, the gate lag characteristic also indicates that the device is slower to respond to the voltage

978-1-6654-4817-8/21 $31.00 © 2021 IEEE

after HTRB stress. Through this experiment, we further explore the degradation reason of depletion-type AlGaN/GaN HEMTs. By analyzing capacitance-voltage and double-pulse characteristics, we discussed the degradation process of devices in detail in Chapter 3, explained the degradation mechanism and inferred the possible degradation location [8].

II. EXPERIMENT

The device used in the experiment is AlGaN/GaN HEMT grown on SiC substrate. Fig. 1a clearly shows the electrode and the film distribution associated with the process from the cross section of the device. Longitudinal heat dissipation can be accelerated by using a silicon carbide substrate. The 1.5-μm buffer layer above the SiC is in contact with the 21nm barrier layer AlGaN. A thin 2DEG exist in the interface of AlGaN and GaN, which can be driven and generated a current by applying a voltage V_{ds}. The gate, source and drain are protected by a SiNx passivation layer, and the gate to source and drain distances are 2 μm and 4 μm respectively. The gate

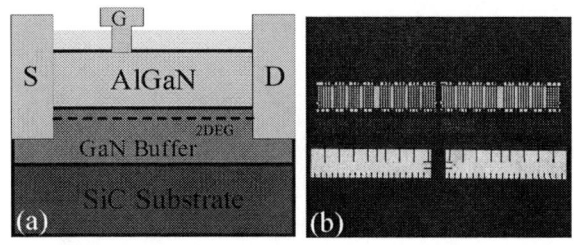

Fig .1 (a) Section view of AlGaN/GaN HEMTs structure and (b) Top view of AlGaN/GaN HEMTs architecture.

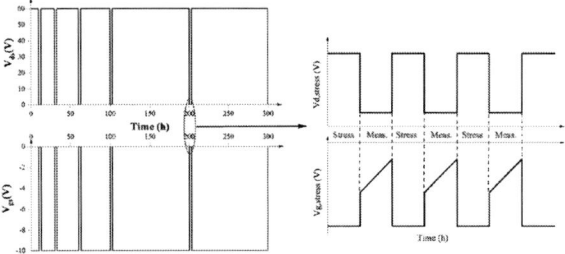

Figure 2. Experimental stress and test waveforms

is a 0.5-μm-length T-gate with width of 1.25 mm.

Fig. 2 shows the schematic diagram of stress and measurement of electrical stress in the experiment. In the past, many experiments were conducted with the same interval time, but the degradation process of the device is often a non-linear process, and the step length setting is very important. Too long step length may miss the time point of device mutation, while too short step length makes the results are affected by the test conditions, making the results inaccurate. In this experiment, we roughly determined the mutation range of the device through preliminary exploration. As we can see, to figure the evolution out, the experiment step time was set to 10h, 30h, 60h, 100h, 200h and 300h with bias stress test ($V_{gstress} = -10$ V, $V_{dstress} = 60$ V). The experiment temperature was set to 150 °C, in order to determine the failure time of the device, we

This work was supported by the National Key Research and Development Plan (2020YFF0218500) and the Key Realm R&D Program of Guangdong Province (2020B010171002,2020B010173001, 2018B010142001 and 2019B010143002).

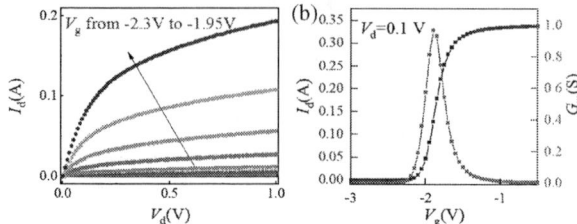

Figure 3. (a) Typical output characteristic and (b) transfer characteristic of the device.

used a computer program to control the applied stress and monitor the real-time current. The I_{ds}-V_{ds} measurement conditions are shown in Figure. 3a, V_{ds}=0.1 V. The device threshold voltage, maximum gate leakage, and peak transconductance can be calculated from the current-voltage characteristics.

III. RESULTS AND DISCUSSION

This work uses three devices, all of which show the same degradation phenomenon, so the following discussion is based on the analysis of a representative device. Fig. 3 shows real-time monitoring diagram of device drain current. For the purpose of clarity, we compare the original device with the device after 100 h and 300 h stress in Fig. 4 to indicate the changes of transfer and output characteristics. The threshold voltage drifted forward 0.27 V before and after the stress, and the current collapsed 68.3 % from the figure, indicating that the device had a certain degree of degradation. However, the threshold voltage was almost unchanged from 100 h to 300 h with the output current collapsed. Transconductance peak was drift forward and down. The presence of surface donor-type states on the G-D surface and in the region under the gate leads to a decrease in peak transconductance.

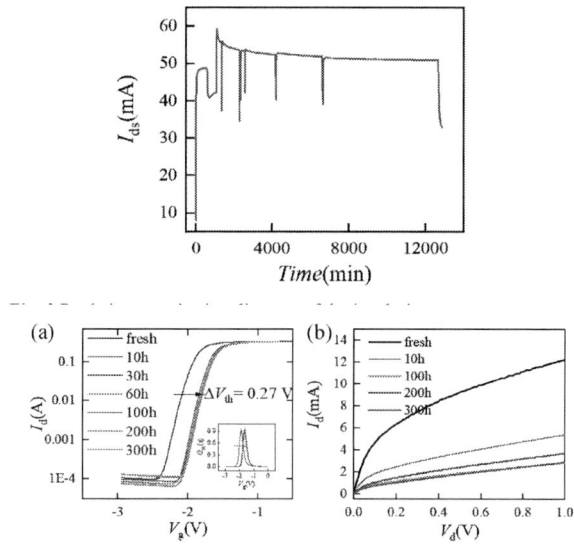

Fig. 4 Fresh to 300h stress transfer curve (a), where the threshold voltage forward drift 0.27V and (b)output curve with calculate the on-resistance at 200h and 300h.

These phenomena indicate that it is possible that the donor traps exist in the place between G-S and /or G-D, or the donor states in interface trap excess electron, leading to the reduction of channel carrier density.

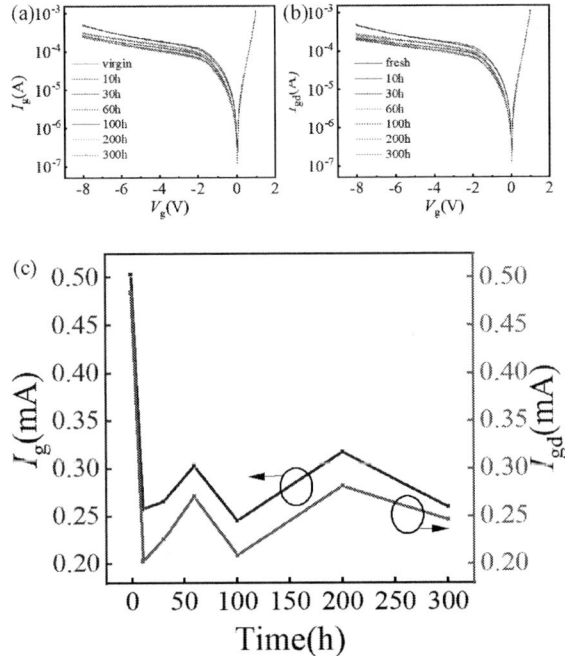

Fig. 5 (a)gate to source leakage current from 10h to 300h stress has decreased by $0.41×10^{-4}$A ,(b)gate to drain leakage current from 10h to 300h stress has decreased by $0.436×10^{-4}$A, (c) Change of leakage current with time at maximum negative gate voltage.

Figure 5 shows the gate and gate-to-drain leakage current at different stress stages. The gate and the gate-to-drain leakage current decreased by 0.041 mA and 0.044 mA before and after the stress, respectively. According to equation (1) and (2), the height of the barrier increased slightly and the ideal factor had a slight decrease.

$$n = \frac{q}{kT} \cdot \frac{1}{Slope} \tag{1}$$

$$\Phi = \frac{kT}{q} ln \frac{AA^*T^2}{Intercept} \tag{2}$$

Figure 6 is the gate-lag (GL)curve before and after the stress. The experimental test conditions are as follows: the gate bias voltage of -10 V is applied in the off mode of the device, the drain voltage of 0.1V is applied, and the source electrode is in the ground state. After maintaining this state for 60 s, the gate voltage changes from -10 V in the off state to 0 V in the full-on state, and the drain voltage remains unchanged. The device needs 30% more time to reach the maximum current. The gate lag is mainly owing to electrons trapping by interface state on the surface and at the transistor barrier[10]. Due to the trap effect in AlGaN/GaN HEMT, electrons will fill the trap on the AlGaN surface, resulting in a large amount of negative charge on the device surface, thus appearing virtual gate effect[11].

Figure 7 shows a 1% duty cycle two-pulse I-V test, including transfer and output characteristics. The static offset point is set to off and can be used to assess current collapse and device sensitivity to the trap capture process. When V_{ds} is

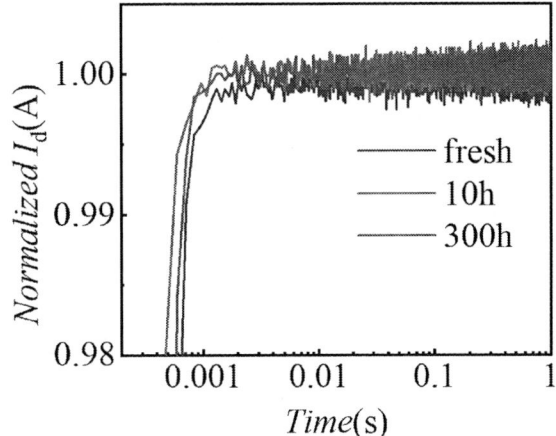

Fig. 6 Gate lag character after 10h to 300h stress. The device takes shorter to recover to 99% of its original value.

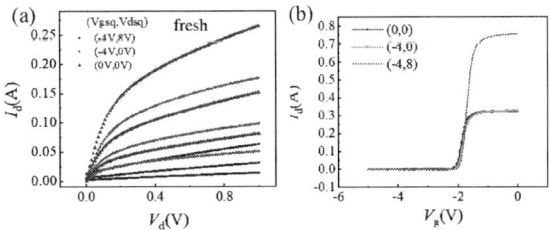

Fig. 7 (a) Pulse output characteristics and (b) transfer characteristics for virgin device. Quasi-static bias points are $(V_{dsq}, V_{gsq}) = (0,0)$, (-4, 0) and (-4, 8) respectively. Pulse width is 1 ms, for duty cycle 1%.

0 V, the pulse output current at $(V_{gsq}, V_{dsq}) = $ (-4 V,8 V) increase to 13 mA, which means the donor states in AlGaN/GaN interface releases electrons then current increases.

Figure 8a is the capacitance curve relative to voltage of the device before and after high temperature deviating stress. It

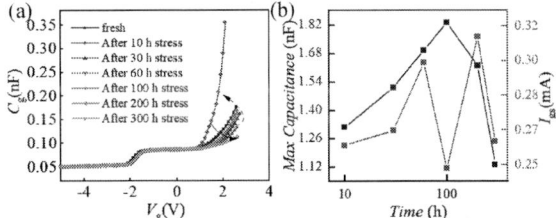

Fig. 8 (a)The capacitance changes during the stress process, and the capacitance changes greatly in the positive pressure region, which goes through three processes: dielectric layer depletion, interface donor state capture of electrons, and barrier layer degradation. (b) The variation trend of capacitance and gate leakage with respect to time is almost the same.

can be seen from the figure that the capacitor appears stretch-in and stretch-out in the forward bias region, which is related to the interface states of metal-semiconductor interface and AlGaN/GaN interface. The slope of capacitance in the forward bias region decreases with the increase of interface state. As the AlGaN potential tends to flatten, electrons are more easily transferred from the AlGaN bottom to AlGaN surface, making the C-V slope steeper [8]. In the early stress stage, at high temperature, the interface state of Schottky

978-1-6654-4817-8/21 $31.00 © 2021 IEEE   169

contact changes, and the dielectric layer under the Schottky gate is consumed, thus forming a tighter metal semiconductor contact [13]. In the middle stress stage, the ionized donor state of AlGaN/GaN interface increases, which increases the net charge of the interface and thus the capacitance (see Fig. 9). With the passage of time, AlGaN barrier degrades at the later stress stage, the dielectric constant decreases and trap-assisted tunneling reduces the capacitance gradually.

We make a comparison with max capacitance (C_{max}) and max reverse gate leakage current (I_g) throughout the stress time in figure7b. It is obvious that their changes are almost identical, suggesting that they had some correlation. The change of I_g has been seen in other experiments. The gate leakage current changes little, indicating that the Schottky

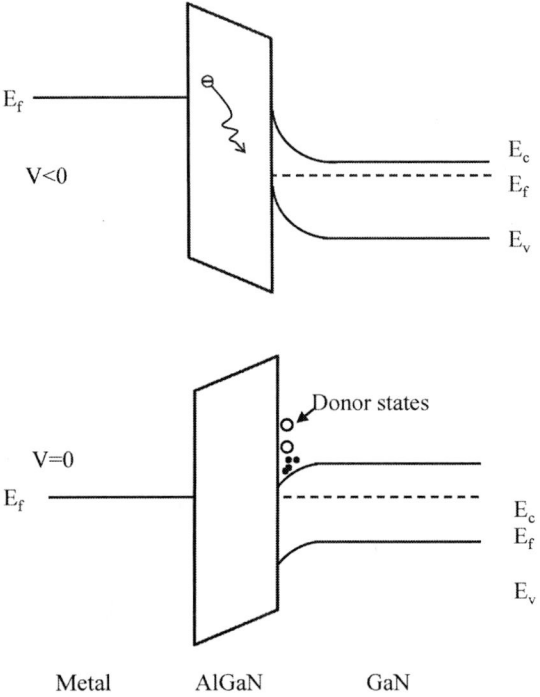

Fig. 9 Schematic diagram of device degradation during HTRB process (a) Reverse gate leakage current may be generated by tunneling effect and screw dislocations in semiconductors may become leakage channels. (b)Donor states exist in AlGaN/GaN interface result in excess electrons

contact of the gate is stable [12]. Of course, some screw dislocations may also exist in the device. The screw dislocation is a leakage channel in semiconductor materials. The conductive continuous states such as dislocation contribute a lot to leakage current[13].

IV. CONCLUSION

The device exhibits different degradation mechanisms at different stages under high temperature reverse bias stress. The medium layer consumption can be improved in the process. The reliability of AlGaN layer and AlGaN/GaN interface has great influence on the device performance. AlGaN/GaN interface donor state causes switching characteristics to be delayed and hinders high-frequency applications.

REFERENCES

[1]   Ambacher O , Smart J . Two-dimensional electron gases induced by spontaneous and piezoelectric polarization charges in N[J]. Journal of Applied Physics, 1999.

[2]   Asubar J T , Kobayashi Y , Yoshitsugu K , et al. Current Collapse Reduction in AlGaN/GaN HEMTs by High-Pressure Water Vapor Annealing[J]. IEEE Transactions on Electron Devices, 2015, 62(8):2423-2428.

[3]   M, Meneghini, I, et al. Negative Bias-Induced Threshold Voltage Instability in GaN-on-Si Power HEMTs[J]. IEEE Electron Device Letters, 2016, 37(4):474-477.

[4]   M. J. Uren, J. Moreke, and M. Kuball, "Buffer design to minimize current collapse in GaN/AlGaN HFETs," IEEE Transactions on Electron Devices, vol. 59, pp. 3327-3333, 2012.

[5]   Esposto, Michele, Krishnamoorthy, et al. Electrical properties of atomic layer deposited aluminum oxide on gallium nitride.[J]. Applied Physics Letters, 2011.

[6]   Son J , Chobpattana V , Mcskimming B M , et al. Fixed charge in high-k/GaN metal-oxide-semiconductor capacitor structures[J]. Applied Physics Letters, 2012, 101(10):063501.

[7]   Minseok Choi, John L. Lyons, Anderson Janotti, et al. Impact of native defects in high-k dielectric oxides on GaN/oxide metal-oxide-semiconductor devices[J]. physica status solidi (b), 2013.

[8]   Yatabe Z , Hori Y , Ma W C , et al. Characterizati- on of electronic states at insulator/(Al)GaN interfaces for improved insulated gate and surface passivation structures of GaN-based transistors[J]. Japanese Journal of Applied Physics, 2014, 53(10):100213.

[9]   Li S , He Z , Gao R , et al. Time-Dependent Threshold Voltage Instability Mechanisms of p-GaN Gate AlGaN/GaN HEMTs Under High Reverse Bias Conditions[J]. IEEE Transactions on Electron Devices, 2020, PP(99):1-4. Robertson J . High dielectric constant gate oxides for metal oxide Si transistors[J]. Reports on Progress in Physics, 2006, 69(2):327-396.

[10] Mitrofanov O . Mechanisms of gate lag in GaN/AlGaN/GaN high electron mobility transistors[J]. Superlattices and Microstructures, 2003, 34(1-2):33-53.

[11] Mitrofanov, O. . "Mechanisms of gate lag in GaN/AlGaN/GaN high electron mobility transistors." Superlattices and Microstructures 34.1-2(2003):33-53.

[12] Ao J P , Yamaoka Y , Okada M , et al. Investigation on current collapse of AlGaN/GaN HFET by gate bias stress[J]. IEICE Transactions on Electronics, 2008, E91-C(7).

[13] L. Y. Yang, "High field degradation effects and thermal problems of gallium nitride based HEMT devices," Ph.D. dissertation, Dept. Microelectron. Eng., Xidian Univ., Xi 'an, Shaanxi, China, 2013

Comprehensive Investigations on Paralleling Operation of SiC MOSFETs based on Subcircuit Model in MATLAB/SIMULINK

Yuqi Wei, Dereje Woldegiorgis, Xia Du, Venkata Samhitha Machireddy, and Alan Mantooth[1]
[1]University of Arkansas, Fayetteville, Arkansas, United States
Email: yuqiwei@uark.edu

Abstract— Wide band gap (WBG) devices have been widely applied in industrial applications owning to their advantages of low switching loss, low on-stage voltage drop, and high operating temperature. Paralleling operation of power devices/modules is attractive due to its cost-effective and high power characteristics. In applications require very high current capability, paralleling operation of off-the-shelf power devices/modules becomes the only choice. However, current balancing operation of individual power device/module becomes difficult due to the differences of circuit parasitics. To investigate the device/module and circuit parasitics influences on the current sharing performance, in this article, a subcircuit model was built in MATLAB. Comprehensive comparisons and analysis are performed, which can provide guidance for engineers when designing the system with paralleling devices/modules. Moreover, the solutions to achieve current balancing operating are proposed with the aid of active gate driver. Experiment results are presented and analyzed to validate the effectiveness of current sharing solutions.

Keywords—Wide band gap, current sharing, circuit parasitics, active gate driving

I. INTRODUCTION

Silicon carbide (SiC) metal–oxide–semiconductor field-effect transistor (MOSFET) enables the improvement of power converter efficiency and power density [1, 2]. It is also advantageous in paralleling operation owning to its positive temperature characteristic of on-state resistance. Nevertheless, paralleling operation is still challenging since the dynamic current sharing will become worse due to its negative temperature characteristic of threshold voltage [3]. In most of the papers, only the influences of on-state resistance and threshold voltage on current sharing performance are investigated. In practice, all circuit and device parasitics would have certain effect on the current sharing operation between paralleled power devices.

To mitigate the current unbalance operation, different solutions have been proposed in the literature. The passive components can be added into the circuits to alleviate the current unbalance issue. Although the static current unbalance can be mitigated with addition series resistor in the parallel branches, no improvement on the dynamic current balancing performance can be achieved with this method. Investigations can also be made on the symmetric parasitics design of the power module with multiple paralleling dies [4, 5]. The development cycle is long and the cost is high. The differential mode choke is inserted to achieve current balance operation [3]. However, the size and power loss of the inserted choke would be the major concerns especially at high power applications. Active gate driver is another promising

technology to achieve current balancing operation among paralleled power devices [6]-[10]. As categorized in [1], the active gate driver for paralleling switches mainly include variable gate resistance, variable gate current, variable delay, and variable gate voltage. A multi-level active gate driving strategy is proposed for the paralleling operation of SiC power modules, where the individual module current information are obtained by detecting the voltage across the Kelvin source parasitic inductance [9, 10]. A complicate circuitry is required to achieve the functions. Moreover, this technique is limited for power devices/modules with Kelvin source terminal. In most of the literature, the static current unbalance is usually neglected due to positive characteristic of on-state resistance. Nevertheless, as analyzed in [3], the on-state resistance is less sensitive to the temperature. Therefore, it is also important to tackle the static current unbalance to improve the system reliability.

The device and circuit parasitics, including on-state resistance, threshold voltage, common source stray inductance, and switching loop stray inductance, effects on the current balancing operation are investigated in [12, 13]. However, there are other circuit parasitics can also affect the current balancing performance of paralleling operation of devices/modules. To fully understand the circuit and device parasitics influences on current sharing performance, all possible circuit and device parasitics are investigated and compared with the aid of the subcircuit model in MATLAB/SIMULINK. To make comparisons among different circuit parasitics influence on the current sharing operation, 20% mismatches for all parasitcis are assumed and simulated. The current error is plotted to demonstrate the effect of each parasitic. The theoretical analysis results can provide insights on the current sharing between paralleling devices. Moreover, to mitigate the current unbalance between paralleling devices/modules, a simple two-level active gate driving solution with adjustable turn-on voltage is proposed. The experiment results demonstrate the effectiveness of using adjustable turn-on voltage for current balancing operation. To further achieve both static and dynamic current sharing operation, a hybrid turn-on gate volatge and gate delay control strategy is discussed and the feasibility is approved by using simulation and subcircuit model.

The rest of the paper is organized as follows. The MATLAB based subcircuit model is briefly discussed in Section II. With the aid of the subcircuit model, the possible device and circuit parasitics influences on current sharing operation are investigated in Section III. The two-level active gate driving solutions are discussed in Section IV, and the experimental results are presented to validate the effectiveness. The simulation studies for a hybrid turn-on gate volatge and

978-1-6654-4817-8/21 $31.00 © 2021 IEEE

gate delay for SiC MOSFETs with paralleling operation are presented. Finally, conclusions are drawn.

II. SUBCIRCUIT MODEL FOR SIC MOSFET

In this work, the SiC MOSFET is modelled by using the conventional method, and its operation can be divided into off region, linear region and saturated region. Fig. 1 shows the subcircuit model of a SiC MOSFET [14].

$$I_{DS} = 0, V_{GS} < V_{th}, \text{off region}$$

$$I_{DS} = \beta((V_{GS} - V_{th})V_{DS} - V_{DS}^2 / 2)(1 + \lambda |V_{DS}|),$$

$$0 < V_{DS} < V_{GS} - V_{th}, \text{linear region}$$

$$I_{DS} = \frac{\beta}{2}(V_{GS} - V_{th})^2 (1 + \lambda |V_{DS}|), 0 < V_{GS} - V_{th} < V_{DS},$$

$$\text{saturated region} \qquad (1)$$

where β is the transistor gain, V_{th} is the threshold voltage, and λ is the channel modulation. These parameters can be extracted based on the transfer characteristic curve (I_{DS}-V_{GS}) provided by the manufacture datasheet.

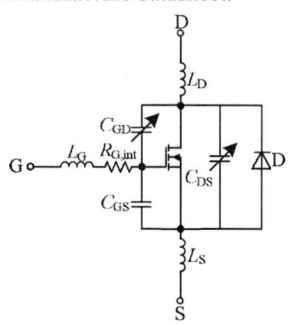

Fig. 1. Model of SiC MOSFET.

Fig. 2 shows the subcircuit model based double pulse test (DPT) circuit with two paralleling devices. Table 1 summarizes the circuit and device parasitics for two devices and assuming 20% difference for investigations, which can be caused by manufacture process or printed circuit board (PCB) layout. The device parasitics are obtained based on the manufacture datasheet. The device parasitic capacitances are modelled by using linearization method, and multiple capacitance values under different drain-to-source voltages are selected.

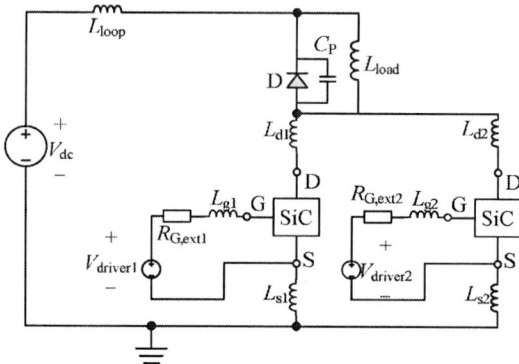

Fig. 2. DPT circuit model with two paralleling devices.

TABLE I. CIRCUIT PARASITICS AND THEIR VALUES

Variable	Definition	Values for device 1	Values for device 2
$R_{G,ext1}$, $R_{G,ext1}$	Gate driver resistor	24 Ω/6 Ω	20 Ω/5 Ω
$R_{G,int1}$, $R_{G,int}$	Device internal gate resistor	1.2 Ω	1 Ω
L_{g1}, L_{g2}	Gate loop parasitic inductance introduced by gate driver	12 nH	10 nH
L_{G1}, L_{G2}	Gate terminal parasitic inductance due to package	11.04 nH	9.2 nH
L_{s1}, L_{s2}	Parasitic inductance from source terminal to ground	6 nH	5 nH
L_{S1}, L_{S2}	Source terminal parasitic inductance due to package	12 nH	10 nH
L_{d1}, L_{d2}	Parasitic inductance from drain terminal to freewheeling diode	6 nH	5 nH
L_{D1}, L_{D2}	Drain terminal parasitic inductance due to package	7.32 nH	6.1 nH
V_{th1}, V_{th2}	Threshold voltage	6.24 V	5.2 V
C_{iss1}, C_{iss2}	Input capacitance	[960 840 624 600 576] pF for [0 1 10 100 800] V	[800 700 520 500 480] pF for [0 1 10 100 800] V
C_{rss1}, C_{rss2}	Reverse transfer capacitance	[504 372 120 36 21.6] pF for [0 1 10 100 800] V	[420 310 100 30 18] pF for [0 1 10 100 800] V
C_{oss1}, C_{oss2}	Output capacitance	[1020 780 312 86.4 43.2] pF for [0 1 10 100 800] V	[850 650 260 72 36] pF for [0 1 10 100 800] V
β_1, β_2	Gain (Related with device on-state resistance)	0.28392	0.2366

III. PARASITICS INFLUENCES ON CURRENT SHARING PERFORMANCE

When investigating the individual parasitic influence on current sharing performance, the other parasitics are kept same for the two paralleled devices. The test condition is fixed at 400 V and 10 A for all the scenarios.

A. Device Parasitic-Internal Gate Resistor

Due to the manufacture process, the internal gate resistor may not consistent for the devices with the same part number. Fig. 3 shows the results with internal gate resistor mismatch. The maximum current difference is 0.6308 A, and the influence of internal gate resistor is not obvious. Meanwhile,

the internal gate resistor will mainly affect the dynamic stages with negligible influence on the steady state. The current sharing performance is still good with the internal gate resistor mismatch.

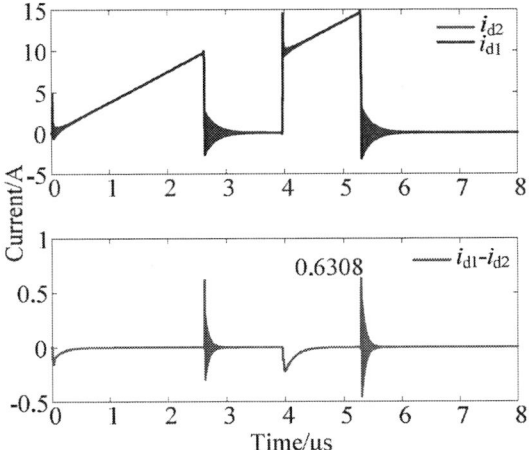

Fig. 3. Current sharing performance with internal gate resistor mismatch.

B. Device Parasitic-Threshold Voltage

The threshold voltage mismatch can also be caused by manufacture process or the degradation of device. Fig. 4 shows the results with threshold voltage mismatch. The maximum current difference is 2.954 A, the influence of threshold voltage is large. The device with a lower threshold voltage will turn-on fast and has higher current. In addition, the threshold voltage has a negative temperature characteristic so that the threshold voltage mismatch should be avoided in the real application. A large current unbalance is observed during the dynamic processes, especially during turn-on process.

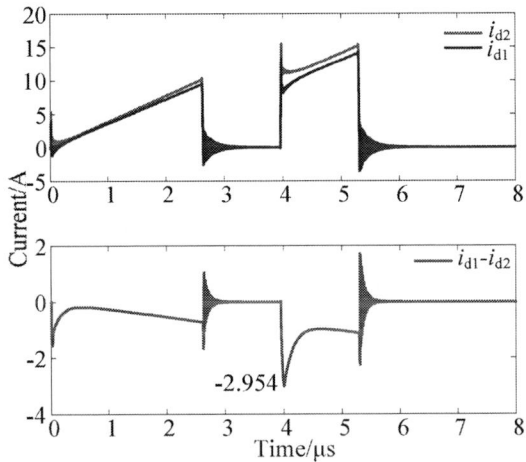

Fig. 4. Current sharing performance with threshold voltage mismatch.

C. Device Parasitic-Gain (Related with Device On-state Resistance)

Fig. 5 shows the current sharing performance with on-state resistance mismatch. From the result one can observe that the on-state resistance will mainly affect the static current sharing performance, while its influence on dynamic current sharing performance is negligible. The maximum current difference is around 2.306 A.

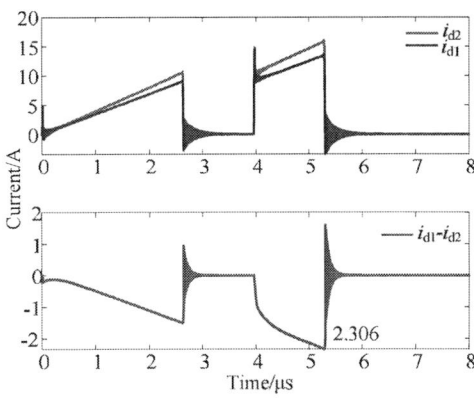

(a) Overall current sharing waveform

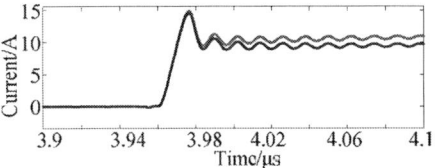

(b) Zoomed-in current sharing waveform during turn-on process

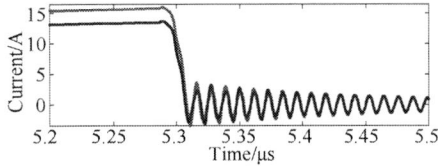

(c) Zoomed-in current sharing waveform during turn-off process

Fig. 5. Current sharing performance with on-state resistance mismatch.

D. Device Parasitic-Gate Terminal Parasitic Inductance

Fig. 6 shows the current sharing performance with gate terminal parasitic inductance mismatch. The maximum current difference is around 0.3178 A. The gate terminal parasitic inductance has small influence on the current sharing performance.

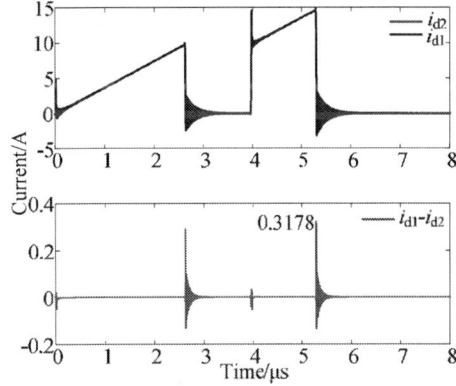

Fig. 6. Current sharing performance with gate terminal inductance mismatch.

E. Device Parasitic-Drain Terminal Parasitic Inductance

Fig. 7 shows the current sharing performance with drain terminal parasitic inductance mismatch. The maximum current difference is around 0.1792 A.

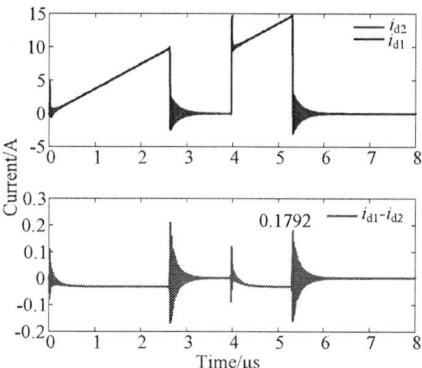

Fig. 7. Current sharing performance with drain terminal inductance mismatch.

F. Device Parasitic-Source Terminal Parasitic Inductance

Fig. 8 shows the current sharing performance with source terminal parasitic inductance mismatch. The maximum current difference is around 2.269 A. Compared with drain terminal parasitic inductance, the source terminal parasitic inductance has large influence on the current sharing performance. The major reason is that the source terminal inductance is also included in the gate loop, which will affect the gate voltages on each device and the dynamic current sharing performance.

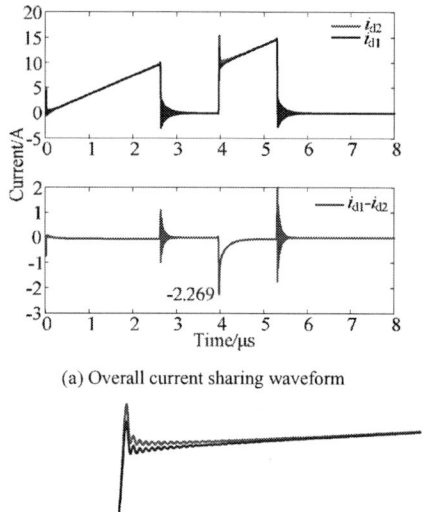

(a) Overall current sharing waveform

(b) Zoomed-in current sharing waveform during turn-on process

Fig. 8. Current sharing performance with source terminal inductance mismatch.

G. Device Parasitic-Input Capacitance

Fig. 9 shows the current sharing performance with input capacitance mismatch. The dynamic switching speed is closely related with the input capacitance. A large input capacitance will lead to slow switching speed. Thus, the maximum current difference of 2.002 A is seen during the turn-on process.

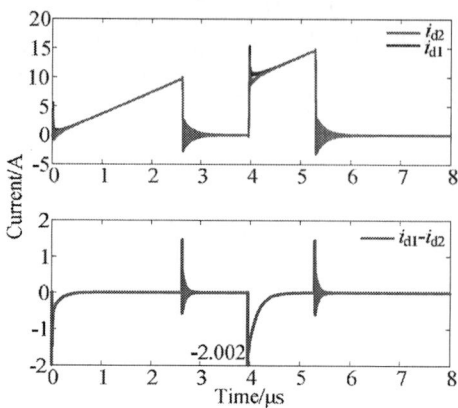

Fig. 9. Current sharing performance with input capacitance mismatch.

H. Device Parasitic-Reverse Transfer Capacitance

Fig. 10 shows the current sharing performance with revere transfer capacitance mismatch, which would also affect the dynamic switching process. The maximum current difference is around 3.783 A.

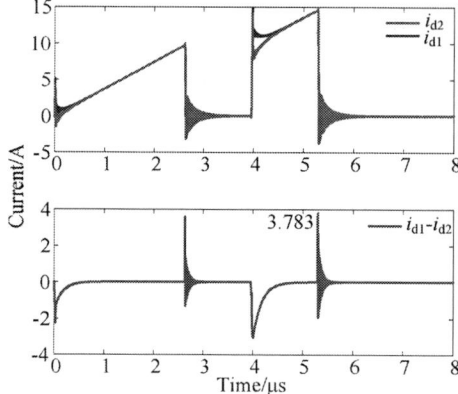

Fig. 10. Current sharing performance with reverse transfer capacitance mismatch.

I. Device Parasitic-Output Capacitance

Fig. 11 shows the current sharing performance with output capacitance mismatch. The device turn-off process is mainly affected by the output capacitance. The maximum current difference is during the turn-off process, which is around 1.476 A.

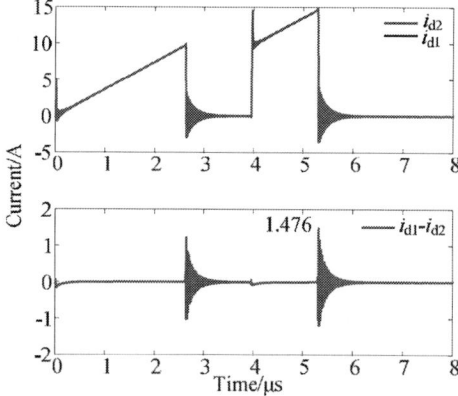

Fig. 11. Current sharing performance with output capacitance mismatch.

978-1-6654-4817-8/21 $31.00 © 2021 IEEE

J. Circuit Parasitic-Turn-on Gate Resistor

Fig. 12 shows the current sharing performance with turn-on gate resistor mismatch. Clearly, the turn-on switching speed is greatly affected by the turn-on gate resistor. The maximum current difference is around 4.154 A, which occurs during the turn-on process.

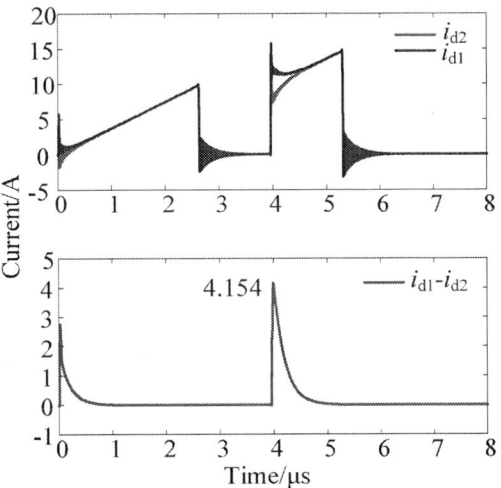

Fig. 12. Current sharing performance with turn-on gate resistor mismatch.

K. Circuit Parasitic-Turn-off Gate Resistor

Fig. 13 shows the current sharing performance with turn-off gate resistor mismatch. It can be seen that mainly the turn-off process is affected by the turn-off gate resistor mismatch. The maximum current difference is around 3.0209 A.

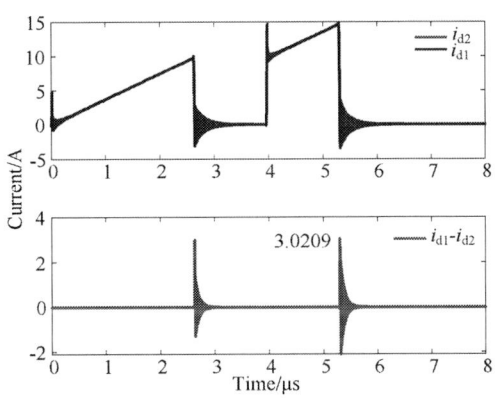

(a) Overall current sharing waveform

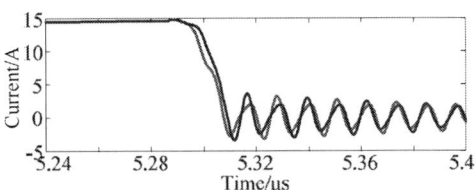

(b) Zoomed-in current sharing waveform during turn-off process

Fig. 13. Current sharing performance with turn-off gate resistor mismatch

L. Circuit Parasitic-Gate Loop Parasitic Inductance

Fig. 14 shows the current sharing performance with gate loop parasitic inductance mismatch. This mismatch can be caused by the asymmetric layout of gate drivers. For the gate

loop parasitic inductance, only the dynamic current sharing performance is affected. The maximum current difference is around 0.6958 A.

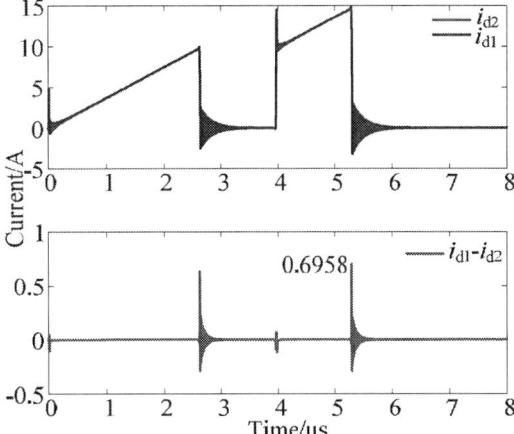

Fig. 14. Current sharing performance with gate loop parasitic inductance mismatch.

M. Circuit Parasitic- Parasitic Inductance from Drain Terminal to Freewheeling Diode

Fig. 15 shows the current sharing performance with circuit drain terminal parasitic inductance mismatch. This mismatch can be caused by the asymmetric layout of PCB. The maximum current difference is around 0.1649 A.

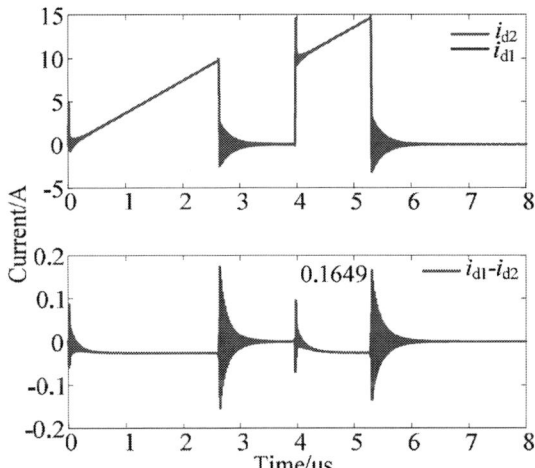

Fig. 15. Current sharing performance with circuit drain terminal parasitic inductance mismatch.

N. Circuit Parasitic- Parasitic inductance from source terminal to ground

Fig. 16 shows the current sharing performance with common circuit source parasitic inductance mismatch. This mismatch can be caused by the asymmetric layout of PCB. The maximum current difference is around 0.1482 A.

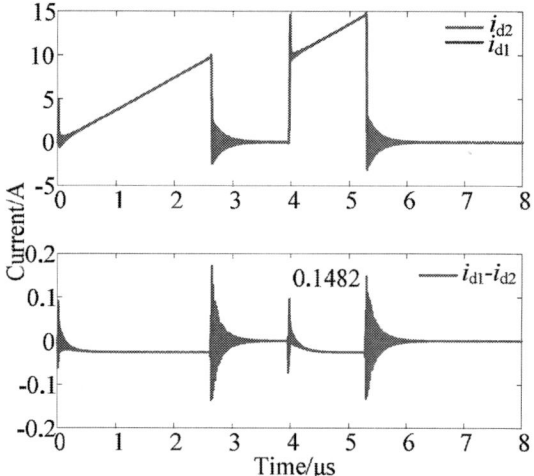

Fig. 16. Current sharing performance with circuit source terminal parasitic inductance mismatch.

Fig. 17 summarizes the parasitics influences on current sharing performance. From Fig. 17, we can observe that to achieve a good current sharing operation, several special considerations should be made: 1) when select the gate resistors, the resistance tolerance should be small so that the variations between each gate driver is minimized. In addition, the gate resistors should also have a stable temperature performance; 2) the common source parasitic inductance should be minimized since it is included in both gate loop and power loop, the device package with kelvin source can alleviate the current unbalance caused by the common source parasitic inductance ; 3) the on-state resistance mismatch and threshold voltage mismatch are the major reasons for the unbalanced device currents. Although efforts can be made to mitigate these parasitics mismatches, it is impossible to achieve identic parasitics for individual devices. In addition, the circuit parasitics are also difficult to be identic when more power devices are in parallel. Therefore, solutions are still required to mitigate the current unbalance.

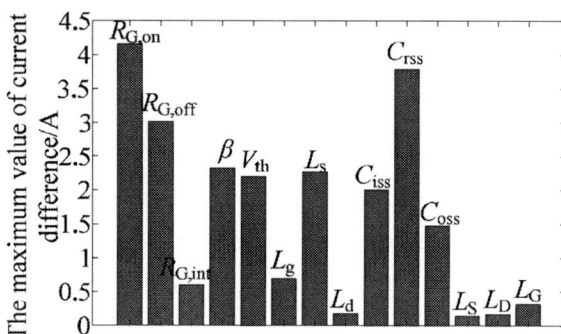

Fig. 17. Current sharing performance comparisons among different circuit parasitics.

IV. Active Gate Driving Method to Mitigate the Current Sharing Performance

According to [10], in the saturation region, the SiC MOSFET current can be modelled as Eq. (2). The mismatches of the dynamic current among different paralleled power devices are caused by the threshold voltage mismatches.

$$i_{dsj} = g_{fs}(v_{gsj} - v_{thj})^2, \quad v_{gsj} > v_{thj} \tag{2}$$

Therefore, to compensate the difference of threshold voltage, different gate voltages can be applied for the individual devices/modules. In addition to the threshold voltage mismatch, the unbalanced transient current caused by other circuit mismatches can also be compensated by the gate volatge. The difficulties of active gate driver with adjustable turn-on voltage including: 1) adjustable turn-on gate voltage circuit design; and 2) current monitor circuit design. This work simply focus on the adjustable turn-on gate voltage circuit design. For medium voltage SiC MOSFET, like 3.3 kV and 10 kV SiC MOSFETs, gate driver with high insulated power supply is required. In this research, a wireless power transfer based high voltage insulated power supply is designed for medium voltage SiC MOSFETs. Fig. 18 shows the circuit topology of the gate driver power supply. The positive turn-on gate voltage is generated by wireless power transfer converter and the negative turn-off gate voltage is generated by a non-isolated dc/dc converter with a regulated -5 V output voltage in a wide input voltage range. Fig. 19 shows the picture of the experiment setup. By adjusting the switching frequency of primary inverter, the turn-on gate voltage can be adjusted accordingly.

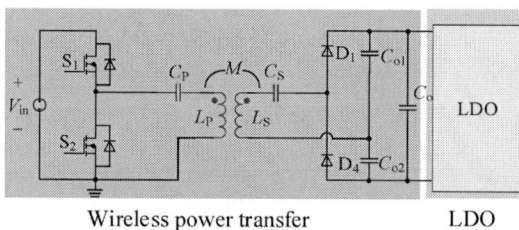

Wireless power transfer   LDO

Fig. 18. Circuit topology implemented for high voltage insulated gate driver.

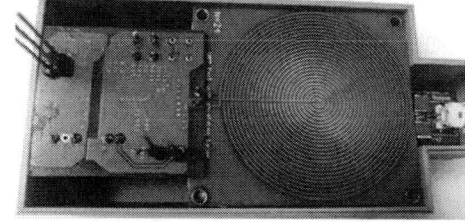

(a) Wireless power transfer converter
Gate driver #1 Gate driver #2

(b) DPT setup with paralleling SiC MOSFETs
Fig. 19. Experiment setup.

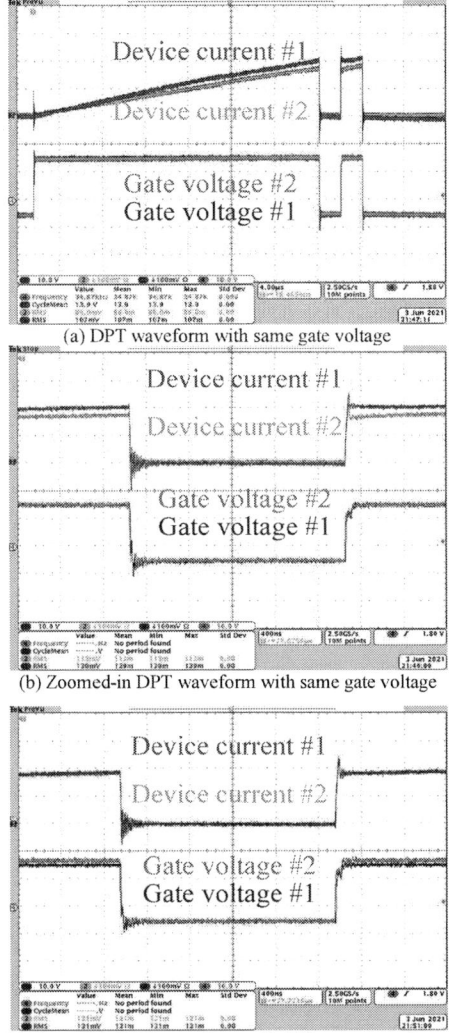

(a) DPT waveform with same gate voltage

(b) Zoomed-in DPT waveform with same gate voltage

(c) Zoomed-in DPT waveform with different gate voltages

Fig. 20. DPT experiment waveforms with same and different gate voltages.

Fig. 20(a) and Fig. 20(b) show the DPT waveforms with same gate voltage for two paralleling devices. Current unbalance exists due to the asymmetric parasitics. Fig. 20(c) shows the DPT waveform with different gate voltages. As can be seen from Fig. 20(b), device 1 turns on faster than device 2. In order to achieve current sharing, the turn-on gate voltage for device 2 can be increased as show in Fig. 20 (c). A well current sharing can be achieved by using adjustable turn-on gate voltage.

V. HYBRID TURN-ON GATE VOLTAGE AND GATE DELAY STRATEGY FOR PARALLELING OPERATION

Although the variable turn-on gate voltage can mitigate the threshold voltage mismatch or other parasitics mismatches, the static current sharing operation is affected at the same time. As discussed in [13], the on-state resistance is sensitive to the gate voltage. To decouple between the dynamic current sharing and static current sharing, a hybrid turn-on gate voltage and gate delay control strategy can be adopted. The basic idea is that during the dynamic process, the gate delays between paralleled devices are adjusted to achieve dynamic current sharing, while the turn-on gate voltages are adjusted

for the purpose of static current sharing caused by the mismatch of on-state resistance. Please note that in the existing literature, the static current sharing is not discussed and only the positive temperature characteristic of on-state resistance is used to achieve current sharing operation. However, as analyzed in [13], when compared with its Si counterpart, the on-state resistance of SiC MOSFET is less sensitive to the temperature. Thus, static current balancing strategy is also required.

Based on the previous discussions, the threshold voltage and on-state resistance are set different between two paralleled power devices. Fig. 21 shows the simulation results with the same gate driver. Both the static and dynamic currents are unbalance. Fig. 22 shows the simulation waveforms with variable gate volatge. Although the dynamic current sharing operation is achieved, the static current unbalance still exists. Fig. 23 shows the hybrid gate voltage and gate delay control strategy, both the turn-on and turn-off delays, and the gate voltage are adjusted to achieve current sharing operation.

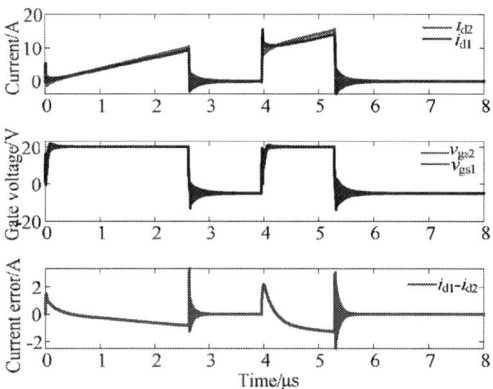

Fig. 21. Simulation t waveforms with the same gate driver.

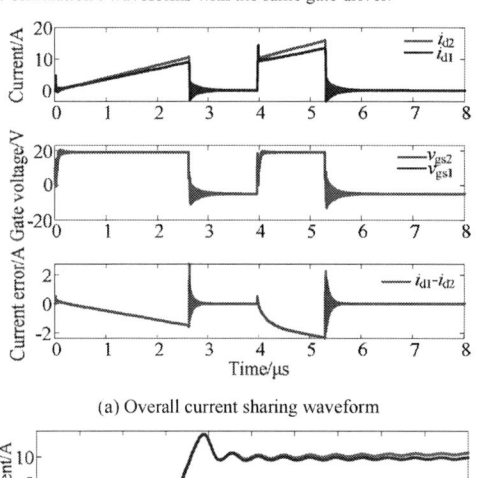

(a) Overall current sharing waveform

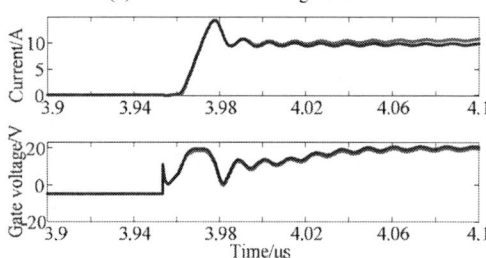

(b) Zoomed-in current sharing waveform during turn-on process

Fig. 22. Simulation t waveforms with the variable gate volatge gate delay control strategy.

978-1-6654-4817-8/21 $31.00 © 2021 IEEE

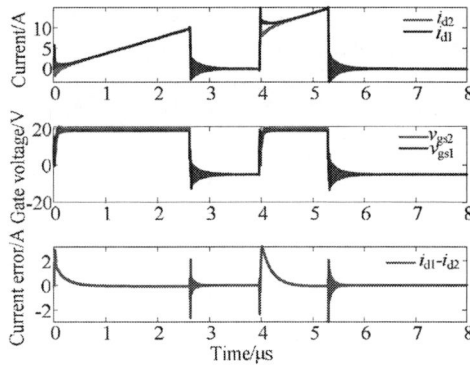

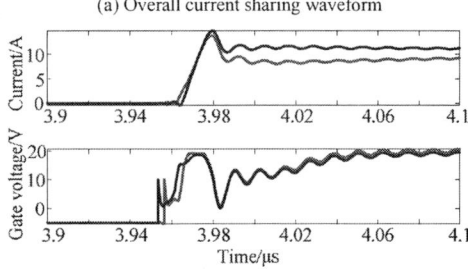

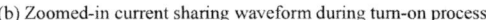

(a) Overall current sharing waveform

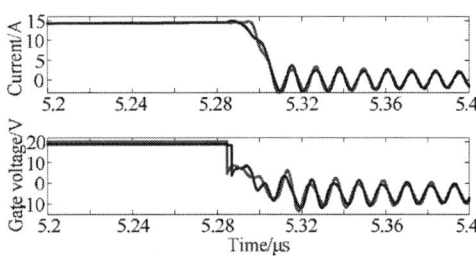

(b) Zoomed-in current sharing waveform during turn-on process

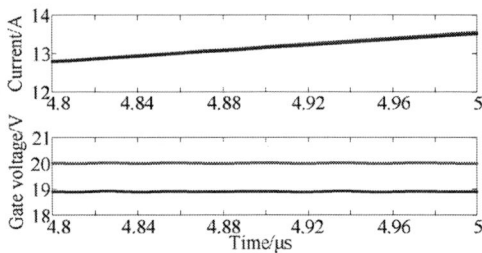

(c) Zoomed-in current sharing waveform during turn-off process

(d) Zoomed-in current sharing waveform during steady state condition

Fig. 23. Simulation t waveforms with the proposed hybrid gate volatge gate delay control strategy.

VI. CONCLUSIONS

In this work, the current sharing performance of paralleling devices/modules are investigated based on the subcircuit model in MATLAB. Each possible device and circuit parasitics are considered and comparisons are presented. To mitigate the current unbalance, an adjustable turn-on gate voltage solution is proposed. By adjusting the switching frequency of the primary inverter on the isolated gate driver power supply, the output volatge is regulated. The experiment results demonstrate the effectiveness of using adjustable turn-on gate voltage for the current sharing operation of paralleling devices. The hybrid gate voltage and gate delay control strategy is proposed and verified by simulation studies.

ACKNOWLEDGMENT

This material is based upon work supported by the National Science Foundation under Grant No. 1939144, GRID Connected Advanced Power Electronics Systems (GRAPES), Project GR-21-06.Any opinions, findings, and conclusions or recommendations expressed in this material are those of the author(s) and do not necessarily reflect the views of the National Science Foundation.

REFERENCES

[1] S. Zhao, X. Zhao, Y. Wei, Y. Zhao and H. A. Mantooth, "A Review on Switching Slew Rate Control for Silicon Carbide Devices using Active Gate Drivers," IEEE Journal of Emerging and Selected Topics in Power Electronics, doi: 10.1109/JESTPE.2020.3008344.

[2] S. Zhao, X. Zhao, A. Dearien, Y. Wu, Y. Zhao and H. A. Mantooth, "An Intelligent Versatile Model-Based Trajectory-Optimized Active Gate Driver for Silicon Carbide Devices," IEEE Journal of Emerging and Selected Topics in Power Electronics, vol. 8, no. 1, pp. 429-441, Mar. 2020.

[3] Z. Zeng, X. Zhang and Z. Zhang, "Imbalance Current Analysis and Its Suppression Methodology for Parallel SiC MOSFETs with Aid of a Differential Mode Choke," IEEE Trans. Ind. Electron., vol. 67, no. 2, pp. 1508-1519, Feb. 2020.

[4] S. K. Singh, N. K. Pilli, F. Guedon, and R. Mcmahon, "PMSM drive using silicon carbide inverter: Design, development and testing at elevated temperature," in Proc. IEEE Int. Conf. Ind. Technol., Mar. 2015, pp. 2612–2618.

[5] H. Li, S. Munk-Nielsen, S. Beczkowski, and X. Wang, "A novel DBC layout for current imbalance mitigation in SiC MOSFET multichip power modules," IEEE Trans. Power Electron., vol. 31, no. 12, pp. 8042–8045, Dec. 2016.

[6] Y. Xue et al., "A compact planar Rogowski coil current sensor for active current balancing of parallel-connected silicon carbide MOS FETs," in Proc. of IEEE Energy Convers. Congr. Expo., Sep. 2015, pp. 4685–4690.

[7] Z. Zhang, J. Dix, F. Wang, B. J. Blalock, D. Costinett, and L. M. Tolbert,"Intelligent gate drive for fast switching and crosstalk suppression of SiC devices," IEEE Trans. Power Electron., vol. 32, no. 12, pp. 9319–9332, Dec. 2017.

[8] S. Kokosis, I. Andreadis, G. Kampitsis, P. Pachos, and S. Manias, "Forced current balancing of parallel connected SiC JFETs during forward and reverse conduction mode," IEEE Trans. Power Electron., vol. 32, no. 2, pp. 1400–1410, Feb. 2017.

[9] Y. Yang, Y. Wen and Y. Gao, "A Novel Active Gate Driver for Improving Switching Performance of High-Power SiC MOSFET Modules," IEEE Transactions on Power Electronics, vol. 34, no. 8, pp. 7775-7787, Aug. 2019.

[10] Y. Wen, Y. Yang and Y. Gao, "Active Gate Driver for Improving Current Sharing Performance of Paralleled High-Power SiC MOSFET Modules," IEEE Transactions on Power Electronics, vol. 36, no. 2, pp. 1491-1505, Feb. 2021.

[11] Y. Mao, Z. Miao, C. Wang and K. D. T. Ngo, "Passive Balancing of Peak Currents Between Paralleled MOSFETs With Unequal Threshold Voltages," IEEE Trans. on Power Electron., vol. 32, no. 5, pp. 3273-3277, May 2017.

[12] H. Li, S. Beczkowski, S. M. Nielsen, R. Maheshwari, and T. Franke, "Circuit mismatch and current coupling effect influence on paralleling SiC MOSFETs in multichip power modules," in Proc. IEEE PCIM Europe, 2015, pp. 1–8.

[13] G. Wang, J. Mookken, J. Rice and M. Schupbach, "Dynamic and static behavior of packaged silicon carbide MOSFETs in paralleled applications," 2014 IEEE Applied Power Electronics Conference and Exposition - APEC 2014, 2014, pp. 1478-1483.

[14] V. Talesara et al., "Dynamic Switching of SiC Power MOSFETs Based on Analytical Subcircuit Model," IEEE Trans. on Power Electron., vol. 35, no. 9, pp. 9680-9689, Sept. 2020.

978-1-6654-4817-8/21 $31.00 © 2021 IEEE

Mode Switchover Strategy for Multi-port Energy Router Based on State Flow Diagram

Jingwen Zheng
State Grid Hubei Electric Power Co.,
Ltd. Electric Power Research Institute
Wuhan, China
1092362588@qq.com

Zhiguo Wei
China University of Geosciences,
Wuhan
Wuhan, China
lacotine@163.com

Zaixun Ling
State Grid Hubei Electric Power Co.,
Ltd. Electric Power Research Institute
Wuhan, China
543762801@qq.com

Yu Guo
State Grid Hubei Electric Power Co.,
Ltd. Electric Power Research Institute
Wuhan, China
511149215@qq.com

Ping Xiong
State Grid Hubei Electric Power Co.,
Ltd. Electric Power Research Institute
Wuhan, China
px_joey@163.com

Yiqun Kang
State Grid Hubei Electric Power Co.,
Ltd. Electric Power Research Institute
Wuhan, China
376523814@qq.com

Abstract—**Up to now, the control strategies of most energy routers are only designed for normal operating conditions, and not designed for failure of a certain module of the system. This paper proposes a switching control strategy for the energy router based on the sudden failure of a certain module of the energy router. The strategy will detect the operating status of each converter. When a converter fails, the system will shut off the converter in time, so that the remaining converters can continue to work and the system bus voltage can be maintained. The system will restore the module automatically when fault of the module is eliminated. Simultaneously, if the energy storage module's power is too high or too low, the system will halt charging and discharging in order to safeguard the energy storage module. Finally, the validity of the above theory is verified through experiments.**

Keywords—**energy router, mode switching, running with failure**

I. INTRODUCTION

The energy router was invented to overcome the intermittent problem of new energy power generation and diversity of electrical equipment. Most of the energy routers have only grid-connected mode, off-grid mode, error mode and standby mode. When a module of the system fails, the system will shut down the whole system, resulting in inefficient utilization of the energy generated by photovoltaic [1]-[3]. A hybrid bus structure energy router with multiple operation modes is proposed in [4]. This structure is expensive in hardware and the control strategy has not been improved. Otherwise, some scholars have proposed the power allocation strategy for energy routers [5]. And a hierarchical coordinated control strategy for energy routers, which used different control methods for the bus voltage of different voltage segments, is proposed [6],[7]. In [8], the failure of one of the devices is simulated when multiple energy routers are interconnected and the dynamic response of the system is analyzed. The control strategies of the energy router described above do not take into account the control method in case of system failure. Moreover, the working model of the converter is relatively simple and the system cannot continue to run with failure.

Aiming at the shortcomings of the current energy router control strategy, this paper proposes an energy router switching control strategy. This strategy can keep the other parts of the system running normally when some modules of the system fail.

II. SYSTEM TOPOLOGY AND CONTROL METHOD

The energy router has four ports, namely photovoltaic (PV) port, battery port, AC load port and grid port. The physical connection form between each module is shown in Fig. 1.

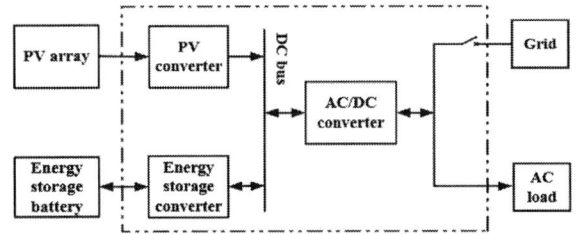

Fig. 1. System topology

The AC/DC converter, PV converter, and energy storage converter terminals are all linked directly to the DC bus. AC/DC converter is responsible for the mutual conversion of AC voltage and DC voltage. The PV converter is responsible for increasing the PV output voltage while completing the maximum power point tracking (MPPT) and the energy storage converter is responsible for the mutual conversion between the energy of the DC bus and the energy of the energy storage battery.

A converter ensures that the bus voltage is maintained at all times. During typical system operation, the energy storage converter maintains the bus voltage. When the energy storage converter fails, the AC/DC converter or PV converter maintains the bus voltage. MPPT mode, constant current mode, and constant bus voltage mode are the three working modes of the PV converter. The PV converter's MPPT and constant current modes use single current loop control, while the constant bus voltage mode uses double loop control. The energy storage converter has only constant voltage charging/discharging mode, and controls the bus voltage. The AC/DC converter has three working modes: grid-connected current mode, rectified voltage mode(control bus voltage) and inverter voltage mode. AC/DC converter adopts decoupling control based on dq coordinate system. The grid-connected current mode adopts single-loop PI control, and the inverter voltage mode and rectified voltage mode adopt dual-loop PI control. The working mode and control quantities of each converter are shown in Table I.

978-1-6654-4817-8/21 $31.00 © 2021 IEEE

TABLE I. CONVERTER MODE

Converter	Converter mode	Control quantity
PV converter	MPPT	PV cell current
	constant current	PV cell current
	constant bus voltage	DC bus voltage
AC/DC converter	grid-connected current	Grid-connected current
	inverter voltage	Inverter AC voltage
	Rectified voltage	DC bus voltage
Energy storage converter	constant voltage charging/discharging	DC bus voltage

III. SYSTEM OPERATION MODE

The battery's service life should be extended as much as possible and the system should prevent overcharging or overdischarging the energy storage battery. When the system is linked to the grid and the energy storage battery is at least 90% charged, which is called overcharge mode, the system adjusts the grid-connected current or PV current to reduce the current of the battery to 0. Conversely, if the battery power is less than 10%, it is called over-discharge mode. The system will reduce the current fed into the grid or take power from the grid to make the current of the battery 0.

When the system is off-grid and the energy storage battery is at least 90% charged, the PV converter exits the MPPT mode and enters the constant current mode. The system makes the PV output power equal to the load power while reducing the battery charging current to 0. On the contrary, if the battery power is less than 10%, the system will cut off part of the load, so that the battery discharge current is 0.

A. Grid-connected mode

The system totally has 5 working modes in grid-connected mode, which are illustrated as follow.

In PV-Bat-Grid-Load mode (mode 1), the energy generated by photovoltaics is given priority to the load. The system maintains the energy storage battery at 50%, and the remaining power will be send to the grid.

In PV-Bat-Load mode (mode 2), the energy produced by photovoltaic is supplied to the load and battery. When the photovoltaic energy is surplus, the photovoltaic converter enters the constant current mode to reduce the photovoltaic output power. When the photovoltaic energy is insufficient, the system cuts off part of the load so that the remaining load of the system can operate normally.

In PV-Grid-Load mode(mode 3), the energy storage converter stops working. If the PV power is higher than the load power, the PV energy is transmitted to the load, and the remaining energy is fed into the grid. If PV power is inadequate to fulfill the load, power is drawn from the grid.

In Bat-Grid-Load mode (mode 4), the photovoltaic converter stops working, the system maintains the energy storage battery power at 50%, and the load energy is provided by the grid.

In the PV-Bat-Grid mode (mode 5), the system maintains the energy storage battery at 90%, and all the energy generated by photovoltaics is fed into the grid. The converter operating mode corresponding to each mode is shown in Table II.

TABLE II. WORKING MODES OF EACH CONVERTER IN GRID-CONNECTED MODE

System mode	AC/DC converter	PV converter	Energy storage converter	AC load
PV-Bat-Grid-Load	Grid-connected current	MPPT	Constant voltage charge/discharge	Running
PV-Bat-Grid	Grid-connected current	MPPT	Constant voltage charge/discharge	Standby
PV-Grid-Load	Rectified voltage	MPPT	Standby	Running
Bat-Grid-Load	Grid-connected current	Standby	Constant voltage charge/discharge	Running
PV-Bat-Load	Inverter voltage	MPPT/Constant current	Constant voltage charge/discharge	Running

B. Off-grid mode

The system totally has 4 working modes in off-grid mode, which are illustrated as follow. PV-Bat-Load mode (mode 2) has been introduced in the previous section.

In PV-Bat mode (mode 6), AC/DC converter is in standby. If the battery power is less than 90%, the PV converter works in MPPT mode to charge the battery. When the battery power reaches 90%, the system enters standby mode.

In PV-Load mode (mode 7), the energy storage converter is currently turned off. The PV converter keeps the bus voltage constant. When the load power is greater than the PV power, the system cuts off part of the load to make the load power less than or equal to the PV output power.

In Bat-Load mode (mode 8), the PV converter is in standby. The energy storage battery provides energy for load. When the battery power is less than 10%, the battery stops discharging and the system enters standby mode. The converter operating mode corresponding to each mode is shown in Table III.

TABLE III. WORKING MODES OF EACH CONVERTER IN OFF-GRID MODE

System mode	AC/DC converter	PV converter	Energy storage converter	AC load
PV-Bat-Load	Inverter voltage	MPPT/Constant current	Constant voltage charge/discharge	Running
PV-Bat	Standby	MPPT/Standby	Constant voltage charge/discharge	Standby
PV-Load	Inverter voltage	constant bus voltage	Standby	Running
Bat-Load	Inverter voltage	Standby	Constant voltage charge/discharge	Running

IV. SYSTEM SWITCHING STRATEGY

A. Grid-connected mode

The system state switching method in grid-connected mode is shown in Fig. 2. The working mode or current command of a certain converter will change when the state is switched.

When the grid voltage is larger than the maximum specified value U_{max} or less than the lowest set value U_{min}, or the frequency is greater than the maximum specified value f_{max}, or the frequency is less than the minimum set value f_{min}, the system will immediately disconnect from the grid and switch the AC/DC converter to inverter voltage mode. The

PV converter chooses to maintain the MPPT mode or enter the constant current mode according to the amount of energy stored. Then the system enters PV-Bat-Load mode (mode 2).

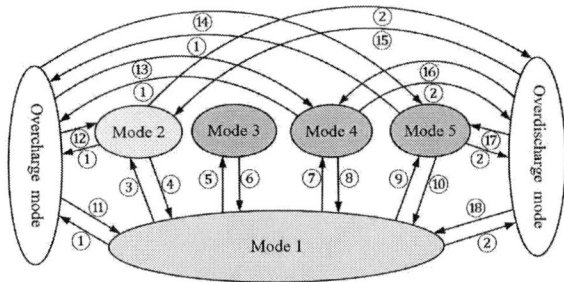

Fig. 2. System state switching diagram

Another situation is to actively disconnect from the grid. When the user issues an off-grid command, the system first reduces the grid-connected current of the AC/DC converter to 0, then disconnects from the grid, and finally switch the AC/DC converter to inverter voltage mode.

If the system detects that the voltage and frequency of the external grid are within the set threshold, or the user issues a grid-connected command, the system stops the AC/DC converter and switches it to grid-connected current mode and closes the relay connected to the grid at the same time. Then system start the phase-locked loop and start the AC/DC converter after the phase-lock is successful. The PV converter enters the MPPT mode, and the system enters the PV-Bat-Grid-Load mode(mode 1).

The energy storage converter and the AC/DC converter will be switched off if the voltage of the energy storage module falls below the minimum setting value U_{bat_min}. And the AC/DC converter's operating mode will be changed to rectified voltage mode. Then the system enters the PV-Grid-Load mode(mode 3). In this mode, the AC/DC converter regulates the bus voltage and the PV converter working mode remains unchanged. The AC/DC converter will boost the grid-connected current to ensure power balance if the bus voltage is higher than the stated reference value. When the system detects that the bus voltage is lower than the given reference value, the grid-connected converter will reduce the grid-connected current to stabilize the bus voltage.

When the system detects that the voltage of the energy storage module is greater than the set value U_{bat_min}, the system stops the AC/DC converter and sets the AC/DC converter working mode to grid-connected current mode, then starts the energy storage converter. The AC/DC converter starts after the bus voltage stabilizes. And the system enters the PV-Bat-Grid-Load mode(mode 1).

When the system detects that the PV voltage is less than the minimum set value U_{pv_min} or higher than the maximum set value U_{pv_max}, the system turns off the PV converter and changes the reference value of the grid-connected current to keep the energy storage discharge current at zero. And the system enters the Bat-Grid-Load mode(mode 4).

On the contrary, when the system detects that the PV voltage is within the set threshold, it sets the PV converter mode to MPPT mode and starts the PV converter, and the system enters the PV-Bat-Grid-Load mode(mode 1).

When the load current is greater than the threshold I_{load_max} set by the user, the system will immediately remove the load, and the system will enter the PV-Bat-Grid mode (mode 5) at the same time. Only after the user clears the error manually, the load relay will close and enter the light-storage-network-load mode. The above switching conditions are shown in Table IV.

TABLE IV.   SWITCHING CONDITION

Serial number	meaning	Serial number	meaning
①	$SOC > 90\%$	②	$SOC < 10\%$
③	$U_{grid} > U_{max}$ or $U_{grid} < U_{min}$ or $f_{grid} > f_{max}$ or $f_{grid} < f_{min}$ or off-grid instruction	④	$U_{grid} < U_{max}$ and $U_{grid} > U_{min}$ and $f_{grid} < f_{max}$ and $f_{grid} > f_{min}$ or grid-connection instruction
⑤	$U_{bat} < U_{bat\_min}$	⑥	$U_{bat} \geq U_{bat\_min}$
⑦	$U_{pv} < U_{pv\_min}$ or $U_{pv} > U_{pv\_max}$	⑧	$U_{pv} > U_{pv\_min}$ and $U_{pv} < U_{pv\_max}$
⑨	$I_{load} > I_{load\_max}$	⑩	Clear errors manually
⑪	$SOC \leq 90\%$ and mode= 1	⑫	$SOC \leq 90\%$ and mode= 2
⑬	$SOC \leq 90\%$ and mode= 4	⑭	$SOC \leq 90\%$ and mode= 5
⑮	$SOC \geq 10\%$ and mode= 2	⑯	$SOC \geq 10\%$ and mode= 4
⑰	$SOC \geq 10\%$ and mode= 5	⑱	$SOC \geq 10\%$ and mode= 1

B. Off-grid mode

The system state switching method in off-grid mode is shown in Fig. 3, and the switching conditions are shown in Table IV.

When the voltage of the energy storage module is lower than the minimum setting value U_{bat_min}, the system will stop the energy storage converter, and the working mode of the PV converter will be switched to the constant bus voltage mode. The working mode of the AC/DC converter remains unchanged, and the system enters the PV-Load mode(mode 7).

On the contrary, when the system detects that the voltage of the energy storage module is larger than the set value U_{bat_min}, the system stops the PV converter and starts the energy storage converter, then sets the working mode of the PV converter to MPPT mode. After the bus voltage stabilizes, the PV converter is started again. And system enter the PV-Bat-Load mode(mode 2).

When the system detects that the PV voltage is less than the minimum setting value U_{pv_min} or higher than the highest setting value U_{pv_max}, the system stops the PV converter, the AC/DC converter working mode remains unchanged. And the system enters the Bat-Load mode(mode 8). If the energy storage capacity is less than 10%, the system stops working and enters standby mode. Instead, when the PV voltage is within the set threshold, the system sets the PV converter

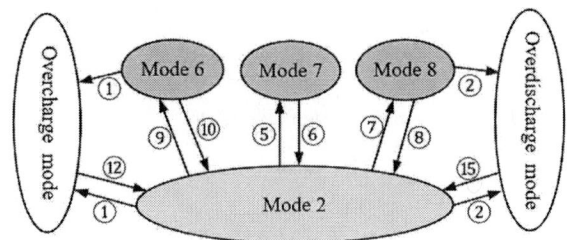

Fig. 3. System state switching diagram in off-grid mode

mode to MPPT mode and restarts the PV converter. And the system enters the PV-Bat-Load mode(mode 2).

When the load current is greater than the threshold I_{load_max} set by the user, the system will immediately cut off the load and enter the PV-Bat mode(mode 6). Only after the user clears the error manually, the load relay will close and the system will enter the PV-Bat-Load mode(mode 2).

V. SIMULATION AND EXPERIMENT

A. Simulation

This paper simulates the switching process between the various modes of the system in MATLAB/Simulink. Battery discharge is positive and the simulation parameters are shown in Table V.

TABLE V. SIMULATION PARAMETERS

parameter	value
Bus voltage	390 V
PV maximum power	4100 W
AC load power	3200 W
Energy storage battery voltage	48 V
Bus capacitance	10 mF
Energy storage converter inductance	1.35 mH
PV converter inductance	1.35 mH
AC/DC converter filter inductance	300 μH
AC/DC converter filter capacitance	20 μF

Fig. 4 shows the simulation results in grid-connected mode. During 0.4s~1s, the system operates in mode 1, and at 1s, it changes to mode 2. Then, the system restores mode 1 at 2s. The energy storage converter exits operation at 2.5s, and the system enters mode 3. The system returns to mode 1 at

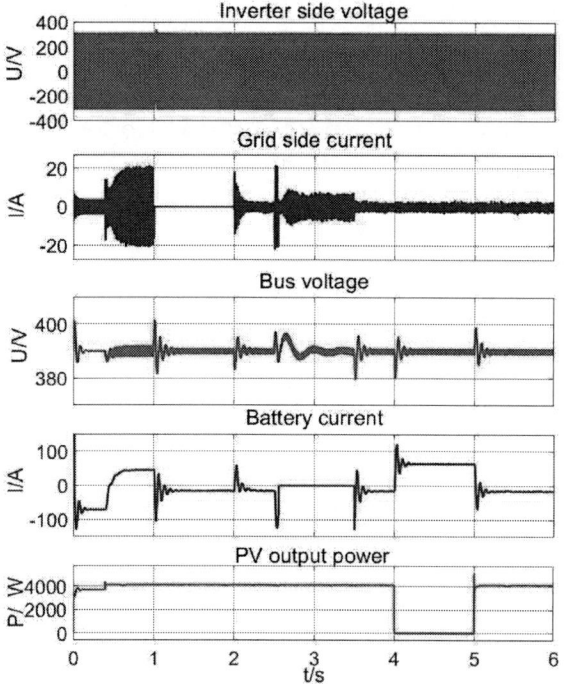

Fig. 4. Simulation results in grid-connected mode

3.5s. The PV converter exits operation at 4s and the system enters mode 4. The PV system resumes operation at 5s and the system enters mode 1.

Fig. 5 shows the simulation results in off-grid mode. The system works in mode 2 during 0.4s~1s, and at 1s, the system switches to mode 7. Then, the system restores mode 2 at 2s. The PV system exits operation at 2.5s, and the system enters mode 8. The system returns to mode 2 at 3.5s. The AC/DC converter exits operation at 4s and the system enters mode 6. The AC/DC converter resumes operation at 5s and the system enters mode 2.

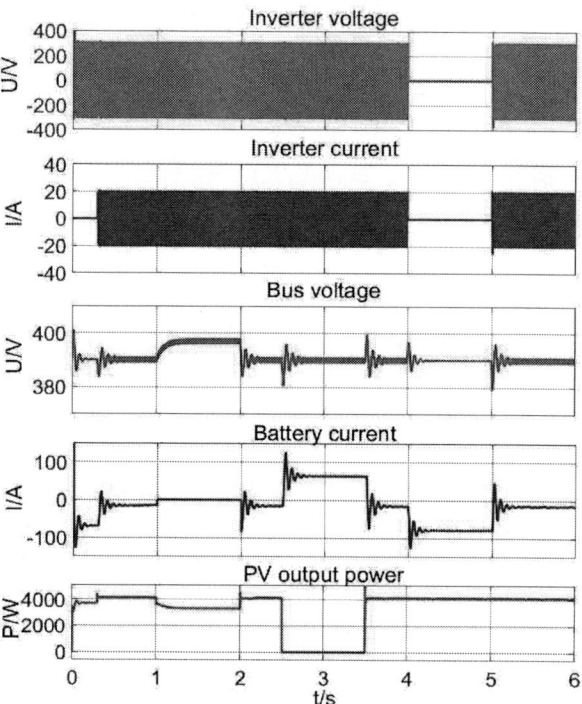

Fig. 5. Simulation results in off-grid mode

It can be seen from the figure that during the switching process of the system, the inverter side voltage remains almost unchanged, and the bus voltage fluctuation value is less than 5%. And the adjustment time is short, the voltage can quickly return to stability. The battery current has a certain overshoot during switching, but it can quickly be adjusted to a stable value.

B. Experimental results

Fig. 6 shows the waveform of switching process of the system mode. The experimental result of the system shifting from mode 1 to mode 2 in grid-connected mode is shown in Fig. 6-(a) and the system turns to off-grid operation. Fig. 6-(b) shows the experimental result of the system when it transitions from mode 1 to mode 5. The system cuts off the load and the grid-connected current increases at the same time. In grid-connected mode, Fig. 6-(c) shows the system's experimental result from mode 1 to mode 4. The PV output power drops to 0 and the battery discharges to feed power to the grid.

It can be seen from these three pictures that the bus voltage fluctuates slightly and the converters work normally during the switching process.

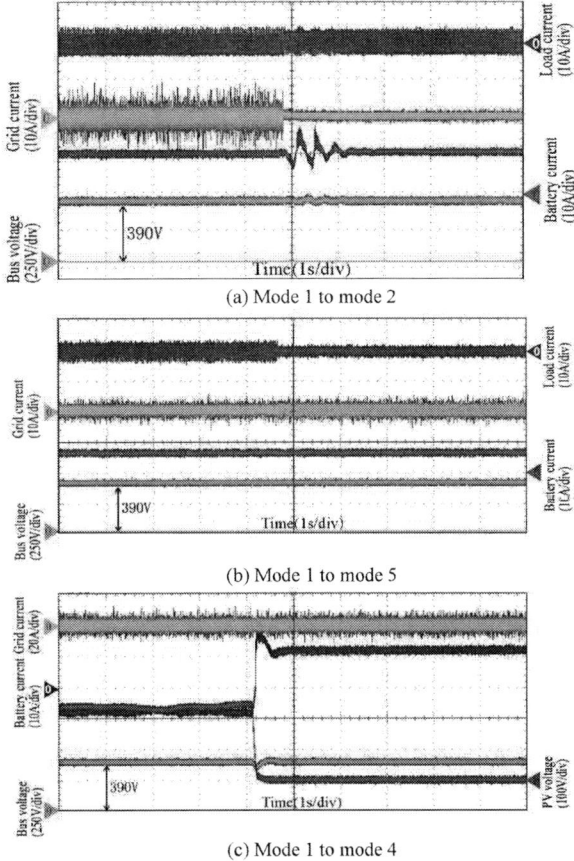

(a) Mode 1 to mode 2

(b) Mode 1 to mode 5

(c) Mode 1 to mode 4

Fig. 6. Mode conversion experiment

VI. CONCLUSION

This paper proposes a switchover strategy for a multi-port energy router, which can make the energy router work in multiple modes. When a module of the system fails or the external working conditions change, the system can continue to run after switching modes. Simulation and experimental results prove the feasibility of the switching strategy. This switching strategy improves the fault-tolerant rate and energy utilization rate of the energy router, and has popularization significance.

REFERENCES

[1] S. Jiao *et al.*, "Cooperative Energy Management for Enterprise Microgrid Using Electrical Energy Router," *2019 IEEE Innovative Smart Grid Technologies - Asia (ISGT Asia)*, Chengdu, China, 2019, pp. 2555-2560.

[2] B. Liu et al., "An AC–DC Hybrid Multi-Port Energy Router With Coordinated Control and Energy Management Strategies," IEEE Access, vol. 7, pp. 109069-109082, 2019.

[3] J. Ahmad, M. Tahir and S. K. Mazumder, "Improved Dynamic Performance and Hierarchical Energy Management of Microgrids With Energy Routing," IEEE Transactions on Industrial Informatics, vol. 15, no. 6, pp. 3218-3229, June 2019.

[4] J. Yu, L. Xiao, Z. Hu, Y. Zhao and J. Nie, "Multi-Terminal Energy Router and Its Distributed Control Strategy in Micro-grid Community Applications," 2020 Asia Energy and Electrical Engineering Symposium (AEEES), Chengdu, China, 2020, pp. 1028-1033.

[5] Y. Ma, X. Wang, X. Zhou and Z. Gao, "An overview of energy routers," 2017 29th Chinese Control And Decision Conference (CCDC), Chongqing, 2017, pp. 4104-4108.

[6] Y. Ma, H. Liu, X. Zhou and Z. Gao, "An Overview on Energy Router Toward Energy Internet," 2018 IEEE International Conference on Mechatronics and Automation (ICMA), Changchun, China, 2018, pp. 259-263.

[7] B. Liu et al., "An AC–DC Hybrid Multi-Port Energy Router With Coordinated Control and Energy Management Strategies," IEEE Access, vol. 7, pp. 109069-109082, 2019.

[8] J. Zhou, Y. Xu, H. Sun and K. Wang, "Flexible Control Strategy of Energy Router for Integration of Local Area Energy Networks with Diverse Services Provision Capability," 2020 IEEE Power & Energy Society General Meeting (PESGM), Montreal, QC, Canada, 2020, pp. 1-5.

A Dynamic Current Sharing Method in Multi-chip SiC Power Module Using Stacked DBC Bridges and Decoupling Capacitors Based on the Original Simple Module Layout

Jianwei Lv
School of Electrical and Electronic Engineering
Huazhong University of Science and Technology
Wuhan,China
jianweil@hust.edu.cn

Chi Zhang
School of Electrical and Electronic Engineering
Huazhong University of Science and Technology
Wuhan,China
chizhang_zach@hust.edu.cn

Cai Chen
School of Electrical and Electronic Engineering
Huazhong University of Science and Technology
Wuhan,China
caichen@hust.edu.cn

Yong Kang
School of Electrical and Electronic Engineering
Huazhong University of Science and Technology
Wuhan,China
ykang@hust.edu.cn

Abstract— **The dynamic current sharing between the parallel SiC chips is important in multi-chip SiC modules. This paper presents a dynamic current balancing approach without changing the original simple module layout in SiC module with multi chips. This method is based on stacked DBC bridges and decoupling capacitors. Stacked DBC bridges soldered with capacitors distribute between the power chips connecting DC+ and DC- electrodes. By this method, distributed decoupling structure is achieved without changing the original simple module layout. The module structure, the design process and the fabrication process are kept simple, and the dynamic current imbalance is reduced. This method is easier to implement than other dynamic current sharing methods. This method is verified by simulations. Compared with the original structure without this method and the concentrated-decoupling structure, dynamic current imbalance of the optimized structure using this method is reduced significantly. Meanwhile, the turn- off overvoltage of the optimized module can be reduced thanks to the decoupling capacitors, and is lower than that of the concentrated-decoupling module due to the lower commutation loop.**

Keywords— *dynamic current sharing, multichip SiC modules, paralleled SiC MOSFETs, distributed decoupling capacitors*

I. INTRODUCTION

Compared to Si power devices, SiC device performs faster switching speed and have lower switching losses [1]. Many SiC power modules uses paralleled SiC MOSFETs to improve the power ratings. However, the dynamic current sharing between paralleled chips is important to ensure the long-time reliability of the multi-chip SiC power modules [2].

There are many methods to achieve dynamic current sharing. In [3], the DBC layout or bonding wires is calculated and changed to achieve dynamic current balancing. In [4], drive-source resistors and power-source inductors are used to improve the transient current sharing. In [5], the driving signals of each paralleled chip are adjusted to achieve dynamic current balancing.

It can be seen that most of the current sharing methods make the structure complex, or are difficult to design and implement. This paper presents a dynamic current sharing

method in multi-chip SiC module using stacked DBC bridges and decoupling capacitors without changing the original simple module layout. By this method, the module structure, the design and fabrication process are kept simple, and the dynamic current imbalance is reduced. To verify this method, the original module, optimized module and the module with concentrated- decoupling capacitors are modeled, simulated and manufactured.

The rest of this paper is organized as follows: Section II presents the baseline SiC module and the dynamic current sharing method. Section III verifies the method by simulation. The fabrication process is presented in Section IV. The conclusions are summarized in section V.

II. THE DYNAMIC CURRENT SHARING METHOD

To be simple and easy to implement, a dynamic current sharing method is presented in multi-chip SiC module without changing the original module layout. In this section, a baseline module is presented firstly. Based on the baseline module, the dynamic current sharing method is presented in detail.

A. The baseline SiC module

The baseline SiC module is a common 2D wire-bonding module. The layout of the module is shown in Fig. 1. It's a commercial EconoDual structure utilizing 8 CREE's 1200V, 98A SiC MOSFETs (CPM212000025B) and 8 1200V, 50A SiC Schottky diodes (CPW51200Z050B). Four MOSFETs and four diodes are paralleled in each switch position to form

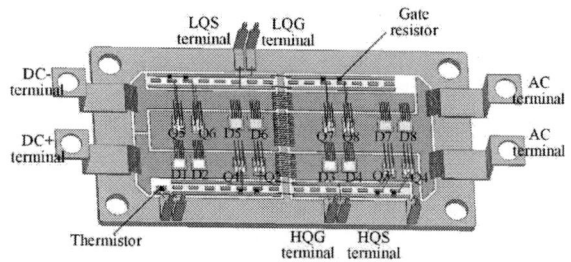

Figure 1: Layout of the original module

978-1-6654-4817-8/21 $31.00 © 2021 IEEE

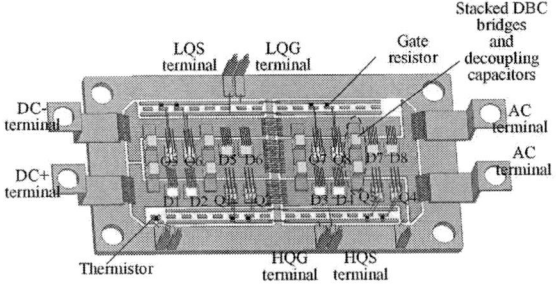

Figure 2: Layout of the improved module

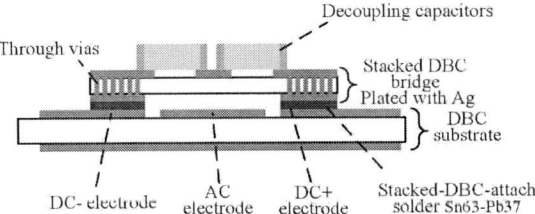

Figure 3: Schematic of the stacked DBC bridge

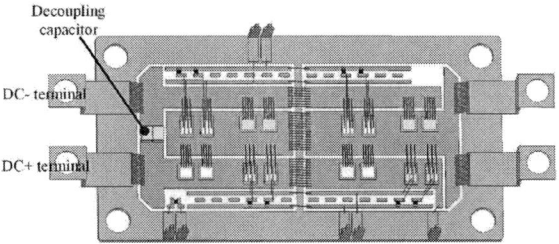

Figure 4: the concentrated-decoupling structure with only one set of capacitors near the terminal

a 1200V, 300A half-bridge module. To ensure the reliability of the DBC substrate in thermal cycling, the DBC is divided into two parts and the two DBC substrates can be connected using bonding wires or copper strips. The gate and source traces are placed on two side of the substrates for Kelvin-type connection, and there is one 1 Ω driving resistor connected to the gate of each MOSFET. There is also an NTC thermistor for temperature measurement.

This structure can realize the balance of the static current between the paralleled MOSFETs and diodes. However, the difference of the transient currents between the paralleled chips is large, caused by the large imbalance of the parasitic inductance between the commutation loops.

B. The proposed current sharing method

To improve the dynamic current sharing, based on this conventional structure, the optimized module adds four stacked DBC bridges on the DBC substrate together with distributed decoupling capacitors to connect the positive and negative DC electrodes, as shown in Fig. 2. In this way, the distributed placement of the decoupling capacitors and the purpose of distributed decoupling are realized. Fig. 3. Shows the schematic of the stacked DBC bridge. The top and the bottom copper layers are connected through copper vias, so that the DC+ and DC- electrodes are led to the top copper layer respectively and then connected through decoupling capacitors. The maximum current of one DBC bridge is 125 A DC which is enough in the 1200V, 300A module. There are 8 decoupling capacitors totally two of which are connected in series to form one branch. There are 4 paralleled branches placing on 4 stacked DBC bridges respectively as seen in Fig. 2. The capacitance value of each capacitor is 47nF, and the equivalent total capacity value is 94nF. It provides each MOSFETs-diodes commutation loop with sufficient decoupling effect during the switching transient [6].

Through the distributed decoupling, the dynamic MOSFETs-diodes commutation loops are almost identical, so that the dynamic current of each MOSFET can be more uniform and the losses are more balanced. During turn-off periods, the overvoltage on the MOSFETs can also be significantly reduced due to the decoupling effect on the terminal inductance. The module structure, the design and fabrication process are kept simple. Meanwhile, using decoupling capacitors, the turn- off overvoltage can be reduced significantly.

III. SIMULATION VERIFICATION

In order to verify the effect of the dynamic current sharing approach, the circuit parameters of the original module and the optimized module are extracted respectively, and double pulse test (DPT) simulations of the bottom MOSFETs is conducted in LTspice. To verify the advantages of the distributed decoupling over common concentrated decoupling, the concentrated-decoupling structure with only one set of capacitors (total capacity value: 94nF) near the terminal shown in Fig. 4 is also modeled and simulated.

The turn-off waveforms at 400V and 100A are presented in Fig. 5. The turn-on waveforms at 400V and 100A are presented in Fig. 6. $I_{DS1\sim4(bottom)}$ is the current of the four parallel MOSFETs. $I_{switch(bottom)}$ is the whole current of the paralleled MOSFETs. I_{Cdec} is the current of the decoupling capacitors.

The overvoltage of the conventional module during turn-off period is 521 V. The peak voltage of the concentrated-decoupling module and the optimized module are both 450 V. The overvoltage of the modules with C_{dec} is reduced by 58% than the conventional module. The current differences between the paralleled MOSFETs in the conventional and concentrated-decoupling modules during turn-on period are both around 140A. Through distributed decoupling, the dynamic current difference is reduced to 50A, which decreases by 63.5% and 65.8% respectively compared to the conventional and concentrated- decoupling modules.

The turn- off overvoltage and turn– on current difference are shown in Fig. 7. It can be concluded that using decoupling capacitors, the turn- off overvoltage is reduced. The overvoltage of the optimized module is lower than the module with concentrated decoupling capacitors due to the lower commutation loop inductance. By the four stacked DBC bridges on the DBC substrate together with distributed decoupling capacitors, the dynamic current difference of the optimized module is reduced than the concentrated-decoupling module and the original module.

978-1-6654-4817-8/21 $31.00 © 2021 IEEE

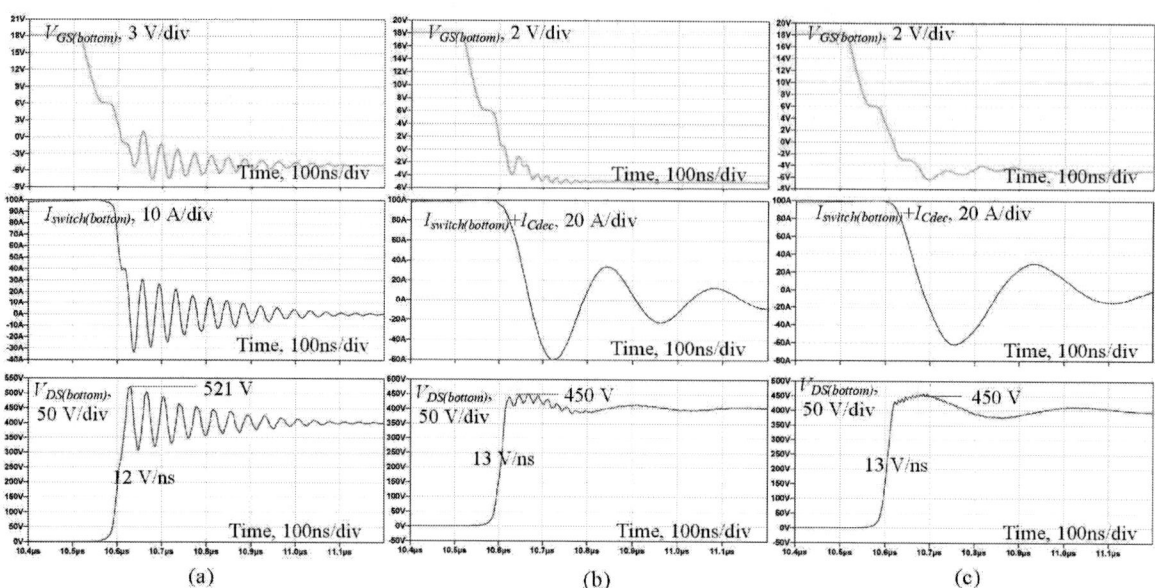

Figure 5: turn-off waveforms of the modules at 400V, 100A. (a) The original module, external $R_{G(off)}$ = 3 Ω, (b) the concentrated-decoupling module, external $R_{G(off)}$ = 3 Ω, (c) the optimized module, external $R_{G(off)}$ = 3 Ω

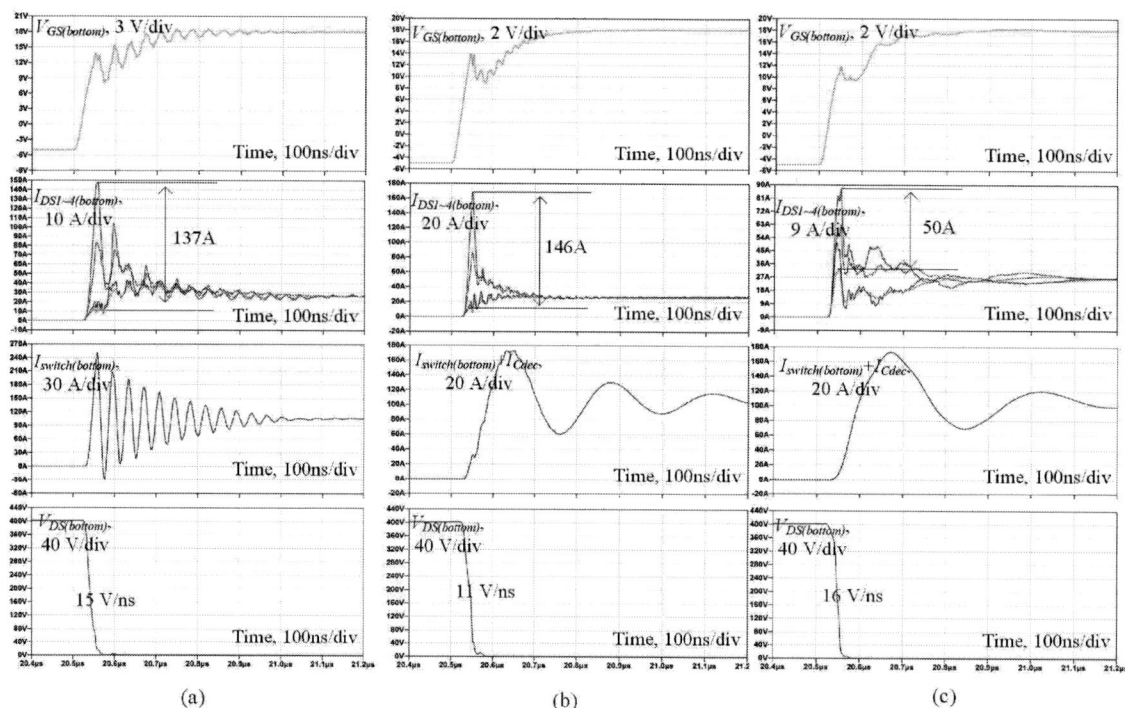

Figure 6: turn-on waveforms of the three modules at 400V, 100A. (a) The original module, external $R_{G(on)}$ = 3 Ω, (b) the concentrated-decoupling module, external $R_{G(on)}$ = 3 Ω, (c) the optimized module, external $R_{G(on)}$ = 3 Ω

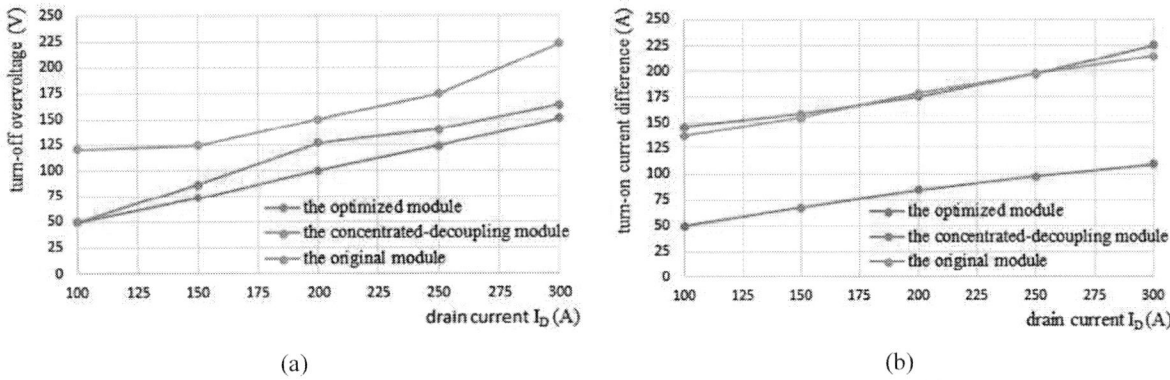

(a)   (b)

Figure 7: turn-off overvoltage and turn-on current difference of the three modules under 400V and different currents with external $R_{G(on)}$ = 3 Ω. (a) The turn-off overvoltage, (b) the turn-on current difference

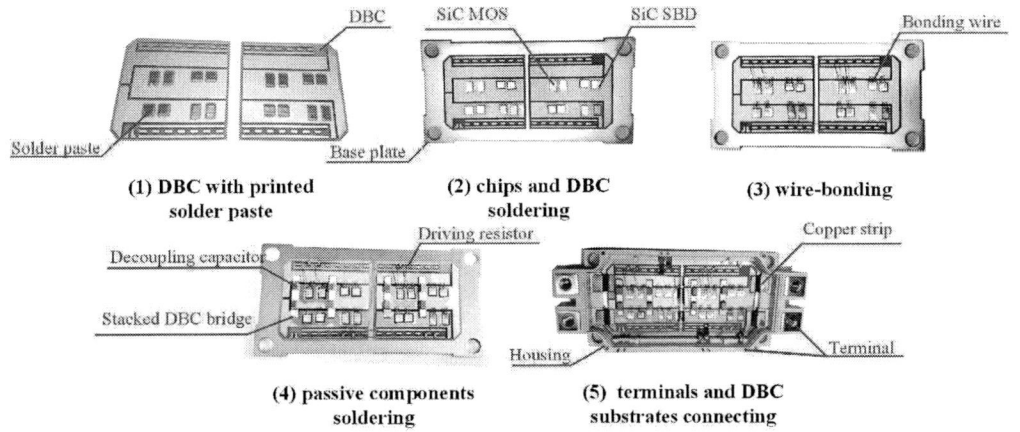

Figure 8: processing steps of the improved power module

IV. FABRICATION

The processing steps of the optimized module with the current balancing approach is depicted in Fig. 8. Firstly, the AlN direct bonded copper substrates is etched to form the desired circuit pattern. Then the SiC power chips are soldered on the direct bonded copper substrates and the the direct bonded copper substrates are soldered on the base plate. Then the top side of the chips and the copper of the two DBC substrates are connected using Al bonding wires. After that, the driving resistors, stacked DBC bridges and the decoupling capacitors are soldered with Sn96.5/Ag3.0/Cu0.5 solder paste. In order to strengthen the conduction capacity, copper strips are soldered using Sn64.7/Bi35/Ag0.3 solder paste to connect the DBC substrates with the terminal and to connect the two DBC substrates. Finally, the module is encapsulated with silicone gel. It can be verified that this method doesn't change the original simple module layout, and the fabrication process is simple.

V. CONCLUSIONS

This paper presents a dynamic paralleling current sharing method with integrated multistage, distributed-decoupling capacitors for a multi-chip commercial-2D-structure SiC power module based on original DBC pattern. The optimized module uses four stacked DBC bridges together with decoupling capacitors to interconnect the DC+ and DC- electrodes to achieve the effect of multistage and distributed decoupling. This method doesn't change the original simple module layout, and the design. The fabrication process is simple.

From the simulation, the optimized module using this method has a 58% dynamic overvoltage reduction over the MOSFETs compared to the conventional module at 400 V, 100 A. Meanwhile, compared to the conventional module and the concentrated-decoupling module, the optimized module has a 63.5% and 65.8% dynamic current difference reduction between parallel MOSFETs during switching periods at 400 V, 100 A. Using decoupling capacitors, the turn- off overvoltage is reduced. The overvoltage of the optimized module is reduced compared to the module with concentrated decoupling capacitors due to the lower commutation loop inductance. By the four stacked DBC bridges on the DBC substrate together with distributed decoupling capacitors, the dynamic current difference of the optimized module is reduced than the concentrated-decoupling module and the original module.

The conventional and optimized modules are fabricated. The fabrication process is simple. So, the method is easy to implement.

REFERENCES

[1] H. Sheng, Z. Chen, F. Wang and A. Millner, "Investigation of 1.2 kV SiC MOSFET for high frequency high power applications," 2010 Twenty-Fifth Annual IEEE Applied Power Electronics Conference and Exposition (APEC), 2010, pp. 1572-1577.

978-1-6654-4817-8/21 $31.00 © 2021 IEEE   187

[2] J. Hu et al., "Robustness and Balancing of Parallel-Connected Power Devices: SiC Versus CoolMOS," IEEE Transactions on Industrial Electronics, vol. 63, no. 4, pp. 2092-2102, April 2016.

[3] H. Li et al., "A novel DBC layout for current imbalance mitigation in SiC MOSFET multichip power modules," IEEE TPEL., vol. 31, no. 12, pp. 8042–8045, 2016.

[4] Y. Mao et al., "Balancing of Peak Currents Between Paralleled SiC MOSFETs by Drive-Source Resistors and Coupled Power-Source Inductors," IEEE Trans. Indus. Electron., vol. 64, no. 10, pp. 8334-8343, 2017.

[5] S. G. Kokosis, I. E. Andreadis, G. E. Kampitsis, P. Pachos and S. Manias, "Forced Current Balancing of Parallel-Connected SiC JFETs During Forward and Reverse Conduction Mode," IEEE Transactions on Power Electronics, vol. 32, no. 2, pp. 1400-1410, Feb. 2017.

[6] Z. Chen, D. Boroyevich, P. Mattavelli and K. Ngo, "A frequency-domain study on the effect of DC-link decoupling capacitors," 2013 IEEE Energy Conversion Congress and Exposition, Denver, CO, 2013, pp. 1886-1893.

978-1-6654-4817-8/21 $31.00 © 2021 IEEE

Power Loop Inductance Extraction with High Order Polynomial Fitting Algorithm for SiC MOSFET Power Module Characterization

Zhikun Wang
Academy for Engineering and Technology
Fudan University
China, Shanghai
20210860095@fudan.edu.cn

Saijun Mao
Academy for Engineering and Technology
Fudan University
China, Shanghai
maosaijun@126.com

Shuhao Yang
Academy for Engineering and Technology
Fudan University
China, Shanghai
20210860083@fudan.edu.cn

Wenyu Li
Academy for Engineering and Technology
Fudan University
China, Shanghai
volcanio@163.com

Yujie Ding
Academy for Engineering and Technology
Fudan University
China, Shanghai
1906626042@qq.com

Keqiu Zeng
Philips Healthcare (Suzhou) Co., Ltd
China, Suzhou
keqiu_zeng@163.com

Abstract—An accurate method for extracting the power loop parasitic inductance with high order polynomial fitting algorithm is proposed for SiC MOSFET power module characterization. Extracting power loop inductance of SiC MOSFET power module characterization is more difficult compared with IGBT power module due to higher switching speed. The relationship between the maximum di/dt and the voltage spike at turn-off transient is revealed. Compared with the four existing methods, high-order fitting method provides accuracy and repeatability. The calculated results are consistent with the Ansys Q3D simulation results. The fitted lines with the calculated stray inductance and di/dt match with the drain-to-source voltage of SiC MOSFET power module at both turn-on and turn-off transient. High-order fitting polynomial method is effective to extract the power loop inductance of SiC MOSFET power module characterization.

Keywords—SiC MOSFET, high order polynomial fitting algorithm, inductance extraction, dynamic characterization test

I. INTRODUCTION

Wide band gap power semiconductor devices such as SiC MOSFET power modules have demonstrated better performance than Si based IGBT for higher voltage rating, faster switching speed and better thermal performance [1]-[3]. It is necessary to acquire the dynamic performance parameters of the power device. An effective method to acquire the dynamic characteristics of the power device is the double pulse test (DPT). Switching performance of SiC MOSFET power module can be characterized with DPT [4], [5].

However, the power loop inductance of DPT platform has affected the dynamic characteristic test accuracy. The parasitic inductance distribution of the power loop is shown in Fig. 1. During the turn-on process, power loss will be generated due to the overlap of the drain-to-source voltage and current of SiC MOSFET power module. The magnitude of the turn-off voltage spike is determined by the power loop inductance and di/dt, and the device can be broken down due to the overvoltage [6]. An accurate method is important to extract the power loop inductance and ensure the accuracy of SiC MOSFET power module switching characterization test.

This work is sponsored by Shanghai Pujiang Program(20PJ1401500) . This work is also sponsored by original research personalized support project by Academy for Engineering&Technology, Fudan University (gyy_yc_2020-6).

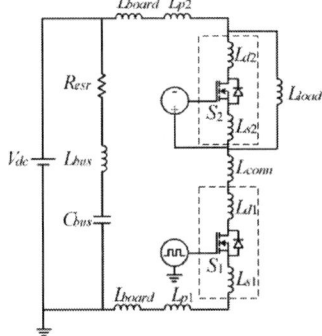

Fig. 1. Power loop inductance distribution for SiC MOSFET power module switching characterization test platform.

Fig. 2. 1200V/600A SiC MOSFET power module for double pulse switching characterization.

In industrial application, differential methods have been used to calculate the power loop inductance [7]-[9]. It is proposed to obtain the parasitic inductance of the power loop based on the least squares fitting method in [10], [11]. In literature [12]-[14], an improved integration method based on the differential method is used to extract the parasitic inductance of the power loop. The sliding mean filter algorithm is often used in nonlinear control and image processing, and it can also be used to fit voltage and current waveforms [15].

However, the mismatch between the delay of the current probe and the voltage probe affects the calculation accuracy of the power loop inductance. In addition, current ripples from the oscilloscope sampling rate also influence the calculation

978-1-6654-4817-8/21 $31.00 © 2021 IEEE

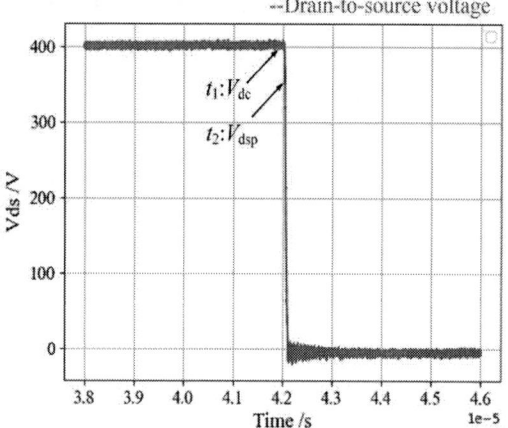

Fig. 3. Drain-to-source voltage waveforms of 400V/600A at turn-on transient

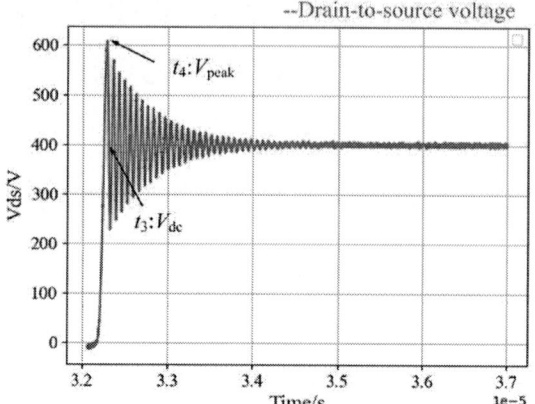

Fig. 4. Turn-off drain-to-source waveforms at 400V/600A

results. In this paper, a method to extract power loop inductance with high order polynomial fitting algorithm is proposed. At turn-off transient, the drain-to-source voltage reaches its spike when the di/dt is the largest. In this method, only the maximum di/dt and voltage spike at turn-off transient is needed to calculate the power loop inductance. Therefore, no attention should be paid to the influence on the probe delay. Besides, the impact of oscilloscope current ripples is minimized by high order polynomial fitting algorithm.

This paper proposes a power loop inductance calculation method based on high order polynomial fitting algorithm, and ANSYS Q3D is used to validate the calculation results. The SiC MOSFET module is shown in Fig. 2. The drain-to-source voltage of SiC MOSFET power module and fitted lines with stray inductance and di/dt match with each other in both turn-on and turn-off transient.

II. METHODS FOR EXTRACTING POWER LOOP INDUCTANCE

A. Extraction of Parasitic inductance of Power Loop Based on Differential Method

The differential method uses the voltage change value and the di/dt during the turn-on and turn-off processes to extract the parasitic inductance of the power loop.

In the turn-on stage, The parasitic inductance of the stray power loop causes a voltage plateau when the drain-to-source voltage is falling [11]. As shown in Fig. 3, The parasitic inductance of the power loop is calculated as:

$$L_S = \frac{V_{dc} - V_{dsp}}{|di_{on}/dt_{on}|} \qquad (1)$$

where V_{dc} is the DC voltage supply; V_{dsp} is the drain-to-source plateau voltage at turn-on stage; di_{on} is the current rise value in this stage; and dt_{on} is the time interval, L_S is the power loop inductance of the DPT platform.

In the turn-off stage, voltage spikes are generated due to the effect of di/dt and power loop inductance. As shown in Fig. 4, the parasitic inductance of the power loop can be calculated as:

$$L_S = \frac{V_{peak} - V_{dc}}{|di_{off}/dt_{off}|} \qquad (2)$$

where V_{peak} is the voltage spike at the turn-off transient, di_{off} is the current drop value at turn-off stage, and dt_{off} is the time interval of this stage.

In the turn-on stage, the voltage plateau is not clearly visible for high switching speed of SiC MOSFET power module, as shown in Fig. 3. Differential method is the simplest method for extracting parasitic inductance of power loop at the turn-off transient. However, calculated errors are yielded due to the mismatch of the current and voltage probes delay. The high frequency current ripples also affect the calculated results.

B. Extract the parasitic inductance of the power loop based on the integral method

The integral method is an improved method of extracting the parasitic inductance of the power loop based on the differential method. This method is to integrate the equations of (1) and (2). In the turn-on stage, the integral method of extracting the parasitic inductance of the power loop can be expressed as follows:

$$L_S = \frac{\int_{t1}^{t2}(V_{dc} - V_{dsp})dt}{|I_{t1} - I_{t2}|} \qquad (3)$$

where I_{t1} is the current value when the voltage is equal to the bus voltage at the turn-on stage, and I_{t2} is the current value when the drain-to-source voltage is equal to the turn-on plateau voltage.

In the turn-off stage, the integral method of extracting the parasitic inductance of the power loop is as follows:

$$L_S = \frac{\int_{t3}^{t4}(V_{peak} - V_{dc})dt}{|I_{t4} - I_{t3}|} \qquad (4)$$

where I_{t3} is the drain current when the drain-source voltage rises to the bus voltage, and I_{t4} is the drain current when the drain-to-source voltage is equal to voltage spike.

978-1-6654-4817-8/21 $31.00 © 2021 IEEE

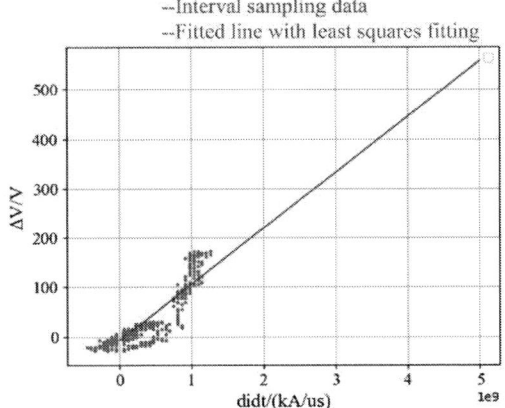

Fig. 5. Extracting the power loop inductance with least squares fitting method.

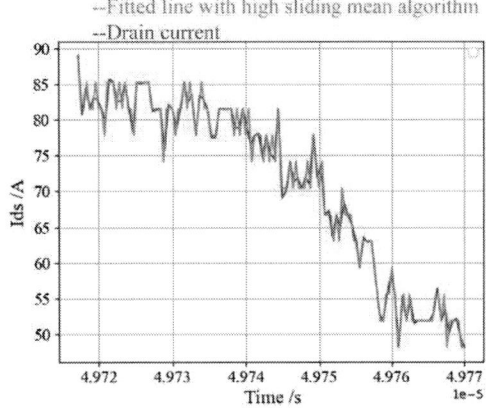

Fig. 6. Extracting the power loop inductance with sliding mean algorithm.

Compared with the differential method, the integral method can decrease the influence of high-frequency ripples due to integration, but it cannot solve the problem caused by the delay of the current probe and voltage probe.

C. Extracting the parasitic inductance of the power loop based on the least squares method

It is proposed to use least squares fitting algorithm to extract the parasitic inductance of the power loop in [10], [11]. In this method, the *di/dt* and voltage change value are obtained by interval sampling method. The least squares method is used to fit a line segment with the *di/dt* and the voltage change value. The slope of the line segment obtained by fitting is the parasitic inductance of the power loop. The curve fitting method of least squares fitting algorithm requires the use of a computer program, and the fitted curve is shown in Fig. 5.

In the turn-on stage, the samples of *di/dt* and voltage change value can be obtained as follows:

$$didt_{on}(i) = \frac{I_d(i+1) - I_d(i-1)}{t(i+1) - t(i-1)} \tag{5}$$

$$dV_{on}(i) = V_{dc} - V_{ds}(i) \tag{6}$$

$$i_{min} = 1 \tag{7}$$

$$i_{onmax} = \frac{t_2 - t_1}{S_{rate}} \tag{8}$$

where $didt_{on}(i)$ is the *di/dt* at the *i* point, $I_d(i+1)$ represents the current of *i*+1 point, $I_d(i-1)$ represents the current of *i*-1 point, $t(i+1)$ represents the time of *i*+1 point, $t(i-1)$ represents the time of i-1 point, $V_{ds}(i)$ represents the drain-source voltage of the *i* point, i_{min} is the minimum value of *i*, i_{max} means the maximum value of *i*, and S_{rate} is the oscilloscope current sampling rate. Power loop inductance can be obtained by least squares fitting with $didt_{on}(i)$ and $dV_{on}(i)$.

In the turn-off phase, the samples of the *di/dt* and the voltage change value can be obtained as follows:

$$didt_{off}(i) = \frac{I_{off}(i+1) - I_{off}(i-1)}{t_{off}(i+1) - t_{off}(i-1)} \tag{9}$$

$$dV_{off}(i) = V_{dsoff}(i) - V_{dc} \tag{10}$$

$$i_{offmax} = \frac{t_4 - t_3}{S_{rate}} \tag{11}$$

where $didt_{off}(i)$ is the *di/dt* at the i point at turn-off transient, $I_{off}(i+1)$ represents the current of *i*+1 point, $I_{off}(i-1)$ represents the current of *i*-1 point, $t_{off}(i+1)$ represents the time of *i*+1 point, $t_{off}(i-1)$ represents the time of *i*-1 point, $V_{dsoff}(i)$ represents the drain-source voltage of *i* point, i_{offmax} is the maximum value of *i* at turn-off transient.

A large number of sampling data can reduce the error of calculation. Calculating the parasitic inductance of the power loop with least squares fitting method has considerable repeatability. However, extracting the parasitic inductance of the power loop based on the least square method is still affected by the high-frequency current ripples and the probe delay.

D. Extracting parasitic inductance of power loop based on sliding mean filtering algorithm

The sliding mean filtering algorithm is usually used in image processing. Fig. 6 shows the current waveform fitted by the sliding mean filter algorithm. The least squares fitting is also used to fit a line segment in Fig. 5. The line segment can be expressed as the following formula:

$$\Delta V = L_S * didt + r(L_S) \tag{12}$$

where *ΔV* is the change value of drain-to-source voltage at turn-off transient, $r(L_S)$ is the relative error of the fitting model.

Fitting the current curve with sliding mean filtering algorithm can greatly decrease the influence of high-frequency current ripples. But the negative impact of mismatch of delay between voltage probe and current probe can not be ignored.

E. The proposed high order polynomial fitting method for extracting power loop inductance

High order polynomial fitting algorithm is used to fit the current curves. For the smooth curve after fitting, the interval sampling method with equations (9), (10) and (11) is used to obtain *di/dt* at turn-off transient. The waveform of the current curve after high order fitting is shown in Fig. 7. The parasitic

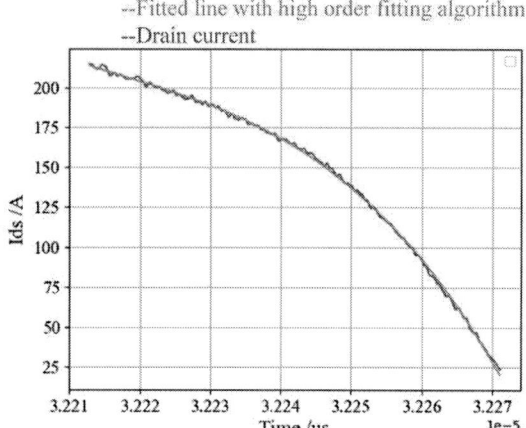

Fig. 7. Extracting the power loop inductance with high order polynomial fitting algorithm.

TABLE. I. Definitions of power loop inductance

Inductance parameters	Definitions
L_{p1}, L_{p2}	Module external inductance
L_{bus}	Bus capacitor inductance
L_{d2}	High side drain inductance
L_{s2}	High side source inductance
L_{board}	Busbar inductance

inductance of the power loop can be expressed by following formula at turn-off transient:

$$didt_{offmax} = \max(didt_{filter}) \quad (13)$$

$$L_S = \frac{V_{max} - V_{dc}}{didt_{offmax}} \quad (14)$$

where $didt_{offmax}$ is the maximum di/dt at turn-off transient, V_{max} is the voltage spike at turn-off transient, $didt_{filter}$ is the di/dt of the current after high order fitting.

The method to extract the parasitic inductance of the power loop based on the high-order fitting algorithm can overcome the problem of the mismatch of the probe delay. At the same time, the high-frequency current ripple is suppressed with high order polynomial fitting algorithm. The accuracy and repeatability of the proposed method is greatly improved.

III. EXPERIMENTAL RESULTS AND VERIFICATION

A. Experimental results

The parasitic inductance distribution of the power loop of the DPT platform is shown in Fig. 1. The components of the power loop parasitic inductance is shown in TABLE. I.

The DPT platform includes 2000V DC power supply, bus capacitor, bus bar, gate driver and the module under test. Fig. 8 and Fig. 9 show the turn-on and turn-off waveforms at 400V/400A. TABLE. II illustrates the calculated results for the stray inductance using different algorithms.

In different test conditions, the high-order fitting algorithm is used to extract the distribution of parasitic inductance of the power loop, and the results ranges from 30.42nH to 32.54nH. The calculated inductance is equal to:

$$L_S = L_{board} + L_{d2} + L_{s2} + L_{bus} + L_{p2} + L_{p1} \quad (15)$$

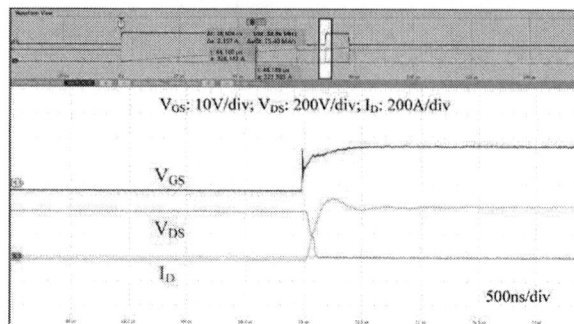

Fig. 8. Turn-on waveform at 400V/400A, turn-on resistance of 1.2 Ohm.

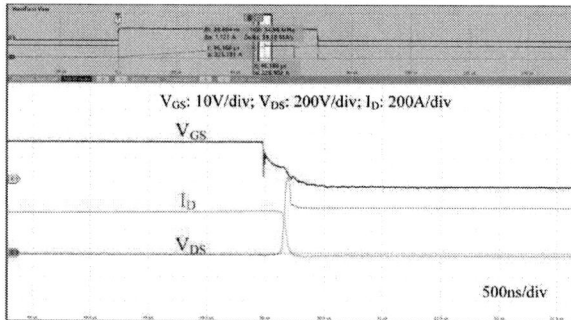

Fig. 9. Turn-off waveform at 400V/400A, turn-off resistance of 2.3 Ohm.

TABLE. II. Inductance extraction by different algorithms at 400V/210A-570A

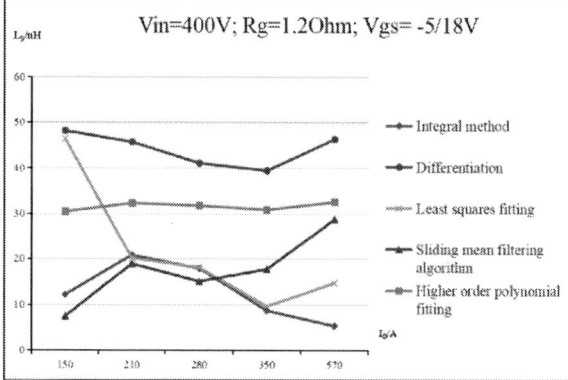

B. ANSYS Q3D simulation

ANSYS Q3D is used to evaluate the parasitic inductance of the power loop. The simulation result of busbar is 8.26nH, and the simulation results are shown in Fig. 10 and Fig. 11.

The bus capacitor of the DPT platform consists of 32 capacitors connected in parallel. The detailed parameters of the bus capacitors are shown in the TABLE. IV. The equivalent inductance value of the bus capacitor is equal to:

$$L_{bus} = \frac{L_{cap}}{n} \quad (16)$$

where L_{cap} is the inductance of a bus capacitor, n is the quantity of bus capacitors; and L_{bus} is the total inductance of bus capacitors.

978-1-6654-4817-8/21 $31.00 © 2021 IEEE

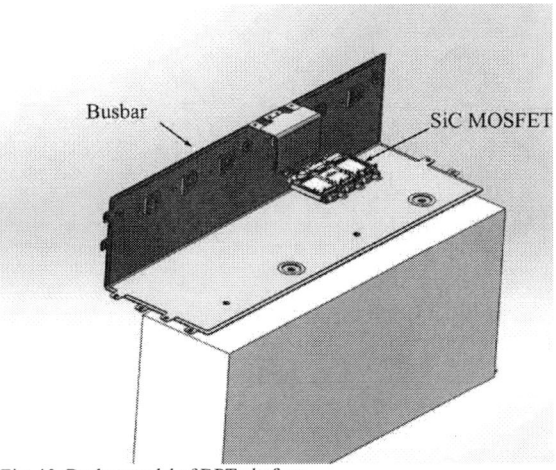

Fig. 10. Busbar model of DPT pltaform.

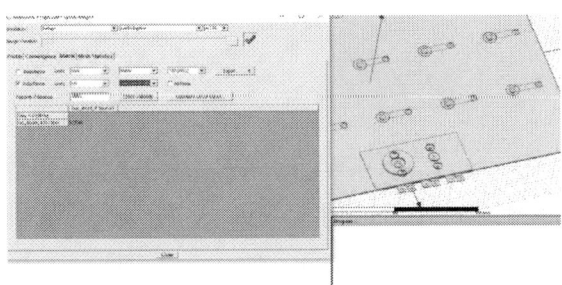

Fig. 11. Busbar inductance simulation for double pulse switching characterization platform.

TABLE. III. Definitions of parasitic parameters

Model name	Inductance /nH
Busbar	8.26
Bus capacitance	1.88
Shunt probe	4.00

TABLE. IV. Bus capacitance parameters

Parameters	Numerical value
capacitance	180 uF
inductance	60 nH
Maximum voltage	2000 V
Maximum current	3660 A

According to the datasheet, the parasitic inductance of the shunt current probe used in the DPT platform is about 4.00nH. The inductance parameters of the test loop are shown in TABLE. III. The value of the internal inductance of the module is determined by the wire length and connection method. The power loop of the test platform is connected by cylindrical wires, and the parasitic inductance can be ignored. The internal drain and source inductance of the SiC power module is:

$$L_\sigma = \frac{1}{N} \frac{\mu_0 l}{2\pi} [\ln(\frac{4l}{d}) + 0.75] \qquad (17)$$

where N is the number of inductors in parallel, u_0 is the vacuum permeability, l is the length of the wire, and d is the diameter of the wire. The simulation result of the power loop inductance of the SiC MOSFET power module DPT platform is about 28.21nH.

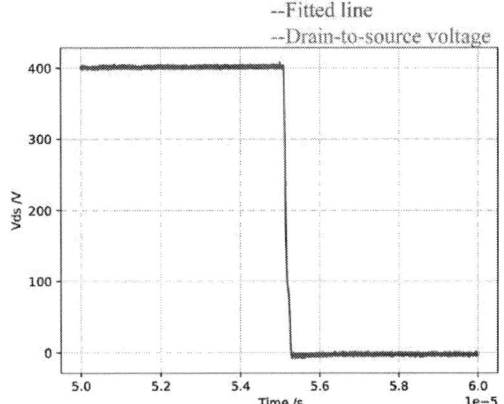

Fig. 12. Comparison between drain-to-source voltage and fitted line with di/dt and stray inductance at turn-on transient.

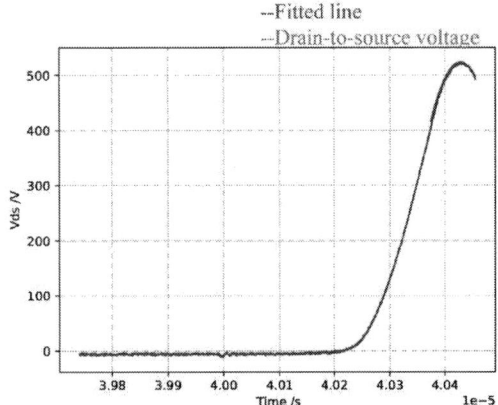

Fig. 13. Comparison between drain-to-source voltage and fitted line with di/dt and stray inductance at turn-off transient

C. Experimental Validation

The interval sampling method is used to obtain the di/dt with equations (5), (7) and (8) in the turn-on stage. A waveform is obtained with the calculated inductance L_S, V_{dc} and $didt_{on}(i)$. The calculated equation can be expressed as follows:

$$V_{dson}(i) = V_{dc} - L_S * didt_{on}(i) \qquad (18)$$

where $V_{dson}(i)$ is the rebuilded voltage with di/dt and L_S at turn-on transient.

Similarly, the rebuilded voltage waveform at turn-off transient can be calculated as follows:

$$V_{dsoff}(i) = V_{dc} + L_S * didt_{off}(i) \qquad (19)$$

where $V_{dsoff}(i)$ is the rebuilded voltage with di/dt and L_S at turn-off transient.

The waveform obtained in the turn-on and turn-off stages is shown in Fig. 12 and Fig. 13. In the turn-on, the turn-off stages, the fitted voltage curve with di/dt and L_S basically matches with the actual drain-to-source voltage curve. This illustrates the feasibility of the high-order fitting algorithm to extract the parasitic inductance of the power loop.

978-1-6654-4817-8/21 $31.00 © 2021 IEEE

IV. Conclusion

Extracting the power loop parasitic inductance of SiC MOSFET DPT platform is of great significance to maximize the performance. A new method of the power loop parasitic inductance extraction based on high order fitting algorithm is proposed. This paper compares the repeatability and accuracy between the proposed method with the four existing test methods. Extracting power loop inductance with high order polynomial fitting method is not affected by high frequency current ripples and probe delay mismatch. The calculation results are basically consistent with the ANSYS Q3D simulation results. The parasitic inductance of the power loop is extracted in different conditions. Compared with the other four methods, extracting parasitic inductance of the power loop based on high order fitting algorithm behaves better accuracy and repeatability.

References

[1] Y. Liu and H. Ye, "Investigation on stray inductance of SiC MOSFET module," in *Conf. 2017 14th China International Forum on Solid State Lighting: International Forum on Wide Bandgap Semiconductors China (SSLChina: IFWS)*, 2017, pp. 193-194.

[2] Z. Zhao et al., "Extraction of Loop Inductances of SiC Half-Bridge Power Module Using An Improved Two-port Network Method," in *Conf. IECON 2018 - 44th Annual Conference of the IEEE Industrial Electronics Society*, 2018, pp. 1204-1208.

[3] J. Milln, P . Godignon, X. Perpi, A. Prez-Toms, and J. Rebollo, "A survey of wide bandgap power semiconductor devices," *IEEE Transactions on Power Electronics.*, vol. 29, no. 5, pp. 2155-2163, May. 2014.

[4] R. O. Lyra, B. J. Cardoso Filho, V . John, and T. A. Lipo, "Coaxial current transformer for test and characterization of high-power semiconductor devices under hard and soft switching," *IEEE Transactions on Industry Applications.*, vol. 36, no. 4, pp. 1181-1188, Jul/Aug. 2000.

[5] S. S. Ahmad and G. Narayanan, "Double pulse test based switching characterization of SiC MOSFET," in *Conf. 2017 National Power Electronics Conference (NPEC)*, 2017, pp. 319-324.

[6] R. S. Krishna Moorthy et al., "Estimation, Minimization, and Validation of Commutation Loop Inductance for a 135-kW SiC EV Traction Inverter," *IEEE Journal of Emerging and Selected Topics in Power Electronics*, vol. 8, no. 1, pp. 286-297, March. 2020.

[7] V olke A, Hornkamp M. IGBT modules[M]. Munich : Infineon Technologies AG, 2012 : 306-307.

[8] Z. Lounis, I. Rasoanarivo and B. Davat, "Minimization of wiring inductance in high power IGBT inverter," *IEEE Transactions on Power Delivery*, vol. 15, no. 2, pp. 551-555, April. 2000.

[9] K. Wada and M. Ando, "Switching Loss Analysis of SiC-MOSFET based on Stray Inductance Scaling," in *Conf. 2018 International Power Electronics Conference (IPEC-Niigata 2018-ECCE Asia)*, 2018, pp. 1919-1924.

[10] M. Ruff and H. Grotstollen, "Identification of the saturated mutual inductance of an asynchronous motor at standstill by recursive least squares algorithm," in *Conf. 1993 Fifth European Conference on Power Electronics and Applications*, 1993, pp. 103-108.

[11] Chen Na, "Switching characteristics testing and modeling of medium and high voltage IGBT power module [D]. " Hangzhou : Zhejiang University, 2012(in Chinese).

[12] X. Tang et al., "Novel method to extract the stray inductance from igbt switching curves," in *Conf. 2017 Sixth Asia-Pacific Conference on Antennas and Propagation (APCAP)*, 2017, pp. 1-3.

[13] X. Li, Y. Luo, Y. Duan, B. Liu, Y. Huang and F. Sun, "Stray Inductance Extraction of High-Power IGBT Dynamic Test Platform and Verification of Physical Model," in *Conf. 2018 IEEE International Power Electronics and Application Conference and Exposition (PEAC)*, 2018, pp. 1-6.

[14] L. Yuan, Q. Gu, G. Feng, Z. Zhao, R. Yang and W. Wang, "Experimental research on stray inductance extraction of planar bus bars based on HVIGBT dynamic characteristics," *in Conf. 2014 17th International Conference on Electrical Machines and Systems (ICEMS)*, 2014, pp. 1957-1962.

[15] M. Basin and P. Rodriguez-Ramirez, "Sliding mode filtering for polynomial systems," in *Conf. 2012 12th International Workshop on Variable Structure Systems*, 2012, pp. 355-360.

Automated SiC MOSFET Power Module Switching Characterization Test Platform

Shuhao Yang
Academy for Engineering & Technology
Fudan University
Shanghai, China
wherearehow@163.com

Saijun Mao
Academy for Engineering & Technology
Fudan University
Shanghai, China
maosaijun@126.com

Zhikun Wang
Academy for Engineering & Technology
Fudan University
Shanghai, China
20210860095@fudan.edu.cn

Xi Lu
UniSiC Technology (Shanghai) Co.,Ltd.
Shanghai, China
xi.lu@unisic.tech

Hansen Chen
UniSiC Technology (Shanghai) Co.,Ltd.
Shanghai, China
Jason.chen@unisic.tech

Keqiu Zeng
Philips Healthcare (Suzhou) Co.,Ltd.
Suzhou, China
keqiu_zeng@163.com

Abstract—To address the problems of large test error and low efficiency in switching characterization test of wide-bandgap (WBG) power semiconductor devices, an automated high accuracy switching performance test platform for SiC MOSFET power module is proposed. This paper focuses on optimizing the design of hardware and software which can efficiently, accurately and safely obtain switching characteristics parameters of SiC MOSFET power module by double pulse test (DPT), as well as short-circuit test and body diode reverse recovery test to acquire more comprehensive switching behaviors of the devices under test (DUT). Real-time test results can be achieved. The multi-process test is serial, and the test data can be compared and managed using a database system. The design principles are experimentally verified by testing the Cree 1.2kV SiC MOSFET power module.

Index Terms—switching characterization, SiC MOSFET, automated test platform, double pulse test (DPT)

I. INTRODUCTION

With the increasing demand for high efficiency and high power density of electrical energy, power semiconductor devices are constantly being innovated and advanced. Silicon carbide (SiC) as one of the representatives of wide-bandgap (WBG) semiconductor can performance lower losses, higher efficiency and higher switching speed compared with traditional Si-based devices. In recent years, SiC MOSFET power modules have been widely used in the fields of photovoltaic, medical, and new energy vehicles due to their excellent properties [1]–[3].

The dynamic characteristics of the power module at the time of switching characterize its own performance and switching losses, but also determine the switching frequency and efficiency of the device. Therefore the module can be tested to obtain the dynamic characteristics of the device at the moment of switching, as a basis for the subsequent evaluation of the project [4], [5].

This work is sponsored by Shanghai Pujiang Program(20PJ1401500). This work is also sponsored by original research personalized support project by Academy for Engineering & Technology, Fudan University (gyy_yc_2020-6).

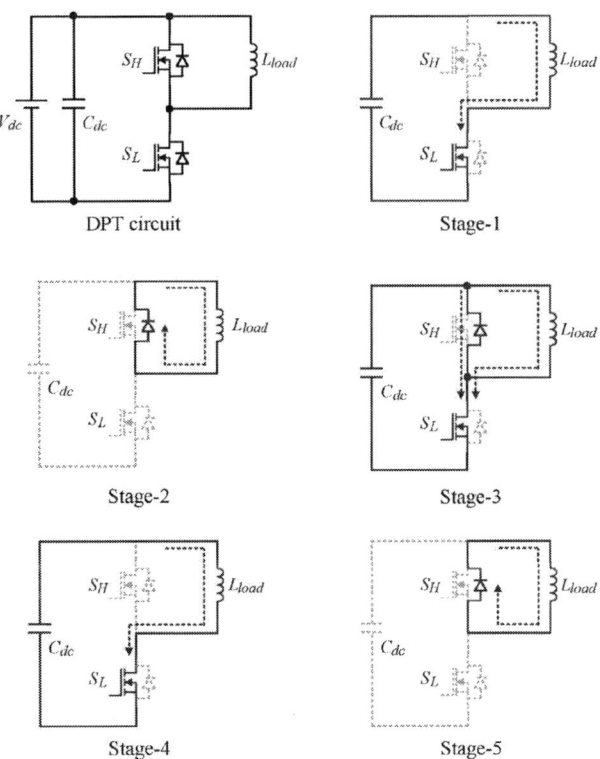

Fig. 1. Double pulse test schematic and equivalent circuit of switching stages.

Among the methods for testing the switching dynamic characteristics of power semiconductor devices, the double pulse test (DPT) is one of the most commonly used and effective [6], [7]. As shown in the Fig.1, the principle of DPT as follows: two pulses are sent to the device under test

978-1-6654-4817-8/21 $31.00 © 2021 IEEE

(DUT) in a clamped inductive load circuit. V_{dc} is the dc power supply, C_{dc} is the bus capacitors, S_H and S_L are the upper and lower switches of the half bridge of the MOSFET module, L_{load} is the load inductor. Take DPT for S_L as an example, by adjusting V_{dc} and pulse width of the first pulse sent to S_L, the switching transients of S_L can be captured at any desired voltage and current conditions at the end of the first pulse and the beginning of the second pulse. The DPT results characterize the switching performance of power semiconductor devices and provide reference indicators for the subsequent design of power electronic systems [8].

However, the high speed switching characteristics of WBG make the devices extremely sensitive to parasitic parameters and very prone to occur voltage overshoot, oscillation, EMI and other problems, which make the test range small and even unable to test properly [7], [9], [10]. At the same time, the accuracy of the measurement probe, oscilloscope channel delay, improper processing of measurement data and other problems also lead to inaccurate test results, limiting the evaluation of WBG device losses [11].

In recent years, researchers have done a lot of work to improve the dynamic characterization of power semiconductor devices. In order to make the test more efficient, researchers have designed automated platforms for testing dynamic characteristics parameters of power devices [12]–[14]. There is also an increasing focus on improving the measurement equipment and hardware layout for enhancing the accuracy of test [11], [15]–[17].

In this study, an automated high-precision switching performance test platform for SiC MOSFET power module is proposed in order to solve the problems of large test errors and low efficiency in the switching characteristics test of WBG power semiconductor devices. The remainder of this paper is organized as follows. The general design of the test platform will be detailed in Section II, the detailed processing of the test automation will be described in Section III, the experimental results of Cree 1.2kV SiC MOSFET will be shown in Section IV, and the Section V summarizes the design of the automated test platform.

II. OVERALL DESIGN OF THE TEST SYSTEM

As shown in the Fig.1, the test system includes parts such as dc bus capacitors, load inductor and dc bus, in addition to SiC MOSFET driver board, measurement equipment, etc. The physical interconnection design and parasitic parameters of these components significantly affect the switching behavior of the DUT. This section will discuss some key techniques for test platform design.

A. Design considerations for each part of the system

1) Busbar Capacitors: Busbar capacitor bulks play an important role in dynamic characteristics test, acting as energy storage devices to provide the energy needed for testing the entire load circuit. Currently, dc bus capacitors are often chosen from film capacitors and electrolytic capacitors, the differences of which are shown in Table I. Especially the safety

and parasitic parameters of capacitor need to be considered to ensure safe and reliable test, therefore film capacitor is the recommended choice.

The number of capacitors in series and parallel is determined by considering the voltage range of the SiC MOSFET under test. It is also necessary to consider that the bus capacitor charges the load inductor at the first pulse, and some of the energy on the bus capacitor will be transferred to the load inductor, causing a drop in the bus voltage, the energy E of the drop is expressed as follows:

$$E = \frac{1}{2}C_{dc}\left(U_1{}^2 - U_2{}^2\right) = \frac{1}{2}L_{load}I_L{}^2 \tag{1}$$

where U_1 is the voltage of the busbar capacitors C_{dc} before charging the load inductor L_{load}, and U_2 is the voltage of the busbar capacitors after charging L_{load}, I_L is the operating current.

In order to prevent busbar voltage drops, pre-charging measures are usually needed to compensate for the drop voltage, so the capacitor withstand voltage needs to be left with sufficient margin. According to [8], it should usually be chosen that:

$$C \geq \frac{L_{load}I_L{}^2}{2k_v V_{dc}{}^2} \tag{2}$$

where k_v is the voltage ripple coefficient, V_{dc} is the operating voltage.

2) Load Inductor: The load inductor affects the voltage drop of the bus capacitor and the current rise rate of the inductor during charging. From the perspective of the bus capacitors transferring energy to the load inductor, it is hoped that the inductance is small, because the smaller the inductance the smaller the energy consumed by the bus capacitors and the smaller the bus voltage drop. However, if the inductance is too small, the current rise rate is too large and the charging time is too short when test small current, which is not easy to control. Therefore, the compromise is to design inductors with adjustable inductance. Meanwhile, the parasitic capacitance of the load inductor affects the current overshoot when the SiC MOSFET is turned on and the damped oscillation when the current overshoot is followed by a drop. The smaller the parasitic capacitance, the better [18].

3) Laminated Busbar: The switching characteristics of high-power devices are closely related to the parasitic inductance of the test circuit [10]. In order to obtain accurate data to reflect the switching characteristics of the devices themselves and to ensure safe and reliable operation of the devices, the parasitic inductance of the test circuit must be very small. Therefore, it is necessary to use laminated busbar as the power circuit of the test circuit. The laminated busbar structure can reduce the equivalent inductance of the circuit. Fig.2 provides the busbar module with optimized design. The simulation result for the parasitic inductance of the busbar is 14.13nH, and the result of impedance analyzer is 17.01nH.

4) SiC MOSFET Gate Driver Board: The gate driver circuit is another essential component of the DPT circuit and has been discussed extensively [8], [16]. SiC MOSFET

TABLE I
COMPARISON OF THE MAIN PARAMETERS OF ELECTROLYTIC CAPACITOR AND FILM CAPACITOR

Parameter	Capacitance Range	Maximum operating voltage	Media	Volume	Price	Lifespan	Parasitic parameters
Electrolytic capacitor	Larger, μF or F	400-600V	Alumina	Larger	Cheap	Short	Larger
Film capacitor	Smaller, μF or mF	Several thousand	Metallized film	Smaller	Expensive	Long	Smaller

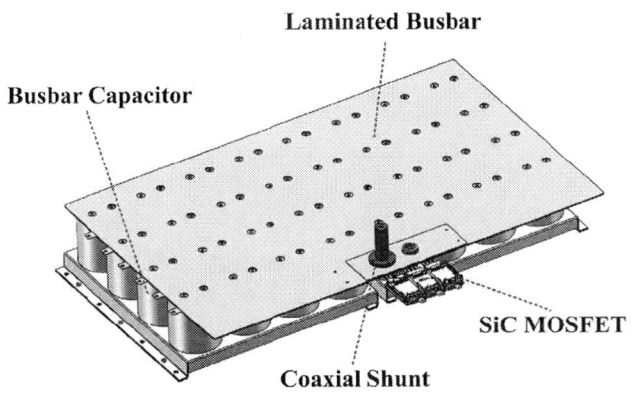

Fig. 2. Laminated busbar model.

Fig. 3. SiC MOSFET power module gate driver board model (left) and prototype photo (right).

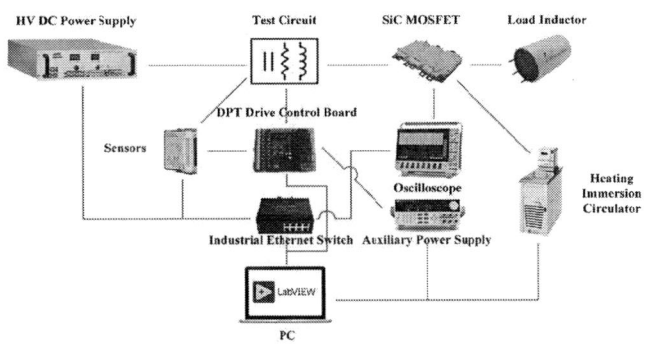

Fig. 4. Hardware structure of diagram for the test platform.

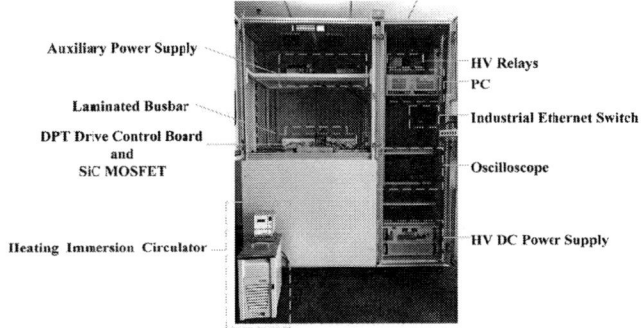

Fig. 5. The photo of the test platform (front side).

has ideally inherent low parasitic capacitances of C_{gd}, C_{gs} and C_{ds}. These low parasitic capacitances help to increase the switching frequency of the device, but they are also inevitably more susceptible to resonance caused by parasitic inductance. Therefore, special attention is needed in the design to keep the power loop away from the gate loop, shorten the distance between the gate driver and the MOSFET, and keep the PCB layout symmetrical with minimum trace length. At the same time, the drive capability of the gate driver IC, the common mode transient immunity of the gate driver isolation and the crosstalk suppression should not be neglected. The driver board designed in this study is shown in Fig.3.

B. Overall Design

The structural connection of the proposed test platform is shown in the Fig.4, where PC is the core part controlling the whole system and establishing communication with the rest hardware parts of the system through VISA, RS232 and other communication protocols. The whole test control and data acquisition monitoring activities are mainly done through the software developed by LabVIEW in the PC.

In order to ensure the accuracy of SiC MOSFET DPT results, measurement equipment that meet the high-speed measurement performance were selected as much as possible. Among them, the oscilloscope is Tektronix MSO64B (1GHz), the voltage probe is Tektronix THDP0200 (200MHz) , and the current probes are coaxial shunt with optical isolation probe Tektronix TIVP02 (200MHz), Rogowski coil PEM CWT Ultra Mini (30MHz) and CP9600S (20MHz), the differences of three current probes will be illustrated in Section IV. A heating immersion circulator is added to the test platform in order to measure the dynamic characteristics of the module at different operating temperatures. The complete test platform setup is shown in Fig.5.

III. TEST CONTROL AND DATA ANALYSIS

In the proposed automated test platform, the control system plays a key role in automating the test. With the established software graphic user interface (GUI), the operator can easily complete automated test for SiC MOSFET module. Real-time

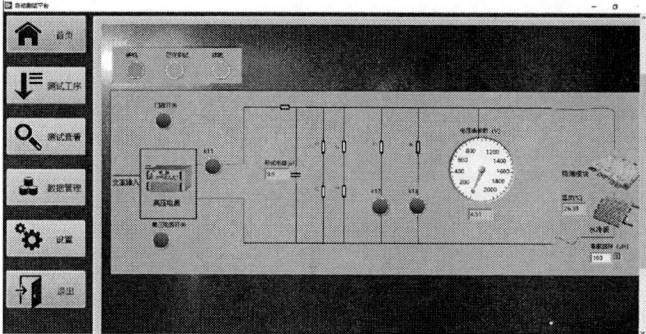

Fig. 6. LabVIEW GUI for the test platform.

test results are available at the end of the test, multi-process test serial, and the test data are compared and managed using a database system. The software GUI of the proposed automated test platform is shown in Fig.6.

In the design of the software should pay attention to several points, first of all, the GUI should be intuitive and easy to operate, the user can easily get started. Secondly, the logic of the test should be considered comprehensively, not only to ensure the safety of the DUT, but also to ensure the safety of the operator, therefore real-time equipment monitoring and audit of the operation logic are indispensable, and the test should be stopped in time when a fault occurs, and the operator should be prompted in time when an operation error occurs.

At the same time, how to process and calculate the data acquired by the oscilloscope is also very critical, which directly affects the results of dynamic switching characteristics parameters such as switching losses of the device. Taking the two parameters of turn-on losses E_{on} and turn-off losses E_{off} as an example, the equations of E_{on} and E_{off} are:

$$
\begin{cases}
E_{on} = \int_{t_1}^{t_2} V_{ds}(t) * I_d(t)dt \\
E_{off} = \int_{t_3}^{t_4} V_{ds}(t) * I_d(t)dt
\end{cases}
\tag{3}
$$

where t_1, t_2, t_3 and t_4 are defined differently, as shown in the TABLE II, for the selection of the time points of the switching losses calculation, each manufacturer has its own set of standards. For example, the E_{on} of Infineon module calculates the integration time starting at 10% of the operating current I_d and ending at 2% of the operating voltage V_{dc}, and the rest of the items are similar to this. The standard of Power Integrations is chosen in this design.

Regardless of the standard, the voltage and current alignment and a smoother curve are the prerequisites so that an accurate integration region can be obtained. In the actual data obtained, the current and voltage waveforms are curves with noise information and each set of data has different delay characteristics due to the limitations of the measurement equipment and other factors. Traditional manual processing of these data would be a tedious and inaccurate process, especially for SiC MOSFET, where the dramatically changing switching waveform data makes it difficult to select the exact point for calculation operations [14]. For the voltage and

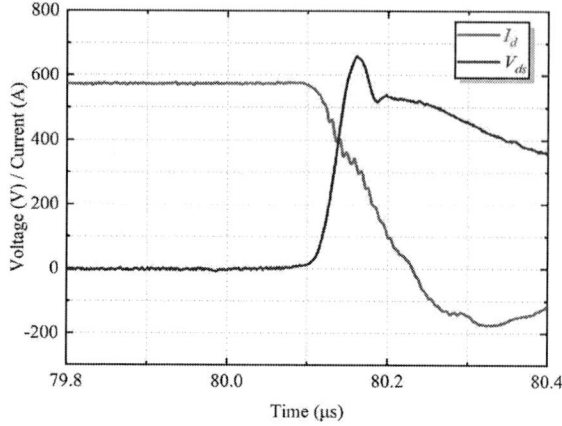

(a) Original turn-off waveforms of SiC MOSFET at 400V/570A.

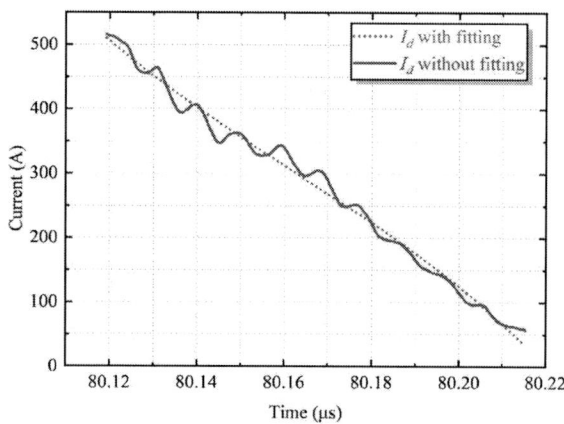

(b) Fitting the current I_d.

Fig. 7. Fitting the current I_d when SiC MOSFET is turned off.

current calibration, some methods are proposed in the literature [8], [19], [20]. As for the processing of waveform noise, it is accurate and simple and convenient to process it by fitting [21], [22], and in this design, the high-order fitting method is chosen for the processing of current noise. The result of the current fitting process for SiC MOSFET when it is turned off is shown in Fig.7. After the fitting process, the current waveform becomes smooth and undistorted, and it is easy to perform the operation of sampling points.

IV. EXPERIMENTAL VALIDATION AND COMPARISON

In order to verify the design principle and evaluate the performance of the automated test platform, the Cree 1.2kV SiC MOSFET module is characterized for DPT. The accuracy of V_{dc}, V_{gs} and I_d is first verified, characterized in terms of the average relative error $\bar{\delta}$, that is:

TABLE II
DEFINITION OF THE TIME POINTS SELECTION FOR SWITCHING LOSSES CALCULATION BY SOME MANUFACTURERS

Parameters	Infineon module	Infineon discrete	STMicroelectronics	IEC	Power Integrations
E_{on}	$0.1I_d \sim 0.02V_{ds}$	$0.2V_{gs} \sim 0.03V_{ds}$	$0.1I_d \sim 0.1V_{ds}$	$0.1V_{gs} \sim 0.02V_{ds}$	$0.1I_d \sim 0.1V_{ds}$
E_{off}	$0.1V_{ds} \sim 0.02I_d$	$0.9V_{gs} \sim 0.01I_d$	$0.1V_{ds} \sim 0I_d$	$0.9V_{gs} \sim 0.02I_d$	$0.1V_{ds} \sim 0.1I_d$

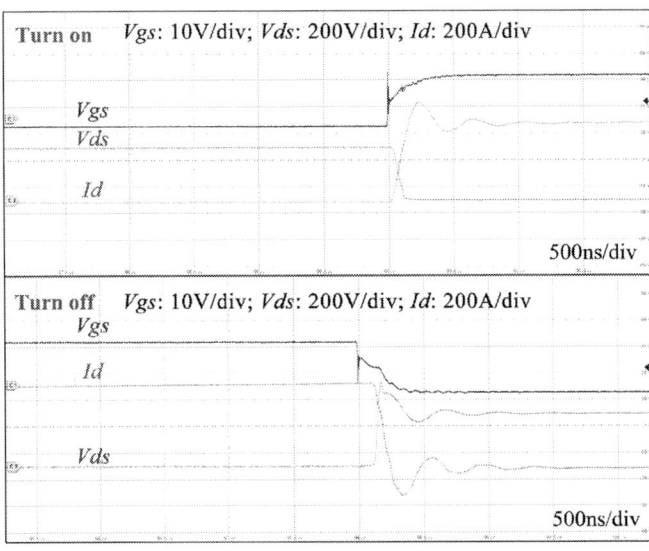

Fig. 8. Turn-on and turn-off waveforms of SiC MOSFET power module with Cree chips at 400V/570A.

$$\bar{\delta} = \frac{1}{n} \sum_{n}^{i} \frac{|x_i - x_i^*|}{x_i} * 100\% \qquad (4)$$

where x is the target value, x^* is the actual test value of x, n is the number of tests and is taken to be 20 here. The relative error of each data is calculated, including 4.3% for positive gate drive voltage, 0.6% for negative gate drive voltage, 1.2% for I_d and 3.6% for V_{dc}.

Fig.8 gives the typical turn-on and turn-off waveforms of SiC MOSFET power module with Cree chips tested at 400V/570A, 25 degrees Celsius. The turn-on resistance is 1.2 Ohm, and the turn-off resistance is 2.3 Ohm. The switching losses of SiC MOSFET with different drive resistances is tested at 25 degrees Celsius at 600V operating voltage. As shown in Fig.9, the switching losses increases as the operating current increases.

In order to accurately capture high-speed switching transients, probes and oscilloscope with sufficient bandwidth are required. There are several viable ways for current measurement, including using coaxial shunt, current transformer, split-core current probe, or Rogowski coil. Among them, the test differences between Rogowski coil, coaxial shunt and optical isolation probe are verified. The information and test waveforms of the three test probes are shown in Fig.10 and Table III.

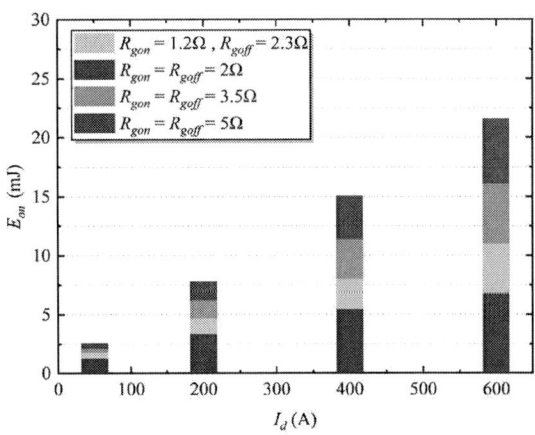

(a) Turn-on losses.

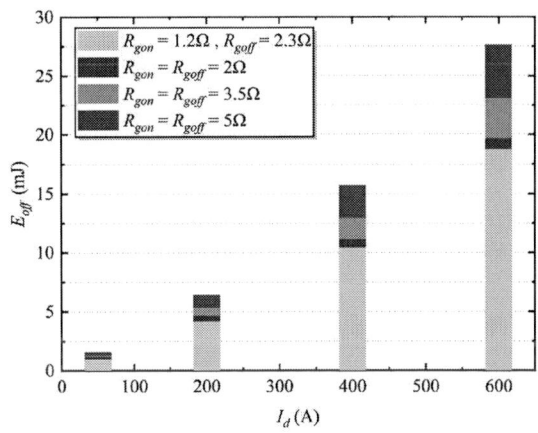

(b) Turn-off losses.

Fig. 9. Turn-on losses(a) and turn-off losses(b) of SiC MOSFET power module with Cree chips at 25°C, 600V, different operating currents and different drive resistances.

The results show that the bandwidth of the probe significantly affects the rise and fall times of the waveforms, and also leads to different calculation results of the switching losses. The higher the bandwidth of the current probe, the faster the waveform of I_d obtained on the oscilloscope changes. In the selection of the probe, it is necessary to consider not only the bandwidth of the DUT, but also the introduction

978-1-6654-4817-8/21 $31.00 © 2021 IEEE

TABLE III
PROBES FOR CURRENT MEASUREMENT

Current probe	Models	Bandwidth	Rise time
Rogowski Coil	CP9600S	20MHz	N/A
Rogowski Coil	PEM CWT Ultra Mini	30MHz	N/A
Coaxial Shunt + Optical Isolation Probe	MMCX50X + TIVP02	200MHz	2ns

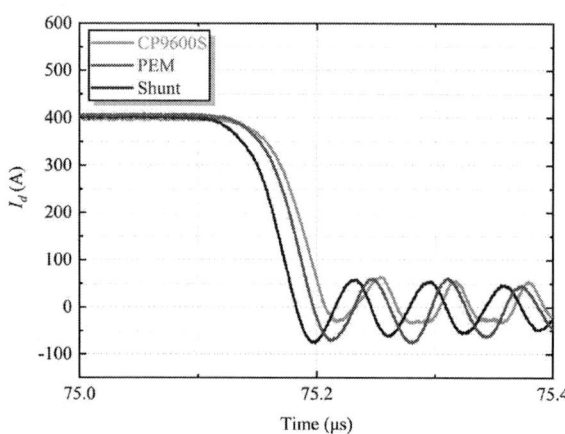

(a) Turn-off waveforms.

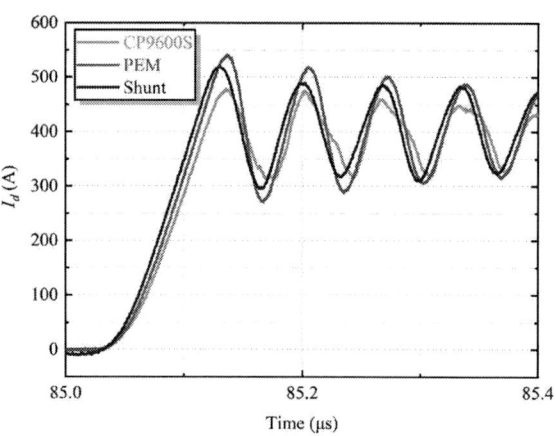

(b) Turn-on waveforms.

Fig. 10. Turn-off (a) and turn-on (b) current waveforms of SiC MOSFET power module with Cree chips tested by three different current probes at 400V/400A.

of parasitic parameters that can interfere with the transient characteristics of the SiC MOSFET [23]. The results of different current probe measurements also have a direct impact on the calculation of switching losses E_{sw} as shown in Fig.11, where $E_{sw} = E_{on} + E_{off}$.

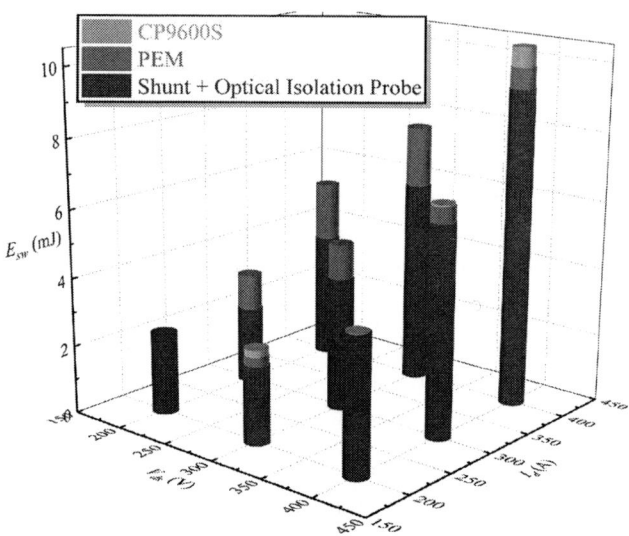

Fig. 11. The effect of three different current probes on swithing losses E_{sw} measurement results for SiC MOSFET power module with Cree chips.

V. CONCLUSION

Switching characteristics are critical for power semiconductor device. The characteristics of WBG semiconductor devices make test difficult. In this paper, an automated SiC MOSFET power module switching characterization test platform is designed, and the construction of the hardware part, the logic processing of the GUI software, the computational processing of the data, and the selection of the measurement equipment are discussed. In the design of the hardware part, the influence of parasitic parameters should be minimized, and high bandwidth, high speed measurement equipment should be used and calibrated to ensure the accuracy of the test. In the software part of the design, the GUI should be simple and easy to operate, fitting or filtering can be considered to reduce the noise of the test data captured from oscilloscope, and the control logic of GUI should ensure safety. The proposed test platform is tested and validated for Cree SiC MOSFET 1.2kV module, and the system has good measurement accuracy and efficiency, which can provide a useful reference for device evaluation.

REFERENCES

[1] S. Ji, Z. Zhang, and F. Wang, "Overview of high voltage SiC power semiconductor devices: development and application," *CES Transactions on Electrical Machines and Systems*, vol. 1, no. 3, pp. 254–264, 2017.

978-1-6654-4817-8/21 $31.00 © 2021 IEEE

[2] J. Watson and G. Castro, "A review of high-temperature electronics technology and applications," *Journal of Materials Science: Materials in Electronics*, vol. 26, no. 12, pp. 9226–9235, 2015.

[3] C. Chen, Y. Chen, Y. Li, Z. Huang, T. Liu, and Y. Kang, "An SiC-based half-bridge module with an improved hybrid packaging method for high power density applications," *IEEE Transactions on Industrial Electronics*, vol. 64, no. 11, pp. 8980–8991, 2017.

[4] J. Millán, P. Godignon, X. Perpiñà, A. Pérez-Tomás, and J. Rebollo, "A survey of wide bandgap power semiconductor devices," *IEEE Transactions on Power Electronics*, vol. 29, no. 5, pp. 2155–2163, 2014.

[5] T. Liu, T. T. Y. Wong, and Z. J. Shen, "A survey on switching oscillations in power converters," *IEEE Journal of Emerging and Selected Topics in Power Electronics*, vol. 8, no. 1, pp. 893–908, 2020.

[6] P. Tu, P. Wang, X. Hu, C. Qi, S. Yin, and M. A. Zagrodnik, "Analytical evaluation of IGBT turn-on loss with double pulse testing," in *2016 IEEE 11th Conference on Industrial Electronics and Applications (ICIEA)*, 2016, pp. 963–968.

[7] L. Zhang, X. Yuan, X. Wu, C. Shi, J. Zhang, and Y. Zhang, "Performance evaluation of high-power SiC MOSFET modules in comparison to Si IGBT modules," *IEEE Transactions on Power Electronics*, vol. 34, no. 2, pp. 1181–1196, 2019.

[8] Z. Zhang, B. Guo, F. F. Wang, E. A. Jones, L. M. Tolbert, and B. J. Blalock, "Methodology for wide band-gap device dynamic characterization," *IEEE Transactions on Power Electronics*, vol. 32, no. 12, pp. 9307–9318, 2017.

[9] B. Zhang and S. Wang, "A survey of emi research in power electronics systems with wide-bandgap semiconductor devices," *IEEE Journal of Emerging and Selected Topics in Power Electronics*, vol. 8, no. 1, pp. 626–643, 2020.

[10] J. Chen, X. Du, Q. Luo, X. Zhang, P. Sun, and L. Zhou, "A review of switching oscillations of wide bandgap semiconductor devices," *IEEE Transactions on Power Electronics*, vol. 35, no. 12, pp. 13 182–13 199, 2020.

[11] Z. Zeng, J. Wang, L. Wang, Y. Yu, and K. Ou, "Inaccurate switching loss measurement of SiC MOSFET caused by probes: Modelization, characterization, and validation," *IEEE Transactions on Instrumentation and Measurement*, vol. 70, pp. 1–14, 2021.

[12] G. P, M. R, R. K. R, and R. J. Vijayan, "Characterisation of 1200V, 35A SiC mosfet using double pulse circuit," in *2016 IEEE International Conference on Power Electronics, Drives and Energy Systems (PEDES)*, 2016, pp. 1–6.

[13] X. Cai, J. Bian, Y. Wang, and Y. Ning, "Design of IGBT parameter automatic test system based on LabVIEW," in *2019 IEEE Symposium Series on Computational Intelligence (SSCI)*, 2019, pp. 2807–2812.

[14] A. Ghosh, C. N. Man Ho, and J. Prendergast, "A cost-effective, compact, automatic testing system for dynamic characterization of power semiconductor devices," in *2019 IEEE Energy Conversion Congress and Exposition (ECCE)*, 2019, pp. 2026–2032.

[15] K. Zou, L. Qi, X. Cui, G. Zhao, and B. Zong, "The design and measurement of test system for dynamic characteristics of IGBT," in *2014 International Conference on Power System Technology*, 2014, pp. 2209–2216.

[16] S. S. Ahmad and G. Narayanan, "Double pulse test based switching characterization of SiC MOSFET," in *2017 National Power Electronics Conference (NPEC)*, 2017, pp. 319–324.

[17] J.-Z. Fu, G. Kapino, and W.-T. Franke, "Effect investigations of double pulse test on the wide bandgap power devices," in *PCIM Europe digital days 2020; International Exhibition and Conference for Power Electronics, Intelligent Motion, Renewable Energy and Energy Management*, 2020, pp. 1–6.

[18] H. Luo, W. Li, X. He, F. Iannuzzo, and F. Blaabjerg, "Uneven temperature effect evaluation in high-power IGBT inverter legs and relative test platform design," *Microelectronics Reliability*, vol. 76, no. sep., pp. págs. 123–130, 2017.

[19] Z. Chen, "Characterization and modeling of high-switching-speed behavior of SiC active devices," Master's thesis, Virginia Tech, 11 2021.

[20] E. A. Jones, F. Wang, D. Costinett, Z. Zhang, B. Guo, B. Liu, and R. Ren, "Characterization of an enhancement-mode 650-V GaN HFET," in *2015 IEEE Energy Conversion Congress and Exposition (ECCE)*, 2015, pp. 400–407.

[21] Z. Li, M. Shuang, S. Wang, and Y. Fang, "The review of IGBT module switching loss calculation method," *Application of Electronic Technique*, 2016.

[22] A. Rajapakse, A. Gole, and P. Wilson, "Electromagnetic transients simulation models for accurate representation of switching losses and thermal performance in power electronic systems," *IEEE Transactions on Power Delivery*, vol. 20, no. 1, pp. 319–327, 2005.

[23] A. Dutta and S. S. Ang, "Effects of parasitic parameters on electromagnetic interference of power electronic modules," in *2017 IEEE Applied Power Electronics Conference and Exposition (APEC)*, 2017, pp. 2706–2710.

978-1-6654-4817-8/21 $31.00 © 2021 IEEE

LLC Resonant Converter Based on Trench Gate SiC MOSFET

Yuming Zhou
School of Electrical Engineering
Anhui University of Technology
Maanshan, China
16688138@qq.com

Jinkun Chu
School of Electrical Engineering
Anhui University of Technology
Maanshan, China
2159918782@qq.com

Jiahui Zhou
School of Electrical Engineering
Anhui University of Technology
Maanshan, China
409241790@qq.com

Abstract—LLC resonant converter is a popular topology for DC-DC converters because of its soft switching characteristics. It has many strong points for example high efficiency, high power density, and lower noise. The traditional Si MOSFET LLC resonant converter cannot continue to improve due to the limitations of the Si characteristics in terms of frequency and efficiency. This article explores the advantages of SiC MOSFET in LLC resonant converter, and compares the conversion efficiency for trench-gate SiC MOSFET, planar-gate SiC MOSFET and Si MOSFET.

Keywords—*LLC resonant converter; trench-gate SiC MOSFET; efficiency*

I. INTRODUCTION

LLC resonant converter is a popular topology for DC-DC converters because of its soft switching characteristics. It has many strong points for example high efficiency, high power density. With the development of converters, the shortcomings of LLC resonant converter use silicon materials have become increasingly prominent, such as low switching frequency and large losses. In order to overcome the shortcomings of Si MOSFET, people began to develop SiC MOSFET converter. Compared with Si MOSFET, SiC MOSFET has many advantages.[1]-[3]. As the second-generation SiC MOSFET, the trench gate SiC MOSFET has a lower resistance and a faster switching rate, which can enable the LLC resonant converter to achieve high frequency while ensuring high efficiency [4]-[5].

After years of development and application of traditional silicon (Si) materials, it has basically reached the limit performance that its material properties can exert. Therefore, device development based on new materials is put on the agenda. Compared with silicon (Si) material devices, silicon carbide (SiC) material devices are more suitable for high-pressure, high-temperature, and high-frequency working environments, have more excellent characteristics, and can develop toward high power density and high efficiency. In the traditional DC-DC conversion circuit, the voltage and current will cross when the circuit is turned on and off. The integral of this cross area is the switching loss. This feature restricts the improvement of efficiency[6]. To reduce the loss in the DC-DC conversion process, it is necessary to explore the DC-DC topology. The LLC resonant converter is the most popular DC-DC topology by virtue of its soft switching and high efficiency characteristics.

The authors would like to thank the Natural Science Foundation of the Colleges and Universities in Anhui [grant number KJ2020A0247]; and Anhui Provincial Natural Science Foundation [grant number 2008085ME157] for their financial support.

II. LLC RESONANT CONVERTER

The circuit topology of the half-bridge LLC resonant converter shows in Fig. 1. There are two switching S_1 and S_2 on the primary side. D_1 and D_2 are the body diodes of the two devices. A resonant cavity in the transformer consists of the capacitor C_r, the resonant inductance L_r, and the magnetizing inductance L_m. The driving signals of the two devices are complementarily turned on, and a very small dead time is added to prevent the switch device from being turned on[7]. The resonant inductance can be replaced by the leakage inductance of the transformer, and made into a magnetic integrated type to increase the power density. The resonant inductance can also be independent to facilitate parameter adjustment. The excitation inductance is provided by the transformer. The secondary side of the transformer has a center tapped structure, and the rectifying part is full-wave rectification connected to a large capacitor filter.

The structure of the resonant network was introduced above, which is composed of three resonant components. According to the different working conditions of the resonant components, it is divided into two different working frequencies. When the secondary side clamps the magnetizing inductance L_m, the voltage across L_m is unchanged in this working condition. At this time, the frequency is only related to C_r and L_r. The resonance frequency f_r is expressed by

$$f_r = \frac{1}{2\pi\sqrt{L_r C_r}} \qquad (1)$$

When the primary and secondary sides of the transformer are disconnected, that is, when no energy transmit between the primary and secondary sides, the secondary side no longer clamps the magnetizing inductance L_m. The resonance frequency is related to L_r, C_r, and L_m. The resonance frequency is

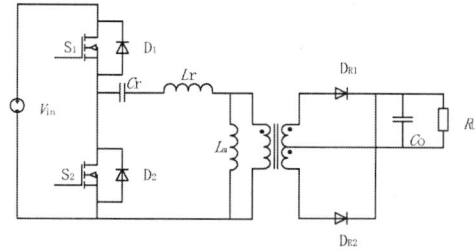

Fig. 1. Topological diagram of half-bridge LLC resonant converter

$$f_{\mathrm{m}} = \frac{1}{2\pi\sqrt{(L_{\mathrm{r}} + L_{\mathrm{m}})C_{\mathrm{r}}}} \tag{2}$$

According to formulas (1) and (2), f_{r} is always greater than f_{m}. When the LLC switching frequency f_{s} starts to change, it can be divided into three situations according to the working principle. (1) When $f_{\mathrm{s}} < f_{\mathrm{m}}$, the reactance of the LLC resonant network is capacitive. In this working state, the current phase leads the voltage. The switching loss is relatively large, and the working state at this time cannot achieve ZVS, the working efficiency is not high and the converter will be damaged. (2) When $f_{\mathrm{m}} < f_{\mathrm{s}} < f_{\mathrm{r}}$, the reactance of the LLC resonant network is inductive. In this working state, the phase of the voltage is ahead of the current, and the LLC can realize ZVS and ZCS. (3) When $f_{\mathrm{s}} > f_{\mathrm{r}}$, LLC can achieve ZVS. In this working state, the secondary side current flows the diode, and the secondary side ZCS cannot be achieved. The magnetizing inductance is always in a clamped state, and does not resonate with other components. There is another case when $f_{\mathrm{s}} = f_{\mathrm{r}}$, which is a special case of the second working state. However, if there is frequency fluctuation, the soft switching state may be lost.

III. EQUIVALENT FHA CIRCUIT DESIGN

When studying LLC resonant converters, parameter design is key to whether LLC resonant converter can achieve good performance. Before parameter design, it is necessary to use the first harmonic approximation (FHA) method to model and analyze LLC resonant converters. The main principle of this modeling method is to ignore the primary resonance. It is assumed that the energy of the converter from input to output is used the component of the voltage and current Fourier transform values, and the output capacitor and transformer are ignored. The influence of leakage inductance on the secondary side greatly simplifies the circuit model of the converter[8]. It is convenient for people to do research on LLC resonant converter.

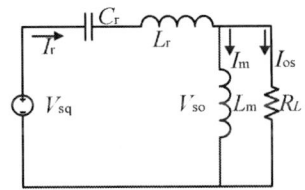

(a) Non-linear non-sinusoidal circuit

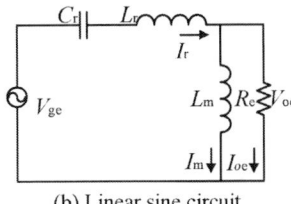

(b) Linear sine circuit

Fig. 2. Half-bridge LLC resonant converter model

When the switching frequency is same as resonance frequency, the AC equivalent circuit can be equivalented to a pure resistive load circuit, as shown in Fig. 2. Among them,

V_{sq} is the square wave voltage, V_{so} is the output voltage, C_{r} is the resonant capacitor, L_{r} is the resonant inductance, L_{m} is the magnetizing inductance, and V_{ge} and V_{oe} are the fundamental components of V_{sq} and V_{so}, respectively. Before the equivalent circuit design, the fundamental wave component is used to represent the single-pole square wave voltage and current of the primary input. The influence of all high-order harmonics, the corresponding parasitic capacitance and transformer leakage inductance are ignored. In Fig. 2(b), V_{ge} is the basic component of V_{sq}. The nonlinear circuit in Fig. 2(a) is approximately transformed into the linear circuit in Fig. 2(b). The voltage V_{ge} and the output voltage V_{oe} are both sinusoidal. The frequency is equal, that is, the component of the square wave voltage, and the formula is derived according to the FHA method.

From the input end, the fundamental component of the square wave voltage is

$$V_{\mathrm{ge}}(t) = \frac{2}{\pi} \times V_{\mathrm{DC}} \times \sin(2\pi f_{\mathrm{s}} t) \tag{3}$$

where V_{DC} is the DC input voltage.

The root mean square (RMS) value is

$$V_{\mathrm{ge}} = \frac{\sqrt{2}}{\pi} \times V_{\mathrm{DC}} \tag{4}$$

From the output point of view, since V_{so} is approximately a square wave, the voltage is

$$V_{\mathrm{oe}}(t) = \frac{4}{\pi} \times n \times V_{\mathrm{o}} \times \sin(2\pi f_{\mathrm{s}} t - \varphi_{\mathrm{v}}) \tag{5}$$

Among them, V_{o} is the output voltage, n is the turn ratio, and Φ_{v} is the phase angle between V_{oe} and V_{ge}.

Its output voltage RMS value is

$$V_{\mathrm{oe}} = \frac{2\sqrt{2}}{\pi} \times n \times V_{\mathrm{o}} \tag{6}$$

With same way, I_{oe} can be obtained by

$$I_{\mathrm{oe}}(t) = \frac{\pi}{2} \times \frac{1}{n} \times I_{\mathrm{o}} \times \sin(2\pi f_{\mathrm{s}} t - \varphi_{\mathrm{i}}) \tag{7}$$

$$I_{\mathrm{oe}} = \frac{\pi}{2\sqrt{2}} \times \frac{1}{n} \times I_{\mathrm{o}} \tag{8}$$

Among them, I_{o} is the output current, and Φ_{i} is the phase angle between I_{oe} and V_{oe}.

The AC equivalent resistance R_{ac} is

$$R_{\mathrm{ac}} = \frac{V_{\mathrm{oe}}}{I_{\mathrm{oe}}} = \frac{8 \times n^2}{\pi^2} \times \frac{V_{\mathrm{o}}}{I_{\mathrm{o}}} = \frac{8 \times n^2}{\pi^2} \times R_{\mathrm{L}} \tag{9}$$

Where R_{L} is the load.

The magnetizing current RMS value is

$$I_{\mathrm{m}} = \frac{V_{\mathrm{oe}}}{\omega L_{\mathrm{m}}} = \frac{2\sqrt{2}}{\pi} \times \frac{n \times V_{\mathrm{o}}}{\omega L_{\mathrm{m}}} \tag{10}$$

After establishing the relationship between the electrical variables, the voltage gain function can be obtained. The link between input and output voltage can be described by the gain

$$M_{\mathrm{g.DC}} = \frac{n \times V_{\mathrm{o}}}{V_{\mathrm{in}}/2} = \frac{n \times V_{\mathrm{o}}}{V_{\mathrm{DC}}/2} \tag{11}$$

$$M_{\mathrm{g.DC}} = \frac{n \times V_{\mathrm{o}}}{V_{\mathrm{in}}/2} \approx M_{\mathrm{g.sw}} = \frac{V_{\mathrm{so}}}{V_{\mathrm{sq}}} \tag{12}$$

In order to simplify the formula, replace $M_{\mathrm{g.DC}}$ with M_{g}, and the variable relationship between V_{oe} and V_{ge} is expressed by L_r, L_m, C_r, and R_{ac}, so the voltage gain formula can be expressed as

$$M_g = \frac{V_{oe}}{V_{ge}} = \left| \frac{jX_{Lm}\,\mathrm{P}\,R_{ac}}{(jX_{Lm}\,\mathrm{P}\,R_{ac}) + j(X_{Lr} - X_{Cr})} \right|$$

$$= \left| \frac{(j\omega L_m)\,\mathrm{P}\,R_{ac}}{(j\omega L_m)\,\mathrm{P}\,R_{ac} + j\omega L_r + \dfrac{1}{j\omega C_r}} \right| \qquad (13)$$

Here, we define the ratio between the two inductances L_r and L_m as

$$K = \frac{L_m}{L_r} \qquad (14)$$

The quality factor is defined as

$$Q_e = \frac{\sqrt{L_r/C_r}}{R_e} \qquad (15)$$

The relationship between the voltage V_{in}, the voltage V_o and the gain M_g can be approximated from the formula (11) – (13) and is expressed by

$$V_o = M_g \times \frac{1}{n} \times \frac{V_{in}}{2} \qquad (16)$$

When the inductance ratio, quality factor and maximum normalized frequency are confirmed, the voltage gain function can be expressed as

$$M_g = \left| \frac{K \times f_n^2}{\left[(K+1) \times f_n^2 - 1 \right] + j\left[(f_n^2 - 1) \times f_n \times Q_e \times K \right]} \right| \qquad (17)$$

At this time, the relationship between output voltage and input voltage can also be obtained from equations (14) – (16).

$$V_o = M_g \times \frac{1}{n} \times \frac{V_{in}}{2} = M_g(f_n, K, Q_e) \times \frac{1}{n} \times \frac{V_{DC}}{2} \qquad (18)$$

IV. DC GAIN ANALYSIS

After getting the DC gain calculation formula, DC gain curve can plot to pave the way for the selection of the inductance ratio K and the quality factor Q_e.

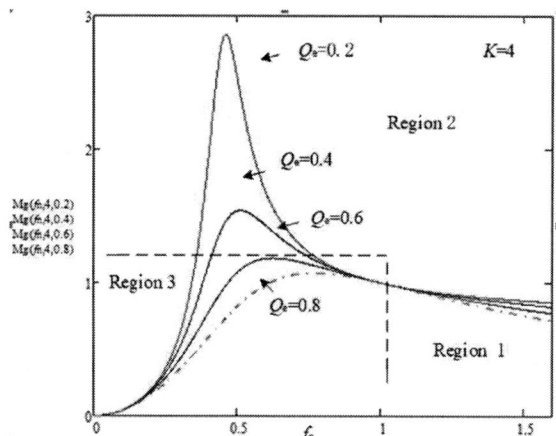

Fig. 3. DC voltage gain curve

MathCAD software was used to draw DC voltage gain curve according to the voltage gain function. The voltage gain curve can help to explore the design requirements and working status of the converter, and help to design LLC resonant converter. Fig. 3 shows the DC gain curve. K is taken as 4, Q_e is 0.2, 0.4, 0.6, 0.8.

The DC gain presents a convex curve around the resonance frequency. According to the reactance characteristics of the resonant network, the voltage gain curve can divided into capacitive. The inductive region can divided into Region 1 and Region 2 as shown in the Fig. 3. When the working state is in Region 1, $f_s > f_r$, the primary side switching device of LLC resonant converter can achieve ZVS, but the secondary side cannot achieve ZCS. When the working state is in Region 2, $f_m < f_s < f_r$, the primary side switching devices of LLC resonant converter can realize ZVS, and the secondary side diodes can also realize ZCS, and the efficiency is the highest at this time. When the resonant network is capacitive, the resonant converter works in Region 3. At this time, $f_s < f_m$, the converter cannot realize ZVS. Generally, the converter does not work in this state.

V. EXPERIMENTAL RESULT

The circuit simulator of LTspice was firstly employed to simulate the LLC resonant converter. In the simulation, the

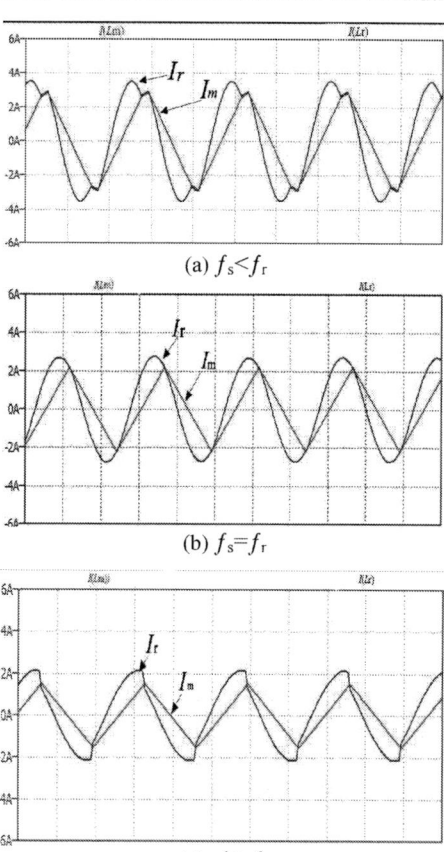

Fig. 4. Resonant current and excitation current waveform from LTspice

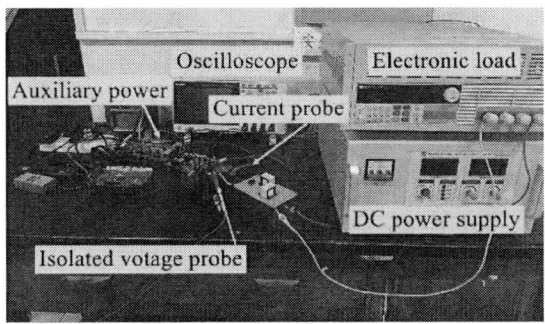

Fig. 5. Experimental test platform

switching device is SCT3060AL (650V/39A), a trench gate SiC MOSFET from ROHM Semiconductor, the rectifying device is C4D10120 (1200V/10A).

The simulation results of the operating current are shown in Fig. 4. The current waveforms of the resonant current I_r and the excitation current I_m at the three operating frequencies. When the operating frequency amounts the resonant frequency, the resonant current I_r is approximately sine wave. The converter has the highest working efficiency, as the frequency decreases, the resonant current and the excitation current have an approximately equal platform.

Simulation can test the rationality of theoretical inferences and design, and experiment can further verify the correctness of simulation and theoretical analysis. LLC resonant converter experiment platform was built based on the LTspice simulation parameters, and the picture is shown in Fig. 5. The platform includes TI DSP (TMS320F28335), a 15V auxiliary power supply, and isolated voltage probes and current probes. The input source is adjustable DC power supply, and the output end is the electronic load from Merno Electronics Inc.

The working current waveform of LLC resonant converter based on SCT3060AL is shown in Fig. 6. In the experiment, the input voltage is 380V, output voltage is 24V, and the load resistor is 2.4Ω. The resonant inductance is 75.4uH, the resonant capacitance is 33.6nF, and the magnetizing inductance is 301.6uH. The switching frequency f_s of SiC device is 100kHz. As comparison, planar-gate SiC MOSFET (C3M0065090D, 900V/36A) and Si MOSFET (IXFX32N90P, 900V/32A) were also employed in the LLC resonant converter. The converter efficiency for three power devices were plotted in Fig. 7. From the figure, we can see that the efficiency for two SiC

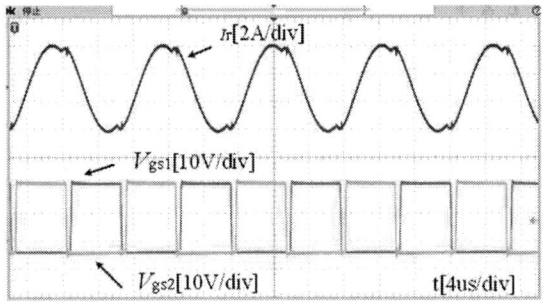

Fig. 6. Resonant current working waveform

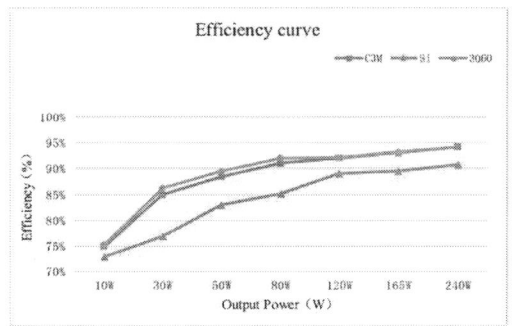

Fig. 7. Efficiency comparison for C3M0065090D, IXFX32N90P, SCT3060AL LLC resonant converter

devices is higher than that of Si device, and trench-gate SiC MOSFET is slightly higher than that of planar-gate SiC MOSFET.

VI. CONCLUSIONS

High frequency and high efficiency are hot topics of today's DC-DC converters. LLC resonant converters can achieve high efficiency because of their soft switching characteristics. Traditional LLC resonant converters mostly use Si MOSFET. Si devices cannot meet the needs of modern technology due to their own material characteristics. Trench-gate SiC MOSFET as the second-generation SiC MOSFET can make LLC resonant converters more efficient.

REFERENCES

[1] M. Gildersleeve, H.P. Forghani-zadeh and G.A. Rincon-Mora, "A comprehensive power analysis and a highly efficient mode-hopping DC-DC converter", ASIC 2002. Proceedings. 2002 IEEE Asia-Pacific Conference on, pp. 153-156, 6-8 Aug. 2002.

[2] Y. Hu, J. Shao and T. S. Ong, "Application of SiC MOSFETs in 6.6kW High-Frequency High-Power-Density Power Converter," PCIM Europe digital days 2020; International Exhibition and Conference for Power Electronics, Intelligent Motion, Renewable Energy and Energy Management, pp. 1-5, 2020.

[3] L. Cao, Q. Guo and K. Sheng, "Comparative Evaluation of the Short circuit Capability of SiC Planar and Trench Power MOSFET," 2018 IEEE 2nd International Electrical and Energy Conference, pp. 653-656, 2018.

[4] Z. Chen, Yao, D. Boroyevich, et al, "A 1200-V, 60-A SiC MOSFET multichip phase-leg module for high-temperature, high-frequency applications," IEEE Trans. Power Electron., vol. 29, no. 5, pp. 2307-2320, May 2014.

[5] M. Paolucci, "Improving Power Density and Efficiency in Servers and Telecom," Power Systems Design, Nov. 2015.

[6] Soheil Khosrogorji, Sepehr Soori, Hossein Torkaman. A new design strategy for DC/DC LLC resonant converter: Concept, modeling, and fabrication. International Journal of Circuit Theory and Applications, 2019, 47(10).

[7] Jianguang Ma, Xueye Wei, Liang Hu, Junhong Zhang. Small-Signal Modeling of the LLC Half-Bridge Resonant Converter. Journal of Circuits, Systems and Computers, 2019, 28(4).

[8] Zhang Lei, Qiu Yafeng, Liu Yixi. Modeling analysis and verification of digitally controlled LLC resonant converter. Electronic Product World, 2015, 22(07): 59-62.

978-1-6654-4817-8/21 $31.00 © 2021 IEEE

A Review of the Crosstalk Suppression Methods for SiC MOSFETs in the Phase-leg Circuit Configuration

Yujie Ding
Academy for Engineering &
Technology
Fudan University
Shanghai, China
yujieding1999@163.com

Saijun Mao
Academy for Engineering &
Technology
Fudan University
Shanghai, China
saijunmao@fudan.edu.cn

Zhikun Wang
Academy for Engineering &
Technology
Fudan University
Shanghai, China
20210860095@fudan.edu.cn

Shuhao Yang
Academy for Engineering &
Technology
Fudan University
Shanghai, China
20210860083@fudan.edu.cn

Wenyu Li
Academy for Engineering &
Technology
Fudan University
Shanghai, China
volcanio@163.com

Keqiu Zeng
Philips Healthcare (Suzhou) Co.,Ltd.
Suzhou, China
keqiu_zeng@163.com

Abstract—SiC (silicon carbide) device as one of the wide-bandgap power semiconductor devices have a wide range of applications in the fields of high frequency, high voltage and high temperature. However, due to lower gate threshold voltage and higher allowable negative gate threshold voltage, SiC MOSFETs in the phase-leg circuit configuration are more sensitive to the voltage spikes caused by crosstalk during the fast-switching process. To effectively suppress crosstalk and exploit the excellent performance of SiC MOSFETs, this paper analyzes the mechanism of crosstalk for SiC MOSFETs. The state-of-the-art modeling methods of crosstalk for SiC MOSFETs are summarized based on gate resistance, common source inductance, Miller capacitor, body diode reverse recovery, et al. The state-of-the-art suppression methods are classified and evaluated from complexity, feasibility and validity. Finally, an effective passive suppression scheme is proposed for SiC MOSFETs with Kelvin package.

Keywords—SiC MOSFETs, crosstalk, modeling, suppression, common source inductance, body diode reverse recovery

I. INTRODUCTION

Power semiconductor devices are the key components of electrical energy conversion [1], [2]. Si devices have approached the material limit in terms of switching loss, power density and switching frequency [3], [4]. With the development of electric vehicles, datacenters, intelligent electronic products and other emerging technologies, SiC semiconductor devices develop rapidly with higher breakdown voltage, higher thermal conductivity and lower switching loss [5]-[10]. At the same time, the application of SiC devices is beneficial to the alleviation of the global energy crisis [11], [12].

With the higher switching speed and operating voltage, the dv/dt and di/dt generated increase accordingly in the phase-leg circuit configuration with SiC MOSFETs [13], [14]. Higher dv/dt and di/dt will cause worse false triggering pulse due to the effect of parasitic parameters [15]-[17]. SiC devices have smaller interelectrode capacitances, lower gate threshold voltage V_{th} and higher negative allowable gate threshold voltage, which make SiC MOSFETs more sensitive to the

This work is sponsored by Shanghai Pujiang Program(20PJ1401500).
This work is also sponsored by original research personalized support project by Academy for Engineering & Technology, Fudan University (gyy_yc _2020-6).

positive and negative gate voltage spikes generated in the switching transient [18], [19]. If the positive false triggering pulse exceeds V_{th}, there will be a risk of shoot-through and increasing switching loss. If the negative false triggering pulse exceeds the negative threshold voltage, there will be a risk of device breakdown. Crosstalk in the phase-leg circuit configuration seriously affect the efficiency and reliability of SiC MOSFETs [20].

The equivalent model of SiC MOSFETs is shown in Fig. 1. The internal parasitic parameters of SiC MOSFETs include gate-source capacitance C_{gs}, gate-drain capacitance C_{gd}, drain-source capacitance C_{ds}, gate resistance R_g, common source inductance L_s, drain inductance L_d, gate inductance L_g and body-diode D. The simplified equivalent model of the phase-leg circuit with SiC MOSFETs is shown in Fig. 2. The output current I_o remains constant during the switching transient since the load is inductive normally. The input voltage source V_{DC} is constant. The gate inductance includes the internal gate inductance L_{g_in} of and the driver loop inductance L_{g_ex}. The gate resistance includes the internal gate resistance R_{g_in} and the driver resistance L_{g_ex} [21]-[24].

If the lower device Q_L is the action one, crosstalk will occur on the complementary device Q_H which should be in off state. In the turn-on transient of Q_L, the driver current i_{g_H} charges C_{gs_H} firstly. Crosstalk will not occur until the channel of Q_H

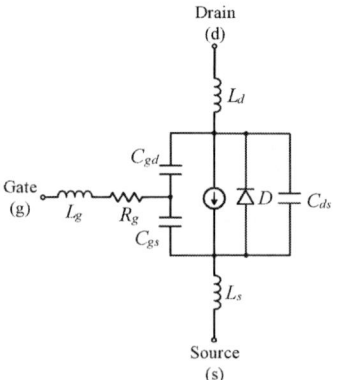

Fig. 1. Equivalent model of SiC MOSFETs.

978-1-6654-4817-8/21 $31.00 © 2021 IEEE

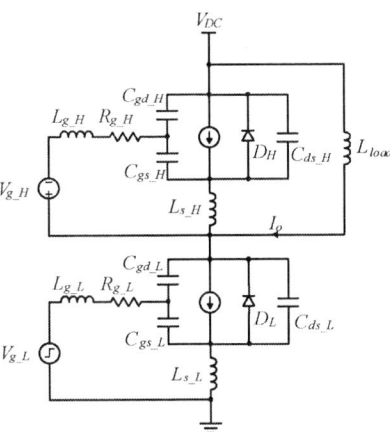

Fig. 2. Equivalent model of the phase-leg circuit with SiC MOSFETs.

is open. After channel opening, the output current I_o commutes from the body diode of Q_H to the channel of Q_L which generates large di/dt. After the commutation, the reverse recovery current of the body diode of Q_H will generate large di/dt as well [25]. di/dt generates an induced voltage drop on the L_{s_H} [26]-[28]. Since the common source inductance belongs to both the driver loop and power loop, the induced voltage drop will be fed back to the driver circuit causing positive false triggering pulse. Then, the gate-source voltage V_{gs_L} is maintained at the Miller platform. The i_{g_H} charges C_{ds_H} which causes V_{ds_H} to rise and generates large dv/dt [29]-[31]. dv/dt generates a displacement current through the Miller capacitor C_{gd_H} which causes negative false triggering pulse. In the turn-off transient of Q_L, the trends of voltage and current are opposite to the turn-on transient and there is no body diode reverse recovery [32]-[34].

Accordingly, this paper provides detailed classification and summary of the state-of-the-art research on the crosstalk in the phase-leg circuit configuration with SiC MOSFETs, based on the analysis of crosstalk formation mechanism.

1) Compare the advantages and disadvantages of different simplified methods for modeling of crosstalk formation mechanism. This will be useful to analyze the key factors of crosstalk for SiC MOSFETs which can assist in the crosstalk suppression and selection of driver parameters.

2) Classify different crosstalk suppression methods from the principle, complexity, feasibility and impact on the circuit cost.

3) Based on the situation that SiC MOSFETs with Kelvin packages are widely used, it is believed that crosstalk can be suppressed effectively with appropriate passive suppression.

II. STATE-OF-THE-ART CROSSTALK MECHANISM MODELING

Commonly, by modeling the phase-leg circuit configuration with SiC MOSFETs, the voltage and current variations in the switching transients are analyzed and the crosstalk mathematical model is developed. The main considerations for crosstalk modeling include dv/dt, di/dt, inter-pole capacitances, gate resistance and common source inductance. Some references also consider the body diode reverse recovery and parasitic inductances. Different modeling approaches have different focus and simplification methods.

The model in Reference [30] considers the effects of inter-pole capacitances of SiC MOSFETs and gate resistance, which shows the dv/dt causes displacement current through the inter-pole capacitances and induces voltage drop through the gate resistance. Reference [35] considers the influences of parasitic inductances and body diode reverse recovery current which indicates that di/dt causes crosstalk feedback to the driver loop through parasitic inductances. Reference [36] establishes an equivalent model of the phase-leg circuit and obtains a fourth-order differential equation considering the loop stray inductances, inter-pole capacitances and gate resistance. However, due to the complexity of the mathematical model, it is difficult to carry out further practical applications.

Reference [37] and Reference [38] further simplify the phase-leg circuit by equating dv/dt and di/dt to a voltage source or a current source which are the fundamental source of crosstalk. The effect of dv/dt can be equated to a current source, since dv/dt induces a displacement current through the Miller capacitor C_{gd}. The effect of di/dt is equated to a voltage source, since di/dt generates an induced voltage drop through the common source inductor L_s. The equivalent circuits are shown in Fig. 3 [37]. Reference [38] analyzes the main source charging C_{gs} at different stages in the switching transient and models the crosstalk voltage with the capacitor charging equation $v=V_x\,(1-e^{-t/\tau})$. This method can sort out the formation mechanism of crosstalk clearly from the root. However, in practical application, the variation of equivalent voltage source and current source cannot be obtained directly.

Reference [39] establishes an equivalent circuit which connects the output capacitances of the upper and lower SiC MOSFETs in parallel shown in Fig. 4. At the same time, the

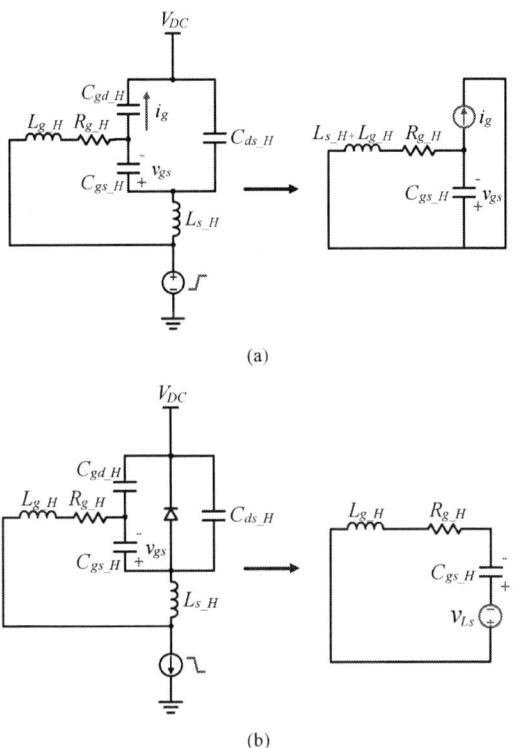

Fig. 3. Crosstalk source equivalent circuit. (a) dv/dt inducing crosstalk equivalent to a current source. (b) di/dt inducing crosstalk equivalent to a voltage source.

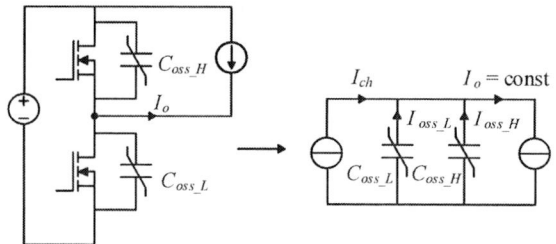

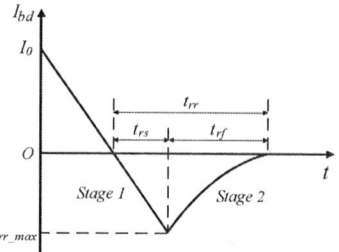

Fig. 4. Equivalent circuit of upper and lower output capacitances in parallel.

Fig. 5. Waveform of the reverse recovery current of the body diode.

voltage and current variations in the switching transients are expressed by the output charge Q_{oss} and the output current I_{oss}. Using this model, time of each phase in the switching transient can be calculated. However, compared to dv/dt and di/dt, Q_{oss} and I_{oss} are difficult to measure and adjust in simulation and circuit measurements.

Reference [40] considers the effect of the gate resistance R_g and output current I_{load}. The smaller the R_g is, the smaller the voltage drop will be generated. However, a smaller resistance causes larger gate current which increases the voltage drop generated across R_g. The two opposite effects make it theoretically possible to have a R_g that minimizes the crosstalk voltage spike. I_{load} affects the di/dt generated in the switching transient and directly determines the crosstalk spikes. This modeling is able to represent the approximate relationship between I_{load}, R_g and crosstalk while considering fewer parameters.

Reference [41] and Reference [42] considers the effect of the reverse recovery current of the body diode in the turn-on transient shown in Fig. 5. At the first stage of the reverse recovery, body diode current increases from 0 to maximum value I_{rr_max}. In this phase, reference [41] expresses the crosstalk voltage with the transfer characteristics of SiC MOSFETs: $V_{gs\_max} = I_{d\_max}/g_{fs} + V_{th}$. However, reference [42] neglects the effect since the common source inductor is in the energy storage stage. At the second stage of the reverse recovery, the body diode current drops from its maximum value to 0 generating di/dt. At the same time, the drain-source voltage of the action device falls generating dv/dt which causes crosstalk in conjunction with the di/dt.

Reference [42] and Reference [43] simplifies the crosstalk formation mechanism model by considering dv/dt and di/dt as constants in the switching transient. Reference [43] considers that the dv/dt can be approximated as v_{ds_max}/T_m. T_m is the time for v_{ds} rising to its maximum value v_{ds_max}. The final result is

$v_{gs\_max} \cong C_{gd}/(C_{gd}+C_{gs}) \times v_{ds\_max}$. This modeling quantifies the crosstalk voltage and provides some guidance for the design of SiC MOSFETs at the device level that reducing the ratio of C_{gd}/C_{gs} can reduce the effect of crosstalk. However, the accuracy of crosstalk voltage calculation cannot be guaranteed since dv/dt and di/dt are variable in the switching transients.

Accordingly, the root causes of crosstalk are the dv/dt and di/dt generated in the switching transient. The parameters that have large influence on crosstalk include: gate resistance, common source inductance, body diode reverse recovery, and inter-pole capacitances. The modeling of the crosstalk formation mechanism needs to be appropriately simplified according to the study purpose. If all the influencing factors are considered, the model will be complex and poorly guided. If fewer factors are considered, the accuracy of the model is not guaranteed.

III. STATE-OF-THE-ART OF CROSSTALK SUPPRESSION

The state-of-the-art crosstalk voltage suppression methods can be divided into three main categories. The first category is adding suitable auxiliary branches to suppress crosstalk voltage which is the main research direction in most related literature. The other two suppression methods are summarized in this paper from application point of view. One is choosing appropriate type and package (such as Kelvin package) of SiC MOSFETs to achieve low crosstalk voltage, the other is debugging on the existing driver circuit to improve the crosstalk.

A. Auxiliary Branches for Crosstalk Suppression

There are two ways to classify the auxiliary branches for crosstalk suppression shown in Fig. 6. Based on the type of auxiliary devices used, they can be classified as active or passive suppression. Passive suppression suppresses crosstalk with auxiliary capacitances, diodes and resistances while active suppression also makes use of triodes, MOSFETs and

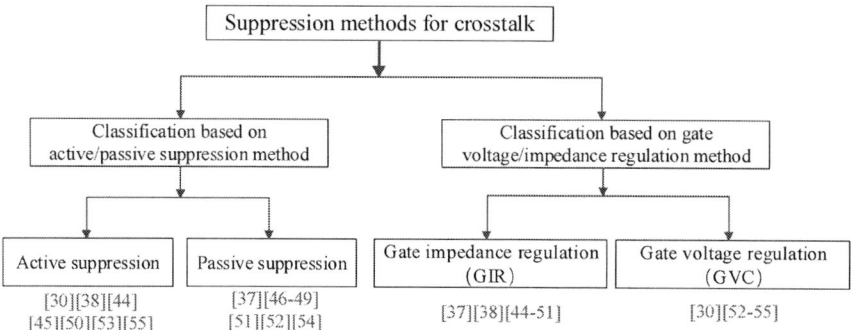

Fig. 6. Classification of the crosstalk suppression methods.

978-1-6654-4817-8/21 $31.00 © 2021 IEEE

voltage sources. Passive suppression has much smaller impact on the complexity and cost of circuit. Depending on the suppression principle, they can be classified as gate impedance regulation (GIR) and gate voltage regulation (GVR).

GIR uses Miller clamping circuit which contains an auxiliary switching device and a capacitance in parallel with the gate and source terminals of SiC MOSFETs as shown in Fig. 7 [38], [44], [45]. The auxiliary capacitance is much larger than the gate-source capacitance to provide a low impedance loop for the displacement current generated in the switching transient. The conventional Miller clamp circuit increases the input capacitance and tends to increase the switching loss [46, 47]. Reference [48] suggests that the design of driver loop impedance and power loop impedance should be coordinated by connecting capacitors in parallel at the gate and source terminals of SiC MOSFETs and gives a selection method for the auxiliary capacitors. Reference [49] adds an auxiliary triode and an auxiliary capacitor to improve the conventional active clamp circuit. This method effectively suppresses the positive spikes and prevents the SiC MOSFETs from turning off too fast. However, it requires a large auxiliary capacitor which cannot be integrated on-chip. Since this method will greatly slow down the rising edge time of the power devices, it cannot exploit the high-speed characteristics of SiC MOSFETs. Reference [50] connected an auxiliary MOSFET in parallel with an auxiliary capacitor. The access or disconnection of the auxiliary capacitor is controlled by an additional drive signal which increases the complexity of the driver loop. Reference [51] proposes an improved gate driver with active crosstalk suppression. An auxiliary capacitor is connected in parallel between the gate and source terminals of SiC MOSFET with an auxiliary transistor turned on when crosstalk occurs. Meanwhile an auxiliary resistor is connected in parallel with the driver resistor through an auxiliary diode. The method increases the gate capacitance and reduces the gate impedance to ensure high speed performance of SiC MOSFETs. Reference [37] adds a parallel diode with junction capacitance in pF level. When the v_{gs} is not reached the diode threshold voltage, the auxiliary diode performs as a junction capacitor. When the diode turns on, the displacement current will be bypassed. This method does not need large auxiliary capacitances or resistances and has less impact on the switching speed. However, only negative crosstalk spikes can be suppressed.

GVR is used to ensure that the gate-source voltage is below the positive gate threshold voltage V_{th} when positive crosstalk occurs and above the negative allowable gate threshold voltage when negative crosstalk occurs [30]. Reference [52] proposes an RCD (resistor-capacitor-diode) level converter that applies a suitable negative voltage to suppress the voltage spikes. However, the selectable range of the negative voltage is small since SiC MOSFETs can tolerate

lower negative threshold voltage. Reference [53] proposes a multi-step active gate driver that varies the gate-source voltage and capacitance in the switching transient. A negative voltage and an auxiliary capacitor are applied at the gate and source terminals of SiC MOSFET when positive voltage spikes occur. A zero voltage is applied when negative voltage spikes occur. So the gate-source voltage is always in the region and the switching speed will not be changed. Reference [54] proposes a gate driver that generates a negative turn-off voltage without using a negative source and adds a negative offset to the gate-source voltage without adding a resistor. Reference [55] proposes a less complex gate driver based on negative bias turn-off voltage containing an auxiliary capacitor and two passive transistors with no additional control signals.

B. Effect of Different Packages on Crosstalk Suppression

Currently, two types of packages are used for discrete SiC MOSFETs. Fig. 8 shows the non-Kelvin package (TO-247-3 of Infineon's IMW120R030M1). Fig. 9 shows the Kelvin package (TO-247-4 of Infineon's IMZ120R030M1). Since the common source inductor L_s in the non-Kelvin package belongs to both the driver loop and power loop, the driver loop is more sensitive to di/dt generated in the power loop. The Kelvin package divides the source terminal into driver-source terminal and power-source terminal, reducing the effect of L_s. SiC MOSFETs with Kelvin package are more effective for crosstalk suppression.

To compare the differences of crosstalk between SiC MOSFETs with non-Kelvin package and Kelvin package , the switching characteristics of C3M0021120D with non-Kelvin package and C3M0021120K with Kelvin package are simulated at 650V/60A. The driver resistance is 3Ω and driver voltage is -3V/18V. The results shown in Fig. 10 prove that Kelvin package effectively suppresses crosstalk by reducing the influence of the common source inductor L_s. At the same time, the negative feedback effect of L_s on the driver loop is reduced

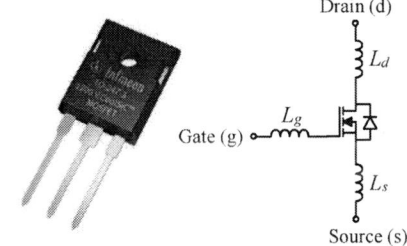

Fig. 8. SiC MOSFET with non-Kelvin package (TO-247-3).

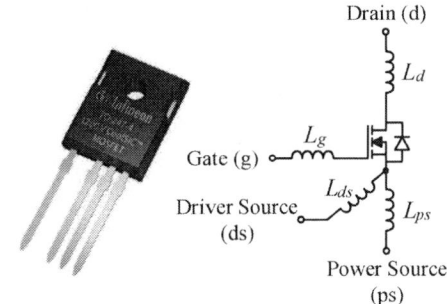

Fig. 9. SiC MOSFET with Kelvin package (TO-247-4).

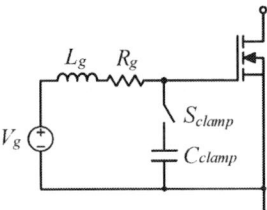

Fig. 7. Equivalent circuit of Miller clamping circuit.

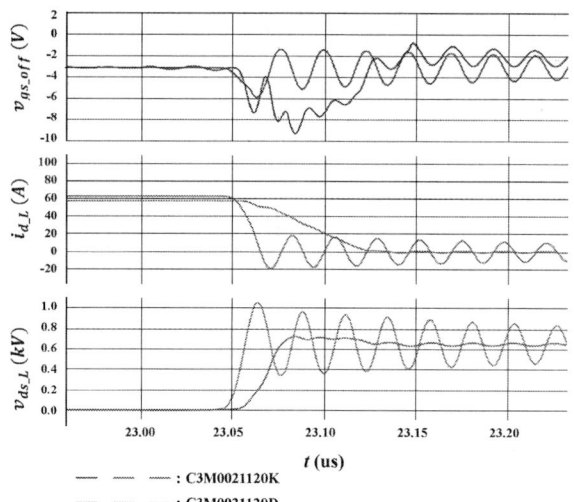

Fig. 10. Comparison of the key waveforms of C3M0021120K and C3M0021120D in the turn-off transient.

in Kelvin package, which accelerates the switching speed beneficial to the high frequency application of SiC MOSFETs.

For SiC MOSFETs with same on-resistance and voltage withstand capability of the same manufacturer, the cost difference between the two packages is not significant. The price provided from Infineon is about $1 more per unit for a Kelvin package than for a non-Kelvin package. Therefore, the SiC MOSFETs with Kelvin package dominate the current application market. The crosstalk suppression for SiC MOSFETs with Kelvin packages is also different from that of Si MOSFETs and SiC MOSFETs with non-Kelvin packages.

C. Effect of Appropriate Driver Parameters for Crosstalk Suppression

Debugging on the existing driver circuit is direct and effective for crosstalk suppression, especially for SiC MOSFETs with Kelvin package. This method will not increase the cost or complexity of the circuit.

Choosing suitable negative turn-off voltage can limit crosstalk within the tolerable range although the negative voltage range that can be selected is small since SiC MOSFETs have lower gate threshold voltage and higher negative breakdown threshold voltage [38].

Adjusting the driver resistance to reasonably control switching speed can suppress crosstalk [56]. According to equation (1), it is generally believed that increasing the driver resistance can reduce dv/dt and suppress crosstalk. However, as the switching speed decreases, the switching loss increases. Many researchers believe it is not a good way for crosstalk suppression [38], [40].

$$v_{gs} = R_g C_{gd} \frac{dv_{ds}}{dt} \left(1 - e^{\frac{-t}{R_g(C_{gd}+C_{gs})}} \right) \qquad (1)$$

In fact, crosstalk and driver resistance are not monotonically related. Larger R_g can suppress the crosstalk caused by dv/dt. However, the larger the R_g is, the larger the induced voltage is. Therefore, proper adjustment of the driver resistance can improve crosstalk.

The driver parameters of C3M0021120K are adjusted, including the driver voltage and driver resistance shown in Fig. 11. Finally, the crosstalk was suppressed within the allowable range which illustrates that adjusting driver parameters of SiC MOSFETs with Kelvin package is an effective means to suppress crosstalk. The experimental waveforms of crosstalk suppression for C3M0021120K is shown in Fig. 12.

The currently proposed driver circuits for crosstalk suppression are complex which require auxiliary devices and complex control signals. On the one hand, it will increase the cost and complexity. On the other hand, auxiliary branches have impacts on the switching speed and loss. Many studies are aimed at SiC MOSFETs with non-Kelvin package which have harsh crosstalk and require relatively complex suppression means. However, with the widespread use of SiC MOSFETs with Kelvin package, effective passive means of crosstalk suppression can be performed.

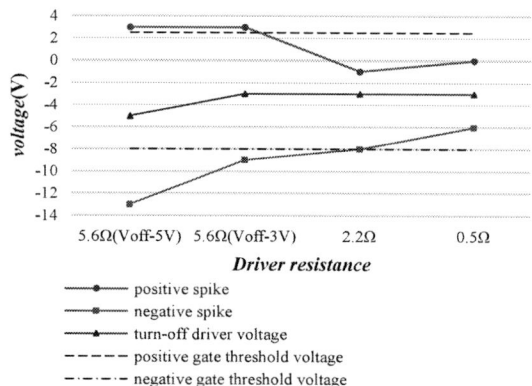

Fig. 11. Crosstalk suppression for C3M0021120K.

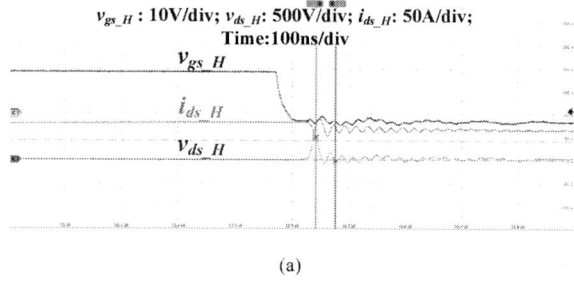

(a)

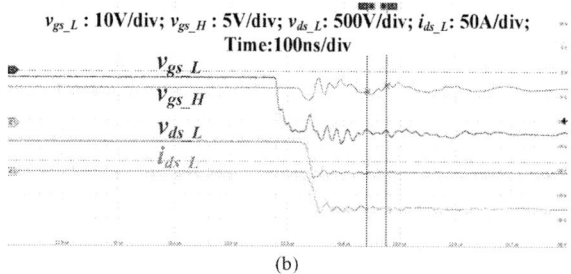

(b)

Fig. 12. Experimental waveforms of crosstalk suppression for C3M0021120K. (a) Turn-on transient. (b) Turn-off transient.

IV. CONCLUSION

This paper reviews the state-of-the-art modeling methods of crosstalk mechanisms and suppression approaches. The modeling considering different key factors influenced are evaluated. The crosstalk suppression methods are classified in terms of complexity, cost, and exploitation of the excellent performance of SiC MOSFETs. The main conclusions obtained are summarized as follows.

1) The crosstalk formation in the phase-leg circuit for SiC MOSFETs is mainly related to inter-pole capacitances, gate resistance, common source inductance, and body diode reverse recovery current. The SiC MOSFETs with Kelvin package weakening the effect of the common source inductance and mitigate the crosstalk caused by di/dt.

2) Compared to active suppression, passive suppression for crosstalk has less impact on circuit complexity and cost. Crosstalk suppression can also be classified as gate impedance regulation (GIR) and gate voltage regulation (GVR). In addition, appropriate type of SiC MOSFETs and driver parameters can suppress crosstalk effectively, especially for SiC MOSFETs with Kelvin package.

REFERENCES

[1] D. Garrido-Diez and I. Baraia, "Review of wide bandgap materials and their impact in new power devices," *IEEE ECMSM*, 2017, pp. 1-6.

[2] Z. Liang, P. Ning and F. Wang, "Development of Advanced All-SiC Power Modules," *IEEE Trans. Power Electron.*, vol. 29, no. 5, pp. 2289-2295, May. 2014.

[3] F. Wang and Z. Zhang, "Overview of silicon carbide technology: Device, converter, system, and application," *CPSS Trans. Power Electron. Appl.*, vol. 1, no. 1, pp. 13-32, Dec. 2016.

[4] A. Marzoughi, R. Burgos and D. Boroyevich, "Active Gate-Driver With dv/dt Controller for Dynamic Voltage Balancing in Series-Connected SiC MOSFETs," *IEEE Trans. Industrial Electron.*, vol. 66, no. 4, pp. 2488-2498, April. 2019.

[5] J. Millán, P. Godignon, X. Perpiñà, A. Pérez-Tomás and J. Rebollo, "A Survey of Wide Bandgap Power Semiconductor Devices," *IEEE Trans. Power Electron.*, vol. 29, no. 5, pp. 2155-2163, May. 2014.

[6] Z. Chen, D. Boroyevich, R. Burgos and F. Wang, "Characterization and modeling of 1.2 kv, 20 A SiC MOSFETs," *in Proc. IEEE Energy Convers. Congr. Expo.*, 2009, pp. 1480-1487.

[7] Z. Wang, X. Shi, Y. Xue, L. M. Tolbert, F. Wang and B. J. Blalock, "Design and Performance Evaluation of Overcurrent Protection Schemes for Silicon Carbide (SiC) Power MOSFETs," *IEEE Trans. Industrial Electron.*, vol. 61, no. 10, pp. 5570-5581, Oct. 2014.

[8] Z. Chen, Y. Yao, D. Boroyevich, K. D. T. Ngo, P. Mattavelli and K. Rajashekara, "A 1200-V, 60-A SiC MOSFET Multichip Phase-Leg Module for High-Temperature, High-Frequency Applications," *IEEE Trans. Power Electron.*, vol. 29, no. 5, pp. 2307-2320, May. 2014.

[9] J. Wang et al., "Characterization, Modeling, and Application of 10-kV SiC MOSFET," *IEEE Trans. Electron. Devices*, vol. 55, no. 8, pp. 1798-1806, Aug. 2008.

[10] M. Imaizumi and N. Miura, "Characteristics of 600, 1200, and 3300 V Planar SiC-MOSFETs for Energy Conversion Applications," *IEEE Trans. Electron. Devices*, vol. 62, no. 2, pp. 390-395, Feb. 2015.

[11] A. Lemmon, M. Mazzola, J. Gafford and C. Parker, "Stability Considerations for Silicon Carbide Field-Effect Transistors," *IEEE Trans. Power Electron.*, vol. 28, no. 10, pp. 4453-4459, Oct. 2013.

[12] A. J. Lelis, R. Green, D. B. Habersat and M. El, "Basic Mechanisms of Threshold-Voltage Instability and Implications for Reliability Testing of SiC MOSFETs," *IEEE Trans. Electron. Devices*, vol. 62, no. 2, pp. 316-323, Feb. 2015.

[13] T. Liu, R. Ning, T. Wong and Z. Shen, "Modeling and Analysis of SiC MOSFET Switching Oscillations," *IEEE J. Emerg. Sel. Topics Power Electron.*, vol. 4, no. 3, pp. 747-756, Sept. 2016.

[14] P. Nayak and K. Hatua, "Parasitic Inductance and Capacitance-Assisted Active Gate Driving Technique to Minimize Switching Loss of SiC MOSFET," *IEEE Trans. Industrial Electron.*, vol. 64, no. 10, pp. 8288-8298, Oct. 2017.

[15] P. Sochor, A. Huerner and R. Elpelt, "A Fast and Accurate SiC MOSFET Compact Model for Virtual Prototyping of Power Electronic Circuits," *PCIM Europe 2019; International Exhibition and Conference for Power Electronics, Intelligent Motion, Renewable Energy and Energy Management*, 2019, pp. 1-8.

[16] S. Yin, K. Tseng, C. Tong, R. SimanJorang, C. Gaianayake, A. Nawawi, et al., "Gate driver optimization to mitigate shoot-through in high-speed switching SiC half bridge module," *IEEE PEDS*, 2015, pp. 484-491.

[17] A. P. Camacho, V. Sala, H. Ghorbani and J. L. R. Martinez, "A Novel Active Gate Driver for Improving SiC MOSFET Switching Trajectory," *IEEE Trans. Industrial Electron.*, vol. 64, no. 11, pp. 9032-9042, Nov. 2017.

[18] H. Li, Y. Jiang, Z. Qiu, Y. Wang and Y. Ding, "A Predictive Algorithm for Crosstalk Peaks of SiC MOSFET by Considering the Nonlinearity of Gate-Drain Capacitance," *IEEE Trans. Power Electron.*, vol. 36, no. 3, pp. 2823-2834, March. 2021.

[19] L. Zhang, X. Yuan, X. Wu, C. Shi, J. Zhang and Y. Zhang, "Performance Evaluation of High-Power SiC MOSFET Modules in Comparison to Si IGBT Modules," *IEEE Trans. Power Electron.*, vol. 34, no. 2, pp. 1181-1196, Feb. 2019.

[20] S. Ji, S. Zheng, F. Wang and L. M. Tolbert, "Temperature-Dependent Characterization, Modeling, and Switching Speed-Limitation Analysis of Third-Generation 10-kV SiC MOSFET," *IEEE Trans. Power Electron.*, vol. 33, no. 5, pp. 4317-4327, May. 2018.

[21] A. Huerner, P. Sochor and R. Elpelt, "Method for extracting internal gate resistance of SiC MOSFETs from double-pulse measurements," *PCIM Europe 2020; International Exhibition and Conference for Power Electronics, Intelligent Motion, Renewable Energy and Energy Management*, 2020, pp. 1-7.

[22] A. P. Arribas, F. Shang, M. Krishnamurthy and K. Shenai, "Simple and Accurate Circuit Simulation Model for SiC Power MOSFETs," *IEEE Trans. Electron. Devices*, vol. 62, no. 2, pp. 449-457, Feb. 2015.

[23] K. Chen, Z. Zhao, L. Yuan, T. Lu and F. He, "The Impact of Nonlinear Junction Capacitance on Switching Transient and Its Modeling for SiC MOSFET," *IEEE Trans. Electron. Devices*, vol. 62, no. 2, pp. 333-338, Feb. 2015.

[24] D. Christen and J. Biela, "Analytical Switching Loss Modeling Based on Datasheet Parameters for mosfets in a Half-Bridge," *IEEE Trans. Power Electron.*, vol. 34, no. 4, pp. 3700-3710, April. 2019.

[25] D. Martin, P. Killeen, W. A. Curbow, B. Sparkman, L. Kegley and T. McNutt, "Comparing the switching performance of SiC MOSFET intrinsic body diode to additional SiC schottky diodes in SiC power modules," *IEEE WiPDA*, 2016, pp. 242-246.

[26] S. Yin and Y. Liu, "A Reliable Gate Driver with Desaturation and Over-Voltage Protection Circuits for SiC MOSFET," *PCIM Asia 2018; International Exhibition and Conference for Power Electronics, Intelligent Motion, Renewable Energy and Energy Management*, 2018, pp. 1-5.

[27] J. Fabre, P. Ladoux and M. Piton, "Characterization and Implementation of Dual-SiC MOSFET Modules for Future Use in Traction Converters," *IEEE Trans. Power Electron.*, vol. 30, no. 8, pp. 4079-4090, Aug. 2015.

[28] D. Han and B. Sarlioglu, "Comprehensive Study of the Performance of SiC MOSFET-Based Automotive DC–DC Converter Under the Influence of Parasitic Inductance," *IEEE Trans. Ind. Appl.*, vol. 52, no. 6, pp. 5100-5111, Nov.-Dec. 2016.

[29] M. R. Ahmed, R. Todd and A. J. Forsyth, "Predicting SiC MOSFET Behavior Under Hard-Switching, Soft-Switching, and False Turn-On Conditions," *IEEE Trans. Industrial Electron.*, vol. 64, no. 11, pp. 9001-9011, Nov. 2017.

[30] F. Gao, Q. Zhou, P. Wang and C. Zhang, "A Gate Driver of SiC MOSFET for Suppressing the Negative Voltage Spikes in a Bridge Circuit," *IEEE Trans. Power Electron.*, vol. 33, no. 3, pp. 2339-2353, March. 2018.

[31] H. Huang, X. Yang, Y. Wen and Z. Long, "A switching ringing suppression scheme of SiC MOSFET by Active Gate Drive," *IEEE IPEMC-ECCE Asia*, 2016, pp. 285-291.

[32] C. Ma and P. Lauritzen, "A simple power diode model with forward and reverse recovery," *IEEE PESC*, 1991, pp. 411-415.

[33] D. Yuan , Y. Zhang , and X. Wang . "An Improved Analytical Model

for Crosstalk of SiC MOSFET in a Bridge-Arm Configuration," *Energies*, vol. 14, no. 3, pp. 1-30, 2021.

[34] X. Li, L. Zhang, S. Guo, Y. Lei, A. Huang and B. Zhang, "Understanding switching losses in SiC MOSFET: Toward lossless switching," *IEEE WiPDA*, 2015, pp. 257-262.

[35] A. Nishigaki, H. Umegami, F. Hattori, W. Martinez and M. Yamamoto, "An analysis of false turn-on mechanism on power devices," *in Proc. IEEE Energy Convers. Congr. Expo.*, 2014, pp. 2988-2993.

[36] S. Jahdi, O. Alatise, J. A. Ortiz Gonzalez, R. Bonyadi, L. Ran and P. Mawby, "Temperature and Switching Rate Dependence of Crosstalk in Si-IGBT and SiC Power Modules," *IEEE Trans. Industrial Electron.*, vol. 63, no. 2, pp. 849-863, Feb. 2016.

[37] J. Zhao, L. Wu, Z. Li, Z. Chen and G. Chen, "Analysis and Suppression for Crosstalk in SiC MOSFET Turn-off Transient," *IEEE IPEMC-ECCE Asia*, 2020, pp. 1145-1150.

[38] M. Liang, Y. Li, Q. Q. Zheng, H. Zhao, "Analysis for Crosstalk of SiC MOSFET with Different Packages in a Phase-Leg Configuration and a Low Gate Turn-Off Impedance Driver," *Transactions of China Electrotechnical Society* (in Chinese), vol. 32, no. 18, pp. 162-174, 2017.

[39] Y. Chen, C. Li, Z. Lu, H. Luo, C. Li, W. Li, et al., "Modeling of SiC MOSFET Crosstalk Voltage in Half Bridge Circuit," *Proceedings of the CSEE* (in Chinese), vol. 40, no. 6, pp. 1775-1786, 2020.

[40] I. Laird and X. Yuan, "Analysing the Crosstalk Effect of SiC MOSFETs in Half-Bridge Arrangements," *in Proc. IEEE Energy Convers. Congr. Expo.*, 2019, pp. 367-374.

[41] J. Wang, H. S. Chung and R. T. Li, "Characterization and Experimental Assessment of the Effects of Parasitic Elements on the MOSFET Switching Performance," *IEEE Trans. Power Electron.*, vol. 28, no. 1, pp. 573-590, Jan. 2013.

[42] B. Zhang, S. Xie, J. Xu, Q. Qian, Z. Zhang and K. Xu, "A Magnetic Coupling Based Gate Driver for Crosstalk Suppression of SiC MOSFETs," *IEEE Trans. Industrial Electron.*, vol. 64, no. 11, pp. 9052-9063, Nov. 2017.

[43] T. Wu and S. Regulator, "Cdv/dt induced turn-on in synchronous buck regulators," International Rectifier (2007).

[44] S. Yin, K. Tseng, C. Tong, R. Simanjorang, C. Gajanayake and A. Gupta, "A novel gate assisted circuit to reduce switching loss and eliminate shoot-through in SiC half bridge configuration," *IEEE APEC*, 2016, pp. 3058-3064.

[45] C. Li et al., "High Off-State Impedance Gate Driver of SiC MOSFETs for Crosstalk Voltage Elimination Considering Common-Source Inductance," *IEEE Trans. Power Electron.*, vol. 35, no. 3, pp. 2999-3011,

March. 2020.

[46] S. Liu, H. Lin, T. Wang and C. Liu, "Design of Drive Parameters Considering Crosstalk Suppression for SiC MOSFET Applications," *in Proc. IEEE Energy Convers. Congr. Expo.*, 2019, pp. 3281-3286.

[47] Y. Li, M. Liang, J. Chen, T. Zheng and H. Guo, "A Low Gate Turn-OFF Impedance Driver for Suppressing Crosstalk of SiC MOSFET Based on Different Discrete Packages," *IEEE J. Emerg. Sel. Topics Power Electron.*, vol. 7, no. 1, pp. 353-365, March. 2019.

[48] Z. Lu, C. Li, H. Wu, W. Li, X. He and S. Li, "Design of Active SiC MOSFET Gate Driver for Crosstalk Suppression Considering Impedance Coordination between Gate Loop and Power Loop," *IEEE APEC*, 2019, pp. 986-990.

[49] Z. Zhong, H. Qin, Y. Yuan, Z. Zhu, H. Xie, "Crosstalk suppression method for silicon carbide MOSFET in phase-leg circuit," *Advanced Technology of Electrical Engineering and Energy* (in Chinese), vol. 34, no. 5, pp. 8-12, 2015.

[50] Z. Zhang, F. Wang, L. M. Tolbert and B. J. Blalock, "Active Gate Driver for Crosstalk Suppression of SiC Devices in a Phase-Leg Configuration," *IEEE Trans. Power Electron.*, vol. 29, no. 4, pp. 1986-1997, April. 2014.

[51] P. Wang, L. Zhang, X. Lu, H. Sun, W. Wang and D. Xu, "An Improved Active Crosstalk Suppression Method for High-Speed SiC MOSFETs," *IEEE Trans. Ind. Appl.*, vol. 55, no. 6, pp. 7736-7744, Nov.-Dec. 2019.

[52] J. Wang and H. S. Chung, "A Novel RCD Level Shifter for Elimination of Spurious Turn-on in the Bridge-Leg Configuration," *IEEE Trans. Power Electron.*, vol. 30, no. 2, pp. 976-984, Feb. 2015.

[53] H. Li, Y. Jiang, Z. Qiu, T. Shao and Y. Wang, "A Multi-step Active Gate Driver for Suppressing Crosstalk of SiC MOSFET," *IEEE IPEMC-ECCE Asia*, 2020, pp. 1868-1873.

[54] H. Zaman, X. Wu, X. Zheng, S. Khan, H. Ali, "Suppression of Switching Crosstalk and Voltage Oscillations in a SiC MOSFET Based Half-Bridge Converter," *Energies*, vol. 11, no. 11, pp. 1-19, 2018.

[55] H. Li, Y. Zhong, R. Yu, R. Yao, H. Long, X. Wang, et al., "Assist Gate Driver Circuit on Crosstalk Suppression for SiC MOSFET Bridge Configuration," *IEEE J. Emerg. Sel. Topics Power Electron.*, vol. 8, no. 2, pp. 1611-1621, June 2020.

[56] Z. Zhang, W. Zhang, F. Wang, L. Tolbert and B. Blalock, "Analysis of the switching speed limitation of wide band-gap devices in a phase-leg configuration," *in Proc. IEEE Energy Convers. Congr. Expo.*, 2012, pp. 3950-3955.

Single Pulse Short-Circuit Failure Mechanism of 1200V Asymmetric Trench SiC MOSFETs

Zhaoxiang Wei, Jiaxing Wei, Xiaowen Yan, Hua Zhou, Hao Fu, Siyang Liu*, Weifeng Sun

National ASIC System Engineering Research Center, School of Electronic Science and Engineering, Southeast University, Nanjing, 210096, China, *E-mail: _liusy2855@163.com_

Abstract- The failure phenomenon and the mechanism for asymmetric trench silicon carbide metal-oxide-semiconductor field-effect transistor (MOSFETs) under single pulse short-circuit (SC) condition are studied in detail. After enduring an ultimate single pulse SC stress, the insulation of the gate oxide of the device fails while the forward and reverse characteristics of the body diode remain intact. Based on the analysis of the gate current of the failed device, it can be inferred that the failure point of the asymmetric trench device is located in the gate oxide close to the channel region. Moreover, TCAD simulations are performed to further demonstrate that the gate damage in the gate-source region, which is resulted from the severe electrical stress there, is meant to be the dominant single pulse SC failure mechanism for the device.

Keywords – Short-circuit; Trench SiC MOSFET; Failure mechanism.

I. INTRODUCTION

Compared with traditional planar-gate power MOSFET devices, trench-gate silicon carbide (SiC) MOSFETs have smaller cell pitches, lower on-resistance and faster switching speed. They are widely used in photovoltaics, automotive and aerospace, which can increase the power densities and reduce losses of these systems [1-3]. The trench SiC MOSFET produced by Infineon uses only one trench sidewall paralleled to the $<1\,1\,\overline{2}\,0>$ crystal plane as the channel, which is targeted for improving the conducting characteristic and the robustness of the device [4]. Nevertheless, very few studies focus on analyzing the reliability of the asymmetric trench device at present.

Under short-circuit (SC) state, the high current and voltage attack the SiC MOSFET leading to malfunction, which is worthy of being studied in depth [5-7]. However, several articles investigated the SC induced failure for the asymmetric trench device, mainly from the perspective of thermal adaptation [8,9]. In this paper, the single pulse SC failure

mechanism of 1200V asymmetric trench SiC MOSFETs are studied in detail. The electrical properties before and after the failure are compared. In addition, relying on TCAD simulations, the physical status of the device is evaluated under the SC condition. It shows that the serious stress, mainly the impact ionization rate, appeals at the channel region, leading to the gate-source failure of the device.

II. DEVICE STRUCTURE AND STRESS CONDITION

A schematic cross section view of asymmetric trench MOSFET produced by Infineon is shown in Fig. 1. The channel of the device is parallel to the $<1\,1\,\overline{2}\,0>$ crystal plane, whose mobility is much higher than that of other crystal planes. Meanwhile, the deep P-well limits the electric field at the bottom and corners of the gate oxide. In this work, a 1200V Infineon product (IMW120R220M1H) is adopted as the targeted device. The normal on-state resistance (R_{on}) and the DC drain current (I_D) of the device are 220mΩ and 13A, respectively.

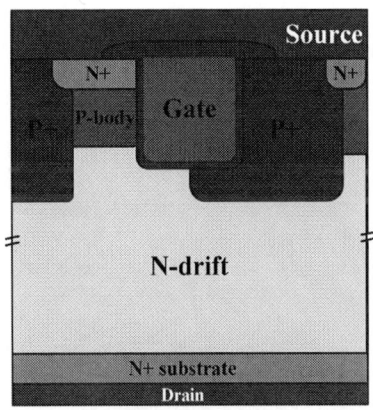

Fig. 1: Schematic cross section of the asymmetric trench SiC MOSFET.

With the help of the test circuit in Fig. 2, SC stress is added to the asymmetric trench SiC MOSFET device under test. The pulse width is gradually increased until the device fails. The waveforms are shown in Fig. 3. The gate-source voltage (V_{gs}) is set from -1 to 15V; the drain-source voltage (V_{ds}) of the device is kept at 400V. Once the gate is on, the drain-source current (I_{ds}) rises quickly to the saturated current ($I_{ds(sat)}$) about 37A during a short time. After that, the I_{ds} continues to drop with the increasing SC time, which is caused by the rise of the junction temperature. At the same time, the excessively high junction temperature causes the tail current and a slight decrease in V_{gs}.

Fig. 2: SC stress system.

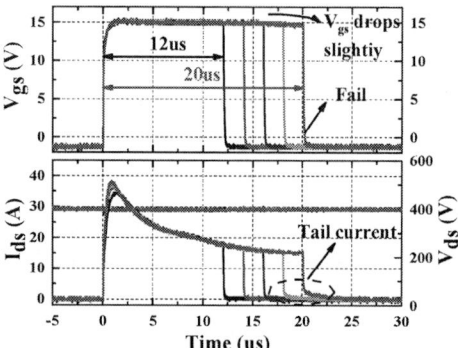

Fig. 3: Oscilloscope waveforms of the asymmetric trench SiC MOSFET under different SC pulse time.

After enduring a 20µs-long SC stress, gate and source electrodes of the asymmetric trench device are shorted. The impedances after the stress are $R_{gs} = 0\Omega$, $R_{gd} = 17.48k\Omega$, $R_{ds} = 31.73k\Omega$. Fig. 4 shows the view of decapping of the failed device. There is no obvious damage on the surface of the failed device, which is quite different from the failure phenomenon caused by the thermal runaway [10]. It can be inferred that the damaged location of the chip is inside the device structure.

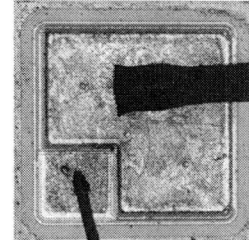

Fig. 4: Top view of the failed asymmetric trench SiC MOSFET.

III. RESULTS AND DISCUSSIONS

In order to verify the internal damage of the device, the changes in electrical characteristics of the asymmetric trench device before and after the SC failure are monitored. As shown in Fig. 5, the threshold voltage (V_{th}) of the device rises slightly with the increase of the SC time. Meanwhile, the body diode is undamaged since it still maintains normal forward and blocking characteristics, as presented in Fig. 6. Before failure, the forward and reverse electrical characteristics hardly degrade, indicating that the asymmetric trench SiC MOSFET has good SC robustness.

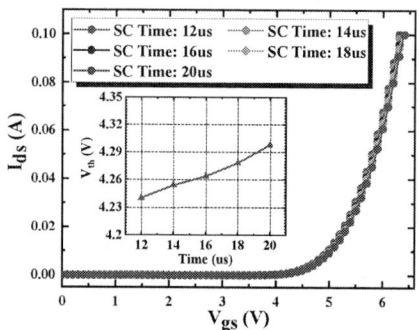

Fig. 5: Extracted V_{th} of the asymmetric trench device suffered from different SC time.

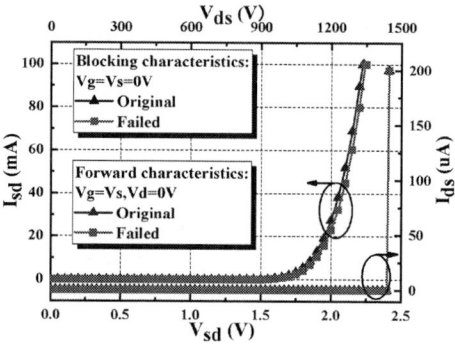

Fig. 6: Comparison of forward and reverse characteristics of the body diode before and after damage.

Since the gate damage occurs while the device fails, the gate leakage current (I_{gss}) is monitored under different pulse width SC stress in Fig. 7. It is clear that the I_{gss} of the device remains stable when the pulse width is shorter than 20µs, proving that the gate oxide of asymmetric trench SiC MOSFET maintains a good insulation property. However, the I_{gss} of the device suddenly rises to 10mA (limited by the test instrument) after enduring a 20µs-long SC stress pulse [11].

978-1-6654-4817-8/21 $31.00 © 2021 IEEE

To analyze the current path of I_{gss} of the failed asymmetric trench device, this work measures the gate-source leakage current (I_{gs}) and the gate-drain leakage current (I_{gd}) under the conditions of drain floating and source floating respectively, which are plotted in Fig. 8. Whatever the gate voltage (V_g) is, the I_{gs} and the I_{gss} are consistent, illustrating that there is a conductive path between the gate and the source of the device. But the I_{gd} presents a completely different curve. On the one hand, The I_{gd} keeps at around 100pA under the negative V_g bias condition. On the other hand, when the V_g is higher than 1V, the I_{gd} increases to 10mA (limited by the test instrument) at an exponential rate. The I_{gd} shares the same curve with the forward conducting characteristic of the body diode as shown in Fig. 8, which will be discussed in the following.

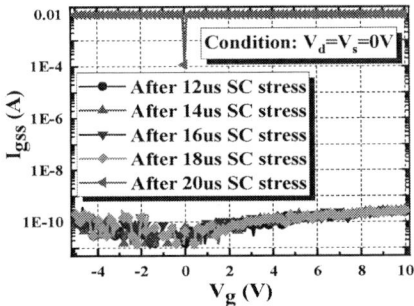

Fig. 7: The I_{gss} of the asymmetric trench device after enduring different pulse widths SC stress.

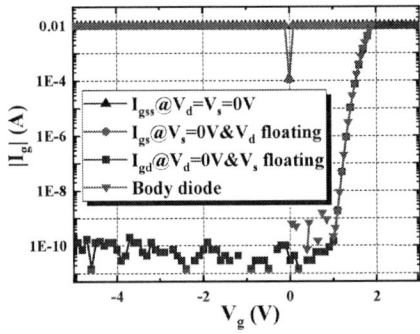

Fig. 8: the comparison among I_{gss}, I_{gs} and I_{gd} of the device after enduring the SC stress.

The damaged location of the device under the single pulse short-circuit stress is summarized in Fig. 9. After enduring a 20μs-long SC stress, the oxide between the gate and the source gets injured, thus an additional conducting path appears in the gate oxide, which means that the source and gate electrodes are shorted. Meanwhile, when the V_{gd} is positive, the body diode consisting of the P+ well and the N drift provides the obvious

I_{gd} current. When the V_{gd} is negative, due to that the gate oxide layer between the gate and the drain is intact, I_{gd} remains at about 100pA, as shown in Fig. 8.

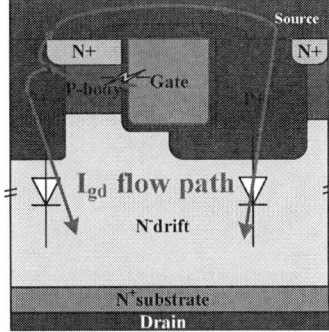

Fig. 9: The damaged location of the asymmetric trench device under the single pulse SC stress.

To further verify the single pulse SC failure mechanism of the asymmetric trench device, Silvaco simulations are performed to extract the physical characteristics of the device during SC condition. The distribution of the current density (J) is shown in Fig. 10. Different from the double-trench device [12] produced by Rohm, other than flowing through the corners and the bottom of the trench, the SC current of the asymmetric trench device passes between the two P+ well, which can more effectively protect the gate oxide from electrical stress.

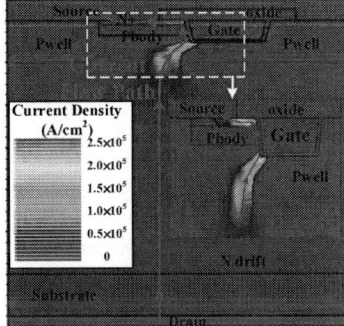

Fig. 10: The Current density distribution of the asymmetric trench device under SC condition.

The impact ionization rates (I.I.) and the electric fields in the gate oxide (Eox) along the SiC/SiO$_2$ interface are extracted in Fig. 11. The peak values of the J and the I.I. appeal in the channel region. And they both decrease gradually along the direction from the channel to the trench corner and the bottom. It demonstrates that the channel of the device bears electrical stress during the SC condition, which causes the gate oxide to be damaged, eventually. Compared with double-trench devices, the stress at the trench corner of the asymmetric trench device is much less serious, suggesting that the oxide at the trench corner is well protected.

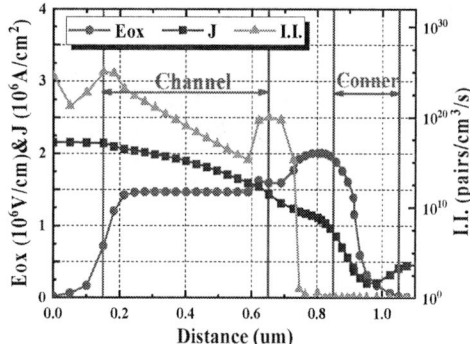

Fig. 11: Extracted E_{ox}, J and I.I. along the SiC/SiO$_2$ interface of the asymmetric trench device

IV. CONCLUSION

This article studies and verifies the single-pulse short-circuit failure mechanism of the asymmetric trench SiC MOSFET. The gate oxide in the channel region of the device is damaged after enduring a 20µs-long SC stress, and there is an additional conductive path. Combined with TCAD simulations, it is verified that there are extremely high impact ionization rates and current density in the channel region, and the strong electrical stress continuously impacts the gate oxide at the channel, leading to damage in the gate oxide of the device. The changes in electrical characteristics before and after the failure of the device also confirm this suspicion. Different from the failure mechanism of the double-trench SiC MOSFET, the gate oxide at the corner of the trench of the asymmetric trench device is well protected.

ACKNOWLEDGMENT

This work was supported in part by the National Key R&D Program of China under Grant 2020YFF0218501, in part by the National Natural Science Foundation of China under Grant 62004037, in part by the Fund for Transformation of Scientific and Technological Achievements of Jiangsu Province under Grant BA2020027, and in part by the Distinguished Young Scholars Program of Southeast University.

REFERENCES

[1] R. Rupp, "CoolSiC™ and major trends in SiC power device development," 2017 47th European Solid-State Device Research Conference (ESSDERC), 2017, pp. 118-121.

[2] J. Wei, S. Liu, J. Fang, S. Li, T. Li and W. Sun, "Investigation on degradation mechanism and optimization for SiC power MOSFETs under long-term short-circuit stress," 2018 IEEE 30th International Symposium on Power Semiconductor Devices and ICs (ISPSD), 2018, pp. 399-402.

[3] L. Cao, Q. Guo and K. Sheng, "Comparative Evaluation of the Short circuit Capability of SiC Planar and Trench Power MOSFET," 2018 IEEE 2nd International Electrical and Energy Conference (CIEEC), 2018, pp. 653-656.

[4] D. Peters et al., "Performance and ruggedness of 1200V SiC - Trench − MOSFET," 2017 29th International Symposium on Power Semiconductor Devices and IC's (ISPSD), Sapporo, Japan, 2017, pp. 239-242.

[5] H. Qin et al., "A comprehensive study of the short-circuit characteristics of SiC MOSFETs," 2017 12th IEEE Conference on Industrial Electronics and Applications (ICIEA), 2017, pp. 332-336.

[6] G. Romano et al., "A Comprehensive Study of Short-Circuit Ruggedness of Silicon Carbide Power MOSFETs," IEEE Journal of Emerging and Selected Topics in Power Electronics, vol. 4, no. 3, pp. 978-987, Sept. 2016.

[7] X. Zhou, H. Su, Y. Wang, R. Yue, G. Dai and J. Li, "Investigations on the Degradation of 1.2-kV 4H-SiC MOSFETs Under Repetitive Short-Circuit Tests," IEEE Transactions on Electron Devices, vol. 63, no. 11, pp. 4346-4351, Nov. 2016.

[8] X. Deng et al., "Short-Circuit Capability Prediction and Failure Mode of Asymmetric and Double Trench SiC MOSFETs," IEEE Transactions on Power Electronics, vol. 36, no. 7, pp. 8300-8307, July 2021.

[9] K. Yao, H. Yano, H. Tadano and N. Iwamuro, "Investigations of SiC MOSFET Short-Circuit Failure Mechanisms Using Electrical, Thermal, and Mechanical Stress Analyses," IEEE Transactions on Electron Devices, vol. 67, no. 10, pp. 4328-4334, Oct. 2020.

[10] Z. Wang et al., "Temperature-Dependent Short-Circuit Capability of Silicon Carbide Power MOSFETs," in IEEE Transactions on Power Electronics, vol. 31, no. 2, pp. 1555-1566, Feb. 2016.

[11] M. Okawa, R. Aiba, T. Kanamori, H. Yano, N. Iwamuro and S. Harada, "Experimental and Numerical Investigations of Short-Circuit Failure Mechanisms for State-of-the-Art 1.2kV SiC Trench MOSFETs," 2019 31st International Symposium on Power Semiconductor Devices and ICs (ISPSD), 2019, pp. 167-170.

[12] J. Wei, S. Liu, J. Tong et al., "Understanding Short-Circuit Failure Mechanism of Double-Trench SiC Power MOSFETs," IEEE Transactions on Electron Devices, vol. 67, no. 12, pp. 5593-5599, Dec. 2020.

978-1-6654-4817-8/21 $31.00 © 2021 IEEE

Modeling and Analysis of the Switching Characteristics Difference for Paralleling SiC MOSFETs in Multichip Power Modules

Wenyu Li
Academy for Engineering & Technology
Fudan University
Shanghai, China
volcanio@163.com

Saijun Mao
Academy for Engineering & Technology
Fudan University
Shanghai, China
maosaijun@126.com

Zhikun Wang
Academy for Engineering & Technology
Fudan University
Shanghai, China
20210860095@fudan.edu.cn

Shuhao Yang
Academy for Engineering & Technology
Fudan University
Shanghai, China
20210860083@fudan.edu.cn

Yujie Ding
Academy for Engineering & Technology
Fudan University
Shanghai, China
yujieding1999@163.com

Keqiu Zeng
Philips Healthcare (Suzhou) Co.,Ltd.
Suzhou, China
keqiu_zeng@163.com

Abstract—**Parallelism applies widely in high power modules to raise the current capacity. This paper models and analyzes a 1200V/600A silicon carbide (SiC) MOSFET power module with multichip in parallel. The equivalent circuits of both current commutation power loop and driver loop are derived. The parasitic parameters are extracted by ANSYS Q3D. The parasitic inductance in the power loop is analyzed considering the mutual influence. The unbalanced layout of the driver loop would cause different switching characteristics of the paralleled chips. Mathematical models are built to analyze current distribution in power loop and the relationship between gate resistor and driver loop inductance. The unbalanced current sharing during the transient can be mitigated by adjusting the gate resistor of each chip. The switching characteristics of the SiC MOSFET power module are measured in the double pulse tester. The experimental results validate the analysis of parasitic parameters.**

Keywords—*SiC MOSFET power module, parasitic inductance, parallelism, ANSYS Q3D, current sharing, switching characteristics.*

I. Introduction

Silicon carbide (SiC) MOSFET as a kind of wide-bandgap (WBG) semiconductor devices offers notorious advantages such as high electron mobility, fast switching speed, low on-state resistance, and high temperature capability. In high-power application scenarios, it often requires SiC MOSFET discrete devices connected in parallel or multichip power modules to achieve higher current rating. However, the connection between discrete SiC MOSFET devices would introduce severe stray inductance mismatch on both drain trace and source trace. It would lead to negative influences such as current unbalance, large current overshoot and high oscillation amplitude.

For multichip power modules, the module package determines the stray inductance. The inductance of the current commutation loop (CCL) reduces to 10% compared to paralleled devices. According to $V=L \cdot di/dt$, high di/dt can result in a large voltage overshoot on a small inductance,

which would lead to device overvoltage with this voltage peak superimposed on the gate source during the turn-off process. At the same time, due to the existence of parasitic parameters, electromagnetic interference will be generated. Since silicon carbide devices have higher switching speeds and higher power densities than silicon-based power modules, the parasitic parameters brought by the package structure have a greater impact on device performance. Therefore, it is necessary to model and decrease the stray inductance from the perspective of a module. References [1] and [2] propose a construction as P-cell and N-cell to reduce the commutation path by diodes rearrangement for IGBT. In [3], a printed circuit board (PCB) is used for the current loop. The chips are connected to the PCB board through bonding wires. And the laminated structure in PCB board can reduce the parasitic inductance. Reference [4] applies the flexible circuit board to lower the inductance with the same principle as magnetic field cancellation. In order to minimize the blanking time, reference [5] proposes a kelvin drain-to-source connection design. References [6] and [7] proposed an improved layout of an existing module by rearranging the DBC layout. In [8], a current-bunch concept is proposed to mathematically model the parasitic inductance.

But at the same time, the current sharing and switching characteristics of individual chips in parallel bring the challenges for SiC power module. To address the current evenness, DBC layout and chip arrangement of the power module result in circuit mismatch, which can further lead to unbalanced thermal coupling and inconsistent electromagnetic affect. Asymmetry circuit layout takes the main responsibility for different current sharing of each chip on steady-state, while the properties of individual chips such as on-state resistance (R_{on}) and gate threshold voltage (V_{th}) determine the unbalance level of transient state. In [9]-[11], the authors analyze the current sharing situation of parallel-connected modules. Devices mismatch could be eliminated by adopting the same manufacturing products. Asymmetry CCL are analyzed to evaluate the influence of the common source stray inductance with external inductance attached. However, the unbalanced driver loop layout in multichip SiC power modules has not been studied in detailed yet.

This work is sponsored by Shanghai Pujiang Program(20PJ1401500). This work is also sponsored by original research personalized support project by Academy for Engineering & Technology, Fudan University(gyy_yc_2020-6).

978-1-6654-4817-8/21 $31.00 © 2021 IEEE

In this paper, modeling and analysis of a 1200V/600A SiC MOSFET power module with multichip in parallel is investigated. And the different switching characteristics of paralleled chips are analyzed in detailed. Section II builds up the circuit model and electromagnetic model of both CCL and driver loop based on the multichip power module structure. In section III, the parasitic inductance is extracted to build the mathematical model. The relationship between gate resistor and driver loop inductance is revealed. The switching characteristics difference for paralleling SiC MOSFETs is illustrated. Finally, the SiC power module is measured with double pulse test, which verifies the effectiveness of the modeling and analysis of the switching characteristics.

II. ELECTROMAGNETIC MODELING FOR MULTICHIP POWER MODULE

Half-bridge structure is a basic switching cell for the power conversion circuit topology. A 1200V/600A half-bridge SiC MOSFET power module is used for the modeling and analysis.

A. Package Structure Modeling

Since each phase of this three-phase module has the symmetrical layout, U phase is selected as a case study in this paper. Its layout and topological structure are shown in Fig. 1. The current flows from the blue squares to the red squares through paralleled SiC MOSFETs. Separate DBCs are connected with boding wires from source of the MOSFETs.

In order to increase the current capability, this module adopts multichip parallel connection in both the upper and lower bridge. It contains 5 SiC MOSFETs from Cree CPM3-1200-0013A in parallel to achieve the rated current of 600A. With kelvin source connection, it avoids the influence of common source inductance. As a result, the voltage overshoot triggered by the large di/dt in CCL will not affect the driver signal.

B. Parasitic Inductance Modeling In CCL For Multichip Power Module

The equivalent circuit of CCL is shown in Fig. 2. In [6], each parallel branch can be regarded as an independent conductive path while regarding the power loop as an open circuit. However, from the perspective of electric potential, this assumption only fits the ideal situation that the performance of each parallel chip is completely consistent. In the actual layout, due to the asymmetry placement of parallel chips, the parasitic parameter varies. Therefore, it is

unreasonable to take the adjacent chips in upper and lower bridge as a parallel commutation loop assuming the current passing through S_6 comes from S_1. The current through the upper bridge will be redistributed at the equipotential of the lower bridge. Fig. 3 shows the balanced current distribution of CCL, which verifies the connections between upper bridge bonding wires and DBC attached to lower chips are equipotential.

In order to accurately analyze the parasitic inductance in the power module, it is necessary to evaluate the self-inductance and mutual inductance of each part. Simplifying the bonding wire to a copper wire with a diameter of 0.36mm and a height of 15mm, the self-inductance is 12.37nH. With a copper wire of the same size arranged in parallel, the mutual inductance between the two wires is 6.72nH with the distance of 1.28mm. Three different arrangements are shown in Fig. 4 to explore the effect of spatial location on electromagnetic coupling. And the direction of current determines the sign of the mutual inductance. When two copper wires are arranged vertically, the mutual inductance is 0, which indicates no electromagnetic interaction.

$$ L_{ij} = \frac{\mu_0}{4\pi} \frac{1}{a_i a_j} \cos\theta \iiint_{b_i}^{c_i} \int_{b_j}^{c_j} \frac{dI_i \, dI_j}{r_{ij}} da_i \, da_j \qquad (1) $$

a_i and a_j are the infinite element of two conductor cross-sections; I_i and I_j are the current vector in the micro-element; (b_i, c_i) and (b_j, c_j) are the start and end positions of the two wires; r_{ij} is the distance between micro-elements.

As for the power module, since the current path shown in Fig. 5 illustrates that the current flows in parallel, mutual inductance of each part should be taken into considerations.

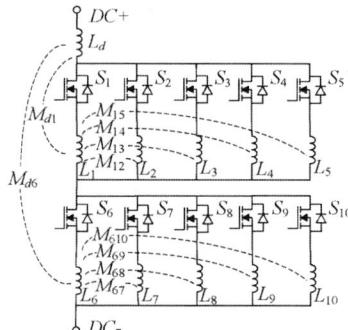

Fig. 2. Equivalent circuit of the commutation loop with mutual inductance influence on S_1 path and S_6 path

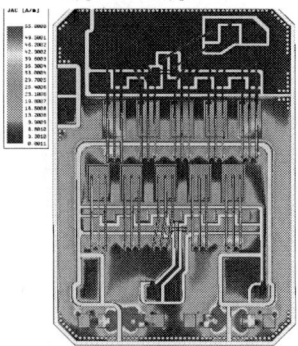

Fig. 3. Current distribution of CCL

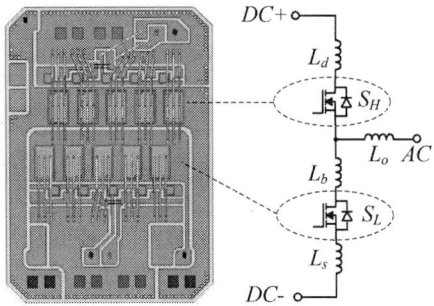

Fig. 1. 1200V/600A SiC MOSFET power module layout and circuit structure

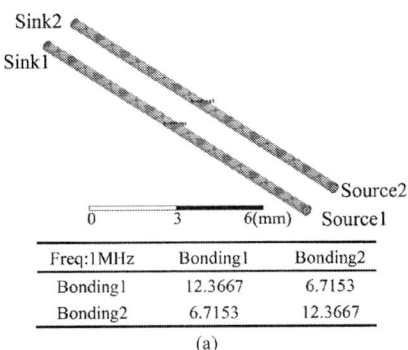

Freq:1MHz	Bonding1	Bonding2
Bonding1	12.3667	6.7153
Bonding2	6.7153	12.3667

(a)

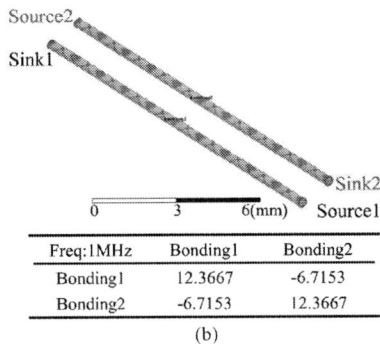

Freq:1MHz	Bonding1	Bonding2
Bonding1	12.3667	-6.7153
Bonding2	-6.7153	12.3667

(b)

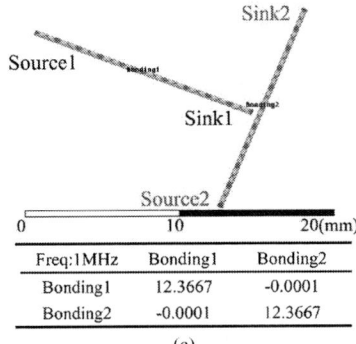

Freq:1MHz	Bonding1	Bonding2
Bonding1	12.3667	-0.0001
Bonding2	-0.0001	12.3667

(c)

Fig. 4. The influence of space position and current direction on mutual inductance: (a) same direction current; (b) reversed direction current; (c) vertical current.

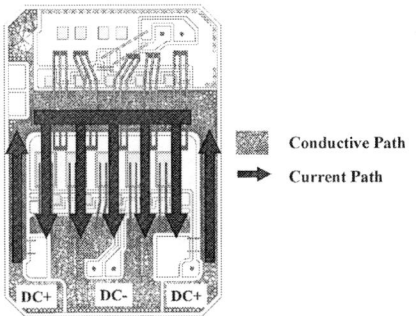

Fig. 5. CCL current path.

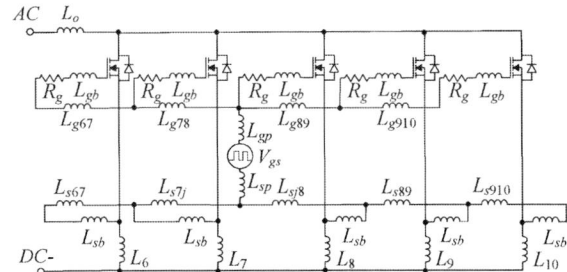

Fig. 6. Unbalanced inductance distribution in the low-bridge drive circuit.

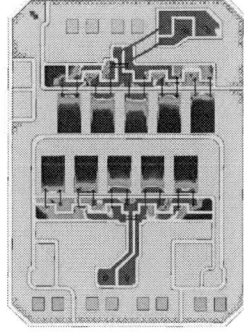

Fig. 7. Current distribution of driver loop

C. Parasitic Inductance Modeling In Driver Loop For Multichip Power Module

Despite the stray inductance in the driver loop has little influence on steady-state current performance, it would result in the signal propagation delays and trigger oscillation. It also has negative feedback on v_{gs}. The equivalent circuit of the lower bridge driver loop circuit derived from the DBC layout is shown in Fig. 6.

It is obvious that the driver path of S_8 is shorter than any other parallel chips. The largest driving inductance difference from S_6 and S_8 is close to 4nH. This would cause different switching characteristics of individual chips, which is not conducive to parallel current sharing in transient time. Fig. 7 shows the current distribution of the driver loop. In spite of ignoring the CCL mutual inductance and the potential switching interference at the AC terminal, the unbalanced phenomenon is conspicuous.

III. MATHEMATICAL MODELING

A. CCL Parasitic Inductance Mathematical Modeling

The power module can be divided into three parts: Drain_track, Bonding_track and Source_track. Meshing of each part is shown in Fig. 8. The stray inductance in the module can be described as:

$$L_P = L_{terminal} + L_{CCL} \qquad (2)$$

$$L_{CCL} = L_d + L_b + L_s \qquad (3)$$

Where, L_P refers to the total stray inductance of the power module; $L_{terminal}$ is the inductance of the DC terminals; L_{CCL} is the total stray inductance of CCL; L_d, L_b and L_s refer to the inductance of drain_track, bonding_track and source_track, respectively.

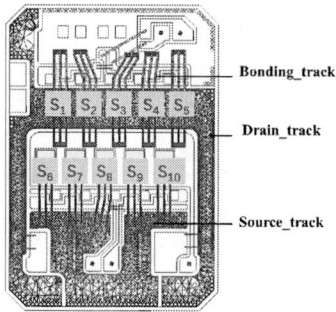

Fig. 8. current path meshing divided in CCL

ANSYS Q3D is used to extract inductance parameters. In order to characterize the mutual inductance influence of each part, the total parasitic inductance of CCL can be expressed as a matrix:

$$
L_{CCL}{}' = \begin{bmatrix} L_d & M_{db} & M_{ds} \\ M_{bd} & L_b & M_{bs} \\ M_{sd} & M_{sb} & L_s \end{bmatrix}
$$

$$
= \begin{bmatrix} 3.006 & -0.093 & -0.578 \\ -0.093 & 0.678 & 0 \\ -0.578 & 0 & 1.780 \end{bmatrix} \text{nH} \quad (4)
$$

Where, M_{ij} refers to the mutual inductance between i_track and j_track. Taking 5 MOSFETs chips paralleled into consideration, it is necessary to expand the paralleled parts to evaluate the influence of parasitic parameters, which can be modified as:

$$
L_b = \begin{bmatrix} L_{b1} & M_{b12} & M_{b13} & M_{b14} & M_{b15} \\ M_{b21} & L_{b2} & M_{b23} & M_{b24} & M_{b25} \\ M_{b31} & M_{b32} & L_{b3} & M_{b34} & M_{b35} \\ M_{b41} & M_{b42} & M_{b43} & L_{b4} & M_{b45} \\ M_{b51} & M_{b52} & M_{b53} & M_{b54} & L_{b5} \end{bmatrix}
$$

$$
= \begin{bmatrix} 2.8603 & 0.2771 & 0.0653 & 0.0653 & 0.0170 \\ 0.2771 & 2.8376 & 0.2695 & 0.0621 & 0.0271 \\ 0.0653 & 0.2695 & 2.8365 & 0.2695 & 0.0659 \\ 0.0653 & 0.0621 & 0.2695 & 2.8390 & 0.2777 \\ 0.0170 & 0.0271 & 0.0659 & 0.2777 & 2.8638 \end{bmatrix} \text{nH} \quad (5)
$$

$$
L_s = \begin{bmatrix} L_{s6} & M_{s67} & M_{s68} & M_{s69} & M_{s60} \\ M_{s76} & L_{s7} & M_{s78} & M_{s79} & M_{s70} \\ M_{s86} & M_{s87} & L_{s8} & M_{s89} & M_{s80} \\ M_{s96} & M_{s97} & M_{s98} & L_{s9} & M_{s90} \\ M_{s06} & M_{s07} & M_{s08} & M_{s09} & L_{s10} \end{bmatrix}
$$

$$
= \begin{bmatrix} 5.7249 & 1.3395 & 0.8313 & 0.4832 & 0.4752 \\ 1.3395 & 5.2951 & 1.3589 & 0.5484 & 0.4774 \\ 0.8313 & 1.3589 & 5.6254 & 0.9498 & 0.5624 \\ 0.4832 & 0.5484 & 0.9498 & 5.3496 & 1.3544 \\ 0.4752 & 0.4774 & 0.5624 & 1.3544 & 5.7377 \end{bmatrix} \text{nH} \quad (6)
$$

Where, M_{bij} refers to the mutual inductance between the current path of S_i and S_j in bonding_track; M_{sij} refers to the mutual inductance between the current path of S_i and S_j in source_track. As is shown from (5) (6), there is little difference between the parasitic inductance of parallel MOSFETs. Therefore, it verifies the balanced current sharing shown in Fig. 3.

The voltage distribution of the single-phase Layout in the module can be expressed as:

$$
\begin{bmatrix} V_d \\ V_b \\ V_s \end{bmatrix} = L_{CCL}{}' \cdot d \begin{bmatrix} i_d \\ i_b \\ i_s \end{bmatrix} / dt \quad (7)
$$

Where, V_d, V_b and V_s refer to the voltage of drain_track, bonding_track and source_track, respectively; i_b refers to the current of L_b ; i_s refers to the current of L_s.

$$
i_b = \begin{bmatrix} i_{b1} \\ i_{b2} \\ i_{b3} \\ i_{b4} \\ i_{b5} \end{bmatrix}; i_s = \begin{bmatrix} i_{s6} \\ i_{s7} \\ i_{s8} \\ i_{s9} \\ i_{s10} \end{bmatrix} \quad (8)
$$

B. Driver Loop Parasitic Unbalanced Circuit Modeling And Switching Characteristics Analysis

In order to study the effect of unbalanced parasitic inductance distributed in driver loop, the current paths of S_6 and S_8 are selected to analyze this asymmetry. Fig. 9 depicts the double-pulse circuit schematic built in Saber to validate the current sharing difference. The parameters of S_6 and S_8 in power loop remain the same. L_{ge} and L_{se} refer to parasitic inductance brought by the additional driver path of S_6.

$$
L_{ge} = L_{g67} + L_{g78} \quad (9)
$$

$$
L_{se} = L_{s67} + L_{s7j} \quad (10)
$$

With simultaneous application of double pulse test for S_6 and S_8, the transient waveforms can be obtained as Fig. 10. It can be seen that the unevenness of the parallel MOSFETs during transient time is particularly severe when MOSFETs turn off. Compared with S_6, S_8 turns off earlier. S_8 reaches the Miller stage first, the voltage drop on S_8 will trigger the current oscillation of the other chip. Since S_6 and S_8 are not switching at the same time, Discharge of gate-source capacitor is out of synchronization. As the initial state of switching process is controlled by v_{gs}, it is necessary to investigate v_{gs} behaviors during the transient time. As the driver circuit depicted in Fig. 6, v_{gs} of S_6 and S_8 can be described as:

$$
v_{gs6} = V_{gs} + \left(L_{gp} + L_{sp} \right) \frac{di_g}{dt} + L_{g78} \frac{di_{g78}}{dt}
$$
$$
+ R_g i_{g6} + \left(L_{gb} + L_{g67} + L_{sb} + L_{s67} \right) \frac{di_{g6}}{dt} \quad (9)
$$

$$
v_{gs8} = V_{gs} + \left(L_{gp} + L_{sp} \right) \frac{di_g}{dt} + R_g i_{g8}
$$
$$
+ \left(L_{gb} + L_{sb} + L_{s8j} \right) \frac{di_{g8}}{dt} \quad (10)
$$

Equation (9) and (10) reveal the relationship between the gate resistor and driver loop inductance of S_6 and S_8. After receiving a low-level signal, the first behavior of the turn-off process is gate-source capacitance (C_{gs}) discharge. When V_{gs} reaches the Miller voltage, the voltage of C_{gs} remains unchanged, and the Miller capacitance begins to increase. During this process, MOSFET remains conductive. Drain current (I_D) would not change rapidly until Miller capacitance is fully charged. The voltage generated by the

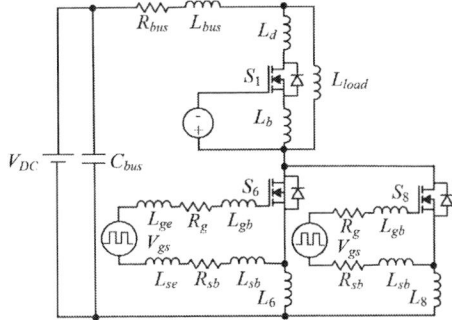

Fig. 9. Schematic diagram of the double pulse tester

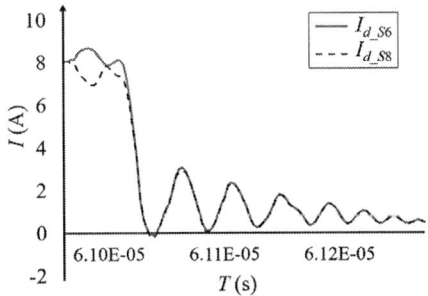

Fig. 10. Turn-off and turn-on current waveforms of S_6 and S_8

asymmetry layout of the driver loop would superimpose to v_{gs}. As the parasitic inductance on S_6 path is larger than S_8 path, there is $v_{gs6} > v_{gs8}$. Since the relationship between I_D and v_{gs} is positive, Fig. 10 verifies this trend with the drain current of S_6 and S_8.

Due to the difference of parasitic inductance on driver loop is around 4nH, it is reasonable to assume $di_{g6}/dt \approx di_{g8}/dt$. Thus, the above can be satisfied as:

$$L_{g78}\frac{di_{g78}}{dt} + R_g i_{g6} + \left(L_{g67} + L_{s67} - L_{s8j}\right)\frac{di_{g6}}{dt} > R_g i_{g8} \quad (12)$$

Apparently, $L_{g67} + L_{s67} - L_{s8j}$ remains constant determined by the physical structure layout. In order to balance the current distribution during transient time, v_{gs} of each MOSFETs needs to be consistent. v_{gs6} decreases by reducing the value of R_{g6}. At the same time, the smaller the R_g, the faster the di/dt. Therefore, the time delay caused by the DBC layout can be corrected by changing the R_g of the parallel chip. After reducing the gate resistance of S_6 from 5Ω to 4.7Ω, the unbalanced signal triggering problem in parallel chips has diminished, which is shown in Fig. 11. As a result, it is necessary to consider the gate inductance influence and adjust the gate resistance to achieve the consistency of the parallel chips.

IV. EXPERIMENTS

Since the parallelled SiC MOSFET chips are packaged inside the power module, the switching characteristics of a single chip cannot be evaluated. In order to further verify the dynamic performance of the half-bridge topology SiC MOSFET power module, a power semiconductor module dynamic performance test platform is built for testing and analysis. The double pulse test mechanism is used to extract the switching characteristics of the power module. Several

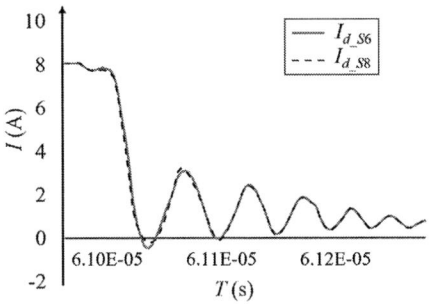

Fig. 11. Turn-off waveforms of S_6 and S_8 after adjustment

tests have been taken to evaluate the performance of the module. The connections during module testing are shown in Fig. 12.

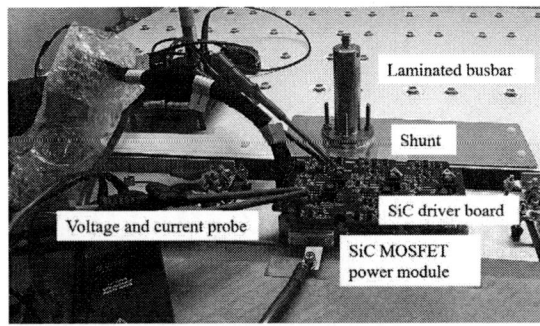

Fig. 12. SiC MOSFET power semiconductor module test experiment setup

A. Parasitic Inductance Validation

In section II B, by extracting the stray inductance in ANSYS Q3D, the theoretical total inductance of the module is 14.12nH through analysis and calculation. From the data in TABLE I, it can be seen that the measured stray inductance under different test conditions is around 31nH. Excluding the stray inductance of 18.21nH introduced by the experimental platform, the parasitic inductance of the commutation circuit plus the terminal section is within 10% of the error of the theoretical analysis, which is acceptable in practical application. This result validates the electromagnetic analysis of the power module in section II B.

TABLE I. COMPARISON OF MODULE PARASITIC PARAMETERS UNDER DIFFERENT TEST CONDITIONS

Test Conditions	Stray Inductance		
	L_{ex}(nH)	$L_{CCL}+L_{terminal}$(nH)	Error
330V-400A	32.69	14.38	1.81%
350V-400A	32.06	13.75	-2.69%
370V-400A	31.57	13.26	-6.49%
390V-400A	31.36	13.05	-8.20%
400V-250A	32.57	14.26	0.98%
400V-400A	30.33	13.57	-4.05%

a. L_{ex} refers to the experimental inductance of the total setup.

B. Voltage Spikes Validation

When characterizing the dynamic performance of power modules, voltage overshoot is important, as both sides of the

TABLE II.   COMPARISON OF VOLTAGE SPIKES UNDER DIFFERENT TEST CONDITIONS

Test Conditions	Voltage Spikes				
	V_{max}(v)	L_{ex}(nH)	$DiDtOFF$(kA/us)	$V_{max(calculate)}$(v)	Error
330V-400A	667.91	32.69	10.26	665.3994	0.38%
350V-400A	688.03	32.06	10.44	684.7064	0.49%
370V-400A	703.25	31.57	10.44	699.5908	0.52%
390V-400A	719.72	31.36	10.43	717.0848	0.37%
400V-250A	663.66	32.57	7.7	650.789	1.98%
400V-400A	726.22	30.33	10.07	721.0316	0.72%

stray inductor generate large additional voltages during fast switching, and these voltage applied to the SiC MOSFET will lead to device overvoltage or even damage.

During the test, the V_{max} measured by the voltage probe is the drain-source voltage V_{DC} of SiC MOSFETs. The voltage spike is the additional voltage generated on the parasitic inductor due to the rapid change of the switching process current, while V_{DC} reflects the superposition of this voltage and the bus voltage

$$V_{max} = V_{peak} + V_{DC} = L_p \frac{di}{dt} + V_{DC} \qquad (13)$$

By testing the effect of different input voltages on voltage spikes at the same current level, it can be visualized that the voltage spikes will increase with the voltage level from Fig. 13.

It can be seen from this data that the current change rate of 10.4 A/ns at the same current level of 400A is kept constant during the shutdown stage, the additional voltage spike V_{peak} is basically unchanged, and the change of voltage peak V_{max} is mainly from the change of voltage level of DC voltage source V_{DC}. When the voltage levels are all 400 V, the main factor affecting the voltage peak is the rate of change of the turn-off current. When the turn-off speed increases, the voltage spike increases, so for the high-speed switching characteristics of SiC power semiconductor devices, the switching speed of the device is usually reduced to prevent voltage overcharge. The experimental results match the theoretical analysis of (13), characterize the influencing factors of turn-off overvoltage, and provide a reference for the design of silicon carbide power devices.

V. CONCLUSIONS

In this paper, the switching characteristics of the 1200V/600A SiC MOSFET power semiconductor module is investigated. The parasitic parameter model of the power module package structure is proposed. The parasitic inductance of the power circuit and the signal circuit is analyzed in detail. The source of each parasitic inductance is analyzed, the mutual inductance is included in the electromagnetic modeling, and the parallel unevenness caused by the asymmetry of the signal circuit of the lower bridge arm is modeled and analyzed. Mathematical derivation establishes the influence of gate resistance and miscellaneous inductance on switching characteristics. A method is proposed to solve the problem of inconsistent switching characteristics of parallel chips by adjusting the gate resistance. The dynamic performance of the power semiconductor module is characterized by experimental

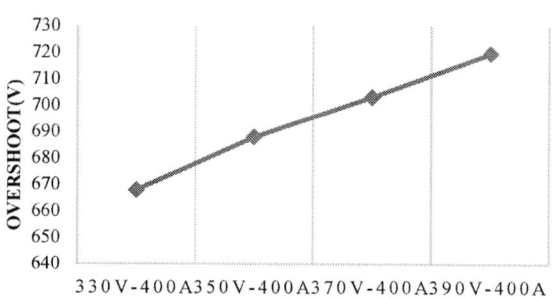

Fig. 13. Voltage overshoot under different voltage test conditions at the same current level

testing, which provides the guidelines for package structure, material selection, layout design and circuit application.

REFERENCES

[1]   S. Li, L. M. Tolbert, F. Wang and F. Z. Peng, "Stray Inductance Reduction of Commutation Loop in the P-cell and N-cell-Based IGBT Phase Leg Module," in *IEEE Transactions on Power Electronics*, vol. 29, no. 7, pp. 3616-3624, July 2014.

[2]   K. Takao and S. Kyogoku, "Ultra low inductance power module for fast switching SiC power devices," 2015 IEEE 27th International Symposium on Power Semiconductor Devices & IC's (ISPSD), Hong Kong, 2015, pp. 313-316,.

[3]   Z. Chen, Y. Yao, D. Boroyevich, K. Ngo and W. Zhang, "An ultra-fast SiC phase-leg module in modified hybrid packaging structure," 2014 IEEE Energy Conversion Congress and Exposition (ECCE), Pittsburgh, PA, 2014, pp. 2880-2886.

[4]   P. Beckedahl, M. Spang, and O. Tamm, "Breakthrough into the third dimension—Sintered multi layer flex for ultra low inductance power modules," in *Proc. 8th Int. Conf. Integr. Power Syst. (CIPS)*, Nuremberg, Germany, pp. 1–5, Feb. 2014.

[5]   F. Yang, Z. Wang, Z. Liang, and F. Wang, "Electrical performance advancement in SiC power module package design with Kelvin drain connection and low parasitic inductance," in *IEEE Journal of Emerging and Selected Topics in Power Electronics*, vol. 7, no. 1, pp. 84–98, Mar. 2019.

[6]   B. Zhang and S. Wang, "Parasitic Inductance Modeling and Reduction for a Wire Bonded Half Bridge SiC MOSFET Multichip Power Module," *2019 IEEE Applied Power Electronics Conference and Exposition (APEC)*, Anaheim, CA, USA, 2019, pp. 656-663.

[7]   J. Ke, S. Huang, Z. Yuan, Z. Zhao, X. Cui, S. S. Ang, and Chen, Z. , "Investigation of Low-Profile, High-Performance 62-mm SiC Power Module Package." in *IEEE Journal of Emerging and Selected Topics in Power Electronics*, 2020.

[8]   L. Wang, Z. Zeng, P. Sun, et al. "Current-Bunch Concept for Parasitic-Oriented Extraction and Optimization of Multi-Chip SiC Power Module." in *IEEE Transactions on Power Electronics*, vol. 36, no. 8, pp. 8593-8599, Aug. 2021.

[9] H. Li et al., "Influences of Device and Circuit Mismatches on Paralleling Silicon Carbide MOSFETs," in *IEEE Transactions on Power Electronics*, vol. 31, no. 1, pp. 621-634, Jan. 2016.

[10] Y. Xue, J. Lu, Z. Wang, L. M. Tolbert, B. J. Blalock, and F. Wang, "Active current balancing for parallel-connected silicon carbide MOSFETs," in *Proc. IEEE Energy Convers. Congr. Expo.*, 2013, pp. 1563–1569.

[11] A. Jauregi, D. Garrido, I. Baraia-Etxaburu, A. Garcia-Bediaga, and A. Rujas, "Static Current Unbalance of Paralleled SiC MOSFET Modules in the Final Layout." *2020 IEEE Vehicle Power and Propulsion Conference (VPPC)* IEEE, 2020.

Multiple UIS Ruggedness of 1200V Asymmetric Trench SiC MOSFETs

Jiayue Liu
School of Electronic Science and
Engineering, University of Electronic
Science and Technology of China
Chengdu, China

Xiaochuan Deng*
School of Electronic Science and
Engineering, University of Electronic
Science and Technology of China
xcdeng@uestc.edu.cn

Xu Li
School of Electronic Science and
Engineering, University of Electronic
Science and Technology of China
Chengdu, China

Xuan Li
School of Electronic Science and
Engineering, University of Electronic
Science and Technology of China
Chengdu, China

Zhiqiang Li
Microsystem and Terahertz Research
Center, China Academy of Engineering
Physics
Mianyang, China

Hongling Lu
China Aerospace Science and
Technology Corporation, The No. 771
Institute
Xi'an, China

Abstract—In this paper, 1200V commercial asymmetric trench silicon carbide (SiC) metal–oxide–semiconductor-field-effect-transistors (MOSFETs) are investigated by experiment under multiple unclamped inductive switching (UIS) events. The degradation of electrical characteristics and failure mode are evaluated under various avalanche energy ratio. Meanwhile, cross-section of damage point of failure device is identified using Focus Ion Beam (FIB). Under low energy ratio condition, the on-resistance (R_{on}) increases ~11% whereas the threshold voltage (V_{th}) remains almost constant, which indicates that thermal fatigue is main reason of failure. Under high energy ratio condition, burn out of field oxide and metal Al is observed, which is caused by high thermal stress.

Keywords—asymmetric trench, SiC MOSFET, multiple UIS, failure mode

I. INTRODUCTION

Due to their better material characteristics, SiC MOSFET shows its desirable performances in high power electronics fields. Its excellent features, high thermal conductivity, low on-state resistance, low switching loss, fast switching speed can bring the superior efficiency and power density of power systems [1]. However, high power density also causes reliability problems in devices. In addition the poor quality of the gate oxide limits the long-term reliability of SiC MOSFETs [2,3]. The interface state between SiC and SiO_2 seriously affects the electrical characteristics of devices [4]. The effect becomes more serious in trench SiC MOSFETs. Therefore, it's important to assess the reliability of SiC MOSFETs under extremely stress conditions.

While using the MOSFETs in the circuits with inductive load, energy saved in the load will force the off-state MOSFET breakdown to form a current path. This is so-called unclamped inductive switching (UIS) [5]. The Si-based power MOSFETs usually failure under avalanche breakdown which is attributed to the activation of parasitic bipolar transistors and subsequent secondary breakdown. After switching off the power, due to the suddenly change of the drain voltage, the avalanche current flows through the reversed PN junction[7]. With the increase of the current, the parasitic bipolar junction transistor (BJT) turns on, causing thermal failure of the MOSFET. Owing to the significant difference between Si and SiC, avalanche breakdown mechanism of SiC MOSFETs can be different greatly with that of Si-based MOSFETs.

With repeated switching, the device will suffer repetitive UIS stress. Under avalanche conditions, devices failure may be due to the high electro-thermal stress. Up to now, there are many reports of UIS ruggedness about planar and double trench SiC MOSFETs [8,9]. Compared with planar SiC MOSFETs, trench SiC MOSFETs have lower on-resistance and higher power density, therefore, trench SiC MOSFETs can make better use of the characteristics of silicon carbide materials. However, the trench gate SiC MOSFETs are suffer from serious gate oxide reliability trouble [10]. Above all, the critical strength of electric field SiC MOSFETs is approximately ten times larger than that of Si MOSFETs, which makes the gate oxide layer of SiC MOSFETs is easy to get to its reliability limit (~3MV/cm) in accordance the Gauss Law [11,12]. Moreover, the SiC/SiO_2 interface is poor, and this matter causes the electrical characteristics of trench SiC MOSFETs to be unreliable [13,14]. Last, the unevenness of oxidation process of SiC MOSFETs leads to a thinner SiO_2 layer thickness at the bottom compare with the thickness at the sidewalls of the trench, that could worsen gate oxide stability of trench SiC MOSFETs. Thence, in order to solve many problems that cause the instability of the gate oxide layer, such as, the gate oxide electric field crowding at the trench bottom, an asymmetric structure is employed to solve [15].

The few investigations of asymmetric trench structure about single UIS event are reported [4,16]. In fact, the device will not failure after single UIS event, if the single avalanche energy does not exceed the maximum value [17]. And then, it is essential to study the electrical degradation of asymmetric trench MOSFETs caused by multiple UIS events.

In this work, the multiple UIS ruggedness of 1200V asymmetric trench SiC MOSFET is investigated. Meanwhile, the electrical degradation and failure modes are evaluated under various avalanche energy ratio. In addition, the cross-section of damage point is performed by FIB cut, which reveals the failure mechanism in the device.

II. STRUCTURE AND TEST METHODPLOGY

The device under test (DUT) is a 1200V asymmetric trench SiC MOSFET. Fig. 1(a) demonstrates the cross-sectional structure of the MOSFET. The specific on-resistance ($R_{on} \times A$) is 4.2m$\Omega \cdot$cm^2 (V_{gs}=15V). The threshold voltage (V_{th}) and drain-source breakdown voltage(BV) are 4.2V and 1375V, respectively. The test temperature is 25 °C. V_{th} is defined as the gate-source voltage when I_{ds}=10mA with V_{gs} =V_{ds}. The breakdown voltage is measured under the condition that gate-source voltage (V_{gs}) and drain-source current (I_{ds}) equals to 0V and 1mA, respectively. The schematic diagram of the equivalent circuit and test platform are shown in Fig. 1(b) and (c). To provide energy for the circuit, two parallel connected 470μF capacitors are coupled with a high voltage power source. The inductor L in the load loop is selected to be 3.6mH. A 1700V Si IGBT module is applied for protect the whole circuit during repetitive UIS test.

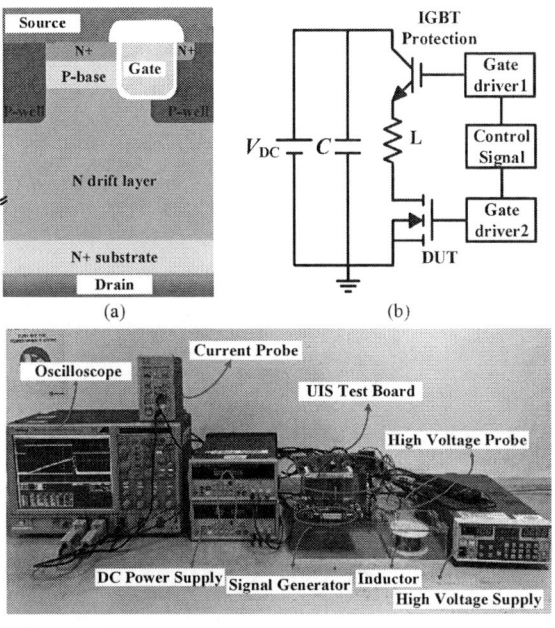

(a)   (b)

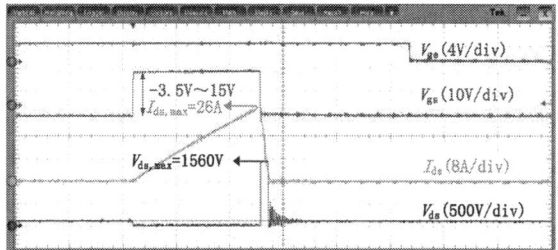

(c)

Fig. 1. (a) Schematic cross-section of asymmetric trench SiC MOSFET, (b) Diagram of equivalent circuit and (c) Test bench.

In the circuit, the DUT is used as a switch to control the charging and discharging of the inductor. When the IGBT turns on, the DUT turns on and the current in the inductor gradually increases. After the device turns off, due to the continuity of the current in the inductor, the current in the circuit cannot reduces to zero immediately, the energy saved in the inductor flows into the DUT, and the avalanche occurs. The avalanche state will continue until the energy saved in the inductive load is completely dissipated. If the energy goes beyond the maximum value that the device can withstand, the DUT will completely fail. After the energy is completely released by the DUT, the IGBT module turns off, and assures the safety of the test bench in off-state, particularly after the avalanche failure of the MOSFET [9,10].

III. EXPERIMENT RESULT

A. UIS Waveforms of DUTs

The avalanche energy (E_{AV}) is calculated by the V_{ds} and I_{ds} of the MOSFET during avalanche regime time (t_{AV}) as

$$E_{AV} = \int_0^{t_{AV}} V_{ds} I_{ds} dt \qquad (1)$$

where I_{ds} and V_{ds} are the current and terminal drain–source voltage of DUT, respectively.

To determine the maximum single avalanche energy (E_{AV}) of the DUT, a single UIS test is implemented at the beginning. The DC power supply voltage (V_{DD}) is 100V. The resistance

between the gate driver and the DUT is set to 20Ω. Turn-off voltage of the device is set to -3.5V. The turn-on voltage of the trench MOSFET is set to 15V.

The typical experimental results of the last test before avalanche failure for the SiC MOSFET is shown in Fig. 2. With the increase of pulse width, the peak avalanche current is gradually increases until the MOSFET reaches its point of avalanche failure. As can be summarize from Fig. 2, during avalanche, the maximum drain current and the voltage between the drain and source is 26A and 1560V, respectively. The t_{AV} is around 64μs.

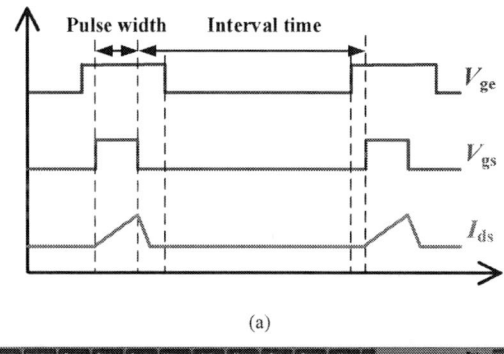

Fig. 2. Single UIS waveforms at the last test before failure.

Fig. 3 shows the schematic and actual results of multiple UIS test of the asymmetric trench SiC MOSFET. The V_{gs} is set from -3.5 to +15V for the DUT. The avalanche energy of each cycle under repetitive UIS test is set to 20% and 40% of E_{av} in single UIS. The interval time between each cycle is set to 500ms, which is enough for the device to dissipate heat to room temperature.

The peak avalanche current gradually increases until the MOSFET get its avalanche failure point. It can be concluded from Fig. 3(b), the maximum drain current and the voltage between the drain and source are 16A and 1550V, respectively. The maximum V_{ds} is almost same with the single UIS test. During repeated UIS tests, the heat sink and fan are used to cool the DUT.

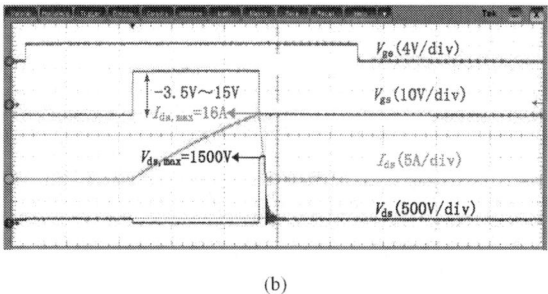

(a)

(b)

Fig. 3. Waveforms of multiple UIS for asymmetric trench SiC MOSFET: (a) schematic and (b) experimental.

B. Electrical Parameters Degradation under Low Energy Ratio

The energy saved in the inductor will be dissipated when the device occurs the avalanche breakdown. In the process of the UIS test, the electrical characteristics of the device change significantly. The energy dissipation leads to avalanche failure which damages the internal structure of the device.

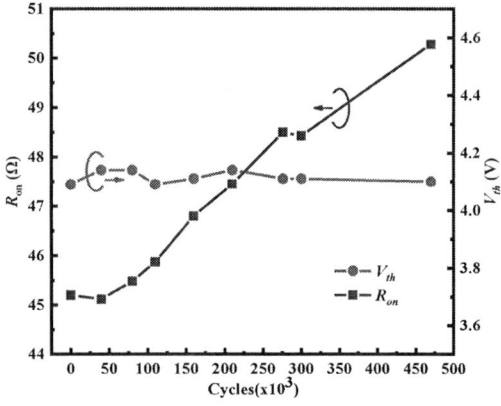

Fig. 4. Degradation of electrical parameters with the increasing of UIS cycles under 20% energy ratio.

Fig. 4 shows the degradation of electrical parameters under multiple UIS test of relative low energy ratio (20%). The R_{on} increases ~11% while V_{th} keeps almost constant with the increasing of UIS cycles. This means that the failure mechanism of devices is thermal fatigue under low energy ratio. During the repeated UIS test, the temperature suddenly increases, causing damage to the wire bonds and pads [10]. Although low energy will produce a relatively low temperature in the MOSFET, the repetitive fluctuation of temperature can also cause thermal fatigue between the wire bonds and source pad.

C. Electrical Parameters Degradation under High Energy Ratio

TABLE I shows the three terminals' impedance of the failed devices. Such a low impedance between the gate and source indicates that the gate-source terminal has failed in the MOSFET. The failure mode is the short connect between gate and source terminal.

TABLE I. THREE TERMINALS' IMPEDANCE OF ASYMMETRIC TRENCH SIC MOSFET AFTER MULTIPLE UIS

Energy Ratio	R_{gs} (Ω)	R_{gd} (Ω)	R_{ds} (Ω)
40%	1.1k	4.2M	4.2M
60%	308	9M	9M

Fig. 5 shows the relation between energy ratio and cycles under repetitive UIS tests for the DUT. Obviously, the repetitive UIS test cycles decrease with the increasing of energy ratio. When energy ratio is set to 20%, the degradation of devices is caused by thermal fatigue. At 40% energy ratio, the DUTs failed with gate rupture. When energy ratio rises to 60%, the failure mode is same with 40% energy ratio.

Fig. 6 shows the cross-sectional view of the decapsulation SiC MOSFET and FIB cut. It shows the thermal point determination of devices under 40% energy ratio. There are

obvious burn out regions between metal Al and field oxide on polysilicon gate, and the thickness of part of the field oxide is even reduced by 50%.

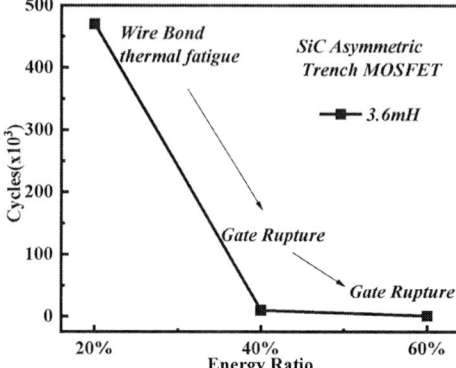

Fig. 5. Relation between energy ratio and cycles under repetitive UIS tests of the DUT

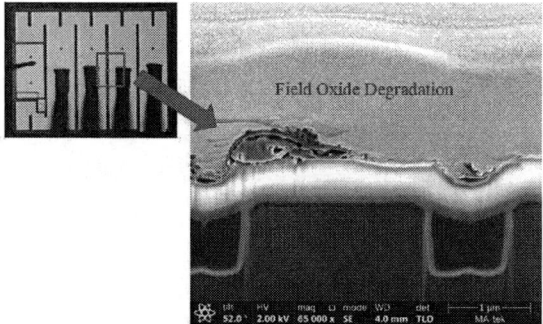

Fig. 6. Thermal point determination of asymmetric trench SiC MOSFET under 40% energy ratio

There is no abnormality on the surface of the die. EMMI and thermal analysis show that the damage point is located in the bottom area of the source finger. Obviously, the failure of SiC MOSFET is caused by thermal stress. Due to the heat sink and fan, the temperature caused by the avalanche drops quickly to room temperature after each UIS test. Because of different expansion coefficients of the two materials, temperature fluctuations can cause a thermal mismatch between the Al and the field oxide on the gate. The thermal stress causes field oxide degradation.

IV. CONCLUSION

Avalanche ruggedness is a very important feature of power MOSFETs. This article has investigated the avalanche capacity of the 1200V asymmetric trench SiC MOSFET through repeated UIS test. Using FIB cut and failure analysis methods, the static electrical characteristics and failure mechanism of the MOSFET are analyzed in detail. Under repetitive avalanche stress, the main failure mode has been verified in the asymmetric trench SiC MOSFETs. The degradation or failure of the SiC MOSFET is attributed to the thermal stress, which causes thermal fatigue between the wire bond and the source pad and burn out of the Al metal on the field oxide.

ACKNOWLEDGMENT

This work was supported in part by the National Key Research and Development Program of China under Grant 2017YFB0102302, in part by the Key Area Research and

978-1-6654-4817-8/21 $31.00 © 2021 IEEE

Development Project of Guangdong Province under Grant 2019B010127001, and in part by the Science Challenge Project under Grant TZ2018003-1-201.

REFERENCES

[1] X. Li, J. Jiang, A. Q. Huang, S. Guo, X. Deng, B. Zhang, and X. She, "A SiC Power MOSFET Loss Model Suitable for High-Frequency Applications," *IEEE Transactions on Industrial Electronics*, vol. 64, no. 10, pp. 8268-8276, 2017.

[2] S. Harada, Y. Kobayashi, K. Ariyoshi, T. Kojima, J. Senzaki, Y. Tanaka, and H. Okumura, "3.3-kV-Class 4H-SiC MeV-Implanted UMOSFET With Reduced Gate Oxide Field," *IEEE Electron Device Letters*, vol. 37, no. 3, pp. 314-316, 2016.

[3] Y. Wang, Y. Ma, Y. Hao, Y. Hu, G. Wang, and F. Cao, "Simulation Study of 4H-SiC UMOSFET Structure With p+-polySi/SiC Shielded Region," *IEEE Transactions on Electron Devices*, vol. 64, no. 9, pp. 3719-3724, 2017.

[4] X. Deng, X. Li, X. Li, H. Zhu, X. Xu, Y. Wen, Y. Sun, W. Chen, Z. Li, and B. Zhang, "Short-circuit Capability Prediction and Failure Mode of Asymmetric and Double Trench SiC MOSFETs," *IEEE Transactions on Power Electronics*, pp. 1-1, 2020.

[5] Trentin, Andrew, et al. "Study of a Silicon Carbide MOSFET Power Module to Establish the Benefits of Adding Anti-parallel Schottky Diodes." *2018 IEEE Energy Conversion Congress and Exposition (ECCE)* IEEE, 2018.

[6] S. Nida, B. Kakarla, T. Ziemann, U. Grossner, "Analysis of Current Capability of SiC Power MOSFETs Under Aval anche Conditions." *IEEE Trans. Electron Devices*, vol. 68, no. 9, pp. 4587–4592, Sep. 2021.

[7] S. Liu, X. Tong, J. Wei, and W. Sun, "Single-Pulse Avalanche Failure Investigations of Si-SJ-mosfet and SiC-mosfet by Step-Control Infrared Thermography Method," *IEEE Transactions on Power Electronics*, vol. 35, no. 5, pp. 5180-5189, 2020.

[8] J. Wei, S. Liu, H. Zhao, H. Fu, X. Zhang, S. Li, and W. Sun, "Verification of Single-Pulse Avalanche Failure Mechanism for Double-Trench SiC Power MOSFETs," *IEEE Journal of Emerging and Selected Topics in Power Electronics*, pp. 1-1, 2020.

[9] X. Deng, H. Zhu, B. Zhang , et al. "Investigation and Failure Mode of Asymmetric and Double Trench SiC MOSFETs under Avalanche Conditions." *IEEE Transactions on Power Electronics*, vol. 35, no. 8, pp. 8524-8531, 2020.

[10] T. Nguyen, A. Ahmed, T. V. Thang, and J. Park, "Gate Oxide reliability issues of SiCMOSFETs under short-circuit operation," *IEEE Trans. Power Electron*, vol. 30, no. 5, pp. 2445–2455, May 2015.

[11] Yao, Kailun , H. Yano , and N. Iwamuro . "Investigations of UIS Failure Mechanism in 1.2 kV Trench SiC MOSFETs Using Electro-Thermal-Mechanical Stress Analysis." *2021 33rd International Symposium on Power Semiconductor Devices and ICs (ISPSD)* 2021.

[12] M. A. Anders, P.M. Lenahan,C. J.Cochrane, andA. J. Lelis, "Relationship between the 4H-SiC/SiO2 interface structure and electronic properties explored by electrically detected magnetic resonance," *IEEE Trans. Electron Devices*, vol. 62, no. 2, pp. 301–308, Feb. 2015.

[13] M. Sagawa, H. Miki, Y. Mori, H. Shimizu, and A. Shima, "Evaluation of gate oxide reliability in 3.3 kV 4H-SiC DMOSFET with J-Ramp TDDB methods," in Proc. *IEEE 30th Int. Symp. Power Semicond. Devices ICs*, 2018, pp. 363–366.

[14] A. Agarwal, K. Han, and B. J. Baliga, "Analysis of 1.2 kV 4H-SiC trench-gate MOSFETs with thick trench bottom oxide," in Proc. *IEEE 6th Workshop Wide Bandgap Power Devices Appl.*, 2018, pp. 125–129.

[15] R. Siemieniec et al., "A SiC Trench MOSFET concept offering improved channel mobility and high reliability," in Proc. *19th Eur. Conf. Power Electron. Appl.*, 2017, pp. P.1–P.13.

[16] Fu, Hao , et al. "Degradation Investigations on Asymmetric Trench SiC Power MOSFETs Under Repetitive Unclamped Inductive Switching Stress." *2021 33rd International Symposium on Power Semiconductor Devices and ICs (ISPSD)* 2021.

[17] Z. Bai; X. Tang; S. Xie; Y. He; et al, "Investigation on Single Pulse Avalanche Failure of 1200-V SiC MOSFETs via Optimized Thermoelectric Simulation." *IEEE Trans. Electron Devices*, vol. 68, no. 3, pp. 1168–1175, Jan. 2021.

Design, Fabrication and Characterization of 6.5kV/100A 4H-SiC PiN power rectifier

Mengling Tao
School of Electronic Science and
Engineering, University of Electronic
Science and Technology of China
Chengdu, China

Xiaochuan Deng*
Institute of Electronic and Information
Engineering of UESTC in Guangdong
Dongguan, China
xcdeng@uestc.edu.cn

Rui Hu
Metrology Testing Center, China
Academy of Engineering Physics
Mianyang, China

Xuan Li
School of Electronic Science and
Engineering, University of Electronic
Science and Technology of China
Chengdu, China

Zhiqiang Li
Microsystem and Terahertz Research
Center, China Academy of Engineering
Physics
Mianyang, China

Hongling Lu
No. 771 Institute, China Aerospace
Science and Technology Corporation,
Xi'an, China

Abstract—**Silicon carbide (SiC) materials have excellent physical properties and a wide range application prospects in the high-voltage devices field. A high-voltage 6.5kV/100A 4H-SiC PiN rectifier is designed, fabricated and characterized in this paper. The termination adopts a combination of mesa structure and multi-ring auxiliary modulation junction termination extension (MAM-JTE). The MAM-JTE expands the ion implantation dose window and increases the breakdown voltage. Compared with the traditional double-zone JTE (DZ-JTE), MAM-JTE does not require additional process steps and masks. The two-dimensional numerical simulation tool TCAD Silvaco is used to design and optimize the termination structure. The 6.5kV SiC PiN power rectifier adopts high-voltage metal ceramic package and the active area of chip is 4.6×4.6mm^2. The device test results show that the breakdown voltage(BV) reaches 7.4kV, corresponding to 93% of an ideal parallel plane junction, the forward voltage drop(V_F) is 3.78V (corresponding to 100A forward current), the differential specific on-resistance($R_{ON,SP}$) is 2.5mΩ·cm^2 (corresponding to 100A/cm^2 forward current density), and the Baliga's figure of merit (BFOM) is 21.9GW/cm^2. In addition, the reverse recovery time is 166ns and softness factor is 1.23 at room temperature. The high termination efficiency shows that the termination structure combining the mesa and MAM-JTE is of great significance to the development of high-voltage SiC devices.**

Keywords—*4H-SiC, PiN power rectifier, MAM-JTE, fabrication and characterization, high-voltage device*

I. INTRODUCTION

Traditional silicon(Si) materials are limited to low critical breakdown electric field strength, low thermal conductivity. It is hard to be applied in high-voltage, high-frequency, and high-energy efficiency fields. Silicon carbide (SiC) has superior characteristics than Si in high-voltage fields [1-2]. Therefore, SiC has broad application prospects such as photovoltaic power generation, electric vehicles and aerospace [3-4].

Due to junction curvature effect, the edge termination design for high voltage and high current SiC devices is a difficult problem [5-7]. The JTE structure has been extensively used for power devices. However, the diffusion constants of the implanted aluminum ions in the silicon carbide is far less than silicon, so it is difficult for SiC devices to achieve a gradual change doping concentration in the lateral direction, thus, many previous studies have adopted many new structures to solve this problem[8-10]. Besides, the JTE

structure is strongly sensitive to the ion implantation dose. In order to overcome above shortcomings, previous literature have been reported spatial-modulation JTE and double reduced surface field JTE(DR-JTE) etc to increase the withstand voltage [11-14]. However, multiple mask ion implantation and multiple etching steps are required, which increases the difficulty and complexity of the process. Huang C et al proposed counter-doping JTE (CD-JTE) [15]. Its most important feature is the addition of n-type doping in the p-type JTE region. However, this cannot effectively increase the breakdown voltage in the low-implantation doses. Zheng-Xin Wen et al designed and fabricated the 10kV SiC p-IGBT with stepped spatial modulation JTE (SSM-JTE) termination [16]. The internal auxiliary ring can share the charge and alleviate the curvature effect, and the external auxiliary ring can reduce peak electric field at the end of the terminal, but SSM-JTE termination efficiency is only 77.4%. The FLRs termination requires numerous rings and large termination area [17-19], and it is necessary to design carefully rings spacing and number.

In this paper, 6.5kV 4H-SiC PiN rectifier is successfully fabricated. The edge termination adopts a new structure combined mesa and MAM-JTE. Compared with DZ-JTE structure, MAM-JTE structure improves the junction termination curvature effect and extends the implantation dose window. The fabricated high-voltage SiC PiN rectifier with MAM-JTE can achieve 7.4kV corresponding to 93% terminal efficiency. The BFOM is 21.9GW/cm^2. The differential specific on-resistance($R_{ON,SP}$) is only 2.5mΩ·cm^2 at 20A forward current (corresponds to 100A/cm^2 forward current density), the forward voltage is 3.78V (corresponds to 100A forward current). Additionally, the dynamic characteristics show that the reverse recovery time is 166ns and softness factor is 1.23 with 20A DC stress at room temperature. The design and fabrication of the 6.5kV/100A 4H-SiC PiN rectifier can be used as a reference for the research and development of high-voltage and high-current silicon carbide modules.

II. DESIGN AND SIMULATION

A. Device structure

There is schematic diagram of the 6.5kV 4H-SiC PiN rectifier with MAM-JTE structure in Fig. 1. The substrate is (0001) 4H-SiC with small off-angles of 4°, then three epitaxial layers are grown. The P++ ohmic layer concentration is

978-1-6654-4817-8/21 $31.00 © 2021 IEEE

$2\times10^{19}cm^{-2}$, and the thickness is 0.5μm, the 1μm P+ injection layer concentration is $5\times10^{18}cm^{-2}$, and the 55μm drift region concentration is 1×10^{15} cm^{-2}.

MAM-JTE includes three floating rings (MAM1) at the bottom of the mesa, and the other floating rings (MAM2 and MAM3) are placed at the junction of JTE1 and JTE2 and the end of JTE2, respectively. The each ring width is 3μm, and the ring spacing from the cell center to the termination direction is 10μm, 8μm and 6μm. As shown in Fig.1, the MAM-JTE termination is divided into five effective JTE dose doping regions. The termination doping concentration shows a downward trend, which can well alleviate the electric field in thethe main junction depletion region. This lateral variable doping (VLD) effect reduces the peak electric field at the edge to enhance the protection of the main junction and the transition region.

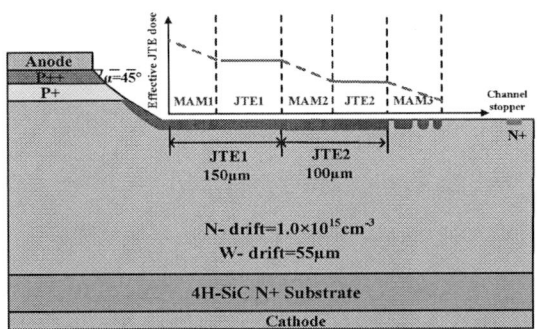

Fig. 1. Schematic diagram of the 6.5kV SiC PiN rectifier with MAM-JTE.

B. Simulation and Optimization

The structure design and dose optimization of the 6.5kV 4H-SiC PiN rectifier are optimized using semiconductor simulator TCAD Silvaco. Fig. 2 compares the dose window of DZ-JTE and MAM-JTE structure. It should be noted that the MAM-JTE dose window ($0.91\sim1.52\times10^{13}cm^{-2}$) to obtain the same target value (6.5kV) is more than twice that of DZ-JTE ($1.05\sim1.32\times10^{13}cm^{-2}$).

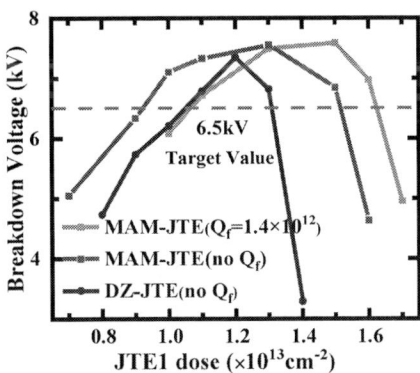

Fig. 2. Influence of interface charge on the implantation dose window.

Moreover, the MAM-JTE breakdown voltage is higher at multiple doses. The dose window of MAM-JTE with 1.4×10^{12} cm^{-2} interface charge (Q_f) is also shown in Fig. 2. With adding positive interface charge is equivalent to neutralize a part of the negative interface charge, and a higher dose is required to achieve the corresponding breakdown voltage. The dose ratio and length ratio of two implantions are 3:2. The JTE1 and JTE2 length is 150μm and 100μm, respectively. The mesa termination structure is used as an auxiliary mean to enhance electric field. The mesa height is 2μm and the angle is 45°.

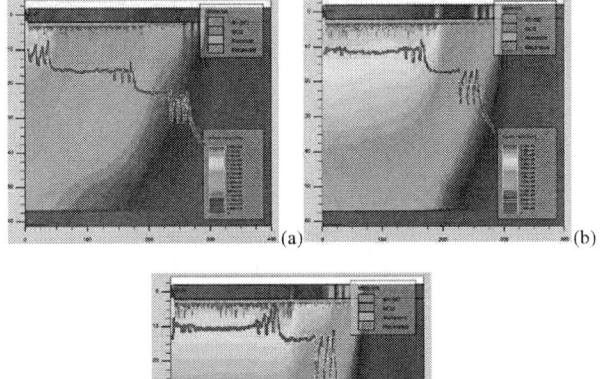

Fig. 3. Breakdown electric field in MAM-JTE with JTE1 dose of (a) $1.1\times10^{13}cm^{-2}$ (b) $1.3\times10^{13}cm^{-2}$ (c) $1.5\times10^{13}cm^{-2}$.

The breakdown electric field in MAM-JTE with different JTE1 doses in Fig. 3. The optimal JTE1 dose is $1.3\times10^{13}cm^{-2}$ and the surface electric field is uniformly distributed compared with the JTE1 doses of $1.1\times10^{13}cm^{-2}$ and $1.5\times10^{13}cm^{-2}$, there is no obvious electric field spike. The simplified fabrication process for 6.5kV SiC PiN rectifier is shown in Fig. 4. It is found that the process for MAM-JTE structure is the same with DZ-JTE structure.

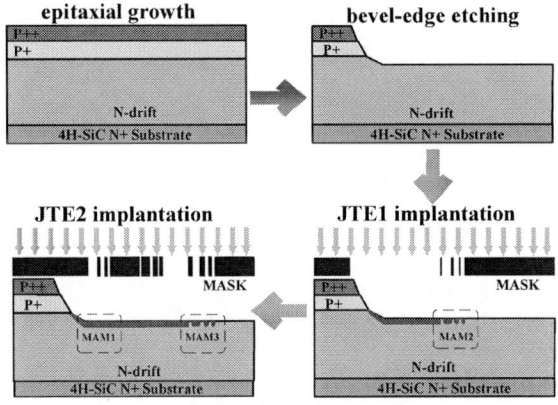

Fig. 4. Main process steps of MAM-JTE termination.

III. FABRICATION AND CHARACTERIZATION

A 6.5kV 4H-SiC PiN device with MAM-JTE is carried out on 4 inches N+ substrate with 0001 crystal plane. The parameters of the N− drift region, the P+ injection layer and the P++ ohmic contact layer are the same as the simulation parameters.

In the process experiment, the mesa is etched by inductively coupled plasma (ICP) technology, and a 3μm SiO$_2$ layer as the etching mask material is deposited by plasma enhanced CVD (PECVD). The multi-step implantation maximum energy is 500 keV at the 500°C temperature. The MAM-JTE total length is 288μm. According to the simulation results, the optimal JTE1 dose is from 1.1×10^{13}cm^{-2} to 1.3×10^{13}cm^{-2}. The implant dose of the MAM1 ring and the MAM3 ring dose is the same as JTE2, and the MAM2 ring implant dose is same as the JTE1, so that the process implantation step is not increased. After the implantation step completed, annealing activation is performed at 1800°C in an Ar environment for 10 minutes. In order to reduce the interface state and suppress the effect of surface leakage current, the first passivation layer adopts a 500Å thermally grown oxide and a 1μm thick SiO$_2$ film. The ohmic contact layer is made by annealing Ni/Ti/Al over a thousand degrees for two minutes. Last, a second passivation layer is formed by deposition of Si$_3$N$_4$ and polyimide. The critical breakdown characteristics of high-voltage devices depend on the surface passivation layer quality. A better quality passivation layer can greatly reduce the device surface defects, reduce leakage current during reverse bias and improve durability.

Fig. 5 (a) and (b) show pictures of the fabricated 6.5kV 4H-SiC PiN rectifier chip and wafer. The rectifier adopts a high-voltage metal ceramic package, and the metal package picture is shown in Fig. 5(c).

Fig. 5.   6.5kV SiC PiN rectifier: (a) chip; (b) wafer; (c) package

A. Static characteristics

The static parameters of the 6.5kV SiC PiN device are characterized by semiconductor parameter test equipment Agilent's B1505. Fig. 6 shows the test results of the device's forward characteristics. The chip active area is 4.6×4.6mm^2, and the forward voltage is 3.2V (current density is 100A/cm^2) at room temperature; The V$_F$ is 3.78V at 100A forward current. The differential specific on-resistance ($R_{ON,SP}$) is 2.5mΩ·cm^2 at 100A/cm^2 current density.

As shown in Fig. 7, the 4H-SiC PiN rectifier reverse characteristic display that the BV is up to 7.4kV@10μA, which achieving 93% of ideal parallel plane junction. It exhibits excellent reverse characteristic and high BFOM, corresponding to 21.9GW/cm^2.

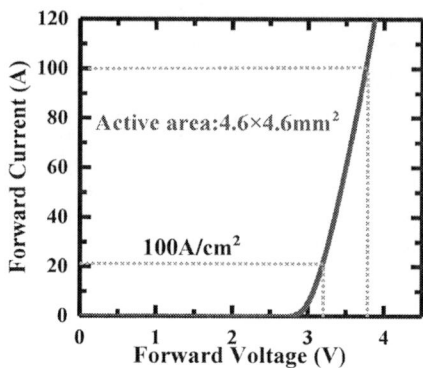

Fig. 6.   6.5kV 4H-SiC PiN power rectifier forward characteristic

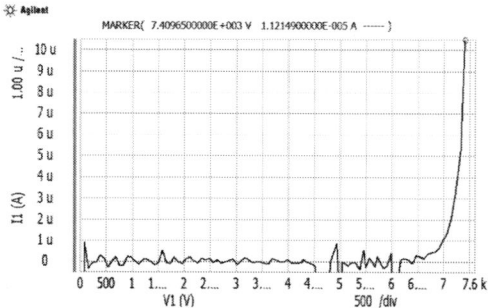

Fig. 7.   6.5kV 4H-SiC PiN power rectifier reverse characteristic

B. Dynamic characteristics

The reverse recovery characteristic is characterized by semiconductor dynamic parameter analyzer Tektronix ITC57220, and the capacitance characteristic is characterized by Tektronix ITC57260.

The dynamic reverse recovery characteristic at different temperatures with 20A direct current (DC) stress is shown in Fig. 8. As the temperature rises, the reverse recovery time increases from 166ns to 263ns (from 25°C to 200°C), and the softness factor decreases from 1.23 to 0.79. Table I shows the softness factor (S) changes, reverse peak current (I_{RM}), extraction time (T_A), recombination time (T_B) and reverse recovery time (T_{rr}) at different temperatures. As the temperature rises, the softness factor shows a downward trend.

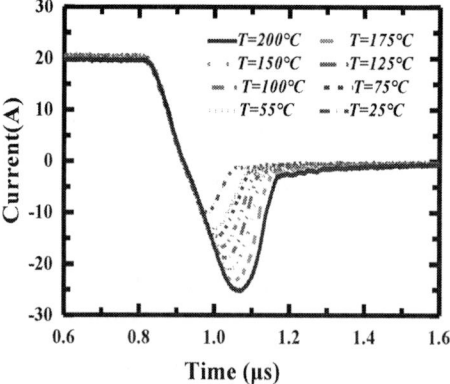

Fig. 8.   Reverse recovery characteristic at variable temperatures

TABLE I. SOFTNESS PARAMETERS AT DIFFERENT TEMPERATURES

$T(°C)$	$I_{RM}(A)$	$T_A(ns)$	$T_B(ns)$	$T_{rr}(ns)$	S
25	10.9	74	91	166	1.23
55	12.5	87	98	186	1.13
75	14.4	96	106	202	1.11
100	16.6	110	101	210	0.92
125	18.4	115	105	220	0.92
150	19.1	118	114	233	0.96
175	23.0	134	123	257	0.92
200	25.0	146	116	263	0.79

Fig. 9 shows the parasitic capacitance versus the reverse voltage at different temperature with 1 MHz frequency. At different temperatures, the capacitance characteristics remain stable. The parasitic capacitance value decreases with the increase of the reverse voltage, and the capacitance is 85pF under the 500V reverse voltage.

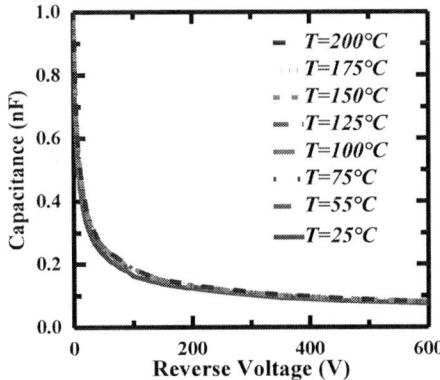

Fig. 9. Capacitance characteristic at variable temperatures

IV. CONCLUSION

In this paper, a high-voltage 6.5kV/100A 4H-SiC PiN device with MAM-JTE is designed, fabricated and characterized. The MAM-JTE dose window is enlarged more than twice compared with DZ-JTE. MAM-JTE adopts ordinary DZ-JTE fabrication process to reduce process steps and fabrication complexity. The device package is in the form of metal shell. The static characteristics show that a forward current of 100A corresponds to V_F of 3.78V, and a low $R_{ON,SP}$ is 2.5mΩ·cm^2 at 100A/cm^2. The reverse voltage reaches 7.4kV@10µA and the high BFOM is 21.9GW/cm^2. The leakage current at 7kV reverse voltage is less than 1µA. The reverse recovery time increases from 166ns to 263ns (from 25°C to 200°C). While the softness factor decreases from 1.23 to 0.79, and the capacitance characteristics hardly changes with temperature increases. These results show that the designed and fabricated high-voltage 6.5kV SiC PiN power rectifier has excellent static and dynamic characteristics. The MAM-JTE widens the JTE ion implantation dose window and reduces the complexity of the process. This research provides a good reference for the development prospects of high-power SiC devices.

ACKNOWLEDGMENT

This work was supported in part by the Science Challenge Project under Grant TZ2018003-1-201, in part by Key Area Research and Development Project of Guangdong Province under Grant 2019B010127001 and in part by Natural Science Foundation of Guangdong under Grant 2019A1515012085.

REFERENCES

[1] J Millan, P Godignon, and X Perpina, "A Survey of Wide Bandgap Power Semiconductor Devices," IEEE Trans. Power Electronics, vol. 29, pp. 2155-2163, 2014.

[2] B. Jayant Baliga, "SiC power devices: From conception to social impact," Lausanne, Switzerland, ESSDERC, pp. 192-197, 2016.

[3] Jiaxi Hu, Wenye Liu, and Jinfeng Yang, "Application of power electronic devices in rail transportation traction system,". Hong Kong, China, International Symposium on Power Semiconductor Devices & IC's, pp. 7-12, 2015.

[4] Palmour J W, "Silicon carbide power device development for industrial markets," International Electron Devices Meeting Technical Digest, vol. 1, pp.1-8, 2014.

[5] Nakayama K, Mizushima T, Takenaka K, "27.5 kV 4H-SiC PiN diode with space-modulated JTE and carrier injection control," Chicago USA: ISPSD, 2018.

[6] Y. Liu et al."Trench Field Plate Engineering for High Efficient Edge Termination of 1200 V-class SiC Devices," International Symposium on Power Semiconductor Devices and ICs, pp. 143–146, 2019.

[7] Tao M L, Deng X C, Wu H, et al, "Design, Fabrication and Characterization of 10kV/100A 4H-SiC PiN Power Rectifier,". Materials Science Forum, pp. 115-119, 2020.

[8] Deng X C, Li L, Wu J, et al, "A Multiple-Ring-Modulated JTE Technique for 4H-SiC Power Device with Improved JTE-Dose Window," IEEE Transactions on Electron Devices, vol. 64, pp. 5042-5047, 2017.

[9] Rui Hu, Xiaochuan Deng, Xiaojie Xu, "An Improved Composite JTE Termination Technique for Ultrahigh Voltage 4H-SiC Power Devices," SSLChina: IFWS, 2019.

[10] C. Xiao, W. Yang, Y. Liu, X. Zhou, "A Trench-Field-Plate High-Voltage Power MOSFET," IEEE Transactions on Electron Devices,vol. 67, pp:2482–2488, 2020.

[11] C. N. Zhou, Y. Wang, R. F Yue, al. Step JTE, an edge termination for UHV SiC power devices with increased tolerances to JTE dose and surface charges[J]. IEEE Trans.Electron Devices, vol.64, pp:1193-1196, 2017.

[12] Y. Saitoh, T. Masuda, H. Michikoshi, "V-groove trench gate SiC MOSFET with a double reduced surface field junction termination extensions structure," Japanese Journal of Applied Physics, vol. 58, pp. SBBD11,2019.

[13] M. Shurrab and S. Singh, "Implantation-free edge termination structures in vertical GaN power diodes," Semiconductor Science and Technology, vol. 35, no. 6, pp. 065005, 2020.

[14] Z. Wang, L. Liang and L. Zhang, "Study of Passivation Layer on Bevel Edge Termination for SiC RSD," 2019 IEEE Workshop on Wide Bandgap Power Devices and Applications in Asia, Taipei, Taiwan, pp. 1-4, 2019.

[15] Huang C, Hsu H, Chu K, "Counter-Doped JTE, an Edge Termination for HV SiC Devices With Increased Tolerance to the Surface Charge," IEEE Transactions on Electron Devices, vol. 62, pp. 354-358, 2015.

[16] Z.-X. Wen et al., "Design and fabrication of 10-kV silicon–carbide p-channel IGBTs with hexagonal cells and step space modulated junction termination extension," Chinese Physics B, vol. 28, pp. 068504, 2019.

[17] Yi Wen, Xiaojie Xu, and Hao Zhu, "Design and Characteristics of an Etching Field Limiting Ring for 10kV SiC Power Device," IFWS, Shenzhen, China, pp. 37-41, 2019.

[18] H. Takashi, O. Hidekatsu, and Y. Kan, "Edge Termination With Enhanced Field-Limiting Rings Insensitive to Surface Charge for High-Voltage SiC Power Devices," IEEE Trans. Electron Devices, vol. 67, pp. 2850–2853, May 2020.

[19] Deng X C, et al,"Experimental study and characterization of an ultrahigh-voltage Ni/4H–SiC junction barrier Schottky rectifier with near ideal performances," Superlattices and Microstructures, 2020.

Dynamic gate leakage current of p-GaN Gate AlGaN/GaN HEMT under positive bias Conditions

Yu Sun
Institute of Microelectronics
Peking University
Beijing, China
1801213572@pku.edu.cn

Maojun Wang
Institute of Microelectronics
Peking University
Beijing, China
mjwang@pku.edu.cn

Wen Lei
Institute of Microelectronics
Peking University
Beijing, China
1801213292@pku.edu.cn

Chun Han
Institute of Microelectronics
Peking University
Beijing, China
1275557821@qq.com

Abstract—**Among the methods to implement E-mode AlGaN/GaN HEMT, the p-type gate GaN HEMT, which has controllable process and good reliability, has drawn a lot of attention now. A Schottky contact on the p-GaN would further reduce the forward leakage current comparing with the ohmic one. However, the p-GaN region enclosed by two barrier layers, the Schottky barrier and AlGaN barrier, will introduce some problems in normal operation. In the problem we will discuss in this article, for commercial 100 V p-GaN AlGaN/GaN HEMTs, we observe the drift of threshold voltage (V_{th}) and dynamic change of gate current when the gate is under positive voltage stress. As the voltage and stress time increases, the V_{th} shifts more to the positive direction, and the change appears to be permanent up to more than 1000 s. On the other hand, a bell shaped gate leakage current with stress time is observed for stress bias larger than 6 V. A model based on dynamic distribution of electron trap during stress is proposed to explain the phenomenon.**

Keywords—GaN HEMT, p-GaN, stress, reliability

I. INTRODUCTION

Compared with traditional semiconductor materials such as Si, GaN based materials have larger band gap width and breakdown field strength. Moreover, benefited by the unique polarization effect, two-dimensional electron gas (2DEG) with high concentration and high mobility will be generated in AlGaN/GaN heterojunction. Because of these advantages, GaN HEMT has good ability to withstand high voltage and low on-resistance, performing excellently in power electronics and RF devices. Conventional GaN HEMT structure is normally on and has potential safety risks in power-related applications. Therefore, enhancement mode GaN HEMT devices need to be developed.

At present, among the existing enhancement mode GaN HEMT structures and process methods, the best performing one is the p-GaN cap GaN HEMT[1]. The p-GaN gate structure not only achieves a positive and adjustable threshold voltage, but also retains the 2DEG channel, which ensures a low on-resistance. In order to reduce the gate current for p-GaN gate, the widely used technology is the Schottky gate. The reverse biased Schottky barrier blocks and reduces the gate current. However, voltage swing of the gate with this structure is very small, and there are many other reliability problems that need to be discussed[2, 3].

In this work, we measured the V_{th} and gate forward current of GaN HEMT with p-GaN Schottky gate after high gate voltage stress. As the voltage and stress time increases, the threshold voltage become larger, and the drift appears to be permanent up to 1000 s. On the other hand, a bell shaped gate leakage current with stress time is observed for stress bias larger than 6 V. Through the analysis of the results and the

comparison with the existing degradation models, a gate degradation model under forward stress is obtained.

II. DEVICE STRUCTURE AND TEST METHOD

A. Device structure

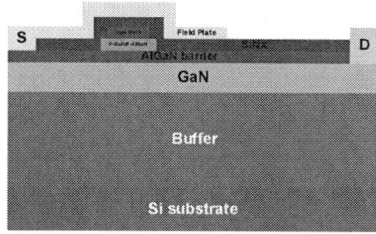

Fig. 1. Device Structure. The p–GaN gate ensures a positive threshold voltage for the device.

The device under test is EPC2036, a commercial device from Efficient Power Conversion (EPC) Corporation. The structure of the device is shown in Figure 1. The substrate of the device is silicon substrate, on which a unintentionally doped GaN layer and a AlGaN barrier layer grown by MOCVD forming AlGaN/GaN heterojunction and the 2DEG channel. P-GaN and schottky contact are used to realize the positive threshold voltage, and the field plate extending from the metal at the source end plays the role of modulating the electric field, weakening the electric field concentration and improving the breakdown voltage of the device.

B. Basic electrical test

We used Agilent B1505A power device analyzer to measure the basic electrical characteristics and gate stress characteristics of the device. The normal gate operating voltage is 5 V, and the gate BV is about 8.5 V. Therefore we chose a gate stress voltage range from 5 V to 7.5 V to monitor the changes of the device. We measured the Vth and gate leakage current before and after the gate stress. The dynamic current during the gate forward stress is also measured.

C. Pulsed electrical stress test

The normal operation status of GaN HEMT is constantly switching to achieve power dependent applications. Simply DC stress test cannot fully explain the degradation process of the device. Measuring the device degradation under the pulsed stress is more in line with the application scenarios of the device[4]. In order to complete the pulsed stress test, one pulse generator to provide the pulsed stress and one ammeter to detect the current are needed. Using the Pulsed IV system provided by AMCAD engineering company, both conditions can be met. The test circuit is shown in Fig.2. The input of the system is connected to the device gate to provide the test pulse, the source and drain are contacted together then connect to the

978-1-6654-4817-8/21 $31.00 © 2021 IEEE

output of the system for current measurement. In order to ensure that the signal applied to the device is with good shape, it is necessary to monitor the real time voltage at the gate and input terminals with an oscilloscope. Because of the gate capacitance and inductance of the test cable, the circuit oscillates under a step signal. At the input end, a small resistance is inserted to increase the rise and fall time of the pulse and reduce the resonance. The observed rise and fall time is 200ns and the test frequency are 1 kHz and 10 kHz.

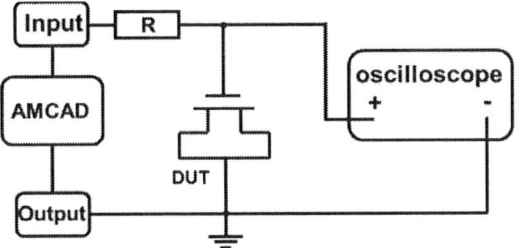

Fig. 2. Setup of the pulsed stress measurement. The resistor in series with the gate reduces the oscillations, and the oscilloscope is used to monitor the pulse waveforms.

III. RESULT AND DISCUSSION

A. Changes of electrical properties after gate stress

We measured the V_{th} of the device with different stress and recovery time, as shown in Fig.3. Under different gate stress, the V_{th} increases with time. But the recovery process is slow. This result indicates that negative charges are constantly injected into the gate area during the static forward stress process. The increase of threshold voltage is positively correlated with stress time and stress voltage.

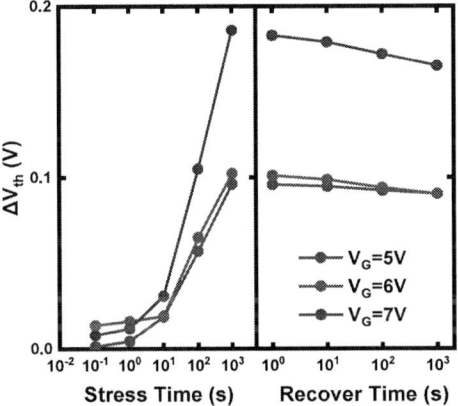

Fig. 3. The shift of threshold voltage with stress time and recover time.

Stress is applied for 100 seconds at different gate voltages from 5.5 V to 7.5 V and the measured current of the device is shown in Fig.4. At the beginning of the application of stress, the behavior is consistent. The gate leakage current gradually increases. But there is a significant difference after that. When the gate voltage is less than 6 V, the gate current continues increasing. When the voltage applied to the gate is greater than 6 V, the gate current increases first and then decreases. With the increase of the gate voltage, the transition time becomes shorter. This phenomenon, however, does not correspond to changes in the device threshold voltage. It is suggested that the variation of gate current is related to the degradation of the gate structure.

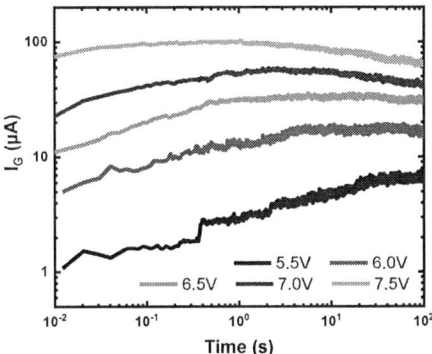

Fig. 4. The variation of gate leakage current with stress time. From 0 s to 100 s, the current rises at the beginning, and when the voltage applied to the gate is greater than 6V, the current will decrease after the rising.

After the forward gate stress, the gate leakage current is measured again. The results are drawn together with the original leakage current as shown in Fig. 5. The current is obviously divided into two sections before and after stress, and the variation trend is similar. When the gate voltage is low, there is a strong correlation between the electric field and temperature, while the gate leakage current is weak temperature dependent when the gate voltage is higher. According to previous reports on gate leakage current of p-GaN gate[5], the mechanism of leakage in the low voltage region relies on the trap-assisted tunneling, and when the gate voltage is greater than 5.5V, the main mechanism of leakage is FN tunneling current. After gate stress, the injection of negative charges changes the electric field near the gate Schottky junction and the dominant mechanism cannot be determined by simple theoretical fitting.

In different voltage stress tests, it is observed that the gate leakage increases first and then decreases under high stress. The repeatability of this change can be determined by repeated measurements. At stress voltage of 7 V, two phases of changes could be observed. We performed repeated stress tests for 100 seconds, with 10 minute intervals between each test. The results are shown in Fig. 6. A significant increase in current was observed in the first 1 second of each test, suggesting that the process could be repeated. But the maximum current during each measurement can only reach the value at the end of the last sweep. The gate current decreased significantly after several times of repeated sweep, which suggests that in a short release period of time (10 minutes), the current decline is not able to restore.

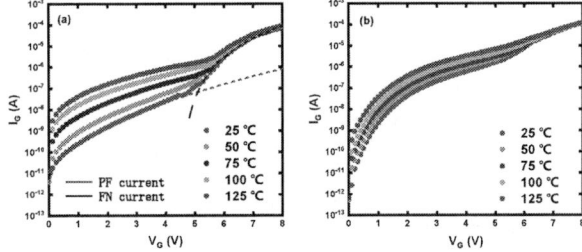

Fig. 5. (a) Gate forward current of the device at different temperature. (b) Temperature dependence of the gate leakage current after 100 s stress at 5 V. Current still has two distinct conduction mechanisms.

978-1-6654-4817-8/21 $31.00 © 2021 IEEE

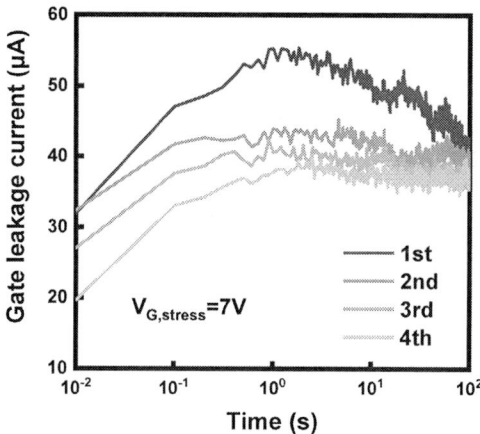

Fig. 6. Repeated gate stress tests at V_G=7 V. There was an interval of 10 minutes between each tests.

B. Degradation model under gate forward stress

Combining the changes of threshold voltage and gate leakage current, we tried to explain the unusual dynamic change of gate current under static stress. As shown in Fig. 7, there are two leakage paths in the gate structure at positive bias. In the initial period of stress, when the gate leakage keeps a low level, the electrons injected by the P-I-N junction are captured by the defects near the Schottky junction interface[6]. The captured electrons enhance the threshold voltage of the device. In the meantime, the electric field of the reverse biased Schottky junction is enhanced and the reverse leakage current increases. When the high bias stress continues to be applied to the gate terminal, electrons may be captured in the passivation layer close to the field plates, which weakens the electric field at the edge of the Schottky junction and reduces the forward leakage current from the gate to p-GaN layer.

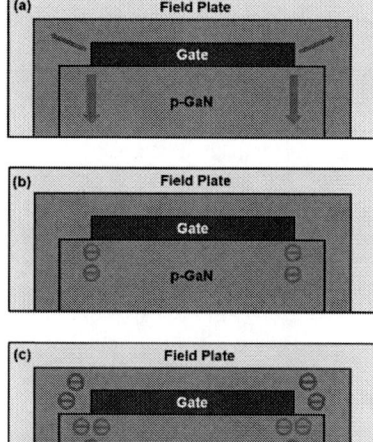

Fig. 7. Degradation model for p-GaN gate GaN HEMT. (a) Gate leakage current distribution. (b) Initial defect states accumulation in the current increase stage. (c) Defect states accumulation in the current decrease stage.

C. Pulsed stress measurement

The gate forward current is measured at a pulsed gate stress of 7.75 V and we got the results in Fig. 8. The gate leakage change could still be divided into two parts under the pulsed test condition. This result indicates that the defects causing the gate current variation are generated and captured faster than the pulse width (50 μs) used in the test.

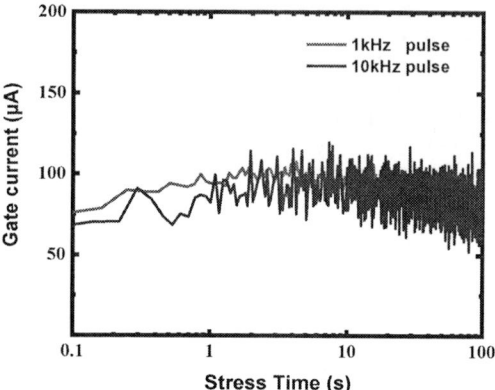

Fig. 8. Time dependent gate leakage current under 1 kHz and 10 kHz pulsed gate stress test.

We also measured the device failure time under different pulsed gate stress. The effective stress time $T_{effective}=T_{total} \times duty\ cycle$ was taken to compare the results under three conditions: DC stress, 1 kHz square wave and 10 kHz square wave stress. The breakdown time and Weibull distribution during pulsed stress have no obvious difference with the dc stress. In addition [7], holes will be trapped at the p-GaN/AlGaN interface states at the positive stress, which increases the gate leakage current. In the pulsed stress test, negative quiescent bias voltage was applied before the pulse to study the effect of hole de-trapping, which may extend the device failure time. Measurement results indicate that the device's failure time and Weibull distribution did not change with quiescent bias during pulsed stress. The results suggest that the gate degradation of the device under positive stress is an irreversible and a rapid process.

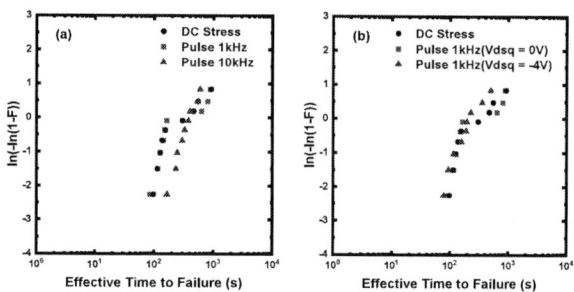

Fig. 9. (a) Gate stress failure test at different frequencies (b) Gate stress failure test at different drain quiescent biases.

IV. CONCLUSION

We measured the electrical characterization and gave a failure analysis of Schottky p-GaN gate HEMT in this work. By static stress test, we obtained the change of the gate leakage current under forward stress and supposed that carrier trapping at different locations in the p-GaN gate structure is responsible for the evolution of gate leakage current with stress time. Similar degradation behavior is observed in pulsed stress test with different quiescent biases comparing with DC stress. This indicates the degradation of p-GaN gate structure is a rapid process which is not affected by relaxation.

REFERENCES

[1] Y. Uemoto *et al.*, "Gate injection transistor (GIT)—A normally-off AlGaN/GaN power transistor using conductivity modulation," *IEEE Transactions on Electron Devices,* vol. 54, no. 12, pp. 3393-3399, 2007.

[2] J. Wei, H. Xu, R. Xie, and K. J. Chen, "Principles and impacts of dynamic threshold voltage in a p-GaN gate high-electron-mobility transistor," *Semiconductor Science and Technology,* vol. 36, no. 2, p. 024006, 2021.

[3] T.-L. Wu *et al.*, "Forward bias gate breakdown mechanism in enhancement-mode p-GaN gate AlGaN/GaN high-electron mobility transistors," *IEEE Electron device letters,* vol. 36, no. 10, pp. 1001-1003, 2015.

[4] J. He, J. Wei, S. Yang, Y. Wang, K. Zhong, and K. J. Chen, "Frequency-and Temperature-Dependent Gate Reliability of Schottky-Type p-GaN Gate HEMTs," *IEEE Transactions on Electron Devices,* vol. 66, no. 8, pp. 3453-3458, 2019.

[5] A. Stockman *et al.*, "Gate conduction mechanisms and lifetime modeling of p-gate AlGaN/GaN high-electron-mobility transistors," *IEEE Transactions on Electron Devices,* vol. 65, no. 12, pp. 5365-5372, 2018.

[6] B. Li *et al.*, "Impact of carrier injections on the threshold voltage in p-GaN gate AlGaN/GaN power HEMTs," *Applied Physics Express,* vol. 12, no. 6, p. 064001, 2019.

[7] G. Meneghesso *et al.*, "GaN HEMTs with p-GaN gate: field-and time-dependent degradation," in *Gallium Nitride Materials and Devices XII,* 2017, vol. 10104: International Society for Optics and Photonics, p. 1010419.

An Integrated GaN-Based Converter Based on Cooling-System-Inductor Structure for point-of-load converters

Longyang Yu, Wei Mu, Huaqing Li, Chengzi Yang, Chenya Wang, Laili Wang
State Key Laboratory of Electrical Insulation and Power Equipment
Xi'an Jiaotong University
Xi'an China
382856693@qq.com

Abstract—The paper proposes an integrated GaN-based power converter, which has these features of a 3-D integrated magnetic component with alloy materials, improving the power density. Besides, the 3-D integrated magnetic component with alloy materials has high thermal conductivity, which can be regarded as a heat sink for an integrated GaN-based power module and further increase the power density. A comprehensive analysis of the thermal and parasitic characteristics of the GaN-based power converter, and the characteristics and design of 3-D integrated magnetic component with heat sink function by finite element software are discussed in detail. Two 90 W, 12-1.8 V, experimental prototype of four-phase interleaving POL converter based ferrite materials and alloy materials on are built and tested, respectively. The experimental results demonstrate that GaN-based power module based on 3-D integrated magnetic component with alloy materials has low parasitic parameters, good thermal performance, high efficiency and high power density than that of ferrite materials.

Keywords—*GaN-based power module, 3-D integrated magnetic component, heat sink.*

I. INTRODUCTION

The POL converters based on GaN devices face many challenges. One of the biggest challenges is the reliability of GaN devices applied in industrial field. Although GaN devices have lower switching loss and smaller package compared to silicon devices, the performance of GaN devices is more fragile. Due to faster switching speed the GaN devices are very sensitive to the parasitic inductor caused by PCB layout. Meanwhile, the gate-to-source voltage of GaN devices from EPC company is 5V under normal operation, but the breakdown voltage of is only 6V. Therefore, the 1V margin indicates that the gate loop parasitic inductance must be reduced very low, which allows the overshoot to be less than 1V or even less. The parasitic inductance inside GaN devices from EPC company is very small or even less than 1nH due to the LGA or BGA package. Hence, the parasitic inductance caused by the external PCB layout is the main influence on the GaN devices. Reducing parasitic inductance in PCB copper traces is a necessary ways to improve reliable operation of GaN devices. Many methods have been proposed to reduce parasitic inductance. An optimal power loop is reported in [1] to minimize parasitic inductance compared with conventional vertical power loop and lateral power loop layout. A multiloop method [2] can further reduce parasitic inductance compared with [1], in which the parasitic inductance can be reduced to be less than 0.5nH. These

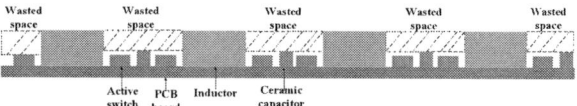

Fig. 1 Side view diagram of the converter with discrete inductors.

methods [1]-[2] have a common characteristic of compact PCB layout. Active integration is a very effective solution for improving the reliability of GaN devices. Its concept is that driver circuit, GaN bare dies and decoupling capacitors can be integrated into a PCB board [3]. The GaN-based power module [4] employs active integration technique to integrate half-bridge circuit, which can minimize the parasitic inductances and can improve power density. Meanwhile, The GaN-based power module can expand N half-bridge circuits according to the POL converters output current level.

The 3-D integrated magnetic component technique can efficiently utilize the entire GaN-based power module space and reduce wasted space caused by the planar magnetic component, increasing the power density. Low temperature co-fired ceramic technology has been proposed to fabricate 3-D integrated inductors employed in high-frequency POL converters. The LTCC inductor can be regarded as a substrate [5], [6], which is attached to the bottom of the buck converter. The performance of the LTCC inductor is evaluated and reported in [7], which demonstrates that the LTCC inductor is very nonlinear and the inductance value decreases rapidly with the increase of the output current. The performance comparison of PCB and DBC using LTCC inductor is reported in [8] and [9], which shows that PCB board has higher efficiency than DBC board in a buck converter, however, thermal performance of DBC board is better than PCB board. A designed multiphase LTCC inductor [10] is applied into small portable electronics, which is attached to the bottom of the multiphase converter. However, the inductance value is very low (<15nH), which is only suitable for VHF converter. The LTCC inductor features low inductance value and non-linearity. The literature utilize the characteristic of the LTCC inductor to design multi-permeability LTCC inductor [11], which shows a large inductance value under light-load condition and very small inductance value under heavy-load condition, improving transient performance under heavy load and the efficiency under light load. The two-phase inverse coupled inductor using LTCC technique have designed and employed into a two-phase buck converter, which has excellent performance.

978-1-6654-4817-8/21 $31.00 © 2021 IEEE

The paper aim at proposing an integrated GaN-based power converter based on cooling-system-inductor structure. A 3-D integrated magnetic component with inverse coupling technique is designed for POL converter, meanwhile, it is multifunctional integrated magnetic component, in which its magnetic core can be employed as heat sink for the entire converter. This paper can be arrangement as follows. Section II describes topologies of POL converters and the integrated GaN-based power converter. Experimental results have been implemented in Section III. Finally, our conclusions has been drawn in Section IV.

II. INTERGRATED GAN-BASED POWER MODULE

It is well known that the GaN devices are very sensitive to parasitic inductances due to very fast switching speed compared with silicon devices. Therefore, PCB layout of GaN devices should be very compact to minimize parasitic inductors from PCB copper traces, which affect steady operation of GaN devices. There are two kinds of parasitic inductances from PCB copper traces. namely, power loop parasitic inductance and gate loop parasitic inductance. Taking a half-bridge as an example, the distribution of parasitic inductances can be presented in Fig. 2.

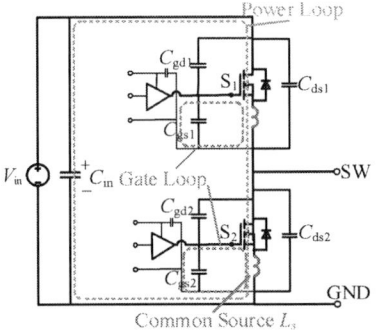

Fig. 2 Distribution of parasitic inductances at half-bridge circuit.

The power loop parasitic inductance can cause oscillations of the drain-source voltage when the GaN devices is turned-on or turned-off. The larger power loop inductance, the more severe the oscillation. The gate loop inductance can result in a large spike of gate-source voltage, which is fatal to GaN devices. The maximum gate-source voltage of GaN devices from EPC company is only 6V, but the gate-source voltage in normal operation is 5V. There is only a margin of 1V, and minimizing gate loop parasitic inductance is very critical for GaN devices. The compact layout is a solution, reducing the power loop and the gate loop inductances from PCB copper traces. Hence, a GaN-based power module is designed by active integration, which is composed of half-bridge drivers, GaN devices and decoupling capacitors [12].

A. Topologies of POL Converter

The power module can minimize power loop and gate loop parasitic inductances. Taking four-phase POL converter as an example, the schematic of a four-phase interleaving POL converter is shown in Fig. 3.

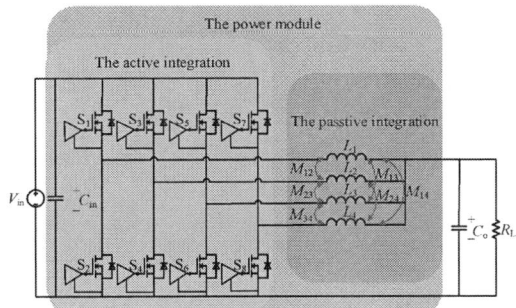

Fig. 3 Topologies of a four-phase buck converter.

B. Active Integration of Power Module

A designed GaN power module in [12] includes active integration. However, the passive integration about magnetic component is not considered. An integrated GaN-based power converter is presented, in which both active and passive integration are included. The active integration of power converter is presented in Fig. 4.

Fig. 4 Active integration in the integrated power module.

It is clear that the GaN devices, driver circuits and decoupling capacitors can be integrated into a PCB board. Thus, these parasitic inductances including the power loop and gate loop inductances can be reduced effectively in the integrated GaN-based power module. The power loop parasitic inductance depends on the loop area, the larger the loop area is, the larger the power loop parasitic inductance is. PCB layout for the power loop have horizontal and vertical structures [9]. The vertical structure is validated to minimize the parasitic inductance. The power loop in the power converter marked by Fig. 2 is made up of an upper switch, a lower switch and ceramic capacitors at a half-bridge circuit. The power loop employs vertical layout, as presented in Fig. 5. Current each half-bridge circuit starts from positive bus decoupling capacitors in the top layer, and then flows through top and bottom switches on the left, and then flows into the bottom layer by via hole. Finally, the current flows into negative bus decoupling capacitors on the right. The note of "SW" per half-bridge circuit is partially slotted so that the integrated inductor structure winding can be fixed to the power module.

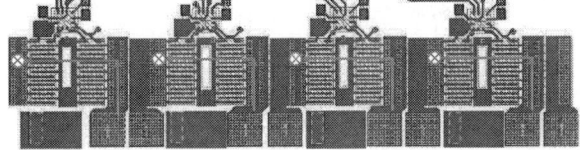

(a)

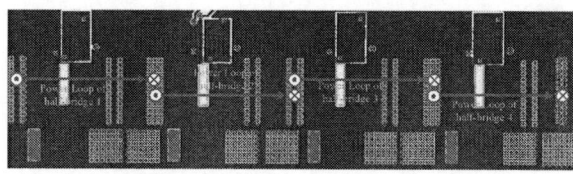

(b)

Fig. 5 The paths of power loop each half-bridge circuit. (a) Top layer. (b) Bottom layer.

Gate loop parasitic inductance depends on area and the distance between the driver and GaN power devices. Hence, the distance can be reduced to minimize the gate loop parasitic inductance. Meanwhile, the area formed by the gate and source paths from the driver circuit to GaN devices must be reduced to make the gate loop parasitic inductance minimized. Fig. 6 shows half-bridge driver circuit can be used to the GaN-based power converter, and gate loop has been marked by the white dotted line.

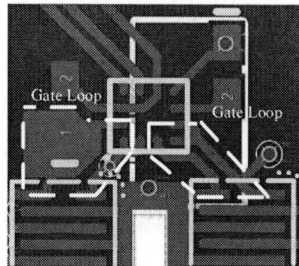

Fig. 6 The paths of gate loop each half-bridge circuit.

C. Possive Integration of Power Module

A novel 3-D magnetic core structure is proposed, as shown in Fig. 7, in which the four-phase inductors in four-phase POL converter can be integrated into the integrated magnetic component by inverse coupling technique. Also, the integrated magnetic component employs alloy material to dissipate heat for the power module. Therefore, the multifunctional integrated magnetic component can act as four-phase inductors and a heat sink. The proposed multifunctional 3-D integrated magnetic component is called cooling-system-inductor.

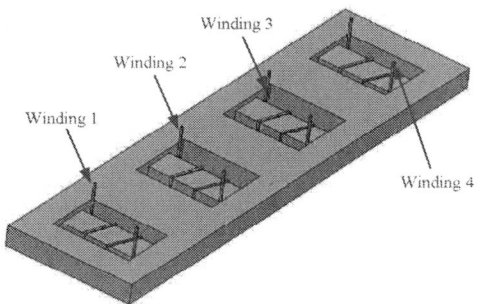

Fig. 7 Multifunctional 3-D integrated magnetic component.

It is seen from Fig. 7 that the windings of four-phase inductor are wound around the respective center legs in the magnetic component. The side legs have two functions, which can not only adjust the inverse coupling coefficient, but also heat GaN power devices and driver circuits.

D. Proposed 3D integrated GaN-Based Power Converter

The schematic of proposed 3D integrated GaN-Based power converter is presented in Fig. 8. It is clear that the multifunctional 3-D integrated magnetic component is arranged to the bottom layer of the power converter. The

windings are implemented by using copper foil, and the windings can be soldered by the respective slot on the power module. Compared with conventional planar magnetic shown in Fig. 1 the proposed 3-D integrated GaN-Based power module has higher space utilization, resulting in higher power density.

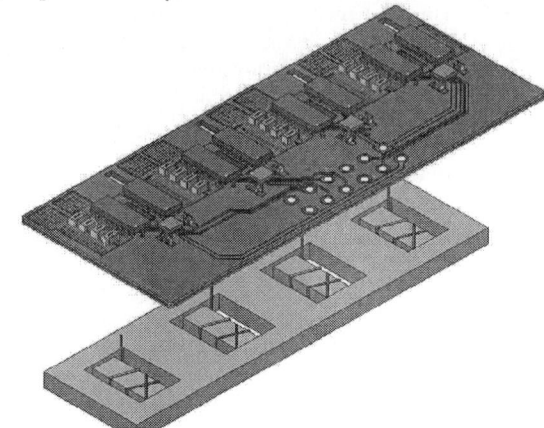

Fig. 8 Proposed integrated power module.

III. Experimental Results

Both alloy and ferrite materials are used to make the multifunctional 3-D integrated magnetic component as shown in Fig. 1. The four-phase POL converter is operated in switching frequency of 1MHz. Therefore, ferrite material chooses DTT-F4 from Shandong Dongtai Electronic Science and Technology company, and the alloy material chooses NPX from POCO HOLDING company. The thermal conductivity of the ferrite material is 3 $Wm^{-1}K^{-1}$, and that of the alloy material is 20 $Wm^{-1}K^{-1}$. Besides, Due to the high saturation density and low permeability, alloy magnetic components do not need to open additional air gaps, while ferrites need to open air gaps to avoid magnetic saturation. For low voltage and high current POL converters, the fringe effect caused by opening air gap is very serious, resulting in higher loss. Hence, the loss of alloy materials is lower than that of ferrite materials. A 90W, 12-to-1.8V, GaN-based experimental prototype shown in Fig. 2 of four-phase POL converters is built and tested, and the detailed parameters are listed in Table I. Fig. 2(a) shows GaN-based power converter, which is composed of four half-bridge circuits, drivers and decoupling capacitors. The power density of GaN-based power module is 540W/inch³. Noted that power density can be calculated without any control,

TABLE I
PARAMETERS OF THE CONVERTER

Parameters	Values
Input voltage V_{in}	12V
Output voltage V_o	1.8V
Rated power P_o	90W
Switching frequency f_s	1MHz
switches	EPC2023
Output capacitor C_o	22µF (ceramic capacitor)
Driver	LMG1205
Controller	DSP28335

driver or auxiliary power supply circuitry. Fig. 2(b) shows that the magnetic component using ferrite material is fixed in bottom of the power module. It is clear that the magnetic component need to open air gap to avoid magnetic saturation. Fig. 2(c) shows that the magnetic component using alloy material is fixed in bottom of the power module. The magnetic component doesn't need to open air gap compared with ferrite material. Fig. 2(d) shows that the whole experimental prototype, which is composed of the power module and motherboard. The motherboard contains output capacitors and input bus capacitors.

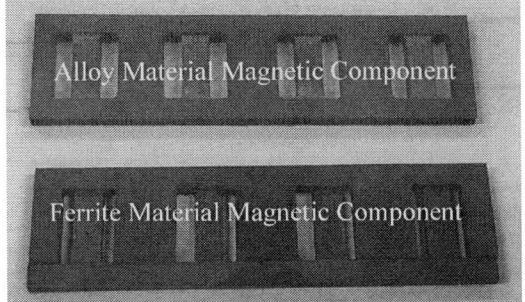

Fig. 1 Multifunctional 3-D integrated magnetic component based on alloy material and ferrite material.

(a)

(b)

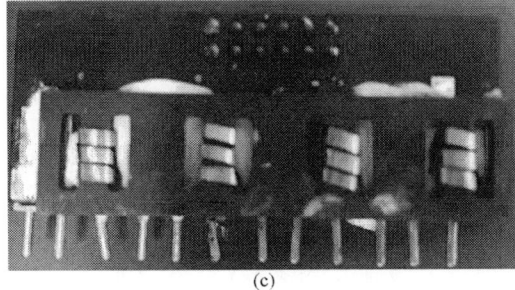

(c)

(d)

Fig. 2 The experimental prototype. (a) GaN-based power module. (b) Magnetic component based on ferrite material. (c) Magnetic component based on alloy material. (d) The whole prototype.

Fig. 3 shows efficiency curve of the four-phase POL converter when magnetic component employs ferrite and alloy materials. The measured peak efficiency using ferrite material is 87.2% when the output current reaches 32A, and the measured efficiency of 84.6% at full load. The measured peak efficiency using alloy material is 89.6% when the output current reaches 32A, and the measured efficiency of 87.1% at full load. It should be noted that the measured efficiency using alloy material is higher than that of using ferrite material regardless of light, medium or heavy load. Fig. 4 shows the thermal performance of the prototype operating at full load, which is measured by FLIR T630SC. It is seen that the maximum temperature is focused on the first-phase half-bridge circuit.Fig. 4(a) shows the thermal performance using ferrite material, and the highest temperature observed on this power module reaches 104 °C, and the hot spot on the first-phase synchronous rectifier is 93.2°C. Fig. 4(b) shows the thermal performance using alloy material, and the highest temperature observed on this power module reaches 90.1 °C, and the hot spot on the first-phase synchronous rectifier is 81.5°C. The thermal performance using alloy material is 13.9 °C cooler than that of ferrite material.

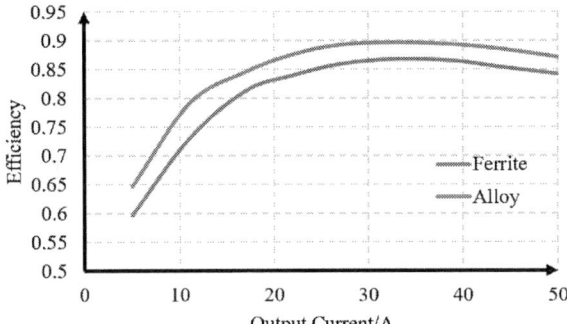

Fig. 3 The efficiency curve.

(a)

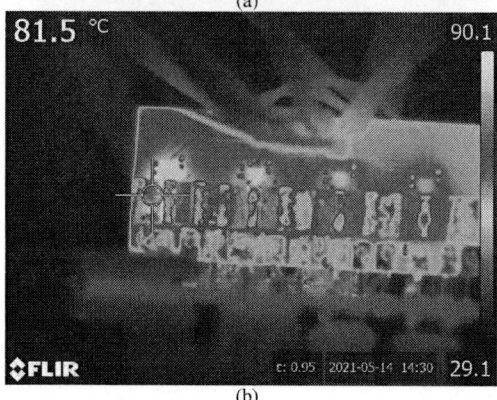

(b)

Fig. 4 Thermal performance of GaN-based power converter at full load. (a) Magnetic component based on ferrite material. (b) Magnetic component based on alloy material.

IV. CONCLUSION

A multifunctional 3-D integrated magnetic component is proposed forGaN-based power converters in this paper. The proposed multifunctional 3-D magnetic component can be pasted to bottom of the GaN-based power module to achieve 3-D integrated GaN-based power module, resulting in increase of power density. In addition, the proposed magnetic component can used as a heat sink to cool the GaN-based power module. Finally, a 90W, 12-1.8V 3-D integrated GaN-based power module is built. The experimental results demonstrate that the proposed multifunctional 3-D integrated magnetic component not only has lower loss but also better heat dissipation. The peak efficiency of the integrated GaN-based power module reaches 89.6% and the full-load efficiency reaches 84.6%. The thermal performance using multifunctional 3-D integrated magnetic component is 13.9 °C cooler than that of ferrite material. The experimental results verify the effectiveness of multifunctional 3-D integrated magnetic component.

ACKNOWLEDGMENT

The authors thank POCO HOLDING CO.,LTD. They provide a very excellent performance of alloy materials called NPX. In addition, they used this alloy material to make the 3-D magnetic component designed by the authors, which provided a very good support for the authors' experiments so that the authors could successfully finish this paper.

REFERENCES

[1] D. Reusch and J. Strydom, "Understanding the Effect of PCB Layout on Circuit Performance in a High-Frequency Gallium-Nitride-Based Point of Load Converter," *IEEE Transactions on Power Electronics*, vol. 29, no. 4, pp. 2008-2015, April 2014.

[2] K. Wang, L. Wang, X. Yang, X. Zeng, W. Chen and H. Li, "A Multiloop Method for Minimization of Parasitic Inductance in GaN-Based High-Frequency DC–DC Converter," *IEEE Transactions on Power Electronics*, vol. 32, no. 6, pp. 4728-4740, June 2017.

[3] J. D. van Wyk, F. C. Lee, Zhenxian Liang, Rengang Chen, Shuo Wang and Bing Lu, "Integrating active, passive and EMI-filter functions in power electronics systems: a case study of some technologies," *IEEE Transactions on Power Electronics*, vol. 20, no. 3, pp. 523-536, May 2005.

[4] K. Wang, B. Li, H. Zhu, Z. Yu, L. Wang and X. Yang, "A Double-Sided Cooling 650V/30A GaN Power Module with Low Parasitic Inductance," *2020 IEEE Applied Power Electronics Conference and Exposition (APEC)*, 2020, pp. 2772-2776.

[5] M. H. F. Lim, J. D. van Wyk and F. C. Lee, "Hybrid Integration of a Low-Voltage, High-Current Power Supply Buck Converter With an LTCC Substrate Inductor," *IEEE Transactions on Power Electronics*, vol. 25, no. 9, pp. 2287-2298, Sept. 2010.

[6] Z. Qi, C. Zhao, L. Wang, F. Yang, Y. Pei and Z. Zheng, "Three-Dimensional Integrated GaN-based DC-DC Converter with an Inductor Substrate," *2019 IEEE Energy Conversion Congress and Exposition (ECCE)*, 2019, pp. 832-838.

[7] Y. Su, Q. Li and F. C. Lee, "Design and Evaluation of a High-Frequency LTCC Inductor Substrate for a Three-Dimensional Integrated DC/DC Converter," *IEEE Transactions on Power Electronics*, vol. 28, no. 9, pp. 4354-4364, Sept. 2013.

[8] D. Reusch, D. Gilham, Y. Su and F. C. Lee, "Gallium Nitride based 3D integrated non-isolated point of load module," *2012 Twenty-Seventh Annual IEEE Applied Power Electronics Conference and Exposition (APEC)*, 2012, pp. 38-45.

[9] S. Ji, D. Reusch and F. C. Lee, "High-Frequency High Power Density 3-D Integrated Gallium-Nitride-Based Point of Load Module Design," *IEEE Transactions on Power Electronics*, vol. 28, no. 9, pp. 4216-4226, Sept. 2013.

[10] D. Hou, F. C. Lee and Q. Li, "Very High Frequency IVR for Small Portable Electronics With High-Current Multiphase 3-D Integrated Magnetics," *IEEE Transactions on Power Electronics*, vol. 32, no. 11, pp. 8705-8717, Nov. 2017.

[11] L. Wang, Y. Pei, X. Yang and Z. Wang, "Design of Ultrathin LTCC Coupled Inductors for Compact DC/DC Converters," *IEEE Transactions on Power Electronics*, vol. 26, no. 9, pp. 2528-2541, Sept. 2011.
[12] Wenkang Huang, and Brad Lehman, "A Compact Coupled Inductor for Interleaved Multiphase DC–DC Converters," *IEEE Trans. Power Electron.*, vol. 31, no.10, pp. 6770–6776, OCT. 2016.

978-1-6654-4817-8/21 $31.00 © 2021 IEEE

A Survey on Modeling of SiC IGBT

Yuwei Wu
Department of Electrical Engineering
Xi'an Jiaotong University
Xi'an, China
wywei99552017@stu.xjtu.edu.cn

Laili Wang
Department of Electrical Engineering
Xi'an Jiaotong University
Xi'an, China
llwang@mail.xjtu.edu.cn

Jianpeng Wang
Department of Electrical Engineering
Xi'an Jiaotong University
Xi'an, China
wangjackmvp@stu.xjtu.edu.cn

Feng Zhang
Department of Physics
Xiamen University
Xiamen, China
fzhang@xmu.edu.cn

Abstract—**SiC IGBT is an emerging device in ultrahigh-voltage power electronic field. With the development of the device manufacture, the circuit models for it have emerged since 2012. First, the behavior of SiC IGBT is briefly introduced from material characteristics, static characteristic and dynamic characteristics. Then, the SiC IGBT circuit models proposed in public are reviewed. All the models are classified into three types, which are the derived models with modified material parameters, the derived models with modified special structure and other new models. The classification guides the scientists who intend to develop models for SiC IGBT or other promising devices. Each of the models is introduced in detail from its mechanism, specialty and simulation performance. Finally, the paper is concluded and the outlook in SiC IGBT modeling is discussed.**

Keywords—**circuit models, SiC IGBT, review.**

I. INTRODUCTION

SiC semiconductor material has fantastic material characteristics, such as wider bandgap and higher thermal conductivity than Si [1], so it becomes a research hotpot nowadays. Especially, SiC IGBT is considered to be the most promising device in ultrahigh-voltage electrical power switches in the future, due to its excellent flow capacity and low on-state resistance [2]. Several SiC models have been proposed to study the behavior of the devices in the past 10 years, which used different methods and focused on different aspects [3-12]. In this paper, the unique behavior of SiC IGBT is studied in comparison with Si IGBT and all the SiC IGBT models are reviewed and compared systematically.

II. THE BEHAVIOR OF SiC IGBT

A. Material Characteristics

The fundamental characteristics of semiconductor materials determine the operating characteristics of power devices. TABLE I summarizes the basic material characteristics of Si and 4H-SiC. The table only contains the 4H-SiC, because its feature is better than other crystalline form of SiC.

TABLE I.   A LIST OF EXISTING SiC IGBT MODELS [13]

feature	Si	4H-SiC
Band gap (eV)	1.11	3.26
Relative permittivity	11.7	9.7
Thermal conductivity (W/cm·K)	1.5	3.7
Electronic affinity (eV)	4.05	3.7

It is important that the band gap of 4H-SiC is two times higher than Si. This results in the intrinsic carrier concentration of SiC is much lower than Si at the same temperature and the collisional ionization coefficient is smaller. Moreover,

the thermal conductivity of 4H-SiC is 2~3 times higher than Si, which means SiC devices performs better in heat dissipation.

B. Static Characteristics

The static characteristics reflect on the conductivity and conduction loss of SiC IGBT.

When a IGBT is on, the voltage drop mainly decided by internal PN junction. The build-in voltage of a PN junction is

$$V_{bi} = \frac{kT}{q} \ln(\frac{N_A^- N_D^+}{n_i^2}) \tag{1}$$

where N_A^- and N_D^+ is the concentration of ionized impurities on both sides of the PN junction, the value of which is almost the same for Si and SiC. However, because the intrinsic carrier concentration (n_i) of SiC is much smaller, the PN junction of SiC IGBT has higher build-in voltage than Si IGBT, which leads to higher on-state resistance.

The output characteristics of IGBT is closely related to the carrier mobility. Fig.1 shows the electron mobility in different temperature and doping concentration.

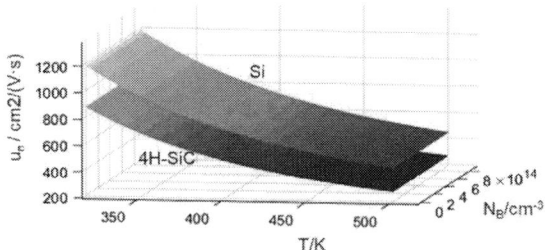

Fig. 1.   The electron mobility in different temperature and doping concentration

As Fig.1 shows, the electron mobility of 4H-SiC is always lower than Si in the same temperature and doping concentration. Therefore, the static current I_c of SiC IGBT is lower than Si IGBT with the same chip design in one gate-emitter voltage.

C. Dynamic Characteristics

Dynamic characteristics determines the switching frequency and switching loss of IGBT. As for SiC IGBT, the turn-off characteristics differs a lot from Si IGBT.

Fig.2 shows the experimental turn-off waveform of a SiC IGBT in double pulse test [14]. The whole process of turn-off transient can be abstracted as five individual process [15], as Fig.3 shows.

Phase I and II are known as sweep-out phases [15]. In phase I, the collector current decreases sharply, which results

978-1-6654-4817-8/21 $31.00 © 2021 IEEE

from displacement current flowing through the parasitic capacitors of SiC IGBT and freewheel diode.

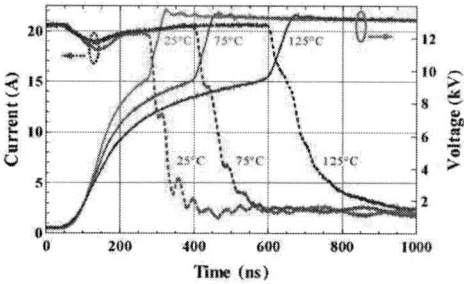

Fig. 2. Double Pluse Test waveform for IGBT with inductive load in 20 A load current and 13 kV bus voltage at different temperatures [14]

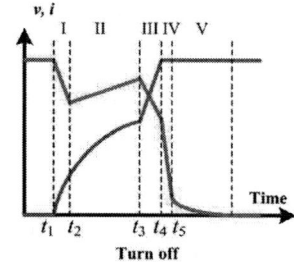

Fig. 3. Typical turn-off waveform of SiC IGBT during hard switch [15]

However, when dealing with conventional Si IGBT this current drop is ignored because of the low *dv/dt*. While in SiC IGBT, the sweep-out phases are notable due to high *dv/dt*. Moreover, this effect weakens as temperature rises, because there is less displacement current due to the increase of the remaining charge in the buffer layer and the decrease of *dv/dt*.[14].

In phase V, the tail current leads to the increase of turn-off time. The length of tail current is positive correlated to the quantity and lifetime of excess carrier. Usually, 4H-SiC IGBT is designed with higher doping concentration and thinner width of the drift region than Si IGBT of the same voltage class, because 4H-SiC has much higher band gap than Si. Therefore, SiC IGBT has shorter turn-off time and thus faster switching speed.

III. REVIEW OF EXISTING MODELS

Since 2012, several researchers have dedicated efforts to find models which can accurately characterize the electrical and terminal behaviors of SiC IGBT, in order to push the path forward for circuit design and packaging structure optimization. All the models are listed in TABLE II with its core arithmetic, foremost contribution, limitation and simulator.

The concept and structure of SiC IGBT is derived from Si IGBT, so some early SiC IGBT models are derived from existing Si IGBT models. Researchers have taken different way to describe the distinct behavior of SiC IGBT which is classified as modifying material parameters and modifying special structures. Moreover, some researchers built their new SiC IGBT models with their own arithmetic. Therefore, 3 types of SiC IGBT models are concluded in this article, and all the models in TABLE II are classified in it and introduced in detail.

A. The Derived Models with Modified Material Parameters

Some SiC IGBT models are based on well-established Si IGBT models with some model parameters modified. The Si IGBT models suitable for extending should be complete in physical mechanism, to ensure that it is convergent when dealing with SiC IGBT. The most famous physical model are Hefner model and Kraus model, on which the SiC IGBT models in this section are based.

Hefner model is the first complete 1-D analytical IGBT model. It can predict the behavior of IGBT under both static and dynamic operation. The model is based on the equivalent circuit that a MOSFET drives a low-gain, high-level injection bipolar transistor, as Fig.4 shows. It also considers the Non-quasistatic effects and nonlinear capacitances for the transient analysis [11]. The Hefner model was further extended to buffer layer IGBTs [16] and a dynamic electro-thermal model [17].

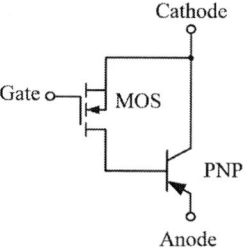

Fig. 4. The equivalent circuit of a n-channel IGBT

In 2012, M. Saadeh *et al.* developed the first SiC IGBT model in the world [9]. All physical equations of the model are derived from Hefner IGBT model [18] and the features of SiC material is reflected on the model's parameters. The procedure of parameter extraction is actualized by curve fitting and look-up table using the software tool Certify. The model parameters are fitted from various characteristic curves, such as transfer characteristics, output characteristics and switching characteristics from datasheet or experimental measurement. Fig.5 shows which parameters need to fit which regions of characteristic curves. Moreover, this model can also describe the behavior of Si IGBT as long as users load the characteristic curves of Si IGBT into Certify. Therefore, it is the first model that can predict the behavior of both Si and SiC IGBT, whether n or p-channel. The ability to represent both static and dynamic behavior was experimentally validated by Cree's 15kV p-channel and n-channel SiC IGBTs.

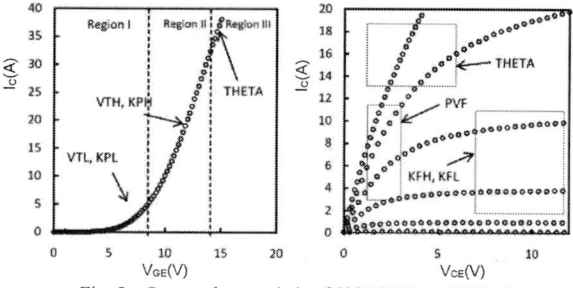

Fig. 5. Output characteristic of SiC IGBT at 150℃ [9]

T.H.Duong *et al.* proposed a model for n-channel SiC IGBTs with a field-stop layer [3]. The physical equations of this model are based on Hefner's dynamic electro-thermal model [17] and Hefner's buffer layer's model [16]. The authors extended the model to SiC IGBT by updating the

TABLE II. A LIST OF EXISTING SiC IGBT MODELS

Paper	[9]	[6] [10]	[7]	[3]	[5]	[8]	[12]
Year	2012	2013-2019	2013	2014	2015	2019	2020
arithmetic	Hefner	HiSIM	Kraus and Elmore	Hefner	equivalent circuit	Hefner	Behavior model
contribution	the first SiC IGBT model	considering the punch-through effect	the first electro-thermal model	a wide range of circuit conditions	give an insight to instan-taneous charge	unified model for all IGBTs	high accuracy and fast convergence
limitation	not suitable for multi-condition simulation	lack of experimental verification	lack of dynamic performance	low accuracy	low accuracy	lack of experimental verification	not suitable for commercial simulator
Simulator	Saber & Pspice	SPICE	SIMULINK	Saber	SIMULINK	Saber, Hspice, Spectre	/

device and material properties from Si to SiC. A software tool was proposed to extract parameters of the SiC IGBT model automatically. It was extended from the software package called IMPACT for Si IGBTs [19] and SiC-IMPACT for SiC power MOSFET [20], which can extract 20 parameters of the Hefner IGBT model for both Si and SiC. The model was implemented in Saber and the validity of the model was confirmed by the static and dynamic characteristics of two SiC N-IGBTs with buffer layer.

The latest physical model of SiC IGBT was presented in 2019 [8], it is the first unified model that is fit to any commercially available IGBT, containing Si/SiC, n-channel and p-channel, no-punch-through and field-stop type. The physical equations of it is also based on the Hefner model [13], but two additional parameters and two approximations were added to make the model more flexible and convergent [8]. In order to build a SiC IGBT model, the material properties were modified by SiC parameters, which contain the intrinsic carrier, dielectric constants [13] and the temperature-dependent hole and electron SiC mobility of SiC[4]. Additionally, a parameter extraction process was presented, which is based on find-up table. It can be easily used by circuit design engineers. Fig.6 shows the simulation result of turn-off transient of the same device at 25°C and 125°C, which fits well to the measured results.

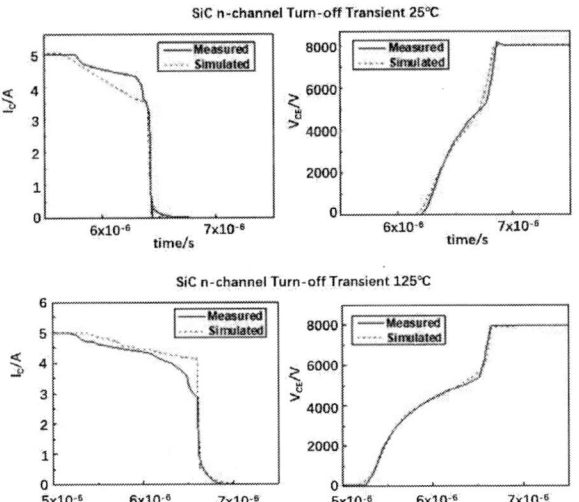

Fig. 6. Turn-off characteristic of SiC IGBT at 25°C and 150°C [9]

The unified model was validated by static and dynamic characteristics of three published Si or SiC IGBTs and the

results are not shown here for the sake of brevity. Moreover, the model has excellent convergence capability and can be applied in system level simulations, which has been validated by a complex simulation circuit in [8].

Another classical analytical IGBT model is the Kraus model [21]. Its distribution is to consider the influence of electron injection from P+ substrate which is aimed at NPT-IGBT with thin emitter. Moreover, the time-dependent ambi-polar diffusion equation for solving the carrier distribution in this model is simplified by a polynomial instead of linear assumption, which is more precise than Hefner model.

In 2013, Arash Nejadpak et al. proposed the first analytical electro-thermal model of SiC IGBT [7]. The electrical part of it is based on Kraus IGBT electrical model. However, from the comparison of simulation and experiment results, Arash Nejadpak et al. found that the Kraus IGBT electrical model cannot predict the behavior of SiC IGBT at higher temperature accurately. So they introduced Elmore thermal model into their model, in order to describe the temperature change in each layer of the IGBT structure. Then, some temperature-dependent parameters were modified, so as to improve the simulation accuracy in high temperature condition. As Fig. 7 shows, the similarity of simulation results and waves from datasheet increases a lot after adding thermal part into the whole model.

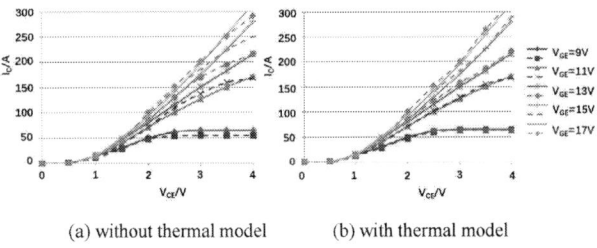

(a) without thermal model   (b) with thermal model

Fig. 7. Output characteristic of SiC IGBT at 150°C [7]

The main contribution of this model is to represent the electric heating effect and describe the temperature rise of the IGBT chip when it is working, in order to guide the design of the new packaging structure.

B. The Derived Models with Modified Special Structure

In order to achieve high efficiency in circuit simulation, there are many kinds of compact models for Si IGBT. These models use various simplification method in modeling to reduce the complexity of physical equations. When expended to SiC IGBT, special structures should be modified or added to the compact model, in order to describe some unique

978-1-6654-4817-8/21 $31.00 © 2021 IEEE

features of SiC IGBTs and improve the simulation accuracy from physical mechanism.

In 2013, [6] proposed a compact SiC IGBT model based on HiSIM-IGBT. Apart from updating the material parameters of the model, the authors added the base carrier distribution into their model in order to describe the punch-through effect of SiC IGBT.

The punch-through effect is an important feature in FS-IGBT switching operation [2]. When a FS-IGBT device is working on a high bus voltage, the depletion width of the internal reverse biased PN junction increases greatly, even punching through the base region. The width of depletion region follows:

$$W_n = \sqrt{\frac{2\varepsilon V_a}{qN_D}} \qquad (2)$$

where ε is the dielectric constant; V_a is the bus voltage; q is the charge unit; N_D is the doping concentration.

In the switching transient of FS-SiC IGBT, the punch-through effect is more likely to occur, the reasons are as follows. First, both of the material have almost same dielectric constant, which means the depletion width of SiC FS-IGBT increases as fast as Si FS-IGBT with the increase of bus voltage. However, the operating voltage of SiC IGBT is always higher than Si IGBT because of higher breakdown voltage [13], so that the depletion region is more easy to reach the field-stop layer. The turn-off characteristic differs a lot between non-punch-through condition and punch-through condition, as show in Fig.8.

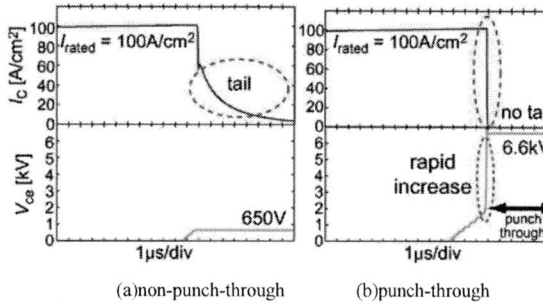

(a)non-punch-through (b)punch-through

Fig. 8. Output characteristic of SiC IGBT at 150℃[6]

In Fig.8, working under the punch-through condition causes abrupt change of I_C and V_{CE}, which has significant effects to the device characteristics. In order to take this effect into consideration, Masataka Miyake *et al.* redefined the R_{base} and Q_{base} in the HiSIM-IGBT model, as shown in Fig.9.

In the developed SiC HiSIM-IGBT model, the carrier distribution is divided into three regions: the depletion region, the quasi-neutral region and the buffer-layer region, which follows the piecewise function:

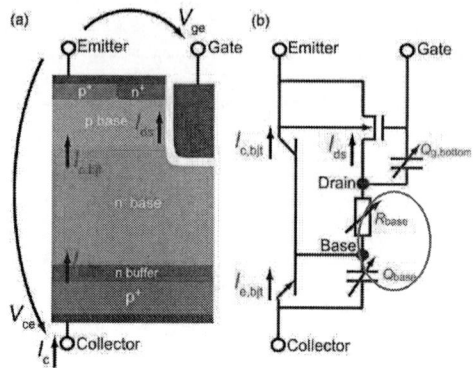

Fig. 9. (a) A half-cell unit of a trench-gate filed-stop IGBT. (b) Equivalent circuit of HiSIM-IGBT.The red circle marked the position of R_{base} and Q_{base} [22]

$$n_{base}(x)$$
$$= \begin{cases} n_{min} & \text{(depletion region)} \\ n_{inj} \cdot \exp(-x/L_{b,bjt}) & \text{(quasi-neutral region)} \\ n_{inj,buff} \cdot \exp[-(x+W_{buff})/L_{buff}] & \text{(buffer region)} \end{cases}$$
$$(3)$$

where n_{min} is the carrier density in the depletion region; n_{inj} is the injected carrier density from P+ substrate; $L_{b,bjt}$ is the diffusion length in the quasi-neutral region; $n_{inj,buff}$ is the injected carrier density in the buffer region and L_{buff} is the diffusion length in the buffer region [6]. While in the original HiSIM-IGBT model, there is no difference between the carrier distribution of depletion region and quasi-neutral region.

Based on the carrier distribution, R_{base} and Q_{base} are modified as the sum of resistance and charge in depletion region and quasi-neutral region respectively. With the new R_{base} and Q_{base}, the SiC HiSIM-IGBT model can explicitly distinguish the punch-through condition. This model has been implemented in SPICE3F5 and the validity of the model is proved by comparison between circuit simulation and 2-D device simulation results. This model is further improved by adding temperature influence in 2019 [10].

C. Some New Models

When building a derived models for SiC IGBT, the change of structure and material parameters makes the model more difficult to converge [15]. In order to build a more efficient SiC IGBT model or be used in special applications, some scientists developed their new SiC IGBT on their own.

Meng-Chia Lee *et al.* implemented their new circuit model of SiC IGBT in Simulink in 2015 [5]. It is an equivalent circuit model derived from charge continuity equation, as shown in Fig.10.

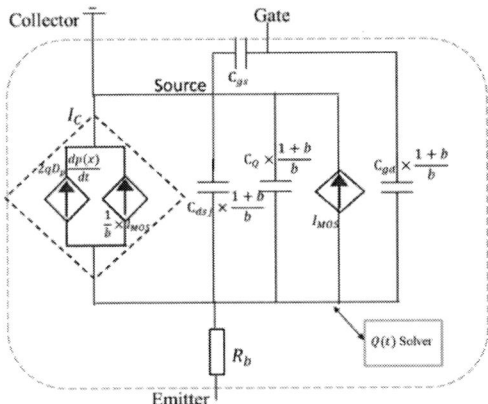

Fig. 10. Equivalent circuit of the IGBT model [5]

In order to describe the carrier distribution in the drift region, several lumped charges were put in the region which were calculated by continuity equation. The dynamic carrier distribution between the lumped charges is simply described by an exponential expression, which is shown in Fig.11. Although some simplification was made to calculate the base carrier distribution, it can also describe the punch-through effect to a degree. Moreover, the model is capacitance-based which can be easily understood by circuit designers and give an insight to instantaneous charge within the device. This model is implemented in SIMULINK environment and the model shows good fitting to the measurement of both static and dynamic characteristics.

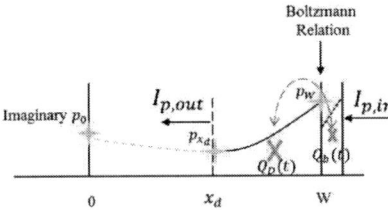

Fig. 11. Distribution of lumped charger [5]

Behavioral models have a better astringency and efficiency than physical models in system level circuit simulation. In 2020, Lubin Han *et al.* proposed the first behavioral model of SiC IGBT, which can be used in circuit simulation for power electronic system [12]. The internal physics of IGBT device was neglected and five controlled current sources is placed to describe the behavior of SiC IGBT. The equivalent circuit of the behavioral model is shown in Fig.12.

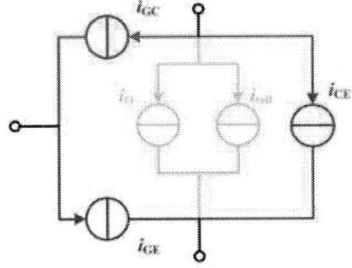

Fig. 12. Equivalent circuit of the behavioral model [12]

In the equivalent circuit, i_{GC}, i_{CE}, i_{GE} represent the behavior of parasitic capacitors and i_O, i_{tail} is for representing the static and transient characteristics of SiC IGBT.

As for the static model, the output current i_O is controlled by v_{GE} and v_{CE}. Therefore, the function of i_O (v_{GE}, v_{CE}) can be obtained by interpolation calculation on discrete data of I-V curves from datasheet.

As for the dynamic model, the authors proposed a method to extract the three capacitors from dynamic switching waveforms. The tail current during turn-off simulation is fitted by a two-stage exponential attenuation equation, which is put into a controlled current source. The result of numerical simulation and the behavior model simulation were compared to verify this model.

IV. CONCLUSION AND OUTLOOK

In this paper, all the existing SiC IGBT circuit models have been classified into three types, and then be introduced in detail. The classification guides the scientists who intend to develop new models for SiC IGBT or other promising devices.

Based on the research above, some trends for future research are put forward as follows:

1) Use some other physical modeling method in SiC IGBT modeling.

Except for Hefner model and Kraus model, there are other physics-based IGBT model with has potentiality to be extend to SiC IGBT, such as physics-based SPICE-model in [23] and the new lumped-charge model in [24].

2) Do proper modify and simplify in physics-based model of SiC IGBT.

Semiconductor physics equations is hard to converge because of wide numeric variation [25]. Moreover, when dealing with SiC material, the problem of convergence is even worse [15]. Therefore, proper modify and simplify to improve the model's efficiency is needed to further research.

3) Broaden the research of the other classes of models such as semimathematical models and seminumerial models.

As for various application nature, the requirement for model accuracy and efficiency differs. Physical model has high accuracy but low efficiency, while behavior model has easier development procedure but cannot forecast the device's behavior. The other types of SiC IGBT models can be a trade-off between accuracy and efficiency, fit for diverse simulation requirements.

4) The electro-thermal model is necessary to improve continuously.

The SiC IGBT is designed to be used in ultra-high voltage condition, so it brings great challenge to packaging structure and insulating material . Using the co-simulation of electrical and thermal model to predict the junction temperature accurately can guide the design of new packaging structure and the prediction of device reliability.

5) Do more research on the modeling for new SiC IGBT design.

SiC IGBT is under continuous development in order to improve its electrical performance and reliability. There are

978-1-6654-4817-8/21 $31.00 © 2021 IEEE

several new structures and parameters optimizations which have been proposed to achieve lower V_f and fast switching [26], such as TC-IGBT [27], DCS-IGBT [28], SC-IGBT [29], BD-IGBT [30], SJ-IGBT [31], AC-IEGT [32] and so on. Moreover, [33] introduced a step space modulated junction termination extension(SSM-JTE) structure to improve the blocking performance of SiC IGBT. These optimized designs can exploit the properties of SiC material better, so they have brighter application potential than conventional IGBT structure. However, there are few models focusing on these new design and it will be a trend in SiC IGBT modeling field.

REFERENCES

[1] J. Millan, P. Godignon, X. Perpina, and A. Perez-Tomas, "A Survey of Wide Bandgap Power Semiconductor Devices," *IEEE Transactions on Power Electronics,* vol. 29, pp. 2155-2163, 2014.

[2] S. H. Ryu, C. Capell, L. Cheng, C. Jonas, A. Gupta, M. Donofrio, J. Clayton, M. O'Loughlin, A. Burk, and D. Grider, "High performance, ultra high voltage 4H-SiC IGBTs," in *Energy Conversion Congress & Exposition,* 2012.

[3] T. H. Duong, A. R. Hefner, J. M. Ortiz-Rodríguez, S. H. Ryu, and J. W. Palmour, "Physics-based electro-thermal Saber model and parameter extraction for high-voltage SiC buffer-layer IGBTs," in *IEEE Energy Conversion Congress & Exposition,* 2014.

[4] T. Hatakeyama, K. Fukuda and H. Okumura, "Physical Models for SiC and Their Application to Device Simulations of SiC Insulated-Gate Bipolar Transist

[5] M. C. Lee, G. Wang and A. Q. Huang, "4H-SiC 15kV n-IGBT physics-based sub-circuit model implemented in Simulink/Matlab," *IEEE,* 2015.

[6] M. Miyake, M. Ueno, U. Feldmann, and H. J. Mattausch, "Modeling of SiC IGBT turn-off behavior valid for over 5-kV circuit simulation," IEEE Transactions on Electron Devices, vol. 60, pp. 622-629, 2013.

[7] A. Nejadpak, A. Nejadpak and O. A. Mohammed, "A physics-based, dynamic electro-thermal model of silicon carbide power IGBT devices," in *Applied Power Electronics Conference & Exposition,* 2013.

[8] S. Perez, R. M. Kotecha, A. Rashid, M. M. Hossain, T. Vrotsos, A. M. Francis, A. H. Mantooth, E. Santi, and J. L. Hudgins, "A Datasheet Driven Unified Si/SiC Compact IGBT Model for N-channel and P-channel Devices," *IEEE Transactions on Power Electronics,* p. 1-1, 2018.

[9] M. Saadeh, H. A. Mantooth, J. C. Balda, J. L. Hudgins, and A. Agarwal, "A unified silicon/silicon carbide IGBT model," in *2012 Twenty-Seventh Annual IEEE Applied Power Electronics Conference and Exposition (APEC),* 2012.

[10] K. Matsuura, Y. Tanimoto, A. Saito, Y. Miyaoku, T. Mizoguchi, M. Miura-Mattausch, and H. J. Mattausch, "Analysis and compact modeling of temperature-dependent switching in SiC IGBT circuits," *Solid-State Electronics, vol. 153,* pp. 59-66, 2019.

[11] R. H. A. and M. D. D., "An experimentally verified IGBT model implemented in the Saber circuit simulator," *IEEE Transactions on Power Electronics,* vol. 9, pp. 532-542, 1994-01-01 1994.

[12] H. L., L. L. and K. Y., "A SiC IGBT Behavioral Model with High Accuracy and Fast Convergence," in 2020 IEEE Workshop on Wide Bandgap Power Devices and Applications in Asia (WiPDA Asia), 2020, pp. 1-6.

[13] B. J. Baliga, Fundamentals of Power Semiconductor Devices: Fundamentals of Power Semiconductor Devices, 2008.

[14] E. V. Brunt, L. Cheng, M. O'Loughlin, C. Capell, and C. Scozzie, "22 kV, 1 cm2, 4H-SiC n-IGBTs with improved conductivity modulation," in *2014 IEEE 26th International Symposium on Power Semiconductor Devices & IC's (ISPSD),* 2014.

[15] L. Han, L. Liang, Y. Kang, and Y. Qiu, "A Review of SiC IGBT," IEEE Transactions on Power Electronics, vol. PP, p. 1-1, 2020.

[16] A. R. Hefner, "Modeling buffer layer IGBTs for circuit simulation," Power Electronics IEEE Transactions on, 1995.

[17] R. H. A., "A dynamic electro-thermal model for the IGBT," IEEE Transactions on Industry Applications, vol. 30, pp. 394-405, 1994-01-01 1994.

[18] A. Jr and D. L. Blackburn, "An analytical model for the steady-state and transient characteristics of the power insulated-gate bipolar transistor - ScienceDirect," Solid-State Electronics, vol. 31, pp. 1513-1532, 1988.

[19] A. Hefner and S. Bouche, "Automated parameter extraction software for advanced IGBT modeling," in Workshop on Computers in Power Electronics, 2000.

[20] T. H. Duong, A. R. Hefner and D. W. Berning, "Automated Parameter Extraction Software for High-Voltage, High-Frequency SiC Power MOSFETs," in Computers in Power Electronics, Compel 06 IEEE Workshops on, 2006.

[21] R. Kraus and K. Hoffmann, "An analytical model of IGBTs with low emitter efficiency," in International Symposium on Power Semiconductor Devices & Ics, Ispsd, 1993.

[22] M. M., N. D., F. U., J. M. H., K. T., O. T., and U. T., "HiSIM-IGBT: A Compact Si-IGBT Model for Power Electronic Circuit Design," IEEE Transactions on Electron Devices, vol. 60, pp. 571-579, 2013-01-01 2013.

[23] M. Cotorogea, "Physics-Based SPICE-Model for IGBTs With Transparent Emitter," IEEE Transactions on Power Electronics, vol. 24, pp. 2821-2832, 2009.

[24] Y. Duan, F. Ia Nn Uzzo and F. Blaabjerg, "A New Lumped-Charge Modeling Method for Power Semiconductor Devices," IEEE Transactions on Power Electronics, vol. 35, pp. 3989-3996, 2020.

[25] K. Sheng, B. W. Williams and S. J. Finney, "A review of IGBT models," IEEE Transactions on Power Electronics, vol. 15, pp. p.1250-1266, 2000.

[26] H. L., L. L., K. Y., and Q. Y., "A Review of SiC IGBT: Models, Fabrications, Characteristics, and Applications," IEEE Transactions on Power Electronics, vol. 36, pp. 2080-2093, 2021-01-01 2021.

[27] K. G. Menon and E. M. S. Narayanan, "Numerical Evaluation of 10-kV Clustered Insulated Gate Bipolar Transistor in 4H-SiC," IEEE Transactions on Electron Devices, vol. 60, pp. 366-373, 2013.

[28] W. Jin, Z. Meng, H. Jiang, S. To, and K. J. Chen, "SiC trench IGBT with diode-clamped p-shield for oxide protection and enhanced conductivity modulation," in 2018 IEEE 30th International Symposium on Power Semiconductor Devices and ICs (ISPSD), 2018.

[29] Y. Liu, Y. Wang, Y. Hao, J. Fang, C. Shan, and F. Cao, "A Low Turn-Off Loss 4H-SiC Trench IGBT With Schottky Contact in the Collector Side," IEEE Transactions on Electron Devices, vol. 64, pp. 4575-4580, 2017.

[30] S. Chowdhury, C. W. Hitchcock, Z. Stum, R. P. Dahal, I. B. Bhat, and T. P. Chow, "Operating Principles, Design Considerations, and Experimental Characteristics of High-Voltage 4H-SiC Bidirectional IGBTs," IEEE Transactions on Electron Devices, vol. 64, pp. 888-896, 2017.

[31] Z. Shen, Z. Feng, L. Tian, G. Yan, and Y. Zeng, "Optimized P-emitter doping for switching-off loss of superjunction 4H-SiC IGBTs," in China International Forum on Solid State Lighting: International Forum on Wide Bandgap Semiconductors China, 2017.

[32] Z. Wen, F. Zhang, Z. Shen, L. Tian, G. Yan, X. Liu, L. Wang, W. Zhao, S. Guosheng, and Y. Zeng, "A Novel Silicon Carbide Accumulation Channel Injection Enhanced Gate Transistor With Buried Barrier Under Shielding Region," IEEE Electron Device Letters, p. 1-1, 2017.

[33] Z. Wen, F. Zhang, Z. Shen, J. Chen, W. He, G. Yan, X. Liu, W. Zhao, L. Wang, and G. Sun, " Design and fabrication of 10-kV silicon-carbide p-channel IGBTs with hexagonal cells and step space modulated junction termination extension," *Chinese Physics B, 2019.*

A Layout Optimization Method to Reduce Commutation Inductance of Multi-Chip Power Module Based on Genetic Algorithm

Yu Zhou
College of Electrical Engineering
Zhejiang University
Hangzhou, China
yzhou16@zju.edu.cn

Yu Chen
College of Electrical Engineering
Zhejiang University
Hangzhou, China
chenyu_ncepu@163.com

Hongyi Gao
College of Electrical Engineering
Zhejiang University
Hangzhou, China
hongyi.gao@zju.edu.cn

Chengmin Li
College of Electrical Engineering
Zhejiang University
Hangzhou, China
lichengmin@zju.edu.cn

Haoze Luo
College of Electrical Engineering
Zhejiang University
Hangzhou, China
haozeluo@zju.edu.cn

Wuhua Li
College of Electrical Engineering
Zhejiang University
Hangzhou, China
woohualee@zju.edu.cn

Xianging He
College of Electrical Engineering
Zhejiang University
Hangzhou, China
hxn@zju.edu.cn

Abstract—This paper describes an automatic optimization method for multi-chip power module (MCPM) layout based on template generation and evolutionary computation techniques. The method is developed with particular emphasis on reducing the commutation inductance and balancing the branch inductances among paralleled chips of the module. An automatic layout generation from netlist and design constraints to ready-to-fabricate prototype is carried using two-step graph-based template generation and genetic algorithm-based sizing approach. The method employs the built-in multi-port discrete circuit model for fast evaluation of layout inductance for parasitics extraction. An analytical model is also established to assist the design of embedded snubber in the post-process stage. The layout design of a 4-chip SiC module is demonstrated. Both the simulations and the experiments are conducted to illustrate the advantage of performing the automatic optimization from the initial stage of the design process.

Keywords—multi-chip power module, layout optimization, parasitic inductance, genetic algorithm, snubber design

I. INTRODUCTION

In the past ten years, applications as electric automobiles and renewable energies have driven the growth of the power semiconductor market from less than $10 billion in 2010 to over $20 billion in 2020. Along with the market prosperity, there is an incremental demand for high-performance multi-chip power modules (MCPMs), which play an essential role in compact system integration and energy saving [1]. To release the potential of semiconductor devices, one of the significant works for MCPM design is dealing with the parasitic inductances, which would severely degrade the device capability by introducing voltage overshoot, ringing, and current imbalance during the switching transient.

Novel structures and design concepts are introduced in MCPMs to alleviate these issues. These techniques include embedding snubber capacitors, stacking a second layer conducting substrate, and using flip-chip packages. Among these techniques, heterogeneous integration of snubbers is the most cost-efficient one thanks to its good compatibility with the conventional packaging technology, i.e., the single-layer

This work was supported by a grant from the National Science Fund for Distinguished Young Scholars of China (No. 51925702), and the Natural Science Foundation of Zhejiang Province of China (No. LQ21E070006).

soldered substrate with wire-bonding interconnections. Furthermore, the embedded snubber is capable of closing the commutation loop in the layout and exclude the impact of power terminals. However, due to the loop and the branch inductances still sensitive to the layout design, the effects of manual works are inevitably limited by the designers' expertise and the project period. One viable way to deal with this bottleneck is to develop and adopt automatic optimization tools as in [2]–[8].

The early approach in [2] proposed an optimization framework for layout traces using the nonlinear programming algorithm. Though trace inductance is considered in the framework, it is treated as a design constraint with an upper bound rather than an objective for optimization, thus limits the output performance. The work in [3] takes the power trace inductance as a design objective. However, only the predefined sizing parameters are changed during the optimization, thus limit the design space. Additionally, due to the employment of finite element method (FEM) software as the parasitics extractor, computing time would be severely extended when dealing with complex module structures. In order to expand the design scope, the work in [4] proposed a two-step method to generate layouts with variable element positions and trace sizes. However, due to the routings are initially generated by random strings in a dense physical grid, there are considerable possibilities to generate unconnected traces. Thus, the connectivity and design rule checks (DRC) must be performed, resulting in a reduced optimization efficiency. Further research in [5] replaces the physical grids with abstract ones, which decreases the complexity. However, the check process may still be required due to the lack of connectivity info in the sequencing loop. Besides, since an integrated fitness function is defined as the design objective in this work, the loop inductance is merely one of the dividers. The output solution is not always the optimal one for parasitic inductances. To deal with the multiple targets in the layout design, e.g., electrical parasitics and thermal resistances, the work in [6] introduces the multi-objective optimization routings for the MCPM layout. In this work, a set of optimal solutions non-dominated to each other, i.e., Pareto-optimal front (POF), are served as design outcomes, providing flexible choices for designers. Meanwhile, the micro-strip inductance model was integrated into the routings to speed up the

TABLE I.   Comparison Among State-of-The-Art Works on MCPM Inductance Optimization

Work	Layout Generation	Inductance Evaluation		Layout Optimization		Snubber Design	
		Method	*Ind. Included*	*Algorithm*	*Outcome*	*Positioning*	*Selection*
Hingora et al. [2], 2010	not included	Invoking extractor	Loop	Nonlinear programming (fmincon)	Solution satisfying the condition	not included	
Hammadi et al. [3], 2011	Template-based sizing	Invoking extractor	Loop	Nonlinear programming (NLPQL)	Sorted solutions	not included	
Ning et al. [4], 2013	Placement & routing	Magnetic vector potential model	Loop	Evolution (GA)	Solution with the highest fitness	not included	
Ning et al. [5], 2015	Element sequencing & placement	Discrete model (PEEC)	Loop	Evolution (GA)	Solution with the highest fitness	not included	
Shook et al. [6], 2013	Template-based sizing	Micro-strip model	Loop	Evolution (NSGA-II)	Pareto-optimal front	Obtained from template	not included
Evans et al. [7], 2019	Template-based sizing	Response surface model	Loop	Evolution (NSGA-II)	Pareto-optimal front	Obtained from template	not included
Razi et al. [8], 2021	Template-based sizing	Discrete model (PEEC)	Loop	Evolution (NSGA-II & Randomization)	Pareto-optimal front	Obtained from template	not included
This work	Template generation & sizing	Discrete model (multi-port PEEC)	Loop & branches	Evolution (NSGA-II)	Pareto-optimal front	Generated with template	Analytical model-based

extraction process. Further works in [7] and [8] build the response surface model and the partial element equivalent circuit (PEEC) model for balancing the speed and accuracy. Finally, it is worth noting that the work in [8] introduces the constraint graphs (CG) to process the design constraints, thus avoid the DRC process and improve the success rate of layout generation. However, it is still a layout sizing tool with limited design space and does not include the branch inductances and the snubber selection in the optimization phase.

TABLE I summarized the characteristics of each work overviewed and the contributions of the approach proposed in this paper. For the first time in the literature, the proposed MCPM layout optimization method combines the following:

1) A MCPM-specific layout generator with template generation and automatic evolutionary sizing, supporting the generation with variable relative positions and no DRC required.

2) A integrated parasitic inductance extractor based on the multi-port PEEC method with an optimal discretization strategy for fast evaluation of loop and branch inductances.

3) An analytical model-based design for layout embedded snubber joining the post-process phase, enabling the close-loop development of heterogeneous modules.

This paper is organized as follows. In Section II, the architecture of the proposed method is described. Then the details of the layout generation and the parasitic inductance extraction methods are discussed in Section III and Section IV. In Section V, implementation results are presented, and finally, the conclusions are drawn in Section VI.

II. Proposed Optimization Flow

The detailed flow of the proposed optimization method is shown in Fig. 1. The process is divided into three parts: construction of the basic layout templates from module setup and design rule inputs, evolutionary extension and sizing the layout according to the evaluation results, and post-processing with Pareto-optimal front (POF) outputs and snubber designs.

A. Layout Template Generation

The layout template describes the geometric topology of the layout. A graph-based method is adopted here to generate the template. In this method, four kinds of basic blocks are defined to represent the areas with chips, snubbers, pads, and pure traces, respectively. The layout template is represented as a grid graph, where nodes correspond to the blocks, and edges represent the interconnection relationships between the blocks. According to the input circuit topology of the module, the grid graph is generated by randomly placing the component blocks and connecting them with trace blocks. Once a grid graph is built, the CG pairs introduced in [8] are generated from the horizontal and the vertical gridlines. Afterward, they are transferred to the extension and sizing loop to create tentative layouts for optimization. Thanks to the introduction of grid graphs, which is aware of the block connectivity during the generation, and the predefined constraints in blocks, the generated layouts are always connectivity check and DRC clean. As a consequence, both the success rate and the generation efficiency are improved. Further details on block definitions, layout representations, and the generation algorithm are described in Section III.

B. GA-based Sizing Loop

In this part, the genetic algorithm is employed to search the front sizing solutions for the input template. The major steps are sketched in order:

1) Optimizer: The layout optimization is carried out by the elitist non-dominated sorting genetic algorithm (NSGA-II) kernel. The kernel is set to solve the multi-objective problem defined as

$$\text{find } w_i, h_j \text{ that minimize } L_s \text{ \& } \Delta L_b$$

$$\text{subject to } w_i \geq w_i^L, \sum w_i = W, i = 1, 2, ..., M \ , \quad (1)$$

$$h_j \geq h_j^L, \sum h_j = H, j = 1, 2, ..., N$$

where w and h are the vectors of CGs edge weights for layout sizing, the objective L_s is the commutation loop inductance, and the objective ΔL_b is the variance of commutation inductances among paralleled chips.

978-1-6654-4817-8/21 $31.00 © 2021 IEEE

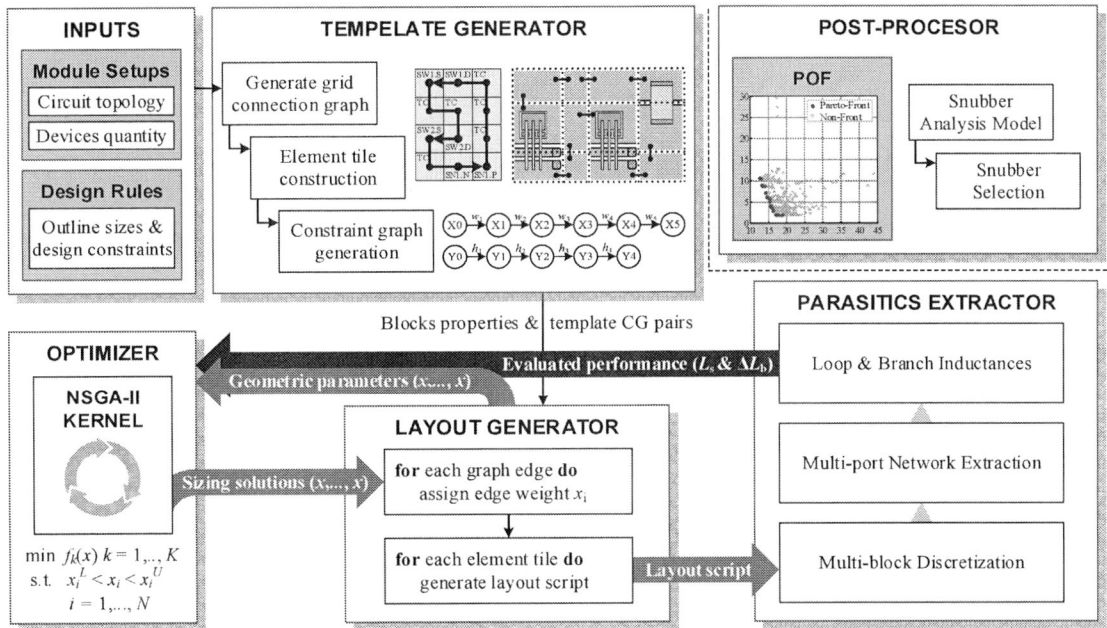

Fig. 1. Proposed optimization flow.

Two sets of lower bounds w^L and h^L are determined by block constraints. The layout outline constraints define the summation limits W and H.

2) Layout Generator: In the layout generator, the extension and sizing solutions from the optimizer are firstly assigned to the CGs edges. Then the geometries of each block are determined, and so is the tentative layout. Afterward, the layout is exported as a customized structure script. The conductors in the script are hierarchically described under the chip-separated nets for multi-port modeling and then passed to the integrated extractor.

3) Parasitics Extractor: The extractor is developed based on the PEEC method, which discrete all conductors into equivalent RL circuits to avoid the time-consuming electromagnetic field calculation. To speed up the extraction of the loop and branch inductances, the MCPM is modeled as a multi-port network, in which the chip electrons are treated as net ports. Then the LU decomposition is employed for fast port solving. As a result, the calculation time of this method is reduced by 85% compared to the Q3D software with 5% errors. Further details on the multi-port PEEC-based extractor are described in Section IV.

C. Post-Processing and Snubber Design

In the post-processing phase, the POF is served as the design outcome. For every solution on the front, the snubber capacitance is calculated by the analytical model in [9] as

$$\left| j\omega_R L_{Ext} \, / / \, \frac{1}{j\omega_R C_{Snb}} \right| \le 20 \left| j\omega_R L_s \right| \qquad (2)$$

where C_{Snb} is the snubber capacitance, L_{Ext} and L_s are respectively the external and internal inductance of the layout, and ω_R is the internal resonance frequency of L_s and device output capacitance C_{oss}.

The proposed method is flexible to modules with different circuit topologies, e.g., half-bridge modules, 3-level ANPC modules, etc. The inputs only need module setups and design rules. Thus, few designers' expertise is required to perform the optimization.

III. AUTOMATIC GENERATION OF LAYOUT TEMPLATE AND IN-THE-LOOP PROTOTYPE

This section provides further details on layout generation: first, the block definition and layout representation, followed by the generation algorithm.

A. Block Definition and Layout Representation

The grid graph is employed to describe the geometric topology of blocks. To build the heterogeneous layout in the graph, four types of blocks are defined in TABLE II. All components in the block are placed on the geometric midpoint. Particularly, paralleled chips in the switch block are placed in a row and perpendicular to the bonding wires to obtain uniform current paths. Input component sizes and design rules determine the minimum block sizes. The chip distance, trace length, and trace width are taken as the zoom variables for block sizing. As for the graph elements, since the switch block and the snubber block contains two adjacent trace areas, their graph elements employ two nodes and a directed edge. On the other hand, the terminal and the trace block employ one node; thus, edges can be constructed from the adjacent four nodes.

Fig. 2 illustrates an exemplary layout with this method. First, to form a commutation loop, the grid graph in Fig. 2(a) consists of a loop whose path starts from the snubber block SN1, then passes the switching blocks SW1 and SW2 through trace blocks, finally ends up in SN1. Next, the graph-constructed template is presented in Fig. 2(b). The graph edges are retained here to illustrate the connectivity between blocks and would be used to merge them in the end. Besides, since the block boundaries are aligned on gridlines, the horizontal and the vertical grid lines are used to build the CG pairs in Fig.

TABLE II.   BLOCK DEFINITION FOR LAYOUT GENERATION

Type	Block Geometry	Graph Element
Switch Block	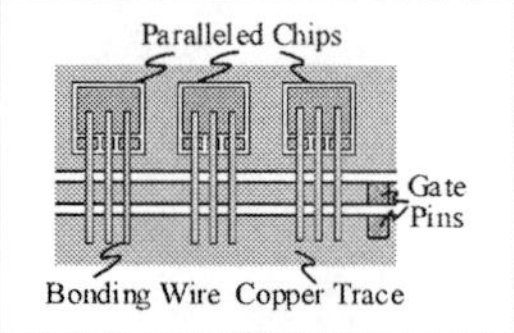 Paralleled Chips / Gate Pins / Bonding Wire Copper Trace	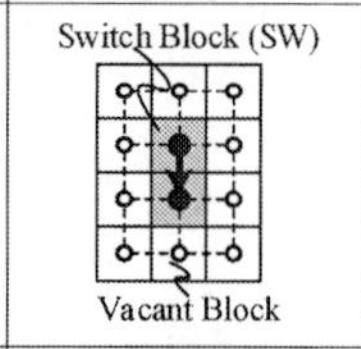 Switch Block (SW) / Vacant Block
Snubber Block	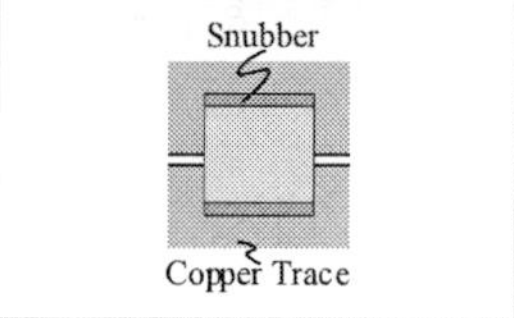 Snubber / Copper Trace	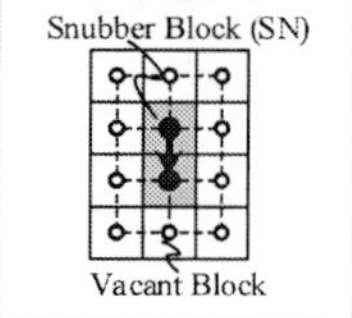 Snubber Block (SN) / Vacant Block
Terminal Block	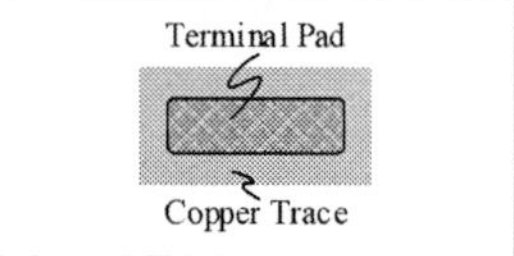 Terminal Pad / Copper Trace	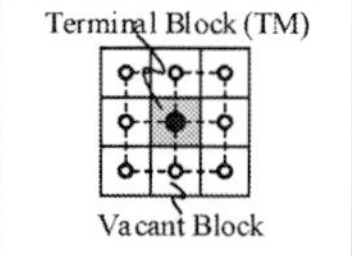 Terminal Block (TM) / Vacant Block
Trace Block	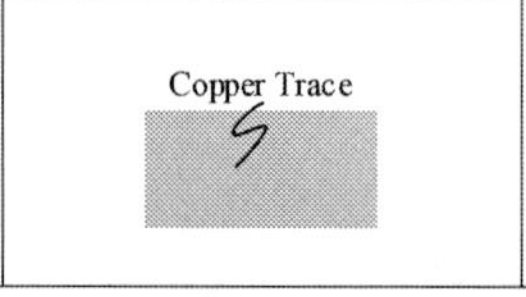 Copper Trace	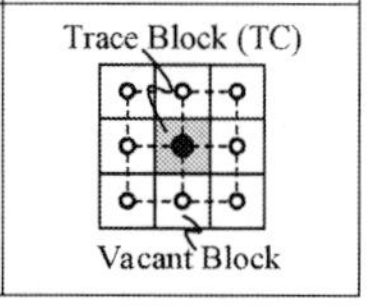 Trace Block (TC) / Vacant Block

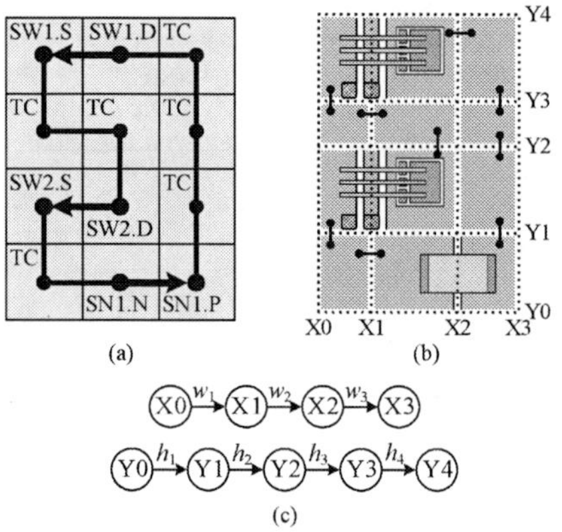

Fig. 2. Layout representations. (a) Block-based gridgraph. (b) Block-generated layout. (c) Constraint graph pairs of blocks.

2 (c). Thus, each node in CGs stands for a gridline, and the edge weight describes the block sizes in between.

B. Layout Generation Routine

The two-step generation routine is shown in Algorithm 1: line 1-7 for template generation and line 8-13 for layout extension and sizing.

In template generation, the circuit topology of the module is inputted in netlist format. A blocklist is firstly created by the components (switches, snubbers, and terminals) in the netlist. Meanwhile, the design constraints are saved in blocklist objectives for subsequent CG creations. Afterward, a grid graph is created as the plant with each block element placed randomly. To form a commutation loop, trace block elements and corresponding edges are created using the shortest-path algorithm. Once all nets are connected, a template is generated after excluding the empty nodes and creating CGs.

According to the GA solution, the CGs are modified by two processes: mirror extension to generate multi-loop layouts

---

Algorithm 1 Layout Generator

input: *Nelist, Constraints, SizingSolutions*
output: *Blocklist, LayoutScript*

1   Create *Blocklist* for all components in the *Netlist*
2   Create an empty *GridGraph*
3   **for** *each block* in the *Blocklist* **do**
4   Place block element randomly
5   **for** *each net* in the *Netlist* **do**
6   Create and place trace blocks using the shortest path algorithm
7   Exclude empty nodes and create *CGs* from gridlines
8   **for** *each variable* in the *SizingSolutions* **do**
9   Extend *CGs* and assign edge weights
10 Find components coordinates for all blocks
11 **for** *each edge* in the *GridGraph* **do**
12   Merge traces for edge-connected blocks
13 **return** convert *Blocklist* to *LayoutScript*

---

and assigning weight for sizing the templates. Consequently, the coordinates of blocks can be determined by evaluating the CGs and thus the component positions. Before the layout output, the edges of the grid graph are completely traversed to merge the connected traces. Then the layout structures are saved as the layout script and transferred to the parasitics extractor.

With this procedure, layout templates and their sizing prototypes can be generated without the back connectivity check and DRC. While, strictly speaking, there is still a possibility to produce failed template due to the stepwise trace blocks placing, it is easy to handle with the rip-up and replace strategy.

IV. LOOP AND BRANCH INDUCTANCES EVALUATION BASED ON MULTI-PORT PEEC METHOD

Due to the prohibitive execution time of commercial extractors, a build-in tool based on the discrete circuit method is used to evaluate the layout parasitics. Furthermore, to increase the efficiency for loop and branch inductances, a multi-port PEEC model is built considering the nature of MCPMs. Thus the matrix decomposition method is employed to accelerate the solving procedure.

By treating the chip electrons as net ports, a half-bridge module can be divided into five subnets, as illustrated in Fig. 3. Since the network is irrelevant to chips status, the calculation of port impedances is conducted through the decomposition of the circuit matrix, as

$$\begin{bmatrix} Z & -A^T \\ A & 0 \end{bmatrix} = LU \tag{3}$$

$$Ly = \begin{bmatrix} 0 & I_S \end{bmatrix}^T \tag{4}$$

$$U\begin{bmatrix} I_b & V_n \end{bmatrix}^T = y \tag{5}$$

where Z and A are the impedance matrix and the incidence matrix of the discrete circuit, I_S is the source vector, in which the port element is assigned as a unit excitation for each

978-1-6654-4817-8/21 $31.00 © 2021 IEEE

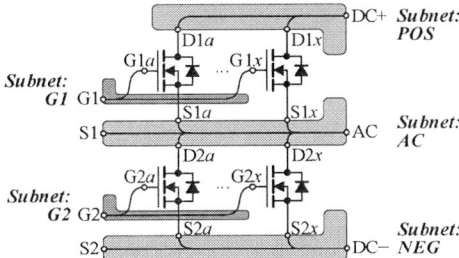

Fig. 3. Multi-port network model of a half-bridge module.

calculation, and $\boldsymbol{I}_b$ and $\boldsymbol{V}_n$ are unknowns. Thanks to the upper and the lower triangle matrix $\boldsymbol{L}$ and $\boldsymbol{U}$, the port impedance calculation is much faster than direct solving the circuit matrix.

Based on the network impedance, the commutation loop inductance is obtained by solving the DC port response with modified port status, which is represented by the network incidence matrix as

$$A_m = \left[diag(A_{POS}, A_{AC}, A_{NEG}) \mid A_{DS} \right] \tag{6}$$

where the left part is the incidence matrix of the power subnets, and the right part A_{DS} describes the new branch introduced by semiconductor chips.

Similarly, the network impedance is modified with new elements as

$$Z_m = \begin{bmatrix} \boldsymbol{Z}_{D1x} & \boldsymbol{Z}_{D1x\text{-}S1x} & \boldsymbol{Z}_{D1x\text{-}D2x} & \boldsymbol{Z}_{D1x\text{-}S2x} & 0 \\ \boldsymbol{Z}_{S1x\text{-}D1x} & \boldsymbol{Z}_{S1x} & \boldsymbol{Z}_{S1x\text{-}D2x} & \boldsymbol{Z}_{S1x\text{-}S2x} & 0 \\ \boldsymbol{Z}_{D2x\text{-}D1x} & \boldsymbol{Z}_{D2x\text{-}S1x} & \boldsymbol{Z}_{D2x} & \boldsymbol{Z}_{D2x\text{-}S2x} & 0 \\ \boldsymbol{Z}_{S2x\text{-}D1x} & \boldsymbol{Z}_{S2x\text{-}S1x} & \boldsymbol{Z}_{S2x\text{-}D2x} & \boldsymbol{Z}_{S2x} & 0 \\ \hline 0 & 0 & 0 & 0 & 0 \end{bmatrix} \tag{7}$$

where the element subscripts correspond to the port in 0.

As for the branch inductances, the variance for output is calculated by finding the maximum difference among power ports inductances for two switches as

$$\Delta L_b = \max(\Delta L_{D1x}, \Delta L_{S1x}, \Delta L_{D2x}, \Delta L_{S2x}) \tag{8}$$

The numerical results show that the calculation error of the method is less than 5%. And the calculation time is reduced by over 85% compared with the ANSYS Q3D software, made it suitable for fast optimization of module layouts.

V. IMPLEMENTATION AND VALIDATIONS

The framework of the proposed method has been implemented in MATLAB and is here tested on an Intel® Xeon™ E5-1650 v4 CPU with 32 GB RAM. To show its potential, a half-bridge module is considered. The input layout outline is 60 × 45 mm, and the module uses four chips (CPM2-1200-0040B, 1200V/40mΩ SiC MOSFET) for a switch. Two kinds of design constraints are defined as clearances for DBC etching (0.7 mm) and component soldering (1 mm).

A. Generated Templates

For the test, 100 layout templates are generated, and the execution time is within 12 s. With the minimum block size, part of the templates is shown in Fig. 4. The interspaces in templates come from the empty nodes inside graphs and would be filled by nearby traces before sizing.

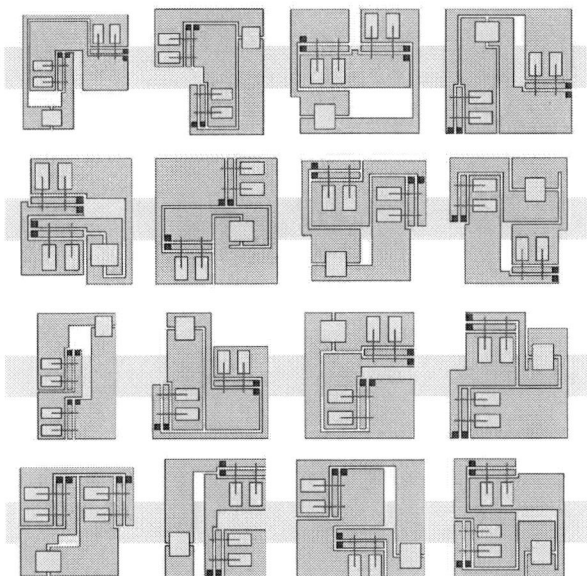

Fig. 4. Part of the generated layout templates.

B. Sizing Optimization Results

To further validate the optimization output, a front solution is selected and illustrated in Fig. 5. In this solution, two commutation loops are formed by mirror extension of the template for inductance reduction and balancing. According to the integrated extractor, the loop inductance is 5.72 nH, and the maximum variance for paralleled chips is 1.09 nH, indicating balanced paths for chip current. The calculated minimum snubber capacitance is 17.23 nF. Accordingly, the design uses two 15 nF/1000 V MLCC capacitors as snubbers.

C. Validations

The extraction results of layout inductances are validated by Ansys Q3D software and the impedance analyzer. The setups of the impedance analyzer are shown in Fig. 6. The loop inductances extracted by the software and the measurement are respectively 5.61 nH and 5.59 nH at the frequency 10 MHz. The relative errors of the results by the integrated extractor are 2.0% and 2.3%. Furthermore, the maximum difference among branch inductance calculated by Q3D software is 1.14 nH. Therefore, the relative error of the extractor result is 4.3%, showing a good accuracy of the integrated extractor.

To varify the performance of the optimized result, the switching behavior is validated by circuit simulation in Pspice. The parasitic parameters, including resistances, self-inductance, and mutual inductance, are extracted by Q3D software and saved as a multi-port network Spice model. And the device model is provided by the semiconductor manufacturer. The simulation is conducted under the 800 V DC voltage and 240 A load current with the drive resistance set to 2 Ω for fast switching. Fig. 7 presents the switching behavior measured on signal terminals for the lower switch. It shows a very low voltage overshoot of less than 50 V and no high-frequency oscillation during turn-off, thus validating the effects of snubber design. Besides, Fig. 8 shows the current distribution of lower devices in switching transient. The maximum current imbalance is 10% appears in the turn-on process.

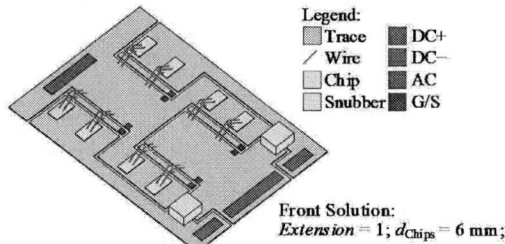

Legend:
- ▨ Trace
- ╱ Wire
- ▢ Chip
- ▨ Snubber
- ▨ DC+
- ▨ DC−
- ▨ AC
- ▨ G/S

Front Solution:
Extension = 1; d_{Chips} = 6 mm;

Fig. 5. Overviews of the selected front solution for validation.

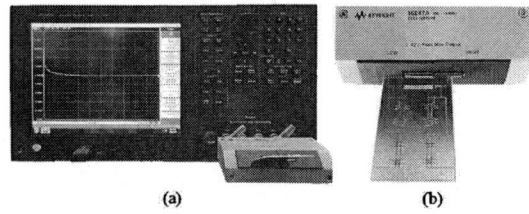

(a)   (b)

Fig. 6. Measurement of the commutation loop inductance. (a) Test setup. (b) Test kit.

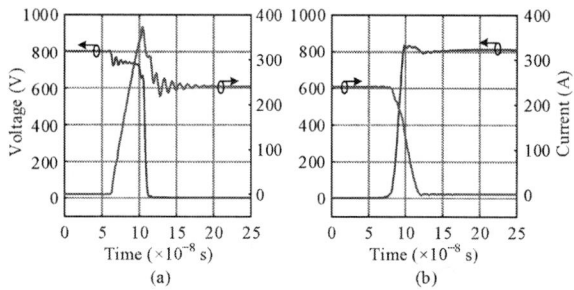

(a)   (b)

Fig. 7. Simulation waveform of drain-source voltage and drain current of the lower switch at switching transient. (a) Turn-on. (b) Turn-off.

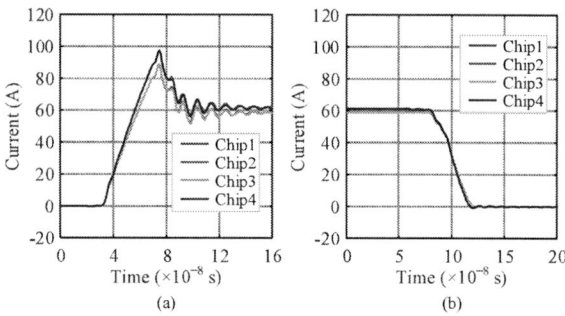

(a)   (b)

Fig. 8. Simulation waveform of drain current of lower chips at swithcing transient. (a) Turn-on. (b) Turn-off.

VI. CONCLUSIONS

Automatic optimization methods for MCPMs layout inductances are essential to speed up the development process and improve the module performance. This paper overviewed the major efforts proposed in this area in the last ten years and proposed a systematic approach by combining the template generation, layout sizing, and inductance extraction. This approach targeted to expand the design scope for MCPMs layout and close the development loop for current balancing and overshoot snubbing. Promising results validated the capability of the method.

In the future work, aiming to complete the MCPMs design flow, the chip planning and vertical structures optimization method with the the power capacity evaluation will be investigated and integrated as a former process before the layout optimization.

REFERENCES

[1] Yole Développement, "Status of the Power Module Packaging Industry 2020: Market and Technology Report," Lyon, 2020.

[2] N. Hingora, X. Liu, Y. Feng, and B. Mcpherson, "Power-CAD : A Novel Methodology for Design , Analysis and Optimization of Power Electronic Module Layouts," 2010 IEEE Energy Convers. Congr. Expo., pp. 2692–2699, 2010.

[3] M. Hammadi, J. Y. Choley, O. Penas, J. Louati, A. Rivière, and M. Haddar, "Layout optimization of power modules using a sequentially coupled approach," Int. J. Simul. Model., vol. 10, no. 3, pp. 122–132, 2011.

[4] P. Ning, F. Wang, and K. D. T. Ngo, "Automatic layout design for power module," IEEE Trans. Power Electron., vol. 28, no. 1, pp. 481–487, 2013.

[5] P. Ning, X. Wen, Y. Mei, and T. Fan, "A fast universal power module layout method," 2015 IEEE Energy Convers. Congr. Expo. ECCE 2015, pp. 4132–4137.

[6] B. W. Shook, A. Nizam, Z. Gong, A. M. Francis, and H. A. Mantooth, "Multi-objective layout optimization for multi-chip power modules considering electrical parasitics and thermal performance," 2013.

[7] T. M. Evans et al., "PowerSynth: A Power Module Layout Generation Tool," IEEE Trans. Power Electron., vol. 34, no. 6, pp. 5063–5078, 2019.

[8] I. Al Razi, Q. Le, T. M. Evans, S. Mukherjee, H. A. Mantooth, and Y. Peng, "PowerSynth Design Automation Flow for Hierarchical and Heterogeneous 2.5-D Multichip Power Modules," IEEE Trans. Power Electron., vol. 36, no. 8, pp. 8919–8933, 2021.

[9] Z. Chen, D. Boroyevich, P. Mattavelli, and K. Ngo, "A Frequency-Domain Study on the Effect of DC-Link Decoupling Capacitors," in 2013 IEEE Energy Conversion Congress and Exposition, 2013, pp. 1886–1893.

A GaN-based High Power Density Power Optimizer for Solar-powered Aircraft Applications

Peng Chen[1], Tao Liu[2], Yujie Cheng[3], Hongfei Wu[1], and Jianxin Zhu[4]

[1] College of Automation, Nanjing University of Aeronautics and Astronautics, Nanjing, China, wuhongfei@nuaa.edu.cn
[2] Shanghai Institute of Space Power-Sources, Shanghai, China, leotau@foxmail.com
[3] Nanjing Electronic Devices Institute, Nanjing, China, zhaolumuzhu@163.com
[4] College of Automation & College of Artificial Intelligence, Nanjing University of Posts and Telecommunications, Nanjing, China, zhujianxin@njupt.edu.cn

Abstract—The energy management system is the control center of a solar-powered aircraft. The most important concerns of the energy system are conversion efficiency and power density and. Considering the requirements of power density and efficiency, a high frequency planar inductor for a GaN-based four-switch buck/boost (FSBB) converter is designed. At the same time, the parameters of the converter are optimized to determine switching frequency. Considering the MPPT of solar panel and battery charging management, the control strategy for the FSBB converter is presented. Working modes can be switched smoothly with the dynamic switching frequency changing. Finally, a 400W solar power optimizer is built. The power density of the power optimizer is 750W/inch³ and the highest efficiency is up to 97.6%.

Keywords—*GaN, planar inductor, four-switch buck/boost, high power_-density, solar optimizer*

I. INTRODUCTION

Solar-powered aircraft, which uses clean and sufficient solar energy as an energy source, can achieve uninterrupted flight day and night by storing the excess energy in the onboard battery. It has the advantages of long-dead time and high altitude. It is widely used in military, commercial, meteorological, and other fields, such as border patrol, geographic mapping, communication relay, meteorological detection, and disaster field command and communication .

In high-altitude environments, the only energy source of UAV is solar cells. At present, the energy conversion efficiency of photovoltaic cells is low, the energy density of energy storage cells is low, and the energy storage technology cannot achieve a qualitative leap in the short term. To complete the task requirements, UAV also has to carry a variety of electronic equipment, to carry more equipment, maximize the use of the output power of solar cells, and reduce the volume and weight of photovoltaic management system, which has become the most critical problem of high-altitude long-endurance UAV airborne energy system [1].

The airborne photovoltaic power optimizer system is mainly divided into a centralized system and distributed system. Although the structure of the centralized system is simple, it cannot achieve the maximum power output of each module, which leads to low power generation efficiency. Compared with the centralized structure, the distributed structure can achieve the maximum power output of each photovoltaic module [2]. As shown in Fig. 1, the distributed structure can be divided into the parallel structure and series

The work is supported by Foundation of SAST.

structures. Parallel structure needs to use a transformer to obtain the high step-up ratio, so it has low efficiency However, in the series structure, the output side of multiple modules is connected in series in order to obtain a high DC voltage, which does not need a high voltage rise, so the efficiency is higher [3].

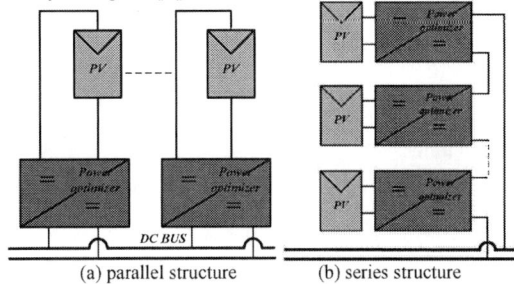

(a) parallel structure   (b) series structure

Fig. 1. The distributed system of power optimizer.

UAV airborne photovoltaic optimizers have been facing the problem of difficulty in balancing efficiency and volume. With the emergence of wide bandgap semiconductor devices such as GaN, the switching frequency of the converter has been greatly improved, and the loss has been greatly reduced. Relying on the excellent switching and conduction performance of GaN, the switching frequency of photovoltaic converter can reach the level of megahertz. This makes the power switching device not a main factor that restricts the power density and efficiency of the converter any more. [4].

However, the megahertz switching frequency brings difficulties to the design of devices such as inductors. The skin effect and proximity effect of high-frequency windings make the AC resistance and loss of traditional wound windings increase greatly; Due to the limitation of the shape and size of the magnetic core, the height of the magnetic component is significantly higher than that of other devices, which restricts the improvement of the power density of the photovoltaic DC module. PCB winding and planar magnetic technology are proved to be effective ways to solve the above problems. The power density and efficiency can be greatly improved by optimizing the design of a planar magnetic core by simulation software [5].

In this paper, a photovoltaic optimizer for solar aircraft is studied. Considering the requirements of efficiency and power density, the optimal switching frequency and GaN device are selected by analyzing the converter; Through simulation analysis, the optimal planar inductor is designed.

978-1-6654-4817-8/21 $31.00 © 2021 IEEE

In addition, considering the control of the converter itself and MPPT, a simple and effective control strategy is analyzed

II. CIRCUIT TOPOLOGY AND CONTROL PRINCIPLE

The input energy source of the photovoltaic optimizer is from the solar photovoltaic array. Due to the wide range of environmental factors such as illumination conditions, temperature, and flight height of airborne photovoltaic panels, the output voltage range of photovoltaic cells fluctuates more than twice. The output terminal of the optimizer is the airborne high-voltage lithium battery, and the output voltage fluctuates more than twice as much as the battery is charged and discharged. The characteristics of wide input and output voltage range require that the photovoltaic optimizer of the solar vehicle has the ability of voltage lifting and lowering at the same time.

A. Topology selection

In the basic non-isolated DC / DC converter topology, Buck-Boost, Cuk, Zeta, and Sepic have the function of voltage step-up and down. The polarity of input and output voltage of Buck-Boost and Cuk is opposite. And there are more inductors and capacitors to Zeta and Sepic, which is not conducive to improving power density. To achieve the function of stepping up and down the voltage simultaneously, a dual-switch Buck/Boost converter can be obtained by cascading Buck converter and Boost converter. A four-switch Buck / Boost (FSBB) converter is obtained by replacing two diodes with synchronous rectifiers as shown in Fig. 2. The polarity of the input and output of FSBB is the same. In addition, the voltage stress of the switch is the input voltage and output voltage respectively, which is lower than that of Buck-Boost [6].

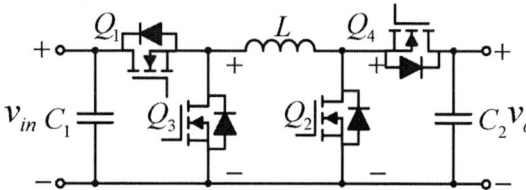

Fig. 2. Schematic diagram of FSBB.

B. Control strategy

Since the input and output voltages of photovoltaic optimizers vary in a wide range, it will inevitably lead to the switching problem of Boost and Buck modes [7].

As shown in Fig. 3, taking into account the MPPT of the system and the control of the converter itself, this paper adopts the control block diagram. The output voltage and current of photovoltaic are sampled, and the maximum output power point of photovoltaic is automatically tracked in the controller by using the perturbation and observation method. The output voltage is sampled to realize the charging control of the airborne battery. The MPPT loop and the constant voltage loop adopt the competition mechanism. Only one control loop works at any time. The battery is always charged with the maximum power before it is fully charged. After the battery is fully charged, it will automatically enter the constant voltage mode and exit the MPPT mode.

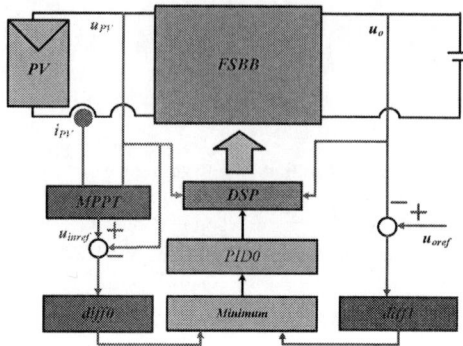

Fig. 3. Converter control block diagram.

As shown in Fig. 4 is a schematic diagram of modulation strategy of the converter. The single modulation wave dual-carrier strategy is used. At any time, the modulation wave is only combined with one carrier, that is, the converter will only work in the Boost or Buck mode at any time, which helps to reduce the loss.

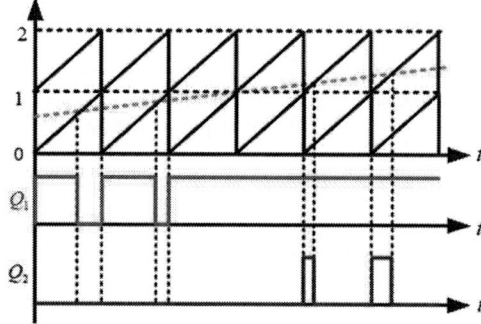

Fig. 4. Schematic diagram of single modulation wave dual carrier.

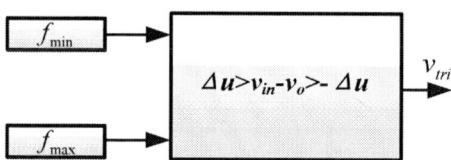

Fig. 5. Frequency switching diagram

Near the switching point of step-up and step-down modes, due to high switching frequency, small duty cycle and the limited resolution of the switch driver chip, the problem of the digital limit cycle will be caused [8]. In turn that will cause the output voltage to fluctuate violently. Therefore, as shown in Fig. 5, a dynamic changing method of switching frequency is adopted in this paper When the difference between input voltage v_{in} and output voltage v_o is less than Δu, the carrier frequency is reduced to reduce the switching frequency so that the two modes can switch smoothly. Otherwise, reducing the switching frequency is also good for reducing the power loss.

In actual work, a half-bridge drive circuit is a bootstrap circuit. Therefore, the upper switch of the bridge arm cannot maintain the straight-through mode.

978-1-6654-4817-8/21 $31.00 © 2021 IEEE

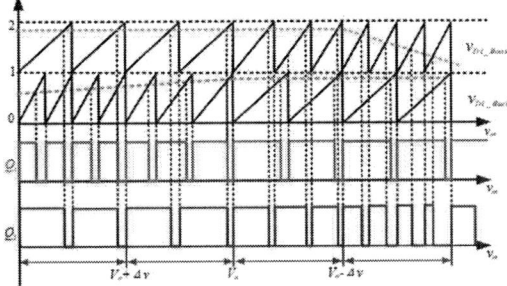

Fig. 6. Actual modulation strategy.

So, the actual modulation method is shown in Fig. 6. As shown in the figure, the modulation has been divided into four periods. 1)$v_{in}>V_o+\Delta v$, the carrier wave of buck is high frequency, while the carrier of boost is low frequency, the modulation wave of boost is constant, the duty of boost is 0.99, FSBB works at buck mode. 2)$V_o+\Delta v>v_{in}>V_o$, while the carrier of boost is still at low frequency, the carrier wave of buck drop from high frequency to low frequency slowly the modulation wave of boost is constant, the duty of boost is 0.99, FSBB works at buck mode. 3)$V_o>v_{in}>V_o-\Delta v$, the carrier wave of buck is low frequency, while the carrier of boost rise from low frequency to high frequency slowly, the modulation wave of buck is constant, the duty of buck is 0.99, FSBB works at boost mode. 4)$V_{in}<V_o+\Delta v$, the carrier wave of buck is low frequency, while the carrier of boost is high frequency, the modulation wave of buck is constant, the duty of buck is 0.99, FSBB works at boost mode. During the whole process, the switching is smooth and there will be no sudden changes.

III. FREQUENCY SELECTION AND INDUCTOR DESIGN

Due to the limitation of structure and dynamic performance of solar-powered aircraft, high efficiency, lightweight, small volume, and high-density photovoltaic optimizer is of great value to improve the performance of solar-powered aircraft. Next, through the determination of the frequency, the selection of the GaN device, and the design of the core of the inductor to optimize the design FSBB. These designs are mainly determined by the parameters of the converter, the main parameters are:
V_{in}:30-70V, V_o:30-70V, power:400W, efficiency:>97%, power-density:>200W/inch³.

A. Frequency determination and switch selection

GaN can increase the switching frequency to the megahertz level. The switching frequency of the converter is determined by the required target efficiency, and the selection of the switch is mainly from the perspective of loss.

Considering the safety margin, the appropriate GaN devices on the market are shown in the TABLE I.

TABLE I. PARAMETERS OF DIFFERENT TYPE OF GAN

type	V_{ds}(V)	R_{ds_on}(Ω)	C_{iss}(pF)	V_{th}(V)	Q_{gd}(nc)
GS61008T	100	0.07	590	1.3	1.5
EPC2034C	200	0.06	1166	1.1	2.1
EPC2034	200	0.05	950	1.4	1.8
EPC2033	150	0.05	1160	1.4	3.2

Through the above-mentioned parameters, the losses of different types of GaN under different switching frequencies can be calculated under V_{in}=70V, V_o=50V. As shown in Fig. 7.

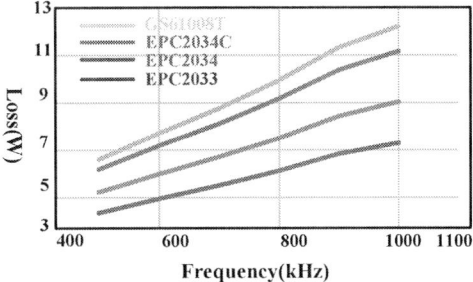

Fig. 7. Switch loss curve.

It can be seen from the figure that GST1008T has the best loss performance. Considering the efficiency requirement, the switching frequency is determined as 800kHz.

B. Inductor design

The design of the inductor is the key factor to determine the efficiency of the converter. By optimizing the design inductor by the simulation software ANSYS, the optimal design parameters of the inductor with consideration of size and loss can be found.

The first is the selection of inductance value, which is determined by the ripple demand of the inductor current.

The inductance value of FSBB is determined by the following equation (1) and (2).

$$L=\begin{cases} \dfrac{V_{in}}{2k\%\cdot I_L}\cdot D_{boost}\cdot T_s (30\le V_{in}\le V_o) \\ \dfrac{V_o}{2k\%\cdot I_L}\cdot (1-D_{buck})\cdot T_s (V_o\le V_{in}\le 70) \end{cases} \quad (1)$$

$$I_L=\begin{cases} 10(30\le V_{in}\le 40\le V_o) \\ \dfrac{P_o}{V_{in}}(40\le V_{in}\le V_o) \\ 10(V_o\le V_{in}\le 70 \&\&V_o\le 40) \\ \dfrac{P_o}{V_{in}}(V_o\le V_{in}\le 70 \&\&V_o\ge 40) \end{cases} \quad (2)$$

Where $2k\%$ is the peak-to-peak ratio of the ripple current of the inductor, P_o is the output power, T_s is period, V_{in} is input voltage, V_o is the output voltage, and I_L is the average ripple current of the inductor.

When V_o=50V, P_o=400W, f=800kHz, and k=20, the inductance changes with the input voltage as shown in Fig. 8.

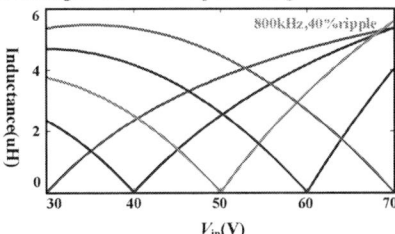

Fig. 8. Inductance under different input voltage.

978-1-6654-4817-8/21 $31.00 © 2021 IEEE

Considering that FSBB works in a continuous current state, the DC bias of current is large, and the copper loss of winding is much larger than that of the magnetic core. Therefore, choosing a larger ripple requirement can effectively reduce the size of the magnetic core without causing too much additional loss. Which is important to an inductor.

When designing the inductor, limit the maximum magnetic density $B_{lim}=300mT$ at the maximum current, and select the different peak-to-peak values of the magnetic flux ΔB_{max}. According to equation (3), the corresponding ripple size $2k$ is obtained. Similarly, the inductance value under different ripples can also be obtained in TABLE II.

$$2k = \frac{\Delta B_{max}}{B_{lim} - \frac{\Delta B_{max}}{2}} \qquad (3)$$

TABLE II. PARAMETERS OF RIPPLE AND L

$B_{lim}(mT)$	$\Delta B_{max}(mT)$	ripple($2k\%$)	$L(uH)$
300	80	30.7%	3.6
300	100	40%	3
300	120	50%	2.5
300	140	60.9%	2

Considering that the switching frequency reaches the megahertz level, the magnetic core adopts DMR51W, and its saturation magnetic density is 350mT.

The inductor structure adopts the magnetic core of the EI structure, and its structure and main parameters are shown in Fig. 9.

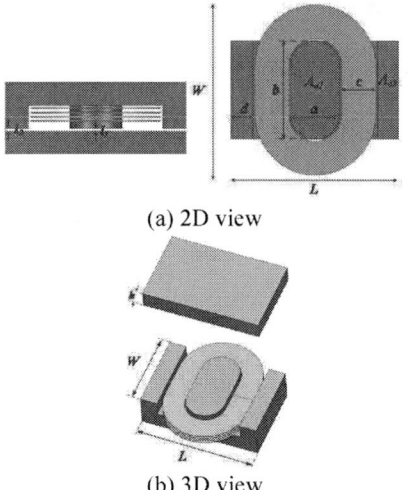

(a) 2D view

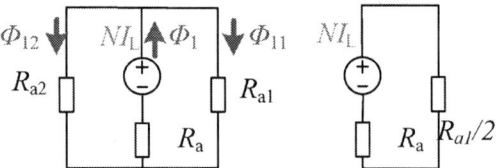

(b) 3D view

Fig. 9. Dimension parameters diagram

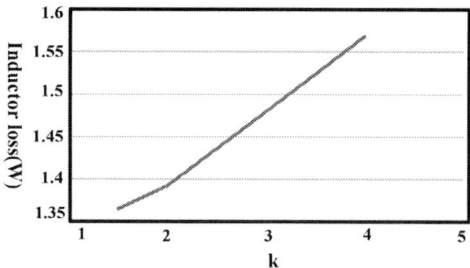

Fig. 10. Reluctance model of the inductor

Its reluctance model is shown in Fig. 10. According to the parameters in the figure, combined with the reluctance model, equation (4) can be obtained.

$$\begin{cases} A_{e1} = 2A_{e2} \\ R_a = \frac{\mu A_{e1}}{l_1} \\ R_{a2} = \frac{\mu A_{e2}}{l_2} \\ l_1 = l_2 \end{cases} \qquad (4)$$

From the above equation, the inductance is obtained in equation(5). N is the turns of winding, and μ is permeability.

$$L = \frac{2N^2 \mu A_{e1}}{l_1} \qquad (5)$$

The relationship between these size parameters is determined by equation(6).

$$\begin{cases} A_{e1} = 0.25\pi a^2 + ba - a^2 \\ A_{e2} = bd \\ k = \frac{b}{a} \\ F_p = L \cdot W = (2d + 2c + a) \cdot (b + 2c) \end{cases} \qquad (6)$$

Combining the equation(5)(6), we can see that under the determined inductance value L, footprint F_p, and k, the entire size of the inductance can be determined. Fix $k=1.5$,$N=4$ scanning different inductance values and footprints to get the loss of different inductances under different footprints as shown in Fig. 11.

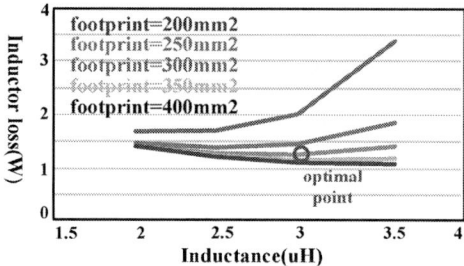

Fig. 11. Loss curves of different inductors under the different footprint

As shown in the figure, it can be seen that under the same inductance value, as the footprint increases, the loss reduction becomes less and less significant. Under the same footprint, the inductance loss is the smallest at 3uH. Finally, select the inductance value of 3uH, footprint 300mm². Then by scanning different k values, different loss values can be obtained under ANSYS simulation as shown in Fig. 12. Finally, $k=1.5$ has been chosen.

Fig. 12. Inductor loss under different k.

ANSYS simulation results are shown in Fig. 13 under I=10A,$2k\%$=40%, as can be seen from the figure, the

978-1-6654-4817-8/21 $31.00 ©2021 IEEE

magnetic density distribution of the magnetic core is relatively balanced.

Fig. 13. Magnetic density distribution.

IV. EXPERIENMT VERIFICATION AND ANALYSIS

A. Experimental waveform

To verify the correctness of the above converter design and the effectiveness of the converter control strategy, the prototype is shown in Fig. 14.

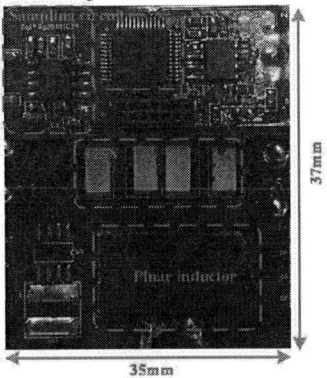

Fig. 14. Prototype of FSBB

Fig. 15 shows the mode switching at $P_o = 400W$, $V_o = 50V$, (a) V_{in} switches from 30V to 70V (b) V_{in} switches from 70V to 30V.

The experimental waveform shows that during the switching process, the current and voltage overshoot is very small and the dynamic adjustment is rapid.

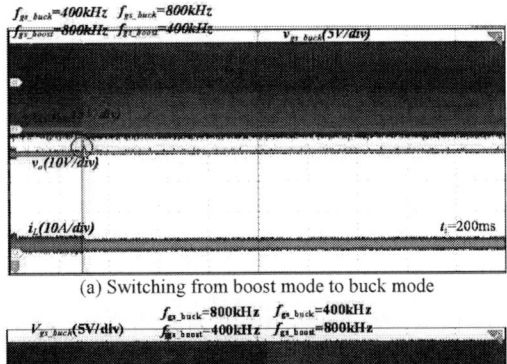

(a) Switching from boost mode to buck mode

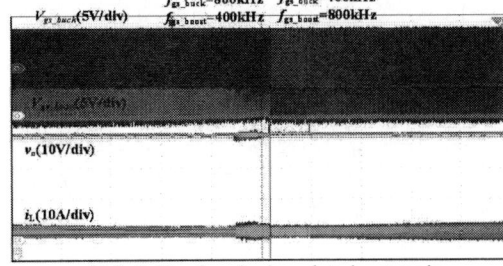

(b) Switching from buck mode to boost mode

Fig. 15. Mode switching between buck and boost.

Fig. 16 shows switching waveform of FSBB in MPPT mode and constant voltage mode. The reference value of the constant voltage loop is the battery charging voltage of 60V. After adjusting the load and increasing the output voltage from 50V to 60V, FSBB exits MPPT mode. When the output voltage is reduced, the converter returns to MPPT mode. The experimental results show the effectiveness of the control strategy.

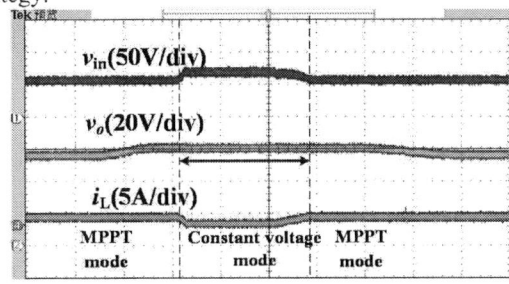

Fig. 16. Mode switching between constant voltage mode and MPPT mode.

B. Efficiency and temperature test

The full load efficiency curve of the FSBB photovoltaic optimizer under input voltage of 30V-70V and the different output voltage is shown in Fig. 17. It can be seen from the test results that the FSBB achieves high conversion efficiency in the full range, and the highest efficiency is achieved when input and output voltage are close, the final test results meet the requirements.

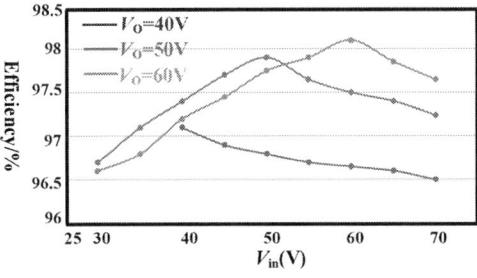

Fig. 17. Efficiency curves of different input and output voltages at full load.

Fig. 18 below shows the efficiency under different power when V_{in} is 70V and V_{out} is 50V. The efficiency is the highest at 300W, reaching 97.6%.

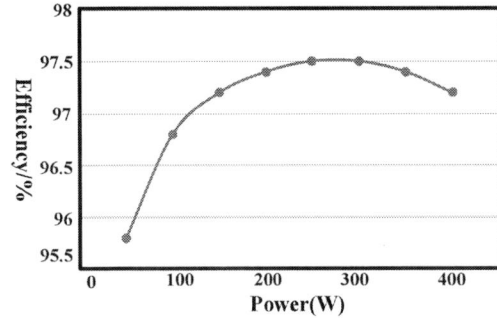

Fig. 18. Efficiency curves of different load at V_{in}=70V, V_o=50V.

Fig. 19 below shows the temperature rise when the ambient temperature is 12.1°C and the power is 200W for 1 minute. It can be seen that the temperature rise meets the requirements and the structure design is reasonable.

978-1-6654-4817-8/21 $31.00 © 2021 IEEE   257

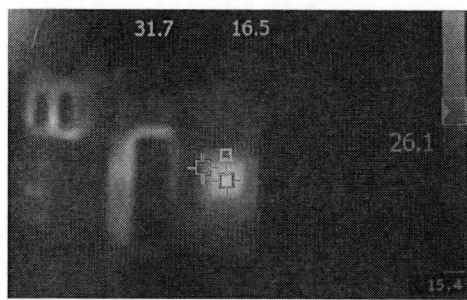

Fig. 19. Diagram of temperature rise.

V. CONCLUSION

In this paper, comprehensively considering the conversion efficiency and power density requirements of the photovoltaic optimizer, a GaN-based FSBB has been developed.

The working principle, control scheme, and modulation strategy of the photovoltaic optimizer are analyzed.

Targeting with the conversion efficiency and power density requirements, the switching frequency is determined, the optimal type of GaN device is selected, and the planar inductor is optimized.

A principle prototype with an input of 30-70V maximum power of 400W was built to verify the rationality of the design. Its power density reaches 750W/inch^3.

REFERENCES

[1] Oettershagen P，Melzer A，Mantel T，et al．A solar-powered hand-launchable UAV for low-altitude multi-day continuous flight[C]．Robotics and Automation(ICRA)，2015 IEEE International Conference. IEEE，2015：3986-3993.

[2] H. Li, D. Yang, W. Su, J. Lu, X. Yu. An overall distribution particle swarm optimization MPPT algorithm for photovoltaic system under partial shading[J]. IEEE Trans.Ind. Electron., vol. 66, no. 1, pp. 265-275, Jan. 2019.

[3] S.-M. Chen, T.-J. Liang, K.-R. Hu. Design, Analysis, and Implementation of Solar Power Optimizer for DC Distribution System[J]. IEEE Trans. Power Electronics, vol. 28, no. 4, pp. 1764-1772, Apr. 2013.

[4] Zheyu Zhang, Ben Guo, Fred Wang, et al. Methodology for Wide Band-Gap Device Dynamic Characterization[J]. IEEE Trans. Power Electronics, 2017, 32(12): 9307-9318.

[5] Ziwei ouyang, Michael A. E. Andersen. Overview of Planar Magnetic Technology—Fundamental Properties[J]. IEEE Transactions on Power Electronics, 2014, 29(9): 4888-4900.

[6] Y. Liu, H. G. Wu, J. Zou, Y. Tai and Z. Ge, "CLL Resonant Converter with Secondary Side Resonant Inductor and Integrated Magnetics," in *IEEE Transactions on Power Electronics*, doi: 10.1109/TPEL.2021.3074646.

[7] Y. Jia, T. Liu, Y. Tai, H. Wu and Y. Xing, "A SiC-Based Dual-Input Buck-Boost Converter with Independent MPPT For Photovoltaic Power Systems," *IECON 2018 - 44th Annual Conference of the IEEE Industrial Electronics Society*, 2018, pp. 1640-1645, doi: 10.1109/IECON.2018.8592687.

[8] Joel Y S，Saikumar H V，Patange S S R．Design &performance analysis of Fuzzy based MPPT control using two-switch non-inverting Buck-Boost converter[C]．Electrical Power and Energy Systems(ICEPES)，International Conference on. IEEE，2016：

Influence of Al/CucorAl wire bonding on reliability of SiC devices

Chao Fang[a,b], Xiang Tang[a,b], Guangyuan Qin[a,b], Haotao Ke[a,b], Yibo Wu[a,b], Jing Zhang[c], Guiqin Chang[a,b] and Haihui Luo[a,b]

[a]State Key Laboratory of Advanced Power Semiconductor Device, Zhuzhou, Hunan 412001, China
[b]Zhuzhou CRRC Times Semiconductor Co., Ltd. Zhuzhou, Hunan 412001, China
[c]Heraeus Materials Technology Shanghai Ltd. Shanghai 201108, China

Abstract— Power semiconductor devices are developing in the direction of miniaturization, high power density and high operating temperature, which puts forward higher requirements for device packaging technology, especially the wire bonding technology which to realize the internal electrical interconnection of devices. Compared with traditional silicon-based semiconductors, the SiC devices have higher requirements for wire bonding reliability due to higher chip junction temperature. In this paper, the influence of aluminum wire and aluminum clad copper wire on the reliability of SiC devices is studied by means of power cycling test and finite element numerical simulation. The experimental results show that the power cycling lifetime of SiC devices bonded with aluminum clad copper wire is 26% higher than that of devices bonded with aluminum wire. The results of finite element simulation show that under the same current load condition, the junction temperature of aluminum clad copper wire bonding device is lower, the current carrying capacity is stronger, and the maximum stress of bonding wire appears at the Al-Cu interface, rather than the bonding wire-chip interface, which leads to longer power cycling lifetime.

Keywords— *bonding wire, reliability, SiC devices, power cycling test, simulation*

I. INTRODUCTION

With the continuous development of power electronics technology, the performance requirements for high-power semiconductor devices are also increasing, such as improving device rated power, reducing device volume, increasing power density and increasing working temperature. In recent years, the application of silicon-based semiconductor devices has been close to the material limit, and silicon-based semiconductor devices have gradually been unable to meet the market demand. Silicon carbide (SiC), as a representative of wide bandgap semiconductor materials, has become a research hotspot due to its good material properties, such as higher withstand voltage, lower conduction loss, higher working temperature (over 200°C, even up to 500°C)[1-3]. It is expected to replace silicon-based semiconductor devices in the future. According to the prediction of French Yole, the market share of SiC devices is expected to reach 1.2 billion US dollars in 2022[4].

In order to satisfy the requirements of high junction temperature and improve the reliability of SiC devices in long-term service, the research institutions continue to develop new packaging technologies, such as using low-temperature silver sintering technology to replace the traditional brazing process[5], using Direct-Lead-Bonding (DLB) technology[6], and the SKiN packaging structure[7]. Comparing with the traditional aluminium wire bonding technology, aluminum clad copper (CucorAl) wire bonding has better mechanical properties, longer fatigue lifetime and stronger current carrying capacity， make it become one of the key technologies to solve the internal chip interconnection of SiC devices[8].

In this paper, the effects of Al bonding wire and CucorAl bonding wire on the power cycling capability of SiC devices are studied by experiment and numerical simulation. The power cycling fatigue lifetime of different bonding wires was obtained through the power cycling test of SiC substrate bonded by Al wires and CucorAl wires. The experimental result shows that the power cycling ability of CucorAl wires is higher than that of Al wires. The simulation result shows that the junction temperature of CucorAl wires bonded chip is lower than that of Al wires bonded chip, and CucorAl wires has stronger current carrying capacity at the same operating temperature. The maximum stress of CucorAl bonding wires is greater than that of Al bonding wires, but the maximum stress of CucorAl bonding wires appears at the Cu/Al interface, and the stress between bonding wires and SiC chip is smaller.

II. POWER CYCLING TEST

A. Experimental sample

In order to eliminate the influence of the later packaging process on the experimental results, the power cycling test was carried out by using the SiC substrate after wire bonding. The experimental sample is shown in Figure 1. Firstly, the drain of a SiC MOSFET chip is sintered on the upper copper of the substrate by silver sintering process, and then the gate and source of the chip are interconnected with the corresponding copper pad by wire bonding technology. Al wire and CucorAl wire are used in wire bonding, and two kinds of SiC MOSFET substrate samples with different bonding wires are obtained. The wire bonded substrate is installed on the power cycling experimental platform, and the conductive column is used to contact the upper copper of the substrate to form the loop between the power supply and the sample, as shown in Figure 2. Six samples were used in the power cycling test, including three samples of Al bonding wire and three samples of CucorAl bonding wire.

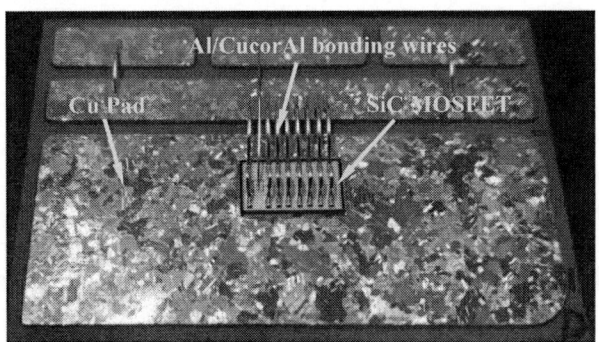

Fig. 1.   SiC substrate for PC test

This work was supported by the National Key Research and Development Program of China under Grant 2017YFB0102303.

Fig. 2.   PC test setup

B. Experimental method

According to the ECPE AQG 324 standard, the power cycling test scheme is developed. The experimental conditions are shown in Figure 3. DC power cycling test is adopted, with each cycle period for 3 seconds, of which the turn-on time is 1 second and the turn-off time is 2 seconds. The conduction stage is to input a 90 A direct current into the sample, and the case temperature of the sample is kept at 25°C~30°C by adjusting the flow rate and water temperature of radiator under the substrate. Chip junction temperature difference is maintained at 120°C.

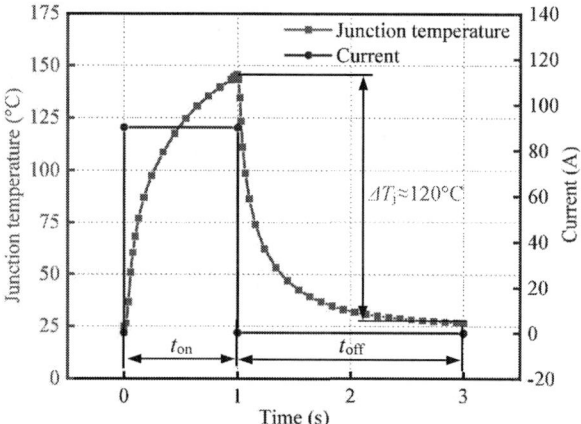

Fig. 3.   Current and Junction temperature of PC test

Before the test, the K-curve of the sample is calibrated. The specific method is to place the sample in a incubator, set the temperature of the incubator, and load a reverse voltage between the gate and source of the chip to ensure that the chip is in the off state. Then, a small current is applied to the sample to record the voltage drop V_f of the chip body diode under this temperature. Next, the temperature of the incubator is adjusted to measure the voltage drop of the chip body diode at different temperatures. The relationship between the junction temperature T_j and the voltage drop V_f of the chip body diode is obtained, which is the K-curve of the chip. The calibration results are shown in Figure 4.

During the power cycling test, the body diode voltage drop, forward conduction voltage drop and case temperature of SiC substrate are collected in real time. The junction temperature of the chip is calculated by K-

curve, and the thermal resistance between the junction and the case is calculated by power loss, junction temperature and case temperature. When the forward conduction voltage drop increases by 5%, or the thermal resistance of the device increases by 20%, the device is judged to be invalid, and the number of power cycles of the device at this time is recorded as the power cycling lifetime of the device. In addition, by monitoring the forward conduction voltage drop of the chip, and using the mutation time of the forward conduction voltage drop to judge whether the bonding wire lift-off, the failure of the device can also be monitored.

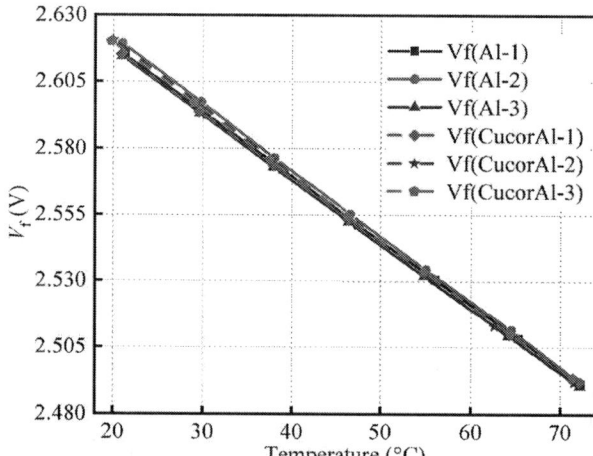

Fig. 4.   Calibration K curve

III. FINITE ELEMENT ANALYSIS

A. Finite element analytical model

According to the structure of the experimental sample, a three-dimensional finite element analysis model is established. The model consists of upper copper pad, ceramic layer, bottom copper pad, silver sintered layer, SiC MOSFET chip and Al/CucorAl bonding wires. The size of the model is consistent with the experimental sample, as shown in Fig. 5.

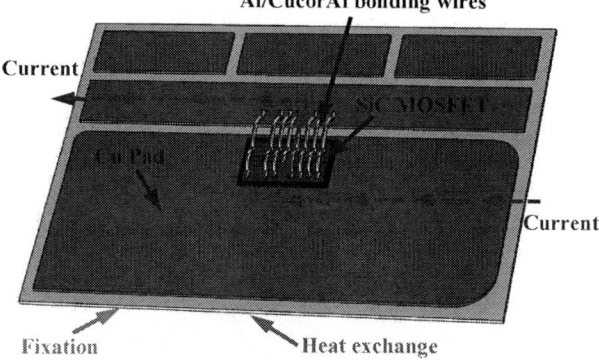

Fig. 5.   3D model for Finite element analysis

B. Material property

In the finite element numerical simulation, a total of five materials, Al, Ag, Cu, SiC and Si_3N_4, are used. The material parameters are listed in Table I. The elastic-plastic model of the material is used for Al, Ag and Cu, the stress-strain data of Al is obtained by tensile test, and the bilinear elastic-plastic constitutive model is used for Ag and Cu. The stress-strain curves of the three materials are shown in Figure 6. The linear elastic model is used for SiC and Si₃N₄. The conductivity of SiC changes with temperature, as shown in Figure 7.

Table I. Material Parameter

Material	Coefficient of heat transfer (mW/(mm·K))	Specific conductance (mS/mm)	Density (tonne/mm³)	Specific heat (mJ/(tonne·K))	Elastic modulus (MPa)	Poisson's ratio	CTE (1/°C)
Al	237	37.74×10^6	2.7×10^{-9}	900×10^6	70.6×10^3	0.33	21×10^{-6}
Ag	240	4.76×10^6	8.4×10^{-9}	234×10^6	10×10^3	0.37	19×10^{-6}
Cu	400	58.82×10^6	8.96×10^{-9}	385×10^6	127×10^3	0.33	16.5×10^{-6}
SiC	370	Temperature dependence	3.21×10^{-9}	750×10^6	430×10^3	0.17	4.1×10^{-6}
Si₃N₄	16.7	1×10^{-12}	3.44×10^{-9}	710×10^6	300×10^3	0.20	3×10^{-6}

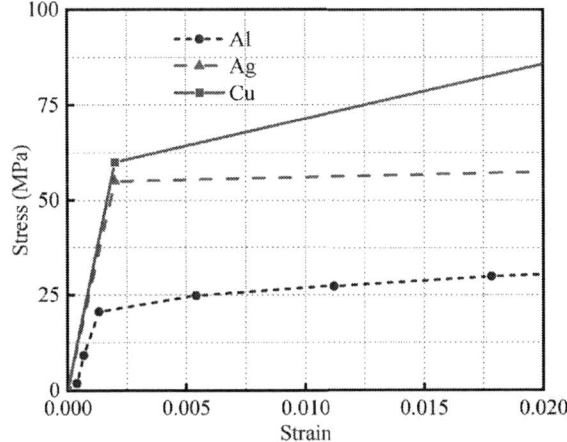

Fig. 6. The stress-strain of Al/Ag/Cu

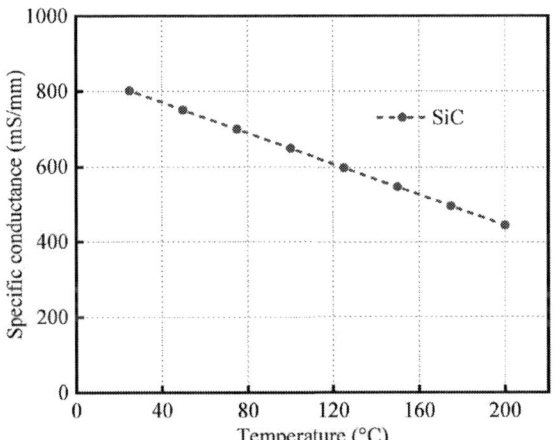

Fig. 7. The specific conductance of SiC

C. Loads and boundary conditions

The finite element numerical simulation is carried out in the way of electrical-thermal-mechanical multiphysics coupling. The Cu pad connected with the drain of the chip is loaded with DC current load, and the Cu pad connected with the source is set as the zero potential area, thus forming the current path. The waveform of DC current load in the finite element numerical simulation is the same as that in the power cycling test, the turn on time is 1 s, the turn off time is 2 s, and the peak current is 90 A. In the power cycling test, the bottom copper of SiC substrate contacts the radiator for heat dissipation, and other surfaces contact the air and copper conductive column. Therefore, the convective heat transfer coefficient of the bottom copper of SiC substrate is set to 4000 W/(m²·K), and that of other surfaces is set to 40 W/(m²·K). The initial temperature of the model is set to 25°C. The ambient temperature is set to 25°C. In addition, in order to prevent unnecessary displacement and rotation of the model, a fixed constraint is imposed on the bottom surface of the bottom copper under the SiC substrate.

D. Mesh verification

In order to give consideration to both calculation efficiency and accuracy, it is necessary to control the number of elements. By setting different element sizes, four kinds of mesh generation methods with different element numbers are obtained. Calculate the models under four grid generation methods, extract the maximum junction temperature of the chip at the time of current turn-off, and get the relationship between the maximum junction temperature of the chip and the number of elements, as shown in Figure 8. It can be seen that when the number of elements is 568,732, the error of the chip junction temperature has met the requirements. At this time, the improvement of the calculation accuracy is no longer obvious by increasing the number of elements, so the 568,732 elements model is used for finite element simulation.

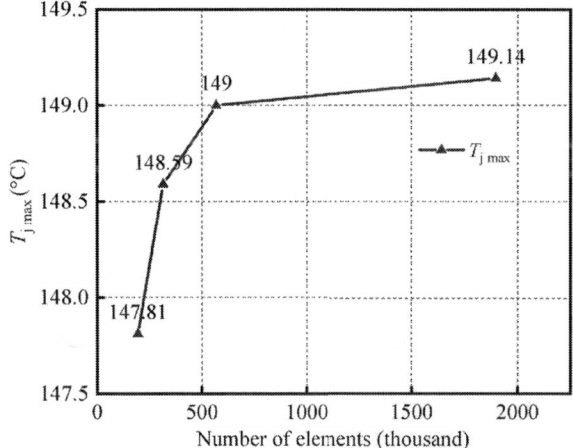

Fig. 8. The relationship between the junction temperature of chip and the element numbers

IV. Result And Discussion

A. PC test results

In the power cycling test, when the first device fails, the power cycling number N_0 is recorded, and based on N_0, the power cycling number when other devices fail is normalized, and the relative lifetime N_f/N_0 is obtained, so that the comparative analysis of device lifetime can be more intuitive. Figure 9 shows the relative lifetimes N_f/N_0 of the six samples. It can be seen that the power cycling lifetime of the device with CucorAl bonding wire is significantly higher than that of the device with Al bonding wire. The average power cycling lifetime of Al wire bonded device is about $1.23N_0$, and that of CucorAl wire bonded device is about $1.55N_0$, which is 26% higher than that of Al wire.

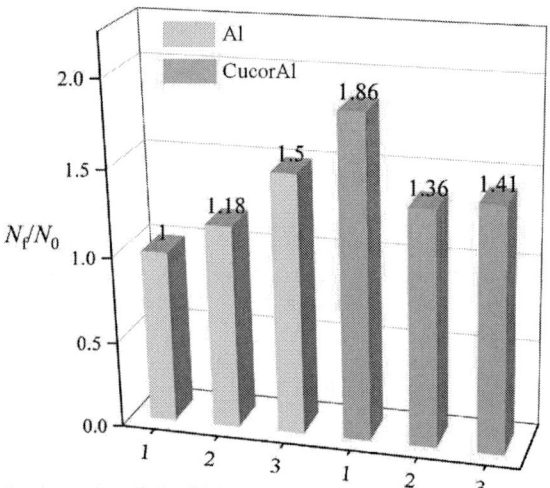

Fig. 9.   The N_f/N_0 of PC test

Figure 10 shows an optical microscope image of the detached CucorAl wire bonded sample. In order to clearly show the morphology of the bonding wires, the bonding wires that lift-off are artificially lifted up. It can be clearly observed that there are a lot of residual Al chips at the bonding areas on the chip surface, and the fracture mainly occurs in the inner region of CucorAl bonding wire, while no obvious cracks are found at the interface between chip and bonding wires.

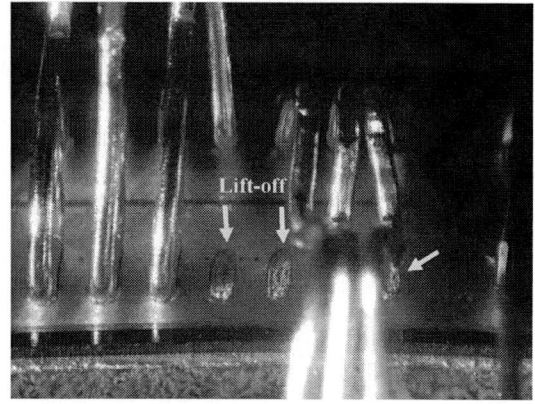

Fig. 10.   The CucorAl bonding wire lift-off

B. Finite element analysis results

Figure 11 shows the temperature distribution of Al bonding wire and CucorAl bonding wire devices at turn-off time. It can be seen that under the same conditions, the chip junction temperature of devices bonded with Al wire is 149.0°C. It is higher than that of the device using CucorAl bonding wire at 147.3°C. For high-power electronic devices,

the increase of temperature will lead to a significant reduction of device lifetime, so the CucorAl bonding wire with relatively low temperature is helpful to improve the power cycling lifetime of the device. In addition, it can be seen that the highest temperature of the device appears on the bonding wires, and the junction temperature of the chip is lower than the temperature of the bonding wires.

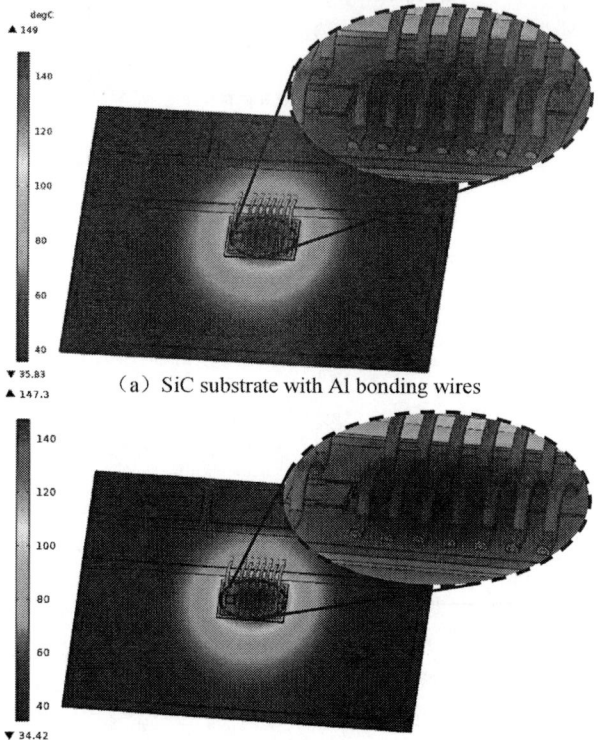

（a）SiC substrate with Al bonding wires

（b）SiC substrate with CucorAl bonding wires

Fig. 11.   The temperature distribution of SiC substrate

Figure 12 shows the Von Mises stress distribution of CucorAl bonding wires. The high stress region of the bonding wire appears at the Al-Cu interface, and the stress at the interface between the bonding wires and the chip is relatively small, which leads to the damage first appearing at the Al-Cu interface. The strength of the Al-Cu interface has a significant impact on the reliability of the bonding wires.

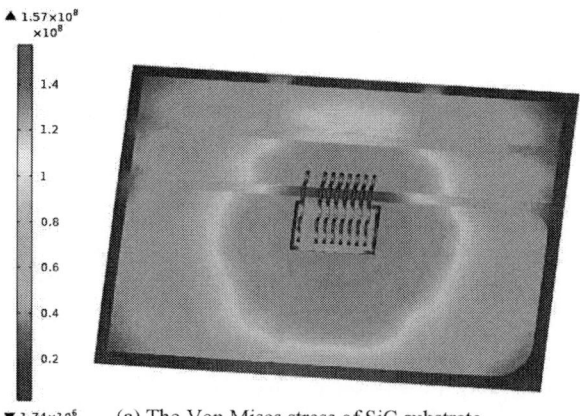

(a) The Von Mises stress of SiC substrate

978-1-6654-4817-8/21 $31.00 © 2021 IEEE

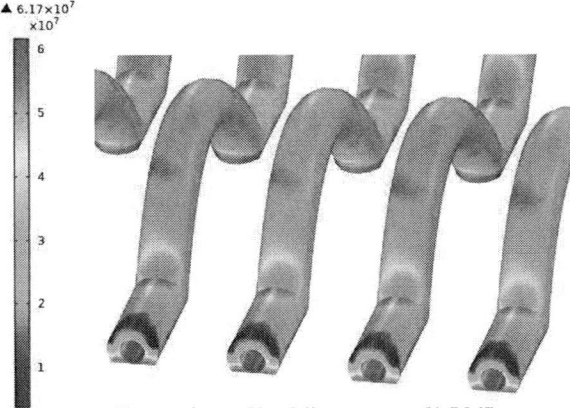

▲ 6.17×10⁷
x10⁷

The maximum Von Mises stress ≈ 61.7 MPa

▼ 3.6×10⁶

(b) The Von Mises stress of Al surface (CucorAl bonding wires)

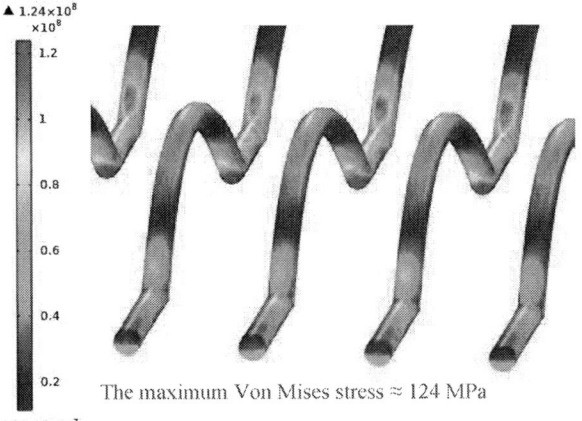

▲ 1.24×10⁸
x10⁸

The maximum Von Mises stress ≈ 124 MPa

▼ 1.13×10⁷

(c) The Von Mises stress of Cu wires (CucorAl bonding wires)

Fig. 12. The Von Mises stress of SiC substrate with CucorAl bonding wires

V. CONCLUSION

In this paper, the influence of Al bonding wire and CucorAl bonding wire on the power cycling reliability of SiC devices is compared by power cycling test and finite element numerical simulation. In order to eliminate the influence of the later packaging process on the experimental results, the sintered and bonded substrate was used as the experimental sample in the power cycling test. As the experiment goes on, the device is aging and degradation, and the junction temperature and thermal resistance of the chip increase in varying degrees. When the bonding wires lift-off, the voltage drop will increase abruptly due to the increase of contact resistance. Comparing the experimental results of the two kinds of bonding wires, the power cycling lifetime of the device with CucorAl bonding wire is 26% higher than that with Al bonding wire. The failure of Al bonding wire device occurs at the interface between bonding wire and chip, while the failure of CucorAl bonding wire device occurs at the Al-Cu interface of bonding wire. The results of finite element simulation also confirm that the junction temperature of CucorAl bonding wire device is lower than that of Al bonding wire device, and the maximum stress of CucorAl bonding wire appears at the Al-Cu interface of bonding wires, which leads to the Al-Cu interface of the bonding wire is the first to be damaged and degraded.

ACKNOWLEDGMENT

This work was supported by the National Key Research and Development Program of China under Grant 2017YFB0102303.

REFERENCES

[1] P. Ning, T. G. Lei, F. Wang, "A novel high-temperature planar package for SiC multichip phase-leg power module," IEEE Transactions on Power Electronics, 2010, 25(8): 2059-2067.

[2] R. A. Wood, T. E. Salem, "Evaluation of a 1200 V, 800 A all-SiC dual module," IEEE Transactions on Power Electronics, 2011, 26(9): 2504-2511.

[3] S. Chowdhury, T. P. Chow, "Performance Tradeoffs for Ultra-High Voltage (15 kV to 25 kV) 4H-SiC n-Channel and p-Channel IGBTs." ISPSD, 2016: 75-78.

[4] Power SiC: Materials, Devices, Modules, and Application report, Yole Development, 2017.

[5] X. P. Dai, Y. B. Wu, Y. M. Zhao, Y. G. Wang, "Packaging Consideration and Development for Fully Sintered SiC Power Module," High power converter technology, 2016(5): 36-40.

[6] E. R. Motto, "Transfer Molded IGBT Module for Electric Vehicle Propulsion," Motor, Driver& Automation System, 2012.

[7] P. Beckedahl, S. Buetow, A. Maul, M. Roeblitz, M. Spang, "400 A 1200 V SiC Power Module with 1nh Commutation Inductance," CIPS 2016, 9th International Conference on Integrated Power Electronics Systems, 2016.

[8] F. Naumann, J. Schischka, S. Koetter, E. Milke, M. Petzold, "Reliability characterization of heavy wire bonding materials," Electronic System-Integration Conference (ESTC), 2012.

An Efficient Voltage Step-up/down Partial Power Converter (SUD-PPC) using Wide Bandgap devices

Chao Liu
Department of Electrical Engineering
Technical University of Denmark
Kgs. Lyngby, Denmark
chali@elektro.dtu.dk

Zhe Zhang
Department of Electrical Engineering
Technical University of Denmark
Kgs. Lyngby, Denmark
zz@elektro.dtu.dk

Michael A. E. Andersen
Department of Electrical Engineering
Technical University of Denmark
Kgs. Lyngby, Denmark
ma@elektro.dtu.dk

Abstract—This paper proposed a novel topology for step-up/down converter based on the concept of partial power processing, which improves efficiency as well as power density. The unified modulation strategy achieves the auto transition between two operating modes without requiring any additional control. The operating principle is analyzed in detail, from which it is found that diode reverse recovery of Si MOSFETs reduces the voltage regulation range in step-down operation mode. Therefore, SiC MOSFET is employed in the high voltage side to avoid duty cycle loss. The SPICE simulations based on the SiC and Si devices have been built, and the simulation results verify the effect of using SiC devices. A 400V prototype based on SiC MOSFET has been built. Measurement results confirm the high efficiency in overall voltage range, and the maximum efficiency exceeds 99%.

Keywords— Step-up/down voltage regulation, partial power processing, unified modulation strategy, SiC devices

I. INTRODUCTION

Recently, Partial Power Processing (PPP) has presented significant advantages in power converter downsizing and efficiency improvement [1]. Compared to conventional full power converters (FPC), partial power converters (PPC) reduce power rating of power electronic devices and systems in different applications, such as solar photovoltaic systems[2], energy storage systems (ESS) [3] and electric vehicle (EV) fast charging stations [4]. The PPC category contains the subcategories of series-connected PPC (S-PPC) and parallel-connected PPC (P-PPC) [5]. The P-PPCs are usually employed in PV module strings, also widely called as differential power processing, which achieve maximum power point tracking (MPPT) at fractional currents regulation. [6]. S-PPCs achieve voltage difference regulation by connecting the PPC, input source and load in series, which was first proposed for photovoltaic applications in the spacecraft technology [7], as shown in Fig.1. Ref [8] and [9] summarized and reviewed the various PPP topologies for PV applications. An analysis and comparison between Dual Active Bridge (DAB) and Isolated Full Bridge Boost (IFBB) topologies based on component stress factor (CSF) is performed in [10]. Depending on the output voltage gain, S-PPCs can be further divided into three types: step-up PPC (SU-PPC), step-down PPC (SD-PPC), and step-up and down PPC (SUD-PPC). A SUD-PPC topology was presented in [11], and the results indicate that the SUD-PPC processes the least active power over the same voltage variation range compared to the other two PPCs, resulting in higher efficiency and power density. However, the unified modulation strategy for the SU and SD operating modes of the SUD-PPC is still a challenge.

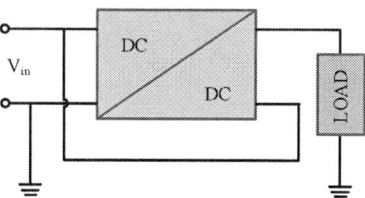

Fig.1 Schematic diagram of the S-PPC structure.

On the other hand, high switching frequency also offers the possibility of increasing the power density due to the reduced size of passive components. Compared to Silicon (Si) MOSFET, wide bandgap (WBG) devices have smaller on-resistance, lower parasitic capacitance and higher operating temperature, thereby are considered as the promising candidates for high frequency power conversion applications [12], [13].

This paper presented a novel SUD PPC topology and the proposed unified modulation strategy achieves the auto switching between two operation modes in absence of any additional control. The rest of this paper is organized as follows. Section II introduces the operating principle of the proposed topology and modulation strategy. In Section III, duty cycle loss caused by the diode reserve recovery has been analysed. And the simulations results verify that SiC devices can be employed in high voltage side to solve this issue. Section IV presents the experimental results of a 400V prototype to verify the feasibility and practicality of the topology and modulation strategy. Finally, Section V concludes this paper.

II. OPERATING PRINCIPLE

Fig.2 shows the proposed SUD-PPC, where V_{c2} is the voltage at the port in series with the power source and load in series; I_{load} is the current across the load. It can be observed the system has two connection configurations. One is that the load is placed between points A and B, while the points C and D are connected by wires. Conversely, the load can be also placed between points C and D, while the points A and B is shorted. Two configurations have the same topological characteristics. Similarly to other S-PPCs, the power source, load and one port of the converter are connected in series. Therefore, the converter only processes a portion of the full power, i.e. $V_{c2} \times I_{load}$. Alternatively, a bidirectional topology can be obtained by replacing the D_{1-4} with active switches. The converter has two operating modes: step-down mode and step-up mode. Fig.3 shows the driver signals and theoretical waveforms of the two operating modes, where Q_{1-4} and S_{1-4} are the drive signals for the corresponded switches in Fig.2; d_q and d_s are the duty cycle for the Q_{1-4} and S_{1-4}, respectively, $d_q = d_s - 1$. The following is the analysis of the operational states of

978-1-6654-4817-8/21 $31.00 © 2021 IEEE

two modes, and assume all devices are ideal and ignore power losses.

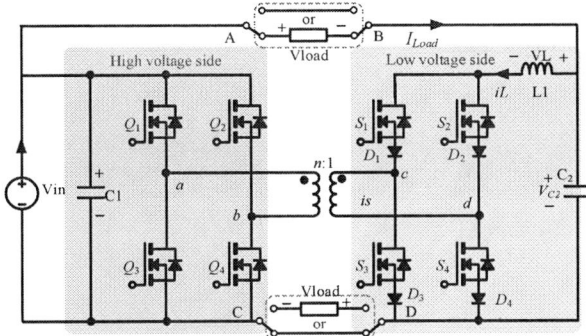

Fig.2 Configuration of the proposed SUD-PPC.

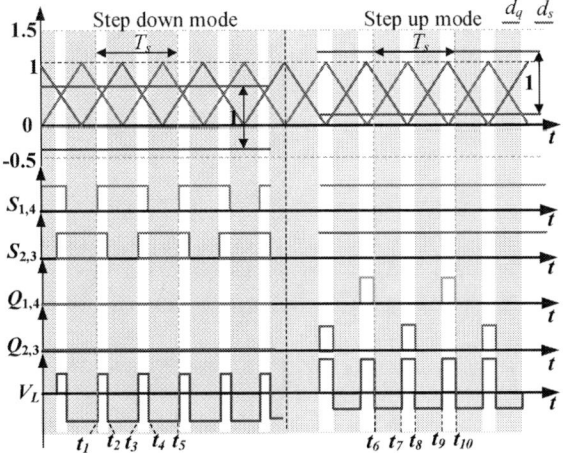

Fig.3 Driver signals and theoretical waveforms

A. Step-down operating mode

In the step-down mode, the range of d_s is from 0 to 1, thus d_q is always smaller than 0. Q_{1-4} maintain off in this operating mode. As shown in Fig.3, one switching period can be divided into 4 states or subintervals. The first two states are symmetrical to the last two states. Fig.4 shows the equivalent circuits of the first two operating states.

1) State 1, t_1-t_2: $(d_s$-$0.5) \times T_s$

In this subinterval, all low-voltage switches are turned on, power source charges to load directly and no power flows through the transformer and high voltage side, as shown in Fig.4 (a). The inductor current equals to load current, which uniform flows through both low-voltage side bridge arms. The inductor voltage v_L equals the difference between input voltage V_{in} and load voltage V_{load}. The inductor voltage for this subinterval is given by

$$v_L = L\frac{di_L}{dt} = V_{in} - V_{load} \qquad (1)$$

2) State 2, t_2-t_3: $(1-d_s) \times T_s$

At t_2, $S_{2,3}$ are switched off, i_L is forced to flow through S_1 and S_4. Therefore, the power is transferred from low-voltage side to high-voltage side through the transformer, as shown in Fig.4(b). In the high-voltage side, the current flows through the body diode of Q_1 and Q_4. The inductor voltage for this subinterval is given by

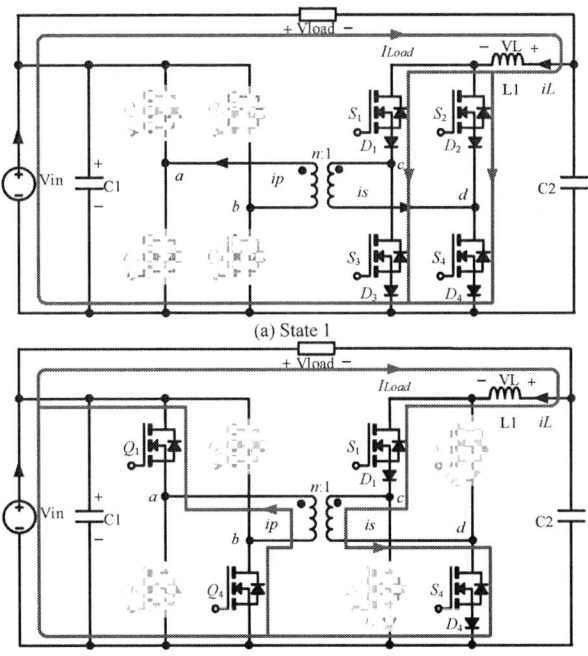

(a) State 1

(b) State 2

Fig.4 Equivalent circuit of the first two operating states in SD mode.

$$v_L = L\frac{di_L}{dt} = \frac{(n-1) \cdot V_{in}}{n} - V_{load} \qquad (2)$$

At t_3 S_{2-3} are turned on again, the operating state is the same as the State 1. At t_4, $S_{1,4}$ are switched off, i_L is forced to flow through S_2 and S_3. Therefore, the last two operating states of the step-down mode are symmetrical to the former two states, and no need to elaborate. The converter works as the isolated boost converter in this mode.

Applying volt-seconds to the inductor over one switching period, the output voltage is calculated by (3).

$$V_{load} = \frac{(n + 2\,d_s - 2)\,V_{in}}{n} \qquad (3)$$

B. Step-up operating mode

In the step-up mode, all low voltage side switches are on state. And same to step-down mode, one switching period can also be divided into 4 parts. Fig.5 shows the equivalent circuit of the first two operating modes.

1) State 1, t_6-t_7: $(0.5-d_q) \times T_s$

In this interval, Q_{1-4} are turned off while S_{1-4} are turned on, the power source V_{in} charges to the load directly, as shown in Fig.5 (a). No power flows through the transformer and high voltage side switches. Therefore, the inductor voltage for this subinterval is the same with that of State 1 in the step-down mode, which is given by

$$v_L = L\frac{di_L}{dt} = V_{in} - V_{load} \qquad (4)$$

2) State 2, t_7-t_8: $d_q \times T_s$

At t_7, $Q_{1,4}$ are switched on. V_{ab} equals to V_{in}, thereby V_{cd} equals to V_{in}/n. Power is transferred from high-voltage side to low-voltage side, as shown in Fig.5(b). The inductor voltage for this subinterval is given by (15).

978-1-6654-4817-8/21 $31.00 © 2021 IEEE

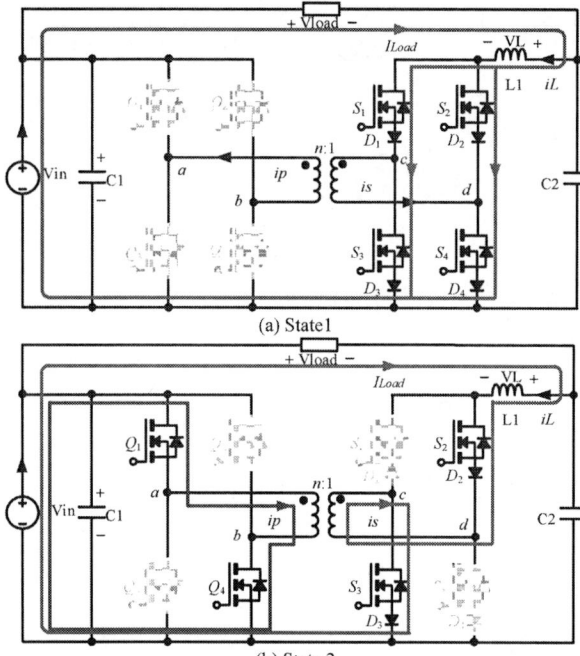

(a) State1

(b) State 2

Fig.5 Equivalent circuit of the first two operating states in SU mode.

$$v_L = L\frac{di_L}{dt} = \frac{(n+1)\cdot V_{in}}{n} - V_{load} \quad (5)$$

At t_8, Q_1 and Q_4 are turned off, the rest two states begin. It can be observed that the converter works as the inverting isolated buck converter in this mode. Applying volt-seconds principle again, the output voltage in step-up mode is obtained by

$$V_{load} = \frac{(n+2\,d_q)\,V_{in}}{n} \quad (6)$$

C. Small-signal modeling

d_q and d_s can be expressed by a unified modulation variable u in(7) and (8), as shown in Fig.6.

$$\begin{cases} d_s = u & 0.5 \le u < 1 \\ d_s = 1 & 1 \le u < 1.5 \end{cases} \quad (7)$$

$$\begin{cases} d_q = 0 & 0.5 \le u < 1 \\ d_q = u-1 & 1 \le u < 1.5 \end{cases} \quad (8)$$

Substituting (7) and (8) into (3) and (6), the output voltage can be expressed

$$V_{load} = \frac{(n+2\,u-2)\,V_{in}}{n} \quad (9)$$

Introducing (7) into (1) and (2), the averaged equations of inductor voltage in step-down mode is given

$$L\frac{d\langle i_L\rangle_{Ts}}{dt} = \frac{(n+2\,u-2)\,\langle v_{in}\rangle_{Ts}}{n} - \langle v_{load}\rangle_{Ts}, \ 0.5 \le u < 1 \ (10)$$

Introducing (8) into (4) and (5), the averaged equations of inductor voltage in step-up mode is given

$$L\frac{d\langle i_L\rangle_{Ts}}{dt} = \frac{(n+2\,u-2)\,\langle v_{in}\rangle_{Ts}}{n} - \langle v_{load}\rangle_{Ts}, \ 1 \le u < 1.5 \ (11)$$

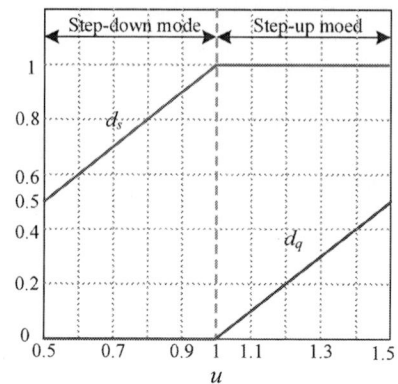

Fig.6 d_q and d_s versus the unified modulation variable u.

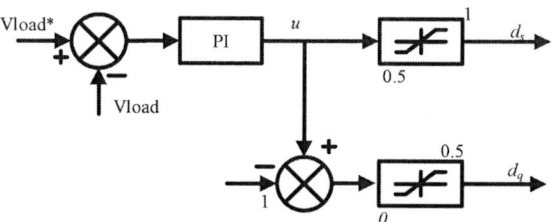

Fig.7 Block diagram of the unified controller for two operating modes.

i_{C2} equals the difference between the RMS value of the load current and i_L, which can be expressed by

$$i_{C2} = C_2\frac{dv_{C2}}{dt} = \frac{v_{load}}{R} - i_L \quad (12)$$

Therefore, both modes have the same averaged equation of output capacitor current, which is given in (13).

$$C_2\frac{d\langle v_{C2}\rangle_{Ts}}{dt} = \frac{\langle v_{load}\rangle_{Ts}}{R} - \langle i_L\rangle_{Ts} \quad (13)$$

Consequently, both operating modes of the converter have the same averaged state equations. The small signal equations for the system can be obtained by introducing perturbation around the steady state value for the state variables, such that $i_L = I_L + \hat{i}_L$.

Then neglecting the second-order terms, the linearized small-signal equations in Laplace domain, containing only the first-order ac terms, are given by the following equations.

$$sL\,\hat{i}_L = \frac{\hat{v}_{in}}{n}(n+2\,u-2) + \frac{2\,V_{in}}{n}\hat{u} - \hat{v}_{load} \quad (14)$$

$$sC_2\big(\hat{v}_{in} - \hat{v}_{load}\big) = \frac{\hat{v}_{load}}{R} - \hat{i}_L \quad (15)$$

Rearranging (15) results in the following equation

$$\hat{i}_L = \frac{(1+sRC_2)\,\hat{v}_{load}}{R} - sC_2\,\hat{v}_{in} \quad (16)$$

Introducing (16) to (14), the control-to-output transfer function is obtained in (17).

$$\left.\frac{\hat{v}_{load}}{\hat{u}}\right|_{\hat{v}_{in}=0} = \frac{2\,RV_{in}}{nR + s^2nLC_2R + snL} \quad (17)$$

Both operating modes have the same control-to-output transfer function, which means that only one common and unified controller is needed for both operating modes, as shown in Fig.7.

978-1-6654-4817-8/21 $31.00 © 2021 IEEE

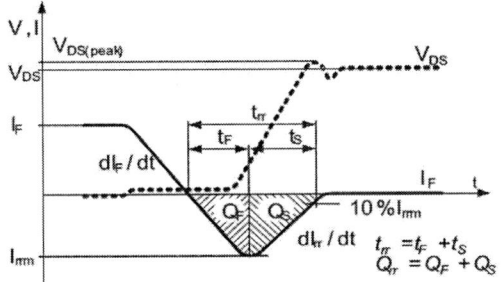

Fig.8 Diode recovery waveform [14]

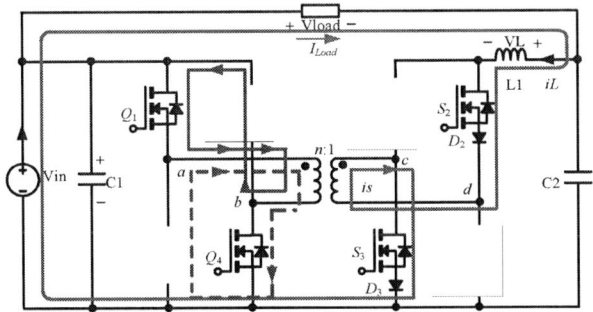

Fig.9 Equivalent circuit of the operating state between state 2 and 3

III. DUTY CYCLE LOSS

In practice, Si MOSTE has long diode reverse recovery time, as shown in Fig.8 [14]. Therefore, in step-down mode, after the low-voltage side switches have been turned on, the transformer voltage can not drop to 0 immediately due to the reverse recovery of the high-voltage switch, meaning that the new operating sates take place between State 2 and 3, and States 4 and 1. As an example, Fig. 9 shows the equivalent circuit of the added operation state occurred between State 2 and 3.

At t_3, S_2 and S_3 are turned on, the capacitors of Q_2 and Q_3 start charge to Q_1 and Q_4, and Q_1 and Q_4 start to reverse recovery. As a result, the voltage of Q_2 and Q_3 start to decrease, while voltage of Q_1 and Q_4 start to increase until the voltage of each high-voltage switch is same and equal to half of V_{in}. In this period, i_L flows through S_2 and S_3. Assuming V_{cd} equals V_{in}/n in this period, the inductor voltage equation can be expressed by (18)

$$ v_L = L \frac{di_L}{dt} = \frac{(n+1) \cdot V_{in}}{n} - V_{load} \quad (18) $$

Define d_{rr} as the division of t_{rr} and the switching period. After this subinterval, the low-voltage side is short again and State 1 begins. As a result, the duration of the Sate1 and 3 reduce to $(u-0.5-d_{rr}) \times T_s$. Define $G(u)$ as the gain of the output voltage and input voltage. Considering the diode reverse recovery, $G(u)$ in the step-down mode is obtained

$$ G(u) = \frac{(n + 2u - 2 + 2d_{rr})}{n} \quad (19) $$

The converter gain considering diode reverse recovery is increased while the regulation range is reduced. Since the u must be larger than $d_{rr}+0.5$, the minimum value of the output voltage is given in (20).

TABLE I. SIMULATION SYSTEM SPECIFICATION

Input voltage	High-voltage switch	Low-voltage switch	Low-voltage diode	Switching frequency
400V	IPW65R095C7	IRF200P223	V30200C	100kHz

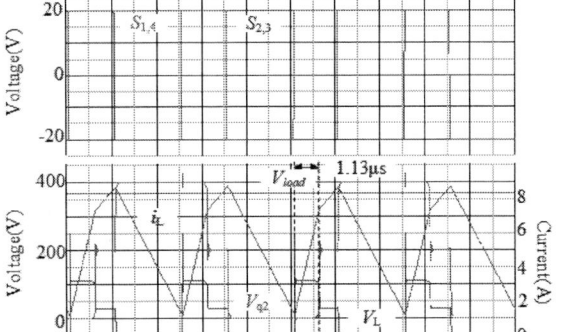

Fig.10 Si MOSFET based simulation waveforms at u=0.7

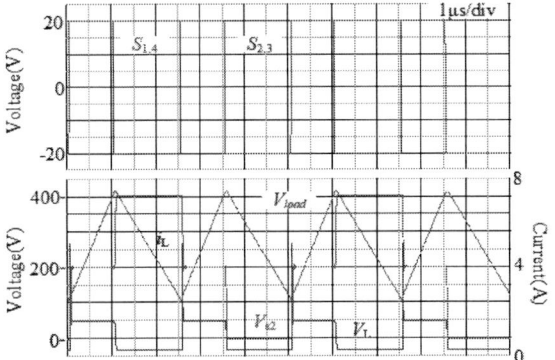

Fig.11 SiC MOSFET based simulation waveforms at u=0.7

$$ V_{Load,min} = \frac{(n - 1 + 4d_{rr}) \, V_{in}}{n} \quad (20) $$

A simulation based on the SPICE model was built in SIMetrix to verify the analysis above. Table I shows the semiconductor specifications. Level1 model for semiconductors was selected and the default temperature is 27 degrees. Fig.10 shows the simulation waveforms at u=0.7, V_{in} is 400V and load is 70Ω. It can be observed that V_{load} is 368V and t_{rr} is 1.13 µs. Since the carrier wave frequency is 100kHz, d_{rr} is 0.113. Introducing d_{rr} = 0.113 into (19). The theoretical output voltage is 369.6V. The simulation result shows an error of less than 0.5% compared to theoretical value. In order to overcome this disadvantage, SiC devices can be employed in high-voltage side due to the shorter reverse recovery time. Fig.11 show the simulation waveforms based on the SiC MOSFET IMW650R072M at u=0.7. It can be observed that the SiC MOSFET has much shorter reverse recovery time, result in the negligible duty cycle loss. The simulation result of V_{load} is 350 with an error of about 0.5% compared to the theoretical value 352V calculated in (9).

Considering the losses of semiconductor, inductor and transformer, the proposed presents high efficiency, as shown in Fig.12. It can be observed that the proposed SUD-PPC has the maximum efficiency operating at u=1, as the all power is transmitted directly to the load. There is only the conductive loss of the low-voltage side, and the maximum efficiency is approx.99.5%.

978-1-6654-4817-8/21 $31.00 © 2021 IEEE

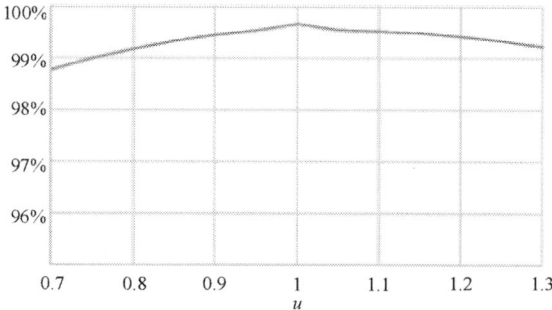

Fig.12 Simulation efficiency versus u at Vin=400V and load=70Ω

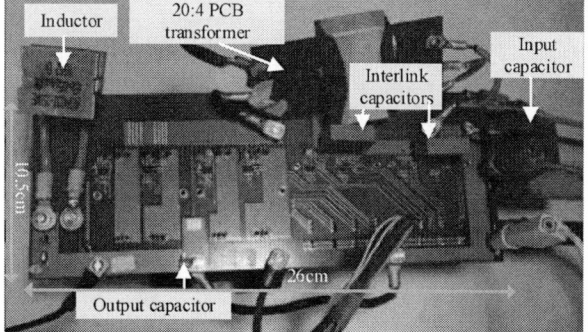

Fig.13 Experimental prototype of the SUD-PPC.

TABLE. II PASSIVE COMPONENTS

Component	Input capacitor	Output capacitor	Interlink capacitor	Inductor
Name	B3291 X1 MKP/SH	CKG57KX 7S2A106M	WIMA FKP1	AGP423 3-153ME
Value	6.8 μF	10 μF	10 nF	15 μH

IV. EXPERIMENTAL RESULTS

A 400V prototype has been built to verify the proposed topology and modulation strategy, as shown in Fig.13. The same semiconductor devices were used for the prototype as for the SiC based simulation. The experimental setup is also the same as the simulation. Input voltage is 400 V, carrier wave frequency is 100kHz and load is 70 Ω. Table II shows the other main components. The interlink capacitors are placed at the ports of Q_{1-4} to reduce voltage oscillation caused by hard switching. For the same reason, a 20:5 PCB transformer was designed to reduce the leakage inductance. Since the blocking voltage of the output capacitor is the difference of V_{in} and V_{load}, A 10μF capacitor with 100V blocking voltage is employed as the output capacitor, resulting in a much smaller volume.

Fig.14 (a) and (b) shows the steady state experimental waveforms at u=0.7 or 1.2, respectively. As it can be observed, the output voltage in the two conditions is close to the theoretical value, which is 353.8V and 431 V, respectively. And duty cycle loss in the step-down mode can be neglected after using SiC MOSFET in high-voltage side.

Fig.15 shows the transient sate waveform of the operation mode switching. In this condition, the reference value of V_{load} is stepped from 360 to 435V. As it can be observed, the switching process is smooth, which verifies the feasibility of the proposed modulation strategy.

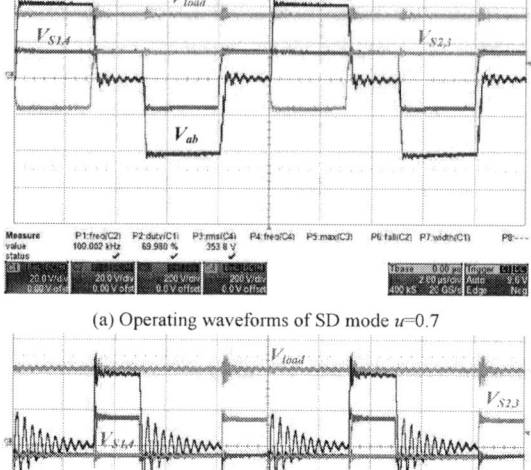

(a) Operating waveforms of SD mode u=0.7

(b) Operating waveforms of SU mode at u=1.2

Fig.14 Steady state waveforms of the prototype

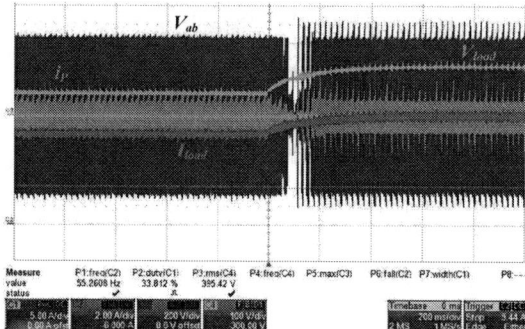

Fig.15 Transient waveforms of the operation mode switching

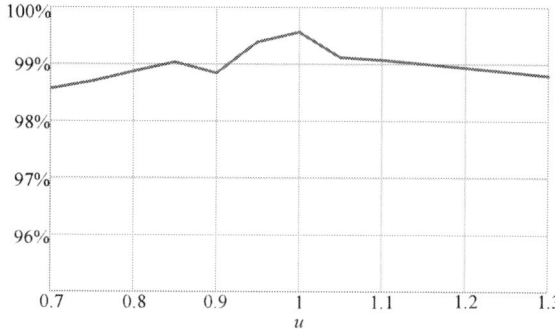

Fig.16 Experimental efficiency versus u at Vin=400V and load=70Ω

Fig.16 shows the experimental efficiency versus u. Two 1mΩ current shunt are placed into the circuit to measure the input and output voltage. The voltages of the current shunts and the input souce and load are measured by 34465A Digit Multimeter. The experimental results of the system efficiency are close to the simulation results. The proposed topology has high efficiency in overall regulation range, and the maximum efficiency exceeds 99.5%.

V. CONCLUSION

This paper proposed a SUD-PPC topology with high efficiency. An unified modulation strategy has been presented for the two operating modes. The detailed analysis of the operating principles has been shown. Moreover, the cause of duty cycle loss is analysed. As the solution, the SiC MOSFETs have been employed in high voltage side. Finally, a 400V prototype has been built and the experimental results verify the feasibility and practicality of the topology and modulation strategy.

In the future, the power losses of the topology will be analysed. And the advantages and disadvantages of GaN and SiC devices in the proposed topology will be compared.

REFERENCES

[1] J. Anzola et al., "Review of Architectures Based on Partial Power Processing for DC-DC Applications," in IEEE Access, vol. 8, pp. 103405-103418, 2020.

[2] H. Zhou, J. Zhao and Y. Han, "PV Balancers: Concept, Architectures, and Realization," in IEEE Transactions on Power Electronics, vol. 30, no. 7, pp. 3479-3487, July 2015.

[3] M. C. Mira, Z. Zhang, K. L. Jørgensen and M. A. E. Andersen, "Fractional Charging Converter With High Efficiency and Low Cost for Electrochemical Energy Storage Devices," in IEEE Transactions on Industry Applications, vol. 55, no. 6, pp. 7461-7470, Nov.-Dec. 2019.

[4] T. Kanstad, M. B. Lillholm and Z. Zhang, "Highly Efficient EV Battery Charger Using Fractional Charging Concept with SiC Devices," 2019 IEEE Applied Power Electronics Conference and Exposition (APEC), 2019, pp. 1601-1608, doi: 10.1109/APEC.2019.8722191.

[5] J. Anzola et al., "Review of Architectures Based on Partial Power Processing for DC-DC Applications," in IEEE Access, vol. 8, pp. 103405-103418, 2020, doi: 10.1109/ACCESS.2020.2999062.

[6] K. A. Kim, P. S. Shenoy and P. T. Krein, "Converter Rating Analysis for Photovoltaic Differential Power Processing Systems," in IEEE Transactions on Power Electronics, vol. 30, no. 4, pp. 1987-1997, April 2015, doi: 10.1109/TPEL.2014.2326045.

[7] M. Button, "An Advanced Photovoltaic Regulator ModuleArray, " in IECEC 96. Proceedings of the 31st IntersocietyEnergy Conversion Engineering Conference, 1996, pp.519–524.

[8] M. Kasper, D. Bortis and J. W. Kolar, "Classification and Comparative Evaluation of PV Panel-Integrated DC–DC Converter Concepts," in IEEE Transactions on Power Electronics, vol. 29, no. 5, pp. 2511-2526, May 2014, doi: 10.1109/TPEL.2013.2273399.

[9] J. R. R. Zientarski, M. L. da Silva Martins, J. R. Pinheiro and H. L. Hey, "Evaluation of Power Processing in Series-Connected Partial-Power Converters," in IEEE Journal of Emerging and Selected Topics in Power Electronics, vol. 7, no. 1, pp. 343-352, March 2019, doi: 10.1109/JESTPE.2018.2869370.

[10] M. C. Mira, Z. Zhang and A. E. Michael Andersen, "Analysis and Comparison of dc/dc Topologies in Partial Power Processing Configuration for Energy Storage Systems," 2018 International Power Electronics Conference (IPEC-Niigata 2018 -ECCE Asia), 2018, pp. 1351-1357, doi: 10.23919/IPEC.2018.8507937.

[11] J. R. R. Zientarski, M. L. d. S. Martins, J. R. Pinheiro and H. L. Hey, "Series-Connected Partial-Power Converters Applied to PV Systems: A Design Approach Based on Step-Up/Down Voltage Regulation Range," in IEEE Transactions on Power Electronics, vol. 33, no. 9, pp. 7622-7633, Sept. 2018.

[12] B. Sun, Z. Zhang and M. A. E. Andersen, "A Comparison Review of the Resonant Gate Driver in the Silicon MOSFET and the GaN Transistor Application," in IEEE Transactions on Industry Applications, vol. 55, no. 6, pp. 7776-7786, Nov.-Dec. 2019, doi: 10.1109/TIA.2019.2914193.

[13] K. Kruse, M. Elbo and Z. Zhang, "GaN-based high efficiency bidirectional DC-DC converter with 10 MHz switching frequency," 2017 IEEE Applied Power Electronics Conference and Exposition (APEC), 2017, pp. 273-278, doi: 10.1109/APEC.2017.7930705.

[14] Data sheet of 650V CoolMOS™ C7 Power Transistor IPW65R095C7

The Method for Decoupling the Parasitic Inductance of the Laminated Busbar with SiC MOSFETs in Parallel

Shaolin Yu
School of Electrical Engineering and Automation
Hefei University of Technology
Hefei, China
hajcysl@163.com

Jianing Wang
School of Electrical Engineering and Automation
Hefei University of Technology
Hefei, China
jianingwang@hfut.edu.cn

Xing Zhang
School of Electrical Engineering and Automation
Hefei University of Technology
Hefei, China
honglf@ustc.edu.cn

Yuanjian Liu
School of Electrical Engineering and Automation
Hefei University of Technology
Hefei, China
liuyuanjian1230@163.com

Zhaoyang Wei
School of Electrical Engineering and Automation
Hefei University of Technology
Hefei, China
zywei0515@163.com

Abstract—Discrete Silicon Carbide (SiC) MOSFETs are usually used in parallel to enhance the current capability for medium and high-power application. However, the rough design of the busbar structure may lead to an asymmetric parasitic inductance of the parallel branches., which can further cause the current imbalance. Thus, a symmetrical busbar design is critical for the parallel application of the discrete SiC MOSFETs. While, mutual coupling between the parallel branches, and with the increase in the parallel devices can lead to a very complicated inductance network. It is not convenient for the prediction of the current balance and estimating the symmetry of the busbar. Based on it, this paper proposes a decoupling method to acquire the equivalent inductance that can easily evaluate the current balance performance. Meanwhile, it can further provide a basis for the busbar structure design. A specific laminated busbar with six devices in parallel in each bridge arm for an electrical vehicle inverter is used as a case study. And the prosed decoupling method is verified by simulation and experimental tests.

Keywords—SiC MOSFETs, laminated busbar, parasitic inductance networks, decoupling, the equivalent inductance

I. INTRODUCTION

Parallel application of discrete SiC MOSFETs is a common approach to increase the current capacity [1]. And the symmetric design of the laminated busbar that used to connect the devices is critical for the parallel application of SiC MOSFETs. This is because the difference in parasitic inductance of each parallel branch may lead to current imbalance [2]. As a result, the unbalanced current can cause unequal power loss, which can further lead to the thermal imbalance that challenge the device safety [3].

In order to predict the current performance among the parallel devices and to enable the symmetric design of the busbar, the primary target is to accurately extract the parasitic inductance of each parallel branch on the busbar. In fact, the busbar inductance has been investigated by many literatures, and currently the partial self-and mutual inductances are the dominant way used to analyze the busbar inductance. Paper [4]-[5] concentrate on the busbar structure optimization to have a minimal loop inductance based on the obtained self-and mutual inductance. However, the paralleling is not the topic discussed in these papers. Actually, for the parallel application, the self-and mutual inductance cannot be directly used to evaluate the symmetry of the busbar structure due to the complex mutual couplings existing in the parallel branches. Therefore, a method is needed to decouple the inductance network into the equivalent inductance that can be used to predict the current balance performance. And meanwhile, the equivalent inductance can also utilized to guide the busbar design. Paper [6] derives different simplified inductance networks based on the current paths relevant to the switching states. However, the simplified network cannot be used in paralleling applications because it still has the mutual inductance parameters. Paper [7] obtain the equivalent inductance of each parallel branch through experimental tests. The method is not efficient and lacks accuracy for its fussy experimental steps. In paper [8], a normal method that measuring *LC* resonant frequency to extract the parasitic loop inductance is presented This is a good method to obtain the total loop inductance. However, it cannot be used in our case to obtain the equivalent inductance of each parallel branch. In paper [9], a method to calculate the equivalent inductance is given. However, the busbar structure only consists of two conducive layers, the approach cannot be utilized for the structure with multiple current paths. In summary, there is no good method to extract the equivalent inductance of each branch in a parallel application.

This paper proposes a method to decouple the parasitic inductance network of the busbar with multiple discrete devices in parallel. It can quickly obtain the equivalent inductance of each parallel branch from the complexed inductance network. Firstly, the inductance model of the busbar is developed according to the real current path. Then, the decoupling method is presented based on the established mathematical model to derive the equivalent inductance from the complex network. Moreover, for one application case, the method is applied to acquire the equivalent inductance of each parallel branch. Finally, the method is validated by the simulation and experiments results.

II. THE PARASITIC INDUCTANCE MODELING FOR THE BUSBAR WITH PARALLEL BRANCHES

For the sake of brevity, the modeling process is illustrated with two devices connected in parallel. For the multiple

This work was supported in part by the Chinese National Natural Science Foundation Program 52077051.

parallel devices, the principle is also applicable. The simplified connection forms of the DC capacitors and the parallel devices for a half-bridge topology is shown in Fig. 1. The laminated busbar used to connect them has three conducting layers, which are the positive layer marked in red, the negative layer marked in green and the AC layer marked in blue. The corresponding equivalent circuit is shown in Fig. 1 as well.

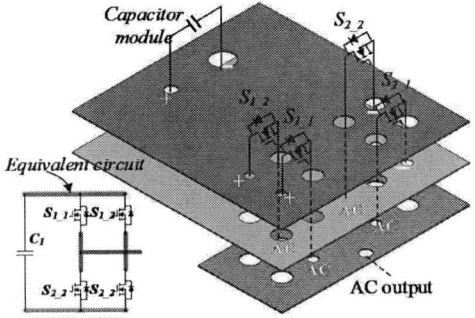

Fig. 1. The laminated busbar with two devices in parallel.

The inductance model that incorporating the self-and mutual inductance is developed based on the real current path that depended on the switching state. Here, it is assumed that the upper bridge arm device is in on state and the lower bridge arm device is in off state at a certain time. And the current flows through the devices in upper bridge arm to the load side, as shown in Fig. 2 (a). Then, the switching devices in upper bridge arm are turned off, while in lower arm is turned on. Due to the inductive load, the load current cannot be changed abruptly. The current is commutated from the upper arm to the lower arm with same rate change, as shown in Fig. 2 (b). Thus, the high frequency component of the commutation current will form loops between capacitors, busbars and devices.

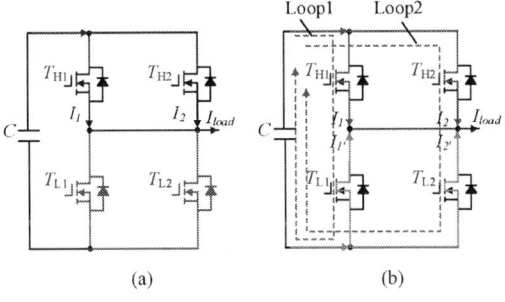

Fig. 2. The analysis of the current commutation loop. (a) Before current commutation. (b) In current commutation transient.

The current paths flowing through the busbar among the capacitor and the devices during the current commutation transient is illustrated in Fig. 3 (a). On the busbar, the current flows from the capacitor through the positive busbar to the drain of upper devices, and then flows from the source terminal to the AC busbar. Then, the current flows from the AC busbar to the drain of the devices in lower bridge arm, and from the source to the negative busbar to the negative terminal of the capacitor. Each current path of the conducting segment can represent a self-inductance. And mutual inductances presents between each two current paths. The inductance model of the busbar with two SiC MOSFETs in parallel can be established in Fig. 3 (b). For clarity, only the mutual inductance between L_{p1p1} and the other self-inductance is shown.

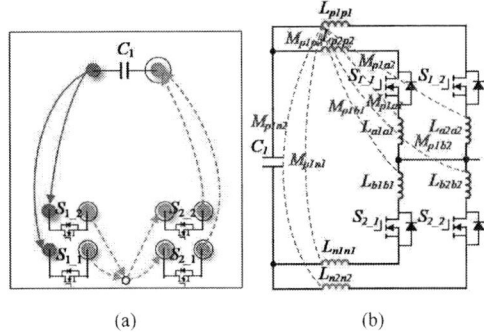

Fig. 3. The modeling of the inductance for the form. (a) Current path on the busbar. (b) The Inductance network incorporates the elements of self-inductance and mutual inductance.

III. DECOUPLING OF THE INDUCTANCE NETWORK

Fig. 4(a) shows the inductance model of the circuit that includes the self-and mutual inductance. The equivalent inductance, as illustrated in Fig. 4(b), rather than the inductance network can be used to evaluate the symmetry of each parallel branch due to the complexed mutual coupling effect. While, the equivalent inductance should be decoupled from the network. The decoupling method shown next is an example of two devices in parallel, and the same principle applies for multiple devices in parallel.

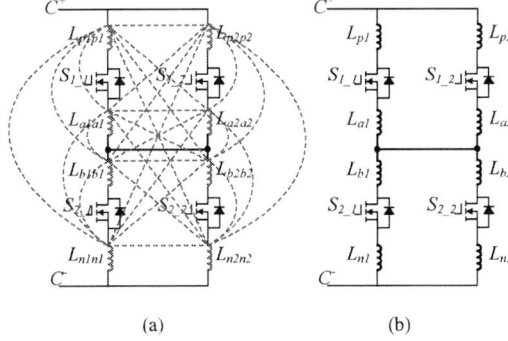

Fig. 4. The inductance model for two parallel devices. (a) Complex inductance model containing the self-inductance and mutual inductance. (b) The equivalent inductance.

The mathematical expression of the inductance network is established to derive the equivalent inductance based on Fig. 5. The voltage across the inductances, the current flowing through the inductance and the inductance network can be met the conditions in (1) in time domain.

$$u=pL \cdot i \qquad (1)$$

where, the p is the differential operator: d/dt, u represents the induced voltage across each inductance, the i is the current through each inductance, and L is the original inductance network including the self- inductance and mutual inductance.

$$u = \begin{bmatrix} u_{p1} & u_{p2} & u_{a1} & u_{a2} & u_{b1} & u_{b2} & u_{n1} & u_{n2} \end{bmatrix}^T$$
$$i = \begin{bmatrix} i_{H1} & i_{H2} & i_{H1} & i_{H2} & i_{L1} & i_{L2} & i_{L1} & i_{L2} \end{bmatrix}^T$$

$$L = \begin{bmatrix} L_{p1p1} & M_{p1p2} & M_{p1a1} & M_{p1a2} & M_{p1b1} & M_{p1b2} & M_{p1n1} & M_{p1n2} \\ M_{p2p1} & L_{p2p2} & M_{p2a1} & M_{p2a2} & M_{p2b1} & M_{p2b2} & M_{p2n1} & M_{p2n2} \\ M_{a1p1} & M_{a1p2} & L_{a1a1} & M_{a1a2} & M_{a1b1} & M_{a1b2} & M_{a1n1} & M_{a1n2} \\ M_{a2p1} & M_{a2p2} & M_{a2a1} & L_{a2a2} & M_{a2b1} & M_{a2b2} & M_{a2n1} & M_{a2d2} \\ M_{b1p1} & M_{b1p2} & M_{b1a1} & M_{b1a2} & L_{b1b1} & M_{b1b2} & M_{b1n1} & M_{b1n2} \\ M_{b2p1} & M_{b2p2} & M_{b2a1} & M_{b2a2} & M_{b2b1} & L_{b2b2} & M_{b2n1} & M_{b2n2} \\ M_{n1p1} & M_{n1p2} & M_{n1a1} & M_{n1a2} & M_{n1b1} & M_{n1b2} & L_{n1n1} & M_{n2n2} \\ M_{n2p1} & M_{n2p2} & M_{n2a1} & M_{n2a2} & M_{n2b1} & M_{n2b2} & M_{n2n1} & L_{n2n2} \end{bmatrix}$$

The equivalent inductance is the network that have same voltage-current (*V-I*) characteristics with the original inductance network. That means it can produce the same voltage drops across segmented conductor at the same rate of current change. Hence, the inductance network L in (1) can be replaced by the equivalent inductance network L_e, as presented in (2). The L_e should not have the mutual coupling, which is easy to evaluate the current sharing.

$$u = pL_e \cdot i \qquad (2)$$

$$L_e = \begin{bmatrix} L_{p1} & 0 & 0 & 0 & 0 & 0 & 0 & 0 \\ 0 & L_{p2} & 0 & 0 & 0 & 0 & 0 & 0 \\ 0 & 0 & L_{a1} & 0 & 0 & 0 & 0 & 0 \\ 0 & 0 & 0 & L_{a2} & 0 & 0 & 0 & 0 \\ 0 & 0 & 0 & 0 & L_{b1} & 0 & 0 & 0 \\ 0 & 0 & 0 & 0 & 0 & L_{b2} & 0 & 0 \\ 0 & 0 & 0 & 0 & 0 & 0 & L_{n1} & 0 \\ 0 & 0 & 0 & 0 & 0 & 0 & 0 & L_{n2} \end{bmatrix}$$

Based on (1) and (2), then the L_e can be derived by (3). The L can be extracted by the software Ansys Q3D. If the currents in each branch are known, then the L_e can be obtained.

$$\begin{cases} L_{p1} = L_{p1p1} + M_{p1ap}\dfrac{i_{H2}}{i_{H1}} + \dots + M_{p1n1}\dfrac{i_{L1}}{i_{H1}} + M_{p1n2}\dfrac{i_{L2}}{i_{H1}} \\ L_{p2} = M_{p2p1}\dfrac{i_{H1}}{i_{H2}} + L_{p2p2} + \dots + M_{p2n1}\dfrac{i_{L1}}{i_{H2}} + M_{p2n2}\dfrac{i_{L2}}{i_{H2}} \\ \vdots \\ L_{n2} = M_{n2p1}\dfrac{i_{H1}}{i_{L2}} + M_{n2p2}\dfrac{i_{H2}}{i_{L2}} + \dots + M_{n2n1}\dfrac{i_{L1}}{i_{L2}} + M_{n2n2} \end{cases} \quad (3)$$

The derivation of the currents is shown as below. Firstly, according to the circuit relationship in Fig. 5, the 8×1 current and voltage matrix can be converted to a 4×1 matrixes as (4) and (5).

$$A^T \begin{bmatrix} i_{H1} \\ i_{H2} \\ i_{L1} \\ i_{L2} \end{bmatrix} = i \qquad (4) \qquad A u = \begin{bmatrix} u_{H1} \\ u_{H2} \\ u_{L1} \\ u_{L2} \end{bmatrix} \qquad (5)$$

where the u_{H1} represents the voltage of the first branch in upper bridge arm, and the u_{H2}, u_{L1}, u_{L2} have the similar meaning, as shown in Fig. 5. The matrix A is described as follow:

$$A = \begin{bmatrix} 1 & 0 & 1 & 0 & 0 & 0 & 0 & 0 \\ 0 & 1 & 0 & 1 & 0 & 0 & 0 & 0 \\ 0 & 0 & 0 & 0 & 1 & 0 & 1 & 0 \\ 0 & 0 & 0 & 0 & 0 & 1 & 0 & 1 \end{bmatrix}$$

According to (4) and (5), the (1) can be further simplified as:

$$pA L A^T \begin{bmatrix} i_{H1} \\ i_{H2} \\ i_{L1} \\ i_{L2} \end{bmatrix} = \begin{bmatrix} u_{H1} \\ u_{H2} \\ u_{L1} \\ u_{L2} \end{bmatrix} \qquad (6)$$

Further, based on Fig. 5, the 4×1 current and voltage matrix can be transferred as 2×1 matrixes, as shown below.

$$\begin{bmatrix} 1 & 1 & 0 & 0 \\ 0 & 0 & 1 & 1 \end{bmatrix} \begin{bmatrix} i_{H1} \\ i_{H2} \\ i_{L1} \\ i_{L2} \end{bmatrix} = \begin{bmatrix} i \\ i \end{bmatrix} \qquad (7)$$

$$\begin{bmatrix} 1 & 0 \\ 1 & 0 \\ 0 & 1 \\ 0 & 1 \end{bmatrix} \begin{bmatrix} u_H \\ u_L \end{bmatrix} = \begin{bmatrix} u_{H1} \\ u_{H2} \\ u_{L1} \\ u_{L2} \end{bmatrix} \qquad (8)$$

where the u_H and u_L represent the upper and lower bridge voltage respectively and the i represents the total currents of the upper or lower bridge arm, which are shown in Fig. 5.

Based on (6), (7) and (8), the proportional relationship between the current in each parallel branch and the sum current i can be deduced, as shown in:

$$\begin{bmatrix} i_{H1} \\ i_{H2} \\ i_{L1} \\ i_{L2} \end{bmatrix} = L_s^{-1} \begin{bmatrix} 1 & 0 \\ 1 & 0 \\ 0 & 1 \\ 0 & 1 \end{bmatrix} \left(\begin{bmatrix} 1 & 1 & 0 & 0 \\ 0 & 0 & 1 & 1 \end{bmatrix} L_s^{-1} \begin{bmatrix} 1 & 0 \\ 1 & 0 \\ 0 & 1 \\ 0 & 1 \end{bmatrix} \right)^{-1} \begin{bmatrix} 1 \\ 1 \end{bmatrix} i \quad (9)$$

where L_s signifies the three matrixes after the operator p in (6). Further, the currents can be used in (3) to calculate the equivalent matrix L_e.

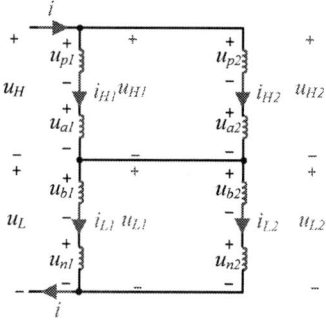

Fig. 5. Scheme for the derivation of the equivalent inductance.

IV. CASE STUDY

To demonstrate the validity of the above method, this paper analyze a laminated busbar that connected with six SiC MOSFETs in parallel for an inverter.

A. The inductance model

The EV inverter has a power rating of approximately 120 kW and a peak current pf 350 A per phase. Six discrete SiC MOSFETs that can handle 75 A at $T_c=100°C$ is required in parallel in each bridge arm. The physical structure of the laminated busbar which is applied to the main circuit of the inverter is shown in Fig. 6 (a). The busbar is composed of three same single-phase bridge arms, thus, only the inductance model of a single phase is analyzed. Fig. 6(b) shows the different conductive layers of the busbar, which include positive layer, the negative layer, and the AC output layer. Each layer of busbar extends out the corresponding terminals to connect with the DC-link capacitors and the power device, which is also shown in this picture

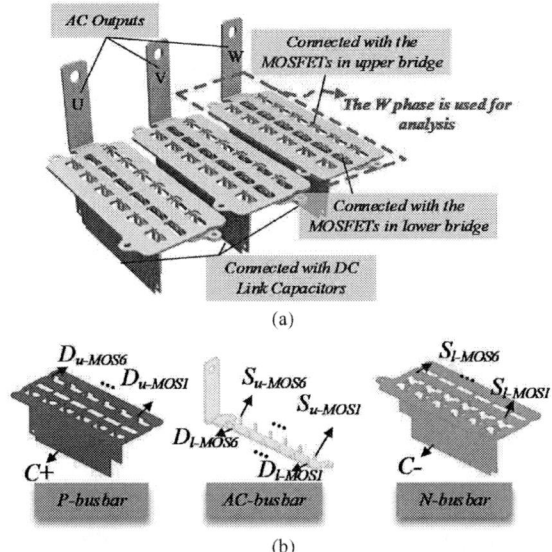

(a)

(b)

Fig. 6. The busbar structure. (a) The overall structure. (b) The connections for the devices and busbar in the W phase.

As analyzed in Section II, the high frequency component of the commutation current forms a loop among the capacitors, busbars, and devices. Thus, for the busbar in this case, there are two current paths. One path is that the current flows from one DC+ terminal to 6 drains of the upper MOSFETs, and then from the 6 MOSFET sources to the AC terminal. The other reversed path is that the current flows from one DC- terminal to 6 sources of the lower MOSFETs, and then from the 6 MOSFET drains to the AC terminal. According to the modeling method in Section III, the inductance model of the busbar that incorporating the self-and mutual inductance can established, as illustrate in Fig. 7(a). The bridge arm inductances are divided into four parts according to inductance modeling approach. For the upper bridge arm, it includes the self-inductances L_{pp}, and L_{aa}. The L_{pp} is formed by the segmented conductor from the positive terminal of the capacitor to the drain terminal of the devices, and the L_{aa} is the segmented conductor from the AC busbar to the source terminal of devices. For the lower arm, it contains the self-inductances L_{bb} and L_{nn}. The L_{bb} is formed by the path from the AC busbar to the drain terminal of devices, and L_{nn} is formed by the path from the negative terminal of the capacitor to the source terminal of devices. There are mutual inductances, such as M_{pa}, M_{pb}, and M_{pn}, among any two self-inductances. Thus, for the six parallel branches, the inductance network can contain three hundred elements, including twenty-four self-inductances and two hundred and seventy-six mutual inductances coupling between them. The complex inductance network cannot be easily used to predict the current sharing performance, thus the network should be simplified as equivalent inductances, as shown in Fig. 7(b), where six MOSFETs are separately shown.

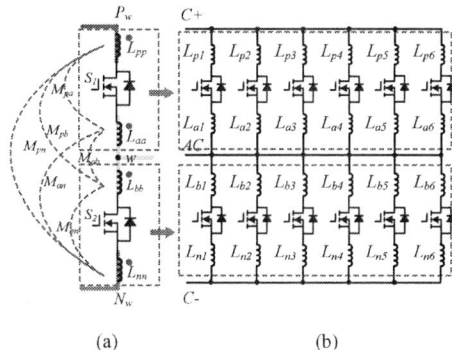

(a)   (b)

Fig. 7. The parasitic inductance model of the busbar. (a) The inductance model including self-inductance and mutual inductance. (b) The equivalent inductance network.

B. The extraction and decoupling of the inductance network

Ansys Q3D is usually used to simulate the *RLC* parameters of the conductors. The laminated busbar used in this case can be imported in this software to extract the inductance network. According to the parallel circuit state, the multiple current excitations on the busbars terminals should be assigned simultaneously. In Q3D, the current path for a conductor is defined by the form of the source to the sink. For this case, the current may flow from one terminal (source) to multiple terminals (sinks) due to the parallel connection. However, for the solver of Q3D, only one sink can be assigned to a conductor, while with multi sources [10]. According to the principle that the inductance of the conductor will not be affected by the current, the excitation source and sink can switch places. Taking an example, for the P busbar in Fig. 8, the original current flows from the DC+ terminal to six parallel devices, while the excitation sink can be allocated to the DC+ terminal, and the sources can be allocated to the drains of the device, which results in the opposite of the actual current path. The current density distribution clouds obtained from the Q3D simulation are shown in Fig. 8. It can be observed that the point with the highest current density distribution corresponds to the excitation source. Because the excitations assigned to the positive busbar have the opposite directions of the actual current, the polarity of the corresponding mutual inductance should be reversed. Another key issue in the simulation is how to determine the point where to assign the AC terminal. The six MOSFET branches physically connect to one point that further connects to the AC line. By the simulation, along the AC busbar the current density becomes stronger and then with unified distribution, as shown in AC-busbar in Fig. 8. Thus, the interface can be set as the current sink.

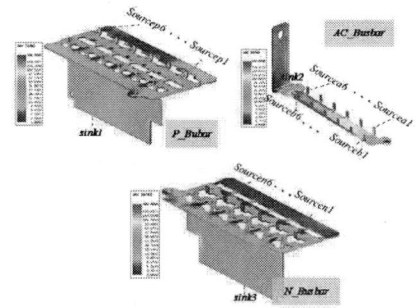

Fig. 8. The excitation source settings and current density of the busbar for extracting the inductance network.

978-1-6654-4817-8/21 $31.00 © 2021 IEEE   273

By the simulation, a 24 multiplied by 24 multi-dimensional inductance matrix can be extracted. However, it cannot be utilized to accurately determine the symmetrical design of the busbar. In addition, the matrix cannot be used to predict the current sharing performance of the parallel devices. By using the proposed decoupling method, the equivalent inductances can be calculate, as presented in Table I.

TABLE I
CALCULATED THE EQUIVALENT INDUCTANCE IN DYNAMIC DURATION

Inductance (nH)	Branch					
	1	2	3	4	5	6
L_p	4.47	4.87	5.02	5.09	5.20	5.42
L_a	0.66	0.65	0.65	0.65	0.65	0.67
L_b	1.70	1.35	1.24	1.17	1.07	0.81
L_n	4.90	4.77	4.78	4.79	4.77	4.79

C. The validation of the method by the simulation

The accuracy of the decoupling method can be validated by contrasting the *V-I* characteristics of the original inductance network and the derived equivalent inductance network. Fig. 9 shows the corresponding simulated circuit of the two inductance networks in Ansys Twin Builder. Fig. 9 (a) is the simulated model that comprising of the self-and mutual inductance. Fig. 9 (b) is the equivalent inductance model decoupled from the original inductance network by the proposed method. The voltage source of the two circuits is same, which is a step signal with 200mV for 15us. The AM stands for ammeter, which used to extract the current of each parallel branches.

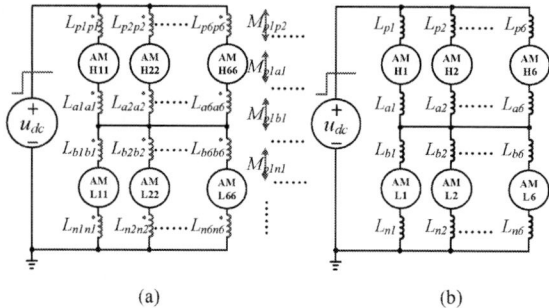

(a)   (b)

Fig. 9. The verification of the decoupling method. (a) The network including the self- inductances and mutual inductance; (b) The network including 24 equivalent inductances.

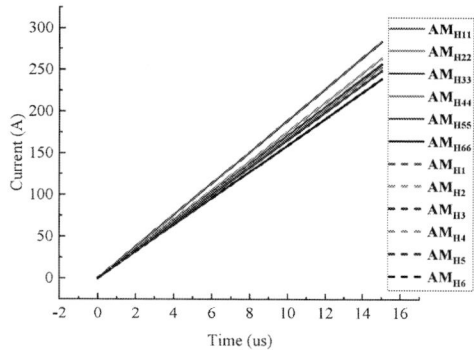

Fig. 10. Comparison of current distribution of devices in upper bridge arm in the above two circuits.

The simulated current of the parallel devices in the upper bridge arm is shown in Fig. 10. The solid lines represent the currents in the complex network that including the self-and mutual inductance, and the dotted lines represent the current

in the derived equivalent inductance network. It can be seen that the solid and dashed lines of the corresponding branches in the figure overlap, meaning that the current of each branch is the same in both networks. It proves that the two inductance networks have the same current at the same voltage excitation. For lower bridges of the two circuit, it also have the same conclusion, which is not shown for clarity. The simulation results demonstrate the equivalence of the two networks, and thus the accuracy of proposed decoupling method can be further proved.

V. EXPERIMENT RESULTS

As shown in Fig. 11, a double-pulse testing platform is established to verify the accuracy of the proposed decoupling method. The prototype in the platform is a full- scale SiC EV inverter that mentioned in Section IV. In the switching transient, the overvoltage induced by the parasitic inductance can be applied to the turned-off devices. Thus, the loop inductance can be obtained by (10)

$$L_{loop} = \frac{\Delta V}{\Delta i} \Delta t \tag{10}$$

where ΔV is the voltage drop caused by the parasitic inductance in specific loop when the MOSFET is in turn off transient. The Δi is the drop in the current flowing through the MOSFET during the period.

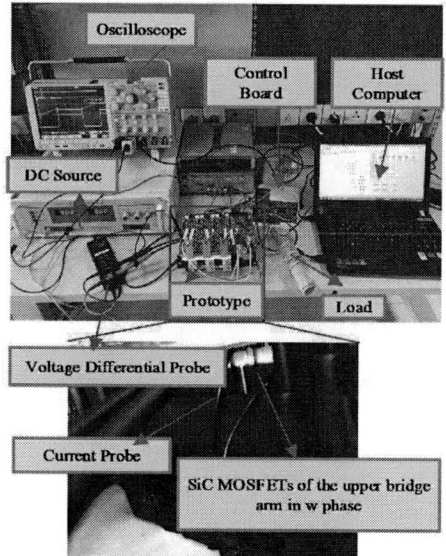

Fig. 11. The experimental platform.

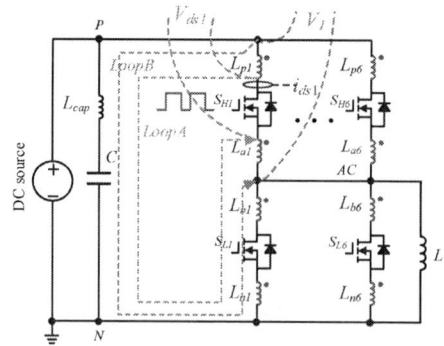

Fig. 12. The test circuit for the equivalent inductance.

Fig. 12 shows the test circuit for obtaining the equivalent inductance of each branch. Here, taking the MOSFET in first branch as an example. The voltage probe 1 clamps the drain and source terminals of this device to test the voltage V_{ds1}. The voltage probe 2 is clamped at the P-busbar and AC-busbar to test the voltage V_1 of the first upper bridge. The current flexible probe is wound around the drain terminal of the first MOSFET for acquiring the current.

The double-pulse test waveforms of the first MOSFET are shown in Fig. 13. The second turn off waveform is selected for calculating the equivalent inductance. The overvoltage ΔV_{d1} is induced by L_{loopB}, which may consist of the device package inductance (L_{MOS}), the capacitor inductance (L_{cap}), and other parasitic inductances (L_{other}). While, the overvoltage ΔV_{ds1} is induced by L_{loopA}, which is composed of the equivalent inductance of the upper branch, the device package inductance (L_{MOS}), the capacitor inductance (L_{cap}), and other parasitic inductances (L_{other}).Compared with L_{loopB}, Ll_{oopA} is larger than it by $L_{p1}+L_{a1}$, which can be expressed in (11). The L_{loopA} and L_{loopB} can be calculated by (10), further, the total busbar equivalent inductances in the upper bridge of the first branch could be calculated by (11).

$$L_{loopA} - L_{loopB} = L_{p1} + L_{a1} \qquad (11)$$

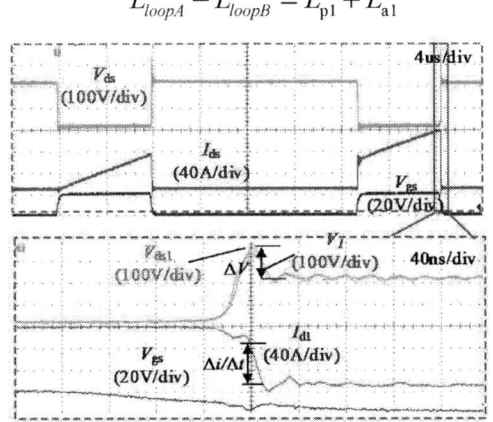

Fig. 13. The experimental double pulse waveform.

TABLE II
THE TESTED INDUCTANCE OF THE FIRST BRANCH

V_{dc} (V)	L_{loopA} (nH)	L_{loopB} (nH)	$L_H=L_{p1}+L_{a1}$ (nH)
200	35.58	29.36	6.22
250	37.51	30.99	6.52
300	36.20	29.59	6.61
Average	36.43	29.98	6.45

Multiple sets of tests with different DC voltages are conducted by averaging to reduce the test errors b. The tested L_{loopA}, L_{loopB} of the first branch are shown in Table II. The L_H represent the sum of $L_{p1}+L_{a1}$.Similarly, by the same testing steps, the L_H of other branches can be acquired e, as presented in Table III. For clarity, the loop inductances are not shown. The error between the tested and calculated results shown in TABLE III is defined as $(L_{tested}-L_{calculated})/L_{calculated}\times100\%$. It can be seen that the tested results are very comparable to the

calculated results that obtained by the decoupling method but with some errors. Considering the small base value of the inductance and the measurement errors during the experiment, the error is in an acceptable range. Thus, the accuracy of the decoupling method can be validated by the experiments.

TABLE III
COMPARISON OF TESTED AND CALCULATED EQUIVALENT INDUCTANCE
(L_P+L_A)

L_H (nH)	Branch					
	1	2	3	4	5	6
Tested	6.45	6.04	6.85	6.66	6.45	6.78
Calculated	5.13	5.52	5.67	5.74	5.87	6.14
Error (%)	25.73	9.42	20.81	16.02	9.88	10.42

VI. CONCLUSION

In terms of the extracted inductance matrix cannot used to evaluate the symmetry of the busbar structure in parallel application. This paper proposes the method to decouple the equivalent inductance that can easily evaluate the current balance performance. Meanwhile the derived equivalent inductance can be further used to estimate the symmetry design of the busbar. It is meaningful to advance the parallel application of the discrete SiC MOSFETs.

REFERENCES

[1] Z. Zeng, X. Zhang and Z. Zhang, "Imbalance Current Analysis and Its Suppression Methodology for Parallel SiC MOSFETs with Aid of a Differential Mode Choke," in IEEE Transactions on Industrial Electronics, vol. 67, no. 2, pp. 1508-1519, Feb. 2020.

[2] H. Li et al., "Influences of Device and Circuit Mismatches on Paralleling Silicon Carbide MOSFETs," in IEEE Transactions on Power Electronics, vol. 31, no. 1, pp. 621-634, Jan. 2016.

[3] C. Zhao, L. Wang and F. Zhang, "Effect of Asymmetric Layout and Unequal Junction Temperature on Current Sharing of Paralleled SiC MOSFETs With Kelvin-Source Connection," in IEEE Transactions on Power Electronics, vol. 35, no. 7, pp. 7392-7404, July 2020.

[4] M. Khan, P. Magne, B. Bilgin, S. Wirasingha and A. Emadi, "Laminated busbar design criteria in power converters for electrified powertrain applications," 2014 IEEE Transportation Electrification Conference and Expo (ITEC), Dearborn, MI, 2014, pp. 1-6.

[5] Z. Huang et al., "A novel low inductive 3D SiC power module based on hybrid packaging and integration method," 2017 IEEE Energy Conversion Congress and Exposition (ECCE), Cincinnati, OH, 2017, pp. 3995-4002.

[6] N. Zhang, S. Wang and H. Zhao, "Develop Parasitic Inductance Model for the Planar Busbar of an IGBT H Bridge in a Power Inverter," in IEEE Transactions on Power Electronics, vol. 30, no. 12, pp. 6924-6933, Dec. 2015.

[7] C. Geng, F. He, J. Zhang, and H. Hu, "Partial stray inductance modeling and measuring of asymmetrical parallel branches on the bus-bar of electric vehicles," Energies, vol. 10, 2017, Art. no. 1519.

[8] A. Shahabi and A. N. Lemmon, "Multi-Branch Inductance Extraction Procedure for Multi-Chip Power Modules," in IEEE Journal of Emerging and Selected Topics in Power Electronics. It is important to emphasize the difference between the previous research and the proposed inductance model.

[9] M. Khan, P. Magne, B. Bilgin, S. Wirasingha and A. Emadi, "Laminated busbar design criteria in power converters for electrified powertrain applications," 2014 IEEE Transportation Electrification Conference and Expo (ITEC), Dearborn, MI, 2014, pp. 1-6.

[10] Q3D Extractor Help. ANSYS, Inc. 2018

EMI Noise Reduction in GaN-based Full-bridge LLC Converter

Yue Cao
Department of Electrical Engineering
Xi'an Jiaotong University
Xi'an, China
yue1403300584@stu.xjtu.edu.cn

Yuxuan Chen
Department of Electrical Engineering
Xi'an Jiaotong University
Xi'an, China
cyx1998@stu.xjtu.edu.cn

Xingwei Huang
Department of Electrical Engineering
Xi'an Jiaotong University
Xi'an, China
huangxw867046635@stu.xjtu.edu.cn

Pengyuan Ren
Department of Electrical Engineering
Xi'an Jiaotong University
Xi'an, China
3120104235@stu.xjtu.edu.cn

Wenjie Chen
Department of Electrical Engineering
Xi'an Jiaotong University
Xi'an, China
cwj@mail.xjtu.edu.cn

Xu Yang
Department of Electrical Engineering
Xi'an Jiaotong University
Xi'an, China
yangxu@mail.xjtu.edu.cn

Abstract—In medium-to-high power converters, full-bridge LLC converter has been a common choice due to its high efficiency and power density. However, today, the EMI noise remains a significant obstacle due to high dv/dt and large parasitic capacitors in transformers and MOSFETs. Through eliminating displacement currents, the presented technique can help to maintain low-profile converter by avoiding the bulky CM chokes, so as to further increase power density. This paper analyses main sources of EMI noise in FB LLC converter and presents a cancellation technique to eliminate EMI noise without other passive components. A lumped simulation circuit shows that the optimized converter generated above 50dB less EMI noise than a traditional converter.

Keywords—*EMI naked-noise, LLC converter, EMI noise cancellation, Symmetric resonant tank.*

I. INTRODUCTION

The full-bridge topology is widely used for the medium-to-high power supplies, such as sever powers, data centers and electric vehicle battery chargers[1]. Today, with the demand of high efficiency and low profile in power supplies, the full-bridge LLC resonant converter is an excellent option, due to soft-switching operation, especially when a high step-down converter with high turn ratio transformer is needed in on-board applications[2, 3].

Today, there are mainly two ways to further increase power density, including increasing switching frequency and high turn ratio transformers to achieve a low-profile power converter. Many efforts have been done to design GaN-based LLC resonant converters with MHz switching frequency. Recently, high turn ratio transformers has been widely studied and applied in power converters, due to its low-profile, parameters repeatability, labor-free, and low thermal resistance. However, high switching frequency and larger dv/dt in switching nodes and larger parasitic capacitors can result in severe CM naked-noise problems in power supplies.

A CM filter is needed in a commercial power converter product to reduce EMI noise, so as to pass the electromagnetic interference(EMI) test. In a CM filter, a CM choke is usually unavoidable to maintain sufficient attenuation for CM noise and usually takes about ¼ of the whole power supply. However, the CM choke is typically larger when facing much larger CM noise, which can significantly deviate the power density. When original CM noise is eliminated, the volume of CM choke can be shrunk while keeping the CM noise lower than the EMI standard. Previously, a shielding technique is normally chosen to avoid displacement current pathing from the primary to the secondary in transformers [3-6]. However, the extra windings for shielding will increase the winding area of transformers and conduction loss due to the eddy current, which will deviate low-profile and efficiency of converter Recently, due to the benefits of no extra windings, paired-layer winding method has been widely applied in fly-back and half-bridge LLC converter[7, 8]. A winding layout has been proposed to main low displacement current flowing through parasitic capacitors between layers. But the paired-layer method is also limited due to the requirement of static points at both the primary and secondary windings, especially in conventional full-bridge topology due to no static point at the primary winding.

In[9, 10], to apply paired-layer method in full-bridge LLC, static-point connection are proposed to establish the static points at the primary winding, so as to build the paired-layer layout. Though a better EMI performance can be achieved, conduction losses of windings in transformers will increase, especially in the primary winding. On the one hand, a rearrangement of primary windings is needed to comply with paired-layer method, which will increase conduction loss due to distortion of winding distribution, especially in high turn ratio transformer. On the other hand, when paired-layer method is adopted, in both half-bridge and full-bridge LLC converter, only partially interleaved structure can be built or paralleled to reduce conduction loss in transformers at the expense of increasing winding area[7]. Especially, in low output voltage and large current power supplies, when a high turn ratio transformers are needed to achieve large step-down converter, those methods are not sufficient enough.

In this article, to eliminate CM noise, a combined method by properly utilizing the dv/dts and transformers to achieving CM noise reduction is proposed. Detailed CM noise models of the full-bridge LLC resonant converter and distributed parasitic capacitances model of transformers are analyzed.

This paper contains five sections. Section II explains CM noise reduction and the CM lumped-circuit model is analysed and developed. Section III analyses the harmonics reduction in this converter. In Section IV, a 1.5kW full-bridge LLC circuit model is simulated in LTSPICE to verify noise reduction method. Finally, conclusion has been made in Section V.

978-1-6654-4817-8/21 $31.00 © 2021 IEEE

II. SUPPRESING CM NOISE IN FULL-BRIDGE LLC CONVERTER

This Section explains how to achieve near zero CM current converter by displacement current sources cancellation. Referring to Fig. 3, these noise paths are I_{A-g} and I_{B-g} (which are due to parasitic capacitors between MOSFETS and heatsink), I_{p-s-g1} and I_{p-s-g2} (which are parasitic capacitors of transformer). A symmetric cancellation method has been analyzed to achieve zero CM noise.

A. Main dv/dts in this converter

Fig. 1 presents the circuit of the LLC converter with symmetric resonant tank. The operational waveforms of this converter under below regions are shown in Fig. 2. The voltage potentials at two mid-points v_A and v_B are also presented. Both v_A and v_B fluctuates between V_{in} and 0. During switching time, dv/dt characteristics of v_A and v_B have the same magnitude and opposite polarity, which are the main CM noise voltage sources in the LLC converter. The waveforms of the two primary winding terminals in transformer have also been presented in Fig.2, which are V_N and V_1.

However, the dv/dt of primary windings in the transformer fluctuate between v_A and v_B. The dv/dts of other layers of the primary winding are distributing linearly due to the repeatability of PCB and characteristics of transformer.

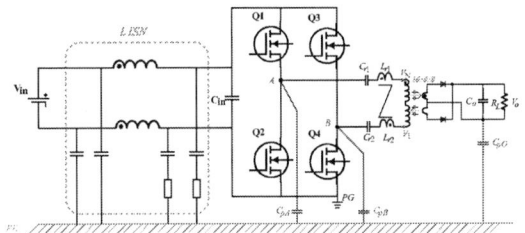

Fig. 1. FB LLC topology and the main parasitic capacitors in CM noise coupling path.

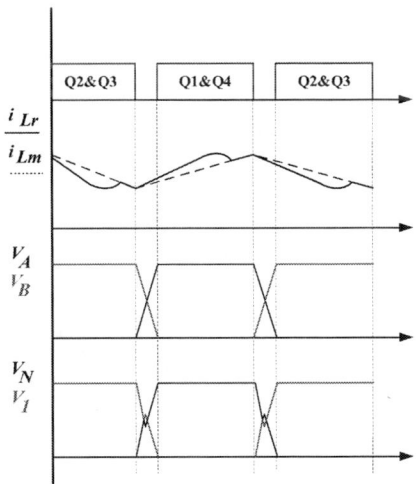

Fig. 2. Critical waveforms in this converter .

B. Corresponding Equivalent CM Circuit

The CM noise currents are pathing through main parasitic capacitors in high dv/dt nodes. In Fig. 3, the main CM noise currents are I_{A-g} and I_{B-g} (which are due to parasitic capacitances C_{pA} and C_{pB} between midpoint of half bridge and protection ground), I_{p-s-g1} and I_{p-s-g2} (which are in transformer).

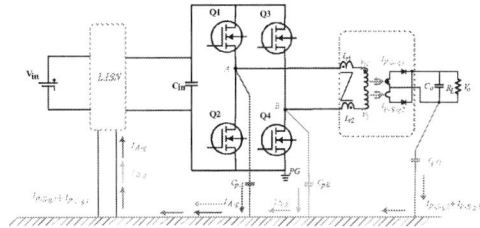

Fig. 3. CM noise currents and their coupling paths.

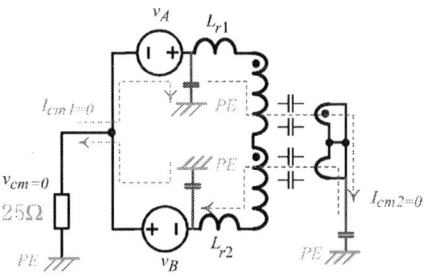

Fig. 4. CM noise lumped circuit model and main current paths for the converter.

Fig. 4 presents the CM noise circuit model. The parasitic capacitors between the midpoints of the two half bridge and PE are usually or can be designed to the same due to the repeatability of PCB technology, then the associated CM noise current can cancel each other ($I_{cm1} = 0$).

The parasitic capacitors between PCB windings in transformers can also be designed to the same, therefore CM noise current can be eliminated when dv/dts of overlapping layers have the same magnitude and opposite polarity. As mentioned in Section II-A, these overlapping layers with these dv/dt characteristic can be easily found, then a near zero CM noise transformer can be designed through appropriate primary and secondary windings arrangement regarding CM noise cancellation method.

As the two divided resonant capacitors will not hold the same value, the dv/dt of the two terminal nodes at the primary winding will not hold the same magnitude and opposite polarity, the CM current reduction in the converter is affected, which will be illustrated in Section III.

III. DIFFERENTIAL-MODE NOISE REDUCTION

In Fig. 5, we can see that the voltage of the mid-point of the primary winding is static, which is the half of the input voltage. At this case, this converter can be considered as two interleaved half bridge LLC resonant converter, which can be investigated to minimize the differential-mode (DM) EMI noise.

978-1-6654-4817-8/21 $31.00 © 2021 IEEE

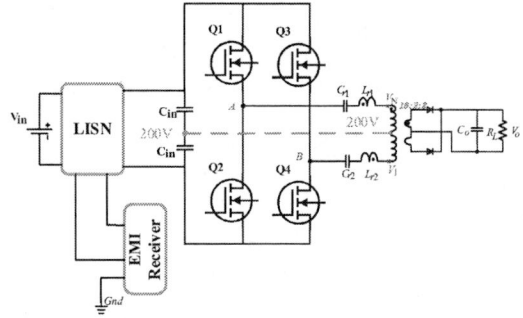

Fig. 5. Circuit of this converter including impedance stabilizing network(LISN) and EMI receicer.

Fig.6 and Fig.7 shows the simulated DM current in frequency domain and in time domain. It can be seen in Fig.7, the harmonic current in each half-bridge are almost symmetric. And in Fig. 6, it can be seen the odd harmonic are eliminated.

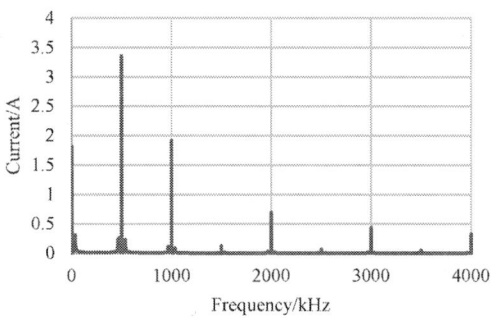

Fig. 6. Simulated DM current in frequency domain

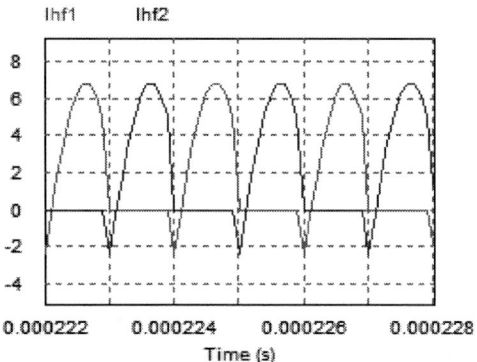

Fig. 7. Simulated DM current in time domain

IV. THE VERIFICATION BY SIMULATION AND EXPERIMENT

A lumped-circuit model is simulated in LTSPICE, so as to verify the method, regarding CM noise currents cancellation. A 1.5kW 400V input and 50V output LLC converter with 500kHz switching frequency is designed. The primary switches are GaN-based devices. A distribution parameter circuit model of transformer is built considering parasitic capacitances and inductances in the simulation. The parasitic parameters of the transformer is achieved by ANSYS

simulation. It demonstrates the LLC converter is working under perfect resonant condition with the best efficiency. As seen in Fig. 8(b), the electric potential versus PG of the primary windings is corresponding to the analysis mentioned in Section II.

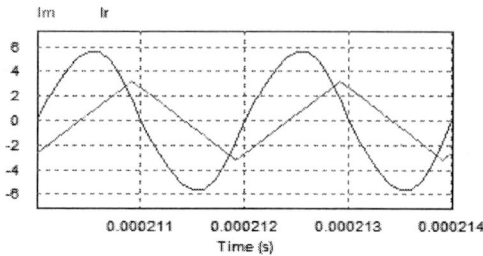

(a)

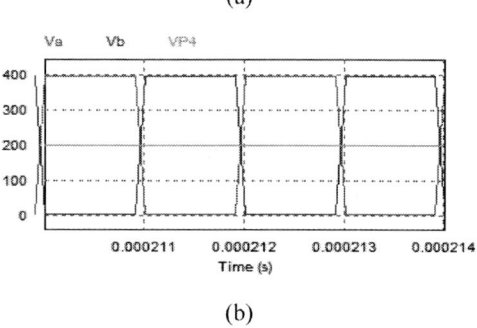

(b)

Fig. 8. Simulational waveforms of this converter. (a)currents of the resonant tank. (b)voltage waveforms of the primary winding.

Based upon the proper working condition of the GaN-based converter, the frequency domain analysis are put forward. In Fig. 9(a), the CM noise of the conventional converter with conventional transformer is shown. From the figure, the first spike of the EMI spectrum is located at 500kHz. It is exactly the same as LLC converter switching frequency. The main challenge in this GaN-based converter is the switching frequency has already increased to EMI standard frequency range. Without the proposed method, the EMI noise level is very high and it can not pass the strict EMI test standard.

Fig. 9(b) presents the CM noise of the method, which is shown in green lines. And the blue dashed line refers to the conventional EMI noise envelope. 50dB CM noise has been reduced, which verifies the cancellation method in this converter.

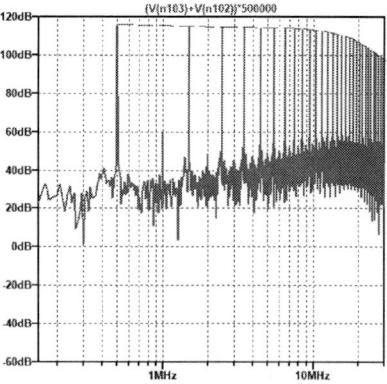

(a)

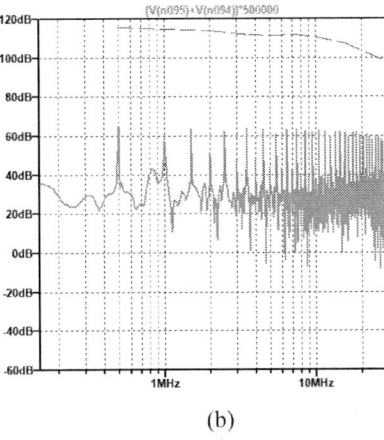

(b)

Fig. 9. The comparison of CM noise of the converter with conventional converter. (a) original converter with conventional transformer. (b) the new converter.

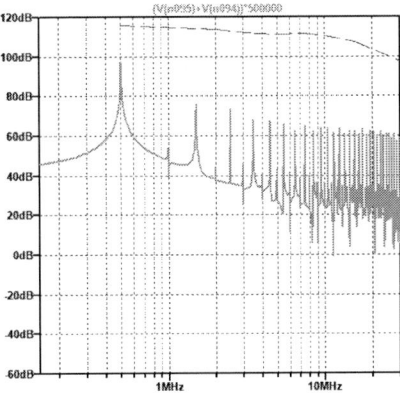

Fig. 10. Comparison of CM noise when divided capacitors are slightly different.

In order to verify the affects of unbalanced resonant capacitances, a tolerance is considered. Fig. 10 compares the CM noise of the formal converter with the novel FB LLC converter with 5% resonant capacitance drift. As can be seen, over 20dB CM noise reduction can be achieved, which is acceptable and verifies that the proposed method is practical.

A prototype is built and the circuit parameters are listed in Table I.

TABLE I. THE PARAMETERS OF THE CONVERTER

Parameters of the converter		
	Main parameters	Value
1	Input/output voltage	400V/48V
2	Switching frequency	500kHz
3	Resonant frequency	500kHz
4	Output power	1500W

TABLE II. TEST EQUIPMENTS

Test Equipments	
LISN	AMN20210125-01
Spectrum Analyzer	R&S ESL3

Fig.11 shows the suggested platform for the conducted EMI tests, and the test equipments are given in Table II.

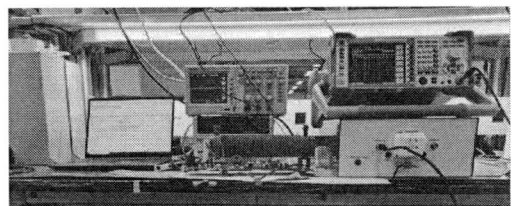

Fig. 11. Test setup for measuring conducted EMI noise.

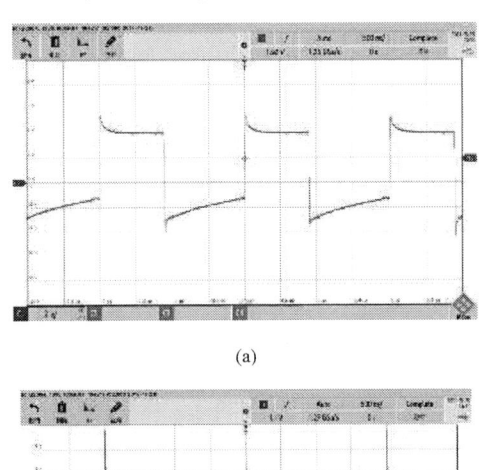

(a)

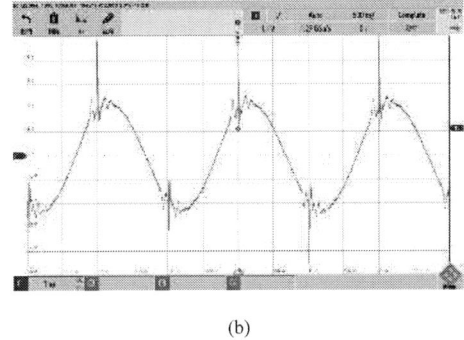

(b)

Fig. 12. Experimental waveforms of the converter. (a) v_{GS} of the choosen GaN devices. (b) The voltage across resonant capacitor.

Fig. 12(a) shows the gate drive voltage waveform of IGT60R070D1. A nagative voltage must be applied to gate-to-source of this GaN device, the waveform shows that this GaN device has been turned-off effectively, misconduction will not happen in this case. Fig. 12(b) verifies that the FB LLC resonant converter is operating properly.

978-1-6654-4817-8/21 $31.00 © 2021 IEEE

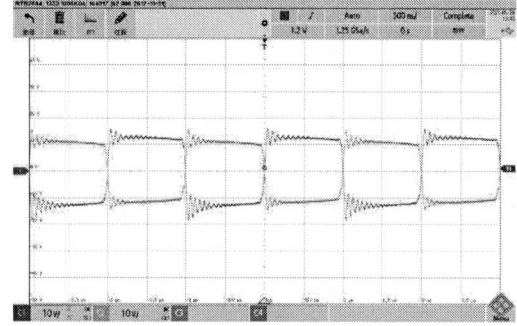

Fig. 13. The experimental voltages of the primary winding.

As can be seen in Fig. 13, the electric potential versus PG of the two terminals of primary windings is corresponding to the theoretical analysis and simulation results.

V. CONCLUSION

In this paper, to further reduce EMI noise without sacrificing efficiency, a novel converter is designed. By dividing resonant components into two parts and designing winding transformer regarding characteristics of dv/dt, the EMI noise can be reduced. A 1.5 kW FB LLC converter is designed and simulated. All the odd harmonics of DM current can be eliminated and CM noise can be reduced by over 50dB compared with the formal converter, and tolerance for the resonant capacitances is verified.

REFERENCES

[1] J.-W. Kim, D.-Y. Kim, C.-E. Kim, and G.-W. Moon, "A Simple Switching Control Technique for Improving Light Load Efficiency in a Phase-Shifted Full-Bridge Converter with a Server Power System," IEEE Transactions on Power Electronics, vol. 29, no. 4, pp. 1562-1566, 2014.

[2] A. Pratt, P. Kumar, T. V. Aldridge, and Ieee, "Evaluation of 400V DC distribution in telco and data centers to improve energy efficiency," in Intelec 07 - 29th International Telecommunications Energy Conference, Vols 1 and 2(International Telecommunications Energy Conference-INTELEC, 2007, pp. 32-39.

[3] C. Fei, Y. Yang, Q. Li, and F. C. Lee, "Shielding Technique for Planar Matrix Transformers to Suppress Common-Mode EMI Noise and Improve Efficiency," IEEE Transactions on Industrial Electronics, vol. 65, no. 2, pp. 1263-1272, 2018.

[4] Y. Yang, D. Huang, F. C. Lee, Q. Li, and Ieee, "Transformer Shielding Technique for Common Mode Noise Reduction in Isolated Converters," in 2013 Ieee Energy Conversion Congress and Exposition(IEEE Energy Conversion Congress and Exposition, 2013, pp. 4149-4153.

[5] Y. Yang, D. Huang, F. C. Lee, Q. Li, and Ieee, "Analysis and Reduction of Common Mode EMI Noise for Resonant Converters," in 2014 Twenty-Ninth Annual Ieee Applied Power Electronics Conference and Exposition(Annual IEEE Applied Power Electronics Conference and Exposition (APEC), 2014, pp. 566-571.

[6] D. Fu, P. Kong, S. Wang, F. C. Lee, M. Xu, and Ieee, "Analysis and Suppression of Conducted EMI Emissions for Front-end LLC Resonant DC/DC Converters," in 2008 Ieee Power Electronics Specialists Conference, Vols 1-10(Ieee Power Electronics Specialists Conference Records, 2008, pp. 1144-1150.

[7] M. A. Saket, M. Ordonez, M. Craciun, and C. Botting, "Improving Planar Transformers for LLC Resonant Converters: Paired Layers Interleaving," IEEE Transactions on Power Electronics, vol. 34, no. 12, pp. 11813-11832, 2019.

[8] M. A. Saket, M. Ordonez, and N. Shafiei, "Planar Transformers With Near-Zero Common-Mode Noise for Flyback and Forward Converters," Ieee Transactions on Power Electronics, vol. 33, no. 2, pp. 1554-1571, Feb 2018.

[9] K.-W. Kim, Y. Jeong, J.-S. Kim, and G.-W. Moon, "Low Common-Mode Noise LLC Resonant Converter With Static-Point-Connected Transformer," IEEE Transactions on Power Electronics, vol. 36, no. 1, pp. 401-408, 2021.

[10] K.-W. Kim, Y. Jeong, J.-S. Kim, and G.-W. Moon, "Low Common-Mode Noise Full-Bridge LLC Resonant Converter With Balanced Resonant Tank," IEEE Transactions on Power Electronics, vol. 36, no. 4, pp. 4105-4115, 2021.

978-1-6654-4817-8/21 $31.00 © 2021 IEEE

The influence of hydrogen annealing on minority carrier lifetimes in 4H-SiC

1st Ruijun Zhang
Department of physics
Xiamen university
Fujian, China
rjzhangxy@163.com

2nd Rongdun Hong
Department of physics
Xiamen university
Fujian, China
rdhong@xmu.edu.cn

3rd Jiafa Cai
Department of physics
Xiamen university
Fujian, China
jfcai@xmu.edu.cn

4th Xiaping Chen
Department of physics
Xiamen university
Fujian, China
xpchen@xmu.edu.cn

5th Dingqu Lin
Department of physics
Xiamen university
Fujian, China
lindq@xmu.edu.cn

6th Mingkun Zhang
Department of physics
Xiamen university
Fujian, China
mkzhang@xmu.edu.cn

7th Shaoxiong Wu
Department of physics
Xiamen university
Fujian, China
wsx@xmu.edu.cn

8th Yuning Zhang
Department of physics
Xiamen university
Fujian, China
zhangyuning@xmu.edu.cn

9th Jingrui Han
Dongguan Tianyu Semiconductor
Technology Co., Ltd
Guangdong, China
han.raye@sicty.com

10th Zhengyun Wu
Department of physics
Xiamen university
Fujian, China
zhywu@xmu.edu.cn

11th Feng Zhang
Department of physics
Xiamen university
Fujian, China
fzhang@xmu.edu.cn

Abstract—In this study, 4H-SiC with different thickness and doping types were annealed in hydrogen at low pressure, the minority carrier lifetimes of which were investigated by microwave photoconductive decay (μ-PCD). It can be observed that the minority carrier lifetime of 4H-SiC will increase and decrease repeatedly after each hydrogen annealing. The phenomenon is due to the surface states and hydrogen passivation instability. After ten times of hydrogen annealing, the minority carrier lifetimes of all the samples showed an increasing trend. This may be due to the dual mechanism of hydrogen passivation and selective etching. The surface roughness of the sample before and after hydrogen annealing were also measured and discussed by atomic force microscopy(AFM).

Keywords—4H-SiC, carrier lifetime, hydrogen annealing.

I. INTRODUCTION

Silicon Carbide (SiC), as one of the most representative wide bandgap semiconductors, is featured by plenty of excellent characteristics such as good thermal conductivity ($4.9\ Wcm^{-1}K^{-1}$) , high carrier saturation rate ($1200\ cm^2v^{-1}s^{-1}$) and wide band gap (3.26 eV), which can be widely applied in power devices with high voltage and high temperature endurance[1][2]. The unipolar power devices can not meet the requirements of ultra-high voltage due to their high resistance. Thus, the bipolar power devices with conductance modulation effect are needed to reduce the forward on-resistances[3]. However, excellent conductance modulation requires suitable minority carrier lifetimes[4][5]. Improving the minority carrier lifetime of 4H-SiC has become a researching focus. At present, the N-type SiC minority carrier lifetimes killer is well known as the $Z_{1/2}$ center[6], which is considered to be the defect center related to carbon vacancies (Vc).[7] The $Z_{1/2}$ defect centers can be eliminated by high temperature oxidation or carbon ion implantation with the introduction of excess carbon atoms.[8][9] Very long minority carrier lifetime of 20-30 μs can be obtained by eliminating V_c through these three methods.[10] However, as for P-type SiC epitaxial wafers, it is reported by Hayashi et al. that the minority carrier lifetime increases by only 1-2 μs after the same high temperature oxidation or carbon ion implantation.[11] Therefore, it can be concluded that the effect of eliminating the carbon vacancy on increasing minority carrier lifetime of N-type 4H-SiC epitaxies is more significant than that of P-type ones.[12][13] Takafumi Okuda et al. further improved minority carrier lifetime of P-type 4H-SiC epitaxial wafers by hydrogen annealing.[14] With or without thermal oxidation, hydrogen annealing will increase minority carrier lifetime of P-type 4H-SiC epitaxial wafers significantly. It is proved that the mechanism of hydrogen annealing to improve minority carrier lifetime is different from that of thermal oxidation. But the mechanism of hydrogen annealing is still not clear.

In this paper, low-pressure hydrogen annealing experiments at 1000 °C were repeatedly carried out for 4H-SiC epitaxial wafers with different thicknesses and doping types. The carrier lifetime and the surface roughness in the epilayer was measured at room temperature by microwave photo conductance decay (μ-PCD) measurements and atomic force microscopy(AFM). The mechanism of increasing minority carrier lifetime of 4H-SiC by low-pressure hydrogen annealing is discussed.

978-1-6654-4817-8/21 $31.00 © 2021 IEEE

II. Experiment

The samples which we used were grown on 4° off-axis

TABLE I. The Sample Parameters

Label	Thickness (μm)	Doping type	Doping concentrations (cm^{-3})
A	30	P	2.5×10^{14}
B	30	N	3×10^{14}
C	60	P	2×10^{14}
D	60	N	2×10^{14}
E	100	P	2×10^{14}
F	100	N	2×10^{14}

4H-SiC (0001) substrates through low pressure chemical vapor deposition (LPCVD). Table.1 shows the parameters of all samples in this experiment.

The hydrogen annealing process was completed by using the LPE-1O9 equipment of Dongguan Tianyu Semiconductor Technology Company. After standard RCA cleaning, all samples were annealed at 1000 ℃ and 100 mbr for 10 times with a cumulative time of 190 min. The time of each hydrogen annealing is 10 min during the first seven hydrogen annealing. To investigate the minority carrier lifetime by a long-time hydrogen annealing, the eighth and the ninth hydrogen annealing time were set to 30 min, and the tenth hydrogen annealing time was set to 1 h.

Both room temperature and μ-PCD measurement of Semilabs WT-2000 instrument were provided to measure the carrier lifetimes in the epilayer. In the μ-PCD measurement, an yttrium lithium fluoride (YLF) laser with a wavelength of 349nm and a pulse width of 15ns was used to excite these samples. During measurements, the decay of the photoconductance was monitored using microwave reflectivity of 26 GHz, which indicated the variation of excess carrier concentration . At last, atomic force microscopy (AFM) was used for the measurement of the surface roughness .

III. Results and Discussion

Fig.1 shows the minority carrier lifetime of 30 μm N-type and P-type 4H-SiC epitaxies after hydrogen annealing. After the first hydrogen annealing, the minority carrier lifetime of sample A and B are increased obviously. But it is decreased after the second hydrogen annealing. The minority carrier lifetime of the two types of SiC epitaxies increased and decreased repeatedly after the first six annealing, and then continuously increased from the seventh to the tenth annealing. The minority carrier lifetime has doubled from the initial value after ten times hydrogen annealing. Simultaneously, the minority carrier lifetime of 60 μm N-type and P-type 4H-SiC epitaxies after hydrogen annealing are showed in Fig.2. It is worth noting that the 60 μm N-type SiC epitaxies keeps the same variation trend as 30 μm epitaxies, but 60 μm P-type SiC epitaxies shows no obvious change after hydrogen annealing. As for Fig.3, it demonstrates the minority carrier lifetime of 100 μm N-type and P-type 4H-SiC epitaxies after hydrogen annealing. The variation trend of minority carrier life of 100 μm P-type SiC epitaxies is basically the same as that of Fig. 1. However, the minority carrier lifetime of 100 μm N-type SiC epitaxies increased abnormally after the fourth and fifth annealing, except which the variation trend of minority carrier lifetime is basically the same as that of 30 μm N-type SiC epitaxies. From the above, it can be concluded that the minority carrier lifetime for both N-type and P-type SiC epitaxial wafers (Except for the 60 um P-type 4H-SiC) could be improved through hydrogen annealing. As Tokuda.et.al reported [15], an unknown defect was passivated by hydrogen annealing, which could be a reason of the enhancement of the minority carrier lifetime caused by hydrogen annealing. However, the effect of minority carrier lifetime increasing which was due to hydrogen annealing was not stable at the beginning, as is shown in Fig.1. The passivation effect of the first hydrogen annealing is eliminated by the second hydrogen annealing and repassivated by the third hydrogen annealing. With the increase of the cumulative annealing time, the phenomenon of repeated passivation effect still exists , but the subsequent continuous increase of minority carrier life is due to another

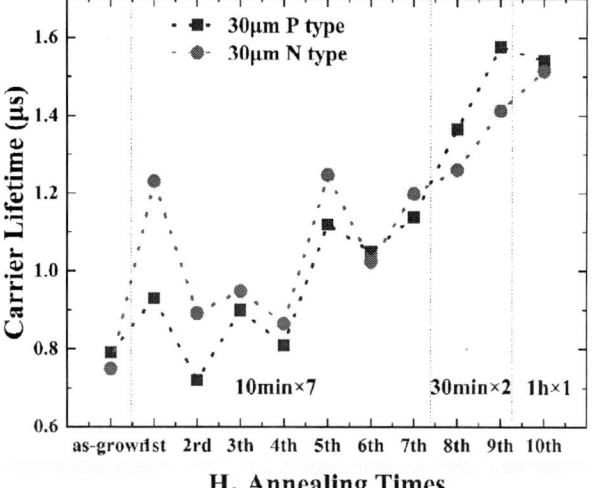

Fig. 1. Measured carrier lifetimes in 30 μm N- and P-type 4H-SiC epitaxies after H₂ annealing (ND=2.5E14cm⁻³、 NA=3E14cm⁻³).

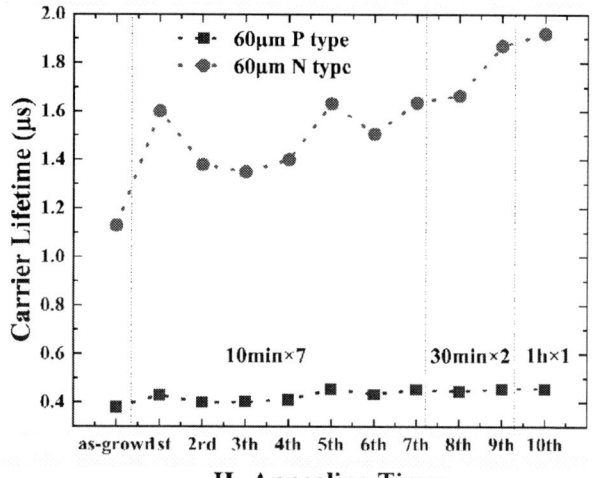

Fig. 2. Measured carrier lifetimes in 60 μm N- and P-type 4H-SiC epitaxies after H₂ annealing (ND=2E14cm⁻³、 NA=2E14cm⁻³).

possible mechanism: the reason for the increase of minority carrier lifetime after long time hydrogen annealing (after 1h) might be the selective Si etching of SiC caused by low pressure hydrogen annealing, during which the surface was gradually enriched with carbon. After a long time of annealing, the excess carbon atoms diffused into the SiC epitaxies to eliminate part of Vc, thus the minority carrier lifetime was continuously improved. This mechanism may play a dominant role after a long period of hydrogen annealing.

pressure were investigated and demonstrated. It is effective for hydrogen annealing to increase the minority carrier lifetimes of n-type and p-type 4H-SiC . At the same time, hydrogen passivation is unstable and can be eliminated by next annealing, leading to the reduction in the minority carrier lifetime. After repeatedly hydrogen annealing, although hydrogen passivation is still unstable, the minority lifetime begins to increase. The reason for the increase of minority carrier lifetime after long-time hydrogen annealing (after 1h) might be the selective Si etching of SiC caused by low pressure hydrogen annealing, during which the surface was gradually enriched with carbon. After a long time of annealing, the excess carbon atoms diffused into the SiC epitaxies to eliminate part of Vc, thus the minority carrier lifetime was continuously improved. The large variation of surface roughness during hydrogen annealing played a role in increasing the minority carrier lifetime, and Hydrogen annealing has little effect on surface defects.

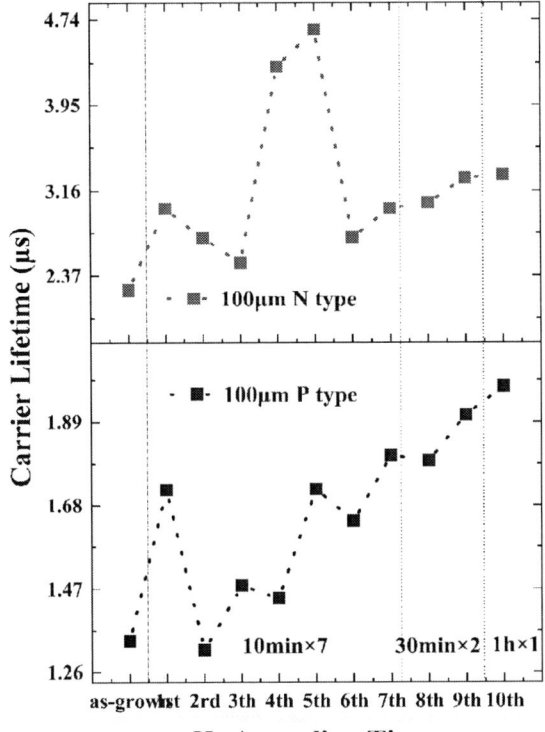

Fig. 3. Measured carrier lifetimes in 100 μm N- and P-type 4H-SiC epitaxies after H$_2$ annealing (N$_D$=2E14cm^{-3}、 N$_A$=2E14cm^{-3}).

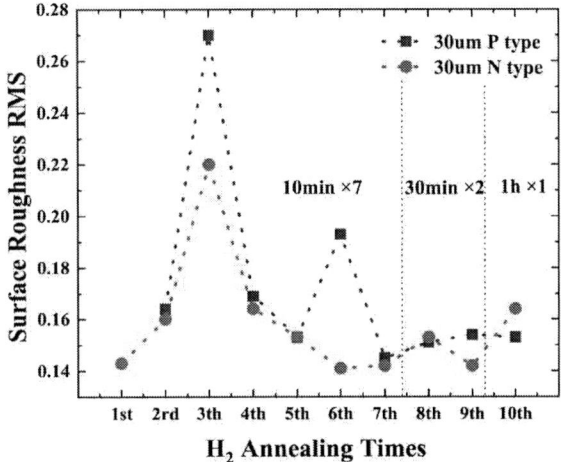

Fig.4. Surface roughness RMS in 30um N- and P-type 4H-SiC epitaxies after H$_2$ annealing.

As shown in Fig. 4, the surface roughness increases dramatically after the third hydrogen annealing, leading to the insignificant increase of the minority carrier lifetime. The third hydrogen repassivation is much less effective than the first because the surface roughness increases dramatically after the third hydrogen annealing, which can be found that the large variation of surface roughness during hydrogen annealing played a negative role in increasing the minority carrier lifetime.

In addition, it can also be measured the number of all samples' surface defects after each hydrogen annealing and found that it had hardly change. Therefore, The change of minority carrier lifetime after hydrogen annealing is independent of surface defects. So the number of all samples' surface defects was not shown here.

IV. CONCLUSION

The minority carrier lifetimes of 4H-SiC with different thickness and doping types annealed in hydrogen at low

ACKNOWLEDGMENT

This work was supported by Key Area Research and Development Project of Guangdong Province (2019B010127001),the Natural Science Foundation of Fujian Province of China for Distinguished Young Scholars (Grant No. 2020J06002), the Science and Technology Project of Fujian Province of China (Grant No. 2020I0001), the Fundamental Research Funds for the Central Universities (Grants No. 20720190049 and 20720190053), and the Science and Technology Key Projects of Xiamen (Grant No. 3502ZCQ20191001).

REFERENCES

[1] H.Matsunami and T.Kimto, "Step-controlled epitaxial growth of SiC: High quality homoepitaxy", Mater. Sci. Eng. R 20(1997) 125.
[2] J. A. Cooper, Jr., M. R. Melloch, R. Singh, A. Agarwal, and J. W. Palmour, "Status and prospects for SiC power MOSFETs",IEEE Trans. Electron Devices 49, 658 (2002).
[3] H. Lendenmann et al., "Operation of a 2500V 150A SiC IGBT / SiC diode module", Mater. Sci. Forum 338 – 342 (2000) 1423.
[4] R. Singh et al., "High-Temperature performance of 10 Kilovolts, 200 amperes (Pulsed) 4H-SiC PiN Rectifiers", Mater. Sci. Forum 389 – 393 (2002) 1265

[5] Y. Sugawara et al.: Proc. Int. Symp. Power Semiconductor Devices and ICs, 2004, p. 365.

[6]T. Tawara et al., "Evaluation of free carrier lifetime and deep levels of the thick 4H-SiC epilayers", Mater. Sci. Forum 457 – 460 (2004) 565.

[7] K. Danno, D. Nakamura, and T. Kimoto, "Investigation of carrier lifetime in 4H-SiC4H-SiC epilayers and lifetime control by electron irradiation", Appl. Phys. Lett. 90, 202109 (2007).

[8] Storasta L and Tsuchida H, "Enhanced annealing of the Z1/2 defect in 4H–SiC epilayers", Appl. Phys. Lett. 90 062116 Page (2007)

[9] Hiyoshi T and Kimoto T, "Reduction of deep levels and improvement of carrier lifetime in n-Type 4H-SiC by thermal oxidation", Appl. Phys. Express 2 041101 (2009)

[10] T. Miyazawa, M. Ito, and H. Tsuchida, "Carrier lifetime and breakdown phenomena in SiC power device material", Appl. Phys. Lett. 97, 202106 (2010).

[11] T. Hayashi, K. Asano, J. Suda, and T. Kimoto, "Enhancement and control of carrier lifetimes in p-type 4H-SiC epilayers", J. Appl. Phys. 112 064503(2012).

[12] W. Shockley and T. Read, "Statistics of the Recombinations of Holes and Electrons", Jr., Phys. Rev. 87, 835 (1952).

[13] T. Hayashi, K. Asano, J. Suda, and T. Kimoto, "Impacts of reduction of deep levels and surface passivation on carrier lifetimes in p-type 4H-SiC epilayers", J. Appl. Phys. 109, 014505 (2011).

[14] T. Okuda, T. Kimoto, and J. Suda, "Improvement of Carrier Lifetimes in Highly Al-Doped p-Type 4H-SiC Epitaxial Layers by Hydrogen Passivation", Appl. Phys. Express 6, 121301 (2013).

[15] Okuda T, Miyazawa T, Tsuchida H, Kimoto T, and Suda J, "Carrier Lifetimes in Lightly-Doped p-Type 4H-SiC Epitaxial Layers Enhanced by Post-growth Processes and Surface Passivation", Journal of ELECTRONIC MATERIALS, Vol. 46, No. 11, (2017)

Comparative Study of Thermal Performance of a SiC MOSFET Power Module Integrated with Vapor Chamber for Traction Inverter Applications

Wei Mu
[1]School of Electrical Engineering
Xi'an Jiaotong University
Xi`an, China
ORCID: 0000-0002-6582-5551

Binyu Wang
[1]School of Electrical Engineering
Xi'an Jiaotong University
Xi`an, China
Email: wangbinyu@stu.xjtu.edu.cn

Shenghe Wang
[2]State Grid Anhui Electric Power Co.,
State Grid Co., Ltd.
Hefei, China
Email: 13605691967@139.com

Haoyuan Jin
[1]School of Electrical Engineering
Xi'an Jiaotong University
Xi`an, China
Email: jinhaoyuan@stu.xjtu.edu.cn

Huaqing Li
[1]School of Electrical Engineering
Xi'an Jiaotong University
Xi`an, China
Email: lihuaqing@stu.xjtu.edu.cn

Laili Wang
[1]School of Electrical Engineering
Xi'an Jiaotong University
Xi`an, China
Email: llwang@mail.xjtu.edu.cn

Abstract—Silicon carbide metal-oxide-semiconductor field-effect transistors (SiC MOSFETs) has the potential to replace silicon-based insulated gated bipolar transistors (IGBT) in medium voltage range for the superiority of wide bandgap (WBG) devices. However, SiC MOSFET has a much smaller die size than Si IGBT, making it has a high thermal resistance and small heat capacitance. In real applications, the shrinkage in die size will lead to higher junction temperature and more significant junction temperature swing, reducing the lifetime and the reliability of the SiC power module. This paper proposes a new packaging design based on vapor chamber (VC) and compares its thermal performance with a conventional DBC module. Utilizing similar working principles as the heat pipes, VCs have advantages including high thermal conductivity, low weight and low cost. In this paper, the VCs are customized and integrated into the SiC power module with a new fabrication process. The FEM simulations along with the experiment demonstrated significant improvements on both steady-state and transient thermal performance compared with conventional direct bonded copper (DBC) substrate. The maximum junction temperature decreased by more than 31.7°C and the junction temperature swing deopped by 46%. By comparing the thermal performance between the conventional module and VC integrated module, this paper reveals the effectiveness of integrating phase-change cooling components into SiC power modules.

Keywords—SiC MOSFET, vapor chamber, packaging, thermal management, phase-change cooling. thermal impedance.

I. INTRODUCTION

In recent years, The demand for high-power-density and high-efficient inverters is increasing due to the application like electric vehicles (EV) amd more elevtric aircraft (MEA) [1]. Increasing the switching frequency can not only reduce the size of passive devices, but also suppress the motor's ripple current and then reduce the power losses on the motor and the motor's working temperature [2]. However, the switching frequency of silicon-based IGBTs is limited due to their relatively high switching losses [3]. So, the conventional silicon devices in traction inverters are being substituted by wide bandgap (WBG) devices.

SiC MOSFETs are drawing more attention in motor drive applications because of their low on-resistance, high blocking voltage, and fast switching process [4]. With the implementation of WBG devices, a 160-kW photovoltaic (PV) inverter based on 4 paralleled SiC MOSFET modules have more than 99.1% peak efficiency [5]. A 50-kW voltage source inverter with 97.91% peak efficiency and 26 kW/kg power density for MEA applications based on SiC devoces is reported in [6].

For the same voltage and current ratings, SiC MOSFETs have a smaller size than their Si counterparts [7]. On the one hand, the smaller die size opens the possibility to further increase the power density. On the other hand, the reduced die size increases the thermal resistance and decreases the heat capacitance, which deteriorates the thermal performance. More advanced cooling solutions have been extensively researched in recent years. Micro and mini-channel liquid cooled heatsinks [8], thermal electric cooling (TEC) [9], jet impingement [10] and phase-change cooling [11] are adopted in thermal management of power devices. Meanwhile, thermally enhanced packaging designs such as double-sided cooling (DSC) [12], 3D flip-chip package [13], Sandwiched Press-Pack [14] and insulated metal substrate (IMS) [15] have been proposed. Vapor chamber (VC) is a two-phase cooling component with very high effective thermal conductivity [16], making them promising in SiC power module packing. This paper evaluates the thermal performance of a SiC power module directly integrated with VCs and compares it with a conventional module using DBC substrate, revealing the potential of integrating phase-change cooling components inside power modules.

II. WORKING PRINCIPLES AND SIMULATION MODLE OF VC

The cross-section view of VC is illustrated in Fig.1. It is a copper shell with wicks at the innder side of the wall. The vacuum space inside VC contains water as the working fluid. Because the air pressure inside the VC is extremely low (around 10^{-4} atm), the boiling point of liquid water inside is also low. The VC is mainly composed of evaporator and condenser sides. When the concentrated heat flux generated by the heat source (SiC MOSFETs in this study) enters the evaporation side, the working fluid will absorb the heat

---

This work was supported by the science and research technology project from Headquarters of State Grid Co., LTD. Project Name: "Research on multi-chip parallel current sharing technology of power electronic devices based on electric-thermal optimization" Number: SGAH0000KJJS1900437

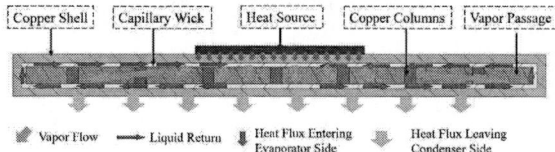

Fig. 1. Cross-section of a vapor chamber.

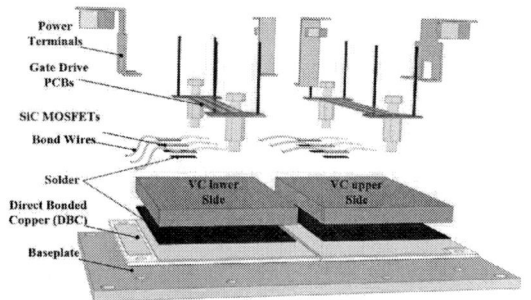

Fig. 2. Structure of the SiC power module integrated with vapor chamber

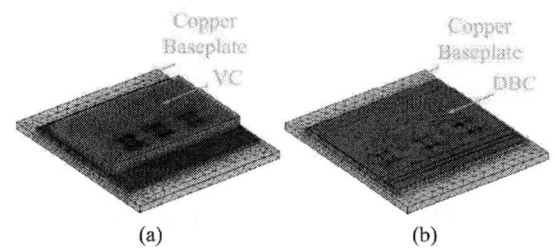

Fig. 3. FEM model of: (a) VC integrated module (b) Conventional module using DBC substrate.

instantly evaporate and fill in the vapor core. When the water vapor comes to a cooler surface, it condenses and releases the latent heat it absorbed from the heat source. In this way, the highly concentrated heat flux is spread to a much larger domain. The condensed working fluid goes back to the heat source using capillary force. Thanks to the high heat transfer coefficient of the phase change process, a VC's effective thermal conductivity reaches 20kW/m·K during optimal working conditions, which is about 50 times higher than red copper [17].

The exploded view of the SiC power module integrated with VC is demonstrated in Fig. 2. Because of the symmetry of the overll structure and the thermal coupling between two VCs are negligible, only the lower side of the module is included in the finite element (FEM) simulation for the simplification of the problem. To compare the thermal performance improvement by integrating VCs directly into SiC power module, a model of a conventional power module with the same layout using a DBC substrate is also built. Both FEM models are shown in Fig. 3. The cooling performance provided by the cold plate based on liquid cooling can be equivalent to a heat transfer coefficient of 1500W/(m²K) at the bottom of the baseplate [15]. The coolant temperature is set to 20°C, consistent with the experiment. All other boundaries except for the baseplate are assumed adiabatic. This is because

TABLE I PARAMETERS OF THE VC AND SIMULATION MODEL

Parameters	Values
VC Size (L)×(W) ×(H)	60mm×40mm ×5mm
Thickness of shell h_{wall}	1 mm
Thickness of wick h_{wick}	0.5 mm
DBC Cu layer thickness	0.3mm
DBC Al₂O₃ thickness	0.635mm
Vacuum Degree	9.31Pa (9.19×10⁻⁵atm)
Thermal conductivity of wick	50W/(m·K)
Thermal conductivity of vapor core	40000 W/(m·K)
Heat transfer coefficient of bottom side	1500W/(m²K)
Coolant temperature	20°C

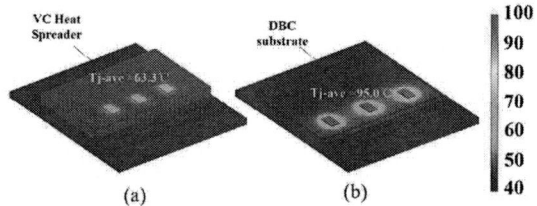

Fig. 4. Steady-state temperature distribution of: (a) VC integrated module (b) Conventional module with DBC substrate.

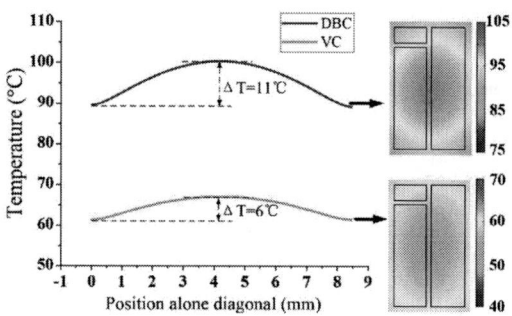

Fig. 5. Comparison of junction temperature distribution along the diagonal of a SiC MOSFET die.

the upper side of the power module is encapsulated with silicone gel for electrical insulation and this encapsulation has poor thermal conductivity. Detailed parameters of the vapor chamber and the simulation model are given in Table I.

III. STEADY-STATE THERMAL SIMULATION

The steady-state simulation results under the power losses of 60 W per die are demonstrated in Fig.4 and Fig.5. Compared with conventional DBC module, the VC integrated module has a 31.7 °C reduction in average junction temperature. Due to the high equivalent thermal conductivity of the vapor chamber, the concentrated heat flux generated by SiC MOSFETs can be spread much further so that the power devices can dissipate heat through a much larger area.

The improvements on thermal conductivity can result in a much uniform junction temperature distribution on each single die. As illustrated in Fig.5, The VC integrated module has a much more uniform junction temperature distribution. The difference between the highest junction temperature and the lowest junction temperature on the die is 11 °C for the conventional module and 6 °C for VC integrated module. This

978-1-6654-4817-8/21 $31.00 © 2021 IEEE

brings two benefits: First, the thermal stress inside the chip is reduced. Second, under the same average junction temperature, the highest junction temperature on the die is lower when the temperature distribution is more uniform.

IV. TRACTION INVERTER MODEL AND TRANSIENT THERMAL SIMULATION

In a motor drive inverter system, the fundamental output frequency has a direct influence on junction temperature variation. Operation in high-torque low-speed region can result in huge junction temperature swing which is a dominant factor for power device failures [18]. To validate the thermal performance improvement of VC integrated modules in real applications, a traction inverter is built in PLECS. In this model, a permanent magnet synchronous motor (PMSM) is fed with a 3-phase inverter. The PMSM operates in high-torque low-speed region, representing the mission profile under the worst-case scenario. The inverter is under direct current control to have a high torque output. The reference 3-phase current i_{abc}^{ref} is calculated using Park transformation. The 3-phase current i_{abc} is captured and compared with i_{abc}^{ref} in a hysteresis current controller to generate the PWM signal for the MOSFETs. The model of the SiC MOSFET and its body diode are imported to the simulation software for transient power loss calculation.

The PMSM operates at a low speed of 120 rpm. The current of each phase and the transient power losses are shown in Fig.7. The transient power losses are pulsating in such working conditions due to the switching of the switches. The peak value of the power losses on the body diode is relatively large for its relatively large on-resistance. The transient power losses are then export to COMSOL to calculate the junction temperature profile using the time-dependent solver.

The comparison of average junction temperature curves of the VC module and conventional DBC module is shown in Fig.8. Under the same cooling condition, the VC can reduce the peak junction temperature by 25% and the maximum junction swing by 48%. In metro traction applications, power modules usually suffer from junction temperature variations caused by the change of the metro's operating state, such as start acceleration, idling, and breaking. A subway system mission profile is chosen as the case study for this paper [19]. This mission profile consists of an acceleration of 40 s, a high-speed cruise phase of 37 s, breaking for 10 s, a low-speed cruise of 40 s, breaking for 22 s, and finally stops at the station for 30 s. Such a mission repeats during the operation of the

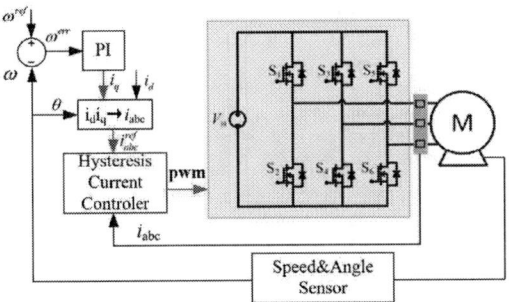

Fig. 6   A 3-phase inverter feeding an Permanent magnet synchronous motor (PMSM)

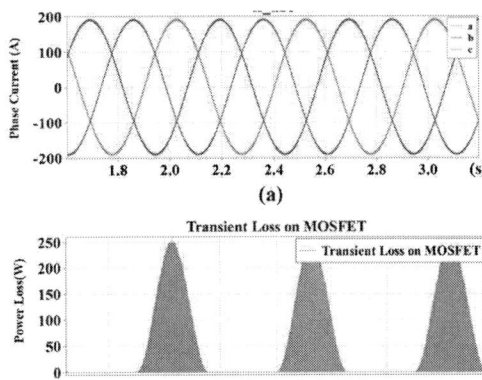

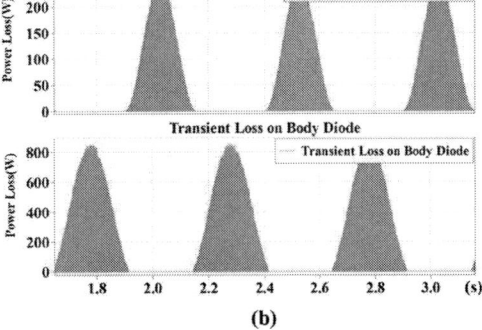

Fig. 7.   (a) Phase current of the inverter (b) Transient power losses on the SiC MOSFETs and their body diodes.

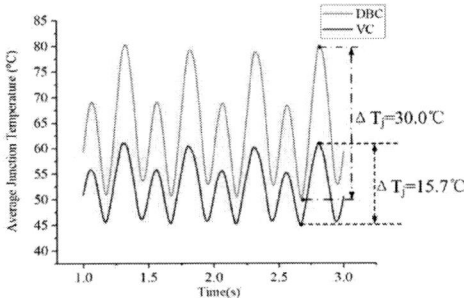

Fig. 8.   Junction temperature swing of SiC MOSFET die in high-

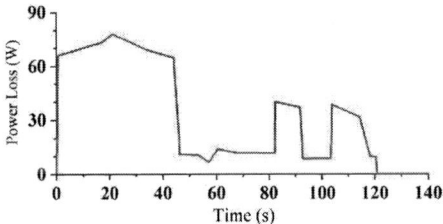

Fig. 9.   Scaled-down transient power loss per SiC die in a typical metro mission profile

metro. The original mission profile is scaled-down and converts to transient power losses shown in Fig. 9 to fit the capacity of the fabricated modules. The transient power loss profile is used as a time-variant heat source in COMSOL, while all other boundary conditions are kept the same as in previous simulations. The result is shown in Fig. 10. The SiC dies reach their maximum junction temperature during the acceleration phase. The VC integrated SiC module can reduce the maximum junction temperature during operation by 36% and the junction temperature swing by 43%.

978-1-6654-4817-8/21 $31.00 © 2021 IEEE   287

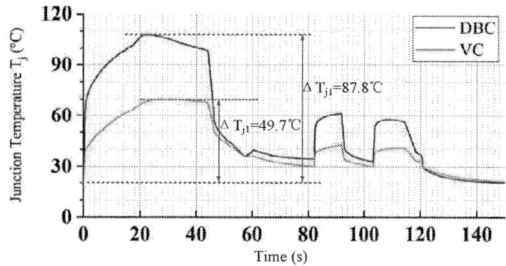

Fig. 10. Comparison of Simulated junction temperature variation during a typical mission profile for metro system.

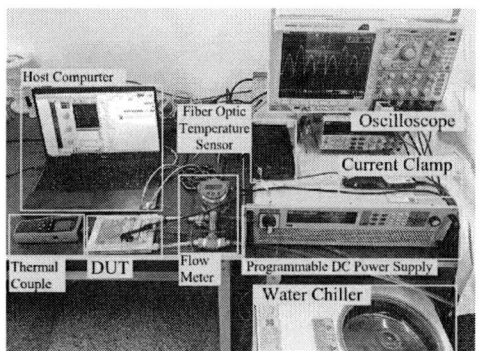

Fig. 11. Comparison of junction temperature variation during a typical mission profile for traction inverter

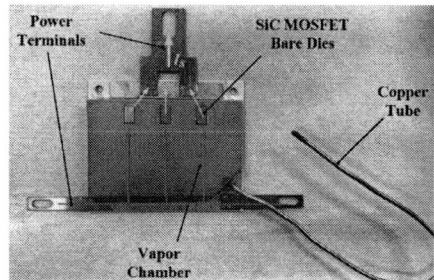

Fig. 12. A prototype of SiC power module integrated with VC (before wire bonding and degassing) and indication of current flow.

V. EXPERIMENT

The designed SiC power modules integrated with VCs are fabricated to validate the thermal performance improvements by directly integrating a phase-change cooling component into the power module. The parameters of VCs are given in Table I, which is consistent with the simulation model. In order to validate the thermal performance improvement by directly integrating VCs, conventional modules with exactly the same SiC MOSFET and layout are made.

The experiment setup is shown in Fig. 11. A programmable DC power supply is connected to the AC and DC- terminal on the substrate. This power source is controlled by a host computer via RS232 serial commination interface to generate the programmed output current. The current waveform is monitored by an oscilloscope using a current clamp. The device under test (DUT) is mounted on a customized cold plate. A water chiller is connected to the cold

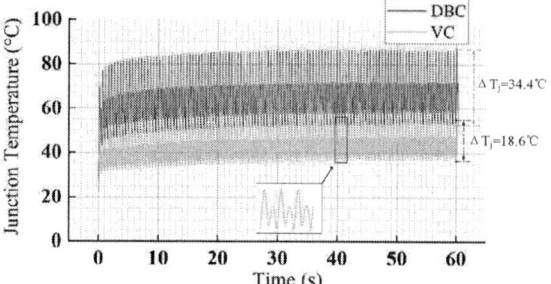

Fig. 13. Comparison of measured junction temperature swing when the inverter works in high-torque low-speed region.

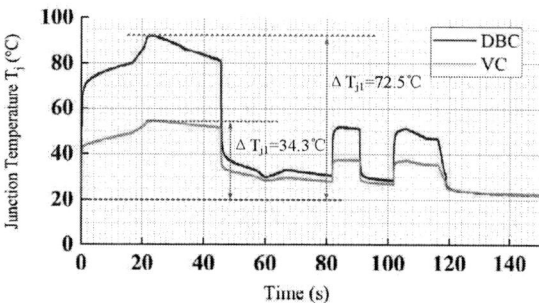

Fig. 14. Comparison of measured junction temperature curve under the metro mission profile.

plate, providing cooling water of 20 °C at the flow rate of 6 L/min. A thermal couple which is extremely thin, is inserted between the substrate and the coold plate to measure the case temperature. A fiber optic temperature sensor with a sampling rate of 1000 samples/s is placed at the center of the SiC MOSFET bare die to measure the transient junction temperature.

Due to the relatively large on-resistance of the body diod of SiC MOSFET, it can be used to heat up the device with a reduced current value. The programmable power supply outputs a 2Hz half-wave sinusoidal heating current to heat up the body diode of the SiC power MOSFETs in the module to imitate the transient power loss when the inverter drives a PMSM at 120 rpm. The result of the experiment is shown in Fig. 13. The VC integrated module provides not only a lower mean junction temperature, but also a significantly lower junction temperature swing. The average junction temperature operating at high-torque low-speed region is 67.1°C for DBC module and 43.8 °C for VC integrated module. Meanwhile, the junction temperature swing during operation is reduced from 34.4 °C to 18.6 °C, which is a reduction of 46%.

The metro mission profile is also tested in the scaled-down experiment to validate the VC's capability of suppressing the junction temperature swing due to the change of output power. The MOSFET is also heated using its body diode. The result of the experiment is given in Fig. 14.. The VC module has a lower junction temperature during the whole mission profile. The mean junction temperature is 48 °C for the DBC module and 36°C for the VC module. The reduction of junction temperature swing is also very significant. The maximum junction temperature ΔT_{j1} is reduced by 38.2 °C, which is a

978-1-6654-4817-8/21 $31.00 © 2021 IEEE   288

reduction of 53%. During the braking phase, the junction temperature swings are reduced by 11.6 °C and 11.1 °C respectively.

VI. CONCLUSION

This paper presents the direct integration of vapor chambers into SiC power modules to improve their thermal performance especially in traction inverter applications. Simulation result reveals that the VC integrated module can reduce the steady-state junction temperature by 31.7 °C and the junction temperature distributes more evenly along the diagonal of the SiC MOSFET die. A traction inverter model is adopted to derive the transient power losses when the inverter drives a PMSM working in high-torque low-speed region. A metro mission profile is also considered. The experiment results validate the transient simulation. In equivalent traction inverter applications, the VC integrated module not only has a 35% lower mean junction temperature, but also a 46% reduction in junction temperature swing. This paper validates the significant improvement in thermal performance of directly integrating VC into SiC power modules by comparing it with conventional power modules. The VC integrated module solves the problem brought by the smaller die size of SiC devices, which is helpful to improve the reliability and the ecpected lifetime of SiC power semiconductor modules.

REFERENCES

[1] K. Yamaguchi, "Design and evaluation of SiC-based high power density inverter, 70kW/liter, 50kW/kg," *2016 IEEE Applied Power Electronics Conference and Exposition (APEC)*, Long Beach, CA, USA, 2016, pp. 3075-3079, doi: 10.1109/APEC.2016.7468302.

[2] K. Yamaguchi, K. Katsura and T. Jikumaru, "Motor loss and temperature reduction with high switching frequency SiC-based inverters," *2017 IEEE Workshop on Wide Bandgap Power Devices and Applications (WiPDA)*, Albuquerque, NM, USA, 2017, pp. 127-131, doi: 10.1109/WiPDA.2017.8170534.

[3] L. Zhang *et al.*, "Evaluation of Different Si/SiC Hybrid Three-Level Active NPC Inverters for High Power Density," in *IEEE Transactions on Power Electronics*, vol. 35, no. 8, pp. 8224-8236, Aug. 2020.

[4] J. Millan, et al. "A Survey of Wide Bandgap Power Semiconductor Devices,", *Ieee Transactions on Power Electronics*, vol. 29, no. 5, pp. 2155-2163, May 2014.

[5] A. Hatanaka, H. Kageyama and T. Masuda, "A 160-kW high-efficiency photovoltaic inverter with paralleled SiC-MOSFET modules for large-scale solar power," *2015 IEEE International Telecommunications Energy Conference (INTELEC)*, Osaka, Japan, 2015, pp. 1-5, doi:

10.1109/INTLEC.2015.7572351.

[6] S. Yin, K. J. Tseng, R. Simanjorang, Y. Liu and J. Pou, "A 50-kW High-Frequency and High-Efficiency SiC Voltage Source Inverter for More Electric Aircraft," in *IEEE Transactions on Industrial Electronics*, vol. 64, no. 11.Nov. 2017.

[7] H. Lee, V. Smet and R. Tummala, "A Review of SiC Power Module Packaging Technologies: Challenges, Advances, and Emerging Issues," in *IEEE Journal of Emerging and Selected Topics in Power Electronics*, vol. 8, no. 1, pp. 239-255, March 2020.

[8] E. Laloya, Ó. Lucía, H. Sarnago and J. M. Burdio, "Heat Management in Power Converters: From State of the Art to Future Ultrahigh Efficiency Systems," in *IEEE Transactions on Power Electronics*, vol. 31, no. 11, pp. 7896-7908, Nov. 2016.

[9] L. Cong, J. Da, J. Jizhou, G. Feng, and W. Jin, "Thermoelectric cooling for power electronics circuits:Modeling and active temperature control," *IEEE Trans. Ind. Appl.*, vol. 50, no. 6, pp. 3995–4005, Nov./Dec. 2014.

[10] I. Silverman, A. L. Yarin, S. N. Reznik, A. Arenshtam, D. Kijet, and A. Nagler, "High heat-flux accelerator targets: Cooling with liquid metal jet impingement," *Int. J. Heat Mass Transf.*, vol. 49, nos. 17–18, pp. 2782–2792, 2006

[11] I. Aranzabal, I. M. de Alegría, N. Delmonte, P. Cova and I. Kortabarria, "Comparison of the Heat Transfer Capabilities of Conventional Single- and Two-Phase Cooling Systems for an Electric Vehicle IGBT Power Module," in *IEEE Transactions on Power Electronics*, vol. 34, no. 5, pp. 4185-4194, May 2019.

[12] Z. Liang, "Integrated double sided cooling packaging of planar SiC power modules," *2015 IEEE Energy Conversion Congress and Exposition (ECCE)*, Montreal, QC, Canada, 2015, pp. 4907-4912.

[13] S. Seal, M. D. Glover, and H. A. Mantooth, "3-D wire bondless switching cell using flip-chip-bonded silicon carbide power devices," *IEEE Trans. Power Electron.*, vol. 33, no. 10, pp. 8553–8564, Oct. 2018.

[14] Y. Chang *et al.*, "Compact Sandwiched Press-Pack SiC Power Module With Low Stray Inductance and Balanced Thermal Stress," in *IEEE Transactions on Power Electronics*, vol. 35, no. 3, pp. 2237-2241, March 2020

[15] E. Gurpinar, S. Chowdhury, B. Ozpineci and W. Fan, "Graphite-Embedded High-Performance Insulated Metal Substrate for Wide-Bandgap Power Modules," in *IEEE Transactions on Power Electronics*, vol. 36, no. 1, pp. 114-128, Jan. 2021.

[16] Chen, Z. , et al. "Design, fabrication and thermal performance of a novel ultra-thin vapour chamber for cooling electronic devices." *Energy Conversion and Management* 187.MAY(2019):221-231.

[17] Avram, Bar-Cohen, Kaiser, Matin, Nicholas, Jankowski, Darin, and Sharar,"Two-Phase Thermal Ground Planes: Technology Development and Parametric Results," *Journal of Electronic Packaging*, vol. 137, pp. 010801-9,2015

[18] R. Bayerer, T. Herrmann, T. Licht, J. Lutz, and M. Feller, "Model for power cycling lifetime of IGBT modules—Various factors influencing lifetime," in *Proc. 5th Int. Conf. Integr. Power Syst.*, 2008, pp. 1–6.

[19] M. Musallam, C. Yin, C. Bailey, and M. Johnson, "Mission profile-based reliability design and real-time life consumption estimation in power electronics," *IEEE Trans. Power Electron.*

978-1-6654-4817-8/21 $31.00 © 2021 IEEE

Optimized Parameter Selection Method of Driving Circuit for SiC MOSFET

Haihong Qin
College of Automationl Engineering
Nanjing University of Aeronautics and
Astronautics
Nanjing, China
qinhaihong@nuaa.edu.cn

Sixuan Xie
College of Automationl Engineering
Nanjing University of Aeronautics and
Astronautics
Nanjing, China
1020098684@qq.com

Feifei Bu
College of Automationl Engineering
Nanjing University of Aeronautics and
Astronautics
Nanjing, China
pufeifei@nuaa.edu.cn

Shishan Wang
College of Automationl Engineering
Nanjing University of Aeronautics and
Astronautics
Nanjing, China
wangshishan@nuaa.edu.cn

Wenming Chen
College of Automationl Engineering
Nanjing University of Aeronautics and
Astronautics
Nanjing, China
wmnuaa@nuaa.edu.cn

Dafeng Fu
College of Automationl Engineering
Nanjing University of Aeronautics and
Astronautics
Nanjing, China
fdf_nuaa@nuaa.edu.cn

Abstract—In order to reduce the switching time and on-resistance of SiC MOSFET, it is usually recommended to use higher driving voltage and lower gate resistance, but this will affect the reliability of gate. Therefore, a parameter selection method which can reduce the total loss and ensure the safety of the gate is adopted in this paper. Firstly, the double pulse test circuit considering parasitic parameters has been analyzed. Then, the mathematical model of the main circuit affecting the drive circuit has been obtained. On this basis, considering the safety of the driving circuit, the parameters of the driving circuit have been selected based on the "optimal comprehensive loss" criterion under the condition of limiting the maximum gate voltage. Simulation and experimental results validate the correctness of the derived analytical model and the theoretical analysis.

Keywords—SiC MOSFET, gate driving circuit, parasitic parameters, driving parameter selection

I. INTRODUCTION

Silicon Carbide (SiC) Metal-Oxide-Semiconductor Field-Effect Transistors (MOSFETs) provide promising potentials for high efficiency, high power density converter design due to their superior performance over Si counterparts [1]. A proper driving circuit needs to have enough driving capability to reduce power losses of SiC MOSFET, as well as maintain the reliable operation of SiC MOSFET under high di/dt (slew rate of current) and high dv/dt (slew rate of voltage) conditions [2]-[3].

For SiC MOSFET, higher driving voltage (V_{DRV}) could minimize on-resistance and switching loss, but it puts greater voltage stress on the gate which can affect the long term reliability of the device. Going with the minimum value for gate resistance (R_G) will minimize the switching losses; however, dv/dt will increase and EMI as well [4]-[5]. Device companies have given recommended driving voltage in datasheets, but these values are based on approximate analysis and cannot exploit the potentials of SiC MOSFET. For the suitable selection of driving parameters for SiC MOSFET, literatures only mentioned that higher V_{DRV} and smaller R_G are recommended [6]. Due to the high switching speed, di/dt and dv/dt of the device are too large during the switching process, which will generate obvious voltage and current oscillations, and increase the switching loss of the device [7]-[8].

Supported by State Key Laboratory of Wide-Bandgap Semiconductor Power Electronic Devices (2019KF001).

Most of the current researches are aimed at studying the influence of the driving circuit on the switching characteristics of SiC MOSFET. But in fact, the main circuit will also affect the driving circuit during the switching process, including the introduction of oscillation to endanger the gate reliability [9]. This shows that the parameters that can affect the switching characteristics, including the driving parameters, can affect the driving circuit by affecting the switching characteristics. The influence of parasitic inductance on switching characteristics has been analyzed in reference [10]; The influence of parasitic capacitance and gate driving resistance on switching-on speed has been studied in reference [11], but other performances has not been mentioned; The analytical expression of SiC MOSFET switching-on current spike in SiC MOSFET-diode commutation circuit has been derived in reference [21], but the derivation process is based on the ideal switching process of the device, without considering the influence of stray inductance in the driving circuit.

The gate voltage stress determines whether the SiC MOSFET can work safely, so it is very important to analyze the source of gate voltage oscillation and the influence of circuit parameters on the oscillation. However, in the current research, there is still a lack of analysis on the coupling effect of the main circuit on the drive circuit. In this paper, based on the double pulse test circuit model considering parasitic parameters, the mechanism of current oscillation in main circuit and its coupling to drive circuit have been analyzed, and the corresponding mathematical model has been established firstly. Then, the influence factors of gate voltage oscillation have been analyzed and studied through simulation. Finally, in the case of limiting the maximum gate voltage, different combinations of driving parameters have been selected, and the driving parameters have been optimized through experimental research.

II. ANALYSIS OF EQUIVALENT CIRCUIT

Figure 1 shows the equivalent circuit of double pulse experimental platform considering stray parameters. In the figure, V_{DC} is the DC bus voltage, and the load inductance is equivalent to the ideal current source I_L in parallel with the equivalent capacitance C_L. For SiC SBD, the simplified model is equivalent to the parallel connection of ideal SBD and parasitic junction capacitance C_J. For SiC MOSFET, $R_{G(int)}$ is the gate internal resistance, C_{GS}, C_{GD}, C_{DS} are the gate-source capacitance, gate-drain capacitance, drain-source capacitance respectively, $L_{D(int)}$ and $L_{S(int)}$ are stray inductances, the gate stray inductance is generally small, and the current change rate of gate loop during the switching transient is much smaller than the current change rate of main circuit, so that the influence of the gate stray inductance is not significant and can be ignored. In the driving circuit, $R_{G(ext)}$ is the external gate resistance, L_G is the gate inductance; in the main circuit, $L_{D(ext)}$ and R_{loop} are the equivalent total strays inductance and resistance between the positive terminal of the DC bus and the drain of the SiC MOSFET, $L_{S(ext)}$ is the stray inductance between the negative terminal of the DC bus and ground.

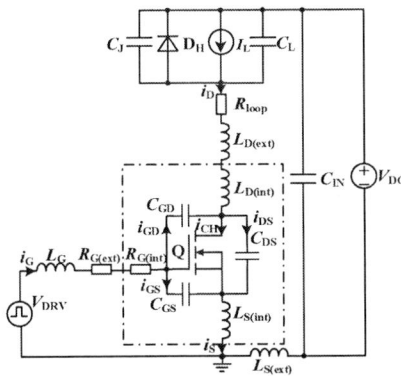

Fig. 1.   Equivalent circuit considering stray parameters

The typical switching-on transient waveform of SiC MOSFET is shown in Figure 2. Since the equivalent junction capacitance C_J of the SiC SBD bears back-pressure charging, the i_D overshoots.

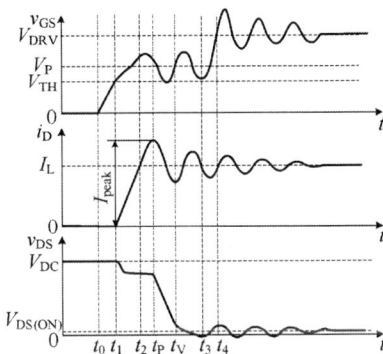

Fig. 2.   SiC MOSFET switching-on transient waveform

At t_V, the drain-source voltage v_{DS} continues to drop, and the SiC MOSFET enters the amplification region. The equivalent junction capacitance C_J of the SiC SBD is connected in parallel with the load equivalent capacitance C_L, and oscillates in series with the stray inductance L_{stray} of the main circuit. Since the stray resistance is generally small, the circuit works in under damping state, and the expression of i_D is shown in (1):

$$i_D(t) = I_L + e^{-\delta_{(1)}(t-t_V)}\left\{ K_{1(1)}\cos\left[\omega_{(1)}(t-t_V)\right] + \right.$$
$$\left. K_{2(1)}\sin\left[\omega_{(1)}(t-t_V)\right]\right\} \tag{1}$$

$$\delta_{(1)} = \frac{R_{loop} + R_{DS(ON)}}{2L_{stray}} \tag{2}$$

$$\omega_{0(1)} = \frac{1}{\sqrt{L_{stray}(C_J + C_L)}} \tag{3}$$

$$\omega_{(1)} = \sqrt{\omega_{0(1)}^2 - \delta_{(1)}^2} \tag{4}$$

$$L_{stray} = L_{D(ext)} + L_{D(int)} + L_{S(ext)} + L_{S(int)} \tag{5}$$

$$K_{1(1)} = i_D\left(t_V^-\right) - I_L \tag{6}$$

$$K_{2(1)} = \frac{\left. di_D(t)/dt\right|_{t=t_V^-} + \delta_{(1)}K_{1(1)}}{\omega_{(1)}} \tag{7}$$

At this time, the current ringing induces a voltage ringing on the common source parasitic inductance $L_{S(int)}$, which is superimposed on the driving circuit and becomes an excitation source, causing the voltage v_{GS} to change. The oscillation part of v_{GS} can be approximately equivalent to (8):

$$\Delta v_{GS}(s) = K_{(2)}\frac{s\sin\psi + \omega_{(1)}\cos\psi}{\left[(s+a)^2 + b^2\right]\left[(s+\delta_{(1)})^2 + \omega_{(1)}^2\right]} \tag{8}$$

$$K_{(2)} = \frac{\omega_{0(1)}\sqrt{K_{1(1)}^2 + K_{2(1)}^2}\,L_{S(int)}}{L_{GS}C_{GS}} \tag{9}$$

$$a = \frac{R_G}{2L_{GS}} \tag{10}$$

$$b = \sqrt{\frac{1}{L_{GS}C_{GS}} - \frac{R_G^2}{4\left(L_G + L_{S(int)}\right)^2}} \tag{11}$$

$$L_{GS} = L_G + L_{S(int)} \tag{12}$$

$$\psi = \text{arctg}\frac{K_{1(1)}}{K_{2(1)}} - \text{arctg}\frac{\omega_{(1)}}{\delta_{(1)}} \tag{13}$$

Solve the inverse Laplace transform:

$$\Delta v_{GS}(t) = 2\left|K_{1(3)}\right|e^{-a(t-t_V)}\cos\left[b(t-t_V)+\alpha\right] + $$
$$2\left|K_{2(3)}\right|e^{-\delta_{(1)}(t-t_V)}\cos\left[\omega_{(1)}(t-t_V)+\beta\right] \tag{14}$$

$$K_{1(3)} = \Delta v_{GS}(s)\cdot(s+a-jb)\Big|_{s=-a+jb} = \left|K_{1(3)}\right|e^{j\alpha} \tag{15}$$

$$K_{2(3)} = \Delta v_{GS}(s)\cdot(s+\delta_{(1)}-j\omega_{(1)})\Big|_{s=-\delta_{(4)}+j\omega_{(4)}} = \left|K_{2(3)}\right|e^{j\beta} \tag{16}$$

If the driving circuit works under damping state, the gate-source voltage expression is:

$$v_{GS}(t) = V_{DRV}\left\{1 + 2K_{(4)}e^a\cos\left[b(t-t_V)+\theta\right]\right\} + \Delta v_{GS}(t) \tag{17}$$

$$K_{(4)} = \frac{V_{DRV}/L_{GS}C_{GS}}{s\cdot\left[(s+a)^2+b^2\right]}\cdot(s+a-jb)\Bigg|_{s=-a+jb} = \left|K_{(4)}\right|e^{j\theta} \tag{18}$$

III.   SIMULATION ANALYSIS OF PARAMETERS

In this chapter, numerical method has been used to study the influence of each parameter on the grid oscillation. The method is to leave an adjustable parameter, and set the other parameters according to the actual situation and experimental conditions, as shown in Table 1.

TABLE I.   SIMULATION CIRCUIT PARAMETERS

Parameter	Value
L_{stray}	80nH
R_{loop}	1Ω
C_J	80pF
L_{GS}	40nH
$L_{S(int)}$	10nH
R_G	12Ω
$R_{DS(ON)}$	0.078Ω
V_{DRV}	18.6V
V_{DC}	400V
I_L	20A

A. Stray Inductance of Main Circuit

The stray inductance L_{stray} is respectively taken as 80nH, 140nH, 200nH, and simulation waveform is shown in Figure 3. It can be seen that as L_{stray} increases, the oscillation frequency decreases. However, as to the oscillation amplitude, L_{stray} is the inductance of the second-order oscillation of the main circuit, as L_{stray} increases, the oscillation amplitude increases; and it is also affect the di_D/dt, as L_{stray} increases, di_D/dt decreases, and the oscillation amplitude decreases. The interaction of two factors causes the gate oscillation amplitude to increase firstly and then decrease during the increase of L_{stray}. Generally, in order to ensure the switching speed, the L_{stray} should be as small as possible.

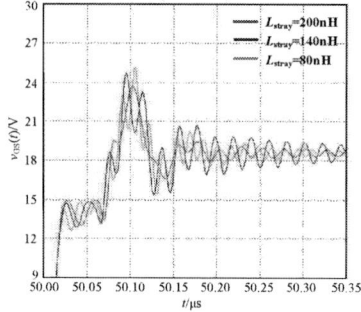

Fig. 3.   The effect of stray inductance L_{stray} on v_{GS}

B. Stray Resistance of Main Circuit

The stray resistance R_{loop} is respectively taken as 1Ω, 0.1Ω, 0.01Ω, and simulation waveform is shown in Figure 4. It can be seen that as the stray resistance R_{loop} decreases, the gate oscillation frequency remains unchanged, while the gate oscillation amplitude increases, and the oscillation recovery stability time increases. As R_{loop} decreases, the damping ratio decreases, while the suppression effect of di_D/dt decreases and the oscillation amplitude increases. Usually in order to reduce power loss, the R_{loop} should be as small as possible.

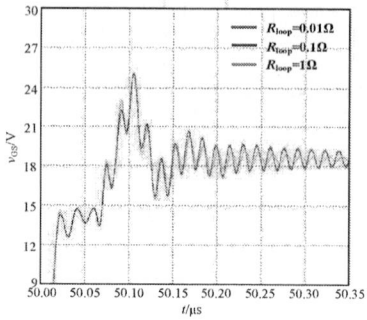

Fig. 4.   The effect of stray inductance R_{loop} on v_{GS}

C. SiC SBD Junction Capacitance

The junction capacitance is respectively taken as 80pF, 140pF, 200pF, and simulation waveform is shown in Figure 5. It can be seen that as the junction capacitance C_J increases, the oscillation frequency decreases, while the gate oscillation amplitude increases from 25.18V to 28.02V. As C_J increases, the damping ratio decreases, and the oscillation amplitude increases. In order to reduce the switching-on ringing, SiC SBD with small junction capacitance should be selected, and air-core inductor should be used to reduce C_L.

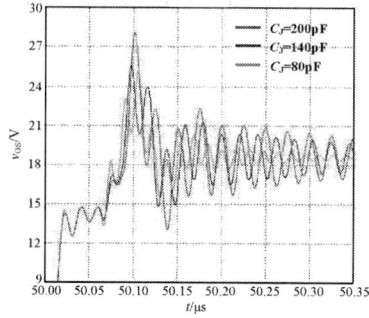

Fig. 5.   The effect of junction capacitance C_J on v_{GS}

D. Gate-Source Inductance

The gate-source inductance is respectively taken as 60nH, 40nH, 20nH, and simulation waveform is shown in Figure 6. It can be seen that as the gate-source inductance L_{GS} increases, the gate oscillation amplitude increases from 22.33V to 25.39V. As L_{GS} increases, the damping ratio of the driving circuit is decreases, and the oscillation amplitude increases. In order to ensure gate reliability and switching speed, L_{GS} should be less than 20nH as much as possible.

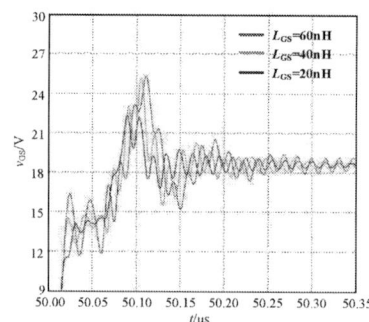

Fig. 6.   The effect of gate-source inductance L_{GS} on v_{GS}

E. Common Source Inductance

The common source inductance is respectively taken as 5nH, 10nH, 20nH, and simulation waveform is shown in Figure 7. It can be seen that as the gate-source inductance $L_{S(int)}$ increases, while the suppression effect of di_D/dt increases, making the voltage drop on the $L_{S(int)}$ not change in one direction. In order to ensure gate reliability and switching speed, $L_{S(int)}$ should be selected as about 10nH.

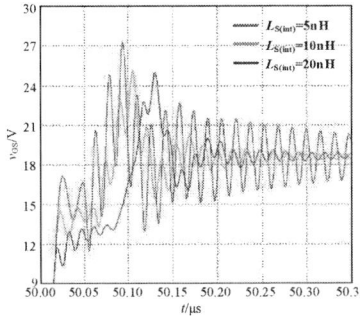

Fig. 7. The effect of common source inductance $L_{S(int)}$ on v_{GS}

F. Gate Driving Resistance

The gate driving resistance is respectively taken as 12Ω, 14Ω, 16Ω, and simulation waveform is shown in Figure 8. It can be seen that as the gate driving resistance R_G decreases, the gate oscillation amplitude increases from 18.91V to 25.18V. As R_G decreases, the damping ratio of the driving circuit is decreases, and di_D/dt increases, and the oscillation amplitude increases. In order to ensure gate reliability and switching speed, the damping ratio of the driving circuit should be selected between 0.77~0.88.

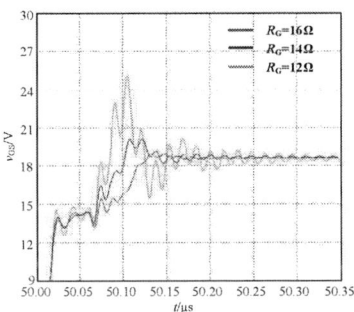

Fig. 8. The effect of gate driving resistance R_G on v_{GS}

G. Bus Voltage

The bus voltage is respectively taken as 200V, 300V, 400V, and simulation waveform is shown in Figure 9. It can be seen that as the bus voltage V_{DC} increases, the gate oscillation amplitude increases from 23.14V to 25.18V. In order to ensure gate reliability, the circuit parameters should be designed under the highest bus voltage.

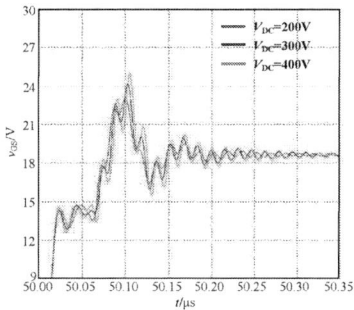

Fig. 9. The effect of bus voltage V_{DC} on v_{GS}

H. Load Current

The load current is respectively taken as 5A, 10A, 20A, and simulation waveform is shown in Figure 10. It can be seen that as the load current I_L decreases, the gate oscillation amplitude increases from 25.18V to 25.98V. In order to ensure gate reliability, the circuit parameters should be designed under the lowest load current.

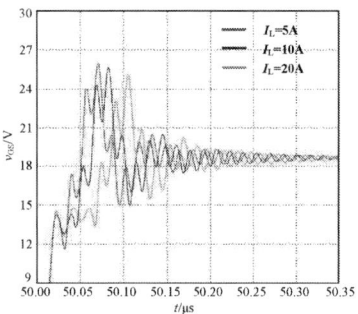

Fig. 10. The effect of load current I_L on v_{GS}

IV. EVALUATION OF OPIMIZED PARAMETERS FOR DRIVER CIRCUIT

Optimizing the selection of the driving circuit parameters effectively reduces the switching loss while ensuring the reliability of the gate. SiC MOSFET switching characteristics changed by adjusting the driving voltage V_{DRV} and the driving resistance R_G is the most simple and effective method.

In order to evaluate the effect of different driving parameters combinations (V_{DRV}, R_G) on the gate reliability and switching energy loss of SiC MOSFET(SCT3060AL, Rohm), this paper has built a double-pulse experimental platform as shown in Figure 11, and the switching characteristics of SiC MOSFET has been tested through double-pulse experiment. The test conditions are shown in Table 2.

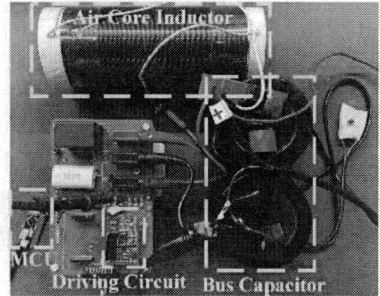

Fig. 11. SiC MOSFET double-pulse experimental platform

978-1-6654-4817-8/21 $31.00 © 2021 IEEE

TABLE II.   SIMULATION CIRCUIT PARAMETERS

Parameter	Value
V_{DC}	400V
I_L	20A
L	1.2mH
(V_{DRV}, R_G)	(20V, 22Ω), (20V, 20Ω), (18.8V, 18Ω), (17.6V, 16Ω), (16.8V, 14Ω), (15.2V, 12Ω)

The switching waveforms of SiC MOSFET under different driving parameter combinations (V_{DRV}, R_G) are shown in the figure 12. In this paper, different driving parameter combinations (V_{DRV}, R_G) are selected to limit the maximum gate voltage between 21V and 22V. For the driving circuit, the gate-source voltage oscillation includes the second-order oscillation of the driving circuit and the high-frequency oscillation of the main circuit coupling when the driving circuit is under damped; When the driving circuit works in the over damping state, the gate-source voltage oscillation only comes from the high frequency oscillation. Obviously, the smaller R_G is, the larger the gate source voltage oscillation is. As R_G decreases from 22Ω to 12Ω, the gate source oscillation increases from 1.2V to 7.2V. At the same time, the turn-on current peak decreases from 24.5A to 23.0A, which is the result of the decrease of R_G and the increase of V_{DRV}; V_{DRV} has little effect on the turn off process, the voltage spike increases with the decrease of R_G, the turn off drain-source voltage spike increased by 22.68% from 485V to 595V.

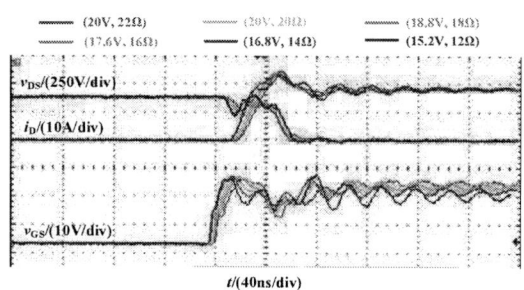

(a) Switching-on waveform

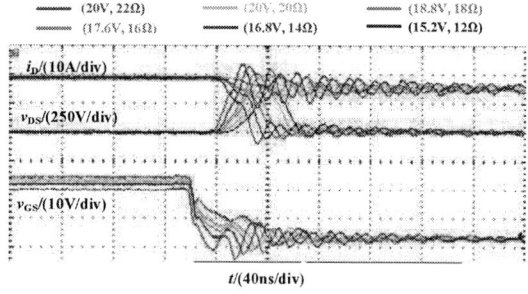

(b) Switching-off waveform

Fig. 12. Switching waveforms of SiC MOSFET driven by different driving parameter combinations (V_{DRV}, R_G)

In order to select the best combination of drive circuit parameters, this paper has adopted the principle of "optimal comprehensive loss". The comprehensive loss includes switching energy loss and conduction loss.

The switching energy loss of SiC MOSFET under different driving parameter combinations (V_{DRV}, R_G) is shown

in Figure 13. The smaller R_G is, the larger gate oscillation is, which leads to the limitation of V_{DRV}, the turn-on energy loss increases, When R_G is reduced from 22Ω to 12Ω, the turn-on energy increased by 41.22% from 153.8μJ to 217.2μJ. V_{DRV} has little effect on the turn off process, the smaller R_G is, the smaller the turn-off energy loss is, When R_G is reduced from 22Ω to 12Ω, the turn-off energy decreased by 35.59% from 104μJ to 66.99μJ.

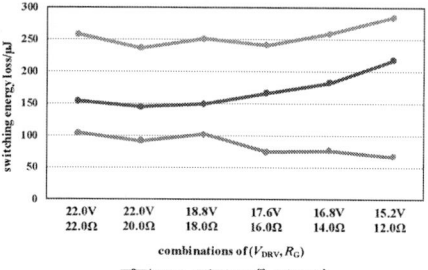

Fig. 13. Switching energy loss of SiC MOSFET driven by different driving parameter combinations (V_{DRV}, R_G)

The comprehensive loss of different combinations of (V_{DRV}, R_G) under different switching frequenciey is shown in Figure 14. It can be seen that at 200kHz, the comprehensive loss of (22.0V, 22.0Ω) is 30.175W less than that of (15.2V, 12.0Ω); At 600kHz, the loss of (22.0V, 22.0Ω) is 49.89W less than that of (15.2V, 12.0Ω). With the increase of frequency, the combination of driving parameters with small switching energy loss has more advantages.

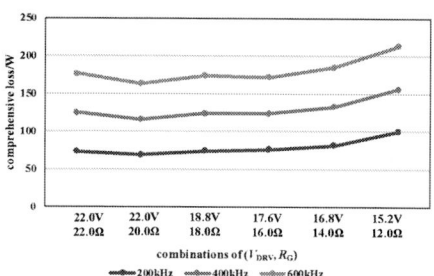

Fig. 14. Comprehensive loss of SiC MOSFET driven by different driving parameter combinations (V_{DRV}, R_G)

In conclusion, considering the trade-off between voltage overshoot and switching loss, the recommended driving parameters of SiC MOSFET are: V_{DRV} =20.0V, R_G =20.0Ω when L_{GS} =42.52nH. If the gate-source inductance is further optimized, R_G can be smaller to reduce the switching energy loss at 20.0V

V. CONCLUSION

The high switching speed of SiC MOSFET significantly increases the nonideal characteristics of its transient process, and makes SiC MSOFET be more sensitive to stray parameters, so it is easy to excite high-frequency oscillation and overshoot. In this paper, starting from the series circuit of SiC MOSFET and SiC SBD, the mathematical model of device turn-on oscillation has been established, the mechanism of device overshoot and oscillation has been revealed, and the optimized driving circuit parameters have been selected through experiments. After the above research, the conclusions are as follows:

978-1-6654-4817-8/21 $31.00 © 2021 IEEE

1) Due to the junction capacitance C_J of SiC SBD and the stray parameters of the main circuit, it is easy to cause overshoot and oscillation in the switching process of SiC MOSFET, which affects the voltage and current stress of the SiC MOSFET.

2) Because of the parasitic parameters, the gate-source voltage v_{GS} is not equal to the driving voltage V_{DRV}. This paper has explored the influence of L_{stray}, R_{loop}, C_J, L_{GS}, $L_{S(int)}$, R_G, V_{DC}, I_L; and has provided guidance for the circuit optimization design.

3) When the parasitic parameters of the circuit and device remain unchanged, increasing the gate resistance R_G can effectively suppress the turn off drain-source voltage spike and gate oscillation voltage.

4) After limiting the maximum of gate voltage, the combination of driving parameters will be selected according to the principle of "optimal comprehensive loss". The experimental results suggest that the R_G should be around the critical damping. If the gate-source inductance L_{GS} is further optimized, at the same V_{DRV}, R_G can be designed smaller to reduce the switching energy loss.

REFERENCES

[1] W. Zhou and X. Yuan, "Experimental Evaluation of SiC Mosfets in Comparison to Si IGBTs in a Soft-Switching Converter," in IEEE Transactions on Industry Applications, vol. 56, no. 5, pp. 5108-5118, Sept.-Oct. 2020.

[2] H Qin, C Ma, D Wang, H Xie and Z Zhu, et al, "An overview of SiC MOSFET gate drivers," 2017 12th IEEE Conference on Industrial Electronics and Applications (ICIEA), Siem Reap, Cambodia, 2017, pp. 25-30.

[3] P Li, X Guo, H Zhou, Q Zhao and L Jia, "A Drive Circuit Design Based on SiC MOSFET and Analysis of Problems," 2019 22nd International Conference on Electrical Machines and Systems (ICEMS), 2019, pp. 1-5.

[4] X. Peng, J. Wang, Z. Peng, Z. Liu, M. Li and Y. Dai, "Study on the CM EMI Characteristics of Si/SiC Hybrid Switch Based Converter," 2020 IEEE Applied Power Electronics Conference and Exposition (APEC), 2020, pp. 2565-2569.

[5] Y. Wu, S. Yin, Z. Liu, H. Li and K. Y. See, "Experimental Investigation on Electromagnetic Interference (EMI) in Motor Drive Using Silicon Carbide (SiC) MOSFET," 2020 International Symposium on Electromagnetic Compatibility - EMC EUROPE, 2020, pp. 1-6.

[6] J. Rice and J. Mookken, "SiC MOSFET gate drive design considerations," 2015 IEEE International Workshop on Integrated Power Packaging (IWIPP), 2015, pp. 24-27.

[7] P. Anthony, N. McNeill and D. Holliday, "High-Speed Resonant Gate Driver With Controlled Peak Gate Voltage for Silicon Carbide MOSFETs," in IEEE Transactions on Industry Applications, vol. 50, no. 1, pp. 573-583, Jan.-Feb. 2014.

[8] W. Zhang, L. Zhang, P. Mao and Y. Hou, "Characterization of SiC MOSFET switching performance," 2018 1st Workshop on Wide Bandgap Power Devices and Applications in Asia (WiPDA Asia), 2018, pp. 100-105.

[9] W. Zhang, X. Wang, M. S. A. Dahidah, G. N. Thompson, V. Pickert and M. A. Elgendy, "An Investigation of Gate Voltage Oscillation and its Suppression for SiC MOSFET," in IEEE Access, vol. 8, pp. 127781-127788, 2020.

[10] H. Qin, Z. Zhu, W. Dai, K. Xu and D. Fu, et al, "Influence of parasitic inductance on switching characteristics of SiC MOSFET," in Journal of Nanjing University of Aeronautics & Astronautics, vol. 49, no. 4, pp. 531-539, Aug. 2017.

[11] D. Cittanti, F. Iannuzzo, E. Hoene and K. Klein, "Role of parasitic capacitances in power MOSFET turn-on switching speed limits: A SiC case study," 2017 IEEE Energy Conversion Congress and Exposition (ECCE), 2017, pp. 1387-1394.

[12] K. Chen, Z. Zhao, L. Yuan, T. Lu and F. He, "The Impact of Nonlinear Junction Capacitance on Switching Transient and Its Modeling for SiC MOSFET," in IEEE Transactions on Electron Devices, vol. 62, no. 2, pp. 333-338, Feb. 2015.

Improved One Cycle Control for Three-Phase Three-Wire VIENNA Rectifier

1st Junnan Gu
School of Mechatronic Engineering and Automation
Shanghai University
Shanghai, China
2857499037@qq.com

2nd Xikun Chen
School of Mechatronic Engineering and Automation
Shanghai University
Shanghai, China
chenxk@shu.edu.cn

3rd Ruiying Li
School of Mechatronic Engineering and Automation
Shanghai University
Shanghai, China
18800202752@163.com

4th Borui Liu
School of Mechatronic Engineering and Automation
Shanghai University
Shanghai, China
lbrshu@163.com

5th Ni Zheng
School of Mechatronic Engineering and Automation
Shanghai University
Shanghai, China
jennyzcnc@163.com

Abstract—**Aiming at the problem of the coupling relationship between the three phases in the traditional one cycle control of the three-phase three-wire VIENNA rectifier, an improved one cycle control method is proposed. The VIENNA rectifier is equivalent to two Boost circuits formed in series, and the power factor correction is realized by controlling the maximum input phase current and the minimum input phase current to be sine waves. In addition, the theoretical analysis demonstrates that the method has self-balancing characteristics of the midpoint potential on the DC side. Finally, a VIENNA rectifier simulation model was built through MATLAB/Simulink, which proved the feasibility of the proposed method and has good dynamic performance.**

Keywords—*one cycle control, VIENNA rectifier, Power factor correction, Midpoint potential equalization*

I. INTRODUCTION

DC charging pile is the core energy source of electric vehicles. It is necessary to design a DC charging pile power module with high efficiency, high power factor and low harmonics, which can improve the performance and life of electric vehicles and reduce pollution. The DC charging pile usually consists of two-stage structure, the front stage AC-DC and the rear stage DC-DC. This paper mainly studies the VIENNA rectifier in the front stage of the charging pile.

Compared with the traditional two-level rectifier, the voltage stress of each switch tube is only half of the two-level rectifier under the same output voltage. It does not have the problem of bridge arm through, so it has caused extensive research by scholars at home and abroad.

[1-4] studied the characteristics of the VIENNA rectifier under space vector modulation. The VIENNA rectifier under this modulation method can achieve power factor correction, but its control is complex and requires complex coordinate transformation, which is not suitable for digital control systems. [5] minimizes the average neutral current in the entire switching cycle by introducing additional switching states to achieve the balance of the midpoint potential. But in this method, the higher harmonic is injected into the input current essentially, which will introduce higher harmonics to the entire system. [6-8] uses the synovial membrane controller as the external DC voltage loop, which improves the system dynamic response and anti-interference ability. But this

control strategy relies on accurate Mathematical model and it is hard to tune parameter.

Because other control methods have their own shortcomings, this article uses one cycle control. The one cycle control has a constant switching frequency, and does not require a multiplier. The control scheme is simple. The traditional one cycle control has the problem of coupling between the three phases in the three-phase three-wire system. Therefore, this paper proposes an improved one cycle control method and analyzes the control strategy in detail. This method decouples a three-phase circuit into two phases, and the decoupled circuit uses one cycle control. Different from the traditional one cycle control, only two switch tubes operate in each cycle, which greatly reduces the switching loss. In addition, this method is the same as the traditional one cycle control, which has the characteristics of self-equalization of the DC side capacitor voltage, which is demonstrated in this paper. Finally, a MATLAB/Simulink simulation model was built, and the simulation results proved the feasibility and effectiveness of the control strategy.

II. SIMPLIFIED VIENNA CIRCUIT AND ITS WORDKING PRINCIPLE

A. Simplified VIENNA rectifier working principle analysis

The main topology of the VIENNA rectifier is shown in Fig. 1. The topology is a three-phase three-wire structure, and the two-way switch of each phase is composed of two common-source switch tubes, so as to realize the two-way flow of current. The output capacitor is composed of two capacitors in series, and the voltage that each capacitor bears is half of the output voltage. The system works mainly by controlling the turn-on and turn-off of the three groups of power switch tubes to control the charging and discharging of the two capacitors at the output terminal to achieve the stability of the output terminal voltage. The control of three groups of bidirectional switches can achieve the purpose of controlling the current of each phase, so as to realize the power factor correction of the VIENNA rectifier.

For the three-phase three-wire VIENNA rectifier, in each power frequency cycle, there is mutual coupling between the three-phase input voltage. Therefore, the topology of Fig. 1 can be further simplified to improve the efficiency of the entire circuit. The simplified equivalent circuit is shown in Fig. 2, which consists of a forward Boost and a reverse Boost in series. In the equivalent circuit, in a voltage range, only the two-phase switches are turned on and off at high frequency, so that the current tracks their respective phase voltages, and the other phase switch remains normally on. Because in the three-wire star connection method, the neutral point current must satisfy Kirchhoff's current law, and the normally-on phase current must also automatically track its phase voltage.

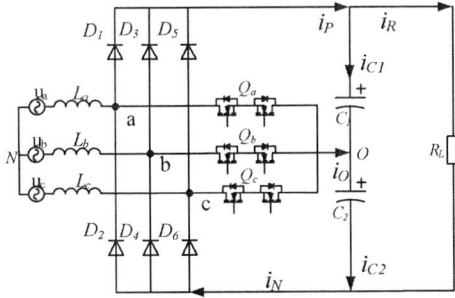

Fig. 1. VIENNA rectifier topology

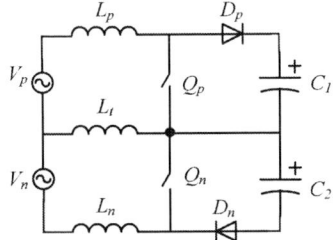

Fig. 2. Target equivalent circuit

Since the equivalent two boost circuits need to achieve a positive voltage and a negative voltage DC output, the input voltage of the two boosts needs to be in both positive and negative states, so the normally-on phase is used as the common phase of the two phases of high-frequency operation. The phase voltage must be between the other two phases. Therefore, a power frequency cycle can be divided into voltage intervals as shown in Fig. 3. In the same interval, the three-phase voltage relationship is fixed and there is no crossover. Taking the 30°~90° interval as an example, as shown in Figure 3(b), in this interval, $V_a > V_b > V_c$ is always present, so the bidirectional switch Q_b is always on. The other two-phase switches Q_a and Q_c work at high frequency, and the two-phase inductor current follows its phase voltage, so the B phase current also follows its phase voltage. At this time, $V_p = V_a - V_c$, $V_n = V_c - V_b$, $L_p = L_a$, $L_t = L_c$, $L_n = L_b$, $Q_p = Q_a$, $Q_n = Q_c$, $D_p = D_1$, $D_n = D_4$. By analogy, in other intervals, the corresponding relationship between the equivalent circuit and the original circuit can also be derived.

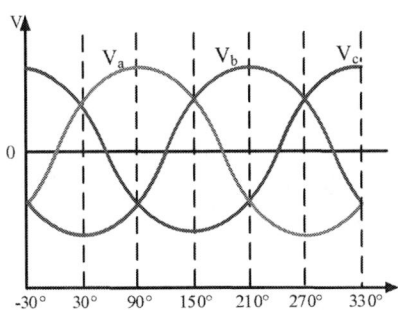

Fig. 3. Grid side three-phase input voltage interval division diagram

The equivalent circuit has four operating modes in each voltage interval, as shown in Fig. 4.

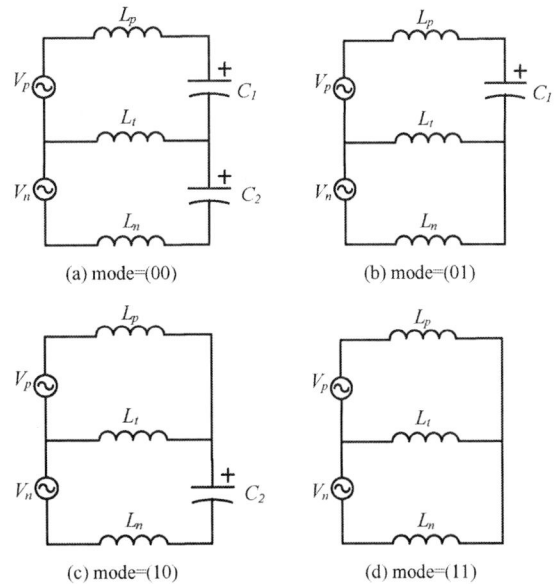

(a) mode=(00) (b) mode=(01)

(c) mode=(10) (d) mode=(11)

Fig. 4. Four working modes of equivalent circuit

B. The mathematical modeling of Simplified VIENNA rectifier

Now we take mode=(10), that is, Q_p turns on and Q_n turns off as an example for analysis. According to the loop current method and the node voltage method, we can get:

$$
\begin{cases}
V_p = L\dfrac{di_{L_p}}{dt} - L\dfrac{di_{L_t}}{dt} \\
V_n = L\dfrac{di_{L_t}}{dt} + \dfrac{1}{2}\cdot V_o + L\dfrac{di_{L_n}}{dt} \\
i_{L_t} = i_{L_n} - i_{L_p}
\end{cases}
\tag{1}
$$

where V_o is the output DC voltage.

For a three-phase VIENNA circuit, the inductor voltage of each phase is

$$
\begin{cases}
V_{L_p} = L\dfrac{di_{L_p}}{dt} \\
V_{L_n} = L\dfrac{di_{L_n}}{dt} \\
V_{L_t} = L\dfrac{di_{L_t}}{dt}
\end{cases}
\tag{2}
$$

Substituting (2) into (1), the relationship between the two boost circuit power supplies V_p and V_n is

$$
\begin{cases}
V_p = 2V_{L_p} - V_{L_n} \\
V_n = 2V_{L_n} - V_{L_p} + \dfrac{1}{2}V_o
\end{cases}
\tag{3}
$$

Substituting (3) into (2), the inductor voltage of each phase is

978-1-6654-4817-8/21 $31.00 © 2021 IEEE

$$\begin{cases} V_{L_p} = \dfrac{2V_p + V_n - \dfrac{1}{2}V_o}{3} \\[4mm] V_{L_n} = \dfrac{V_p + 2V_n - V_o}{3} \\[4mm] V_{L_t} = \dfrac{V_n - V_p - \dfrac{1}{2}V_o}{3} \end{cases} \tag{4}$$

By analogy, the inductance of each phase of the remaining three switching states can also be deduced, and the inductor voltage as shown in Table I can be obtained.

TABLE I. THE VOLTAGE ON THE INDUCTANCE OF THE TARGET EQUIVALENT CIRCUIT

Mode	V_{L_p}	V_{L_n}	V_{L_t}
00	V_p^*	V_n^*	V_t^*
01	$V_p^* - \dfrac{2}{3}V_{c1}$	$V_n^* - \dfrac{1}{3}V_{c1}$	$V_t^* + \dfrac{1}{3}V_{c1}$
10	$V_p^* - \dfrac{1}{3}V_{c2}$	$V_n^* - \dfrac{2}{3}V_{c2}$	$V_t^* - \dfrac{1}{3}V_{c2}$
11	$V_p^* - \dfrac{2}{3}V_{c1} - \dfrac{1}{3}V_{c2}$	$V_n^* - \dfrac{1}{3}V_{c1} - \dfrac{2}{3}V_{c2}$	$V_t^* + \dfrac{1}{3}V_{c1} - \dfrac{1}{3}V_{c2}$

where

$$\begin{cases} V_p^* = \dfrac{2}{3}V_p + \dfrac{1}{3}V_n \\[3mm] V_n^* = \dfrac{1}{3}V_p + \dfrac{1}{3}V_n \\[3mm] V_t^* = \dfrac{1}{3}V_n - \dfrac{1}{3}V_p \\[3mm] V_{c1} = V_{c2} = \dfrac{1}{2}V_o \end{cases} \tag{5}$$

Since the average value of the inductor voltage is zero in one cycle, so

$$\begin{cases} V_p^* d_n + (V_p^* - \dfrac{1}{3}V_{c2})(d_p - d_n) \\ \quad + (V_p^* - \dfrac{2}{3}V_{c1} - \dfrac{1}{3}V_{c2})(1 - d_p) = 0 \\[3mm] V_n^* d_n + (V_n^* - \dfrac{2}{3}V_{c2})(d_p - d_n) \\ \quad + (V_n^* - \dfrac{1}{3}V_{c1} - \dfrac{2}{3}V_{c2})(1 - d_p) = 0 \\[3mm] V_t^* d_n + (V_t^* - \dfrac{1}{3}V_{c2})(d_p - d_n) \\ \quad + (V_t^* + \dfrac{1}{3}V_{c1} - \dfrac{1}{3}V_{c2})(1 - d_p) = 0 \end{cases} \tag{6}$$

where d_p is the time that Q_p is turned on in one cycle, and d_n is the time that Q_n is turned on in one cycle.

In the three-phase symmetric system, satisfy

$$V_p^* - V_n^* + V_t^* = 0 \tag{7}$$

Substituting (7) into (6), we can get

$$\begin{bmatrix} 1-d_p \\ 1-d_n \end{bmatrix} = \frac{2}{V_o}\begin{bmatrix} 2 & -1 \\ -1 & 2 \end{bmatrix}\begin{bmatrix} V_p^* \\ V_n^* \end{bmatrix} \tag{8}$$

In the interval of 30°~90°, as shown in Figure 3, we can get

$$\begin{cases} V_p = V_a - V_c \\ V_n = V_c - V_b \end{cases} \tag{9}$$

For a three-phase balanced and symmetrical system, there are

$$V_a + V_b + V_c = 0 \tag{10}$$

Substituting (9) and (10) into (5), we can get

$$\begin{cases} V_p^* = V_a \\ V_n^* = -V_b \end{cases} \tag{11}$$

Therefore, the PFC control target can be written as

$$\begin{cases} V_p^* = R_e * i_{L_p} \\ V_n^* = R_e * i_{L_n} \end{cases} \tag{12}$$

where R_e is the equivalent input resistance.

Substituting (12) into (8), we can get

$$\begin{bmatrix} 1-d_p \\ 1-d_n \end{bmatrix} = \frac{R_e}{V_o R_s} R_s \begin{bmatrix} 2 & -1 \\ -1 & 2 \end{bmatrix}\begin{bmatrix} i_{L_p} \\ i_{L_n} \end{bmatrix} \tag{13}$$

where R_s is the equivalent current detection resistance.

Letting $V_m = \dfrac{V_o R_s}{R_e}$, where V_m represents the output of the voltage error compensator, then

$$V_m\begin{bmatrix} 1-d_p \\ 1-d_n \end{bmatrix} = R_s \begin{bmatrix} 2 & -1 \\ -1 & 2 \end{bmatrix}\begin{bmatrix} i_{L_p} \\ i_{L_n} \end{bmatrix} \tag{14}$$

Equation (14) is the core equation for the improved one cycle control. It can be seen that the simplified circuit can also achieve power factor correction using one cycle control.

III. THE LOGIC REALIZATION OF IMPROVED ONE CYCLE CONTROL

Based on the above analysis, the improved one cycle control principle block diagram shown in Fig. 5 can be obtained.

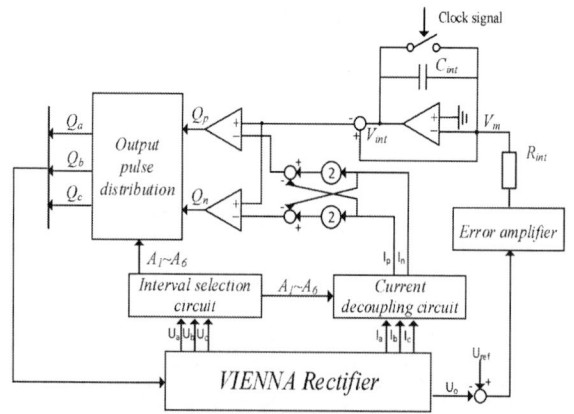

Fig. 5. Block diagram of improved one cycle control principle

First, we divide an entire power frequency cycle into 6 intervals, as shown in Fig. 3. In each interval, the relative magnitude of the three-phase voltage is fixed. The i_{Lp} and i_{Ln} in the simplified circuit is also equal to the two-phase current. By judging the magnitude of the three-phase voltage, the interval in which this working state is located can be obtained, which is represented by $A_1 \sim A_6$. Then the current is decoupled according to the voltage interval, and the three-phase current sampled at this time is transformed into the simplified control current i_{Lp} and i_{Ln} of the double series boost circuit. Then, according to the one cycle core control equation, the output driving duty ratios Q_p and Q_n of the dual-series boost circuit in one switching cycle are calculated. Finally, Q_p and Q_n are inversely transformed into actual three-phase drive pulses Q_a, Q_b, and Q_c according to the previous interval judgment flag bit. This completes the entire control process.

A. Voltage range selection

In each 60° voltage interval, the relative magnitude of the three-phase voltage is determined, so it can be defined as follows: when the voltage of phase A is greater than the voltage of phase B, then a-b=1, otherwise a-b=0, and so on for the remaining two phases. Then in the voltage interval of 30° to 90°, A2=(a-c)(c-b)=1, and the remaining intervals are all 0. Then according to the three-phase voltage relationship in Fig.3, the logic relationship of the flag bits of the six voltage phase intervals can be obtained as

$$\begin{cases} A_1 = (a-b)(c-a) \\ A_2 = (a-c)(c-b) \\ A_3 = (a-b)(b-c) \\ A_4 = (b-a)(a-c) \\ A_5 = (b-c)(c-a) \\ A_6 = (c-b)(b-a) \end{cases} \quad (15)$$

B. Current decoupling

In each 60° voltage interval, the current relationship between the two equivalent currents and the original circuit is shown in Table II.

TABLE II. THE RELATIONSHIP BETWEEN EQUIVALENT CURRENT AND ORIGINAL THREE-PHASE CURRENT

Interval marker	i_p	i_n
$A_1 = 1$	i_c	$-i_b$
$A_2 = 1$	i_a	$-i_b$
$A_3 = 1$	i_a	$-i_c$
$A_4 = 1$	i_b	$-i_c$
$A_5 = 1$	i_b	$-i_a$
$A_6 = 1$	i_c	$-i_a$

Summarizing the relationship between the flag bit and the current i_p, i_n in Table 2, we can get

$$\begin{cases} i_p = i_a(A_2 + A_3) + i_b(A_4 + A_5) + i_c(A_1 + A_6) \\ i_n = -i_a(A_5 + A_6) - i_b(A_1 + A_2) - i_c(A_3 + A_4) \end{cases} \quad (16)$$

C. Driving signal generation

According to the current i_p, i_n and the improved one cycle control core control equation, two drive signals Q_p and Q_n can be obtained. To generate the final three-phase driving signal, it needs to be inversely transformed, as shown in Table III.

TABLE III. THE RELATIONSHIP BETWEEN THE EQUIVALENT CIRCUIT DRIVE SIGNAL AND THE ORIGINAL THREE-PHASE SWITCH DRIVING SIGNAL

Interval marker	Q_a	Q_b	Q_c
$A_1 = 1$	1	Q_n	Q_p
$A_2 = 1$	Q_p	Q_n	1
$A_3 = 1$	Q_p	1	Q_n
$A_4 = 1$	1	Q_p	Q_n
$A_5 = 1$	Q_n	Q_p	1
$A_6 = 1$	Q_n	1	Q_p

Summarizing the relationship between the flag bit and the three-phase drive signal in Table III, we can get

$$\begin{cases} Q_a = (A_1 + A_4) + Q_p(A_2 + A_3) + Q_n(A_5 + A_6) \\ Q_b = (A_3 + A_6) + Q_p(A_4 + A_5) + Q_n(A_1 + A_2) \\ Q_c = (A_2 + A_5) + Q_p(A_1 + A_6) + Q_n(A_3 + A_4) \end{cases} \quad (17)$$

In summary, (15) ~ (17) are the logic realization of improved one cycle control.

IV. ANALYSIS OF AUTOMATIC BALANCE ON DC BUS OUTPUT CAPACITOR VOLTAGE

For three-level rectifiers, the midpoint balance problem is common, so the midpoint potential is analyzed.

Combining vertical (8) and (14), we can get

$$\begin{cases} \dfrac{R_s V_{c1}}{V_m} \cdot (2i_{L_p} - i_{L_n}) = 2V_p^* - V_n^* \\ \dfrac{R_s V_{c2}}{V_m} \cdot (-i_{L_p} + 2i_{L_n}) = -V_p^* + 2V_n^* \end{cases} \quad (18)$$

Simplifying (15) into a function of i_{Lp} and i_{Ln}, we can get

$$\begin{cases} i_{L_n} = \dfrac{V_m}{3R_s} \cdot \left[\dfrac{2V_{c2} - 2V_{c1}}{V_{c1}V_{c2}} V_p^* + \dfrac{-V_{c2} + 4V_{c1}}{V_{c1}V_{c2}} V_n^* \right] \\ i_{L_p} = \dfrac{V_m}{3R_s} \cdot \left[\dfrac{4V_{c2} - V_{c1}}{V_{c1}V_{c2}} V_p^* + \dfrac{-2V_{c2} + 2V_{c1}}{V_{c1}V_{c2}} V_n^* \right] \end{cases} \quad (19)$$

When the circuit is in a steady state, that is, $V_{c1}=V_{c2}$, the average value of the inductor current changes linearly with the expected voltage, so power factor correction can be achieved; when $V_{c1} \neq V_{c2}$, let

$$\begin{cases} V_{c1} = V_c + \Delta u \\ V_{c2} = V_c - \Delta u \end{cases} \quad (20)$$

Then, (16) can be simplified to

$$\begin{cases} i_{L_p} = \dfrac{V_m}{R_s V_c} \cdot V_p^* + \dfrac{V_m \cdot \Delta u}{3R_s V_c^2} \left[-5V_p^* + 4V_n^* \right] \\ i_{L_n} = \dfrac{V_m}{R_s V_c} \cdot V_n^* + \dfrac{V_m \cdot \Delta u}{3R_s V_c^2} \left[-4V_p^* + 5V_n^* \right] \end{cases} \quad (21)$$

Fig. 2 shows the current difference flowing into the two capacitors: $\Delta i = i_{c2} - i_{c1} = (1 - d_n)i_{L_n} - (1 - d_p)i_{L_p}$, then the average value of the current difference in a power frequency cycle is

$$\Delta i = \frac{2V_m}{R_s V_c^2} \cdot \frac{\Delta u}{3V_c} \cdot K \quad (22)$$

where $K = \dfrac{1}{T} \int_0^T (V_p^* + V_n^*)^2 \, dt$.

978-1-6654-4817-8/21 $31.00 © 2021 IEEE

At steady state, $\Delta u = 0$, $\Delta i = 0$, so there is no voltage difference between the two capacitors. When $\Delta u < 0$, $\Delta i < 0$, the current will charge C_1, so V_{c1} rises and V_{c2} falls until $V_{c1} = V_{c2}$. In the same way, the system will eventually stabilize to $V_{c1} = V_{c2}$ when $\Delta u > 0$.

From the above derivation, it can be seen that the improved one cycle control has a self-equalizing characteristic of the DC bus output capacitor voltage.

V. SIMULATION RESULTS

In order to verify the correctness and feasibility of the above control strategy, this paper builds an improved one cycle control simulation model of VIENNA rectifier based on the MATLAB/Simulink simulation platform. Simulation parameters: AC line voltage 380V, input filter inductance 2mH, DC bus voltage 1000V, output power 10kW, DC side capacitance 2000uF, switching frequency 20KHz.

Fig. 6 shows the A phase input voltage and input current waveforms. Through this figure, it can be seen that the improved one cycle control can achieve power factor correction well, and the sine of the input current is relatively high. Using Simulink's built-in FFT module to perform Fourier analysis on the phase A current, as shown in Fig. 7, we can see that the phase A input current total harmonic distortion factor THD=3.94%. If the fundamental wave displacement factor is assumed to be 1, the factor value can be calculated that PF is 0.9992, which can be seen that the one cycle control can control the harmonics well and make the power factor close to 1. Fig. 8 shows the driving waveform of the switching tube. It can be seen from the figure that there is a switching tube that is always on in every 60° interval, and the driving signal continues to output a high level. The other two switching tubes are switched off at high frequency. Therefore, improved one cycle control has less loss. Fig. 9 is the efficiency comparison curve between traditional one cycle control and improved one cycle control. The orange is the improved one cycle control efficiency curve, and the blue is the traditional one cycle control efficiency curve. It can be seen from the curve that the control method proposed in this paper has higher efficiency no matter it is in a light load state or a heavy load state.

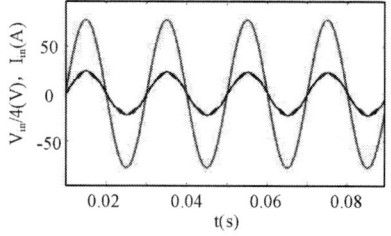

Fig. 6. Simulation waveforms of phase A input voltage and input current

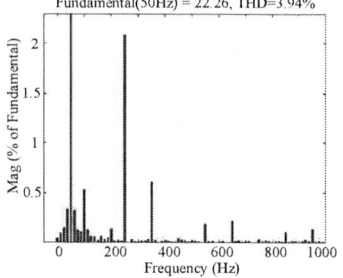

Fig. 7. FFT analysis waveform of AC side input current

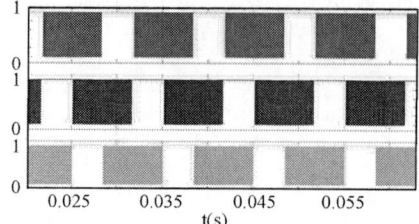

Fig. 8. Switch tube driving simulation waveform

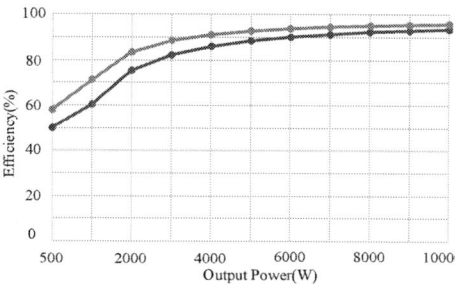

Fig. 9. Improved one cycle control and traditional one cycle control efficiency comparison chart

Fig. 10 shows the simulation waveform of the DC bus voltage when the load is suddenly reduced from 10kW to 5kW. It can be seen from the figure that at 0.3s, the load is suddenly reduced, and the DC bus voltage rises at the moment of the sudden load reduction. Under the negative feedback regulation of the system, approximately after 50ms, it reaches 1000V stabilization again. Fig.11 shows the simulation waveform of the A-phase input current when the load is suddenly reduced from 10kW to 5kW. It can be seen from the figure that the input current decreases after load shedding, and the current stabilizes after two mains cycles. This shows that the improved VIENNA rectifier system under one cycle control has better rapidity and stability.

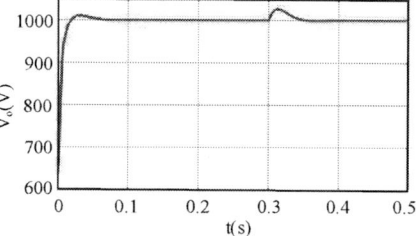

Fig. 10. The simulation waveform of the DC bus voltage when the load is suddenly reduced from 10kW to 5kW

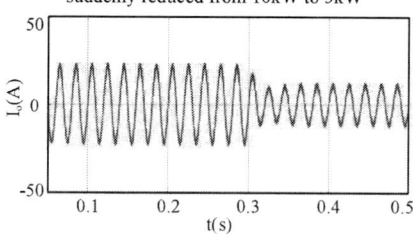

Fig. 11. A phase input current simulation waveform when the load is suddenly reduced from 10kW to 5kW

Fig. 12 shows the voltage waveforms at both ends of the capacitor when V_{c1}=600V and V_{c2}=300V. It can be seen that although their initial voltages are different, after 0.2s, their voltages tend to be the same, both are 500V, which meets the voltage self-balancing characteristics of capacitances analyzed above.

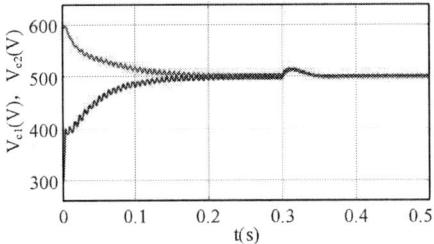

Fig. 12. Simulation waveform of the voltage across the capacitor when V_{c1}=600V, V_{c2}=300V

VI. CONCLUSION

This paper is aimed at the VIENNA rectifier. We redivided the interval of the power frequency cycle and the VIENNA rectifier in each interval after the redivision can be simplified into a circuit formed by two boost series. For this circuit, an improved one cycle control is established. The mathematical model derives the core control equations of the improved one cycle control in detail. At the same time, theoretical analysis demonstrates that the control method has the characteristics of capacitor voltage self-equalization. Finally, a simulation model was built through MATLAB/ Simulink. The simulation results show that the VIENNA rectifier grid-side current waveform under this control mode has high sinusoidality, high power factor, low loss. It has the characteristics of capacitor voltage self-equalization, and has good dynamics and stability. This paper verifies the rationality and feasibility of the proposed method, and provides a good theoretical basis for the design of the actual system.

REFERENCES

[1] T. Wang, C. Cheng, S. Duan, et al, "Space Vector Modulation Strategy of Vienna rectifier for improving current zero-crossing distortion," Transactions of China Electrotechnical Society, 2019, 34(18):3854-3864.

[2] S. Liu, J. Jiang and G. Cheng, "Research on Vector Control Strategy of Three Phase VIENNA Rectifier Employed in EV Charger," 2019 Chinese Control And Decision Conference (CCDC), 2019, pp. 4914-4917.

[3] K. Siriphan and P. Khamphakdi, "Analysis of Center-Aligned Space Vector Pulse Width Modulation Realization for Three-Phase Vienna Rectifier," 2019 International Conference on Power, Energy and Innovations (ICPEI), 2019, pp. 44-47.

[4] H. Cheng and J. Huang, "Research on SVPWM Control Strategy of Three Phase VIENNA Rectifier," 2018 5th International Conference on Systems and Informatics (ICSAI), 2018, pp. 166-170.

[5] H. Cheng, Y. Ma and C. Wang, "A Modified One Cycle Control of VIENNA Rectifier for Neutral Point Voltage Balancing Control Based on Cycle-by-cycle Correction," 2018 IEEE International Power Electronics and Application Conference and Exposition (PEAC), 2018, pp. 1-6.

[6] C. Bing, M. Hui, X. Yunxiang and W. Yingpin, "Sliding mode and predictive current control for vienna-type rectifiers," 2015 18th International Conference on Electrical Machines and Systems (ICEMS), 2015, pp. 340-344.

[7] X. Huang, D. Ma and X. Chen, "Sliding Mode Variable Structure Control on Vienna Rectifier," 2020 Chinese Control And Decision Conference (CCDC), 2020, pp. 2997-3002.

[8] M. Hui, X. Yunxiang, L. Wenjing and C. Bing, "Research on direct power control based on sliding mode control for vienna-type rectifier," IECON 2015 - 41st Annual Conference of the IEEE Industrial Electronics Society, 2015, pp. 25-30.

Research on the strategy of parallel wide range bidirectional DC-DC converter

Zehui Peng
School of Mechatronic Engineering and Automation
Shanghai University
Shanghai, China
came1lia@shu.edu.cn

Xikun Chen
School of Mechatronic Engineering and Automation
Shanghai University
Shanghai, China
chenxk@shu.edu.cn

Borui Liu
School of Mechatronic Engineering and Automation
Shanghai University
Shanghai, China
chenxk@shu.edu.cn

Yongjian Chen
School of Mechatronic Engineering and Automation
Shanghai University
Shanghai, China
chenxk@shu.edu.cn

Junnan Gu
School of Mechatronic Engineering and Automation
Shanghai University
Shanghai, China
chenxk@shu.edu.cn

Ruiying Li
School of Mechatronic Engineering and Automation
Shanghai University
Shanghai, China
chenxk@shu.edu.cn

Abstract—To improve energy shortage and environmental pollution, the research of new energy power generation technology and microgrid has become a research hotspot in the field of electrical engineering. The energy storage system is an important part of the microgrid. In the process of realizing the energy interaction between the microgrid and the energy storage system, the wide voltage range bidirectional DC-DC converter plays a significant role. Based on the requirements of modular wide voltage range bidirectional DC-DC converters, in order to improve the output power level, modular parallel connection is adopted,and CLLLC resonant converter and its parallel current sharing technology have been studied in detail. To demonstrate the feasibility and coherence of the hybrid control of the bidirectional CLLLC resonant converter based on virtual impedance,experiments were carried out on an experimental platform with two 3kw outputs in parallel.

Keywords—DC-DC converter, CLLLC resonant converter, hybrid control, parallel flow

I. INTRODUCTION

With the increasing tension of energy and environmental problems, the development of new energy has been widely concerned. As the speedy advancement of renewable energy applications, distributed generation and DC microgrid technologies emerge as the times require[1]. Renewable energy such as photovoltaic and wind power generation is intermittent and unstable, which requires a storage system composed of batteries or supercapacitors to store excess energy. For the energy storage system in the DC distribution network, a bidirectional DC-DC converter with a large capacity and wide input voltage gain range is demanded.

A bidirectional LLC resonant converter is widely used because of its unique soft switching technology. In reference [2], the converter only uses constant frequency control. When the output voltage range is wide, there will be greater return power, which will reduce the efficiency of the system. In reference[3], the CLLLC resonant converter only uses frequency conversion control. When the voltage transfer is wide, a large variation range should be needed by switching frequency, which is not conducive to the optimization of resonant parameters. In reference[4], a hybrid control based on full-bridge LLC is proposed, but it can only realize one-way power transmission. In reference[5], LLC cascaded topology is proposed, in which the front stage is LLC

topology ,while the backstage is interleaved buck/boost topology to achieve high dynamic performance and wide range voltage regulation.This cascaded topology increases the size and cost of the converter. In reference[6], a full-bridge three-level LLC resonant converter with control policy combination is introduced. The control strategy and topology are complex and difficult to analyze. In reference[7], an LLC resonant converter with a semiactive variable structure rectifier is developed, which can worbadk in a broad output voltage range.Nonetheless, in the mode switching, the converter is not smooth switching, which plays a crucial role in the dynamic characteristics and stability of the circuit.

This paper proposes a two-way CLLLC resonant converter based on a hybrid control strategy to achieve electrical isolation,wide-range voltage regulation, and two-way power transmission. To advance the output power, two identical converters are associated in parallel with input parallel and output parallel, and the virtual impedance method is adopted to solve the problem of parallel current sharing. Finally, two 3kW experimental prototypes were prepared. The experimental outcomes exhibit that the proposed control method can reduce the unbalance of the output current with high effeciency, which verifies the feasibility of the method.

II. CLLLC RESONANT CONVERTER TOPOLOGY

A. Topology analysis of CLLLC resonant converter

The main circuit topology of the CLLLC is shown in Figure 1. Compared with the traditional LLC topology , the secondary side has two more resonant components, and the secondary side rectifier diode is replaced by a power switch tube, which can realize bidirectional energy flow.

Define that the converter transfers power from V_1 to V_2 just as a forward working mode.Determined transfers power from V_2 to V_1 as a reverse working mode. The structure of the CLLLC is symmetrical, which can obtain the two-way energy transmission. The working principles of the forward working mode and the reverse working mode are symmetrical. In this essay, the forward direction mode is taken as an example. And the converter can reach the soft-switching characteristics of the traditional LLC topology whether it works in the forward or reverse direction, that is, the zero-voltage-switching(ZVS) turn-on of the primary side

978-1-6654-4817-8/21 $31.00 © 2021 IEEE

switch and the zero-current-switching(ZCS) turn-off of the secondary side rectifier diode.

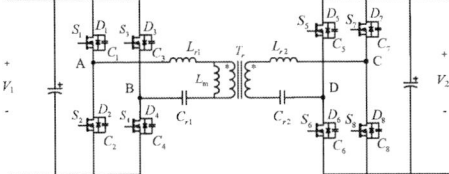

Fig. 1. CLLLC resonant converter circuit topology

B. The working principle and characteristic analysis of frequency conversion control

The frequency conversion control rule of the CLLLC is the same as the control principle of the full-bridge LLC , that is, the output voltage is adjusted by changing the switching frequency of the switch tube. Define two resonant frequencies here, which are the resonant frequency f_r of the resonant capacitor C_{r1} and the resonant inductance L_{r1}; the resonant frequency f_m of the C_{r1}, the L_{r1} and the magnetizing inductance L_m.

$$f_r = \frac{1}{2\pi\sqrt{L_r C_r}} \quad (1)$$

$$f_m = \frac{1}{2\pi\sqrt{(L_r + L_m)C_r}} \quad (2)$$

In the region of $f_m < f_s \leq f_r$, the resonant converter can not only accomplish the ZVS turn-on of the primary side switch tube, but also realize the ZCS turn-off of the secondary side rectifier diode,achieving the highest efficiency. Fig.2 presents the main waveform of the CLLLC resonant converter when $f_m < f_s < f_r$.

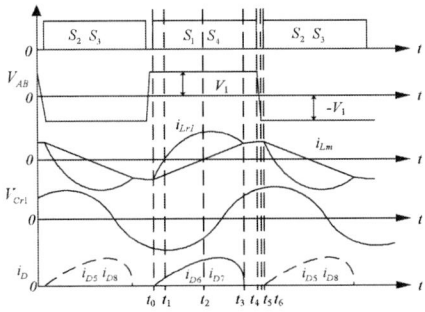

Fig. 2. The main working waveform of the converter when $f_m < f_s < f_r$

The equivalent circuit of the CLLLC can be obtained by using the fundamental wave analysis method. The parameters of the secondary side of the converter are converted to the primary side, and assuming that $L'_{r2} / L_{r1} = C'_{r2} / C_{r1} = 1$, the gain of the CLLLC is determined as M, and the expression can be obtained as follow:

$$M = \frac{1}{\sqrt{(1 + \frac{1}{K} - \frac{1}{K f_n^2})^2 + \frac{Q^2}{K^2}\left[f_n(2K+1) - \frac{2K+2}{f_n} + \frac{1}{f_n^3}\right]^2}} \quad (3)$$

Among them: K denotes the ratio of the L_m to the L_{r1}, $K = L_m / L_{r1}$; f_n denotes the normalized frequency, f_s denotes the switching frequency, $f_n = f_s / f_r$, Q denotes the quality factor.

Figure 3 shows the curve of voltage gain M and the normalized frequency f_n under different Q values. It can be seen from the figure that, when f_n is the same, the voltage gain of the converter decreases as the quality factor Q increases.

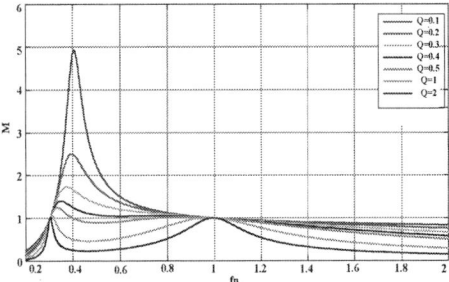

Fig. 3. The curve of voltage gain varies from normalized frequency under different Q values when $K=5$

C. Working principle and characteristic analysis of fixed frequency control

When using the phase shift control, the switching frequency of the switching tube is constant as the resonance frequency f_r, and the output voltage is controlled by changing the phase shift angle of the driving signals of the switching tubes S_1 and S_4. Assuming that the switching tubes S_1 and S_4 are turned on concurrently, the duty cycle is D, and its transient analysis waveform is shown in Fig. 4.

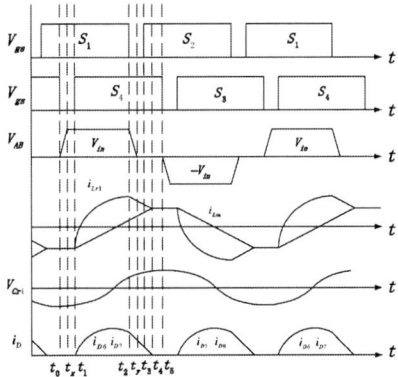

Fig. 4. Main working waveforms under phase shift control

When the phase shift control is selected in the system, for the sake of suppressing the harmonic components, the time domain analysis method is usually used for analysis. Because it is hard to solve the equations directly, the numerical solution of M is usually solved by assigning D to the value, and the relationship between M and D is obtained. curve. When $K=5$, the curve of voltage gain ratio M and duty cycle D under different Q values is shown in Fig. 5:

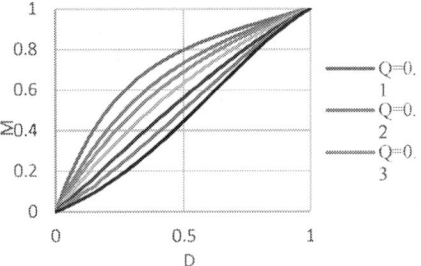

978-1-6654-4817-8/21 $31.00 © 2021 IEEE

Fig. 5. The gain curve of phase shift control when K=5

According to Fig. 5 , when the value of Q is constant, as D decreases, the gain of the converter decreases monotonously. The converter under phase shift control works in step-down mode.

III. CLLLC RESONANT CONVERTER CONTROL STRATEGY

A. Hybrid control strategy

For a single CLLLC , if the frequency conversion control is adopted, when the normalized frequency f_n is greater than 1, the anti-parallel diode on the secondary side is hardly turned off, and there is a reverse recovery problem; if the converter is working in the step-down mode, the control strategy turns to phase shift control. Phase shift control can achieve the ZCS turn-off of the anti-parallel diode on the secondary side and advance the conversion efficiency of the system. The control block diagram is displayed in Figure 6. If the adjusted frequency exceeds the f_r, phase shift control is adopted; if the frequency is less than the f_r, frequency conversion control is adopted.

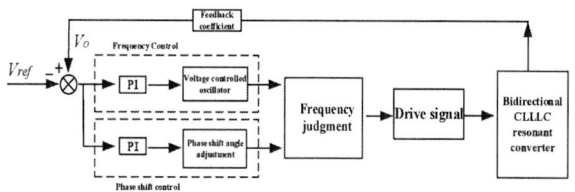

Fig. 6. Hybrid control block diagram

B. Parallel current sharing method based on virtual impedance

When two converters are connected in parallel, as the inequality between the parameters of the resonant network, the corresponding parameter of the two modules are not equal. Under the premise of equal gain, it can be known that the output load resistance of each module is not equal, and under the condition of natural current sharing, the unbalance of the output current of the two converters is extremely large, and it is inevitable to introduce a current-sharing loop to control the module current.

According to Thevenin's theorem, the AC equivalent circuit can be equivalent to a circuit structure composed of an equivalent input voltage V_{eq} and an equivalent impedance Z_{eq}. The equivalent output current of each channel of the multi-module parallel system can be obtained.

Determined the output circulating current variable I_{ci} to measure the current sharing effect of the parallel connection of the modules, and the output circulating current expression can be obtained as follows:

$$I_{ci} = \frac{\left|V_{eqi} - V_i\right| - \dfrac{1}{n}\sum_{i=1}^{n}\left|V_{eqi} - V_i\right|}{\left|Z_{eqi}\right|} \quad (4)$$

Among them, V_{eqi} denotes the equivalent input voltage, V_i denotes the equivalent output voltage, and Z_{eqi} denotes the equivalent impedance.

Since Z_{eqi} is determined by the parameters of the circuit itself and cannot be changed, to reduce the impact on the system, the virtual impedance Z_s can be serialized in equation (4), and the output loop current I_{ci} can be reduced by increasing Z_s to achieve the inter-module Parallel current sharing.

The transfer function block diagram of the CLLLC resonant converter controlled by the single voltage closed-loop is shown in Fig. 7.

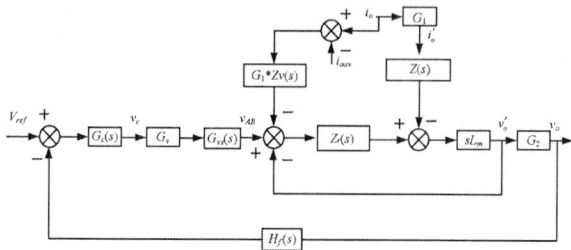

Fig. 7. Transfer function block diagram of CLLLC resonant converter controlled by single voltage loop

Assuming that the resonant network parameters of each converter are the same, according to the distinction between the reference value of each output voltage and the average voltage of the n-channel parallel system and the ratio of the transfer function in Fig.7, the expression of the loop output current $I_{ci}(s)$ and the loop impedance $Z_{ci}(s)$ can be obtained, the virtual impedance $Z_v(s)$ is introduced into the expression of the loop impedance $Z_{ci}(s)$, and the output characteristic expression after introducing the virtual impedance can be obtained as follows:

$$V_o(s) = \frac{G_c(s)G_v G_{vs}(s)G_2}{C_0(s)} V_{refi}(s) - \frac{Z_{ro}(s)G_1 G_2}{C_0(s)} I_{oi}(s)$$
$$- \frac{Z_v(s)G_1 G_2}{C_0(s)}(I_{oi}(s) - I_{av}(s)) \quad (5)$$

Where $Z_{ro}(s) = Z(s)Z_r(s)$.

According to the expression (5), the output transfer function which introduced the virtual impedance can be regarded as the introduction of a current feedback loop in a single voltage loop closed-loop control system.and the introduction of virtual impedance in the feedback loop, which can be adjusted by regulating the virtual impedance. The circulating current can be controlled by adjusting the virtual impedance.Introduce the current feedback loop and migrate the circulating feedback point to the output of the transfer function G_v, and the equivalent transfer function block diagram is shown in Figure 8. When the n-way parallel system is working, the output current of each converter is unbalanced, which will produce output circulating current. The output circulating current passes through the virtual impedance link of the feedback channel, and the frequency or phase shift angle is corrected to diminish the output circulating current, thereby realizing the current sharing between the modules.

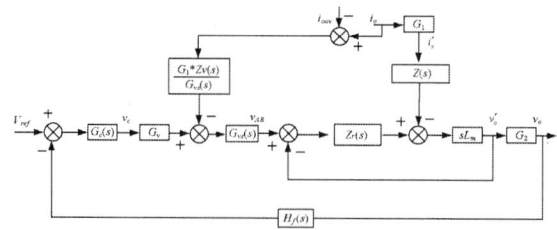

978-1-6654-4817-8/21 $31.00 © 2021 IEEE   304

Fig. 8. The adjusted transfer function block diagram of the CLLLC resonant converter with virtual impedance introduced

IV. EXPERIMENTAL VERIFICATION

A. Experimental Research on Single Module System

1) The waveform of the primary side input voltage is 300V, the secondary side output voltage is 600V.

Taking the forward working state as an example, the experimental results are compared and analyzed. As shown in Fig.9(a) , in responce to the constant reference, the output voltage is stable at 600V, and the output is 3kW.The converter works in the under-resonant frequency range is depicted in Fig.9(b), that is, $f_s < f_r$. The primary side switch tube can realize ZVS turn-on, the current rectified on the secondary side is intermittent, and the diode on the secondary side which can get to switch off with no current.

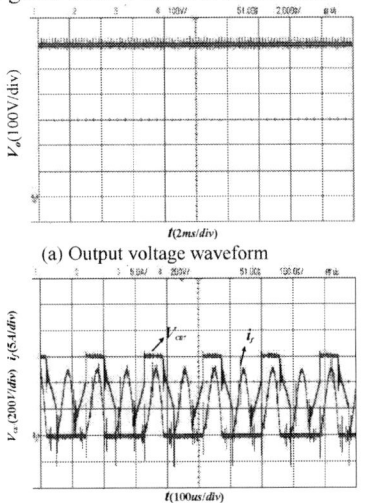

(a) Output voltage waveform

(b) Tube voltage drop V_{CE7}, current if waveform after secondary side rectification

Fig. 9. The main waveforms with reference input voltage of 300V

2) The waveform of the reference input voltage on the primary side is 450V, and the output voltage on the secondary side is 300V.

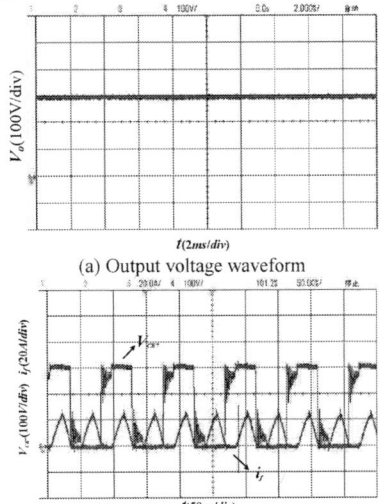

(a) Output voltage waveform

(b) waveform of tube voltage drop V_{CE7}, and rectifid current on the secondary side

Fig. 10. The main waveforms with reference input voltage of 450V

As illustrated in Fig.10(a), the referene input voltage is 450V, and the output voltage tracks its references as a constant value of 300V, while the output power is 3kW. Fig.10 (b) shows the voltage drop V_{CE7} of the secondary side switch tube S_7 and the current i_f rectified on the secondary side. The experiment outcomes of Fig.10 shows the system works in the phase shift control mode, and the ZVS turn-on of the primary side can still be carried out. The ZCS turn-off is consistent with the theoretical analysis of previous parts.

B. Experimental Research on Parallel System

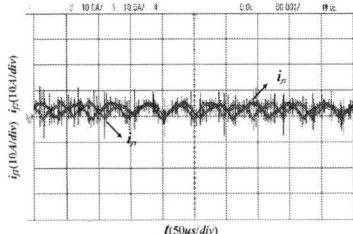

(a) The waveforms of rectified current i_{f1} and i_{f2} in module1 and module2

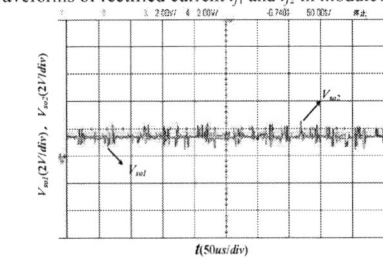

(b) The sampling voltage waveform of output current in module1 and module2

Fig. 11. the main current waveforms when the converters are working in 1.4kW in forwarding parallel operation

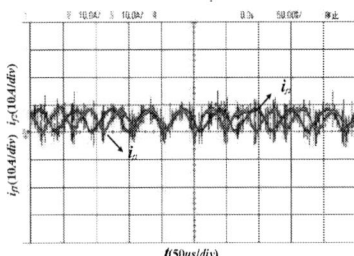

(a) The waveforms of rectified current i_{f1} and i_{f2} in module1 and module2

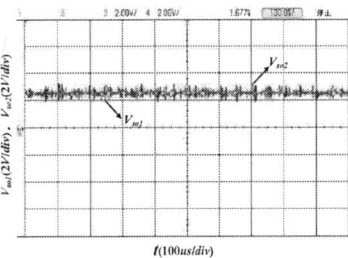

(b) The sampling voltage waveform of output current in module1 and module2

Fig. 12. The main current waveforms when the converters are working in 1.4kW in forwarding parallel operation

978-1-6654-4817-8/21 $31.00 © 2021 IEEE

As shown in Figs.11 and 12, under the adjustment of the current-sharing loop, the output currents of the module1 and module2 maintain dynamic balance when the two modules are working in the forward direction, while the average of the output currents tends to be the same, and the current imbalance is small. With the module1 and module 2,when the outcomes of output power is stable at 1.4kW, the average current of module 2 is 3.29A, while the current of module 1 is 3.15A, and the current imbalance of the module is 2.2%, which meets the design requirements.When the output power is 2.5kW, the module2 average output current is 5.4A, and the module 1 average output current is 5.2A. The current imbalance of the module at this time can be obtained as 1.8%, which meets the design requirements.

CONCLUSION

On the basis of the analysis to the working principle and characteristics of the two-way CLLLC resonant converter under variable frequency control and phase shift control, this paper proposes a hybrid control strategy which can adjust dynamiclyfrom frequency control mode to phase shift control mode and adopts a parallel current sharing strategy based on virtual impedance. Converters applied to the strategy can achieve a wide voltage gain range, wide load range bidirectional operation,and power balance between modules. Finally, two 3kW experimental prototypes were formed to corroborate the feasibility and rationality of the hybrid control and single voltage control of the bidirectional CLLLC based on virtual impedance.

REFERENCES

[1] SONG Qiang, ZHAO Biao, LIU Wenhua, ZENG Rong. An Overview of Research on Smart DC Distribution Power Network [J]. Proceedings of the CSEE, 2013, 33(25): 9-19+5.

[2] Safaee A , Karimi-Ghartemani M , Jain P K , et al. Time-Domain Analysis of a Phase-Shift-Modulated Series Resonant Converter with an Adaptive Passive Auxiliary Circuit[J]. IEEE Transactions on Power Electronics, 2016, 31(11):7714-7734.

[3] Schneider T , Kratz S , Wegener R , et al. Symmetrical Bidirectional CLLC-Converter with Simplified Synchronous Rectification for EV-Charging in Isolated DC Power Grids[C]// 2019 IEEE 28th International Symposium on Industrial Electronics (ISIE). IEEE, 2019.

[4] Li Ju, Ruan Xinbo. Hybrid Control Strategy of Full Bridge LLC Converters [J]. TRANSACTIONS OF CHINA ELECTROTECH-NICAL SOCIETY, 2013, 28(04): 72-79+94.

[5] YAO Hongtao, SU Mei, DAN Hanbing, XIONG Wenjing, WU Sisheng, HU Ziheng. Input-parallel-output-parallel Wide-range Bidirectional Isolated DC/DC Converter [J]. Journal of Power Supply, 2020, 18(03): 13-21.

[6] JIN Ke, RUAN Xin-bo. Hybrid Full Bridge Three-level LLC Resonant Converter [J]. Proceedings of the CSEE, 2006(03): 53-58.

[7] H. Wu, Y. Li and Y. Xing, "LLC Resonant Converter With Semiactive Variable-Structure Rectifier (SA-VSR) for Wide Output Voltage Range Application," in IEEE Transactions on Power Electronics, vol. 31, no. 5, pp. 3389-3394, May 2016, doi: 10.1109/TPEL.2015.2499306.

Analysis of an Output Series High Voltage Gain Impedance Source Circuit Based on SiC Switch

Qing Cheng
Department of Electrical Engineering
Harbin Institute of Technology
Harbin, China
c_qing07@163.com

Wei Wang
Department of Electrical Engineering
Harbin Institute of Technology
Harbin, China
wangwei602@hit.edu.cn

Yueshi Guan
Department of Electrical Engineering
Harbin Institute of Technology
Harbin, China
hitguanyueshi@163.com

Tingting Yao
Department of Electrical Engineering
Harbin Institute of Technology
Harbin, China
hityaott@163.com

Dianguo Xu
Department of Electrical Engineering
Harbin Institute of Technology
Harbin, China
xudiang@hit.edu.cn

Abstract—In order to further improve the voltage gain of the traditional Y-source inverter, a series output type high voltage-gain impedance source inverter topology is proposed. The topology is a two-stage circuit. The front-stage circuit is composed of two symmetrically designed switches in front of an improved Y-source DC-DC topology, which optimizes the switch stress and expands the duty cycle adjustment range; the second-stage circuit is full bridge inverter circuit. Through the symmetrical design of the pre-stage circuit, a higher DC bus voltage is obtained by connecting the outputs in series. To further improve the power density, the planar magnetic components and silicon carbide (SiC) switches are adopted. Establish a theoretical model of planar inductance, analyze the advantages and disadvantages of different winding arrangements, and determine the inductance parameters with the help of finite element simulation software. In order to verify the theoretical analysis, an experimental platform is built, and a Y-source coupled planar inductor is designed and manufactured. The experimental results verifies: under the rated power, the boost ratio can reach 5, which effectively improves the voltage gain.

Keywords—Y-source converter; output series; planar inductance; winding optimization.

I. Introduction

Due to its non-renewability, fossil energy has begun to fail to meet the further development of the global economy. As an important part of renewable energy, photovoltaic power generation has developed rapidly. As a key part of the photovoltaic power generation system, the inverter has also received more attention and research from scholars [1-3]. The Y source impedance network inverter is a commonly used photovoltaic inverter. The basic topology is shown in Fig. 1(a), Among them, the impedance source adopts Y source coupled inductor, which can obtain a higher DC bus boost ratio, and the design of the turns ratio of the coupled inductor is more flexible. However, the traditional Y-source inverter still has some problems: 1) The starting impulse current is too large; 2) The input current is discontinuous; 3) The leakage inductance of the coupled inductor affects the DC bus voltage spike. In order to solve the above problems, some scholars have proposed an improved Y-source inverter [4-8].

The improved Y source structure topology is shown in Fig. 1(b). A diode and a capacitor are added on the basis of the traditional Y source structure, which effectively overcomes some of the shortcomings of the above-mentioned traditional

This work was supported by the National Natural Science Foundation of China under Grant 51977045.

Y source topology, but the boost ratio is still limited.

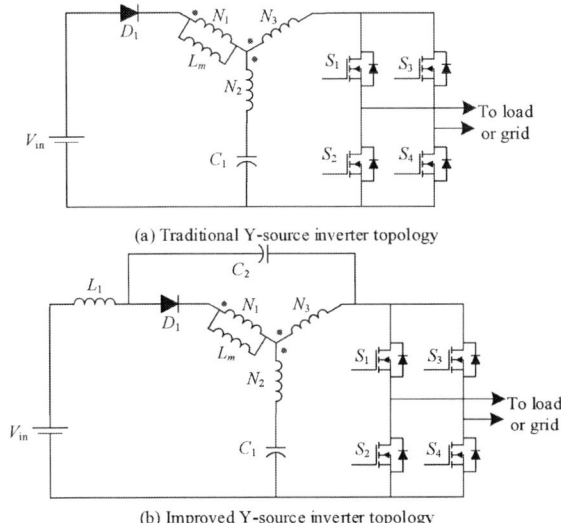

(a) Traditional Y-source inverter topology

(b) Improved Y-source inverter topology

Fig. 1 Two Y source inverter topologies

On the basis of the improved Y source topology, in order to further improve the voltage gain of the inverter, this paper proposes a two-stage Y source topology circuit. The front stage circuit is an improved Y source DC-DC topology with the switch in front. By transferring the switch from the output side to the input side, the voltage and current stress of the switch is effectively reduced, and the duty cycle adjustment range of the switch is enlarged [9]; The latter circuit is a bridge inverter circuit, which adopts The two-stage topology separates the boost from the inverter, and the switch modulation of the inverter circuit part can be performed separately, which simplifies the control strategy [10-11].

By symmetrically designing a single Y source DC topology, the loads of two symmetric Y source circuits can be connected in series to form an effective current path, and then the outputs of the two symmetric circuits are superimposed in series, which further improves the boost ratio.

Y-source coupled inductor is a key component in the circuit. In order to overcome the shortcomings of traditional winding inductors such as high height, large volume, and complicated winding, this paper uses planar inductors to make Y-source coupled inductors, which form inductor windings through multilayer printed circuit boards. It has the characteristics of high coupling coefficient, good thermal characteristics, and high process repeatability. Different winding arrangements have a greater impact on the parameters of the planar inductor, such as winding loss and leakage inductance. This paper makes a detailed comparative analysis of the different arrangements of multilayer windings, and determines an optimal winding arrangement way [12-16].

Based on the above aspects, this paper builds a hardware experimental platform for output series high voltage gain Y-source inverters using planar inductors, which effectively achieves higher voltage gain output.

II. CIRCUIT ANALYSIS

The proposed complete topology is shown in Fig. 2, It is divided into two levels: DC-DC and DC-AC. The DC-DC circuit is composed of two improved Y-source DC-DC circuits I and II with symmetrical design in front of the switch.

As shown in Fig. 2, circuit I and circuit II adopt a symmetrical structure design. On the basis that the Kirchhoff voltage and current equations of the two circuits are consistent, that is, to ensure that the working principles of the two circuits are consistent, a symmetrical topology is designed to connect the outputs in series.

It is this symmetrical circuit that connects the negative output terminal of circuit I to the negative pole of the input power supply, and the positive output terminal of circuit II connects to the positive pole of the input power supply, which can form an output series loop, and the inverter circuit III can obtain a higher DC bus voltage input. The input of DC bus voltage finally increases output V_o.

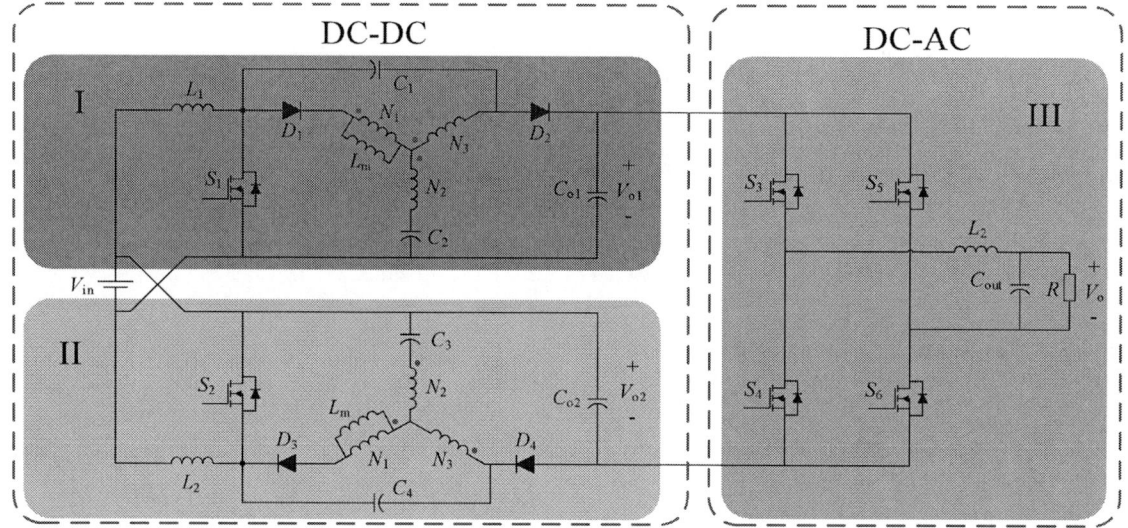

Fig. 2 Output series Y source inverter topology

A. Boost Ratio Analysis

The front-stage switch adopts the alternate conduction mode, the circuit is divided into two modes: a and b, As shown in Fig. 3.

1) Mode a:

At this time, S_1 is turned on and S_2 is turned off. According to the coupling relationship between the three-phase inductance:

$$V_\text{N1} = V_\text{Lm}, V_\text{N2} = \frac{N_2}{N_1}V_\text{Lm}, V_\text{N3} = \frac{N_3}{N_1}V_\text{Lm} \quad (1)$$

Analyze the two loops of circuit I:
$$V_\text{L1} - V_\text{in} = 0 \quad (2)$$

$$-V_\text{C2} - V_\text{N2} + V_\text{N3} + V_\text{C1} = 0 \quad (3)$$

2) Mode b:

At this time, S_1 is turned off and S_2 is turned on. Analyze the three loops of circuit II:

$$-V_\text{in} + V_\text{L1}' - V_\text{C1} + V_\text{o1} = 0 \quad (4)$$

$$V_\text{C2} + V_\text{N2}' - V_\text{N3}' - V_\text{o1}' = 0 \quad (5)$$

$$-V_\text{in} + V_\text{L1}' + V_\text{N1}' + V_\text{N2}' + V_\text{C2} = 0 \quad (6)$$

According to the volt-second balance between the input inductance L_1 and the magnetizing inductance L_m:

$$D_1 V_\text{L1} + (1 - D_1)V_\text{L1}' = 0 \quad (7)$$

$$D_1 V_\text{Lm} + (1 - D_1)V_\text{Lm}' = 0 \quad (8)$$

The voltage gain expression can be obtained:

$$G_1 = \frac{V_\text{o1}}{V_\text{in}} = \frac{1 + D_1 K}{1 - D_1} \quad (9)$$

Among them K is the turns ratio of Y source coupling inductor:

$$K = \frac{N_3 + N_1}{N_3 - N_2} \quad (10)$$

Analyzing the loop formed by V_in, C_o1, C_o2 and V_dc:
$$V_\text{dc} = V_\text{o1} + V_\text{o2} - V_\text{in} \quad (11)$$

the DC bus voltage can be obtained:

978-1-6654-4817-8/21 $31.00 © 2021 IEEE   308

$$V_{dc} = \frac{1 + K(D_1 + D_2) - (2K+1)D_1 D_2}{(1-D_1)(1-D_2)} V_{in} \quad (12)$$

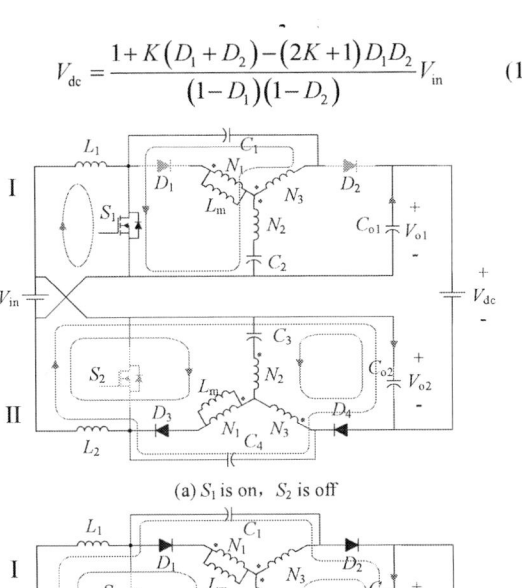

(a) S_1 is on, S_2 is off

(b) S_1 is off, S_2 is on

Fig. 3 The working mode diagram of the pre-circuit

B. Switch Stress Analysis

1) Switch front

According to the above deduction, it can be concluded that:

$$V_{sw1} = V_o - V_{C2} = \frac{K+G}{K+1} V_{in} \quad (13)$$

2) Switch rear

The Y source DC-DC topology behind the switch is shown in Figure 5:

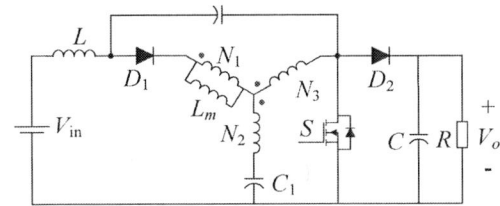

Fig. 4 Y source DC-DC topology after switch tube

Analyze the circuit to get:

$$V_{sw2} = V_o = G V_{in} \quad (14)$$

Because it is a boost circuit, the boost ratio $G>1$, we can get:

$$\frac{K+G}{K+1} < G$$

That is $V_{sw1} < V_{sw2}$, the voltage stress of the switch under the front of the switch is less than the voltage stress of the rear switch. The topology in front of the switch tube effectively reduces the voltage stress of the switch tube.

III. OPTIMAL DESIGN OF PLANAR MAGNETIC COMPONENTS

A. Optimized Design of Winding Arrangement

After analysis and calculation, the selected three-phase winding turns are: $N_1=6$, $N_2=3$, $N_3=6$. Considering the area and height of the magnetic core window comprehensively, five PCB are used to form the entire winding, and three-turn coils are arranged on each PCB.

When the alternating current flows through the conductor, the induced eddy current inside the conductor will have a proximity effect, which will increase the high-frequency AC loss between two adjacent conductors. The maximum magnetomotive force under different arrangements of multilayer planar windings is very different, and the winding losses caused are also different

The current relationship between the three-phase windings is: $I_1 : I_2 : I_3 = 1 : 4 : -3$ (positive and negative refer to the direction of the current). The three arrangements are marked as: N_1-N_1-N_2-N_3-N_3 (arrangement a), N_1-N_3-N_2-N_3-N_2 (arrangement b), N_3-N_1-N_2-N_1-N_3 (arrangement c). It can be seen from Fig. 8, that the windings are arranged according to the mode b, and the maximum magnetomotive force in the winding space is the smallest.

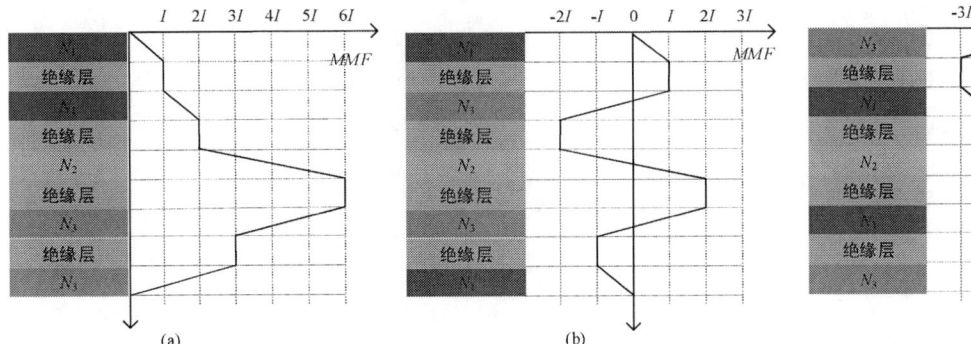
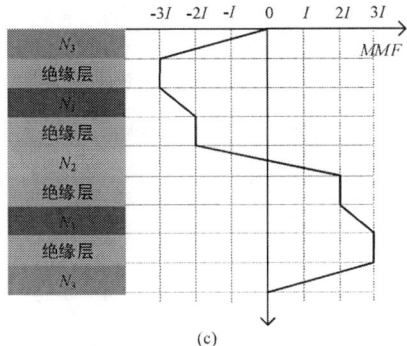

Fig. 5 Maximum magnetomotive force under different winding arrangements

B. Winding Leakage Inductance Analysis

The leakage inductance of the coupled inductor can be analyzed and defined from the perspective of energy: by solving the energy of each winding layer, the leakage inductance of the winding can be obtained:

$$E_{lk} = \frac{\mu_0}{2} \sum \int_0^h H^2 \cdot l_w \cdot b_w \cdot dx \qquad (15)$$

Where H is the field strength in the winding layer. The value of the field strength is constant along the plane of the winding layer and changes along the longitudinal direction of the winding layer.

$$H = \frac{I}{b_w} \cdot \frac{x}{h} \qquad (16)$$

According to (15) and (16), the leakage inductance of three different arrangements can be obtained:

$$L_{lka} = \mu_0 \cdot \frac{l_w}{b_w} \left(\frac{5h_1 + 28h_2 + 45h_3}{3} + 50h_\Delta \right) \qquad (17)$$

$$L_{lkb} = \mu_0 \cdot \frac{l_w}{b_w} \left(\frac{2h_1 + 16h_2 + 18h_3}{3} + 10h_\Delta \right) \qquad (18)$$

$$L_{lkc} = \mu_0 \cdot \frac{l_w}{b_w} \left(\frac{50h_1 + 16h_2 + 18h_3}{3} + 26h_\Delta \right) \qquad (19)$$

Among them, h_1, h_2 and h_3 are the thickness of the three-phase winding conductor, h_Δ is the thickness of the insulating layer, l_w is the length of each layer of winding conductor, and b_w is the width of each layer of winding conductor. Substituting the parameters, the calculation can be obtained: L_{lka} = 1285.13nH, L_{lkb} = 261.5nH, L_{lkc} = 677.7nH.

From the calculated data, it can be seen that the leakage inductance is the smallest under the arrangement b. Compared with the non-cross arrangement a, the leakage inductance is greatly reduced. Therefore, from the perspective of electromagnetic field energy, the total leakage energy of the windings under arrangement b is the smallest, and the leakage inductance is the smallest.

The following is a further simulation to verify the leakage inductance comparison of the three winding arrangements through ANSYS software, and the built ANSYS simulation model is shown in Fig. 6:

Fig. 6 Y source coupled inductor ANSYS simulation model

In the simulation model, different colored coils are used to simulate different copper windings. The copper thickness of each layer of winding is 1 ounce; the green plane layer is the simulated PCB with a thickness of 1mm;

the magnetic core model is a 1:1 EIW58.4 Magnetic core, the material is set to ferrite.

In the simulation under the 50KHz eddy current field, the leakage inductance data obtained under the three winding arrangements are shown in Table I:

Table I
ANSYS SIMULATION LEAKAGE INDUCTANCE DATA

	a	b	c
L_{lk1}/nH	623.7	273.74	423.3
L_{lk2}/nH	14.8	70.5	50.5
L_{lk3}/nH	595.7	15.97	232.2

L_{lk1}, L_{lk2}, and L_{lk3} in Table I are the leakage inductances of winding N_1, winding N_2, and winding N_3, respectively.

The simulation data shows that different winding crossing modes have obvious effects on the leakage inductance of the coupled inductor. Using arrangement a, the smallest leakage inductance of the N_2 winding can be obtained, but the value is only a slight difference of tens of nanohenries; for winding N_1 and winding N_3, the arrangement b has the smallest leakage inductance, because windings N_1 and N_3 are The leakage inductance accounts for a relatively large proportion, so the arrangement b can be used to obtain the best leakage inductance optimization results compared to the other two methods.

Based on the above theoretical analysis calculations and simulation verification results, the final selection of the winding arrangement of the planar Y source coupled inductor is N_1-N_3-N_2-N_3-N_2.

IV. EXPERIMENTAL VERIFICATION

In order to verify the feasibility of the output series high voltage gain impedance source circuit proposed in this paper, a system prototype was built, as shown in Fig. 7: (1), (2) and (3), (4) are two Y-source DC-DC conversion circuits and Y-source coupled inductors respectively, (5) is a full-bridge inverter circuit.

Fig. 7 Output series impedance source inverter prototype

Table II lists the models and parameter values of the main components of the experimental prototype.

Table II
The Main Component Models and Parameter Values of The Prototype

Parameters or Components	Parameter value or Model
L_1-L_3	1mH
S_1、S_2	12N80L-TF2-T
S_3-S_6	STF4N80K5
C_1-C_4	220μF
D_1-D_4	C3D02060A
C_{out}	2.2μF
R	200Ω

Table III shows the detailed design parameters of the Y-source planar coupled inductor, including the measured inductance and leakage inductance values of each phase winding.

Table III
Plane Y Source Coupled Inductor Parameters

	Winding N_1	Winding N_2	Winding N_3
Number of winding layers	2	1	2
Number of turns per layer	3	3	3
Winding Thickness /μm	35	35	35
Winding width /mm	5.08	5.08	5.08
Distance between layers /mm	1.365	1.365	1.365
Inductance value /mH	115.9	29.3	116.25
Leakage inductance /nH	344.7	34.3	22.7

Fig. 8 shows the output waveform of a single improved Y-source DC-DC circuit. The working parameters are: input voltage V_{in}=10V, switching frequency f=50KHz, switching duty cycle D=1/3.

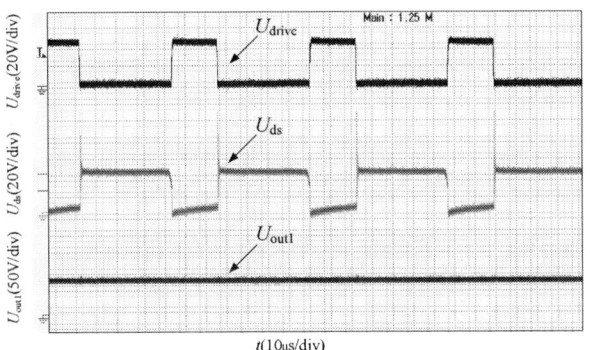

Fig. 8 Single Y source DC-DC circuit output waveform

The purple waveform is the output voltage U_{out1}=32.4V measured on the output capacitor C_{o1}, and the theoretical value U_{cal1}=34.96V calculated by (9). Considering the various losses of the components in the circuit, it can be considered that the measured value is basically equal to the theoretical value; The switch turn-off voltage $U_{ds\text{-}off}$=10.4V, which is basically equal to the input voltage V_{in}, and the voltage stress of the front switch has been effectively improved.

When the Y source coupling inductor turns ratio K=4, the switching frequency of the switch remains unchanged, and the duty cycle is fixed at D=1/3, change the input voltage V_{in}, measure the output voltage of the single Y source DC-DC circuit, and plot the output characteristic curve is shown in Fig. 9. It can be seen that within a reasonable input voltage range, the output voltage is in good agreement with the theoretical value.

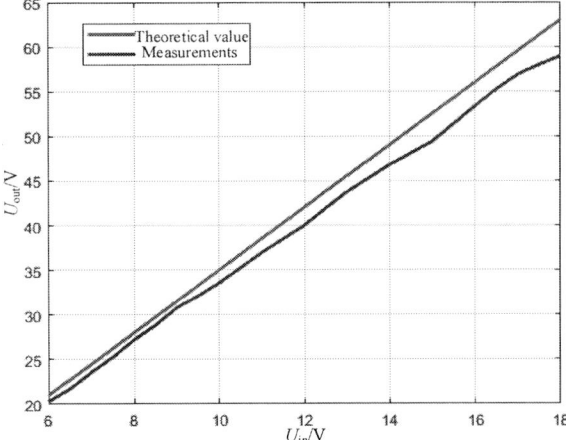

Fig. 9 Single Y source DC-DC circuit output characteristics(D=1/3)

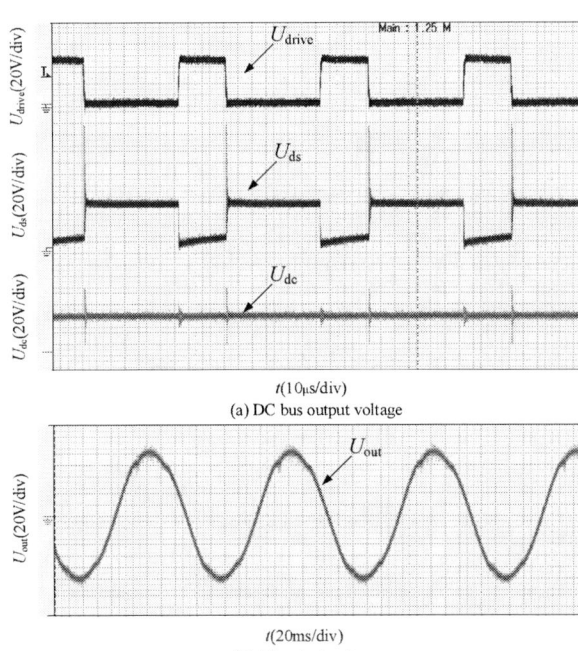

Fig. 10 Converter DC bus voltage and inverter output waveform

Fig. 10(a) is the voltage on the DC bus, that is the output series voltage of two single Y-source DC-DC topology, and Fig. 10(b) is the final AC output voltage U_{out}. At this time, the input voltage is V_{in}=10V, the duty cycle of the upper and lower switches are both D=1/3, and the voltage V_{dc}=51.2V on the DC bus is measured. The DC-DC output voltage of a single Y source measured in Fig. 8 is V_{out1}=32.4V. According to (12), it can be calculated that the theoretical value of the DC bus voltage at this time should be U_{cal2}=54.8V, Taking into account the loss factors such as the internal resistance of the planar inductor, the experimentally measured values are basically consistent with the theoretical values.

Therefore, in the proposed symmetrical design topology, it is feasible to obtain a higher voltage gain by connecting the outputs of two symmetrical Y-source DC-DC topologies in series.

V. CONCLUSION

This paper presents an output series type high voltage gain impedance source inverter circuit. The topology is a two-stage circuit. The output is connected in series through the two improved Y-source DC-DC circuits of the previous stage, which effectively improves the output voltage gain. And by adjusting the position of the switch, the stress of the switch is effectively reduced.

At the same time, a planar Y source coupled inductor is used, and the leakage inductance and winding loss are reduced after the winding arrangement is optimized. The produced experimental prototype has a DC bus voltage of 51.2V when the input is 10V, and it can maintain high efficiency within a reasonable range of the switch duty cycle.

REFERENCES

[1] J. M. Carrasco *et al.*, "Power-Electronic Systems for the Grid Integration of Renewable Energy Sources: A Survey," in *IEEE Transactions on Industrial Electronics*, vol. 53, no. 4, pp. 1002-1016, June 2006.

[2] S. M. Moosavian, N. A. Rahim and J. Selvaraj, "Photovoltaic power generation: A review," *2011 IEEE Conference on Clean Energy and Technology (CET)*, 2011, pp. 359-363.

[3] Y. P. Siwakoti, P. C. Loh, F. Blaabjerg, S. J. Andreasen and G. E. Town, "Y-Source Boost DC/DC Converter for Distributed Generation," in *IEEE Transactions on Industrial Electronics*, vol. 62, no. 2, pp. 1059-1069, Feb. 2015.

[4] M. Forouzesh, A. Baghramian and N. Salavati, "Improved Y-source inverter for distributed power generation," *2015 23rd Iranian Conference on Electrical Engineering*, 2015, pp. 1677-1681.

[5] Y. P. Siwakoti, P. C. Loh, F. Blaabjerg and G. E. Town, "Effects of Leakage Inductances on Magnetically Coupled Y-Source Network," in *IEEE Transactions on Power Electronics*, vol. 29, no. 11, pp. 5662-5666, Nov. 2014.

[6] H. Liu, Z. Zhou, K. Liu and W. Wang, "An Improved Y-Source Inverter with Continuous Input Current," *2018 IEEE 27th International Symposium on Industrial Electronics (ISIE)*, 2018, pp. 346-351.

[7] F. Xupeng, D. Xiaokang, L. Fengzhao and T. Yingying, "Improved Y-source Inverter," *2019 IEEE 10th International Symposium on Power Electronics for Distributed Generation Systems (PEDG)*, 2019, pp. 565-569.

[8] F. M. Shahir, E. Babaei and M. Farsadi, "Extended Topology for a Boost DC–DC Converter," in *IEEE Transactions on Power Electronics*, vol. 34, no. 3, pp. 2375-2384, March 2019.

[9] J. Yuan, A. Mostaan, Y. Yang, Y. P. Siwakoti and F. Blaabjerg, "A Modified Y-Source DC–DC Converter With High Voltage-Gains and Low Switch Stresses," in *IEEE Transactions on Power Electronics*, vol. 35, no. 8, pp. 7716-7720, Aug. 2020.

[10] Y. Ran, Y. Wang, W. Wang and H. Liu, "Energy-Shaping Control Strategy of the Improved Y-Source Inverter," *2018 21st International Conference on Electrical Machines and Systems (ICEMS)*, 2018, pp. 1082-1087.

[11] M. Mohammadi, J. S. Moghani and J. Milimonfared, "A Novel Dual Switching Frequency Modulation for Z-Source and Quasi-Z-Source Inverters," in *IEEE Transactions on Industrial Electronics*, vol. 65, no. 6, pp. 5167-5176, June 2018.

[12] Z. Ouyang, O. C. Thomsen and M. A. E. Andersen, "Optimal Design and Tradeoff Analysis of Planar Transformer in High-Power DC–DC Converters," in *IEEE Transactions on Industrial Electronics*, vol. 59, no. 7, pp. 2800-2810, July 2012.

[13] W. G. Hurley, E. Gath and J. G. Breslin, "Optimizing the AC resistance of multilayer transformer windings with arbitrary current waveforms," *30th Annual IEEE Power Electronics Specialists Conference. Record. (Cat. No.99CH36321)*, 1999, pp. 580-585 vol.1.

[14] Wei Chen, Yipeng Yan, Yuequan Hu and Qing Lu, "Model and design of PCB parallel winding for planar transformer," in *IEEE Transactions on Magnetics*, vol. 39, no. 5, pp. 3202-3204, Sept. 2003.

[15] A. Reatti and M. K. Kazimierczuk, "Comparison of various methods for calculating the AC resistance of inductors," in *IEEE Transactions on Magnetics*, vol. 38, no. 3, pp. 1512-1518, May 2002.

[16] W. G. Hurley, E. Gath and J. G. Breslin, "Optimizing the AC resistance of multilayer transformer windings with arbitrary current waveforms," in *IEEE Transactions on Power Electronics*, vol. 15, no. 2, pp. 369-376, March 2000.

A Compact Model for Si/SiC IGBT Implemented in LTspice

Md Maksudul Hossain
Electrical Engineering, University of Arkansas
Fayetteville, AR 72701, U.S.A
mh080@uark.edu

Arman Ur Rashid
Electrical Engineering, University of Arkansas
Fayetteville, AR 72701, U.S.A
aurashid@uark.edu

Yuqi Wei
Electrical Engineering, University of Arkansas
Fayetteville, AR 72701, U.S.A
yuqiwei@uark.edu

H. Alan Mantooth
Electrical Engineering, University of Arkansas
Fayetteville, AR 72701, U.S.A
mantooth@uark.edu

Abstract— A well-established unified physically-based compact model for IGBT has been implemented in LTspice, a notable free SPICE circuit simulator. A 12.5 kV n-channel SiC IGBT, a p-channel (13 kV) SiC IGBT, and a 1200 V/60 A field stop Si IGBT from IXYS have been used in this paper for model verification. The parameters have been extracted with ICCAP software using LTspice as an external simulator. This model combines both Si and SiC materials for n and p-channel IGBT that add to the flexibility of a circuit designer. This paper shows an LTspice implementation of a model published in [3]. It describes the MOSFET channel, internal bipolar transistor, and nonlinear capacitances to a high degree of accuracy. It also has temperature scaling capabilities, as well as a reliable yet straightforward datasheet-driven parameter extraction mechanism.

Keywords—Simulator, compact model, LTspice, IGBT, SiC

I. INTRODUCTION

Integrated gate bipolar transistors (IGBTs) are still the device of choice in high power applications thanks to their lower on-state power loss and higher blocking capability. Since IGBT is a 4-layer structure with a cascade of MOSFET and BJT, it shares the ease of gate control scheme that a MOSFET offers while harnesses the benefit of conductivity modulation that a bipolar device offers. Although the current turn-off is slower than MOSFET due to its longer recombination time, newer wide bandgap materials e.g., silicon carbide (SiC) offer better performance to overcome the silicon (Si) limitations. With the advent of newer materials, the need for compact modeling is becoming more compelling to aid the designer with the capability to parametric analysis before building an actual prototype. Despite the fact that compact modeling of silicon power IGBTs has been an active research area, SiC IGBT modeling for both n and p-channel is comparatively new. Reference [1] demonstrated finite element-based TCAD models considering most of the physical effects, however, these models are uniquely recommended for device studies, not for complex circuit simulations. The paper in [2] presented a physics-based SiC compact model in Saber®. However, a proprietary software package has been used to extract the parameters which limit the flexibility to some extent. Although most device manufacturers make SPICE models for their products, they often lack physical parameters and are less accurate in predicting device behavior. Moreover, sometimes the models are encrypted due to the non-disclosure of proprietary information makes the use limited to the model user in terms of the parametric study. In addition, the non-physical nature of the parameters makes the parameter extraction more tedious. This paper implements an LTspice version of the compact model in [3]. Both the dc and dynamic behavior of Si and SiC can be accurately predicted using this approach. The previously published paper in [3] was implemented in Verilog-A and MAST platform. Unfortunately, each of them requires commercial simulation software like Saber, HSpice, or Spectre which limits the broader reach of the user base. Since LTspice is a free simulator and very popular for fast simulation capability, the work presented in this paper will help the designers validate their design with greater confidence.

II. PHYSICAL STRUCTURE AND OPERATION PRINCIPLE

The device structure of a field stop IGBT is illustrated in Fig. 1. An extra P$^+$ layer distinguishes this from the power MOSFET structure. Instead of creating a body diode, it forms an internal PNP bipolar junction transistor (BJT) structure. As can be observed, the input side resembles the MOSFET while output will consist of a BJT controlled by the MOSFET channel current. That is the reason it is called a MOSFET driving a BJT. The IGBT operates in different modes as described below:

A. Reverse Blocking Mode

The junction between the P+ injection region - N- base region and the N$^+$ source region - P$^+$ body region becomes reverse biased, whereas the junction between the P$^+$ body region and the N$^-$ base region becomes forward biased during the reverse blocking mode. The majority portion of the blocking voltage stems from the epitaxial drift layer due to its low doping density. Punch-through (PT) IGBTs are distinguished from non-punch through (NPT) IGBTs by the existence of a N$^+$ buffer layer. The N$^+$ buffer zone in the PT structure serves to lower minority carrier lifespan and shorten the tail current during turn-off. However, it cannot withstand higher reverse blocking voltage while the NPT structure has a symmetry in both blocking operations.

978-1-6654-4817-8/21 $31.00 © 2021 IEEE

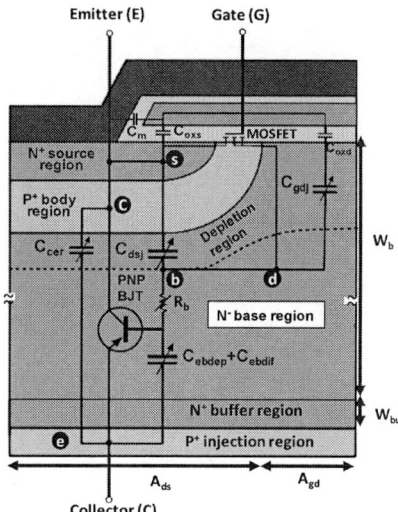

Fig. 1. Structure of a field stop (FS) IGBT with equivalent circuit elements.

B. Forward Blocking Mode

When the gate-emitter are shorted in forward blocking operation, the N^+ source – p^+-body junction and P^+ injection-N^- buffer junction are forward biased, while the P^+ body and N^- drift junction becomes reverse biased, and the latter blocks the forward voltage.

C. Conduction Mode

The P^+ body region causes the inversion as gate-emitter voltage rises beyond the threshold voltage, and that channel connects the MOSFET source and the N^- drift region. The drain current causes a voltage drop across the drift zone, which causes the P^+-N^+ junction to be forward biased. The drift region current is the internal PNP BJT's base current, resulting in a greater collector current. These holes in the drift region attract electrons from the N^+- source. Conductivity modulation is the result of both carriers creating a low resistance base region. This is why IGBTs do not suffer from the $R_{ds(on)}$ vs. breakdown voltage constraint that power MOSFETs do, because the carriers of both polarities effectively reduce the on resistance even if the drift resistance grows at higher breakdown voltage devices.

III. IGBT MODELING APPROACH

A. MOSFET Component

As described in section II, the input of the IGBT is a MOSFET. The MOSFET operating regions are described with (1) as described in [3].

$$I_{mos_{V_{ds} \geq 0}} = \begin{cases} 0, V_{gs} < vt \\ \frac{kp \cdot kf \cdot \left(V_{gs} - vt - kf \frac{V_{ds}}{2}\right) \cdot V_{ds}}{mufact}, V_{ds} \leq \frac{V_{gs} - vt}{kf} \\ \frac{0.5 \cdot kp \cdot (V_{gs} - vt)^2}{mufact}, V_{ds} \geq \frac{V_{gs} - vt}{kf} \end{cases} \quad (1)$$

Here *mufact* accounts for the vertical field effect on mobility while *kp*, *kf* are transconductance parameters, *vt* is the threshold voltage. The circuit depicted in Fig. 1 illustrates the inter-electrode capacitances i.e., gate-source (C_{gs}), Miller (C_{gd}), and drain-source (C_{ds}) capacitance. C_{gs} consists of the oxide overlap portion of the source in parallel with the metal electrode capacitance. Cgd is created by combining the oxide

capacitance and the depletion capacitance in series. Cds is the same as internal base-collector capacitance. After taking the derivative of the computed charges, the current contribution has been calculated in the model.

B. BJT Component

The BJT has three current components e.g., base, collector, and emitter current. They can be expressed with (2) [3].

$$ibp = \frac{qceb}{tauhl} + \frac{4 \cdot qceb^2 \cdot nb^2 \cdot isne}{qb^2 \cdot ni^2}$$

$$rb = \begin{cases} \frac{w}{mun \cdot a \cdot q \cdot nb} + rs, qceb \leq 0 \\ \frac{w}{mueff \cdot a \cdot q \cdot neff} + rs, qceb > 0 \end{cases}$$

$$icp = \frac{irb}{(1+b)} + \frac{b}{(1+b)} \cdot \left(\frac{4 \cdot dp \cdot qceb}{w^2}\right) \quad (2)$$

Here, *qceb* represents the emitter-base charge, *qb* represents the intrinsic doping concentration, *w* is the width of the base, *rs* is the series resistance parameter, *mun* is the electron mobility, *mueff* is the bipolar mobility, and *neff* is the base donor concentration. The effective mobility is calculated using on the equation in reference [3] which results from conductivity modulation in the base.

The emitter-base charge was solved such that expression (3) holds true:

$$V_{ebj} = V_{eb} + voff \quad (3)$$

where V_{ebj} is given as the piecewise relation in (4) for reverse conduction, forward conduction and strong on-state regions [3]:

$$V_{ebj} = \begin{cases} V_{ebdep}, qceb < 0 \\ \frac{qceb \cdot V_{ebdif}}{qceb0}, 0 \leq qceb \leq qceb0 \\ V_{ebdif}, qceb \geq qceb0 \end{cases} \quad (4)$$

Here V_{ebdep} is the emitter-base depletion voltage, V_{ebdif} is the emitter-base diffusion voltage, *qceb0* is the zero bias base charge.

C. Field-stop Layer Component

The field-stop layer has been implemented by the equations described in [4]. The base's excess minority carrier was divided into two sections for the base and filed-stop layer. They are related with *qceb* and *ql* in a steady state.

D. SiC Modeling

The material properties of SiC have been incorporated in the equations. The temperature-dependent intrinsic carrier concentration is as follows:

$$ni_{SiC} = 1.7 \times 10^{16} \cdot \frac{templim^{1.5}}{e^{(20800/templim)}} \quad (5)$$

For temperature-dependent mobility, equations in [5-6] have been utilized. The dielectric constant has also been added.

E. Model Formulation

The model was implemented in LTspice following the equivalent circuit presented in Fig. 1. In the code, conditional statements dictate the use of p or n-channel IGBT or Si and

SiC material selection. The common nodes like b and d are combined inside the model. The equations are implemented in the same way described in [7]

IV. IMPLEMENTATION IN LTSPICE

A. Voltage-dependent Current Source

Implementing the unified IGBT in LTspice is not very similar to MAST (modeling language for Saber) implementation. Since this model includes a buffer layer, there are two implicit equations involved for calculating the emitter-base charge ($qceb$) which requires an iterative solution. This charge is responsible for the tail current during IGBT switching turn-off. This iterative solution is implemented in MAST the following way:

$$qceb: V_{ebj} = V_{eb} \qquad (6)$$

Equation (6) states "Find qceb so that the predicted value of emitter-base junction voltage equals the emitter-base terminal voltage". In LTspice this function is not directly available. As a result, the implicit equation was implemented in the following manner:

$$
\begin{aligned}
Evqceb \; qceb \; 0 \; value \qquad (7)\\
= \{vied + voff \\
- vebj(vde, v(qceb), vcie)\}
\end{aligned}
$$

It can be noticed that an E source named Evqceb has been declared which represents $qceb$. This is a floating branch and connected to the ground. The E source is an alternative to the behavioral source in LTspice. An E source in LTspice can be both a voltage-dependent voltage source and an alternative of the behavioral BV source depending on the syntax. Expressions can contain the through and across variables of circuit elements, conditionals, parameters, and functions. The node voltage is v(qceb) which needs to be solved. However, the expression on the right also depends on this node voltage v(qceb). Therefore, these two are solved simultaneously. This implementation is LTspice-equivalent implementation of an "implicit equation". This approach is very useful in any bipolar device e.g., a power diode whenever the reverse recovery current is implemented. [8]

The capacitance has been implemented using the special syntax as follows:

$$
\begin{aligned}
Cgs \; gate \; emitter \; Q = if(channel = \qquad (8)\\
= 1, CapON * cgs \\
* x, -CapON * cgs * x)
\end{aligned}
$$

This equation is ideally suitable for models that considers computing speed as the key benchmark [9]. It can be noted that this is not a built-in SPICE feature, but an additional feature available within LTspice.

The base resistance was modeled using "voltage-dependent variable resistances", a useful component in compact modeling in LTspice. [10] Rb has been implemented with the following equation:

$$
\begin{aligned}
Rrb \; collector \; internal_{emitter} \; R = if(rb(drain, emitter), v(qceb) \qquad (9)\\
\leq 0.0, v(collector, internal_{emitter}), rb(v(drain, emitter)), v(qceb))
\end{aligned}
$$

The MOSFET drain current and BJT collector current have been described by a G-source by using the "value=<expression>" syntax. Based on the gate voltage, and drain-source volatge, a conditional statement defines the device's operational region.

$$
\begin{aligned}
Gmos \; drain \; Emitter \; value \qquad (10)\\
= \{imos(vde, vg) \\
+ imult(vde, vge, vqceb, vcie, viee)\}
\end{aligned}
$$

The capacitance has been implemented using the special syntax as follows:

$$
\begin{aligned}
Cgs \; gate \; emitter \; Q = if(channel = \qquad (11)\\
= 1, CapON * cgs \\
* x, -CapON * cgs * x)
\end{aligned}
$$

V. PARAMETER EXTRACTION

The parameter extraction was performed with IC-CAP software from Keysight. Although LTspice is not a native simulator in IC-CAP, it can be added as a spice3 simulator. Minimax and gradient algorithms have been selected for the dc output and transfer curves, respectively. An RMS error of 1×10^{-3} has been selected as the stopping criterion. The extraction procedure in [3] has been followed.

VI. EXPERIMENTAL VALIDATION

The model in [3] was implemented in Verilog-A. In this paper, the LTspice version implements the same physical equations in SPICE specific manner. Fig. 2 and 3 demonstrate the equivalence of each model prediction.

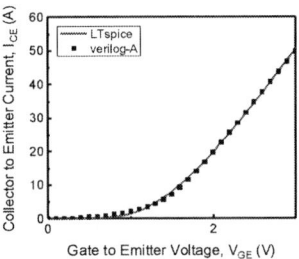

Fig. 2. Equivalence of Verilog-A [3] and the LTspice model predicted transfer function with a Si NPT IGBT from IXYS.

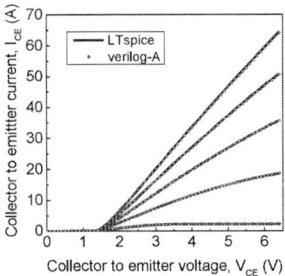

Fig. 3. Equivalence of Verilog-A [3] and the LTspice model predicted dc output characteristics.

For model validation, a 1.2 kV field-stop Si IGBT, a 12.5 kV SiC n-IGBT [11], a SiC p-IGBT (13 kV) [12] were employed. Fig. 4 shows the static fit of the model for both SiC n and p-IGBT. The simulated and empirical capacitance values are compared in Fig. 5 while Fig. 6 presents the gate charge comparison of a silicon n-IGBT. The clamped load switching test for gate charge measurement used $V_{CE} = 600$ V and $I_{CE} = 25$ A [3]. The charge was calculated by integrating the gate current. Fig. 7 presents the turn-off transient of a SiC n-IGBT at 125°C. The two humps in the curve originate from the ceasing of holes in the first stage and then the "reaching through" to the field stop layer which reshapes the electric field and increases the voltage faster. This behavior has been

978-1-6654-4817-8/21 $31.00 © 2021 IEEE

captured by incorporating the buffer equations described in reference [4]. The resulting empirical match suggests the accuracy of the proposed model.

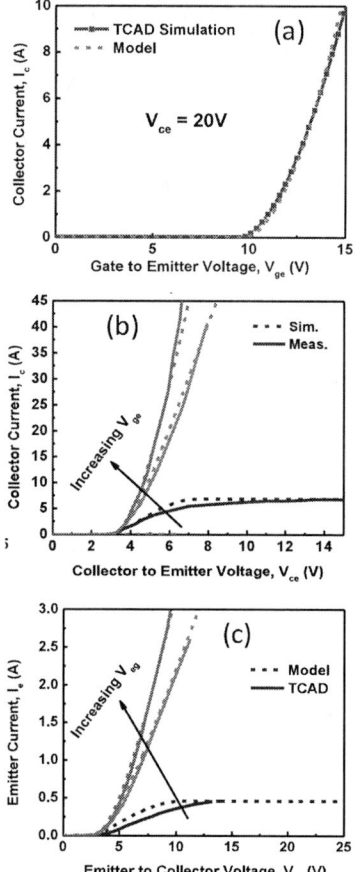

Fig. 4 Model fit results for (a) SiC IGBT (n-channel) transfer curve (b) SiC IGBT (n-channel) dc output curve (c) p-channel SiC IGBT dc output characteristics.

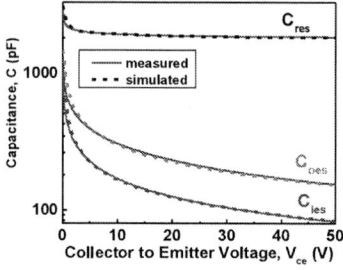

Fig. 5 Capacitance-voltage fit for an n-channel Si-IGBT.

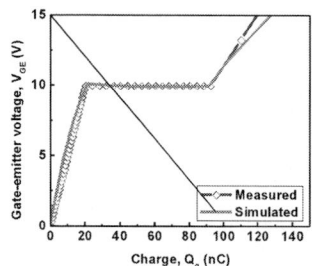

Fig. 6 N-channel Si gate charge model vs. measured data comparison.

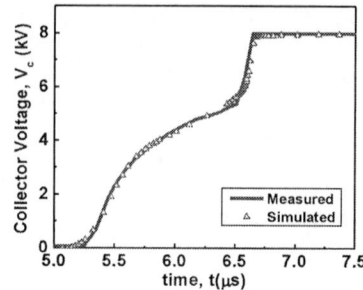

Fig. 7 N-channel SiC turn off transient at 125°C.

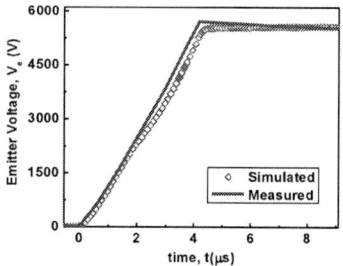

Fig. 8 P-channel SiC turn off transient at 250°C.

VII. APPLICATION SIMULATION

A typical double pulse test (DPT) is simulated with the model. Fig. 8 demonstrates the collector current and voltage waveforms. The switching power and energy can be calculated from (12). The steps in the energy waveform correlates to the turn-on and turn-off switching loss, while the linear region indicates conduction loss, as seen in Fig. 9.

$$E_{loss}(t) = \int I_{CE}(t) V_{CE}(t) dE \qquad (12)$$

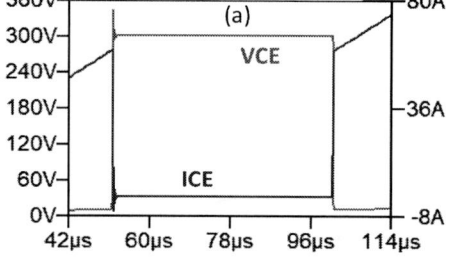

Fig. 9 Double pulse test simulation waveforms.

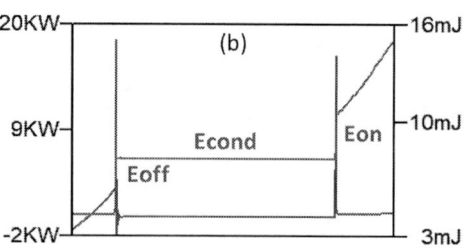

Fig. 10 Transient power and energy loss calculation.

VIII. CONCLUSION

This work extends the prior work in [3] by implementing an IGBT model in LTspice, a free and popular simulator. The parameters have been extracted using LTspice as a simulator in IC-CAP software. The model has been validated against a practical case study of both n and p-type SiC and an n-channel Si-IGBTs which will give the designer more flexibility in analyzing circuits before prototyping. Furthermore, the physics-based equations will provide intuitive predictions across a broad range of operating areas without the need for precise device design and fabrication data.

ACKNOWLEDGMENT

This material is based upon work supported in part by the NSF S-STEM grant number DUE- 0728636. Any opinions, findings, and conclusions, or recommendations expressed in this material are those of the authors and do not necessarily reflect the views of the National Science Foundation.

REFERENCES

[1] T. Hatakeyama, K. Fukuda and H. Okumura, "Physical Models for SiC and Their Application to Device Simulations of SiC Insulated-Gate Bipolar Transistors," in *IEEE Transactions on Electron Devices*, vol. 60, no. 2, pp. 613-621, Feb. 2013, doi: 10.1109/TED.2012.2226590.

[2] T. H. Duong *et al.*, "Physics-based electro-thermal Saber model and parameter extraction for high-voltage SiC buffer-layer IGBTs," *2014 IEEE Energy Conversion Congress and Exposition (ECCE)*, Pittsburgh, PA, USA, 2014, pp. 460-467, doi: 10.1109/ECCE.2014.6953430.

[3] S. Perez *et al.*, "A Datasheet Driven Unified Si/SiC Compact IGBT Model for N-Channel and P-Channel Devices," in *IEEE Transactions on Power Electronics*, vol. 34, no. 9, pp. 8329-8341, Sept. 2019, doi: 10.1109/TPEL.2018.2889263.

[4] A. R. Hefner, "Modeling buffer layer IGBTs for circuit simulation," *Power Electronics, IEEE Transactions on*, vol. 10, no. 2, pp. 111–123, 1995.

[5] S. Kagamihara, H. Matsuura, T. Hatakeyama, T. Watanabe, M. Kushibe, T. Shinohe, and K. Arai, "Parameters required to simulate electric characteristics of SiC devices for n-type 4H–SiC," *Journal of Applied Physics*, vol. 96, no. 10, p. 5601, 2004.

[6] T. Hatakeyama, K. Fukuda, and H. Okumura, "Physical Models for SiC and Their Application to Device Simulations of SiC Insulated-Gate Bipolar Transistors," *IEEE Transactions on Electron Devices*, vol. 60, no. 2, pp. 613–621, Feb. 2013.

[7] M. M. Hossain, L. Ceccarelli, A. U. Rashid, R. M. Kotecha and H. A. Mantooth, "An Improved Physics-based LTSpice Compact Electro-Thermal Model for a SiC Power MOSFET with Experimental Validation," *IECON 2018 - 44th Annual Conference of the IEEE Industrial Electronics Society*, 2018, pp. 1011-1016, doi: 10.1109/IECON.2018.8592522.

[8] A. Rashid, M. M. Hossain, A. Emon and H. E. C. A. Mantooth, "Datasheet-driven Compact Model of Silicon Carbide Power MOSFET Including Third Quadrant Behavior," in *IEEE Transactions on Power Electronics*, doi: 10.1109/TPEL.2021.3062737.

[9] B. W. Nelson *et al.*, "Computational Efficiency Analysis of SiC MOSFET Models in SPICE: Dynamic Behavior," in *IEEE Open Journal of Power Electronics*, vol. 2, pp. 106-123, 2021, doi: 10.1109/OJPEL.2021.3056075.

[10] M. M. Hossain, L. Ceccarelli, A. U. Rashid, R. M. Kotecha and H. A. Mantooth, "An Improved Physics-based LTSpice Compact Electro-Thermal Model for a SiC Power MOSFET with Experimental Validation," *IECON 2018 - 44th Annual Conference of the IEEE Industrial Electronics Society*, 2018, pp. 1011-1016, doi: 10.1109/IECON.2018.8592522.

[11] H. Ryu, C. Capell, C. Jonas, L. Cheng, M. O'Loughlin, A. Burk, A. Agarwal, J. Palmour, and A. Hefner, "Ultra high voltage (>12 kV), high performance 4H-SiC IGBTs," in *2012 24th International Symposium on Power Semiconductor Devices and ICs (ISPSD)*, 2012, pp. 257–260.

[12] T. Deguchi, T. Mizushima, H. Fujisawa, K. Takenaka, Y. Yonezawa, K. Fukuda, H. Okumura, M. Arai, A. Tanaka, S. Ogata, T. Hayashi, K. Nakayama, K. Asano, S.-I. Matsunaga, N. Kumagai, and M. Takei, "Static and dynamic performance evaluation of >13 kV SiC p-channel IGBTs at high temperatures," in *2014 IEEE 26th International Symposium on Power Semiconductor Devices IC's (ISPSD)*, 2014, pp. 261–264.

Temperature-Dependent Current Collapse and Gate Leakage in AlGaN/GaN HEMTs With Si-rich SiN Interlayer

Jielong Liu, Yuwei Zhou
School of Advanced Materials and Nanotechnology, Xidian
University
Xi'an 710071, People's Republic of China
liujielong@stu.xidian.edu.cn

Minhan Mi, Jiejie Zhu, Siyu Liu, Qing zhu, Pengfei
Wang, Hong Wang, Xiaohua Ma, Yue Hao
School of Microelectronics, Xidian University
Xi'an 710071, People's Republic of China

Abstract—In this paper, SiN passivation with Si-rich SiN interlayer has been studied in AlGaN/GaN HEMTs. The temperature-dependent TLM revealed that sheet resistance for HEMTs with Si-rich SiN increases from 310 Ω/sq at 300 K to 794 Ω/sq at 420 K (without Si-rich SiN increases from 341 Ω/sq at 300 K to 911 Ω/sq at 420 K). The Si-rich SiN interlayer resulted in reduced off-state current of approximately two orders of magnitude. At pulse (-8 V, 40 V) measurement, HEMTs without Si-rich SiN passivation exhibit significant current collapse (~48.1%), while HEMTs with Si-rich SiN passivation show small current collapse (~7.5%) at 420 K. The reverse gate leakage of AlGaN/GaN HEMTs is dominated by Poole-Frenkel (PF) emission mechanism. The extracted trap energy level (ϕ_t) for HEMTs without Si-rich SiN and with Si-rich SiN are 0.65 eV and 0.26 eV, respectively. This paper has shown that the capability of bilayer SiN to effectively suppress gate leakage and current collapse in AlGaN/GaN HEMTs.

Keywords—AlGaN/GaN HEMTs, Si-rich SiN, current collapse, temperature-dependent TLM, Poole-Frenkel (PF) emission.

I. INTRODUCTION

Gallium nitride (GaN) based high-electron-mobility transistors (HEMTs) are one of the most expected candidates for next generation power and microwave applications due to their large bandgap, high breakdown voltage, high electron mobility, and high electron density [1]-[3]. However, GaN based HEMTs encounter several obstacles of realibility and stability. For example, the gate leakage and the drain current collapse are badly in need of immediate solution [4]. Previous studies have demonstrated that current collapse was attributed to the high density of surface traps under a negative gate voltage, cause a virtual gate, which deplete channel electrons and increase on-state resistance simultaneously [5], [6]. Miscellaneous approaches were reported to suppress current collapse in AlGaN/GaN HEMTs, such as surface passivation, surface pretreatment, and field-plate structures [7].

Surface passivation technology using insulator films such as SiO_2, SiN, AlN, Al_2O_3 and others, are significant process for achieving high stability and reliability. Among all kinds of passiation materials to suppressing current collapse, SiN passivation is the most popular scheme because of low leakage current and high stability [8]-[10]. T. Huang has demonstrated that bilayer LPCVD SiN delivers smooth appearance and suppressed current collapse. Bilayer SiN passivation has been demonstrated that pulsed I-V and RF load-lines are greatly reduced owing to Si-rich SiN [11]. Bilayer SiN passivation has been applied in W band devices and achieved excellent output power density of 3 W/mm in W band [12]. However, they only researched surfacce

passivation at room temperature. The passivation effect in GaN HEMTs is significantly affected by temperature. Because the trap has different energy levels at different temperatures [13], so it is necessary to analyse the temperature-dependent of current collapse and gate leakage for RF power devices. In this study, we report a comparative investigation of HEMTs with and without Si-rich SiN interlayer as surface passivation. We find that Si-rich SiN is an effective passivation technology, suppressed current collapse and reduced gate leakage at high temperature.

II. DEVICE FABRICATION

Fig. 1(a) exhibits the cross section diagram of $Al_{0.25}Ga_{0.75}N$/GaN HEMTs with Si-rich SiN. The epitaxial layers were grown on 3-inch sapphire substrate by metal organic chemical vapor deposition, which consists of a 2 um undoped GaN buffer layer, a 1 nm AlN spacer and a 20 nm $Al_{0.25}Ga_{0.75}N$ barrier layer. The electron sheet charge density and mobility of this structure were $8.9\times10^{12}cm^{-2}$ and 1820 $cm^2 \cdot V^{-1} \cdot s^{-1}$ by room temperature Hall measurement, respectively. A pair of ohmic contacts formation using an alloyed Ti/Al/Ni/Au (=20/160/55/45 nm) metal stack was fabricated, followed by rapid thermal annealing (RTA) at 830 °C for 30 s in N_2 gas. The ohmic contact (R_c)=0.4 $\Omega \cdot mm$ using transmission line measurement (TLM). 110 nm mesa etch depth was carried out by inductively coupled plasma (ICP) . Then, SiN was deposited by plasma-enhanced chemical vapor deposition (PECVD). The pressure, platen power and deposition temperature were maintained at 600 mToor, 22 W and 250 ℃, respectively. The only step which varied in the process was the stoichiometry of the PECVD SiN passivation layer. For conventional SiN structure, SiN structure was passivated with an 60 nm standard stoichiometric Si_3N_4 (STO-SiN, 2%SiH_4:NH_3=100:2 sccm). For bilayer SiN structure, the first layer of Si-rich SiN (2%SiH_4:NH_3=450:2 sccm) with a thickness of 15 nm and the second layer of standard stoichiometric Si_3N_4 (STO-SiN) (2%SiH_4:NH_3=100:2 sccm) with a thickness of 45 nm. The first thin Si-rich SiN film was applied in order to improve the SiN/GaN interface properties, because it contacts device surface and affects the density of surface dangling bond and trap. A T-shaped gate structure was formed on the opening in the SiN film. The Ni/Au/Ni (=45/200/20 nm) multilayer gate electrode was formed by electron beam evaporation with footprint of 1 um, stem of 60 nm high and head of 1.5 um. The gate-source, gate-drain distance, and gate width are 3 um, 3 um, and 100 um.

978-1-6654-4817-8/21 $31.00 © 2021 IEEE

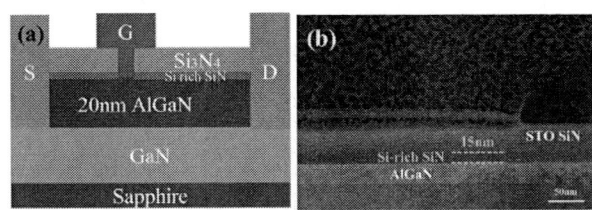

Fig. 1. (a) cross section diagram of AlGaN/GaN HEMTs with Si-rich SiN Interlayer. (b) SEM of HEMTs with Si-rich SiN Interlayer.

III. RESULTS AND DISCUSSION

As can be seen from Fig. 1(b), 15 nm Si-rich SiN has obvious stratification with 45 nm STO SiN using scanning electron microscopy (SEM). The refractive index values of STO SiN and Si-rich SiN films are 1.9 and 2.4 by spectroscopic ellipsometer at a wavelength of 632 nm, respectively. In order to assess the composition of the SiN and analyse SiN film quality, Fourier transform infrared (FTIR) spectroscopy was measured using a Nicolet iS 50 FTIR from 400 cm^{-1} to 4000 cm^{-1} as shown in Fig. 2(a). The relative content of Si–H in Si-rich SiN is approximately three times that in STO-SiN. It has been reported that excess Si in the high refractive index SiN film reacts with Ga-O in the interface of SiN/GaN, which reduces the density of Ga-O at the interface, essentially reducing the O impurity at the interface and suppressing current collapse [12].

Keithley 4200 semiconductor characterization system was utilized for dc and pulse measurements. In order to determine the mechanism of SiN passivation, the temperature-dependent sheet resistance was carried out by transmission line model, as shown in Fig. 2(b). At 300K, the sheet resistances are 310 Ω/sq and 341 Ω/sq for HEMTs with and without Si-rich SiN, respectively. The slope of the sheet resistance versus temperature are 4.73 Ω/(sq·K) and 4.05 Ω/(sq·K) for HEMTs with and without Si-rich SiN, respectively.

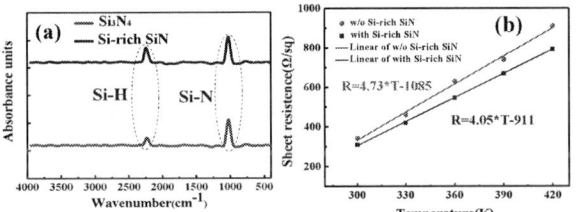

Fig. 2. (a) FTIR absorption spectrum of Si_3N_4 and Si-rich SiN films. (b) Temperature-dependent R_{sh} extracted with 100 μm metal pad width by TLM.

The measured transfer characteristics of the fabricated transistors, for a gate bias (V_g) swept from -10 V to +2 V at drain-source bias (V_{ds}) of +10 V, are shown in Fig. 3(a). Defining threshold voltage (V_{th}) as gate voltage with drain current of 1 mA/mm, the V_{th} of HEMTs with and without Si-rich SiN are both -3.9 V. The off-state current (V_g=-10 V) of HEMTs with Si-rich SiN is 3.8×10^{-4} mA/mm and that of HEMTs without Si-rich SiN is 7.9×10^{-3} mA/mm, respectively. It is worth noting that the off-state current of HEMTs without Si-rich SiN increases exponentially with the increase of reverse gate voltage, in contrast, the increment of bilayer SiN HEMTs remains small. The extrinsic peak transconductance (G_m) are 172 mS/mm and 168 mS/mm for HEMTs with and without Si-rich SiN, respectively. The

maximum drain current density is 727mA/mm for HEMTs with Si-rich SiN as shown in Fig. 3(b), which is a bit of larger than 692 mA/mm for HEMTs without Si-rich SiN due to its reduced sheet resistance.

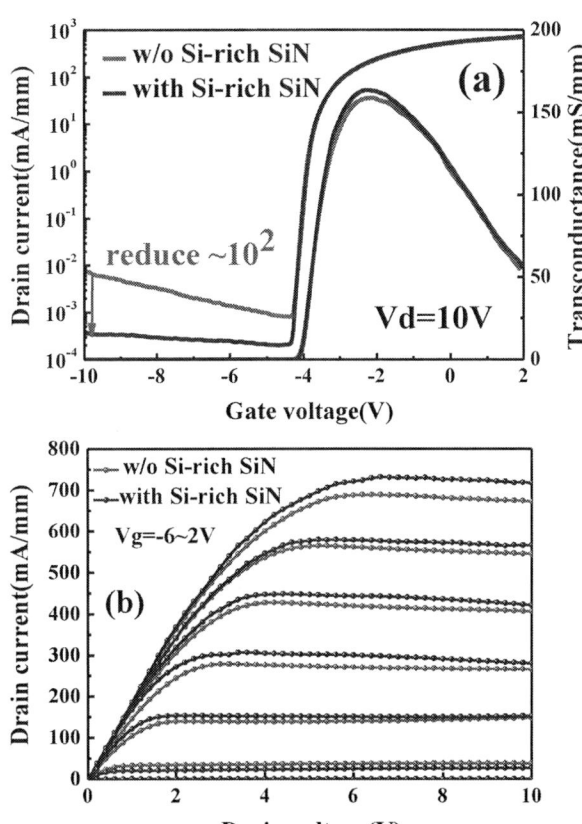

Fig. 3. (a) Transfer curves of the AlGaN/GaN HEMTs without and with Si-rich SiN. (b) Comparison of output characteristics between HEMTs without and with Si-rich SiN.

For the double pulse measurement, different quiescent bias points were carried out. The pulse period and pulse width were 1 ms and 500 ns, respectively. Compared to devices without Si-rich SiN, devices without Si-rich SiN exhibit small current collapse at 300 K. Compared to pulse (0 V, 0 V), saturated drain current density of HEMTs with Si-rich SiN shows a small decline (~9.4%), while saturated drain current density of HEMTs without Si-rich SiN exhibits an obvious decline (~17.5%) at ($V_{gs.q}$, $V_{ds.q}$) = (-8 V, 40 V).

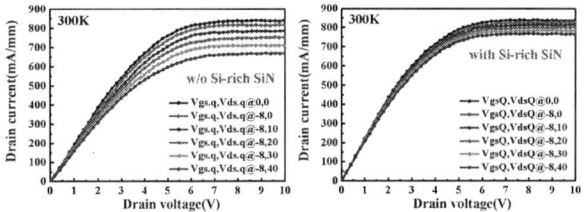

Fig. 4. Pulsed I-V curves at V_{gs} = 2 V tested under different quiescent bias conditions for AlGaN/GaN HEMTs (a) without and (b) with Si-rich SiN.

To confirm the impact of Si-rich SiN on the device temperature-dependent current collapse, the dc and ($V_{gs.q}$, $V_{ds.q}$) = (-8 V, 40 V) pulsed characteristics of two samples were measured at 300 K to 420 K temperature range as shown in Fig. 4 and Fig. 5. We evaluated the influence of SiN passivation on current collapse by comparing maximum current. The maximum current of both sample decreases with the temperature increasing. At 420 K temperature, it is observed that the current collapse of HEMTs with Si-rich SiN is 7.5%, whereas that of HEMTs without Si-rich SiN is 48.1%. The temperature-dependent current collapse phenomenon can be effectively suppressed by utilizing Si-rich SiN interlayer for AlGaN/GaN HEMTs. It is demonstrated that the current collapse of HEMTs with Si-rich SiN is better than that of HEMTs without Si-rich SiN when the devices operate at high temperature.

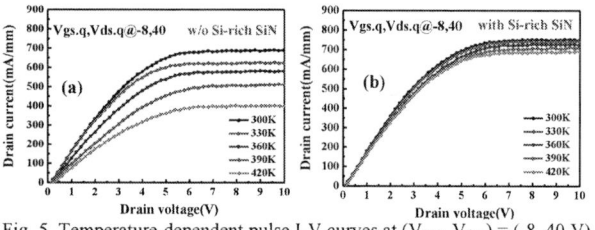

Fig. 5. Temperature-dependent pulse I-V curves at ($V_{gs.q}$, $V_{ds.q}$) = (-8, 40 V) for AlGaN/GaN HEMTs (a) without and (b) with Si-rich SiN.

The AlGaN/GaN schottky gate characteristics at different temperatures are shown in Fig. 6. The forward current of ring schottky diodes with Si-rich SiN is similar to that of ring schottky diodes without Si-rich SiN. The characteristics of the reverse voltage are quite different, the reverse leakage current of ring schottky diodes without Si-rich SiN exibits a strong temperature-dependent and voltage-dependent relationship, whereas that of ring schottky diodes with Si-rich SiN exibits a weak temperature-dependent and voltage-dependent relationship. So, we consider Poole- Frenkel (PF) emission as the dominant mechanism of reverse leakage current.

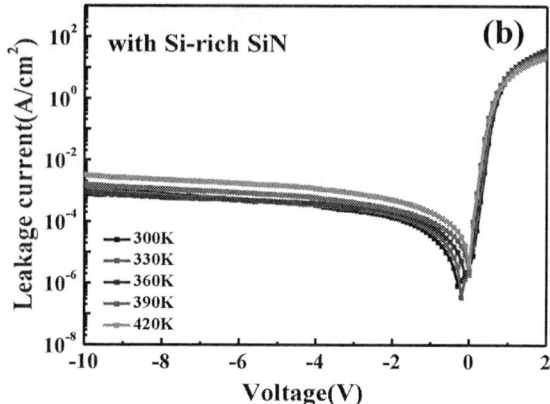

Fig. 6. Temperature-dependent I-V curves of AlGaN/GaN ring schottky diodes (a) without and (b) with Si-rich SiN.

The current density (J_{PF}) can be decribed by the following expression: [14], [15]

$$In(\frac{J_{PF}}{E}) = m(T)\sqrt{E} + c(T) \qquad (1)$$

$$m(T) = \frac{q}{kT}\sqrt{\frac{q}{\pi \varepsilon_i}} \qquad (2)$$

$$c(T) = -\frac{q\phi_t}{kT} + In(C) \qquad (3)$$

where C is a constant, ϕ_t is the trap energy level and ε_i is the permittivity of the AlGaN barrier at high frequency, E is the electric field. From the plots of c(T) versus q/kT in Fig. 7(a) and (b), the extracted values of ϕ_t are 0.65 eV and 0.26 eV for AlGaN/GaN HEMTs without and with Si-rich SiN passivation, respectively. Evidently, the deep trap level of surface traps changed to shallow trap level, making it easier to detrap electrons. It is consistent with significant difference in current collapse for two samples.

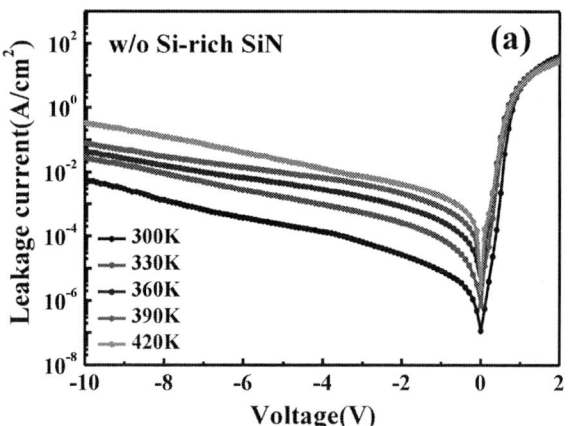

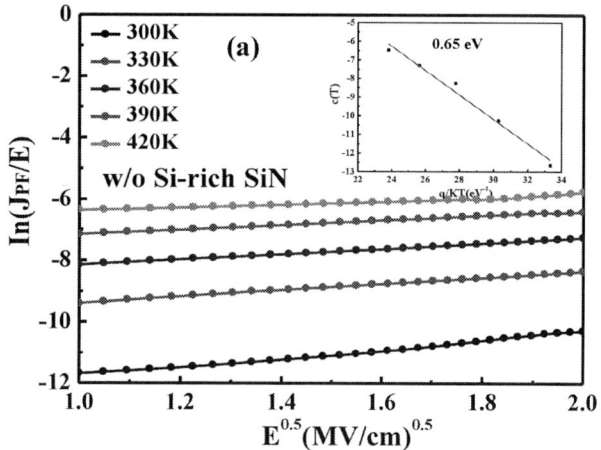

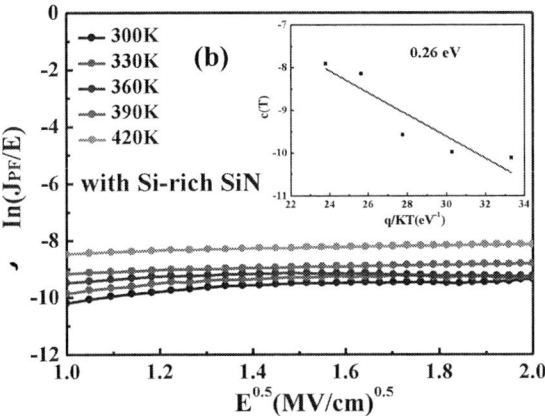

Fig. 7. PF mission plots for reverse I-V characteristics of AlGaN/GaN ring schottky diodes (a) without and (b) with Si-rich SiN passivation. The inset shows dependence of c(T) on q/kT.

IV. CONCLUSION

AlGaN/GaN HEMTs with Si-rich SiN and without Si-rich SiN passivation were fabricated in this paper. After SiN interlayer was adopted, the sheet resistance at 300 K was decreased from 341 Ω/sq to 310 Ω/sq and the slope of Rsh versus T was reduced from 4.73 Ω/(sq·K) to 4.05 Ω/(sq·K). At pulsed (-8 V, 40 V), the HEMTs with Si-rich SiN exhibit serious current collapse (~48.1%), while the HEMTs with Si-rich SiN only shows small current collapse (~7.5%). The devices gate leakage was mitigated by inserting an Si-rich SiN, which can be ascribed to the reduced electron trap level (from 0.65 eV to 0.26 eV).

ACKNOWLEDGMENT

This work was supported by the National Key R&D Program of China under Grant 2020YFB1804902 and Grant 11690042, in part by the National Natural Science Foundation of China under grant 61904135, in part by the China Postdoctoral Science Foundation under Grant 2018M640957 and BX20200262, in part by the Nature Science Foundation of Shaanxi Province under Grant 2020JQ-316.

REFERENCES

[1] S. Ho, C. Lee, A. Tzou, H. Kuo, Y. Wu, and J. Huang, "Suppression of Current Collapse in Enhancement Mode GaN-Based HEMTs Using an AlGaN/GaN/AlGaN Double Heterostructure," *IEEE Trans. Electron Devices*, vol. 64, no. 4, pp. 1505-1510, April 2017, doi: 10.1109/TED.2017.2657683.

[2] M. Asif Khan, A. Bhattarai, J. N. Kuznia, and D. T. Olson, "High electron mobility transistor based on a GaN-AlxGa1-xN heterojunction," *Appl. Phys. Lett.*, vol. 63, no. 12, Jun. 1993, Art. no. 1214. doi: 10.1063/1.109775.

[3] Y. F. Wu, M. Moore, A. Saxler, T. Wisleder, and P. Parikh, "40-W/mm Double Field-plated GaN HEMTs," in *Proc. Device Res. Conf.*, Jun. 2006, pp. 151-152, doi: 10.1109/DRC.2006.305162.

[4] T. Watanabe, A. Teramoto, Y. Nakao, S. Sugawa and T. Ohmi, "Low Interface Trap Density and High Breakdown Electric Field SiN Films on GaN Formed by Plasma Pretreatment Using Microwave-Excited Plasma-Enhanced Chemical Vapor Deposition," *IEEE Trans. Electron Devices*, vol. 63, no. 4, pp. 1795-1801, April 2016, doi: 10.1109/TED.2016.2525766.

[5] R. Vetury, N. Q. Zhang, S. Keller and U. K. Mishra, "The impact of surface states on the DC and RF characteristics of AlGaN/GaN HFETs," *IEEE Trans. Electron Devices*, vol. 48, no. 3, pp. 560-566, March 2001, doi: 10.1109/16.906451.

[6] Gao, F, Chen, D. , Tuller, H. L. , Thompson, C. V. , & Palacios, T. . (2014). On the redox origin of surface trapping in algan/gan high electron mobility transistors. *Journal of Applied Physics*, 115(12), 132-1140.

[7] Y. F. Wu et al., "30-W/mm GaN HEMTs by field plate optimization," *IEEE Electron Device Lett.*, vol. 25, no. 3, pp. 117-119, March 2004, doi: 10.1109/LED.2003.822667.

[8] W. M. Waller et al., "Control of Buffer-Induced Current Collapse in AlGaN/GaN HEMTs Using SiNx Deposition," *IEEE Trans. Electron Devices*, vol. 64, no. 10, pp. 4044-4049, Oct. 2017, doi: 10.1109/TED.2017.2738669.

[9] H. Jiang, C. Liu, Y. Chen, X. Lu, C. W. Tang and K. M. Lau "Investigation of In Situ SiN as Gate Dielectric and Surface Passivation for GaN MISHEMTs," *IEEE Trans. Electron Devices*, vol. 64, no. 3, pp. 832-839, March 2017, doi: 10.1109/TED.2016.2638855.

[10] K. Makiyama, T. Ohki, N. Okamoto, M. Kanamura, S. Masuda, Y. Nakasha, K. Joshin, K. Imanishi, N. Hara, S. Ozaki, N. Nakamura, and T. Kikkawa, "High-Power GaN-HEMT for Millimeter-Wave Amplifier," Phys. Status Solidi C, vol. 8, no. 7-8, pp. 2442-2444, Mar. 2011, doi: 10.1002/pssc.201001034.

[11] T. Huang et al., "Suppression of Dispersive Effects in AlGaN/GaN High-Electron-Mobility Transistors Using Bilayer SiNx Grown by Low Pressure Chemical Vapor Deposition," *IEEE Electron Device Lett.*, vol. 36, no. 6, pp. 537-539, June 2015, doi: 10.1109/LED.2015.2427294.

[12] K. Makiyama et al., "Collapse-free high power InAlGaN/GaN-HEMT with 3 W/mm at 96 GHz," *2015 IEEE International Electron Devices Meeting (IEDM)*, Washington, DC, 2015, pp. 9.1.1-9.1.4, doi: 10.1109/IEDM.2015.7409659.

[13] J. Zhu et al., "Impact of Recess Etching on the Temperature-Dependent Characteristics of GaN-Based MIS-HEMTs With Al2O3/AlN Gate-Stack," *IEEE Trans. Electron Devices*, vol. 64, no. 3, pp. 840-847, March 2017, doi: 10.1109/TED.2017.2657780.

[14] Turuvekere S , Karumuri N , Rahman A A , et al. "Gate Leakage Mechanisms in AlGaN/GaN and AlInN/GaN HEMTs: Comparison and Modeling," *IEEE Trans. Electron Devices*, vol. 60, no. 10, pp. 3157-3165, Oct. 2013, doi: 10.1109/TED.2013.2272700.

[15] Dutta G , Dasgupta N , Dasgupta A, "Gate Leakage Mechanisms in AlInN/GaN and AlGaN/GaN MIS-HEMTs and Its Modeling," *IEEE Trans. Electron Devices*, vol. 64, no. 9, pp. 3609-3615, Sep. 2017, doi: 10.1109/TED.2017.2723932.

Design and Verification of Gate Driver for 6.5 kV SiC MOSFET Module

Yijian Wang
State Key Laboratory of Advanced
Electromagnetic Engineering and
Technology, School of Electrical and
Electronic Engineering, Engineering
Research Center of Power Safety and
Efficiency, Ministry of Education
Huazhong University of Science and
Technology
Wuhan, China

Lin Liang
State Key Laboratory of Advanced
Electromagnetic Engineering and
Technology, School of Electrical and
Electronic Engineering, Engineering
Research Center of Power Safety and
Efficiency, Ministry of Education
Huazhong University of Science and
Technology
Wuhan, China
lianglin@hust.edu.cn

Hai Shang
State Key Laboratory of Advanced
Electromagnetic Engineering and
Technology, School of Electrical and
Electronic Engineering, Engineering
Research Center of Power Safety and
Efficiency, Ministry of Education
Huazhong University of Science and
Technology
Wuhan, China

Lubin Han
State Key Laboratory of Advanced
Electromagnetic Engineering and
Technology, School of Electrical and
Electronic Engineering, Engineering
Research Center of Power Safety and
Efficiency, Ministry of Education
Huazhong University of Science and
Technology
Wuhan, China

Abstract—Gate driver for 6.5kV/25A Silicon Carbide (SiC) MOSFET module, which features low coupling capacitance, is proposed in this paper. In contrast to 6.5 kV Silicon (Si) IGBT module, 6.5 kV SiC MOSFET module captures the ability to switch at a much higher frequency, improving the power density greatly. However, the common mode current, which flows through the coupling capacitance between primary and secondary sides of the isolation transformer of gate driver produced by high dv/dt during quick switching, may produce annoying electromagnetic interference (EMI). Firstly, the coupling capacitance of single ring transformer and double ring transformer is simulated based on electrostatic Finite Element Method (FEM). Then, these two schemes are adapted to make isolation transformer of the gate driver for 6.5kV SiC MOSFET module. In the end, double pulse test circuit is built, and thus, double pulse test is carried out so as to assess the designed gate driver. Experimental results show that, compared with single ring transformer, common mode current, which flows by the coupling capacitance between primary and secondary sides of the double ring isolation transformer, has been reduced by 13.1%. In addition, the 6.5kV SiC MOSFET module is switched on and switched off by the designed gate driver at 4500V supply voltage and 15A current with low oscillation at gate-source voltage. To be specific, voltage overshoot during turn off is no more than 200V, about 4.5% of the bus voltage, and, the amplitude of reverse recovery current is about 3A during turn on process, all of which verify the gate driver's effectiveness.

Keywords—SiC MOSFET, gate driver, coupling capacitance, common mode current, isolation transformer

I. INTRODUCTION

As a kind of wide bandgap (WBG) material, compared with conventional semiconductor material—Si, SiC material

This project is supported by the National Key Research and Development Program of China (2018YFB0905700, 2018YFB0905705).

has many advantages, such as higher thermal conductivity, higher saturation drift velocity, higher breakdown electric field, and higher melting point, etc [1], [2]. Owing to these advantages of SiC material, it is possible for SiC MOSFET to operate at higher voltage level, shorter turning-on and turning-off time, i.e., faster switching speed, higher temperature and so on, with much lower loss, and thus, help power electronic (PE) systems to achieve higher performance, e.g., higher power density, higher efficiency as well as smaller volume [1]. SiC MOSFETs get much attention and rapid development in the last decade or so. To date, semiconductor manufacturers such as Cree and Infineon have introduced SiC MOSFETs of different voltage levels, such as 600V, 1200V and 1700V, etc. What's more, SiC MOSFET of even higher voltage levels are demonstrated in laboratory [3]-[12]. For instance, SiC MOSFET with 15kV blocking voltage is reported in [12].

Owing to its excellent electrical performance, SiC MOSFETs have been adapted in some industrial applications, such as solar inverter, uninterruptible power supply (UPS), traction, electric vehicles (EVs), and induction heating (IH), etc [13]-[20]. In solar inverter applications, the adaption of SiC MOSFETs has achieved higher weight power density and higher efficiency [13], [14]. For instance, as reported in [13], by using 1.2kV SiC MOSFETs, the 50kW PV string inverter has nearly triple the weight power density versus traditional Si based PV inverters, from 0.38 kW/kg to 1 kW/kg. As for higher voltage applications, e.g., solid state transformer (SST), a 15kV SST is developed based on 15kV SiC MOSFET and it has achieved high efficiency while obtaining only 50% the size as well as 25% the weight of conventional transformer.

As reported in [21], SiC MOSFET of 10kV is compared with Si IGBT of 6.5 kV/25A in SST application, and the result shows that 10kV SiC MOSFET has the capability of switching at a frequency 7-10 times higher than that of 6.5 kV/25A Si IGBT. However, there is another side. Fast switching of SiC MOSFETs may result in issues in high voltage applications, such as crosstalk in the bridge circuit, surge voltages, spike currents, common-mode current which flows by the isolation transformer of gate driver, etc [22], [23]. In respect of crosstalk in the bridge circuit, it occurs at upper SiC MOSFET when the device at lower leg is switched on and switched off.

receives optic signal from fiber optic emitter HFBR-1521Z and transfer it to the fiber optic receiver HFBR-2521Z.

A. Isolation Transformer Design

The common mode current that we care most, which flows by the coupling capacitance, is calculated by (1).

$$i_{CM} = C \cdot \frac{dv_{CM}}{dt} \tag{1}$$

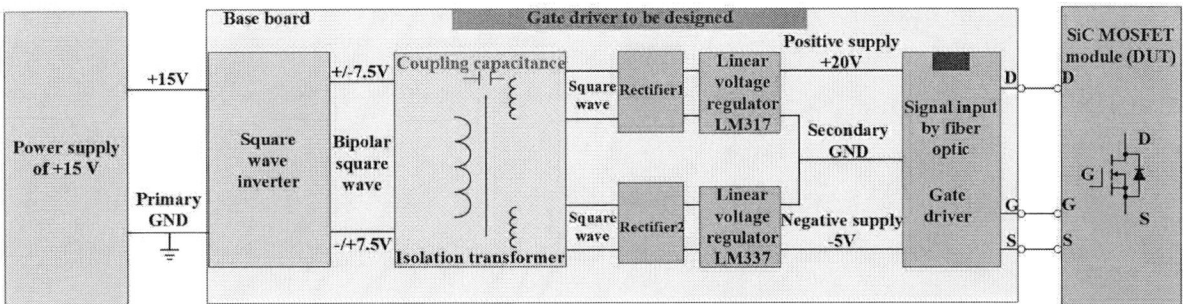

Fig. 1 Diagram of the gate driver.

Also, crosstalk will also be generated at the lower device during switching transition of upper one semiconductor device. For the purpose of suppressing the crosstalk, several circuit topologies are proposed, such as adding capacitors in parallel with C_{gs} (the parasitic capacitance which exists in the gate-to-source terminals of SiC MOSFETs) [24], adapting variable voltage and variable gate resistance during turning-on and turning-off process [25], [26], and setting extra discharge path between gate and source terminals [27], [28]. Common mode current, which flows by coupling capacitance between primary and secondary sides of the isolation transformer, is generated by high dv/dt during quick switching and may result in annoying EMI, and thus, disturb the safe operation of the gate driver as well as micro-controller (MCU). For the purpose of suppressing common mode current, it is practical to reduce coupling capacitance of isolation transformer while keeping enough isolation. In [29] and [30], several design schemes of isolation transformer are proposed, such as square core transformer, ferrite toroidal core transformer, as well as transformer with PCB winding, etc. When it comes to double galvanic isolation transformer implemented in [29], it obtains lower coupling capacitance than single ring isolation transformer. However, its effectiveness has not been verified by actual experiment. Therefore, the design of gate driver for SiC MOSFET module of 6.5 kV including the parameter design of isolation transformer, coupling capacitance comparison between these two schemes, design of driver output stage, in attached with its verification by double pulse test circuit, is implemented.

II. DESIGN OF GATE DRIVER

Fig. 1 presents diagram of the gate driver for SiC MOSFET module of 6.5 kV. As presented in Fig. 1, it is obvious that it consists of several parts, such as square wave inverter, isolation transformer, rectifier, as well as driver output stage, etc. It has to be noted that the power isolation is achieved by the good design of isolation transformer, while the signal isolation is implemented by fiber optic, which

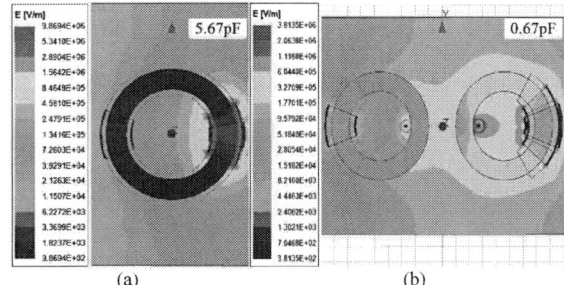

Fig. 2 Simulation results of two schemes (a) single ring. (b) double ring.

As presented in formula (1), i_{CM} is the common mode current to be calculated while C and dv_{CM}/dt represent coupling capacitance of isolation transformer and common-mode voltage change ratio of the second side of the gate driver for SiC MOSFET module of 6.5 kV during switching process, respectively. In order to suppress i_{CM} flowing through the

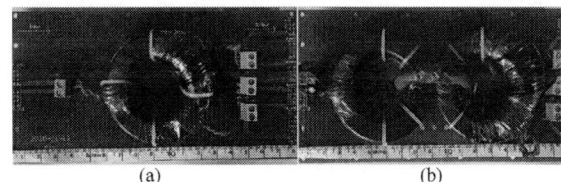

Fig. 3 Isolation transformer. (a) single ring. (b) double ring.

isolation transformer of gate driver, the coupling capacitance needs to be decreased, which exists between primary and secondary sides of the isolation transformer.

Based on the isolation voltage, taking robust into consideration, the design of isolation transformer is determined. To obtain low magnetizing current, the primary and secondary windings should have enough turns. Here, the turn of the primary winding is 4 and there are 16 and 12 turns on secondary windings, respectively. Ten layers of high voltage and high temperature Kapton tape are wrapped around the core to get good insulation between windings and the core,

978-1-6654-4817-8/21 $31.00 © 2021 IEEE

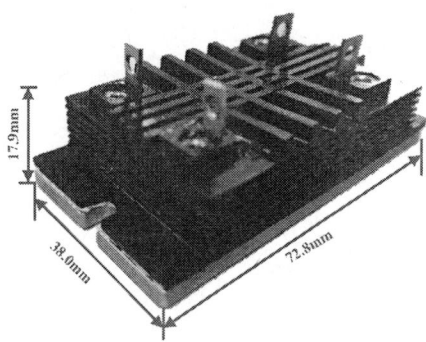

Fig. 4 6.5kV/25A SiC MOSFET module developed by our group.

while leaving enough space for selected turns of windings. Finally, ferrite toroidal core produced by TDK, HF60T62X13X39, is used to make isolation transformer of these two schemes.

Then, electrostatic FEM is carried out so as to simulate the coupling capacitance of these two schemes. The coupling capacitance of single ring transformer and double ring transformer are 5.67 pF and 0.67 pF, respectively, which is presented in Fig. 2. What's more, Fig. 3 presents the actual isolation transformer of these two schemes.

B. Driver Output Stage Design

Negative voltage of -5 V is adapted instead of 0 V so as to suppress the gate-source voltage spikes, prevent from mis-conduct as well as reduce energy loss during turning-off transition of the 6.5 kV SiC MOSFET module. What's more, Sub Miniature version A (SMA) connectors and RF coaxial

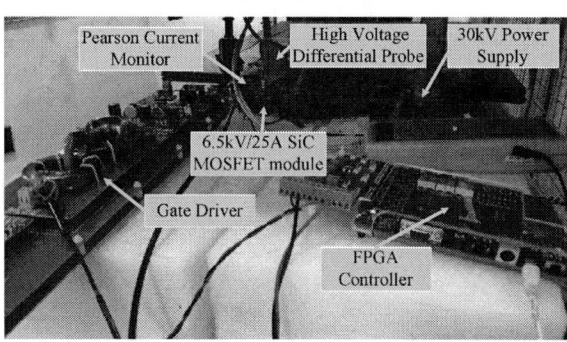

Fig. 5 Double pulse test circuit.

line around 10cm in length with very low lead inductance and resistance, rather than twisted-pair, are used to decrease the signal distortion during the transfer path, i.e., the path between output stage and gate terminal of DUT. In addition, wires on PCB (printed circuit board) are widened to decrease lead inductance, which will lead to gate oscillation during switching transitions.

III. VERIFICATION OF THE DESIGNED GATE DRIVER

As presented in Fig. 4, the SiC MOSFET module of 6.5 kV/ 25 A, is developed by our group based on 6.5 kV/ 25 A SiC MOSFET and 6.5 kV/ 25 A SiC JBS (junction barrier schottky) diode bare dies.

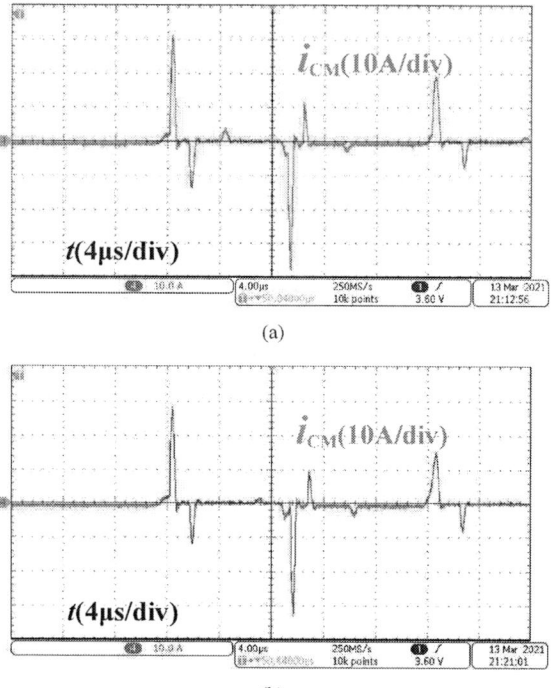

(a)

(b)

Fig. 6 Comparison of common mode current of two schemes (a) single ring. (b) double ring.

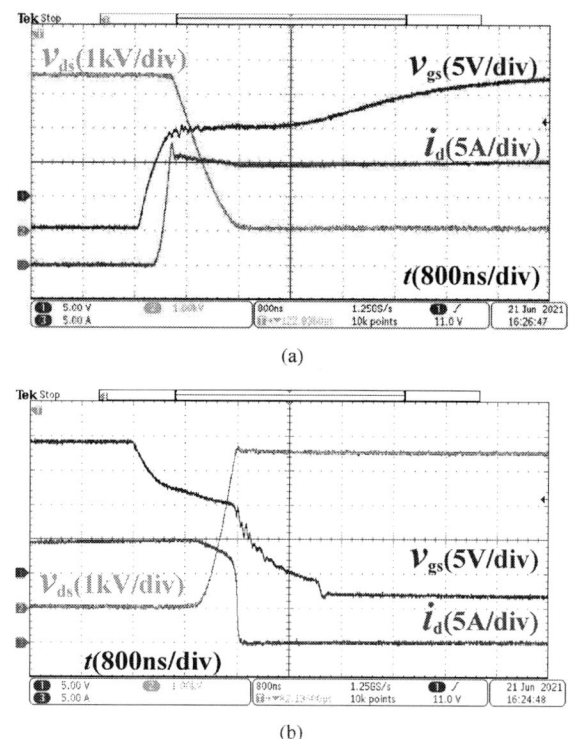

(a)

(b)

Fig. 7 Waveforms of (a) switching-on @ 4.5 kV/ 15 A. (b) switching-off @ 4.5 kV/ 15A with R_g of 100Ω.

As presented in Fig. 5, for the purpose of evaluating the designed gate driver for 6.5kV SiC MOSFET module, double

978-1-6654-4817-8/21 $31.00 © 2021 IEEE   324

pulse test circuit is built. To be more specific, the DC bus voltage is supported by four high voltage film capacitors, which are connected in parallel. Three air-core inductors with same inductance of 8.25 mH are connected in series, and it is selected as the load inductor, and thus, total inductance of approximately 25 mH is achieved. The gate signal is produced by altera FPGA (Field Programmable Gate Array), EP4CE10F17C8N. To be more specific, the first pulse generated by FPGA is made up of high level for $80\,\mu s$ followed by low level for 40µs, while the second pulse made up of high level for 20µs followed by low level for 20µs, which is called double pulse. In order to capture V_{ds}, high voltage differential probe manufactured by Keysight, N2891A, is connected to the drain terminal and source terminal of 6.5 kV SiC MOSFET module. In addition, pearson current monitor model 3972 and oscilloscope model MDO3054 with a bandwidth of 500MHz are selected to capture i_d and record

TABLE I. DYNAMIC PARAMETERS OF 6.5kV SiC MOSFET MODULE

Parameters	Values	Unit
t_{on} (10% of i_d to 90% of i_d)	166.4	ns
E_{on} (10% of i_d to 5% of v_{ds})	20.2	mJ
t_{off} (90% of i_d to 10% of i_d)	319.2	ns
E_{off} (10% of v_{ds} to 2% of i_d)	15.0	mJ

waveforms, respectively.

Firstly, the common mode current that we care most, which flows through the coupling capacitance between primary and secondary sides of the isolation transformer of two schemes, is measured by current monitor. As shown in Fig. 6, with respect to single ring transformer, peak common mode current that flows by the primary side of double ring transformer has reduced by 13.1%, which verifies the effectiveness of the isolation transformer design.

What's more, Fig. 7 shows that the developed gate driver can safely switch on and switch off 6.5 kV SiC MOSFET module at 4.5 kV/ 15 A with low gate-source voltage oscillation. In detail, during switching-on transition, the peak current produced by reverse recovery process of SiC JBS diode is around 3 A, which is within limits of acceptability. During switching-off transition, the voltage overshoot caused by di/dt is no more than 200 V, about 4.5% of the bus voltage. Additionally, Table I presents the switching characteristics of the DUT, 6.5 kV SiC MOSFET module, which is also obtained at 4500 V supply voltage and 15 A drain-source current.

IV. CONCLUSION

In this paper, design process of gate driver for 6.5kV SiC MOSFET module, which mainly includes isolation transformer design and driver output stage design, is proposed. Firstly, simulation results obtained by electrostatic FEM show that the coupling capacitances of single ring transformer and double ring transformer are 5.67pF and 0.67pF, respectively. Then, negative voltage -5V is selected to suppress the gate-source voltage spikes during turning-off transition, while SMA connectors and RF coaxial line with very low lead inductance and resistance are used to decrease

the distortion of gate signal during the transfer path. After that, double pulse test circuit is built, and in order to assess the designed gate driver, double pulse test is done. Experimental results indicate that the common mode current which flows by the double ring transformer is 13.1% lower than that flowing through the single ring transformer. In addition, the effectiveness of the gate driver is verified by low reverse recovery current during turning-on at 4.5kV/15A and low drain-source voltage overshoot during turning-off at 4.5kV/15A.

REFERENCES

[1] X. She, A. Q. Huang, Ó. Lucía and B. Ozpineci, "Review of Silicon Carbide Power Devices and Their Applications," in IEEE Transactions on Industrial Electronics, vol. 64, no. 10, pp. 8193-8205, Oct. 2017.

[2] J. Biela, M. Schweizer, S. Waffler and J. W. Kolar, "SiC versus Si—Evaluation of Potentials for Performance Improvement of Inverter and DC–DC Converter Systems by SiC Power Semiconductors," in IEEE Transactions on Industrial Electronics, vol. 58, no. 7, pp. 2872-2882, July 2011.

[3] W. Ni et al., "Design and Fabrication of 3300V 100mΩ 4H-SiC MOSFET with Stepped p-body Structure," 2019 16th China International Forum on Solid State Lighting & 2019 International Forum on Wide Bandgap Semiconductors China (SSLChina: IFWS), 2019, pp. 50-53.

[4] S. Li, Y. Chen, H. Liu, R. Huang, Q. Liu and S. Bai, "Simulation, Fabrication and Characterization of 3300V/10A 4H-SiC Power DMOSFETs," 2018 15th China International Forum on Solid State Lighting: International Forum on Wide Bandgap Semiconductors China (SSLChina: IFWS), 2018, pp. 1-4.

[5] S. Li et al., "Simulation, fabrication and characterization of 6500V 4H-SiC power DMOSFETs," 2017 14th China International Forum on Solid State Lighting: International Forum on Wide Bandgap Semiconductors China (SSLChina: IFWS), 2017, pp. 144-147.

[6] D. Yujie et al., "Fabrication and dynamic switching characteristics of 6.5kV400A SiC MOSFET module," 2020 17th China International Forum on Solid State Lighting & 2020 International Forum on Wide Bandgap Semiconductors China (SSLChina: IFWS), 2020, pp. 67-70.

[7] Q. Xiao, Y. Yan, X. Wu, N. Ren and K. Sheng, "A 10kV/200A SiC MOSFET module with series-parallel hybrid connection of 1200V/50A dies," 2015 IEEE 27th International Symposium on Power Semiconductor Devices & IC's (ISPSD), 2015, pp. 349-352.

[8] J. B. Casady et al., "New Generation 10kV SiC Power MOSFET and Diodes for Industrial Applications," Proceedings of PCIM Europe 2015; International Exhibition and Conference for Power Electronics, Intelligent Motion, Renewable Energy and Energy Management, 2015, pp. 1-8.

[9] C. DiMarino, I. Cvetkovic, Z. Shen, R. Burgos and D. Boroyevich, "10 kV, 120 a SiC MOSFET modules for a power electronics building block (PEBB)," 2014 IEEE Workshop on Wide Bandgap Power Devices and Applications, 2014, pp. 55-58.

[10] S. Sabri et al., "New generation 6.5 kV SiC power MOSFET," 2017 IEEE 5th Workshop on Wide Bandgap Power Devices and Applications (WiPDA), 2017, pp. 246-250.

[11] A. Q. Huang, Q. Zhu, L. Wang and L. Zhang, "15 kV SiC MOSFET: An enabling technology for medium voltage solid state transformers," in CPSS Transactions on Power Electronics and Applications, vol. 2, no. 2, pp. 118-130, 2017.

[12] V. Pala et al., "10 kV and 15 kV silicon carbide power MOSFETs for next-generation energy conversion and transmission systems," 2014 IEEE Energy Conversion Congress and Exposition (ECCE), 2014, pp. 449-454, doi: 10.1109/ECCE.2014.6953428.

[13] J. Mookken, B. Agrawal and J. Liu, "Efficient and Compact 50kW Gen2 SiC Device Based PV String Inverter," PCIM Europe 2014; International Exhibition and Conference for Power Electronics, Intelligent Motion, Renewable Energy and Energy Management, 2014, pp. 1-7.

[14] M. H. Todorovic et al., "SiC MW PV Inverter," PCIM Europe 2016; International Exhibition and Conference for Power Electronics,

978-1-6654-4817-8/21 $31.00 © 2021 IEEE

Intelligent Motion, Renewable Energy and Energy Management, 2016, pp. 1-8.

[15] S. Buschhorn and K. Vogel, "Saving money: SiC in UPS applications," PCIM Europe 2014; International Exhibition and Conference for Power Electronics, Intelligent Motion, Renewable Energy and Energy Management, 2014, pp. 1-7.

[16] J. Casarin, P. Ladoux and P. Lasserre, "10kV SiC MOSFETs versus 6.5kV Si-IGBTs for medium frequency transformer application in railway traction," 2015 International Conference on Electrical Systems for Aircraft, Railway, Ship Propulsion and Road Vehicles (ESARS), 2015, pp. 1-6.

[17] M. Su, C. Chen, S. Sharma and J. Kikuchi, "Performance and cost considerations for SiC-based HEV traction inverter systems," 2015 IEEE 3rd Workshop on Wide Bandgap Power Devices and Applications (WiPDA), 2015, pp. 347-350.

[18] H. Sarnago, Ó. Lucía and J. M. Burdío, "A Comparative Evaluation of SiC Power Devices for High-Performance Domestic Induction Heating," in IEEE Transactions on Industrial Electronics, vol. 62, no. 8, pp. 4795-4804, Aug. 2015.

[19] F. Wang, G. Wang, A. Huang, W. Yu and X. Ni, "Design and operation of A 3.6kV high performance solid state transformer based on 13kV SiC MOSFET and JBS diode," 2014 IEEE Energy Conversion Congress and Exposition (ECCE), 2014, pp. 4553-4560.

[20] X. She, A. Q. Huang and R. Burgos, "Review of Solid-State Transformer Technologies and Their Application in Power Distribution Systems," in IEEE Journal of Emerging and Selected Topics in Power Electronics, vol. 1, no. 3, pp. 186-198, Sept. 2013.

[21] G. Wang, X. Huang, J. Wang, T. Zhao, S. Bhattacharya and A. Q. Huang, "Comparisons of 6.5kV 25A Si IGBT and 10-kV SiC MOSFET in Solid-State Transformer application," 2010 IEEE Energy Conversion Congress and Exposition, 2010, pp. 100-104.

[22] L. F. S. Alves, P. Lefranc, P. Jeannin and B. Sarrazin, "Review on SiC-MOSFET devices and associated gate drivers," 2018 IEEE International Conference on Industrial Technology (ICIT), 2018, pp. 824-829.

[23] Y. Liu and Y. Yang, "Review of SiC MOSFET Drive Circuit," 2019 IEEE International Conference on Electron Devices and Solid-State Circuits (EDSSC), 2019, pp. 1-3.

[24] S. Yin, K. J. Tseng, C. F. Tong, R. Simanjorang, C. J. Gajanayake and A. K. Gupta, "A 99% efficiency SiC three-phase inverter using synchronous rectification," 2016 IEEE Applied Power Electronics Conference and Exposition (APEC), 2016, pp. 2942-2949.

[25] A. Paredes, V. Sala, H. Ghorbani and L. Romeral, "A novel active gate driver for silicon carbide MOSFET," IECON 2016 - 42nd Annual Conference of the IEEE Industrial Electronics Society, 2016, pp. 3172-3177.

[26] J. Dix, Z. Zhang and B. J. Blalock, "CMOS gate drive IC with embedded cross talk suppression circuitry for SiC devices," 2016 IEEE Applied Power Electronics Conference and Exposition (APEC), 2016, pp. 684-691.

[27] S. Yin, K. J. Tseng, C. F. Tong, R. Simanjorang, C. J. Gajanayake and A. K. Gupta, "A novel gate assisted circuit to reduce switching loss and eliminate shoot-through in SiC half bridge configuration," 2016 IEEE Applied Power Electronics Conference and Exposition (APEC), 2016, pp. 3058-3064.

[28] Z. Zhang, F. Wang, L. M. Tolbert and B. J. Blalock, "Active Gate Driver for Crosstalk Suppression of SiC Devices in a Phase-Leg Configuration," in IEEE Transactions on Power Electronics, vol. 29, no. 4, pp. 1986-1997, April 2014.

[29] K. Mainali, S. Madhusoodhanan, A. Tripathi, K. Vechalapu, A. De and S. Bhattacharya, "Design and evaluation of isolated gate driver power supply for medium voltage converter applications," 2016 IEEE Applied Power Electronics Conference and Exposition (APEC), 2016, pp. 1632-1639.

[30] T. Batra, G. Gohil, A. K. Sesham, N. Rodriguez and S. Bhattacharya, "Isolation design considerations for power supply of medium voltage silicon carbide gate drivers," 2017 IEEE Energy Conversion Congress and Exposition (ECCE), 2017, pp. 2552-2559.

978-1-6654-4817-8/21 $31.00 © 2021 IEEE

Soft Precharging Method for Four-Level Hybrid-Clamped Converter

Yihui Zhao
School of Electrical Engineering
Chongqing University
Chongqing, China
zhaoyihui@cqu.edu.cn

Jianyu Pan
School of Electrical Engineering
Chongqing University
Chongqing, China
panjianyu@cqu.edu.cn

Yao Luo
School of Electrical Engineering
Chongqing University
Chongqing, China
luoyao@cqu.edu.cn

Jian Li
School of Electrical Engineering
Chongqing University
Chongqing, China
lijian@cqu.edu.cn

Abstract—A four-level hybrid clamped converter(4L-HCC) is one of the next-generation multilevel converters for medium voltage motor drives without transformers. However, dc-link and flying capacitors in 4L-HCCs must be precharged to their nominal voltage. Due to the distinctive structure of capacitors in a 4L-HCC, few papers have made an explicit solution on the safe start-up process without inducing large inrush currents. This paper presents an effective soft precharging method using the selected switching states. A straightforward algorithm is developed to control the switching states to build different precharging circuits properly. All capacitors are precharged through internal elements rapidly with suppressed and controllable charging current. The simulation results verify the effectiveness of the method.

Keywords—*Precharging method, capacitor voltage control, hybrid clamped converter.*

I. INTRODUCTION

The four level hybrid clamped converter (4L-HCC) combing two parts of multilevel topologies have attracted extensive attention due to its simple structure and strong control flexibility [1-3]. Compared with three classic multilevel converters (e.g., NPC [4], FC [5], and CHB [6]), the 4L-HCC features better current and voltage harmonic characteristics. Additionally, in contrast to topology known as MMC [7], the 4L-HCC requires only half number of switches and capacitors of that in MMC at the same voltage level. Furthermore, the redundancy of eight switching states gives more control freedoms for a 4L-HCC to limit capacitor voltage fluctuation at low frequencies, which is still a technical challenge for MMC [8]. In view of above salient advantages, the 4L-HCC shows great potential for applications of medium-voltage dc-ac power conversion, like motor drives and renewable energy integration.

For the last several years, numerous studies have been done to address challenge issues related to the operation and control of the 4L-HCC [9-11], which mainly focus on modeling, voltage balancing, integrated control method, and modulation strategies, etc. But there are few reports on another technical key issue: the precharging of the dc-link and flying capacitor of 4L-HCC. These capacitors in the 4L-HCC are need to be precharged to their nominal voltage at the start-up process. However, if existing control methods of 4L-HCC are applied, the internal capacitors cannot be charged to the rated voltage, and a large inrush current during the charging process could be induced.

Recently, different precharging methods for other multilevel converter topologies have been published to precharge the capacitors of their systems to nominal voltage before getting into operation [12-14]. Nevertheless, due to the distinctive structure of capacitors in a 4L-HCC, their method cannot apply to the HCC without inducing large inrush currents.

Hence, this paper proposed a soft precharging method for the 4L-HCC, including the proper precharging circuit configuration and switch state selection. A 4L-HCC is connected to the dc source through a current limiting resistor and a bypass breaker in parallel to establish the precharging circuit. This method takes advantage of the existing switches and capacitors in 4L-HCC and it needs no auxiliary power supply. By controlling the state of the switches, all capacitors can be precharged to the nominal voltage while ensuring the 4L-HCC system works without inrush currents. Verified by the single-phase simulation with MATLAB/Simulink, the proposed method is simple, efficient and economical. Besides, the precharging method can switch over to the mode of output voltage/current control smoothly with no instability issue.

II. 4L-HCC CIRCUIT REVIEW

The one-phase circuit topology of a 4L-HCC is presented in Figure 1. Each phase has eight power switches and one flying capacitor C_{fx}. It consists of two two-level half-bridges (S_{1x}, S_{1x}'), (S_{4x}, S_{4x}') and one three-level flying capacitor unit (S_{2x}, S_{2x}', S_{3x}, S_{3x}'). And the V_{dc} is divided into three equal parts by dc-link capacitors, C_{d1}, C_{d2}, and C_{d3}.

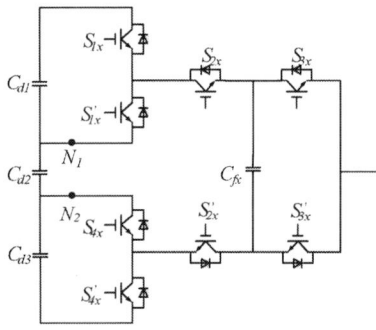

Fig. 1. The 4L-HCC converter.

978-1-6654-4817-8/21 $31.00 © 2021 IEEE

The operating rules of 4L-HCC are described below.

1) Each dc-link and flying capacitor need to be precharged, in order to reach a third of the total dc power supply voltage $V_{dc}/3$. Where E refers to the nominal value of each capacitor voltage, $E=V_{dc}/3$.

2) Each stage is comprised of four pairs of complementary switches: (S_{1x}, S_{1x}'), (S_{2x}, S_{2x}'), (S_{3x}, S_{3x}') and (S_{4x}, S_{4x}'). For each of them, if one of the switches is on (such as $S_{1x}=1$), the complementary switch must be off ($S_{1x}'=0$) to prevent short circuits from happening, and vice versa.

3) S_{1x}, S_{2x}, S_{3x} operate independently. The switches S_{4x} has the same state as S_{1x}, in order to keep switches S_{4x}' and S_{1x}' operate in phase. If S_{1x} and S_{4x}' are switched on concurrently, either S_{1x}' or S_{4x} will bear the voltage stress up to 2E.

Once the states of S_{1x}, S_{2x}, S_{3x} are identified, the states of the residual switches will also be determined. Different combination of switching states enables each phase to output four voltage levels. Taking one phase as an instance, figure 2 shows eight current flows in the single-phase circuit, which matches up with the eight switch states respectively. Assuming that the output current i_o is positive, the purple curve represents the current flow loop of the 4L-HCC. As can be seen, for the charging and discharging of flying capacitor, four different current flows b, c, f and g can be used.

III. PROPOSED PRECHARGE METHOD

A. Precharge Circuit

This paper presents the precharging circuit using the dc voltage source. A single-phase 4L-HCC with external precharing circuits is displayed in Fig.3. DC power source in series with the circuit breaker S_0 and current limiting resistor R_0 is connected with the dc bus of 4L-HCC.

Particularly, dc power-supply voltage is triple the normal operating voltage of capacitors. The dc power supply used for precharging is the same as the dc power supply voltage when the system works normally, which avoids the trouble of changing the dc voltage source. The current limiting resistor R_0 in the precharging circuit is used to avoid large impulsive current during precharge. When the dc-link capacitors precharging is completed, cut off the current limiting resistor. However, how to choose the basis for judging the completion of dc-link capacitors precharging and the value of current limiting resistance is the key to avoid large impulsive current. It will be explained in the next section. What's more, in normal operation, the precharging circuit is not needed be removed or disconnected manually after precharging, because they can also be used when the 4L-HCC system works normally.

B. The proposed precharging strategy

Fig.4 indicates the control flowchart of proposed precharging method. This section describes the precharging method based upon the single-phase 4L-HCC structure.

Firstly, the initial driving signals of all switches in the whole system are set to low level. A third of the voltage of dc power-supply is taken as the reference value of capacitors. Then, the internal switches S_{1x}, S_{2x} and S_{3x}' of the 4L-HCC are given the high-level driving signal, and the other switches are given the low-level signal. At this time, by connecting them to the source of direct current, there will be two precharging circuits, as presented in the Fig.5. The

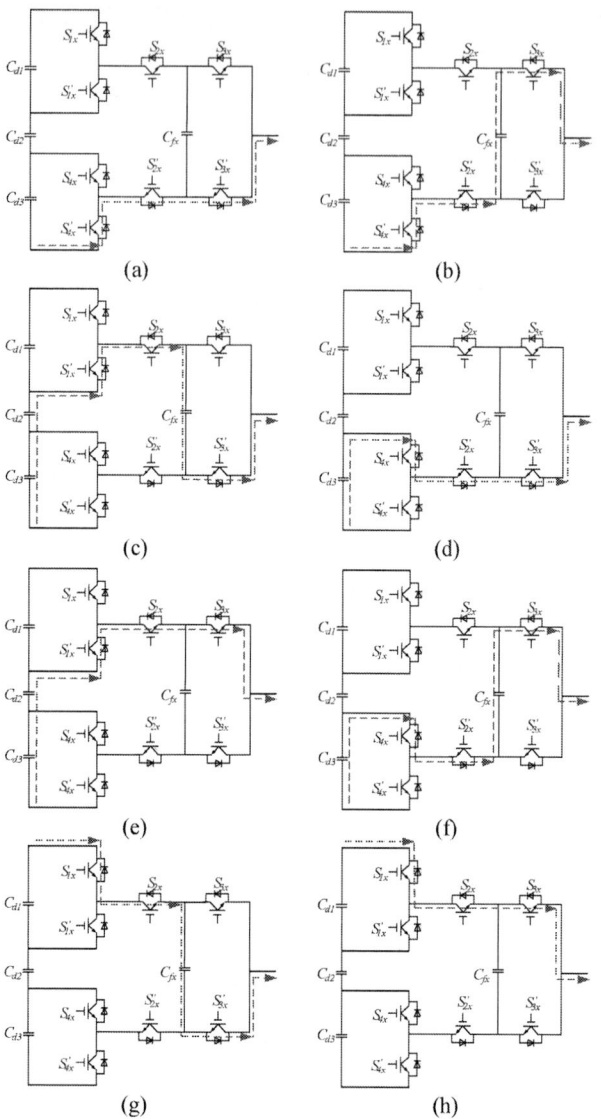

Fig. 2. Eight current flows of the 4L-HCC: (a) V_1=0, (b) V_2=E, (c) V_3=E, (d) V_4=E, (e) V_5=2E, (f) V_6=2E, (g) V_7=2E, and (h) V_8=3E.

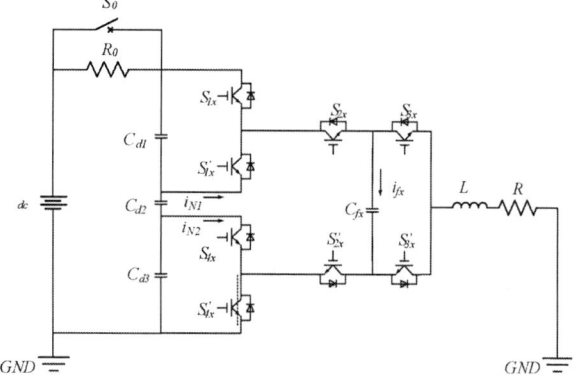

Fig. 3. Diagram of 4L-HCC with the precharging circuit

978-1-6654-4817-8/21 $31.00 © 2021 IEEE   328

precharging circuit I is composed of dc power-supply, resistor R_0 and three dc-link capacitors. And the precharging circuit II is composed of dc power-supply, current limiting resistor, switches S_{1x}、S_{2x} and S_{3x}', flying capacitor and load. All capacitors start charging. As mentioned above, the crucial factor to avoid large current is how to select the proper time to cut off the current limiting resistor. Based on the analysis of the circuit structure, the current in precharging circuit I and circuit II can be written as (1) and (2), respectively.

$$i_{c1} = \frac{v_{dc} - (v_{cap\_1} + v_{cap\_2} + v_{cap\_3})}{R_0} \qquad (1)$$

$$i_{c2} = \frac{v_{dc} - v_{cap\_fx}}{\sqrt{(R_0 + R)^2 + (XL)^2}} \qquad (2)$$

It can be easily concluded from formula (1) and (2) that current in the precharging process can be controlled by adjusting the resistance of current limiting resistor R_0, this is, if the resistance of R_0 is big enough, the precharging current will be regulated to very small.

And the current when the R_0 is cut off can be obtained through formula (3) according to the switching states and internal resistance R_r of the switch S_0.

$$i_{c1\_inrush} = \frac{v_{dc} - k*v_{dc}}{R_r} = \frac{(1-k)v_{dc}}{R_r} \qquad (3)$$

Due to the existence of the resistor R_0, the dc-link capacitors obviously cannot be fully charged to 100% of the dc power supply. The value of k is closer to 1, the longer the precharging time of dc-link capacitors is and the smaller the current $i_{c1\text{-}inrush}$ is. Considering the balance of precharging time and current, the value of k is selected as 99.99%. It can control the impulse current within 10 A when the dc power supply is less than or equal to 10 kV. Of course, the value of k can be adjusted to meet different precharging current and time requirements, which makes the proposed precharging method flexible and suitable for various applications of 4L-HCC.

To prevent the influence of precharging circuit II on the precharge of dc-link capacitor, after the first charge of flying capacitor is finished, switch off the precharging circuit II. And the low-level drive signal is given to the internal switches S_{1x}、S_{2x} and S_{3x}', which stop flying capacitor charging. And when the values of dc-link capacitors voltage reach 99.99% of the nominal voltage, throw the cut-off switch. After cutting off the current limiting resistance, the dc-link capacitors and dc power-supply are directly connected in parallel, and dc-link capacitor voltage is compensated to maintain the nominal voltage.

What's more, before getting into operation, when it's detected that voltage of flying capacitor(FC) drops below the nominal voltage, give a high-level driving signal to the switches S_{1x}、S_{2x} and S_{3x}'. Thus, as to sustain the voltage of FC at nominal voltage. The state of switches is continuously controlled to compensate the voltage of FC in real time, which realize the active control of flying capacitor's precharge.

In this way, the general precharging process can be divided into three parts. During the stage 1, dc-link capacitors and flying capacitor are charged simultaneously. When it comes to stage 2, the charging of FC is completed and dc-link capacitors keep charging. Stage 3 is dynamic compensation control, which control the state of internal switches to

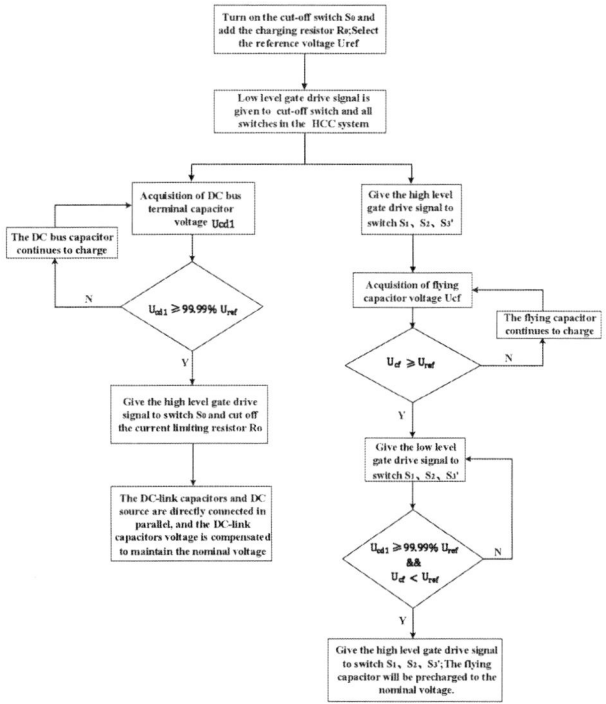

Fig. 4. Flow chart of precharging control.

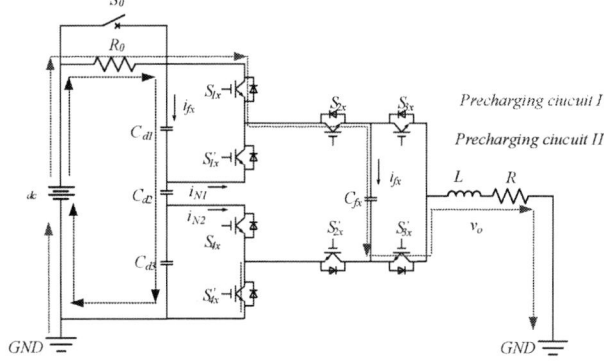

Fig. 5. Two precharging ciucuits.

maintain the voltage of all capacitors at the nominal working voltage.

When it comes to the single load and two leg 4L-HCC system, which is applied in actual situation, the principle of proposed precharging method is also applicable. To charge capacitors of one-phase 4L-HCC, the dc-link capacitors and flying capacitors are supposed to be inserted into precharging circuit. As shown in Fig. 2 and Fig. 6, obviously, all these capacitors meet this requirement. By combining states b, c, f and g, the suitable charging circuit can be formed to charge the flying capacitor, which means that this method can also be applied to multi-phase Structure of 4L-HCC. Therefore, the capacitors in one-phase 4L-HCC system also can be charged by proposed precharging method. The precharging procedure for the single load and two leg 4L-HCC can be seperated into three steps below:

978-1-6654-4817-8/21 $31.00 © 2021 IEEE   329

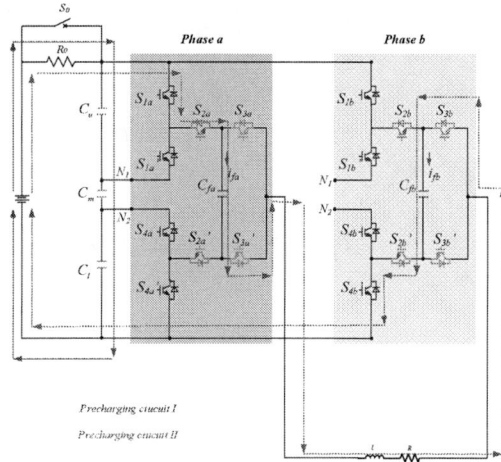

Fig. 6. Two precharging ciucuits of one-load and two-leg 4L-HCC.

Step 1: Before the precharging process, all internal switches of the 4L-HCC and the switch S_0 are set to the low-level gate signal. When the precharging process begins, the dc power supply is connected to the single phase 4L-HCC through a limiting resistor R_0. Switches S_{1a}, S_{2a}, S_{3a}', S_{3b}, S_{2b}' and S_{4b}' are set to 1 to turn on those switches. As shown in Fig.6, there will be two precharging circuits used to charge the dc-link capacitors and FC respectively.

Step 2: When the voltages of the flying capacitors reach the nominal voltage, switches S_{1a}, S_{2a}, S_{3a}', S_{3b}, S_{2b}' and S_{4b}' are given the low level driving signal. Simultaneously, the charge of flying capacitor is stopped and dc-link capacitors keep charging.

Step 3: When the voltages of dc-link capacitors reach 99.99% of the nominal voltage, throw the switch S_0. After cutting off the current limiting resistance, the dc-link capacitors and dc power-supply are directly connected in parallel, and the dc-link capacitors' voltage is compensated to maintain the nominal voltage. At the same time, if the flying capacitors' voltages drop below the nominal value, the switches S_{1a}, S_{2a}, S_{3a}', S_{3b}, S_{2b}' and S_{4b}' will be set to 1 to maintain it at the nominal voltage before getting into operation.

The previous analyses manifest that proposed method is suitable for single-phase 4L-HCC system in actual application.

IV. SIMULATION RESULTS

To demonstrate proposed precharging method, this paper builds a 50 Hz, 1 MW 4L-HCC system prototype with a 6 kV dc-bus voltage in Matlab/Simulink. Table I lists the simulation parameters for precharging 4L-HCC with the proposed method. In the 4L-HCC, each phase has three dc-link capacitors and a flying capacitor. All capacitances are 500 μF. Nominal dc-link and flying capacitors voltages are 2 kV, which match with the dc-bus voltage. The current limiting resistance is 2 k. The load inductance and resistance are 5mH and 15Ω.

Fig. 7 presents the results of precharging the single-phase 4L-HCC system by the proposed precharging method. The dc-

TABLE I SIMULATED SYSTEM PARAMETERS

Parameter	Description	Simulated values
V_{dc} (V)	DC input voltage	6000
C_x (μF)	DC-link capacitance	500
C_f (μF)	Flying capacitance	500
R_0 (Ω)	Current limiting resistance	2000
R_r (Ω)	Switch S_0 resistance	0.1
f_c (kHz)	Carrier frequency	4
f_{equ} (kHz)	Equivalent system frequency	12
L_{load} (mH)	Load Inductance	5
R_{load} (Ω)	Load Resistance	15

link capacitors voltage and the FC voltage are shown in Fig. 7, respectively. the voltage value of the FC is rapidly charged from 0 to the rated working value. After 3.48 s, the three dc bus capacitors are charged synchronously, and the voltage is rapidly charged from 0 to about 2000 v. Then, the capacitor voltage is monitored and compensated in real-time, and the capacitor voltage is maintained at the ideal working voltage of 2000 v. The whole charging process is completed within 4 s, which proves that the precharging circuit and the proposed precharging method can quickly and reliably realize the precharge of the capacitors in 4L-HCC system. Through the precharging method, the precharging process is divided into three parts: dc bus capacitor voltage charging, flying capacitor voltage charging and dynamic balance control of flying capacitor voltage. Due to the existence of inductance, the current of flying capacitor won't change abruptly during the precharging.

Fig.8 shows that the current of flying capacitor increases from 0 a to 2.2 A and then decreases to 0 A after 0.55 s. The dc bus capacitor current decreases slowly from the maximum value of 3 A. During the precharging process, the charging of flying capacitor is completed at 0.55 s. The charging speed of

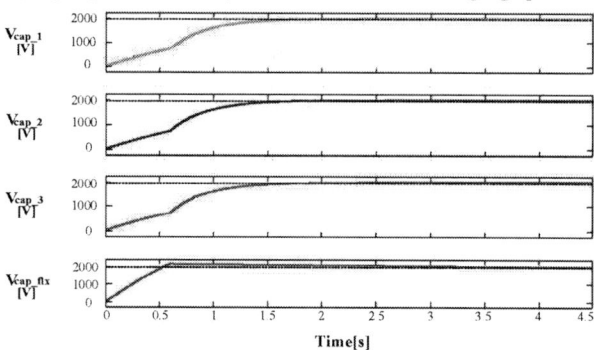

Fig. 7. Simulated result at Vdc=6000 V: dc-link capacitors voltage, flying capacitor voltage.

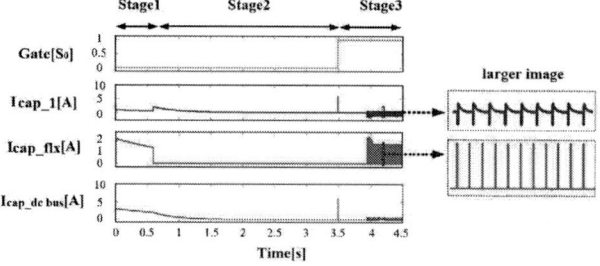

Fig. 8. Simulated result at Vdc=6000 V: Gate driving signal of switch S_0, dc-link capacitor current, flying capacitor current and dc bus current.

dc bus capacitor is accelerated and the current reduction speed is also accelerated. When the charging is completed in 3.48 s, there is a small impulse current of 6 A when the current limiting resistor is cut off.

Fig. 9 shows the load current is not exceed 3 A, which is safe for circuit components. During the stage 3, with the dynamic balance control of flying capacitor voltage, the precharging circuit II is continuously formed with the on-off states of the switches. At the same time, the voltage of resistance-inductance load varies regularly between 0 and 4000 V and the load current oscillation occurs, which is correspond to the expected analysis. Hence, the proposed precharging method can effectively and rapidly charge all capacitors in the 4L-HCC system to the nominal voltage without large current.

Furthermore, to verify the convenience of the precharging method, switching from charging state to the state of output voltage/current control when the precharging process is completed have been taken into consideration, too. This article choose the PS-PWM as the control strategy during normal work. As shown in Fig. 10, when V_{ref_x} compared to a set of three 120° phase-shifted carriers, three channels of gate signals will be obtained to determine the logic values of S_{1x}, S_{2x}, and S_{3x}. Fig. 11 shows that the 4L-HCC system can directly switch from the precharging state to the normal working state without changing the precharging circuit.

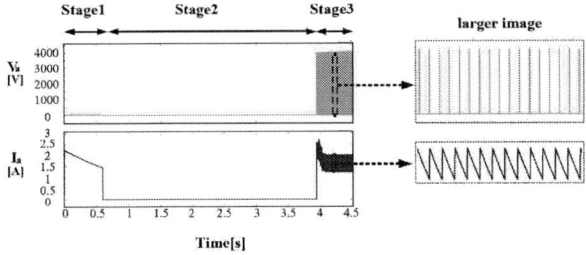

Fig. 9. Simulated result at Vdc=6000 V: Resistance-inductance Load voltage, Resistance-inductance Load current.

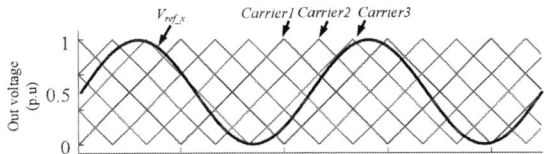

Fig. 10. PS-PWM for one-phase 4L-HCC.

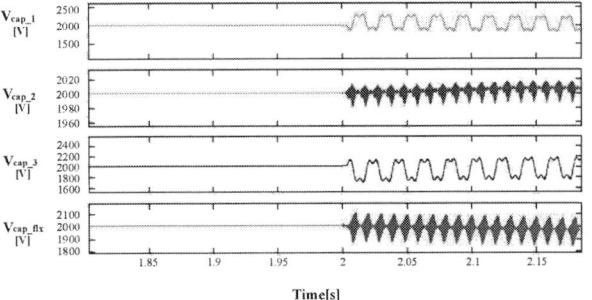

Fig. 11. Simulated results of switching from charging state to standby state: dc-link capacitors voltage, flying capacitor voltage.

Meanwhile, each capacitor voltage remains stable. The simulation results verify the proposed method.

V. CONCLUSION

This paper proposes a novel and soft precharging method with a straightforward algorithm for the 4L-HCC. Using the selected switching states, the proposed method can precharge the dc-link and flying capacitors voltages for single-phase 4L-HCCs rapidly. Simulation is used to verify the effectiveness of the proposed method with the following advantages:

1) The dc-link and flying capacitors voltages can be precharged to the nominal value quickly and safely from the dc voltage source. The precharging process does not induce any large inrush currents.

2) Realize the active control of the flying capacitors' precharge. All capacitors can be controlled at the normal voltage entirely.

3) No additional operation is required between the normal operation and precharging process. When the precharging is completed, it can be smoothly switched to the normal operating state without any instability issue.

ACKNOWLEDGMENT

This research was funded by National Natural Science Foundation of China (52007021), and China Postdoctoral Science Foundation (2020M673133).

REFERENCES

[1] J. Pan, R. Na, Y. Yang, H. Cai, and L. Xu, "Capacitor Voltage Balancing and Stabilization for 4-Level Hybrid-Clamped Converter Using Selected Switching States," *IEEE Transactions on Power Electronics*, vol. 34, no. 12, pp. 12453-12463, Dec 2019.

[2] J. Pan, Y. Yang, H. Cai, and L. Xu, "Capacitor Voltage Fluctuation Minimization for Four-Level Hybrid Clamped Converter Using Improved Common-Mode Voltage Injection," *IEEE Transactions on Power Electronics*, vol. 35, no. 7, pp. 7563-7573, Jul 2020.

[3] K. Wang, L. Xu, Z. Zheng and Y. Li, "Voltage Balancing Control of a Four-Level Hybrid-Clamped Converter Based on Zero-Sequence Voltage Injection Using Phase-Shifted PWM," in *IEEE Transactions on Power Electronics*, vol. 31, no. 8, pp. 5389-5399, Aug. 2016.

[4] A. Nabae, I. Takahashi and H. Akagi, "A New Neutral-Point-Clamped PWM Inverter," in *IEEE Transactions on Industry Applications*, vol. IA-17, no. 5, pp. 518-523, Sept. 1981.

[5] J. A. Dickerson and G. H. Ottaway, "Transformerless power supply with line to load isolation," U.S. Patent 3 596 369, Aug. 3, 1971.

[6] W. McMurray, "Fast response stepped-wave switching power converter circuit," U.S. Patent 3 581 212, May 25, 1971.

[7] R. Marquardt, "Stromrichterschaltungen mit verteilten energiespeichern," German Patent DE10103031A1, Jan. 24, 2001.

[8] A. J. Korn, M. Winkelnkemper and P. Steimer, "Low output frequency operation of the Modular Multi-Level Converter," *2010 IEEE Energy Conversion Congress and Exposition*, Atlanta, GA, 2010, pp. 3993-3997.

[9] J. Jung, H. Lee and S. Sul, "Control Strategy for Improved Dynamic Performance of Variable-Speed Drives With Modular Multilevel Converter," in *IEEE Journal of Emerging and Selected Topics in Power Electronics*, vol. 3, no. 2, pp. 371-380, June 2015.

[10] Y. Yang *et al.*, "An Optimized Model Predictive Control for Three-Phase Four-Level Hybrid-Clamped Converters," *IEEE Transactions on Power Electronics*, vol. 35, no. 6, pp. 6470-6481, Jun 2020.

[11] J. Pan, Y. Yang, J. Li, F. Wang, Y. Abdullah, and L. Xu, "Control of High-Performance Drive Feeding by Four-Level Hybrid Clamped Converter for Transportation Electrification," *IEEE Transactions on Transportation Electrification*, vol. 6, no. 2, pp. 568-577, Jun 2020.

[12] Y .Xue, Z. Xu, and G. Tang, "Self-start control with grouping

sequentially precharge for the C-MMC-based HVDC system," *IEEE Trans. Power Del.*, vol. 29, no. 1, pp. 187–198, Feb. 2014.

[13] K. Shi, F. Shen, D. Lv, P Lin, M. Chen, and D. Xu, "A novel start-up scheme for modular multilevel converter," in *Proc. IEEE Energy Convers. Congr. Expo.*, 2012, pp. 4180–4187.

[14] Z. Liu, J. Lu, Z. Ou, M. Ma, C. Yuan, X. Xiao, and H. Wang, "The start control strategy design of unified power quality conditioner based on modular multilevel converter," in *Proc. IEEE Int. Elect. Mach. Drives Conf.*, 2013, pp. 933–937.

Design and Research on Package Insulation of Highvoltage Silicon Carbide Module

Yang Zhou, Ling Sang, Xinling Tang, Hao Shi

State Key Laboratory of Advanced Power Transmission Technology, Global Energy Interconnection Research Institute Co.,Ltd, Beijing 102209, China

ABSTRACT

In order to give full play to the performance advantages of silicon carbide (SiC) modules under high temperature and high pressure, it is necessary to carry out targeted design from the perspectives of module structure design, high temperature resistance and material insulation based on SiC characteristics. This paper relies on the 6.5 kV SiC single-chip module package scheme to carry out high voltage SiC module package design and material insulation research. First of all, a variety of package insulation materials were explored, and the significant correlation between high temperature leakage current and long-term withstand voltage of devices was analyzed. After optimizing the internal insulation, the internal insulation of the device has obviously been improved by comparing different design structures. The reverse leakage current of the device was reduced from 380 μA to 4 μA at 150 °C and the working peak reverse voltage of 6.5 kV. The pass rate of 800h 6kV high-temperature reverse-bias test reached 100%, of which the reliability was also significantly improved. Furthermore, it is believed that charge behaviors such as partial discharge and flashover discharge are the main factors leading to the failure of module high-temperature insulation materials.

Index Terms —**Highvoltage Silicon Carbide Module, Package Insulation, Insulation failure mechanism.**

1 INTRODUCTION

The silicon carbide as the representative of the third generation of wide band gap semiconductor devices, shows the advantages of high voltage grade, large flow capacity, small loss, fast heat dissipation [1], so it is widely used in solid-state transformers, solid-state circuit breakers, electric vehicle controllers, other new equipment, as well as traditional FACTS and DC transmission equipment and others [2]. Currently, SiC devices are increasingly becoming a research and development hotspot at home and abroad, among which commercial SiC devices are mainly concentrated between 600V and 1700V, and the studies on high-voltage SiC devices is still in the laboratory sample development stage [3-4]. However, the package schemes of most SiC modules are not targeted based on SiC characteristics at present, meanwhile, some of them still follow the traditional silicon-based package form, thus severely limiting the development of SiC power electronic devices [5-7].

With the increasing voltage level of SiC modules (up to 6kV or even more than 15kV), the trend of miniaturization and the operation requirements under high temperature conditions, higher requirements are put forward for the high temperature insulation performance of SiC modules. Therefore, it is necessary to fully study the package structure parameters and package material performance [8]. However, there are relatively few theoretical and experimental studies on insulation package of high-voltage SiC devices at present, so it is of great significance to carry out research in this field.

In this paper, 6.5 kV/25A SiC diode and 6.5 kV/20A SiC MOS chips designed independently are selected to study the package insulation of high voltage SiC module from the perspectives of insulation material selection, package design and device testing.

2 INSULATION ANALYSIS OF PACKAGE MATERIAL

The insulating materials in the SiC module mainly involves copper coated ceramic plate, insulating shell and potting material. Copper coated ceramic plate is used to carry electronic interconnection lines and need to have outstanding electrical insulation performance and excellent thermal conductivity. For high-voltage SiC modules above 6.5 kV, AlN ceramic plates with a thickness of about 1mm are generally selected, with the thermal conductivity as high as 170W/(m·K) and the insulation strength as high as

978-1-6654-4817-8/21 $31.00 © 2021 IEEE

20mV/mm. On the other hand, the AMB-AlN is finally selected, hence the Active Metal Bonding process (AMB) reveals obvious advantages in porosity and reliability compared with the Direct Bonding Copper process (DBC). The potting materials, which mainly contain epoxy resin and silicone gel, play a role in increasing electrical insulation between potentials in the module, and hold certain moisture and gas isolation functions. Due to the elastic modulus of the silicone gel is much smaller than that of epoxy resin, the temperature impact damage to the chip and the bonding wire can be effectively avoided, furthermore, the breakdown electric field of the silicone gel is above 23mV/mm, and the long-term temperature resistance can reach above 210 °C, so the high-temperature resistant silicone gel becomes an ideal material for module potting. The shell insulating material mainly plays the role of mechanical protection, mechanical support as well as strengthen the voltage discharge distance between terminals, therefore, it is necessary to have excellent high temperature resistance and insulating property. At present, commonly used insulating polymer materials include PA66, PPA, PEEK, PPS, and others, among which, PPA has higher bending modulus and high-temperature resistance than PA66 due to the introduction of aromatic rings in the main chain. Furthermore, its CTI can reach about 600V (more than twice that of PPS), and its price is far lower than that of other high temperature materials such as PEEK. In addition, the introduction of glass fiber can effectively improve the mechanical strength and temperature resistance of the body. When it comes to the insulating performance of high temperature SiC module, besides considering the dielectric strength of the material, we study the influence of leakage current of the material under different temperature and voltage conditions in this paper. As shown in Fig. 1, the leakage current of shell insulating material raises significantly with the increase of temperature and electric field, especially at medium and high voltage, the relationship between leakage current and temperature is exponential. Ramu model in multi-stress accelerated test believes that the insulation life of the material conforms to the power exponential relationship with the electric field strength, conforms to Arrhenius formula with the temperature, and has the same change trend with the leakage current of the material [9]. Therefore, it can be seen that the high temperature leakage current of the material is an important factor affecting the long-term high temperature insulation reliability of the material.

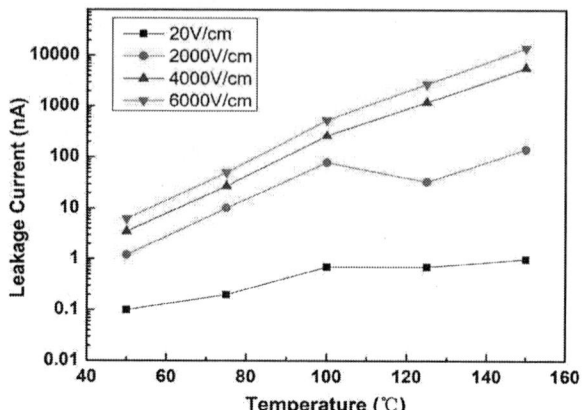

Figure 1. The temperature dependence of leakage current of the insulating shell under different electric field , and it shows the linear fitting of the experimental data.

3 PACKAGE DESIGN OF SIC MONOLITHIC MODULE

In order to further study the influence of package design on module insulation performance, as shown in Fig. 2, we have designed several SiC single-chip welding modules. An AMB-ALN board with a thickness of 1mm was used in the module. The AMB was designed with four regions: drain region, power source region, signal source region and gate region. Therefore, it can be applied to the package of SiC diodes or MOS (IGBT) monolithic chips. The chip was electrically interconnected with the copper clad board through brazing and aluminum wire bonding. The module shell frame was made of PPA containing 30% glass fiber. In order to increase the external creepage distance, four square sawtooth structures with 2mm in width were added around the frame. Three square sawtooth structures with 2mm in width were also designed for high-potential terminals and low-potential terminals. Finally, silicone gel was filled to further enhance insulation protection.

Figure 2. The image of 6.5kV silicon carbide module.

We have designed three kinds of schemes to study the internal insulation of the device. In Scheme A, we directly filled the whole device with silicone gel, while in Scheme B, the isolation between power terminals was mainly realized by T-shaped insulation board, while the chip and bonding wire were only coated with silicone gel, the T-shaped insulation board of Scheme B is shown in Fig. 3, while the process design scheme of T-shaped insulation board and silicone gel coating metal terminals was adopted in Scheme C.

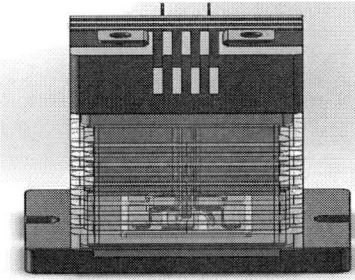

Figure 3. The schematic plot of the Scheme B, reinforcing ribs witch is between the power terminals are used to increase the creepage distance, and the lower part is protected by silicone gel potting.

4 TEST AND EVALUATION OF MODULE INSULATION PERFORMANCE

First of all, the reverse leakage current of three modules were measured at different temperatures. As shown in Table 1, with the increase of temperature, the reverse leakage current of the module of Scheme B increased significantly, reaching about 380uA at 150 °C, while the optimized Module C, its reverse leakage current at 150 °C, was still very low. In addition, we tested the reliability of three modules through HTRB. Under the constant voltage of 150 °C and 6kV, the module of Scheme A and the module of Scheme B failed after being pressurized for just a few hours, while the result of the module of Scheme C is shown in Fig. 4, the leakage current is stable during the HTRB test for 800h, and the pass rate of all modules in this scheme is 100%.

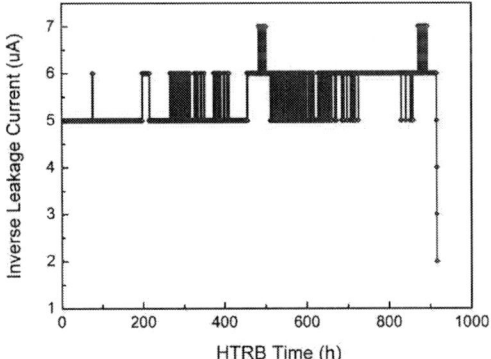

Figure 4. The result of HTRB reliability assessment for 1000 hour.

Table 1. Summary of the Id of three package scheme modules at different temperature (6500V)

Id(μA) @6500V	Module A	Module B	Module C
25°C	0	0	0
100°C	0	7	0
150°C	24	380	4

Afterwards, failure analysis of the failed devices was performed. The results showed that the failure positions were all located inside the insulating shell, and the withstand voltage of their chip was still normal. As shown in Fig. 5, since the module of Scheme A was designed no internal creepage design, it presented surface breakdown, while the module of Scheme B produced breakdown holes in the T-shaped insulating board. It is believed that charge behaviors such as partial discharge and flashover discharge are the main factors leading to the failure of insulating materials. The T-type insulating board of the Module B and the power terminal had a long parallel distance in the air, and there will be some pollution points or bulges on their surface. Above-mentioned abnormal structures is generally believed to easily induce corona discharge [10], thus causing the high-temperature insulating performance of the polymer material to be significantly lower than its theoretical value. The module of Scheme C increased internal creepage through the T-shaped insulation board structure, and at the same time, the surface of the metal terminal and the insulating board were all coated with silicone gel, these improved process effectively avoided flashover discharge.

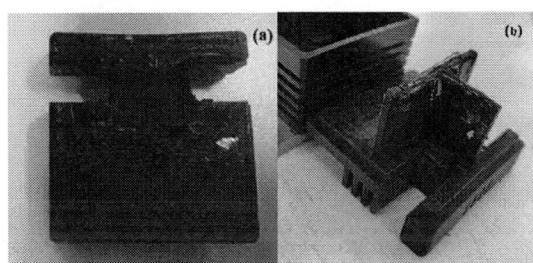

Figure 5. The morphology of plastic shells of the failed modules devices at the HTRB, (a) is the module of Scheme A, (b) is the module of Scheme B.

Subsequently, we used the method of partial discharge test to compare and verify the insulating performance of the three package forms. Two power terminals of the module were connected to the test circuit, and the signal waveform including sinusoidal, square wave, pulse and DC was output through the function signal generator (Tektronix AFG3022C), and finally the partial discharge pulses were collected by the partial discharge instrument (MPD-600).The test equipment is shown in Fig. 6, the output frequency range of the function signal generator is from 0 Hz to 1 MHz, the peak-to-peak output voltage range is 10mV-20Vp-p, and the output waveform types include sine, square wave, pulse, DC. Meanwhile, the high-voltage amplifier has a magnification of 4000 and the maximum output voltage is ±40kV, its input voltage range of the high voltage amplifier is from -10 V to 10 V, the maximum output current is 20 mA, and the output voltage frequency range is from 0 kHz to 5 kHz.

We measured the partial discharge starting voltages of the three samples at positive DC voltages at different temperatures respectively. The results are shown in Table 2 and Fig 7. The partial discharge starting voltages of the module of Scheme A and the module of Scheme C exceed 6.5 kV, and the partial discharge starting voltages of the module of the module of Scheme B at normal temperature are 6.31 kV. Then the temperature of the device was raised to 150 °C. Although the initial voltages of partial discharge of the three modules decreased obviously, the partial discharge voltage of the module of Scheme A was obviously higher than that of the module of Scheme C, and the devices were not damaged. Scheme B device experienced overcurrent trip during the experiment, which basically indicated that it had been damaged at high temperature. The test conclusion of the above partial discharge experiment is consistent with the HTRB results, which proves that the structural optimization of the module of Scheme C can significantly improve the partial discharge voltage, and then improve the withstand voltage performance of the module.

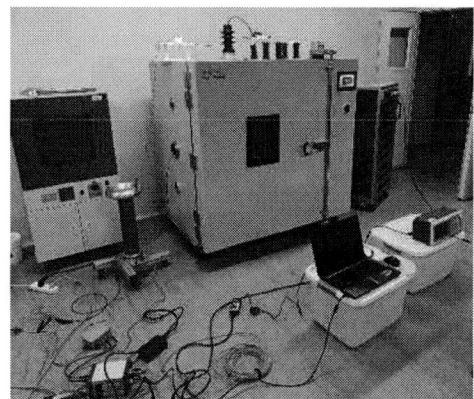

Figure 6. The image of the device used by partial discharges

Table 2. The partial discharge initial voltage of three package scheme modules.

Samples	Partial discharge initial voltage , kV	
A	8.17	2.0
B	6.31	0
C	8.02	4.5

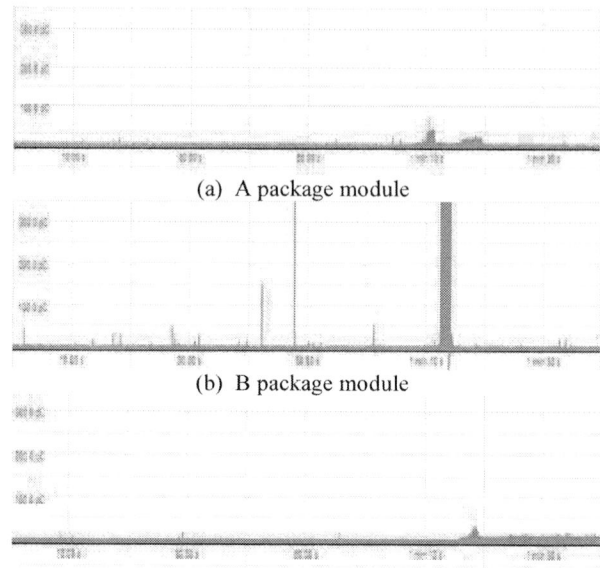

(a) A package module

(b) B package module

(c) C package module

Figure 7. The partial discharge pulse diagram of three package modules at room temperature.

4 CONCLUSIONS

In summary, the insulation studies on high voltage SiC modules suitable for high temperature conditions were carried out. First of all, through the test of material leakage current at high temperature and high voltage and the T-Ramu model, we believe that the leakage current of insulation materials is an important factor affecting the insulation life of materials. Secondly, by comparing three package schemes of single-chip SiC module, long-term high temperature reliability (HTRB test) and partial discharge test, charge behavior such as partial discharge and flashover discharge are the main factors leading to insulation material failure. By optimizing the creepage distance inside the module and improving the potting method, its insulation reliability can be significantly improved, and the passing rate of 800h 6kV high temperature reverse bias test of this module at 150 °C reaches 100%.

ACKNOWLEDGMENT

This work is supported by the National Key Research and Development Program (2016YFB0400500). At the same time, thanks for the support of State Key Laboratory of Advanced Power Transmission Technology.

REFERENCES

[1] Fengze H, Wenbo W, et al. Review of Packaging Schemes for Power Module [J]. IEEE Journal of Emerging and Selected Topics in Power Electronics, 2020, 8(1):223-238.

[2] Siergiej R R, Clarke R C. Advances in SiC Materials and Devices: An Industrial Point of View [J]. Material Science and Engineering, 1999, 61-62: 9-17.

[3] Bhatnagar M, Baliga B J. Comparison of 6H-SiC, 3C-SiC, and Si for power devices [J]. Electron Devices, IEEE Transactions on, 1993, 40(3): 645-655.

[4] Chen C,Luo F, Kang Y. A Review of SiC Power Module Packaging: Layout, Material System and Integration [J]. Cpss Transactions On Power R Electronics And Application, 2017, 2(3):170-187.

[5] Siyang L, Baliga B J, Yifan J, et al. Electrical Performances and Physics Based Analysis of 10kV SiC Power MOSFETs at High Temperatures, Materials Science Forum [J].2018,924:719-722.

[6] Wolfgang B, Swen G, et al. Comparison of Si- and SiC-powerdiodes in 100A-modules, 2007 European Conference on Power Electronics and Applications [J]. 2007, 1(10): 705.

[7] Guoyou L, Yongdian P, Guiqin C, Wei Y, Development of 3300V/500A SiC Hybrid for Traction Applications, [J].Power Electronics,.2017, 951(8):4-7.

[8] Coppola L, Huff D, Wang F, et al. Survey on High-temperature Packaging Materials for SiC-based Power Electronics Modules [J]. IEEE Power Electronics Specialists Conference Records, 2007, 1: 2234-2240.

[9] Srinivas M B, Ramu T S. Multifactor Aging of HV Generator Stator Insulation Including Mechanical Vibrations [J]. IEEE Transactions on Electrical Insulation, 1992, 27(5): 1009.

[10] Lebey T, Malec D, et al. Partial Discharges Phenomenon in High Voltage Power Modules[J]. Transactions on Dielectrics and Electrical Insulation, 2006, 13(4),810-819.

Adaptive Digital Technique Assisted Hard Switching Fault Detection for SiC MOSFETs

Saravanan Dhanasekaran
Electrical Engineering
Indian Institute of Technology Madras
Chennai, India
ee18s075@smail.iitm.ac.in

Vamshi Krishna Miryala
Electrical Engineering
Indian Institute of Technology Madras
Chennai, India
ee14d010@ee.iitm.ac.in

Kamalesh Hatua
Electrical Engineering
Indian Institute of Technology Madras
Chennai, India
kamalesh@ee.iitm.ac.in

Abstract—A Hard Switching Fault (HSF) detection technique for SiC MOSFETs is presented in this paper. In fault circumstances, the dependability of SiC MOSFETs is critical and can lead to failure. To detect the HSF situation, the suggested technique solely requires device voltage sensing. Instead of the desaturation method's set blanking time, the proposed method adaptively alters the blanking time during each switching cycle. This allows for speedier identification of the shoot-through occurrence, as well as a reduction in the amplitude of the fault peak current. A discrete 1kV, 32A SiC MOSFET was used in the experimental verification of the proposed approach. The results reveal that the proposed technique works well, with the HSF event being detected in a few tens of nanoseconds.

Index Terms—HSF, Short-Circuit, Active Gate Driver, Adaptive time, SiC MOSFETs, Fault Current, Fault Detection.

I. INTRODUCTION

Wide bandgap semiconductor devices are becoming increasingly popular as a result of their improved performance over their Si counterparts. Greater breakdown voltage, higher operating temperature, reduced loss, and faster switching speed are all advantages of SiC MOSFETs. In the 600V-1.7kV range, SiC MOSFETs are emerging as next-generation switching devices for constructing small and high-power density converters [1]. However, ensuring the dependability of SiC MOSFET-based converters requires safeguarding the SiC MOSFET during a fault situation [2]. Because of their quick switching speed, short-circuit protection becomes even more difficult, and the short-circuit withstanding time of SiC MOSFETs is shorter than that of Si IGBTs, and it decreases further as the junction temperature rises [3]. As a result, for widespread adoption of SiC MOSFETs, a reliable short-circuit detection and quenching mechanism is critical.

For SiC MOSFET short-circuit protection, several protective systems have been developed. The gate charge characteristics reveal the variety in distinguishing the normal and short-circuit events, and the protective circuit is built utilising the characteristics [4]. [5] The current transformer (CT) is utilised to sense the drain current and a quick comparator is employed to identify the short circuit event in this current transformer based short-circuit protection. To detect the fault event, the Rogowski coil is utilised to measure the di/dt, and the drain current is reconstructed using the integrator and compared to the specified fault current reference [6]. In [7], [8] the fault

current is detected using the stray inductance between the gate return and power supply terminal of SiC MOSFETs.All of the strategies discussed above improve short-circuit performance, but they are difficult to execute.

The desaturation technique is commonly used for short-circuit protection of IGBTs. It has the advantage of easy implementation, less complexity, and high noise immunity. But it cannot be directly used for SiC MOSFETs due to the fixed blanking time [9]- [11]. The blanking time is provided to differentiate between the normal turn-on and the short-circuit event. [12] discuss self-adaptive blanking time implemented in IGBTs, where the blanking time variation is brought by monitoring the saturation voltage, and the comparator reference is varied. However, this method cannot be directly implemented in SiC MOSFETs owing to the fast switching speed, the signal integrity issues will be dominant and makes it difficult to implement. [13] presents a modified desaturation circuit, Where an additional resistor is added between the gate and desaturation sensing circuit to minimise the charging time of the blanking capacitor. But the blanking time reduction is less due to the limited range for resistor selection.

Short-circuit protection for SiC MOSFETs can be achieved using a variety of methods. Protective circuitry can be constructed utilising gate charge characteristics that show the variation in identifying whether a normal or short-circuit event occurred. [5] It demonstrates short-circuit protection using a current transformer (CT) and a quick comparator to sense drain current and identify the short circuit occurrence. By using the Rogowski coil, di/dt is measured, and the integrator reconstructs the drain current, which is then compared with the set fault current to detect the fault occurrence. SiC MOSFETs are being used in [7], [8] to detect the fault current by using the stray inductance between the gate return and power supply terminals. Although all of the above-mentioned techniques improve short-circuit performance, they are difficult to execute.

II. WORKING PRINCIPLE OF THE PROPOSED METHOD

This section describes the difference between the typical turn-on and HSF conditions. By looking at the HSF transient characteristics, the drain to source voltage of the device alone is sufficient to differentiate HSF event. Based on this, the

978-1-6654-4817-8/21 $31.00 © 2021 IEEE

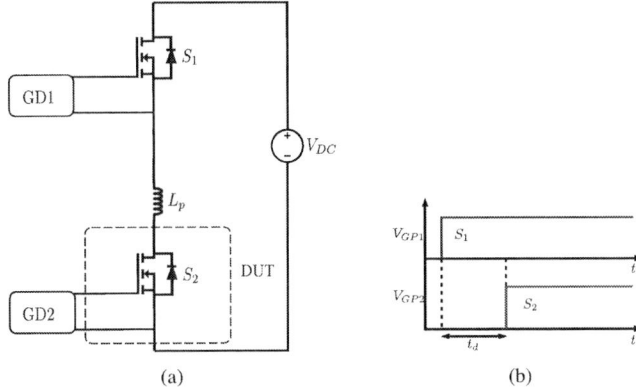

(a)   (b)

Fig. 1: Protection method Testing circuit (a) S_2 switch subjected to HSF (b) Gate signal for switch S_1 and S_2

working principle of the proposed method is explained. Fig.1a shows the test circuit in which S_1 and S_2 are two active switches. L_p represents the layout inductance of the power circuit. The switch S_1 is turned on first and after a time delay t_d the switch S_2 is turned on. Thus short-circuit event during turn-on of switch S_2 is created. Fig.1b shows the Gate Pulse signal applied to the two switches.

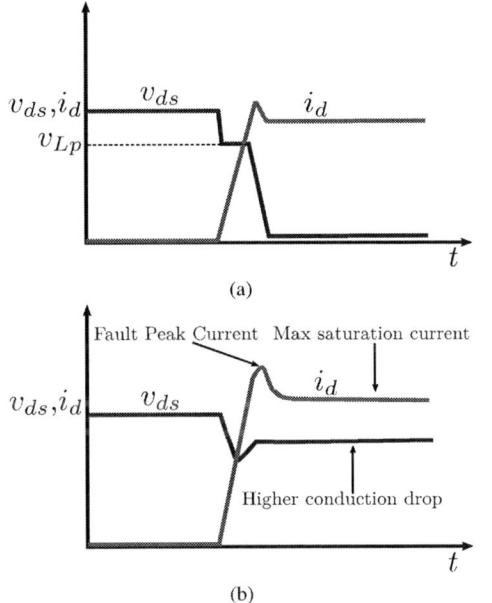

(a)

(b)

Fig. 2: Typical waveform for (a) Conventional Turn-On (b) HSF Event

Fig.2a and Fig.2b shows the waveforms under the conventional turn-on and HSF event respectively. During HSF event, the device conducts in saturation, and the drain current i_d increases. Due to this, the junction temperature increases and considerable power loss happens. The increase in temperature further reduces the channel carrier mobility charges, which reduces the increase in device current and stays at the maximum

saturation current. It can be seen, the device voltage alone is enough to differentiate the HSF event from normal turn-on.

The proposed method consists of v_{ds} sensing circuit with comparator, Adaptive time generation block, Fault generation block, and Turn off switching stage signal generation block as shown in Fig.3. The v_{ds} sensing circuit consists of a resistive divider with a lead compensated capacitor. The gain of the sensing stage can be adjusted by choosing proper value of R_1, R_2 and C_1 , C_2. The sensed v_{ds} signal is compared with V_{ref1}. The V_{ref1} selection will be discussed in the subsequent sections. Based on the comparator signal, the adaptive time generation circuit is used to compute the turn-on switching time T_{on} at each switching cycle and the reference time T_{ref}. The T_{ref} time contains the information of the previous turn-on time as well as the maximum allowable load current change.

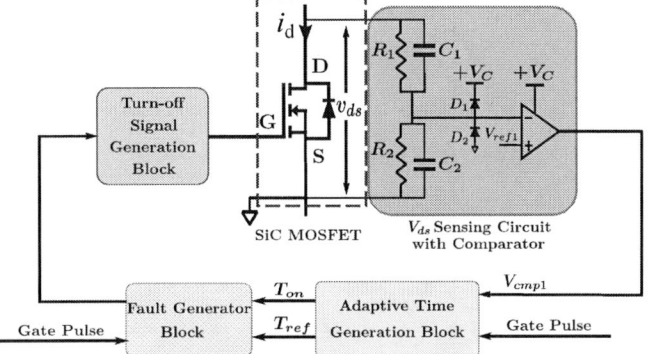

Fig. 3: Proposed HSF detection method

In each switching cycle, T_{ref} gets updated with the new value of turn-on switching time T_{on}. In fault generation block, the turn-on switching time T_{on} is compared with reference time T_{ref} at the same switching cycle to detect the Hard Switching Fault (HSF) event. In normal turn on condition, the computed turn-on time T_{on} at that switching cycle will always be less than the reference time T_{ref}. In the case of the HSF event, the computed turn-on switching time T_{on} will be higher than the reference time T_{ref}. Hence, the fault is generated and the turn off stage is initiated.

A. Adaptive time generation block

The on board CPLD controller is used to implement the Adaptive time generation block. Fig.4 shows the computation of turn-on time T_{on} and the reference time T_{ref}.

Fig.5 shows the typical turn-on transient under normal condition. Upon receiving the Gate signal V_{GP}, the current rise i_d starts after a turn-on delay time (time interval t_1 to t_2) $t_{d(on)}$. The turn-on delay time $t_{d(on)}$ can be expressed by (1) [14].

$$t_{d(on)} = \tau \ln(\frac{V_{GS}}{V_{GS} - V_{TH}}) \qquad (1)$$

Where τ represents time constant $R_g C_{iss}$. The delay1 block shown in Fig.4 contains the information of pre calculated turn-

978-1-6654-4817-8/21 $31.00 © 2021 IEEE   339

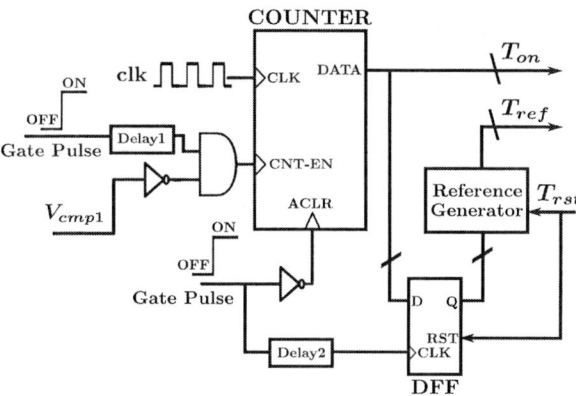

Fig. 4: Detailed Adaptive time generation circuit

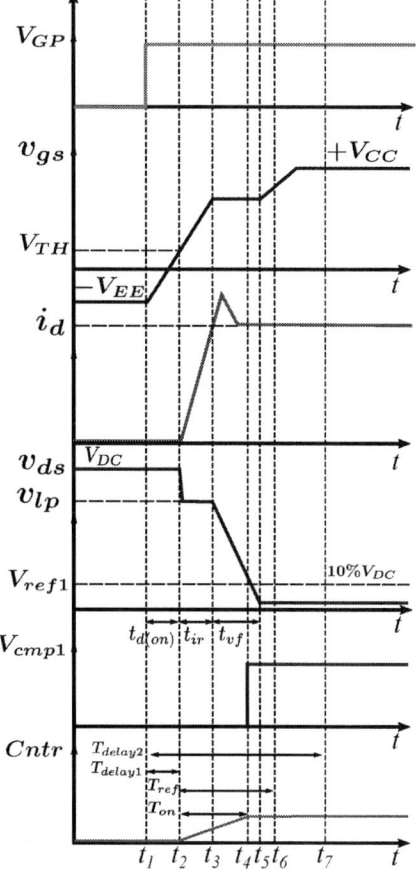

Fig. 5: Turn-on switching waveform with proposed method

The counter module is enabled at the positive edge of the gate pulse V_{GP} and disabled at the positive edge of the comparator signal V_{cmp1}. By this, the turn-on time T_{on} is measured at the each turn-on switching cycle. The counter is cleared at the negative edge of the gate pulse V_{GP}. This ensures the proper T_{on} measurement in the next switching cycle (t_2 to t_4). The reference time T_{ref} can be realised as following (2).

$$T_{ref}(N) = T_{on}(N-1) + T_{sf} \qquad (2)$$

At any instant, the reference time of the current switching cycle T_{ref} contains the information of previous turn-on time T_{on} and the safety factor time T_{sf}. T_{sf} represents maximum allowable change in turn on switching time from previous switching cycle to current switching cycle. This accounts for maximum possible variation in drain current that occurs during consecutive turn-on switching cycles. The computed time T_{on} should be updated to the reference time T_{ref}. In normal turn-on condition, the measured T_{on} time at each switching cycle remains constant after completion of the turn-on switching transient. The D-Flip Flop (DFF) is used to sample the T_{on} time with delay 2 block. The delay 2 block contains the delay time T_{delay2} (in Fig.5) higher than the switching transient time (i.e higher than the reference time). The higher delay time T_{delay2} is provided for the proper update of T_{on} switching time to generate T_{ref} time for next switching cycle. In case of HSF event, the T_{rst} signal is used to reset to the initial conditions. During the first switching cycle (N=1 in (2)), where $T_{on}(0)$ represents turn-on switching time at 0A load current and it can be pre calculated using data sheet values or it can be measured experimentally at 0A load current.

B. Fault generation Block

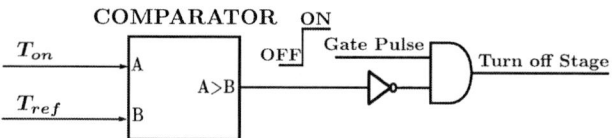

Fig. 6: Fault generation Block

Fig.6 consist of a digital comparator in which the T_{on} is compared with the reference time T_{ref}. T_{on} is lower than the reference time T_{ref}, the output of the comparator becomes lower (i.e normal turn-on event).The comparator's output is set to High if the T_{on} time exceeds the T_{ref} reference time. However, the current is still increasing and the Hard Switching Fault (HSF) has taken place.

Fig.7 shows the turn-on switching waveform during Hard Switched Fault (HSF) condition. In normal turn-on switching cycle the drain to source voltage v_{ds} falls below V_{ref1} and the comparator output goes from low to high, the T_{on} time is estimated (as shown in Fig.6 V_{cmp1} signal). The measured T_{on} time is instantly compared with T_{ref} time. During HSF event the drain current i_d increases and the drain to source voltage v_{ds} will have a higher conduction drop and never reaches the

on delay time T_{delay1}(in Fig.5). This block is used to deduct the turn-on delay time $t_{d(on)}$ from the instant the gate pulse V_{GP} is received. The time interval t_2 to t_4 shows the current rise time t_{ir} and voltage fall time t_{vf}. The V_{ref1} is selected such that the v_{ds} falls below 10% of the DC bus voltage (V_{DC}). When v_{ds} falls below the V_{ref1}, the output of the comparator signal V_{cmp1} (as shown in Fig.3) becomes high.

978-1-6654-4817-8/21 $31.00 © 2021 IEEE

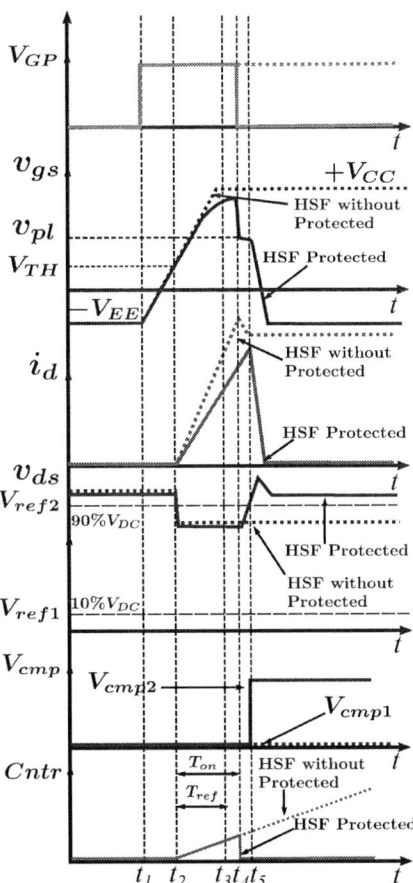

Fig. 7: Turn-on waveform with HSF event

$$i_g = -(I_{f1} + I_{f2}) \qquad (3)$$

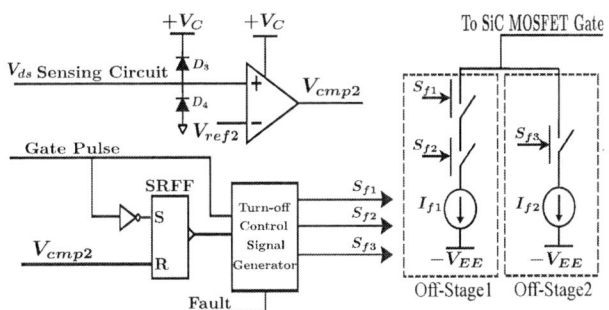

Fig. 8: Fault generation circuit

Here, I_{f1} of higher gate current and I_{f2} of lower gate current is chosen such that fault peak current and voltage overshoot is reduced. The off-stage1 as shown in Fig.8 consists of two series switches S_{f1} and S_{f2}. The switch S_{f1} is biased to normally on condition. The sensed voltage from v_{ds} sensing circuit is compared with V_{ref2} voltage. The V_{ref2} value is set at 90% of V_{DC}. When the turn-off gate pulse arrives, the switch S_{f2} and S_{f3} is turned on. The switch S_{f1} is already turned on, which makes the complete turn-on of the off-stage1. The turn-off is initiated, as the device voltage v_{ds} reaches the V_{ref2} comparator output becomes high, which turns off the switch S_{f1} and eventually off-stage1 is turned off. During this time, the gate voltage v_{gs} quickly reaches the pleatue voltage and the remaining I_{f2} current in (3) will bring down the fault current and the device will be turned off. By this, fast turn-off is achieved and eventually the fault peak current is also reduced.

set reference voltage V_{ref1}. Due to this, the comparator output stays at low and the counter used to measure T_{on} time will not be disabled. As the T_{on} time increases above the the reference time T_{ref}, the comparator output becomes high and the turn-off is initiate. From the Fig.7 the time interval t_2 to t_3 shows the computed T_{ref} and at t_3 instant the T_{on} time exceeds the T_{ref}. The turn-off is initiated at t_4 instant. Due to the faster detection, the turn-off delay time is reduced and fast turn-off can be achieved.

C. Turn-Off switching stage

The turn-off stage of the HSF event is designed to protect the device from voltage and current stress. In order to reduce the voltage stress, the di/dt of the fault current to be brought down must be increased at a higher rate, which results in an overshoot. Suppose the turn-off is made slower the fault current peak increases, which further increases the junction temperature. To achieve optimum turn-off process, the voltage stress and current stress to be controlled within the safe limit. The proposed turn-off stage consists of two stages. The purpose of the off-stage1 is to reduce the voltage overshoot by actively controlling the gate current i_g and off-stage2 is to bring down the fault current and clamp the gate voltage to $-V_{EE}$. The gate current i_g can be represented as (3).

III. SIMULATION RESULTS

The proposed method is verified with LTspice simulation environment. Fig.9 shows the HSF result taken at 0A load current condition. The practical delays (Process delay time and Comparator delay) are not included in the simulation. The current limit is set at 25A. The total fault handling time is 60ns and the fault peak current is limited at 50A. Fig.10 shows the

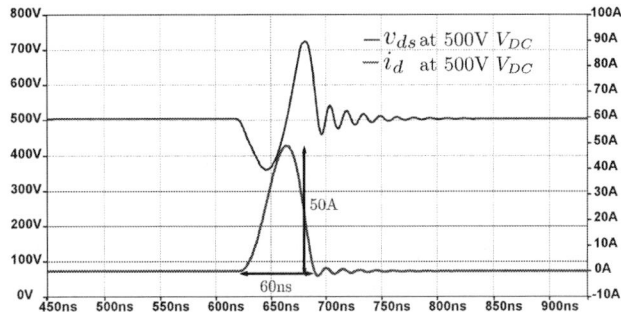

Fig. 9: Hard Switched Fault (HSF) Result at 0A Load Current

Hard Switching Fault result taken with varied DC bus voltage.

With a lower DC bus voltage, the peak current of a failure will be reduced. The DC bus voltage increases the peak current will be increased. When determining the reference time, the change in DC bus voltage will be considered as well. The safety factor time T_{sf} can be varied to include the maximum deviation in the DC bus voltage as well as Junction temperature. Fig.11 shows the HSF event at different Junction temperature.

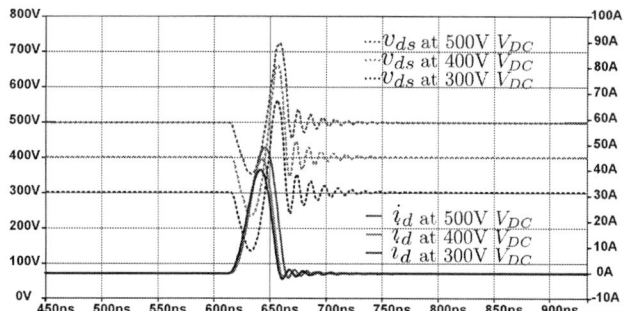

Fig. 10: HSF Result at varied DC bus voltage

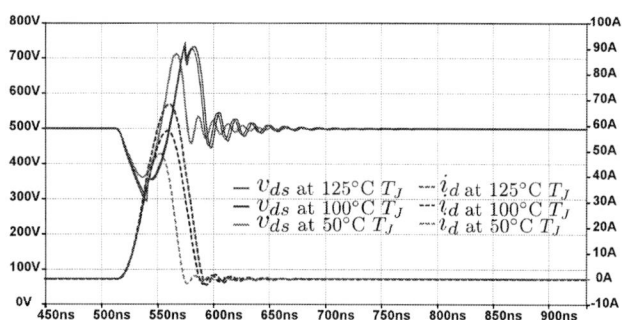

Fig. 11: HSF Result at varied Junction temperature

IV. EXPERIMENTAL RESULTS

The proposed method is tested in Double Pulse Test (DPT) setup at 500V DC bus voltage shown in Fig.12. CREE make part number C3M0065100K (1000V, 32A) SiC MOSFET is used with 130nH layout inductance. The CPLD controller is operating with a clock frequency of 100MHz. The experimental evaluation of Hard Switched fault (HSF) is presented here.

The turn-on switching time T_{on} is pre-calculated either by datasheet values or by measuring experimentally. The calculated value of $T_{on}(0)$ at 0A load current is around 20ns. Fig.13 shows the measured turn-on time for 20A load current is 40ns. From this, the safety factor time T_{sf} can be estimated, which is around 40ns. The safety factor time includes the maximum change in the load current and the comparator signal delay time.

Hence the total reference time T_{ref} can be calculated as 60ns. Fig.14 shows the Hard Switched Fault (HSF) Result taken at 0A load current. The Fault current limit is set at 15A. T_{doff} represents the turn-off delay time which includes the comparator delay time and process delay time for generating the turn-off control signals. Due to the finite process delay

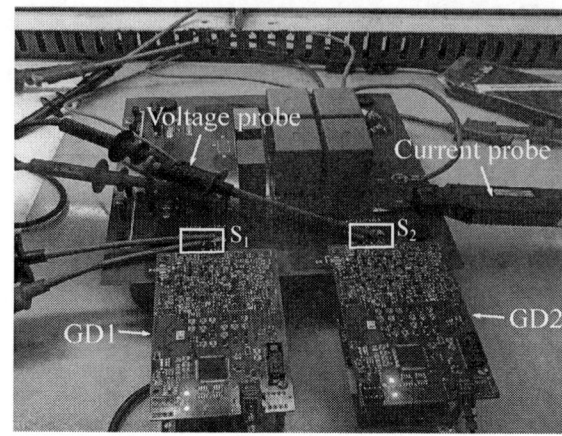

Fig. 12: Photo of Hardware test setup

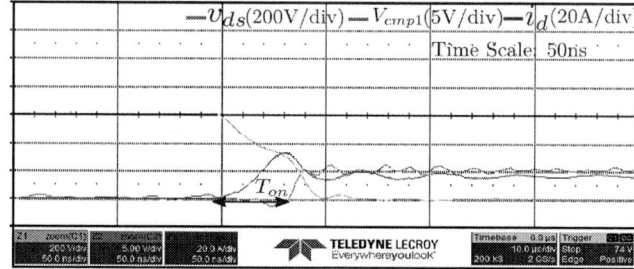

Fig. 13: Estimation of Turn-on time

time, the fault peak current is limited to 45A, and the total HSF time is 130ns (detection time, process time, and total turn off delay time)

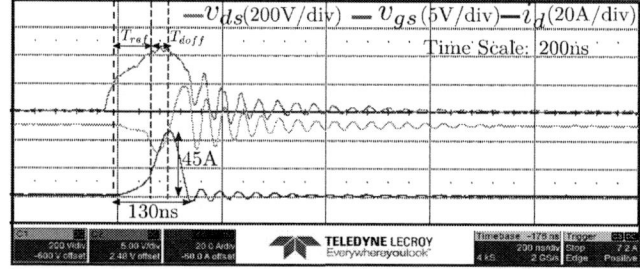

Fig. 14: Hard Switched Fault (HSF) Result at 0A Load Current

The Hard Switching Fault (HSF) event can happen at any load current. HSF event is created at 15A load current. Fig.15 shows the HSF event at 15A load current. The Fault current limit is changed to 25A. It can be seen, the maximum fault peak current is limited to 60A (as shown in Fig.16), which is within the safe limit of the device. The results also show that the proposed method does not trigger false tripping during the normal turn-on switching. Fig.17 shows the proposed method with conventional 2- level turn-off (2LTO) for 0A load current. The total HSF time is 240ns which is longer than the suggested two-stage turn-off, which greatly reduces the turn-off delay time.

978-1-6654-4817-8/21 $31.00 © 2021 IEEE

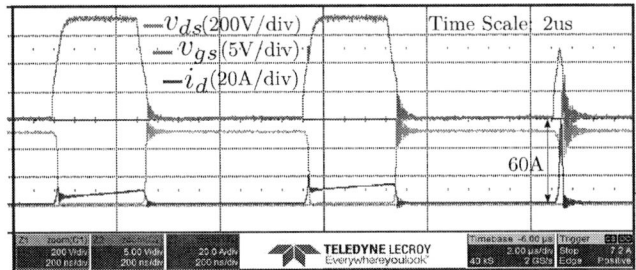

Fig. 15: Hard Switched Fault (HSF) Result at 15A Load Current

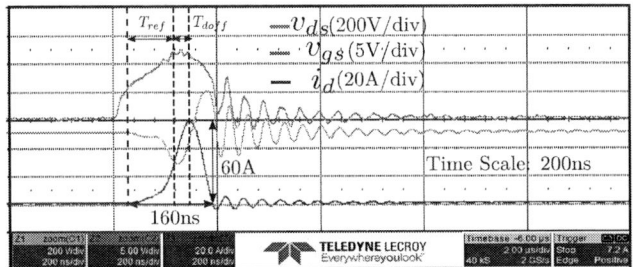

Fig. 16: Transient of Hard Switched Fault (HSF) Result at 15A Load Current

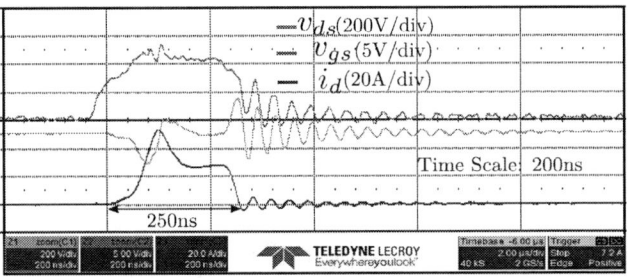

Fig. 17: Conventional 2-Level Turn-off

V. CONCLUSION

The paper presents an adaptive time-based Hard Switching Fault (HSF) detection method, and a two-stage turn-off method is also discussed. The proposed method shows by only sensing the device voltage, the HSF event can be estimated. The detection method reduces the HSF detection time, and also a faster turn-off can be achieved by reducing the fault peak current, thus improving the reliability of SiC MOSFETs.

REFERENCES

[1] X. She, A. Q. Huang, Ó. Lucía and B. Ozpineci, "Review of Silicon Carbide Power Devices and Their Applications," in IEEE Transactions on Industrial Electronics, vol. 64, no. 10, pp. 8193-8205, Oct. 2017, doi: 10.1109/TIE.2017.2652401.

[2] A. Fayyaz, L. Yang and A. Castellazzi, "Transient robustness testing of silicon carbide (SiC) power MOSFETs," 2013 15th European Conference on Power Electronics and Applications (EPE), 2013, pp. 1-10, doi: 10.1109/EPE.2013.6634645.

[3] P. D. Reigosa, F. Iannuzzo, H. Luo and F. Blaabjerg, "A Short-Circuit Safe Operation Area Identification Criterion for SiC MOSFET Power Modules," in IEEE Transactions on Industry Applications, vol. 53, no. 3, pp. 2880-2887, May-June 2017, doi: 10.1109/TIA.2016.2628895.

[4] S. Yano, Y. Nakamatsu, T. Horiguchi and S. Soda, "Development and Verification of Protection Circuit for Hard Switching Fault of SiC MOSFET by Using Gate-Source Voltage and Gate Charge," 2019 IEEE Energy Conversion Congress and Exposition (ECCE), 2019, pp. 6661-6665, doi: 10.1109/ECCE.2019.8912618.

[5] D. Rothmund, D. Bortis and J. W. Kolar, "Highly compact isolated gate driver with ultrafast overcurrent protection for 10 kV SiC MOSFETs," in CPSS Transactions on Power Electronics and Applications, vol. 3, no. 4, pp. 278-291, Dec. 2018, doi: 10.24295/CPSSTPEA.2018.00028.

[6] S. Mocevic et al., "Comparison between desaturation sensing and Rogowski coil current sensing for shortcircuit protection of 1.2 kV, 300 A SiC MOSFET module," 2018 IEEE Applied Power Electronics Conference and Exposition (APEC), 2018, pp. 2666-2672, doi: 10.1109/APEC.2018.8341393.

[7] J. Xue, Z. Xin, H. Wang, P. C. Loh and F. Blaabjerg, "An Improved di/dt-RCD Detection for Short-Circuit Protection of SiC mosfet," in IEEE Transactions on Power Electronics, vol. 36, no. 1, pp. 12-17, Jan. 2021, doi: 10.1109/TPEL.2020.3000246.

[8] K. Sun, J. Wang, R. Burgos and D. Boroyevich, "Design, Analysis, and Discussion of Short Circuit and Overload Gate-Driver Dual-Protection Scheme for 1.2-kV, 400-A SiC MOSFET Modules," in IEEE Transactions on Power Electronics, vol. 35, no. 3, pp. 3054-3068, March 2020, doi: 10.1109/TPEL.2019.2930048.

[9] B. Yu, X. Guo, X. Bu and J. Wu, "Research on the SiC MOSFETs Short Circuit Detection and Protection optimization Method," 2020 IEEE Vehicle Power and Propulsion Conference (VPPC), 2020, pp. 1-7, doi: 10.1109/VPPC49601.2020.9330860.

[10] Z. Wang, X. Shi, Y. Xue, L. M. Tolbert, F. Wang and B. J. Blalock, "Design and Performance Evaluation of Overcurrent Protection Schemes for Silicon Carbide (SiC) Power MOSFETs," in IEEE Transactions on Industrial Electronics, vol. 61, no. 10, pp. 5570-5581, Oct. 2014, doi: 10.1109/TIE.2013.2297304.

[11] S. Yin and Y. Liu, "A Reliable Gate Driver with Desaturation and Over-Voltage Protection Circuits for SiC MOSFET," PCIM Asia 2018; International Exhibition and Conference for Power Electronics, Intelligent Motion, Renewable Energy and Energy Management, 2018, pp. 1-5.

[12] M. Chen, D. Xu, X. Zhang, N. Zhu, J. Wu and K. Rajashekara, "An Improved IGBT Short-Circuit Protection Method With Self-Adaptive Blanking Circuit Based on V CE Measurement," in IEEE Transactions on Power Electronics, vol. 33, no. 7, pp. 6126-6136, July 2018, doi: 10.1109/TPEL.2017.2747587.

[13] J. Kim and Y. Cho, "Overcurrent and Short-Circuit Protection Method using Desaturation Detection of SiC MOSFET," 2020 IEEE PELS Workshop on Emerging Technologies: Wireless Power Transfer (WoW), 2020, pp. 197-200, doi: 10.1109/WoW47795.2020.9291267.

[14] Y. Sukhatme, V. K. Miryala, P. Ganesan and K. Hatua, "Digitally Controlled Gate Current Source-Based Active Gate Driver for Silicon Carbide MOSFETs," in IEEE Transactions on Industrial Electronics, vol. 67, no. 12, pp. 10121-10133, Dec. 2020, doi: 10.1109/TIE.2019.2958301.

978-1-6654-4817-8/21 $31.00 © 2021 IEEE

Short-circuit Protection Circuit of SiC MOSFET Based on Drain-source Voltage Integral

Hong Li
School of Electrical Engineering
Beijing Jiaotong University
Beijing, China
hli@bjtu.edu.cn

Yuting Wang
School of Electrical Engineering
Beijing Jiaotong University
Beijing, China
yutingwang@bjtu.edu.cn

Zhidong Qiu
School of Electrical Engineering
Beijing Jiaotong University
Beijing, China
18121484@bjtu.edu.cn

Zuoxing Wang
School of Electrical Engineering
Beijing Jiaotong University
Beijing, China
19117029@bjtu.edu.cn

Xiaofei Hu
School of Electrical Engineering
Beijing Jiaotong University
Beijing, China
21121428@bjtu.edu.cn

Jia Zhao
Infineon Integrated Circuit (Beijing)
Co., Ltd.
Beijing, China
Jia.Zhao@infineon.com

Abstract—**The short-circuit withstand time of SiC MOSFET is only 2-7μs, which is much shorter than that of Si IGBT, therefore, the short-circuit detection circuit must respond fast and have high bandwidth. According to the short-circuit characteristic of SiC MOSFET that the short-circuit withstand time decreases with the increase of DC bus voltage, a short-circuit protection circuit based on drain-source voltage integral is proposed in this paper to realize the self-adapting short-circuit protection, namely, the higher the DC bus voltage is, the faster the short-circuit protection action is. Further, the experimental platform is built and the experimental results under different DC bus voltage levels are given to verify the rapidness and effectiveness of the proposed short-circuit protection circuit. Finally, the conclusion is drawn. This paper provides a new choice for short-circuit protection of SiC MOSFETs.**

Keywords—*SiC MOSFET, short-circuit characteristic, short-circuit protection, self-adapting*

I. INTRODUCTION

As a typical SiC device, SiC MOSFET has been developed rapidly in recent years. Compared with silicon devices, SiC MOSFET has higher breakdown voltage, higher thermal conductivity and lower conduction loss, these excellent physical characteristics make SiC MOSFET have broad development prospects in high temperature, high voltage, high frequency and other application.

Although SiC MOSFET has many advantages, its short-circuit reliability cannot be ignored. The short-circuit current of IGBT is about 4-6 times of the rated current, while that of SiC MOSFET can reach about 10 times, as shown in Fig. 1. Moreover, the chip area of the SiC MOSFET is smaller than that of the IGBT of the same level, so the short-circuit current density of SiC MOSFET is larger. Therefore, the short-circuit withstand time of SiC MOSFET is much shorter than that of IGBT. Taking the discrete SiC MOSFET of Infineon as an example, the datasheet points out that the short-circuit withstand time of the SiC MOSFET is only 3μs when the DC bus voltage is 800V and the gate-source voltage is 15V. In addition, many researchers have studied the short-circuit characteristic of SiC MOSFET, although the short-circuit withstand time of different SiC MOSFET is not the same, the experimental test results show that the short-circuit withstand

time of SiC MOSFET is shorter with the increase of drive voltage and bus voltage[1-7]. Therefore, the short-circuit protection of SiC MOSFET needs to be self-adapting.

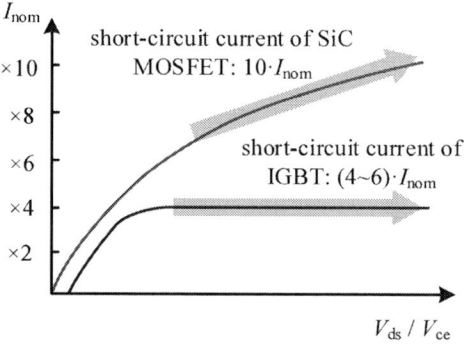

Fig. 1. The relationship between the short-circuit current and the rated current of SiC MOSFET and IGBT.

In recent years, some researchers have designed short-circuit protection circuits for SiC MOSFET. The traditional desaturation detection circuit is adopted to carry out short-circuit protection for SiC MOSFET in [5], which is the most common short-circuit protection method for IGBT application, and whether a short circuit occurs is determined by detecting the conduction voltage of the device. Unlike IGBT, the conduction voltage of SiC MOSFET is more susceptible to temperature than that of Si IGBT, so it is necessary to design the reference voltage and blanking time of desaturation detection reasonably. In [8], the drain current of SiC MOSFET is detected by Rogowski coil. In order to prevent magnetic saturation phenomenon at high frequency, the bandwidth of Rogowski coil needs to be increased. In [9], the hard-switching fault (HSF) is determined by detecting the gate voltage and gate charge of SiC MOSFET, but this short-circuit detection circuit is ineffective for fault under load (FUL), and is affected by the parasitic parameters of the gate loop.

According to the short-circuit characteristic of SiC MOSFET that the short-circuit withstand time decreases with the increase of DC bus voltage, this paper proposes a short-circuit detection circuit based on conduction voltage integral, and the reaction speed of the short-circuit protection increases with the increase of DC bus voltage.

This work was partly supported by the Excellent Youth Scholars of National Natural Science Foundation of China(51822701), the Key Project of National Natural Science Foundation of China(U1866211), and partly supported by Infineon Integrated Circuit (Beijing) Co., Ltd.

II. TRADITIONAL DESATURATION PROTECTION CIRCUIT

In IGBT-based power electronic devices, desaturation detection is usually used for short-circuit protection. According to the output characteristic of IGBT, when short-circuit occurs, the collector current increases rapidly, and the collector-emitter voltage rises rapidly to trigger the action of short-circuit protection circuit, i.e., the device are operated from saturation region to amplification region.

Fig. 2 is the common desaturation detection circuit, and the voltage waveforms is shown in Fig. 3. When the drive signal PWM of SiC MOSFET is low level, SiC MOSFET is in cutoff region, and the transistor T_1 is in on-state, so the voltage at point M is very small. When the drive signal PWM changes to high level, SiC MOSFET is turned on gradually. At the beginning, the drain-source voltage V_{ds} remains large, the diode D_{sat} is reverse blocked, and the transistor T_1 is in off-state, so the voltage VCC charges the capacitor C_{blk} through the resistor R_{blk}, and the voltage at point M gradually increases. when the voltage V_M exceeds the value of V_{ds}, the diode D_{sat} is forward biased, and V_M will change with the forward bias voltage of D_{sat} and V_{ds}. If short circuit occurs, V_{ds} increases rapidly, and VCC charges the capacitor C_{blk}. When V_M exceed the reference voltage V_{ref}, the output voltage of the comparator changes, then the short-circuit protection starts to work, so the drive signal PWM goes low, and the SiC MOSFET is turned off.

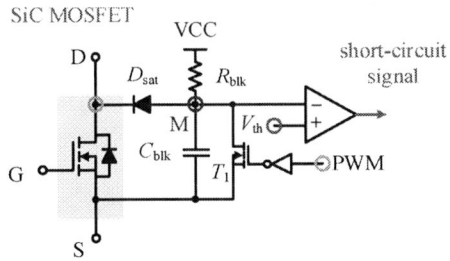

Fig. 2. The desaturation detection circuit.

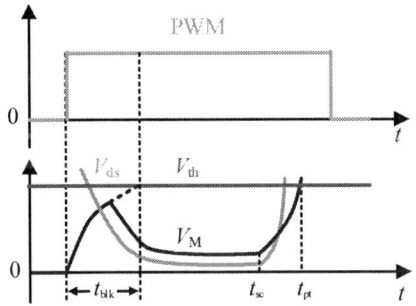

Fig. 3. The voltage waveforms of desaturation detection circuit.

In order to prevent the desaturation detection circuit from triggering by mistake during normal turned on process of SiC MOSFET, it is necessary to reserve enough blanking time T_{blk}, which is given by (1).

$$T_{blk} = R_{blk} \cdot C_{blk} \cdot \ln\left(\frac{VCC}{VCC - V_{th}}\right) \qquad (1)$$

As can be seen from (1) that the desaturation detection circuit is the integral of the DC voltage VCC. When the circuit

parameters are set to a certain value, the blanking time T_{blk} in different operating conditions is basically the same, so it does not have adaptability and needs to meet the requirements of rapid protection in the condition of ensuring that no false trigger occurs.

In view of the short-circuit characteristic of SiC MOSFET that when DC bus voltage is larger, the short-circuit withstand time is shorter, a self-adapting short-circuit protection circuit based on the integral of drain-source voltage is proposed, and the working principle and parameter design method are introduced below.

III. SHORT-CIRCUIT DETECTION CIRCUIT BASED ON DRAIN-SOURCE VOLTAGE INTEGRAL

The voltage and current waveforms of SiC MOSFET under normal operating condition are shown in Fig. 4, and the waveforms under HSF condition and FUL condition are shown in Fig. 5 and Fig. 6, respectively. The conduction voltage $V_{ds(on)}$ of SiC MOSFET is only a few volts, and when a short circuit fault occurs, the drain-source voltage of SiC MOSFET is almost close to the DC bus voltage. Therefore, the integral of conduction voltage $V_{ds(on)}$ of SiC MOSFET can be taken as the short-circuit detection signal, that is the area of the light green shaded area in Fig. 5 and Fig. 6. When the integral of $V_{ds(on)}$ exceeds the reference voltage, a short-circuit signal is output. And the higher the DC bus voltage, the faster the integral value of drain-source voltage reaches the reference value, that is the faster the short-circuit protection speed, which is consistent with the short-circuit characteristic that the short-circuit withstand time of SiC MOSFET decreases with the increase of DC bus voltage, indicating that the proposed short-circuit detection circuit can realize the self-adapting short-circuit protection.

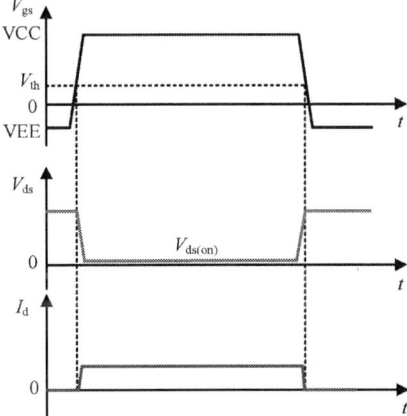

Fig. 4. The voltage and current waveforms of SiC MOSFET under normal operating condition.

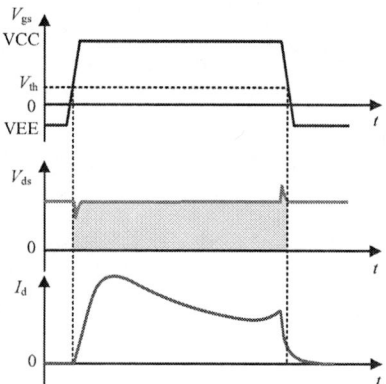

Fig. 5. The voltage and current waveforms of SiC MOSFET under HSF condition.

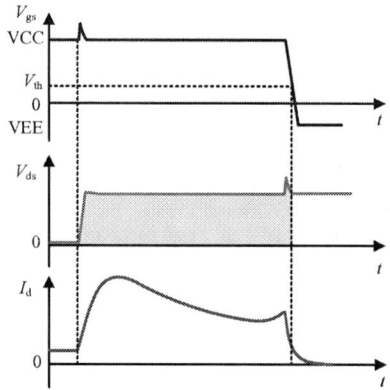

Fig. 6. The voltage and current waveforms of SiC MOSFET under FUL condition.

According to the principles, a short-circuit detection circuit is proposed as shown in Fig. 7, the drain-source voltage V_{ds} of SiC MOSFET is sampled by resistor divider circuit, and the voltage at point A in Fig. 7 is expressed by(2).

$$V_A = \frac{R_2}{R_1 + R_2} \cdot V_{ds} \qquad (2)$$

The active integrator is used to achieve higher bandwidth, and the output voltage V_B of the integral circuit is expressed as (3).

$$V_B = -\frac{1}{R_{int}C_{int}} \int \frac{R_2}{R_1 + R_2} \cdot V_{ds} dt \qquad (3)$$

Since the impedance of integral circuit is far less than that of the voltage divider, the voltage follower is adopted for isolation and buffering. The drive signal of the reset switch S_1 is inverted from the drive signal of the SiC MOSFET, so that the drain-source voltage V_{ds} is integrated only when SiC MOSFET is in on-state, and a discharge circuit is provided by S_1 for the integral capacitor C_{int}. The output of the integral circuit is sent to the non-inverting input of the low delay comparator. Once the integral value exceeds the reference value V_{ref}, the output of the comparator changes from high level to the low one, thereby a short-circuit fault signal is output.

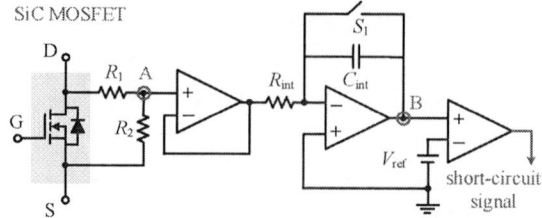

Fig. 7. Short-circuit detection circuit based on drain-source voltage integral.

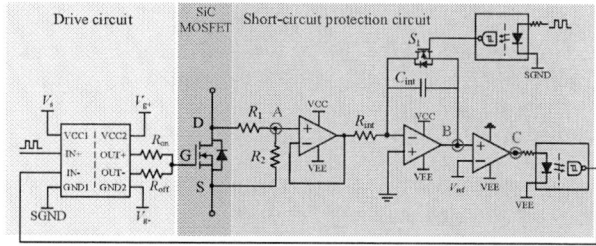

Fig. 8. The proposed short-circuit protection circuit based on drain-source voltage integral.

According to (3), the short-circuit detection time of SiC MOSFET under FUL condition can be expressed as (4).

$$t_{det} = \frac{R_1 + R_2}{R_2} \cdot \frac{R_{int}C_{int}}{V_{ds}} \cdot V_{ref} \qquad (4)$$

Assuming that the delay time caused by the parasitic capacitance of the sampling resistor is T_d, the short-circuit detection time of SiC MOSFET is expressed as (5).

$$t_{det} = T_d + \frac{R_1 + R_2}{R_2} \cdot \frac{R_{int}C_{int}}{V_{ds}} \cdot V_{ref} \qquad (5)$$

Based on the proposed short-circuit detection circuit based on drain-source voltage integral, a short-circuit protection circuit is designed as shown in Fig. 8. The SiC MOSFET IMW120R030M1H of Infineon is selected as the device under test, and the single channel isolation driver IC 1EDI60H12AH of Infineon is used to drive the SiC MOSFET, and the drive voltage is 15/-5V. For the voltage follower and the integrator, the two-channel operational amplifier THS3062DDA of TI is selected, and the supply voltage is VCC=5V, VEE=-5V. The high speed comparator TLV3501AIDBVR of TI is selected, and the reset switch S_1 is a high speed and low power N-channel MOSFET BSS214N of Infineon. According to (5) and the maximum DC bus voltage, the circuit parameters are reasonably selected. The circuit parameters in this paper are selected as R_2/R_1=1/1200, R_{int}=220Ω, C_{int}=330pF, V_{ref}=-2.5V.

An optocoupler is connected to the comparator to achieve electrical isolation, which sends the short-circuit protection signal to the inverted driver input (IN-) of the driver IC. The type of the optocoupler is Broadcom's HCPL-2430, and the input signal is inverted from the output signal, and another optocoupler is used to drive S_1. The Logic input to output switching behavior of the driver IC is shown in the fig. 9. It can be seen that the IN- input of the driver IC is equivalent to the enable pin of the driver IC. Then, the waveforms of the proposed short-circuit protection circuit of SiC MOSFET under normal operation and short circuit conditions are depicted in the fig. 10 and fig. 11, respectively.

978-1-6654-4817-8/21 $31.00 © 2021 IEEE

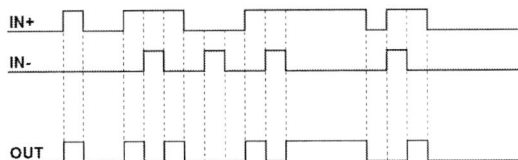

Fig. 9. Logic input to output switching behavior of the driver IC[10].

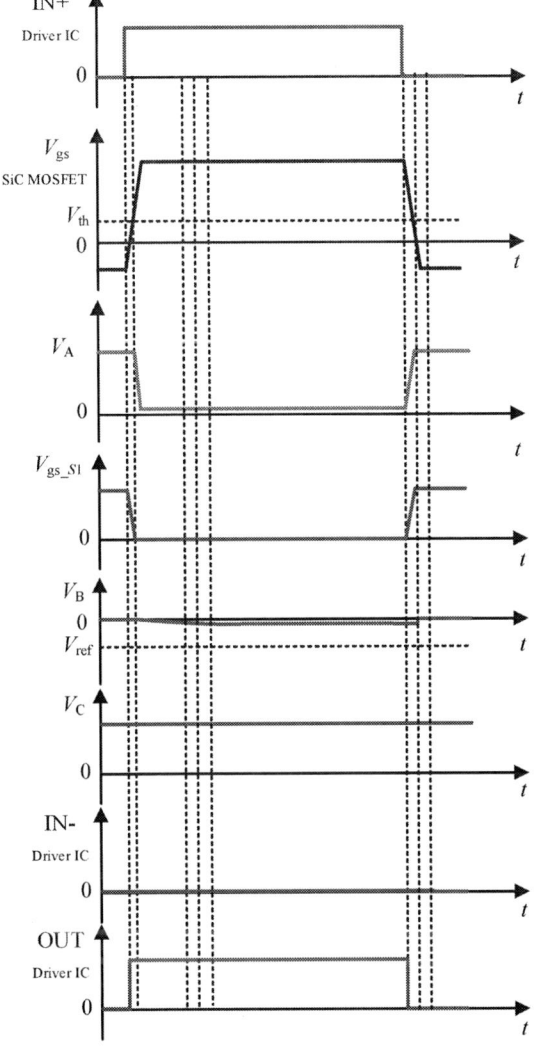

Fig. 10. The waveforms of the short-circuit protection circuit under normal operating condition.

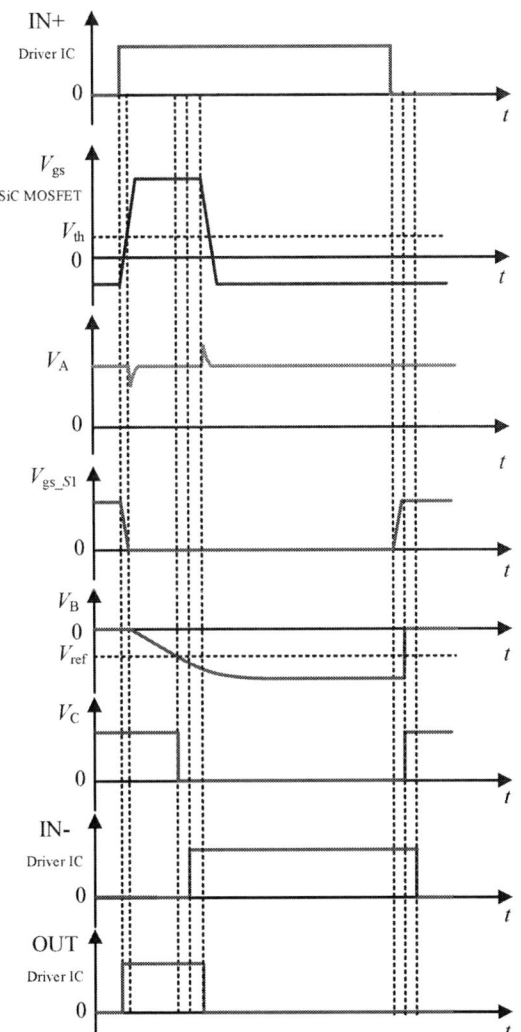

Fig. 11. The waveforms of the short-circuit protection circuit under HSF condition.

IV. EXPERIMENTAL VERIFICATION

In this section, hard-switching short-circuit fault is taken as an example to verify the rapidness and effectiveness of the proposed short-circuit protection circuit.

The short-circuit protection test is carried out by double pulse circuit, and the schematic diagram is shown in fig. 12. The experimental platform is shown in fig. 13, and the circuit parameters are shown in Table I. In short circuit protection experiment, the drain and source electrode (point E and F in fig. 12) of the SiC MOSFET Q_1 is shorted with a wire.

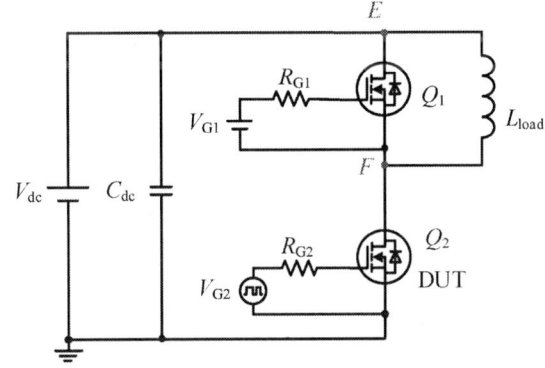

Fig. 12. Double pulse circuit schematic diagram.

TABLE I.   DOUBLE PULSE CIRCUIT PARAMETERS

Parameters	L_{load}	$R_{G1,2}$	V_{G1}	V_{G2}
Values	160µH	10Ω	-5V	-5/15V

978-1-6654-4817-8/21 $31.00 © 2021 IEEE   347

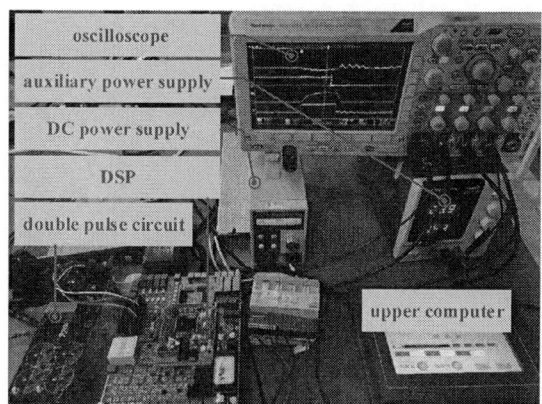

Fig. 13. The short-circuit protection experimental platform.

The short-circuit protection experimental results under different DC bus voltages are shown in Fig. 14 - Fig. 17, as is observed, SiC MOSFET is turned off approximately 163ns after the short-circuit signal from the comparator. The total time for short-circuit protection at different DC bus voltages is presented in Table II, and it shows that SiC MOSFET can be turned off within 2μs of the occurrence of short-circuit. And as the DC bus voltage increases, the short-circuit protection reaction is faster. The experimental results indicate that the short-circuit protection circuit based on drain-source voltage integral can achieve the self-adapting short-circuit protection.

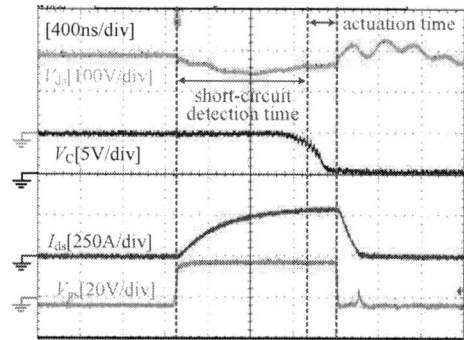

Fig. 14. Short-circuit protection test result at V_{dc}=200V.

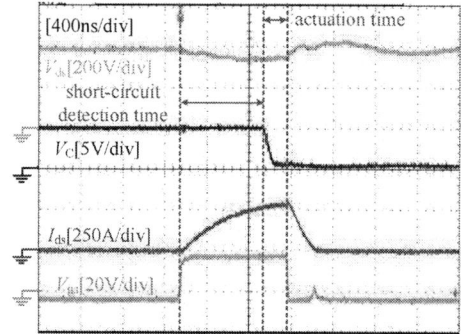

Fig. 15. Short-circuit protection test result at V_{dc}=400V.

TABLE II.   SHORT-CIRCUIT PROTECTION TIME AT DIFFERENT DC BUS VOLTAGES

DC bus voltage	200V	400V	600V	800V
Short-circuit protection time	1.68μs	1.12μs	1.06μs	743ns

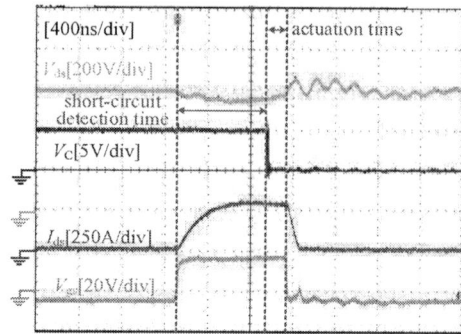

Fig. 16. Short-circuit protection test result at V_{dc}=600V.

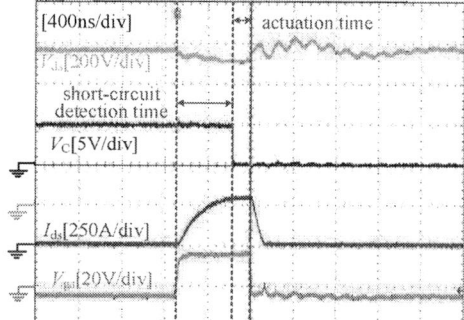

Fig. 17. Short-circuit protection test result at V_{dc}=800V.

Furthermore, in order to verify the noise immunity of the proposed short-circuit protection circuit, double pulse experiments are carried out with the short-circuit protection test circuit, and the experimental circuit parameters are the same as Table I. The double pulse experimental results at different DC bus voltages are shown in Fig. 18 - Fig. 21. It shows that the output voltage V_C of the comparator always maintains a high level, and no short-circuit fault signal is output. Therefore, SiC MOSFET can work normally under different operating conditions, which indicates that the short-circuit protection circuit based on drain-source voltage integral has good anti-noise performance.

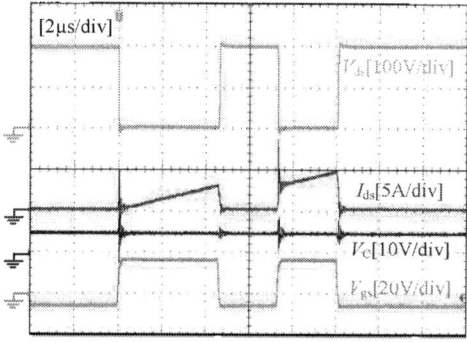

Fig. 18. Double pulse test result at V_{dc}=200V.

978-1-6654-4817-8/21 $31.00 © 2021 IEEE

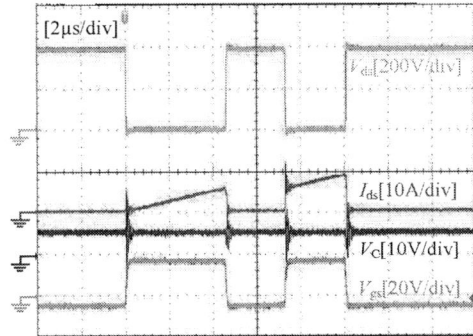

Fig. 19. Double pulse test result at V_{dc}=400V.

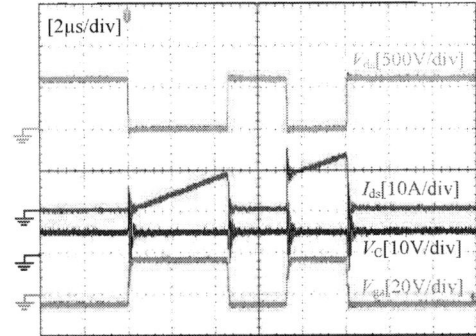

Fig. 20. Double pulse test result at V_{dc}=600V.

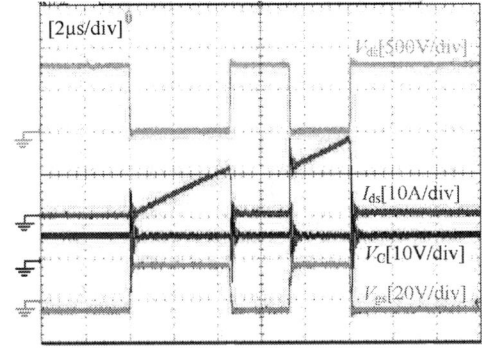

Fig. 21. Double pulse test result at V_{dc}=800V.

V. CONCLUSION

In this paper, a short-circuit protection circuit based on drain-source voltage integral is proposed to realize the self-adapting short-circuit protection of SiC MOSFET, and the rapidness and effectiveness of the short-circuit protection circuit are verified by experiments. As the DC bus voltage increases, the short-circuit protection action response faster, when DC bus voltage is 800V, the short-circuit protection circuit can act within 743ns. Therefore, this paper provides an effective short-circuit protection circuit for SiC MOSFETs.

REFERENCES

[1] H. Qin et al., "A comprehensive study of the short-circuit characteristics of SiC MOSFETs," 2017 12th IEEE Conference on Industrial Electronics and Applications (ICIEA), Siem Reap, Cambodia, 2017, pp. 332-336.

[2] S. Ji et al., "Short circuit characterization of 3rd generation 10 kV SiC MOSFET," 2018 IEEE Applied Power Electronics Conference and Exposition (APEC), San Antonio, TX, USA, 2018, pp. 2775-2779.

[3] H. Chen et al., "Investigation on Short Circuit Test of 3300V SiC MOSFET," 2019 IEEE International Conference on Electron Devices and Solid-State Circuits (EDSSC), Xi'an, China, 2019, pp. 1-3.

[4] G. Romano et al., "Short-circuit failure mechanism of SiC power MOSFETs," 2015 IEEE 27th International Symposium on Power Semiconductor Devices & IC's (ISPSD), Hong Kong, China, 2015, pp. 345-348.

[5] L. Cao, Z. Gao, Q. Guo and K. Sheng, "Experimental Investigations of SiC MOSFETs under Short-Circuit Operations," 2019 31st International Symposium on Power Semiconductor Devices and ICs (ISPSD), Shanghai, China, 2019, pp. 227-230.

[6] J. Kim and Y. Cho, "Overcurrent and Short-Circuit Protection Method using Desaturation Detection of SiC MOSFET," 2020 IEEE PELS Workshop on Emerging Technologies: Wireless Power Transfer (WoW), Seoul, Korea (South), 2020, pp. 197-200.

[7] C. Lin, J. Wu, H. Xu, N. Ren, Q. Guo and K. Sheng, "Comparison and Analysis of Short Circuit Performance of 1200V SiC MOSFETs," 2020 17th China International Forum on Solid State Lighting & 2020 International Forum on Wide Bandgap Semiconductors China (SSLChina: IFWS), Shenzhen, China, 2020, pp. 81-84.

[8] J. Wang, Z. Shen, C. DiMarino, R. Burgos and D. Boroyevich, "Gate driver design for 1.7kV SiC MOSFET module with Rogowski current sensor for shortcircuit protection," 2016 IEEE Applied Power Electronics Conference and Exposition (APEC), Long Beach, CA, USA, 2016, pp. 516-523.

[9] . T. Horiguchi, S. Kinouchi, Y. Nakayama and H. Akagi, "A fast short-circuit protection method using gate charge characteristics of SiC MOSFETs," 2015 IEEE Energy Conversion Congress and Exposition (ECCE), Montreal, QC, Canada, 2015, pp. 4759-4764.

[10] Infineon. 1EDI60H12AH_datasheet[EB/OL]. https://www.infineon.com/dgdl/Infineon-1EDIxxy12AH-DataSheet-v02_00-EN.pdf?fileId= 5546d46253f6505701543843c049027b.

978-1-6654-4817-8/21 $31.00 © 2021 IEEE

Failure Analysis of 200V p-GaN HEMT under Unclamped Inductive Switching Conditions

Junjie Ye
State Key Laboratory of Electronic Thin Film and Integrated Devices University of Electronic Science and Technology of China
Chengdu, China
junjiey0303@gmail.com

Li Xuan*(corresponding author)
State Key Laboratory of Electronic Thin Film and Integrated Devices University of Electronic Science and Technology of China
Chengdu, China
andrew_xuanli@foxmail.com

Yangyang Wu
State Key Laboratory of Electronic Thin Film and Integrated Devices University of Electronic Science and Technology of China
Chengdu, China
wuyangyang_uestc@163.com

Xiaochuan Deng
State Key Laboratory of Electronic Thin Film and Integrated Devices University of Electronic Science and Technology of China
Chengdu, China
xcdeng@uestc.edu.cn

Zhiqiang Li
Microsystem and Terahertz Research Center China Academy of Engineering Physics
Mianyang, China

Bo Zhang
State Key Laboratory of Electronic Thin Film and Integrated Devices University of Electronic Science and Technology of China
Chengdu, China
zhangbo@uestc.edu.cn

Abstract—Power electronic systems based on gallium nitride (GaN) devices are expected to significantly reduce the power loss and increase the operating frequency. This makes GaN devices very promising for next-generation medium-low voltage power electronics, like applications in electric vehicles, 5G communication, lasers, and fast charge. However, reliability issues are still an obstacle to the widespread application of p-GaN high electron mobility transistor (HEMT). It is of great significance to study the reliability of p-GaN HEMT which suffers unclamped inductive switching (UIS) shock. This work reveals the UIS failure mechanism of commercial 200V p-GaN HEMT. During entire UIS process, the p-GaN HEMT does not consume the energy stored by the load inductance. The energy transferred from the load inductance makes the electric field of the passivation layer under the source field plate (SFP) rise rapidly, causes the catastrophic breakdown of the passivation layer. Once the maximum drain-source voltage ($V_{DS,max}$) exceeds the critical peak voltage (V_{peak}), the UIS failure occurs. The ionized traps in passivation layer and hot carrier injection caused by high electric field lead to the formation of a penetration current path between SFP and AlGaN barrier layer. This work provides a new UIS failure mechanism of 200V p-GaN HEMT.

Keywords—200V p-GaN HEMT, UIS, failure mechanism, SFP

I. Introduction

The current commercial gallium nitride (GaN) high electron mobility transistors (HEMTs) are mainly lateral structures and have no parasitic body diodes. They provide extremely low gate charge and no reverse recovery charge, which is suitable for high-efficiency power conversions and high power density [1]-[5]. Due to the polarization effect, GaN HEMTs are usually normally-on devices. The GaN HEMTs with the p-GaN gate create a diode-like characteristic to deplete the two-dimensional-electron-gas (2DEG) in the channel, which achieves normally-off characteristic[6]-[8]. When working in power electronics circuit with inductive loads (e.g., DC-AC converter), the p-GaN HEMT regularly suffers unclamped inductive switching (UIS) stress. It is necessary to study the UIS tolerance of p-GaN HEMT.

There are some researches involved the UIS tolerance and its failure mechanism of p-GaN HEMT. Saito et al. conclude that the p-GaN/AlGaN/GaN gate structure can remove holes so that the p-GaN HEMT has certain "avalanche" capability [9]-[12]. Martínez et al. claim that impact ionization is the predominant breakdown mechanism of 600V p-GaN HEMT under UIS test [13]. Siyang Liu et al. believe that high electric field during the UIS test enhances the inverse-piezoelectric effect, leading to the increase of leakage current and thermal runaway [14]. Yuhao Zhang et al. demonstrate that the failure mechanism of 650V p-GaN HEMT is electrical breakdown and dielectric fatigue related to the peak electric field [15].

Different from the previous researches, we conduct UIS test on commercial 200V p-GaN HEMT for the first time and reveal its UIS failure mechanism. The UIS behaviors of p-GaN HEMT are obviously distinct from those of SiC MOSFET [16]-[18].

II. UIS Test

A. Experimental Background

The subjects used in the UIS test are p-GaN HEMTs (EPC2010C) with the rated voltage of 200V and the rated current of 22A [19]. Fig. 1 shows the schematic diagram of the structural cross-section of the device under test (DUT).

The UIS test circuit topology and the test bench are shown in Fig. 2. The DUT is driven by the gate driver (LM5114) supplied by 4V DC power supply for safe operation. The added gate resistance in the drive loop is 12.5Ω, and the load inductance in the power loop is 0.3mH.

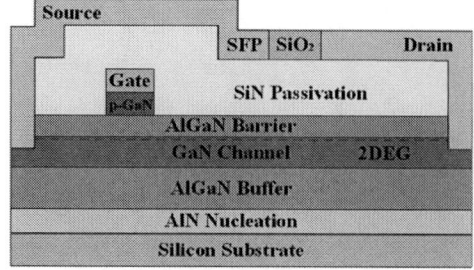

Fig. 1. Schematic diagram of the cross-sectional structure of the DUT.

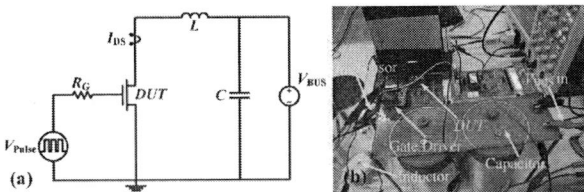

Fig. 2.   (a) Topology of the UIS test circuit and (b) test bench.

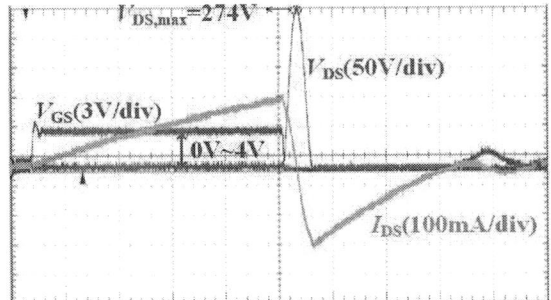

Fig. 3.   Typical waveform of the UIS without failure.

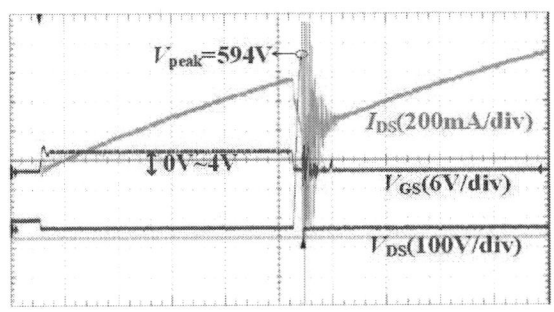

Fig. 4.   Typical failure waveform of the UIS.

B. UIS Stress Process

When the DUT is turned on, the bus voltage (V_{BUS}) charges the load inductance. When the DUT is turned off, the energy of load inductance is released through the DUT. Fig. 3 shows the typical waveform of the UIS without failure. Different from SiC MOSFET [20-21], the drain-source voltage (V_{DS}) exhibits a shape of half sine waveform instead of being clamped to a certain value. It is observed that the maximum drain-source voltage ($V_{DS,max}$) is greater than the rated voltage of the DUT, indicating that the DUT withstands transient electrical stress during the UIS test. It is worth noting that when the V_{DS} reaches $V_{DS,max}$, the drain-source current (I_{DS}) is exactly zero. In other words, the DUT does not consume energy during the UIS test.

The load inductance current can be changed by changing the load inductance charging time or the power supply voltage. As the load inductance current increases, the DUT fails when V_{DS} reaches the critical peak value (V_{peak}, i.e., the $V_{DS,max}$ in the critical failure case) of 594V as shown in Fig. 4. After that, the I_{DS} starts to oscillate harshly and run away. The resistances R_{gs}, R_{gd}, and R_{ds} between the three terminals of the failure device are 5MΩ, 5MΩ, and 0.5kΩ, respectively. There is short-circuit case with the drain and source of the DUT.

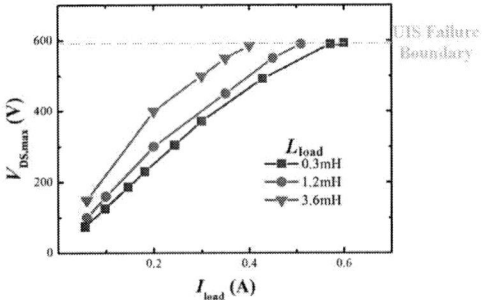

Fig. 5.   UIS failure boundary of the DUT under different load inductors.

C. Failure Boundary and Withstanding Capability

To further investigate whether the failure of DUT is governed by the critical V_{peak}, the UIS capability of DUT is tested with different load inductance. Fig. 5 shows the $V_{DS,max}$ varying with load inductance current (I_{load}) with different load inductances. With the load inductance unchanged, the $V_{DS,max}$ increases with the increase of I_{load}. It can be seen that DUT fails at same V_{peak} with different load inductances. The value of the V_{peak} is defined as a UIS failure boundary of the device, which of the 200V p-GaN HEMT is approximately 600V. The smaller the load inductance is, the higher the I_{load} requires for the $V_{DS,max}$ of the DUT to reach the failure boundary. The UIS tolerance of p-GaN HEMTs is limited by the UIS failure boundary-V_{peak}.

III. UIS FAILURE ANALYSIS

A. Simulation Analysis

For the purpose of understanding the physics process of p-GaN HEMT during UIS test, the device model of p-GaN HEMT is constructed using physics-based technology computer-aided design (TCAD) simulation. The UIS waveform of p-GaN HEMT is achieved by TCAD mixed-mode as shown in Fig. 6. The simulated waveform is almost the same with the typical experimental test waveform as shown in Fig. 3, which suggests reliable analysis in the following parts.

Fig. 7 shows the electric field distribution in p-GaN HEMT when $V_{DS,max}$ is equal to the failure boundary. Due to existence of the source field plate (SFP) inside the device, the new peak electric field appears in the passivation layer at the edge of the SFP. During UIS test, the p-GaN HEMT withstands V_{peak} exceeding the static blocking voltage, which

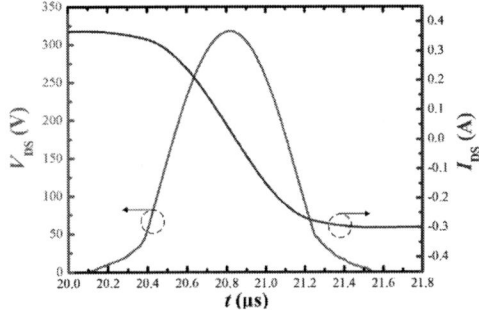

Fig. 6.   Simulated UIS waveform of p-GaN HEMT.

978-1-6654-4817-8/21 $31.00 © 2021 IEEE

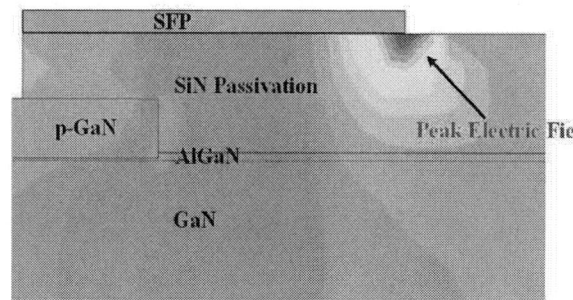

Fig. 7. Electric field distribution in p-GaN HEMT.

makes the passivation layer under the SFP withstand the peak electric field. This indicates that p-GaN HEMT is mainly affected by electrical stress during UIS test.

B. Failure Mechanism

According to the simulation results, the intrinsic physics of p-GaN HEMT during UIS test is revealed. For p-GaN HEMT, the acceptor traps inside the passivation layer are the intrinsic cause of the current collapse. This indicates that the acceptor traps capture electrons from the surface under high electric field.

Before the UIS test, the energy band diagram of the SiN layer and the AlGaN layer under SFP is shown in Fig. 8(a). The acceptor traps in the passivation layer are almost empty. During UIS test, the energy band diagram of the SiN layer and the AlGaN layer under SFP is shown in Fig. 8(b). On one hand, the peak electric field under the SFP makes the empty traps begin to capture electrons to become ionized traps. On the other hand, part of carriers is accelerated to become hot carrier under the high electric field and injected into the AlGaN barrier layer surface to form more traps. With the enhancement of the peak electric field, plenty of ionized traps are accumulated in the passivation layer. Eventually, the ionized traps overlap to form a conduction path, as shown in Fig. 8(c), which causes dielectric breakdown and makes the p-GaN HEMT drain and source short.

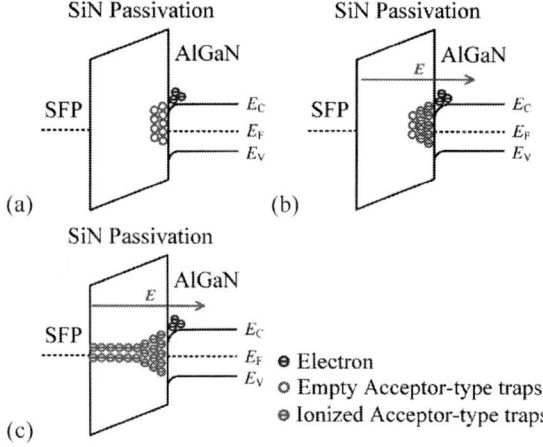

Fig. 8. Schematic diagram of UIS failure mechanism. (a)Before UIS. (b)During UIS before failure. (c)UIS failure.

IV. Conclusion

We conduct UIS test and simulation analysis on commercial 200V p-GaN HEMT, revealing its UIS failure mechanism. Compared with the avalanche features of SiC MOSFETs, p-GaN HEMT without avalanche capability mainly relies on its inherent overvoltage capability to withstand transient electric stress. The V_{peak} plays an important indicator to determine p-GaN HEMT UIS tolerance. The device fails, as long as the $V_{DS,max}$ exceeds its V_{peak} during UIS test. The UIS failure boundary of 200V p-GaN HEMT is approximately 600V. The ionized traps in the passivation layer and hot carrier injection caused by high electric field lead to the formation of a penetration current path between SFP and AlGaN barrier layer, resulting in p-GaN HEMT drain-source short circuit. This research provides an in-depth understanding for the UIS reliability of 200V p-GaN HEMTs.

Acknowledgment

This work was supported in part by the National Natural Science Foundation of China under Grants 62004033, the China Postdoctoral Science Foundation under Grant 2020M683287, and the Science Challenge Project under Grant TZ2018003-1-201.

References

[1] Q. Huang, R. Yu, Q. Ma and A. Q. Huang, "Predictive ZVS Control With Improved ZVS Time Margin and Limited Variable Frequency Range for a 99% Efficient, 130-W/in3 MHz GaN Totem-Pole PFC Rectifier," in IEEE Transactions on Power Electronics, vol. 34, no. 7, pp. 7079-7091, July 2019, doi: 10.1109/TPEL.2018.2877443.

[2] W. Qian, J. Lu, H. Bai and S. Averitt, "Hard-Switching 650-V GaN HEMTs in an 800-V DC-Grid System With No-Diode-Clamping Active-Balancing Three-Level Topology," in IEEE Journal of Emerging and Selected Topics in Power Electronics, vol. 7, no. 2, pp. 1060-1070, June 2019, doi: 10.1109/JESTPE.2018.2873156.

[3] X. Zhou, Y. Wang, L. Wang, Y. -F. Liu and P. C. Sen, "A Soft-Switching Transformerless DC−DC Converter With Single-Input Bipolar Symmetric Outputs," in IEEE Transactions on Power Electronics, vol. 36, no. 8, pp. 8640-8646, Aug. 2021, doi: 10.1109/TPEL.2020.3048230.

[4] X. Zhu et al., "A Sensorless Model-Based Digital Driving Scheme for Synchronous Rectification in 1-kV Input 1-MHz GaN LLC Converters," in IEEE Transactions on Power Electronics, vol. 36, no. 7, pp. 8359-8369, July 2021, doi: 10.1109/TPEL.2020.3042340.

[5] J. Gareau, R. Hou and A. Emadi, "Review of Loss Distribution, Analysis, and Measurement Techniques for GaN HEMTs," in IEEE Transactions on Power Electronics, vol. 35, no. 7, pp. 7405-7418, July 2020, doi: 10.1109/TPEL.2019.2954819.

[6] E. A. Jones, F. F. Wang and D. Costinett, "Review of Commercial GaN Power Devices and GaN-Based Converter Design Challenges," in IEEE Journal of Emerging and Selected Topics in Power Electronics, vol. 4, no. 3, pp. 707-719, Sept. 2016, doi: 10.1109/JESTPE.2016.2582685.

[7] R. Xie et al., "Switching transient analysis for normally-off GaN transistors with p-GaN gate in a phase-leg circuit," 2017 IEEE Energy Conversion Congress and Exposition (ECCE), 2017, pp. 399-404, doi: 10.1109/ECCE.2017.8095810.

[8] Y. Q. Chen et al., "Degradation Behavior and Mechanisms of E-Mode GaN HEMTs With p-GaN Gate Under Reverse Electrostatic Discharge Stress," in IEEE Transactions on Electron Devices, vol. 67, no. 2, pp. 566-570, Feb. 2020, doi: 10.1109/TED.2019.2959299.

[9] T. Naka and W. Saito, "UIS withstanding capability and mechanism of high voltage GaN-HEMTs," 2016 28th International Symposium on Power Semiconductor Devices and ICs (ISPSD), 2016, pp. 259-262, doi: 10.1109/ISPSD.2016.7520827.

[10] T. Naka and W. Saito, "Relation between UIS withstanding capability and gate leakage currents for high voltage GaN-HEMTs," 2017 29th International Symposium on Power Semiconductor Devices and IC's (ISPSD), 2017, pp. 199-202, doi: 10.23919/ISPSD.2017.7988922.

978-1-6654-4817-8/21 $31.00 © 2021 IEEE

[11] W. Saito and T. Naka, "UIS test of high-voltage GaN-HEMTs with p-type gate structure," Microelectronics Reliability, vol. 64, pp. 552–555, Sep. 2016, doi: 10.1016/j.microrel.2016.07.066.

[12] W. Saito and T. Naka, "Relation between UIS withstanding capability and I-V characteristics in high-voltage GaN-HEMTs," Microelectronics Reliability, vol. 76–77, pp. 309–313, Sep. 2017, doi: 10.1016/j.microrel.2017.07.009.

[13] Martínez, P. J., et al. "Failure analysis of normally-off GaN HEMTs under avalanche conditions." Semiconductor Science and Technology 35.3 (2020): 035007.

[14] S. Liu et al., "Single Pulse Unclamped-Inductive-Switching Induced Failure and Analysis for 650 V p-GaN HEMT," in IEEE Transactions on Power Electronics, vol. 35, no. 11, pp. 11328-11331, Nov. 2020, doi: 10.1109/TPEL.2020.2988976.

[15] R. Zhang, J. P. Kozak, M. Xiao, J. Liu and Y. Zhang, "Surge-Energy and Overvoltage Ruggedness of P-Gate GaN HEMTs," in IEEE Transactions on Power Electronics, vol. 35, no. 12, pp. 13409-13419, Dec. 2020, doi: 10.1109/TPEL.2020.2993982.

[16] J. Wei, S. Liu, S. Li, J. Fang, T. Li and W. Sun, "Comprehensive Investigations on Degradations of Dynamic Characteristics for SiC Power MOSFETs Under Repetitive Avalanche Shocks," in IEEE Transactions on Power Electronics, vol. 34, no. 3, pp. 2748-2757, March 2019, doi: 10.1109/TPEL.2018.2843559.

[17] J. Wei, S. Liu, X. Zhang, W. Sun and A. Q. Huang, "Modeling Avalanche Induced Degradation for 4H-SiC Power MOSFETs," in IEEE Transactions on Power Electronics, vol. 35, no. 11, pp. 11299-11303, Nov. 2020, doi: 10.1109/TPEL.2020.2984650.

[18] J. Qi et al., "Avalanche Capability Characterization of 1.2 kV SiC Power MOSFETs Compared with 900V Si CoolMOS," 2019 IEEE 10th International Symposium on Power Electronics for Distributed Generation Systems (PEDG), 2019, pp. 55-59, doi: 10.1109/PEDG.2019.8807671.

[19] EPC: EPC2010C datasheet. [Online]. Available: EPC2010C_datasheet.pdf (epc-co.com)

[20] X. Li et al., "Failure Mechanism of Avalanche Condition for 1200-V Double Trench SiC MOSFET," in IEEE Journal of Emerging and Selected Topics in Power Electronics, vol. 9, no. 2, pp. 2147-2154, April 2021, doi: 10.1109/JESTPE.2020.2965002.

[21] X. Deng et al., "Investigation and Failure Mode of Asymmetric and Double Trench SiC mosfets Under Avalanche Conditions," in IEEE Transactions on Power Electronics, vol. 35, no. 8, pp. 8524-8531, Aug. 2020, doi: 10.1109/TPEL.2020.2967497.

978-1-6654-4817-8/21 $31.00 © 2021 IEEE

Low Roughness SiC Trench Formed by ICP Etching with Sacrificial Oxidation and Ar Annealing Treatment

Changwei Zheng
State key Laboratory of Advanced Power Semiconductor Devices
Zhuzhou CRRC Times Semiconductor Company, Ltd.
Zhuzhou, China
zhengcw@csrzic.com

Zhicheng Wang
State key Laboratory of Advanced Power Semiconductor Devices
Zhuzhou CRRC Times Semiconductor Company, Ltd.
Zhuzhou, China
wangzc1@csrzic.com

Shasha Jiao
State key Laboratory of Advanced Power Semiconductor Devices
Zhuzhou CRRC Times Semiconductor Company, Ltd.
Zhuzhou, China
jiaoss@csrzic.com

Qijun Liu
State key Laboratory of Advanced Power Semiconductor Devices
Zhuzhou CRRC Times Semiconductor Company, Ltd.
Zhuzhou, China
liuqj2@csrzic.com

Yehui Luo
State key Laboratory of Advanced Power Semiconductor Devices
Zhuzhou CRRC Times Semiconductor Company, Ltd.
Zhuzhou, China
luoyh5@csrzic.com

Jieqin Ding
State key Laboratory of Advanced Power Semiconductor Devices
Zhuzhou CRRC Times Semiconductor Company, Ltd.
Zhuzhou, China
dingjq@csrzic.com

Chengzhan Li
State key Laboratory of Advanced Power Semiconductor Devices
Zhuzhou CRRC Times Semiconductor Company, Ltd.
Zhuzhou, China
licz@csrzic.com

Abstract—**A simple post-trench treatment of SiC trench was developed to improve the trench morphology. The post-trench treatment was comprised of sacrificial oxidation at 1150℃ and high temperature annealing in Ar ambient at 1500℃ where the sacrificial oxidation was carried out before the high temperature Ar annealing. The sacrificial oxidation can eliminate most of the step bunching in the trench sidewall which resulted in a relatively smooth surface. The high temperature Ar annealing can further reduce the roughness of the trench sidewall as well as round the trench corner. An ideal morphology of the SiC trench without obvious degradation in the SiC top surface has been realized by the proposed post-trench treatment.**

Keywords—*SiC trench, ICP, low roughness, sacrificial oxidation, Ar annealing*

I. INTRODUCTION

The SiC trench MOSFET (UMOSFET) has been proved to have lower on-state resistance than planar MOSFET due to it's smaller cell pitch and higher electron mobility in the trench sidewall. The quality of SiC trench plays a very critical role in manufacturing UMOSFET because the inversion channel is established at the trench sidewall whose topography has strong effect on the interface characteristic of trench gate oxide layer. In addition, the high electric field tends to crowd at the gate oxide near the trench corner especially when the corner is angular. This may cause reduction in breakdown voltage[1]. However it's very hard to improve the trench quality by only optimizing the dry etching conditions[2].

A post-trench treatment of high temperature annealing in various atmosphere is proven to be necessary and effective[3]. During high temperature annealing the transformation of SiC trench leads to smooth trench sidewall and rounded trench corner. However the high temperature annealing will cause the evaporation of Si or SiC[4] which may result in the top surface roughness increased and with the temperature increased the trench surface morphology degraded as well. Therefore the annealing temperature should be as low as possible. In this paper, a simple post-trench treatment was developed and a procedure based on this process was evaluated. Through the process optimization, the SiC trench with low roughness sidewall and rounded corner was successfully achieved.

II. EXPERIMENT DETAILS

A procedure of SiC tench etching with post trench treatment is shown in Fig.1. The experiments were performed on 4 inch n-type 4H-SiC substrates with 4°-off orientation. After standard RCA cleaning, the SiO_2 films with a thickness of 2μm were deposited on the SiC substrates by plasma enhanced chemical vapor deposition (PECVD). Photoresist masks with trench width of 1μm were formed on SiO_2. Then the SiO_2 films were etched by reactive ion etching (RIE) to form trench pattern. Since the SiO_2 films were served as SiC etching masks, the SiO_2 trenches with vertical sidewall were desired. The photoresist was removed by O_2 plasma ashing and H_2SO_4/H_2O_2 cleaning to make sure that there were no photoresist residue as well as etching residuals. The SiC substrates were then etched by inductively coupled plasma (ICP) etching after which the SiO_2 masks were removed by wet etching using buffered oxide etchant (BOE).

978-1-6654-4817-8/21 $31.00 © 2021 IEEE

After SiC trench formation, a post-trench treatment comprised of sacrificial oxidation and high temperature annealing in Ar ambient was performed. The sacrificial oxidation at 1150℃ followed by diluted HF rinsing was carried out before high temperature Ar annealing. The thickness of the sacrificial oxide layer was more than 20nm. At last, the Ar annealing was applied with a temperature of 1500℃ for 5min which is insufficient to cause significant sublimation of SiC. The SiC trench morphology was observed by scanning electron microscopy (SEM) and the surface roughness was measured by atomic force microscopy (AFM).

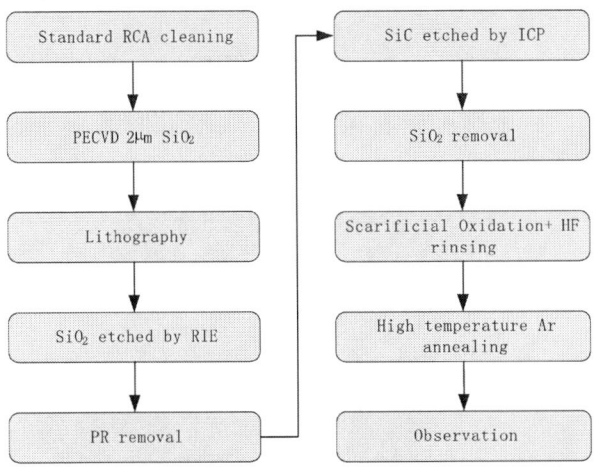

Fig. 1. SiC tench etching procedure with post trench treatment

III. RESULTS AND DISCUSSION

Fig.2(a) shows the cross section of as-etched SiC trench without any post-trench treatment. The SiC etching process has been optimized by adjusting the process conditions such as ICP power, bias power, gas flow and working pressure. It can be seen that the trench is free of microtrench and the sidewall is approximately vertical. However the corners in the trench opening and bottom are angular. That is because the physical etching is stronger than chemical etching in SiC etching process. The SEM image of the trench sidewall is shown in Fig.2(b) where serious step bunching caused by dry etching can be observed. These step bunching can not be eliminated by just optimizing SiC etching process conditions. The RMS roughness of the trench sidewall measured by AFM is 5.25nm which is too high to fabricate UMOSFET. All the experiments were carried out based on this kind of trench.

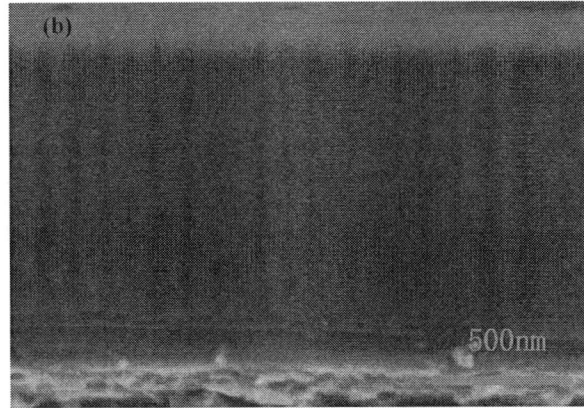

Fig. 2. (a) The cross section and (b) trench sidewall view of the as-etched SiC trench

Sacrificial oxidation doesn't have obviously effect on the trench shape but it is helpful to remove the etching residuals and smooth the trench surface. Fig.3 shows the SEM and AFM image of trench sidewall after sacrificial oxidation at 1150℃. As can be seen the step bunching are eliminated mostly which is probably because the convex part of the stripe feature has a faster oxidation rate than the concave part. Meanwhile, the sacrificial oxidation can remove the graphitic carbon on the surface. Through controlling the thickness of oxide layer above 20nm a low RMS roughness of 0.92nm is obtained. Besides the sacrificial oxidation is also beneficial for reducing the Ar annealing time so that decreasing the negative effect of high temperature annealing on SiC top surface.

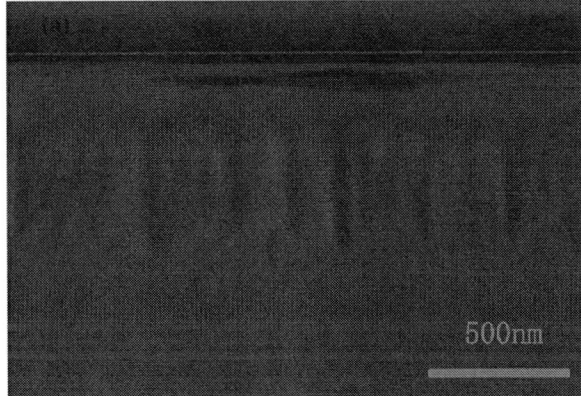

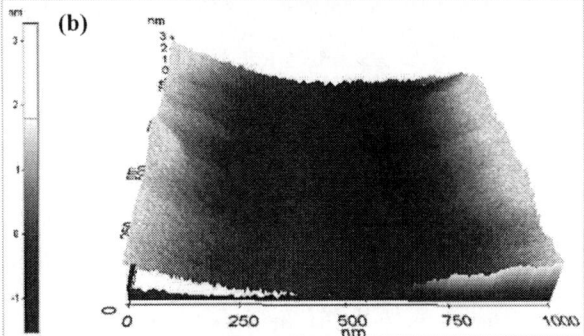

Fig. 3. (a) The SEM and (b) AFM image of SiC trench sidewall after sacrificial oxidation at 1150℃

Fig.4(a) shows the SEM image of trench sidewall after Ar annealing at 1500℃. Benefited from the sacrificial oxidation the Ar annealing time can be decreased to less than 10min. It can be seen that the trench sidewall is extremely smooth without any step bunching which suggests that a reconstitution of SiC on the surface takes place during the high temperature annealing. The AFM image of trench sidewall is shown in Fig.4(b). Compared with the trench before Ar annealing, a much lower RMS roughness of 0.46nm is obtained. This value is close to the original roughness of the SiC wafer surface which is suitable for fabrication of UMOSFET.

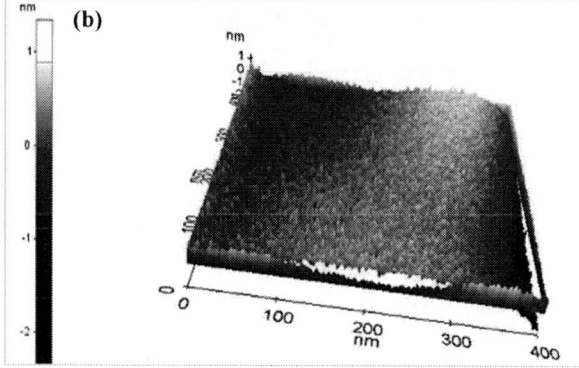

Fig. 4. (a) The SEM image and (b) AFM image of SiC trench sidewall after Ar annealing at 1500℃

Fig.5(a) shows the cross sectional view of SiC trench after Ar annealing. The trench opening and bottom corners are both rounded with a radius of curvature of 0.18μm and 0.22μm respectively which are equal to approximately 1/5 width of the trench. Excluding the corners the trench sidewall is keeping perpendicular to the top surface while the bottom is keeping parallel to the top surface. On this basis the subsequent gate oxidation process with considerable performance can be expected. Fig.5(b) shows the AFM image of SiC top surface in which very shallow steps are observable. It is because that the Si sublimation is unavoidable under high temperature annealing. However when the annealing time is short enough, the steps height can be reduced to a negligible point. The measured RMS roughness is 0.53nm regardless of the white particles which may caused by AFM sample preparation. It is indicated that the post-trench treatment of sacrificial oxidation and high temperature Ar annealing is promising for formation of high quality SiC trench without significant degradation in the top surface.

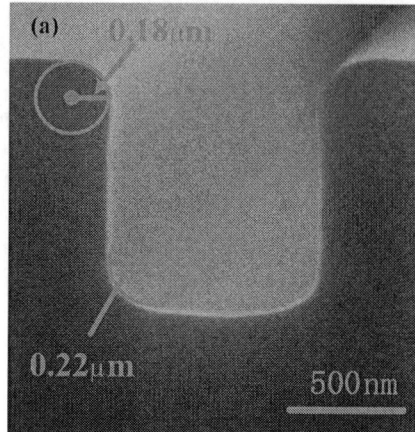

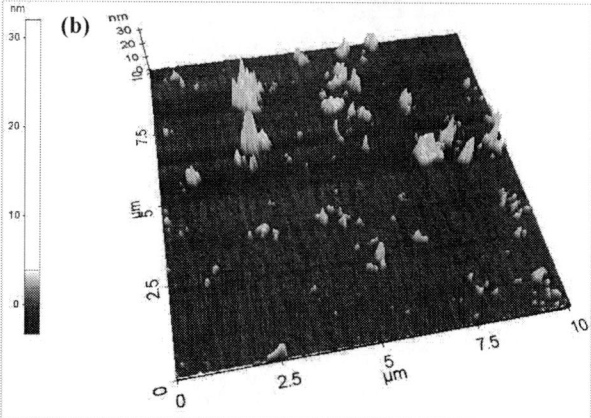

Fig. 5. (a) The cross sectional view of SiC trench and (b) AFM image of SiC top surface after Ar annealing at 1500℃

During the post-trench treatment optimization, we found that the results were quite different when the sacrificial oxidation sequence changed. Fig.6(a) shows the SEM image of trench sidewall after Ar annealing followed by sacrificial oxidation. It can be seen that the topography of the trench sidewall is restored to the state before post-trench treatment. A RMS roughness of 6.6nm was measured by AFM as shown in Fig.6(b). The reappearance of the step bunching with much sharper features reveals that the interaction of sacrificial oxidation and Ar annealing is complementary when sacrificial oxidation carried out first. However when the sacrificial oxidation carried out after Ar annealing, those beneficial results of Ar annealing will be offset.

978-1-6654-4817-8/21 $31.00 © 2021 IEEE

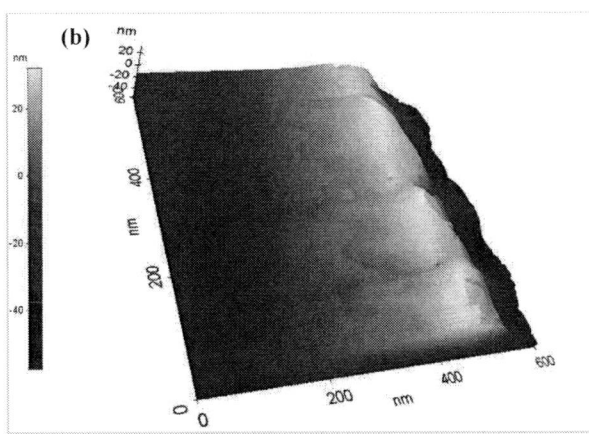

Fig. 6. (a) The SEM image and (b) AFM image of SiC trench sidewall after Ar annealing followed by sacrificial oxidation

IV. CONCLUSION

A simple post-trench treatment of SiC trench was developed to improve the trench morphology. The post-trench treatment of sacrificial oxidation at 1150℃ followed by high temperature annealing in Ar ambient at 1500℃ was effective to reduce the roughness of the trench sidewall and round the trench corner without significant degradation in the SiC top surface. The process sequence was also critical, when the sacrificial oxidation was carried out after the Ar annealing, the topography of the trench sidewall became even worse than the trench as-etched.

ACKNOWLEDGMENT

This work was supported by the National Basic Research Program (Grant No. 2016YFB0400403).

REFERENCES

[1] O. Seok, Y. J. Kim, and W. Bahng, "Micro-trench free 4H-SiC etching with improved SiC/SiO_2 selectivity using inductively coupled $SF_6/O_2/Ar$ plasma," Physica Scripta, vol. 95, p. 045606, 2019.

[2] Y. Kawada, T. Tawara, S. I. Nakamura, T. Tamori, and N. Iwamuro, "Shape Control and Roughness Reduction of SiC Trenches by High-Temperature Annealing," Japanese Journal of Applied Physics, vol. 48, pp. 116508.1-116508.6, 2009.

[3] C. T. Banzhaf, M. Grieb, A. Trautmann, A. J. Bauer, and L. Frey, "Investigation of Trenched and High Temperature Annealed 4H-SiC," Materials Science Forum, vol. 778-780, pp. 742-745, 2014.

[4] H. Naik, K. Tang and T. P. Chow, "Effect of graphite cap for implant activation on inversion channel mobility in 4H-SiC MOSFETs," Materials Science Forum, vol. 615-617, pp. 773-776, 2009.

Research on Threshold Voltage Hysteresis of D-mode and Fully Recessed E-mode AlGaN/GaN MIS-HEMTs with HfO$_2$ Dielectric

Zicheng Yu
School of Materials Science and Engineering
Shanghai University
Shanghai, China
zcyu2020@sinano.ac.cn

Chi Sun
Nanofabrication facility
Suzhou Institute of Nano-Tech and Nano-Bionics
Suzhou, China
csun2017@sinano.ac.cn

Xiaoyu Ding
School of Materials Science and Engineering
Nanjing University of Science and Technology
Nanjing, China
xyding2017@sinano.ac.cn

Xing Wei
Nanofabrication facility
Suzhou Institute of Nano-Tech and Nano-Bionics
Suzhou, China
xwei2018@sinano.ac.cn

Weining Liu
Nano Science and Technology Institute
University of Science and Technology of China
Suzhou, China
wnliu2020@sinano.ac.cn

Li Zhang
Nanofabrication facility
Suzhou Institute of Nano-Tech and Nano-Bionics
Suzhou, China
lizhang2017@sinano.ac.cn

Zhang Chen
School of Materials Science and Engineering
Shanghai University
Shanghai, China
chenzhang@shu.edu.cn

Guohao Yu
Nanofabrication facility
Suzhou Institute of Nano-Tech and Nano-Bionics
Suzhou, China
ghyu2009@sinano.ac.cn

Baoshun Zhang
Nanofabrication facility
Suzhou Institute of Nano-Tech and Nano-Bionics
Suzhou, China
bszhang2006@sinano.ac.cn

Abstract—**D-mode and fully recessed E-mode GaN-based MIS-HEMTs using atomic layer deposition (ALD) hafnium oxide (HfO₂) as gate dielectric were fabricated on Si substrates. Threshold voltage hysteresis (ΔV_{th}) of the D-mode sample was reduced from 1.11V to 0.42V after post-gate-annealing (PGA) at the temperature of 400°C in the nitrogen atmosphere for 5 minutes. This phenomenon can be explained by the reduction of interface-state density (from 8.96×10^{12}-1.2×10^{14}eV^{-1}·cm^{-2} to 2.6×10^{12}-7.6×10^{13}eV^{-1}·cm^{-2}). By contrasting the ΔV_{th} of the D-mode sample and the E-mode sample after PGA, it was found that AlGaN/AlN barrier of D-mode sample could limit electron flux to the interface, leading to its threshold voltage hysteresis being smaller than that of the E-mode sample.**

Keywords—*GaN MIS-HEMTs, HfO₂, interface-state density, threshold voltage hysteresis*

I. INTRODUCTION

Combining with the excellent material and device structure properties such as wide direct energy gap, high 2-D electron gas (2DEG) and high electron saturation velocity, Gallium Nitride (GaN) based transistors are particularly well suited for multiple applications like microwave frequency and high-power, which have been the focus of much research during the last few years [1]. According to whether the threshold voltage (V_{th}) is positive, GaN high electron mobility transistors (HEMTs) were divided into two categories, one of which is depletion-mode (D-mode) and the other is enhancement-mode (E-mode). Compared with D-mode HEMTs, the E-mode HEMTs on Silicon (Si) substrate has normally-off channel that it can greatly improve reliability issues and circuit safety, thus, it is highly preferred. Among the methods for the preparation of E-mode HEMTs, the recessed-gate technique is a common approach for fabrication. To further enlarge the gate swing and decrease the high gate leakage current (I_G), metal-insulator-semiconductor structure

MIS-HEMTs were fabricated by inserting gate dielectric between the AlGaN barrier epilayer or GaN epilayer and the gate metal.

In the numerous gate dielectric, atomic layer deposition (ALD) hafnium oxide (HfO₂) is promising because of its high quality, relatively high dielectric (high-k) constant and wide bandgap properties. Nevertheless, the additional dielectric/semiconductor interface exists interface-defect state that can reduce the stability of the V_{th}. Among the effects, threshold voltage hysteresis (ΔV_{th}) can be monitored in bi-directional transfer sweeps characteristics from below V_{th} to high positive bias and then backward sweep again. Surface treatment was applied to reduce the interface defect states density in several reports, such as pre-gate nitridation and post-deposition annealing [2]. However, there has been less study on the ΔV_{th} of MIS-HEMTs with ALD HfO₂ dielectric and comparison of the difference of ΔV_{th} between the D-mode and the E-mode MIS-HEMTs. Thus, in this study, ΔV_{th} of D-mode and barrier layer completely etched (fully recessed) E-mode HfO₂ MIS-HEMTs with and without post-gate annealing (PGA) was compared by analyzing HfO₂/semiconductor interface traps. Moreover, with the calculation of interface-state density, 400°C PGA in the nitrogen atmosphere for 5 minutes was found that it can lead to reduction of it so that the ΔV_{th} is reduced. In addition, the observation of bi-directional gate transfer sweeps and the trapped electron density at the interface suggests that the AlGaN/AlN barrier of D-mode MIS-HEMTs could suppress the ΔV_{th} at small positive gate bias, which is the fully recessed E-mode devices do not have.

978-1-6654-4817-8/21 $31.00 © 2021 IEEE

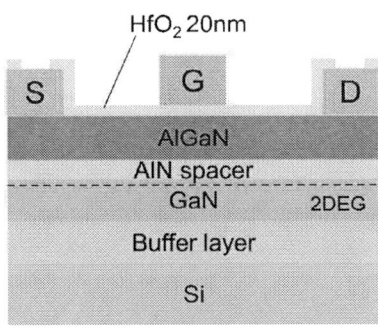

HfO₂ 20nm

D-mode MIS-HEMTs

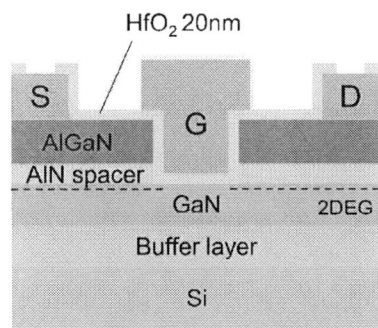

E-mode MIS-HEMTs

Fig. 1. The cross-section structure of HfO₂ D-mode GaN MIS-HEMTs and fully recessed HfO₂ E-mode GaN MIS-HEMTs.

II. STRUCTURE DESCRIPTION

Fig. 1 shows the cross-sections structure of the D-mode HfO₂ MIS-HEMTs sample and the fully recessed E-mode HfO₂ MIS-HEMTs sample. The multiple epilayers grown on silicon substrate consist of a 4.24μm GaN and multiple Al$_x$Ga$_{1-x}$N buffer layer, a GaN channel layers which thickness is 265nm, a 1nm AlN as the pace layer, and a 20nm AlGaN barrier layer. Two kinds of MIS-HEMTs preparation were performed as follows. First, the mesa area was formed by F ion implantation. Second, source area and drain area were deposited Ti/Al/Ni/Au, which the thickness is 20nm/130nm/50nm/50nm followed by annealing at 875 °C in N₂ environment for 30s to obtain the ohmic contacts. In the next step, to achieve E-mode operation, 2μm gate-recess area was defined, and the AlGaN/AlN barrier layer was completely removed by ICP etch process. Then, 20nm HfO₂ dielectric layer was deposited on D-mode sample and E-mode sample by using atomic layer deposition (ALD). Next, Ni/Au (50nm/150nm) as the gate electrodes were deposited by using e-beam evaporation. Then, the electrical properties of D-mode (W/O PGA D-mode) sample and E-mode sample were characterized by using Agilent B1505A measurement system. Finally, post-gate annealing at the temperature of 400°C in N₂

atmosphere for 5min was implemented to improve performance and followed by characterizing the electrical properties again of each kind of HfO₂ MIS-HEMTs (W/ PGA D-mode sample and W/ PGA E-mode sample). The gate length (L_G) = 2μm, gate-to-source spacing (L_{GS}) = 5μm, gate-to-drain spacing (L_{GD}) = 16μm for D-mode devices and 4μm, 4μm, 16μm respectively for E-mode devices.

III. RESULTS AND DISCUSSION

The typical dc transfer curve measurements characteristics of W/O PGA D-mode sample and W/ PGA D-mode sample are shown in Fig. 2. The gate voltage (V_{GS}) of both samples were swept from -10 to 6V and back to -10V, with the drain voltage (V_{GS}) was 10V. The V_{th} was extracted from the corresponding gate voltage when the drain current (I_D) was 10μA/mm. According to the backward voltage sweep curve, the I_{OFF} current values decreased about one order of magnitude after PGA. Furthermore, the V_{th} of the W/ PGA D-mode sample positively shifts by 1.89V compared with the W/O PGA D-mode sample. The decrease of the positive fixed oxide charge, which at the HfO₂/AlGaN interface, result in V_{th} moving toward the positive direction after PGA. It also can be seen from Fig. 2, there was a considerable threshold voltage hysteresis (ΔV_{th} =1.11V) of the W/O PGA D-mode sample and reduced to 0.42V after PGA.

For further probing of the reason for the reduction of ΔV_{th} after PGA, the interface-state density of the W/O PGA D-mode sample and the W/ PGA D-mode sample was analyzed by a C-V method [3]. Different frequencies (f_m varying from 400kHz to 10kHz) at 25°C were measured for both samples and demonstrated in Fig. 3a-b. Based on the onset voltage (V_{on}) of the second slope of f-dispersions, using formulas shown below, the interface-state density of these samples was calculated.

$$\Delta E_T(f_m, T_m) = kT_m ln(\frac{v_{th}\sigma_n N_c}{2\pi f_m}) \tag{1}$$

$$\Delta E_{T\_AVG} = \frac{\Delta E_T(f_1, T) + \Delta E_T(f_2, T)}{2} \tag{2}$$

$$\Delta E_{dis} = \Delta E_T(f_1, T) - \Delta E_T(f_2, T) \tag{3}$$

$$\Delta V_{on} = V_{on}(f_2, T) - V_{on}(f_1, T) \tag{4}$$

$$D_{it}(E_c - E_T = \Delta E_{T\_AVG}) = \frac{C_{ox} \cdot \Delta V_{on}}{q \cdot \Delta E_{dis}} - \frac{C_{ox} + C_B}{q^2} \tag{5}$$

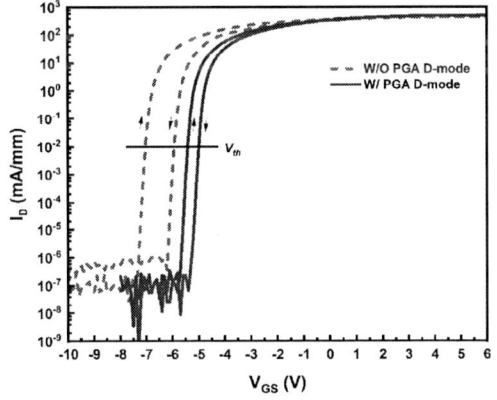

Fig. 2. Transfer characteristics (I_D-V_{GS}) of W/O PGA D-mode sample and W/ PGA D-mode sample in log scale at V_{DS}=10V.

978-1-6654-4817-8/21 $31.00 © 2021 IEEE

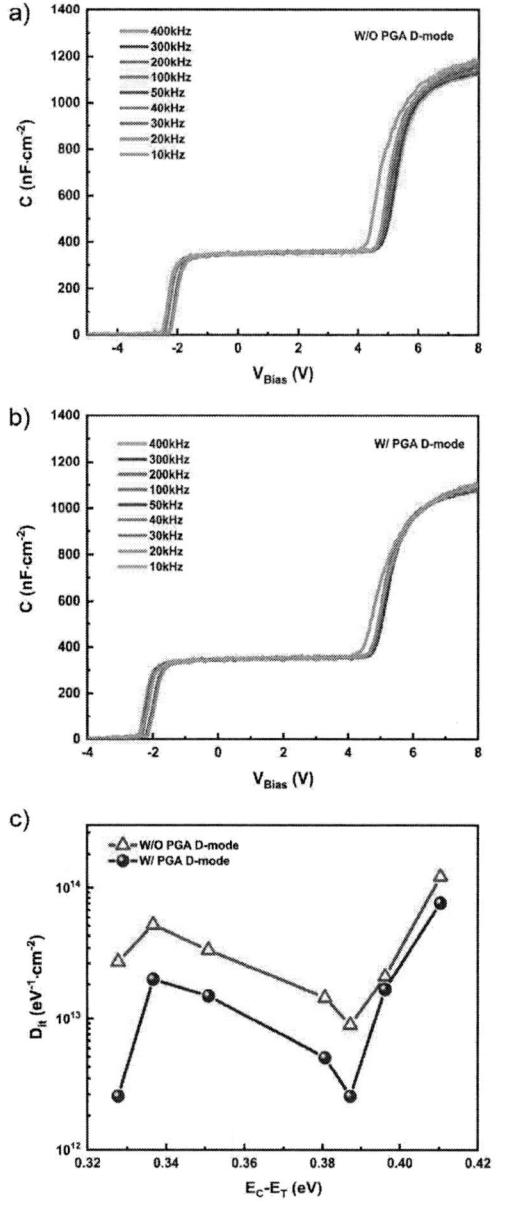

measurement, respectively. In (5), C_{ox} is gate dielectric capacitance and C_B is barrier capacitance. It can be seen from Fig. 3c, the minimum interface-state density is $10^{12} eV^{-1} \cdot cm^{-2}$, and the maximum is $10^{14} eV^{-1} \cdot cm^{-2}$ in the measured energy levels (0.33 eV to 0.41 eV), which is corresponded to the results of the previous studies. In addition, as can also be seen that the interface-state density of the W/ PGA D-mode sample ($2.6 \times 10^{12} - 7.6 \times 10^{13} eV^{-1} \cdot cm^{-2}$) was lower than the W/O PGA D-mode sample ($8.96 \times 10^{12} - 1.2 \times 10^{14} eV^{-1} \cdot cm^{-2}$) in this energy levels range, which is the reason for the decrease of ΔV_{th}.

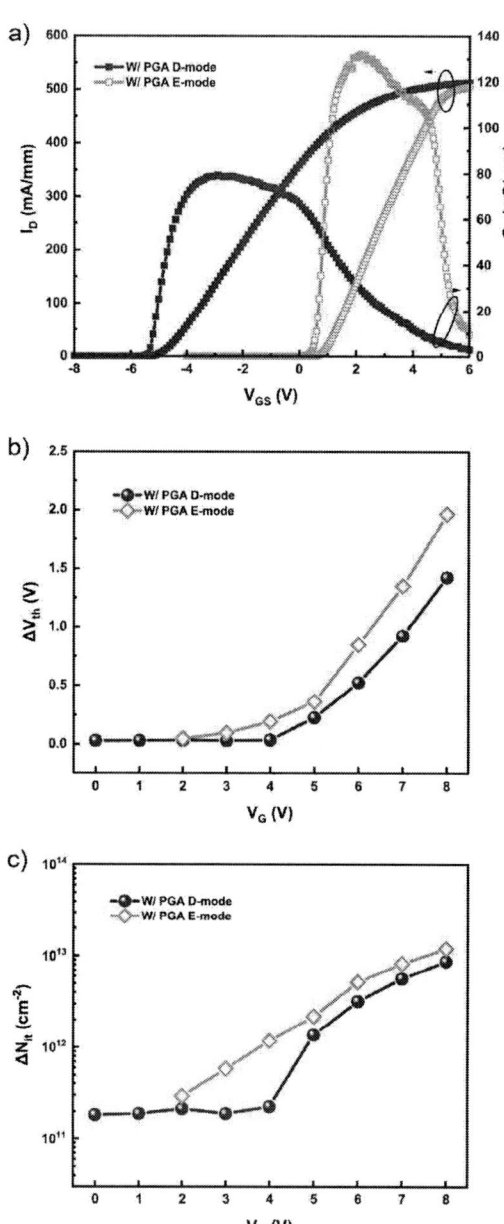

Fig. 3. C-V characteristics with different measurement frequencies and interface-state density for W/O PGA D-mode sample and W/ PGA D-mode sample: a) f_m varying from 400kHz to 10kHz at 25°C of W/O PGA D-mode sample; b) f_m varying from 400kHz to 10kHz at 25°C of W/ PGA D-mode sample; c) The relationship between interface-state density and energy levels for W/O PGA D-mode sample and W/ PGA D-mode sample.

In (1), ΔE_T is the energy range of the dielectric/semiconductor interface states that can be measured under conditions of frequency f_m and temperature T_m. Moreover, N_c is the effective density of states, σ_n is the electron capture cross-section, and v_{th} is the electron thermal velocity in (1). ΔE_{T_AVG} and ΔE_{dis} are the average and difference between $\Delta E_T(f_1,T)$ and $\Delta E_T(f_2,T)$ in (2) and (3), respectively. In (4), $V_{on}(f_1,T)$ and $V_{on}(f_2,T)$ are the second onset voltage with the condition of f_1 and f_2 during the C-V

Fig. 4. a) I_{DS}-V_{GS} characteristics and transconductance at V_{DS}=10V of W/ PGA D-mode sample and W/ PGA E-mode sample; b) Comparison of ΔV_{th} between W/ PGA D-mode sample and W/ PGA E-mode sample; c) The corresponding variation of the trapped electron density at the HfO$_2$/semiconductor interface of W/ PGA D-mode sample and W/ PGA E-mode sample.

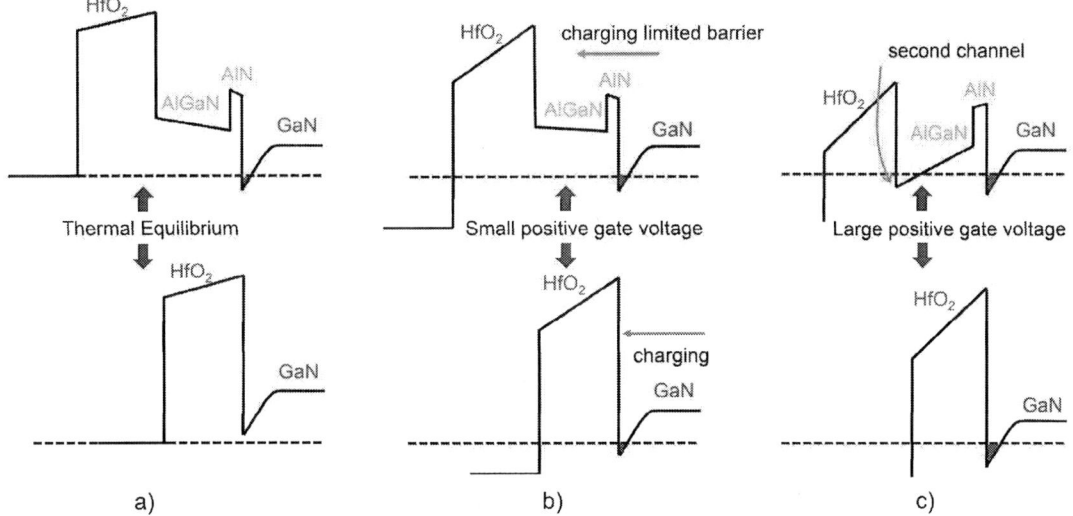

Fig. 5. Schematic band diagrams of W/ PGA D-mode sample and W/ PGA E-mode sample: a) Conduction band edge of W/ PGA D-mode sample and W/ PGA E-mode sample under thermal equilibrium; b) Conduction band edge of W/ PGA D-mode sample and W/ PGA E-mode sample at small positive gate bias; c) Conduction band edge of W/ PGA D-mode sample and W/ PGA E-mode sample at large positive gate bias.

Transfer characteristics of W/ PGA D-mode sample and W/ PGA E-mode sample are shown in Fig. 4a, and both samples were treated with post-gate annealing at the temperature of 400°C in N_2 ambient for 5min. For the W/ PGA E-mode sample, the threshold voltage was 0.6V. The small positive threshold voltage can be attributed to the positive fixed charge at the HfO_2/GaN surface. The maximum I_D values of the W/ PGA E-mode sample is 505.9 mA/mm, which comparable to the W/ PGA D-mode sample. Furthermore, the maximum transconductance (G_m) of the W/ PGA E-mode sample increases to 131.5 mS/mm. The ΔV_{th} is extracted from bi-directional gate transfer sweeps by applying gate voltage from -10V to kinds of forward gate voltage and then back to -10V. Each test used a new device. The evolution of ΔV_{th} with the maximum positive gate voltage in bi-directional gate transfer sweeps measurement are demonstrated in Fig. 4b. The corresponding variation of the trapped electron density at the interface (Fig. 4c), ΔN_{it}, can be obtained from ΔV_{th}, which was measured above with this formula

$$\Delta N_{it} = \frac{C_D \Delta V_{th}}{q} = \left(\frac{\varepsilon_0 \varepsilon_D \Delta V_{th}}{q t_D} \right) \quad (6)$$

where q, C_D, ε_0, ε_D, t_D is the unit charge, the dielectric capacitance, vacuum permittivity, dielectric constant, and the thickness of the dielectric layer in (6), respectively. As can be seen from Fig. 4b, the ΔV_{th} of W/ PGA D-mode sample maintained a small value of about 0.03V when the maximum positive gate voltage was below 4V and was beginning to increase from around V_G of 4V. However, for the W/ PGA E-mode sample, ΔV_{th} was increasing from the V_G=2V, which just exceeded the V_{th} of E-mode sample, and it was larger than the ΔV_{th} of the W/ PGA D-mode sample in the range of 2 to 8V for the V_G. It is shown in Fig. 4c, ΔN_{it} of these samples range from 10^{11} cm^{-2} to 10^{12} cm^{-2}. ΔN_{it} of the W/ PGA E-mode samples increased linearly with the increase of V_G in the range of 2-6V and then tended to be saturated. However, for the W/

PGA D-mode sample, ΔN_{it} increased significantly from V_G=4V and was gradually approaching saturation.

The difference of ΔV_{th} and ΔN_{it} changed with V_G between the W/ PGA D-mode sample and the W/ PGA E-mode sample can be interpreted with the following mechanisms. It is demonstrated in Fig. 5, the process of applying V_G from -10V to various forward gate sweep voltage can be divided into three conditions. At thermal equilibrium, which the V_G=0V, there is a common Fermi-level, and no net current flux existed for both samples. At small positive gate bias condition, there is a small ΔV_{th} for the W/ PGA D-mode sample, and it is expected that few net electron fluxes from electronic channel to HfO_2/AlGaN interface, which was limited by AlGaN/AlN barrier. So that the Fig. 4c blue line shows that the ΔN_{it} of W/ PGA D-mode samples keep around at 10^{11} cm^{-2} when the V_G<4V. However, at this condition (V_G>2V), for the W/ PGA E-mode sample, the channel electrons can easily move to the HfO_2/semiconductor interface on account of lacking AlGaN/AlN barrier which was completely removed during the fully recessed gate etch process. Thus, the increasing of ΔV_{th} and ΔN_{it} can be observed from the beginning of V_G=2V. At large positive gate bias condition (V_G>4V), a second electron channel of W/ PGA D-mode sample at HfO_2/AlGaN interface is formed, electron can flux to interface directly and the AlGaN/AlN barrier can be considered as transparent so that the ΔN_{it} increased obviously to 10^{12} cm^{-2} and ΔV_{th} began to increase.

IV. CONCLUSION

This paper shows the effect of PGA on the ΔV_{th} of the HfO_2 MIS-HEMTs and the difference of ΔV_{th} between the W/ PGA D-mode sample and the W/ PGA E-mode sample. The ΔV_{th} of the W/ PGA D-mode sample was reduced from 1.11V to 0.42V after post-gate annealing at the temperature of 400°C in N_2 ambient for 5min, which is due to the decrease of HfO_2/AlGaN interface-state density in the measured energy levels from 0.33 eV to 0.41 eV (shallow energy level).

Further, the ΔV_{th} of W/ PGA D-mode sample maintained a small value when the V_G<4V is attributed to the AlGaN/AlN barrier which is the W/ PGA E-mode sample does not exist. And second electron channel was formed when V_G=5V led to the increase of ΔV_{th} for the W/ PGA D-mode sample.

ACKNOWLEDGMENT

This paper was supported by the Youth Innovation Promotion Association CAS (No.2020321). The authors would grateful for the technical support from Nano-X, Platform for Characterization and Test, Chinese Academy of Sciences and Nano Fabrication Facility.

REFERENCES

[1] K. J. Chen *et al.*, "GaN-on-Si Power Technology: Devices and Applications," *IEEE Trans. Electron Devices,* vol. 64, no. 3, pp. 779-795, 2017, doi: 10.1109/TED.2017.2657579.

[2] Z. H. Zaidi *et al.*, "Effects of surface plasma treatment on threshold voltage hysteresis and instability in metal-insulator-semiconductor (MIS) AlGaN/GaN heterostructure HEMTs," *J. Appl. Phys.*, vol. 123, no. 18, p. 184503, 2018/05/14 2018, doi: 10.1063/1.5027822.

[3] S. Yang, S. Liu, Y. Lu, C. Liu, and K. J. Chen, "AC-Capacitance Techniques for Interface Trap Analysis in GaN-Based Buried-Channel MIS-HEMTs," *IEEE Trans. Electron Devices*, vol. 62, no. 6, pp. 1870-1878, 2015, doi: 10.1109/TED.2015.2420690.

978-1-6654-4817-8/21 $31.00 © 2021 IEEE

Investigation of the Insulation Failure of Power Modules by Observation of Electrical Trees

Kaixuan Li
State Key Lab of Electrical Insulation and Power Equipment
Xi'an Jiaotong University
Xi'an China
lkx@stu.xjtu.edu.cn

Xingwen Li
State Key Lab of Electrical Insulation and Power Equipment
Xi'an Jiaotong University
Xi'an China
xwli@xjtu.edu.cn

Boya Zhang
State Key Lab of Electrical Insulation and Power Equipment
Xi'an Jiaotong University
Xi'an China
zhangby@xjtu.edu.cn

Haotao Ke
State Key Lab of Advanced Power Semiconductor Devices
Zhuzhou CRRC Times Electric Co., Ltd.
Zhuzhou China
keht@csrzic.com

Abstract—With the increasing of voltage level, the electric field has been distorted more severely in insulated gate bipolar transistor (IGBT). When the distorted electric field lasts for a long time, it will lead to insulation failures in the modules. Triple junction is the area of copper, ceramic and silicone gel in IGBT modules. Partial discharges (PDs) usually initiate from triple junctions, spread on the interface of ceramic and gel, finally lead to insulation failure. In fact, there are processing defects at the edge of triple junctions, such as hollows and protrusions. These defects lead to insulation failure in reality. The relationship between the behaviors of discharge and defects is still unclear. This paper presents an experimental study of the origin of insulation failure in power modules. The influence of defects on the origin of the electrical trees is analyzed. The results show that protrusion defects are rare but awfully dangerous. Because electrical trees can grow from these points easily. The hollow defects are frequent but they not likely to trigger electrical trees compared with protrusion.

Keywords—IGBT modules, triple junctions, defects, electrical trees, insulation failure

I. INTRODUCTION

The power industry is building the system which consists of high proportion of clean power and electronic devices to meet the increasing energy demand. It requires electronic devices to maintain a long-term stable state under complex conditions. The insulation failure is hidden danger in insulated gate bipolar transistor (IGBT) modules. Therefore, it is necessary to conduct research on the insulation problem in IGBT modules[1].

Fig. 1 shows the IGBT module which adopted a direct bonded copper (DBC) ceramic substrate. The chips and diodes solder to the upper surface of the DBC. After that, silicone gel is poured into the module. The encapsulation structure composed of silicone gel and ceramic prevent electrical discharges. The encapsulation structure sustains high electric stress between the upper copper layer of DBC and grounded when the IGBT chip is on blocked state. The distribution of electric field is extremely abnormal at the triple junctions, which has been recorded in red in Fig. 1. These are defined as the junctions of the ceramic, copper layer and silicone gel[2]. PDs initiate at the triple junctions, spread on the surface of ceramic and failure at the end[3].

At present, several studies have been tested and the discharge phenomena has been analyzed at triple junctions in IGBT modules. The electrical detection and optical detection

are common detection methods. The first method detects the partial discharge inception voltage (PDIV) and partial discharge phase distribution (PRPD) of DBC by the partial discharge test instrument such as pulse current detector [4,5,6]. This method can obtain information accurately which characterizes the intensity of discharge, such as the apparent charge and phase. However, it cannot observe the discharge phenomena directly at the triple junctions. The second method can capture the light phenomena by using charge coupled device (CCD) camera, intensified charge coupled device (ICCD) camera, and photomultiplier tub. It positions the discharge locations to the triple junctions [7,8]. However, it is usually required to be carried out in a dark room. And the light phenomena disappear after the process of discharge ends. It is difficult to confirm the relationship between defects and discharges. This is a great obstacle to analyze the influence of defects on insulation fault.

Obviously, it is not enough to position the insulation fault location to the triple junctions through electrical and optical method. In most cases, there are defects at the edge of triple junctions, such as hollows and protrusions. These defects usually are formed by processing and make the modules insulation failure through electrical trees. In this regard, this paper presents an experimental study on the origin of insulation failure in power modules. Insulation failure is equivalent to the growth of electrical trees in this work. And the electrical trees are observed by using a stereo-microscope. The influence of defects on the origin of the electrical trees is analyzed.

II. EXPERIMENTAL SYSTEM

A. Test Sample and Cell

The test sample in this experiment consisted of ceramic, copper and silicone gel. The copper layer was attached to the ceramic by directed bonding processing. This processing caused defects. It could been seen from Fig. 2, several defects

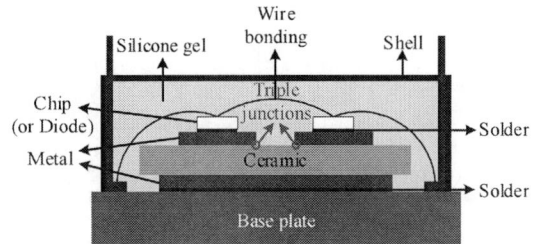

Fig. 1. IGBT module.

Project supported by the National Key Research and Development Program of China (2020YFA0710500).

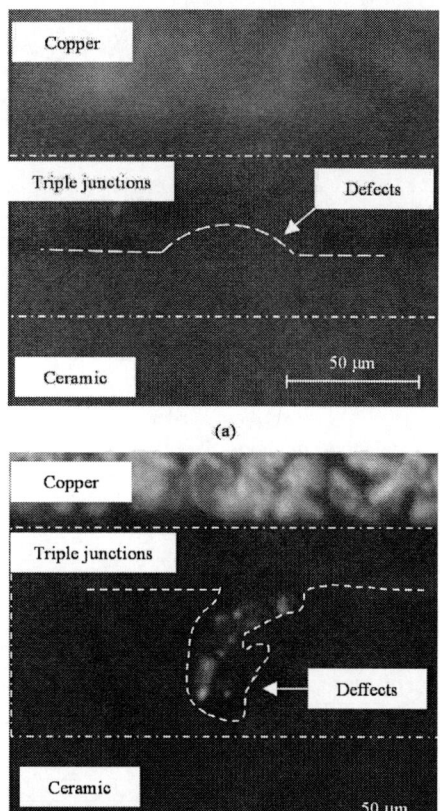

(a)

(b)

Fig. 2. Typical defects caused by processing around the edge of copper layer. (a) hollow, (b) protrusion.

were around the edge of copper layer. All of them may grow electrical trees and lead to insulation failure. As for the size of sample, copper was 24 mm × 25 mm × 0.3 mm and ceramic was 30 mm × 35 mm × 0.63 mm.

As showed in Fig. 3, the upper copper layer was set as high voltage electrode and the lower copper layer was grounded. The DBC substrates were positioned in test cell, which was made of the PTFE. The creepage distance along ceramic from high voltage electrode to ground electrode was long enough to ensure the growth of electrical trees.

Silicone gel in IGBT modules belonged to two component addition-cured gelatinous polymer. The gel consisted of two components: the base silicone oil A and the crosslinking agent B. The preparation process in this study was as follows.

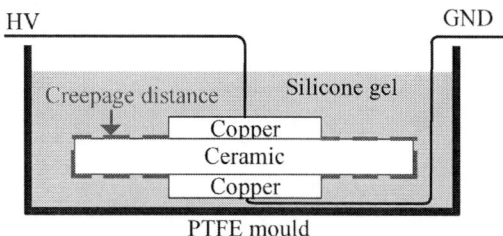

Fig. 3. Test cell.

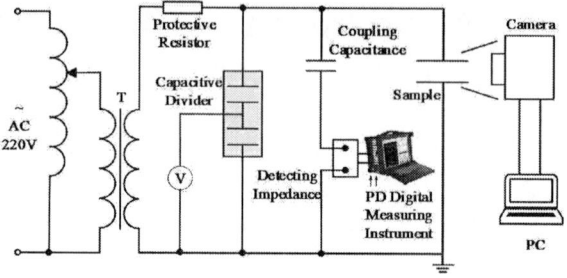

Fig. 4. Test circuit.

1) Mixing. Mixed the part A and part B at a ratio of 10:1, and stirred the mixed liquor for 5 minutes.

2) Degassing. Purred the mixed liquor into test cell and degassed the mixed liquor under a vacuum (at 0.02 MPa) environment for 20 minutes.

3) Curing. Heated the degassed mixed liquor in oven at 100 °C for 60 minutes to cure the gel.

B. Test Circuit and Procedure

The test circuit was shown in Fig. 4. The supply was AC voltage, which consisted of a voltage regulator (GDYD-55D) and a transformer (YDQW). The range of supply was 0~50kV rms. A protection resistor (GR-50kV/5kΩ) was connected in circuit to limit the current. A voltage divider (FVM) was used to measure the voltage. As showed in Fig. 5, a great attention was to capture the electrical trees, a complementary metal oxide semiconductor (CMOS) camera (1200wSONY) was used as microscope eyepiece to record the growth of electrical trees. The magnification of microscope objective was from 0.35 to 9 times. A computer was connected to CMOS in order to analysis the image.

Before testing, all samples were pre-scanned and defects were recorded. After pre-scanning, the samples were aged at 5 kV rms for 2.5 h to achieve condition of operation. The voltage was applied in from 0 V to 10 kV continuously and kept for 15 minutes to grow the electrical trees. After growth, the triple junctions was scanned carefully and each electrical tree was recorded with the defect where it grew from.

III. RESULTS AND ANALYSIS

A. Discussion of Electrical Trees Growing Points

The hollows and protrusions are typical defects which appeared around triples junctions. Electrical trees grew from these defects in most cases and Fig. 6 illustrated electrical trees grown from different cases.

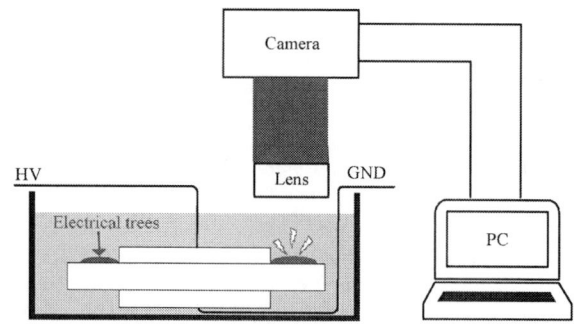

Fig. 5. Electrical trees record.

978-1-6654-4817-8/21 $31.00 © 2021 IEEE

Fig. 6(a) presented that an electrical tree grew from a protrusion defect. The growing point of electrical tree was at the tip of protrusion. There was no other electrical tree near this tree, even if there were hollow defects nearby. This showed that protrusion defects were more likely to generate electrical trees than hollow defects, and the risk of insulation failure was higher.

Fig. 6(b) showed that two electrical trees grew from the corner of two hollow defects respectively. They started to grow separately from two hollows, extended outwards and eventually grew crosswise. Both of electrical trees came from one of two corners in a hollow. This case was the most common in this experiment. It could be concluded that the corners of hollows were more easier to cause discharge than other locations in a hollow. However, there was no tendency about which one corner between two corners of a hollow was more likely to be the growth point. It was determined by electrical field distortion at corner point.

Fig. 6(c) gave an image that an electrical tree grew from the common corner of two hollow defects. As the improper processing, two hollows were close together and formed three corners. Finally, the electrical tree grew from the middle corner. It was a special case and could be classified in cases above mentioned. In one hand, the middle corner could be called protrusion because it was higher than either side of it. In the other hand, the middle corner should be defined as a corner of hollow(s) as either side of it was hollow. Thus, it was no doubt that this complex position became the growing point of electrical tree.

Fig. 6(d) drew that an electrical tree grew from two corners of a hollow defect and Fig. 6(e) drew that an electrical tree grew from a corner and center of a hollow defect. Both of these cases had a common feature that one electrical tree had two growing points. As it was hard to record the process of growth from beginning, it was not clear that how these two cases were formed. However, there were two hypothesizes to explain these cases. The first statement was that two electrical trees grew from two points independently at beginning, but met soon and grew together. The second explanation was that an electrical tree grew from one point, when it grew nearby second point, the charges carried by branches or bubbles of electrical tree changed the electric field around second point and the second electrical tree started to grow, then two electrical trees met soon and grew together. Whether which statement was true, it was no doubt that both of these cases appeared around hollow defects. This meant it was necessary to pay more attention on hollow, even if it was safer than protrusion.

Fig. 6(f) recorded that an electrical tree grew from smooth edge. This case appeared at the round of the copper layer edge, where was the weakness area in triple junctions [9]. And there were no defects at this round and this may be the reason why smooth edge could be growing point. This case also proved that electrical trees not only grew from defect, but also from smooth edge of triple junctions.

B. Statistics of Electrical Trees Growing Points

It can be observed that the probability of electrical tree growth in different defects from Fig. 7. Defects were classified

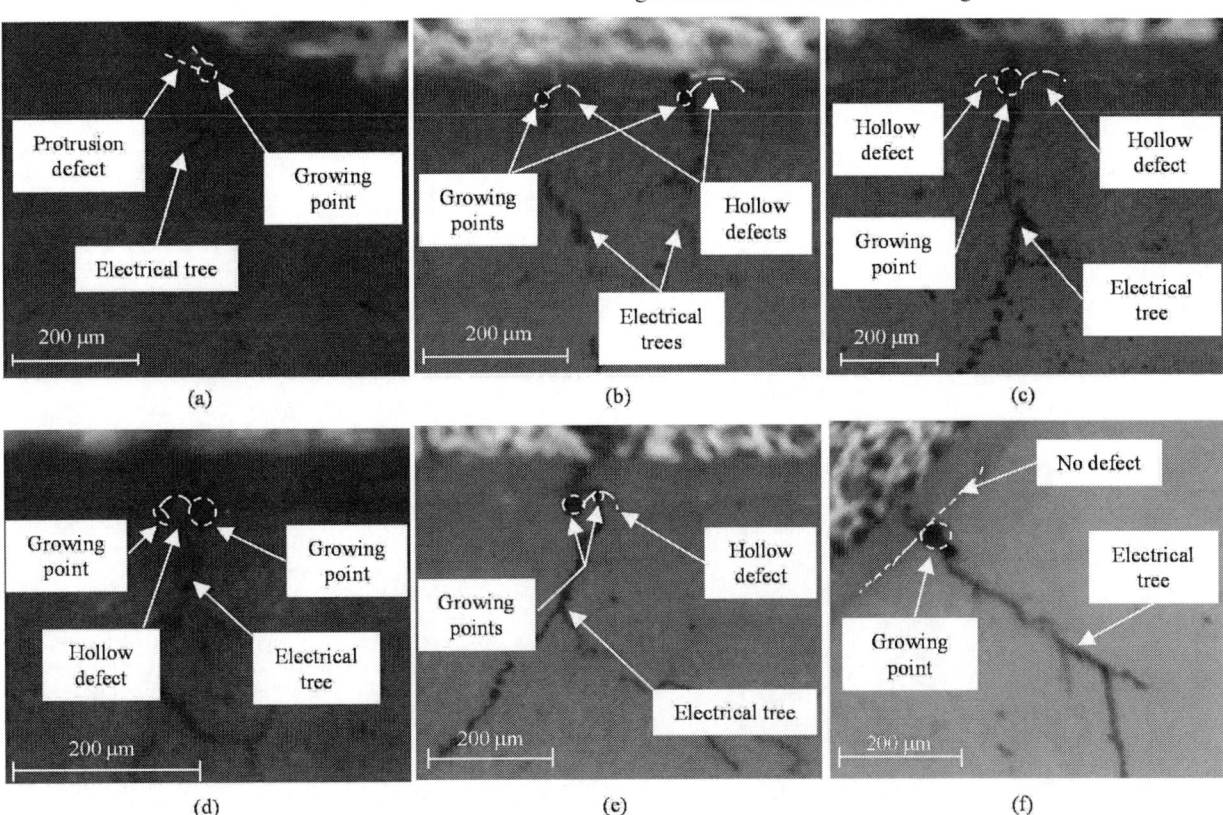

Fig. 6. Typical growing points of electrical trees in triple junctions. (a) an electrical tree grew from the tip of a protrusion defect, (b) two electrical trees grew from the corner of two hollow defects respectively, (c) an electrical tree grew from the common corner of two hollow defects, (d) an electrical tree grew from two corners of a hollow defect, (e) an electrical tree grew from a corner and center of a hollow defect, (f) an electrical tree grew.

into three types, namely protrusion defect, hollow defect with two corners and hollow defects with three corners. According to the statistics, there were 95 % chances that the protrusion defect became the growing point of electrical tree. This data meant that protrusion was the most danger defect which may lead IGBT modules to insulation failure. As for hollow defect, the hollows with three corners were easier to reach the critical conditions for electric tree growth than the hollow with two corners. It was because that the middle corner have the characteristic of protrusion. Even if the hollow with two corners was the safest defect than other defects, we also should pay more attention on it, because this defect was the most common defect. Fig.8 demonstrated this defect had great potential for insulation failure.

Fig. 8 showed the type ratio of electrical tree growing points. There were four type points where grew electrical tree. The smooth edge represented the remained points except defects. In real cases, most growing points located in defects. The ratio of growing points occurring at the defect was around 94 %. This showed that it was necessary to study the relationship between defects and electrical trees. It was not enough to clear out the danger of triple junctions. The hollow with two corners played an important role in growing electrical trees. There were many hollows around the copper layer edge. This caused that a half ratio of growing points located at this defect. As for protrusion defect, its' ratio was lower than hollows' as the number of protrusion in a sample was rare.

IV. CONCLUSION

This paper presented an experimental study of the origin of insulation failure in power modules. Insulation failure was equivalent to the growth of electrical trees in this work. The relationship of defects on the origin of the electrical trees was analyzed. The main conclusions are as follows.

1) It is no doubt that triple junctions are the most danger area leading IGBT modules to insulation failure. However, the defects around triple junctions caused by processing play an important role in the process of breakdown. The effect of defects on insulation failure should not be ignored.

2) There are two typical defects around triple junctions. The protrusion defect is rare and awfully dangerous. It is pretty easy to grow electrical tree at the tip of protrusion defect. Fortunately, the number of protrusion is less in a sample. It

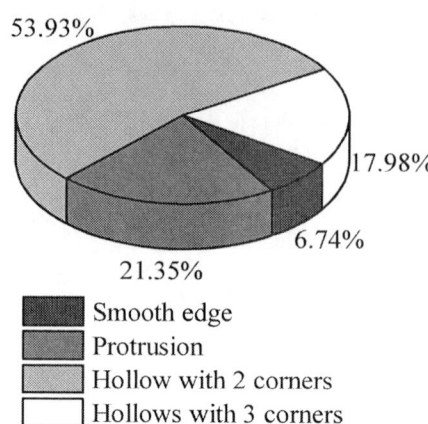

Fig. 8. The type ratio of electrical tree growing points.

leads to a phenomenon that the ratio of electrical trees growing from protrusion is low. Compared with protrusion, the hollow defect is frequent but safe. It can be divided into two kinds of hollows, hollow with two corners and hollows with three corners. The former is the most frequent defect. And one corner of this defect usually becomes the growing point of electrical tree. The later has a similar feature with protrusion. The electrical trees related this defect always start from the middle corner of it. In summary, all of defects should be avoid to form during processing in order to avoid the insulation failure in power modules.

ACKNOWLEDGMENT

This work was supported by the National Key Research and Development Program of China (2020YFA0710500).

REFERENCES

[1] B.Y. Zhang, M. Ghassemi, and Y.X. Zhang, "Insulation materials and systems for power electronics modules: a review identifying challenges and future research needs," IEEE Trans. Dielectr. Electr. Insul., vol. 28 no. 1, pp. 290-302, February 2021.

[2] G. Mitic and G. Lefranc, "Localization of electrical-insulation and partial-discharge failures,". IEEE Trans. Ind. Appl., vol. 38, no. 1, pp. 175-180, January 2002.

[3] U. Waltrich, C. F. Bayer, M. Reger, A. Meyer, X. Tang, and A. Schletz, "Enhancement of the partial discharge inception voltage of ceramic substrates for power modules by trench coating," in Proc. ICEP, Hokkaido, Japan, 2016, pp. 536-541.

[4] J.-L. Augé, O. Lesaint, and A.T. Vuthi, "Partial discharges in ceramic substrates embedded in liquids and gels," IEEE Trans. Dielectr. Electr. Insul., vol. 20, no. 1, pp. 262-274, January 2013.

[5] T. Lebey, D. Malec, S. Dinculescu, V. Costan, F. Breit, and E. Dutarde, "Partial discharges phenomenon in high voltage power modules," IEEE Trans. Dielectr. Electr. Insul., vol. 13 no. 4, pp. 810-819, August 2006.

[6] P.Y. Fu, R. Jin, Z.B. Zhao, T. Wen, H.Y. Wang, and L. Li, "A high sensitivity partial discharge current measurement method for the high voltage IGBT," in Proc. 2017 Sixth Asia-Pacific Conference on Antennas and Propagation (APCAP), Xi'an, China, 2017, pp. 1-3.

[7] A. A. Abdelmalik, A. Nysveen, and L. Lundgaard, "Influence of fast rise voltage and pressure on partial discharges in liquid embedded power electronics,". IEEE Trans. Dielectr. Electr. Insul., vol. 22 no. 5, pp. 2770-2778, October 2015.

[8] M. T. Do, J.-L. Augt I and O. Lesaint, "Optical measurement of partial discharges in silicone gel under repetitive pulse voltage," in Proc. 2005 International Symposium on Electrical Insulating Materials, Kitakyushu, Japan, 2005, pp. 1-17.

H.Y. You, Z. Wei, B.X. Hu Z. Zhao, R. Na and J. Wang, "Partial discharge behaviors in power modules under square pulses with ultrafast dv/dt,". IEEE Trans. Power Electron., vol. 36, no. 3, pp. 2611-2620, Mar. 2021.

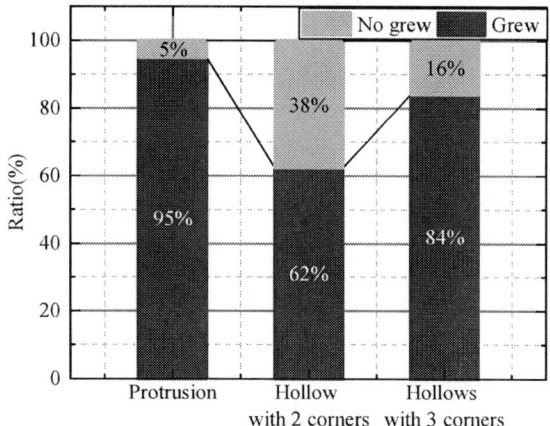

Fig. 7. The probability of electrical tree growth in different defects.

978-1-6654-4817-8/21 $31.00 © 2021 IEEE

A Novel GaN MIS-HEMT with a Source-connected Clamp Electrode for Suppressing Short-channel effect

Yijun Shi
CEREI
Guangzhou, China
syj20094870@sina.com

Shan Wu*
Guangdong Eco-Engineering Polytechnic
Guangzhou, China
785136202@qq.com

Hongyue Wang
CEREI
Guangzhou, China
1090145826@qq.com

Peng Zhao*
CEREI
Guangzhou, China
zhaopeng@ceprei.com

Yiqiang Chen
CEREI
Guangzhou, China
chenyiqiang@ceprei.com

Abstract—To suppress short-channel effect (SCE) in short-channel GaN MIS-HEMT, we have inserted a Source-connected clamp (SC) electrode near to the Drain-side Gate edge of the device, which can clamp the transverse potential at Gate edge to less than 2V. That low transverse potential cannot reduce the barrier of the Gate channel, then SCE can be suppressed. With the conventional GaN MIS-HEMT as comparison, the GaN MIS-HEMT with SC electrode possesses much lower off-state electron density (n_{off}) under the Drain voltage of 10V (reduced by 7 orders of magnitude), and the device also exhibits much reduced off-state leakage current (I_{off}, reduced by 6 orders of magnitude). Meanwhile, the proposed GaN MIS-HEMT also exhibits much reduced reverse Gate-to-Drain capacitor (RC_{GD}). Both increase in the length and depth of SC electrode can further reduce n_{off}, I_{off}, and RC_{GD}. For the GaN MIS-HEMT, the voltage drop in the on state is often less 2V, so the SC electrode will not obviously affect the forward conduction characteristics. The excellent characteristics of the proposed SC GaN MIS-HEMT show its promising future in power application.

Keywords—*GaN MIS-HEMT, short-channel effect, Clamp electrode, Gate-to-Drain capacitor.*

I. INTRODUCTION

GaN-based power device has attracted many attentions in power application, in which the device with high breakdown voltage, high temperature operating capability, and high conversion efficiency is highly desired [1-4]. Despite the promise, the *E*-mode GaN MIS-HEMTs have suffered from high resistance in Gate region (R_g) [5-6]. To decrease R_g, many researchers are devoted to improve the electron mobility of Gate channel, namely the Self-Terminated layer technology, thin AlGaN barrier layer technology and AlN inserted layer technology [7-9]. The other simplest method to decrease R_g is to downsize the length of Gate region (L_g) [10-14]. However, the SCE will inevitably occur with the decrease of L_g. The SCE affects the performance in such a way as the increase of I_{off} deterioration of Gate controllability, and decrease of threshold voltage (V_{TH}). Because the SCE adversely affects GaN MIS-HEMT performance, the effect should be minimized or eliminated.

In this work, we propose a novel GaN MIS-HEMT with SC electrode, which can clamp the transverse potential at Drain-side Gate edge to less than 2V, subsequently suppressing the SCE. Compared with the Con. short-channel GaN MIS-HEMT, both I_{off}, n_{off} and RC_{GD} of SC GaN MIS-HEMT are greatly reduced. Furthermore, the dependences of

I_{off}, n_{off}, and RC_{GD} on the SC electrode's length (L_{SC}) and depth (T_{SC}) are also investigated.

II. SHORT-CHANNEL EFFECT AND DEVICE STRUCTURE

A. SHORT-CHANNEL EFFECT

The cross-sections of Con. GaN MIS-HEMT and SC GaN MIS-HEMT are exhibited in Fig. 1. Before introducing SC GaN MIS-HEMT, there is a necessary to describe SCE in Con. device. In this part, the TCAD Sentaurus is used to study the SCE in Con. device [15-17]. The Gate-to-Source length (L_{gs}) and Gate-to-Drain length (L_{gd}) of Con. device are 2 and 5µm, respectively. L_g is 0. 25 µm or 0.5 µm.

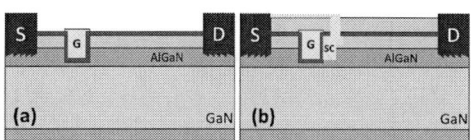

Fig.1 The cross-sections of (a) Con. GaN MIS-HEMT and (b) SC GaN MIS-HEMT.

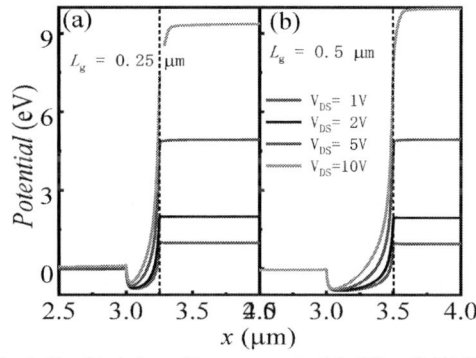

Fig. 2. Simulated channel transverse potential of Con. GaN MIS-HEMT with L_g of 0.25/0.5 µm and V_G of 0V

Fig. 2 plots the simulated channel transverse potential of the Con. GaN MIS-HEMT with L_g of 0. 25/0.5 µm. The barrier height in Gate region is lowered with V_{DS} increasing from 1V to 10V, also with the significant reduction in barrier thickness, which makes the electron more easily to punch through that barrier, and n_{off} will be greatly increased. As shown in Fig. 3, n_{off} increases from 10^{11} cm^{-3} to 10^{18} cm^{-3} (V_G=0V) for the Con. 0. 25 µm GaN MIS-HEMT when with V_{DS} increasing from 1V to 10V, which will lead to the weaken breakdown

---

This research was supported by the National Natural Science Foundation of China (NSFC) under Grant No. 62004046.

978-1-6654-4817-8/21 $31.00 © 2021 IEEE

characteristics, increased I_{off}, deteriorated Gate controllability, and decreased V_{TH} (Fig. 4). From the figure, it can be seen that I_{off} increases from 10^{-8} A/mm to 10^{-1} A/mm ($V_G = 0$V) for the Con. 0.25 μm GaN MIS-HEMT when with V_{DS} increasing from 1V to 10V, which indicates that V_{TH} is less than 0V under V_{DS} of 10V. And the subthreshold slops of Con. GaN MIS-HEMT with L_g of 0. 25/0.5 μm are also greatly degenerated with the increase of V_{DS}.

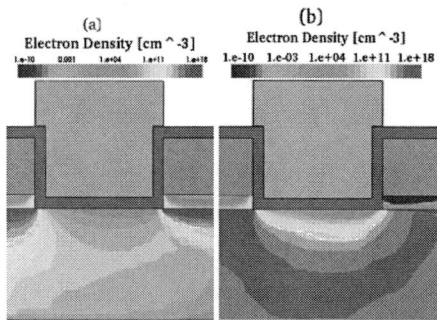

Fig. 3. n_{off} of Con. 0.25 μm GaN MIS-HEMT with V_G of 0V. (a) V_{DS} of 1V. (b) V_{DS} of 10V.

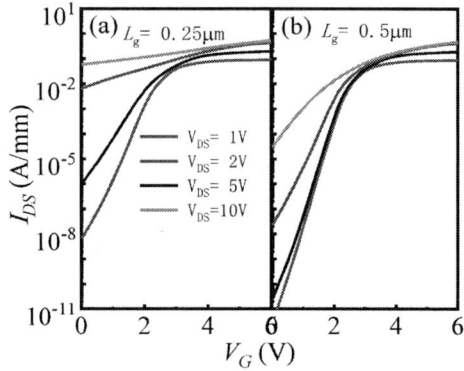

Fig. 4. The transfer characteristic of Con. GaN MIS-HEMT with (a) L_g of 0. 25 μm. (b) L_g of 0.5 μm

B. THE SC GAN MIS-HEMT

In Fig. 2, it can be also found that the high transverse potential at Gate edge may be the immediate cause for the barrier reduction in Gate region. To suppress SCE, clamping the transverse potential to small value is an effective method. To achieve this goal, we propose a novel GaN MIS-HEMT featuring a SC electrode (Fig. 1(b)). The SC electrode is embedded into AlGaN layer, which can clamp the transverse potential to a small value, then avoiding high transverse potential and suppressing SCE. For GaN MIS-HEMT, the voltage drop in the on state is often less 2V, so the SC electrode will not obviously affect the forward conduction characteristics. The SC electrode's working mechanism is explained in Fig. 5.

When with the Drain voltage (V_{DS}) smaller than clamping potential (CP), there is still many 2DEG under the SC region (Fig. 5 (a) and (b)). The resistance of Gate-region channel still occupies the main part of channel resistance. Then, the Gate channel region bear the mainly transverse voltage drop. So, in this time, the SC electrode nearly has no effect on the transverse potential, and transverse potential at this point is

close to V_{DS}. With the increase of V_{DS}, 2DEG under the SC region will be gradually depleted. When with V_{DS} equal to CP, 2DEG channel will be pinched off at the right side of SC electrode (Fig. 5 (c)). With the further increase of V_{DS}, the depletion area will come into being under the SC electrode (Fig. 5 (d)). Then, the clamping electrode will bear the transverse voltage drop higher than CP. So the transverse potential at Gate edge will be clamped to be a fixed and small value, which is much less than V_{DS}. As a result, the SCE will be suppressed.

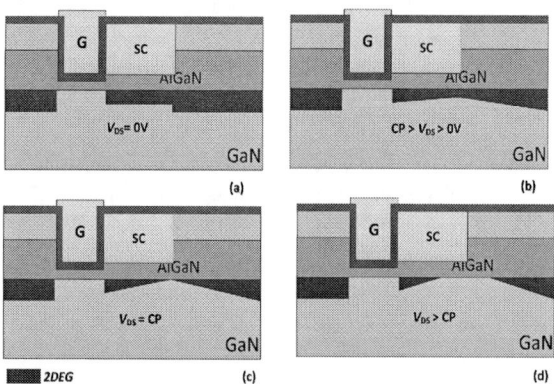

Fig. 5 The working mechanism of SC electrode.

The change in the length and depth of SC electrode can change the clamping potential, which will yield to different suppressing effect. Among them, changing the depth of SC electrode has obvious influence on the clamping potential. The increase in the depth of SC electrode will reduce 2DEG under the SC electrode. On this occasion, 2DEG channel under the SC electrode will be pinched off at a smaller value. That is to say, the increase in the depth of SC electrode will decrease the clamping potential. In this work, the influence of L_{SC} and T_{SC} on I_{off}, n_{off}, and RC_{GD} will be investigated by the TCAD Sentaurus. L_{gs}, L_{gd} and L_g of the proposed SC GaN MIS-HEMT are same to the Con. short-channel GaN MIS-HEMT.

III. RESULT AND DISUSSION

Fig. 6 gives the transfer characteristics of the SC GaN MIS-HEMTs with T_{SC} of 12 nm and different L_{SC}. Compared with the Con. device, the SC GaN MIS-HEMT exhibits much smaller I_{off} and improved subthreshold characteristic. When with $L_{SC} = 0.5$ μm, I_{off} of SC GaN MIS-HEMT decreases from about 10^{-1} A/mm to 10^{-5} A/mm (at V_G=0V and V_{DS}=10V) for the device with $L_g = 0. 25$ μm, and decreases from 10^{-5} A/mm to 10^{-9} A/mm (at V_G=0V and V_{DS}=10V) for the device with $L_g = 0. 5$ μm. I_{off} of SC GaN MIS-HEMT is decreasing with increase of L_{SC}. And SCE can be effectively suppressed for the devices with long enough L_{SC} (larger than 0.5μm for the device with $L_g = 0.25$ μm and larger than 0.25μm for the device with $L_g = 0.5$ μm). When with L_{SC} increasing from 0.25 μm to 1 μm, I_{off} of SC GaN MIS-HEMT decreases from about 10^{-3} A/mm to 10^{-7} A/mm (at V_G=0V and V_{DS}=10V) for the device with $L_g = 0.25$ μm, and decreases from 10^{-8} A/mm to 10^{-10} A/mm (at V_G=0V and V_{DS}=10V) for the device with $L_g = 0. 5$ μm. In addition, when with $V_{DS} = 1$V, I_{off} and subthreshold characteristic of SC GaN MIS-HEMT are almost same as that of the Con. device. The reasons will be explained in the following part.

978-1-6654-4817-8/21 $31.00 © 2021 IEEE

The decrease in I_{off} (at V_G=0V and V_{DS}=10V) of the SC GaN MIS-HEMT is because that the SC electrode can clamp the transverse potential at Gate edge to be a small value (Fig. 7). It can be found that the transverse potential at Gate edge of SC GaN MIS-HEMT is less 2V under V_{DS} of 10V, which is much smaller than that of the Con. device. That low transverse potential at Gate edge cannot reduce the barrier in the Gate region of SC GaN MIS-HEMT. The high barrier will prevent the electron from punching through, and SCE of the short-channel device can be avoided. And because the SC electrode can clamp the transverse potential at Gate edge to less 2V, I_{off} and subthreshold characteristic of SC GaN MIS-HEMT will exhibit no obvious change with V_{DS} less than 2V (Fig. 6).

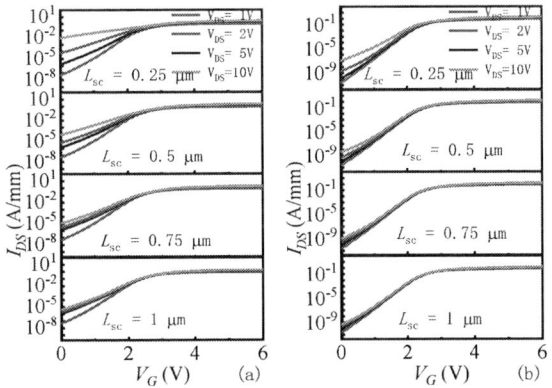

Fig. 6. The transfer characteristic of SC GaN MIS-HEMT with different L_{SC} (T_{SC} of 12 nm). (a) L_g of 0. 25 μm. (b) L_g of 0.5 μm

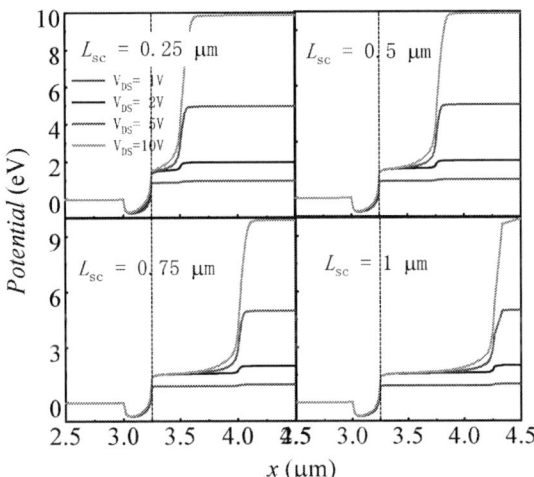

Fig. 7. Simulated channel potential of SC GaN MIS-HEMT with different L_{SC}. (L_g of 0. 25 μm and T_{SC} of 12 nm)

Fig. 8 exhibits n_{off} under Gate region (at V_G=0V and V_{DS}=10V) for the SC GaN MIS-HEMT with T_{SC} = 12 nm and different L_{SC}. It can be seen that the SC electrode can decrease n_{off}. Compared with the Con. short-channel device with L_g = 0.25 μm, n_{off} decreases from 10^{18} cm^{-3} to 10^{11} cm^{-3} (at V_G=0V and V_{DS}=10V) for the SC GaN MIS-HEMT with L_g = 0.25 μm and L_{SC} = 0.5 μm. It can also be seen, when only with L_{SC} = 0.25 μm, the proposed SC GaN MIS-HEMT with L_g = 0.25 μm still exhibits much high off-state electron density. So, to

make the SC electrode to effectively suppress the SCE for the SC GaN MIS-HEMT, the length of SC electrode should be long enough. According to Fig. 6 and Fig. 8, the length of SC electrode should be larger than 0.5μm for the device with L_g = 0.25 μm and larger than 0.25μm for the device with L_g = 0.5 μm. In summary, introducing the SC electrode to the short-channel device can enable the device with low n_{off}, low I_{off}, improved subthreshold characteristic and high V_{TH}.

The switching characteristic is also important for GaN power device, which is related to RC_{GD}. Fig. 9 gives RC_{GD} for the Con. short-channel device and the SC GaN MIS-HEMT. The SC electrode can help to decrease RC_{GD} about one order for the SC GaN MIS-HEMT, which may result from the electrostatic shielding effect of the SC electrode. And RC_{GD} is decreasing with the increase in L_{SC}. The decrease of RC_{GD} can help to improve the device switching characteristics. So, the SC GaN MIS-HEMT possesses both the outstanding static characteristics and excellent switching characteristics, showing its promising for future power applications.

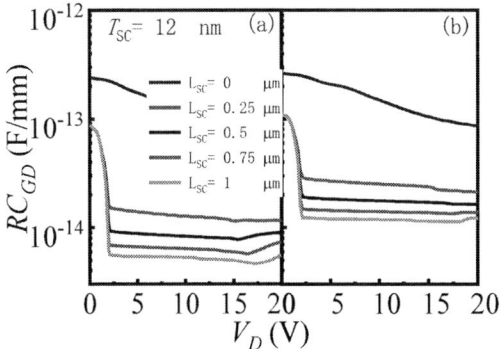

Fig. 9. Simulated RC_{GD} of the SC GaN MIS-HEMT with different L_{SC}: (a) L_g of 0.25 μm. (b) L_g of 0.5 μm.

In this paragraph, the influences of T_{SC} on I_{off} and n_{off} under the Gate region, and RC_{GD} are investigated. Fig. 10 plots the transfer characteristics of SC GaN MIS-HEMTs with L_{SC} = 0.5 μm and different T_{SC}. I_{off} of SC GaN MIS-HEMT is decreasing with the increase in T_{SC}. When with T_{SC} increasing from 8 nm to 14 nm, I_{off} decreases from 10^{-4} A/mm to 10^{-6} A/mm (at V_G=0V and V_{DS}=10V) for the proposed SC GaN MIS-HEMT with L_g = 0.25 μm, and decreases from 10^{-8} A/mm to 10^{-10} A/mm (at V_G=0V and V_{DS}=10V) for the device with L_g = 0.5 μm. It is because that the increase in T_{SC} can decrease the clamped potential at the Gate edge (Fig. 11). In addition, I_{off} (at V_G=0V and V_{DS}=1V) of SC GaN MIS-HEMT with L_g = 0. 5 μm exhibits no obvious change with the increase in T_{SC}, and are almost same as that of the Con. short-channel device. But I_{off} (at V_G=0V and V_{DS}=1V) of SC GaN MIS-HEMT with L_g = 0.25 μm and T_{SC} = 14 nm exhibits a slight decrease. It is because that when with T_{SC} = 14 nm, the transverse potential at the Drain-side Gate edge for the proposed SC GaN MIS-HEMT with L_g = 0.25 μm is clamped to about 0.9V, as shown in Fig. 10. While for the proposed SC GaN MIS-HEMT with L_g = 0.5 μm, the transverse potential at Gate edge is still larger than 1V. Fig. 12 plots RC_{GD} of SC GaN MIS-HEMT with different T_{SC}. The increase in T_{SC} may not affect RC_{GD} of the proposed SC GaN MIS-HEMT with same L_{SC}, when with V_{DS} larger than 5V. So, the change in T_{SC} will not affect the switching characteristics of the SC GaN MIS-HEMT.

978-1-6654-4817-8/21 $31.00 © 2021 IEEE

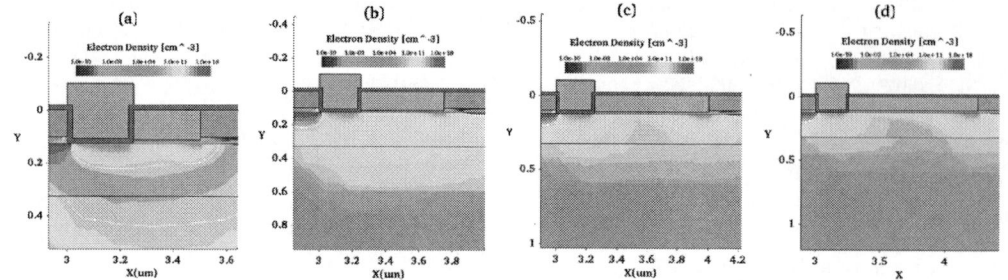

Fig. 8. Simulated n_{off} of SC GaN MIS-HEMT with different L_{SC}: (a) L_{SC} =0. 25 μm. (b) L_{SC} =0. 5 μm. (c) L_{SC} =0. 75 μm. (d) L_{SC} =1.0 μm.

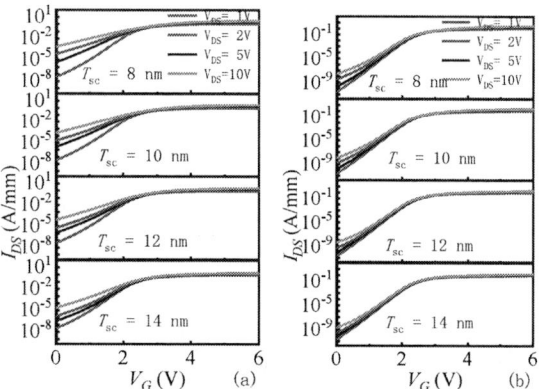

Fig. 10. The transfer characteristic of the SC GaN MIS-HEMT with different T_{SC} (L_{SC} = 0.5 μm). (a) L_g of 0.25 μm. (b) L_g of 0.5 μm.

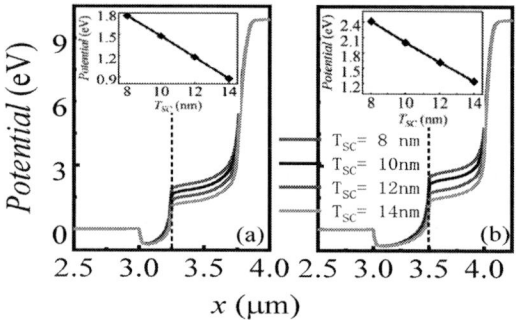

Fig. 11. Simulated channel potential of the SC GaN MIS-HEMT with different T_{SC}. (a) L_g of 0.25 μm. (b) L_g of 0.5 μm.

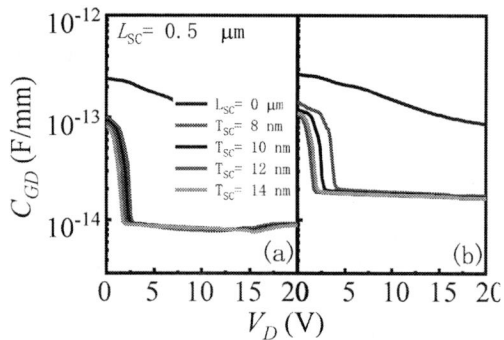

Fig. 12. Simulated RC_{GD} of the SC GaN MIS-HEMT with different T_{SC}: (a) L_g of 0.25 μm. (b) L_g of 0.5 μm

IV. CONCLUSION

To suppress SCE in short-channel GaN MIS-HEMT, we have inserted a SC electrode near to the Drain-side Gate edge of the device, which can clamp the transverse potential at Gate edge to less than 2V. That low transverse potential cannot reduce the barrier of the Gate channel, then SCE can be suppressed. With the conventional GaN MIS-HEMT as comparison, the GaN MIS-HEMT with SC electrode possesses much lower n_{off} under the Drain voltage of 10V (reduced by 7 orders of magnitude), and the device also exhibits much reduced I_{off} (reduced by 6 orders of magnitude). Meanwhile, the proposed GaN MIS-HEMT also exhibits much reduced RC_{GD}. Both increase in the length and depth of SC electrode can further reduce n_{off}, I_{off}, and RC_{GD}. For the GaN MIS-HEMT, the voltage drop in the on state is often less 2V, so the SC electrode will not obviously affect the forward conduction characteristics. The excellent characteristics of the proposed SC GaN MIS-HEMT show its promising future in power application.

REFERENCES

[1] K. J. C., O. H., and A. L., et al. *IEEE TED*, 2017, 64 (3): 779-795.

[2] N. M. S., Y. L., and T. S., et al. *IEEE TED*, 2019, 66(4): 1694-1698.

[3] Y. S. W. C., and X. C., et al. *IEEE TED*, 2018, 65(12): 5322-5328.

[4] Y. S., S. H., and Q. B, et al. *IEEE TED*, 2016, 63(2): 614-619.

[5] J. Z., X. M., and B. H., et al. *IEEE TED*, 2018, 65(12):5343-5349.

[6] Z. T., S. H., and X. T., et al. *IEEE TED*, 2014, 61(8): 2785–2792

[7] S. L., M. W., and F. S., et al. *IEEE EDL*, 2016, 37(4).

[8] S. H., X. L., and X. W., et al. *IEEE EDL*, 2016, 37(12): 1617-1620.

[9] J. W., S. L., and B. L., et al. *IEEE EDL*, 2015, 36(12): 1287-1290.

[10] H. S., K. B. L., and L. Y., et al. *IEEE RFIT*, Singapore, 2012: 204-206.

[11] M. J. U., K. J. N., and R. S. B., et al. *IEEE TED*, 2006, 53(2): 395-398.

[12] P. S. P. and S. R. *IEEE TED*, 2011, 58(3): 704-708.

[13] C. G., A. G., and A. K. B., et al. *EDSSC*, Hsinchu, 2017: 1-2.

[14] M. U., D. G. H., and R. S. B., et al. *EMICC*, Manchester, 2006: 65-68.

[15] J. S., J. W., and W. C. B., et al. *IEEE TED*, 2013, 60(10): 3223-3229.

[16] Z. W., B. Z., and W. C., et al. *IEEE TED*, 2013, 60(15): 1607-1612

[17] Synopsys, Inc.: Sentaurus device user guide.

Resonant Gate Driver with Wide Range Adjustment of Driving Speed

Hao Peng
Power Electronics and Energy
Management Key Laboratory, Ministry of
Education of China
Huazhong University of Science and
Technology
Wuhan, China
iridescent@hust.edu.cn

Han Peng
Power Electronics and Energy
Management Key Laboratory, Ministry of
Education of China
Huazhong University of Science and
Technology
Wuhan, China
pengh@hust.edu.cn

Qiaozhi Yue
Power Electronics and Energy
Management Key Laboratory, Ministry of
Education of China
Huazhong University of Science and
Technology
Wuhan, China
george_yue@hust.edu.cn

Abstract—**Resonant gate driver (RGD) is the more advanced solution to drive SiC MOSFETs in high frequency applications with driver loss reduction and power density improvement. The driving speed of RGD is mainly determined by the resonance mode and resonant inductor, the options of which are relatively limited and make the smooth adjustment of driving speed not quite easy. In this paper, a new driving speed adjustment approach is proposed by changing the average driving current in driving transition. The acceleration of driving speed is achieved through inductor current pre-charging and the reduction of driving speed is realized through multi-pulse resonant driving. A full-bridge 1/4 period RGD prototype is built for verification. The baseline of driving time is 250 ns when initial inductor current is zero, and the measured driving times using the proposed method vary from 80 ns to 510 ns. Different driving speeds for turn-on and turn-off transients are also realized.**

Keywords—*Resonant gate driver, driving speed adjustment, inductor current pre-charging, multi-pulse resonant driving, SiC MOSFET*

I. INTRODUCTION

With the applications of wide bandgap semiconductor such as SiC and GaN transistors, power converters are using higher switching frequencies for smaller size and higher power density. More advanced gate drivers are needed to drive power transistors effectively and efficiently. Conventional gate driver (CGD) based on RC first-order circuit is shown in Fig. 1. The driving loss is calculated by $P_{CGD}=Q_g(V_{cc}+|V_{ee}|)f_{sw}$ [1], where Q_g is the total gate charge of M, V_{cc} and V_{ee} are the driving voltages, and f_{sw} is the switching frequency. Gate driver loss increases linearly with switching frequency, which requires large isolated power supply and large gate driver board [2]. Hence, the increase of switching frequency will decrease the overall power density of the system.

To reduce the driving loss, resonant gate driver (RGD) is developed to employ an inductor to resonantly charging and discharging the gate capacitor C_{iss}. Driver loss reduction is

realized by energy recovery after resonant driving [3], or by energy recycling during resonant driving transients [4]. RGD was first applied in high-frequency and lower-power converters to improve system efficiency [5],[6]. With the wide applications of SiC devices, RGD modules with high efficiency and small size become more and more important.

In CGD, driving speed is determined by the gate resistor, and different turn-on or turn-off speeds can be realized by changing gate resistors. As shown in Fig. 1, a resistor of 22 Ω is used for turn-on process to suppress oscillation and overshoot, while 2 Ω (22 Ω in parallel with 2.2 Ω) is employed to realize fast turn-off for crosstalk suppression [7].

The driving speed of RGD is mainly determined by the resonance mode and resonant inductor. RGD is usually set with 1/2 period or 1/4 period resonance mode [8]. Once the resonant inductor is selected, the driving speed will be determined. The options of inductor value and package are relatively limited compared with resistor, which make the smooth adjustment in resonant inductor not quite easy. Furthermore, since inductor current cannot change suddenly, it is not practical to parallel multiple inductors. Hence, it is important to explore a possible solution to adjust driving speed in RGD.

In this paper, a new driving speed adjustment approach is proposed by changing the average driving current in driving transition. The acceleration of driving speed is achieved through inductor current pre-charging and the reduction of driving speed is realized through multi-pulse resonant driving. The content of this article is arranged as follows. Section II gives the analysis of the proposed driving speed adjustment method. Section III analyzes the prototype design and experimental results. Section IV summarizes the work of this article.

II. PROPOSED DRIVING SPEED ADJUSTMENT METHOD

As shown in Fig. 2, a full-bridge RGD topology is used for analysis, where M is the power transistor, S_1-S_4 are switches and L_r is the resonant inductor. When it is operated without the initial inductor current, the driving time is given as

$$T_{ref} = \pi\sqrt{L_r C_{iss}}/2 \text{ [3].}$$

A. Driving speed acceleration

The driving process can be accelerated by inductor current pre-charging. The signals of S_1-S_4, along with waveforms of the resonant inductor current i_L, the driving current i_c and the driving voltage v_{gs} are illustrated in Fig. 3. With S_1 and S_4 turning on, the pre-charging loop is formed. L_r is charged by V_{cc} and V_{ee} until i_L rises to I_{L0}. Then S_4 shuts off and the

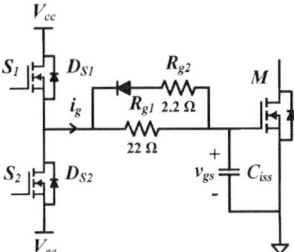

Fig. 1 CGD topology with different driving resistors

This work was supported by the National Natural Science Foundation of China under Grant 52007077.

978-1-6654-4817-8/21 $31.00 © 2021 IEEE

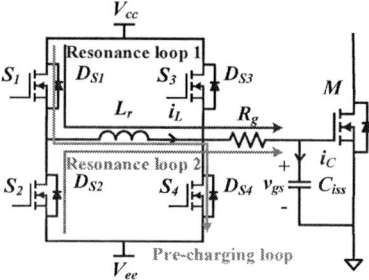

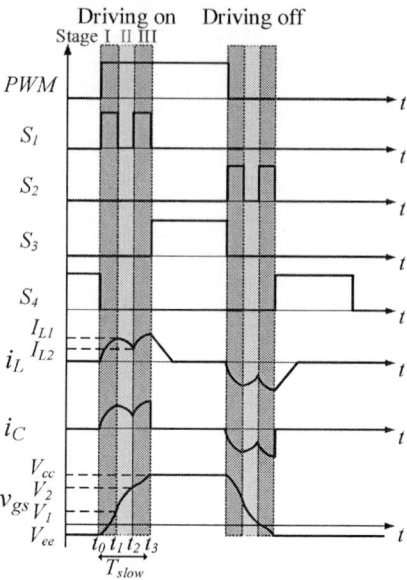

Fig. 2 Full-bridge RGD topology

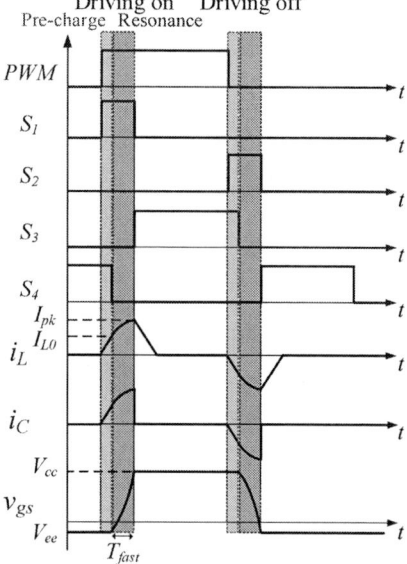

Fig. 3 Critical driving waveforms for driving speed acceleration

resonant driving process starts. V_{cc}, S_1, L_r and C_{iss} form the resonance loop 1 to charge C_{iss}. During this process, i_L and v_{gs} can be expressed as:

$$\begin{cases} i_L(t) = I_{L0}\cos\omega_0 t + \dfrac{V_{cc} - V_{ee}}{Z_0}\sin\omega_0 t \\ v_{gs}(t) = V_{cc} - (V_{cc} - V_{ee})\cos\omega_0 t + I_{L0}Z_0\sin\omega_0 t \end{cases} \quad (1)$$

where $\omega_0 = 1/\sqrt{L_r C_{iss}}$ and $Z_0 = \sqrt{L_r / C_{iss}}$.

The driving time is the time interval between v_{gs} rising from V_{ee} to V_{cc}, as:

$$T_{fast} = \frac{1}{\omega_0}\left(\frac{\pi}{2} - \arctan\frac{I_{L0}Z_0}{V_{cc} - V_{ee}}\right) \quad (2)$$

The higher initial current I_{L0} is, the shorter driving time and the faster driving speed will be. The fastest driving speed is realized when the driving current is approximate to constant (e.g. peak current of i_L is no more than $(1+10\%)\times I_{L0}$), because the driving current cannot increase infinitely. The peak current can be calculated as:

$$I_{pk} = i_L(T_{fast}) = \frac{\sqrt{(V_{cc} - V_{ee})^2 + I_{L0}^2 Z_0^2}}{Z_0} \quad (3)$$

The ratio of T_{fast} to T_{ref} is given as:

Fig. 4 Critical driving waveforms for driving speed reduction

$$\frac{T_{fast}}{T_{ref}} = 1 - \frac{2}{\pi}\arctan\frac{I_{L0}Z_0}{V_{cc} - V_{ee}} \quad (4)$$

It can measure the degree of driving speed acceleration through inductor current pre-charging.

B. Driving speed reduction

The driving speed can be reduced by multi-pulse resonant driving in one driving transient. A two-pulse resonant driving is used for the following analysis. As shown in Fig. 4, the driving process is divided into three resonance sub-stages.

Resonance stage I (t_0-t_1): The first pulse starts to turn on S_1. V_{cc}, S_1, L_r and C_{iss} form the resonance loop 1 to charge C_{iss}. Both i_L and v_{gs} increase resonantly, as:

$$\begin{cases} i_L(t) = \dfrac{V_{cc} - V_{ee}}{Z_0}\sin\omega_0(t - t_0) \\ v_{gs}(t) = V_{cc} - (V_{cc} - V_{ee})\cos\omega_0(t - t_0) \end{cases} \quad (5)$$

Resonance stage I ends when v_{gs} gets to V_1 at t_1. The duration of this stage is the first pulse width, as:

$$\Delta t_1 = t_1 - t_0 = \frac{1}{\omega_0}\arccos\frac{V_{cc} - V_1}{V_{cc} - V_{ee}} \quad (6)$$

And i_L reaches to:

$$I_{L1} = i_L(t_1) = \frac{\sqrt{(V_{cc} - V_{ee})^2 - (V_{cc} - V_1)^2}}{Z_0} \quad (7)$$

Resonance stage II (t_1-t_2): Turn off S_1 at t_1 and the resonance loop 2 composed of V_{ee}, D_{S2}, L_r and C_{iss} is formed to charge C_{iss}. v_{gs} increases while i_L decreases, as:

$$\begin{cases} i_L(t) = I_{L1}\cos\omega_0(t - t_1) \\ \qquad - \dfrac{V_1 - V_{ee}}{Z_0}\sin\omega_0(t - t_1) \\ v_{gs}(t) = V_{ee} + I_{L1}Z_0\sin\omega_0(t - t_1) \\ \qquad + (V_1 - V_{ee})\cos\omega_0(t - t_1) \end{cases} \quad (8)$$

This stage ends when v_{gs} rises to V_2 at t_2. The duration of this stage is the time interval between two pulses, as:

$$\Delta t_2 = t_2 - t_1 = \frac{1}{\omega_0}\arcsin\frac{V_2 - V_{ee}}{\sqrt{2(V_{cc} - V_{ee})(V_1 - V_{ee})}}$$
$$- \frac{1}{\omega_0}\arctan\frac{V_1 - V_{ee}}{\sqrt{(V_{cc} - V_{ee})^2 - (V_{cc} - V_1)^2}} \quad (9)$$

And i_L drops to:

$$I_{L2} = i_L(t_2) = \frac{\sqrt{2(V_{cc} - V_{ee})(V_1 - V_{ee}) - (V_2 - V_{ee})^2}}{Z_0} \quad (10)$$

Resonance stage III (t_2-t_3): The second pulse starts to turn on S_1 again. C_{iss} is charged by the resonance loop 1. Both i_L and v_{gs} rise up, as:

$$\begin{cases} i_L(t) = I_{L2}\cos\omega_0(t - t_2) \\ \qquad + \dfrac{V_{cc} - V_2}{Z_0}\sin\omega_0(t - t_2) \\ v_{gs}(t) = V_{cc} + I_{L2}Z_0\sin\omega_0(t - t_2) \\ \qquad - (V_{cc} - V_2)\cos\omega_0(t - t_2) \end{cases} \quad (11)$$

This stage ends when v_{gs} rises to V_{cc} at t_3. The duration of this stage is the second pulse width, as:

$$\Delta t_3 = t_3 - t_2$$
$$= \frac{1}{\omega_0}\arctan\frac{V_{cc} - V_2}{\sqrt{2(V_{cc} - V_{ee})(V_1 - V_{ee}) - (V_2 - V_{ee})^2}} \quad (12)$$

So the total driving time of the two-pulse resonant driving is the summation of stage I-III, as:

$$T_{slow} = \frac{1}{\omega_0}\arctan\frac{V_1 - V_{ee}}{\sqrt{(V_{cc} - V_{ee})^2 - (V_{cc} - V_1)^2}}$$
$$+ \frac{1}{\omega_0}\arctan\frac{\sqrt{2(V_{cc} - V_{ee})(V_1 - V_{ee}) - (V_2 - V_{ee})^2}}{2V_1 - V_2 - V_{ee}} \quad (13)$$

According to the above analysis, the longer Δt_2 is, the smaller I_{L2} and the longer driving time will be. The slowest driving speed is realized when i_L drops to zero at the end of stage II (i.e. I_{L2}=0).

The ratio of T_{slow} to T_{ref} is given as:

$$\frac{T_{slow}}{T_{ref}} = \frac{2}{\pi}\arctan\frac{V_1 - V_{ee}}{\sqrt{(V_{cc} - V_{ee})^2 - (V_{cc} - V_1)^2}}$$
$$+ \frac{2}{\pi}\arctan\frac{\sqrt{2(V_{cc} - V_{ee})(V_1 - V_{ee}) - (V_2 - V_{ee})^2}}{2V_1 - V_2 - V_{ee}} \quad (14)$$

It can measure the degree of driving speed reduction through inductor current pre-charging.

It can be seen that driving speed adjustment is realized by changing the average driving current I_{avg}: I_{avg} can be increased by inductor current pre-charging to accelerate the driving process; I_{avg} can be decreased by multi-pulse (e.g. two-pulse) resonant driving to slow down the driving process. Fig. 5 shows the rule of driving speed adjustment approach.

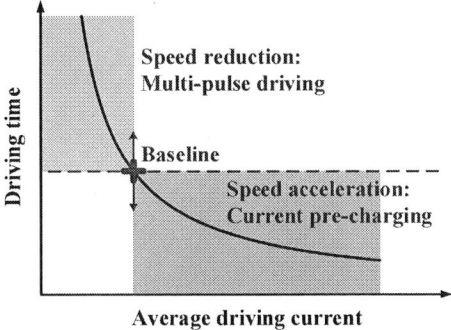

Fig. 5 The rule of driving speed adjustment approach

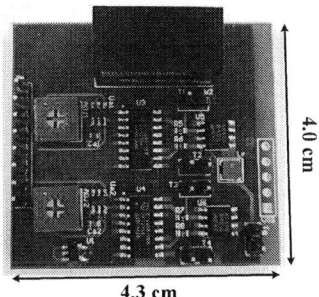

Fig. 6 The full-bridge RGD prototype

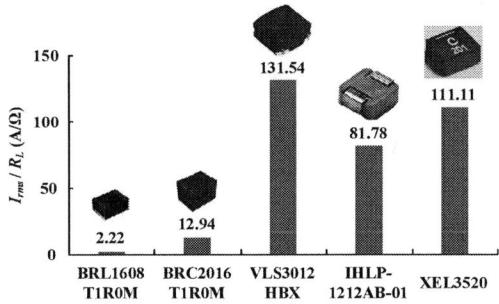

Fig. 7 Inductors comparison and selection

III. PROTOTYPE DESIGN AND EXPERMENTAL RESULTS

A prototype with adjustment driving speed is designed to verify the proposed driving speed adjustment method, as shown in Fig. 6. The driven power transistor uses SiC MOSFET C3M0016120K with 1.2kV/115A from Cree [9] for this verification. The figures of merit (I_{rms}/R_L) of five types of chip inductors are listed in Fig. 7. The higher I_{rms}/R_L is, the bigger current handling capability with less loss will be. Furthermore, the resonant quality factor ought to be greater than 3 to realize decent resonant driving performances [2]. Therefore, a 1500 nH inductor VLS3012HBX from TDK [10] is selected in this design. Two Si4946CDY with dual n-channel MOSFETs in one package are used to form the full bridge, which are controlled by 2EDF7275F from Infineon. Four driving signals are provided by signal generators. V_{cc} and V_{ee} is generated by a 2 W isolated power supply from Murata.

The turn-on transient of RGD with 1/4 cycle resonance is shown in Fig. 8 as the baseline. Without the initial inductor current, the turn-on process lasts 1/4 resonance period of 250 ns, and the peak current reaches 1.6 A. The accelerated driving processes by inductor current pre-charging are illustrated in Fig. 9 and Fig. 10. In Fig. 9, the initial current I_{L0} is 1.6 A after 100 ns of pre-charging interval. The turn-on time is shortened to 130 ns. In Fig. 10, turn-on time is further reduced to 80 ns with the initial current I_{L0} of 2.8 A. This driving process can

978-1-6654-4817-8/21 $31.00 © 2021 IEEE

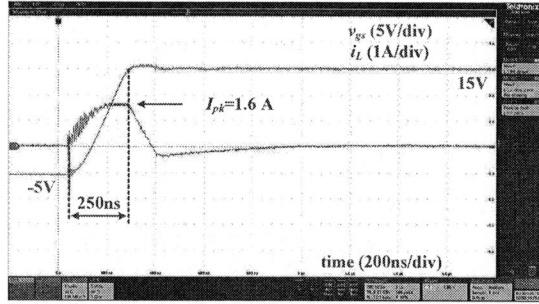

Fig. 8 Resonant turn-on process when $I_{L0}=0$

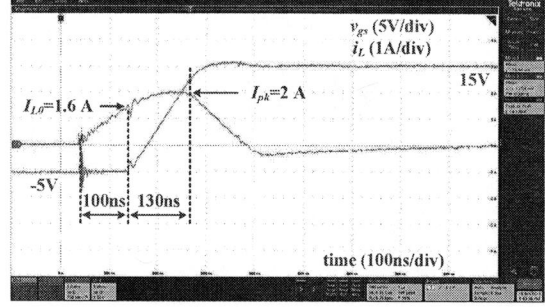

Fig. 9 Resonant turn-on process when $I_{L0}=1.6$ A

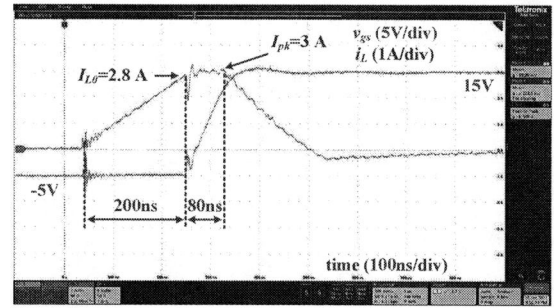

Fig. 10 Resonant turn-on process when $I_{L0}=2.8$ A

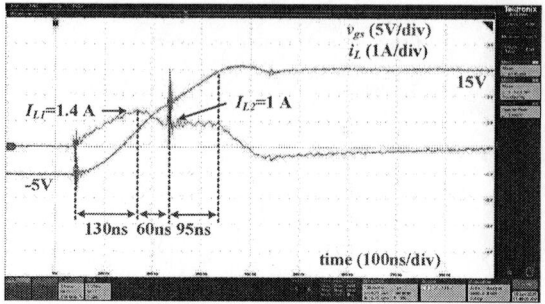

Fig. 11 Resonant turn-on process with three resonance sub-stages

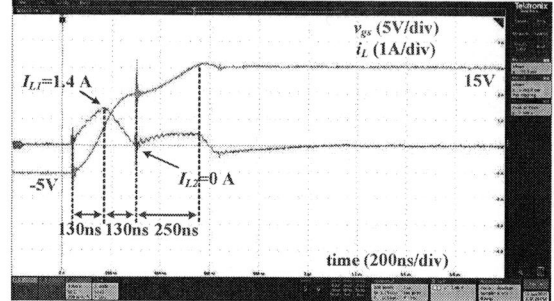

Fig. 12 Resonant turn-on process when $I_{L2}=0$

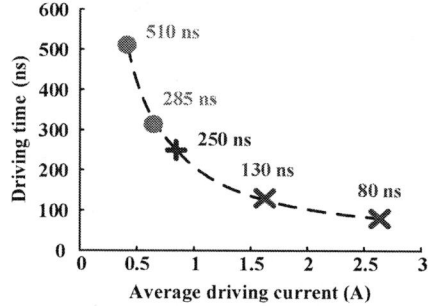

Fig. 13 The measured adjustment range of driving time

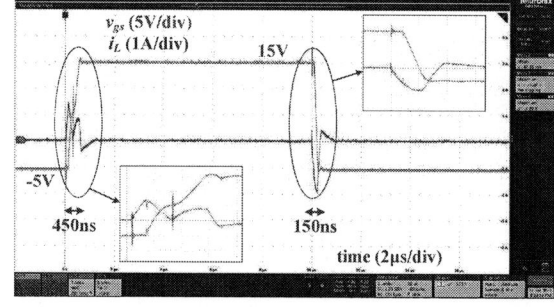

Fig. 14 Driving transients with different speeds in one switching cycle

be considered as constant current driving because the peak current I_{pk} is 3 A, no more than (1+10%)×2.8 A.

The reduction of driving speed by two-pulse resonant driving is depicted in Fig. 11. The turn-on time lasts 285 ns, where the durations of three resonance sub-stages are 130 ns, 60 ns and 95 ns respectively. In Fig. 12, turn-on time is further extended to 510 ns, where the durations of three sub-stages become 130 ns, 130 ns and 250 ns respectively. The driving process is the slowest under this experimental condition since i_L decreases to zero at the end of stage II. The above experimental results show that the driving time can be adjusted from 0.32 to 2.04 times of the baseline, as summarized in Fig. 13.

The proposed method can also realize different driving speeds in turn-on and turn-off transients. Fig. 14 gives the waveforms of i_L and v_{gs} in one switching cycle. The two-pulse resonant driving is used for slow down the turn-on process, which lasts 450 ns, while the inductor current pre-charging is used for accelerate the turn-off process, which lasts 150 ns.

IV. CONCLUSIONS

A novel driving speed adjustment method is proposed in this paper to realize wide range adjustment of driving speed in RGD, where the acceleration of driving speed is realized through inductor current pre-charging and the reduction of driving speed is achieved through multi-pulse resonant driving.

A full-bridge 1/4 period RGD topology is used for analysis and the rule of driving speed adjustment is concluded. In the prototype verification, the baseline of driving time is 250 ns without initial inductor current, and the measured driving times using the proposed method vary from 80 ns to 510 ns. Different driving speeds for turn-on and turn-off transients are also realized.

REFERENCES

[1] J. Zhang, H. Wu, J. Zhao, Y. Zhang, and Y. Zhu, "A resonant gate driver for silicon carbide MOSFETs," *IEEE Access*, vol. 6, pp. 78394-78401, 2018.

[2] H. Peng *et al.*, "A time-segmented resonant gate driver analysis for loss, speed and SiC switching performance tri-optimization," *IEEE Journal*

978-1-6654-4817-8/21 $31.00 © 2021 IEEE

of Emerging and Selected Topics in Power Electronics, vol. 9, no. 2, pp. 2212-2226, Apr. 2021.

[3] Y. Chen, F. C. Lee, L. Amoroso, and H. Wu, "A resonant MOSFET gate driver with efficient energy recovery," *IEEE Transactions on Power Electronics*, vol. 19, no. 2, pp. 470-477, Mar. 2004.

[4] H. Fujita, "A resonant gate-drive circuit capable of high-frequency and high-efficiency operation," *IEEE Transactions on Power Electronics*, vol. 25, no. 4, pp. 962-969, Apr. 2010.

[5] Y. Panov and M. M. Jovanovic, "Design considerations for 12-V/1.5-V, 50-A voltage regulator modules," *IEEE Transactions on Power Electronics*, vol. 16, no. 6, pp. 776-783, Nov. 2001.

[6] K. Yao and F. C. Lee, "A novel resonant gate driver for high frequency synchronous buck converters," *IEEE Transactions on Power Electronics*, vol. 17, no. 2, pp. 180-186, Mar. 2002.

[7] Infineon, "Design guidelines and application example in the Infineon 800 W ZVS PSFB evaluation board," Version 3.0-7, Jun. 15, 2020. [Online]. Available: https://www.infineon.com/dgdl/Infineon-GateDriverICs_EiceDRIVER_2EDi_Using_the_EiceDRIVER_2EDi_family-ApplicationNotes-v03_00-EN.pdf?fileId=5546d46267354aa001675a431da84a41

[8] H. Peng, H. Peng, Q. Tong, X. Ding and Y. Kang, "Review of Resonant Gate Driver from the Perspective of Driving Energy and Time," *IEEE Journal of Emerging and Selected Topics in Power Electronics*, to be published. DOI: 10.1109/JESTPE.2020.3044151.

[9] (Apr. 2019). *Cree C3M0016120K, Datasheet*. [Online]. Available: https://www.wolfspeed.com/downloads/dl/file/id/1483/product/643/c3m0016120k.pdf

[10] (Jan. 2019). *TDK VLS3012HBX, Datasheet*. [Online]. Available: https://product.tdk.com/system/files/dam/doc/product/inductor/inductor/smd/catalog/inductor_commercial_power_vls3012hbx_en.pdf

A Predictive Method for Switching Time of Nanosecond Pulsed Power System of Ohmic Loads Using SiC MOSFETs

Yifei Luo
National Key Laboratory of Science
and Technology on Vessel Integrated
Power System
Naval University of Engineering
Wuhan, China
yfluo16@163.com

Xin Li*
National Key Laboratory of Science
and Technology on Vessel Integrated
Power System
Naval University of Engineering
Wuhan, China
l.x04@mail.scut.cn

Fei Xiao
National Key Laboratory of Science
and Technology on Vessel Integrated
Power System
Naval University of Engineering
Wuhan, China
xfeyninger@qq.com

Zenan Shi
State Key Laboratory of Electrical
Insulation and Power Equipment
Xi'an Jiaotong University
Xi'an, China
szn961021@gmail.com

Ruitian Wang
National Key Laboratory of Science
and Technology on Vessel Integrated
Power System
Naval University of Engineering
Wuhan, China
wangrt4321@163.com

Feng Xie
National Key Laboratory of Science
and Technology on Vessel Integrated
Power System
Naval University of Engineering
Wuhan, China
449093346@qq.com

Abstract—In this paper, a predictive method for accurately determining the switching time of nanosecond pulsed power system is proposed, which is important to evaluate and better design higher-grade pulse power system. Compared to the traditional predictive method of switching time, the ohmic loads and nonlinearity of capacitances are considered in the proposed method. In detail, the expressions of the switching time are deduced, in which, the main parasitic elements affecting the switching transient are considered. In order to verify the validity of the method, an ohmic-load double pulse circuit is established. The experimental device is a CREE SiC MOSFET module CAS325M12HM2 (1200V/325A). The predictive results of the proposed method are compared with experiments and the traditional predictive method under different voltages and currents. The comparison results show that the proposed method has a good agreement with experiments and has better predictive accuracy than the traditional method, which verifies the effectiveness and accuracy. The method of prediction the switching time of nanosecond pulsed power system of ohmic loads using SiC MOSFETs offers a theoretical reference to evaluate and improve pulse power systems.

Keywords—*SiC MOSFETs, nanosecond pulsed power system, ohmic loads, switching time, predictive method.*

I. INTRODUCTION

The Wide Band-Gap (WBG) power semiconductor devices provide a new opportunity for the high-speed and high-voltage pulse power systems [1, 2]. Among the WBG devices, silicon carbide (SiC) devices can achieve higher switching speeds (<100ns) and switching frequencies (> 100kHz) due to the special material characteristics, such as higher electrical breakdown field, wider bandgap, higher thermal conductivity and faster carrier saturation velocity [3, 4]. In the past several years, some traditional converters, such as the pulse generator, have been greatly improved because of the use of the SiC devices. In pulsed generators, one of the most concerned characteristic is the rising and

falling speed of the voltage, which has reached several nanoseconds to tens of nanoseconds [5–7]. In order to achieve such fast switching speeds in high-voltage devices, the SiC MOSFET becomes one of the most promising semiconductor devices.

At present, the pulse power generators based on SiC MOSFETs have the advantages of short switching speeds and high voltage outputs. For example, the switching transients of up to 43V/ns [8, 9] of the new SiC MOSFETs have been measured and the latest pulse power system has reached the output voltage of ±50kV. Such fast switching speeds of the SiC MOSFETs are generally measured by inductive load circuits because the inductive-load is the commonly used load in the PES (power electronic systems). However, in pulse generators, the pulse output is generally connected to a load with resistance characteristics such as the gas, water or cells. Therefore, in the pulse generator design process, the load type is generally designed as an ohmic load [10–13]. The ohmic loads often slow down the overall switching speed of the device. In addition, for pulse power systems, modular topologies are commonly used, such as the Marx pulse generator, which composed of many modules cascaded to achieve a very high output voltage. For the system with clamped inductive loads, many analytical models have been developed [14-16]. But for the SiC MOSFETs with a large resistance, there is currently no accurate predictive method to calculate the transient switching time. The performance evaluation of these systems with ohmic loads is still based on empirical methods or experimental measurements, which makes it difficult for the design optimization and evaluation of the systems. Thus, it is important to develop a predictive method for accurately determining the switching performance of nanosecond pulsed power system with considering the ohmic loads.

In this paper, an accurate predictive method for the switching time of the nanosecond pulsed power system with ohmic loads using SiC MOSFETs is proposed. The predictive method is based on the physical mechanism of the SiC MOSFET switching transient and relevant parasitic elements are also considered. In the previous Refs, the Laplace transform method [15] and state equation method are often adopted to obtain analytical solutions of higher-

---

This work was supported by National Key Research and Development Project (2019YFC0119101) and JCJQ program (2020-JCJQ-ZD-105 and 2020-JCJQ-JJ-139).

Corresponding author: Xin Li.

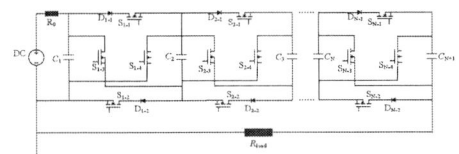

Fig. 1. Circuit diagram of nanosecond Marx pulsed power system.

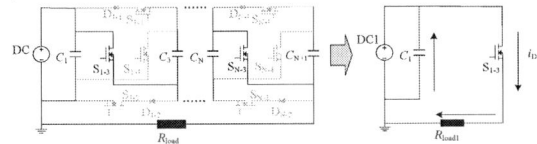

(a) Positive pulse circuit   (b) Simplified circuit
Fig. 2. Investigated positive Marx generator and its simplified circuit.

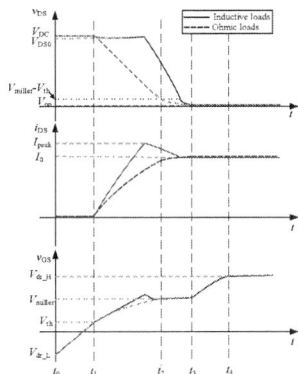

Fig. 3. Comparison of turn-on transient of the inductive and ohmic load.

order circuit equations. But the analytical solutions becomes more difficult to obtain as the circuits become more and more complex. Hence the segment capacitances and partial linearization method are used in this paper to get analytical solutions. In Section II, a simplified circuit of the Marx pulse generator is proposed based on its operation principles. Then the output transient characteristics of target circuit with a ohmic load are modeled using the piecewise capacitance and partial linearization methods in Section III. In Section IV, the experimental results are compared with the calculated results using the proposed method. Finally, the conclusions are drawn in Section V.

II. CIRCUIT DESCRIPTION AND SIMPLIFICATION

As shown in Fig. 1, the Marx pulse power system is composed of SiC MOSFETs, SiC MPS diodes and non-inductance capacitors. The mode of parallel charging and series discharging is adopted, so as to achieve the target output voltage. Unlike traditional circuits that use inductive clamping loads, in the Marx pulse power systems in this paper, the output load is a large resistance of 300Ω. Nowadays, new power module design and advanced semiconductor packaging techniques are proposed. For example, CREE SiC MOSFET chips have been embedded in a printed circuit board (PCB), and its layout is very compact, so very low parasitic inductance can be obtained. Furthermore, the optimization of the manufacturing techniques have reduced the differences between devices and make the differences of the switching transient within a very small range. Therefore, as shown in Fig. 2, the switching transient process of multiple devices in series in the pulse power system can be equivalent to that of a single device.

The pulse power system in this paper is a 7-level Marx generator with the goal to achieve the output voltage of 5kV and output current of 16A. In the positive discharge state, the topology diagram of the Marx generator is shown in Fig. 2. It can be found that the pulse generator discharges through the multiple capacitors in series. The simplified circuit diagram of this process is shown in Fig. 2(b), which consists of a capacitor, a resistor and a SiC MOSFET. The equivalent voltage and current of this simplified circuit are 720V and 16A, respectively.

III. MODELLING OF SWITCHING TRANSIENT AND PREDICTIVE METHOD FOR SWITCHING TIME

The transient characteristics of SiC MOSFETs are mainly determined by the charge and discharge of the nonlinear capacitances, which include gate-drain capacitance, source-

drain capacitance and gate-source capacitance. These non-linear capacitances have a great influence on the dynamic and static characteristics of SiC MOSFETs. There have been some methods for analyzing and modeling these capacitances, and they are mainly divided into two categories. In the first category of model [9], the average values are often used for modeling of the capacitances, whose accuracy is low. The second category of model fits the capacitance curves using some complex nonlinear equations [18, 19], whose accuracy is high. However, the use of these model makes it difficult to have analytical solutions for the SiC MOSFET switching transient. In this paper, the switching transient of the SiC MOSFET with the ohmic load is divided into different stages, and different capacitance values are used (see Fig. I). Using this method, the analytical results of the complex differential ADE equations can be avoided and it can achieve a good trad-off between calculation efficiency and accuracy.

TABLE I
SEGMENTED VALUE OF NONLINEAR CAPACITANCES

Criterion	$v_{GS}>V_{th}$, $v_{DS}\leq V_{GS}-V_{th}$	$v_{GS}>V_{th}$, $v_{DS}>V_{GS}-V_{th}$	$v_{GS}<V_{th}$
Operation region	Ohmic	Saturation	Cutoff
C_{GD}	C_{GDH}	C_{GDA}	C_{GDL}
C_{DS}	C_{DSH}	C_{DSA}	C_{DSL}
C_{GS}	C_{GS}	C_{GS}	C_{GS}

In addition, g_{fs} and V_{th} also have large non-linearities under different working conditions, which has a significant impact on the output dynamic characteristics of SiC MOSFETs. The function fitting method [9] is used in this paper for the modeling of g_{fs} and V_{th} to improve the model accuracy. The qualitative switching waveforms of the target circuit are depicted in Fig. 3 and Fig. 4: the waveforms of drain-source voltage and current and gate-source voltage between the ohmic load and inductive load of a double-pulse test with SiC MOSFETs are compared. It is obvious that the switching time of ohmic load is longer than that of inductive loads. The modeling of the circuit with inductive loads has been investigated in [15, 16]. For the ohmic loads, The detailed analysis and modeling process is discussed below.

A. Turn-on process

As shown in Fig. 3, the turn-on transient process is divided into four sub-stages. The detailed descriptions and analysis of each stage are as follows.

Stage 1 (t_0-t_1): This is the turn-on delay stage. The input capacitance is charged when the gate voltage starts to increases from V_{dr_L}. The SiC MOSFET will remain in the cutoff-region until the gate–source voltage v_{GS} reaches the threshold voltage V_{th}. Hence, the SiC MOSFET drain–source voltage v_{DS} is equal to V_{DC}, the drain current i_D is equal to 0. During this stage, only v_{GS} rises exponentially over time, it can be expressed as follows.

$$V_{GS} = V_{drive\_H}[1 - e^{(t-t_0)/\tau_{on}}] \tag{1}$$

Here, $\tau_{on} = R_G(C_{GS} + C_{GDL})$. Then the delay time can be derived:

$$t_{on\_delay} = R_G(C_{GS} + C_{GDL})\ln[(V_{dr\_H} - V_{dr\_L})/(V_{dr\_H} - V_{th})] \tag{2}$$

Stage 2 (t_1-t_2): During this stage, v_{GS} exceeds V_{th}, then the drain current i_D gradually increases. Because the load is a large resistance, the voltage across the ohmic load starts to decrease as the current rises. The v_{Lstray} induced by the rising SiC MOSFET current across the stray inductances L_{stray} will decrease slowly as the current rate di/dt decreases. When the voltage v_{DS} drops to ($V_{millier}$-V_{th}), the current has not yet increased to the steady-state current, but the MOSFET will enter the ohmic region from the saturation region. In this process, the average gate current can be expressed as

$$i_{G(av)} = (C_{GS} + C_{GDA})\frac{dv_{GS}}{dt} - C_{GDA}\frac{dv_{DS}}{dt} \tag{3}$$

During this period, the gate voltage is approximately considered to increase linearly, then the average current can be expressed as

$$i_{G(av)} = \left[V_{dr\_H} - 0.5(V_{miller} + V_{th}) - L_m\alpha I_0/(t_2 - t_1)\right]/R_G \tag{4}$$

where L_m represents the parasitic inductance of the SiC MOSFET module. As shown in Fig. 3, the change rate of gate–source voltage v_{GS} of this process can be expressed using Eq. (5) which is as follows.

$$dv_{GS}/dt = (V_{miller} - V_{th})/(t_2 - t_1) \tag{5}$$

The source-drain current is assumed as αI_0 when the gate voltage reaches the miller voltage, then the current change rate during this stage can be expressed as:

$$di_D/dt = \alpha I_0/(t_2 - t_1) \tag{6}$$

Hence, the change rate of the MOSFET voltage is obtained:

$$dv_{DS}/dt = -R_{load}(di_D/dt) - v_{Lstray}/(t_2 - t_1) \tag{7}$$

Here, v_{Lstray} represents the voltage of the stray inductance induced by the current rate di/dt, which can be expressed as:

$$v_{Lstray} = L_{stray}\alpha I_0/(t_2 - t_1) \tag{8}$$

As the SiC MOSFET is in the saturated region during this stage, its current can be expressed as:

$$i_D = g_{fs}(v_{GS} - V_{th}) \tag{9}$$

During the MOSFET switching process, the current i_D at t_2 is αI_0, and it can be obtained the following expression:

$$i_D = g_{fs}(V_{miller} - V_{th}) = \alpha I_0 \tag{10}$$

Therefore, the miller voltage can be approximated be derived:

$$V_{miller} = \alpha I_0/g_{fs} + V_{th} \tag{11}$$

The resulting equation for the time (t_2-t_1) derived from Eqs. (3)-(11) are of quadratic order, for which the analytical solution exists. Hence, the positive solution can be obtained:

$$t_2 - t_1 = \left(-B_0 + \sqrt{B_0^2 - 4A_0C_0}\right)/(2A_0) \tag{12}$$

Here, A_0, B_0 and C_0 are expressed as follows.

$$\begin{cases} A_0 = g_{fs}V_{dr\_H} - (0.5\alpha I_0 + g_{fs}V_{th}) \\ B_0 = -g_{fs}R_{load}C_{GDA}R_G\alpha I_0 - g_{fs}L_m\alpha I_0 - \alpha I_0(C_{GS} + C_{GDA}) \\ C_0 = -g_{fs}R_GC_{GDA}L_{srtay}\alpha I_0 \end{cases} \tag{13}$$

Stage 3 (t_2-t_3): During this stage, the voltage continues to fall. The SiC MOSFET will switch into the ohmic region when v_{DS} drops to the boundary voltage (V_{miller}-V_{th}). Hence, $C_{GD} = C_{GDH}$. As $v_G = V_{miller}$, the voltage continues to decrease and the device current continues to rise while the gate voltage remains at the miller voltage. At the end of this stage ($t=t_3$), v_{DS} reaches its steady-state voltage V_{on}. In this process, the average gate current of the MOSFET can be obtained.

$$i_{G(av)} = -C_{GDH}\frac{dv_{DS}}{dt} \tag{14}$$

Similarly, the gate current is also related to the gate voltage and resistance, so the average current of the process can be approximately expressed as:

$$i_{G(av)} = \left(V_{dr\_H} - V_{miller} - L_m(di_D/dt)\right)/R_G \tag{15}$$

During this stage, in order to obtain the change characteristics of the voltage and current, the changes of the drain-source voltage and current is also assumed to be linear, so the following Eqs. (16) and (17) can be obtained

$$di_D/dt = (1-\alpha)I_0/(t_3 - t_2) \tag{16}$$

$$dv_{DS}/dt = (V_{miller} - V_{th} - I_0R_{on})/(t_3 - t_2) \tag{17}$$

By substituting Eqs. (11), (16) and (17) into Eqs. (14) and (15) and making an arrangement, the time of this stage can be expressed as follows.

$$t_3 - t_2 = \frac{g_{fs}L_m(1-\alpha)I_0 - R_GC_{GDH}(\alpha I_0 - g_{fs}I_0R_{on})}{g_{fs}(V_{dr\_H} - V_{th}) - \alpha I_0} \tag{18}$$

Stage 4 (t_3-t_4): The gate capacitance continues to be charged. Once v_{DS} reaches V_{on}, it remains at V_{on} and i_D keeps constant at I_0, no longer controlled by v_{GS}. v_{GS} will reach V_{dr_H} at t_4. Therefore the time can be calculated as

$$t_4 - t_3 = 2R_G(C_{GS} + C_{GDH}) \tag{19}$$

B. Turn-off process

As shown in Fig. 4, the turn-off transient process is divided into five sub-stages. Some detailed process descriptions and analysis are as follows.

Stage 1 (t_5-t_6): The gate voltage is set to the low level, then the v_{GS} drops exponentially from V_{dr_H} to V_{miller}. SiC MOSFET works in the ohmic region, the drain–source voltage v_{DS} and current i_D can be expressed as $v_{DS}= V_{on}$ and $i_D =I_0$. Input capacitance C_{iss} is discharged through gate resistance. The time of this stage can be derived in the similar way as the turn-on delay stage. It can be approximately expressed as:

$$t_{delay\_off} = R_G(C_{GS} + C_{GDH})[\frac{V_{dr\_H} - V_{dr\_L}}{V_{miller} - V_{dr\_L}}] \tag{20}$$

Here, the miller voltage can be approximated by Eq. (10).

Stage 2 (t_6-t_7): The drain-source voltage v_{DS} starts to rise. During this process, the SiC MOSFET operates in the ohmic region, hence $C_{GD}=C_{GDH}$. After v_{GS} drops to V_{miller}, the drain-source voltage v_{DS} rises from V_{on} to V_{miller}-V_{th} with a smaller slope. Therefore, the drain current i_D also decreases slightly during this stage. Under the condition of ignoring the influence of stray inductance of the loop (L_{stray}), the voltage change is approximately considered linear, so the gate current can be expressed as

$$i_{G(av)} = C_{GS}\frac{dv_{GS}}{dt} + C_{GDH}\frac{d(v_{GS} - v_{DS})}{dt} \tag{21}$$

During this process, the voltage v_{GS} remains constant, so the following equation for the gate current can be given

$$i_{G(av)} = -C_{GDH}\frac{dv_{DS}}{dt} = -C_{GDH}\frac{V_{miller} - V_{th} - R_{on}I_0}{t_7 - t_6} \tag{22}$$

The gate voltage in this process is approximately regarded to be constant, then the average current in this process can also be expressed as:

$$i_{G(av)} = \frac{|V_{dr\_L} - V_{miller}|}{R_G} \tag{23}$$

Combining the Eqs. (22) and (23), the voltage rise time of this stage can be calculated as follows.

$$t_7 - t_6 = \frac{R_G C_{GDH}(V_{miller} - V_{th} - R_{on}I_0)}{V_{miller} - V_{dr\_L}} \tag{24}$$

At t=t_7, v_{DS} reaches to V_{miller}-V_{th} and the MOSFET starts to work in the saturation region.

Stage 3 (t_7-t_8): During this stage, v_{DS} continues to rise. As the SiC MOSFET works in the saturation region, hence C_{GD} changes to a small value C_{GDA} and a larger voltage change rate than that of the previous stage is observed. At the same time, the drain-source current i_D continues to drop to i_{DS1} with a large rate. At t=t_7, the drain-source voltage is given as follows.

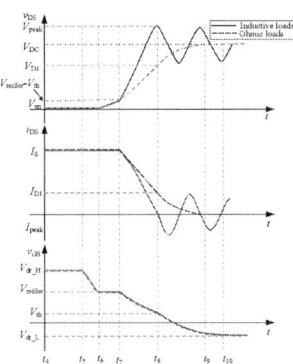

Fig. 4. Comparison of turn-off transient of the inductive and ohmic load.

$$V_{DS(t_7)} = I_0 / g_{fs} = V_{miller} - V_{th} \tag{25}$$

In this process, the gate voltage V_{GS} is expressed as:

$$v_{GS} = \begin{cases} I_0 / g_{fs} + V_{th} & t = t_7 \\ V_{th} & t = t_8 \end{cases} \tag{26}$$

The gate-source voltage, drain-source voltage, gate-drain voltage and the load voltage are assumed to change linearly during this stage and it can be expressed as follows:

$$\frac{dv_{GS}}{dt} = \frac{dv_{DS}}{dt} - \frac{dv_{GD}}{dt} \tag{27}$$

$$\frac{dv_{load}}{dt} = \frac{dv_{DS}}{dt} \tag{28}$$

Similarly, the current of the process has the following expression:

$$\frac{di_{load}}{dt} = \frac{di_{DS}}{dt} = g_{fs}\frac{dv_{GS}}{dt} \tag{29}$$

The resulting equation for the time (t_8-t_7) derived from Eqs. (25)-(29) are of quadratic order, for which the analytical solution exists. Hence, the positive solution (t_8-t_7) can be expressed as [9]:

$$t_8 - t_7 = \left(-B_1 + \sqrt{B_1^2 - 4A_1C_1}\right)/(2A_1) \tag{30}$$

Here, A_1, B_1, C_1 are given as

$$\begin{cases} A_1 = V_{DC} + 2R_{load}g_{fs}(V_{th} - V_{dr\_L}) \\ B_1 = V_{DC}(C_x - 2C_{GD}R_G g_{fs})R_{load} - 2V_{DC}(C_{GS}R_G + g_{fs}L_m) + 2(V_{th} - V_{dr\_L})R_{load}^2 g_{fs}C_x \\ C_1 = -2V_{DC}[R_{load}(R_G(C_y C_{GD} + C_{GS}(C_{DS} + C_L)) + L_m C_x g_{fs}) + R_G g_{fs}C_{GD}(L_{stray} + L_m)] \end{cases} \tag{31}$$

where $C_x=C_{DS}+C_L+C_{GD}$, $C_y=C_{DS}+C_L+C_{GS}$, $C_{GD}=C_{GDA}$ and $C_{DS}=C_{DSA}$.

Stage 4 (t_8-t_9): During this stage, the gate–source voltage v_{gs} decrease from V_{th}. As the capacitor in the loop continues to discharge, the drain current i_D continues to decrease to 0. The fall time of this process for the ohmic load can be calculated using the RC circuit discharging. The time of this stage can be expressed as follows [9].

978-1-6654-4817-8/21 $31.00 © 2021 IEEE

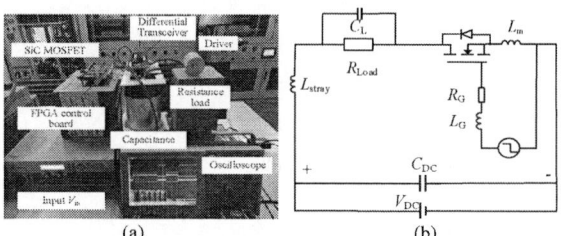

Fig. 5. Double-pulse test experiment circuit. (a) Experimental platform. (b) Experimental equivalent circuit diagram.

$$t_9 - t_8 = R_{\text{load}}(C_{\text{DSL}} + C_{\text{GDL}} + C_{\text{L}}) \ln\left[\frac{V_{\text{load1}}}{0.1 V_{\text{DC}}}\right] \qquad (32)$$

where V_{load1} is the voltage of the R_{load} at t_8.

Stage 5 (t_9-t_{10}): After t_9, the MOSFET works in the cut-off region and the drain-source voltage v_{DS} and drain current i_{D} enter the oscillation stage because of the stray inductance. However, as the load of the circuit is a large resistance, the oscillation may be small and it decays quickly. In the operation of very large resistances, the oscillation phenomenon may no longer be observed. During this stage, the gate-source voltage v_{GS} drops to $V_{\text{drive\_L}}$ exponentially, The time of this stage can be approximately expressed as:

$$t_{10} - t_9 = 2R_{\text{G}}(C_{\text{GS}} + C_{\text{GDL}}) \qquad (33)$$

From the above analysis, it can be concluded that the turn-off switching transient is an approximate reverse symmetrical process of the turn-on transient.

IV. SIMULATION AND EXPERIMENTAL RESULTS

The experimental platform is shown in Fig. 5, an ohmic-load double-pulse-switching test is used for verification. Fig. 5 (a) is a schematic diagram of the entire experimental platform and its equivalent circuit diagram are shown in Fig. 5 (b). Considering the high switching speed of SiC MOSFETs, it is necessary to minimize the parasitic inductances of the test circuit. Therefore, a low-inductance double-pulse test bus is designed and some low-inductance components are used. The load is a non-inductive resistor, the capacitance is non-inductive capacitor with the value of 2000μF, and the charging source is a high-voltage and high-power DC source of 1500V/0-30A(15000W). In addition, the turn-on and turn-off of the whole test platform is controlled by FPGA. The experimental device is a CREE SiC MOSFET module (1200V/325A, CAS325M12HM2). The type of gate driver for the SiC MOSFET is CGD15HB62P1. The measuring equipment should also meet the requirements of the fast switching speed of the SiC devices. TABLE II shows the types of measuring instruments used in the experimental platform.

In order to accurately predict the switching time, the model parameters need to be extracted from the SiC MOSFET module. In the development process of SiC devices, many parameter extraction methods have been proposed. In this section, the parameters used in the method are divided into two categories: the device physical parameters and parasitic parameters. The reverse engineering experiments and impedance analyzer are used to get these parameters and the extraction procedures adopted in this section can be found in [14, 20, 21].

TABLE II
TYPES AND PARAMETERS OF MEASURING INSTRUMENTS

Instruments	Type Parameters	Parameters
Oscilloscope	DPO5104B	1GHz
Voltage probe	TEK THDP0200	200MHz/1500V
Current probe	CWT_Mini50HF	50MHz/300A

Fig. 6 shows the switching transient waveforms of v_{DS} and i_{D} of measurement under the voltages from 100V to 800V. The ohmic load of the test circuit is about 39Ω, which results in a load current of about 2.5A-20.5A. Under the voltage of 800V, a rising and falling switching time of T_{on}=76ns, respectively T_{off}=322ns is measured. Although the turn-on and turn-off speeds of the SiC MOSFET are fast, the output pulse has almost no peak voltage due to the low parasitic inductance of the experimental platform and the large load resistance. Besides, it can be found from Fig. 6 that the turn-on time clearly decreases with the bus voltage, whereas the turn-off time increases with the voltage. The detailed reasons for this characteristics are as follows. The on-state load current increases with the bus voltage due to the ohmic load. The charging speed of the gate driver for the gate capacitance can be seen constant under the same gate resistance and voltage, so the current rise time is basically linear with the increase of the steady-state current. Hence, the voltage rise time increases linearly with the bus voltage. The voltage fall time gradually decreases with the increase of the voltage. This is because the on-state current increases with the bus voltage. Then the ohmic load consumes more energy in the initial stage of the turn-off transient, which makes the voltage fall faster. Due to some errors in the linear assumption and parameter extraction, the accuracy of the prediction method is affected a little. The worst error is around 10% which can meet the requirements of the time prediction of the power pulse system, hence verifying the effectiveness of the proposed prediction method.

In order to further verify the effectiveness and accuracy of the prediction method, the calculated switching time of the output voltage using the proposed method is compared with the traditional model [9] and experiments. The comparison results are shown in Fig. 7 and Fig. 8. The traditional model did not consider the nonlinearity of the parasitic capacitance and piecewise linearization of the switching process, resulting in a low accuracy with the maximum error of nearly 40%. For the method in this paper, the prediction results are in good agreement with the experiments. The comparison results show that the average error is reduced from about 40% of the traditional method to about 10% using the proposed method under different voltage and current conditions. This method can accurately predict the transient performance of the SiC MOSFET with an ohmic load, and the prediction results can be used for the evaluation, design and improvement of nanosecond pulsed power systems.

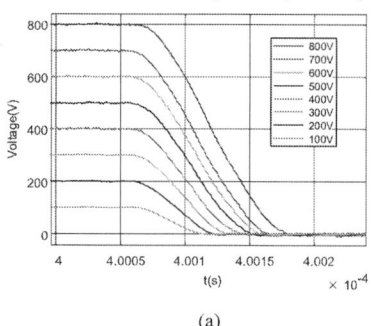

(a)

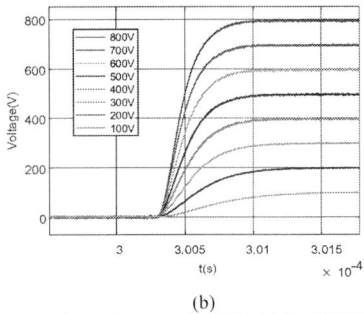

(b)

Fig. 6. Switching transient characteristics of the SiC MOSFET under different working conditions. (a) Turn-on; (b) Turn-off.

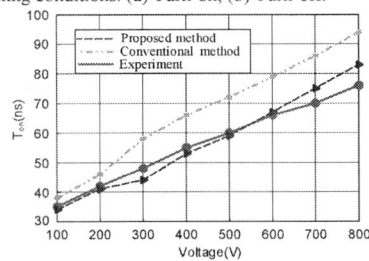

Fig. 7. Experiment and simulation comparison of turn-on speed.

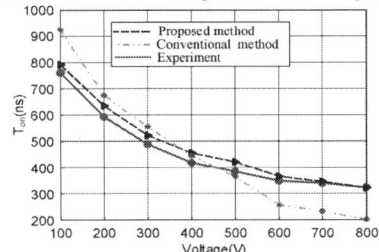

Fig. 8. Experiment and simulation comparison of turn-off speed.

V. CONCLUSION

SiC MOSFETs offer a great opportunity for very fast and high voltage pulsed power system. However, there is a lack of accurate mathematical methods, that can quantitatively predict the switching time of pulsed power system of ohmic loads. This paper proposed a predictive method for the SiC MOSFET switching transient time. For pulse power systems composed of the modular topology, the equivalent simplification is performed and the basic switching cell is obtained. Based on the nonlinear capacitance and partial linearization assumptions, the mathematical models of the switching transient time are established. In order to verify the theoretical analysis and the proposed prediction method, a double-pulse circuit using the SiC MOSFET and an ohmic load is established. Experiments are conducted under different voltages and load currents, and the results show that the turn-on time increases significantly with the bus voltage, while the turn-off time decreases with the bus voltage. The prediction method can accurately describe this trend. Errors between the analysis method and the experimental results under different voltages and currents are within 10%, which verifies the effectiveness and accuracy of the proposed prediction method.

REFERENCES

[1] B. Jayant Baliga, *Wide Bandgap Semiconductor Power Devices: Materials, Physics, Design, and Applications*. UK: Woodhead, 2019, pp. 12-38.

[2] L. Pang, T. Long, K. He, Y. Huang and Q. Zhang, "A Compact Series-Connected SiC MOSFETs Module and Its Application in High Voltage Nanosecond Pulse Generator," IEEE Transactions on Industrial Electronics, vol. 66, no. 12, pp. 9238-9247, Dec. 2019.

[3] X. She, A. Q. Huang, Ó. Lucía and B. Ozpineci, "Review of Silicon Carbide Power Devices and Their Applications," IEEE Transactions on Industrial Electronics, vol. 64, no. 10, pp. 8193-8205, Oct. 2017.

[4] X. Li, F. Xiao, Y. Luo and Y. Duan, "Analysis and Modeling of SiC MPS Diode and Its Parasitic Oscillation," *IEEE Journal of Emerging and Selected Topics in Power Electronics*, vol. 8, no. 1, pp. 152-162, March 2020.

[5] L. Collier, T. Kajiwara, J. Dickens, J. Mankowski and A. Neuber, "Fast SiC Switching Limits for Pulsed Power Applications," in IEEE Transactions on Plasma Science, vol. 47, no. 12, pp. 5306-5313, Dec. 2019.

[6] M. Azizi, J. J. van Oorschot and T. Huiskamp, "Ultrafast Switching of SiC MOSFETs for High-Voltage Pulsed-Power Circuits," IEEE Transactions on Plasma Science, vol. 48, no. 12, pp. 4262-4272, Dec. 2020.

[7] Y. He, X. You, J. Ma, L. Yu, W. Zeng, S. Dong, Z. Zhang and C. Yao, "A Polarity-Adjustable Nanosecond Pulse Generator Suitable for High Impedance Load," IEEE Transactions on Plasma Science, vol. 48, no. 10, pp. 3409-3417, Oct. 2020.

[8] G. Regnat, P. Jeannin, J. Ewanchuk, D. Frey, S. Mollov, and J. Ferrieux, "Optimized power modules for silicon carbide MOSFET," IEEE Trans. on Industry Applications, vol. 54, no. 2, pp. 1634–1644, Apr. 2018.

[9] R. Risch and J. Biela, "Nanosecond switching of ohmic loads using SiC MOSFETs in ultra-low inductive PCB-packages," in *Proc.* EPE '19 ECCE Europe, Genova, Italy, 2019, pp. P.1-P.10.

[10] E. G. Cook et al., "Solid-State Modulator R&D at LLNL," Int. Workshop on Recent Progress of Induction Accelerators, Dec. 2002.

[11] J. Rao, Y. Lei, S. Jiang, Z. Li, and J. F. Kolb, "All Solid-State Rectangular Sub-Microsecond Pulse Generator for Water Treatment Application," IEEE Trans. on Plasma Science, vol. 46, no. 10, pp. 3359–3363, Oct. 2018.

[12] W. L. Waldron, J. E. Galvin, W. B. Ghiorso, and C. Pappas, "The design and testing of an inductive voltage adder for ALS-U kicker magnets," in in Proc. IPMHVC, Jul. 2016, pp. 176–178.

[13] J. Holma and M. J. Barnes, "The Prototype Inductive Adder with Droop Compensation for the CLIC Kicker Systems," IEEE Trans. on Plasma Science, vol. 42, no. 10, pp. 2899–2908, Oct. 2014.

[14] X. Wang, Z. Zhao, K. Li, Y. Zhu and K. Chen, "Analytical Methodology for Loss Calculation of SiC MOSFETs," IEEE Journal of Emerging and Selected Topics in Power Electronics, vol. 7, no. 1, pp. 71-83, March 2019.

[15] J. Wang, H. S. Chung and R. T. Li, "Characterization and Experimental Assessment of the Effects of Parasitic Elements on the MOSFET Switching Performance," IEEE Transactions on Power Electronics, vol. 28, no. 1, pp. 573-590, Jan. 2013.

[16] D. Christen and J. Biela, "Analytical Switching Loss Modeling Based on Datasheet Parameters for mosfets in a Half-Bridge," IEEE Transactions on Power Electronics, vol. 34, no. 4, pp. 3700-3710, April 2019.

[17] M. Liang, T. Zheng, and Y. Li, "An Improved Analytical Model for Predicting the Switching Performance of SiC MOSFETs," Journal of Power Electronics, vol. 16, no. 1, pp. 374–387, Jan. 2016.

[18] A. Alhoussein, H. Alawieh, Z. Riah and Y. Azzouz, "A New SiC Power MOSFET Model with a Parameter Optimization Procedure," in Proc. EPE '19 ECCE Europe, Genova, Italy, 2019, pp. P.1-P.11.

[19] A. Stefanskyi, Ł. Starzak and A. Napieralski, "Review of commercial SiC MOSFET models: Topologies and equations," in Proc. 2017 MIXDES, Bydgoszcz, Poland, 2017, pp. 484-487.

[20] X. Li, F. Xiao, Y. Luo, and Y. Duan, "Parameter extraction method for a physics-based lumped-charge SiC MPS diode model," IET Power Electronics, vol. 13, no. 14, pp. 2992-3000, 2020.

[21] W. Jouha, A. E. Oualkadi, P. Dherbécourt, E. Joubert and M. Masmoudi, "A new extraction method of SiC power MOSFET threshold voltage using a physical approach," in Proc. ICEIT, Rabat, Morocco, 2017, pp. 1-6.

Active Magnetic Bearing Amplifier Design based on SiC Devices

Gang Cao
School of Electrical and Electronic Engineering
Huazhong University of Science and Technology
Wuhan, China
cao_gang@hust.edu.cn

Hongbo Sun
School of Electrical and Electronic Engineering
Huazhong University of Science and Technology
Wuhan, China
hongbo_sun@hust.edu.cn

Gao Yang
Science and Technology on Ship Integrated Power System Technology Laboratory
Wuhan, China
lamb1234@sina.com

Dong Jiang
School of Electrical and Electronic Engineering
Huazhong University of Science and Technology
Wuhan, China
jiangd@hust.edu.cn

Abstract—**Active Magnetic Bearing (AMB) levitate rotor by magnetic force without friction. The purpose of this paper is to design a SiC-based power amplifier for the AMB controller. Firstly, the structure and control principle of the magnetic bearing system are introduced. Secondly, the improvement of power amplifier performance of SiC device relative to traditional Si device is analyzed. Device selection and digital PI control circuit are also introduced. In addition, step response experiments at 20kHz and 40kHz switching frequencies are designed for the power amplifier of SiC devices. Then the AMB at two switching frequencies is evaluated from response speed. Finally, based on the experimental results under two different working conditions of AMBS, the corresponding conclusions are drawn.**

Keywords—*Active Magnetic Bearing (AMB), SiC , Digital PI controller, Power amplifier*

I. INTRODUCTION

Active Magnetic Bearing (AMB) supports the rotor with un-contact magnetic force, and it has been applied in high-speed motor, compressor, and machine tool etc[1]. Different from traditional machine bearing, AMB is a control system including sensors, amplifier, and controller etc. The structure of radial AMB system is shown in Fig.1. The sensors detect the displacement of rotor, and the controller generates control instruction based on displacement signals and control algorithm. Amplifier produces real current in the coils, which creates magnetic force in the coils. Generally, the amplifier is a current control system.

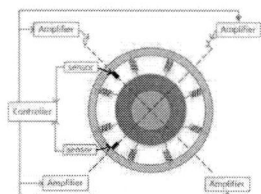

Fig. 1 Structure of radial AMB system

In the design of traditional magnetic bearing controller, due to the limitation of the upper limit of switching speed and bandwidth of the hardware circuit, the designed controller has some defects in performance. Moreover, the voltage withstand value of the device is low, so the magnetic bearing controller can only be designed under low voltage and low bandwidth. In recent years, the third generation semiconductor devices represented by SiC have been developed rapidly. Compared with traditional silicon devices, SiC devices have the characteristics of high temperature resistance, high voltage resistance, high switching frequency and so on, which has a great application prospect[2].

There are great differences between SiC devices and IGBT devices in switching characteristic driving and performance testing. By comparing the performance of SI-IGBT and SiC MOSFET power modules in DC/DC power conversion circuit, Maximilian Slawinski [3] etc found that SiC MOSFET modules had better conversion performance and power density. Junhui Ao[4] etc proposes a master-slave gating driver for current sharing control of parallel SiC MOSFETs, in order to extend current capability, which will be used in transportation sector. Xudong Wang [5] etc proposes an analytical method to evaluate the THD of the high power CDAs, using Silicon Carbide (SiC) MOSFETs, in order to improve the fidelity of the CDAs. Karsten Haehre [6] etc applies SiC MOSFETs to Class-E Amplifier, which works at a switching frequency of 2.5 MHz with an output power up to 830 W environment. Cheng-Po Chen [7] etc using the high temperature resistance of SiC MOSFETs, they designed an integrated circuit with a wide temperature range, highlight. Their research is to apply SiC devices to integrated circuits. Chen Peng [8] etc applied the SiC switch device to the battery charge and discharge system. The voltage and current oscillations in the fast switching process are analyzed by establishing the circuit model of the half-bridge SiC MOSFET switching process. Henry Barth [9] etc studied the performance of SiC-MOSFET and Si-IGBT in AC motor driven inverters. Kazuaki Mino [10] etc developed and designed a high-power photovoltaic inverter(Mega solar)boost chopper using all-SiC modules, which reduced input voltage fluctuations and increased power density by 25%. The purpose of this paper is to apply the superior performance of SiC devices in power amplifier circuit of the AMB system, through the selection of hardware, analysis and design, so that the system can meet the basic functions of AMB and have more superior performance.

978-1-6654-4817-8/21 $31.00 © 2021 IEEE

In part II, the structure and basic control principle of AMB are analyzed, the influence of switching frequency and DC bus voltage on the control performance is discussed, and the influence of discrete PI control on current loop control is also analyzed. In part III, the advantages that SiC devices can bring to AMB control system compared with traditional silicon devices are introduced, and the process and results of device selection are analyzed. Finally, the points for attention when drawing SiC hardware PCB are introduced. The experimental results on the test bench are shown in part IV, the experimental results show that the designed SiC-based magnetic bearing controller is effective in basic functions and performance improvement. Conclusions are summarized in part V.

II. AMB CONTROL PRINCIPLE

A. AMB control principle

From [1] , We can draw the overall structure of the AMB system, as shown in Fig.2, It has two parts, the Control board and the circuit, including two Control frames, the position outer loop and the current inner loop. The real-time position signal of the rotor is collected by the position sensor, and the current signal is collected by the current sensor. After the conditioning circuit, the signal is fed back to the DSP processor, and then the DSP outputs the appropriate PWM wave to control the on-off of the switch tube, so as to realize the suspension of the rotor.

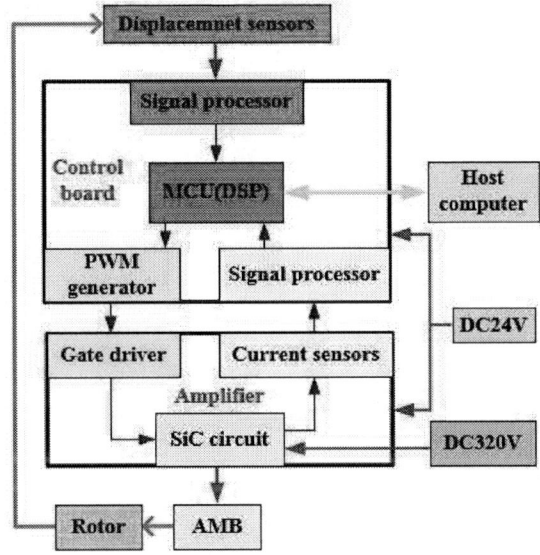

Fig. 2 Structure of AMB system

In the AMB system, the outer control loop is displacement control loop, and the inner control loop is current control loop. In each degree of freedom (DOF), two currents are applied to generate opposite magnetic force. Many types of topology of amplifier have been proposed, and the basic topology for current control is H bridge. Two H bridges with PI controllers are applied to control the two currents in one DOF. The structure amplifier in one DOF is shown in Fig.3.

Existing amplifier designs are mainly based on Si devices, for each unipolar magnetic bearing coil, a unipolar H-bridge topology is used for power electronic converters, which is shown in Fig.4, the corresponding PWM wave is generated by the controller, controlling the duty cycle of main circuit board switches, so as to realize the switch of main circuit different

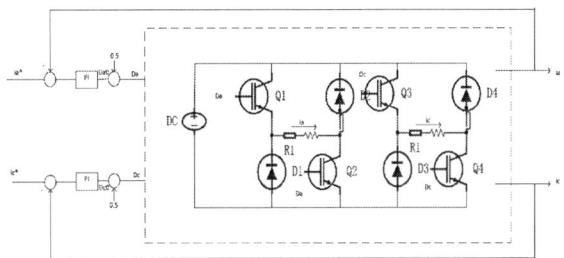

Fig. 3 Structure of amplifier for one DOF

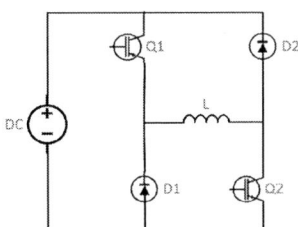

Fig. 4: Unipolar H-bridge topology

work mode. By changing the state of the two switch tubes to change the working mode of the circuit, the current through the coil can be changed. Fig.5 shows the four operating modes of the circuit :(a) Two switching devices are turned on at the same time, current flows through the two switching devices, and the positive DC bus voltage at both ends of the coil is loaded, at which point the coil current rises rapidly; In (b) and (c), only one of the two devices is turned on and the other is turned off. At this time, the winding current continues through the diode of the on-switching device and the other bridge arm. Due to the coil resistance and the voltage drop of the device, the coil current at this time drops slowly;(d) When two switching devices are turned off at the same time, the coil current can only be continued through the diodes on the two bridge arms. When the negative DC bus voltage at both ends of the coil is loaded, the coil current will drop rapidly.

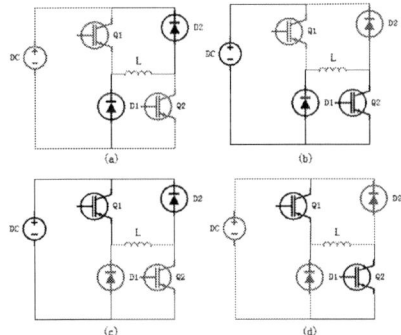

Fig. 5: The four operating modes of the circuit

Through the combination of these four operating modes, the coil current can be controlled.

The current i_L flowing through the magnetic bearing coil can be expressed as:

$$i_L = \frac{v_L}{j\omega L + R} = \frac{s_1 V_{dc} - (1-s_2)V_{dc}}{j\omega L + R} \qquad (1)$$

Where, s1 and s2 are the state equations of two switching devices Q1 and Q2 respectively. When the switch tube is on, s =1, and when it is off, s=0. The duty cycle d1 and d2 of the driving signal of the switching device can be substituted, and the Laplace transform can be carried out to obtain the expression of the current under the average model:

$$i_L(s) = \frac{(d_1 + d_2 - 1)V_{dc}}{sL + R} \quad (2)$$

It can be concluded that the circuit is a first-order model, in which the current is controlled by the duty ratio of the two switching devices, and the values of d1 and d2 range from 0 to 1. Let d1 = d2 = dc + 0.5, then (2) can be written as:

$$i_L(s) = \frac{2d_c V_{dc}}{sL + R} \quad (3)$$

At this time, the value range of the average voltage applied at both ends of the coil is changed to $-V_{DC} \sim +V_{DC}$, and the voltage utilization ratio is two. The control system block diagram of a single DOF power amplifier can be drawn as shown in Fig.6.

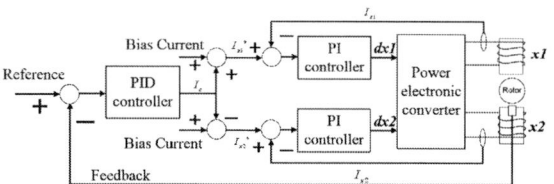

Fig. 6: Control block diagram of the one DOF amplifier

Both switches have the same duty cycle, which is the instruction generated by PI control plus 0.5. The two devices also use opposite carriers to generate PWM drive signals to achieve unipolar modulation. The transfer function of PI controller is:

$$G_c(s) = K_p + \frac{K_i}{s} \quad (4)$$

According to the model of coil current, it can be set:

$$K_p = \frac{\omega_c L}{V_{dc}} \quad K_i = \frac{\omega_c R}{V_{dc}} \quad (5)$$

In which ωc is a variable coefficient. Then the open loop transfer function of the system becomes:

$$G(s) = (K_p + \frac{K_i}{s})\frac{2V_{dc}}{sL + R} = \frac{2\omega_c}{s} \quad (6)$$

It can be concluded that the bandwidth of the first-order system is just the coefficient ωc. Therefore, by changing the parameters of the controller, the bandwidth of the current loop can be changed.

B. Digital PI control

The current loop modeling described above is carried out under the continuous time model. When PI control is adopted in the continuous time model, the characteristic equation of the system can be obtained as follows:

$$\frac{K_p s + K_i}{s} \frac{2V_{dc}}{Ls + R} + 1 = 0 \quad (7)$$

In the formula, the control parameters are taken in the (5), and the following formula can be obtained:

$$(s + 2\omega_c)(Ls + R) = 0 \quad (8)$$

It is easy to obtain that the characteristic roots of the system are all on the negative real axis, and the system is always stable no matter what the control parameter ωc is. In this paper, we use digital control, so we need to discretize the control system. The discretization of the system introduces the sampling time Ts, and the difference of sampling time will affect the overall performance and stability of the system. Analyze the system open-loop Bode diagram with the same control parameters and different sampling times, as shown in Fig.7. Different sampling frequencies truncate the open loop Bode diagram of the system at different frequency points. If the sampling time is too long and the truncation is before the zero-crossing point of the amplitude-frequency characteristic curve, the system will be unstable. The shorter the sampling time is, the closer the zero-crossing point of the system at the amplitude-frequency characteristic curve will be to the continuous system, and the greater the phase margin will be. However, the high sampling frequency has higher requirements on the system hardware in the actual implementation. Therefore, we use SiC device to replace the original Si device, so as to improve the switching frequency and sampling frequency, so that the performance of the system is more superior. Combining with mathematical PI control algorithm, the controller only the amount of control is calculated according to the deviation of the sampling times, when the sampling frequency is high enough, the output is basically fitted with the analog PI[11].

III. METHODS TO IMPROVE POWER AMPLIFIER

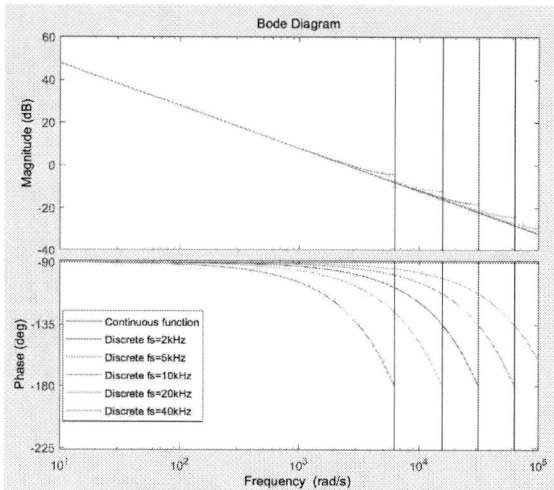

Fig. 7: Current loop discrete model with different sampling time system open loop Bode diagram

The SiC switch device is used to replace the traditional Si device, and the diode is also replaced by SiC device to avoid the phenomenon of diode breakdown caused by excessive voltage. It is necessary to build an appropriate drive circuit for SiC devices, and then select the appropriate drive chip and power module to ensure the normal operation of SiC devices. Through the improvement of hardware circuit, the control cycle of the whole controller can be reduced, making the system respond faster. Because the SiC device has the ability of high switch frequency and high pressure, the hardware circuit can be based on the traditional AMB power circuit, greatly improve the dc bus voltage, because the dc bus voltage will affect current control effect, and improve the dc bus voltage, can eliminate the limit, it will bring to the current control by II analysis can know. The increase of switching frequency can indirectly improve the bandwidth of the controller, which can greatly improve the performance of the system.

A. Advantages of SiC device in power amplifier

Withstand voltage of switching devices based on Silicon Carbide (SiC) is larger than which are based on Silicon (Si). As a result, switching frequency and voltage of DC bus can both be increased, which helps improve stability and increase bandwidth of amplifier. This paper puts forward a kind of H half bridge amplifier based on SiC MOSFET device, by SiC device to replace the original Si device, improving the ability of the main circuit board of the high pressure and high temperature resistance, improving the switching frequency, reducing filter capacitance and inductance, making the current ripple smaller, increasing the controller bandwidth.

B. Components Selection

1) Power electronic devices: SiC devices have higher voltages and switching frequencies than Si devices. Therefore, they can be applied to magnetic suspension bearing system to improve controller bandwidth. Due to the high switching frequency of SiC devices, they are particularly sensitive to parasitic capacitors and stray inductors. Therefore, the driver chip selected in this paper has protection and control functions, which can support efficient and fast design of high-performance systems. Moreover, the SiC device uses a track-rail separated driver, which greatly enhances dV/dt control. The emitter of SiC device has the characteristics of positive voltage on and negative pressure off, so the driver chip in this paper adopts bipolar power supply. Compared with the emitter of a traditional Si device, a positive voltage is usually applied to the SiC driver pole to turn on and a negative voltage to turn off. The negative-power turn-off setting helps prevent the Mosfet switch tube from opening dynamically due to the extra charge generated by the input capacitance.

2) Current sensors: In general, a single current sensor, such as current transformer, Hall iron core sensor, always occupies a considerable space in the design of power amplifier circuit. In the AMB system, hole-core Hall sensors are used to optimize the volume of the controller, so that all measurement and processing steps are integrated into the circuit board.

C. Power circuit design

The SiC MOSFET device selected in this paper is a PCB (printed board) welding device, and the SiC device is placed on the bottom side of the PCB. In the power circuit, the switch tube can choose the same type of device, and the reverse diode also chooses SiC diode. The following considerations are made in the layout and design of the PCB board:

1) The DC energy storage capacitor is arranged as close to the SiC device as possible, and the capacitor is placed near the pin of the SiC device, the purpose is to absorb the interference voltage caused by the high di/dt change rate of the SiC device, so as to avoid the turnoff over-voltage spike generated on the parasitic inductance parameters of the circuit and reduce EMI (electromagnetic interference).

2) In order to reduce the parasitic inductance of the circuit, the positive and negative busbar and AC busbar copper sheet between DC energy storage capacitor and SiC device on PCB are laminated.

3) The drive circuit is as close to the SiC device as possible. In this paper, the drive circuit and the SiC device are arranged on the same PCB board, in order to reduce the influence of parasitic parameters such as inductance from the drive circuit to the SiC MOSFET on the gate. At the same time, on the PCB surface where the SiC device is located, the device that is susceptible to high frequency should be avoided.

IV. THE EXPERIMENTAL RESULTS

In order to verify the proposed structure and method, the main circuit of magnetic bearing based on SiC was designed and completed. The hardware circuit of AMB consists of a SiC MOSFET main circuit board and a DSP controller board, as shown in Fig.8.The bus part of the whole controller adopts 320V DC power supply, and the control circuit adopts 24V power supply.

Fig. 8: Hardware platform of AMB controller

A. Rotor static suspension experiment

In order to verify the function of the designed SiC power amplifier, the static suspension experiment of the magnetic suspension rotor was first designed. The experimental results are shown in Fig.9. It can be seen from the experimental results that when the rotor is successfully suspended, it can be stabilized near the central axis and the fluctuation range is very small, only $\pm 10\mu m$, which indicates that the controller has normal function, good control performance and good suspension effect.

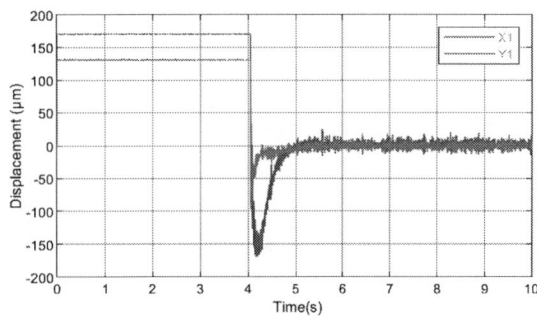

Fig. 9: Rotor static suspension results

B. Low-frequency rotation experiment of the rotor

Furthermore, in order to test the performance of the controller, we used the platform to complete the low-frequency rotation experiment of the rotor after suspension. The experimental results are shown in Fig.10. It can be seen from the experimental results that when the rotor is suspended successfully and rotates at a low speed, the rotor can still remain near the central axis and fluctuate within the range of ±50μm, indicating that the controller has a good function.

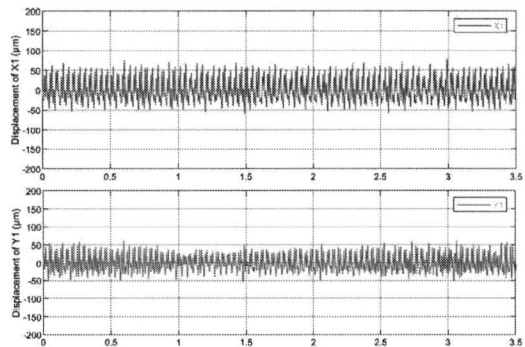

Fig. 10: Low frequency rotor rotation results

C. Step experiments based on SiC devices at switching frequencies of 20kHz and 40kHz

To verify that improving switch frequency can improve system performance, and verify power amplifier based on SiC device does help to improve the performance of magnetic suspension bearing system, we also design in different switching frequency of the rotor step response experiment, change of the switch tube by the controller PWM signal cycle, make its in 20 KHZ and 40 KHZ, through contrast different switching frequencies were used to verify the superior performance of high switching frequencies. The experimental

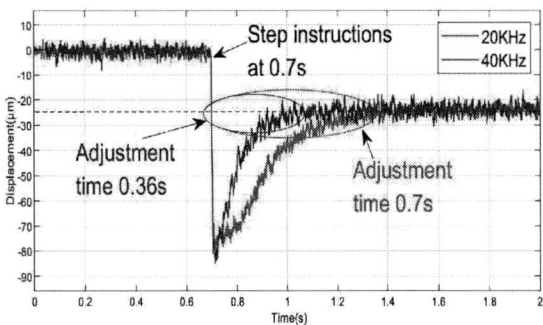

Fig 11: Stepping response of rotors at different switching frequencies

results were shown in Fig.11. The experiment begins by suspending the rotor from the protective bearing. When the rotor is stable in suspension, a step signal is given to the rotor to make it jump from one stable suspension point to the next one. It can be seen that at 40kHz, the steady-state recovery time is shorter than that at 20kHz, indicating that the control cycle of the whole system is shorter and the response is more rapid at 40kHz.

V. CONCLUSION

This paper discusses the function of the proposed AMB power amplifier based on SiC device and its improvement to the performance of the magnetic bearing system, then introduces the design process of the power amplifier and device selection, and finally verifies the function of the designed controller. The analysis shows that the application of SiC in AMB power amplifier circuit can effectively improve the DC voltage and switching frequency of the system, and significantly improve the controller bandwidth. In addition, a SiC device power amplifier is designed at 20kHz and 40kHz switching frequencies to verify the effectiveness of improving the system performance.

ACKNOWLEDGMENT

The authors thank the Science and Technology on Ship Integrated Power System Technology Laboratory for supporting the research in this paper

REFERENCES

[1] Schweitzer G, Maslen E H. Magnetic bearings: theory, design, and application to rotating machinery[M]// Magnetic Bearings—Theory, Design and Application to Rotating Machinery. 2009.

[2] Jun Wang, Longxiang Xu. Analysis and modeling of a switching power amplifier for magnetic bearing[C]// 2009 4th IEEE Conference on Industrial Electronics and Applications, Xi'an, 2009, pp. 2257-2261.

[3] Maximilian Slawinski, Daniel Heer, Tim Villbusch and Marc Buschkuehle, "System study of SiC MOSFET and Si IGBT power module performance using a bidirectional buck-boost converter as evaluation platform," 18th European Conference on Power Electronics and Applications (EPE'16 ECCE Europe), 5-9 Sept. 2016, pp. 1-8.

[4] Junhui Ao, hao Wang, Jin Chen, Li Peng and Yu Chen, "The Cost-Efficient Gating Drivers with Master-Slave Current Sharing Control for Parallel SiC MOSFETs," IEEE Transportation Electrification Conference and Expo, Asia-Pacific, 6-9 June 2018, pp. 1 – 5.

[5] Xudong Wang, Zhengming Zhao, Kainan Chen, Kai Li and Liqiang Yuan, "An Analytical Methodology to Evaluate the THD of High Power Class D Amplifiers," 21st International Conference on Electrical Machines and Systems (ICEMS), 7-10 Oct. 2018, pp. 2327 – 2332.

[6] Karsten Haehre, Nicolai Hildebrandt, Rainer Kling and Wolfgang Heering, "Class-E Amplifier with SiC-MOSFET switching at 2.5 MHz," International Exhibition and Conference for Power Electronics, Intelligent Motion, Renewable Energy and Energy Management, 20-22 May 2014..

[7] Cheng-Po Chen and Reza Ghandi, "Designing Silicon Carbide NMOS Integrated Circuits for Wide Temperature Operation," IEEE International Symposium on Circuits and Systems (ISCAS), 24-27 May 2015,pp. 109-112

[8] Chen Peng, Guo chun Xiao, Shuai Zhang, Chun He,Zhihao Zhai,Xinwei Wang,Qilei Wang and Xudong Du, "SiC MOSFET switching characteristic optimization and application in battery charging/discharging," IEEE 10th International Symposium on Power Electronics for Distributed Generation Systems (PEDG), 3-6 June 2019, pp. 13 – 18.

[9] Henry Barth and Wilfried Hofmann, "Potentials and Boundaries of Discrete SiC-Transistors in AC Drives," 20th European Conference on Power Electronics and Applications (EPE'18 ECCE Europe), 17-21 Sept. 2018,pp. 1-8

[10] Kazuaki Mino, Ryuji Yamada, Hiroshi Kimura and Yasushi Matsumoto, "Power electronics equipments applying novel SiC power semiconductor modules," International Power Electronics Conference (IPEC-Hiroshima 2014 - ECCE ASIA), 18-21 May 2014,pp. 1920 - 1924

[11] Chip Rinaldi Sabirin, Andreas Binder. Rotor Levitation by Active Magnetic Bearing Using Digital State Controller[C]// 2008 13th International Power Electronics and Motion Control Conference (EPE-PEMC), pp. 1625 – 1632.

An Integrated Buck-Boost Converter with SRC for Wide Input Voltage

1st Yanqing Wang
China-EU Institute for Clean and Renewable Energy
Huazhong University of Science and Technology
Wuhan, China
m201971309@hust.edu.cn

4th Changle Xu
State Key Laboratory of Advanced Electromagnetic Engineering and Technology, School of Electrical and Electronic Engineering
Huazhong University of Science and Technology
Wuhan, China
xcl_1995@hust.edu.cn

2nd Yutao Lou
Shanghai Institute of Satellite Engineering
Shanghai, China
286490277@qq.com

5th Xudong Zou
State Key Laboratory of Advanced Electromagnetic Engineering and Technology, School of Electrical and Electronic Engineering
Huazhong University of Science and Technology
Wuhan, China
xdzou@mail.hust.edu.cn

3rd Xiang Guo
State Key Laboratory of Advanced Electromagnetic Engineering and Technology, School of Electrical and Electronic Engineering
Huazhong University of Science and Technology
Wuhan, China
guoxiang@hust.edu.cn

6th Yong Kang
State Key Laboratory of Advanced Electromagnetic Engineering and Technology, School of Electrical and Electronic Engineering
Huazhong University of Science and Technology
Wuhan, China
ykang@hust.edu.cn

Abstract—The series resonant converter has attracted more and more attentions because of its resonant characteristic. However, the gain of series resonant converter is less than 1, which is not applicable for wide input voltage. In this paper, an innovative DC/DC converter integrated Buck-Boost and series resonant converter (IBBSRC) was put forward. The IBBSRC integrates intermediate leg of Buck-Boost converter and SRC to diminish amount of switches and switching losses. Besides, with the modulation strategy of two-edge-modulation + single-phase-shift (TEM+SPS), constant output voltage and zero voltage switching of switches are realized. Six operation modes for above resonant frequency operation were analyzed in detail. The parameter design and zero voltage switching of the converter were studied. Simulation results with an IBBSRC model were indicated to support the vlegality and viability of the propounded circuit topology.

Keywords—four-switch buck-boost, series resonant converter, dual-edge-modulation, single-phase-shift, wide input voltage, ZVS

I. INTRODUCTION

In order to lessen transistor losses and enhance operation of the converter, soft switching technology has become an important method of DC/DC converter. Accordingly, resonant technology is a conventional way to realize soft switching, which utilizes resonance between the capacitor and inductor. As a consequence, the switches are able to switch on or off at the zero-crossing of the voltage or current, and soft switching is naturally realized. The series resonant converter (SRC) is the most typical topology in resonant converters since a resonant inductor cascaded with a resonant capacitor to form a resonant structure.

Compared with other DC/DC converters, SRC shows more superiorities. Specially, a filter capacitor, instead of a bulky filter inductor, is required for SRC at the output side, which cuts down the expense and volume of the converter. Further, the resonant capacitor of SRC could effectively block the DC component of the resonant current, and substantially prevent output transformer core from saturation. Moreover,

the SRC possesses the short-circuit self-protection. More importantly, smooth changing waveforms, low degree of fluctuations of voltage and current with time bring in better electromagnetic compatibility (EMC)[1-5]. Unfortunately, the switching frequency of SRC suffers from significant variations to keep the demanded output voltage under wide input voltage situation. The circumstance will be worse at light loads since the voltage gain can be hardly changed, which will pose great challenges to the design and majorization of magnetic devices and filters, along with fall short in suppressing some specific harmonics[3-6]. In short, the SRC has difficulty when applied to the wide input voltage.

To deal with the issue, several countermeasures were proposed. In [7-9], the three-element LLC series resonant converter was presented through attaching an inductor to the basic SRC. Although the optimized LLC converter can work under both boost and buck mode which could ensure wide input voltage, wide gain range means small magnetizing inductance and high circulating current losses. In [10], a quasi-Z-source series resonant converter (qZSSRC) was come up with. The converter added a quasi-Z-source converter to SRC, thus extending the scopes of input voltage, while the mode process and parameters design were difficult. In [11], a three-level series resonant bidirectional DC/DC converter was put forward. This converter could run in a extensive range of voltage gain by adjusting zero potential duty cycle, but the converter has twelve switches so that the control strategy would be more complicated.

As can be seen, existing researches have some shortcomings in power conversion and topology structures. For this point, this article proffers a fresh isolated DC-DC converter integrated Buck-Boost converter and SRC (IBBSRC). The IBBSRC integrates the wide voltage gain characteristics of four switch Buck-Boost (FSBB) converter and resonant characteristics of SRC, and thus, the converter can not only operate normally under wide input voltage condition, but also zero voltage switching (ZVS) can be fulfilled. In addition, with the modulation strategy of TEM +

978-1-6654-4817-8/21 $31.00 © 2021 IEEE

SPS, the serious problem of EMI caused by the wide range of frequency variation could be avoided and desired output voltage and zero voltage switching of transistors would be obtained.

The rest of this essay is indicated as below. The IBBSRC topology, DC gain and modulation strategy are discussed in Section II. In Section III, the working principles of IBBSRC are discussed comprehensively. The parameters calculations of components are presented in Section IV. Then, the simulation consequences are displayed and researched in Section V to demonstrate the theoretical analysis. Finally, conclusions are made in Section VI.

II. DERIVATION OF IBBSRC

A. Topology of IBBSRC

The schematic of SRC is exhibited in Fig. 1. The SRC adjusts the output voltage by changing switching frequency then altering the impedance of the resonant circuit. Unfortunately, the DC gain of SRC is less than 1, and this will be explained in part *B*. In other words, the SRC is equal to a buck converter, which has poor regulation under wide input voltage. To utilize the wide gain characteristic of FSBB, the proposed IBBSRC is obtained by combining the boost leg of FSBB and the leading leg of SRC, which is shown in Fig. 2. The IBBSRC includes six switches ($Q_1 \sim Q_6$). $Q_1 \sim Q_4$, L_b and C_b form a Buck-Boost converter; and $Q_3 \sim Q_6$, $D_1 \sim D_4$, L_r, C_r and T form a SRC. Main voltages and currents are marked in Fig. 2.

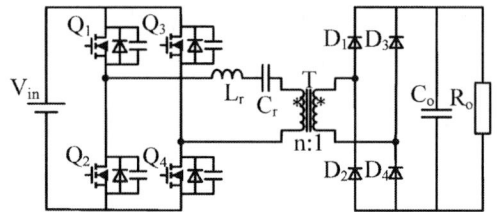

Fig. 1. The schematic of SRC

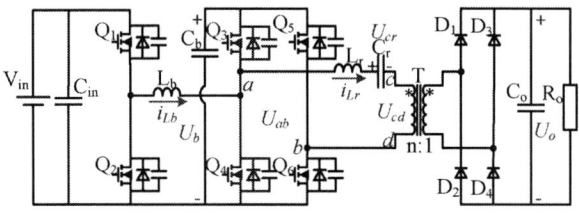

Fig. 2. The topology of IBBSRC

B. DC gain of IBBSRC

Since the shared intermediate leg has no effect on the gain of FSBB converter and SRC, the gain of cascaded converter could be obtained by analyzing the gain of two-stage converter separately. The gain of FSBB converter is calculated by

$$M_1 = \frac{d_1}{1 - d_2} \qquad (1)$$

where d_1 and d_2 are the duty cycle of Q_1 and Q_4. Fig. 3 shows the DC gain of FSBB converter in relation to d_1 and d_2. The DC gain range of FSBB is 0.11~9. The gain of SRC is written as

$$M_2 = \frac{1}{\sqrt{1 + Q^2 \left(k - \frac{1}{k}\right)^2}} \qquad (2)$$

where Q means load factor, and k is the deviation degree of switching frequency f_s and resonant frequency f_r. Fig. 4 exposes to view the characteristics of gain of SRC at different values of load factor Q. As is shown in Fig. 4, the gain of SRC will be the highest and close to 1 when k is equal to 1, and the slight the Q is, the greater the voltage gain will be.

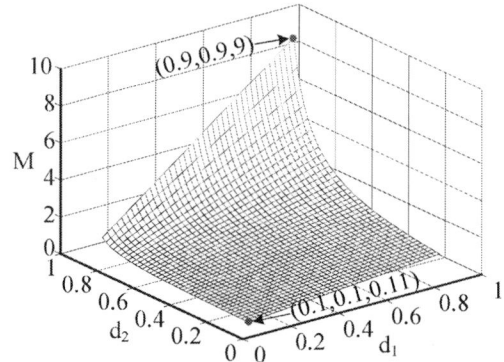

Fig. 3. The gain of FSBB converter

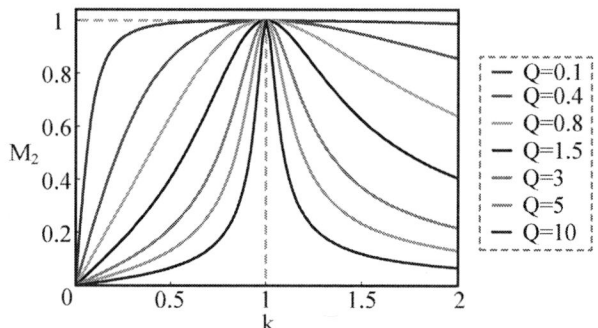

Fig. 4. The gain of SRC

Therefore, the DC gain of IBBSRC could be expressed as

$$M = M_1 \cdot M_2 = \frac{d_1}{(1 - d_2)\sqrt{1 + Q^2 \left(k - \frac{1}{k}\right)^2}} \qquad (3)$$

Fig. 5(a) demonstrates the DC gain of IBBSRC as a function of d_1 and k when supposing d_2=0.5, Q=1.5. Fig. 5(b) indicates the gain of IBBSRC when supposing d_2=0.5, k=1.1 under different Q (the value of Q increases from top to bottom). From Fig.5, the value of DC gain is from 0 to 1.8, which is preferable to SRC. Moreover, the gain will be larger if k is close to 1. In addition, the gain of IBBSRC decreases as Q increases. In this paper, the value of d_2, k and Q are set to 0.5, 1.1, 1.5 separately.

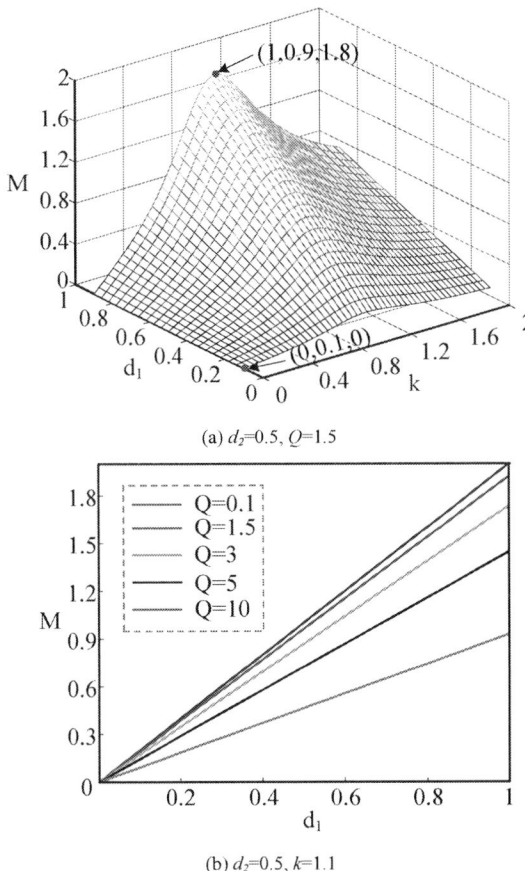

(a) d_2=0.5, Q=1.5

(b) d_2=0.5, k=1.1

Fig. 5. The gain of IBBSRC

C. The modulation scheme of IBBSRC

At present, there are two main modulation strategies of FSBB converter: two-edge modulation (TEM) and constant frequency ZVS modulation. The latter requires negative current detection and the inductance is difficult to design. The DEM strategy can effectively reduce the effective value and fluctuations of inductor current, and this modulation strategy is easy to realize by software. Therefore, DEM is applied in the front-stage converter in this paper.

The modulation strategy of SRC concludes variable frequency (VF) modulation and phase-shift (PS) control. VF modulation is not conducive to the optimal design of magnetic components. The post-stage SRC selects constant frequency PS control correspondingly in this treatise, and adjusts the voltage by changing the phase difference between the leading leg and lagging leg.

In short, the IBBSRC chooses the two-edge-modulation + single-phase-shift (TEM+SPS) strategy. All switches operate at the same switching frequency, the duty cycle of $Q_3 \sim Q_6$ is 0.5, the switch of the same leg is complementary to each other, the intermediate voltage of IBBSRC is altered by modifing the duty cycle of front stage converter, and then desired output voltage is achieved through altering the phase difference of the second-stage converter.

III. ANALYSIS OF IBBSRC MODES

To make the study easier, assumptions are suggested:

- Every transistor and diode are ideal.

- The dead time is short enough.

- The leakage inductance and parasitic capacitance of the transformer are omitted, and the ratio is presupposed to be 1.

- There is no power loss in the conversion.

- The output capacitive filter is large enough.

- The input voltage is constant in one switching period.

The proposed IBBSRC has six operation modes for different duty cycles and different loads, as shown in Fig. 6. Modes 1 to 3 are the front-stage FSBB converter functioning in the step-up mode, and modes 4 to 6 are the FSBB converter operating in the buck status. When Q_1 and Q_4 are switched on concurrently, the current i_{Lb} of the intermediate inductor rises. When Q_1 and Q_4 are switched off at the same time, the current i_{Lb} of the intermediate inductor remains unchanged. When Q_1 and Q_3 are turned on, if d_1 is larger than 0.5, then the current i_{Lb} of the inductor increases (mode1 ~ mode3); if d_1 is smaller than 0.5, then the current i_{Lb} of the inductor decreases (mode4 ~ mode6). The operation characteristics of the rest part of modes 4 to 6 are the same as mode 1 to 3, thus the operational principles of modes 1 to 3 are only analyzed in this paper.

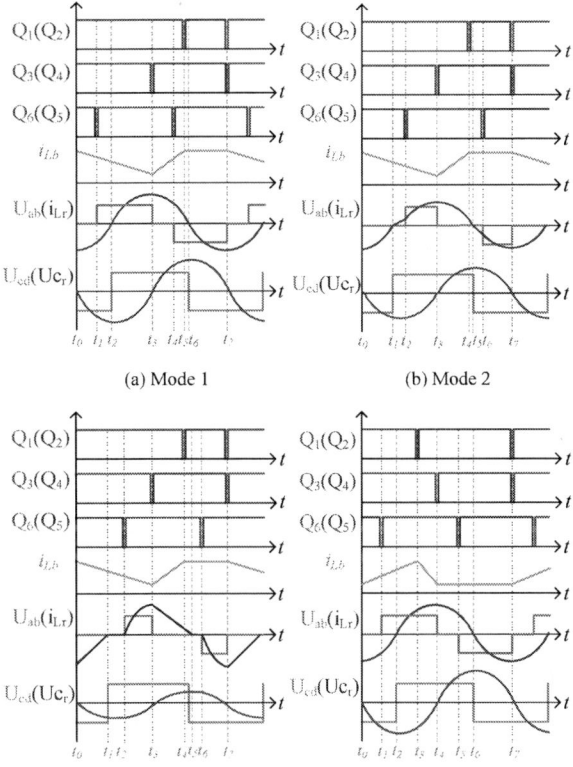

(a) Mode 1   (b) Mode 2

(c) Mode 3   (d) Mode 4

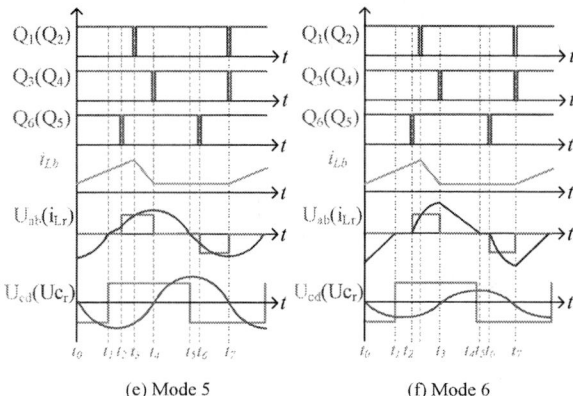

|(e) Mode 5 | (f) Mode 6|

Fig. 6. The operation modes of IBBSRC

A. Mode 1

The basic waveform of mode 1 is depicted in Fig. 6(a).

At $t_0 \sim t_1$, Q_1, Q_3 and Q_5 are conducted, the current i_{Lb} of the intermediate inductor begins to drop from positive, the voltage U_{ab} is zero, and the current i_{Lr} of the resonant inductor is negative;

At $t = t_1$, Q_5 is turned off, the resonant current continues through the antiparallel body diodes of Q_3 and Q_6, and the voltages at both ends of Q_3 and Q_6 are clamped at zero potential. During this period, Q_3 and Q_6 are turned on to realize ZVS/ZCS;

At $t_1 \sim t_2$, Q_1, Q_3 and Q_6 are conducted, the current i_{Lb} falls further, the voltage U_{ab} is positive, and the resonant current i_{Lr} drops to zero at $t = t_2$;

At $t_2 \sim t_3$, the current i_{Lb} continues to decrease, the voltage U_{ab} is positive, and the resonant current i_{Lr} flows forward and increases gradually;

At $t_3 \sim t_4$, Q_1, Q_4 and Q_6 are conducted, the current i_{Lb} starts to rise, the voltage U_{ab} drops to zero, and the resonant current i_{Lr} starts to decline. During this period, Q_4 and Q_5 are turned on to realize ZVS/ZCS;

At $t_4 \sim t_5$, Q_2, Q_4 and Q_5 are conducted the current i_{Lb} continues to rise, the voltage U_{ab} is negative, and the resonant current i_{Lr} decreases until it drops to zero at $t = t_6$;

At $t_6 \sim t_7$, Q_1, Q_4 and Q_5 are conducted the current i_{Lb} remains unchanged, the voltage U_{ab} is negative, and the resonant current i_{Lr} begins to flow in reverse;

At $t = t_7$, the voltage U_{ab} drops to zero, and the antiparallel diode of Q_3 turns on, and Q_4 achieves soft turn-off switching under the effect of the parasitic output capacitance. Then Q_3 is turned on, and the above cycle is repeated.

From the above analysis, it can be seen that the voltage U_{cd} is reversed at the same time when the resonant current i_{Lr} is reversed, while the voltage of the resonant capacitor is 90° different from the phase of the bus current, and the voltage of the capacitor is the maximum when the resonant current is 0.

B. Mode 2

The basic waveform of mode 2 is indicated in Fig. 6(b). The operational principle of the current i_{Lb} and ZVS of Q_3, Q_4 are the same as mode 1. The difference is that the resonant

current i_{Lr} has dropped to zero at $t = t_5$ before Q_6 turns off and during the voltage U_{ab} is equal to zero. At this time, in that the voltage of the resonant capacitor is greater than the voltage U_{cd}, the resonant current i_{Lr} increases reversely and rises along another resonant curve. Moreover, the antiparallel diode of Q_6 is conducted and acts with Q_4 to short circuit the resonant network, and then Q_6 is turned off to realize its ZCS. At $t = t_6$, Q_5 is conducted, the voltage U_{ab} is negative, the resonant current i_{Lr} rises sinusoidally, and Q_5 is hard switching, so this mode is unfitted for long-term work.

C. Mode 3

The basic waveform of mode 3 is demonstrated in Fig. 6(c). The resonant current of this mode is discontinuous. What is different from mode 2 is that the voltage of the resonant capacitor is lower than the voltage U_{cd} when the resonant current is zero, so that the resonant current remains at zero and does not increase reversely. In the meantime, due to no current charge and discharge, the voltage of the resonant capacitor remains at the maximum value, and the voltage U_{cd} is zero. in this mode, when Q_5 and Q_6 are turned on, the switching losses are really high because of the existence of the parasitic capacitors, and ZVS is completely lost. Meanwhile, there are disadvantages of non-sinusoidal resonant current waveform and low energy transfer rate.

IV. PARAMETERS DESIGN OF IBBSRC

A. Intermediate Inductor L_b

Lowering the RMS current of the intermediate inductor can reduce the conduction loss of the power semiconductor devices and the core loss of the inductor, thus improving the efficiency of the converter. according to the waveform of the current i_{Lb} of the intermediate inductor in Fig. 6(a), the expression of i_{Lb} in a period is

$$
i_{Lb}(t) = \begin{cases} I_m - \dfrac{U_b - U_{in}}{L}t, & t_0 \le t \le t_3 \\[2mm] I_m - \dfrac{U_b}{L}(t_3 - t_0) + \dfrac{U_{in}}{L}t, & t_3 \le t \le t_4 \\[2mm] I_m, & t_4 \le t \le t_7 \end{cases} \quad (4)
$$

where I_m is the maximum value of L_b. Thus the RMS current of L_b is calculated from[15,17]

$$
i_{Lb\_rms} = \begin{cases} \sqrt{\dfrac{U_b}{2U_{in}}\left[4I_b^2 + \dfrac{1}{12}\left(\dfrac{U_b - U_{in}}{2L_b f_s}\right)^2\right]+} \\ \qquad\qquad \sqrt{\dfrac{U_{in} - \frac{1}{2}U_b}{U_{in}}(2I_b - \dfrac{U_{in}-U_b}{4L_b f_s})^2}, \quad U_{in} \ge U_b \\[3mm] \sqrt{\dfrac{1}{2}\left\{4I_b^2 + \dfrac{1}{12}\left[\dfrac{U_b(U_{in}-U_b)}{2U_{in}L_b f_s}\right]^2\right\}+} \\ \qquad\qquad \sqrt{\dfrac{1}{2}\left[2I_b - \dfrac{U_b(U_{in}-U_b)}{4U_{in}L_b f_s}\right]^2}, \quad U_{in} \ge U_b \end{cases} \quad (5)
$$

where I_b is the intermediate output current. Supposed U_b=220V, f_s=10kHz, I_b=0.45A(P=1kW), the variation of RMS current with inductance under wide input voltage situation is shown in Fig. 7.

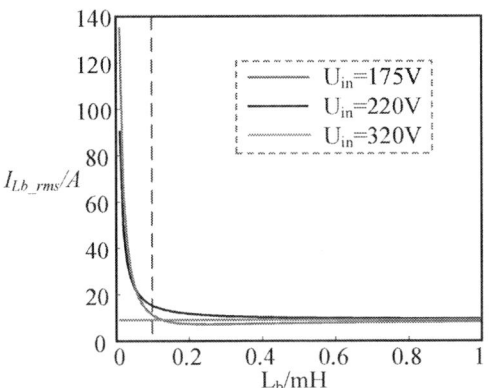

Fig. 7. RMS of inductor current i_{Lb} as a function of L_b

It can be seen from the above figure that for wide input voltage range, when $L_b \approx 100uH$, the RMS current of the L_b reaches the lowest value, and the conduction loss of the converter is the lowest, so L_b is equal to 100uH in this paper.

B. Resonant Inductor L_r and Resonant capacitor C_r

Equations (6) and (7) are the expressions of resonant capacitor voltage gain and impedance gain of the proposed IBBSRC.

$$M_c = \frac{1}{\sqrt{(1-k^2)^2 + (\frac{k}{Q})^2}} \tag{6}$$

$$M_z = \sqrt{1 + Q^2(m - \frac{1}{m})^2} \tag{7}$$

where k is the deviation degree between switching frequency and resonant frequency, Q is the quality factor, $m=2n-1(n=1,2,3,\dots)$, and n is the harmonic order. Fig. 8 and Fig. 9 are characteristics of resonant capacitor voltage and impedance.

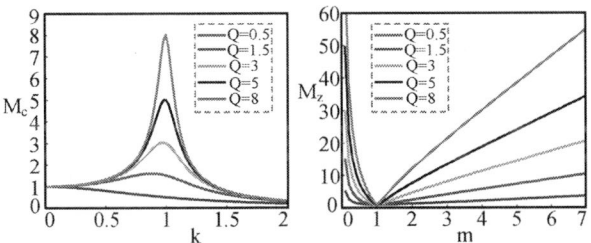

Fig. 8. Characteristics of resonant capacitor

Fig. 9. Characteristics of impedance

From above figures, the larger the Q is, the bigger the M_c is, and the capacitor with larger withstand voltage value is needed. When the f_s is identical with the f_r, the resonant network impedance reaches the minimum, the smaller the Q is, the lower the impedance gain of high-order harmonic is. In other words, the influence of high-order harmonic on IBBSRC could be ignored if smaller value of Q is selected. In addition, if Q is too small, the IBBSRC will operate at mode 3 or mode 6. Thus, $Q=1.5$.

When the resonant frequency is close to the switching frequency, the maximum output power of IBBSRC is obtained.

In this paper, the value of k is 1.1. Based on $k = \frac{f_s}{f_r}$,

$$Q = \frac{\sqrt{L_r / C_r}}{R_o}, \quad f_r = \frac{1}{2\pi\sqrt{L_r C_r}}, \text{ the resonant inductance and}$$

capacitor is expressed as

$$C_r = \frac{k}{2\pi Q f_s R_o} \tag{8}$$

$$L_r = C_r (Q R_o)^2 \tag{9}$$

In this paper, $k=1.1$, $Q=1.5$, f_s=10kHz, R_o=0.576Ω, and then C_r=0.41uF, L_r=741uH.

V. SIMULATION RESULTS OF IBBSRC

A. Simulation Description

To demonstrate the correctness of the proposed topology, a 1kW, 24VDC IBBSRC simulation platform is built by matlab/simulink and the simulation waveform in a broad input voltage scope is analyzed. The simulation parameters of the model are manifested in Table I.

TABLE I. PARAMETERS OF IBBSRC

parameters	value	parameters	value
V_{in}/V	200	C_r/μF	0.41
C_{in}/μF	4700	f_s/kHz	10
L_b/μH	400	n	7
C_b/μF	2000	C_o/μF	100
L_r/μH	741	V_o/V	24

B. Steady Waveforms

The closed loop simulation results of IBBSRC topology are displayed in Fig. 10. Fig. 10(a) indicates the steady waveforms of 175V voltage input while Fig. 10(b) shows the 320V voltage input. As can be seen, for V_{in}=175V, the converter operates at mode 1; for V_{in}=320V, the converter operates at mode 4. The simulation waveforms accord with the academic waveforms, and the output voltage can be stabilized at about 24V, which proves the usefulness of the propounded circuit topology. The peak current of the resonant inductor is far less than the voltage U_{ab}, which greatly reduces the circuit loss, indicating that the circuit parameters are properly configured. In short, with DEM+SPS modulation strategy, the IBBSRC is applicable for wide input voltage.

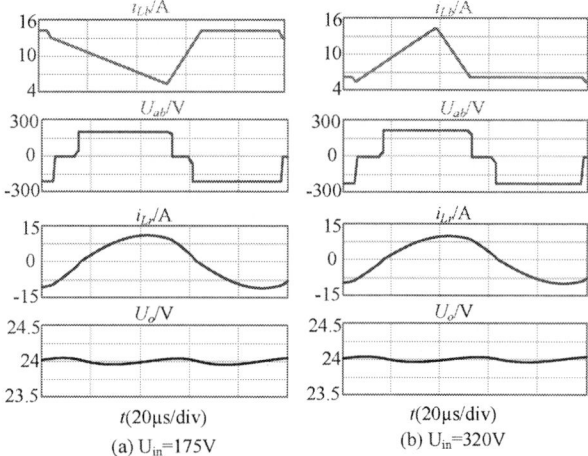

Fig. 10. Steady state waveforms of IBBSRC

C. ZVS Analysis

When the intermediate voltage is constant, the soft switching analysis of the second-stage switches is consistent. The following is only analysis on soft switching of $Q_3 \sim Q_6$ for $V_{in}=175V$, $V_o=24V$. The main simulation waveforms of switches are drawn in Fig. 11, in which V_{ds_Q*} appears for the drain source voltage, V_{gs_Q*} represents 100 times the switching driving signal. From Fig. 11, when the drain source voltage of the switch is reduced to zero, the driving signal begins to act on the switch, that is, $Q_3 \sim Q_6$ all realize ZVS, which is consistent with the theoretical analysis.

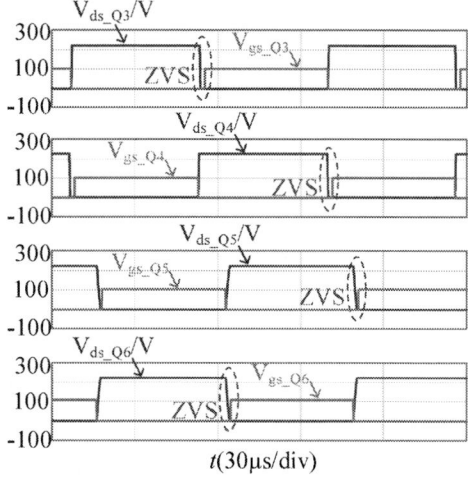

Fig. 10. The ZVS veryfication of IBBSRC

VI. CONCLUSION

The integrated Buck-Boost+SRC proffered in this essay has characteristics of wide input voltage range. By reusing the intermediate leg, the number of switches is reduced and the efficiency of the converter is improved. With DEM+SPS modulation strategy, switches operate at a constant frequency, which effectively decreases the impact of EMI, and some switches can achieve ZVS, reducing the switching losses.

REFERENCES

[1] I. Batarseh, "Resonant converter topologies with three and four energy storage elements," in *IEEE Transactions on Power Electronics*, 2002, 9(1):64-73.

[2] Steigerwald, L. R., "A comparison of half-bridge resonant converter topologies," in *IEEE Transactions on Power Electronics*, 2002, 3(2):174-182.

[3] J. Hong, K. Li, "Analysis and design of half-bridge series-parallel resonant DC-DC converter," in *Modern Electronics*, 2000(1):48-54.

[4] X. Guo, D. Zhu, and et al., "Dynamic inertia evaluation for type-3 wind turbines based on inertia function," in *IEEE J. Emerg. Sel. Topics Circuits Syst.*, vol. 11, no. 1, pp. 28-38, Mar. 2021.

[5] J. Zhu, F. Luo, "Power resonant converter and its development trend," in *Advanced Technology of Electrical Engineering and Energy*, 2004(01):55-59.

[6] H. Lv, Y. Huang, "Zhang Zhongchao. Research on control methods of series resonant single-phase full bridge inverter," in *Power Supply Technologies and Applications*, 2002(05):216-218.

[7] J. F. Lazar, R. Martinelli, "Steady-state analysis of the LLC series resonant converter," in *Proc. IEEE Appl. Power Electro. Conf. Expo.*, 2001, vol. 2, pp. 728-735.

[8] B. Yang, F. Lee, A. Zhang, G. Huang, "LLC resonant converter for front end DC/DC conversion," in *IEEE APEC*, 2002:1108-1112.

[9] K. Jin, X. Ruan, "Hybrid full bridge three-level LLC resonant converter," in *Proceeding of the CSEE*, 2006, 26(3): 53-58.

[10] D. Vinnikov, A. Chub, I. Roasto, L. Liivik, "Multi-mode quasi-Z-source series resonant dc/dc converter for wide input voltage range applications," in *Proc. Appl. Power Electron. Conf. Expo.*, Mar. 2016, pp. 2533-2539.

[11] Y. Ma, Q. Li, H. Li, X. Zhou, H. Liu, "Wide gain three-level series-resonant bidirectional dc-dc converter," in *Proceedings of the CSU-EPSA*, 2020, 32(10):17-27.

[12] Y. Lo, C. Lin, M. Hsieh, and et al., "Phase-Shifted Full-Bridge Series-Resonant DC-DC Converters for Wide Load Variations," in *IEEE Transactions on Industrial Electronics*, 2011, 58(6):2572-2575.

[13] Y. Mohamed, "Control and modeling of high frequency resonant DC/DC converters for powering the next generation microprocessors," in *Canada: Queen's university*, 2005.

[14] J. Hayes, M. Egan, "A comparative study of phase-shift, frequency, and hybrid control of the series resonant converter supplying the electric vehicle inductive charging interface," in *IEEE Applied Power Electronics Conference and Exposition*, 1999:450-457.

[15] X. Ren, X. Ruan, M. Li, and et al., "Dual edge modulated four-switch buck-boost converter," in *Proceedings of the CSEE*, 2009(12):16-23.

[16] X. Sun, J. Chou, X. Li, and et al., "An integrated buck-boost LLC cascaded converter with wide input voltage range," in *Proceedings of the CSEE*, 2016(6):1667-1673.

[17] C. Xu, S. Liu S, X. Guo, and et al., "A novel converter integrating buck-boost and DAB converter for wide input voltage," in *IECON 2020 The 46th Annual Conference of the IEEE Industrial Electronics Society*, Singapore, 2020.

Design of a High Power Density Bidirectional AC/DC Converter Based on GaN

Jiajia Guan
School of Electrical and Electronic Engineering
Huazhong University of Science and Technology
Wuhan,China
jiajiaguan@hust.edu.cn

Zhiwei Wang
School of Electrical and Electronic Engineering
Huazhong University of Science and Technology
Wuhan,China
D201980470@hust.edu.cn

Ziyan Tang
School of Electrical and Electronic Engineering
Huazhong University of Science and Technology
Wuhan,China
tang_ziyan@hust.edu.cn

Jianwei Lv
School of Electrical and Electronic Engineering
Huazhong University of Science and Technology
Wuhan,China
jianweil@hust.edu.cn

Cai Chen
School of Electrical and Electronic Engineering
Huazhong University of Science and Technology
Wuhan,China
caichen@hust.edu.cn

Yong Kang
School of Electrical and Electronic Engineering
Huazhong University of Science and Technology
Wuhan,China
ykang@hust.edu.cn

Abstract—Traditional Si-based semiconductors have a large reverse recovery charge, which brings great challenges to the application of totem-pole bridgeless PFC. With the rise of the third-generation semiconductor device SiC/GaN, totem-pole bridgeless PFC has gradually attracted people's attention. This paper designs a high-power density, high-efficiency bidirectional AC/DC converter based on GaN HEMTs. When working in the forward direction, the totem-pole bridgeless PFC is connected to a synchronous Buck to realize 220V@AC input, and 300V to 400V@DC 1A constant current output; When working in the reverse direction, Boost is connected to a unipolar inverter to achieve 300V to 400V@DC input, 220V 1.5A@AC output, and the output frequency and phase can follow the reference sinusoidal signal. The converter can realize 400W power conversion, with a peak efficiency of 97.2% and power density of 85W/inch³.

Keywords—bidirectional AC/DC converter, high power density, power factor correction

I. INTRODUCTION

Totem-pole power factor correction (PFC) can be regarded as a Boost converter with synchronous rectification. For the Boost converter with synchronous rectification, if it works in continuous conduction mode (CCM), the reverse recovery charge of the MOSFET body diode is a big problem. This means that the totem-pole bridgeless PFC can only work in discontinuous conduction mode (DCM) or boundary conduction mode (BCM) when using traditional Si MOSFETs, but both are challenging [1]. The PFC in the DCM mode is generally used in low-power applications, and the PFC in the BCM mode has a wide range of operating frequency. In addition, the peak current of the PFC in the BCM mode will be twice that of the PFC in the CCM mode, which increases the difficulty of EMI filter design and efficiency optimization. New semiconductor switching devices based on wide band gap materials, such as silicon carbide (SiC) and gallium nitride (GaN), have ultra-small reverse recovery charges and other advantages. This allows the totem-pole bridgeless PFC to work in CCM mode to achieve higher efficiency and higher power [2,3].

GaN devices have been widely used in a variety of topologies due to their excellent performance, such as Buck,

Boost, LLC, etc. Based on GaN HEMTs, this paper designs a 400W bidirectional AC/DC converter (Fig1), which has the characteristics of high efficiency and high power density. This article first analyzes the working mode and control principle of the totem-pole bridgeless PFC when working in the forward direction; then analyzes the control principle of the unipolar SPWM modulation inverter when working in the reverse direction. After calculation of key component parameters and estimation of loss and heat generation, a compact layout is adopted, which not only ensures the heat dissipation conditions but also reduces the size of the prototype to achieve high power density. Finally, the working waveforms of the prototype is given. The power of the prototype is 400W, the efficiency is greater than 94% when working in both directions, and the power density reaches 85W/inch³.

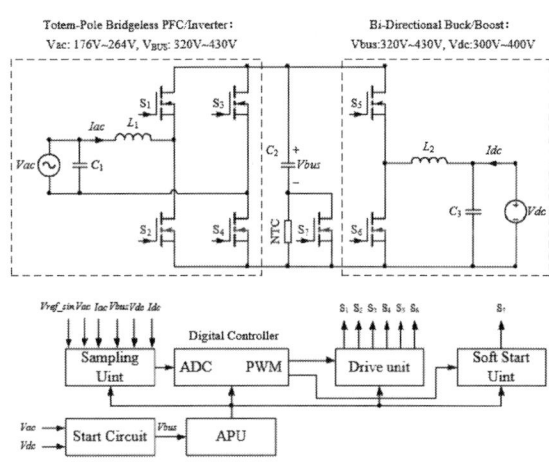

Fig.1 The schematic of bidirectional AC/DC converter.

II. DESIGNER OF THE BI-DIRECTIONAL AC/DC CONVERTER

A. AC to DC Mode

The working mode of totem-pole bridgeless PFC is shown in Figure 2. Where, Q1 and Q2 are turned on complementary, Q3 and Q4 are turned on complementary. Q1 and Q2 are high-frequency switching devices, so GaN HEMTs are used to reduce switching losses, Q3 and Q4 are low-frequency

978-1-6654-4817-8/21 $31.00 © 2021 IEEE

switching devices, and Si MOSFETs are used to ensure low conduction losses. Its working principle is as follows: During the positive half cycle of AC input, Q4 is turned on, and Q1 and Q2 are switched on and off with high frequency under the control of the controller. When Q1 is off and Q2 is on, the grid-side voltage charges the inductor, and the inductor current rises approximately linearly and positively. During the Q2 off period, the inductor current flows through Q1 and the inductor charges the output capacitor. In the negative half cycle of AC input, Q3 is turned on,and Q1 and Q2 are switched at high frequency under the control of the controller. When Q2 is off and Q1 is on, the grid-side voltage charges the inductor, and the inductor current rises approximately linearly and negatively. During the off period of Q1, the inductor current flows through Q2 and the inductor charges the output capacitor.

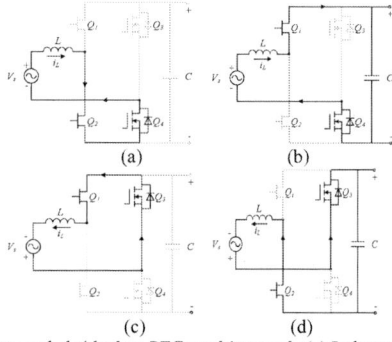

Fig 2 Totem-pole bridgeless PFC working mode.(a) Inductance charging during positive half cycle. (b) Inductance discharging during positive half cycle. (c) Inductance charging during negative half cycle. (d) Inductance discharging during negative half cycle.

The control block diagram of the totem-pole bridgeless PFC is shown in Figure 3. Three circuit parameters are need to be obtained: DC bus voltage Vbus, AC input voltage Vac, and AC input current Iac. The reference value of the voltage outer loop is the expected bus voltage, and the reference value of the current inner loop is the product of the output value of the voltage outer loop and the absolute value of the AC input voltage. The voltage loop ensures that Vbus fluctuates around 450V, and the current loop ensures that the phase of Iac follows the phase of Vac to achieve power factor correction.

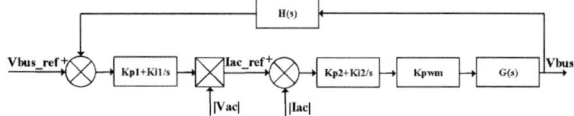

Fig.3 The control block diagram of Totem-pole bridgeless PFC.

B. DC to AC Mode

In reverse operation, under unipolar SPWM modulation, the circuit structure is the same as the forward totem pole rectification, and the bridge arm composed of Q1 and Q2 is still a high-frequency bridge arm, working at a high-frequency switching frequency; The bridge arm composed of Q3 and Q4 is still a low-frequency bridge arm, working in the state of power frequency switching. As shown in Figure 4, the modulation wave of unipolar bilateral SPWM is a complete sine wave with positive and negative half waves. The corresponding carrier is a positive triangle wave when the modulation wave is a positive half wave, and negative when the modulation wave is a negative half wave. When the modulating wave is a positive half-wave, the low-frequency arm lower tube Q2 is turned on, and when it is a negative half-

wave, Q1 is turned on. When the modulating wave is larger than the carrier, Q3 is turned on, otherwise, Q3 is turned off, and the high-frequency arm lower tube Q4 is the complementary switch of Q3 [4-6]. Compared with the bipolar SPWM, the unipolar SPWM modulation has only two high-frequency switches, so the switching loss is small, the electromagnetic interference is less, and it is matched and consistent with the high-frequency bridge arm and the low-frequency bridge arm of the totem pole rectifier.

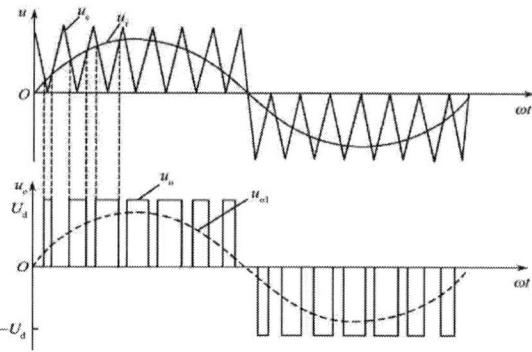

Fig.4 Unipolar SPWM modulation

When working in reverse direction, the inverter circuit is similar to the rectifier circuit, and it is also divided into high-frequency bridge arms and low-frequency bridge arms. The sinusoidal reference signal is multiplied by the given amplitude signal after the phase-locked link to obtain the given Iac value Iref, and then compared with the actual value of Iac. Through the current closed-loop control, the unipolar SPWM modulation is adopted to realize the output constant current and the phase and frequency follow the reference sinusoidal signal.

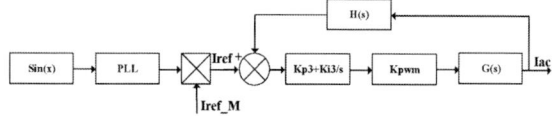

Fig.5 The control block diagram of inverter circuit.

C. Converter Parameters Designer

The PFC inductance is designed when the boost ratio is the largest, and is comprehensively considered based on factors such as input voltage and current, maximum ripple current, duty cycle, and switching frequency, among which the current ripple rate r is generally 0.4. According to (1), the minimum value of PFC inductance can be obtained.

$$ L_{min} = \frac{V_{in}^{2} \eta PF(V_{bus,min} - V_{in})}{\sqrt{2} r P_o V_{bus,min} f_s} \tag{1} $$

The value of the bus capacitor is mainly determined by the bus voltage ripple. According to the output voltage ripple, the required value of bus capacitance can be calculated by (2).

$$ C_{bus} = \frac{P_0}{2\pi f_{bus} V_{bus,min} V_{ripple}} \tag{2} $$

TABLE I. PARAMETERS OF THE PROPOSED CONVERTER

Circuit	Description	Chip	Amount
Power stage	PFC high frequency switchs	GS66502B	2
	PFC low frequency switchs	IPL60R065C7	2
	Buck/Boost switchs	GS66502B	2
	PFC inductor core	CH229060	1
	Buck/Boost inductor core	77314A7	1
	DC link capacitor	47 uF/450V	4
	Output capacitor	2.2 uF/450V	2
Driver circuit	Gate driver	Si8271GB-IS	6
Control circuit	Micro controller	STM32F334R8T6	1
Sampling circuit	Current shunt monitor	INA286	1
	Dual low temperature drift precision op amp	OPA2187	1
	Hall sensor	ASC724LLCTR-05AB-T	1

Buck/Boost inductance is designed when the duty cycle is the smallest, while considering the switching frequency and current ripple rate, the ripple rate r is generally 0.4, and the minimum value of Buck/Boost inductance can be calculated by (3).

$$L_{min} = \frac{V_o(1-D_{min})}{rI_L f_s} \tag{3}$$

III. EXPERIMENTAL VERIFICATION

In this work, STM32F334R8T6 is used as micro controller. Si8271GB-IS is used as the gate driver of GaN devices. The detail loss breakdown of the converter is shown in Fig 6. Switching devices, PFC inductance and other losses (auxiliary power supply, sampling circuit, etc.) account for 82% of the total losses. By optimizing these aspects, the efficiency can be further improved. The main heating components are four high-frequency switching devices and auxiliary power transformers. Therefore, when drawing the circuit board, pay special attention to the layout of these components to ensure good heat dissipation.

TABLE II. LOSS ASSESSMENT OF PROPOSED CONVERTER

Category	Loss/W
Switching devices	4.4107
PFC Inductance	2.5121
Buck/Boost Inductance	1.09
ESR	1
Others	2.5
Total	11.5128

Total Loss 11.5128 W

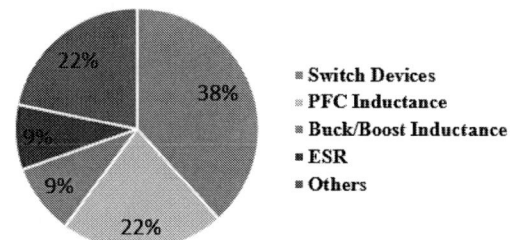

Fig.6 Loss Breakdown of proposed converter.

After evaluating the loss of each part, consider that the heat dissipation condition is natural cooling, so the power devices should not be placed in the same area, which is not conducive to heat dissipation. The PCB three-dimensional layout of the converter is shown in Figure 7, including: AC port, DC port, PFC/Inverter power unit, Buck/Boost power unit, sampling unit, control unit, auxiliary power supply unit, soft start unit, etc. In addition, the DC bus capacitor, flyback transformer, input differential mode filter inductance, PFC inductance, and Buck/Boost inductance are all installed on the bottom of the PCB, as shown in Figure 8(b). High-frequency switching devices are placed in the upper right and lower right corners, the auxiliary power transformer is placed in the lower left corner, and the three main heat sources are placed in three corners. At the same time, a large amount of copper is deposited on the bottom of the switching device to enhance its heat dissipation capacity.

In this design, due to the large PFC inductance value, if a single magnetic ring is used to make the inductor, the overall thickness will be greatly increased, which will reduce the power density of the prototype. Therefore, double magnetic rings are used to respectively wind the inductors and then connect them in series to achieve the required inductance value. Finally, the size of the prototype is 85 mm×53 mm×17mm, and the power density is 85W/inch³.

Fig.7 PCB board layout.

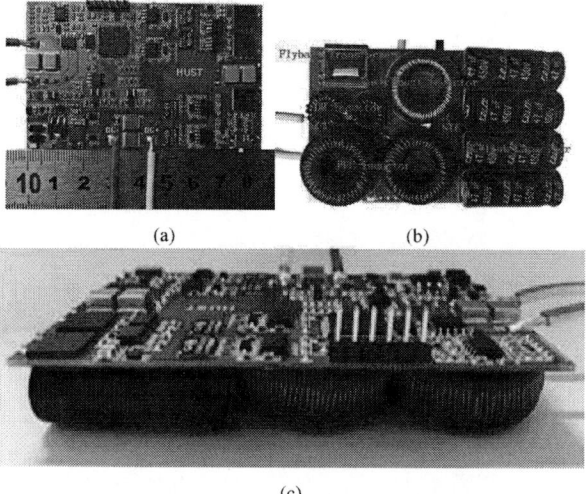

(a)   (b)

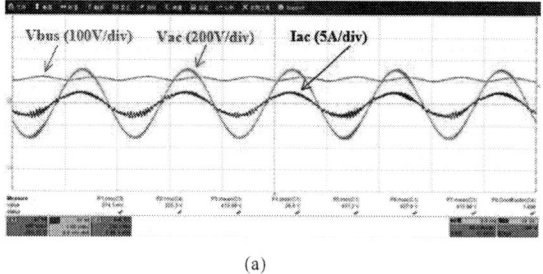

(c)

Fig.8 Prototype of proposed converter. (a) Top. (b) Bottom. (c) Front.

A. AC to DC Mode

When working in the forward direction, the RMS value of the AC side input voltage is 220V, and when the output is 400W, the input voltage, input current and DC bus voltage waveforms are shown in Figure 9(a). When the output power is 400W, the prototype can achieve power factor PF=0.98, THD=9%, and the peak-to-peak value of the bus voltage is about 20V. The prototype has completed a long-term temperature and reliability assessment, and the temperature rise test is shown in Figure 9(b). The maximum temperature of the prototype is 89.4°C, and the temperature rise is about 65°C.

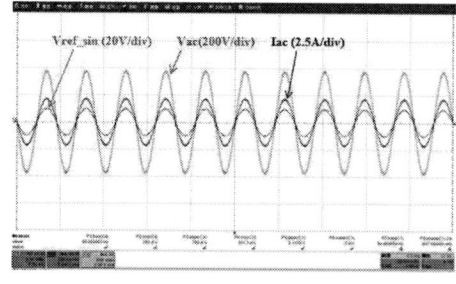

(a)

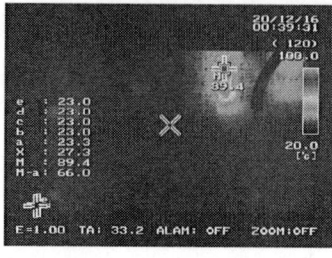

(b)

Fig.9 (a) Totem-pole bridgeless PFC working waveform. (b) Temperature rise test.

B. DC to AC mode

The DC side input voltage is 400V, the reference sinusoidal signal and the output voltage and current are shown in Figure 10(a). When the output power is 400W, the output current (Iac) can be stabilized at 1.5A, and its frequency and phase can be locked to the given signal (Vref_sin). Also, the prototype has completed a long-term temperature and reliability assessment, and the temperature rise test is shown in Figure 10(b). The maximum temperature of the prototype is 82.2°C, and the temperature rise is about 57°C.

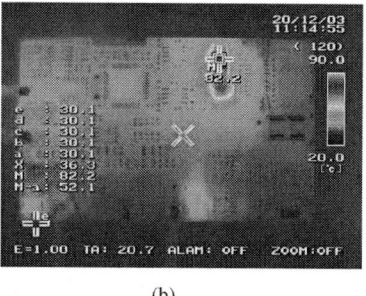

(a)

(b)

Fig.10 (a) Unipolar inverter operating waveform. (b) Temperature rise test.

IV. CONCLUSIONS AND FUTURE WORK

In this paper, a high-power density, high-efficiency bidirectional AC/DC converter is designed based on a new semiconductor GaN HEMTs. When working in the forward direction, the front stage adopts totem pole PFC structure, and the latter stage uses synchronous Buck to realize constant current or constant voltage control. In reverse operation, unipolar inverter control is adopted, which can greatly reduce the loss of switching devices. After the loss evaluation, the main heating elements are evenly laid out, and then a compact layout is adopted, which not only ensures the heat dissipation conditions but also reduces the size of the prototype. It can be seen from the experimental results that the prototype can achieve power factor correction when working in the forward

978-1-6654-4817-8/21 $31.00 © 2021 IEEE

direction, and it also has a phase lock function when working in the reverse direction. The prototype can achieve 400W power conversion, the peak efficiency can reach 97.2%, and the power density can reach 85W/inch^3.

REFERENCES

[1] S. Bin, Z. Junming, and L. Zhengyu, "Totem-pole Boost Bridgeless PFC Rectifier With Simple Zero-Current Detection and Full-Range ZVS Operating at the Boundary of DCM/CCM,", IEEE Transactions on Power Electronics, vol. 26, pp. 427-435, 2011.

[2] Liu C, Salih A, Padmanabhan B. Breakthroughs for 650V GaN power devices: Stable high-temperature operations and avalanche capability. Transactions on Power Electronics, 2015, 2(3): 44-50.

[3] Huang, Qingyun. Review of GaN totem-pole bridgeless PFC[J]. CPSS Transactions on Power Electronics and Applications, 2017, 2(3):187-196.

[4] Zou Y , Guan W G , Zhang Z . Detection and Simulation of Input Voltage of Totem-Pole Bridgeless PFC. Journal of Liaoning University of Technology(Natural Science Edition), 2019.

[5] Rojas J , DMB Leguizamón, Bautista D , et al. Simulation of the Model, Design and Control of a Current Source Inverter with Unipolar SPWM Modulation[C]// 2019 IEEE 15th Brazilian Power Electronics Conference and 5th IEEE Southern Power Electronics Conference (COBEP/SPEC). IEEE, 2020.

[6] Pan W C . Research on driving modes of unipolar SPWM inverter bridge. Journal of Zhejiang University of Science and Technology, 2012.

978-1-6654-4817-8/21 $31.00 © 2021 IEEE

Influence of the Interface Traps Distribution on I-V and C-V Characteristics of SiC MOSFET Evaluated by TCAD Simulations

Yumeng Cai
State Key Laboratory of Alternate
Electrical Power System with
Renewable Energy Sources
North China Electric Power University
Beijing, China
caiyumeng@ncepu.edu.cn

Hao Xu
State Key Laboratory of Alternate
Electrical Power System with
Renewable Energy Sources
North China Electric Power University
Beijing, China
haoxu@ncepu.edu.cn

Peng Sun
State Key Laboratory of Alternate
Electrical Power System with
Renewable Energy Sources
North China Electric Power University
Beijing, China
sunpeng@ncepu.edu.cn

Zhibin Zhao
State Key Laboratory of Alternate
Electrical Power System with
Renewable Energy Sources
North China Electric Power University
Beijing, China
zhibinzhao@126.com

Zhong Chen
Department of Electrical Engineering
University of Arkansas
Fayetteville, USA
chenz@uark.edu

Abstract—The high interface traps density of SiC/SiO_2 interface has always been a major reliability issue of SiC MOSFET. This paper evaluates the influence of the interface traps distribution on the I-V (I_D-V_{GS}) and C-V(C_G-V_G) characteristics of SiC MOSFET by TCAD simulations. First, a TCAD model of SiC MOSFET is established. Then, the impact of the interface traps distribution on the I_D-V_{GS} and C_G-V_G characteristics is studied in detail under different trap types and energy levels. Moreover, the correlation between I_D-V_{GS} and C_G-V_G curves is analyzed. The acceptor traps close to E_V or the donor traps close to E_C cause the overall C_G-V_G curve to shift. Moreover, the impact of the interface traps on V_{TH} reflected in the C_G-V_G curve is consistent with that in I_D-V_{GS}. The analysis in this paper is important in the comprehension of interface traps distribution and the calibration of the TCAD modeling of SiC MOSFET.

Keywords—SiC MOSFET, I-V, C-V, interface traps, TCAD

I. Introduction

Due to material limitations, traditional silicon (Si)-based power devices have approached the intrinsic limit of materials, and it is difficult to meet people's higher demands [1]. As a new type of wide-bandgap semiconductor material, silicon carbide (SiC) has received extensive attention from the industry recently due to its excellent physical properties such as high bandgap, high saturation rate, high thermal conductivity and high breakdown field strength [2]. These features make SiC-based power devices, for example, SiC MOSFETs, widely used in power electronics and other fields with their advantages of high temperature, high voltage, high frequency, and low loss [3].

However, the quality of the SiC/SiO_2 interface has always been a major reliability issue hindering the development of SiC MOSFET. Generally, the interface state density (D_{it}) of SiC/SiO_2 is nearly two orders of magnitude higher than that of Si/SiO_2 [4]. Such a high D_{it} will reduce the channel mobility (μ_{ni}), cause the instability of the threshold voltage (V_{TH}) and increase the leakage current [5]. TCAD is a powerful tool for power semiconductor devices modeling. However, the complicated interface traps distribution that is difficult to

obtain accurately leads to the inaccuracy of TCAD simulation. Therefore, it is very important to evaluate the influence of the interface traps distribution on device characteristics, which helps to establish an accurate TCAD simulation model.

Recently, some scholars have evaluated the influence of SiC/SiO_2 interface traps distribution on the characteristics of SiC MOSFET. For C-V characteristics, literature [6] quantitatively analyzed the influence of interface traps distribution on the C-V characteristics through TCAD simulation. An uneven trap distribution of the SiC/SiO_2 interface was proposed to achieve an accurate calibration of the TCAD model [7]. For I-V characteristics, the effect of interface traps distribution on V_{TH} instability under DC gate bias was investigated in [8]. Moreover, some researchers simulated the phenomenon of V_{TH} hysteresis at different temperatures [9]. Existing studies have evaluated the effect of the interface traps distribution on the C-V and subthreshold I-V characteristics. However, the impact of the interface traps distribution on transfer I-V characteristics has not been reported. Moreover, the comprehensive analysis of the effect of the interface traps distribution on the I-V and C-V characteristics of the device remain to be explored.

This paper comprehensively evaluates the influence of interface traps distribution on the I-V and C-V characteristics of SiC MOSFET. The rest of this paper is arranged as follows. In Section II, the TCAD model of SiC MOSFET is established. Based on this model, Section III investigates the effect of the interface traps distribution on the I-V and C-V characteristics of the device comprehensively. Moreover, the calibration process of TCAD simulation model is given. Section IV summarizes the full text.

II. Simulation Setup

A. Simulated Structure

To evaluate the influence of the interface traps distribution on the characteristics of SiC MOSFET, this paper takes a 1200V planar SiC MOSFET as an example to build a TCAD simulation model. Due to device symmetry, a half elementary cell model is shown in Fig. 1. Although the model does not

978-1-6654-4817-8/21 $31.00 © 2021 IEEE

fully match the characteristics of commercial device, it can reflect the structure and characteristics of an actual device.

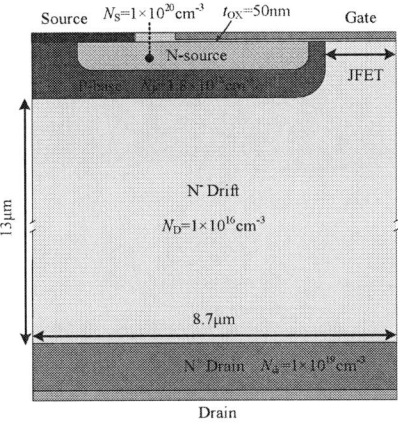

Fig. 1. Simulated structure.

This work focuses on the impact of the interface traps distribution on the I-V and C-V characteristics, therefore the device needs to be calibrated in these two aspects. The calibration involves the selection of suitable models and the appropriate adjustments of the parameters. The doping and structural parameters of the device are from [10]. The mobility model takes into account the influence of interface traps and is considered to be suitable for SiC MOSFET.

In addition, the behavior of device is largely dependent on the characteristics of the SiC/SiO₂ interface. It is necessary to calibrate the interface traps distribution of SiC/SiO₂ interface before evaluating the impact of the interface traps distribution on the device characteristics. This enables the TCAD simulation model to reflect the I-V and C-V characteristics of the actual device. Many literatures have investigated the impact of interface traps on SiC MOSFET [11]. The trap levels are generally considered to be acceptor traps near the conduction band (E_C) for NMOSFET. Therefore, in this paper, the model is calibrated under the condition that only the acceptor traps close to E_C are considered.

B. Simulation Framework

Data from reference [6] reports that interface traps include donor traps and acceptor traps. The former type is electrically neutral when occupied by electrons, and is positively charged when electrons are released. While the latter type is electrically neutral when its energy level is empty, and negatively charged after accepting electrons. To comprehensively evaluate the impact of the interface traps distribution on the device, both the donor and acceptor traps have been considered. In addition, both traps may occupy different energy levels in the forbidden gap of SiC substrate, which depends heavily on the process steps of device manufacturing. Therefore, to consider the complex distribution of the traps at the SiC/SiO₂ interface, the traps in the band gap of SiC are set in two kinds of distribution forms: distributed close to E_C or close to valance band (E_V).

It is impossible to have only a single energy level trap distribution in the band gap of SiC MOSFET, so a complex distribution of multiple energy levels is considered in the simulation. Moreover, the interface state density generally presents an exponential distribution [12]. Therefore, the

distribution of interface traps evaluated in this paper is as follows:

1) Obtain the distribution D_{it}-E of acceptor traps close to E_C by calibrating the interface traps in the model.

2) Maintain the distribution form of D_{it}-E, and investigate the I-V and C-V characteristics of the device under the following four interface traps distribution: acceptor traps close to E_C, acceptor traps close to E_V, donor traps close to E_C, and donor traps close to the E_V.

Taking the donor (acceptor) interface traps as an example, the D_{it} distribution close to E_C or E_V is shown in Fig. 2.

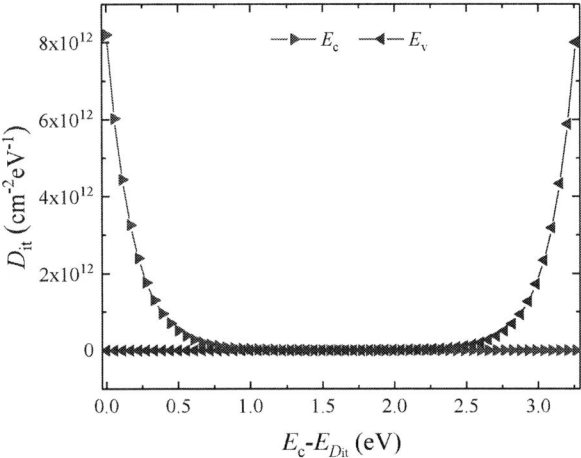

Fig. 2. Interface traps distribution.

Fig. 3(a) (b) show of the I_D-V_{GS} and C_G-V_G characteristics respectively. v_{GS} in Fig. 3(a) is a step-up DC voltage, and v_{GS} in Fig. 3(b) is the superposition of a small sinusoidal AC signal and a step-up DC voltage. A fixed voltage V_{DS} is set between the source and drain of the device when testing I_D-V_{GS}, and the drain and source are shorted when testing C_G-V_G to obtain the total gate capacitance.

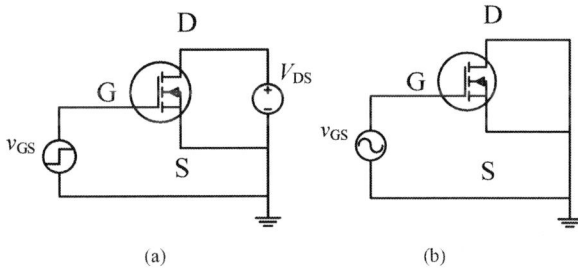

Fig. 3. Schematic diagrams used for the TCAD simulation. (a) I-V. (b) C-V.

III. SIMULATION RESULTS AND ANALYSIS

A. Simulation Results of I-V Characteristics

The effect of type and energy level of the interface traps on the I_D-V_{GS} characteristics is studied and the results are shown in Fig. 4. Moreover, Figs .5 and 6 demonstrate the influence on V_{TH} and μ_{ni}. Note that in Fig. 4(b), the I_D-V_{GS} curves for no traps and donor traps close to E_V are overlapped.

Fig. 4(a) shows that compared with the distribution of acceptor traps close to E_C, the I_D-V_{GS} curve when the acceptor

978-1-6654-4817-8/21 $31.00 © 2021 IEEE

traps located close to E_V almost shifts to the right. Moreover, compared with the I_D-V_{GS} curve for no traps in Fig. 4(b), the turning point of the I_D-V_{GS} curves are all backward whether the acceptor traps are close to E_C or E_V. The overlapped curves in Fig. 4(b) indicates that the donor traps close to E_V do not influence I_D-V_{GS} characteristic. However, the turning point of the I_D-V_{GS} curve is advanced, and then the slope of the curve decreases when the donor traps is close to E_C.

The changes in the I_D-V_{GS} curves under different interface traps distributions reflect the changes in V_{TH} and μ_{ni}. The above results indicate that the V_{TH} increases whether the acceptor traps are close to E_C or E_V. However, it decreases with the donor traps located near the E_C and hardly affected by the donor traps located near the E_V. This is consistent with the extraction data of V_{TH} mentioned in Fig. 5. Moreover, the slope of the I_D-V_{GS} curves from Fig. 4 shows that μ_{ni} almost does not change when the acceptor traps of the same density change from E_C to E_V. While it increases a lot for the donor traps with the same change. This is confirmed in the μ_{ni} shown in Fig. 6, which is calculated from I_D-V_{GS} characteristics simulations.

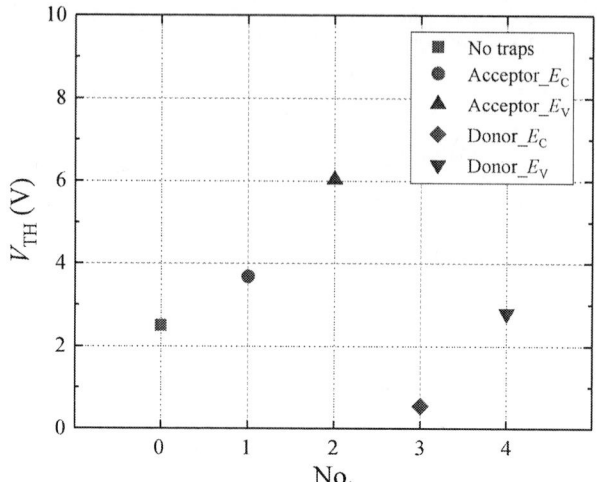

Fig. 5.　Numerical V_{TH}.

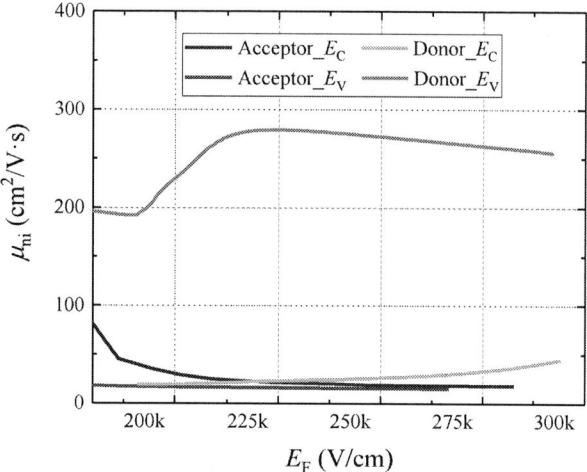

Fig. 6.　Numerical μ_{ni}.

B. Simulation Results of C-V characteristics

The effect of type and energy level of the interface traps on the C_G-V_G is studied and the results are shown in Fig. 7.

As shown in Fig. 7 that the acceptor traps located near the E_C only affect the right half of the C_G-V_G curve, making it shifts to the right. Similarly, the donor traps located near the E_V only affect the left half of the C_G-V_G curve, making it shifts to the left. However, both the acceptor traps located near the E_V and the donor traps located near the E_C affect the entire C_G-V_G curve. The former shifts the entire C_G-V_G curve to the right, and the latter shifts the entire C_G-V_G curve to the left.

Parameters like the flat band voltage (V_{FB}) and the V_{TH} can all be deduced from the study of C_G-V_G curve. It can determine the V_{FB} and the V_{TH} from the C_G-V_G in such a way that V_{TH} corresponds to the increase of the capacitance, and V_{FB} is related to the decrease of the capacitance. The C_G-V_G results from Fig. 7 indicate that V_{TH} increases regardless of whether the acceptor traps near to E_C or E_V. Moreover, V_{FB} increases when the acceptor traps near to E_V. For donor traps, V_{FB} decreases regardless of whether the donor traps near to E_C or E_V, and V_{TH} decreases when it is close to E_C.

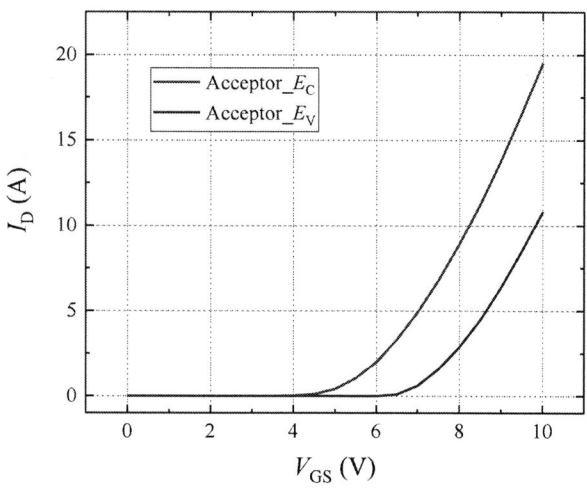

(a)

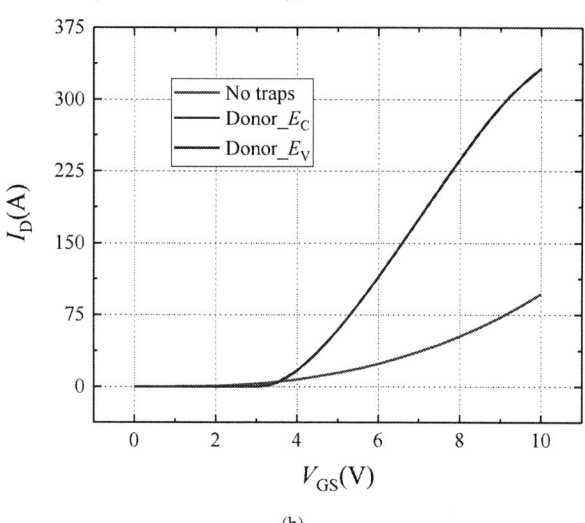

(b)

Fig. 4.　Numerical I-V curves. (a) Acceptors. (b) Donors.

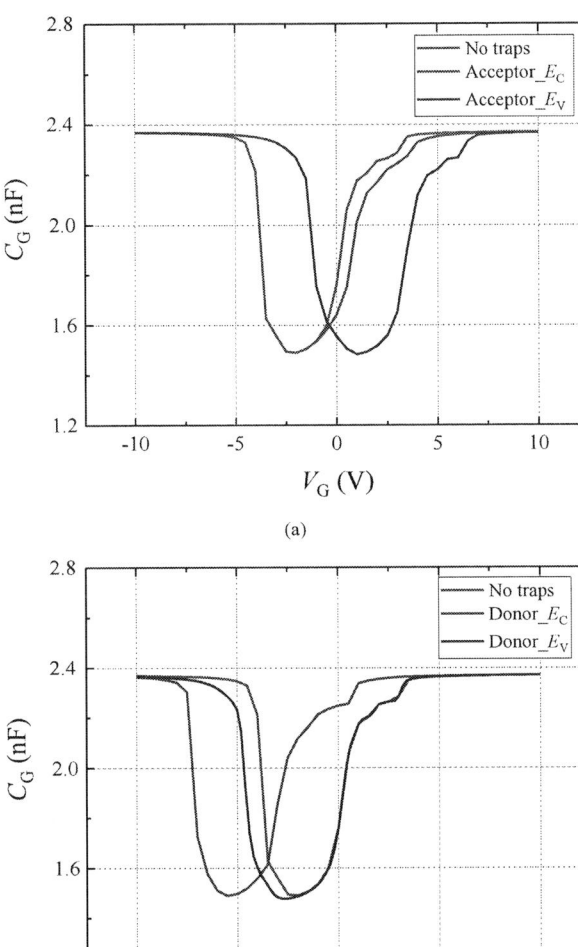

Fig. 7. Numerical C-V curves. (a) Acceptors. (b) Donors.

C. Comprehensive Analysis of I-V and C-V characteristics

The analysis of the I_D-V_{GS} and C_G-V_G characteristics of different types of interface traps with different distributions shows that the I_D-V_{GS} and C_G-V_G characteristics can reflect different device parameters respectively (e.g., μ_{ni} in I_D-V_{GS} and V_{FB} in C_G-V_G). In addition, they also reflect the same device parameters, such as V_{TH}. Moreover, it can be found that the V_{TH} changes reflected in the I_D-V_{GS} and C_G-V_G characteristics are consistent. This shows that the I_D-V_{GS} and C_G-V_G characteristics of TCAD simulation can mutually verify the accuracy of the results. On this basis, it is also effective to investigate other device parameters reflected by I_D-V_{GS} and C_G-V_G characteristics.

This paper mainly analyzes the interface traps distribution on the C-V and I-V characteristics by TCAD simulations. Then an accurate TCAD simulation model can be established through the analysis in this paper. Therefore, the calibration of the TCAD simulation model can be implemented according to the following steps:

1) Measurement of the I_D-V_{GS} and C_G-V_G characteristics of the device.

2) Establishment of a cell simulation model of the device, including the definition of structure and doping parameters, and the choice of mobility model.

3) Numerical simulation of I_D-V_{GS} and C_G-V_G characteristics of the device without interface traps as the reference curves.

4) Adjustment of distribution of the interface traps in the numerical analysis, aiming at the consistency of the I_D-V_{GS} and C_G-V_G curves of numerical simulation and measurement. The interface traps type and the energy level distribution should be considered when acting on the calibration of the interface traps distribution.

IV. CONCLUSION

This paper gives a comprehensive study of the impact of interface traps distribution on the C-V and I-V characteristics of SiC MOSFET by TCAD simulations. From the I_D-V_{GS} numerical results, the V_{TH} increases whether the acceptor traps are close to E_C or E_V. However, it decreases with the donor traps located near the E_C and hardly affected by the donor traps located near the E_V. The μ_{ni} almost does not change when the acceptor traps of the same density change from E_C to E_V. But it increases for the donor traps with the same change. From the C_G-V_G numerical results, the acceptor traps located near the E_C and the donor traps located near the E_V only affect the right half and left half of the C_G-V_G curve respectively. The overall curve shifts when there are acceptor traps near the E_V or donor traps near the E_C. This reflects the change in V_{TH} and V_{FB}, and the change in V_{TH} is consistent with the results of I_D-V_{GS}. A more accurate interface traps distribution and TCAD model of SiC MOSFET can be obtained under the combination of the analysis in this paper and the measurement of the I-V and C-V characteristics.

REFERENCES

[1] H. Sakairi, T. Yanagi, H. Otake, N. Kuroda and H. Tanigawa, "Measurement Methodology for Accurate Modeling of SiC MOSFET Switching Behavior Over Wide Voltage and Current Ranges," IEEE Transactions on Power Electronics, vol. 33, no. 9, pp. 7314-7325, Sept. 2018.

[2] Baliga, B. Jayant. Fundamentals of power semiconductor devices. Springer Science & Business Media, 2010.

[3] M. Bhatnagar and B. J. Baliga, "Comparison of 6H-SiC, 3C-SiC, and Si for power devices," IEEE Transactions on Electron Devices, vol. 40, no. 3, pp. 645-655, March 1993.

[4] C. Raynaud, J. L. Autran, J. B. Briot, B. Balland, N. Becourt, T. Billon, and C. Jaussaud. "Comparison of trapping-detrapping properties of mobile charge in alkali contaminated metal-oxide silicon carbide structures." Applied physics letters 66, no. 18 (1995): 2340-2342.

[5] V. V. Afanasev, M. Bassler, G. Pensl, and M. Schulz. "Intrinsic SiC/SiO2 interface states." physica status solidi (a) 162, no. 1 (1997): 321-337.

[6] L. Maresca, I. Matacena, M. Riccio, A. Irace, G. Breglio and S. Daliento, "Influence of the SiC/SiO2 SiC MOSFET interface traps distribution on CV measurements evaluated by TCAD simulations," IEEE Journal of Emerging and Selected Topics in Power Electronics, 2019, PP(99):1-1.

[7] L. Maresca, I. Matacena, M. Riccio, A. Irace, G. Breglio and S. Daliento, "TCAD model calibration for the SiC/SiO2 interface trap distribution of a planar SiC MOSFET," 2020 IEEE Workshop on Wide Bandgap Power Devices and Applications in Asia (WiPDA Asia), 2020, pp. 1-5.

[8] S. Cascino, M. Saggio, A. Guarnera, "Modeling of Threshold Voltage Hysteresis in SiC MOSFET Device," Materials Science Forum, 2020, 1004:671-679.

[9] A. Vasilev, M. Jech, A. Grill, G. Rzepa, S. Tyaginov, "Modeling the Hysteresis of Current-Voltage Characteristics in 4H-SiC Transistors," IEEE International Integrated Reliability Workshop (IIRW), 2020, pp. 1-4.

[10] G. Romano, A. Fayyaz, M. Riccio, L. Maresca, G. Breglio, A. Castellazzi, "A Comprehensive Study of Short-Circuit Ruggedness of Silicon Carbide Power MOSFETs," IEEE Journal of Emerging and Selected Topics in Power Electronics, vol. 4, no. 3, pp. 978-987, Sept. 2016.

[11] H. Yano, T. Hirao, T. Kimoto, H. Matsunami, "Interface Properties in Metal-oxide-semiconductor Structures on n-type 4H-SiC (0338)", App. Phys. Lett., vol. 81, no. 25, pp. 4772-4774, 2002.

[12] D. Cornigli, A. N. Tallarico, S. Reggiani, C. Fiegna, E. Sangiorgi, L. Sanchez et al., "Characterization and Modeling of BTI in SiC MOSFETs," ESSDERC 2019 - 49th European Solid-State Device Research Conference (ESSDERC), 2019, pp. 82-85.

High Breakdown Voltage AlGaN/GaN HEMT with Graded Fluorine Ion Implantation Terminal in Thick Passivation Layer

Siyu Deng, Xiaorong Luo*, Jie Wei*, Yanjiang Jia, Tao Sun, Lufan Xi, Zhuolin Jiang, Kemeng Yang, Qingfeng Jiang and Bo Zhang

University of Electronic Science and Technology of China, Chengdu, P. R. China

Email: xrluo@uestc.edu.cn ; weijieuestc@uestc.edu.cn

Abstract—A novel high electron mobility transistor (HEMT) with graded fluoride ion (F-) implantation is presented. It features a graded F- implantation into a thick passivation layer as terminal GaN HEMT (GFT HEMT). The shape of the GFT is an isosceles trapezoid in the xz plane, and the injection area decreasing from the gate to drain. Firstly, the gradual reducing F- implantation area from the gate to drain is more conducive to form a uniform E-field distribution and increase *BV*; secondly, a gradual F-distribution is achieved by controlling the F- implantation area through a mask and once implantation process, simplifying the process; finally, F- implantation in the thick passivation layer can greatly reduce the damage to 2DEG. The *BV* and I_d of the GFP HEMT are 955V and 494mA/mm, respectively.

Keywords—*AlGaN/GaN HEMT, Breakdown voltage, SiNx passivation layer, Graded Fluorine ion implantation.*

I. INTRODUCTION

GaN-based high electron mobility transistor (HEMT) has broad prospects due to the characteristics of GaN materials such as high electron mobility and high critical breakdown electric field [1-5]. However, problems such as electric field congestion at the edge of the gate or substrate leakage can cause premature breakdown in the off state, making the *BV* of the GaN HEMT lower than its theoretical limit. F-injection [13-15], polarized super junction [11-12], RESURF technology [9-10] and Field plate (FP) [6-8] are commonly used technologies to solve the congestion of the electric field at the gate edge. FP technology will increase the gate-drain capacitance of AlGaN/GaN HEMTs, which will affect the switching characteristics of the device. However, it is difficult to achieve high concentration of the hole and high-quality P-type buffer layer to RESURF technology. Polarization super junction would introduce the two-dimensional hole gas and reduce the two-dimensional electron gas (2DEG) concentration.

F- implantation in thin passivation layer or AlGaN barrier layer would not introduce the parasitic capacitance. However, the concentration of 2DEG will decreases due to the peak position of the vacancy distribution and fluorine distribution close to the 2DEG channel. Ion implantation would cause physical damage to AlGaN materials and have adverse effects

on the 2DEG mobility, which leads to dynamic and static characteristic degradation significantly.

In this paper, a novel high electron mobility transistor with Graded Fluoride ion (F-) implantation Terminal (GFT) is presented. The *BV* of the GFT HEMT has been improved significantly by GFT. Meanwhile, F- implantation in the thick passivation layer can avoid the physical damage to the AlGaN material and reduce the adverse effects on the mobility of 2DEG. The GFT-HEMT can significantly reduce the electric field congestion at the edge of the gate and improve *BV* by simulation results.

II. DEVICE STRUCTURE AND MECHANISM

Fig. 1(a) shows the structure schematic of the proposed GFT HEMT. It features a GFT in the thick passivation layer instead of in the thin barrier layer. The shape of GFT is an isosceles trapezoid in the xz-plane and the F- injection area decreases from the gate to drain, as shown by the red region in Fig. 1(a). Firstly, the GFT could play a role of variable lateral doping (VLD) technology [16] to assist depleting the 2DEG as shown in Fig. 1(b). Compared with the conventional regular F-implantation terminal, the GFT modulates the E-field distribution more effective of the drift region in the off-state, so as to improve the *BV* dramatically. Moreover, the implantation damage and depleting effect to the 2DEG induced by the GFT in the thick passivation layer is much smaller, since the F-implantation layer is far away from the thin AlGaN barrier layer and 2DEG. So the proposed GFT HEMT could suppress the degradation of $I_{d,sat}$ effectively. F- implantation terminal in passivation layer HEMT in our previous work [13] with different *I* (UFT HEMT-a, *I*=1.2 μm, the same acreage of the F-implantation with GFT HEMT/UFT HEMT-b, *I*=2 μm, the same length of F- implantation with GFT HEMT) and a conventional HEMT used for comparison are shown in Fig.2(a) and Fig.1(b), respectively. The gate length L_G, the gate-drain distance L_{GD}, the gate-source distance L_{GS}, and the passivation layer depth T_{SiN} are 0.8 μm, 5 μm, 1 μm and 0.05μm for all devices, respectively. The F- implantation depth for GFT HEMT and UFT HEMT is T_F=0.02μm.

978-1-6654-4817-8/21 $31.00 © 2021 IEEE

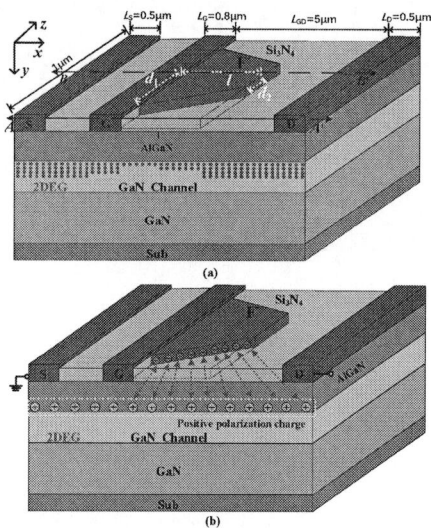

Fig. 1 Structure schematic of (a) GFT HEMT (b) Mechanism of the GFT HEMT at the off state

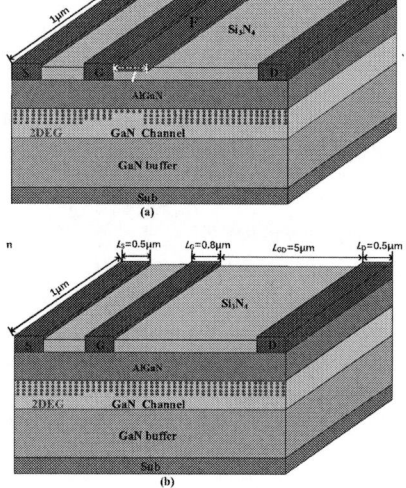

Fig. 2 Structure schematic of (a) UFT HEMT (b) CON HEMT

The AlGaN/GaN heterostructure could be epitaxial grown by metal organic chemical vapor deposition (MOCVD) on 6-in (111) silicon substrate. Then, a 40-nm-thick SiNx passivation layer by low pressure chemical vapor deposition (LPCVD) could be deposited. Fluoride ion can be implanted by using AZ5214 photoresist as a mask, and the window length of fluoride ion implantation is 3 μm, and the energy and dose of fluoride ion implantation are 10 keV and 1×10^{12} cm^{-2}, respectively [13]. Furthermore, through changing the width in the z direction，the GFT could be implemented by F-implantation through one mask at once, simplifying the process.

Synopsys based 2D Sentaurus TCAD has been used for the analysis of physical mechanisms. Several important physical models such as hydrodynamic transport equation, band gap narrowing, doping dependence, high field saturation,

spontaneous polarization and Shockley-Read-Hall are also considered,.

III. RESULTS AND DISCUSSIONS

Fig. 3 compares the lateral electric field (E_x-field) distributions at breakdown. The GFT could realize the same modulation effect as VLD technology [16], then the GFT HEMT exhibits more uniform E_x-field and achieves the highest BV of 955V. But the CON HEMT prematurely breakdown at the gate edge with a BV of 49V. The UFT HEMT-b shows a higher BV of 731V than 407V of UFT HEMT-a, owing to the larger F-implantation area and longer l of the F- implantation terminal. Although GFT HEMT and UFT HEMT-a have the same l of the F- implantation terminal, GFT HEMT shows a higher BV with the likely-VLD technology

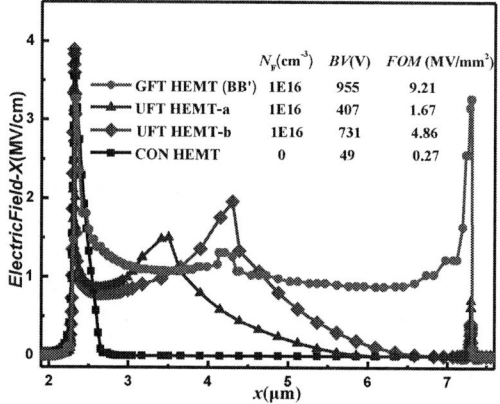

Fig. 3 E_x-field distribution at breakdown

Fig. 4 shows the 2DEG density distributions of the three devices. GFT HEMT shows a smaller 2DEG depletion area, compared with UFT HEMT-b. Furthermore, the total F-implantation dosage for GFT HEMT and UFT HEMT-a are same and less than that of UFT HEMT-b, so the their $I_{d,sat}$ are almost the same and higher than that of UFT HEMT-b, due to the weakened depletion of F- on 2DEG as shown in Fig. 5. Without F- implantation, the $I_{d,sat}$ of the CON HEMT is the highest. Compared with UFT HEMT, GFT HEMT makes more effective use of the F- injection area, which leads to the higher BV and less saturation current reduction. Therefore, the proposed GFT HEMT achieves the highest figure of merit ($FOM = BV^2/R_{on,sp}$), which is 89.5% higher than that of UFT HEMT-b as labeled in Fig. 3.

978-1-6654-4817-8/21 $31.00 © 2021 IEEE

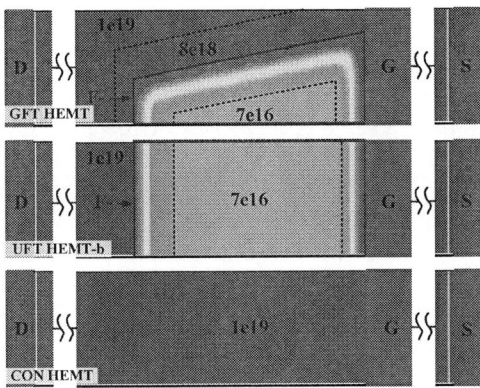

Fig. 4 2DEG density distribution(cm^{-3})

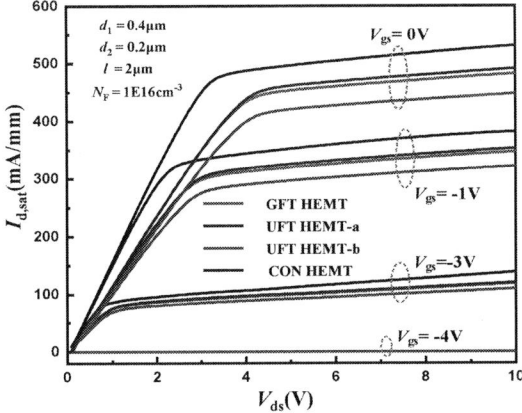

Fig. 5 I-V output characteristics

Fig. 6(a) shows the influences of concentration of F-implantation layer (N_F) on BV and $I_{d,sat}$ for the GFT HEMT. Too low/high N_F leads to premature breakdown. Moreover, the $I_{d,sat}$ decreases as N_F increases, because more 2DEG are depleted by F-. For the GFT HEMT, when the N_F < 1E16cm^{-3}, the drift region cannot be depleted fully, which causes the E-field concentrating at drain edge, as shown in Fig. 6(b). When the N_F > 1E16cm^{-3}, the drift region has been fully depleted, but F- concentration is too high to shield the electric field below the GFT region, which causes the E-field concentrating at drain edge. The equipotential contours are uniformly distributed in the drift region for an optimized N_F = 1E16cm^{-3}, and the BV and $I_{d,sat}$ of the GFP HEMT are 955V and 494mA/mm, respectively.

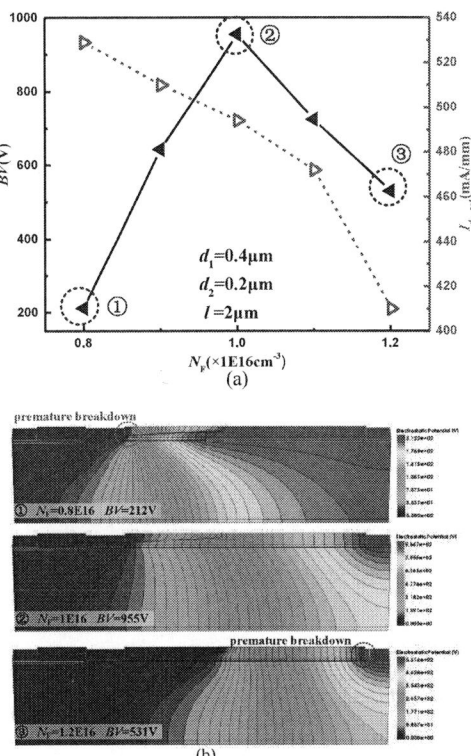

Fig. 6 Influence of N_F on (a) BV and $I_{d,sat}$, (b) equipotential contours at breakdown.

As illustrated in Fig. 7, although there is no F- implantation, the surface electric field (at AA') has been still significantly optimized. Large d_1 value is beneficial to suppress the E_x-field (at AA') peak at the edge of the gate, and increase the electric field in the drift region. The $I_{d,sat}$ increases as d_1 decreases, because the F- injection area increases and more 2DEG are depleted by F-. When $d_1 \geq 0.4$ μm, BV increases slightly as d_1 increases, which means the drift region has been almost completely depleted. However, the $I_{d,sat}$ decreases significantly on the contrary. The optimized d_1 value is 0.4 μm in comprehensive consideration.

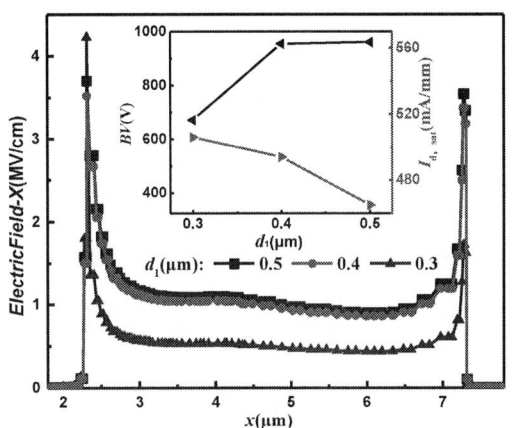

Fig. 7 E_x-field (at AA') distribution, BV and $I_{d,sat}$ with different d_1 (N_F=1E16cm^{-3}, d_2=0.2 μm and l=2μm)

978-1-6654-4817-8/21 $31.00 © 2021 IEEE   405

Fig.8 shows the influences of d_2 and l on BV and $I_{d,sat}$. Obviosly, the $I_{d,sat}$ decreases with increasing d_2 and l values, because more F- are implanted in the passivation layer. However, when d_2 and l exceed certain values, the drift region has been completely depleted. Then, as d_2 and l increase, the BV of the GFT HEMT decreases. Excessive F- injection area will cause premature breakdown in the off state, because the electric field concentrates at the edge of the GFT or at the edge of the drain. To maintain high $I_{d,sat}$ and high BV simultaneously, the optimized values are $d_2 = 0.2\mu m$ and $l = 2\mu m$.

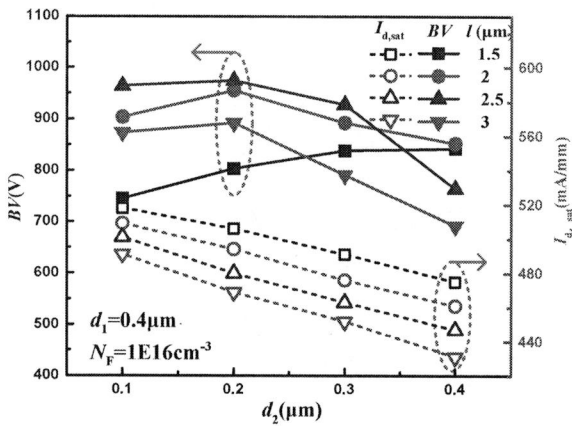

Fig.8 Influences of d_2 and l on BV and $I_{d,sat}$

IV. CONCLUSION

A novel high electron mobility transistor with graded fluoride ion implantation is presented. It features a graded F-implantation into a thick passivation layer as terminal. The shape of the GFT is an isosceles trapezoid in the xz plane, and the injection area decreasing from the gate to drain. GFT in the thick passivation layer can relieve the E-field concentration, increase the E-field in the drift region and improve the BV. Meanwhile, the thick passivation layer avoid the degradation of dynamic and static characteristics. Through changing the width in the z direction，the GFT could be implemented by F-implantation through one mask at once, simplifying the process. In summary, the BV of GFT HEMT increases to 955V from 49V of the CON HEMT, and the $I_{d,sat}$ of the GFT HEMT decreases only 6.5%. Compared with UFT HEMT-b, the GFT HEMT increases the BV by 30.6% and improves the $I_{d,sat}$ by 7%.

REFERENCES

[1] W. E. Newell. Power electronics-emerging from limbo[J]. IEEE Transactions on Industry Applications, 1974, IA-10(1): 7-11

[2] Y. Zhang, A. Dadgar, T. Palacios. Gallium nitride vertical power devices on foreign substrates: a review and outlook[J]. Journal of Physics D: Applied Physics, 2018, 51(27): 1-13

[3] K. J. Chen, O. Häberlen, A. Lidow, et al. GaN-on-Si power technology: devices and applications [J]. IEEE Transactions on Electron Devices, 2017, 64(3): 779-795

[4] U. K. Mishra, P Parikh, YF Wu. AlGaN/GaN HEMTs-an overview of device operation and applications. Proc IEEE, 90: 1022–1031 (2002)

[5] T. L. Wu. Long term stability of enhancement mode GaN power devices[D]. Belgium: University of Leuven, 2016, 6-9

[6] C. Yang, X. R. Luo, T. Sun, et al. High Breakdown Voltage and Low Dynamic ON-Resistance AlGaN/GaN HEMT with Fluorine Ion Implantation in SiNx Passivation Layer, Nanoscale Research Letters, 2019, 14: 191

[7] M. J. Wang, K. J. Chen. Off-state breakdown characterization in AlGaN/GaN HEMT using drain injection technique[J]. IEEE Transactions on Electron Devices, 2010, 57(7): 1492-1496

[8] D. Song, J. Liu, Z. Q. Chen, et al. Normally off AlGaN/GaN low-density drain HEMT (LDD-HEMT) with enhanced breakdown voltage and reduced current collapse[J]. IEEE Electron Device Letters, 2007, 28(3): 189-191

[9] S. Karmalkar, J. Deng, M. S. Shur, et al. RESURF AlGaN/GaN HEMT for High Voltage Power Switching[J]. IEEE Electron Device Letters, 2001, 22(8): 373-375

[10] A. Nakajima, K. Adachi, M. Shimizu, et al. Improvement of unipolar power device performance using a polarization junction[J]. Applied Physics Letters, 2006, 89(19): 193501

[11] H. L. Xing, Y. Dora, A. Chini, et al. High breakdown voltage AlGaN-GaN HEMTs achieved by multiple field plates[J]. IEEE Electron Device Letters, 2004, 25(4): 161-163

[12] W. Huang, T. P. Chow, Y. Niiyama, et al. Experimental Demonstration of Novel High-Voltage Epilayer RESURF GaN MOSFET[J]. IEEE Electron Device Letters, 2009, 30(10): 1018-1020

[13] A. Nakajima, Y. Sumida, M. H. Dhyani, et al. GaN-based super heterojunction field effect transistors using the polarization junction concept[J]. IEEE Electron Device Letters, 2011, 32(4): 542-544

[14] S. Karmalkar, U. K. Mishira. Enhancement of breakdown voltage in AlGaN/GaN high electron mobility transistors using a field plate[J]. IEEE Transactions on Electron Devices, 2001, 48(8): 1515-1521

[15] S. Jiang, K. B. Lee, Z. H. Zaidi, et al. Field plate designs in all-GaN cascode heterojunction field-effect transistors[J]. IEEE Transactions on Electron Devices, 2019, 66(4): 1688-1693

[16] S. Zhang, J. K. O. Sin, T. M. L. Lai, et al., IEEE Trans. Electron Devices, 46(9): 1036–1041 (1999)

978-1-6654-4817-8/21 $31.00 © 2021 IEEE

Homogeneous-Flux Transmitter Coil Design with Improved Position Tolerance

Yunfeng Liu, Yi Dou, Ziwei Ouyang, Michael A. E. Andersen

Department of Electrical Engineering, Technical University of Denmark, Kongens Lyngby, Denmark

Abstract—This paper presents a winding configuration design algorithm for transmitter coils of the wireless power transfer (WPT) system to overcome the power transfer shift from the receiver's position shift. The Algorithm can deal with the winding configuration design for the air-core coils and the coils with a magnetic shield based on the magnetic flux-density modeling. The detailed design considerations and the homogeneous-flux winding configuration evaluation have been investigated. The proposed design algorithm is implemented in the case study, and the results are verified by finite-element-analysis (FEA) simulation and experimental results.

Keywords—Homogeneous-Flux, Magnetic field, Turns distribution, Freedom of placement

I. INTRODUCTION

As shown in Fig.1(a), the WPT system usually consists of two stages. First stage is the input AC-DC (or PFC) satge. As depicted in Fig.1(b), here we use the bridgeless voltage step-down Dual-SEPIC PFC to convert the ac voltage to 100 Vdc. The second stage is the DC-DC stage, consisting of a 6.78 MHz Class E DC-AC high-frequency inverter, two compensation networks, transmitter coil, receiver coil, and secondary 6.78 MHz Class E AC-DC rectifier as shown in Fig.1(c). The power transfer depends on the magnetic field coupling between transmitter and receiver coils. Thus, the relative position between the coils intrinsically influences the power transferring. An improved position tolerance between the transmitter coil and the transmitter coil is normally preferred for applications to build a robust system.

The freedom of the placement for the receiver side of the system has been tricky for the WPT system for many years. To deal with this problem, several attempts have been made to address this challenge previously. In [1]–[3], the coil matrix was implemented on the transmitter side of the system, and the effective power transfer area can be expanded with the increased number of transmitter coils; In [4], the compensation network and the inductive coupler were designed as insensitive with the coupling between the primary coil and the secondary coil to exclude this position impact to the power transfer. There are also several efforts were made by implementing system controlling on inverter/rectifier stages or by output regulation to compensate the inductive coupling shifting [5], [6]; Besides, shaping the distribution of the flux-density generated by the transmitter coil is an alternative to overcome the power transfer shift from the position shift, which is initially introduced in [7].

In this paper, the authors first adopt the concept of shaping the magnetic flux from the transmitter coil to eliminate the coupling shift from the position shift and then present the overall winding configuration design based on the systems'

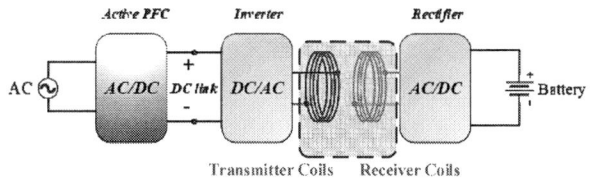

(a)

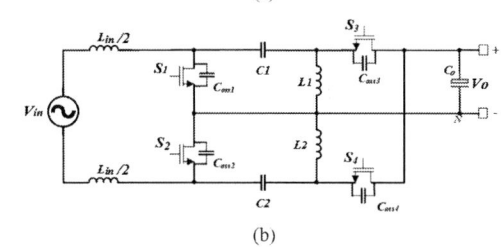

(b)

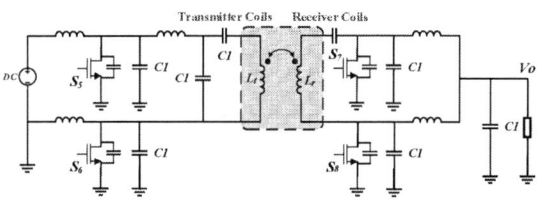

(c)

Fig. 2. Topology of the WPT sytem. (a).The simplified structure of the wireless power transfer system. (b). first stage is SEPIC PFC. (c). Second

GA method. Section II introduces the flux-density modeling both for air-core coil and magnetics shield's coils. Section III the winding configuration design algorithm, and FEA simulation and experimental verification. The experimental results of the hardware prototype are presented in Section IV. Finally, Section V gives the conclusion of the paper.

II. FLUX-DENSITY MODELLING

A. Flux-Density Modelling for Air-Core Coils

During the flux-density modeling process, the wire diameter of the transmitter coil is regarded as zero, and the connection between each turn is eliminated. Thus, the modeled coil can be illustrated in the 2-D cross-section view, as in Fig.2 (b) with a cylindrical coordinate. Based on the Biot-Savart laws, then the flux-density for a circular close-loop source current can be calculated by

$$\overrightarrow{B_q} = \frac{\mu_0}{4\pi} \int_c \frac{I \overrightarrow{dl} \cdot \overrightarrow{R}}{R^3} \tag{1}$$

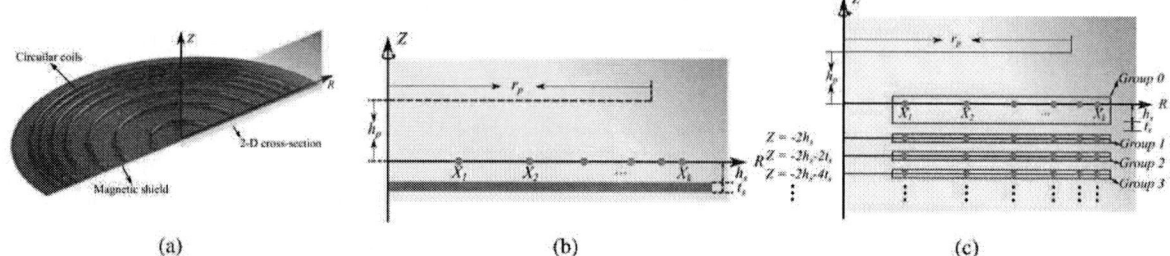

Fig. 2. Illustration of the modeled transmitter coil: (a) 3-D graphic of a coil with magnetic shield (b) Simplified 2-D cross-section of the coil with the dimensions (c) Illustration of the equivalent current sources by the current-mirror method

The elemental magnetic flux density generated by the source current is integrated, and R is the distance vector from the integrated elemental current source to the target position. By implementing 1 on the simplified transmitter coil model. The flux density at the target position $Q(x_q; y_q; z_q)$ can be finally derived as the function of the radius of each turn for the transmitter coil as

$$\overrightarrow{B_q} = \sum_{i=1}^{k} \int_{\theta=0}^{2\pi} \frac{I \, \overrightarrow{dl} \cdot \overrightarrow{R_o}(x_q, y_q, z_q, x_k, \theta)}{R_o^3(x_q, y_q, z_q, x_k, \theta)} \quad (2)$$

where the position of the windings in transmitter coils is simplified in 2-D coordinate as $x_1, x_2 \dots x_k$; I_s is the filamentary current source; R_o is the distance vector, and R_o is the norm of this distance vector. These auxiliary vectors can be calculated as

$$\overrightarrow{I_s} = \begin{bmatrix} -sin(\theta) \\ cos(\theta) \\ 0 \end{bmatrix}, \overrightarrow{R_o} = \begin{bmatrix} x_q \\ y_q \\ z_q \end{bmatrix} - x_k \cdot \begin{bmatrix} cos(\theta) \\ sin(\theta) \\ 0 \end{bmatrix} \quad (3)$$

B. Flux-Density Modelling for Magnetics Sheild's Coils

For the modeling for magnetics shield's coils, the current-mirror method is selected to deal with the magnetic flux shift and distortion from the magnetic shield. The other application of the method can refer to [9]. In Fig.2 ©, the positions of the virtual current sources, which can provide the same flux shift and distortion as the magnetic shield, are illustrated. Consequently, the magnetic flux-density can also be calculated at the upper half-region (beyond the transmitter coil) as

$$\overrightarrow{B_{qs}} = \sum_{i=1}^{k} \int_{\theta=0}^{2\pi} \frac{I \, \overrightarrow{dl} \cdot \overrightarrow{R_o}(x_q, y_q, z_q, x_k, \theta)}{R_o^3(x_q, y_q, z_q, x_k, \theta)} + $$
$$\sum_{n=1}^{\infty} \sum_{i=1}^{k} \int_{\theta=0}^{2\pi} \frac{\lambda_n \overrightarrow{I_s}(\theta) \cdot \overrightarrow{R_o}(x_q, y_q, z_q + 2h_s + 2(n-1)t_s, x_k, \theta)}{R_o^3(x_q, y_q, z_q + 2h_s + 2(n-1)t_s, x_k, \theta)} \quad (4)$$

where h_s is the distance from the transmitter coil to the magnetic shield, t_s is the thickness of the magnetic shield. Λ_n is the factor indicating the ratio between the virtual current sources and can be expressed by

$$\lambda_n = \begin{cases} \eta & , if \ n=0 \\ (\eta^2-1)\eta^{2n-1} & , if \ n>0, \eta = (\mu_r-1)/(\mu_r+1) \end{cases} \quad (5)$$

here μ_r is the relative permeability of the material in the shield. Based on (5), the flux-density generated by the transmitter coil with a magnetic shield can be calculated similarly to that in the homogeneous medium by adding series virtual current sources.

III. HOMOGENEOUS FLUX COIL DESIGN

In [10], the author directly solved the continuous current distribution at the position of the transmitter coil by limiting the induced flux-density is homogenous at the specific distance. And then, the winding configuration is designed to fit the continuous current distribution. However, the error from the discrete current distribution in each winding to the continuous current distribution essentially occurs. It is difficult to get the analytic solution of the continuous current distribution, even harder for the coil with a magnetic shield. In this Section, a Genetic Algorithm (GA) based winding configuration design algorithm is proposed combining with the magnetic-flux density model derived mentioned above.

A. Winding Configuration Algorithm

Genetic Algorithm can solve global optimization problems. A typical GA process is shown in Fig.3, it consists of initialization, selection, genetic operators, and finally, the Algorithm terminates if a solution is found, or another termination condition is fully filled.

In the beginning, the perspective output and the constraints for the winding configuration design algorithm should be defined: the output of the Algorithm should indicate the windings' positions for the transmitter coil, which are expressed as a vector $x_n = [x_1, x_2, ..., x_{nw}]$ (here now is the selected number of turns for the coil); and the constraints

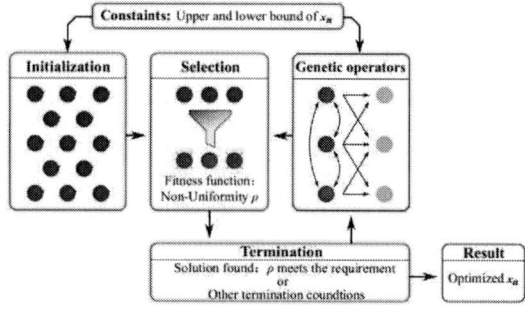

Fig. 3. General illustration of the proposed GA-based algorithm

limit the distribution range of the coil. The fitness function of the GA-based Algorithm given as:

$$\rho(h_p, r_p, h_s, t_s, \mu_r, nw, x_n) =$$

$$\frac{Std.deviation(\overrightarrow{B_{qs1}}, \overrightarrow{B_{qs2}}, ..., \overrightarrow{B_{qsl}})}{Avg.(\overrightarrow{B_{qs1}}, \overrightarrow{B_{qs2}}, ..., \overrightarrow{B_{qsl}})} \times \begin{bmatrix} 0 \\ 0 \\ 1 \end{bmatrix} \quad (6)$$

As illustrated in Fig.2(b), h_p is the distance between the transmitter coil to the target receiver's plate; r_p is the radius of the target receiver's plate, where the induced magnetic flux should be homogeneous. To evaluate the uniformity of the flux-density at the specific distance, the radius of the target receiver's plate is equally divided into l pieces, where the flux-density is defined as the ratio between the standard deviation of the series of the flux-density to the average value of the flux-density, whose minimal value is pursued by search the values in x_n in the Algorithm.

B. Evaluation of Generated Homo-Flux Winding Configurations

A virtual design was conducted to verify the proposed GA-based winding configuration design algorithm. In the case study, the radius of the target receiver's plate r_p is set as 1 to normalize other parameters of the geometry. In Table I, the optimized results when the number of turns n_w equals 7 and 14, and the distance is set as 0.2 are presented. Actually, the authors made a parameter sweeping for the case with/without a magnetic shield for the number of turns from 5 to 15 (where hs and ts are normalized as 0.01 and 0.005).

TABLE I.

TABLE TYPE STYLES

Number of turns n_w	Winding's position X_n
7	1.213 1.214 1.213 1.213 1.012 1.212 0.621
14	0.803 1.478 1.427 0.496 1.206 1.372 1.243 1.451 1.231 1.286 1.398 1.467 1.365 0.984

FEA simulation for the cases in Table I were made to verify the design results: in Fig.4 (a), the simulated flux-density at the target distance ($r_p = 0.2$) is plotted, from which the induced flux-density is found as near homogeneous and the value of the flux-density can be directly read; Besides, Fig.4 (b) illustrates the shape of flux line and the overall flux distribution of the 14-turn configuration.

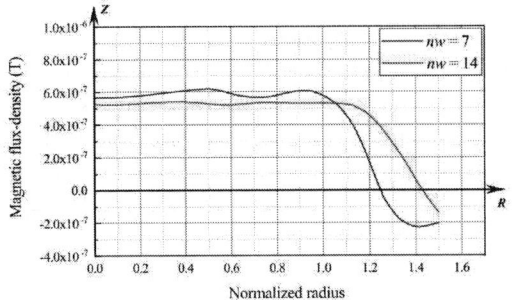

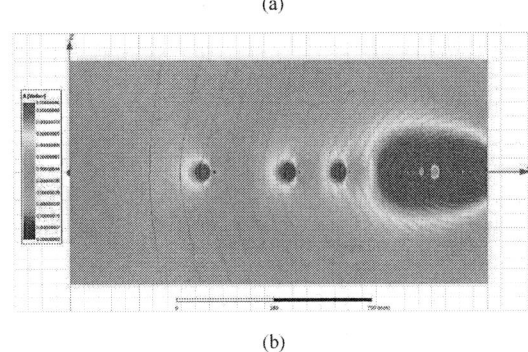

(a)

(b)

Fig. 4. FEA simulation verification for the design results: (a) flux-density distribution at the target position (b) flux distribution and flux line illustration

IV. EXPERIMENTAL RESULTS

A 6.78 MHz experimental prototype was built to verify the homogeneous flux-density transmitter coil. The photo of the system prototype is shown in Fig.5. The Push-Pull class-E inverter is used as the DC-AC high-frequency converter. The input voltage is 48 V, and the design output voltage is 12 V, and the output is around 20 W. The compensation network is LLC-S. For the receiver side, the rectifier bridge has been used.

Fig.6 shows the waveform of the Vds of the Class E inverter and the current of the transmitter coil. In Fig.7, the waveform of the output voltage is given. The x-axis is the positive of the receiver coil. And the zero value is the center of the transmitter coil. It can be seen that the output voltage is maintained at around 12 V. that means the homogeneous flux-density generated by the transmitter coil can maintain a high tolerance to the receiver's position shift.

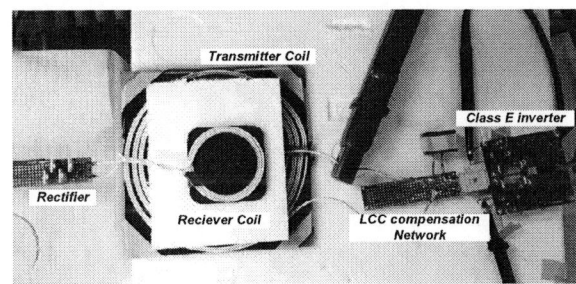

Fig. 5. photo of the 6.78 MHz system prototype

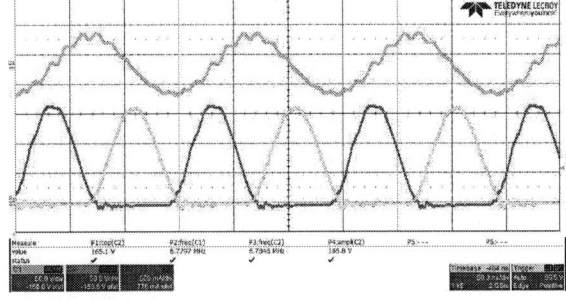

Fig. 6. the waveform of the Vds of the Class E inverter and the current of the transmitter coil.

978-1-6654-4817-8/21 $31.00 © 2021 IEEE

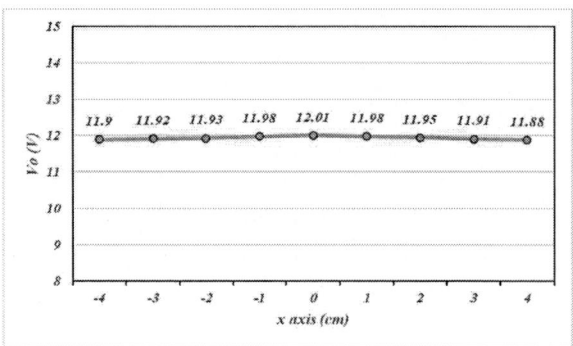

Fig. 7. waveform of the output voltage in the different positions.

V. CONCLUSION

Generally, in this paper, a GA-based winding configuration design algorithm is proposed for transmitter coils with homogeneous flux density. With the improved transmitter coil, the receiver's position tolerance can be enhanced and, finally, a robust and simplified control system with the ability of position-free for the secondary coil is built to verify the proposed method.

REFERENCES

[1] S. Y. R. Hui and W. W. C. Ho, "A new generation of universal contactless battery charging platform for portable consumer electronic equipment," IEEE Transactions on Power Electronics, vol. 20, no. 3, pp. 620–627, 2005.

[2] S.-Y. R. Hui, "Planar inductive battery charging system," Aug. 18 2009, uS Patent 7,576,514.

[3] W. X. Zhong, X. Liu, and S. Y. R. Hui, "A novel single-layer winding array and receiver coil structure for contactless battery charging systems with free-positioning and localized charging features," IEEE Transactions on Industrial Electronics, vol. 58, no. 9, pp. 4136–4144, 2011.

[4] S. Li, W. Li, J. Deng, T. D. Nguyen, and C. C. Mi, "A double-sided lcc compensation network and its tuning method for wireless power transfer," IEEE Transactions on Vehicular Technology, vol. 64, no. 6, pp. 2261–2273, 2015.

[5] J. Kim, D. Kim, and Y. Park, "Free-positioning wireless power transfer to multiple devices using a planar transmitting coil and switchable impedance matching networks," IEEE Transactions on Microwave Theory and Techniques, vol. 64, no. 11, pp. 3714–3722, 2016.

[6] C. Cai, J. Wang, H. Nie, P. Zhang, Z. Lin, and Y. G. Zhou, "Effective-configuration wpt systems for drones charging area extension featuring quasi-uniform magnetic coupling," IEEE Transactions on Transportation Electrification, vol. 6, no. 3, pp. 920–934, 2020.

[7] X. Liu and S. y. Hui, "Optimal design of a hybrid winding structure for planar contactless battery charging platform," in Conference Record of the 2006 IEEE Industry Applications Conference Forty-First IAS Annual Meeting, vol. 5, 2006, pp. 2568–2575.

[8] D. J. Griffiths, "Introduction to electrodynamics," 2005. [9] E. Waffenschmidt, "Design and application of thin, planar magnetic components for embedded passives integrated circuits," in 2004 IEEE 35th Annual Power Electronics Specialists Conference (IEEE Cat. No.04CH37551), vol. 6, 2004, pp. 4546–4552 Vol.6.

[9] E. Waffenschmidt, "Shielding properties of soft-magnetic layers for planar inductors," in Proceedings of14th International Power Electronics and Motion Control Conference EPE-PEMC 2010, 2010, pp. S15–17–S15–24.

[10] E. Waffenschmidt, "Homogeneous magnetic coupling for free positioning in an inductive wireless power system," IEEE Journal of Emerging and Selected Topics in Power Electronics, vol. 3, no. 1, pp. 226–233, 2015.

978-1-6654-4817-8/21 $31.00 © 2021 IEEE

Review of soft-switching high-frequency GaN-based single-phase Bridgeless Rectifier

Yunfeng Liu, Ziwei Ouyang, Michael A. E. Andersen

Department of Electrical Engineering, Technical University of Denmark, Kongens Lyngby, Denmark

***Abstract*— This paper reviewed the topologies and soft-switching (also called Zero Voltage Switching, ZVS) for the high-frequency GaN-based single-phase Bridgeless PFC. First, a brief introduction of single-phase bridgeless PFC has been presented. Then a review of the bridgeless PFC topologies has been given. Next, the theoretical analysis of the ZVS and the state plane trajectory is introduced, and the case study for the voltage step-up two-phase Totem-Pole PFC and voltage step-down Dual-SEPIC PFC are presented. Finally, the prototype for Totem-Pole PFC and Dual-SEPIC PFC are built, and the soft-switching has been achieved with high efficiency based on the ZVS extension strategy.**

Keywords—GaN devices, High frequency, CRM, Bridgeless PFC, soft-switching

I. INTRODUCTION

Power factor correction is widely used in modern power supplies for EV chargers, consumer devices, data centers, and telecommunication equipment. The need for miniaturization of the electronics devices is driving increased power density and also the high efficiency to deal with the thermal management in a smaller bulk [1].

The conventional PFC consists of a front-end diode bridge rectifier and a followed DC-DC converter which is shown in Fig.1(a). the bridge rectifier will first convert the AC lines to the double line frequency DC voltage, and it can easily consume 1% to 2% of the total power. Compared with the conventional PFC shown in Fig.1(a), the bridgeless active PFC reduces the number of power devices that the current flows. Therefore, the conduction losses of the power devices can be greatly reduced [2],[3].

As illustrated in Fig.1(b), the bridgeless PFC only has one stage AC-DC converter. The different polarity input voltage in the positive half-line cycle and negative half-line cycle needs two cells to work, respectively [3],[4]. As shown in Fig.2, each cell can be any discrete DC-DC converter, and the two cells may also be merged.

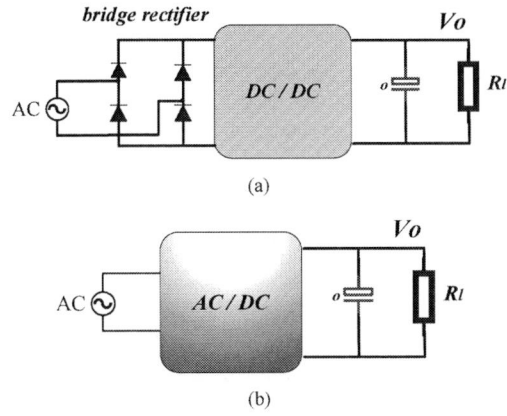

(a)

(b)

Fig. 1. PFC topologies. (a) Conventional PFC. (b) Bridgeless PFC.

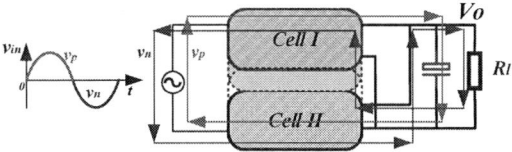

Fig. 2. Bridgeless PFC topologies with two cells.

The power devices, the control strategy, and the topologies dramatically influence the power density and system efficiency of the power converters [5]. With the development of the third semiconductor wide bandgap device (GaN, SiC), it is possible to push the switching frequency to MHz to further improve the power density [6]. However, at the same time, the turn-on power loss of the power device will dominate the power loss, which will greatly reduce the system efficiency. Thus, the turn-on loss increases significantly when the switching frequency increases, limiting the switching frequency around a few hundred kHz.

In order to address the turn-on loss issue, the zero-voltage-switching (ZVS) transaction technique can be employed. The drain-source voltage of the power switch will be discharged to zero before turning it on. Thus, the turn-on loss will be almost eliminated. [6-8]. Therefore, the ZVS transaction can improve the power density and system efficiency in a very high switching frequency.

This paper gives a topologies survey in Section II firstly. Then a comparison of the 600 V MOSFETs is presented. Section III gives a review of the soft-switching of high-frequency GaN-based CRM single-phase bridgeless Rectifiers. The ZVS turn-on of Totem Pole PFC and Dual-Sepic PFC has been analyzed. Section IV gives the experiment results of the bridgeless PFC, and Section V will give the conclusion of the paper.

II. TOPOLOGIES AND GAN DEVICES

A. A Topology survey of bridgeless PFC

According to the DC-bus voltage, the bridgeless active PFC generally can be divided into three different types: Step-up PFC, Step-down-Type, Step-up-down.

The Boost-type step-up topologies have been widely used in industry as depicted in Fig.3(a) to Fig.3(c). For this type of PFC, the output voltage must be higher than the peak input voltage, usually around 400 V. Fig.3(a) is the Dual-Boost PFC. And Fig.3(b) is the Dual-Boost PFC with diode [1],[10]. S_3 will remain turned on, and S_1 is the high switching

978-1-6654-4817-8/21 $31.00 © 2021 IEEE

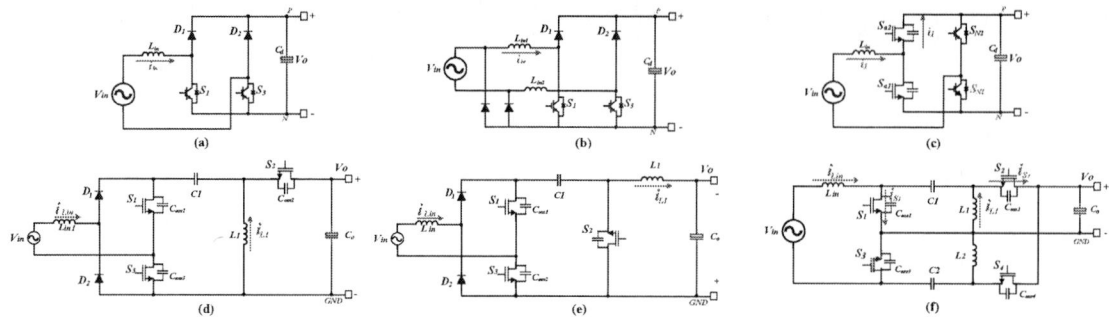

Fig. 3. Bridgeless PFC topologies. (a) Dual Boost PFC. (b) Dual Boost PFC with Diode. (c) Totem Pole PFC. (d) Single SEPIC PFC.
(e) Single CUK PFC. (f) Dual SEPIC PFC.

frequency device during the positive half line cycle. On the contrary, S_1 will become the high switching frequency device in the negative half-line cycle. Fig.3(c) is the Totem-Pole PFC [6]-[8],[15]-[17]. Buck PFC is a step-down PFC. the drawback for Buck PFC is the dead angle issue that will deteriorate the current THD and power factor [9]. Since it is not popular in the industry, it has not been introduced here. The other type is Step-up-down bridgeless PFC. The candidate topologies could be the SEPIC, CUK, Buck-Boost. Fig.3(d) is Single-SEPIC PFC, and Fig.3(e) is single-CUK PFC, the output voltage is reversed. Fig.3(f) is the Dual-SEPIC PFC[13]-[16].

B. GaN Devices

In recent years, due to the superior properties, wide bandgap semiconductors such as SiC and GaN for power electronics applications have been very attractive. Table I gives the key electrical properties of the three semiconductor materials. E_G represents the wide bandgap energy. μ defines Electron mobility. Vs represent the saturation velocity. E_{BR} represents the critical field breakdown voltage. It can be seen that the SiC and GaN have a high Band Gap, higher saturation electron drift velocity, and around ten times critical field than the Si MOSFET [18].

For GaN devices, the high channel mobility, short drain to source spacing, and the great reduction of the substrate conduction resistance make the GaN devices the best power devices in very high-frequency applications. Furthermore, with the advantage of no existing PN junction in the enhancement-mode type of GaN devices, there is no reverse recover power loss for the GaN MOSFETs, making it a perfect fit for Totem-Pole PFC, SEPIC PFC, and other topologies. The performance comparison is given by the figures of merits (FOMs) for the different Si MOSFETs, SiC MOSFETs, and GaN devices are shown in TABLE II.

FOM1 is defined by the R_{ds} x C_{iss}. Lower R_{ds} reduces conduction losses. The smaller FOM1 will contribute to faster gate drives of the power devices. It can be seen that the wide bandgap devices (SiC and GaN) can greatly reduce the switching loss than the SiC MOSFETs. Moreover, FOM1 of the GaN devices is much smaller than SiC MOSFETs. FOM2 defines the switching loss for both hard-switching and soft-switching. If the FOM2 is smaller, it means it has less switching loss. It can be seen that the latest Si MOSFETs from the Infineon can achieve even smaller FOM2 than the wide bandgap devices (SiC and GaN). the reverse recovery loss of the power devices can be presented by the FOM3. The lower FOM3 represents the less reverse recovery loss. It can be seen that the wide bandgap devices (SiC and GaN) have much smaller FOM3 than the conventional Si MOSFETs. The enhancement-mode type of GaN devices IGT60R070D1 and GS66508T have zero reverse recovery loss. Furthermore, TP65H035WSQA is cascode GaN, so the reverse recovery loss is coming from the MOSFET. The wide bandgap devices (SiC and GaN) can greatly reduce the reverse recovery loss.

TABLE I. COMPARISON OF THREE MATERIALS

	E_G (eV)	μ Cm²/V.s	V_S (10^6cm/s)	E_{BR} (10^6)V/cm
Si	1.1	1500	10	0.3
SiC	3.26	650	20	3
GaN	3.44	900-2200	25	3.5

TABLE II.

COMPARISON OF 600 V MOSFETs

MOSFETs	Part Number	V_{ds} (V)	R_{on} (mΩ)	C_{iss} (pF)	FOM1 ($R_{on}*C_{iss}$)	C_{oss} (pF)	Q_{oss} (nC)	FOM2 ($R_{on}*Q_{oss}$)	Q_{rr} (nC)	FOM3 ($R_{on}*Q_{rr}$)
Si	IPT60R028G7	600	28	4820	135000	99	74	2072	8700	24300
	STF28N60M2	600	135	1440	194400	104	78	2072	6500	877500
SiC	IMW65R027M1H	650	27	2131	57537	244	147	3969	239	6453
	C3M0060065D	650	60	1020	61200	80	48	2880	190	11400
GaN	TP65H035WSQA	650	41	1500	61500	196	178	7298	178	7298
	IGT60R070D1	600	70	380	26600	72	41	2870	0	0
	GS66508T	650	50	520	26000	65	57	2850	0	0

978-1-6654-4817-8/21 $31.00 © 2021 IEEE

The superior properties of the GaN devices have made it possible to push the switching frequency higher than MHz. This enables the feasibility of MHz Totem-Pole PFC and SEPIC PFC, which will be discussed in the next section.

C. Work mode

Generally, the operation mode for PFC can be classified into three different work modes: the continuous conduction mode (CCM), the discontinuous mode (DCM), and the critical conduction mode (CRM) [14]. For boost-type PFC, the work mode is only decided by the current waveform of the input side inductor. However, it becomes more complicated when it comes to the SEPIC, CUK PFC since it has two inductors [13].

The advantage of the CCM is that it has a much smaller inductor current ripple and better THD performance. But in some applications, the CRM work mode can obtain the ZVS transaction for the high-switching frequency devices, which makes it possible to work in very high frequency around MHz.

III. ZVS TURN-ON OF PFC

A. ZVS analysis of the Bridgeless PFC

From the above-mentioned bridgeless PFC, it is noticed that all the topologies with the input inductor can be simplified by the two branches. As shown in Fig.4. One is the charge branch. The input inductor is charged by the input voltage. The energy is stored in the input inductor from the input AC voltage source. And the other branch is the discharge branch, and then the input inductor current is discharged. The energy is delivered to the output side.

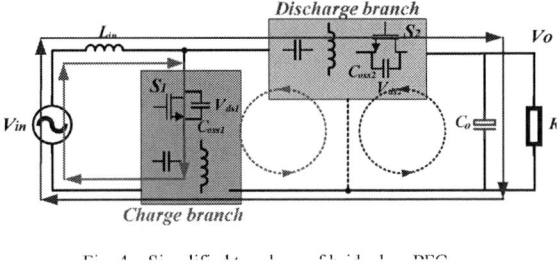

The ZVS turn on for all the switches of the single-phase PFC are occurring in the resonant stage. Two resonant stages occurred during the current switch between the two branches.

When the current goes through the charge branch, the inductor current increase, so when switched to the discharge branch, the output capacitor voltage of the switch S_2 is easy to discharge to zero by the input inductor current. However, when the inductor current switch back to the charge branch, it's a different story because the charge branch needs inverse current to discharge the output capacitor of the S_1. to discharge the output capacitor voltage of the switch, a ZVS extension strategy is needed [6]-[8],[12]-[17]. Then the negative current is provided by the inductor to discharge the voltage of output capacitor of the main switch.

Here taking the Totem Pole PFC and Dual-SEPIC PFC as an example, and the ZVS analysis is given below, respectively.

B. ZVS for Totem-Pole PFC

For Totem-Pole PFC, the topology is shown in Fig. 3(c). S_{a1} and S_{a2} are high switching frequency GaN devices, and S_{N1} and S_{N2} are low switching frequency devices. The inductor current waveform and the drain-source voltage waveform are shown in Fig.5. It can be seen that after the inductor current reduces to zero, it will continue to reduce to the negative value. The ZVS stage for the main switch S_{a1} starts when the synchronous switch S_{a2} is off at t_4. At this moment, the output capacitor of both the main switch and synchronous switch C_{oss1}, C_{oss2} start resonant with the input inductor L_{in}. The drain-source voltage of S_{a1} is changed from Vo to 0. Simultaneously, the drain-source voltage of S_{a2} is changed from 0 to Vo. This resonant stage ends when the main switch S_{a1} turns on. The equation during this resonant stage can be calculated as:

$$
\begin{cases}
I_{dis} \cdot cos(w_0 t) - \dfrac{V_M}{Z_n} sin(w_0 t) = i_{dis}(t) \\
V_{in} + I_{dis} \cdot Z_n sin(w_0 t) + V_M \cdot cos(w_0 t) = V_{ds1}(t) \\
V_M \cdot [1 - cos(w_0 t)] - I_{dis} \cdot Z_n sin(w_0 t) = V_{ds2}(t)
\end{cases} \quad (1)
$$

where $Z_n = \sqrt{L_{eq}/C_{eq}}$, and $w_o = 1/\sqrt{L_{eq}C_{eq}}$, $C_{eq} = C_{oss1} + C_{oss2}$. For Totem Pole PFC, $I_{dis} = I_{Lin}$, $V_M = (V_O - V_{in})$, $L_{eq} = L_{in}$. According to (1), finally, it can be rewritten as:

$$
Z_n^2 \cdot i_{dis}(t)^2 + (V_{ds1}(t) - V_{in})^2 = I_{dis}^2 \cdot Z_n^2 + (V_M)^2 \quad (2)
$$

According to (2), the state plane trajectory of ZVS Totem-Pole PFC is depicted in Fig.4. It can be seen that the state-plate-trajectory of the output voltage V_{ds1} of the main switch S_{a1} from t_5 to t_6 is a partial part of a circle, and the center of the circle is located at $(0, V_{in})$.

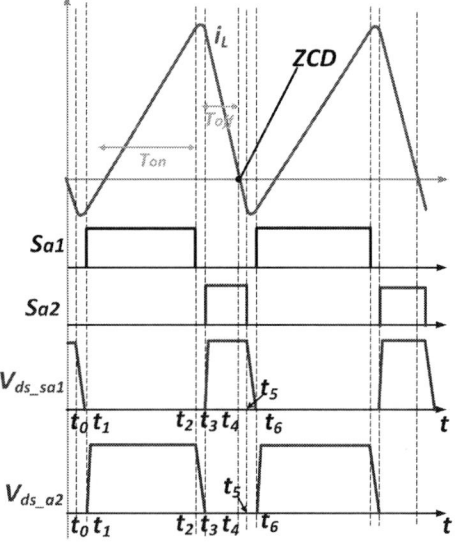

Fig. 5. The current waveform of the Totem-Pole PFC

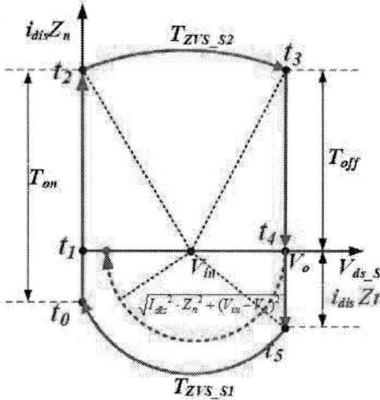

Fig. 6. state plane trajectory of Totem-Pole PFC

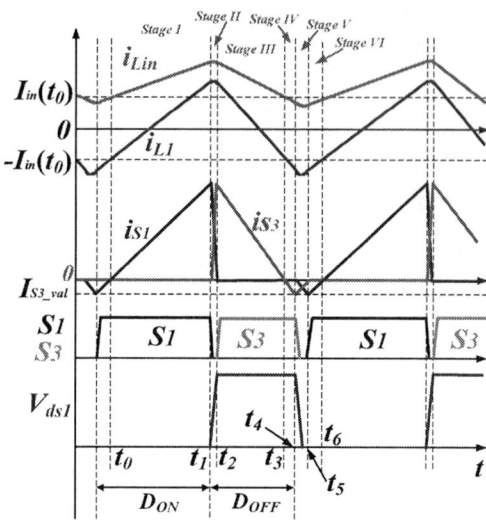

Fig. 7. The current waveform of the Dual-SEPIC PFC

For Totem Pole PFC, the radius is $\sqrt{I_{dis}^2 \cdot Z_n^2 + (V_{in} - V_o)^2}$, as shown in Fig. 6. For the conventional CRM control strategy without ZVS extension, it means $I_{dis} = 0$. The radius is ($Vo - V_{in}$) for Totem Pole PFC. In this condition. When the input voltage V_{in} is larger than $0.5Vo$, the ZVS transaction cannot be obtained, reducing the system efficiency dramatically. Therefore, to solve this issue when Vin > $0.5Vo$, an additional stage from t_4 to t_5 is added to provide an amount of negative discharge current is needed to guarantee the ZVS transaction.

C. ZVS for SEPIC/CUK PFC

For SEPIC or CUK PFC, as shown in Fig.3(d) to Fig. 3(f). The operation mode is quite different because those topologies have the intermediate capacitor. The intermediate capacitor voltage will follow the input ac voltage. So the voltage stress for the switch is ($Vin + Vo$).

The inductor current waveform is shown in Fig.7. The resonant stage for ZVS of the main switch begins with turning the output side synchronous switch S_3 off at t_4. The resonant stage occurs between the output capacitor C_{oss1} of S_1, C_{oss3} of S_3, and the two inductors L_{in} and L_1. Same with the Totem-Pole PFC, the output voltage of C_{oss3} is changed from 0 to ($Vin + Vo$), and simultaneously, the output voltage of C_{oss1} is changed from ($Vin + Vo$) to 0. The equations for this resonant stage are the same as (1). Where $I_{S3} = I_{L1} + I_{Lin}$, $V_M = V_O$, $L_{eq} = L_1 + L_{in}$. and the state plane trajectory of SEPIC or CUK is expressed by

$$Z_n^2 \cdot i_{dis}(t)^2 + (V_{ds1}(t) - V_{in})^2 = I_{dis}^2 \cdot Z_n^2 + V_o^2 \qquad (3)$$

Same with the Totem-Pole PFC, The state-plate-trajectory of the output capacitor voltage, i.e., drain-source voltage V_{ds1} from t_4 to t_5 is also a partial part of a circle. Moreover, the radius is $\sqrt{I_{dis}^2 \cdot Z_n^2 + V_o^2}$, as shown in Fig. 8. For the conventional CRM work mode without ZVS extension, then the $I_{dis} = 0$. Therefore, the radius is Vo for SEPIC PFC. So, when the input voltage is large Vo. The ZVS cannot be obtained. In order to solve this issue when Vin > Vo, same with Totem-Pole PFC, another stage is added to provide some negative current is needed to guarantee the ZVS transaction.

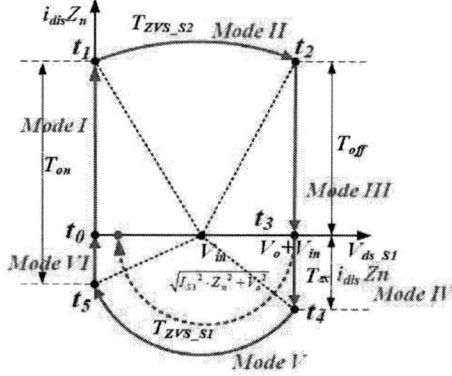

Fig. 8. State plane trajectory of Dual-SEPIC PFC

IV. EXPERIMENTAL RESULTS

To validate the ZVS control strategy, both the hardware prototype of the two-phase interleaved Totem Pole PFC in Fig. 9(a) and Dual-SEPIC PFC in Fig. 9(b) are built.

All the high switching frequency switches use the GaN device (GS6608T, GaN System, 650V, 50 mΩ). And two Si MOSFETs (IPW60R018CFD7XKS, 600 V) from the Infineon are used for the line frequency switches S_{N1} / S_{N2} for Totem-Pole PFC. Moreover, four paralleled electrolysis capacitors are installed in the dc link to buffer the voltage ripple. The DSP microcontroller from the TI was used to design the digital control system. The switching frequency ranges from a few hundred kHz to MHz under CRM mode.

Fig.10~12 shows the waveform for the two-phase interleaved Totem Pole PFC. The output voltage is 400 V_{dc}. Fig.10 shows the input voltage waveform, two inductor currents, and the input current in the line cycle. Moreover, Fig.11 shows the current waveform of each inductor current and the input current in each switching cycle. Furthermore, it can be obtained that after the two phases interleaved, the input current ripple has been significantly reduced. Moreover, the inductor current for each phase of the ripple

978-1-6654-4817-8/21 $31.00 © 2021 IEEE   414

also has been reduced by the negative coupled inductor. Fig.12 shows that the ZVS transaction both for the low-side switch and high-side switch has been achieved.

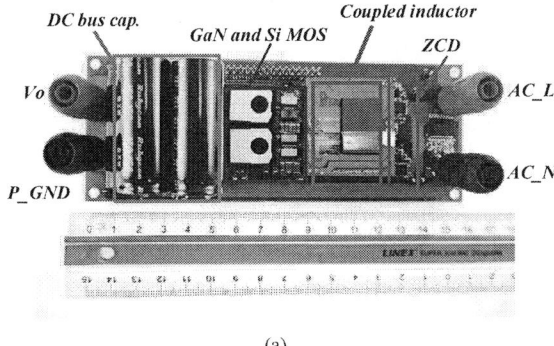

(a)

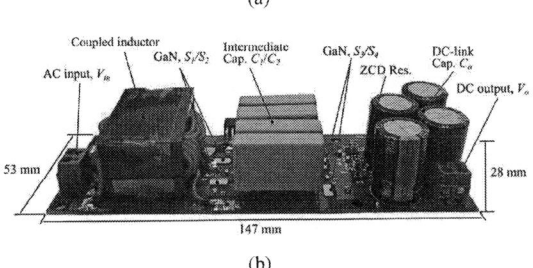

(b)

Fig. 9. Prototype of Bridgeless PFC (a) two-phase Totem-Pole PFC.

(b) Dual-SEPIC PFC

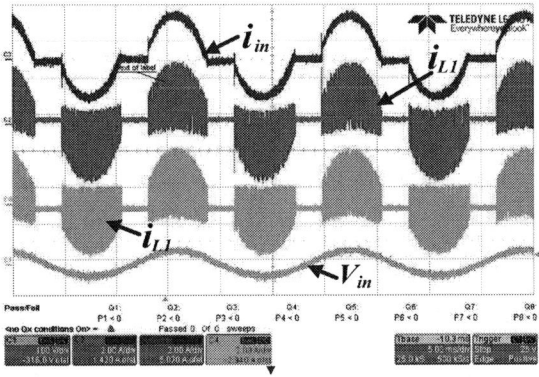

Fig. 10. Current waveform and input voltage waveform.

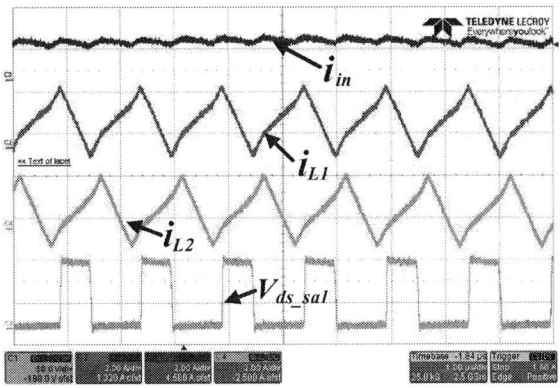

Fig. 11. Zoom in the current waveform in each switching cycle.

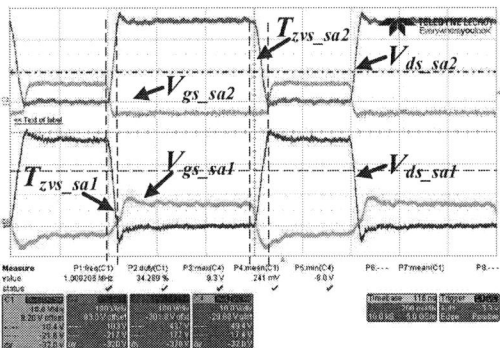

Fig. 12. ZVS achieved for all switches.

And Fig. 13~15 shows the waveforms for the bridgeless Dual-SEPIC PFC. The voltage value in the dc-link for the Dual-SEPIC PFC is 100 V_{dc}. The waveform of input voltage, the input current, and output voltage in the line cycle is shown in Fig.13. The two intermediate capacitor voltages follow the input voltage in each half-line cycle shown in Fig.14. It can be seen that the intermediate capacitor will follow the input voltage. And Fig.15 shows the main switch S_1 achieves the ZVS with the help of the negative current i_{S3}.

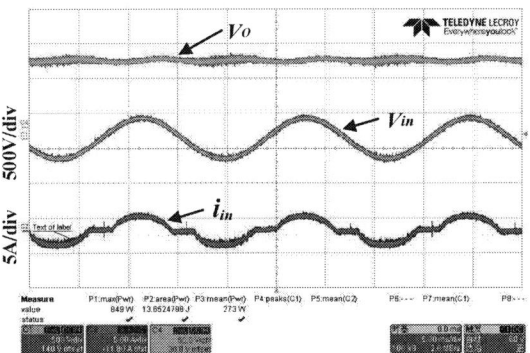

Fig. 13. Input current, input voltage, and output voltage Vo.

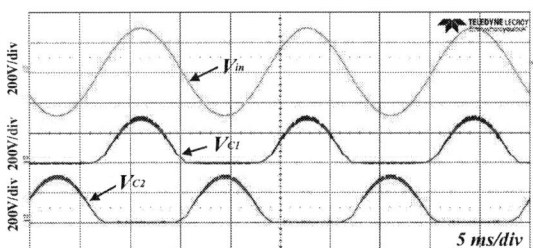

Fig. 14. Bridgeless PFC topologies

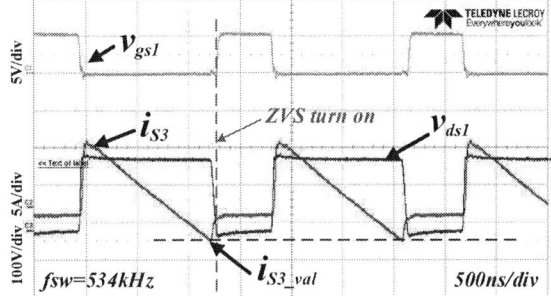

Fig. 15. ZVS achieved for S_1 by the ZVS extension strategy.

V. CONCLUSION

This paper reviewed the topologies and soft-switching for the high-frequency GaN-based single-phase Bridgeless PFC. Then the two cases study for the bridgeless PFC is presented. Theoretical analysis of the ZVS and the state plane trajectory for the bridgeless voltage step-up Totem-Pole PFC and bridgeless voltage step-down Dual-SEPIC PFC are presented. Finally, soft switching, both for Totem-Pole PFC and SEPIC PFC prototypes with high efficiency and high-power density, has been achieved based on the ZVS extension strategy in the author's group to verify the correction of the ZVS analysis.

REFERENCES

[1] L. Huber, Y. Jang, and M. M. Jovanovic, "Performance evaluation of bridgeless PFC boost rectifiers," IEEE Trans. Power Electron., vol. 23, no. 3, pp. 1381–1390, May 2008.

[2] K. Raggl, T. Nussbaumer, G. Doerig, J. Biela, and J. W. Kolar, "Comprehensive design and optimization of a high-power-density single-phase boost PFC," IEEE Trans. Ind. Electron., vol. 56, no. 7, pp. 2574–2587, Jul. 2009.

[3] Z. Chen, B. Liu, Y. Yang, P. Davari and H. Wang, "Bridgeless PFC Topology Simplification and Design for Performance Benchmarking," in IEEE Transactions on Power Electronics, vol. 36, no. 5, pp. 5398-5414, May 2021.

[4] Z. Chen, P. Davari and H. Wang, "Single-Phase Bridgeless PFC Topology Derivation and Performance Benchmarking," in IEEE Transactions on Power Electronics, vol. 35, no. 9, pp. 9238-9250, Sept. 2020.

[5] F. C. Lee, Q. Li, Z. Liu, Y. Yang, C. Fei, and M. Mu, "Application of GaN devices for 1 kW server power supply with integrated magnetics," CPSS Trans. Power Electron. Appl., vol. 1, no. 1, pp. 3–12, Dec. 2016.

[6] Q. Huang, A. Q. Huang, "Review of GaN Totem-Pole Bridgeless PFC," in CPSS Transactions on Power Electronics and Applications, vol. 2, no. 3, 2017.

[7] Q. Huang, R. Yu, A. Q. Huang and W. Yu, "Adaptive zero-voltage switching control and hybrid current control for high efficiency GaN MHz Totem-pole PFC rectifier," 2017 IEEE Applied Power

Electronics Conference and Exposition (APEC), Tampa, FL, 2017, pp. 1763-1770.

[8] Z. Liu, F. C. Lee, Q. Li, and Y. Yang, "Design of GaN-based MHz Totem-pole PFC rectifier," IEEE J. Emerg. Select. Topics Power Electron.,vol.4,no. 3, pp. 799–807, Sep. 2016.

[9] X. Lin and F. Wang, "New Bridgeless Buck PFC Converter with Improved Input Current and Power Factor," in IEEE Transactions on Industrial Electronics, vol. 65, no. 10, pp. 7730-7740, Oct. 2018.

[10] K. K. M. Siu and C. N. M. Ho, "A critical review of Bridgeless PFC boost rectifiers with common-mode voltage mitigation," IECON 2016 - 42nd Annual Conference of the IEEE Industrial Electronics Society, 2016, pp. 3654-3659,

[11] M. Mahdavi and H. Farzanehfard, "Bridgeless SEPIC PFC rectifier with reduced components and conduction losses," IEEE Trans. Ind. Electron., vol. 58, no. 9, pp. 4153-4160, 2011.

[12] Y. Liu, Y. Sun and M. Su, "A Control Method for Bridgeless Cuk/Sepic PFC Rectifier to Achieve Power Decoupling," in IEEE Transactions on Industrial Electronics, vol. 64, no. 9, pp. 7272-7276, Sept. 2017.

[13] Y. Liu, X. Huang, Y. Dou, Z. G. Ouyang and M. A. E. Andersen, "GaN-based ZVS Bridgeless Dual-SEPIC PFC Rectifier with Integrated Inductors," in IEEE Transactions on Power Electronics.

[14] Marxgut, C.; Krismer, F.; Bortis, D.; Kolar, J.W., "Ultraflat Interleaved Triangular Current Mode (TCM) Single-Phase PFC Rectifier," Power Electronics, IEEE Transactions on , vol.29,no.2, pp.873,882, Feb. 2014

[15] Y. Liu, M. Li, Y. Dou, Z. Ouyang and M. A. E. Andersen, "Investigation and Optimization for Planar Coupled Inductor dual-phase interleaved GaN-based Totem-Pole PFC," 2020 IEEE Applied Power Electronics Conference and Exposition (APEC), 2020, pp. 1984-1990.

[16] Y. Liu, X. Huang, Y. Dou, M. Li, O. Ziwei and M. A. E. Andersen, "Adaptive dead time control for ZVS GaN based CRM Totem Pole PFC," 2020 IEEE 9th International Power Electronics and Motion Control Conference (IPEMC2020-ECCE Asia), 2020, pp. 438-442.

[17] Huang, X.; Li, Q.; Liu, Z.; Lee, F.C., "Analytical Loss Model of High Voltage GaN HEMT in Cascode Configuration," Power Electronics, IEEE Transactions on , vol.29, no.5, pp.2208,2219, May 2014.

[18] EPC. Gallium nitride technology overview. https://epc-co.com/epc/campaigns/WhatIsGaN/GaN%20Transistors%20for%20E fficient%20Power%20Conversion-chapter-1.pdf

A Low Winding Loss Magnetic Circuit Structure Design of Planar Inductor for GaN-based Totem-Pole PFC

Pengyuan Ren
Department of Electrical Engineering
Xi'an Jiaotong University
Xi'an, China
3120104235@stu.xjtu.edu.cn

Wenjie Chen
Department of Electrical Engineering
Xi'an Jiaotong University
Xi'an, China
cwj@mail.xjtu.edu.cn

Xingwei Huang
Department of Electrical Engineering
Xi'an Jiaotong University
Xi'an, China
huangxw867046635@stu.xjtu.edu.cn

Yue Cao
Department of Electrical Engineering
Xi'an Jiaotong University
Xi'an, China
yue1403300584@stu.xjtu.edu.cn

Yuxuan Chen
Department of Electrical Engineering
Xi'an Jiaotong University
Xi'an, China
cyx1998@stu.xjtu.edu.cn

Xu Yang
Department of Electrical Engineering
Xi'an Jiaotong University
Xi'an, China
yangxu@mail.xjtu.edu.cn

Abstract—In recent years, power converters are increasingly moving towards high-efficiency, low losses and miniaturization. And the application of the wide-bandgap power semiconductor devices has further improved the operating frequency and efficiency of the power converter. PCB-based planar inductors are suitable for further reducing the volume of the power converter. But the increased operating frequency makes the low AC loss design of passive components at high frequency has become a bottleneck. By changing the path of leakage flux, a method of adding a low-permeability magnetic ring to the magnetic core is proposed to weak the increase in winding AC loss caused by the edge effect of the magnetic core. The results gained from the 3-D FEA simulations show that the AC resistance of the planar PCB inductor can be significantly reduced by 19.89%.

Keywords—high-frequency, AC winding loss, planar inductor, fringing effects, distributed magnetoresistance.

I. Introduction

Power converters are increasingly moving towards high-frequency, modular and miniaturization, and the planar inductors design for high-efficiency AC/DC or DC/DC converters is more important. However, because of their relatively smaller volume and larger winding area perpendicular to the air gap, its AC copper loss due to fringing effect and skin effect is usually more serious than traditional wire wound inductors. Especially for low voltage and large current power supplies of medium and small power used in communication base stations and data centers, the design of low AC losses planar magnetic components while reducing the volume has become more and more important.

In planar PCB inductors and transformers, the PCB winding area perpendicular to the leak magnetic flux is larger, thus the windings' AC loss caused by air gaps' fringing effect is usually serious. In [1][2], a method of cutting windings around air gaps to move away from them to avoid the effect of fringing leakage flux is proposed. But the number of windings layers and total height will be increased. To ameliorate the uneven distribution of AC current caused by the proximity effect and skin effect in the time-varying electromagnetic field, various planar Litz winding structures designed based on vertical and horizontal transposition have been proposed [3][4][5]. To a certain extent, the effectiveness of this method will be greatly influenced by the winding size, winding structure, operating frequency and so on, and it often requires many layers or turns which is difficult to implement.

For windings with multiple turns in the one layer, the optimization of the width and thickness of one turn has also been studied in [6][7]. In order to substantially weaken the fringing magnetic fields around air gaps, orthogonal air gaps in planar inductor structure has been discussed in [8], the leakage flux near air gaps is reduced by magnetic flux offset and the AC loss of the inductor is reduced. A multi-permeability distributed air-gap planar inductor structure is proposed in [9] to optimize the AC loss of inductor windings and improve the material utilization compared with the traditional structure.

In this paper, a method of adding a low-permeability magnetic ring to the magnetic core to change the path of leakage flux around air gaps is proposed, which can help to reduce the increase in AC wingding loss caused by the air gaps' fringing effect. Some basic principles and the analysis and modeling of the magnetic ring method will be given in the section II. Several key parameters used in the design of the magnetic ring will be analyzed in III. 3-D FEA analyses based on ANSYS have been done and the simulation results will be discussed in section IV. Section V discusses the conclusions.

II. Analytical Modeling

In principle, the reasons for the increase in AC winding loss can be summarized into three categories: the skin effect caused by high-frequency, the proximity effect caused by other conductors in the magnetic field, and the fringing effect caused by magnetic leakage. Among them, the eddy current loss in planar rectangular cross-sectional conductors caused by magnetic leakage around the air gap is significant in most cases, especially for the windings used in planar PCB inductors and transformers, which are close to the air gap. Thus, the AC winding loss in planar PCB conductors caused by the air gaps' fringing effect will be mainly discussed.

A. Fringing Fields of Air Gaps and the AC Winding Loss

As shown in Fig. 1, compared with the magnetic field generated around a cylindrical conductor, the biggest difference of the air gap fringing field is that its relative position relationship with the conductor has two-dimensional (2-D) characteristic [10]. Through the orthogonal decomposition method, the magnetic field vector of a point in space can be decomposed into two scalars in the x direction and the y direction, as shown in Fig. 1, which are parallel and perpendicular to the planar inductor windings respectively.

978-1-6654-4817-8/21 $31.00 © 2021 IEEE

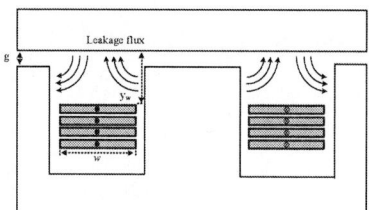

Fig. 1. Planar inductor core and air gap leakage flux

A scalar potential approach based 2-D analytical method described in [11] can be used in describing of how the fringing field changes in 2-D space. As shown in Fig. 1, several key geometrical parameters needed in this calculation is the gap length g, the distance y_w from the upper surface of the top winding to the core segment parallel to it, winding width w and the window width of magnetic core l. And the expresses for components x and y of the fringing field H are:

$$H_x\left(x,y\right)=\frac{H_g}{2\pi}\ln\left[\frac{x^2+\left(y-h\right)^2}{x^2+\left(y+h\right)^2}\right] \quad (1)$$

$$H_y\left(x,y\right)=\frac{H_g}{\pi}\left[\tan^{-1}\left(\frac{2xh}{x^2+y^2-h^2}\right)+k_0\pi\right] \quad (2)$$

Where k_0 equals to 1 if $x^2+y^2<h^2$ and k_0 equals to 0 if $x^2+y^2>h^2$. In actual planar inductors, the windings are generally located below the air gap, thus the equation (2) can be simplified to:

$$H_y\left(x,y\right)=\frac{H_g}{\pi}\tan^{-1}\left(\frac{2xh}{x^2+y^2-h^2}\right) \quad (3)$$

Under the premise that the magnetic core permeability is considered to be large enough, according to the law of Ampere's loos, Hg can be expressed as:

$$H_g=\frac{0.9NI}{2l_g+l_m\big/\mu}\cong\frac{0.9NI}{2l_g} \quad (4)$$

As we all know, no matter what angle of the air gap leakage flux intersecting the winding, the AC copper loss is merely related to the component perpendicular to the winding of the fringing magnetic field. For a thin rectangular conductor, the relationship between AC copper loss per unit length and H_y can be expressed as:

$$P_{ac}=\frac{1}{6\rho}\left(\mu_0\pi H_y f\right)^2 w^3 t \quad (5)$$

Where μ_0 is the magnetic permeability of vacuum; f is the operating frequency; ρ is the resistivity of the conductor material. Thus we can know that the magnitude of the eddy loss caused by the fringing effect is proportional to the square of the magnetic field component which is perpendicular to the PCB windings, which is [12]:

$$P_{ac}\propto F\left(f,w,t\right)\times H_y^2 \quad (6)$$

Therefore, H_y must be reduced if a lower loss planar PCB winding is expected.

B. Equivalent Distributed Magnetoresistance Model

In additional to the main magnetic flux distributed in the magnetic core, there is also a small part of the magnetic flux distributing in core window, around air gaps, windings and outside of the magnetic core in the planar inductor, which can be verified by finite element simulation. In this paper, we collectively refer to this part of the magnetic flux occurring in the air and intersecting with the winding as leakage flux.

Take the magnetic core in Fig. 1 as an example, its equivalent distributed magnetoresistance model is shown in Fig. 2. R_{ii}, R_{gi}, R_{ei} and R_{wi} are the equivalent magnetic resistance of I segment core, air gaps, E segment core and the magnetic core window respectively. The equivalent distributed magnetoresistance model can be established in more detail, but it will increase the complexity of the analysis.

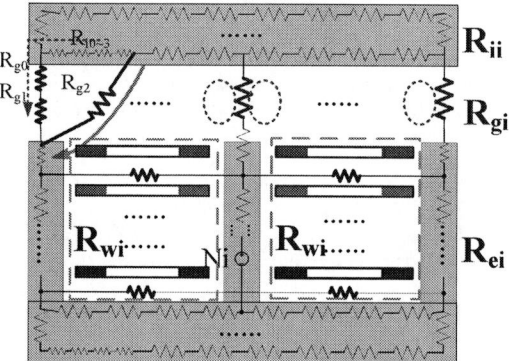

Fig. 2. Distributed Magnetoresistance Model of Planar PCB Inductor

Take the refined magnetoresistance model near the air gap on the left as an example, the principle of minimum reluctance indicates that the flux in space always flows alone the path of least reluctance. In the fringing field near the air gaps, there are two paths for the magnetic flux: one is a right-angle path formed by R_{i0-3}, R_{g0} and R_{g1}, as drawn by the dotted blue line in Fig. 2, and the other is an approximately straight path formed by R_{g2} as drawn by the solid green line in Fig. 2. R_{i0-3}, R_{g0-1} can be respectively expressed as:

$$R_{i\cdot}=\sum_{n=0}^{N_1}R_{in}=\frac{\displaystyle\sum_{n=0}^{N_1}l_{cn}}{\mu_0\mu_r\cdot k_{area\_c}A_c}=\frac{k_l l_c}{k_{area\_c}\mu_0\mu_r A_c} \quad (7)$$

$$R_{g\cdot}=\sum_{\substack{m=0\\m\neq2}}^{N_2}R_{gm}=\frac{\displaystyle\sum_{m=0}^{N_2}l_{gm}}{\mu_0 A_g}=\frac{g}{\mu_0 A_g} \quad (8)$$

where N_1 and N_2 represent the quantity of equivalent magnetoresistance in the I segment core and the air gap; l_{cn} and l_{gm} represents the magnetic flux path length of each equivalent magnetoresistance; k_l is the ratio of the magnetic circuit length that the magnetic leakage in the I-segment magnetic core should have to the total magnetic circuit length l_c; k_{area_c} equals to the cross-sectional area of the leakage flux to the total cross-sectional area of main flux in the I segment magnetic core A_c, usually $k_{area\_c}\ll1$; A_g is the cross-sectional area of the gap; μ_r is the relative permeability. Simplify the actual magnetic leakage path to the triangular relationship as shown in Fig. 2. Due to the divergence of the leakage flux, it is assumed that the magnetic cross-sectional leakage flux in the air is close to A_g. Thus we have $k_{area\_a}=1$ and:

978-1-6654-4817-8/21 $31.00 © 2021 IEEE

$$R_{g2} = \frac{l_{g2}}{\mu_0 k_{area\_a} A_g} \approx \frac{l_{g2}}{\mu_0 A_g} \qquad (9)$$

Where k_{area_a} equals to the cross-sectional area of R_{g2} to A_g. And the magnetic resistance on the two flux paths can be calculated as:

$$\begin{cases} R_{rec} = R_i + R_{g'} \\ R_{air} = R_{g2} \end{cases} \qquad (10)$$

Make the difference between the two and we have:

$$R_{rec} - R_{air} = \frac{1}{\mu_0 A_c}\left(\frac{k_l l_c}{k_{area\_c}\mu_r} + g - \sqrt{k_l^2 l_c^2 + g^2} \right) > 0 \qquad (11)$$

Since $k_{area\_c} \ll 1$, we can assume that $k_{area\_c}\mu_r$ is less than 1, and therefore equation (11) is greater than 0. The above equation indicates that near the air gap of an actual planar magnetic component, there will be a flux path composed entirely of air reluctance whose reluctance is less than $R_{rec}+R_{air}$. Thus, according to the principle of minimum reluctance, the generation of air gap magnetic leakage is proved theoretically. To simplify the analysis, in the derivation of formula (11), we assume that A_c equals to A_g.

Further analysis of the distributed reluctance model reveals the location of three types of flux leakage in the planar magnetic components as shown in Fig. 3. They all occur where the flux path makes an approximate right-angle turn. This phenomenon implies that magnetic flux leakage will occur at the interface of different media where the permeability changes and the magnetic circuit is turned.

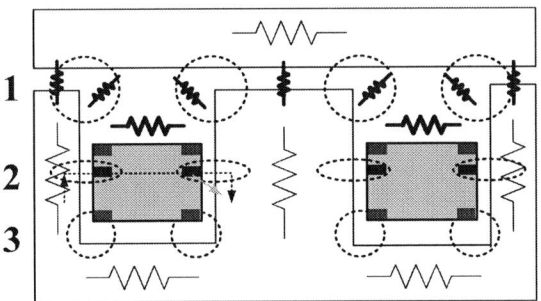

Fig. 3.   Three types of positions where magnetic flux leakage occurs in the planar magnetic component

III. Low Permeability Magnetic Ring Structure Design

A. Principles of Magnetic Ring Design

As shown in Fig. 4 below, the reluctance of the magnetic circuit in the space above the windings can be reduced by placing some magnetic rings with low permeability between the PCB winding and air gaps.

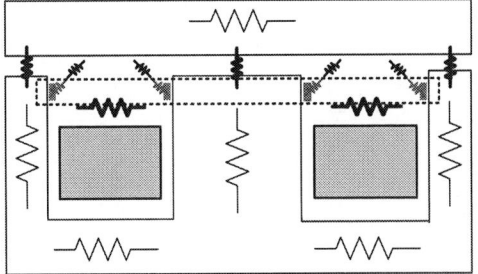

Fig. 4.   Planar magnetic component with a low permeability magnetic ring

Therefore, the flux leakage near the air gap that would have intersected the windings will enter the magnetic core through a path with lower reluctance and away from the windings, as shown in Fig. 5 below.

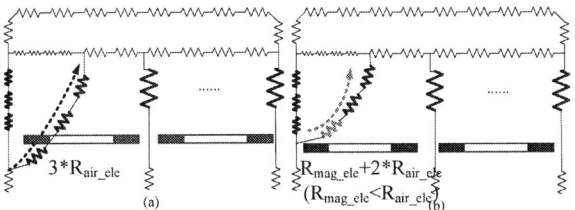

Fig. 5.   The influence of the low permeability magnetic ring on the magnetic path; (a) Without magnetic ring core; (b) With the low permeability magnetic ring

As can be seen in Fig. 5, one air reluctance R_{air_ele} is replaced by a low permeability magnetoresistance R_{mag_ele}, thus the total magnetoresistance above the windings reduce and less leakage flux will intersect with the top windings. That is, part of the leakage flux is transferred to the magnetic flux path avoiding the winding According to equation (5), the decrease of H_y will contribute to the improvement of AC winding loss caused by fringing field.

B. Key Parameters of the Magnetic Ring structure

According to the principle of minimum reluctance, the use of a magnetic ring made of low-permeability materials will help guide air gap leakage to avoid windings. However, two possible consequences of changing the reluctance, as shown in Fig .6 below, which need to be carefully considered:

1)   Magnetic rings placed close to the air gap may lead to the increase of leakage flux near the air gap. For example, if the low permeability material completely encapsulates the air gap, it will increase the area of magnetic flux leakage and cause the increase of leakage flux.

2)   Placing the magnetic ring closer to the winding may cause the leakage flux which is originally parallel to the PCB winding to intersect with PCB winding, resulting in increased loss.

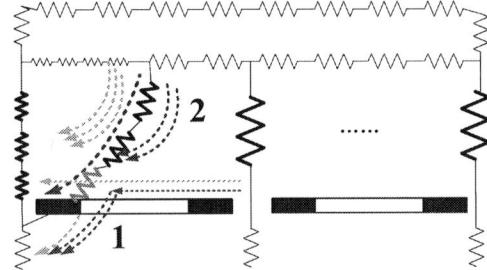

Fig. 6.   Two inappropriate positions of the magnetic rings

Therefore, the choice of the relative position of the magnetic ring between the air gap and the PCB winding is very important. In addition, the relative permeability μ_r, thickness δ and height h_r of the magnetic ring as shown in Fig. 7 will also affect the loss reduction. At the same time, to minimize the effect of magnetization on the magnetic loss of planar magnetic components, low-permeability materials with good high-frequency loss characteristics should also be selected.

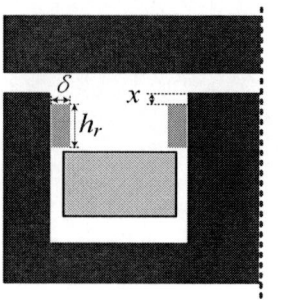

Fig. 7.   Geometric parameters of magnetic rings

IV. VERIFICATION

The designed planar inductor is used for a 1kW totem pole PFC operating at 200~500kHz frequency is shown in Fig. 8. And the 3D structure of the planar magnetic core with a magnetic ring is shown in Fig. 9. In the planar inductor used, the distance between the top PCB winding and air gaps y_w=2.1mm, and the size of the magnetic core is 32mm×27mm×12mm.

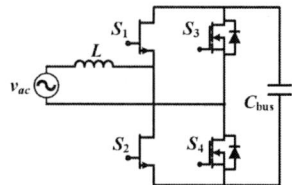

Fig. 8.   Totem pole PFC circuit topology

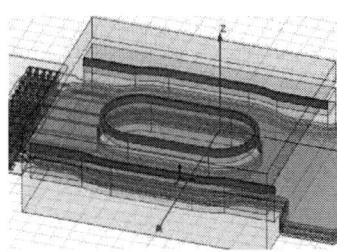

Fig. 9.   Planar PCB inductor with low permeability magnetic ring structure

It can be seen from the analysis in Part III that the magnetic ring should change the magnetoresistance on the set position or the desired path within a small range, so magnetic materials with low permeability and low loss density should be used to design the magnetic ring structure. On the contrary, materials with high magnetic permeability may have a greater impact on the local magnetic field, which will cause the two problems mentioned before to occur more easily. In the current design of the magnetic ring, an iron-silicon metal powder core material NPX is used. Its relative permeability is 26, while its loss is even lower than that of the 95 series ferrites, and its loss characteristics are close to those of the MDR51 series at low and medium frequencies.

A. Optimization of the Key Parameters

On the ANSYS Maxwell 3D simulation platform, parameterized scanning of key parameters δ, h_r and x of the

magnetic ring was carried out for the purpose of minimizing the AC loss. In the simulation, 12A current excitation of 300kHz is applied on the winding, as shown in the Fig. 10, Fig. 11 and Fig. 12.

First, fix the height h_r and thickness δ of the magnetic ring, perform a parametric scan on x, and it can be found that the total loss of the planar inductor and the winding loss both are the lowest at x=0.1mm, as shown in Fig. 10. Negative x means that the magnetic ring moves up and enters the air gap; then fix x equal to 0.1mm , When the height h_r or thickness δ of the magnetic ring is parametrically scanned, another parameter is fixed, and the results obtained are shown in Fig. 11 and Fig. 12.

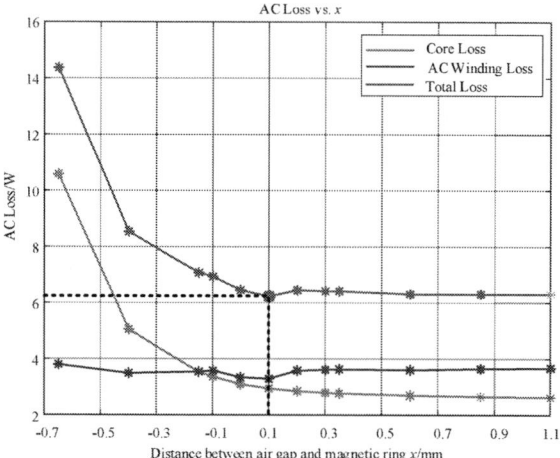

Fig. 10. The curve of the plane inductance AC loss with the distance x between the magnetic ring and air gaps

For the optimization of AC loss, the optimal thickness of the side column magnetic ring and the center column magnetic ring in the current structure is δ_1=0.3mm and $\delta_2=\delta_1$+0.1=0.4mm respectively, as shown in Fig. 11. In the parametric scan of the core height h_r, it is found that the total loss is close to the lowest point when h_r is 0.6, 0.9 and 1.35 respectively.

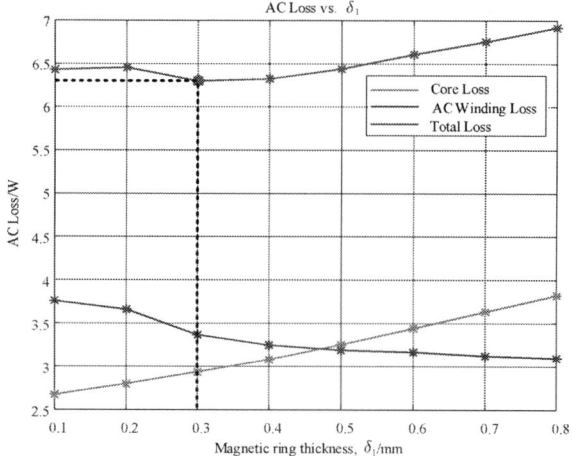

Fig. 11. The curve of the plane inductance AC loss with the magnetic ring thichness δ

978-1-6654-4817-8/21 $31.00 © 2021 IEEE

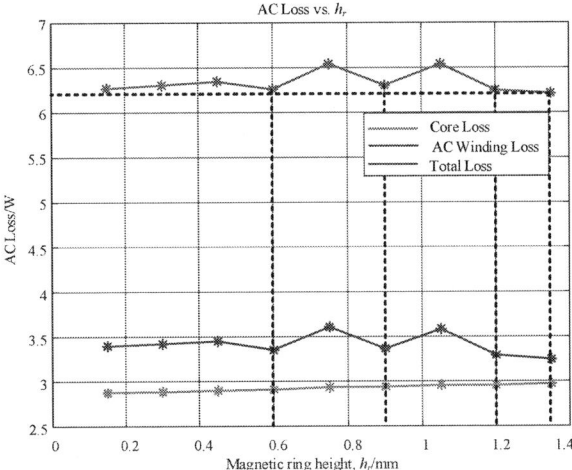

Fig. 12. The curve of the plane inductance AC loss with the magnetic ring high h_r

B. Final Results and Analysis

Finally, the size of the optimized magnetic ring is:

$$\begin{cases} x = 0.1mm \\ \delta_1 = 0.3mm \\ \delta_2 = 0.4mm \\ h_r = 1.35mm \end{cases} \quad (12)$$

Under the premise that neither the core nor the winding need to be changed, the final optimized winding AC resistance R_{ac}, PCB winding loss, iron loss and total loss are compared as shown in the Fig. 13 and Fig. 14.

In Fig. 13, the copper loss can be reduced by 19.89% but the total loss is only reduced by 6.85%, the reason is that the magnetic ring guides more leakage flux through the magnetic ring into the magnetic core, thus the core loss has increased after the magnetic ring is added, as shown in Fig. 14.

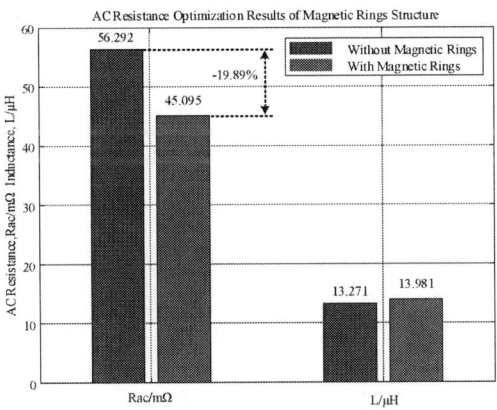

Fig. 13. AC Resistance Optimization Results of Magnetic Rings Structure

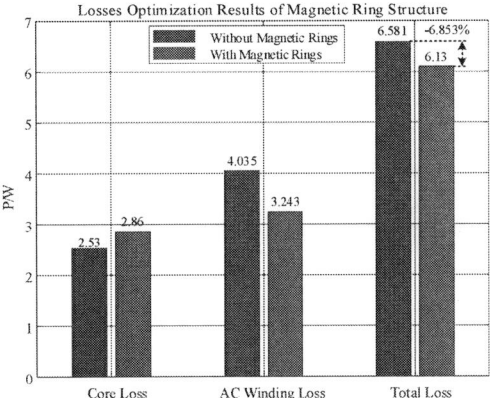

Fig. 14. Losses Optimization Results of Magnetic Rings Structure

Fig. 15 shows the comparison of the current density distribution on the windings with or without a magnetic ring made of low magnetic permeability, which reveals that the phenomenon of current density concentration in the windings near the center column has been significantly improved, that is, the current distribution is more uniform, and the AC winding loss is lower.

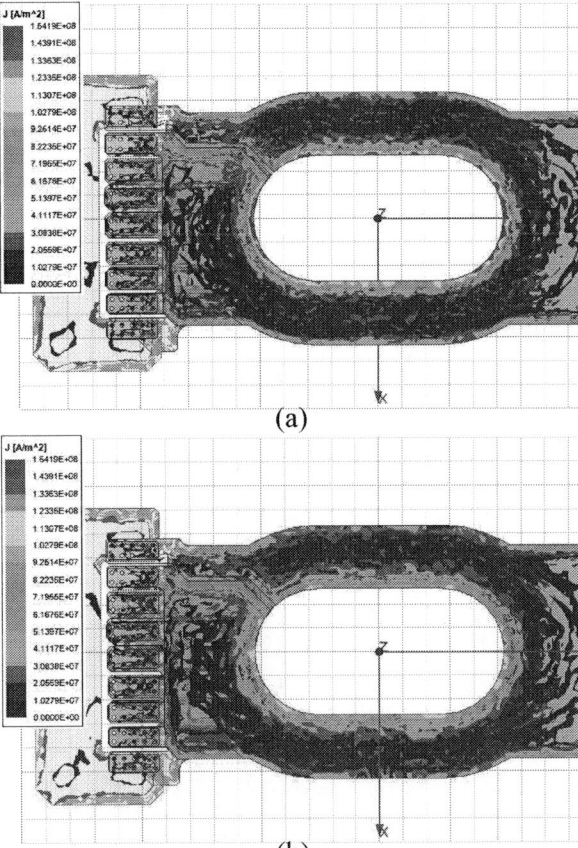

Fig. 15. Current density distribution of PCB winding without and with magnetic ring; (a) without magnetic ring; (b) with magnetic ring

Therefore, the magnetic ring structure can effectively weaken the increase in the copper loss of the winding caused by the edge effect of the air gap, but it is necessary to look for a low-permeability material with lower loss, or to optimize the core structure near the air gap to further reduce the structure The effect on core loss will be something that can be further studied in the future.

V. CONCLUSION

Based on the principle of minimum flux and the an established equivalent distributed magnetoresistance model, this paper has analyzed and proved the generation of air gap edge effect, that is, air gap leakage flux, from the perspective of reluctance. And the general conclusion has been given: the leakage flux is always generated in three types of magnetic core locations under the condition of a certain cross-sectional area of the leakage flux can be satisfied.

To reduce the increase of the winding AC loss caused by the fringing field, a low permeability magnetic ring structure is proposed. It can lessen the AC copper loss of the winding by guiding the magnetic leakage and avoiding the winding. Finally, three key parameters of the magnetic ring were optimized by parameterized scanning on FEA simulation platform. The simulation results show that the AC loss of winding can be significantly reduced by 19.89% if the magnetic ring can be added appropriately.

The advantages of this method are no need to make any changes to the magnetic core and winding, small volume, suitable for the demand of planar magnetic parts, and the winding loss reduction effect is obvious. However, the addition of magnetic ring will lead to a slightly greater magnetic core loss, and the gain of reducing the total loss of planar inductor will be reduced. Therefore, it is worth studying to find low permeability materials with lower loss or further optimize the shape of magnetic core near the air gap in the future.

REFERENCES

[1] D. Fu, F. C. Lee and S. Wang, "Investigation on transformer design of high frequency high efficiency dc-dc converters," 2010 Twenty-Fifth Annual IEEE Applied Power Electronics Conference and Exposition (APEC), Palm Springs, CA, USA, 2010, pp. 940-947.

[2] Y. Yang, M. Mu, Z. Liu, F. C. Lee and Q. Li, "Common mode EMI reduction technique for interleaved MHz critical mode PFC converter with coupled inductor," 2015 IEEE Energy Conversion Congress and Exposition (ECCE), Montreal, QC, Canada, 2015, pp. 233-239.

[3] R. Yu, T. Chen, P. Liu and A. Q. Huang, "A Novel Three Dimensional (3D) Winding Structure for Planar Transformers," 2020 IEEE Applied Power Electronics Conference and Exposition (APEC), New Orleans, LA, USA, 2020, pp. 202-207.

[4] Shen Wang, M. A. de Rooij, W. G. Odendaal, J. D. van Wyk and D. Boroyevich, "Reduction of high-frequency conduction losses using a planar Litz structure," IEEE 34th Annual Conference on Power Electronics Specialist, 2003. PESC '03., Acapulco, Mexico, 2003, pp. 887-891 vol.2.

[5] I. Lope, C. Carretero, J. Acero, R. Alonso and J. M. Burdio, "Frequency-Dependent Resistance of Planar Coils in Printed Circuit Board With Litz Structure," in IEEE Transactions on Magnetics, vol. 50, no. 12, pp. 1-9, Dec. 2014, Art no. 8402409.

[6] Y. Guan, Y. Wang, W. Wang and D. Xu, "Analysis and Design of a 1-MHz Single-Switch DC–DC Converter With Small Winding Resistance," in IEEE Transactions on Industrial Electronics, vol. 65, no. 10, pp. 7805-7817, Oct. 2018.

[7] C. Fei, F. C. Lee and Q. Li, "High-Efficiency High-Power-Density LLC Converter With an Integrated Planar Matrix Transformer for High-Output Current Applications," in IEEE Transactions on Industrial Electronics, vol. 64, no. 11, pp. 9072-9082, Nov. 2017.

[8] S. Mukherjee, Y. Gao and D. Maksimović, "Reduction of AC Winding Losses Due to Fringing-Field Effects in High-Frequency Inductors With Orthogonal Air Gaps," in IEEE Transactions on Power Electronics, vol. 36, no. 1, pp. 815-828, Jan. 2021.

[9] L. Wang, Z. Hu, Y. Liu, Y. Pei and X. Yang, "Multipermeability Inductors for Increasing the Inductance and Improving the Efficiency of High-Frequency DC/DC Converters," in IEEE Transactions on Power Electronics, vol. 28, no. 9, pp. 4402-4413, Sept. 2013.

[10] I. Lope, C. Carretero, J. Acero, R. Alonso and J. M. Burdio, "AC Power Losses Model for Planar Windings With Rectangular Cross-Sectional Conductors," in IEEE Transactions on Power Electronics, vol. 29, no. 1, pp. 23-28, Jan. 2014.

[11] W. A. Roshen, "Fringing Field Formulas and Winding Loss Due to an Air Gap," in IEEE Transactions on Magnetics, vol. 43, no. 8, pp. 3387-3394, Aug. 2007.

[12] E. C. Snelling, Soft Ferrites: Properties and Applications, 2nd ed. London, U.K.: Butterworth, 1988.

Design Methodology of SiC MOSFET Based Bidirectional CLLC Resonant Converter for Wide Battery Voltage Range

Mingjie Liu
State Key Laboratory of Advanced Electromegnetic Engineering and Technology
Huazhong University of Science and Technology
Wuhan, China
liumingjie@hust.edu.cn

Xuehua Wang
State Key Laboratory of Advanced Electromegnetic Engineering and Technology
Huazhong University of Science and Technology
Wuhan, China
wang.xh@hust.edu.cn

Jiangtao Xu
State Key Laboratory of Advanced Electromegnetic Engineering and Technology
Huazhong University of Science and Technology
Wuhan, China
jt_xu@hust.edu.cn

Abstract—The CLLC resonant converter has gained much attention in DC microgrid due to its bidirectional power transfer capability and high power density. When interlinking the DC-link bus and energy storage battery in DC microgrid, the CLLC converter needs to possess wide voltage regulation ability in forward mode (FM) and backward mode (BM) due to the wide battery voltage range. To solve this issue, this paper presents a simple and effective design methodology of bidirectional CLLC resonant converter based on the first harmonic approximation (FHA) model. Considering the voltage gain and zero-voltage-switching (ZVS) region, the design procedures are discussed in detail. Finally, a 10-kW SiC MOSFET based prototype converter was designed and built. Experiment results verify the theoretical expectations and the maximum power conversion efficiency is 97.2%.

Keywords—CLLC resonant converter; wide voltage range; FHA model; design methodology

I. INTRODUCTION

Due to the large potential of renewable energy sources and electric vehicles, the DC microgrid has attracted increasing attention [1]. Bidirectional DC-DC converters (BDCs) play a key role in the DC microgrid, interfacing between DC-link bus and power loads [2]. With the rapid development of wide-bandgap semiconductor devices, such as SiC and GaN, BDCs shows significant advantages comparing to traditional isolation transformers.

Several isolated BDC topologies have been recommended for applications of DC microgrid [3]. Among them, dual-active-bridge (DAB) and CLLC resonant converter have gained widespread attention in the research of isolated BDCs due to their common advantages such as modular symmetrical structure, ZVS feature and high power density [4]-[5]. However, the soft switching region of the DAB converter is unfortunately limited to a narrow voltage range and the DAB still suffers from high turn-off current and high circulating power, which causes a tremendous loss [6]. In contrast, the CLLC converter is able to implement soft switching in full load range and has a relatively small turn-off loss [5]. Therefore, the CLLC converter is a more promising candidate in the future distributed energy system.

Previous studies on the bidirectional CLLC converter can be summarized as topology research [7]–[8], control strategy [9]–[10], and the parameter design [11]–[12]. The CLLC topology was first introduced in [7] and the operating principle and the mathematical model of the circuit are explained in detail. In the applications of the wide input and output voltage, an integrated half-bridge CLLC converter is presented, which is suitable to interlink two DC voltage buses [8]. As for the control strategies, the small signal model for the CLLC converter is derived and a controller design method is proposed in [9]. To further improve efficiency of the CLLC converter, synchronous rectification (SR) control is introduced to reduce the conduction loss of the MOSFETs [10]. According to the FHA method, a design methodology based on voltage gain curve for the CLLC converter in Vehicle-to-Grid (V2G) applications is proposed [11]. In [12], an AI algorithm-based optimal design method to reduce the power loss for the CLLC converter is presented, but this method requires a huge amount of calculation, which is not conducive to promotion.

When the CLLC converter interlinking the DC-link bus and energy storage battery, due to the wide battery voltage range in FM (battery in charging mode) and BM (battery in discharging mode), the operation and design of the CLLC converter face great challenges. Motivated by these challenges, this paper presents a simple and effective design methodology of the CLLC converter for wide battery voltage. Moreover, a CLLC prototype based on SiC MOSFET is built to verify the reliability of the design.

This paper is organized as follows. In Section II, the voltage gain and ZVS region are derived and discussed. In Section III, the design procedures for the CLLC converter is presented in detail. In Section IV, experiments are implemented to verify the theoretical expectations. Section V concludes this paper.

II. ANALYSIS OF CLLC CONVERTER

The CLLC converter illustrated in Fig. 1 has the full-bridge symmetric structure, where the primary inverting and the secondary rectifying stages are isolated by a high-frequency transformer. Meanwhile, the power flow directions are also given in Fig. 1, which are defined as FM and BM. The CLLC resonant tank is composed of five resonant elements: magnetizing inductance L_m, resonant inductance L_1, L_2 and resonant capacitor C_1, C_2, which play an important role in determining the resonance process of the converter.

978-1-6654-4817-8/21 $31.00 © 2021 IEEE

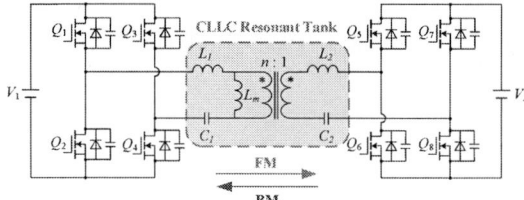

Fig. 1. Configuration of the CLLC converter.

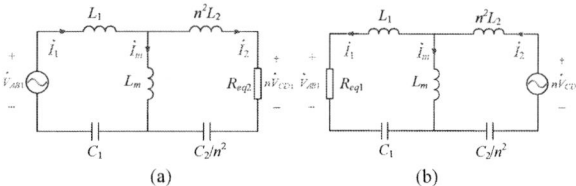

(a)   (b)

Fig. 2. FHA model of CLLC converter. (a) FM. (b) BM.

A. Gain analysis based on FHA Model

The FHA is a common method to analyze the resonant converters. From Fig. 1, the FHA model of the CLLC converter can be obtained shown in Fig. 2. Since the power of resonant converter is approximately transferred by the fundamental component, the quasi-square voltage v_{AB} and v_{CD} can be represented by their fundamental components v_{AB1} and v_{CD1} respectively. In addition, R_{eq2} is the load equivalent impedance referred to the primary side in FM and R_{eq1} is the load equivalent impedance in BM. They are expressed as

$$R_{eq2} = \frac{8n^2}{\pi} \frac{V_2^2}{P_o}; \quad R_{eq1} = \frac{8}{\pi} \frac{V_1^2}{P_o} \quad (1)$$

According to the FHA model, the voltage gain in FM and BM is derived respectively.

$$M_f(\omega_n) = \frac{k\omega_n}{\sqrt{Q_f^2 A(\omega_n)^2 + B(\omega_n)^2}} \quad (2)$$

$$M_b(\omega_n) = \frac{k\omega_n}{\sqrt{Q_b^2 A(\omega_n)^2 + C(\omega_n)^2}} \quad (3)$$

Where

$$\begin{cases} A(\omega_n) = a\omega_n^2 - b + c\frac{1}{\omega_n^2} \\ B(\omega_n) = \omega_n + k\omega_n - \frac{1}{\omega_n} \\ C(\omega_n) = h\omega_n + k\omega_n - \frac{1}{g\omega_n} \end{cases} \quad (4)$$

$$k = \frac{L_m}{L_1}, \quad h = \frac{n^2 L_2}{L_1}, \quad g = \frac{C_2}{n^2 C_1} \quad (5)$$

According to (2) and (3), the voltage gain with respect to the normalized operating frequencies ω_n can be shown in Fig. 3. Fig. 3(a) shows the voltage gain curve of the symmetrical CLLC converter where $h=1$ and $g=1$. It is obvious that the voltage gain curves of the symmetrical CLLC converter are exactly the same in FM and BM. In addition, the CLLC converter always has unit gain when $\omega_n=1$ and the gain decreases as the load increases at other operating frequency. To sum up, the basic characteristics of the CLLC resonant converter are consistent with those of the LLC resonant converter.

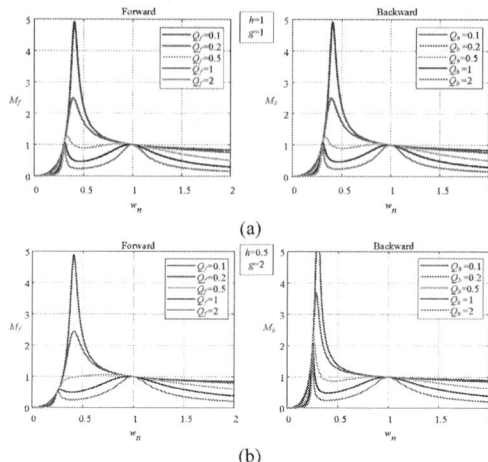

(a)

(b)

Fig. 3. Voltage gain with respect to the switching frequency. (a) Symmetric CLLC. (b) Asymmetric CLLC.

Fig. 3(b) shows the voltage gain curve of the asymmetric CLLC converter where $h=0.5$ and $g=2$. It can be seen that the voltage gain curves of the asymmetric CLLC converter are different, specifically, the voltage gain in FM is greater than that in BM at the same operating frequency. Moreover, the difference between the FM and BM gain curves will change with different h and g, which brings a great challenge to the design of the asymmetric CLLC converter.

B. Analysis of ZVS region

Achieving ZVS in the full load range is of great importance to the design of the resonant converters. For the LLC-type resonant converters, as the load increases, the input impedance of the resonant tank may change from inductive to capacitive, causing the primary side switches losing ZVS. Therefore, the inductive input impedance is a necessary condition for the CLLC converter to achieve ZVS. According to Fig. 2, the input impedance of the CLLC converter operating in FM and BM are derived as follows

$$Z_{fin} = \frac{-A(\omega_n) + B(\omega_n)C(\omega_n) + j\left(Q_f A(\omega_n)C(\omega_n) + \frac{B(\omega_n)}{Q_f} \right)}{Q_f^{-2} + C(\omega_n)^2} R_{eq2} \quad (6)$$

$$Z_{bin} = \frac{-A(\omega_n) + B(\omega_n)C(\omega_n) + j\left(Q_b A(\omega_n)B(\omega_n) + \frac{C(\omega_n)}{Q_b} \right)}{Q_b^{-2} + B(\omega_n)^2} R_{eq1} \quad (7)$$

From (6) and (7), the constraint conditions of the quality factor Q are obtained in FM and BM when the input impedance is purely resistive.

$$Q_f^2 = -\frac{B(\omega_n)}{A(\omega_n)C(\omega_n)} \quad (8)$$

$$Q_b^2 = -\frac{C(\omega_n)}{A(\omega_n)B(\omega_n)} \quad (9)$$

Substituting (8) and (9) into (2) and (3) respectively, the pure resistive line can be drawn in the voltage gain curves to determine the ZVS region, as shown in Fig. 4, where the ZVS regions are marked with green shading. Coincidentally, the pure resistive lines of the symmetric CLLC converter are the

978-1-6654-4817-8/21 $31.00 © 2021 IEEE

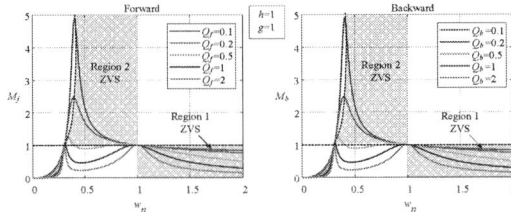

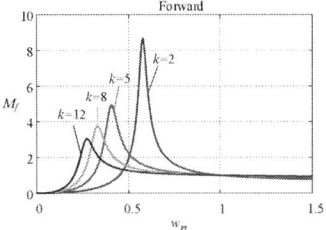

Fig. 4. Voltage gain and ZVS region.

Fig. 5. Voltage gain curves with different k values.

unity gain curves (M_f=1 and M_b=1). ZVS can be realized above the pure resistive line when ω_n <1 as well as below the pure resistive line when ω_n >1. The ZVS regions in FM and BM are identical due to the symmetrical structure. From the perspective of the inductive input impedance, the symmetric CLLC can easily realize ZVS in the full load range.

III. DESIGN PROCEDURES FOR CLLC CONVERTER

A. Proposed design procedures

The parameters to be designed for the CLLC resonant converter are transformer ratio n and the resonant elements (L_m, L_1, L_2, C_1 and C_2). There are many factors that need to be considered in the design procedures, such as the voltage gain, operating frequency range and soft switching, etc. Therefore, the design of the CLLC converter is complicated and there is currently no universal and effective method. On the basis of the voltage gain curves and ZVS region in section II, a design methodology is proposed in this section. The design procedures are listed as follows.

1) Choose a proper resonant frequency f_r.

$$f_r = \frac{1}{2\pi\sqrt{L_1 C_1}} = \frac{1}{2\pi\sqrt{L_2 C_2}} \quad (10)$$

Since the CLLC converter has excellent soft switching feature, which significantly reduces switching loss when the converter is equipped with SiC MOSFETs, thus the converter has the ability to maintain high efficiency under high operating frequency. The choice of a proper resonant frequency should consider the volume of magnetic components, the efficiency of the converter as well as the complexity of the control algorithm.

2) Design the transformer ratio n. The CLLC converter has the highest efficiency at the resonant frequency f_r, so the converter is suggested to operate at f_r under rated condition according to the designed transformer ratio. If the primary and secondary power switches of the converter are assumed to be ideal, then the transformer ratio can be expressed as

$$n = \frac{V_{1nom}}{V_{2nom}} \quad (11)$$

3) Design the magnetizing inductance L_m. In order to achieve ZVS for the primary side switches, the resonant current is supposed to discharge the junction capacitors of the power switches during the dead-time. Studies have proven that the magnitude of resonant current mainly depends on the L_m and the operating frequency f_s. Therefore, the maximum value of the magnetizing inductance is limited by (12).

$$L_m \leq \frac{t_{dead}}{16 f_{s\max} C_{oss}} \quad (12)$$

Where t_{dead} is the dead-time and C_{oss} is the junction capacitance of the switches. A smaller magnetizing

inductance ensure the power switches to achieve ZVS easier in full load range. However, L_m cannot be too low as it would form a very high magnetizing current, resulting in large conduction losses. A larger magnetizing inductance ensure a smaller magnetizing current, but it limits the voltage gain of the converter. So there is a trade-off between the voltage gain and efficiency about the design of L_m.

4) Design k, h and g. In order to ensure that the voltage gain of the CLLC converter at the resonant point is unity gain, hg=1 is a necessary condition. The design result corresponding to h=g=1 is the symmetric CLLC while that corresponding to h>g or h<g is the asymmetric CLLC, and which one to choose depends on the specific application. Fig. 5 shows the voltage gain curves with different k values while L_m is constant. It can be seen that the smaller value of k, the steeper the gain curve, which is beneficial to reduce the operating frequency range. However, if the value of k is too small, the unexpected circulating energy in the resonant tank will increase because L_m is no longer much larger than L_1 and L_2. Empirically, the value of k is recommended to be 3~7 for the CLLC converter operating at variable frequency. The accurate k value can be solved according to (2)-(5) if the expected operating frequency has been determined.

5) Calculate resonant parameters. After determining the transformer ratio n, the L_m and the parameters k, g, h. Combined (5) and (10), the values of resonant inductance L_1, L_2 and resonant capacitor C_1, C_2 can be easily calculated. In the end, it is worth noting that the ZVS region of the CLLC converter needs to be verified according to Fig. 4.

B. Design example

A 10kW bidirectional CLLC converter, which can be used for V2G application to interface the 700~800V DC bus and 500~800V wide battery voltage is employed as a design example.

Considering the size and heat dissipation of the converter, the resonant frequency is designed as 73 kHz. The normal voltage of the DC bus is 750V, and the normal voltage of the battery is 600V. So the transformer turns ratio comes out to be 1.25. For a dead-time of 150 ns and a junction capacitance of 220 pF, the maximum magnetizing inductance is calculated to be 584μH according to (12). Fig. 6 shows the voltage gain curves with different L_m. The voltage gain requirement of the converter in FM is 0.94~1.07 and that in BM is 0.75~1.20. In order to take a larger L_m under the premise of satisfying the voltage gain, the actual L_m is designed to be 160μH.

For the consistency of the FM and BM, the resonant tank is designed to be symmetrical in V2G application, that is h=g=1. In addition, the size of the passive components is determined by the lowest operating frequency. Limited by the

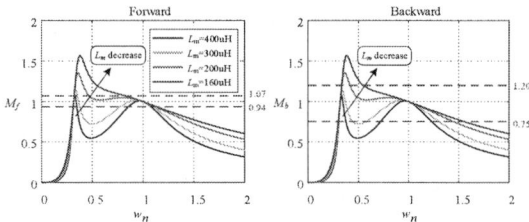

Fig. 6. Voltage gain curves with different L_m values.

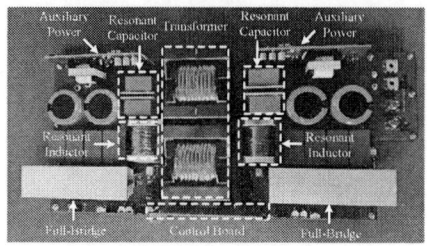

Fig. 7. Prototype of the bidirectional CLLC converter.

minimum operating frequency of 40 kHz, the appropriate value of k is derived as 4.44 according to (2), (3) and (4). On this basis, the values of L_1, L_2 and C_1, C_2 are obtained.

Finally, the ZVS region verification shows that the designed parameters can ensure the converter to realize ZVS in wide input and output voltage range.

IV.   EXPERIMENTAL RESULTS

A 10-kW CLLC prototype based on SiC MOSFET was built to verify the design methodology. The prototype parameters are listed in Table I. The power switches are CoolSiC^{TM} MOSFET discretes IMZ120R0301H from Infineon. The control is implemented in an XMC4400 Infineon microcontroller.

TABLE I.   PROTOTYPE PARAMETERS

Parameters	Symbol	Value	Parameters	Symbol	Value
Input voltage	V_1	700~800V	Resonant inductor 1	L_1	36 μH
Output voltage	V_2	550~800V	Resonant inductor 2	L_2	22 μH
Transformer ratio	n	20:16	Resonant capacitor 1	C_1	132 nF
Magnetizing inductance	L_m	160 μH	Resonant capacitor 2	C_2	216 nF

The prototype of the converter is shown in Fig. 7 and the main components have been marked in the figure. Fig. 8 shows the steady-state waveforms of the prototype under three different input and output voltages in FM. The waveforms of each figure from top to bottom is the output voltage V_2, the primary resonant current i_{L1} and the drive of the primary switch v_{gs_Q1}. It can be seen from i_{L1} and v_{gs_Q1} that the converter can achieve ZVS under the three operating conditions because the negative resonant current has already flowed through the anti-parallel diode of Q_1 before Q_1 turns on.

In Fig. 8(a), the gain of the resonant tank M_f is 1.43 with the input voltage of 700V and the output voltage of 800V, and the operating frequency is 48 kHz, which is higher than the resonant frequency 73 kHz. In Fig. 8(b), the input voltage is 750V and the output voltage is 600V, which respectively correspond to the rated voltages of the DC-link bus and the battery. In this condition, the converter operates at resonant frequency with unit gain and achieves the maximum power conversion efficiency of 97.2%. In Fig. 8(c), the converter operates at 95 kHz with the input voltage of 800V and the output voltage of 550V. Fig. 9 shows the efficiency of the CLLC converter at resonant frequency under different power conditions. It can be seen that the converter achieves high

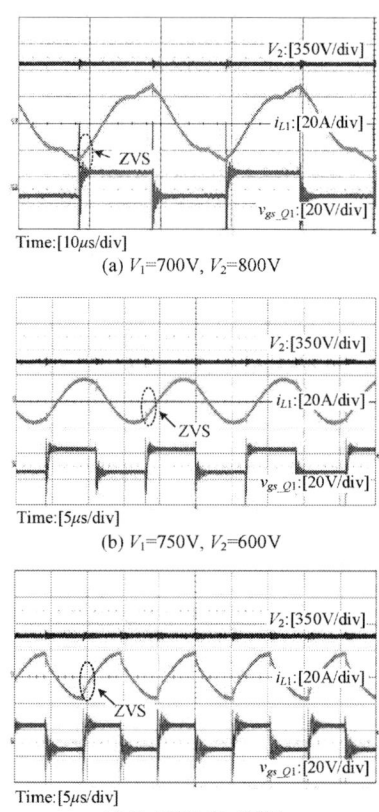

Fig. 8.   Steady state operating waveforms. (a) $f_s>f_r$. (b) $f_s=f_r$. (c) $f_s<f_r$.

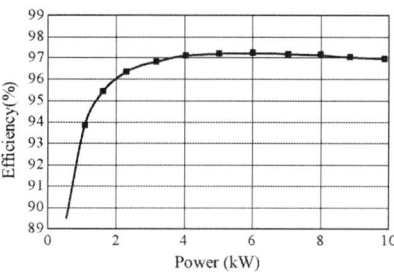

Fig. 9.   Efficiency of the CLLC converter.

power conversion efficiency in full load range. This indicates that the design parameters are perfect, which enable the prototype to achieve soft switching and high efficiency.

V. CONCLUSIONS

The CLLC resonant converter interlinking the DC-link bus and energy storage battery in DC microgrid was briefly introduced in this paper. Then the voltage gain and ZVS region of the CLLC converter are derived and discussed. A simple and effective design methodology of the CLLC converter based on the FHA model is presented. Finally, a 10-kW SiC MOSFET based prototype was designed and built. The prototype interface a 700-800V DC-link bus and a 500-800V battery and the maximum power conversion efficiency is 97.2%.

REFERENCES

[1] T. Dragičević, X. Lu, J. C. Vasquez, and J. M. Guerrero, "DC Microgrids—Part II: A Review of Power Architectures, Applications, and Standardization Issues," *IEEE Trans. Power Electron.*, vol. 31, no. 5, pp. 3528–3549, May 2016.

[2] J. Huang, J. Xiao, C. Wen, P. Wang, and A. Zhang, "Implementation of Bidirectional Resonant DC Transformer in Hybrid AC/DC Micro-Grid," *IEEE Trans. Smart Grid*, vol. 10, no. 2, pp. 1532–1542, Mar. 2019.

[3] Y. Du, S. Lukic, B. Jacobson, and A. Huang, "Review of high power isolated bi-directional dc-dc converters for PHEV/EV dc charging infrastructure," in Proc. *IEEE Energy Convers. Congr. Expo.* , Sep. 2011, pp. 553–560.

[4] K. Wu, C. W. de Silva, and W. G. Dunford, "Stability Analysis of Isolated Bidirectional Dual Active Full-Bridge DC–DC Converter With Triple Phase-Shift Control," *IEEE Trans. Power Electron.*, vol. 27, no. 4, pp. 2007–2017, Apr. 2012.

[5] P. He and A. Khaligh, "Comprehensive Analyses and Comparison of 1 kW Isolated DC–DC Converters for Bidirectional EV Charging Systems," *IEEE Trans. Transport. Electrific.*, vol. 3, no. 1, pp. 147–156, Mar. 2017.

[6] F. Krismer and J. W. Kolar, "Accurate power loss model derivation of a high-current dual active bridge converter for an automotive application," *IEEE Trans. Ind. Electron.*, vol. 57, no. 3, pp. 881–891, Mar. 2010.

[7] J. H. Jung, H. S. Kim, M. H. Ryu, and J. W. Baek, "Design methodology of bidirectional CLLC resonant converter for high-frequency isolation of DC distribution systems," *IEEE Trans. Power Electron.*, vol. 28, no. 4, pp. 1741–1755, Apr. 2013.

[8] C. Zhang, P. Li, Z. Kan, X. Chai, and X. Guo, "Integrated half-bridge CLLC bidirectional converter for energy storage systems," *IEEE Trans. Ind. Electron.*, vol. 65, no. 5, pp. 3879–3889, May 2018.

[9] W. L. Malan, D. M. Vilathgamuwa, and G. R. Walker, "Modeling and control of a resonant dual active bridge with a tuned CLLC network," *IEEE Trans. Power Electron.*, vol. 31, no. 10, pp. 7297–7310, Oct. 2016.

[10] S. Zong, G. Fan, and X. Yang , "Double voltage rectification modulation for bidirectional DC/DC resonant converters for wide voltage range operation, " *IEEE Trans. Power Electron.*, vol. 34, no. 7, pp. 6510–6521, Jul. 2019.

[11] Z. U. Zahid, Z. M. Dalala, R. Chen, B. Chen, and J.-S. Lai, "Design of bidirectional DC–DC resonant converter for Vehicle-to-Grid (V2G) applications," *IEEE Trans. Power Electron.*, vol. 1, no. 3, pp. 232–244, Oct. 2015.

[12] B. Zhao, X. Zhang, and J. Huang, "AI Algorithm-Based Two-Stage Optimal Design Methodology of High-Efficiency CLLC Resonant Converters for the Hybrid AC–DC Microgrid Applications," *IEEE Trans. Ind. Electron.*, vol. 66, no. 12, pp. 9756–9767, Dec. 2019.

Design of a 10kW, High-Frequency Dual Active Bridge Converter Using SiC Devices

Haoyuan Jin
Power Electronics and Renewable
Energy Research Center
Xi'an Jiaotong University
Xi'an, China
jinhaoyuan@stu.xjtu.edu.cn

Huaqing Li
Power Electronics and Renewable
Energy Research Center
Xi'an Jiaotong University
Xi'an, China
lihuaqing@stu.xjtu.edu.cn

Junduo Wen
Power Electronics and Renewable
Energy Research Center
Xi'an Jiaotong University
Xi'an, China
EE_Wjd@stu.xjtu.edu.cn

Chengzi Yang
Power Electronics and Renewable
Energy Research Center
Xi'an Jiaotong University
Xi'an, China
lemonyang@stu.xjtu.edu.cn

Hang Kong
Power Electronics and Renewable
Energy Research Center
Xi'an Jiaotong University
Xi'an, China
konghang@stu.xjtu.edu.cn

Laili Wang
Power Electronics and Renewable
Energy Research Center
Xi'an Jiaotong University
Xi'an, China
LLwang@mail.xjtu.edu.cn

Abstract—**Dual active bridge (DAB) converter is widely used as a dc transformer in high power applications. This paper gives a general design method for high-frequency DAB converter using silicon carbide (SiC) MOSFETs, and gives mathematical analysis for designing the inductor and transformer for the DAB converter. The inductance calculation is critical in the designing process, especially in the high-frequency area, considering the dead-time effect and zero-voltage-switching (ZVS) requirement. It can be divided into four steps, including converter power requirement, inductor energy requirement, dead-time effect requirement and rms of the inductor current. A 10kW, 100kHz DAB converter using SiC MOSFETs is built, achieving ZVS turn-on successfully. The efficiency is up to 97.1% at 10kW, and the power density is 4.18W/cm³.**

Keywords—*dual active bridge (DAB), silicon carbide (SiC), inductance, zero-voltage-switching (ZVS), dead-time.*

I. INTRODUCTION

With the rapid development of wide bandgap power semiconductor devices, dc transmission plays a more and more significant role in electrical power systems because of its low cost, high efficiency and high reliability. The dc transformer is a key link in the dc transmission process, which realizes the power transformation. In general, the transmission power is proportional to the number of power semiconductor switches, so the DAB converter is widely used as a dc transformer in high power applications[1].

DAB was proposed by De Doncker in 1991[2]. It compared three soft-switched dc/dc converters and DAB has the most favorable characteristics. The topology of the DAB converter is shown in Fig. 1. It consists of two full bridges, an inductor, and a transformer. The transformer, with turns ratio n, not only can isolate the voltage between the input side and output side, but also change the amplitude of the voltage to meet the ZVS turn-on condition. The inductor L, including the series inductor L_s and the leakage inductor of the transformer L_σ, is a main component in the designing process, which transfers the energy[2].

There are different control methods for DAB converter, e.g., single-phase-shift (SPS) control[3], dual-phase-shift (DPS) control[4], extended-phase-shift (EPS) control[5] and

PWM control[6]. SPS control is widely used when the working condition is constant. The condition to achieve soft-switching operation considering the parasitics, including the transformer magnetizing inductance and device snubber capacitances, was analyzed in [7]. The conclusion is based on that the device snubber capacitances keep constant during the switching transient. The complete ZVS analysis which considered the nonlinear parasitic MOSFET capacitances was given in [8], including the SPS control and DPS control. It used a constant voltage source for the secondary side. In addition, the dead-time will also influence the ZVS condition. When the dead-time is short, the drain-source voltage will not decrease to zero[8]. When the dead-time is long, or the inductance is small, voltage polarity reversal and voltage sag phenomena may appear[9, 10], causing non-ZVS turn-on.

The traditional high power DAB converter often used silicon devices, which cannot achieve high power density.[3] built a 10kW DAB converter, and the switching frequency is less than 20kHz, which will use a big transformer. With the development of SiC devices, the operating frequency is high, decreasing the volume of the magnetic components. Many recent papers built DAB operating less than 100kHz, the phase-shift-time of which is much longer than the dead-time. However, when the frequency is high, the dead-time effect can't be neglected. The restricted conditions of the inductance considering the soft switching requirement have not been discussed in the previous paper, especially the dead-time effect requirement.

The objective of this work is to design a 10kW DAB converter using SiC devices, which operate in ZVS turn-on condition at 100kHz, and give mathematical analysis for designing the inductor and transformer for a high power density DAB converter. Section I introduced the application background and research status. Section II gives the principle of SPS control. Section III gives the specifications of the designed converter, chooses the SiC MOSFETs, and calculates the conduction and switching loss. Section IV is the major contribution of this paper, which gives the restricted conditions of the inductance and dead-time, as well as the core selection, wire selection, and loss calculation for the inductor and transformer. Section V gives the simulation and experiment results. Finally, Section VI gives the conclusion.

The work was supported by 2018 steadily supports research projects from Key Laboratory of National Defense Science and Technology. Project number: 614221720180401.

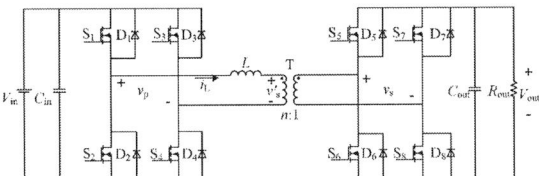

Fig. 1. Topology of DAB converter.

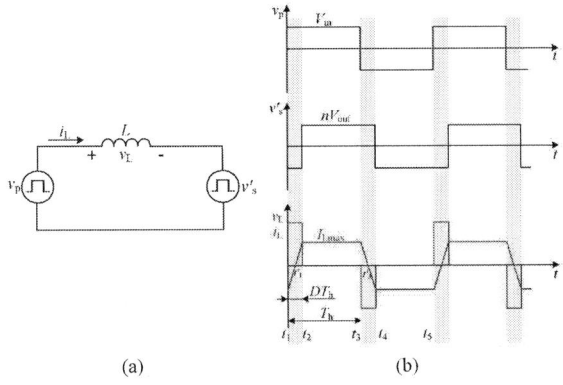

Fig. 2. The basic principle of DAB. (a) Simplified model. (b) Working principle of SPS control.

II. BASIC PRINCIPLE OF THE CONVERTER

Fig. 2(a) shows the simplified model of the DAB converter with phase shift modulation. Fig. 2(b) shows the working principle of SPS control. The gate drive signal is 50% duty cycle for all of the switches. There is no inner phase shift angle in two bridges, but an outer phase difference between them, so-called phase-shift ratio D[3], as shown in Fig. 2(b). If $0<D<1$, the power will flow from the primary side to the secondary side. The operating process is discussed in [3] in detail, including heavy load and light load conditions.

If the input and output voltage matches with the transformer's turns ratio, in other words,

$$k = \frac{V_{in}}{nV_{out}} = 1 \tag{1}$$

the converter can achieve soft switching in the full control range in the ideal case. Hence, $k=1$ is usually chosen as a convenient design point[7]. In this condition, the inductor current is a trapezoid waveform at the steady-state, and there is no dc bias[11] because of the series line resistor in the circuit.

The supplied power of the converter is[3]

$$P = \frac{1}{T_h} \int_0^{T_h} v_p i_L \, dt = \frac{nV_{in}V_{out}}{2f_sL} D(1-D) \tag{2}$$

where T_h is half of the switching cycle, f_s is the switching frequency, and $T_h=1/(2f_s)$. D can be calculated from (2). The maximum and rms value of the inductor current can be derived using Fig. 2. For the condition $V_{in}=nV_{out}$, it is [12]

$$I_{Lmax} = \frac{V_{in}}{4f_sL}\left(1 - \sqrt{1 - \frac{8Pf_sL}{V_{in}^2}}\right) \tag{3}$$

$$I_{Lrms} = \frac{\sqrt{6}}{12f_sL}\sqrt{V_{in}^2 - (V_{in}^2 + 4Pf_sL)\sqrt{1 - \frac{8Pf_sL}{V_{in}^2}}} \tag{4}$$

TABLE I. SPECIFICATIONS OF DAB CONVERTER

Parameter	Value
Input voltage V_{in}	1000V
Output voltage V_{out}	1000V
Output current I_{out}	10A
Switching frequency f_s	100kHz

III. SELECTION OF POWER SEMICONDUCTOR DEVICES

TABLE I shows the specifications for the converter. Compared with Si devices, SiC MOSFETs have lower on-state resistance, especially in the high blocking voltage area. In addition, the switching loss is lower, which makes SiC MOSFETs operate at a higher frequency. To meet the requirement of the 1kV application, 1.7kV SiC MOSFETs C2M0080170P are selected, which have approximately 80mΩ on-state resistance. To achieve high power density, the body diodes of the MOSFETs are used.

In DAB converter, the current of the MOSFETs can flow from source to drain, e.g. S5 endures reverse current during t_2 and t_3'. [13] analyzed current path for SiC MOSFETs in synchronous rectification mode. When the total voltage drop across the device is smaller than the voltage drop of the body diode, the current flows only through the channel of the MOSFET. For this MOSFET, it means the reverse current is less than 51A, which meets this condition. In other words, the body diode will conduct only in the dead-time. As a result, the conduction loss of each MOSFET is

$$P_{con} = 0.5I_{Lrms}^2 R_{ds(on)} \tag{5}$$

where I_{Lrms} is given in (4), $R_{ds(on)}$ is the on-state resistance of the MOSFET. As for the switching loss, the converter only has turn-off loss if $k=1$. In this case, the turns ratio n of the transformer is 1. The switching loss for each MOSFET can be estimated using $E_{off,nom}$, given in the datasheet of the MOSFET.

To design or choose the heatsink, the total loss for each MOSFET, including the conduction loss and switching loss, should be calculated after the inductance L is confirmed.

IV. DESIGN OF THE INDUCTOR AND TRANSFORMER

The design of the magnetic component is a critical process for the converter, including the inductor and transformer design for DAB. This section will give a general process to design them when DAB works in SPS control and $k=1$ mode.

A. Calculation of the Inductance

The inductor should be designed carefully in order to achieve the required power, meet the ZVS turn-on condition, and have lower power loss. It can be divided into four steps.

1) Converter power requirement

The supplied power of the converter is given in (2), when the phase-shift ratio $D=0.5$, it reaches the maximum power

$$P_{max} = \frac{nV_{in}V_{out}}{8f_sL} > V_{out}I_{out} \tag{6}$$

Using the specifications in TABLE I, the maximum value of the inductance $L_{max}=125.0\mu H$. Considering the efficiency, the maximum value should be lower than that.

2) Inductor energy requirement

In order to achieve ZVS turn-on, the inductor energy at the instant of turn-off should be large enough to charge and discharge the output capacitances of the MOSFETs[7]. There

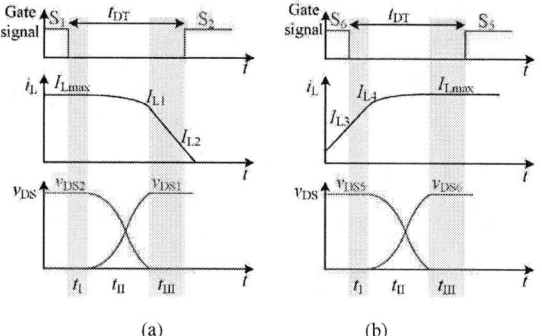

(a)   (b)

Fig. 3. Waveform during the transient process. (a) t_3 moment. (b) t_2 moment.

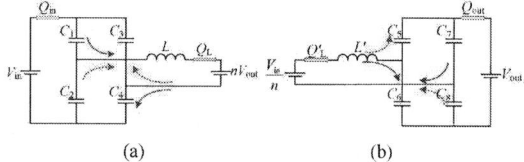

(a)   (b)

Fig. 4. Equivalent circuit during t_{II} process. (a) t_3 moment. (b) t_2 moment.

are four half-bridges and each has two current commutating processes. Because of the symmetry of the converter, it is only necessary to analyze two current commutating processes, one for a half-bridge in the primary side, and the other for the secondary side.

Fig. 3 shows the waveform during the transient switching process. Fig. 3(a) is the current commutating process for the primary side, corresponding to the time t_3 in Fig. 2(b). Fig. 3(b) is for the secondary side, corresponding to the t_2 moment. Such a process can be divided into three periods. t_I indicates the time delay difference between turn-on and turn-off. For the driver and device, it includes the gate delay time and the switching time[14]. After that, the channel of the MOSFET is turned off, and it can be considered as a capacitor paralleled with a diode. In t_{II} period, the output capacitor resonates with the inductor, and the drain-source voltage changes. In t_{III} period, the anti-parallel diode clamps voltage, and the current flows from it. The turn-on signal of the other MOSFET is generated during t_{III}, so it can achieve ZVS turn-on.

The inductor energy requirement means that, during t_{II}, the drain-source voltage of one MOSFET should achieve zero. The equivalent circuit during t_{II} is shown in Fig. 4. The current changes from the blue solid path to the red dotted one. For Fig. 4(a), the total charge flowing through V_{in} is zero, and flowing into the equivalent voltage source nV_{out} is $Q_L = 2Q_{oss}(V_{in})$. The minimum inductor energy for ZVS turn-on is[8]

$$\frac{1}{2}LI_{Lmax}^2 \geq 2Q_{oss}(V_{in}) \cdot nV_{out} \tag{7}$$

where

$$Q_{oss}(V_{in}) = \int_0^{V_{in}} C_{oss}(v)dv \tag{8}$$

For Fig. 4(b), the total charge flowing through V_{out} is zero and flowing out of the equivalent source V_{in}/n is $2Q_{oss}(V_{out})$. Therefore, the inductor energy increases, i.e., the secondary side can always satisfy the inductor energy requirement if $D>0$.

For the selected MOSFET C2M0080170P, using (8), the charge $Q_{oss}(1000V)$ is 181.9nC after curve fitting the C_{oss} plot

in the datasheet. For the case $k=1$, combining (3) and (7), such condition can be derived

$$L > \frac{4Q_{oss}(V_{in})V_{in}}{\left[\dfrac{P}{V_{in}} + 8f_s Q_{oss}(V_{in})\right]^2} \tag{9}$$

Using the specifications in TABLE I, the minimum value of the inductance $L_{min}=7.1\mu H$.

3) Dead-time effect requirement

An appropriate dead-time is necessary for the bridge circuit. During the dead-time, the current will flow through the body diode of the MOSFET instead of the channel, which will increase the switching loss. In addition, there are also many other phenomena caused by the dead-time, such as voltage polarity reversal and voltage sag phenomena, etc[9]. However, if the dead-time is too short, the converter will not work in the ZVS mode, which may even cause the shoot-through problem. In consequence, the deadtime should be ensured first.

a) Set the dead-time

The minimum dead-time can be considered as t_I+t_{II} according to Fig. 3. The time t_I is set to avoid the shoot-through problem in a bridge. To be simplified, it can be considered as the maximum turn-off delay time minus the minimum turn-on delay time, as

$$t_I = (t_{PDHL(max)} + t_{d(off)} + t_f) - (t_{PDLH(min)} + t_{d(on)} + t_r) \tag{10}$$

where $t_{PDHL(max)}$, $t_{PDLH(min)}$ is the gate driver's maximum propagation delay time for falling edge, the minimum time for rising edge, separately. $t_{d(off)}$, t_f, $t_{d(on)}$, and t_r are the MOSFET's turn-off delay time, fall time, turn-on delay time, and rise time, separately. The gate driver is UCC21530, so the time t_I is appropriate 34ns according to (10). It should be noted that this time is related to the gate driver resistance, MOSFET's voltage and current, etc., and this value is considered as the maximum value, in order to choose an adequate dead-time.

The time t_{II} is set to ensure the drain-source voltage of MOSFETs can drop to zero before the turn-on signal. For the primary side, if the inductor energy requirement is satisfied, the inductor current at the end of t_{II} is[8]

$$I_{L1} = \sqrt{I_{Lmax}^2 - \frac{4}{L}Q_{oss}(V_{in}) \cdot nV_{out}} \tag{11}$$

The accurate calculation for t_{II} should consider the C_{oss} and i_L changing with time[8]. In the case of SPS control and $k=1$, i_L won't have severe changes, which can be considered as the average of I_{Lmax} and I_{L1}. t_{II} can be reckoned as

$$t_{II} = \frac{4Q_{oss}(V_{in})}{I_{Lmax} + I_{L1}} \tag{12}$$

For the inductance L ranges from 7.1μH to 125μH, the relation curve of the resonant time t_{II} and inductor current i_L varied with L is shown in Fig. 5. With L increases, the maximum inductor current I_{Lmax} and the current after resonant I_{L1} both increase, while the reduced value during this process decreases. When L is at the minimum value, I_{L1} is nearly zero. In addition, the time t_{II} is shorter with the increase of L. When $L=7.1\mu H$, the period t_{II} is 67ns. The dead-time is supposed to have some margins and can be set as

$$t_{DT} = 1.5(t_I + t_{II}) \tag{13}$$

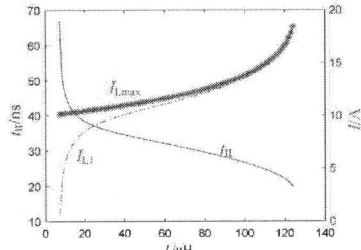

Fig. 5. Relation curve of the resonant time t_{II} and inductor current varied with the inductance L.

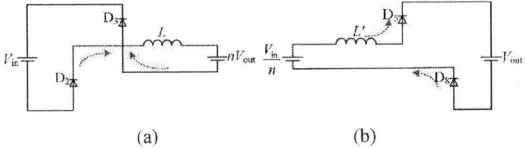

(a)　　　　　　　　　　　(b)

Fig. 6. Equivalent circuit during t_{III} process. (a) t_3 moment. (b) t_2 moment.

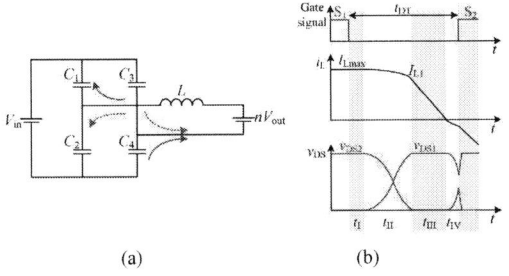

(a)　　　　　　　　　　　(b)

Fig. 7 The condition if the dead-time is too long for t_3 moment. (a) Equivalent circuit in t_{IV} period. (b) Waveform.

For the secondary side, i_L increases during t_{II}, and the waveform is symmetric with the primary side. Thus, the supposed dead-time is the same.

b) Inductance range to avoid the dead-time effect

The resonant process of the inductor and capacitors stops at the end of t_{II}. During t_{III} period, the diodes conduct current. The equivalent circuit during t_{III} process is shown in Fig. 6. In Fig. 6(a), after v_{DS2} and v_{DS3} drop to zero, the diodes D_2 and D_3 conduct, clamping v_{DS2} and v_{DS3} to approximately zero. In this case, v_L is a negative constant value, and i_L is decreasing linearly. For the secondary side, as shown in Fig. 6(b), v_L is zero and i_L keeps constant for the case $k=1$.

For Fig. 6(a), S_2 and S_3 should turn on before i_L changes direction. Otherwise, the ZVS turn-on will not fulfill. Fig. 7 shows the condition that the dead-time is too long at t_3 moment. At the end of t_{III} period, i_L becomes zero. After that, the diodes D_2 and D_3 will not conduct current and clamp the voltage. The equivalent circuit for period t_{IV} is shown in Fig. 7(a). The inductor current changes direction, charging C_2, C_3, and discharging C_1, C_4. The voltage v_{DS2} is rising towards V_{in} during t_{IV}, as shown in Fig. 7(b). Consequently, the MOSFETs in the primary side will be non-ZVS turn-on.

For Fig. 6(b), the inductor current stays constant and keeps positive. Even the dead-time is long, v_{DS5} and v_{DS8} are still clamped to zero by the diodes. Thus, the MOSFETs in the secondary side will always achieve ZVS turn-on.

To avoid the dead-time effect for the primary side, one of the necessary conditions for ZVS turn-on is

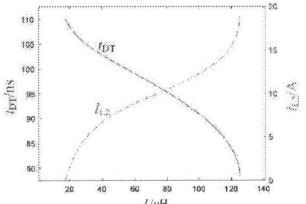

Fig. 8. Relation curve of the dead-time and inductor current after the dead-time varied with the inductance.

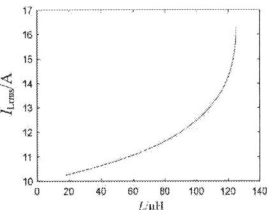

Fig. 9. Relation curve of the inductor's rms current I_{Lrms} varied with the inductance.

$$I_{L2} = I_{L1} - \frac{2V_{in}^{'}}{L} t_{III} > 0 \tag{14}$$

To get the minimum value of the inductance, t_{III} should be considered as the maximum value, i.e., the time delay t_I is considered as zero. Accordingly, the inductance range to avoid the dead-time effect is

$$L > \frac{2V_{in}(t_{DT} - t_{II})}{I_{L1}} \tag{15}$$

Based on the relationship between L various with t_{II} and I_{L1} given in Fig. 5, the inductance range satisfied (15) is 17.5μH<L<125.0μH. The relation curve of the dead-time and inductor current after the dead-time I_{L2} varied with the inductance is shown in Fig. 8. With the increase of L, the supposed dead-time decreases, while I_{L2} increases. When L is at the minimum value 17.5μH, I_{L2} is approximately zero, which is a critical situation for ZVS turn-on. Generally, the minimum value should be larger than it, considering the loss factor.

4) Rms value of the inductor current

The previous limitations give a possible range of the inductance. The optimal solution should consider the loss factor, which is related to the rms value of i_L. The expression of the rms value is given in (4). The relationship between I_{Lrms} and L is shown in Fig. 9. The larger the inductance, the higher the loss. Safe operation area (SOA) design is also important[3], which is according to the I_{Lmax} given in Fig. 5. Finally, the inductance is chosen as 25μH, and the corresponding dead-time is 107ns according to Fig. 8.

B. Design of the Inductor

The magnetic component design includes the wire selection and core selection. For the high-frequency DAB converter, the current has no dc bias, the amplitude for ac current is high, so it is necessary to use Litz wire. The effective frequency for the current is[15]

$$f_{eff} = \frac{\mathrm{rms}\left\{\dfrac{di(t)}{dt}\right\}}{2\pi I_{Lrms}} = \frac{V_{Lrms}}{2\pi L I_{Lrms}} \tag{16}$$

978-1-6654-4817-8/21 $31.00 © 2021 IEEE　　　431

where V_{Lrms} is the rms value of the inductor voltage, which can be derived from Fig. 2(b). For $k=1$,

$$V_{Lrms} = \sqrt{\frac{1}{T_h} \int_0^{DT_h} (2V_{in})^2 dt} = 2V_{in}\sqrt{D} \quad (17)$$

When L=25μH, D=0.0528 according to (2), I_{Lrms}=10.37A according to (4), the effective frequency is 282kHz according to (16). The skin depth is 0.143mm at 100°C. The strand diameter is chosen as 0.05mm, which is approximately 1/3 of the skin depth as a rule. The number of strands is chosen as 1200 according to the total area of the wire and I_{Lrms}.

For the core working in high frequency, ferrite is the most widely used material. In this application, material N87 is used, whose optimum frequency range is 25kHz to 500kHz. The shape of the core should be chosen according to the loss. Through trial and error, ETD59/31/22 is chosen. The gap for the core is 2mm, and the number of turns N is 9. The maximum magnetic flux density is

$$B_{max(L)} = \frac{LI_{Lmax}}{NA_{e(ETD59)}} \quad (18)$$

where I_{Lmax}=10.56A according to (3). $A_{e(ETD59)}$ is the effective area of the core. As a result, $B_{max(L)}$=80mT, and the inductor does not saturate. The shape of the magnetic flux density is the same as the inductor current waveform. For nonsinusoidal waveform, the "improved generalized Steinmetz equation" (iGSE)[16] can be used to calculate the core loss. Using the parameter of N87 material and iGSE equation, the power loss per unit volume is 272mW/cm³, and the core loss for the inductor is approximately 13.9W.

The wire loss of the inductor can be calculated as

$$P_w = F_R I_{Lrms}^2 R_{dc} \quad (19)$$

where F_R is a proportion of ac resistance with dc resistance given in [15], R_{dc} is the dc resistance of the Litz wire. In this application, $F_R \approx 1.14$, $R_{dc} \approx 6.4$mΩ. Thus, $P_w \approx 0.78$W. The total loss of the inductor $P_L \approx 14.7$W.

C. Design of the transformer

The selection of the wire for the transformer is 0.05×1200-Litz wire as well. The selection of the core should also use the trial and error method. Two sets of EE65/32/27 are used for the transformer according to the saturated magnetic flux density and the winding space limitation. The magnetic flux density of the transformer, derived from Faraday's law, is a triangular waveform, and the maximum value is

$$B_{max(T)} = \frac{V_{in}}{4N_1 f_s A_{e(EE65)}} \quad (20)$$

where $A_{e(EE65)}$ is 10.8cm² for 2 set of core. For the turns N_1=N_2=18, $B_{max(T)}$ is 129mT, which does not saturate. By using iGSE, the core loss is 13.7W. The wire loss calculation is similar to the inductor's, and is approximately 10.8W. As a result, the total loss of the transformer $P_T \approx 24.5$W.

V. Simulation and Experimental Results

A. Simulation Results

The simulation results are based on LTspice, using the spice model of MOSFET C2M0080170P. The inductance used in the simulation is 25μH, and the dead-time is 107ns.

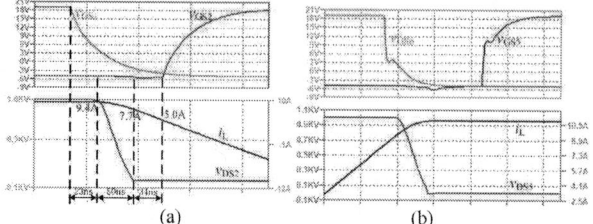

Fig. 10. ZVS transient simulation waveforms. (a) t_3 moment. (b) t_2 moment.

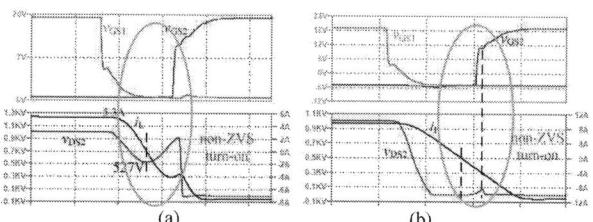

Fig. 11. Transient simulation waveform at t_3 moment when (a) L=5μH. (b) L=13μH.

The phase-shift time is 264ns according to (2). The waveforms for ZVS transient are shown in Fig. 10.

For the t_3 moment, v_{DS2} starts to change after 23ns when the turn-off signal of S_1 is generated, which is smaller than the reckoned maximum value using (10). In the next 50ns, v_{DS2} drops to zero, while i_L decreases from 9.4A to 7.7A. The final value of the inductor current I_{L1} satisfies (11), which uses the initial i_L of this period. The time interval is 50ns, a little longer than the calculated time according to (12), which is 43ns. This is because the premise of the calculation is i_L changes linearly, causing an acceptable error for the dead-time setting. For the remaining part of the dead-time, v_{DS2} keeps zero and i_L continues decreasing. According to (14), i_L after this period is 5.0A, conforming to the simulation value. Then, the turn-on signal of S_2 is generated, accomplishing the ZVS turn-on. For the t_2 moment, i_L increased when v_{DS5} changes, and then it stays constant, which ensures the ZVS turn-on condition no matter with the inductance.

Transient simulation waveform at t_3 moment when L=5μH is shown in Fig. 11(a). In this situation, the phase-shift time is 50ns. The inductance, less than 7μH, is not satisfied with the inductor energy requirement. The inductor current at the beginning of the resonant process is 5.3A. This energy is not enough to charge and discharge the MOSFETs' output capacitors. Therefore, the minimum v_{DS2} value is 527V when the inductor energy is released totally. In this circumstance, the primary MOSFETs can not achieve ZVS turn-on no matter how long the dead-time is.

Transient simulation waveform at t_3 moment when L=13μH is shown in Fig. 11(b). This inductance can satisfy the inductor energy requirement, but not the dead-time effect requirement. The dead-time is set as 114ns according to (13). After i_L drops to zero, the deadtime is not ended, causing reverse charging for v_{DS2}. Therefore, the primary MOSFETs can not achieve ZVS turn-on.

B. Experiment Results

The experimental prototype is shown in Fig. 12. The dimension is 29cm×12.9cm×6.4cm=2394cm³. The calculated total power density of the converter is 4.18W/cm³. The converter is controlled by DSP TMS320F28335, and the

(a)   (b)

Fig. 12. Experimental prototype. (a) Front side. (b) Bottom side.

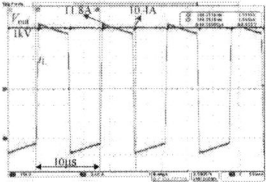

Fig. 13. Steady-state experiment waveform.

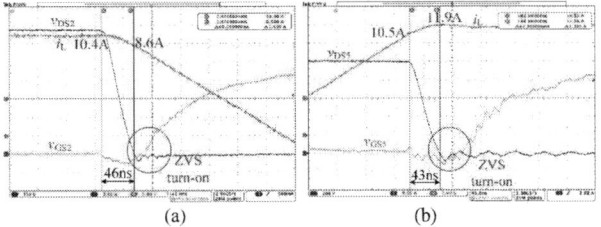

(a)   (b)

Fig. 14. Transient experiment waveforms. (a) Primary side. (b) Secondary side.

phase-shift time is calculated by a digital PI controller. The series inductance is 24.6μH, and the leakage inductance of the transformer is 4.0μH, measured by Agilent E4980A. Thus, the total inductance L=28.6μH. The dead-time is set as 107ns.

The steady-state waveform is shown in Fig. 13. The output voltage is 1kV, and the inductor current is an ac waveform with no dc bias. Different from Fig. 2(b), i_L slightly decreases when v_p and v_s are all positive. There are mainly two reasons. Firstly, the turns ratio of the transformer is not exactly 1. Secondly, due to the voltage drop in the MOSFETs, the amplitude of v_p will be lower than V_{in}, while v_s will be higher than V_{out}. This will result in the inductor voltage v_L being negative at that time, causing i_L to decrease.

The ZVS turn-on process is shown in Fig. 14. Fig. 14(a) is the transient process for the primary side at t_3 moment. The voltage v_{DS2} drops to zero in 46ns, while i_L drops from 10.4A to 8.6A. Then, MOSFET S_2 turns on before i_L drops to zero, achieving ZVS turn-on. Fig. 14(b) is for the secondary side at t_2 moment. i_L increases during the resonant process and stays constant after that, which ensures the ZVS turn-on. However, the calculated variation value for i_L using (11) is a little smaller than in the experiment. This is mainly because the calculation is based on the assumption that the loss during the resonant process is zero, which will cause some small errors. The measured efficiency is up to 97.1% at 10kW.

VI. CONCLUSION

In this paper, a universal design method for high frequency and high power density DAB converter is discussed. The main contribution of this paper is the inductance design method in the high-frequency condition, which can be divided into four steps. The maximum inductance is decided by the converter power requirement, and the minimum inductance is decided

by the inductor energy requirement, as well as the dead-time effect requirement. The current stress and power loss should be considered to choose the optimal inductance within the aforementioned range. An experimental prototype of a 10kW, 100kHz DAB converter is built. The converter can achieve ZVS turn-on successfully. The efficiency is up to 97.1% at 10kW, and the power density is 4.18W/cm^3.

REFERENCES

[1] B. Zhao, Q. Song, W. Liu, and Y. Sun, "Overview of Dual-Active-Bridge Isolated Bidirectional DC–DC Converter for High-Frequency-Link Power-Conversion System," *IEEE Transactions on Power Electronics*, vol. 29, no. 8, pp. 4091-4106, 2014.

[2] R. W. A. A. D. Doncker, D. M. Divan, and M. H. Kheraluwala, "A three-phase soft-switched high-power-density dc/dc converter for high-power applications," *IEEE Transactions on Industry Applications*, vol. 27, no. 1, pp. 63-73, Jan.-Feb. 1991.

[3] C. Mi, H. Bai, C. Wang, and S. Gargies, "Operation, design and control of dual H-bridge-based isolated bidirectional DC–DC converter," *IET Power Electronics*, vol. 1, no. 4, 2008.

[4] B. Hua and C. Mi, "Eliminate Reactive Power and Increase System Efficiency of Isolated Bidirectional Dual-Active-Bridge DC–DC Converters Using Novel Dual-Phase-Shift Control," *IEEE Transactions on Power Electronics*, vol. 23, no. 6, pp. 2905-2914, 2008.

[5] B. Zhao, Q. Yu, and W. Sun, "Extended-Phase-Shift Control of Isolated Bidirectional DC–DC Converter for Power Distribution in Microgrid," *IEEE Transactions on Power Electronics*, vol. 27, no. 11, pp. 4667-4680, 2012.

[6] A. K. Jain and R. Ayyanar, "Pwm control of dual active bridge: Comprehensive analysis and experimental verification," *IEEE Transactions on Power Electronics*, vol. 26, no. 4, pp. 1215-1227, 2011.

[7] M. H. Kheraluwala, R. W. Gascoigne, D. M. Divan, and E. D. Baumann, "Performance Characterization of a High-Power Dual Active Bridge dc-to-dc Converter," (in English), *IEEE Transactions on Industry Applications*, Article vol. 28, no. 6, pp. 1294-1301, Nov-Dec 1992.

[8] Y. Yan, H. Gui, and H. Bai, "Complete ZVS Analysis in Dual Active Bridge," *IEEE Transactions on Power Electronics*, vol. 36, no. 2, pp. 1247-1252, 2021.

[9] B. Zhao, Q. Song, W. Liu, and Y. Sun, "Dead-Time Effect of the High-Frequency Isolated Bidirectional Full-Bridge DC–DC Converter: Comprehensive Theoretical Analysis and Experimental Verification," *IEEE Transactions on Power Electronics*, vol. 29, no. 4, pp. 1667-1680, 2014.

[10] Y. Yan, H. Bai, A. Foote, and W. Wang, "Securing Full-Power-Range Zero-Voltage Switching in Both Steady-State and Transient Operations for a Dual-Active-Bridge-Based Bidirectional Electric Vehicle Charger," *IEEE Transactions on Power Electronics*, vol. 35, no. 7, pp. 7506-7519, 2020.

[11] B. Zhao, Q. Song, W. Liu, and Y. Zhao, "Transient DC Bias and Current Impact Effects of High-Frequency-Isolated Bidirectional DC–DC Converter in Practice," *IEEE Transactions on Power Electronics*, vol. 31, no. 4, pp. 3203-3216, 2016.

[12] B. Zhao, Q. Song, W. H. Liu, and Y. D. Sun, "A Synthetic Discrete Design Methodology of High-Frequency Isolated Bidirectional DC/DC Converter for Grid-Connected Battery Energy Storage System Using Advanced Components," (in English), *IEEE Transactions on Industrial Electronics*, vol. 61, no. 10, pp. 5402-5410, Oct 2014.

[13] Z. Wang and A. Castellazzi, "Device loss model of a fully SiC based dual active bridge considering the effect of synchronous rectification and deadtime," presented at the 2017 IEEE Southern Power Electronics Conference (SPEC), 2017.

[14] J. Li, Z. Chen, Z. Shen, P. Mattavelli, J. Liu, and D. Boroyevich, "An Adaptive Dead-time Control Scheme for High-Switching-Frequency Dual-Active-Bridge Converter," presented at the 2012 Twenty-Seventh Annual IEEE Applied Power Electronics Conference and Exposition (APEC), 2012.

[15] C. R. Sullivan and R. Y. Zhang, "Simplified design method for litz wire," presented at the 2014 IEEE Applied Power Electronics Conference and Exposition - APEC 2014, 2014.

[16] K. Venkatachalam, C. R. Sullivan, T. Abdallah, and H. Tacca, "Accurate prediction of ferrite core loss with nonsinusoidal waveforms using only Steinmetz parameters," presented at the 2002 IEEE Workshop on Computers in Power Electronics, 2002. Proceedings., 2002.

Evaluating Switching Performance of GaN HEMT Using Analytical Modeling

Yingzhe Wu[1]*, Shan Yin[2,3]**, Hui Li[2], Minghai Dong[2], Xi Liu[2], and Yuhua Cheng[1]

[1] School of Automation Engineering, University of Electronic Science and Technology of China, Chengdu, China
[2] School of Aeronautics and Astronautics, University of Electronic Science and Technology of China, Chengdu, China
[3] Innoscience (Zhuhai) Techonolgy Co., Ltd., Zhuhai, China
Email: microuestc@163.com*; shanyin@innoscience.com**

Abstract—Evaluating switching performance of the gallium nitride high electron mobility transistor (GaN HEMT) is critical for its application in power converter/inverter. In this work, we have elaborated an analytical model aims to comprehensively estimate switching characteristics of GaN HEMT. The model has considered nonlinear junction capacitance as well as forward and reverse transconductance, skin effect during oscillation period, and interactions between high- and low-side GaN HEMTs. As a result, the switching losses, switching oscillation, and crosstalk issue during switching transient can be correctly reflected together. Comparisons between measured and calculated waveforms manifest that the model can evaluate switching performance of the GaN HEMT in sufficient accuracy when compared with other models. Consequently, the model can give many valuable design references in applying GaN HEMT.

Index Terms—Gallium nitride high electron mobility transistor (GaN HEMT), switching loss, switching oscillation, crosstalk, analytical modeling.

I. INTRODUCTION

With the development of the wide bandgap semiconductor device, the gallium nitride high electron mobility transistor (GaN HEMT) promises rapid switching speed, lower switching losses as well as higher switching frequency. As a result, the GaN HEMT enables higher density and efficiency compared with the silicon (Si) devices [1]–[3], which is critical for converter design. Despite advantages described above, the GaN HEMT will also cause some detrimental effects, such as aggravated oscillation during switching transient, which may seriously impact reliability and safety of the device in applications [4], [5]. Moreover, the GaN HEMT is more susceptible to crosstalk due to its lower threshold voltage ($v_{gs,th}$) [6]. Therefore, comprehensive evaluating switching performance of the GaN HEMT is required for power converter design.

As an effective method for assessing switching performance of the semiconductor devices, analytical modeling has been widely adopted to investigate switching characteristic of the GaN device in aspects of switching losses, voltage/current oscillation, and crosstalk [7]–[12]. In the works of [7], [8], switching losses have been estimated, which is critical for efficiency and density of the converter. Oscillations during switching transient has been studied in [5], [9], which is responsible for the high-frequency electromagnetic interference (EMI). The crosstalk occurs between phase-legs impacts reliability and safety of the GaN HEMT significantly, and

has been analyzed in [10]–[12]. In [13], the gate overshoot during the turn-on transition has also been discussed using analytical modeling. However, there are few works focused on investigating switching performance of the GaN HEMT by considering switching losses, switching oscillation, and crosstalk together.

In this work, we have elaborated an analytical model aims to comprehensively assess switching characteristics of GaN HEMT. The whole switching process of the device has been well demonstrated with elaborated circuit equation deduction. In order to correctly reflect voltage/current slew rate and switching losses, the model has considered non-linearity of junction capacitance, forward (g_{fs}) and reverse (g_{rs}) transconductance. The skin effect during ringing period has been properly considered, which is critical to high-frequency EMI. Meanwhile, interactions between high- and low-side GaN HEMTs have also been paid sufficient attentions. Therefore, the analytical model can estimate switching characteristic of the GaN HEMT in aspects of switching losses, voltage/current oscillation, and crosstalk.

The rest of this paper is organized as follows. The switching process of the GaN HEMT has been analyzed in Section II, and the modeling of switching process has been elaborated in Section III. Modeling of the nonlinear junction capacitance as well as forward and reverse transconductance are demonstrated in Section IV. In Section V, experiment verification is presented. Finally, the conclusions as well as future works have been summarized in Section VI.

II. SWITCHING PROCESS OF THE GAN HEMT

A typical inductive clamped circuit (see Fig. 1) is adopted to study switching characteristics of the GaN HEMT. In this situation, M_1 operates in hard switching mode, and can be considered as control device, while M_2 switches in freewheeling mode, and can be viewed as synchronous device. Thus, the switching characteristics are mainly determined by switching transition of M_1. In Fig. 1, $C_{gd1,2}$, $C_{gs1,2}$, and $C_{ds1,2}$ refer to gate-drain, gate-source, as well as drain-source capacitances, $L_{d1,2}$, $L_{s1,2}$, and $L_{g1,2}$ represent drain inductance, source inductance, and gate inductance, R_{g12}, $R_{1,2}$ $L_{1,2}$, and L_{loop} denote gate resistance, stray resistance, stray inductance, and loop inductance.

978-1-6654-4817-8/21 $31.00 © 2021 IEEE

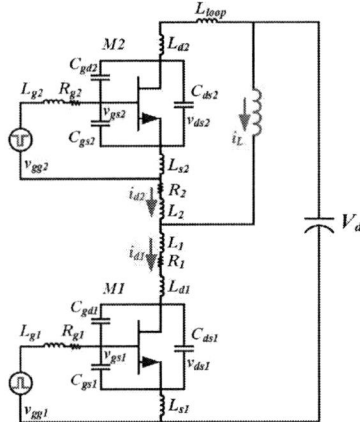

Fig. 1. Typical inductive clamped circuit based on GaN HEMT.

Fig. 2 illustrates typical switching waveforms of M_1 (v_{gs}, v_{ds}, and i_d) during switching transients. Based on operation principle of the device, the switching transients can be divided into 6 stages. During turn-on transient, it should be noticed that the v_{ds1} may fall below on-state voltage ($v_{ds(on)}$) before i_{d1} attains to load current (i_L) when GaN HEMT operates in low voltage condition [14]. During turn-off transient, it should be noted that the sustainable oscillation may occur over the duration from t_{10} to t_{13} (see Fig. 2 (b)) because the damping ratio of the system approaches to 0 [15]. Despite analyzing sustained oscillation is out of scope of this paper, the parasitic parameters (such as $R_{g1,2}$, $L_{s1,2}$ and L_{loop}) should be properly selected to avoid such undesired ringing.

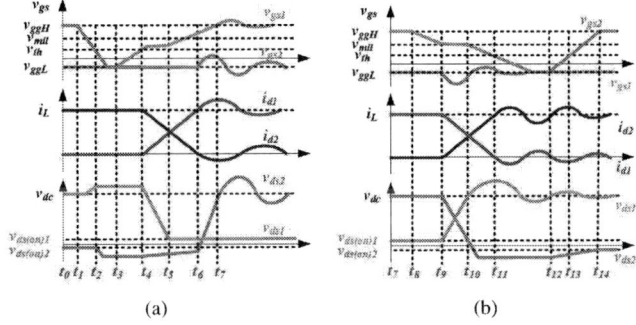

Fig. 2. Typical waveform of the GaN device (M_1) during switching transient, (a) Turn-on, (b) Turn-off.

III. MODELING OF GaN DEVICE DURING ON AND OFF TRANSITIONS

A. Turn-on Transient

1) Stage 0 ($t_0 \sim t_1$): In this stage, M_1 is in cut-off region, and M_2 is in on-state region. As a result, it can be acquired that: $v_{gs1} = v_{ggL}$, $v_{ds1} = v_{dc} + R_{ds(on)} \cdot i_L$, $i_{d1} = 0$, $v_{gs2} = v_{ggH}$, $v_{ds2} = -R_{ds(on)} \cdot i_L$, and $i_{d2} = -i_L$. In this case, $R_{ds(on)}$, v_{ggH} and v_{ggL} are on-state resistance, gate voltage during on-state, and gate voltage during off-state, respectively.

2) Stage 1 ($t_1 \sim t_2$): The gate voltage of M_2 is moved at the beginning of this stage. Thus, the input capacitance ($C_{iss2} = C_{gs2} + C_{gd2}$) discharges. The channel of M_2 conducts until v_{gs2} drops to $v_{gs,th}$. Therefore, the following equations can be obtained.

$$v_{gg2} = \underbrace{(L_{g2} + L_{s2})}_{L_{gs2}} \frac{di_{g2}}{dt} + R_{g2}i_{g2} + v_{gs2} \quad (1)$$

$$i_{g2} = C_{gs2} \frac{dv_{gs2}}{dt} + C_{gd2} \frac{dv_{gd2}}{dt} \quad (2)$$

Since M_2 switches in soft switching mode, v_{ds2} can be considered as $v_{ds(on)}$. Thus, (2) can be rewritten as:

$$i_{g2} \approx C_{iss2} \frac{dv_{gs2}}{dt} \quad (3)$$

Combining (1) and (3), the slew rates of i_{g2} and v_{gs2} can be expressed as:

$$\frac{di_{g2}}{dt} = -\frac{R_{g2}}{L_{gs2}}i_{g2} - \frac{1}{L_{gs2}}v_{gs2} + \frac{1}{L_{gs2}}v_{gg2} \quad (4)$$

$$\frac{dv_{gs2}}{dt} = \frac{1}{C_{iss2}}i_{g2} \quad (5)$$

3) Stage 2 ($t_2 \sim t_4$): In this stage, the load current can still flow through M_2 because of the reverse conduction characteristic. The i_{ch2} is governed by v_{gd2} in this condition. Based on operation principle of the GaN HEMTs, the following equations can be obtained.

$$i_{d1} = C_{oss1} \frac{dv_{ds1}}{dt} - C_{gd1} \frac{dv_{gs}}{dt} \quad (6)$$

$$i_{d2} = i_{d1} - i_L \approx i_{ch2} \quad (7)$$

$$i_{ch2} = -g_{rs}(v_{gs2} - v_{ds2} - v_{gd,th}) \quad (8)$$

$$i_{g1} = \underbrace{(C_{gs1} + C_{gd1})}_{C_{iss1}} \frac{dv_{gs1}}{dt} - C_{gd1} \frac{dv_{ds1}}{dt} \quad (9)$$

$$i_{g2} = C_{iss2} \frac{dv_{gs2}}{dt} - C_{gd1} \frac{dv_{ds2}}{dt} \quad (10)$$

$$v_{gg1} = \underbrace{(L_{g1} + L_{s1})}_{L_{gs1}} \frac{di_{g1}}{dt} + L_{s1} \frac{di_{d1}}{dt} + R_{g1}i_{g1} + v_{gs1} \quad (11)$$

$$v_{gg2} = L_{gs2} \frac{di_{g2}}{dt} + L_{s2} \frac{di_{d1}}{dt} + R_{g2}i_{g2} + v_{gs2} \quad (12)$$

$$v_{ds1} = v_{dc} + R_2 i_L - v_{ds2} - L_p \frac{di_{d1}}{dt} - (R_1 + R_2) i_{d1} \quad (13)$$

Where, g_{rs} and $v_{gd,th}$ are transconductance and threshold voltage of the reverse conduction, and $L_p = L_{loop} + L_1 +$

$L_{d1} + L_{s1} + L_2 + L_{d2} + L_{s2}$. Then, the following relationship can be obtained based on (7) and (8).

$$\frac{dv_{ds2}}{dt} - \frac{dv_{gs2}}{dt} = \frac{1}{g_{rs}}\frac{di_{d1}}{dt} \tag{14}$$

Then, the slew rates of i_{d1}, i_{g1}, i_{g2}, v_{ds1}, v_{gs1}, v_{ds2} and v_{gs2} can be obtained according to (6), (9)~(14).

$$\frac{di_{d1}}{dt} = -\frac{R_{12}}{L_p}i_{d1} - \frac{1}{L_p}v_{ds1} - \frac{1}{L_p}v_{ds2} + \frac{1}{L_p}v_{dc} + \frac{R_2}{L_p}i_L \tag{15}$$

$$\frac{di_{g1}}{dt} = \frac{L_{s1}R_{12}}{L_{gs1}L_p}i_{d1} - \frac{R_{g1}}{L_{gs1}}i_{g1} + \frac{L_{s1}}{L_{gs1}L_p}v_{ds1} - \frac{1}{L_{gs1}}v_{gs1} + \frac{L_{s1}}{L_{gs1}L_p}v_{ds2} + \frac{1}{L_{gs1}}v_{gg1} - \frac{L_{s1}}{L_{gs1}L_p}v_{dc} - \frac{L_{s1}R_2}{L_{gs1}L_p}i_L \tag{16}$$

$$\frac{di_{g2}}{dt} = \frac{L_{s2}R_{12}}{L_{gs2}L_p}i_{d1} - \frac{R_{g2}}{L_{gs2}}i_{g2} + \frac{L_{s2}}{L_{gs2}L_p}v_{ds1} + \frac{L_{s2}}{L_{gs2}L_p}v_{ds2} - \frac{1}{L_{gs2}}v_{gs2} + \frac{1}{L_{gs2}}v_{gg2} - \frac{L_{s2}}{L_{gs2}L_p}v_{dc} - \frac{L_{s2}R_2}{L_{gs2}L_p}i_L \tag{17}$$

$$\frac{dv_{ds1}}{dt} = \frac{C_{iss1}}{C_{iss1}C_{oss1} - C_{gd1}^2}i_{d1} + \frac{C_{gd1}}{C_{iss1}C_{oss1} - C_{gd1}^2}i_{g1} \tag{18}$$

$$\frac{dv_{gs1}}{dt} = \frac{C_{gd1}}{C_{iss1}C_{oss1} - C_{gd1}^2}i_{d1} + \frac{C_{oss1}}{C_{iss1}C_{oss1} - C_{gd1}^2}i_{g1} \tag{19}$$

$$\frac{dv_{ds2}}{dt} = -\frac{m_1 R_{12}}{L_p}i_{d1} + \frac{1}{C_{iss2} - C_{gd2}}i_{g2} - \frac{m_1}{L_p}v_{ds1} - \frac{m_1}{L_p}v_{ds2} + \frac{m_1}{L_p}v_{dc} + \frac{m_1 R_2}{L_p}i_L \tag{20}$$

$$\frac{dv_{gs2}}{dt} = -\frac{m_2 R_{12}}{L_p}i_{d1} + \frac{1}{C_{iss2} - C_{gd2}}i_{g2} - \frac{m_2}{L_p}v_{ds1} - \frac{m_2}{L_p}v_{ds2} + \frac{m_2}{L_p}v_{dc} + \frac{m_2 R_2}{L_p}i_L \tag{21}$$

In which, $C_{oss1,2} = C_{ds1,2} + C_{gd1,2}$, $m_1 = C_{iss1}/g_{rs}(C_{iss2} - C_{gd2})$, $m_2 = C_{gd2}/g_{rs}(C_{iss2} - C_{gd2})$, and $R_{12} = R_1 + R_2$.

4) *Stage 3 (t_4~t_5):* At the beginning of this stage, v_{gs1} arrives to v_{th}, M_1 starts to conduct. At this period, i_L transfers from M_2 to M_1 gradually. M_1 is operating in saturation mode, and the i_{ch1} is governed by v_{gs1}. Meanwhile, v_{ds1} begins to fall because of the positive voltage drops on L_p. As a result, the following equations can be obtained.

$$i_{d1} = i_{ch1} + C_{oss1}\frac{dv_{ds1}}{dt} - C_{gd1}\frac{dv_{gs1}}{dt} \tag{22}$$

$$i_{ch1} = g_{fs}(v_{gs1} - v_{gs,th}) \tag{23}$$

In which, g_{fs} is the transconductance of the GaN HEMT. Since M_2 still works in reverse conduction mode, (17), (20), and (21) are suitable for describing slew rates of i_{g2}, v_{ds2}, and v_{gs2}. It should be noticed that the gate and power circuit of M_1 can be depicted by (17) and (20). Thus, slew rates of i_{d1}

and i_{g1} can be represented by (15) and (16). The slew rates of v_{ds1} and v_{gs1} can be calculated based on (9), (22), and (23).

$$\frac{dv_{ds1}}{dt} = \frac{C_{iss1}}{C_{iss1}C_{oss1} - C_{gd1}^2}i_{d1} + \frac{C_{gd1}}{C_{iss1}C_{oss1} - C_{gd1}^2}i_{g1} - \frac{C_{iss1}}{C_{iss1}C_{oss1} - C_{gd1}^2}i_{ch1} \tag{24}$$

$$\frac{dv_{gs1}}{dt} = \frac{C_{gd1}}{C_{iss1}C_{oss1} - C_{gd1}^2}i_{d1} + \frac{C_{oss1}}{C_{iss1}C_{oss1} - C_{gd1}^2}i_{g1} - \frac{C_{gd1}}{C_{iss1}C_{oss1} - C_{gd1}^2}i_{ch1} \tag{25}$$

5) *Stage 4 (t_5~t_6):* At t_5, M_1 is fully conducted, and v_{ds1} equals $v_{ds(on)}$. As a result, the following equation can be acquired.

$$i_{g1} \approx C_{iss1}\frac{dv_{gs1}}{dt} \tag{26}$$

The slew rate of v_{gs1} can be obtained as:

$$\frac{dv_{gs1}}{dt} = \frac{1}{C_{iss1}}i_{g1} \tag{27}$$

At this moment, M_2 still works in reverse conduction mode because $i_{d1} \le i_L$. It should be noted that (15), (16), (17), (20), and (21) are still validated to describe slew rates of i_{d1}, i_{g1}, i_{g2}, v_{ds2}, and v_{gs2} in this stage.

6) *Stage 5 (t_6~t_7):* At t_6, i_{d1} arrives to i_L. M_2 enters to cut-off region, and begins to block voltage. At this moment, positive spurious voltage on v_{gs2} will be triggered because of the variation of v_{ds2}. Meanwhile, the skin effect should be taken into consideration to dampen the current oscillation. Thus, the following equations can be acquired.

$$i_{d2} = C_{oss2}\frac{dv_{ds2}}{dt} - C_{gd2}\frac{dv_{gs2}}{dt} \tag{28}$$

$$v_{ds1} = v_{dc} - v_{ds2} - L_p\frac{di_{d1}}{dt} - R_1 i_{d1} - (R_2 + R_{ac})i_{d2} \tag{29}$$

Combining (9)~(12), (28) and (29), slew rates of i_{d1}, i_{g1}, i_{g2}, v_{ds2}, and v_{gs2} can be obtained. In this situation, (27) is still suitable for slew rate of v_{gs1}.

$$\frac{di_{d1}}{dt} = -\frac{(R_{12}+R_{ac})}{L_p}i_{d1} - \frac{1}{L_p}v_{ds1} - \frac{1}{L_p}v_{ds2} + \frac{1}{L_p}v_{dc} + \frac{(R_2+R_{ac})}{L_p}i_L \tag{30}$$

$$\frac{di_{g1}}{dt} = \frac{L_{s1}(R_{12}+R_{ac})}{L_{gs1}L_p}i_{d1} - \frac{R_{g1}}{L_{gs1}}i_{g1} + \frac{L_{s1}}{L_{gs1}L_p}v_{ds1} - \frac{1}{L_{gs1}}v_{gs1} + \frac{L_{s1}}{L_{gs1}L_p}v_{ds2} + \frac{1}{L_{gs1}}v_{gg1} - \frac{L_{s1}}{L_{gs1}L_p}v_{dc} - \frac{L_{s1}(R_2+R_{ac})}{L_{gs1}L_p}i_L \tag{31}$$

$$\frac{di_{g2}}{dt} = \frac{L_{s2}(R_{12}+R_{ac})}{L_{gs2}L_p}i_{d1} - \frac{R_{g2}}{L_{gs2}}i_{g2} + \frac{L_{s2}}{L_{gs2}L_p}v_{ds1} + \frac{L_{s2}}{L_{gs2}L_p}v_{ds2} - \frac{1}{L_{gs2}}v_{gs2} + \frac{1}{L_{gs2}}v_{gg2} - \frac{L_{s2}}{L_{gs2}L_p}v_{dc} - \frac{L_{s2}(R_2+R_{ac})}{L_{gs2}L_p}i_L \tag{32}$$

$$\frac{dv_{ds2}}{dt} = \frac{C_{iss2}}{C_{iss2}C_{oss2} - C_{gd2}^2}i_{d1} + \frac{C_{gd2}}{C_{iss2}C_{oss2} - C_{gd2}^2}i_{g2} - \frac{C_{iss2}}{C_{iss2}C_{oss2} - C_{gd2}^2}i_L \tag{33}$$

$$\frac{dv_{ds2}}{dt} = \frac{C_{gd2}}{C_{iss2}C_{oss2}-C_{gd2}^2}i_{d1} + \frac{C_{oss2}}{C_{iss2}C_{oss2}-C_{gd2}^2}i_{g2}$$
$$-\frac{C_{gd2}}{C_{iss2}C_{oss2}-C_{gd2}^2}i_L \quad (34)$$

B. Turn-off Transient

1) Stage 6 ($t_7 \sim t_8$): In this stage, M_1 operates in ohmic region, and M_2 is turned off. It can be figured out that: $v_{gs1}=v_{ggH}$, $v_{ds1}=R_{ds(on)}\cdot i_L$, $i_{d1}=i_L$, $v_{gs2}=v_{ggL}$, $v_{ds2}=v_{dc}+R_{ds(on)}\cdot i_L$, and $i_{d2}=0$.

2) Stage 7 ($t_8 \sim t_9$): At t_8, the gate drive voltage of M_1 is removed, and C_{iss1} begins to discharging. M_1 operates in ohmic region until v_{gs1} drops below v_{mil} ($v_{mil} = \frac{i_L}{g_{fs}} + v_{gs,th}$). Thus, the following equation can be obtained.

$$i_{d2} = C_{oss2}\frac{dv_{ds2}}{dt} - C_{gd2}\frac{dv_{gs2}}{dt} \quad (35)$$

Combining (26) and (35), slew rates of i_{g1} and v_{gs1} can be calculated.

$$\begin{cases} \frac{di_{g1}}{dt} = -\frac{R_{g1}}{L_{gs1}}i_{g1} - \frac{1}{L_{gs1}}v_{gs1} + \frac{1}{L_{gs1}}v_{gg1} \\ \frac{dv_{gs1}}{dt} = \frac{1}{C_{iss1}}i_{g1} \end{cases} \quad (36)$$

3) Stage 8 ($t_9 \sim t_{10}$): As v_{gs1} drops below v_{mil}, M_1 enters to saturation region, and the i_{ch1} is governed by v_{gs1}. At the same instance, the voltage commutation between M_1 and M_2 begins, which can cause negative spurious voltage on v_{gs2} due to the changing of v_{ds2}. In this condition, (9)~(13), (22), (23), and (28) are responsible for depicting switching behaviors of M_1 and M_2. Therefore, the slew rates of i_{d1}, i_{g1}, i_{g2}, v_{ds1}, v_{gs1}, v_{ds2}, and v_{gs2} can be demonstrated by (15)~(17), (24), (25), (33), and (34), respectively.

4) Stage 9 ($t_{10} \sim t_{11}$): At t_{10}, M_2 switches to reverse conduction mode because $v_{gd2} \geq v_{gd,th}$. At this moment, i_L transfers from M_1 to M_2. It can be figured out that this stage is identical to Stage 3 during turn-on transient. As a result, the slew rates of i_{d1}, i_{g1}, i_{g2}, v_{ds1}, v_{gs1}, v_{ds2}, and v_{gs2} can be described by (15)~(17), (24), (25), (20), and (21), respectively.

5) Stage 10 ($t_{11} \sim t_{13}$): At t_{11}, M_1 goes back to cut-off region because $v_{gs1} < v_{gs,th}$. This stage is almost the same with Stage 2 during turn-on transient. As a result, slew rates of i_{g1}, i_{g2}, v_{ds1}, v_{gs1}, v_{ds2}, and v_{gs2} can be represented by (16)~(19). It should be noticed that the skin effect is represented by R_{ac} connected in series with R_1 in this situation. Thus, the flowing relationship can be obtained as:

$$v_{ds1} = v_{dc} - v_{ds2} - L_p\frac{di_{d1}}{dt} - (R_1 + R_{ac})i_{d1} - R_2 i_{d2} \quad (37)$$

According to (10), (14), and (37), slew rates of i_{d1}, v_{ds2}, and v_{gs2} can be calculated as:

$$\frac{di_{d1}}{dt} = -\frac{(R_{12}+R_{ac})}{L_p}i_{d1} - \frac{1}{L_p}v_{ds1} - \frac{1}{L_p}v_{ds2}$$
$$+\frac{1}{L_p}v_{dc} + \frac{R_2}{L_p}i_L \quad (38)$$

$$\frac{dv_{ds2}}{dt} = -\frac{m_1(R_{12}+R_{ac})}{L_p}i_{d1} + \frac{1}{C_{iss2}-C_{gd2}}i_{g2} - \frac{m_1}{L_p}v_{ds1}$$
$$-\frac{m_1}{L_p}v_{ds2} + \frac{m_1}{L_p}v_{dc} + \frac{m_1 R_2}{L_p}i_L \quad (39)$$

$$\frac{dv_{gs2}}{dt} = -\frac{m_2(R_{12}+R_{ac})}{L_p}i_{d1} + \frac{1}{C_{iss2}-C_{gd2}}i_{g2} - \frac{m_2}{L_p}v_{ds1}$$
$$-\frac{m_2}{L_p}v_{ds2} + \frac{m_2}{L_p}v_{dc} + \frac{m_2 R_2}{L_p}i_L \quad (40)$$

6) Stage 11 ($t_{13} \sim t_{14}$): This stage is almost the same with Stage 10. Thus, slew rates of i_{d1}, i_{g1}, i_{g2}, v_{ds1}, and v_{gs1} can be expressed by (38), (16)~(19). The only difference from Stage 10 is that M_2 is in forward conduction mode, and v_{ds2} can be approximated as $v_{ds(on)}$. As a result, slew rate of v_{gs2} is represented by (4) in this condition.

IV. MODELING OF JUNCTION CAPACITANCE AND FORWARD/REVERSE TRANSCONDUCTANCE

A. Modeling of Junction Capacitance

The dv/dt is critical to determine the switching losses of the GaN device. As a consequence, accurate modeling of junction capacitance is critical for acquiring reliable dv_{ds}/dt. In addition, the nonlinear junction capacitance also impacts the crosstalk between phase-legs.

In this work, the modeling of the junction capacitance can be interpreted as C-V curves fitting based on equation presented in [16]. Fig. 3 illustrates the fitted values of C_{gd}, $C_{oss}(=C_{ds}+C_{gd})$, and $C_{iss}(=C_{gs}+C_{gd})$ compared with the measured ones obtained from datasheet. Obviously, a good agreement is acquired, which can ensure more accurate results of the analytical model. The values of C_{gd}, C_{oss}, and C_{iss} should be updated with the latest calculated v_{ds}.

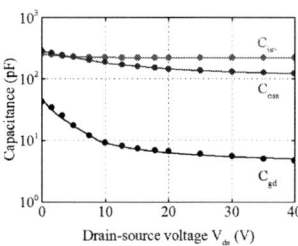

Fig. 3. The measured (solid dots) and fitted (solid lines) values of C_{gd}, C_{iss}, and C_{oss}.

B. Modeling of Forward/Reverse Transconductance

It should be noted that g_{fs} determines i_{ch} when GaN HEMT works in forward conduction, while g_{rs} plays a significant role once the device operates in reverse conduction. The modeling process of g_{fs} and g_{rs} is understood as curve fitting of I-V characteristic, which can be easily obtained with Matlab. Fig. 4 illustrates the measured and calculated I-V curves of the GaN HEMT in forward and reverse conduction, which manifests that the calculated I-V curve is in good accordance to the measured ones acquired directly from datasheet.

V. EXPERIMENT VERIFICATION

A. Experiment Setup

Experiment setup of the low-voltage GaN device is presented in Fig. 5. EPC2014C (40 V, 10 A) is selected as the

978-1-6654-4817-8/21 $31.00 © 2021 IEEE

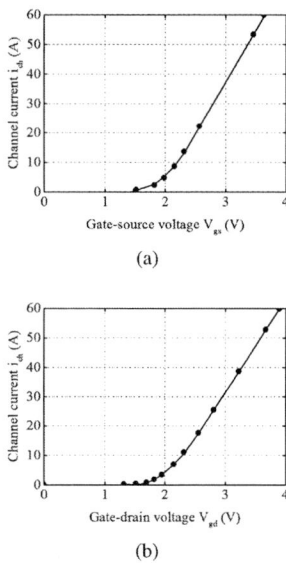

(a)

(b)

Fig. 4. The measured (solid dots) and fitted (solid lines) I-V characteristics of the GaN device, (a) Forward conduction (g_{fs}), (b) Reverse conduction (g_{rs}).

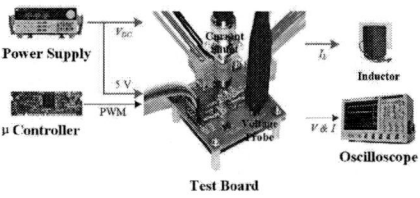

Fig. 5. Experiment setup of the low-voltage GaN device.

GaN HEMT in this study. LMG1205 is selected to design gate drive circuit, which provides 5 V/0 V drive voltage to quickly turn-on/turn-off the GaN devices. A micro-controller (TMS320F28335) is used to generate double-pulse signal. A 31 μH air-core inductor with single layer is chosen as the current load to avoid the magnetic saturation, and ensure minimized interturn capacitances. The switching waveform can be acquired with Tektronix's DPO7354C. The v_{gs} and v_{ds} are measured by differential probes with high-bandwidth (TDP1000, 42 V, 1 GHz), and the i_d can be obtained with current shunt from T&M research (25 mΩ, 1.2 GHz).

B. Measured and Calculated Waveforms of the Low-voltage GaN Device

The DPT is conducted with supply voltage v_{dc}=15 and 18 V and load current i_L=4, 6, 8, and 10 A. The measured and calculated waveforms are shown in Figs. 6~8. The correctness and effectiveness of the analytical model have been effectively validated because of the good agreement of the measured and calculated waveforms.

In addition, Fig. 9 summarizes the calculated and measured switching energies, peak values of the v_{ds}, i_d, and positive/negative spurious voltage generated on v_{gs2} as the i_L varies from 4 to 10 A. It is obvious that the proposed

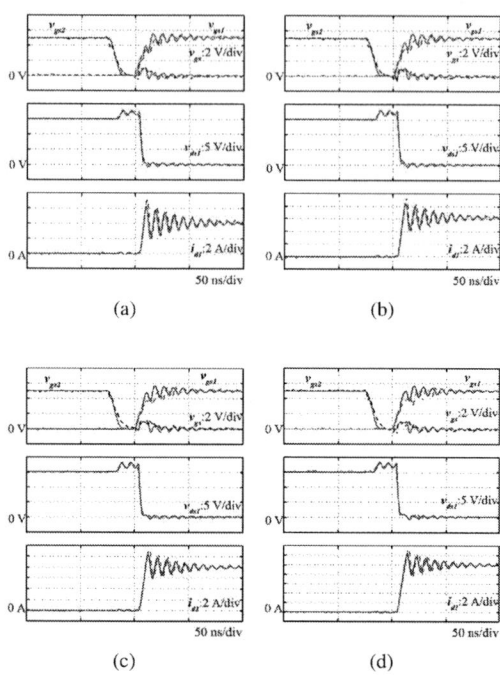

(a)   (b)

(c)   (d)

Fig. 6. Comparisons between measured (dashed line) and calculated (solid line) waveforms when v_{dc}=15 V and i_L varies from 4 to 10 A during turn-on transient, (a) i_L=4 A, (b) i_L=6 A, (c) i_L=8 A, (d) i_L=10 A.

model can demonstrate switching losses, switching oscillation, and crosstalk simultaneously during switching transient of the GaN HEMT in sufficient accuracy when compared with other models presented in [7]–[9], [11], [13].

VI. CONCLUSION AND FUTURE WORKS

Switching characteristics of the GaN HEMT have been comprehensively evaluated using analytical modeling in this work. The switching process of the low-voltage GaN HEMT has been elaborated based on detailed circuit equation deduction. Meanwhile, the model has considered nonlinear junction capacitances as well as forward and reverse transconductances, skin effect during oscillation period, and interactions between high- and low- side GaN HEMTs. According to measured and calculated results, we can figure out that the switching characteristic of the low-voltage GaN device can be well represented in aspects of switching losses, voltage/current oscillation and crosstalk with the proposed model.

In future study, more detailed works will be focused on investigating influences of parasitic elements on switching characteristics of the low-voltage GaN HEMT, especially in aspects of voltage/current oscillation and crosstalk, which is expected to give many valuable design references in practical applications.

REFERENCES

[1] J. L. Hudgins, "Power electronic devices in the future," *IEEE J. Emerg. Sel. Topics Power Electron.*, vol. 1, no. 1, pp. 11–17, 2013.

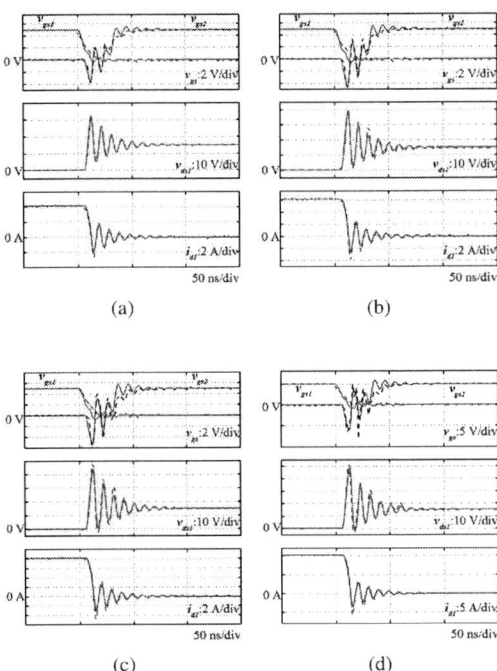

(a)   (b)

(c)   (d)

Fig. 7. Comparisons between measured (dashed line) and calculated (solid line) waveforms when v_{dc}=15 V and i_L varies from 4 to 10 A during turn-off transient, (a) i_L=4 A, (b) i_L=6 A, (c) i_L=8 A, (d) i_L=10 A.

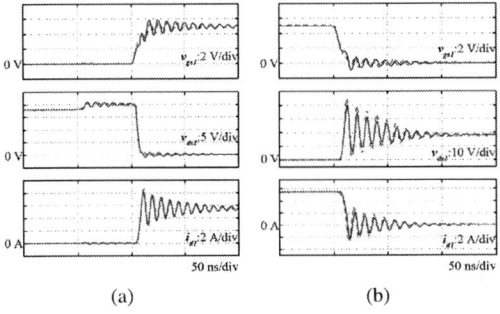

(a)   (b)

Fig. 8. Comparisons between measured (dashed line) and calculated (solid line) waveforms when v_{dc}=18 V and i_L=6 A during switching transient, (a) Turn-on, (b) Turn-off.

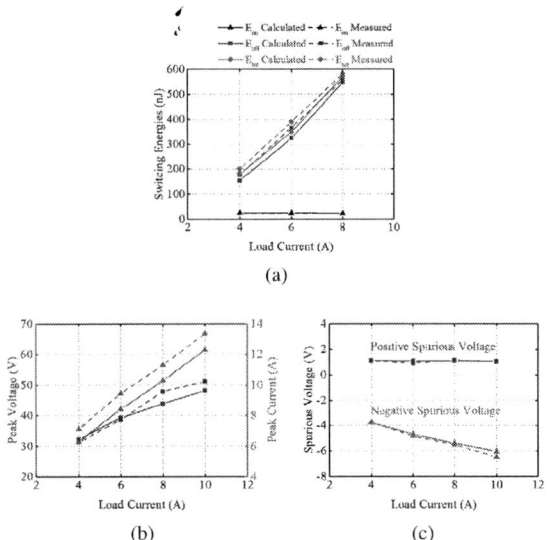

(a)

(b)   (c)

Fig. 9. Comparisons between measured (dashed line) and calculated (solid line) results, (a) Switching energies (v_{dc}=18 V, i_L=4, 6, 8 A), (b) Voltage/current peaks (v_{dc}=15 V, i_L=4, 6, 8, 10 A), (c) Positive/negative spurious voltage (v_{dc}=15 V, i_L=4, 6, 8, 10 A).

phase-leg topology," in *Proc. IEEE Workshop on Wide Bandgap Power Devices and Applications (WiPDA)*, 2015, pp. 98–103.

[7] J. Chen, Q. Luo, J. Huang, Q. He, and X. Du, "A complete switching analytical model of low-voltage eGaN HEMTs and its application in loss analysis," *IEEE Trans. Ind. Electron.*, vol. 67, no. 2, pp. 1615–1625, 2020.

[8] Y. Xin, W. Chen, R. Sun, Y. Shi, C. Liu, Y. Xia, F. Wang, M. Li, J. Li, Q. Zhou, X. Deng, T. Chen, Z. Li, and B. Zhang, "Analytical switching loss model for GaN-based control switch and synchronous rectifier in low-voltage buck converters," *IEEE J. Emerg. Sel. Top. Power Electron.*, vol. 7, no. 3, pp. 1485–1495, 2019.

[9] X. Long, W. Liang, Z. Jun, and G. Chen, "A normalized quantitative method for GaN HEMT turn-on overvoltage modeling and suppressing," *IEEE Trans. Ind. Electro.*, vol. 66, no. 4, pp. 2766–2775, 2019.

[10] R. Xie, H. Wang, G. Tang, X. Yang, and K. J. Chen, "An analytical model for false turn-on evaluation of high-voltage enhancement-mode GaN transistor in bridge-leg configuration," *IEEE Trans. Power Electron.*, vol. 32, no. 8, pp. 6416–6433, 2017.

[11] B. Li, G. Wang, S. Liu, N. Zhao, G. Zhang, X. Zhang, and D. Xu, "Modeling and analysis of bridge-leg crosstalk of GaN HEMT considering nonlinear junction capacitances," *IEEE Trans. Power Electron.*, vol. 36, no. 4, pp. 4429–4439, 2021.

[12] X. Long, Z. Jun, L. Pu, D. Chen, and W. Liang, "Analysis and suppression of high speed dv/dt induced false turn-on in GaN HEMT phase-leg topology," *IEEE Access*, vol. 9, pp. 45 259–45 269, 2021.

[13] J. P. Kozak, A. Barchowsky, M. R. Hontz, N. B. Koganti, W. E. Stanchina, G. F. Reed, Z.-H. Mao, and R. Khanna, "An analytical model for predicting turn-on overshoot in normally-off GaN HEMTs," *IEEE J. Emerg. Sel. Topics Power Electron.*, vol. 8, no. 1, pp. 99–110, 2020.

[14] M. Dong, H. Li, S. Yin, Y. Wu, and K. Y. See, "A postprocessing-technique-based switching loss estimation method for GaN devices," *IEEE Trans. Power Electron.*, vol. 36, no. 7, pp. 8253–8266, 2021.

[15] K. Wang, X. Yang, L. Wang, and P. Jain, "Instability analysis and oscillation suppression of enhancement-mode GaN devices in half-bridge circuits," *IEEE Trans. Power Electron.*, vol. 33, no. 2, pp. 1585–1596, 2018.

[16] M. R. Ahmed, R. Todd, and A. J. Forsyth, "Predicting SiC MOSFET behavior under hard-switching, soft-Switching, and false turn-on conditions," *IEEE Trans. Ind. Electro.*, vol. 64, no. 11, pp. 9001–9011, 2017.

[2] J. Millan, P. Godignon, X. Perpina, A. Perez-Tomas, and J. Rebollo, "A survey of wide bandgap power semiconductor devices," *IEEE Trans. Power Electron.*, vol. 29, no. 5, pp. 2155–2163, 2014.

[3] E. A. Jones, F. F. Wang, and D. Costinett, "Review of commercial GaN power devices and GaN-based converter design challenges," *IEEE J. Emerg. Sel. Topics Power Electron.*, vol. 4, no. 3, pp. 707–719, 2016.

[4] T. Liu, T. T. Y. Wong, and Z. J. Shen, "A Survey on switching oscillations in power converters," *IEEE J. Emerg. Sel. Topics Power Electron.*, vol. 8, no. 1, pp. 893–908, 2020.

[5] J. Chen, Q. Luo, J. Huang, Q. He, P. Sun, and X. Du, "Analysis and design of an RC snubber circuit to suppress false triggering oscillation for GaN devices in half-bridge circuits," *IEEE Trans. Power Electron.*, vol. 35, no. 3, pp. 2690–2704, 2020.

[6] E. A. Jones, F. Wang, D. Costinett, Z. Zhang, and B. Guo, "Cross conduction analysis for enhancement-mode 650-V GaN HFETs in a

A Novel SiC Trench MOSFET Structure with Enhanced Short Circuit Robustness

Chongyu Jiang[a], Hongyi Xu[a], Na Ren[a b*], Qing Guo[a b], Kuang Sheng[a b]
[a] College of Electrical Engineering, Zhejiang University, Hangzhou, China
[b] Hangzhou Global Scientific and Technological Innovation Center, Zhejiang University
Hangzhou, China
* ren_na@zju.edu.cn

Abstract—In this paper, a novel SiC trench MOSFET （TMOS） structure N-implanted TMOS is proposed and compared with previous P-implanted TMOS. Specific on-resistance ($R_{dson,sp}$), breakdown voltage (BV), and maximum electric field in gate oxide (E_{ox_max}) were optimized by using numerical simulation. It was found the optimal N-implanted TMOS has a 390V higher breakdown voltage than optimal P-implanted TMOS under similar $R_{dson,sp}$. However, E_{ox_max} in N-implanted TMOS is higher than P-implanted TMOS under the same blocking voltage. Further, short circuit (SC) simulations were studied under bus voltage of 400 V/800 V. The peak short circuit current (normalized with current rating @V_{ds} = 2 V) and energy dissipations were compared for the two devices with varied designs. Both the peak short circuit current and energy dissipation are lower for the N-implanted TMOS at the same SC pulse width. The N-implanted TMOS device have better short circuit robustness.

Keywords—Silicon Carbide, N-implanted TMOS, specific on-resistance, breakdown voltage, maximum electric field in gate oxide, enhanced short circuit robustness

I. INTRODUCTION

Silicon Carbide(SiC) devices, especially Silicon Carbide Metal Oxide Field Effect Transistor(SiC MOSFET), have proven their technical advantage in power electronic applications with high-speed switching, high breakdown voltage, and high-temperature performance [1].

It is well known that SiC trench-gate MOSFET has lower specific on-resistance with narrower cell pitch, higher channel mobility compared to planar-gate MOSFET. However, the high electrical field in gate oxide at the corner of the trench causes device degradation and even device failure[2][3][4]. To resolve this issue, many novel structures have been proposed. J. Tan introduced a P+ implant region under the gate trench to provide electric field shielding and protect the gate oxide[5] as shown in Fig.1(a). The device structure is designated as P-implanted TMOS in this work. A Thicker N-epi layer is required in the P-implanted TMOS and the on-resistance is increased. Furthermore, the P-implanted region should be well-grounded [6], otherwise, the dynamic on-resistance is founded during the operation due to the depletion of the P-implanted region after blocking status. Qingwen Song, et al presented a novel SiC trench MOSFET structure with an L-shaped gate (LSG) [7].

However, uniformity of the L-shaped channel is hard to produce in this structure. Rohm proposed a double trench structure to eliminate the crowding electrical field at the corner of the trench gate oxide. The manufacturing process is simplified, but the cell pitch of this structure is hard to shrink[8]. Infineon announced an unsymmetric TMOS structure to protect the trench gate. But the ultra-high-energy implantation process is needed in this structure and is hard to fulfill[9].

In this paper, a novel SiC trench MOSFET (named N-implanted TMOS) is proposed, with an introduced N-implant region surrounding the gate trench, as shown in Fig.1(b). The comparison study between the two devices with regard to static characteristic performance and short circuit robustness is carried out in this work.

II. STATIC CHARACTERISTIC SIMULATION

A. Key Parameters of Device

The key structural parameters of the two devices are listed in Table 1. Three indexes for static performance, i.e., device specific on-resistance ($R_{dson,sp}$), breakdown voltage (BV) and maximum electric field in gate oxide (E_{ox_max}) are optimized by tuning the structural parameters.

As shown in Fig.1(a) and (b), for P-implanted TMOS, the distance between the channel and P+ region under gate oxide (L_{p-ch}) and the width of the P+ region under gate oxide (L_{p+}) have significant effects on device performance, while for N-implanted TMOS, the doping concentration N_{sp} and width of N-implanted region under gate oxide L_{sp} are essential parameters.

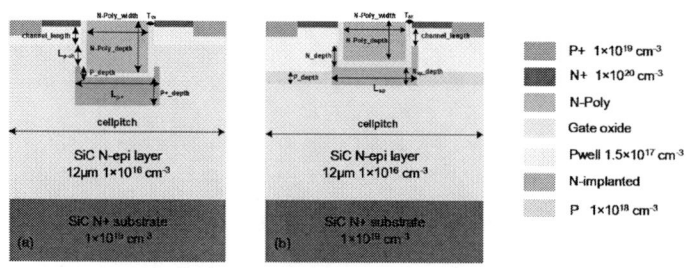

Fig. 1. Schematic of (a) P-implanted TMOS and (b) N-implanted TMOS.

B. Output Characteristic Simulation

The key parameters of the devices are given in Table I.

This research was funded by grants from the Guangdong Key Research and Development Program of China (2019B010143001) and National Natural Science Foundation of China (52007165).

978-1-6654-4817-8/21 $31.00 © 2021 IEEE

The output characteristic curves are shown in Fig.2(a), (b) for P-implanted TMOS with varied L_{p+} (0.7, 0.8, 0.9, 1.0 μm) and L_{p-ch} (0.8, 0.9, 1 μm), Fig.2(c), (d) for N-implanted TMOS with varied N_{sp} (1.1×10^{17}, 1.5×10^{17}, 2×10^{17} cm^{-3}) and L_{sp} (0.7, 0.8, 0.9, 1.0 μm). The resistance of these two devices is displayed in Fig.3.

For P-implanted TMOS, larger L_{p+} increases the $R_{dson,sp}$ shown in Fig.2.(a) due to the increment of R_{p-ch} shown in Fig.3(a). The $R_{dson,sp}$ decreases with an increasing L_{p-ch}, shown in Fig.2.(b) due to the decrement of R_{p-ch} shown in Fig.3(a). For N-implanted TMOS, N_{sp} affects R_{nsp} shown in Fig.3(b). With an increasing L_{sp}, the $R_{dson,sp}$ decreases shown in Fig.2.(d) due to the decrease of R_{nsp} shown in Fig.3(b).

C. Blocking Characteristic Simulation

The breakdown curves are shown in Fig.4(a), (b) for P-implanted TMOS with varied L_{p+} (0.7, 0.8, 0.9, 1.0 μm) and L_{p-ch} (0.8, 0.9, 1.0 μm), Fig.4(c), (d) for N-implanted TMOS with varied N_{sp} (1.1×10^{17}, 1.5×10^{17}, 2×10^{17} cm^{-3}) and L_{sp} (0.7, 0.8, 0.9, 1 μm). The electrical field distribution in oxide at $V_{ds} = 1200$ V of P-implanted TMOS and N-implanted TMOS are shown in Fig.5.(a) and (b). For P-implanted TMOS, P+ provide the protection for trench bottom, the BV increases with longer L_{p+} shown in Fig.4.(a). The BV decreases with increasing L_{p-ch} shown in Fig.4.(b). For N-implanted TMOS, with an increasing N_{sp} and L_{sp}, the BV remains almost the same shown in Fig.4.(c) and Fig.4.(d).

TABLE I.   KEY STRUCTURAL PARAMETERS OF SiC TRENCH MOSFET USED IN THE SIMULATION.

Parameters	P-implanted TMOS	N-implanted TMOS
Cellpitch(μm)	3	
T_{ox}(nm)	50	
L_{p+}(μm)	0.7,0.8,0.9,1	*
L_{p-ch}(μm)	0.8,0.9,1	*
L_{sp}(μm)	*	0.7,0.8,0.9,1
N_{sp}(cm^{-3})	*	1.1×10^{17},1.5×10^{17},2×10^{17}

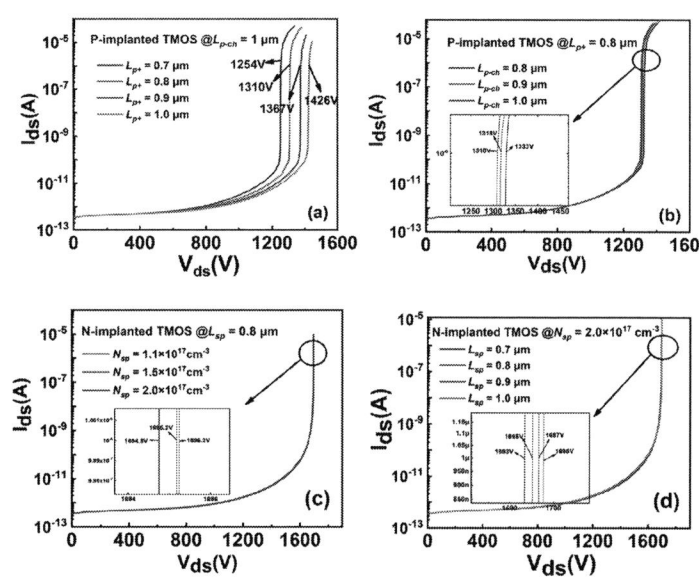

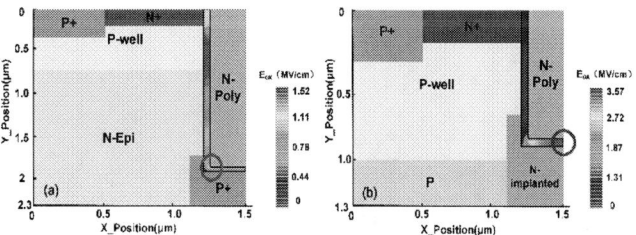

Fig. 4. Breakdown characteristics of (a) P-implanted TMOS with varied L_{p+} (b) P-implanted TMOS with varied L_{p-ch} (c) N-implanted TMOS with varied N_{sp} (d) N-implanted TMOS with varied L_{sp}

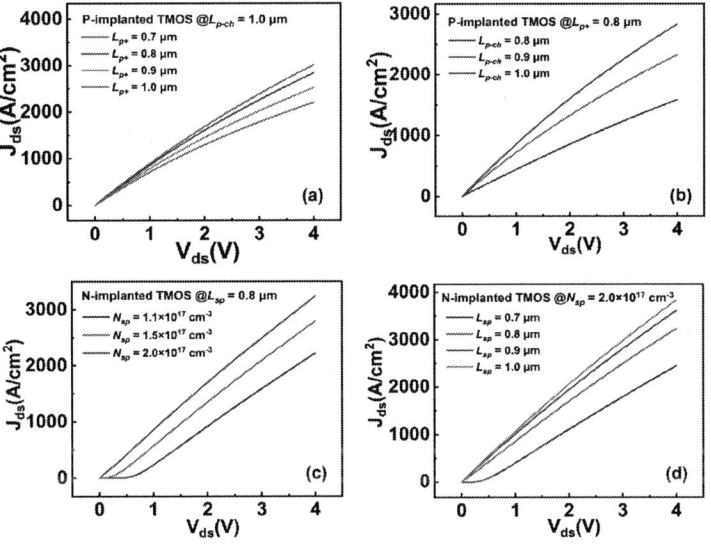

Fig. 2. Output characteristics of a) P-implanted TMOS with varied L_{p+} (b) P-implanted TMOS with varied L_{p-ch} (c) N-implanted TMOS with varied N_{sp} (d) N-implanted TMOS with varied L_{sp}

Fig. 5. The electrical field distribution in oxide at $V_{ds} = 1200$ V of (a) P-implanted TMOS and (b) N-implanted TMOS. Peak electrical field positions are circled respectively.

D. Tradeoff between the $R_{dson,sp,}$ and E_{ox_max}

According to the output and breakdown simulation results, the influence of the key parameters on static performance

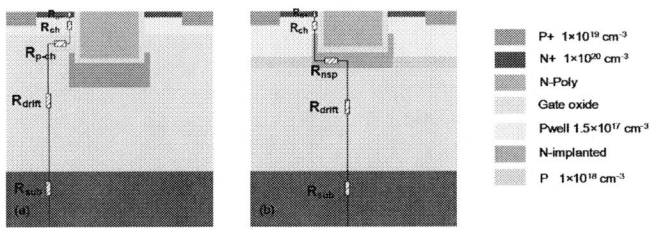

Fig. 3. The resistance of (a) P-implanted TMOS and (b) N-implanted TMOS

(including $R_{dson,sp}$, E_{ox_max}, and BV) is displayed in Fig.6 and Fig.7.

As shown in Fig.6, for P-implanted TMOS, the distance between channel and P+ region under gate oxide (L_{p-ch}) and the width of P+ region under gate oxide (L_{p+}) have significant effects on device performance. With an increasing L_{p+}, the BV increases as well as the $R_{dson,sp}$ shown in Fig.6.(a). With an L_{p-ch} increasing, BV shows a slight decrease, and $R_{dson,sp}$ reduces significantly shown in Fig.6.(a). The influences of L_{p-ch} and L_{p+} on E_{ox_max} are also plotted in Fig.6.(b) which shows E_{ox_max} decreases with an increasing L_{p+} and decreasing L_{p-ch}. An optimal structure (L_{p-ch} = 1 µm and L_{p+} = 0.8 µm) can be concluded (labeled with red star in Fig.6).

As shown in Fig.7, for N-implanted TMOS, the doping concentration N_{sp} and width of the N-implanted region under gate oxide L_{sp} are essential parameters. The $R_{dson,sp}$ and BV of N-implanted TMOS with various N_{sp} and L_{sp} are illustrated in Fig.7.(a). The E_{ox_max} and $R_{dson,sp}$ are also extracted in TMOS with different N_{sp} and L_{sp} illustrated in Fig.7.(b). Thus, an optimal structure with N_{sp} = 2.0×10^{17} cm^{-3} and L_{sp} = 0.8 µm can be concluded (labeled with blue star in Fig.7). When comparing the respective optimal structure for the two devices (P-implanted and N-implanted TMOS), E_{ox_max} are all below 4MV/cm, $R_{dson,sp}$ is similar (1.16mΩ·cm^2 for P-implanted and 1.17mΩ·cm^2 for N-implanted) but N-implanted TMOS shows a 390V higher breakdown voltage.

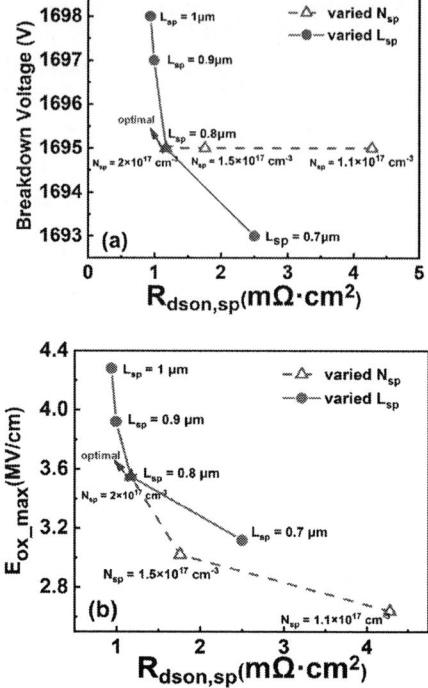

Fig. 7. Simulation results of N-implanted TMOS (a) the relationship between $R_{dson,sp}$ and BV with different N_{sp} and L_{sp}. (b) the relationship between $R_{dson,sp}$ and E_{ox_max} with different N_{sp} and L_{sp}. The point labeled with the blue star was selected with the best static performance $R_{dson,sp}$ =1.17 mΩ·cm^2, BV=1695 V.

III. SHORT CIRCUIT SIMULATION

In addition, short circuit (SC) simulations are conducted. The short circuit schematic is shown in Fig.8. The V_{DC} is selected as 400V and 800V. The turn-on pulse time is 10 µs for V_{DC} = 400 V and 5 µs for V_{DC} = 800 V.

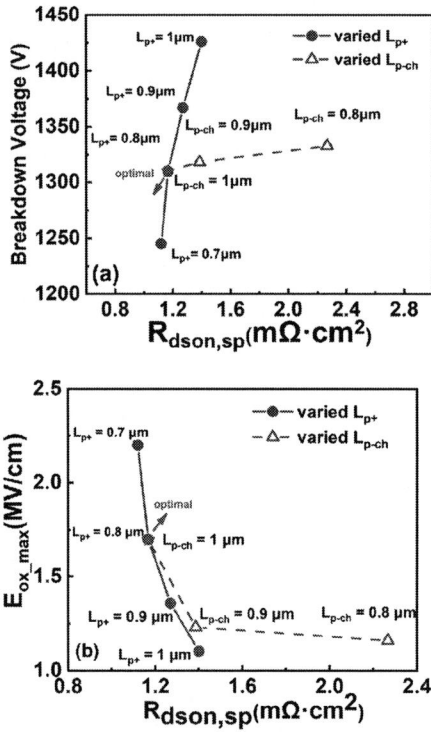

Fig. 6. Simulation results of P-implanted TMOS (a) the relationship between $R_{dson,sp}$ and BV with different L_{p+} and L_{p-ch}. (b) the relationship between $R_{dson,sp}$ and E_{ox_max} with different L_{p+} and L_{p-ch}. The point labeled with the red star was selected with the best static performance $R_{dson,sp}$ =1.16 mΩ·cm^2, BV=1310 V.

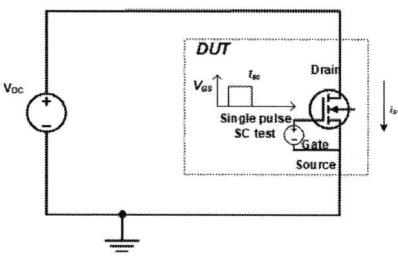

Fig. 8. Short circuit simulation schematic

The short circuit current density waveforms of the two devices under bus voltage 400 V/800 V are shown in Fig.9(a). The peak short circuit current (normalized with current rating @V_{ds} = 2 V at room temperature) and energy dissipations are compared in Fig. 9(b) and (c) for the two devices with varied designs. Both the peak short circuit current and energy dissipation are lower for the N-implanted TMOS at the same SC pulse width. Therefore, the N-implanted TMOS device should have higher short circuit robustness (longer SC withstand time)[10-13].

978-1-6654-4817-8/21 $31.00 © 2021 IEEE

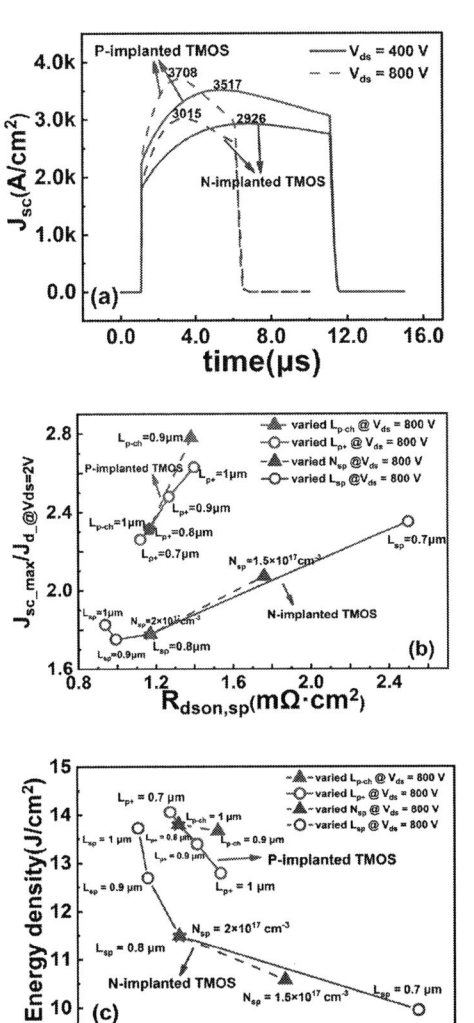

Fig. 9. Simulation results of (a) Short circuit current density waveforms for two optimal structures at $V_{ds} = 400$ V,10 µs and $V_{ds} = 800$ V,5 µs. (b) The peak short circuit current (normalized with current rating @$V_{ds} = 2$ V) and (c) energy dissipations in short circuit of P+ implanted TMOS (red solid and dotted line) and N-implanted TMOS (blue solid and dotted line) at $V_{ds} = 800$ V in varied design. The optimal designs are labeled with star for this two structures respectively.

IV. CONCLUSION

A new trench MOSFET structure N-implanted TMOS was proposed, and compared with P-implanted TMOS. Considering three indexes for static performance, i.e., device-specific on-resistance ($R_{dson,sp}$), breakdown voltage (BV), and maximum electric field in gate oxide (E_{ox_max}), the two structures are both optimized by tunning key parameters. For the two optimal TMOS, E_{ox_max} are all below 4 MV/cm, and $R_{dson,sp}$ is close (1.16 mΩ·cm² for P-implanted and 1.17 mΩ·cm² for N-implanted), but N-implanted TMOS shows a 390 V higher breakdown voltage (1310 V for P-implanted and 1695 V for N-implanted). Short circuit capability of P-implanted and N-implanted TMOS

is also simulated. The results show both the peak short circuit current and energy dissipation are lower for the N-implanted TMOS at the same SC pulse width. Therefore, the N-implanted TMOS device should have better short circuit robustness (longer SC withstand time).

REFERENCES

[1] T. Kimoto and J. A. Cooper, Fundamentals of Silicon Carbide Technology:Growth, Characterization, Devices and Applications: John Wiley & Sons, 2014.

[2] Zeng J Z J , Dolny G , Kocon C , et al. An ultra dense trench-gated power MOSFET technology using a self-aligned process[C]// ISPSD, 2001.

[3] H. Yano, H. Nakao, T. Hatayama, Y. Uraoka, and T. Fuyuki, Increased Channel Mobility in 4H-SiC UMOSFETs Using On-Axis Substrates, Mater. Sci. Forum, Vol.556-557, pp. 807-810, 2007.

[4] T. Nakamura, M. Aketa, Y. Nakano, M. Sasagawa, and T. Otsuka, Novel developments towards increased SiC power device and module efficiency, in Proc. IEEE Energytech. Cleveland, OH, USA, May 2012, pp. 1–6, doi: 10.1109/EnergyTech.2012.6304633.

[5] J. Tan, J. A. Cooper, Jr., and M. R. Melloch, High-voltage accumulation-layer UMOSFETs in 4H-SiC, in 56th Annu. Device Res. Conf. Dig., Charlottesville, VA, USA, Jun. 1998, pp. 88–89, doi: 10.1109/DRC.1998.731133.

[6] J. Wei, M. Zhang, H. Jiang, B. Li, Z. Zheng and K. J. Chen, "Investigations of p-Shielded SiC Trench IGBT with Considerations on IE Effect, Oxide Protection and Dynamic Degradation," 2019 31st International Symposium on Power Semiconductor Devices and ICs (ISPSD), 2019, pp. 199-202, doi: 10.1109/ISPSD.2019.8757642

[7] Q. Song et al., 4H-SiC Trench MOSFET With L-Shaped Gate, in IEEE Electron Device Letters, vol. 37, no. 4, pp. 463-466, April 2016, doi: 10.1109/LED.2016.2533432.

[8] T. Nakamura, et al., IEEE International Electron Devices Meeting, pp. 26.5.1-26.5.3 (2011).

[9] Dethard Peters, Ralf Siemieniec, Thomas Aichinger, et al., IEEE International Symposium on Power Semiconductor Devices and IC's, pp. 239-242 (2017).

[10] M. Namai, J. An, H. Yano, and N. Iwamuro, "Investigation of Short-Circuit Failure Mechanisms of SiC MOSFETs by Varying the DC Bus Voltage, Japanese Journal of Applied Physics, vol.57, 074102 1-10, (2018).

[11] M. Namai, J. An, H. Yano and N. Iwamuro, "Experimental and Numerical Demonstration and Optimized Methods for SiC Trench MOSFET Short-Circuit Capability," in Proceedings of the International Symposium on Power Semiconductor Devices and ICs, 2017, pp. 363-366

[12] X. Jiang, J. Wang, J. Lu, J, Chen, X. Yang, Z. Li, C. Tu, and Z.S. Shen, "Failure mode and mechanism analysis of SiC MOSFET under short-circuit condition," Microelectronics Reliability, vol.89-90, pp.593-597, (2018).

[13] M. Okawa, R. Aiba, T. Kanamori, H. Yano, N. Iwamuro and S. Harada, "Experimental and Numerical Investigations of Short-Circuit Failure Mechanisms for State-of-the-Art 1.2kV SiC Trench MOSFETs," 2019 31st International Symposium on Power Semiconductor Devices and ICs (ISPSD), 2019, pp. 167-170, doi: 10.1109/ISPSD.2019.8757617.

A High Power Density Chip-on-Chip Gan-based Module with Ultra-Low Parasitic Inductance

Yi Zhang
School of Electrical and Electronic
Engineering
Huazhong University of Science and
Technology
Wuhan, China
D201880465@hust.edu.cn

Zongheng Wu
School of Electrical and Electronic
Engineering
Huazhong University of Science and
Technology
Wuhan, China
D201780377@hust.edu.cn

Cai Chen
School of Electrical and Electronic
Engineering
Huazhong University of Science and
Technology
Wuhan, China
caichen@hust.edu.cn

Yong Kang
School of Electrical and Electronic
Engineering
Huazhong University of Science and
Technology
Wuhan, China
ykang@hust.edu.cn

Han Peng
School of Electrical and Electronic
Engineering
Huazhong University of Science and
Technology
Wuhan, China
pengh@hust.edu.cn

Abstract—In this paper, a 650V/120A rated half-bridge chip-on-chip GaN module has been proposed. The proposed module is based on a chip-on-chip structure, which allows the distributing decoupling capacitors integrated very close to each device. These distributing decoupling capacitors can help to equalize the dynamic turn-on current of parallel devices. Based on this structure, the parasitic inductance including the power terminal of the proposed GaN module is only 2.1 nH. The effect of parasitic inductance on parallel devices is analyzed and optimized. By double-sided cooling, the module shows good thermal performance.

Keywords—GaN HEMTs, Packaging, Double-sided cooling.

I. INTRODUCTION

Gallium nitride (GaN) devices have the advantages of lower switching loss, almost no reverse recovery, and lower gate charge, which makes them an excellent candidate for high power density application. However, because of the fast switching transient, the parasitic parameters show a more significant effect on the performance of devices. Because of the large parasitic inductance, A commercial module with a wire-bonding structure is demonstrated only 36% load capacity of the total devices rated current [1]. Literature [2] shows the asymmetrical distributive source parasitic inductance for parallel devices, which is called quasi-common source inductance, will induce severe gate-source voltage resonance. Hence, to make full use of GaN devices, the parasitic parameters need to be optimized during module design.

There has been lots of work about the optimization of parasitic parameters for modules based on vertical SiC devices [3-4], where the concept of a 3D power loop has been proposed to realize mutual inductance cancellation. However, because GaN devices are lateral, these designs are not appropriate for GaN modules. Literature [5] has proposed a GaN module based on direct bond copper (DBC) substrate and printed circuit board (PCB) and Literature [2] has proposed a full PCB GaN module. In these papers, the commutation loop has been optimized to reduce the parasitic inductance (1.5 nH for [5], and 3.2 nH for [2]). However, the power terminals have not been included in the mutual inductance cancellation loop and still have large parasitic inductance. Further, the equalization of turn-on current for parallel transistors has not been considered.

This paper proposes a 650V, 120A rated half-bridge chip-on-chip GaN module based on GS66508T from GaNsystem. A vertical current commutation loop with ultra-low parasitic inductance (2.111nH for power loop) is formed by the chip-on-chip structure. With the integrated distributed decoupling capacitors, the dynamic current distribution of parallel MOSFETs during turn-on transient has been improved. Further, by optimizing the power terminal, the high-side parasitic inductance is designed symmetrically with the low-side, which can improve the EMI performance [6]. The effect of quasi-common source inductance is easily restrained by distributed decoupling resistors. To improve the heat dissipation, both the bottom and top of the module are soldered to a heat sink.

II. STRUCTURE OF PROPOSED GAN MODULE

The cross-section of the proposed half-bridge GaN module is shown in Fig. 1. A chip-on-chip structure is established for lower parasitic inductance. This structure consists of two PCBs, GaN HEMTs GS66508T, and two DBC substrates. To realize the desired 120 A current-carrying capacity, one single leg consists of four GaN HEMTs in parallel. At the bottom and top of the proposed module, the top-side cooling pads of GaN HEMTs are soldered to an aluminum nitride DBC and then soldered to an aluminum heat sink for better heat dissipations. In the middle, two three-layer PCBs are used to connect the electrodes of devices. Four GaN HEMTs are mounted on each PCB to form the upper and lower leg. Accordingly, the positive and negative DC bus is on the top and bottom PCB respectively. Between two PCBs, the upper and lower leg of the bridge is connected via holes, and the output terminal is extracted between two PCBs. The horizontal view of the PCBs is shown in Fig. 2. The drain of the upper devices is soldered on the top PCB with the +DC bus, and the source is connected to the output pad through the via hole. Then, the bottom PCB is soldered with the output pad of the top PCB directly. On the bottom PCB, the drain of lower devices is connected with the output pad through the via

978-1-6654-4817-8/21 $31.00 © 2021 IEEE

hole, and the source is connected with -DC bus. Therefore, the vertical commutation loop is established, which can significantly shorter the length of the power communication loop and reduce the parasitic inductance.

To optimize the parasitic, around the PCBs, decoupling capacitors (C_{de}) are soldered to +DC and -DC bus vertically. The structure allows the decouple capacitors to be distributed very close to each GaN HEMT, which can further reduce the parasitic inductance, and equalize the dynamic turn-on current of parallel devices.

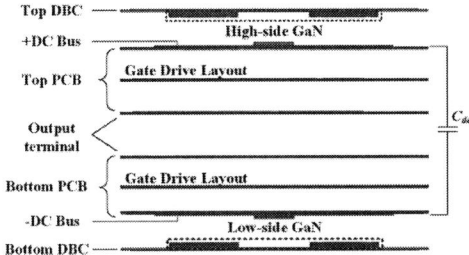

Figure 1: the cross-section of the proposed GaN module.

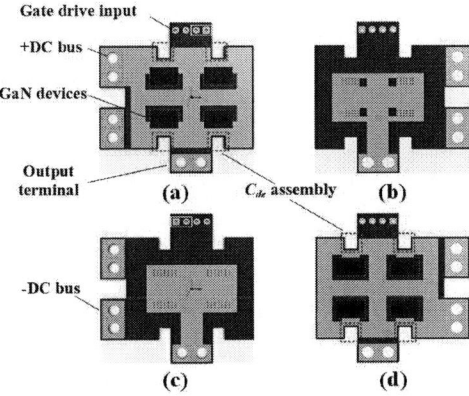

Figure 2: (a) top picture of the top PCB; (b) bottom picture of the top PCB; (c) top picture of the bottom PCB; (d) bottom picture of the bottom PCB:

III. MODELING AND SIMULATION

A. Parasitic parameters extraction

For wide band-gap devices with high switching speed, the parasitic will affect the switching behavior significantly. Therefore, the parasitic inductance is extracted by Maxwell Quick 3D. The power loop and equivalent circuit are illustrated in Fig. 3 (a), and the simulation model is illustrated in Fig. 3 (b). Without considering the integrated distributed decoupling capacitors, the details of the extracted parasitic inductance are shown in Table I. As we can see, both the total inductance (L_{sl} and L_{dh}) and the respectively distributed inductance (L_1 to L_8) of high-side and low-side are very symmetrical. As proposed in the literature [6], the symmetry of parasitic inductance can reduce the radiation EMI. Thanks to the proposed vertical commutation loop structure, the parasitic inductance of the power loop is only 2.111nH (include the power terminal). The ultra-low power loop inductance can reduce the turn-off voltage overshoot, and helps to take full advantage of the high switching speed of the devices. However, as

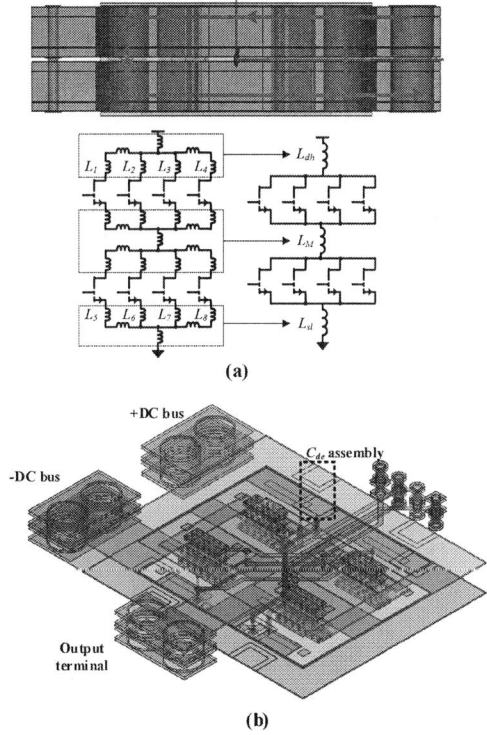

Figure 3: (a) schematic and equivalent circuit of the communication loop; (b) Q3D model.

mentioned in Table I, the respective parasitic inductance from each device to the power terminal varies greatly (from 1.005 nH to 7.073 nH). These differences in distributed parasitic inductance will lead to problems, such as the gate-source voltage oscillation and unbalance in turn-on current distribution, which will increase the total loss and the difficulty of the thermal design. To handle the problems induced by the unsymmetrical parasitic inductance, the distributed decoupling capacitors and distributed gate resistance are used, which are discussed in the following.

TABLE I. EXTRACTED PARASITIC INDUCTANCE

High-side	L_{dh}	L_1	L_2	L_3	L_4	L_M
Value (nH)	0.888	1.005	3.522	5.316	6.820	0.0231
Low-side	L_{sl}	L_5	L_6	L_7	L_8	
Value (nH)	0.992	1.127	3.940	5.579	7.073	

B. Dynamic current equalization with integrated distributed decoupling capacitors

As mentioned above, one problem caused by the asymmetric parasitic inductance for parallel devices is the unbalanced turn-on transient. The unbalanced turn-on transient will lead to a huge difference in turn-on transient current between the parallel devices, which leads to unbalanced turn-on loss. Because the di_d / dt during turn-on transient will induce the inductive voltage on the parasitic inductance, the turn-on voltage of the parallel switches is different. Because the turn-on voltage is different, the turn-on current of the parallel devices is also different. Since the device farther away from the DC terminal suffers the higher inductive voltage, the devices closest to the power terminal

978-1-6654-4817-8/21 $31.00 © 2021 IEEE

will bear the maximum turn-on current. If not optimized, the unbalanced transient will increase the heat burden or even damage the device. Therefore, the distributed decoupling capacitors are integrated close to each GaN HEMTs to reduce the effect of L_1 to L_8. As shown in Fig. 2(b), the distributed decoupling capacitors are mounted around the PCB symmetrically, which is very close to the devices. These distributed decoupling capacitors can help to maintain the turn-off voltage of the parallel devices, and make the commutation loop of each parallel device symmetrical. Besides, these distributed decoupling capacitors can further reduce the communication loop and parasitic inductance, which can further decrease the turn-off overshoot voltage of the devices.

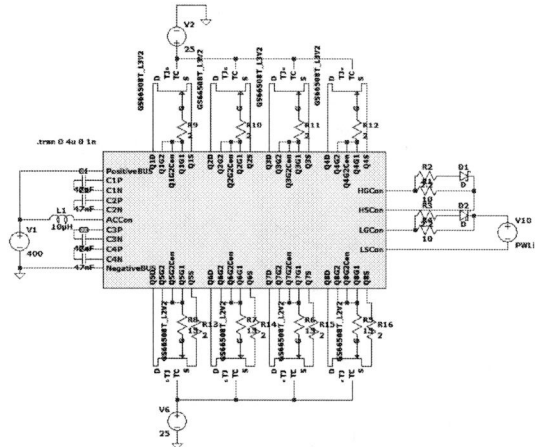

Figure 4: LTspice model

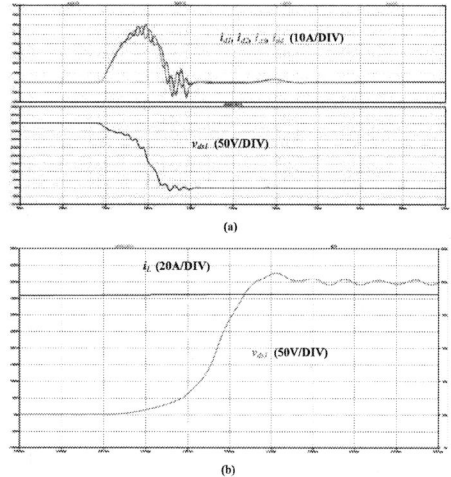

Figure 5: (a) schematic and equivalent circuit of the communication loop; (b) Q3D model.

The improvement of the distributed decoupling capacitors is verified by LTspice simulation. The simulation model is shown in Fig.4, and the simulation results are shown in Fig. 5 (a). As we can see, because the turn-on transient equilibrium is improved by the distributed decoupling capacitors, the turn-on drain current (i_d) of the parallel devices is approximately the same with 100 nF distributed decoupling capacitors. For the turn-off transient,

as shown in Fig. 5 (b), because the parasitic inductance is further reduced, the overshoot of turn-off drain-source voltage is only 26 V, when the devices are turned off at 90 A / 400 V.

C. Decoupling resistors to restrain gate-source resonance caused by GaN devices paralleling

Besides, the asymmetric parasitic inductance will cause severe gate-source resonance, which will cause false turn-on and damage the devices. For simplicity, the analysis is based on two devices in parallel, and Fig. 6(a) shows the equivalent circuit. Although the source pads of each parallel device are connected with the -DC bus or output terminal, the distributed parasitic inductance from the source pad of the devices to the -DC bus is different, which results in the difference in the respective source parasitic inductance (L_{qs1}, and L_{qs2}). Therefore, when the devices are turned off, even though the turn-off current of these devices is the same, the different voltages will be induced on L_{qs1}, and L_{qs2}. This difference in induced voltage will cause serious gate-source voltage v_{gs} oscillation through the loop shown in Fig. 6(b). As mentioned in literature [2], even nH-level differences in the respective source parasitic inductance will cause huge gate-source voltage oscillations,

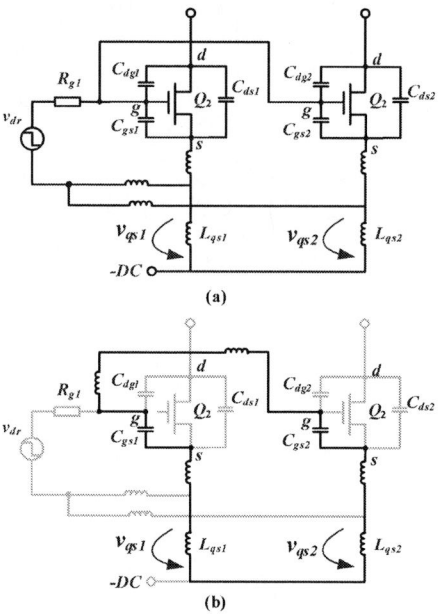

Figure 6: (a) Equivalent circuit of the parallel devieces; (b) gate-source oscillation loop.

which will lead to the false turn-on.

Factors that can cause the different inductive voltage can induce the gate-source voltage oscillations. First, the unbalanced parasitic inductance in the power loop will lead to the unbalanced distribution of the turn-on and turn-off transient current. Therefore, the unbalanced di_d / dt will induce different voltages on the distributed source parasitic inductance. The above problems caused by unbalanced dynamic current equalization can be solved by distributed decoupling capacitors. Secondly, the gate inductance will influence the switching speed, and then affect the dynamic di_d /dt. To solve the gate-source oscillation problem by

978-1-6654-4817-8/21 $31.00 © 2021 IEEE   446

optimizing the layout, it is necessary to maintain a highly symmetrical parasitic inductance both in the gate and power loop. However, when more than two devices are connected in parallel, designing a highly symmetrical layout is very difficult.

Even though the layout is symmetrical, the fabrication difference in devices can also lead to the above problems. Therefore, in the proposed module, the gate-source oscillation is handled by decoupling the oscillation loop. To decouple the gate-source oscillation loop, the gate of each device is connected to a decoupling resistor and then aggregated to the gate driver. The decoupling resistor can greatly increase the impedance of the oscillation circuit to avoid the false turn-on. The improvement of the decoupling resistor is verified by Ltspice simulation. As shown in Fig. 7, only a 2 Ω decoupling resistor can restrain the resonance effectively. In the proposed module, in order not to reduce the switching speed of the device, the original centralized gate resistance R_{g1} is reduced to make the equivalent gate resistance the same.

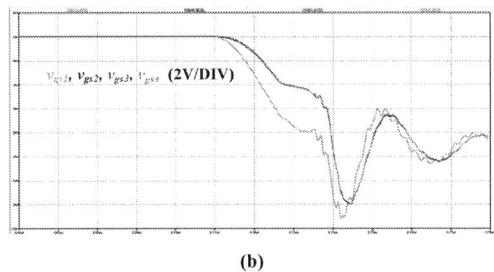

(b)

Figure 7: LTspice simulation result with decoupling resistors.

D. Thermal Simulation

As depicted in Fig. 1, two DCBs are soldered on the top-sided cooling pad for insulation, and then soldered to the heat sink respectively. Soldered to the alumina ceramic substrate, the thermal resistance between the cooling pad of the GaN HEMTs and the heat sink can be reduced. The heat sinks above and below the module provides independent heat dissipation paths for the upper and lower leg. The thermal performance of the proposed modular is evaluated by Flotherm thermal simulation. The simulation results are shown in Fig. 8. In the simulation, the switching frequency is 500kHz, and the turn-off voltage/current is 400V / 70A.

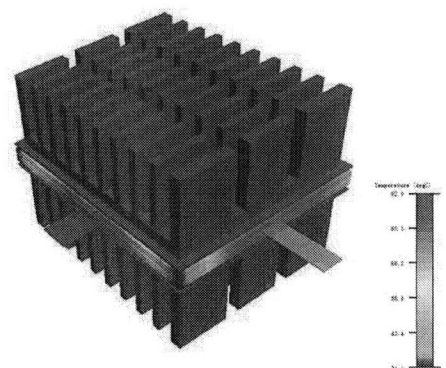

Figure 8: fabrication process of the proposed module.

As seen from the simulation results, the maximum temperature rise is 61.8 ℃ with 3.14m/s forced air cooling. The simulation results show the proposed double-sided cooling structure has a good heat dissipation ability.

IV. FABRICATION AND EXPERIMENTAL VERIFICATION

Finally, the proposed GaN-based module is fabricated in the reflow furnace. As mentioned in Section II, the proposed modular consist of two PCB, and two DBC. To ensure the effective welding between the substrate and devices, solders at different welding temperatures are used to fabricate the proposed module. Fig. 9 shows the pictures of the proposed modules in different fabrication processes. Firstly, 227 ℃ solder (sn99ag0.3cu0.7) was used to soldered GaN HEMTs to PCBs. The upper and lower PCBs were simultaneously welded in the reflow furnace. After the devices are welded, a 183℃ solder (Sn63Pb37) is used to weld the upper and bottom DBC. Then, a 172℃ solder (Sn64.7 bi35ag0.3) is used to weld these PCBs together. Finally, a 138℃ solder (Sn42Bi58) is used to weld the heat sink.

Figure 9: fabrication process of the proposed GaN HEMTs based module.

V. CONCLUSIONS AND FUTURE WORK

A chip-on-chip GaN module has been proposed in this paper. The proposed vertical structure provides ultra-low parasitic inductance (2.111nH for power loop including the terminals). For parallel GaN devices, the asymmetric distributed parasitic inductance will lead to unbalanced turn-on transient and severe gate-source resonance. Therefore, this paper proposes a design of decoupling capacitors and resistors to reduce the effect of asymmetric distributed parasitic inductance. The above design is verified by simulation.

The proposed model has been fabricated. Due to time constraints, the following experimental verification will be done in the future.

REFERENCES

[1] J. A. Brothers and T. Buechner, "GaN Module Design Recommendations Based on the Analysis of a Commercial 3-Phase GaN Module," *2019 IEEE Energy Conversion Congress and Exposition (ECCE)*, Baltimore, MD, USA, 2019, pp. 4109-4116.

[2] J. L. Lu and D. Chen, "Paralleling GaN E-HEMTs in 10kW-100kW systems," *2017 IEEE Applied Power Electronics Conference and Exposition (APEC)*, Tampa, FL, 2017, pp. 3049-3056.

[3] F. Yang, Z. Wang, Z. Liang and F. Wang, "Electrical Performance Advancement in SiC Power Module Package Design With Kelvin Drain Connection and Low Parasitic Inductance," in *IEEE Journal of Emerging and Selected Topics in Power Electronics*, vol. 7, no. 1, pp. 84-98, March 2019.

[4] Z. Chen, Y. Yao, D. Boroyevich, K. Ngo and W. Zhang, "An ultra-fast SiC phase-leg module in modified hybrid packaging structure," *2014 IEEE Energy Conversion Congress and Exposition (ECCE)*, Pittsburgh, PA, 2014, pp. 2880-2886.

[5] J. L. Lu, D. Chen and L. Yushyna, "A high power-density and high efficiency insulated metal substrate based GaN HEMT power module," *2017 IEEE Energy Conversion Congress and Exposition (ECCE)*, Cincinnati, OH, 2017, pp. 3654-3658.

[6] A. Domurat-Linde and E. Hoene, "Analysis and Reduction of Radiated EMI of Power Modules," in *2012 7th International Conference on Integrated Power Electronics Systems (CIPS)*, 2012, pp. 1-6.

978-1-6654-4817-8/21 $31.00 © 2021 IEEE

Modeling and Experimental Verification of Common Mode Crosstalk with Shielded Cables in Power Converter System

Ruizhou Xue
School of Electrical and Electronic Engineering
Huazhong University of Science and Technology
Wuhan, China
m202071568@hust.edu.cn

Xuejun Pei
School of Electrical and Electronic Engineering
Huazhong University of Science and Technology
Wuhan, China
ppei215@hust.edu.cn

Chunyu Yang
School of Electrical and Electronic Engineering
Huazhong University of Science and Technology
Wuhan, China
m201871404@hust.edu.cn

Yi Yu
School of Electrical and Electronic Engineering
Huazhong University of Science and Technology
Wuhan, China
u201711895@hust.edu.cn

Abstract—**With the development of wide bandgap semiconductor technology, the electromagnetic interference (EMI) caused by wide bandgap semiconductor is more complex. In addition, interconnect cable space is becoming smaller due to the trend of systematization and miniaturization of power converters. Therefore, the crosstalk problem between cables cannot be ignored. In order to suppress crosstalk, shielded cables were often selected in practice. However, a lot of interference can still be found in practice applications. In order to estimate the shielding performance of shielded cable in crosstalk suppression, a cascaded active multiport network was proposed for modelling the crosstalk. Firstly, the circuit model was established based on the proposed method. And then, mathematic derivations were carried out for estimating of common mode (CM) crosstalk. Crosstalk can be accurately predicted by the derived formula and the established model. At last, the research was carried out by building an experimental platform. Experimental results verified the correctness of the proposed model.**

Keywords—crosstalk, shielded cables, circuit modeling, matrix

I. INTRODUCTION

With the development of wide bandgap semiconductor technology, the switching rate of wide bandgap power devices such as SiC and GaN is increasing, and the electromagnetic interference (EMI) caused by them is more complex. Now power converter has been widely used in trains and ships. The distance between cables is very close due to the narrow space on ships and trains, so there is a non-negligible CM crosstalk problem on these occasions [1-2]. In order to suppress crosstalk, shielded cables were often selected in practice. However, a lot of interference can still be found in practice applications when shielded cables is selected. Crosstalk may be coupled into weak current circuit or sensitive circuit, such as some control signals, which may cause instability of the control system or even lead to system fault [3-5]. Hence establishing the interference coupling model between shielded cables and explore interference coupling mechanism is very meaningful for practical applications.

Establishing transmission line equations with per-unit-length distribution parameters was the traditional approach for modeling cables. For obtaining the distribution parameters of cables, direct measurement [6], analytical formulas [7] and electromagnetic simulation [8] were utilized. Based on these methods, the modeling of CM conducted EMI in short cables was indicated [9]. For long multi-conductor cables, a recipe aimed to create time-domain and frequency-domain models was presented [10]. However, the above research lacked modeling the crosstalk between two independent cables. Aiming at two sets of shielded cables, most researches utilized the multi-conductor transmission line (MTL) method. On the basis of the MTL, the hybrid solver combining the MTL theory with the method of moments (MoM) was proposed [11]. However, the above study regarded shielded cables as ideal condition, lacking consideration of the influence of shield layer. So a interference coupling model considering the non-ideal characteristics of the shielding layer of shielded cable is proposed in this paper. This paper proposes a cascaded multi-network model to describe the high frequency performance of the shielded cables. The cascaded multi-network model is length scalable, which is suitable for different length of shielded cables. Accurate CM crosstalk prediction can be realized using the method. Besides, establishing the correct crosstalk model can also help to offer effective proposes for avoiding severe crosstalk in the interfered cables.

The paper is organized as follows. Section II introduces the near filed coupling model between shielded cables, and the test bench for modeling is given. In Section III, the cascaded multiport network model of crosstalk is established and the expression of interference is obtained through the equivalent circuit. In Section IV, the correctness of the model is verified by experiments. The last part is the conclusions.

II. A NEAR-FIELD COUPING MODEL BETWEEN SHIELDED CABLES

A. Coupling model between shielded cables

Because of parasitic parameters between cables, when there is high frequency interference in the current flowing

---

Project Supported by National Natural Science Foundation of China (51977091)

978-1-6654-4817-8/21 $31.00 © 2021 IEEE

through one cable, the high frequency current will propagate between cables through the parasitic parameters. In order to establish an accurate crosstalk model, the near-filed coupling mechanism of shielded cables should be studied first. See Fig. 1 below, the 3D model of two dual-core shielded cables was established by Q3D software. In Q3D, different lengths of cables can be modeled and the distribution parameters between cables can be extracted. However, in order to extract the accuracy of parameters, the length of cables should not be too long, otherwise the parasitic parameters of cables cannot be accurately calculated due to the solution method of Q3D. The impedance matrix is extracted by using the finite element analysis, and the distribution parameters of cable are finally obtained after numerical procession.

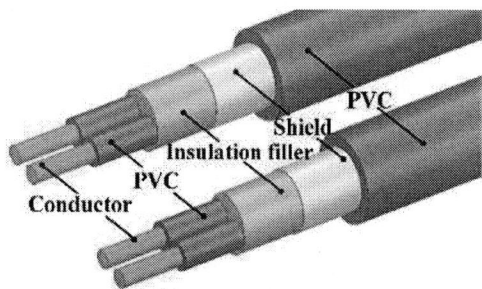

Fig. 1   3D model of shielded cables.

According to the physic structure of the shielded cables and the extraction parameters results of Q3D, the cable coupling model per unit length is drawn, which is illustrated in Fig. 2. The parasitic parameters between conductors that are farther apart are smaller than those that are closer together. Therefore, only the coupling parameters between conductors close to each other are considered. In Fig.2, L_2 and L_1 are the self-inductance of shield layer and conductor respectively. M_1 is mutual inductance between conductor and shield layer and M_2 is between shield layers. C_1 and C_2 are the parasitic capacitors between conductor and shield layer, and between shield layers respectively.

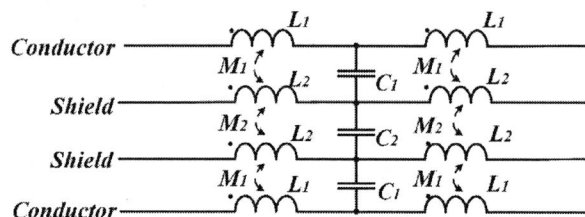

Fig. 2   Cable coupling model per unit length

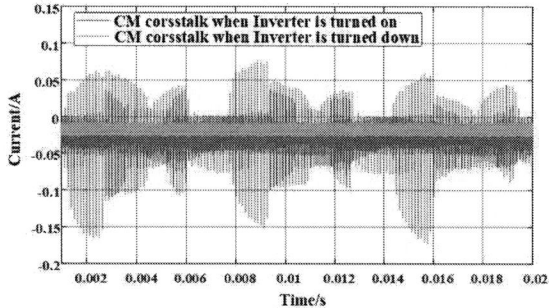

Fig. 3   Crosstalk in the signal circuit

Crosstalk process between shielded cables can be explained as follows. When high frequency electric current flows through a conductor, high frequency current will coupling to the shield layer through the mutual inductances and parasitic capacitors C1 and C2. Because there is mutual inductance and parasitic capacitance between shielding layers, high frequency current will couple from one shield layer to the other shield layer, and then flow into the conductor, resulting crosstalk. In the actual measurement, the phenomenon of crosstalk can be clearly observed, which shows in Fig. 3.

B. The Test Bench for Crosstalk Modeling

In order to study the crosstalk phenomenon, Fig. 4 shows the test bench building in laboratory which consists of two parts. One was the power electronic circuit, containing DC power supply with adjustable voltage, LISNs and PWM inverter powered by shielded cables, and the other was the switching mode power supply power to resistive load through shielded cables. The LISNs and shield layers are connected to the same test ground. The measurement position of crosstalk is in the signal circuit close to LISN side.

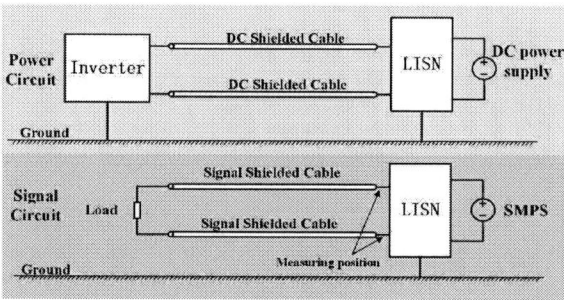

Fig. 4   The test bench for crosstalk modeling.

Meantime, a 2m×2m aluminous plate was used as the reference ground. All the equipment is placed on the aluminum plate with grounding wire connected to the aluminum, and all cables were placed 30 cm above aluminum plate. In the test bench, the PWM inverter will produce a lot of EMI interference in the cable due to the high-speed switching action of the switching device. A large conducted EMI would be generated in the DC cables, which would couple to the interfered circuit through the distribution parameters between the cables. In the next section, the propagation path of interference between shielded cables will be analyzed and the equivalent model of crosstalk will be established.

III. Math Modeliing Of Crosstalk Between Shielded Cables

A. The transmission path of shielded cable crosstalk

Taking the nonideal characteristic of the shield into consideration, the high frequency crosstalk model of the shielded cables is equivalent to a multiport crosstalk model. Fig. 5 shows multiport crosstalk model of the shielded cables. C1 and C2 are the parasitic capacitors between conductor and shield obtained by finite element analysis. C3 is the parasitic capacitance between shielding layers. M1, M2 and M3 are mutual inductance between conductors. The shield layers of the shield cables are single-terminal grounding, which link to the reference ground.

978-1-6654-4817-8/21 $31.00 © 2021 IEEE   450

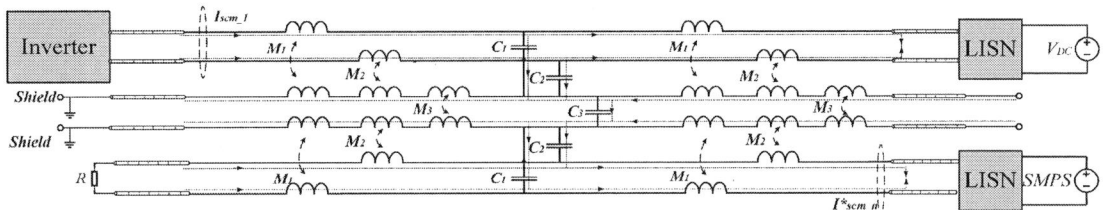

Fig. 5 The model of shielded cables and CM crosstalk's conduction path

The inverter will generate high dv/dt in operation, causing conducted interference flowing in the power circuit. The common-mode interference in the power circuit is coupled to the shielding layer of the power circuit cable through the parasitic parameters. Later, due to the non-ideal characteristics of the shielding layers, the interference will also enter the signal circuit through the parasitic parameter coupling between the shielding layers, and finally generate common-mode interference in the signal circuit.

The conduction path of CM crosstalk can be obtained based on the multiport crosstalk model of the shielded cables. The conduction path of common-mode interference is indicated in Fig.5. Since the common-mode interference is the same on both conductors of a dual-core shielded cable, the common-mode conduction path can be simplified. The coupling model of cable CM interference is indicated in Fig.6. The simplified equivalent circuit is a three-port network. I_{scm_1} is the CM interference generated in the power circuit. Because of mutual inductance and parasitic capacitance between the shielding layers, common-mode interference in the power circuit can be coupled into the signal circuit through distributed parameters. Finally, it forms the CM crosstalk current which $I^*_{scm\_1}$ is in the interfered cables.

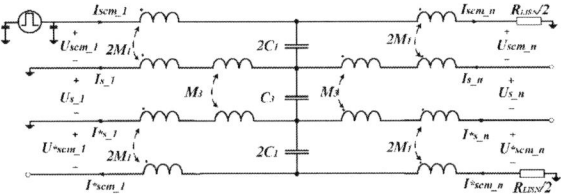

Fig. 6 Coupling model of shielded cable CM interference

B. Mathematical derivation of common mode crosstalk current

From one cell of the CM crosstalk simplified model, establishment of equivalent circuit model is down. One cell of equivalent crosstalk circuit model is shown in Fig. 7, in

which $Z_{L1}= j\omega L_1$, $Z_{L2}=j\omega L_2$, $Z_{C1}=-1/j\omega C_1$ and $Z_{C2}=-1/j\omega C_2$. According to the equivalent circuit in Fig.7, the transfer matrix of interference from the left to the right can be derived. From it, the matrix (1) is obtained through theoretical calculation of circuit, the value of A-T can be seen in the appendix. From the matrix (1), it can be got that there is a recursive relation between the voltage and current in one cell of the CM crosstalk circuit model. It can be got by parity of reasoning:

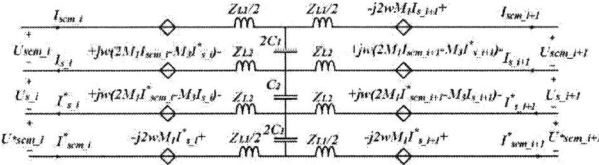

Fig. 7 CM crosstalk circuit model

$$
\begin{bmatrix} I_{scm\_i-1} \\ I_{s\_i-1} \\ I^*_{s\_i+1} \\ I^*_{scm\_i+1} \\ U_{scm\_i+1} \\ U_{s\_i-1} \\ U^*_{scm\_i+1} \end{bmatrix} = \delta_i^* \begin{bmatrix} I_{scm\_i} \\ I_{s\_i} \\ I^*_{s\_i} \\ I^*_{scm\_i} \\ U_{scm\_i} \\ U_{s\_i} \\ U^*_{scm\_i} \end{bmatrix} \Rightarrow \begin{bmatrix} I_{scm\_n} \\ I_{s\_n} \\ I^*_{s\_n} \\ I^*_{scm\_n} \\ U_{scm\_n} \\ U_{s\_n} \\ U^*_{scm\_n} \end{bmatrix} = \delta_1^* \delta_2^* \cdots \delta_n^* \begin{bmatrix} I_{scm\_1} \\ I_{s\_1} \\ I^*_{s\_1} \\ I^*_{scm\_1} \\ U_{scm\_1} \\ U_{s\_1} \\ U^*_{scm\_1} \end{bmatrix} \quad (2)
$$

From the matrix (2), the CM crosstalk current $I^*_{scm\_n}$ is related to U_{scm_1}, I_{scm_1} and I_{scm_n}, which is complex and need to be further simplified. From the coupling model of shielded cable CM interference, boundary conditions can be deduced. There are two boundary conditions, $I_{scm\_1} = I_{s\_n} = I^*_{s\_n} = 0$ and $U_{s\_1} = 0$, respectively. Combining matrix (2) and boundary conditions, it can be deduced as the result of (3).

From the equation (3), It can be seen that the CM crosstalk in the interfered cable is related to the I_{scm_1} and transfer impedance. The CM crosstalk current $I^*_{scm\_n}$ in the interfered cables can be estimated by multiplying the I_{scm_1} in the DC cable and corresponding transfer impedance in the frequency domain. The CM crosstalk current in the interfered cables can

$$
\begin{bmatrix} I_{scm\_i+1} \\ I_{s\_i+1} \\ I^*_{s\_i+1} \\ I^*_{scm\_i+1} \\ U_{scm\_i+1} \\ U_{s\_i+1} \\ U^*_{scm\_i+1} \end{bmatrix} = \begin{bmatrix} \dfrac{2Z_{C3}-j4\omega M_1+Z_{L1}}{2Z_{C3}} & \dfrac{-j2\omega M_1+Z_{L2}}{Z_{C3}} & \dfrac{j\omega M_3}{Z_{C3}} & 0 & -\dfrac{1}{Z_{C3}} & 0 & 0 \\ A & B & C & \dfrac{j\omega M_1}{Z_{C1}} & -\dfrac{2Z_{C1}}{Z_{C3}} & \dfrac{1}{2Z_{C1}} & 0 \\ A & B-\dfrac{j\omega M_3}{Z_{C3}}-1 & D & E & \dfrac{2Z_{C1}}{Z_{C3}} & \dfrac{1}{2Z_{C1}} & \dfrac{1}{Z_{C3}} \\ 0 & -\dfrac{j\omega M_3}{Z_{C3}} & \dfrac{j2\omega M_1+Z_{L2}}{Z_{C3}} & \dfrac{j4\omega M_1+Z_{L1}+2Z_{C3}}{2Z_{C3}} & 0 & 0 & -\dfrac{1}{Z_{C3}} \\ F & G & H & I & J & \dfrac{j2\omega M_1-Z_{L2}-j\omega M_3}{2Z_{C1}} & \dfrac{j\omega M_3}{Z_{C3}} \\ K & L & M & N & O & -\dfrac{Z_{L1}+4Z_{C1}}{4Z_{C1}} & \dfrac{j2\omega M_1+Z_{L2}+j\omega M_3}{Z_{C3}} \\ P & Q & R & S & T & \dfrac{j2\omega M_1+Z_{L2}+j\omega M_3}{2Z_{C1}} & \dfrac{Z_{L1}-2Z_{C3}+2Z_{L3}}{2Z_{C1}} \end{bmatrix} \cdot \begin{bmatrix} I_{scm\_i} \\ I_{s\_i} \\ I^*_{s\_i} \\ I^*_{scm\_i} \\ U_{scm\_i} \\ U_{s\_i} \\ U^*_{scm\_i} \end{bmatrix} \quad (1)
$$

978-1-6654-4817-8/21 $31.00 © 2021 IEEE

be estimated accurately. From equation (3), we can also see that the common mode crosstalk $I^*_{scm\_n}$ can be suppressed by changing the transfer impedance and I_{scm_1}. Therefore, this model can provide guidance for suppressing CM interference generated by crosstalk

$$
\begin{aligned}
I^*_{scm\_n} = [&(\omega^2 L_2 C_1 + 2\omega^2 L_2 C_2 + L_2 C_3 - \omega M_1 C_1 - 2\omega M_2 C_2 - \\
&\omega^4 L_2^2 C_1 C_2 - \omega^4 L_2^2 C_1 C_3 - \omega^4 L_2^2 C_2 C_3 + \omega^2 M_2^2 C_1 C_2 + \omega^2 M_2^2 C_1 C_3 + \\
&\omega^2 M_2^2 C_2 C_3 - 1 + \omega^3 M_1 C_1 C_2 l_2 + \omega^3 M_1 C_1 C_3 l_2 - \omega^2 M_1 M_2 C_1 C_2) \\
&(\omega M_2 + \omega^2 M_2^2 C_2 - \omega^4 L_2^2 C_2) / (\omega^2 C_1 C_3 l_1 - j25\omega C_1 C_3)]^n I_{scm\_1}
\end{aligned}
\tag{3}
$$

IV. EXPERIMENTAL VERIFICATION

As shown in Fig. 8, a test bench is built to study crosstalk. It corresponds to the bench described in Section II B. The main parameters are indicated in Table I. All devices should be placed on the grounding aluminum plate. The distance between the cables and the grounding aluminum is 30cm. The distance between the two cables is 5cm. The inverter in the power circuit is the main device to generate CM interference. Finally, the receiver is used to test the CM crosstalk in the signal circuit.

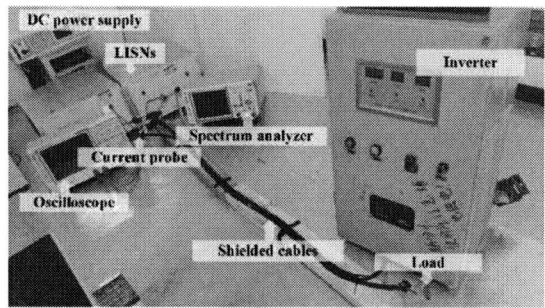

Fig. 8   The test bench for verification

TABLE I.   MAIN PARAMETERS OF THE TEST BENCH

Parameter	Value
Inverter operating frequency	8 kHz
DC voltage of inverter	400 V
Load	5Ω
Voltage of switching power supply	5V
Length of shielded cables	2m

According to the test bench, CM crosstalk current in the signal circuit is obtained through current probe. Then the measured spectrum of CM crosstalk is got through FFT method.

In order to obtain the predicted spectrum, per unit length of the shielded cable' near-field 3d model is built in ANSYS Q3D firstly. Besides, the parameters matrix of the cable near-field coupling is obtained, and then the cascaded crosstalk modeling is built in ANSYS. Finally, the transfer impedance is obtained through frequency sweep in ANSYS, and the predicted spectrum is obtained by multiplying the CM interference in the power circuit with the corresponding transfer impedance in the frequency domain.

Fig. 9 shows the interference measured in the laboratory compared with the interference predicted by the established model. Compared the measured spectrum with the predicted spectrum, It can be seen that the proposed model realize the accurate prediction of CM crosstalk.

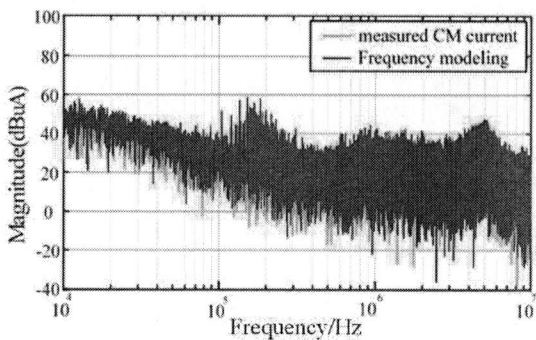

Fig. 9   The predicted spectrum and measured spectrum

V. CONCLUSION

In this paper, the crosstalk coupling model between shielded cables is established considering the non-ideal characteristics of shielded layers. Firstly, the coupling principle between shielded cables is studied. Then by establishing the equivalent circuit of the interference coupling path, the interference of the loop coupling into the signal back is expressed by parasitic parameters through mathematical derivation. Finally, by comparing the experimental results with the simulation results, it can be seen that the CM crosstalk of the shielded cable is accurately predicted.

VI. APPENDIX

$$
A = \frac{Z_{e2}(Z_{e2} - jM_1 + Z_{L1})}{Z_{e2}} - \frac{jM_1}{Z_{e1}}, B = \frac{Z_{e1}(-jM_1 + Z_{L2})}{Z_{e2}} + \frac{Z_{e1} + Z_{e2} - jM_2}{Z_{e1}}
$$

$$
C = \frac{jM_2 Z_{e1}}{Z_{e2}} + \frac{Z_{L2} + jM_2}{Z_{e1}}, D = \frac{jM_2 Z_{e1}}{Z_{e2}} + \frac{Z_{L2} + jM_2}{Z_{e1}} + \frac{Z_{L2} + jM_1}{Z_{e2}} + 1
$$

$$
E = 2 + \frac{jM_1}{Z_{e1}} - \frac{Z_{L1} + jM_1}{Z_{e2}}, F = \frac{(jM_1 - Z_{e2} + Z_{L1})(Z_{e2} - jM_1 + Z_{L1})}{Z_{e2}} + jAM_1 - AZ_{L1} - jAM_2 + Z_{e2}
$$

$$
G = \frac{(-jM_1 + Z_{L1})(jM_1 - Z_{e2} + Z_{L1})}{Z_{e2}} + B(jM_1 - Z_{L2} - jM_2) + jM_2 + \frac{M_2^2}{Z_{e2}}
$$

$$
H = \frac{jM_2(jM_1 - Z_{e2} + Z_{L1})}{Z_{e2}} - jM_2 + (jM_1 - Z_{L2} - jM_2) - \left(\frac{jM_2 Z_{e1}}{Z_{e2}} + \frac{Z_{L2} + jM_2}{Z_{e1}}\right)
$$

$$
I = jM_2 - \frac{(jM_1 + Z_{L1} + Z_{e2})(jM_1 - Z_{L2} - jM_2)}{Z_{e2}} + \frac{jM_1(jM_1 - Z_{e2} + Z_{L1})}{Z_{e2}}
$$

$$
J = -\frac{(jM_1 - Z_{e2} + Z_{L1})}{Z_{e2}} - \frac{Z_{L1}(jM_1 - Z_{L2} - jM_2)}{Z_{e2}}, K = \frac{(Z_{e2} - jM_1 + Z_{L1})(jM_1 + Z_{L1})}{Z_{e2}} - A(2Z_{L1} + Z_{e1}) - Z_{e1}
$$

$$
L = B(2Z_{L1} + Z_{e1}) + \frac{jM_2(jM_1 + Z_{L1} + jM_2)}{Z_{e2}} + Z_{L2} + jM_2 + Z_{e1}
$$

$$
M = \frac{jM_2(jM_1 + Z_{L1})}{Z_{e2}} + jM_1 + jM_2 - \frac{(jM_1 + Z_{L1} + Z_{e2})(jM_1 + Z_{L2})}{Z_{e2}} + \left(\frac{jZ_{e2}M_2}{Z_{e2}} + \frac{Z_{L2} + jM_2}{Z_{e1}}\right)(2Z_{L1} + Z_{e1}) + Z_{L2}
$$

$$
N = -\frac{jM_1}{Z_2}(2Z_{L1} + Z_{e1}) + Z_{L23} + \frac{(jM_1 + Z_{L1} + Z_{e2})(jM_1 + Z_{L2} + jM_2)}{Z_{e2}}
$$

$$
O = -\frac{jM_1 + Z_{e1}}{Z_{e2}} + \frac{Z_{e1}(2Z_{L1} + Z_{e1})}{Z_{e2}}, P = -A(jM_1 + Z_{L1} + jM_2)
$$

$$
Q = Z_{L2} + jM_1 - B(jM_1 + Z_{L2} + jM_2) - \frac{jM_2}{Z_{e2}}(Z_{e2} - Z_{L2} - Z_{L1})
$$

$$
R = -(jM_1 + Z_{L2} + jM_2)\left(\frac{Z_{e2}jM_2}{Z_{e2}} + \frac{Z_{L2} + jM_2}{Z_{e1}}\right) + \frac{(jM_1 + Z_{L2})(Z_{e2} - Z_{L2} - Z_{L1})}{Z_{e2}}
$$

$$
S = jM_1 - \frac{jM_1(jM_1 + Z_{L2} + jM_2)}{Z_{e1}} + \frac{(jM_1 + Z_{L1} + Z_{e2})(Z_{e2} - Z_{L2} - Z_{L1})}{Z_{e2}}, T = \frac{Z_{e1}(jM_1 + Z_{L2} + jM_2)}{Z_{e2}}
$$

REFERENCES

[1] Y. Xiang, X. Pei, and W. Zhou, "A Fast and Precise Method for Modeling EMI Source in Two-Level Three-Phase Converter," IEEE Trans. Power Electron., vol. 34, no. 11, pp. 10650-10664, Jan. 2019.

[2] S. Jahdi, O. Alatise, J. A. Ortiz Gonzalez, R. Bonyadi, L. Ran and P. Mawby, "Temperature and Switching Rate Dependence of Crosstalk in Si-IGBT and SiC Power Modules," IEEE Trans. Ind. Electron., vol. 63, no. 2, pp. 849-863, Feb. 2016.

[3] C. J. Collins and J. R. Bray, "Worst-Case Crosstalk Measurements of Cables-The Multi-Network Analyzer Method," IEEE Trans. Electromagn. Compat., vol. 60, no. 4, pp. 1061-1068, Aug. 2018.

[4] Z. Li, L. L. Liu, J. Yan, A. W. Xu, Z. Y. Niu, and C. Q. Gu, "Anefficient simplification scheme for modeling crosstalk of complex cable bundles

above an orthogonal ground plane," IEEE *Trans. Electromagn. Compat.*, vol. 55, no. 5, pp. 975–978, Oct. 2013

[5] A. Sugiura and Y. Kami, "Generation and Propagation of Common-Mode Currents in a Balanced Two-Conductor Line," IEEE *Trans. Electromagn. Compat.*, vol. 54, no. 2, pp. 466-473, April 2012.

[6] Y. Weens, N. Idir, R. Bausiere, and J. J. Fanchaud, "Modeling and simulation of unshielded and shielded energy cables in frequency and time domains," *IEEE Trans. Electromagn. Compat.*, vol. 42, no. 7, pp. 1876–1882, Jul. 2006.

[7] Y. Tanji and A. Ushida, "Closed-form expression of RLCG transmission line and its application," Electron. Commun. Jpn (Part III: Fundamental Electron. Sci.), vol. 87, no. 4, pp. 1–11, Apr. 2004.

[8] K. B. Smida, P. Bidan, T. Lebey, F. B. Ammar, and M. Elleuch, "Identification and time-domain simulation of the association inverter-cable-asynchronous machine using diffusive representation," *IEEE Trans. Ind. Electron.*, vol. 56, no. 1, pp. 257–265, Jan. 2009.

[9] T. S. Pang, P. L. So, and K. Y. See. "Common-Mode Current Propagation in Power Line Communication Networks Using Multi-Conductor Transmission Line Theory," IEEE Int. Symp. Power-Line Commun. its. Appl., pp. 517–522, Mar. 2007.

[10] I. Stevanović, B. Wunsch, G. L. Madonna and S. Skibin, "High-Frequency Behavioral multi-conductor Cable Modeling for EMI Simulations in Power Electronics," *IEEE Trans. Ind. Informat.*, vol. 10, no. 2, pp. 1392-1400, May 2014.

[11] S. A. Pignari and A. Orlandi, "Long-cable effects on conducted emissions levels," *IEEE Trans. Electromagn. Compat.*, vol. 45, no. 1, pp. 43-54, Feb. 2003.

Analytical Averaged Loss Model of a Three-level NPC-type Converter With SiC Devices

Xinyue Guo
School of Electrical and Electronic Engineering
Huazhong University of Science and Technology
Wuhan, China
xinyueguo@hust.edu.cn

Yue Xie
School of Electrical and Electronic Engineering
Huazhong University of Science and Technology
Wuhan, China
xie_yue@hust.edu.cn

Cai Chen
School of Electrical and Electronic Engineering
Huazhong University of Science and Technology
Wuhan, China
caichen@hust.edu.cn

Yong Kang
School of Electrical and Electronic Engineering
Huazhong University of Science and Technology
Wuhan, China
ykang@hust.edu.cn

Abstract—**This paper proposes an analytical averaged loss model of a three-level Neutral Point Clamped（3L-NPC）converter using SiC devices with sinusoidal based PWM modulation. Both conducting and switching loss have been analyzed. Their relationship with modulation index and power factor angle has also been studied. Finally, simulation has been carried out to verify the accuracy of the model. The result shows that with the parameters only provided by the datasheet, the maximum error of the calculation is 9.3% compared to the simulation, which means the model is an effective and simple way to estimate loss of a three-level NPC-type converter in practical design.**

Keywords—loss model, three-level, npc converter, sic device

I. INTRODUCTION

In the field of aircraft and high-speed drives applications, it is common that the output fundamental frequency is greater than 1kHz, which requires sufficient bandwidths in the control. Therefore, the switching frequency increased to above 50 kHz is often needed to achieve good output performance. However, the efficiency of the system will decrease when switching frequency increases. With loss increasing, the pressure on the cooling system will also increase. Multi-level topology can solve this problem well. Previous studies have shown that the efficiency of a multilevel topology has very low dependence on the switching frequency. Furthermore, when the frequency is higher than 10kHz, the 3L-NPC topology is more efficient than the traditional 2L-VSC [1]. Therefore, the three-level NPC can be well applied in the high-speed field.

With the development of wide-bandgap semiconductor technologies such as silicon carbide, the use of SiC devices can further improve the performance of NPC topology. Compared with Si IGBT, switching loss of SiC MOSFET is smaller even working at a higher switching frequency. SiC Schottky barrier diodes (SBD), basically having no reverse recovery loss, can reduce turn-on losses of IGBT by 35% and 85% compared to ultra-fast and fast Si diodes at 150℃ junction temperature [2]. Therefore, the NPC topology using SiC can greatly improve the efficiency of the converter.

In order to obtain the loss of each device, and its relationship with circuit parameters, such as modulation index and power factor, etc., it is very necessary to build an accurate loss model. Several loss models on multilevel topology have been proposed and studied in previous literatures. Literature [3] introduced a loss model of a 3L T-type converter with Si IGBTs and diodes based on sinusoidal PWM modulation. But it needs extra work to get some parameters because not all parameters can be obtained from datasheet. Literature [4] introduced a loss model of three-level NPC topology based on space vector modulation. In this paper, an analytical averaged loss model of a three-level NPC-type converter using SiC MOSFETs and SiC SBDs with sinusoidal based PWM modulation is proposed. Both the conducting loss and switching loss can be determined with the proposed loss model. And all the parameters needed in the loss model can be obtained from datasheet and design requirement. Their relationship with modulation index and power factor has also been analyzed. Finally, simulation has been carried out to verify the accuracy of the model. The result shows that the maximum error of the calculation is 9.3% compared to the simulation, which means the model is an effective and simple way to estimate loss when selecting devices in the practical applications.

II. LOSS MODEL OF NPC-TYPE CONVERTER WITH SiC DEVICES

A. Device characterization

The three-level NPC-type topology is depicted in Fig. 1. T1-T4 are SiC MOSFETs with anti-paralleled body diode, D5 and D6 are SiC SBDs.

The conducting loss of device can be determined by its forward voltage drop and current while the switching loss can be determined by turn-off and turn-on energy. The forward voltage drop can be modelled by the initial forward voltage drop plus a resistive voltage drop as shown in equation (1).

$$u(t) = V_0 + i(t)R_{on} \qquad (1)$$

Where V_0 is initial voltage drop, i(t) is conducting current, R_{on} is equivalent on-state resistance to represent the linear region of device voltage drop. It has to be noted that the initial voltage drop of SiC MOSFET can be omitted.

The switching energy of SiC MOSFET can be modelled linearly with the device current and switch-off voltage as shown in (2).

$$E_{sw} = E_{test} \frac{i_D}{I_D} \frac{V_{dc}/2}{V_{DS}} \qquad (2)$$

978-1-6654-4817-8/21 $31.00 © 2021 IEEE

where the E_{test} is the switching energy under the test drain-source voltage V_{DS} and test drain current I_D in the datasheet. E_{test} corresponds to E_{on} in the datasheet during turn-on transient and corresponds to E_{off} respectively during turn-off transient. The switch-off voltage of three-level NPC topology is only half of DC-link voltage.

Since SiC SBD has virtually no reverse recovery effect, its switching loss is neglected.

B. Conducting loss model

Assuming the load current as sinusoidal, the conducting loss in one fundamental period can be scaled as equation (3).

$$P_{cond} = \frac{1}{T}\int_0^T u(t)i(t)dt \qquad (3)$$

Replacing the $u(t)$ with equation (1), and using the electric output angle as the integrate variable, the equation (3) can be turned into equation (4).

$$P_{cond} = \frac{1}{2\pi}\int_{\theta_1}^{\theta_2} d(\omega t)[V_0 + I_M \sin(\omega t - \varphi)R_{on}] \bullet \\ I_M \sin(\omega t - \varphi)d(\omega t) \qquad (4)$$

where $d(\omega t)$ is the conducting duty cycle function, φ denotes power factor angle. I_M denotes peak load current.

The integral boundaries in equation (4) depend on the on-time of the device in a fundamental period. The three-level NPC-type converter in this paper is controlled based on sinusoidal PWM modulation. The conducting interval of each device is shown as Fig. 2.

The SPWM wave generated in practice is generally calculated using the regular sampling method. Using triangular carrier, within one carrier cycle, the on-time δ of the device can be obtained as given in equation (5):

$$\delta = T_c M \sin(\omega t) \qquad (5)$$

Where T_c denotes carrier period, M denotes modulation index.

So the duty cycle function can be determined in equation (6):

$$d(\omega t) = \frac{\delta}{T_c} = M \sin(\omega t) \qquad (6)$$

Within one fundamental period, the corresponding duty cycle function of each device in each conducting interval is shown in Table I.

With formula (4) and the duty cycle function in Table I, the conducting loss can be obtained as shown in (7).

$$P_{cond,T1} = P_{cond,T4} = \frac{MI_M^2 R_{DS(on)}}{3\pi}(1 + \cos^2\varphi)$$

$$P_{cond,T2} = P_{cond,T3} = \frac{I_M^2 R_{DS(on)}}{4}$$

$$P_{cond,D5} = P_{cond,D6} = I_M^2 R_f\left[\frac{1}{4} - \frac{1}{3\pi}\left(1 + \cos^2\varphi\right)\right] + \qquad (7)$$

$$\frac{V_0 I_M}{2\pi}\left\{2 - M\left[\left(\frac{\pi}{2} - \varphi\right)\cos\varphi + \sin\varphi\right]\right\}$$

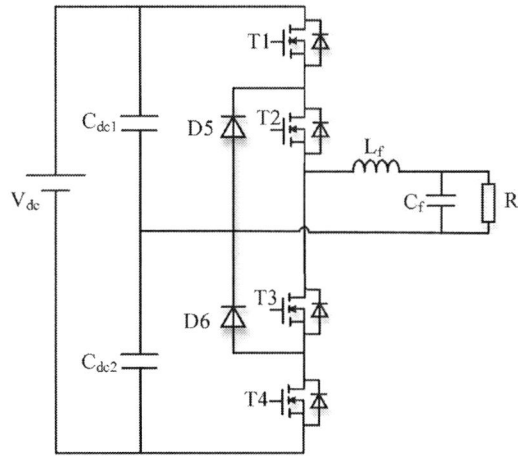

Fig. 1. Single phase three-level NPC-type topology

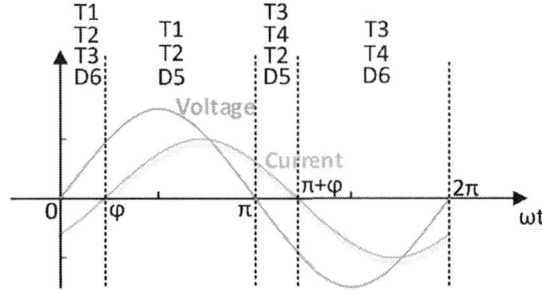

Fig. 2. Conducting interval of each device

TABLE I. DUTY CYCLE FUNCTION OF EACH DEVICE

	(0, φ)	(φ, π)	(π, π+φ)	(π+φ, 2π)
T1	Msinwt	Msinwt		
T2	Msinwt	1	1+ Msinwt	
T3	1- Msinwt		- Msinwt	1
T4			- Msinwt	- Msinwt
D5		1- Msinwt	1+ Msinwt	
D6	1- Msinwt			1+ Msinwt

For SPWM modulation, the loss distribution for the top cell devices (T1, T2, D5) is the same as the bottom cell devices (T4, T3, D6). It can be seen that the conducting loss of T2 and T3 only depends on the peak load current and the on-state resistance, but has nothing to do with modulation index and power factor angle.

C. Switching loss model

The switching loss of SiC MOSFET can be modelled by formula (8):

$$P_{sw} = \frac{f_s}{2\pi}\int_{\theta_1}^{\theta_2} E_{test}\frac{V_{dc}/2}{V_{DS}}\frac{i(t)}{I_D}d(\omega t) \qquad (8)$$

Where f_s represents switching frequency.

The upper and lower limits of the integral in the formula depend on the interval in which the device switches. Take the interval $(0, \pi)$ as an example, T1 and T3 switch complementarily.

TABLE II.   SWITCHING INTERVAL OF EACH DEVICE

	T1	T2	T3	T4
Switching interval	(φ, π)	$(\pi, \pi+\varphi)$	$(0, \varphi)$	$(\pi+\varphi, 2\pi)$

During the period $(0, \varphi)$, due to the anti-parallel body diode of T1, it can achieve ZVS, so the switching loss can be ignored while T3 produces switching loss. During the interval (φ, π), since no current flows through T3, the switching loss caused by T3 can be ignored. During this interval, switching loss occurs on T1. Since T2 remains on during the entire positive half-cycle of load voltage, its switching loss is negligible. The analysis in interval $(\pi, 2\pi)$ is the same. The switching interval of each device is given in Table II. Since SiC SBD has virtually no reverse recovery effect, its switching loss is neglected.

According to equation (8) and the integration boundary in Table II, the switching loss of SiC MOSFET can be obtained:

$$P_{sw,T1} = P_{sw,T4} = \frac{f_s V_{dc} I_M (E_{on} + E_{off})}{4\pi V_{DS} I_D}(1 + \cos\varphi)$$
$$P_{sw,T2} = P_{sw,T3} = \frac{f_s V_{dc} I_M (E_{on} + E_{off})}{4\pi V_{DS} I_D}(1 - \cos\varphi)$$

(9)

It can be seen that T2 and T3 have no switching losses when working at the inverter operation, while T1 and T4 have no switching losses when working at the rectifier operation.

III. CALCULATED LOSS RESULT

According to the loss model, the means conduction loss and switching loss of each device in rectifier mode and inverter mode over one fundamental period can be obtained as shown in Fig. 3. From the figure it can be learnt that the loss distribution is unbalancing. Furthermore, only T2 and T3 produce switching losses in rectifier mode while in inverter mode, T1 and T4 produce switching losses.

Figure 4 shows loss variation with modulation index at a power factor angle of $\pi/4$ and a switching frequency of 50kHz. It can be obtained from the figure that the conduction loss and switching loss of two internal switches, T2 and T3 have nothing to do with the modulation index. The conduction loss of T1 and T4 has a positive relationship with the modulation index; and the conduction loss of the clamp diodes D5 and D6 has a negative correlation with the modulation ratio. When M=0.74, the conduction losses of the four devices are equal.

Figure 5 shows the loss variation with power factor angle when modulation index is 0.4 and the four switches work at frequency of 50kHz. It is able to know from the figure that the conduction losses of T2 and T3 have nothing to do with the power factor angle. The conduction loss of T1 and T4 has a sinusoidal relationship with the power factor angle. When $\varphi=\pi/2$, conduction loss of T1 gets valley, while conduction loss of D5 gets peak. Switching loss of two external switches, T1 and T4 reach maximum in the case of $\varphi=0$.

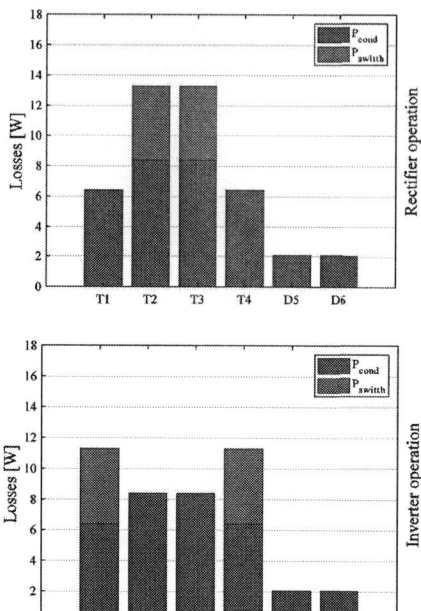

Fig. 3.   Means loss of each device for the rectifier and inverter operation (operation point: M=0.9, I_M=20.5A)

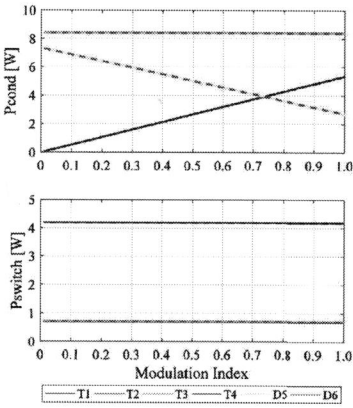

Fig. 4.   Loss variation with modulation index (operation point: $\varphi=\pi/4$, I_M=20.5A)

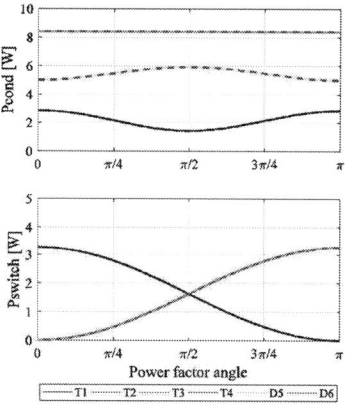

Fig. 5.   Loss variation with power factor angle (operation point: M=0.4, I_M=20.5A)

IV. SIMULATED RESULT

A simulation model has been built in Lt Spice according to the topology shown in Fig. 1. The devices used in the simulation are CPM2-1200-0080B 1200V/24A SiC MOSFETs and CPW5-1200-Z050B 1200V/50A SiC SBDs from CREE. The devices' main parameters needed for calculation are shown in Table III. Working condition of the circuit are shown in Table IV.

In the simulation, the DC-link voltage sets as 1200V, and peak load current sets as 20.5A. Under pure resistive load condition, simulations with different modulation index are carried out. The simulated result at modulation index of 0.5 has been depicted in Fig.6. Load current and load voltage waveforms is depicted in Fig.6 (a), where unity power factor is achieved with sinusoidal current. The output phase voltage which has three level is depicted in Fig.6 (b).

TABLE III. CHARACTERISTICS OF THE MOSFET AND SBD

	V_0/V	R_{on}/mΩ	E_{on}^a/µJ	E_{off}^a/µJ
MOSFET	0	80	265	135
SBD	0.9	14	0	0

ᵃ Test Condition: V_{DS}=800V, I_D=20A

TABLE IV. OPERATING CONDITION OF THE 3L-NPC

Parameter	Description	Value
V_{dc}	DC-link voltage	1200V
I_M	peak load current	20.5A
φ	power factor angle	0
M	modulation index	0.1 ~ 0.9
f_s	switching frequency	50kHz

The simulated loss can be obtained based on the simulation model. Comparison between calculated losses and simulated losses at different modulation index for inverter operation is shown in Fig. 7.

It is able to know from the figure that the simulated loss of diode is very close to the calculated losses. However, it is weird that the simulated loss of SiC MOSFET is a little smaller than the calculated one. Further analysis has found that three reasons lead to this:

1) the actual on-state resistance of SiC MOSFET in simulation (72mΩ) is a little smaller than that in datasheet (80 mΩ),

2) the actual turn-off energy in simulation is smaller than that in datasheet,

3) the actual current on MOSFET is not constant in one carrier period. When MOSFET turns on, the current increases. And it starts to decrease from its peak value in one carrier period when the MOSFET t urns off. So the actual switch-on current is a little smaller than the load current while the actual switch-off current is a little bigger than the load current. Since the turn-on energy of MOSFET used in this paper is nearly two times of turn-off energy, the calculated loss is bigger than the simulated loss.

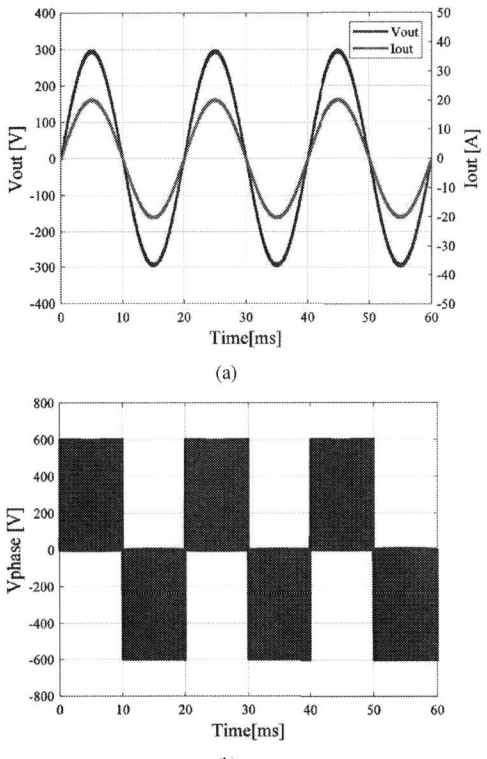

(a)

(b)

Fig. 6. Simulated results (operation point: M=0.5, I_M=20.5A)

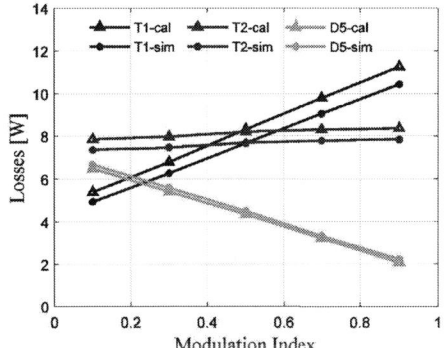

Fig. 7. Simulated loss and calculated loss at different modulation index for inverter operation

Though the data provided by the datasheet are deviated from the actual data, the maximum error is 9.3%. So the loss model is a very simple and effective way to estimate loss with only data from datasheet and design requirements.

V. CONCLUSION

This paper has proposed an analytical averaged loss model of a three-level NPC-type converter with SiC devices. All the parameters needed in the loss model can be gotten in the datasheet and the design requirements, which makes it a very simple way to estimate the loss of each device. Further simulation has verified the loss model in different modulation index. The maximum error is less than 10% although the data used for calculation has deviation from the actual one. So the loss model is effective in practical.

REFERENCES

[1] R. Teichmann and S. Bernet, "A comparison of three-level converters versus two-level converters for low-voltage drives, traction, and utility applications," in IEEE Transactions on Industry Applications, vol. 41, no. 3, pp. 855-865, May-June 2005, doi: 10.1109/TIA.2005.847285.

[2] A. Elasser et al., "A comparative evaluation of new silicon carbide diodes and state-of-the-art silicon diodes for power electronic applications," in IEEE Transactions on Industry Applications, vol. 39, no. 4, pp. 915-921, July-Aug. 2003, doi: 10.1109/TIA.2003.813730.

[3] X. Yuan, "Analytical averaged loss model of a three-level T-type converter," 7th IET International Conference on Power Electronics, Machines and Drives (PEMD 2014), Manchester, UK, 2014, pp. 1-6, doi: 10.1049/cp.2014.0343.

[4] M. Schweizer, T. Friedli and J. W. Kolar, "Comparison and implementation of a 3-level NPC voltage link back-to-back converter with SiC and Si diodes," 2010 Twenty-Fifth Annual IEEE Applied Power Electronics Conference and Exposition (APEC), Palm Springs, CA, USA, 2010, pp. 1527-1533, doi: 10.1109/APEC.2010.5433434.

Dual-Side Three-stage Asymmetric Phase Shift Strategy for Bidirectional Inductive Power Transfer System with SiC Power Module

Haowen Chen, Changsong Chen, Mengjie Jiang, Shuran Jia, Xuezheng Huang
The State Key Laboratory of Advanced Electromagnetic Engineering and Technology, Huazhong
University of Science and Technology, Wuhan, China

Abstract

Power exchange between vehicle and grid by bidirectional inductive power transfer (BIPT) system. This paper proposes a dual-side three-stage asymmetrical phase shift (TAPS) control strategy to achieve a high transmission efficiency with zero voltage switching (ZVS) in a wide range of load. With the proposed dual-side TAPS strategy, asymmetrically adjust the amplitude of the ac fundamental excitation voltage with three adjacent control periods as a waveform period. A wide ZVS range and a high system power factor can be realized simultaneously. In addition, when the load is light, the number of switching actions is significantly reduced. The control scheme for the full bridge of both sides is presented, which realizes both optimization of transmission efficiency and zero-voltage switching operation for all power switches. Simulation are performed to compare the effects of triple-phase-shift (TPS) strategy and proposed TAPS strategy on the transmission efficiency, which shows that the system with proposed TAPS control strategy realizes a higher transmission efficiency in a wide range of load.

Introduction

BIPT system allows power flow wirelessly between grid and vehicle. The power of charging or discharging change in a wide range, so it is necessary for BIPT system to achieve high transmission efficiency in a wide range of transmission power.

In [1], a dual-side phase shift stage is proposed. By changing the width of the voltage square waves on both sides, voltage matching can be achieved, and the overall efficiency can be optimized. However, some power switches operate in hard switching state, which will bring greater switching losses to the system. Paper [2] proposed a triple-phase-shift (TPS) control scheme. In addition to adjusting the width of the square wave voltage on both sides, the phase difference of the primary and secondary voltages is also adjusted. By TPS, a wide range of ZVS is achieved while optimizing efficiency. However, the realization of wide-range ZVS reduces the system power factor.

In this paper, a TAPS control strategy with both a higher system power factor and a wider ZVS range for BIPT system with SiC is proposed. Compared with the traditional control strategy, when adopting TAPS strategy, system transmission efficiency is significantly improved in a wide range of load.

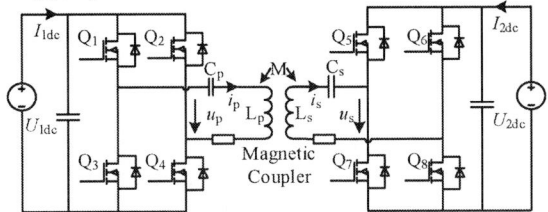
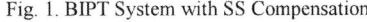

Fig. 1. BIPT System with SS Compensation

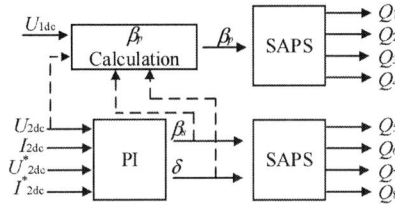

Fig. 2. Block Diagram of the Control Strategy

Proposed Dual-Side Three-stage Asymmetric Phase Shift Strategy

Fig. 1 shows the main circuit of the series-series compensated BIPT system. U_{1dc} and U_{2dc} are the dc bus voltage of both sides respectively. The converters of primary and secondary are composed of MOSFETs $Q_1 \sim Q_4$ and $Q_5 \sim Q_8$. C_p and C_s are the resonant compensation capacitors of both sides respectively. And L_p,

L_s are coil inductance of primary side and secondary side . M is the mutual inductance between the primary coil and secondary coil. The BIPT system works in resonance.

Fig. 2 shows the control method of the BIPT system. The secondary-side controller regulates U_{2dc} or I_{2dc} by adjusting equivalent pulse width β_s. Primary-side equivalent pulse width β_p can be calculated by β_s using (1). Define three adjacent switching periods as a waveform period. In each waveform period, the output square wave voltage of the converter is the same. As shown in the Fig. 3 and Fig. 4, in every waveform period, the positive(negative) pulse widths are β_1, β_2, β_3. When transmission power changes, β_p and β_s will be adjusted by sequentially adjusting β_2, β_1, β_3 in three stages to meet power output requirements and improve transmission efficiency.

$$\beta_p = 2\arcsin(U_{2dc}\sin(\beta_s/2)/U_{1dc}) \tag{1}$$

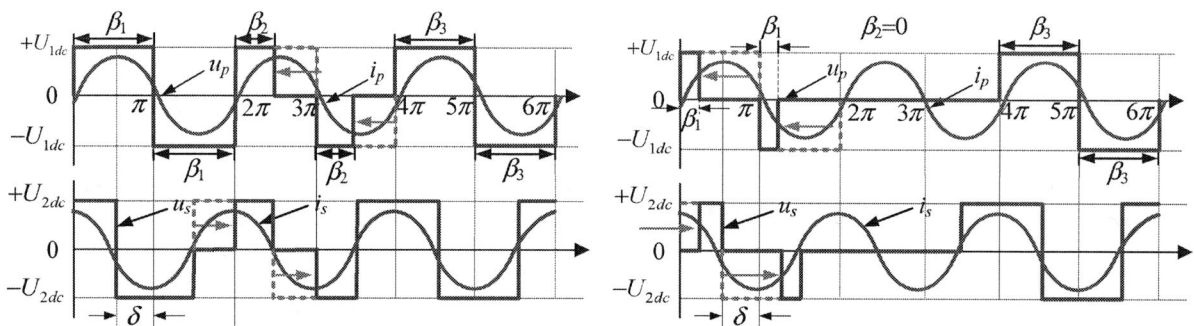

Fig. 3. Voltage and Current waveform of the Stage.1   Fig. 4. Voltage and Current waveform of the Stage.2

At the first stage, it is easy to find Q_1, Q_3 are easier to realize ZVS than Q_2, Q_4, and Q_5, Q_6 are easier to realize ZVS than Q_7, Q_8, so just asymmetrically move the right edge of the positive pulse or negative pulse of the primary side to the left, and the left edge of the positive pulse or negative pulse of the secondary side to the right as the Fig. 3 shows.

The amplitude $|A_m|$ and the phase θ relative to resonant current of fundamental component of u_p and u_s can be obtained by Fourier decomposition (2)~(3). If the primary-side ac current and secondary-side ac current are both at the critical ZVS, then θ is also the power factor angle of the BIPT system.

$$u_s(t) = a_1\sin(\omega_s t) + b_1\cos(\omega_s t) = |V_{sm}|\sin(\omega_s t + \theta + \delta) \tag{2}$$

$$\begin{cases} a_1 = \dfrac{1}{6\pi}\int_0^{6\pi} f(\omega t)\,d(\omega t) = \dfrac{10-2\cos\beta}{3\pi} \\[2mm] b_1 = \dfrac{1}{6\pi}\int_0^{6\pi} f(\omega t)\,d(\omega t) = \dfrac{2\sin\beta}{3\pi} \end{cases}, \quad \begin{cases} |V_m| = \sqrt{a_1{}^2 + b_1{}^2} = \dfrac{\sqrt{104-40\cos\beta}}{3\pi} \\[2mm] \theta = \arctan\dfrac{b_1}{a_1} = \arctan\dfrac{2\sin\beta}{10-2\cos\beta} \end{cases} \tag{3}$$

As shown in the Fig.7, for example, when β_2 decreases from $180°$ to $0°$, the power factor θ first decreases and then increases, and the unit value of fundamental voltage $|A_m|$ gradually decreases from 1 to 0.67. In other words, as the transmission power of BIPT system continues to decrease, the transmission efficiency decreases first and then increases. The system theoretically achieves unity power factor again when β_2 decreases to $0°$.

If the transmission power continues to decrease, the second stage starts, as shown in Fig.4. It is similar to the first stage, the transmission efficiency first decreases and then increases as β_1 is reduced. The number of switching actions is reduced to 2/3 of the original. In the third stage, the transmission efficiency first decreases and then increases, too. The number of switching actions will be reduced to 1/3.

Simulation

PLECS is used to establish a BIPT simulation model. SiC power MOSFET SCT3022AL is used as power switch. The voltage of the double-sided DC side are both 300V, and the rated transmission power of the system is 3.7kW. The primary-side and secondary-side loss resistance r_1 and r_2 are 0.5 Ω, resonance inductance L_p and L_s are 183 μH, resonance capacitor C_p and C_s are 19nF, switching frequency and resonance frequency are 85k.

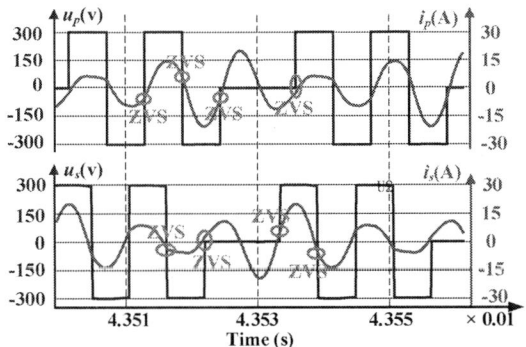

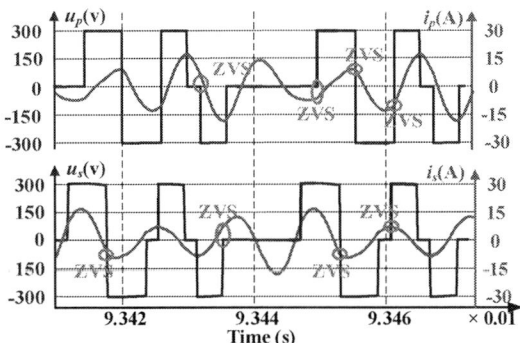

Fig. 5. Voltage and current waveform at 44% Load

Fig. 6. Voltage and current waveform at 35% Load

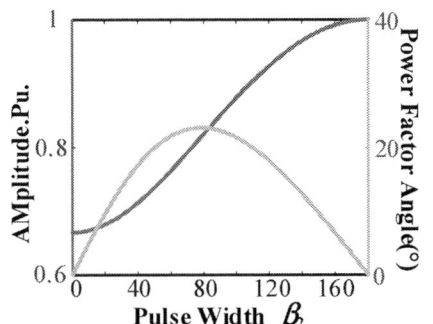

Fig. 7 Parameter relationship

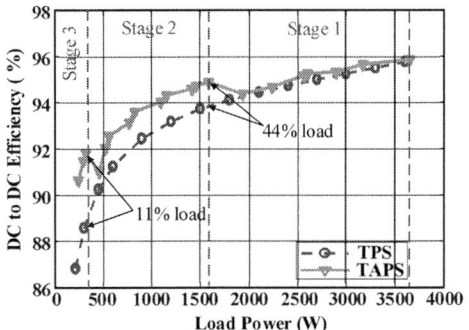

Fig. 8 Transmission Efficiency with TPS and TAPS

The voltage and current waveform of the primary and secondary side at 44% load and 35% load by TAPS strategy is shown in Fig. 5 and Fig. 6. When the system parameters are the same, the system transmission efficiency by TAPS and TPS Strategy is shown in Fig.8. It is found that in the full load range, especially in the 0-4/9 load range, the transmission efficiency by TAPS strategy is significantly higher than that by traditional TPS strategy. It is consistent with the previous discussion that the local extreme points of transmission efficiency probably appear at full load, 4/9 full load and 1/9 full load. In conclusion, the TAPS strategy proposed realizes the better optimization of the system transmission efficiency.

Conclusion

The system power factor and zero-voltage-switching are very important to the improvement of transmission efficiency for BIPT. Especially when the load is light, in order to achieve ZVS, system power factor will be reduced. In this paper, a TAPS strategy is proposed for BIPT system to achieve a higher efficiency in a wide range of transmission power, a higher system power factor and ZVS for all power switches. Comparing with TPS, the transmission efficiency with the proposed TAPS is increased significantly.

[1]　Y. Li, J. Hu, F. Chen, Z. Li, Z. He and R. Mai, "Dual-Phase-Shift Control Scheme With Current-Stress and Efficiency Optimization for Wireless Power Transfer Systems," in IEEE Transactions on Circuits and Systems I: Regular Papers, vol. 65, no. 9, pp. 3110-3121, Sept. 2018.

[2]　X. Zhang et al., "A Control Strategy for Efficiency Optimization and Wide ZVS Operation Range in Bidirectional Inductive Power Transfer System," in IEEE Transactions on Industrial Electronics, vol. 66, no. 8, pp. 5958-5969, Aug. 2019.

A Synchronous Boot-strapping Technique with Increased On-time and Improved Efficiency for High-side Gate-drive Power Delivery

Nathan M. Ellis, Rahul Iyer, Robert C.N. Pilawa-Podgurski
Dept. of Electrical Engineering and Computer Sciences
University of California, Berkeley, U.S.A.
Email: {nathanmilesellis, rkiyer, pilawa} @berkeley.edu

Abstract—In this work we demonstrate a synchronous boot-strap power delivery scheme applied to a high level count GaN-based flying capacitor multi-level (FCML) converter. The proposed approach is well suited for high-frequency operation as it eliminates conventional boot-strap diodes and allows for precise on-time control for a maximized conduction duration. Importantly, we note the synchronous boot-strapping scheme's capability for bi-directional energy transfer: Gate driver power can be injected into the chain from either a ground referenced supply, a high-side line referenced supply, or both simultaneously for further reduced voltage droop. A discrete 6-level FCML hardware prototype switching at 500 kHz (2.5 MHz effective) is designed and constructed to validate this approach. A maximum deviation in supply voltage of 156 mV throughout all 10 series stacked gate drivers for converter duty ratios spanning 15-85% is measured. Subsequently the need for local regulation throughout the gate-drive chain is eliminated, which in turn improves efficiency, simplifies design, and reduces cost.

I. MOTIVATION & BACKGROUND

In the quest towards improved power density, efficiency and cost, industries continue to be receptive to improvements made in micro-assembly and co-packaging, in addition to monolithic solutions, with power electronics modules gaining a significant market share in recent years. To that end, there is a strong push to revisit and improve converter sub-systems that would not have been considered limiting in the recent past.

One essential, yet often trivialized aspect of power converter design is the power delivery scheme for the gate-drivers contained therein. While this circuitry is expected to leave a minimal footprint, improvements in this area have not kept pace with the rapidly developing power converter landscape in which higher order topologies eschew simplicity for dramatic performance improvements [1]–[3]; advanced control schemes and assistive circuitry minimize parasitic effects [4]–[6]; and advanced semiconductor devices, such as Gallium Nitride, consume significantly less area for the same switch conductivity while simultaneously boasting reduced parasitics. Subsequently, switching frequencies have been increased, resulting in the ordinarily dominant volume of passive components reducing considerably [7], [8].

Conventional gate-driver power delivery has historically employed isolated voltage supplies when dealing with topologies with complexity beyond the standard half-bridge structure (e.g. [9]). However, this approach tends to be expensive while consuming considerable converter volume, especially for converters with a high switch count, such as the flying capacitor multi-level (FCML) converter [10]–[12]. In response, in recent years there has been a push in the literature to promote and develop non-isolated gate-driver power delivery techniques with the promise of greatly increased overall converter density [13]–[16].

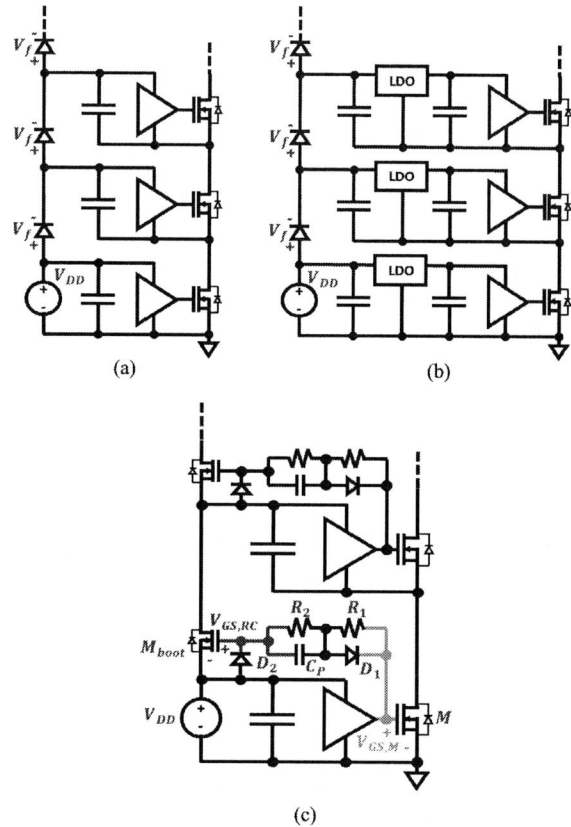

Fig. 1. Prior work on non-isolated gate driver power delivery. (a) Straight-forward cascaded boot-strapping using diodes ([13], [14]); suffers from accumulating voltage droop due to repeated diode forward voltage drops. (b) Cascaded boot-strap with LDO regulation [16]; local regulation ensures correct voltage presented to each gate driver, albeit with poor efficiency. (c) RC-delayed synchronous boot-strapping [15], [17]; removes diodes drops, increasing efficiency and feasible switching frequency, although bootstrap conduction time is still limited by RC delay.

The commonly used boot-strap diode offers a simple and elegant solution for half-bridge structures, with many commercial products even integrating this element on-chip [18]. However, when applied over an increased number of switches, repeated diode drops leads to a significant decrease in voltage provided to each subsequent gate driver (Fig. 1 (a)). While [14] has described a voltage recovery scheme leveraging a finite dead-time duration, this approach is load and timing dependent and only applicable to switches that are subjected to reverse conduction as part of normal converter operation. In GaN-based converters where tight supply tolerances are required, this approach may be considered infeasible in practice.

978-1-6654-4817-8/21 $31.00 © 2021 IEEE

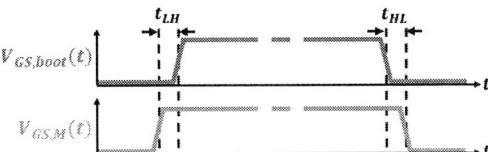

Fig. 2. Timing diagram of ideal V_{GS} waveforms in a synchronous bootstrap solution. Time durations t_{LH} and t_{HL} must be substantial enough to avoid shoot-through currents, but sufficiently short so as to maximize conduction through M_{boot}, reducing voltage droop in a cascaded array of gate drivers.

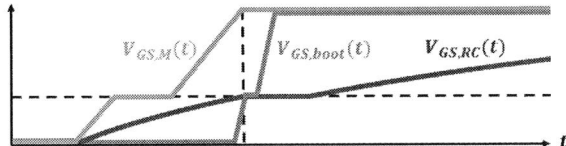

Fig. 3. Example V_{GS} waveforms occurring at the turn-on of primary switching device M (gold). Since the RC-delayed gate signal for M_{boot} (red) begins rising at the same time as $V_{GS,M}$, its slope must be greatly reduced such that M_{boot} does not turn on until $V_{DS,M} = 0V$. Conversely, the proposed buffered signal (green) may turn on sharply, allowing M_{boot} to reach minimum $R_{DS,ON}$ much sooner.

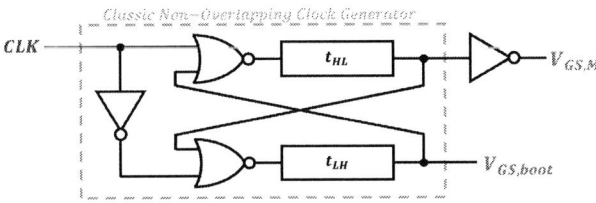

Fig. 4. Appropriate synchronous bootstrap signals may be synthesized locally using modified dead-time control circuitry, and can be easily integrated on-chip within existing gate driving solutions. Time delays t_{HL} and t_{LH} are kept as short as possible while avoiding shoot-through currents within the bootstrapping network.

In [16], this accumulating voltage droop is overcome by increasing the ground-referenced supply voltage (V_{DD}) and introducing local regulation within each gate driver (Fig. 1 (b)). Recognizing that this approach leads to significant LDO-incurred inefficiency, [16] further proposes splitting the power delivery path using gate driven charge pumps, reducing the path impedance and leading to reduced LDO losses. While both of these approaches offer compelling results, they still rely on lossy diodes and ultimately require local regulation.

Alternatively [17] improved on [15] by suggesting a gate-driven synchronous boot-strap using an RC delay network to provide an appropriate gate drive signal to a diode-replacing boot-strap FET (M_{boot} Fig. 1 (c)). Here, M_{boot} must only be turned on within the conduction period of primary FET, M (Fig. 2). To do otherwise risks unintended and potentially damaging large reverse current transients from a high-side bypass capacitor back down to the preceding low-side charge resevoir. In this solution, t_{LH} is defined by the RC time constant of R_1 and the intrinsic gate capacitance, C_{ISS}, of M_{boot}. Here, C_p is large and stores a DC offset equal to V_{DD}, effectively performing voltage level translation. D_1 acts to bypass R_1 during the falling edge and allows M_{boot} to rapidly turn off at the same time as M with a marginally acceptable t_{HL} of zero seconds. D_2 and R_2 are high impedance elements that act to bias C_p at V_{DD} in steady-state.

This approach eliminates diode voltage drops and alludes to the possible omission of local regulation stages [19], [20]. However, the slope of this rising RC-delayed gate signal ($V_{GS,RC}$ in Fig. 3) must be severely limited since it begins rising at the same time as the gate of the primary switching device, M, and must be designed such that M_{boot} does not

turn on until after the V_{DS} of primary switch M has fully discharged to 0 V. As such, M_{boot} does not reach its minimum $R_{DS,ON}$ until much later. For high frequency converters this delayed turn-on can be detrimental and allows the aforementioned voltage droop issues to persist, as recorded in [16].

II. SYNCHRONOUS BOOTSTRAP WITH MAXIMIZED ON-TIME

Here we explore a synchronous boot-strapping technique that instead applies controlled timing delays such that a sharp buffered gate-drive signal can be quickly applied to M_{boot} as soon as $V_{DS,M}$ has reached 0V. This approach avoids the RC-settling observed in $V_{GS,RC}$ (Fig. 3) and enables M_{boot} to conduct for its full allowable duration ($V_{GS,boot}$ in Fig. 3), with minimum $R_{DS,ON}$. In turn, this results in maximum charge being transferred and extends the achievable frequency of operation while minimizing voltage droop on successive cascaded gate drivers.

There are several ways to generate the ideal $V_{GS,boot}$ signal depicted in Figures 2 and 3, such that it maintains precise timing relationships to $V_{GS,M}$, as defined by t_{HL} and t_{LH}. Here, rather than using separate level-shifted clock signals to control M and M_{boot} (with t_{XX} defined within the controller), instead $V_{GS,boot}$ is synthesized locally from the same single level-shifted signal destined for M. To do so, the classic non-overlapping clock generator (Fig. 4) may be modified to include one additional inverter on its output: The result is a $V_{GS,boot}$ waveform that exclusively overlaps with $V_{GS,M}$, thereby providing the desired ideal buffered waveforms depicted in Fig. 2.

To demonstrate this in hardware, existing integrated dead-time circuitry was leveraged, making this approach viable as a fully integrated circuit (IC) gate-drive solution. Figures 5 and 6 depict the gate-drive circuitry used to synthesize appropriate gate signals for both the primary switching device, M, and its associated boot-FET, M_{boot}. The internal dead-time circuitry of a high-speed commercial GaN gate-driver is used to provide tuned delays, with an additional gate-drive IC

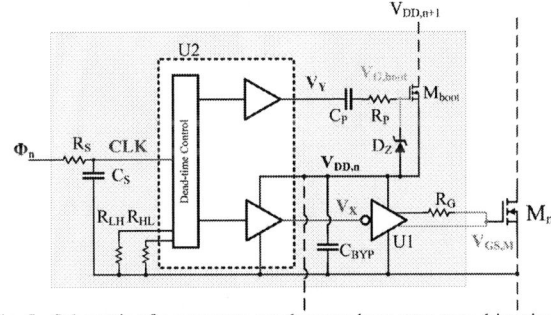

Fig. 5. Schematic of a prototype synchronous boot-strap gate-driver including a dead-time control IC, U_2; an output driver stage, U_1; and a level-shifting charge-pump to drive M_{boot}. The charge-pump may be omitted if M_{boot} is instead made depletion mode.

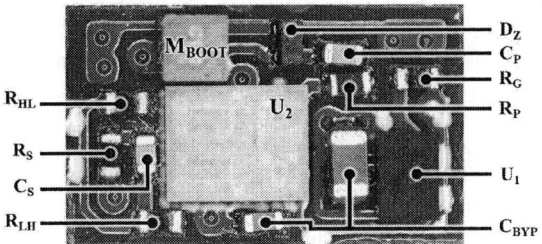

Fig. 6. Photograph of constructed daughter-board proof of concept, containing gate-driver and synchronous boot-strap circuitry.

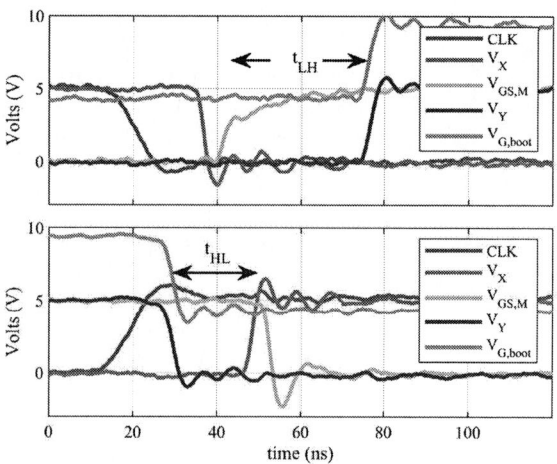

Fig. 7. Measured voltage waveforms demonstrating the correct synthesis of $V_{GS,boot}$ ($= V_{G,boot} - 5V$) with respect to $V_{GS,M}$. Top and bottom figures zoom in on rising and falling edges of $V_{GS,M}$ respectively. $V_{G,boot}$ only goes high after a programmed delay t_{LH}, and goes low before $V_{GS,M}$, avoiding any possibility of shoot-through currents while simultaneously maximizing the conduction window of M_{boot}.

TABLE I
DAUGHTERBOARD COMPONENTS

Component	Description
U_1	Gate Driver, LMG1020
U_2	Dual Gate Driver, PE29102
M_{boot}	100 V 73 mΩ, EPC2036
R_S	200 Ω 0201
C_S	50 pF 0201
R_{HL}	100 kΩ 0201
R_{LH}	750 kΩ 0201
D_z	5.6 V Zener 0201
C_P	10 nF 0201
R_P	10 Ω 0201
R_G	5.1 Ω 0201
C_{BYP}	0.1 μ F (0201), 2.2μF (0402)

providing a required signal inversion. Similar to C_P in Fig. 1 (c), here capacitor C_P maintains a 5V offset allowing a level-shifted drive signal to be applied to M_{boot}. However, we note that a depletion mode device may instead be used, eliminating the need for components C_P, R_P and D_Z (e.g. [21]). R_P is employed here to limit $V_{GS,boot}$ overshoot given that the gate driver producing V_Y is greatly oversized for this application. D_Z ensures that $\sim 5V$ is maintained on C_P and is a discrete alternative to D_2 and R_2 in Fig. 1 (c). R_s and C_s are optional noise filtering components. The time durations of t_{HL} and t_{LH} may be adjusted via R_{LH} and R_{HL}. Measured voltage waveforms validating intended operation are depicted in Figure 7 and the specific component values used are listed in Table I.

III. APPLICATION TO A 6-LEVEL FCML

The described synchronous boot-strapping circuitry was demonstrated within a 6-level FCML converter, depicted in Figures 8 and 9 and in which there are ten GaN switching devices connected in series (M_{1-10}). Consequently, these switches are controlled using ten copies of the daughter-board depicted in Fig. 6. Here, ten much smaller GaN-FETs, serving as M_{boot}, are used to replace the diodes used in a conventional cascaded bootstrap. While validating this boot-strapping approach in hardware, power and voltage levels were kept significantly lower than rated, while the converter was switched at 500 kHz, resulting in the FCML's output inductor seeing an effective 2.5 MHz switching frequency with a line voltage, V_{IN}, of 140V (Fig. 10). Table II lists the components used to construct the power stage.

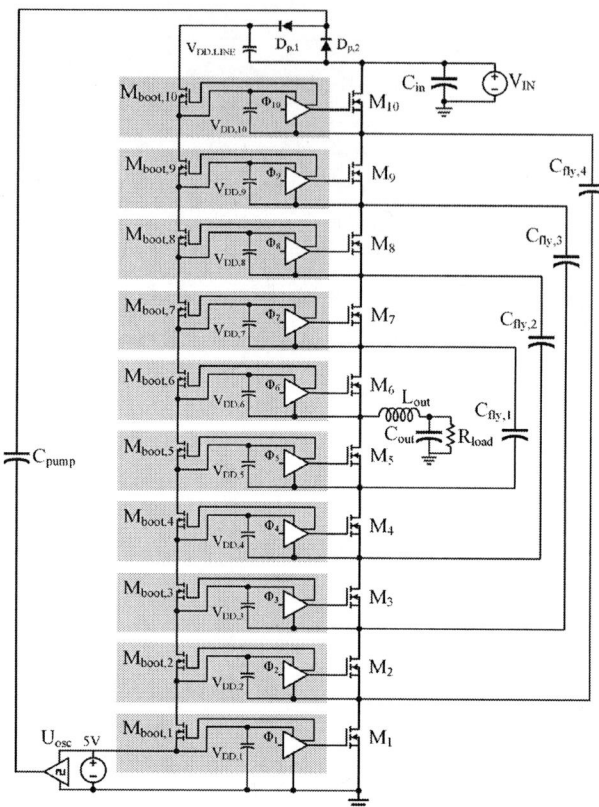

Fig. 8. Simplified schematic of the constructed 6-level FCML prototype using synchronous boot-strapping within each gate-driver. Ten copies of the previously described daughterboards are used to drive and provide power to ten series connected GaN devices, M_{1-10}. An additional charge pump, comprised of U_{osc}, C_{pump}, $D_{p,1}$, and $D_{p,2}$ is included to provide a 5V line referenced supply, $V_{DD,LINE}$, and allows power to be delivered from both ends of the boot-strapping network, further reducing voltage droop.

TABLE II
FCML COMPONENTS

Component	Description
L_{out}	7.92μH, 11A, 10.8mΩ
M_{1-10}	100V 3.2mΩ, EPC2218
$C_{fly,1,2,3,4}$	3.2μF, 4.2μF, 3.6μF, 3.2μF, X7R
C_{in}	4.2 μF, X7R
Level Shift	2EDF7275K
U_{osc}	Gate Driver, 1EDN7550U
$D_{p1,2}$	20V Schottky 0402
C_{pump}	47nF 450V X7T 0805

We note that this work serves as the first demonstration of bi-directional energy delivery throughout a cascaded bootstrapping network: Since M_{boot} can conduct in either direction when turned on, dissimilar to a diode, it follows that a gate driver may receive energy from the high-side driver immediately above it, provided it can access this energy. That is, this gate driver must possess sufficient initial charge to push the gate of its M_{boot} high, thereby enabling charge to flow down onto its local bypass capacitor[1]. This requirement that charge be stored locally in order to facilitate access to energy provided by gate drivers higher up the bootstrapping chain, necessitates an initial upward flow of energy from a ground referenced supply (5V in Fig. 8) at start-up, as is conventional.

[1]This was not the case in [20] where PMOS devices were instead used. In this case, M_{boot} is controlled by the high-side driver and may commence a downward flow of charge irrespective of the charge stored within gate drivers beneath it. However, with this approach appropriate generation of $V_{GS,boot}$ becomes significantly more challenging as it cannot be generally synthesized from the adjacent primary switching device.

978-1-6654-4817-8/21 $31.00 © 2021 IEEE

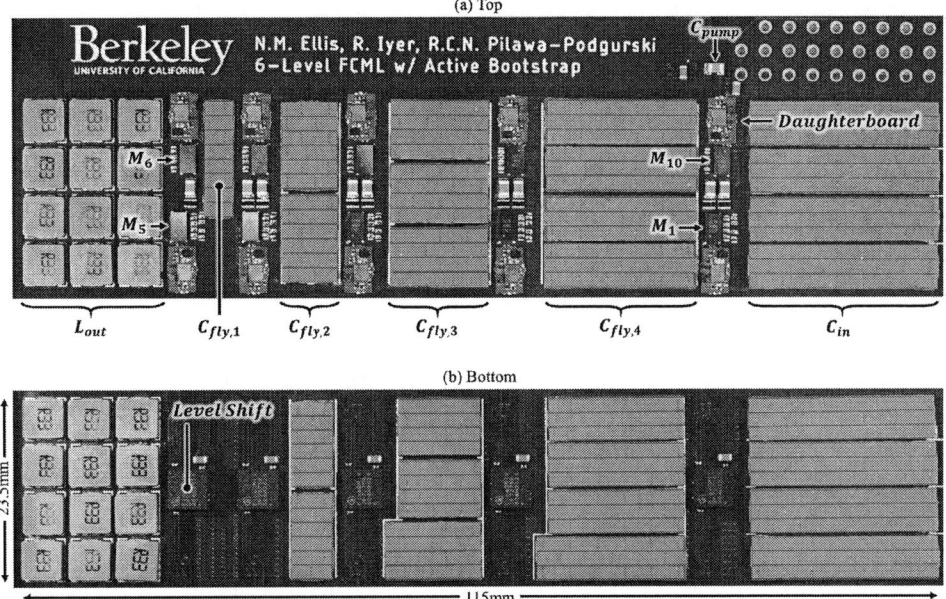

(a) Top

(b) Bottom

Fig. 9. Photograph of the constructed 6-level FCML prototype using the synchronous boot-strap power delivery on all 10 GaN-FETs.

However, once all gate drivers linked by synchronous boot-strapping are sufficiently charged, energy may be derived from a downward flowing line-referenced supply, labelled $V_{DD,LINE}$ in Fig. 8, in addition to the ground referenced upward flowing supply, $V_{DD,1}$. This essentially halves the impedance of the gate-drive power delivery path since charge may be delivered from both ends of the bootstrapping network. Alternatively, once a downward flowing charge path has been established, the ground referenced supply may be disabled with energy delivery from $V_{DD,LINE}$ self-sustaining, although this may not be desirable in practice. In this prototype, a simple charge-pump consisting of an oscillator, a pumping capacitor, and two diodes was used to produce $V_{DD,LINE}$, thereby allowing gate-driver energy to be fed into the chain via $M_{boot,10}$.

To validate the aforementioned concepts, the steady-state DC voltages stored on all 10 of the gate-driver's bypass capacitors was measured for FCML duty cycles of 15%, 50%, and 85% while switching at 500 kHz. For each of these operating points, gate-drive power was supplied either from the 5V low-side supply $V_{DD,1}$, the charge-pump generated high-side supply $V_{DD,LINE}$, or both simultaneously for minimal voltage droop throughout the chain. Figure 11 documents these results and illustrates that when the synchronous boot-strapping network is fed energy from both a ground-referenced and line-referenced supply, a worst-case maximum voltage droop of 156 mV is observed across a wide duty cycle range of 15-85%. This is a significant improvement over the RC-delayed synchronous approach which yielded a ~ 1 V droop at a relaxed 75% duty cycle in [16].

IV. CONCLUSION

To conclude, this work demonstrates a synchronous gate-drive power delivery approach with a greatly reduced volume as compared to conventional fully isolated solutions. It performs significantly better than cascaded diode approaches due to the absence of any forward voltage drops, ultimately eliminating the need for local regulation. Additionally, the boot-strap conduction duration is maximized, making it well suited for high frequency converters. Furthermore, appropriate boot-strap control signals may be generated locally

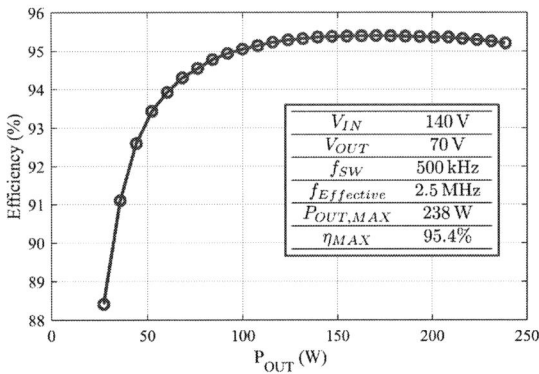

Fig. 10. Measured efficiency versus output power for a 6-level FCML prototype with $V_{IN} = 140V$ and a 50% duty cycle.

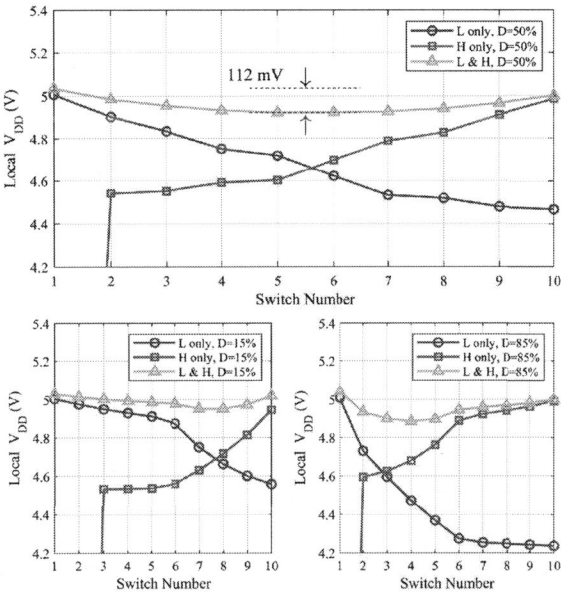

Fig. 11. Measured supply voltages across all 10 gate-drivers with varying duty cycle and three permutations of power delivery: low-side power only, high-side power only, and both simultaneously. At 50% duty cycle, applying power from both ends of the boot-strap network leads to a low voltage droop of 112 mV.

within a given gate-driver, with the majority of components suitable for complete monolithic integration. Moreover, we explore the possibility of bi-directional energy flow through a synchronous boot-strap network and demonstrate that voltage droop can be further reduced by applying power to both ends of the chain, approximately halving its impedance.

This approach was successfully demonstrated as part of a high level count FCML where power is delivered to ten series connected switches. A maximum voltage droop of 158 mV is observed across all switches over a wide conversion range of 15-85%, verifying that this approach is well suited for providing power to high-order GaN-FET networks which demand strict supply tolerances.

REFERENCES

[1] Y. Lei, C. Barth, S. Qin, W.-C. Liu, I. Moon, A. Stillwell, D. Chou, T. Foulkes, Z. Ye, Z. Liao *et al.*, "A 2-kw single-phase seven-level flying capacitor multilevel inverter with an active energy buffer," *IEEE Transactions on Power Electronics*, vol. 32, no. 11, pp. 8570–8581, 2017.

[2] Z. Ye, Y. Lei, and R. C. Pilawa-Podgurski, "A resonant switched capacitor based 4-to-1 bus converter achieving 2180 W/in^3 power density and 98.9 % peak efficiency," in *2018 IEEE Applied Power Electronics Conference and Exposition (APEC)*. IEEE, 2018, pp. 121–126.

[3] N. Ellis and R. Amirtharajah, "A Resonant 1: 5 Cockcroft-Walton Converter Utilizing GaN FET Switches with N-Phase and Split-Phase Clocking," in *2020 IEEE Applied Power Electronics Conference and Exposition (APEC)*. IEEE, 2020, pp. 19–25.

[4] Z. Ye, Y. Lei, and R. C. Pilawa-Podgurski, "A 48-to-12 v cascaded resonant switched-capacitor converter for data centers with 99% peak efficiency and 2500 w/in 3 power density," in *2019 IEEE Applied Power Electronics Conference and Exposition (APEC)*. IEEE, 2019, pp. 13–18.

[5] N. M. Ellis and R. Amirtharajah, "Reducing coss switching loss in a gan-based resonant cockcroft-walton converter using resonant charge redistribution," in *2020 IEEE Energy Conversion Congress and Exposition (ECCE)*. IEEE, 2020, pp. 158–164.

[6] N. Ellis, E. Sousa, and R. Amirtharajah, "A Resonant Gate Driver with Variable Gain and a Capacitively Decoupled High-Side GaN-FET," in *2020 IEEE Energy Conversion Congress and Exposition (ECCE)*. IEEE.

[7] R. C. Pilawa-Podgurski, A. D. Sagneri, J. M. Rivas, D. I. Anderson, and D. J. Perreault, "Very-high-frequency resonant boost converters," *IEEE Transactions on Power Electronics*, vol. 24, no. 6, pp. 1654–1665, 2009.

[8] D. J. Perreault, J. Hu, J. M. Rivas, Y. Han, O. Leitermann, R. C. Pilawa-Podgurski, A. Sagneri, and C. R. Sullivan, "Opportunities and challenges in very high frequency power conversion," in *2009 Twenty-Fourth Annual IEEE Applied Power Electronics Conference and Exposition*. IEEE, 2009, pp. 1–14.

[9] Analog Devices, *Dual-Channel Isolators with Integrated DC-to-DC Converters*, (accessed April 1st, 2021), https://www.analog.com/.

[10] T. Meynard and H. Foch, "Multi-level conversion: high voltage choppers and voltage-source inverters," in *PESC'92 Record. 23rd Annual IEEE Power Electronics Specialists Conference*. IEEE, 1992, pp. 397–403.

[11] A. Stillwell, E. Candan, and R. C. Pilawa-Podgurski, "Active voltage balancing in flying capacitor multi-level converters with valley current detection and constant effective duty cycle control," *IEEE Transactions on Power Electronics*, vol. 34, no. 11, pp. 11 429–11 441, 2019.

[12] N. Pallo, S. Coday, J. Schaadt, P. Assem, and R. C. Pilawa-Podgurski, "A 10-level flying capacitor multi-level dual-interleaved power module for scalable and power-dense electric drives," in *2020 IEEE Applied Power Electronics Conference and Exposition (APEC)*. IEEE, 2020, pp. 893–898.

[13] C. Klumpner and N. Shattock, "A cost-effective solution to power the gate drivers of multilevel inverters using the bootstrap power supply technique," in *2009 Twenty-Fourth Annual IEEE Applied Power Electronics Conference and Exposition*. IEEE, 2009, pp. 1773–1779.

[14] Z. Ye and R. C. Pilawa-Podgurski, "A power supply circuit for gate driver of gan-based flying capacitor multi-level converters," in *2016 IEEE 4th Workshop on Wide Bandgap Power Devices and Applications (WiPDA)*. IEEE, 2016, pp. 53–58.

[15] S. Biswas and D. Reusch, "Gan based switched capacitor three-level buck converter with cascaded synchronous bootstrap gate drive scheme," in *2018 IEEE Energy Conversion Congress and Exposition (ECCE)*. IEEE, 2018, pp. 3490–3496.

[16] Z. Ye, Y. Lei, W.-C. Liu, P. S. Shenoy, and R. C. Pilawa-Podgurski, "Improved bootstrap methods for powering floating gate drivers of flying capacitor multilevel converters and hybrid switched-capacitor converters," *IEEE Transactions on Power Electronics*, vol. 35, no. 6, pp. 5965–5977, 2019.

[17] Efficient Power Conversion, *EPC2108 Datasheet (Figure 14)*, (accessed April 1st, 2021), https://epc-co.com/.

[18] Texas Instruments, *LMG1205 80V, 1.2A to5A, Half Bridge GaN Driver with Integrated Bootstrap Diode*, (accessed June 22nd, 2021), https://www.ti.com/.

[19] M. Choi and D.-K. Jeong, "18.6 A 92.8 %-Peak-Efficiency 60 A 48 V - to- 1 V 3-Level Half-Bridge DC-DC Converter with Balanced Voltage on a Flying Capacitor," in *2020 IEEE International Solid-State Circuits Conference-(ISSCC)*. IEEE, 2020, pp. 296–298.

[20] P. Assem and R. C. Pilawa-Podgurski, "Quad gate-driver controller with start-up and shutdown for cascaded resonant switched-capacitor converter," in *2021 IEEE Custom Integrated Circuits Conference (CICC)*. IEEE, 2021, pp. 1–2.

[21] EPC Efficient Power Conversion, *EPC2152 80V, 15A ePowerTM Stage, Preliminary Datasheet, revision 2.0*, (accessed June 23nd, 2021), https://www.epc-co.com/.

978-1-6654-4817-8/21 $31.00 © 2021 IEEE

Single-Pulse Avalanche Failure Characterization of Single and Paralleled SiC MOSFETs

Hua Mao[1], Huaping Jiang[1*], Guanqun Qiu[1], Yifu Zhang[1], Xiaohan Zhong[1], Hao Feng[1], and Li Ran[1,2]

[1] State Key Laboratory of Power Transmission Equipment and System Safety and New Technology, Chongqing University, Chongqing, 400044, China

[2] School of Engineering University of Warwick Coventry, U.K.

*Email: Stefan.Jiang@foxmail.com

Abstract—**The voltage spikes generated by the turn-off of the high-speed switches can easily drive the devices into an avalanche mode and even failure. In order to study the silicon carbide (SiC) MOSFET's avalanche limit of a single device and the influence of electrical parameters of paralleled devices, an unclamped inductance switching (UIS) test platform for single and paralleled SiC MOSFETs is set up. This paper summarizes the single-pulse avalanche limit of single MOSFET under different inductances and different temperatures through experiments. In addition, the characteristics of parallel connected MOSFETs under different electrical parameters are also analyzed, and the main factors that affect the avalanche failure are shown.**

Keywords— silicon carbide MOSFET, unclamped inductance switching, single-pulse avalanche

I. INTRODUCTION

Compared with Si IGBT, silicon carbide (SiC) power MOSFET shows the characteristics of high switching speed, low switching loss, and low turn-off leakage current, which is suitable in high power density, high frequency and high temperature operating conditions [1]-[2]. However, robustness and reliability remain a major issue, such as SiC/SiO$_2$ interface reliability [3], threshold voltage instability [4] and energy shock reliability under different time scales [5]-[6]. The avalanche is one of the conditions that seriously affects the reliability of the device [8]. Due to the coupling of di/dt and stray inductance, the voltage spikes generated in high-speed switches [7] and automotive applications like antilock braking module [6] can easily drive the device into avalanche mode. Generally, a single avalanche breakdown will not cause the device to fail immediately, only when the avalanche energy is greater than a certain value. In engineering applications, parallel devices are often used to increase the current level, but the current sharing problem in paralleled SiC MOSFET often leads to an unbalanced distribution of energy, even reduce the system reliability [9]. Therefore, studying the avalanche failure performance of single and paralleled SiC MOSFETs is of great significance for improving the device design and operational reliability.

The effects of different temperatures, inductances, and gate resistances for the single-pulse avalanche energy (E_{AS}) and current (I_{av}) have been studied in [6][8], and related studies have also been performed on paralleled devices [10]-[11]. However, most of the studied devices are with planar gates, and few on trench MOSFET [12], meanwhile, there are still some issues in the avalanche failure factors of paralleled devices that need to be studied.

This work was supported by the National Key Research and Development Program of China under Grant 2018YFB0905803.

An unclamped inductive switching (UIS) platform was set up in this paper and analyzed the avalanche process in section II. Single device avalanche test with different inductances and different temperatures are described in section III. Moreover, the avalanche waveforms of the paralleled devices with different electrical parameters are explained. Finally, the paper draws main factors that affect the avalanche failure of the paralleled devices.

II. AVALANCHE TEST SYSTEM SET-UP AND ANALYSIS

The simplified UIS test schematic and its ideal output waveforms are shown in Fig. 1 for parallel connected devices under test (DUTs). The power supply starts to charge the inductance when the potentials of S and DUTs are high, and turns off the DUTs when reaching the preset charging time. At this time, due to the rapid turn-off of the switching device, the coupling of high di/dt and inductance generates an instantaneous high voltage, which is applied to the drain-source terminals of the devices. The gate drive is shared by two paralleled DUTs, which can be easily considered as the current is divided equally. The current continues to flow through the diode until I_{ds} returns to zero, and one single-pulse avalanche is complete. The instantaneous high voltage can be expressed as

$$V_{BR} = L \cdot \frac{di}{dt} + V_{DD} \tag{1}$$

and the avalanche time is:

$$t_{av} = \frac{I_{av}}{V_{BR} - V_{DD}} \cdot L \tag{2}$$

Therefore, the energy of an avalanche can be expressed as:

$$E_{av} = \int_0^{t_{av}} V_{BR} \cdot I_{ds} \cdot dt = \frac{1}{2} \cdot V_{BR} \cdot \frac{V_{BR} - V_{DD}}{L} \cdot t_{av}^2$$

$$= \frac{1}{2} \cdot I_{av}^2 \cdot L \cdot \frac{V_{BR}}{V_{BR} - V_{DD}} \tag{3}$$

Fig. 2 shows the UIS test bench that includes power source, test circuit, controller, auxiliary power source and inductive load, and electrical parameter test equipment, Agilent B1505A. The main power supply is charging for the capacitors and storing energy. The low-power power supply is for gate drivers and current sensor. The test bench also contains a safety protection circuit. When the sampling current exceeds the preset value, the feedback signal will be sent by the controller to cut off S and the main power supply. Here, S is replaced by IGBT.

In this paper, ROHM SiC MOSFET SCT3120AL was used for DUTs, which features a maximum drain–source voltage V_{ds} of 650 V and a dc-drain current I_d of 21A under 25°C. The next experiments are conducted from two aspects. The first is to find the avalanche capability limit of a single SiC MOSFET under different inductive loads and different temperatures, and then to explore the impact of parallel connection devices with different electrical parameters on avalanche waveforms. Table I shows the experimental parameters, in which V_{DD}, L, and t_{on} are selected according to different experiments.

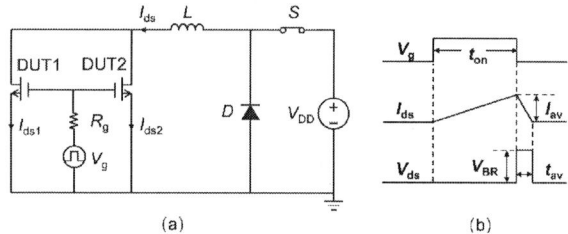

Fig. 1. The UIS (a) test schematic and (b) output waveforms.

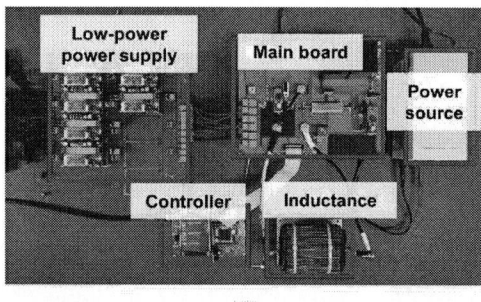

(a)

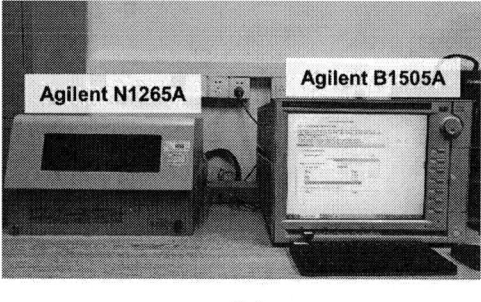

(b)

Fig. 2. Photograph of (a) test rig and (b) electrical parameter test equipment.

TABLE I.   AVALANCHE TEST PARAMETERS

Parameters	Value	
	Series	*Parallel*
V_{DD} / V	[10, 20, …, 100]	60
L / mH	[1, 2, 3]	1
t_{on} / μs	[100, 200, 300]	100
Gate resistance / Ω	10	10
Temperature	[25, 50, …, 175]	Room temperature
Abort condition	Device failure	Device failure

III. EXPERIMENT AND DISCUSSION

For one-pulse avalanche test, perform a single device test on the DUT first. The S in this article are all switched on earlier than DUT making no vibration when DUT is turned on, and turned off later to suppress the resonance rather than these two switches turned off simultaneously. Besides, the current would increase before S is switched off when the device is failure, and the overcurrent signal can be detected by current sensor at this cycle.

In Fig. 3, the gate voltages of S and DUT are shown by purple and dark blue lines, and the DUT's drain-source current and voltage are shown by light blue and green lines, respectively. Fig. 3 (a) describes an overall waveform about one single-pulse avalanche, and more details at the moment of avalanche are shown in Fig. 3 (b). The DUT suddenly turns off when I_{ds} is charged to a certain current, and drops a little during the rapidly rise in V_{ds}, and then returns to normal value, which because when the device is turned off, a high-value drain-source voltage is induced across the device. At the same time, the loop stray capacitance is charged in the reverse direction, causing the current to drop rapidly, and then the current returns to normal value when the drain-source voltage is clamped at breaking voltage (V_{BR}). Then I_{ds} oscillates, which is related to gate resistance. V_{BR} starts to drop when current approaches to 0, and the drain-source capacitance discharges, leading the current to flow in the reverse direction and oscillate around 0 ampere. This is the whole process of an avalanche. As shown in Fig. 3 when the inductance is 1mH, the I_{ds} reaches 9A after 200μs charging process, and the voltage between drain and source can even reach 1460V.

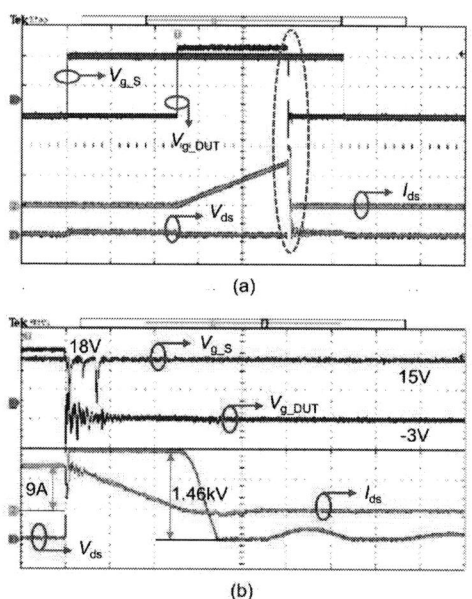

Fig. 3. (a) Single-pulse avalanche process output waveform and (b) the zoom in output waveform at the moment of avalanche.

A. Single Device Avalanche Test with Different Power Source

Fig. 4 (a) shows the current waveform of DUT when V_{DD} changes from 5V to final 62V where device fails under the case of t_{on}=200μs. Here we can find that the current increases with the increased voltage until the DUT cannot be turned off. The corresponding breaking voltages are shown in Fig. 4 (b). As the peak current increases, the corresponding V_{BR} and t_{av} also increase. On the other hand, for one certain avalanche process during t_{av}, as the avalanche energy increases, V_{BR} increases slightly at first because the chip is heated up with the increased single avalanche energy and V_{BR} is positively correlated with temperature; then V_{BR} decreases a little as the

978-1-6654-4817-8/21 $31.00 © 2021 IEEE

current drops much and fast due to the blocking characteristic of the SiC MOSFET; finally, V_{BR} quickly drops to 0. Then oscillate after I_{av} returns to 0 because S is still on.

B. Single Avalanche Energy Limit under Different Inductances

Fig. 5 shows the current (I_{av}) and energy (E_{AS}) limits of a single-pulse avalanche with different inductances. The light blue rectangles represent the maximum current that can pass before device failure, and the dark blue rectangles represent the maximum energy before invalidation. From Fig. 5 we can easily find that the I_{av} decreases and E_{AS} increases as the inductance increases. It can be concluded that the larger the loop inductance, the easier an avalanche happens with high-speed switching devices at low currents. However, the avalanche energy is proportional to the inductance and the square of I_{av}. The larger the capacity of the inductor, the more energy it can withstand, and the large energy stored in it can only be released through the DUT when an avalanche occurs.

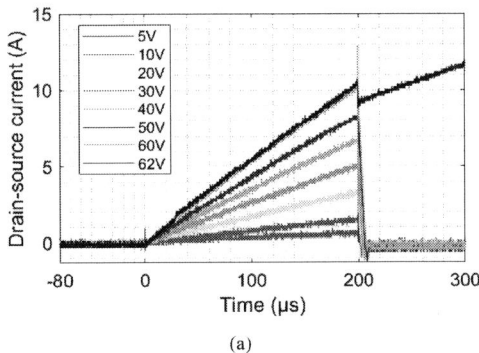

(a)

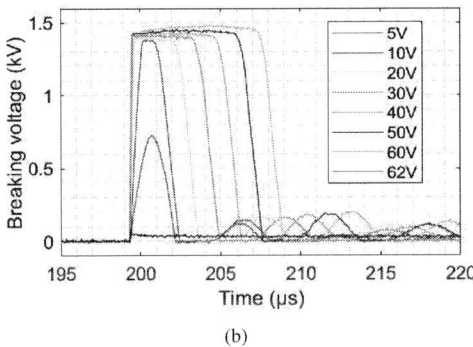

(b)

Fig. 4. (a) Avalanche currents under different V_{DD} and (b) avalanche voltages under different V_{DD}.

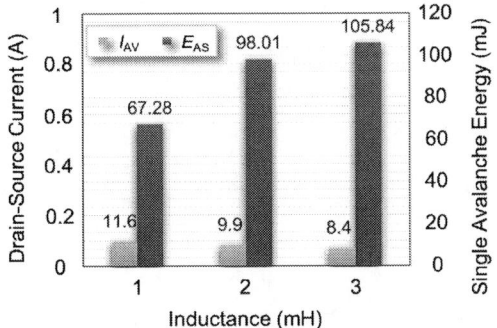

Fig. 5. Avalanche current and energy limits with different inductances.

C. Single Avalanche Energy Limit under Different Temperatures

As shown in Fig. 6, keep the t_{av}=100μs and L=1mH, and change the temperature of SiC MOSFET, exploring the relationship between E_{AS} limit and chip temperature. 25°C~175°C are selected as the temperature range, and increase in steps of 25°C of a total 7 temperatures. The heating plate is controlled by the heating table and connected with the device to ensure that the chip's case temperature fluctuates within 0.5°C. Adjust the charging power supply and increase the avalanche current until it fails. The maximum current before failure will be the limit of the avalanche current at that temperature. Fig. 6 shows that the E_{AS} limit of SiC MOSFET is inversely proportional to the temperature. The higher the temperature, the lower the I_{av} limit and E_{AS} limit, which means the increase in junction temperature will reduce the avalanche robustness of the device.

Fig. 7 shows the V_{BR} waveforms obtained by changing the temperature of DUT with the same E_{AS} = 24.5mJ. We can clearly find that the V_{BR} gradually increases as the temperature increases because of the positive temperature coefficient, while t_{av} gradually decreases, which further verifies that the temperature does not change the total energy, but the device can withstand different maximum energy values under different temperatures.

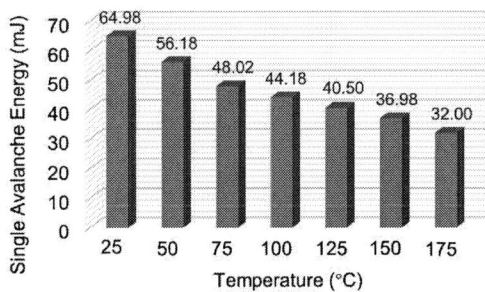

Fig. 6. Avalanche energy limits with different temperatures.

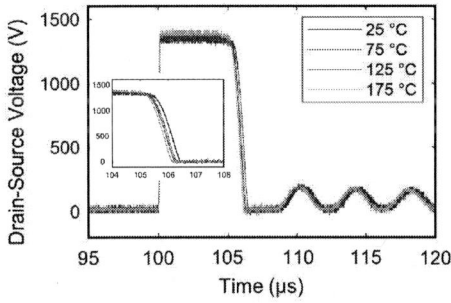

Fig. 7. Avalanche voltages with different temperatures.

There is a parasitic transistor inside the SiC MOSFET. During the avalanche process, the current gain of the parasitic transistor becomes larger due to the temperature rise, and the base current flowing through the parasitic transistor is relatively large. At this time, the parasitic transistor is latched up and the cell is out of control. As for SiC MOSFET, it is composed of multiple cells in parallel, and there will always be relatively weak cells that first enter the out-of-control state due to the above process, and the avalanche current will be concentrated in these weak cells. At the same time, although

978-1-6654-4817-8/21 $31.00 © 2021 IEEE

the avalanche energy is not very high, its avalanche power is extremely large during the relative short t_{av}, which can reach kilowatt level. Such high thermal power cannot be withstood by the chip locally, which further causes the lattice temperature to rise, aggravating the latch-up of the transistor and current concentration of the weak cells, and ultimately lead to the device failure by a positive feedback. Based on the above analysis, it can be considered that there is a temperature limit (T_{lim}) for chips. When the local temperature exceeds this limit, the device fails.

Combining with the experiments in this article, set the initial temperature of the device as T_0, and the average heat capacity C_{av} is introduced to describe the temperature-energy characteristics of the chip.

$$C_{av} \times (T_{lim} - T_0) = Q_{av} \qquad (4)$$

where Q_{av} is the heat generated locally when current is concentrated during the avalanche process.

It is considered that when device fails, the local heat can be regarded as invariable with high avalanche power and short avalanche time. On the other hand, most of the energy is dissipated in the form of heat and generated in local area where the current is concentrated. So we can consider that

$$Q_{av} \approx E_{av} \qquad (5)$$

Then,

$$E_{av} = C_{av} \times (T_{lim} - T_0) \qquad (6)$$

Therefore, when temperature rises, the corresponding avalanche energy limit drops. At the same time, (6) also corresponds to the experimental results, that is, the relationship between the avalanche energy limit and the initial case temperature is close to linear.

D. Parallel Connected Avalanche Test with Different Parameters

The parallel connection of SiC MOSFETs is very common in practical because of the limitation in manufacturing process of SiC power devices. It can only be packaged by chips or multiple discrete devices in parallel to increase the current level. However, the parallel connection of devices with different parameters will cause uneven current sharing, unbalanced energy distribution, and reduced device reliability, which may further lead to system instability. Therefore, it may have a greater impact for avalanche when high-speed switching devices connected in parallel, which requires us to study the effect of various electrical parameters carefully on the avalanche of paralleled devices.

Fig. 8 selects the waveform diagrams of eight parallel-connected devices with different electrical parameters, under the test conditions of V_{DD}=60V, t_{av}=100μs. Four channels separately measured the shared gate voltage in dark blue, and the drain-source currents for paralleled devices in light blue and purple, and voltages of the DUTs in green. The electrical parameters of eight tested devices are listed in Table II. In Fig. 8, three areas are represented by red, blue, and green dashed circles, which show the uneven current distribution caused by R_{ON}, V_{TH}, and V_{BR}. It can be found that the smaller the R_{ON} or the smaller the V_{BR}, the more current is distributed to the charging device. The influence of V_{TH} is not very large, which is covered by the reverse charging of stray capacitance. R_{ON}

and V_{BR} have a greater influence on the paralleled avalanche current waveforms, but V_{BR} dominants the final failure. During an avalanche, the device with smaller V_{BR} withstands more avalanche energy. The greater the V_{BR} difference, the greater the current gap, so the main consideration is the ability of a single device avalanche energy limit. Therefore, parallel-connection does not mean double avalanche-withstand ability.

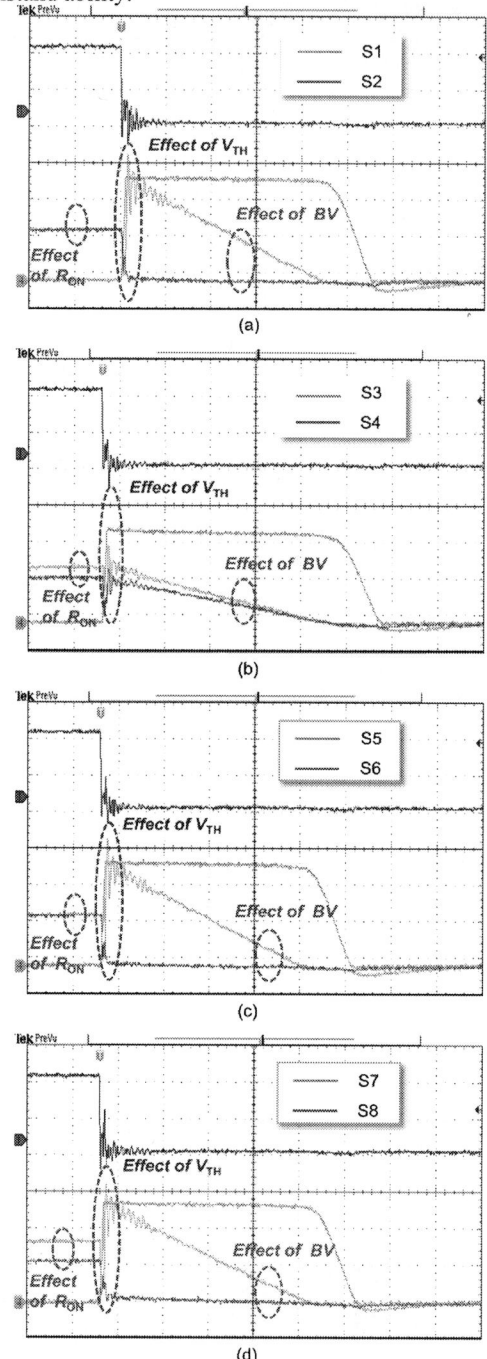

Fig. 8. Avalanche waveform diagrams of the parallel-connected devices with different electrical parameters.

After multiple sets of experiments, it is found that if the device fails due to the occurrence of an avalanche instead of

TABLE I. PARAMETERS FOR PARALLEL AVALANCHE TEST

DUT	S1	S2	S3	S4	S5	S6	S7	S8
V_{TH}	4.611	4.68	4.5	6.45	5.45	4.755	6.3	6.35
R_{ON}	120.4	123.4	116	149.6	131.3	128.4	134.6	144.2
V_{BR}	1239.8	1312	1115	1116	1235	1307	1186	1231

the excessive current during the charging process, one of the paralleled DUTs must fail first, and it is the one with smaller V_{BR}, which is determined by the blocking characteristic curves of the two DUTs. Fig. 10 shows the comparison of the blocking characteristic curves of S1 and S2. The drain-source voltage (V_{DS}) corresponding to the two DUTs at 1mA is V_{BR}, and the V_{BR} of S1 is much smaller than that of S2. When an avalanche occurs, the V_{DS} of the two DUTs are clamped at a certain value. The smaller the V_{BR}, the more current will be obtained. In other words, during an avalanche, the device with smaller V_{BR} will withstand more avalanche energy, and eventually fails. Therefore, multi-chip paralleling does not mean that the withstand energy is also increased by quantity, which is the content that we will study in the future.

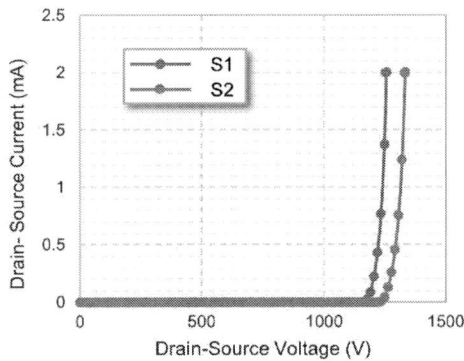

Fig. 9. The comparison of the blocking characteristic curves of S1 and S2.

IV. CONCLUSION

To study the avalanche limit of single SiC MOSFET and avalanche characteristics of paralleled devices with various electrical parameters, a UIS test platform was set up to conduct a single-pulse avalanche. When increasing the power supply with fixed charging time, V_{BR} and t_{av} increase constantly; at the same time, for a single avalanche test with higher energy, V_{BR} rises first because of the heat generated from chip, and then drops slightly caused by the continuous decrease of I_{ds}. Secondly, the larger the loop inductance is, the easier an avalanche happens for high-speed switching device with low current and high energy. Thirdly, the avalanche energy limit decreases with the increase of temperature, which is affected by the temperature limit of the chip. Finally, R_{ON} and V_{BR} have a greater influence on the parallel avalanche current waveforms, but the final failure is mainly determined by V_{BR}. Devices with smaller V_{BR} withstanding the main energy fail earlier.

ACKNOWLEDGMENT

This work was supported by the National Key Research and Development Program of China under Grant 2018YFB0905803.

REFERENCES

[1] M. Treu, R. Rupp and G. Sölkner, "Reliability of SiC power devices and its influence on their commercialization - review, status, and remaining issues," *2010 IEEE International Reliability Physics Symposium*, pp.156-161, 2010.

[2] A. Merkert, T. Krone and A. Mertens, "Characterization and scalable modeling of power semiconductors for optimized design of traction inverters with Si- and SiC-devices," *IEEE Transactions on Power Electronics*, vol. 29, no. 5, pp. 2238-2245, 2014.

[3] B. Asllani, A. Castellazzi, O. A. Salvado, A. Fayyaz, H. Morel and D. Planson, "VTH-hysteresis and interface states characterisation in SiC power MOSFETs with planar and trench gate", *Proc. IEEE Int. Rel. Phys. Symp. (IRPS)*, pp. 1-6, 2019.

[4] H. Jiang, X. Zhong, G. Qiu, L. Tang, X. Qi, L. Ran, "Dynamic gate stress induced threshold voltage drift of silicon carbide MOSFET," *IEEE Electron Device Letters*, vol. 41, no. 9, pp. 1284-1287, 2020.

[5] K. Sun, J. Wang, R. Burgos, D. Borovevich, "Design, analysis, and discussion of short circuit and overload gate-driver dual-protection scheme for 1.2-kV, 400-A SiC MOSFET modules," *IEEE Transactions on Power Electronics*, vol. 35, no. 3, pp. 3054-3068, 2020.

[6] X. Zhou, H. Su, R. Yue, G. Dai, J. Li, Y. Wang, Z. Yu, "A deep insight into the degradation of 1.2-kV 4H-SiC MOSFETs under repetitive unclamped inductive switching stresses," *IEEE Transactions on Power Electronics*, vol. 33, no. 6, pp. 5251-5261, 2018.

[7] M. Nawaz, "Evaluation of SiC MOSFET power modules under unclamped inductive switching test environment," *Microelectronics Reliability*, vol. 63, pp. 97–103, 2016.

[8] L. Yang, A. Fayyaz and A. Castellazzi, "Characterization of high-voltage SiC MOSFETs under UIS avalanche stress," *7th IET International Conference on Power Electronics, Machines and Drives (PEMD 2014)*, pp. 1-5, 2014.

[9] H. Li, S. Munk-Nielsen, X. Wang, R. Maheshwari, S. Beczkowski, C. Uhrenfeldt, W. Franke, "Influences of device and circuit mismatches on paralleling silicon carbide MOSFETs," *IEEE Transactions on Power Electronics*, vol. 31, no. 1, pp. 621-634, 2016.

[10] J. Hu, O. Alatise, J. Gonzalez, R. Bonyadi, L. Ran, P. Mawby, "The effect of electrothermal nonuniformities on parallel connected SiC power devices under unclamped and clamped inductive switching," *IEEE Transactions on Power Electronics*, vol. 31, no. 6, pp. 4526-4535, 2016.

[11] A. Fayyaz, B. Asllani, A. Castellazzi, M. Riccio, A. Irace, "Avalanche ruggedness of parallel SiC power MOSFETs," *Microelectronics and Reliability*, vol. 88-90, pp. 666-670, 2018.

[12] J. Wei, S. Liu, L. Yang, L. Tang, R. Lou, T. Li, J. Fang, S. Li, C. Zhang, W. Sun, "Investigations on the degradations of double-trench SiC power MOSFETs under repetitive avalanche stress," *IEEE Transactions on Electron Devices*, vol. 66, no. 1, pp. 546-552, 2019.

978-1-6654-4817-8/21 $31.00 © 2021 IEEE

DC Transform Circuit Design Based on Multiplier Rectification

Penghui Yin
*State Key Laboratory of
Advanced Electromagnetic
Engineering and Technology
Huazhong University of Science
and Technology*
Wuhan,P.R.China
m202071577@hust.edu.cn

Xuehua Wang
*State Key Laboratory of
Advanced Electromagnetic
Engineering and Technology
Huazhong University of Science
and Technology*
Wuhan,P.R.China
wang.xh@hust.edu.cn

Xinbo Ruan
*State Key Laboratory of
Advanced Electromagnetic
Engineering and Technology
Huazhong University of Science
and Technology*
Wuhan,P.R.China
ruanxb@nuaa.edu.cn

Abstract—**Due to the high reliability and high efficiency, the DC/DC converters are rapidly developing. However, a challenge is faced which is the output voltage range of the DC/DC converter is limited by the withstand voltage of the device in practice. In order to solve this problem, this paper designs a DC/DC circuit based on the voltage double rectifier circuit and half-bridge LLC topology. It realizes high voltage output in the case of low voltage input. Meanwhile, the requirements for the withstand voltage of the device are relaxed, which is convenient for the selection of the device. Furthermore, the zero voltage switch (ZVS) of the primary-side switches and the zero current switch (ZCS) of the secondary-side diodes can be achieved. The correctness and effectiveness of the proposed method are finally verified by simulation.**

Keywords—*multiplier rectification, zero current switching, the withstand voltage of the device, DC/DC converter, LLC resonant circuit*

I. INTRODUCTION

Recently,the rapid development of distributed generation and energy storage has led to the increasing popularity of the switching converters as an ever-lasting key interface [1].As the penetration of renewable DC sources, energy sources, and DC loads (e.g., consumer electronics, motor drives, electric vehicle charges), the DC network attracts more and more attention since it can facilitate the connection of DC sources and loads, avoiding multiple stages of energy conversions.In the DC network, the DC/DC conversion topologies and technology are critical. Several topologies of DC/DC converter have been proposed.Among them, the one proposed in [2] is attractive, which however, require the components with high withstand voltage. Therein, the most important part is the resonant circuit, which transforms the square waveform voltage of the primary side into the sinusoidal voltage of the secondary side [3].Usually, the secondary side adopts synchronous rectification or uncontrolled diode rectification, but these methods call for high withstand voltage of the diode and bring high losses of the rectifier circuit [4].

Inspired by this, many studies are devoted to exploring more efficient and practical DC converters. One of the solutions is to use devices with higher withstand voltage levels [5]. Although this method can solve the withstand voltage problem, the corresponding cost will also increase. A new single-switch quadratic boost DC/DC converter with low voltage stress proposed in [6]. In this topology, the voltage on the device is limited, and a higher voltage gain is obtained without a large duty cycle. However, coupled-inductors(CI) and voltage multipliers(VM) are used in this topology, which increases the complexity of the topology. In [7], A high-boost

DC/DC converter based on coupled inductor and switched capacitor is proposed. The introduction of switched capacitors reduces the voltage of power components, but this topology needs to regulate the voltage gain by changing the number of turns of the coupled inductor and the number of switched capacitor units. This method increases the complexity and volume of the system. A new hybrid non-isolated DC/DC converter suitable for photovoltaic applications is proposed [8]. This topology reduces the withstand voltage on the device, but the application field of this topology is subject to many restrictions.

To solve this problem, this paper proposes a rectifier circuit in which the secondary side uses 14-stage double-voltage rectifier units in series to achieve 28-fold voltage output. The primary side uses an asymmetric half-bridge LLC structure to convert the input DC voltage into an AC square wave. This design method improves the working efficiency of the converter and reduces the requirements on the device. The proposed topology has three merits: (1) achieve high voltage output under low-voltage input conditions, (2) achieve ZVS of the primary side switches and ZCS of the secondary side rectifier diode. (3) reduce the withstand voltage of the components, facilitating device selection.

II. SYSTEM DESCRIPTION

The topology of the DC/DC resonant converter based on multi-voltage rectifier circuit is shown in Fig. 1. The primary side of the isolation transformer Tr is connected to the high-voltage side, and the secondary side is connected to the low-voltage side. The primary side is a half-bridge LLC topology, consisting of two main switches $Q_1 \sim Q_2$ and two DC-link capacitor C_{oss1}&C_{oss2}. The secondary side is fourteen double voltage rectification topologies. The switches work in a complementary working mode with a driving signal duty cycle of 50% and a square wave voltage v_{sq} is generated as the input of the resonant circuit. In order to prevent the upper and lower bridge arms from passing through, the driving signal has a certain dead time T_d. The resonant network of LLC resonant converter is formed of the resonant capacitor C_r, resonant inductance L_r and transformer magnetizing inductance L_m. When the twenty-eight double-voltage rectification circuit works, the working modes of the fourteen double voltage rectification circuits connected in series are basically the same. In order to facilitate the analysis, the twenty-eight times voltage rectification circuit is replaced by a double voltage rectification circuit and the topology is shown in Fig. 2.

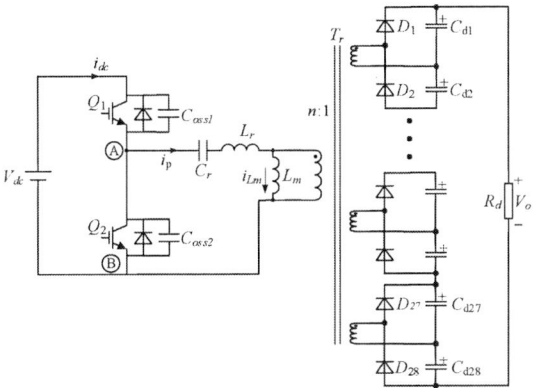

Fig. 1. The topology of the DC/DC resonant converter based on multi-voltage rectifier circuit.

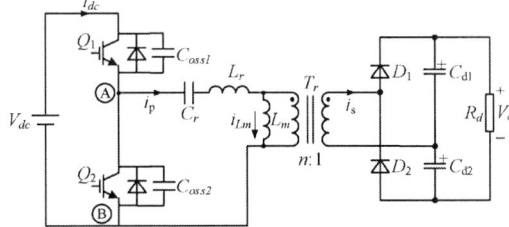

Fig. 2. Circuit diagram of half-bridge LLC resonant converter.

Based on the operating characteristics of the LLC resonant converter, there are two resonant frequencies:

Case 1: When the rectifier diode on the secondary side is turned on, the voltage on the secondary side of the transformer is the same as the voltage of the magnetizing inductance L_m. The voltage on the excitation inductance L_m is constant, and it does not participate in resonance. The resonant frequency f_r can be obtained with

$$f_r = \frac{1}{2\pi \cdot \sqrt{L_r C_r}} \quad (1)$$

Case 2: When the rectifier diode on the secondary side is cut off, the transformer is disconnected from the primary side to the secondary side. At this time, the primary side has no energy transferred to the secondary side, and the magnetizing inductance L_m is no longer clamped by the secondary side voltage. The resonant components are the resonant inductor L_r, the resonant capacitor C_r and the magnetizing inductance L_m. The resonant frequency f_m can be obtained with

$$f_m = \frac{1}{2\pi \cdot \sqrt{(L_r + L_m) C_r}} \quad (2)$$

The LLC resonant converter adopts the method of changing the frequency to control the converter(PFM), and the energy transmission is adjusted by changing the switching frequency f_s. The LLC resonant converter has different operating modes. According to the relationship between the switch's operating frequency fs and resonant frequency fr, as well as factors such as input voltage and load, the LLC resonant converter can be classified into three operating modes, namely $f_s > f_r$, $f_m < f_s < f_r$ and $f_s = f_r$.

Each operating mode has its own operating characteristics. Since this article wants to realize the zero-current shutdown (ZCS) of the secondary side rectifier diode, the main analysis converter operating mode is $f_m < f_s < f_r$, and the converter works under the rated load conditions. Make the following assumptions before the analysis: (1) the power switches $Q_1 \sim Q_2$, rectifier diodes $D_1 \sim D_2$, capacitors, inductors and transformers are all considered to be in an ideal state. (2) the parasitic capacitances $C_{oss1} \sim C_{oss2}$ of the switches $Q_1 \sim Q_2$ do not participate in resonance. (3) the value of the output filter capacitor C_o is very large, and the output voltage is considered to be a constant DC voltage approximately.

TABLE I.   SWITCHING STATES OF THE HALF-BRIDGE CIRCUIT

	$t_0 \sim t_1$	$t_1 \sim t_2$	$t_2 \sim t_3$
Q_1	on	off	off
Q_2	off	off	off
D_1	on	off	off
D_2	off	off	off

When the converter operating mode is $f_m < f_s < f_r$, the switch control sequence is summarized in Table I, and the equivalent circuits of each stage in the working process are shown in Fig. 3. The main working waveform is shown in Fig. 4.

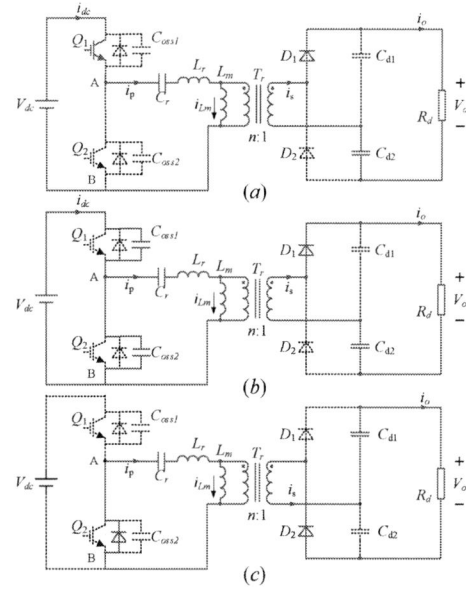

Fig. 3. The equivalent circuits of each stage in the working process.

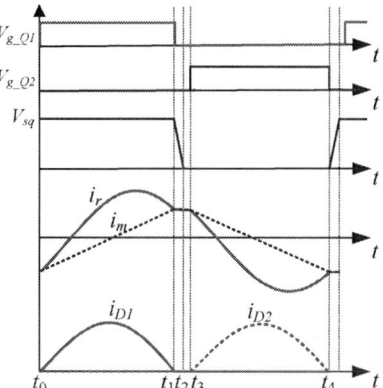

Fig. 4. The main working waveform.

978-1-6654-4817-8/21 $31.00 © 2021 IEEE

According to Fig. 4, it can be seen that the switching frequency f_s is less than f_r, and the LLC primary side switches can achieve zero voltage turn-on, and the secondary side rectifier diode can realize zero current turn-off, so that all power semiconductor devices in the circuit realize soft switching. For the primary side switches, the switching loss is reduced. The loss caused by the reverse recovery of the secondary side rectifier diode is reduced. At this time, there is no voltage spike in the rectifier diode, so a diode with a lower withstand voltage can be used.

III. ANALYSIS OF CIRCUIT PRINCIPLE

A. Fundamental Wave Approximation

Since the resonant current waveform in the LLC resonant tank is similar to a sine wave, the fundamental wave approximation method (FHA) can be used to analyze the resonant circuit. In the case of ignoring the non-linear characteristics of the secondary side rectifier circuit, the secondary side rectifier circuit and the filter circuit are equivalently converted into the AC load on the secondary side of the transformer. When the high-order component in the square wave excitation source of the resonant circuit is ignored on the primary side of the transformer, a linear sinusoidal excitation circuit is formed. A linear circuit converted from a non-linear circuit is shown in Fig. 5.

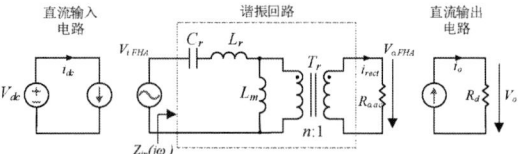

Fig. 5. Equivalent circuit diagram of half-bridge LLC resonant converter.

In the resonant circuit of the dual-port network model, $V_{i.FHA}$ is the input voltage of the equivalent resonant circuit port, and $V_{o.FHA}$ is the output voltage of the equivalent resonant network port. $Z_{in}(j\omega)$ is the input impedance, and $R_{o.ac}$ is the equivalent resistance load on the secondary side of the transformer. The equivalent resistance of the secondary side of the transformer $R_{o.ac}$ can be calculated as

$$R_{o.ac} = \frac{V_{o.FHA}}{I_{rect}} = \frac{8}{\pi^2}\frac{V_{out}^2}{P_{out}} = \frac{8}{\pi^2}R_d \qquad (3)$$

The transformation ratio of the transformer is represented by n. The equivalent AC load on the primary side of the transformer R_{ac} can be expressed as

$$R_{ac} = \frac{8}{\pi^2}n^2 R_d \qquad (4)$$

The gain relationship between the input voltage and the output voltage of the LLC resonant converter M can be obtained with:

$$M = 2n\frac{V_o}{V_{dc}} \qquad (5)$$

The normalized frequency f_n is the ratio of the switching frequency f_s to the resonant frequency f_r, and k is the ratio of magnetizing inductance to resonant inductance. Using the

fundamental wave approximation method for LLC resonant converter, the voltage gain M can be also expressed as

$$M\left(f_n,k,Q\right) = \frac{1}{\sqrt{\left(1+\frac{1}{k}-\frac{1}{kf_n^2}\right)^2 + Q^2\left(f_n - \frac{1}{f_n}\right)^2}} \qquad (6)$$

B. Principle of Voltage Doubler Rectifier Circuit

The topology of the twenty-eight times voltage rectifier circuit is shown in Fig. 6. The transformer has one winding on the primary side and fourteen independent windings on the secondary side. The secondary winding is composed of rectifier diodes $D_1 \sim D_{28}$ and capacitors $C_{d1} \sim C_{d28}$. The mode of the switching process is shown in Fig. 7. There are two cases when the circuit is working:

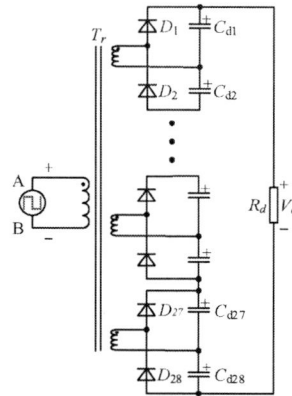

Fig. 6. The topology of the twenty-eight times voltage rectifier circuit.

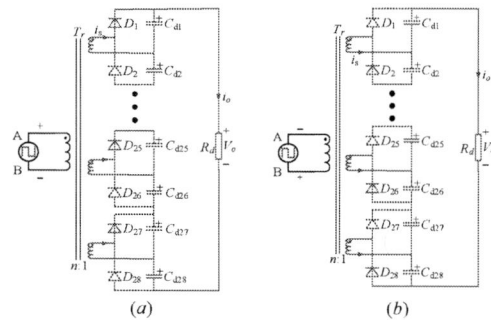

Fig. 7. The mode of the switching process.

Case 1: In the positive half cycle, when the A port of the power supply is positive and the port B of the power supply is negative, the diode D_1 turns on and the diode D_2 turns off. The power supply charges the capacitor C_{d1} through the diode D_1. In an ideal situation, the diode D_1 can be regarded as in a short-circuit state, while the capacitor C_{d1} is charged to $V_o/28$ in half a cycle. At this time, the reverse voltage on the diode D_2 is $V_o/14$.

Case 2: In the negative half cycle, when the A port of the power supply is negative and the port B of the power supply is positive, the diode D_1 turns off, and the diode D_2 turns on. The power supply charges the capacitor C_{d2} through the diode D_2. In an ideal situation, the diode D_2 can be regarded as in a short-circuit state, while the capacitor C_{d2} is charged to $V_o/28$

in half a cycle. At this time, the reverse voltage on the diode D_1 is $V_0/14$.

From the previous analysis of LLC resonant circuit, it can be seen that the transformer secondary current $i_s(t)$ can be approximated as a sine wave. It can be obtained with

$$i_s(t) = \sqrt{2} I_{rect} \sin(2\pi f_s t) \tag{7}$$

Among them, the reference current I_{rect} can be expressed as

$$I_{rect} = \frac{\pi}{2\sqrt{2}} \cdot \frac{P_{out}}{V_{out}} \tag{8}$$

The current i_{D1} flowing through the diode D_1 is half of the current i_s. It can be expressed as

$$i_{D1}(t) = \begin{cases} \sqrt{2} I_{rect} \sin(2\pi f_s t), t < T_s/2 \\ 0, T_s/2 < t < T_s \end{cases} \tag{9}$$

The load current i_o can be obtained with

$$i_o(t) = \frac{P_{out}}{V_{out}} \tag{10}$$

According to Kirchhoff's current law, the current flowing through the capacitor C_{d1} can be expressed as

$$i_{Cd1}(t) = \begin{cases} \sqrt{2} I_{rect} \sin(2\pi f_s t) - \dfrac{P_{out}}{V_{out}}, t < T_s/2 \\ -\dfrac{P_{out}}{V_{out}}, T_s/2 < t < T_s \end{cases} \tag{11}$$

The current waveform is shown in Fig. 8.

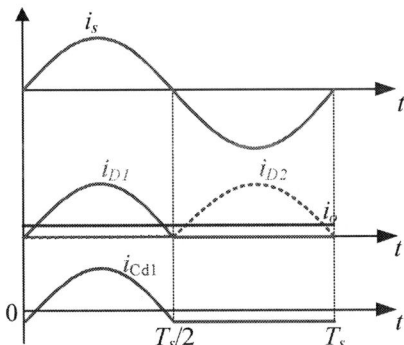

Fig. 8. The current waveform.

Since the current i_o of the load is relatively small, it can be approximately regarded as $i_{Cd1} = i_{D1}$. The voltage ripple on the capacitor C_{d1} can be calculated as

$$\Delta V_{Cd1} \approx \frac{1}{C_d} \int_0^{T_s/2} \sqrt{2} I_{rect} \sin(2\pi f_s t) dt = \frac{P_{out}}{2C_d f_s V_{out}} \tag{12}$$

IV. DESIGN OF LLC CIRCUIT PARAMETERS

The value of the resonance frequency f_r in this article is 200kHz. According to (6), the normalized frequency f_n is

taken as the abscissa, and the voltage gain M is taken as the ordinate. The relationship between the voltage gain M and the frequency f_n is shown in Fig. 9. Regardless of the value of the inductance ratio k and the quality factor Q, the voltage gain curve always passes through the $(f_n, M) = (1, 1)$ point. It can be seen from the Fig. 9 that when the switching frequency f_s of the LLC circuit is L_r and the resonant frequency of the capacitor C_r is f_r, the voltage gain M is always equal to 1. Therefore, most LLC resonant circuits choose to work near this point.

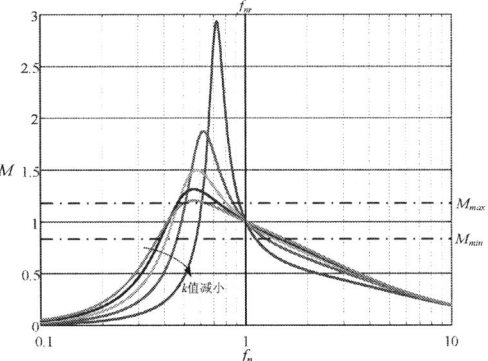

Fig. 9. The relationship between the voltage gain M and the frequency f_n.

The output voltage of the LLC resonant converter is changed by adjusting the switching frequency of the input square wave voltage. When the output power drops or the input DC voltage increases, the output voltage is adjusted by increasing the switching frequency to keep it stable. In order to cope with the load change, the LLC resonant converter should work near the unity gain point. In this case, the output voltage can be adjusted for wide load changes within a relatively small range of switching frequency change.

Since the output rectifier diode is required to have the characteristic of zero-current turn-off in this article, the switching frequency of the LLC converter should be less than the resonance frequency. In this article, the rated value of the input voltage is 100V, and the fluctuation range is ±10%, so as to determine the maximum gain M_{max} and the minimum gain M_{min}. According to fo (5), $M_{max}=1.305$ and $M_{min}=1.026$ are obtained.

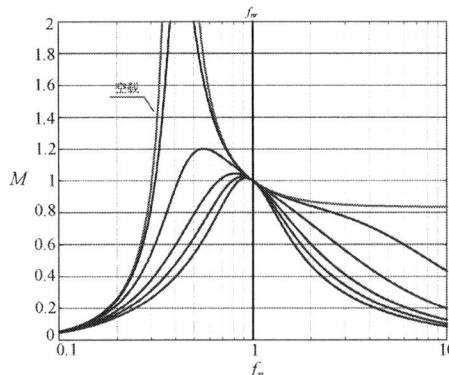

Fig. 10. The LLC resonant converter voltage gain curve.

The LLC resonant converter voltage gain curve is shown in Fig. 10. When the quality factor Q is zero, the gain curve of the no-load voltage M_{OL} can be obtained with

$$M_{OL}(f_n, k) = \frac{1}{\left| 1 + \dfrac{1}{k} - \dfrac{1}{kf_n^2} \right|} \qquad (13)$$

When the normalized frequency f_n tends to infinity, the voltage gain at no-load tends to an asymptotic value M_∞

$$M_\infty = M_{OL}(f_n \to \infty, \lambda) = \frac{k}{1+k} \qquad (14)$$

In order to ensure that the converter can run without load at the required minimum voltage gain M_{min}, it is necessary to ensure

$$M_{min} = 2n\frac{V_{out}}{V_{dc.max}} > \frac{k}{1+k} \qquad (15)$$

It can be seen from the Fig. 9 that as the inductance ratio k decreases, the peak value of the voltage gain curve increases, and the curve gradually becomes steeper and shrinks toward the unity gain point. The second resonance frequency f_{mo} corresponding to the peak also approaches the resonance frequency f_r accordingly.

According to the intersection of the voltage gain curve, the maximum gain curve and the minimum gain curve, the minimum switching frequency and the maximum switching frequency can be determined. Therefore, the appropriate value of the inductance ratio k can be selected to obtain the appropriate range of the switching frequency. The minimum switching frequency f_{s_min} can be determined according to the intersection of the voltage gain curve and the maximum gain curve. Similarly, the maximum switching frequency f_{s_max} can be obtained. Therefore, selecting the appropriate value of the inductance ratio k can obtain the appropriate switching frequency range. Since the value of the inductance ratio k is related to the magnetizing inductance of the transformer, the inductance ratio k=5 in this article.

The relationship between the voltage gain M and the quality factor Q is shown in Fig. 11. It can be seen from the figure that as the value of Q increases, the peak value of the voltage gain curve also decreases. The point corresponding to the second resonant frequency also approaches the unity gain point. It can be seen from the expression of the quality factor Q that when the system is in the no-load state, the value of Q is 0 at this time. When the system is fully loaded, the value of Q is the maximum. Therefore, in order to ensure that the system can work normally under the full load, the LLC resonant circuit parameters should be designed when the value of Q takes the maximum value within its range.

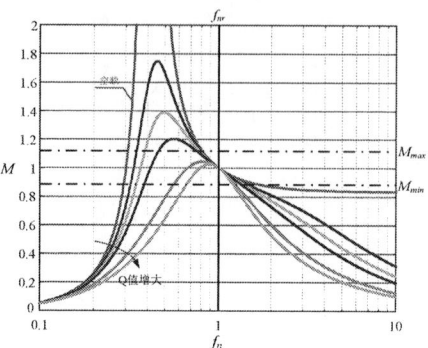

Fig. 11. The relationship between the voltage gain M and the quality factor Q.

The expression of the normalized input impedance Z_n of the resonant circuit is defined as

$$Z_n(f_n, k, Q) = \frac{Z_{in}(f_n, k, Q)}{Z_o} = \frac{jf_n}{\dfrac{1}{k} + jf_n Q} + \frac{1 - f_n^2}{jf_n} \quad (16)$$

By assuming that the imaginary part of $Z_n(f_n, k, Q)$ is zero, that is, the input impedance Z_{in} of the resonant tank has a zero phase angle. Since the output impedance Z_o is a real number, it does not affect the phase. Therefore, the boundary of the capacitive and inductive regions of the resonant tank can be determined as

$$Q_z(f_n, k) = \sqrt{\frac{1}{k(1 - f_n^2)} - \left(\frac{1}{kf_n}\right)^2} \qquad (17)$$

In (17), the parameter Q_z is the maximum value of the quality factor. When it is lower than this value, the impedance of the resonant tank is inductive. According to (6) and (16), the voltage gain curve with purely resistive input impedance can be obtained with

$$M_{res}(f_n, k) = \frac{f_n}{\sqrt{f_n^2\left(1 + \dfrac{1}{k}\right) - \dfrac{1}{k}}} \qquad (18)$$

Therefore, the LLC resonant converter works on the boundary between the inductive and capacitive regions. The inductive and capacitive area of LLC resonant converter is shown in Fig. 12.

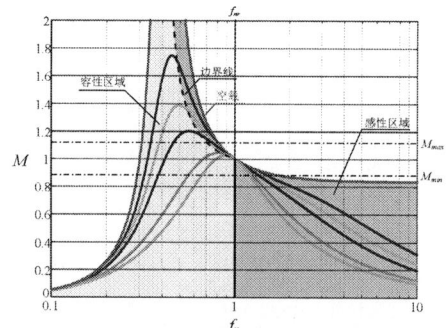

Fig. 12. The inductive and capacitive area of LLC resonant converter.

The intersection of the boundary line and the maximum gain M_{max} is defined as the normalized minimum switching frequency f_{nmin}

$$f_{n\min} = \sqrt{\frac{1}{1+k(1-\frac{1}{M_{max}^2})}} \qquad (19)$$

It allows the LLC resonant converter to obtain the maximum gain at the boundary of the capacitive and inductive regions. The intersection of the no-load voltage gain curve and the minimum voltage gain M_{min} is defined as the normalized maximum switching frequency f_{nmax}

$$f_{n\max} = \sqrt{\frac{1}{1+k(1-\frac{1}{M_{min}})}} \qquad (20)$$

V. SIMULATION VERIFICATION

TABLE II.   SIMULATION PARAMETERS

Parameter	Value
P_{out}	100W
V_{in}	90V~110V
V_o	10kV
f_s	112kHz~188kHz
L_m	30μH
L_r	6μH
C_r	100nF

To verify the correctness of the circuit, simulation is performed according to the main parameters listed in Table II. Fig. 13 shows the resonant inductor current i_{Lr} and excitation current waveform i_{Lm} of the half-bridge LLC resonant converter under the condition of 100V input voltage.

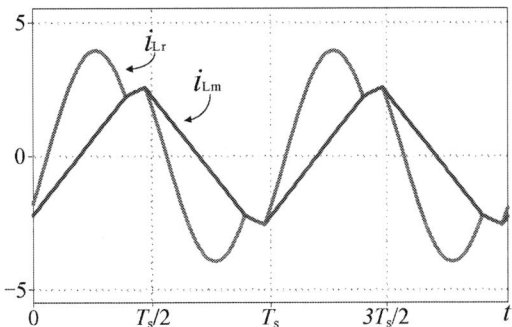

Fig. 13. The current waveform of the resonant inductance L_r and the waveform of the current of the excitation inductance L_m.

It can be seen from the Fig. 13 that the LLC resonant converter can realize the zero voltage turn-on of the switchs on the primary side. Fig. 14 shows the main waveforms of the voltage doubler rectification circuit under the full-load condition. It includes the waveform of the output voltage, the waveform of the rectifier diode voltage, the waveform of the rectifier diode current, the waveform of the voltage of the high-voltage capacitor, and the waveform of the output ripple voltage.

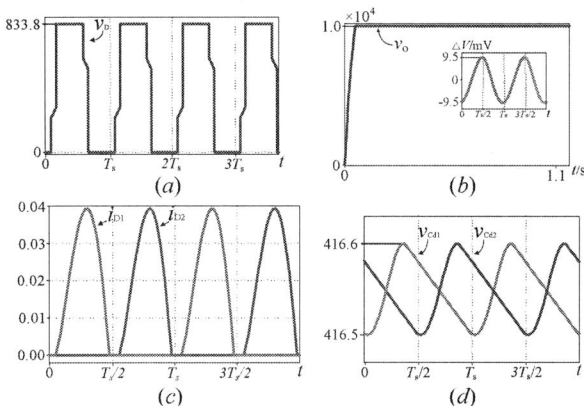

Fig. 14. The main waveforms of the voltage doubler rectification circuit under the full-load condition: (a) v_D; (b) v_o, ΔV (c) i_{D1}, i_{D2} (d) v_{Cd1}, v_{Cd2}.

It can be seen from the Fig. 14 that the average value of the output voltage is 10 kilo volts, and the voltage ripple is less than 0.2 volt. It also can be seen from the waveforms of the current i_{D1} and the current i_{D2} that the rectifier diode can achieve zero-current shutdown.

VI. CONCLUSIONS

This paper proposes a rectifier circuit in which the secondary side uses 14-stage double-voltage rectifier units in series to achieve 28-fold voltage output. The primary side uses an asymmetric half-bridge LLC structure. The converter is controlled by changing the frequency of the switches. This topology can reduce the voltage on the rectifier diodes and the voltage on the key devices. The simulation results demonstrate the effectiveness of the transformer and the feasibility of reducing the voltage on the devices.

REFERENCES

[1] Jialiang Li,Bo Zhu,Mingyu Cheng,Yuan Li,Mingliang Wang. Design and control algorithm of dual switch forward DC/DC converter with DSP digital control[J]. Journal of Physics: Conference Series,2021,1754(1).

[2] Wu Xiaogang,Liu Zhengxin,Du Jiuyu,Yu Boyang. Research on Zero Voltage Switching Non-inductive Current Circulation Control of Bidirectional DC/DC Converter for Hybrid Energy Source System of Electric Vehicle[J]. Journal of Electrical Engineering & Technology,2021,16(2).

[3] Matsushita Yoshinori,Noguchi Toshihiko,Taguchi Noritaka,Ishii Makoto. 2 kW Dual-Output Isolated DC/DC Converter Based on Current Doubler and Step-Down Chopper[J]. World Electric Vehicle Journal,2020,11(4).

[4] Güler Naki,Irmak Erdal. Design, implementation and model predictive based control of a mode-changeable DC/DC converter for hybrid renewable energy systems[J]. ISA Transactions,2020(prepublish).

[5] Cho Min Su,Mun Hye Jin,Lee Sang Ho,An Hee Dae,Park Jin,Jang Jaewon,Bae JinHyuk,Kang In Man. Design and Analysis of DC/DC Boost Converter Using Vertical GaN Power Device.[J]. Journal of nanoscience and nanotechnology,2021,21(8).

[6] Hasanpour Sara,Siwakoti Yam,Blaabjerg Frede. New Single-Switch quadratic boost DC/DC converter with Low voltage stress for renewable energy applications[J]. IET Power Electronics,2021,13(19).

[7] Ding Jie,Zhao Shiwei,Yin Huajie,Qin Ping,Zeng Guanbao. High step-up DC/DC converters based on coupled inductor and switched capacitors[J]. IET Power Electronics,2020,13(14).

[8] António Manuel Santos Spencer Andrade,Tiago Miguel Klein Faistel,Ronaldo Antonio Guisso. Single-switch high-efficiency hybrid boost–Ćuk DC/DC converter with high-voltage gain and low-voltage stress[J]. IET Power Electronics,2020,13(12).

Impact of Gate Resistances on Switching-on Behaviors of Si/SiC Hybrid Switch

Xiaofeng Jiang[1], Huaping Jiang[1,*], Hongyu Yu[1], Jinhong Jiang[1], Hao Feng[1], Hua Mao[1],
Lei Tang[1], Xiaohan Zhong[1], Li Ran[1,2]

[1] State Key Laboratory of Power Transmission Equipment and System Safety and New
Technology, Chongqing University, Chongqing, 400044, China
[2] School of Engineering University of Warwick Coventry, U.K
*Email: Stefan.Jiang@foxmail.com

Abstract—A SiC/Si hybrid switch (HyS) composed of a larger current IGBT and a smaller current MOSFET can provide higher cost performance. Previous research on the hybrid switch mostly focuses on reducing loss and balancing junction temperature through different control strategies, neglecting the reliability issue caused by the load current exceeding the safe operating area (SOA) of SiC MOSFET with a lower current rating. In addition, it remains an unsolved issue that how to design gate resistances for a Si/SiC HyS. This manuscript investigates the impact of gate resistances on switching-on behaviors of a Si/SiC HyS in a double pulse test. Furthermore, the tradeoff between switching loss and the device reliability can be used as a reference for selecting the switching-on gate resistance of a hybrid switch.

Keywords—Hybrid switch, gate resistances, safe operating area, switching loss, SiC

I. INTRODUCTION

Compared to the silicon insulated gate bipolar transistor (IGBT), the silicon carbide power metal-oxide-semiconductor field-effect transistor (MOSFET) has the advantage of lower switching loss, higher breakdown voltage, and heat conductivity. However, the application of the SiC MOSFET is impeded because of the high cost of SiC materials. Si IGBT has the advantage of lower conduction loss under heavy load and the low cost. To achieve a high cost performance, the concept of a Si IGBT/SiC MOSFET hybrid switch is proposed to combine advantages of SiC MOSFET and Si IGBT [2-4]. Compared with all-Si IGBT and all-SiC MOSFET at the same rated power, the HyS has a lower conduction loss in the full-load current range. Therefore, the HyS can be more attractive in power electronic converters due to a high cost performance [5].

Previous research mostly focuses on reducing the device loss and balancing the junction temperature of a Si/SiC HyS [6]. Nevertheless, these switching control methods cannot solve the reliability issue of a Si/SiC HyS due to dynamic current exceeding the SOA of small current SiC MOSFET. A new switching strategy is proposed, to eliminate the effect of dynamic overcurrent stress on the reliability of SiC MOSFET [7]. When the total load current is over the SOA limit current of SiC MOSFET, the low-current MOSFET is protected by high current Si IGBT. The IGBT turns on earlier than SiC MOSFET and turns off at the same time as the MOSFET. However, the active gate sequencing method is more complicated in applications. Except for that, it remains an unsolved issue that how to design gate resistances for a Si/SiC HyS. In this manuscript, we found that the gate resistance can adjust the change rate of voltage and current during the switching-on and switching-off of devices, which can suppress the current overshoot of SiC MOSFET when the gate signals of Si IGBT and SiC MOSFET are triggered at the same

moment. Consequently, total passive dynamic current balancing can be achieved without resorting to the active gate sequencing method.

This paper investigates the impact of gate resistances on switching-on voltage and current of a HyS. We experimentally explore the impact of gate resistances on the dynamic current distribution and switching loss when a hybrid switch turns on. A double pulse test (DPT) with two parallel devices is established to investigate the impact of gate resistances on its reliability and switching loss. Proper gate resistances are extracted and implemented to reach the tradeoff between the device reliability and switching loss.

II. DOUBLE PULSE TEST BASED ON A SI/SIC HYS

A. Gate Control Pattern Selection of a HyS

Generally, a Si/SiC HyS is composed of a large-current rating IGBT and a small-current rating MOSFET to achieve high cost and low cost performance. Infineon 1200 V/50 A Si IGBT (IKW25T120) and 1200 V/25 A SiC MOSFET (IMW120R090M1H) are selected. Key electrical parameters of power devices are illustraed in Table I. Gate driver signals are illustrated in Fig. 1. The minimum switching-on loss of a HyS occurs dur to larger dI/dt when the MOSFET and IGBT are swicthed on at the same moment [7]. To investigate the impact of gate resistance on switch-on behaviors of the HyS, $T_{\text{on\_delay}} = 0$ could be adopted.

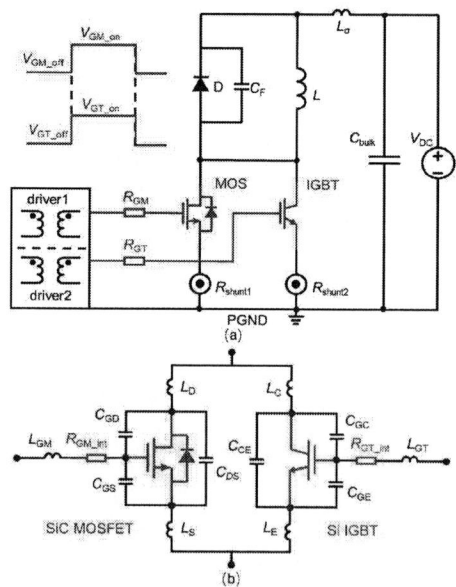

Fig. 1. The schematic diagram of double pulse test and a Si/SiC hybrid switch: V_{DC}, L_σ, and L represent input voltage, the parasitic inductance of the dc bus and load inductance, respectively. (a) double pulse test. (b) hybrid switch.

978-1-6654-4817-8/21 $31.00 © 2021 IEEE

B. Gate Voltage Selection and Static Characteristics Comparison

Conduction characteristic curves of the studied hybrid switch with different gate voltages at 25°C are illustrated in Fig. 2. The condition loss of a Si/SiC hybrid switch decreases with gate voltage of the device increases. When the gate voltage increases from 15V to 20V, the forward voltage V_F decreases from 3.1V to 2.7V under full load 75A at room temperature. Nevertheless, the short circuit withstanding time of power device is shortened under the large positive gate voltage. Meanwhile, a large dv/dt will mislead the device through the Miller capacitor under zero voltage switching-off. Therefore, the gate voltages of SiC MOSFET and Si IGBT can not be too large and the switching-off voltage should be negative. To highlight the effect of gate resistance, the MOSFET and IGBT adopt the same driving voltage. Combined with the recommended gate voltage in the device datasheet, +15V/-5V could be adopted.

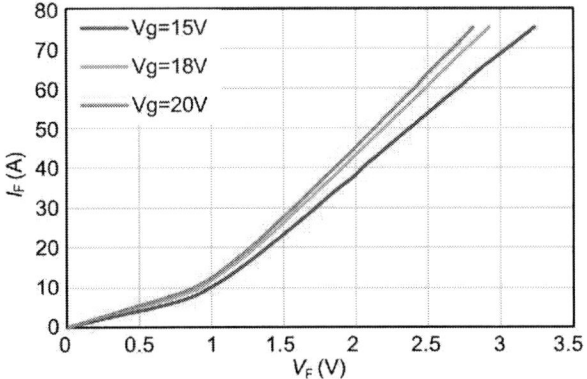

Fig. 2. Conduction characteristic curves of the studied Si/SiC Hys with different gate voltages at room temperature.

TABLE I.   KEY PARAMETERS OF A SI/SIC HYS

Parameter	IKW25T120	IMW120R090M1H
V_{CE} and V_{DS} (V)	1200	1200
I_C and I_D (A)	50	25
I_{SOA} (A)	75	50
R_{Gint} (Ω)	8	9

C. Turn on transient current Analysis of double pulse test

The structure of the double pulse test is indicated in Fig. 1. When a Si/SiC HyS is turned on, the transient current can be described as follows:

$$i_{HyS} = i_L + \Delta i = i_L + i_{rr} + i_F \qquad (1)$$

where i_L is the load current, Δi is the current overshoot, i_{rr} is the diode reverse recovery current, and i_F is the diode parasitic capacitance charging current. The switching-on transient current of the HyS i_{HyS} is greater than the rated load current i_L due to the effect of the diode. In this study, the SiC Schottky Barrier Diode (SBD) is used as a freewheeling diode. Therefore, the diode reverse recovery current can be neglected. When a hybrid switch is turned on for the first time, i_L=0, the diode reverse voltage increases as the HyS voltage decreases. At that moment, the dc bus capacitor C_{bulk} charges the diode parasitic capacitance C_F. The diode parasitic capacitance charging current i_F begins to flow from C_{bulk} to C_F, and then to a Si/SiC hybrid switch. It can be expressed as follows:

$$i_F = C_F \frac{dv_F}{dt} \approx -C_F \frac{dv_{DS}}{dt} \qquad (2)$$

where v_F, and v_{DS} are the diode reverse voltage and the hybrid switch voltage, respectively. Obviously, i_F decreases as the switching-on process of the HyS slows down. In other words, in addition to changing the transient current distribution of the IGBT and the MOSFET, the gate resistance of a HyS can also change the charging current of the diode parasitic capacitance by changing the dv_{ds}/dt of a HyS. The gate resistance of the MOSFET and IGBT can be expressed as follows:

$$R_{GM} = R_{GM\_ext} + R_{GM\_int} \qquad (3)$$

$$R_{GT} = R_{GT\_ext} + R_{GT\_int} \qquad (4)$$

where R_{GM_int} and R_{GT_int} are respectively internal gate resistances of the MOSFET and the IGBT, R_{GM_ext} and R_{GT_ext} are respectively external gate resistances of the MOSFET and the IGBT.

III. EXPERIMENTAL TESTS

A. Experimental test platform

To verify the impact of the gate resistance on the switching-on current distribution of a HyS, as depicted in Fig. 3, the experimental test platform is built, which includes the double pulse test with two parallel devices, DC power, signal generator, inductor and digital oscilloscope. Experimental electrical parameters are illustrated in Table II.

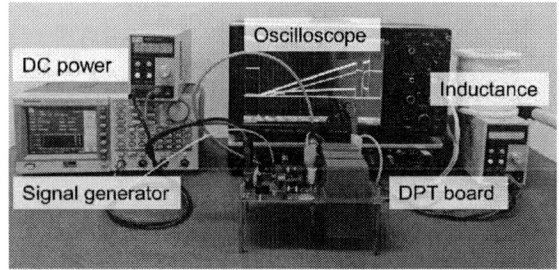

Fig. 3. Experimental platform.

TABLE II.   EXPERIMENTAL ELECTRICAL PARAMETERS

Parameter	Values	Parameter	Values
DC voltage (V)	600	Shunt Resistor (mΩ)	10
Load current (A)	75	Gate voltage (V)	-5/15
DC capacitor (μF)	240	R_{GT} (Ω)	8~48
Load inductor (μH)	568	R_{GM} (Ω)	9~109

B. Switching-on transient waveform of HyS

Fig. 4 and Fig. 5 indicate how the gate resistances R_{GT}, R_{GM} impact swtiching-on behaviors of a HyS. As demonstrated in Fig. 4, the charging speed of C_{GS} and C_{GD} decreases as the value of R_{GM} increases. The peak current of the MOSFET is smaller than its SOA current in the case of R_{GM} = 29 Ω. In the case of R_{GT} = 49 Ω, the load current flows into the IGBT first and then into the MOSFET. The peak current of the IGBT becomes large as the R_{GM} increases.

Fig. 5 illustrates the measured switching-on curves of a HyS when R_{GM} is fixed, R_{GT} is increased. The switching-on speed of Si IGBT decreases when R_{GT} increases from 8 Ω to 48 Ω. Therefore, the MOSFET is forced to conduct most of the HyS transient current, which will exacerbate the reliability issue of SiC MOSFET. Observing the load current i_L curves in Fig. 4 and Fig. 5, the increase of gate resistances reduces the diode parasitic capacitance charging current i_F (i.e. the peak of the transient current i_{HyS}) by slowing down the speed of devices. This agrees with the analysis in Section II. V_{DS} begins to drop when the load current i_L reaches a steady state. However, at that time, the current of the internal Si/SiC HyS has not reached steady-state current. During this period, the on-resistance of the IGBT decreases because of the conductivity modulation and the increase of the gate voltage. Dynamic current distribution could lead the commutation process of internal HyS. Therefore, gate resistances can also affect the commutation process by adjusting the current distribution of a Si/SiC HyS.

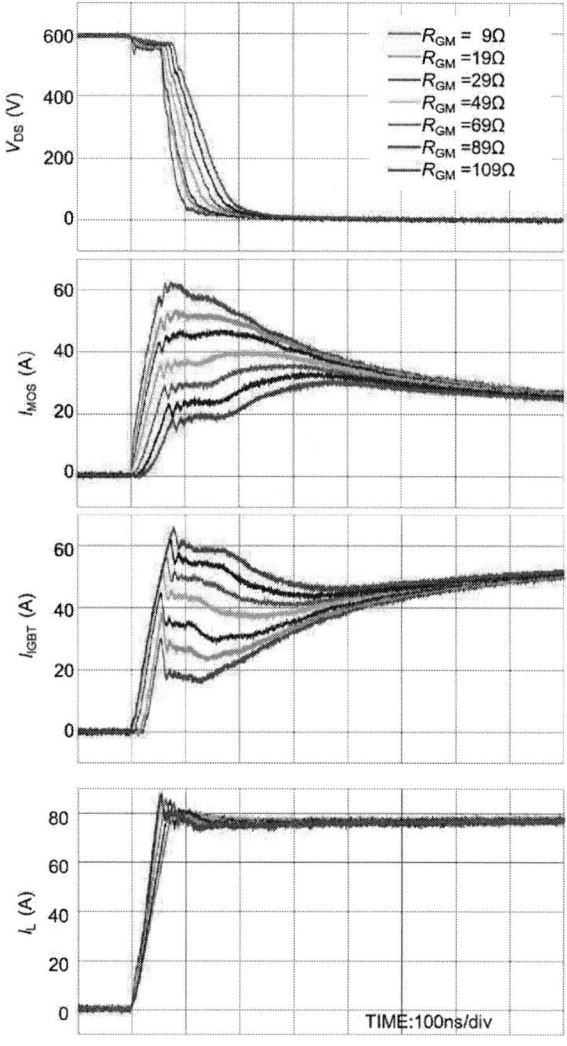

Fig. 4. The swicthing-on behavior of a HyS with different gate resistances R_{GM} of a SiC MOSFET and a fixed gate resistance (8 Ω) R_{GT} of a Si IGBT.

Fig. 6 illustrates the swicthing-on loss of a Si/SiC HyS by adjusting different gate resistances. Small R_{GT} and small R_{GM} can accelerate the switching-on process, resulting in reducing the swicthing-on loss. When the HyS turns on, the overcurrent stress coefficients of the MOSFET and IGBT are determined as:

$$\alpha_{M\_on} = I_{Mp\_on} / I_{MSOA} \tag{5}$$
$$\alpha_{T\_on} = I_{Tp\_on} / I_{TSOA} \tag{6}$$

where I_{Mp_on} and I_{Tp_on} are the peak current of SiC MOSFET and Si IGBT during swicthing-on process, I_{MSOA} and I_{TSOA} are the safe operation area limit current of inernal IGBT and MOSFET. The overcurrent stress coefficients of the MOSFET and IGBT by adjusting different gate resistances are depicted in Fig.7 and Fig.8, respectively.

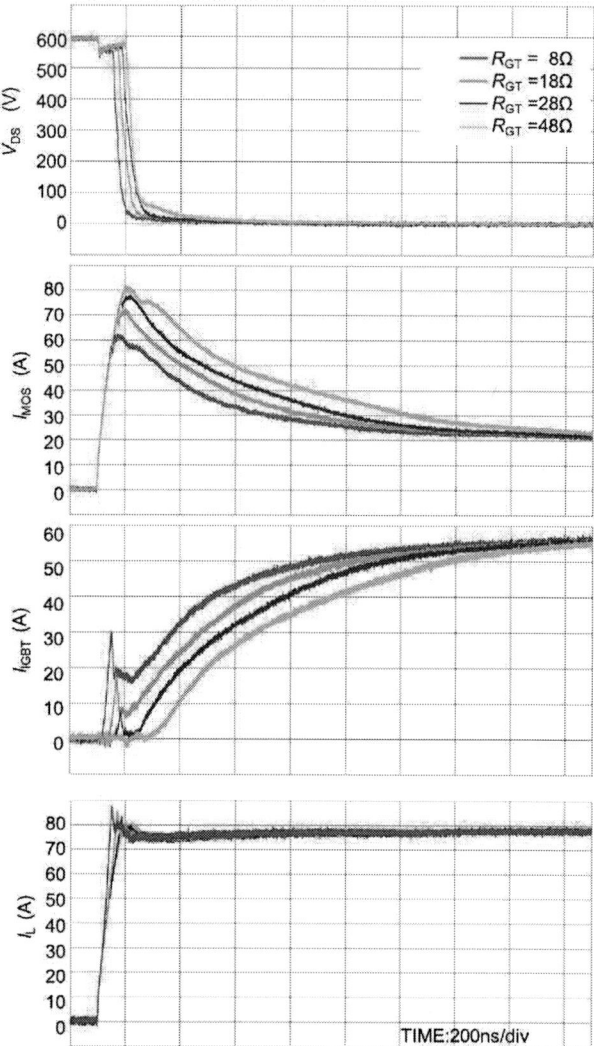

Fig. 5. The swicthing-on behavior of a hybrid switch by adjusting different gate resistances R_{GT} of the Si IGBT and a fixed gate resistance (9 Ω) R_{GM} of SiC MOSFET.

As demonstrated in Fig. 7, the overcurrent stress coefficient of Si IGBT is always less than 1. An increased R_{GM} reduces the charging speed of gate capacitances of the MOSFET, which drives more dynamic current into the IGBT and naturally leading to higher loss. Therefore, the R_{GT} should be small but to be tuned to a lower limit (8 Ω), which meets maximum dv/dt under the actual operating conditions of a HyS. R_{GM} can be adjusted to a moderate level to force the IGBT to carry a higher current so that the low-current SiC MOSFET can ride through the harsh switching-on condition. As illustrated in Fig. 8, all columns with $\alpha_{M\_on} > 1$ are marked

978-1-6654-4817-8/21 $31.00 © 2021 IEEE

yellow. The peak current of the MOSFET is below the device SOA limit when the R_{GT} is small and R_{GM} is large.

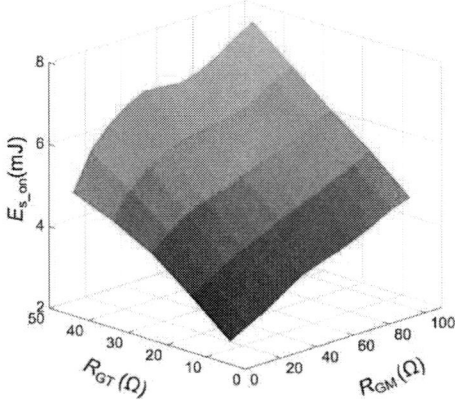

Fig. 6. The switching-on loss of a Si/SiC HyS by adjusting different gate resistances.

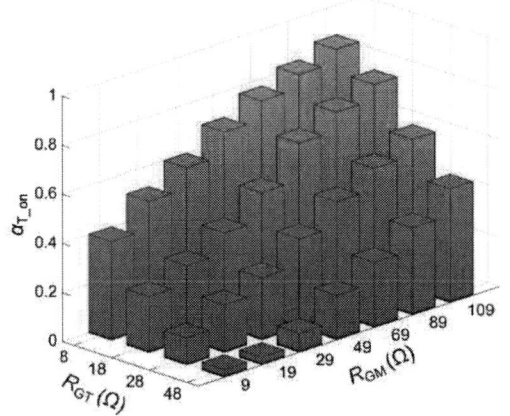

Fig. 7. The overcurrent stress coefficient of the Si IGBT by adjusting different gate resistances.

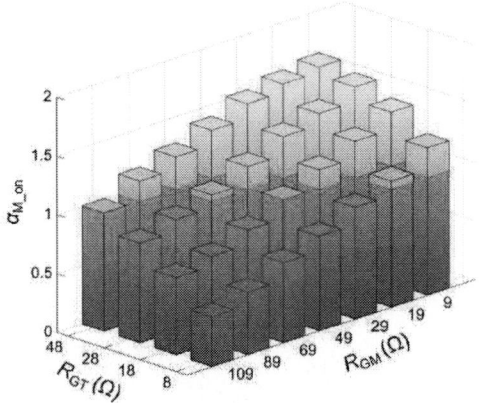

Fig. 8. The overcurrent stress coefficient of the SiC MOSFET by adjusting different gate resistances

Combined with Fig. 6 and Fig. 8, the trade-off between the swithing-on loss of a HyS and the overcurrent stress

coefficient of the MOSFET is indicated in Fig. 9. R_{GM} is adjusted to balance the switching-on loss and switching-on dynamic current redistribution under the premise of meeting the SiC MOSFET SOA. E_{s_on} decreases with the rise of α_{M_on}. The swicthing-on loss of a hybrid switch is reduced by 59.8% when α_{M_on} increases from 0.42 ($R_{GM} = 109\ \Omega$) to 0.95 ($R_{GM} = 29\ \Omega$). The recommended gate resistances in the devices' datasheet are equivalent to $R_{GM} = 9\ \Omega$ and $R_{GT} = 18\ \Omega$ (point B) in Fig. 9. However, a 29 Ω R_{GM} (point A in Fig. 9) and an 8 Ω R_{GT} are finally chosen, reducing switching-on loss by 11.8% compared to point B while improving the reliability of SiC MOSFET.

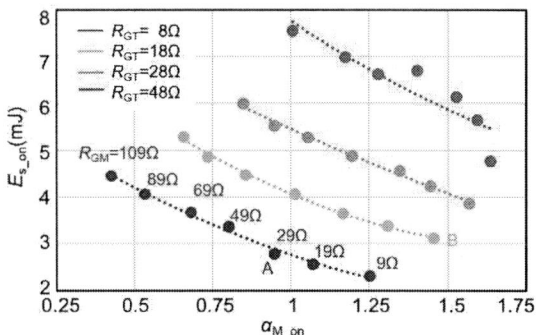

Fig. 9. The trade-off between the swicthing-on loss of a HyS and the overcurrent stress coefficient of the SiC MOSFET.

IV. CONCLUSION

The impact of gate resistances on switching-on behaviors of Si/SiC HyS is investigated by experiment in this paper. Switching loss and switching-on overcurrent of the hybrid switch can be regulated gate resistances. Switching-on loss of a HyS increases as R_{GT} and R_{GM} increase. The dynamic peak current of the IGBT and MOSFET could be below their respective SOA limit current when R_{GT} ranges from 8 Ω to 28 Ω while R_{GM} ranges from 29 Ω to 109 Ω. The switching-on loss of a HyS is reduced by 59.8% when α_{M_on} increases from 0.42 to 0.95. Therefore, tradeoff between switching-on loss and the reliability of device can be used as a reference for selecting switching-on gate resistance of the Si/SiC HyS.

The minimum R_{GT} should be limited to the maximum allowable dv/dt under the actual operating conditions of a HyS. And R_{GM} could reduce the switching loss under the premise of ensuring that the dynamic current is below the SOA limit of SiC MOSFET. In this paper, $R_{GT} = 8\ \Omega$, and $R_{GM} = 29\ \Omega$ can achieve a better performance/reliability tradeoff than other resistors matching. In the future, the impact of gate resistances on the switching-off behaviors and the active driver of the Si/SiC HyS would be studied.

ACKNOWLEDGMENT

This stduy was supported by the National Key Research and Development Program of China under Grant 2018YFB0905803. The authors would like to thank them for financial support.

REFERENCES

[1] M. H. Mohamed Sathik, P. Sundararajan, F. Sasongko, J. Pou, and V. Vaiyapuri, "Short circuit detection and fault current limiting method for IGBTs," *IEEE Transactions on Device and Materials Reliability*, vol. 20, no. 4, pp. 686-693, 2020.

[2] M. Rahimo *et al.*, "Characterization of a silicon IGBT and silicon carbide MOSFET cross-switch hybrid," *IEEE Transactions on Power Electronics,* vol. 30, no. 9, pp. 4638-4642, 2015.

[3] M. Rahimo *et al.*, "The cross switch "XS" silicon and silicon carbide hybrid concept," in *Proceedings of PCIM Europe 2015; International Exhibition and Conference for Power Electronics, Intelligent Motion, Renewable Energy and Energy Management*, 2015, pp. 1-8.

[4] Z. Li, J. Wang, Z. He, J. Yu, Y. Dai, and Z. J. Shen, "Performance comparison of two hybrid Si/SiC device concepts," *Ieee Journal of Emerging and Selected Topics in Power Electronics,* vol. 8, no. 1, pp. 42-53, Mar 2020.

[5] C. Li *et al.*, "Space vector modulation for SiC and Si Hybrid ANPC Converter in Medium-Voltage High-Speed Drive System," *IEEE*

Transactions on Power Electronics, vol. 35, no. 4, pp. 3390-3401, 2020.

[6] J. Wang, Z. Li, X. Jiang, C. Zeng, and Z. J. Shen, "Gate control optimization of Si/SiC hybrid switch for junction temperature balance and power loss reduction," *IEEE Transactions on Power Electronics,* vol. 34, no. 2, pp. 1744-1754, 2019.

[7] Z. Peng *et al.*, "A variable-frequency current-dependent switching strategy to improve tradeoff between rfficiency and SiC MOSFET overcurrent stress in Si/SiC-Hybrid-Switch-Based Inverters," *IEEE Transactions on Power Electronics,* vol. 36, no. 4, pp. 4877-4886, 2021.

The Influence of Dynamic Threshold Voltage Drift on Third Quadrant Characteristics of SiC MOSFET

Lei Tang[1], Huaping Jiang[1,*], Hua Mao[1], Zebing Wu[1], Xiaohan Zhong[1], Xiaowei Qi[1], Li Ran[1,2]

[1] State Key Laboratory of Power Transmission Equipment and System Safety and New Technology,
Chongqing University, Chongqing, 400044, China
[2] School of Engineering University of Warwick Coventry, U.K
*Email: Stefan.Jiang@foxmail.com

Abstract—The conduction characteristics of the third quadrant of SiC MOSFET include the current distribution of the body diode and the MOS channel, which are sensitive to many factors. In this paper, the influence of dynamic threshold voltage drift on third quadrant characteristics of SiC MOSFET is studied. By combining the equivalent potential model, TCAD simulation, and experiment, it is revealed that once the threshold voltage drifts under dynamic gate voltage stress, the static and dynamic characteristics of the third quadrant of SiC MOSFET will change. Furthermore, the current sharing of parallel devices is also affected by dynamic threshold voltage drift. These results provide an effective guide for the third quadrant application of SiC MOSFET.

Keywords—Body diode, SiC MOSFET, body effect, third quadrant characteristics, dynamic threshold voltage drift.

I. INTRODUCTION

Thanks to superior material properties, compared with Si IGBT, SiC MOSFET shows low switching loss and promises potential in high-temperature application, which is widely used in various fields [1]-[3]. As the shorter average carrier lifetime of SiC compared to Si and the manufacturing process of SiC MOSFET is optimized, the reverse recovery performance of the body diode has been improved [4], and the body diode instead of SBD can be used to reduce costs and increase power density. Therefore, the body diode is commonly used for freewheeling. Although SiC MOSFETs are constantly being improved, some problems such as unstable forward voltage (V_F) and the unclear mechanism of the third quadrant characteristics need further research.

At present, research institutions are exploring the third quadrant characteristics of SiC MOSFET [5]-[6]. The influence of temperature and negative gate voltage on the third quadrant characteristics is reported in [7], and the V_F increases with the $|V_{GSOFF}|$ and decreases with the temperature. Furthermore, the third quadrant characteristics under positive gate voltage bias are described in [8]. The body effect is the main factor that affects the third quadrant characteristics [9]. Based on the model of body effect [10], the third quadrant characteristic is used for monitoring PBTI and NBTI are proposed in [11]. The threshold voltage (V_{TH}) is a very important parameter of SiC MOSFET, the switching loss and speed of the device are related to it. It is observed that threshold drifts largely under dynamic gate stress, and the excessive turn-off voltage (V_{GSOFF}) will cause threshold drift [12], and its influence on the third quadrant has not been reported in recent researches.

In this paper, Sentaurus TCAD simulation software and experiments are used to study the influence of dynamic threshold voltage drift on the third quadrant characteristics of SiC MOSFET. Furthermore, the body effect is explained in

This work was supported by the National Key Research and Development Program of China under Grant 2018YFB0905803.

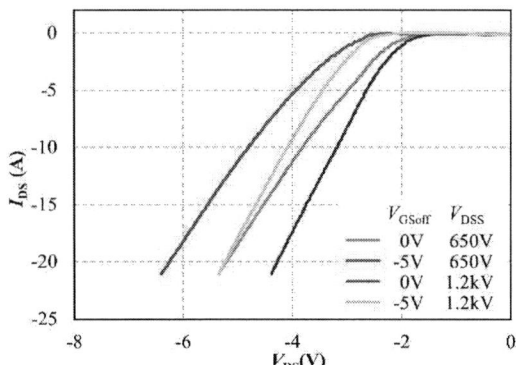

Fig. 1. Third quadrant characteristics of trench SiC MOSFETs with different voltage levels

combination with the equivalent potential distribution, which is the mechanism for the V_F affected by V_{TH}. It is hoped that this research will contribute to a deeper understanding of the third quadrant characteristics of SiC MOSFET.

In this paper, section II presents the third quadrant characteristics, body effect theory, and the Sentaurus TCAD simulation for body effect. Section III illustrates the dynamic threshold voltage drift and its influence on the static and dynamic characteristics of the third quadrant of SiC MOSFET. Finally, section IV concludes this paper.

II. THIRD QUADRANT CHARACTERISTICS AND BODY EFFECT MECHANISM ANALYSIS

This section introduces the third quadrant conduction characteristics and body effect mechanism, which provides a theoretical basis for investigating the relationship between the dynamic threshold drift and the third quadrant characteristics.

A. Third quadrant characteristics

As shown in Fig. 1, the third quadrant performance of 650V and 1.2 kV SiC trench MOSFETs under different V_{GS} conditions are tested by the power device analyzer (Agilent B1505A).

When $V_{GS} \leq$ -10V, the SiC MOSFET channel is almost completely closed. At this time, the third quadrant characteristic can be approximately equivalent to a PiN diode. Equation (1) is used to describe the voltage drop V_F of a PiN diode with a conduction current density of J_T.

$$V_F = \frac{2kT}{q} ln\left(\frac{J_T d}{2qD_a n_i F\left(d / L_a \right)} \right) \tag{1}$$

k is the Boltzmann constant, T is the temperature, q is the elementary charge, d is the length of the drift region, D_a is the

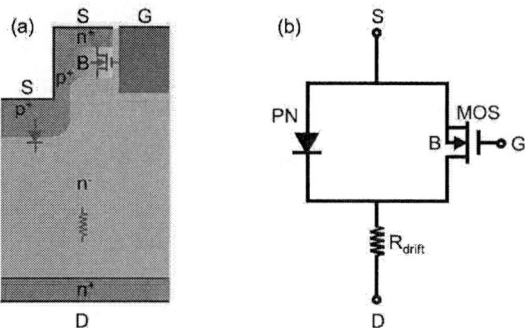

Fig. 2. (a) Trench cell (b) Third quadrant equivalent circuit

diffusion coefficient, n_i is the intrinsic carrier concentration, L_a is the bipolar diffusion length, and F is the composite function depending on d and L_a.

When the V_{GS} is 0V, the SiC MOSFET will be partially turned on under the influence of the body effect. The current will be divided between the channel and the body diode according to Ohm's law. As shown in Fig. 1, under the same V_{DS}, the I_{DS} ($V_{GS} = 0V$) is higher than the I_{DS} ($V_{GS} = -5V$) and the difference between them can be considered as the part that flows through the channel.

According to the above theory, Fig. 2 shows the cell structure and the third quadrant equivalent model which is composed of three parts: PN junction, MOS, drift region resistance R_{drift}, the third quadrant voltage is mostly distributed on the MOS / PN and drift region. It is worth noting that the figure emphasizes the effect of the B region potential on the characteristics of the third quadrant. The threshold voltage is influenced by the potential of the B region.

B. Body effect theory

The body effect theory of the MOSFET is mainly derived from the influence of the bias voltage between the S-B (Source-Bulk) on the V_{TH} of the MOSFET. With the advent of trench gate MOSFETs, the channel density is greatly improved, and the body effect is more obvious. This article will describe the influence of the body effect from the perspective of potential distribution.

What stands out in Fig.3 is the potential distribution of the first and third quadrants of the SiC MOSFET. The I-V characteristic curves of the first and third quadrants are shown in Fig.3 (a). When $V_{GS}=0V$, the SiC MOSFET is not turned on at point F ($V_{DS} = 3V$), but the current is generated at point T ($V_{DS} = -3V$). The above phenomenon is caused by the body effect which induces the threshold voltage of SiC MOSFET to drop. As shown in Fig.3 (b) and (c), the physical electrode is the same as the electrical electrode. However, in Fig.3 (d) and (e), the physical drain corresponds to the electrical source, and the physical source corresponds to the electrical drain, the details are shown in Table I. All electrodes in the picture are electrical electrodes, it can be seen from the potential distribution that the potentials of B and G at point T are higher than those at point F. The increasing potential at point B is the result of body effect, and the increasing potential at point G is caused by reverse connection, the influence of body effect on V_{TH} will be verified by TCAD simulation and formula.

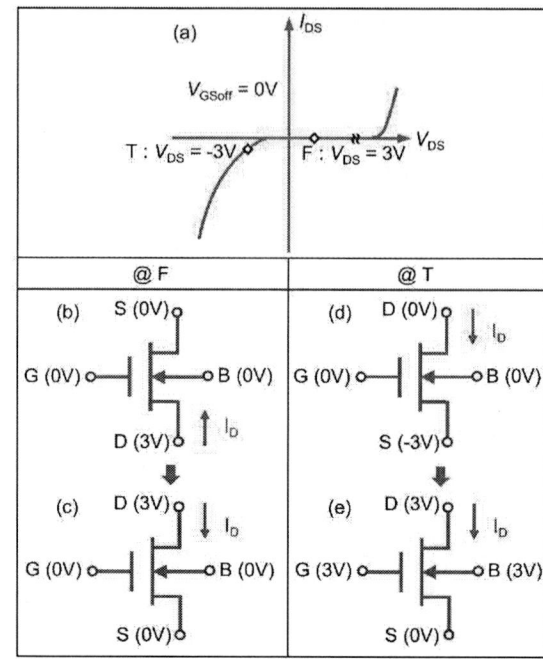

Fig. 3. Equivalent potential distribution of SiC MOSFET under forward and reverse voltage

TABLE I. PHYSICAL ELECTRODES AND ELECTRICAL ELECTRODES

Point	Physical G	Physical D	Physical S	Physical B
@F	Electrical G	Electrical D	Electrical S	Electrical B
@T	Electrical G	Electrical S	Electrical D	Electrical B

Under the effect of the body effect, the V_{TH} depends on the body source potential V_{BS}, which is shown in formula (2)-(3). The V_{Bias} is the equivalent voltage applied to the gate under the influence of the body effect. In the third quadrant of the SiC MOSFET, the V_{Bias} is negative. Therefore, as shown in Fig. 4 (b), the V_{Bias} decreases with the increase of the V_{BS}.

$$V_{TH} = V_{TH0} + V_{Bias} \tag{2}$$

$$V_{Bias} = \gamma \left(\sqrt{2\phi_p - V_{BS}} - \sqrt{2\phi_p} \right) \tag{3}$$

The zero-bias threshold voltage V_{TH0} and the body effect coefficient γ as follow:

$$V_{TH0} = V_{FB} + 2\phi_p + \frac{\sqrt{2\varepsilon_s q N_A (2\phi_p)}}{C_{ox}} \tag{4}$$

$$\gamma = \frac{\sqrt{2\varepsilon_s q N_A}}{C_{ox}} \tag{5}$$

C_{OX} is the oxide capacitance; V_{FB} the flat band voltage; Φ_p the surface potential; ε_s the dielectric constant of silicon carbide; N_A the effective channel doping.

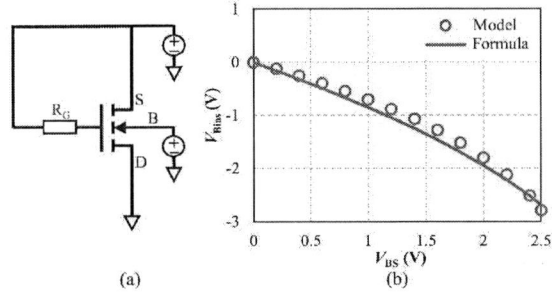

(a)   (b)

Fig. 4. (a) Threshold voltage measurement circuit (b) The relationship between V_{Bias} and V_{BS}.

This is illustrated in Fig 4, the solid line in the figure which uses (2)-(5) to draw the V_{BS} - V_{bias} (t_{ox}=40nm, N_A=2×17cm^{-3}), the scattered points in Fig.4 (b) are drawn using Sentaurus TCAD simulation software, the device cell is shown in Fig.2. In the actual device, the body effect area (B) is always connected to the physical source, but the B area can be connected separately in the simulation software, The simulation circuit diagram is shown in Fig.4 (a). As V_{BS} increases, V_{bias} decreases rapidly. When V_{BS} is 2.5V, V_{bias} = -2.75V. This means that under the influence of the body effect, the threshold voltage will be reduced by 2.75V, and the potential at point G is higher than the potential at point S, these factors induce channel formation.

According to the above theoretical analysis, in the third quadrant of the SiC MOSFET, the channel may turn on due to the body effect. Therefore, when the V_{TH} of SiC MOSFET changes, it will affect the characteristics of the third quadrant.

III. IMPACT OF DYNAMIC THRESHOLD VOLTAGE DRIFT ON THIRD QUADRANT CHARACTERISTICS

This section analyzes the theory through experimental data, and the impact of dynamic threshold drift on the characteristics of the third quadrant is divided into static and dynamic.

A. Dynamic threshold voltage drift

Compared with PBTI and NBTI, dynamic threshold voltage drift is a long-term effect, which can cause lasting drift to SiC MOSFET. Excessive V_{TH} will make the device turn on difficultly. Therefore, it affects the switching loss of the SiC MOSFET, and the threshold voltage drift has a greater impact on the parallel connection of devices.

The dynamic threshold aging test bench is presented in Fig. 5. The SiC MOSFET threshold voltage aging circuit is shown in Fig.6 (a). The trench commercial SiC MOSFETs are tested and their recommended gate-source voltage is -5 V / 18V. Fig.6 (b) shows V_{GS} waveform. In the dynamic threshold voltage drift experiment, the duty cycle of the gate voltage stress is 50%. The power supply needs 2 seconds to realize the voltage scan and the V_{TH} reading of the device under test.

The threshold voltage in this article is the voltage difference between the gate and the source when the drain current reaches 10mA. To shorten the experiment time, this article sets the voltage stress frequency to 150kHz. Aging devices with different V_{TH} can be obtained by changing the

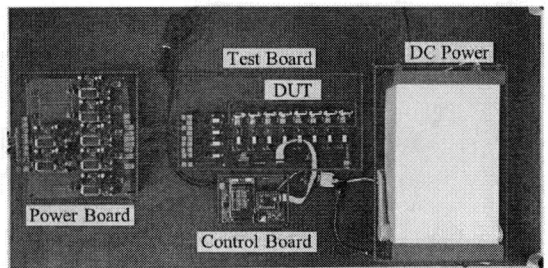

Fig. 5. Test bench for dynamic threshold voltage drift.

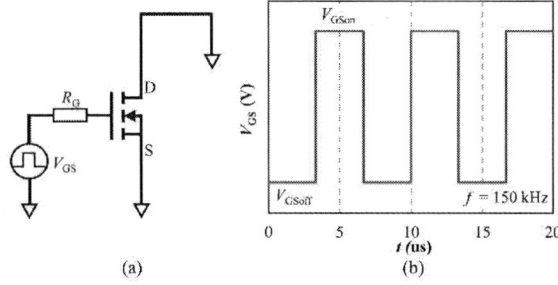

(a)   (b)

Fig. 6. Threshold voltage aging circuit and gate voltage stress waveform.

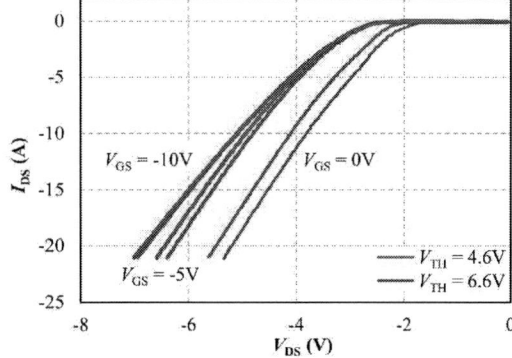

Fig. 7. The influence of the threshold voltage on the static characteristics of the third quadrant.

aging time. Finally, the third quadrant characteristics of aging devices are tested by the B1505A and double pulse test (DPT) bench.

B. The third quadrant Static characteristics

It is important to research the third quadrant static characteristics for power switching applications. Especially those that need reverse freewheeling, such as motor drives, synchronous buck converters, and synchronous rectifiers.

In the previous section, devices with different threshold voltages are obtained by dynamic gate voltage stress aging. As shown in Fig. 7, it can be found that in the case of V_{GS} = 0V and V_{GS} = -5V, $|V_{DS}|$ rises with the increased threshold voltage at the same I_{DS}, this trend is caused by the body effect. The larger the V_{TH} of the SiC MOSFET, the more difficult the channel to be formed. When V_{GS} = -10V, the devices with different threshold voltages, whose third quadrant I-V characteristic curves almost overlap. Because the

978-1-6654-4817-8/21 $31.00 © 2021 IEEE   485

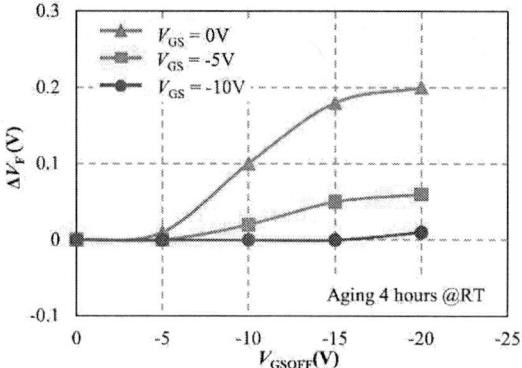

Fig. 8. The influence of gate stress on the static characteristics.

channel is completely closed and the body effect does not exist at this time, all current flows through the body diode.

As shown in Fig. 8, fixed the V_{Gson} to 18V, and the V_{Gsoff} is variable. It is apparent from this picture that the ΔV_F increases with the V_{GSoff}. This is because excessive turn-off voltage will aggravate the dynamic threshold voltage drift. Therefore, according to the previous theory, the ΔV_F increases with the threshold voltage. Furthermore, in the reverse freewheeling circuit, if $|V_{GSOFF}|$ is very large, it will cause an excessive dynamic drift, the circuit loss will increase with the ΔV_F.

C. The third quadrant parallel Dynamic characteristics

The third quadrant dynamic characteristics focus on reverse recovery characteristics. With the promotion of SiC MOSFETs in high-power applications, more and more SiC MOSFET devices are connected in parallel. This section uses DPT to test the influence of different thresholds on the parallel reverse recovery.

As shown in Fig. 9 and TABLE II, in the double-tube parallel test, when $V_{GS} = 0V$, compared with devices with a smaller V_{TH}, the SiC MOSFET with a larger V_{TH} has a smaller static current, and the reverse recovery charge (Q_{RR}) and peak reverse recovery current (I_{RRM}) are also smaller than the MOSFET with a smaller V_{TH}. Due to body effect factors, the channel is turned on, the device with a larger V_{TH} has a larger equivalent resistance, and it gets a smaller current. Furthermore, the Q_{RR} and I_{RRM} are affected by the forward current [7]. Therefore, at the same forward current, devices with a large V_{TH} have a smaller channel current. At room temperature, the channel current will produce larger I_{RRM} and Q_{RR} when the body diode current is the same as the channel current, devices with a small V_{TH} have larger I_{RRM} and Q_{RR}. when $V_{GS} = -5V$, ΔQ_{RR} reduced from 10nC to 5nC, the ΔI_F and the ΔI_{RRM} also dropped. Because the gate negative pressure is so large and the inversion layer cannot be formed, so the body effect is weakened, most of the current flows through the body diode.

Overall, these experiments confirmed that the dynamic threshold drift in the parallel connection will aggravate the current unevenness, and most SiC MOSFET body diodes have a positive current temperature coefficient. Therefore, the reliability of the parallel device is reduced by the problem of the dynamic threshold drift.

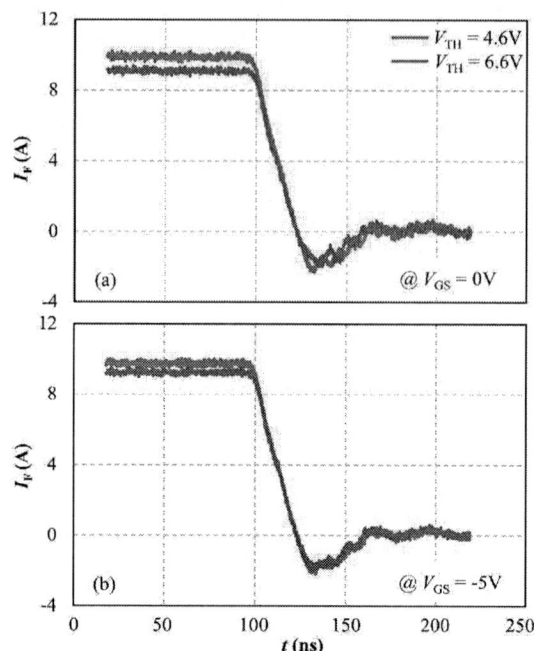

Fig. 9. The effect of different threshold voltages on parallel body diodes.

TABLE II. COMPARISON OF THE THIRD QUADRANT CHARACTERISTICS OF PARALLEL BODY DIODES WITH DIFFERENT THRESHOLD VOLTAGES.

V_{TH}	$V_{GS} = 0V$			$V_{GS} = -5V$		
	I_F	I_{RRM}	Q_{RR}	I_F	I_{RRM}	Q_{RR}
4.6V	10 A	2.3 A	47 nC	9.7 A	2.2 A	46 nC
6.6V	9 A	2 A	37 nC	9.3 A	2 A	41 nC
Δ	1 A	0.3 A	10 nC	0.4 A	0.2 A	5 nC

IV. CONCLUSION

The influence of dynamic threshold voltage drift on the third quadrant characteristics of SiC MOSFET is presented in this paper. Compared with the Si MOSFET, the SiC MOSFET body effect is more obvious. In the third quadrant, due to the SiC MOSFET body effect, most of the current flows through the channel instead of the body diode at $V_{GS} = 0V$. In terms of static characteristics, the higher the threshold voltage, the higher the V_F. In the case of parallel connection, the device with a large threshold has smaller I_{RRM} and Q_{RR}. By using a smaller $|V_{GSOFF}|$ to mitigate threshold drift, benefits like the loss reduction and good parallel current sharing can be achieved. This work will provide effective guidance for the third quadrant application.

ACKNOWLEDGMENT

This work was supported by the National Key Research and Development Program of China under Grant 2018YFB0905803.

REFERENCES

[1] B. J. Baliga, *Silicon Carbide Power Devices*. Singapore: World Scientific, Jan. 2006.

978-1-6654-4817-8/21 $31.00 © 2021 IEEE

[2] T. Kimoto and J. A. Cooper, *Fundamentals of Silicon Carbide Technology: Growth, Characterization, Devices, and Applications.* Singapore: Wiley, Nov. 2014.

[3] S. Sabri et al, "New generation 6.5 kV SiC power MOSFET," *2017 IEEE 5th Workshop on Wide Bandgap Power Devices and Applications(WiPDA)*, pp. 246-250, 2017.

[4] A. Bolotnikov et al, "Utilization of SiC MOSFET body diode in hard switching applications," *Mater. Sci. Forum*, pp. 947-950, 2014.

[5] R. Callanan, J. Rice and J. Palmour, "Third quadrant behavior of SiC MOSFETs," *2013 Twenty-Eighth Annual IEEE Applied Power Electronics Conference and Exposition (APEC)*, pp. 1250-1253, 2013.

[6] Huerner, A., T. Heckel, A. Endruschat et al, "Analytical Model for the Influence of the Gate-Voltage on the Forward Conduction Properties of the Body-Diode in SiC-MOSFETs," *Mater. Sci. Forum*, pp. 901-904, 2018.

[7] K. Peng, S. Eskandari and E. Santi, "Characterization and modeling of SiC MOSFET body diode," *2016 IEEE Applied Power Electronics Conference and Exposition (APEC)*, pp. 2127-2135, 2016.

[8] R. Zhang, X. Lin, J. Liu, S. Mocevic, D. Dong and Y. Zhang, "Third Quadrant Conduction Loss of 1.2–10 kV SiC MOSFETs: Impact of Gate Bias Control," *IEEE Transactions on Power Electronics*, vol. 36, no. 2, pp. 2033-2043, Feb. 2021.

[9] K. Lindberg-Poulsen, L. P. Petersen, Z. Ouyang and M. A. E. Andersen, "Practical investigation of the gate bias effect on the reverse recovery behavior of the body diode in power MOSFETs," *2014 International Power Electronics Conference (IPEC-Hiroshima 2014 - ECCE ASIA)*, pp. 2842-2849, 2014.

[10] G. M. Dolny, S. Sapp, A. Elbanhaway, and C. F. Wheatley, "The influence of body effect and threshold voltage reduction on trench MOSFET body diode characteristics," *Proceedings of the 16th International Symposium on Power Semiconductor Devices and ICs*, pp. 217-220, 2004.

[11] J. A. O. González and O. Alatise, "A novel non-intrusive technique for BTI characterization in SiC MOSFETs," *IEEE Transactions on Power Electronics*, vol. 34, no. 6, pp. 5737-5747, June 2019.

[12] H. Jiang. X. Zhong, G. Qiu et al, "Dynamic gate stress induced threshold voltage drift of silicon carbide MOSFET," *IEEE Electron Device Letters*, vol. 41, no. 9, pp. 1284-1287, Sept. 2020.

An Optimal Design Scheme of Intermediate Bus Voltage for two-stage LLC Resonant Converter Based on SiC MOSFET

Feng Wang
State Key Laboratory of Advanced Electromegnetic Engineering and Technology
Huazhong University of Science and Technology
Wuhan, China
wfd@hust.edu.cn

Xuehua Wang
State Key Laboratory of Advanced Electromegnetic Engineering and Technology
Huazhong University of Science and Technology
Wuhan, China
wang.xh@hust.edu.cn

Xinbo Ruan
State Key Laboratory of Advanced Electromegnetic Engineering and Technology
Huazhong University of Science and Technology
Wuhan, China
ruanxb@nuaa.edu.cn

Abstract—The two-stage LLC resonant converter has numerous superiorities in wide voltage input occasions. The fixed-frequency operating state of LLC converter is conducive to the design of magnetic components and a high efficiency could be achieved. In such a two-stage converter system, the optimal design of the intermediate bus voltage is of great significance. This parameter not only determines the working voltage of tubes, but also affects the efficiency of converter. To optimize the overall performance of the converter, a majorized design scheme of intermediate bus voltage is proposed in this paper. The optimal point can be obtained via building and solving the mathematical model. Finally, a 1-kW converter with the DC-link voltage of 70~120V and the output voltage of 28V is built. The peak efficiency of the prototype is 97.3%.

Keywords—two-stage LLC resonant converter; loss analysis; optimal design

I. INTRODUCTION

The two-stage LLC converter has been widely used in wide voltage input occasions, such as fuel cell power system [1], electric vehicles [2] and onboard chargers [3]. This topology consists of a non-isolated converter and an LLC converter. The non-isolated converter functions as a regulator to maintain the output voltage, whereas the LLC converter operates as a dc transformer (DCX). The pulse width modulation (PWM) converters such as buck converter [4] boost converter [5]-[6], and buck-boost converter [7] is adopted as the regulator according to the wide voltage regulating range and the simple control.

The LLC converter operates at a fixed switching frequency slightly lower than the resonant frequency, which is convenient to the design of magnetic components. Meanwhile, the LLC converter features soft switching for all load conditions, a high efficiency can be harvested with wide range of input voltage [8].

Numerous studies have been made in parameter of the two-stage LLC converter. In [9], a parameter design scheme is proposed to make the LLC work as a DCX to maintain a nearly constant voltage gain even though the parameters drift. Moreover, to reduce the additional component caused by the cascading structure, a two-stage buck-boost integrated LLC converter is proposed and the design procedure is presented in [10].

At present, there are few optimal designs on the power density of this converter. The two-stage converter has inductor, transformer, and capacitor which occupy the most of the volume. In [11], a size reduction method for boost + LLC converter is proposed in a 60W AC-DC adapter. A loss model of boost + LLC converter is established, and the optimal intermediate bus voltage is obtained Correspondingly in [1]. Yet the accuracy of loss model is not high. Besides, with the rapid development of wide-bandgap semiconductor devices, there will be improvements in power density with SiC MOSFET adopted in this topology.

In order to improve the modeling accuracy, an accurate current calculation method is adopted. Basing on the accurate model, an optimal design scheme is proposed to obtain higher efficiency. Further, synchronous rectification is applied to reduce the power loss. Finally, a boost + LLC prototype based on SiC MOSFET is established to verify the reliability of the design method.

This paper is organized as follows. In Section II, the parameter design procedure is illustrated. In Section III, the optimal design scheme of intermediate bus voltage for boost + LLC converter is presented in detail. In Section IV, experimental results are conducted to verify the effectiveness of analysis. Section V concludes this paper.

II. DESIGN FOR TWO-STAGE RESONANT CONVERTER

The two-stage LLC converter consists of a non-isolated converter and an LLC converter. The LLC converter runs as a DCX to realize the function of electrical isolation and voltage matching, whereas the non-isolated converter operates as a pre-regulator to regulate the output voltage.

The non-isolated converters which operates as pre-regulator, are both PWM chopper circuits that regulate the output voltage via changing the duty radio. Here we take boost + LLC converter as an example to implement the optimal design scheme. Fig. 1 shows the topology diagram of boost + LLC converter, where L_b, C_b are the inductor and output capacitor of boost converter, L_r, C_r, L_m and C_o are the series inductor, series capacitor, magnetizing inductor and output capacitor of LLC resonant converter. n is the transformer ratio. i_{Lb}, i_{Lr}, I_{rect}, I_o are the boost inductor current,

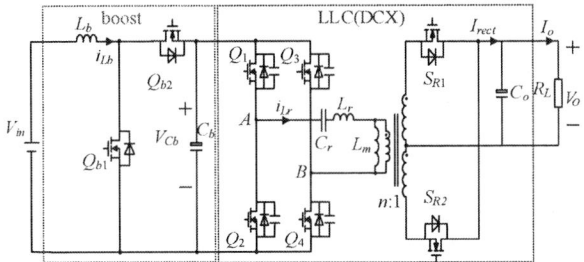

Fig. 1. The topology of boost+LLC converter.

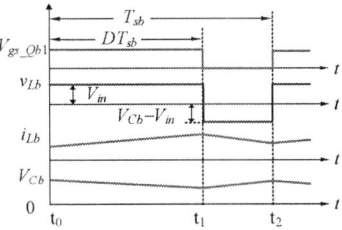

Fig. 2. Waveforms of the boost converter in CCM mode.

resonant current, rectifier current and load current respectively and V_{Cb} represents the intermediate bus voltage. A set of parameters of converter should be determined firstly.

A. Design for Boost Converter

In the boost converter, the design of boost inductor and capacitor is very important. These parameters not only determine the working characteristic, but also affect the efficiency of converter. In Fig. 2, according to the trajectory of i_{Lb} and V_{Cb}, the value of L_b and C_b can be selected to pursue satisfied power quality under CCM mode.

(1) Design for boost inductor L_b: From Fig. 2, in $[t_0, t_1]$, the boost inductor is charged by input voltage V_{in}, and the pulsation of the inductor current ΔI_{Lb} can be deduced as

$$\Delta I_{Lb} = \frac{1}{L_b}\int_{t_0}^{t_1} V_{in}dt = \frac{V_{in}D}{L_b f_{s\_boost}} \tag{1}$$

Where, D and f_{s_boost} represent the duty radio and switching frequency of boost converter respectively.

Due to (1), the value of L_b should be guaranteed as follow

$$L_b \geq \frac{V_{in}D}{\Delta I_{Lb}I_{Lb}f_{s\_boost}} \tag{2}$$

Where I_{Lb} is the DC component of i_{Lb}.

(2) determination of boost capacitor C_b

Likewise, From Fig. 2, in $[t_0, t_1]$, the boost capacitor is discharged through the load, and the pulsation of the capacitor voltage ΔV_{Cb} can be derived as

$$\Delta V_{Cb} = \frac{1}{C_b}\int_{t_0}^{t_0+DT_{sb}} \frac{P_o}{V_{Cb}}dt = \frac{DP_o}{C_b V_{Cb} f_{s\_boost}} \tag{3}$$

Where P_o is the transmitted power of boost converter.

Based on (2) and (3), the proper values of L_b and C_b can be obtained under the specific requirements.

B. Implementation of DCX with LLC Converter

(1) Design for L_m: Basically, the energy of L_m needs to charge and discharge to the junction capacitors of power switches within the dead time t_d. Thus, L_m is constrained to achieve ZVS for the power tubes. Specifically, L_m can be calculated as [9]

$$L_{m\_max} = \frac{t_d}{8 f_{s\_LLC} C_{oss}} \tag{4}$$

Where t_d is the dead time, f_{s_LLC} is the switching frequency of LLC converter and C_{oss} represents the output capacitance of the power switches.

(2) Design for L_r and C_r: In the boost + LLC converter, The LLC converter needs to be designed as a DCX, whose voltage gain approximately remains constant with the load fluctuates.

The voltage gain of LLC converter is as follow[12]

$$M_{LLC} = \frac{nV_o}{V_{Cb}} = \frac{1}{\sqrt{(f_n - \frac{1}{f_n})^2 Q^2 + (1 + \frac{1}{\lambda} - \frac{1}{\lambda f_n^2})^2}} \tag{5}$$

Where, V_o represents the output voltage, n is the turns-ratio of transformer, $f_n = f_{s\_LLC} / f_r$, $f_r = 1/(2\pi\sqrt{L_r C_r})$ is the resonant frequency of LLC converter, $\lambda = L_m / L_r$.

The quality factor Q is

$$Q = \frac{\pi^2 I_o}{8n^2 V_o}\sqrt{\frac{L_r}{C_r}} \tag{6}$$

Suppose the maximum deviation of the voltage gain is σ, $|M_{LLC} - 1| \leq \sigma$. It is noting that $M_{LLC} > 1$ when f_n is slightly less than 1, The value of M_{LLC} up to maximum when $Q = 0$. To reduce the fluctuation of voltage gain, the value of λ should be as large as possible. And the minimum value of λ is [9]

$$\lambda_{min} = (\frac{1}{f_n^2}-1)(1+\frac{1}{\sigma}) \tag{7}$$

Thus, the maximum value of L_r can be determined due to (7) and C_r can be obtained once f_r is selected.

Based on the design procedure above, the parameters of LLC converter can be harvested, which are shown in Table I in Section VI. In this case, M_{LLC} still remains unchanged with the fluctuates of load and resonant parameters. This means that the LLC converter operates as a DCX perfectly.

III. Loss Modeling and Optimal Design

In order to obtain the relationship between the optimization target and V_{Cb}, the loss models of elements regarding the transformer ratio n are derived in this section. Since the LLC resonant functions as DCX, $V_{Cb}=nV_o$. Note that the inductor and transformer here are adopted optimized design method, and their losses involving iron loss and copper loss are also calculated.

As described in Fig. 3, R_{Lb} is the parasitic resistance of boost inductor L_b, R_{on_Qb} represents the on-resistance of tubes Q_{b1} and Q_{b2}, R_{on_Q} is the on-resistance of power switches of

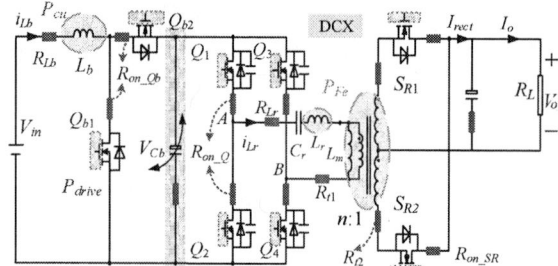

Fig. 3 Loss distribution of the boost+LLC converter.

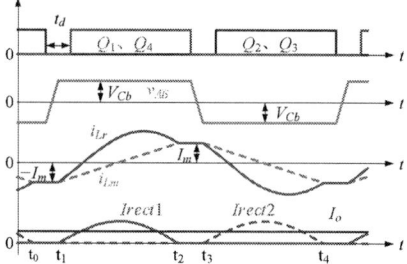

Fig. 4 Working waveforms of the LLC converter.

bridge arm, R_{Lr} is the parasitic resistance of L_r, R_{on_SR} is the on-resistance of synchronous rectifier tubes, R_{t1} and R_{t2} are the resistance of transformer windings. The losses of converter include copper loss of parasitic resistance (red portion), iron loss of magnetic components (blue shade) and loss of driving power switches (orange shade). The optimal value of V_{Cb} can be solved via the loss model to achieve highest efficiency. In order to theoretically calculate the loss of components, working conditions should be solved firstly.

A. Working Condition Analysis of Components

The working trajectory of LLC converter is presented in Fig. 4, and v_{AB} is the output voltage of bridge arm. Due to the LLC converter works as a DCX, the period $[t_2, t_3]$ is small, the resonant current can be approximated as a sine wave. Meanwhile, v_{AB} is a square wave with amplitude V_{Cb}. Further, the phase of i_{Lr} lags v_{AB} to realize ZVS of the switch in primary side.

Based on Fig. 5, the input impedance Z_r of the LLC converter can be deduced as

$$Z_r = \frac{1}{j\omega_s C_r} + j\omega_s L_r + \frac{R_e j\omega_s L_m}{R_e + j\omega_s L_m}$$
$$= \frac{L_m/C_r - \omega_s^2 L_m L_r + j(R_e w_s L_r - R_e/(\omega_s C_r))}{R_e + j\omega_s L_m} \quad (8)$$
$$= \frac{aR_e - b\omega_s L_m + j(a\omega_s L_m + bR_e)}{R_e^2 + (\omega_s L_m)^2}$$

Where $R_e = 8n^2 V_o^2/(\pi^2 P_o)$ is the equivalent load, $\omega_s = 2\pi f_{s\_LLC}$ and

$$\begin{cases} a = L_m/C_r - \omega_s^2 L_m L_r \\ b = R_e w_s L_r - R_e/(\omega_s C_r) \end{cases} \quad (9)$$

Thus, the phase angle i_{Lr} lagging behind the fundamental component of input voltage v_{AB} is

$$\varphi = \arctan(\frac{a\omega_s L_m + bR_e}{aR_e - b\omega_s L_m}) \quad (10)$$

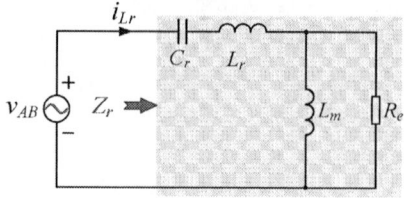

Fig. 5. The equivalent impedance circuit of LLC converter.

For LLC converter, according to the power conservation

$$P_{inL}\eta_L = \frac{2\sqrt{2}}{\pi} V_{Cb} \frac{I_{Lr}}{\sqrt{2}} \cos\varphi = P_o = V_o I_o \quad (11)$$

Where I_{Lr} is the amplitude of i_{Lr}, P_{inL} and η_L are the input power and efficiency of LLC resonant converter respectively. Here $\eta_L = 0.98$.

The intermediate bus voltage V_{Cb} and output voltage V_o satisfy

$$V_o = V_{Cb} M_{LLC}/n \quad (12)$$

Considering $M_{LLC} \approx 1$, and substituting (12) into (11), leads to

$$I_r = \frac{\pi I_o}{2\eta_L N \cos\varphi} \quad (13)$$

As shown in Fig. 4, At t_2, i_{Lr} is equal to i_{Lm}, the secondary side of the transformer is open, i_{Lr} and i_{Lm} maintain approximately constant as I_m, and the express of I_m is

$$I_m = \frac{nV_o}{4f_r L_m} \quad (14)$$

Supposing that the current I_{rect} is full-wave rectified waveform. The relationship between the mean of I_{rect} and the output current I_o is presented as

$$I_{rect\_mean} = \frac{2I_{rect\_peak}}{\pi} = I_o \quad (15)$$

Thus, the RMS of I_{rect1} and I_{rect2} can be solved as

$$I_{rect1\_rms} = I_{rect2\_rms} = \frac{\pi I_o}{4} \quad (16)$$

Likewise, power conservation is applied in boost converter, yields

$$P_{inb}\eta_b = V_{in}I_{Lb}\eta_b = \frac{V_{in}}{(1-D)}I_m \quad (17)$$

Where P_{inb} and η_b are the input power and efficiency of boost converter. Here $\eta_b = 0.97$.

The DC component of input current i_{Lb} is

$$I_{Lb} = \frac{I_o}{\eta N \cos\varphi(1-D)} \quad (18)$$

The inductor current is a triangle wave in CCM mode, then the RMS of i_{Lb} is

$$I_{Lb\_rms} = \sqrt{\frac{10I_{Lb}^2 + 5I_{Lb}\Delta I_{Lb} + \Delta I_{Lb}^2}{3}} \quad (19)$$

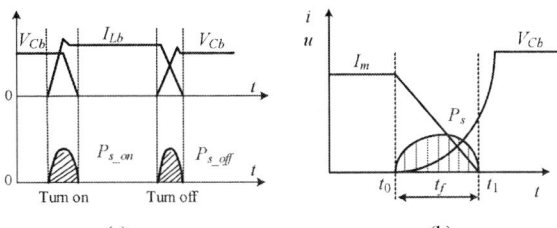

(a)   (b)

Fig. 6. Switching process of power switches. (a) hard switching process of boost. (b) turn-off process of LLC converter.

Where ΔI_{Lb} is the pulsation of the inductor current. Here $\Delta I_{Lb}=20\%$.

B. Loss Model Establishment

(1) Loss of boost converter: Fig .6(a) shows the switching process of the power tubes in boost converter and the switching loss of the tubes is the shaded area. Assuming that the voltage and current change linearly, the switching loss can be approximated as

$$P_{s\_boost} = \frac{nV_o I_{Lb}(t_{r1}+2t_{f1})f_{s\_boost}}{6} \quad (20)$$

Where t_{r1} and t_{f1} are the rise time and fall time of tubes in boost converter respectively.

According to the analysis above, the loss generated by the parasitic resistances of boost inductor and the on-resistances of tubes is

$$P_{on\_boost} = I_{Lb\_rms}^2 R_{Lb} + I_{Lb\_rms}^2 R_{on\_Qb} \quad (21)$$

(2) Loss of LLC converter: The LLC converter operates in region showing inductive characteristic, the ZVS of tubes can be realized in primary side, so only the turn off process should be taken into consideration.

Similarly, the switching loss of tubes as shown in Fig. 6(b) is

$$P_{s\_LLC} = \frac{t_{f2}^2 f_{s\_LLC}}{12 C_{oss}} I_m^2 \quad (22)$$

Where t_{f2} is the fall time of tubes $Q_1 \sim Q_4$.

The copper consumption of LLC components can be concluded as

$$P_{on\_LLC} = I_r^2 \left(R_{on\_Q} + \frac{R_{Lr}+R_{t1}}{2}\right) + 2I_{rect1\_rms}^2(R_{t2}+R_{on\_SR}) \quad (23)$$

Comprehensively, integrating (20)-(23), the loss of converter can be derived as

$$P_{loss\_all} = P_{on\_boost} + P_{s\_boost} + P_{on\_LLC} + P_{s\_LLC} + P_{add} \quad (24)$$

Where P_{add} is the additional power loss being independent of V_{Cb}, which includes iron loss of magnetic components, drive loss of power tubes and loss of capacitors.

C. Optimal Design of V_{Cb}

After obtaining the loss model, the goal is to improve the overall efficiency of the converter while ensuring the voltage meets the tube withstand voltage requirements. The detailed design steps are as follows: firstly, calculate losses of

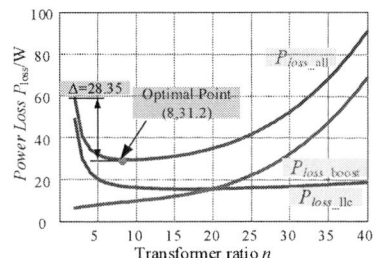

(a)

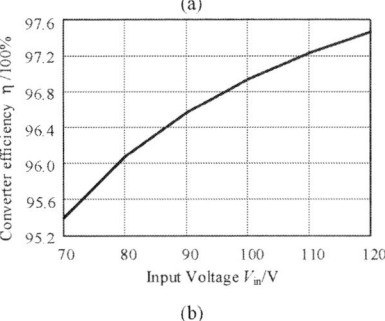

(b)

Fig.7 Theoretical calculation results. (a) The curve of total losses P_{loss} versus transform radio n under $V_{in}=100\text{V}$ and $P_o=1\text{kW}$. (b) The curve of efficiency η versus input voltage V_{in}.

components based on transformer ratio n; these losses are superimposed to form the overall loss model of the converter; finally, draw the curve of loss versus transformation ratio n, then the optimal point can be harvested. If the bus voltage V_{Cb} at the best point is high, a compromise between efficiency and voltage level needs to be considered.

According to the parameters in Table I in Section VI, substituting known parameters into (24), the curve of P_{loss} versus transform radio n and the curve of efficiency η versus input voltage V_{in} are drawn in Fig. 3(a)and (b) respectively. The power loss is the lowest when $n=8$, and as V_{in} changes, the efficiency of the converter can be maintained at a point around 96.5%.

IV. EXPERIMENTAL RESULTS

A 1-kW boost + LLC prototype based on SiC MOSFET was built to verify the proposed design scheme. The prototype parameters are listed in Table I. The power switches of boost converter and full bridge in primary side of LLC converter are CoolSiC™ MOSFETs (IMW65R027M1H). The selected rectifier tube on the secondary side is IPA030N10N3.

TABLE I. PROTOTYPE PARAMETERS

Parameters	Symbols	Value
Input voltage	V_{in}	70~120V
Rated power	P_o	1kW
Output voltage	V_o	28V
Transformer ratio	n	8
Switching frequency of boost	f_{s_boost}	100kHz
Switching frequency of LLC	f_{s_LLC}	100kHz
Boost inductor	L_b	200μH
Boost capacitor	C_b	300μF
Magnetizing inductor	L_m	180μH
Resonant inductor	L_r	6.5μH
Resonant capacitor	C_r	320nF

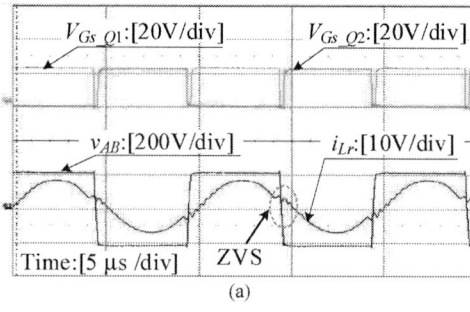

(a)

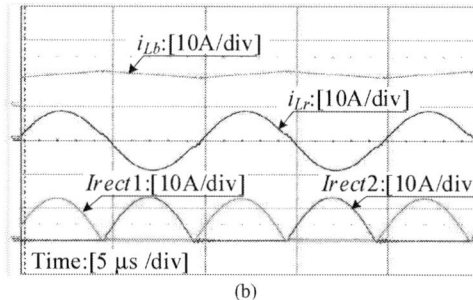

(b)

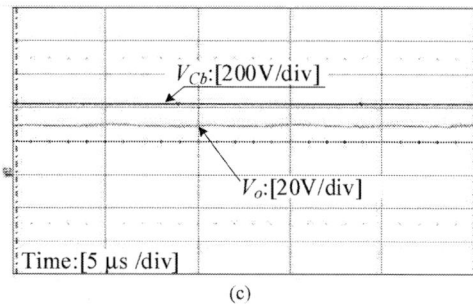

(c)

Fig. 8. Waveform of key voltage and current with V_{in}=100V and full load. (a) Key waveforms to verify ZVS. (b) Currents of converter. (c) Voltages of V_{Cb} and V_o.

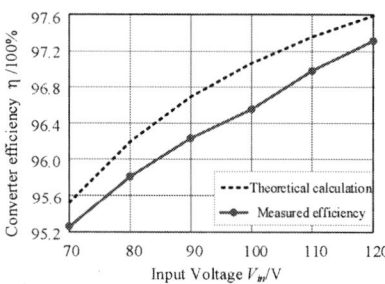

Fig. 9. The curve of efficiency η versus input voltage V_{in}.

In Fig .9, a series of experiments was carried out under the different V_{in} to examine the efficiency. The key waveforms of the full-load experiment are described in Fig. 8. It can be seen from v_{AB} and i_{Lr} that the converter achieves ZVS in Fig. 8(a). In this condition, the switching frequency f_s of LLC converter is slightly lower than the f_r (f_{s_LLC}=0.9f_r) and the converter achieves the maximum power conversion efficiency of 96.5% at the normal input V_{in}=100V.

As presented in Fig. 9, the simulation result is consistent with the measured efficiency of prototype and the difference

is about 0.4%. Besides, it can be observed that a high efficiency could be maintained with wide range of input voltage. Moreover, the measured efficiency reaches 97.3% when V_{in} = 120V. Thus, the boost + LLC converter adopting optimal scheme of V_{Cb} is suitable for the low voltage high current output situation.

V. CONCLUSIONS

The design of boost + LLC converter is discussed in this paper. An accurate loss model is established and an optimal design of the intermediate bus voltage V_{Cb} is presented to achieve higher efficiency. Finally, a 1-kW prototype based on SiC MOSIFET was designed and built. The power conversion efficiency was 96.5% at the normal input V_{in} = 100V and the high efficiency could be maintained with wide range of input voltage.

REFERENCES

[1] J. Y. Lee, Y. S. Jeong, and B. M. Han, "An isolated dc/dc converter using high-frequency unregulated LLC resonant converter for fuel cell applications," IEEE Trans. Ind. Electron., vol. 58, no. 7, pp. 2926–2934, Jul. 2011.

[2] X. Sun, J. Qiu, X. Li, B. Wang, L. Wang, and X. Li, "An improved wide input voltage buck-boost+LLC cascaded converter," in Proc. IEEE Energy Convers. Congr. Expo., 2015, pp. 1473–1478.

[3] C. Shi, H.Wang, S. Dusmez, and A. Khaligh, "A SiC-based high-effificiency isolated onboard PEV charger with ultrawide dc-link voltage range," IEEE Trans. Ind. Appl., vol. 53, no. 1, pp. 501–511, Jan. 2017.

[4] M. Fu, C. Fei, Y. Yang, Q. Li, and F. C. Lee, "Optimal design of planar magnetic components for a two-stage GaN-based DC–DC converter." in IEEE Transactions on Power Electronics, vol. 34, no. 4, pp. 3329–3338, Apr. 2019.

[5] X. Sun, Y. Shen, Y. Zhu and X. Guo,"Interleaved Boost-Integrated LLC Resonant Converter With Fixed-Frequency PWM Control for Renewable Energy Generation Applications," in IEEE Transactions on Power Electronics, vol. 30, no. 8, pp. 4312-4326, Aug. 2015.

[6] Y. Wang, Y. Guan, K. Ren, W. Wang and D. Xu, "A Single-Stage LED Driver Based on BCM Boost Circuit and LLC Converter for Street Lighting System," in IEEE Transactions on Industrial Electronics, vol. 62, no. 9, pp. 5446-5457, Sept. 2015.

[7] X. Sun, J. Qiu, X. Li, B. Wang, L. Wang and X. Li, "An improved wide input voltage buck-boost + LLC cascaded converter," 2015 IEEE Energy Conversion Congress and Exposition (ECCE), pp. 1473-1478, 2015.

[8] F. Musavi, M. Craciun, D. S. Gautam, W. Eberle and W. G. Dunford, "An LLC Resonant DC–DC Converter for Wide Output Voltage Range Battery Charging Applications," in IEEE Transactions on Power Electronics, pp. 5437-5445, 2013.

[9] F. Liu, G. Zhou, X. Ruan, S. Ji, Q. Zhao and X. Zhang, "An Input-Series-Output-Parallel Converter System Exhibiting Natural Input-Voltage Sharing and Output-Current Sharing," in IEEE Transactions on Industrial Electronics, vol. 68, no. 2, pp. 1166-1177, Feb. 2021.

[10] Q. Liu, Q. Qian, B. Ren, S. Xu, W. Sun and L. Yang, "A Two-Stage Buck-Boost Integrated LLC Converter With Extended ZVS Range and Reduced Conduction Loss for High-Frequency and High-Efficiency Applications," in IEEE Journal of Emerging and Selected Topics in Power Electronics, vol. 9, no. 1, pp. 727-743, Feb. 2021.

[11] J. Kim, M. Kim, C. Yeon and G. Moon, "Analysis and design of Boost-LLC converter for high power density AC-DC adapter," 2013 IEEE ECCE Asia Downunder, pp. 6-11, 2013.

[12] H. Huang, "FHA-based voltage gain function with harmonic compensation for LLC resonant converter," 2010 Twenty-Fifth Annual IEEE Applied Power Electronics Conference and Exposition (APEC), pp. 1770-1777, 2010.

Modeling and Design of A 10MHz Class Φ_2 Inverter

Yongzhi Liu
School of Electrical and Electronic Engineering
Huazhong University of Science and Technology
Wuhan, China
yongzhi@hust.edu.cn

Yiyang Yan
School of Electrical and Electronic Engineering
Huazhong University of Science and Technology
Wuhan, China
yanyiyang@hust.edu.cn

JiaJia Guan
School of Electrical and Electronic Engineering
Huazhong University of Science and Technology
Wuhan, China
m202071539@hust.edu.cn

Cai Chen
School of Electrical and Electronic Engineering
Huazhong University of Science and Technology
Wuhan, China
caichen@hust.edu.cn

Yu Chen
School of Electrical and Electronic Engineering
Huazhong University of Science and Technology
Wuhan, China
ayu03@hust.edu.cn

Yong Kang
School of Electrical and Electronic Engineering
Huazhong University of Science and Technology
Wuhan, China
ykang@hust.edu.cn

Abstract—The Class Φ_2 inverter has become a research hotspot for high frequency applications due to low voltage stress and load-independent ZVS operation. However, the previous design procedures based on the Bode diagram of the high order resonant network are inconvenient for parameter tuning during simulation. In this paper, a unit-less model of the Class Φ_2 inverter is presented to facilitate the generalized circuit design. And a new parametric design method is proposed, that achieves decoupling design of the switch voltage stress and reduces design iterations by the constraint lines of the capacitor voltage stress and zero voltage switching. It is verified by simulation results that the optimal parameters can be found in the design area, and the experimental prototype is being built.

Keywords—Class Φ_2 inverter, resonant converter, parametric design

I. INTRODUCTION

High frequency resonant inverters can improve power density and dynamic performance, and achieve high efficiency with soft switching, which are widely used in the resonant dc-dc converter and the plasma generator [1]. And with the development of wide band semiconductor, GaN HEMTs with higher switching frequency and lower parasitic capacitance have been widely used in high frequency resonant converters. At present, resonant inverters operating at frequencies above 10 MHz include Class E, Class DE, Class Φ_2, and so on. The Class E inverter has a simple resonant network consisting of an input inductor and a shunt capacitor, but its voltage stress is up to 4 times the input voltage [2]-[4]. The Class DE inverter has a voltage stress equal to the input voltage, while it requires a more complex high-side gate drive [5], [6]. The Class Φ_2 inverter [7], shown in Fig. 1, adds a second harmonic parallel branch to the Class E inverter so that the drain-source voltage consists of the fundamental and the third harmonic, resulting in voltage stress of about 2.2 times the input voltage.

However, since the Class Φ_2 inverter has a fourth-order resonant network Z_{MR}, its design process is more complex compared to the Class E inverter. The original design method proposed in [7] requires the loading network Z_L to look inductive at the fundamental frequency f_s, which makes the zero voltage switching (ZVS) operation significantly influenced by the load. Reference [8] proposed four improved rules, including making Z_L resonant at f_s to achieve load-independent ZVS operation. But both of these methods

empirically use the impedance Bode diagram to design parameters, which is not conducive to parameter sweep and iteration during simulation.

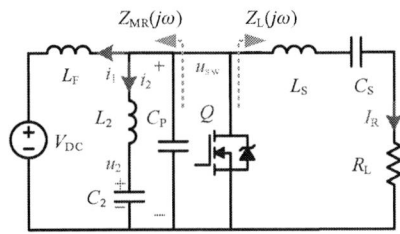

Fig. 1. Class Φ_2 resonant inverter.

In this paper, a unit-less state space model is proposed, which is only related to the two relative poles of the fourth-order resonant network and M_R, the ratio of input inductive reactance to load resistance. Therefore, it can be demonstrated that if two relative poles and M_R remain unchanged, different input voltages, input inductance values and frequencies will not change voltage stress or other unit-less parameters due to the same generalized model.

Then, for the parametric design, a new parameter M_P is proposed that mostly determines V_{pk}^*, the voltage stress of the switching device, decoupling the design variables of V_{pk}^*. And to ensure ZVS and to limit the excessive voltage peak of the 2nd-order resonant capacitor C_2, constraint line models of ZVS and the capacitor voltage stress are established in a certain range of M_R. The final design parameters can be obtained directly in the area enclosed by the constraint lines, which is verified by simulation.

II. MODEL OF THE CLASS Φ_2 INVERTER

A. Basic State-Space Model

According to the switching state of the switch, the circuit has two operation modes: Mode I ($0 < \omega_0 t \leq (1-D)T_S$) with the switch off and Mode II ($(1-D)T_S < \omega_0 t \leq T_S$) with the switch on.

In Mode I, The Class Φ_2 inverter can be equivalent to a fourth-order resonant network under the excitation of input voltage and output sinusoidal current when the switch is off, as shown in Fig. 2.

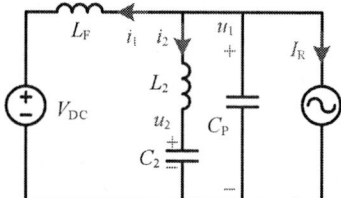

Fig. 2. Class Φ_2 resonant inverter model in Mode I.

With state vector $\boldsymbol{x_t} = [\ i_1\ \ i_2\ \ u_1\ \ u_2\ ]^T$, the state-space model of Mode I can be derived as:

$$\frac{d\boldsymbol{x_t}}{dt} = \begin{bmatrix} 0 & 0 & \dfrac{1}{L_F} & 0 \\[2mm] 0 & 0 & \dfrac{1}{L_2} & \dfrac{-1}{L_2} \\[2mm] \dfrac{-1}{C_P} & \dfrac{-1}{C_P} & 0 & 0 \\[2mm] 0 & \dfrac{1}{C_2} & 0 & 0 \end{bmatrix} \boldsymbol{x_t} + \begin{bmatrix} \dfrac{-V_{DC}}{L_F} \\[2mm] 0 \\[2mm] \dfrac{-i_R}{C_P} \\[2mm] 0 \end{bmatrix} \qquad (1)$$

Because the Z_{MR} is short-circuited to the second harmonic, the drain-source voltage u_1 in one period is mainly composed of the 1st and 3rd harmonic voltages, which approximates a trapezoidal wave. Note that u_1 can be approximated as square wave u_{sq} at the 1st and 3rd harmonic and the ac drain-source voltage $u_{1(ac)}$ can be equivalent to a bipolar square wave with a peak-to-peak value of $2V_{DC}$, as shown in Fig. 3.

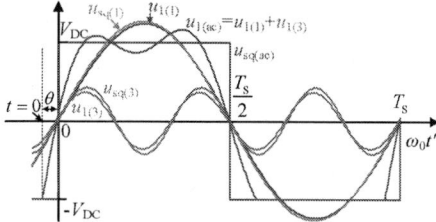

Fig. 3. The ac drain-source voltage $u_{1(ac)}$, square wave u_{sq} and their first and third harmonic components.

Since the load network resonates at f_s, the output voltage u_R is equal to $u_{1(1)}$, the fundamental component of the drain-source voltage, which is equivalent to $u_{sq(1)}$, the first harmonic of the square wave. And the output current i_R is as follows:

$$i_R(t) = \frac{u_R}{R_L} = \frac{u_{sq(1)}}{R_L} = \frac{4V_{DC}}{\pi R_L}\sin(\omega_0 t') = I_R \sin(\omega_0 t - \theta) \qquad (2)$$

Besides, the amplitude of i_R can be obtained, as in:

$$I_R = \frac{U_{sq(1)}}{R_L} = \frac{4V_{DC}}{\pi R_L} \qquad (3)$$

If the rising edge time of u_1 is considered, the fundamental wave will have a phase shift θ relative to $t = 0$, which can be calculated with the Fourier coefficients of the u_1:

$$\tan\theta = -\frac{a_1}{b_1} = -\frac{\dfrac{1}{\pi}\displaystyle\int_0^{2\pi(1-D)} u_1 \cos\omega_0 t\ d(\omega_0 t)}{\dfrac{1}{\pi}\displaystyle\int_0^{2\pi(1-D)} u_1 \sin\omega_0 t\ d(\omega_0 t)} \qquad (4)$$

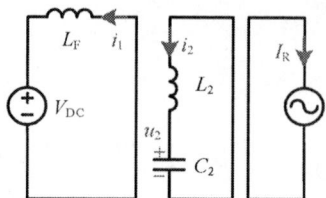

Fig. 4. Class Φ_2 resonant inverter model in Mode II.

In Mode II, the circuit can be divided into three parts: the input part of V_{DC} and L_F, the parallel branch of L_2 and C_2, the output circuit of L_S, C_S and R_L, as shown in Fig. 4.

With state vector $\boldsymbol{x_t} = [\ i_1\ \ i_2\ \ u_1\ \ u_2\ ]^T$, the state-space model of Mode II can be derived as:

$$\frac{d\boldsymbol{x_t}}{dt} = \begin{bmatrix} 0 & 0 & 0 & 0 \\ 0 & 0 & 0 & -1/L_2 \\ 0 & 0 & 0 & 0 \\ 0 & 1/C_2 & 0 & 0 \end{bmatrix} \boldsymbol{x_t} + \begin{bmatrix} \dfrac{-V_{DC}}{L_F} \\[2mm] 0 \\[2mm] 0 \\[2mm] 0 \end{bmatrix} \qquad (5)$$

B. Unit-less State-Space Model

Because Z_{MR} mainly determines the shape of u_1, it is necessary to discuss the zero and poles of Z_{MR}. To begin with, Z_{MR} can be expressed as:

$$Z_{MR}(j\omega) = \frac{j\omega L_F}{1-\omega^2 L_F C_P}\ //\ (j\omega L_2 + \frac{1}{j\omega C_2}) \qquad (6)$$

Thus, the magnitude of the complex impedance Z_{MR}:

$$|Z_{MR}(j\omega)| = \frac{\omega L_F(1-\omega^2 L_2 C_2)}{\omega^4 L_F C_P L_2 C_2 - \omega^2(L_2 C_2 + L_F C_P + L_F C_2)+1} \qquad (7)$$

Assuming that the poles of the resonant network impedance Z_{MR} are $\omega_1 = k_1\omega_0$ and $\omega_3 = k_3\omega_0$, and the zero of Z_{MR} is $\omega_2 = 2\omega_0$, as in:

$$\begin{aligned}|Z_{MR}(j\omega_1)| &= |Z_{MR}(jk_1\omega_0)| = \infty \\ |Z_{MR}(j\omega_3)| &= |Z_{MR}(jk_3\omega_0)| = \infty \\ |Z_{MR}(j\omega_2)| &= |Z_{MR}(j2\omega_0)| = 0 \end{aligned} \qquad (8)$$

Therefore, the parameters of the resonant components can be expressed as:

$$\begin{aligned} C_P &= \frac{4}{L_F k_1^2 k_3^2 \omega_0^2} = \frac{1}{\omega_0 K_P X_L} \\[2mm] C_2 &= \frac{(4-k_1^2)(k_3^2-4)}{4L_F k_1^2 k_3^2 \omega_0^2} = \frac{1}{4\omega_0 K_L X_L} \\[2mm] L_2 &= \frac{L_F k_1^2 k_3^2}{(4-k_1^2)(k_3^2-4)} = K_L L_F \end{aligned} \qquad (9)$$

where parameters K_L, K_P, and reactance X_L are defined as:

$$\begin{aligned} K_P &= \frac{k_1^2 k_3^2}{4} \\[2mm] K_L &= \frac{k_1^2 k_3^2}{(4-k_1^2)(k_3^2-4)} \\[2mm] X_L &= \omega_0 L_F \end{aligned} \qquad (10)$$

To simplify the basic model, a new unit-less state vector $x = [\ i_1{}^* \ \ i_2{}^* \ \ u_1{}^* \ \ u_2{}^*\ ]^T$ can be defined as:

$$x = \left[\frac{i_1 X_L}{V_{DC}} \quad \frac{i_2 X_L}{V_{DC}} \quad \frac{u_1}{V_{DC}} \quad \frac{u_2}{V_{DC}}\right]^T \quad (11)$$

In addition, define a new parameter M_R as follows:

$$M_R = \frac{X_L}{R_L} \quad (12)$$

And the output current can also be expressed as:

$$i_R^* = \frac{i_R X_L}{V_{DC}} = u_{1(1)}^* M_R, \quad I_R^* = \frac{I_R X_L}{V_{DC}} = \frac{4}{\pi} M_R \quad (13)$$

By substituting (9), (11) and (13) into the base model (1) and (5), a more concise unit-less model can be obtained.

When the switch is off, the model of Mode I can be expressed as:

$$\frac{dx}{d(\omega_0 t)} = \begin{bmatrix} 0 & 0 & 1 & 0 \\ 0 & 0 & \dfrac{1}{K_L} & \dfrac{-1}{K_L} \\ -K_P & -K_P & 0 & 0 \\ 0 & 4K_L & 0 & 0 \end{bmatrix} x + \begin{bmatrix} -1 \\ 0 \\ -K_P M_R u_{1(1)}^* \\ 0 \end{bmatrix} \quad (14)$$

When the switch is on, the model of Mode II can be expressed as:

$$\frac{dx}{d(\omega_0 t)} = \begin{bmatrix} 0 & 0 & 0 & 0 \\ 0 & 0 & 0 & -1/K_L \\ 0 & 0 & 0 & 0 \\ 0 & 4K_L & 0 & 0 \end{bmatrix} x + \begin{bmatrix} -1 \\ 0 \\ 0 \\ 0 \end{bmatrix} \quad (15)$$

Since the unit-less model is only related to K_L, K_P and M_R, the generalized circuit model is determined by the model parameters k_1, k_3 and M_R. Thus, the optimal parameters can be designed with the unit-less model. As long as k_1, k_3 and M_R are kept constant, that is, the model remains unchanged, the generalized waveforms are invariant and not affected by the irrelevant parameters, such as V_{DC}, L_F and f_s. For example, for $k_1 = 1.2$, $k_3 = 2.4$ and $M_R = X_L/R_L = 2\pi f_s L_F/R_L = 0.8$, simulation waveforms of generalized voltage $u_1^* = u_1/V_{DC}$ and input current $i_1^* = i_1 X_L/V_{DC}$ are invariant, when $V_{DC} = 30 \sim 70$ V, $L_F = 300 \sim 700$ nH and $f_s = 5 \sim 20$ MHz, as shown in Fig. 5.

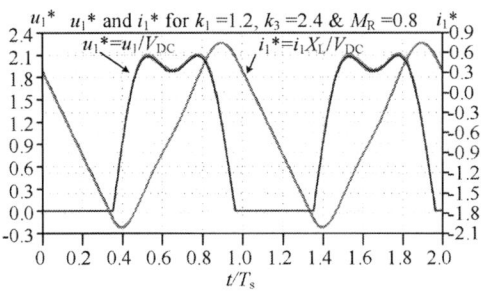

Fig. 5. The generalized drain-source voltage u_1^* and generalized input current i_1^* for $k_1 = 1.2$, $k_3 = 2.4$ and $M_R = 0.8$.

III. PARAMETRIC DESIGN METHOD

A. Constraint of the Switch Voltage Stress V_{pk}^*

The switch voltage stress V_{pk}^* is an important issue, especially for the resonant converter. Since improper parameter design may significantly increase the voltage stress of the Class Φ_2 inverter, it is necessary to discuss the relationship between the three model parameters (k_1, k_3 & M_R) and voltage stress based on the unit-less circuit model.

The plot of the switch voltage stress V_{pk}^* versus k_1 and k_3 at different M_R can be obtained. For example, the case of $M_R = 0.8$ are shown in Fig. 6. It can be seen that the constant V_{pk}^* lines are approximately parallel to each other. Therefore, there may be a linear relationship of V_{pk}^*, k_1 and k_3 and it can be verified by multiple linear regression.

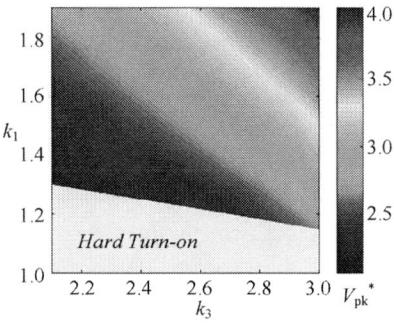

Fig. 6. The switch voltage stress V_{pk}^* versus k_1 and k_3 for $M_R = 0.8$.

Thus, for $M_R = 0.2 \sim 1.0$, under constraints of ZVS and $V_2^* < 10$, the multiple linear regression model of V_{pk}^* can be established as follows:

$$V_{pk}^* = a_M k_1 + b_M k_3 + c_M \quad (16)$$

TABLE I. REGRESSION PARAMETERS OF V_{pk}^*

95% confidence intervals	Fitting Parameters			
	Slope: a_M	Slope: b_M	Intercept: c_M	Determined Coefficient: R^2
Estimated	1.665	1.005	-2.325	0.9714
Lower Bound	1.653	0.997	-2.301	
Upper Bound	1.678	1.012	-2.349	

At 95% confidence level, the coefficients within the confidence interval can be chosen to define a new parameter M_P that mostly determines the voltage stress, denoted as,

$$M_P = V_{pk}^* - c_M = a_M k_1 + b_M k_3 = 1.67 k_1 + k_3 \quad (17)$$

Therefore, by the new parameter M_P, the coupling problem that k_1 and k_3 jointly affect V_{pk}^* at the same time is solved, and the independent decoupling design of V_{pk}^* is realized, which is convenient for parameter tuning in simulation. Taking $M_R = 0.8$ as an example, it can be found that the switch voltage stress V_{pk}^* is determined by the M_P, as shown in Fig. 7.

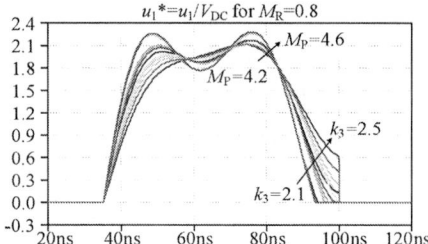

Fig. 7. The generalized drain-source voltage u_1^* at different M_P.

B. Constraint of ZVS

In order to avoid large switching loss caused by hard turn-on at muti-MHz switching frequency, ZVS is a design requirement that the resonant inverter must meet. Notice that in Fig. 7, it tends to transition from ZVS to hard turn-on when k_3 increases and M_P is the same . Therefore, there is a critical constraint line of ZVS, l_{ZVS}, related to M_R in M_P and k_3 coordinates. And when M_P and k_3 are constant, the increase of M_R (e.g. from light load to heavy load) will also lead to the transition from ZVS to hard turn-on, as shown in Fig. 8.

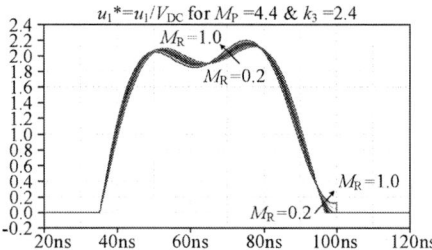

Fig. 8. The generalized drain-source voltage u_1^* at different M_R.

According to (23), M_R is proportional to P_o/V_{DC}^2. Therefore, in order to ensure that ZVS can be realized from no-load to rated power, it is necessary to calculate the maximum M_R corresponding to the minimum input voltage V_{DC} and maximum output power P_o, and design the optimal M_P and k_3 under this value.

The ZVS constraint lines, l_{ZVS} for $M_R = 0.2 \sim 1.0$ are shown in Fig. 9. It can be seen that when the M_R increases, the ZVS constraint lines move upward and the ZVS critical value of k_3 for a certain M_P decreases, so the area of ZVS above l_{ZVS} is reduced.

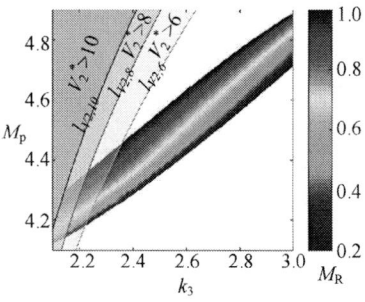

Fig. 9. The ZVS constraint lines, l_{ZVS} for $M_R = 0.2 \sim 1.0$.

For $M_R = 0.2 \sim 1.0$, the fitting model of l_{ZVS} can be given by the following equation:

$$l_{ZVS}: \quad M_P = a_Z k_3 + b_Z M_R + c_Z \tag{18}$$

TABLE II. PARAMETERS OF ZVS CONSTRAINT LINE

95% confidence intervals	Slope: a_Z	Slope: b_Z	Intercept: c_Z	Determined Coefficient: R^2
Estimated	0.723	0.238	2.484	0.991
Lower Bound	0.718	0.233	2.471	
Upper Bound	0.728	0.243	2.498	

C. Constraint of the Capacitor Voltage Stress V_2^*

The resonant branch composed of inductor L_2 and capacitor C_2 is short-circuited to the second harmonic of the drain-source voltage u_1, which generates a second harmonic voltage on C_2. In order to avoid excessive V_2^* (the voltage stress of capacitor C_2) and correspondingly high resonant currents i_2, appropriate design parameters are required.

Define a new parameter N_P proportional to K_P as follows:

$$N_P = k_1^2 k_3^2 = 4K_P \tag{19}$$

Then the plot of the capacitor voltage stress V_2^* versus N_P and k_3 at different M_R can be obtained. For example, the case of $M_R = 0.8$ is shown in Fig. 10.

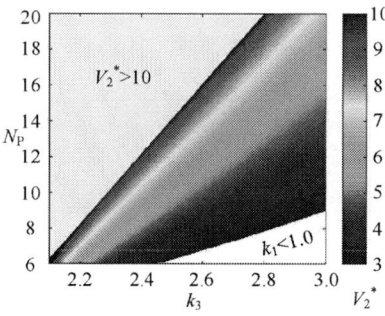

Fig. 10. The capacitor voltage stress V_2^* versus k_3 and N_P for $M_R = 0.8$.

It can be seen that the constant V_2^* line, l_{V2}, is a straight line in the N_P and k_3 coordinates. The constant V_2^* line can be fitted for the three cases of $V_2^* = 6$, 8, and 10 within the range of $M_R = 0.2 \sim 1.0$, as shown in Fig. 11.

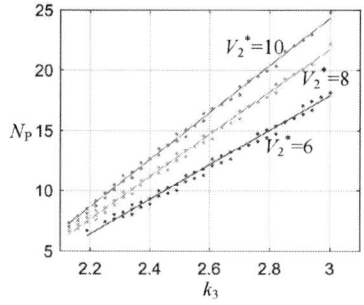

Fig. 11. The fitting constant V_2^* lines, l_{V2} for $V_2^* = 6$, 8 and 10.

Thus, a linear model of the constant V_2^* line is established as shown below:

$$l_{V2}: \quad N_P = k_1^2 k_3^2 = a_N k_3 + b_N \tag{20}$$

TABLE III. PARAMETERS OF CONSTANT V_2^* LINES

V_2^*	Fitting Parameters		
	Slope: a_N	Intercept: b_N	Determined Coefficient: R^2
6	14.3	-25.2	0.9959
8	17.8	-31.6	0.9979
10	19.9	-35.4	0.9985

In order to design the parameters directly in the M_P and k_3 coordinates, it is also necessary to transform the constant V_2^* line, l_{V2}, given in (20) into the following equation:

$$l_{V2}: \quad M_P = \frac{1.67}{k_3}\sqrt{a_N k_3 + b_N} + k_3 \qquad (21)$$

D. Parametric Design Procedure

The parametric design procedure is given in Fig. 12. As u_1 approximates trapezoidal wave and i_1 approximates triangular wave, X_L and L_F can be selected according to the maximum inductance current I_{Lmax}, as in:

$$I_{Lmax} = I_{in} + I_{1m} = \frac{P_o}{V_{DC}} + \frac{\pi V_{DC} D}{X_L} \qquad (22)$$

And M_R can be determined according to the rated output power P_o, as in:

$$P_o = \frac{1}{2} I_R^2 R_L = \left(\frac{4}{\pi}\right)^2 \frac{V_{DC}^2}{2 X_L} M_R \qquad (23)$$

Given $V_{DC} = 30$ V, $P_o = 20$ W, $I_{Lmax} = 3$A, and $D = 0.5$, it can be calculated that $X_L = 10\pi$ Ω, $L_F = 500$ nH and $M_R = 0.86$ at $f_s = 10$ MHz.

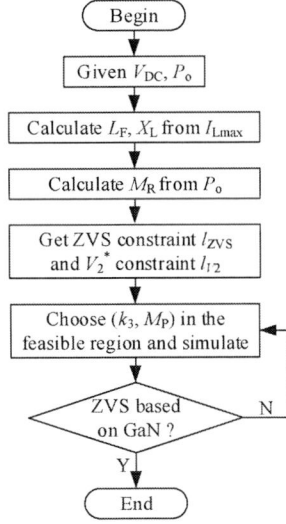

Fig. 12. Parametric design procedure.

Therefore, in this example design, when $M_R = 0.86$, the design areas for voltage stress V_{pk}^* and duty cycle D constrained by l_{ZVS} and l_{V2} in the M_P and k_3 coordinates are shown in Fig. 13 and Fig. 14 respectively. It can be verified from Fig. 13 that V_{pk}^* is mainly dominated by M_P. Meanwhile,

as can be seen from Fig. 14 that the smaller M_P and the larger k_3 correspond to the smaller D.

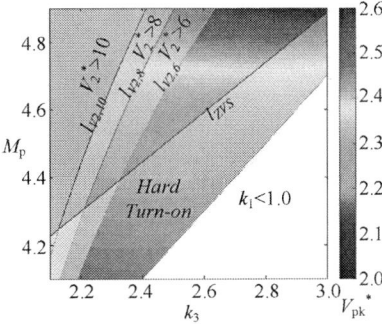

Fig. 13. The switch voltage stress V_{pk}^* versus M_P and k_3 for $M_R = 0.86$.

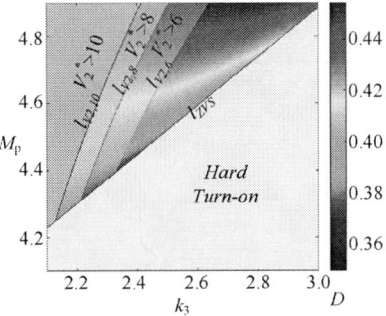

Fig. 14. Duty cycle D versus M_P and k_3 for $M_R = 0.86$.

As a result, the optimal design with minimum voltage stress for $V_2^* < 6$ is at $M_P = 4.40$, $k_3 = 2.34$ with duty cycle $D = 0.39$. The simulation results based on optimal parameters using the ideal switching device are shown in Fig. 15, where it can be seen that the voltage stress is about 2.2 times V_{DC}, and load-independent ZVS operation is achieved.

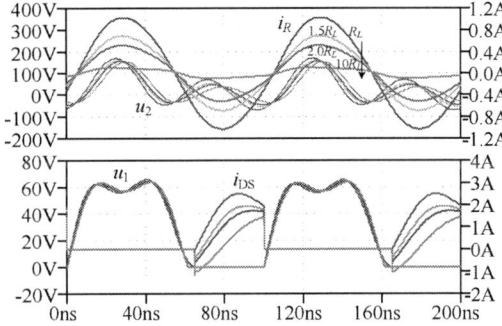

Fig. 15. Simulation results based on the ideal switch.

As for the prototype, EPC2019 is chosen as the GaN HEMTs switching device with $Q_G = 1.8$nC. So the 236 pF capacitance C_P consists of additional capacitance C_{Pext} and parasitic capacitance C_{oss}. The component parameters are shown in Table IV.

The simulation results using EPC2019 as the switching device are shown in Fig. 16. Since the output capacitance of GaN HEMT is nonlinear, the switch voltage stress increases. For example, C_{oss} of EPC2019 drops significantly from 375 to 130 pF over the V_{DS} range of 0 to 60V, and the voltage stress increases by about 0.2 compared with the ideal case.

TABLE IV. PROTOTYPE BOM

Parameter	Value
Q	EPC2019
L_F	500nH
L_2	1460nH
C_2	43pF
C_{Pext}	30pF
L_S	2860nH
C_S	91pF

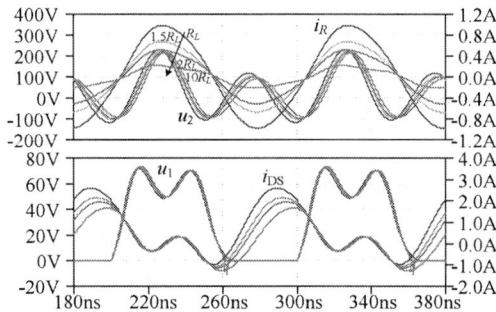

Fig. 16. Simulation results based on EPC2019.

IV. CONCLUSION

Although the voltage stress of the Class Φ_2 inverter is relatively low, the parameter design of its fourth-order resonant network based on Bode diagram proposed by predecessors is empirical and not convenient to find the optimal parameters by parameter sweep. To improve the circuit design method, a unit-less model of the Class Φ_2 inverter is derived, which is determined by three model parameters k_1, k_3 and M_R. And it is illustrated that the waveforms of the unit-less state variables are unchanged for varying irrelevant parameters. Then the decoupling design of V_{pk}^* is realized by the parameter M_P consisting of k_1 and k_3. And in the M_P and k_3 coordinates, the constraint line models of ZVS and V_2^* are given, by which the optimal parameters are selected in the feasible region. Finally, the design method is verified by simulation using the ideal switch and GaN HEMT as the switching device respectively.

REFERENCES

[1] W. Liang, L. Raymond, J. Rivas, C. Charles, and R. Boswell, "Structurally supportive RF power inverter for a CubeSat electrothermal plasma micro-thruster with PCB inductors," in Proc. 2017 IEEE Applied Power Electronics Conf. and Exposition (APEC), 2017, pp. 2141-2145.

[2] Y. Han, O. Leitermann, D. A. Jackson, J. M. Rivas and D. J. Perreault, "Resistance Compression Networks for Radio-Frequency Power Conversion," in IEEE Transactions on Power Electronics, vol. 22, no. 1, pp. 41-53, Jan. 2007, doi: 10.1109/TPEL.2006.886601.

[3] L. S. Mendonça, T. C. Naidon, R. F. Raposo and F. E. Bisogno, "An Unit-less Mathematical Model for Analysis and Design of Class-E Resonant Converters," 2019 IEEE 15th Brazilian Power Electronics Conference and 5th IEEE Southern Power Electronics Conference (COBEP/SPEC), 2019, pp. 1-6.

[4] L. Roslaniec, A. S. Jurkov, A. A. Bastami and D. J. Perreault, "Design of Single-Switch Inverters for Variable Resistance/Load Modulation Operation," in IEEE Transactions on Power Electronics, vol. 30, no. 6, pp. 3200-3214, June 2015, doi: 10.1109/TPEL.2014.2331494

[5] M. P. Madsen, A. Knott and M. A. E. Andersen, "Very high frequency half bridge DC/DC converter," 2014 IEEE Applied Power Electronics Conference and Exposition - APEC 2014, 2014, pp. 1409-1414.

[6] P. Thummala, D. B. Yelaverthi, R. A. Zane, Z. Ouyang and M. A. E. Andersen, "A 10-MHz GaNFET-Based-Isolated High Step-Down DC–DC Converter: Design and Magnetics Investigation," in IEEE Transactions on Industry Applications, vol. 55, no. 4, pp. 3889-3900, July-Aug. 2019.

[7] J. M. Rivas, O. Leitermann, Y. Han, and D. J. Perreault, "A very high frequency dc–dc converter based on a class Φ2 resonant inverter," IEEE Transactions on Power Electronics, vol. 26, no. 10, pp. 2980–2992,2011.

[8] L. Gu, K. Surakitbovorn, and J. Rivas-Davila, "High-frequency resonant converter with synchronous rectification for high conversion ratio and variable load operation," in Proc. Int. Power Electron. Conf. (IPEC-Niigata-ECCE Asia), May 2018, pp. 632–638.

An Automated Electro-Thermal-Mechanical Co-Simulation Methodology Based on PSpice-MATLAB-COMSOL for SiC Power Module Design

Yayong Yang, Yuxin Ge, Zhiqiang (Jack) Wang, Yong Kang
School of Electrical and Electronics Engineering
Huazhong University of Science and Technology
Wuhan, China
zhiqiangwang@hust.edu.cn

Abstract—The design of power module requires an accurate multi-physical fields simulation methodology to simulate the electro-thermal stress accurately. The paper proposed an automated multi-physical fields modeling and co-simulation methodology based on intelligent software interfaces for SiC power module design. Firstly, the multi-physical fields coupling mechanism and basic co-simulation principle of the proposed methodology are described. Then, the co-simulation software interface and indirect coupling strategy are designed for this methodology, which can reflect the evolution of various physical fields from initial condition to steady state through continuous transient simulation. Finally, the proposed co-simulation methodology is verified by direct steady-state thermal simulation of a SiC MOSFET power module. The simulation results from the steady-state thermal simulation demonstrate the feasibility and accuracy of the proposed simulation methodology.

Keywords—Electro-thermo-mechanical, SiC power module, interface, co-simulation.

I. INTRODUCTION

With the superior properties like high switching frequency, low conduction resistance and high temperature tolerance capability, SiC power module is expected to apply to high power density converters [1]. The design of high power density SiC converters remains challenging. One of the challenges is the lack of the precise co-simulation methodology to reveal the stress distribution in the SiC power module which is the core component of converters. The power semiconductor module is a complex coupling system with multiple physical fields and multiple time scales. The interaction of multiple physical fields and multiple time scales must be considered comprehensively in modeling and simulation. The simulation results obtained from a single physical field perspective exist a big difference with the actual situation in most cases. It is of great significance to establish a precise electro-thermal-mechanical simulation platform for power module design.

The key of establishing the simulation platform is to establish the equations describing the coupling relationship. The mathematical partial differential equations describing multi-physical field coupling simulation model come down to complex calculation. Utilizing multi-physics finite-element analysis (FEA) simulation software is a good choice to solve complicated equations. As far as the design of power module is concerned, FEA is necessary but not sufficient for the lack of circuit simulation capability. Circuit simulation software and FEA simulation software are supposed to be implemented on power module design. Circuit simulation software like PSpice, Saber, Simulink are widely applied in the simulation of electrical characteristics of SiC power module. The simulation of thermal and mechanical characteristics is usually carried out in FEA simulation software such as ANSYS and COMSOL. Although commercial software ANSYS and COMSOL have powerful simulation capability for multiple fields, there are still many problems when they are applied to the multi-physical fields co-simulation of SiC power module.

(1) The SiC device spice models from manufacturers are generally incompatible with ANSYS Twin Builder, leading to limited electrical simulation capability with device behavior models. Alternatively, a SiC device behavior model can be rebuilt in ANSYS Twin Builder, which can be time and energy consuming.

(2) The thermal network model used in ANSYS Twin Builder is based on thermo-electric analogy, which is only valid for linear and time invariant (LTI) systems [2]. However, the real thermal systems for SiC power modules are essentially not linear time-invariant systems, resulting in unsatisfactory simulation accuracy by using the thermal network model based on model reduction technique.

(3) Moreover, the field-circuit coupled simulation in ANSYS is based on the traditional thermal network method which can only get the temperature values of a few network nodes, such as junction temperature. It is difficult to obtain a continuous temperature distribution with RC thermal network. The strain distribution in the power module needs to simulate in other software separately.

All of these problems restrict the application of ANSYS in precise electro-thermo-mechanical co-simulation of SiC power module. Another commercial software COMSOL lacks the capability to perform accurate co-simulation in combination with SiC device model either, since the circuit simulation module, i.e.AC/DC Module, does not support device spice models.

In recent years, several multi-physics coupling co-simulation methodologies based on COMSOL or ANSYS have been proposed to overcome the above problems. In [3], a co-simulation model of IGBT based on FEA is established in COMSOL. The loss of IGBT is estimated through mathematical derivation, instead of circuit simulation with temperature-dependent real device model for exact loss. Similar methodologies have also been discussed and explored in [4] in which the electrical models are replaced by simplified behavior models to extract power losses. In addition, a multi time-scale electro-thermal-mechanical co-simulation method is proposed in [5] by exchanging data between COMSOL and PSpice. This methodology combines the strengths of the device spice model and the FEA model, while it does not consider the difference in the temperature distribution of each heat source in the power module during data exchanging process.

All of these methodologies can not reflect the impact of uneven temperature distribution of SiC chips on electrical

978-1-6654-4817-8/21 $31.00 © 2021 IEEE

characteristics, leading to compromised accuracy in practical applications.

To pursue excellent and insightful design, this paper proposes an automated co-simulation methodology which can be used for both steady-state simulation and transient simulation research. This methodology takes into account the uneven junction temperature feedback at the end of each thermal simulation by updating the temperature assignment statement of spice model. The thesis is structured as follows: Section II puts forward the suggested co-simulation methodology. Section III introduces the designed SiC MOSFET power module and comparison of results between co-simulation and direct steady-state thermal simulation. Section IV draws the conclusion of this paper.

II. PROPOSED CO-SIMULATION METHODOLOGY

To predict the electrical, thermal, strain characteristics, and their interactive behaviors within SiC power module accurately, an automated co-simulation methodology with intelligent software interfaces is proposed in this section, with a special focus on the difficulty of coupling analysis between numerical calculation software COMSOL and existing circuit simulation software PSpice.

A. Multi-physical Fields Coupling Mechanism

The interaction of multi-physical fields is demonstrated in Fig. 1. The coupling relationship between different physical fields is established through certain coupling variables. The coupling variable power loss leads to the rise of module temperature. Temperature rise tends to change the power module packaging material characteristics, which is eventually reflected as the change of module electrical characteristics. Variation of packaging material properties also induces module structure deformations because of the thermal expansion. Due to the different thermal expansion coefficients of materials in each layer of the power module, thermal stress will be generated when the temperature changes.

According to the theory of linear thermal stress, the general strain stress can be divided into two parts: ε^T and ε^E. The strain stress can be given in (1)[6].

$$\begin{cases} \varepsilon = \varepsilon^T + \varepsilon^E \\ \varepsilon^T = \alpha(T)(T - T_{\text{ref}}) \\ \varepsilon^E = 0.5[\nabla u + (\nabla u)^T] \end{cases} \quad (1)$$

where u represents the strain displacement; $\alpha(T)$ is the thermal expansion coefficient; T_{ref} and T are the reference temperature and the actual temperature respectively. The coupling between the mechanical field and the electric field is weak, and thus the coupling between them is usually ignored. It may be considered under some special conditions such as high frequency and large pulse current environment.

The accuracy of the model established considering above main coupling variables is sufficient and accurate for the packaging design of SiC power module. The coupling between different physics can be achieved by transferring the corresponding coupling variables through a dedicated coupling interface. The thermal-mechanical coupling interface called thermal expansion is integrated in COMSOL while the electro-thermal coupling interface is lacking. The coupling relationship between temperature and power loss is

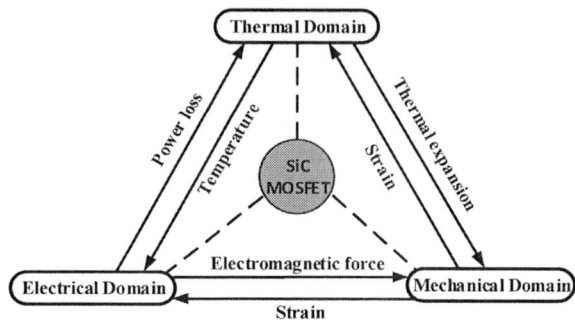

Fig. 1. Multi-domain coupling mechanism of SiC power module.

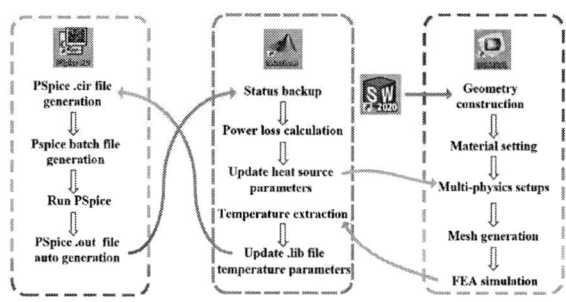

Fig. 2. Coupling interface of proposed co-simulation methodology.

supposed to fully considered for the implementation of electro-thermal coupling interface.

B. Coupling Interface of Proposed Co-simulation Methodology.

The proposed coupling interface is illustrated in Fig. 2, in which the automatic iteration loop, including some essential steps of co-simulation, is built up as a co-simulation interface to support the proposed design methodology. The coupling interface involves three major components: PSpice circuit simulation based on spice model provided by device manufacturers, thermal-mechanical coupling simulation in COMSOL FEA simulation software, and a monitoring script program in MATLAB. The MATLAB control program is implemented to deal with the data and then coordinate data transfer between PSpice and COMSOL at each thermal-mechanical simulation step. The operating principle of this interface is based on integrating MATLAB code with COMSOL and PSpice.

Since each action in the GUI of COMSOL can be realized as a set of construction code, FEA simulation commands executed in COMSOL can be directly invoked from the co-simulation algorithms implemented in MATLAB. For each operation in the COMSOL, there is a corresponding Java command code. MATLAB prompts Java command as lines of pre-coded instructions and imports it into COMSOL to eliminate the involvement of manual operation. The conversion of COMSOL operation steps to Java command code enables the whole simulation process (including geometry construction, material setting, and physics settings, mesh algorithm implementing, applying finite element solver and evaluation of the results) can be controlled by MATLAB. Cooperating with SOLIDWORKS for the construction of the auxiliary components, the thermal-mechanical model is constructed under the control of MATLAB. After the completion of FEA thermal-mechanical coupling simulations, the temperature distribution is generated. Then the maximum junction temperature or the strain are acquired. MATLAB uses the acquired temperature data value to update the corresponding temperature parameters in the SiC MOSFET device model files with the suffix .lib.

MATLAB can extract the temperature of each device through the interface of COMSOL Multiphysics 5.6 with MATLAB. However, the lack of simulation interface between MATLAB and PSpice makes it impossible to directly control and process data for circuit simulation. To realize the process control and data processing, it is necessary to process the files related to the PSpice circuit simulation.

The key to co-simulation interface programming of PSpice and MATLAB is the processing of various input and output format files to realize data exchange. PSpice circuit simulation can also be performed in the PSpice AD by executing input files with the suffix .cir, in addition to using graphical operations in the Capture CIS. After executing the circuit simulation file with the suffix of .cir written in spice language in PSpice, an output file with the suffix of .out will be automatically generated. The device loss data at different times can be extracted from the output file by writing the relevant SPICE command in the input file. MATLAB processes the loss data and calculates the power loss of each heat source. The calculated loss data is then transmitted to COMSOL to update heat source parameters.

C. Indirect Coupling Strategy of Proposed Co-simulation Methodology

Since the electrical model and thermal-mechanical model are constructed in different simulation software, the backhanded coupling strategy need to adopt to achieve co-simulation function. The detailed indirect coupling strategy is shown in Fig. 3. In this coupling strategy, the COMSOL time steps is equal to PSpice time steps. The simulation time of the two software is set to be the same since the action time of each physical field in power module is equal. However, PSpice internal time steps are generally much smaller than COMSOL internal time steps. Specifically, the switching transient of SiC MOSFET is μs level, so nanosecond internal simulation steps in Pspice are required to ensure precise assessment of power loss. On the contrary, the time scale of heat transfer is ms level, millisecond time steps in COMSOL are sufficient for thermal-mechanical simulation. In long time scale thermal-mechanical simulation, second-level time steps may even be adopted in order to achieve good balance between calculation accuracy and efficiency. Due to the significant difference between the time scale of circuit and thermal simulation, the power loss parameter transmitted from PSpice to COMSOL is actually average power loss. The temperature parameter passed from COMSOL to PSpice is actually the maximum junction temperature of SiC power module.

Under co-simulation state, the MATLAB script synchronizes the PSpice and COMSOL simulations and coordinates the data sharing between them. In the first step, the circuit simulation at a given initial temperature is performed in PSpice to obtain the power loss in the heat source area of the SiC module. The circuit simulation is suspended when the preset data iteration step is reached. In the second step, the calculated loss data is transmitted to COMSOL through MATLB. In the third step, the thermal-mechanical simulation is performed by means of the differential equations in COMSOL according to the received loss parameters, and it also pauses when the preset data iteration step is reached. In the fourth step, MATLAB feeds the updated parameters back to the PSpice circuit model which uses them as the new initial

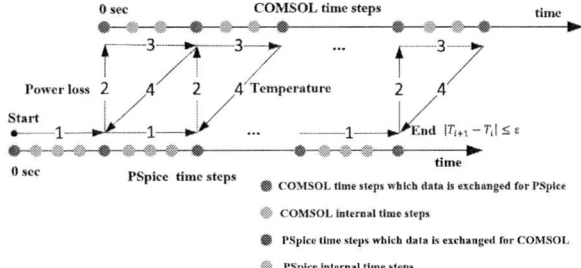

Fig. 3. Indirect coupling strategy of proposed co-simulation methodology.

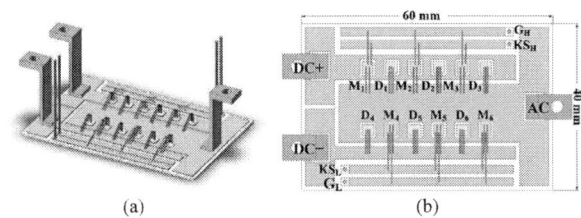

Fig. 4. SiC MOSFET power semiconductor module. (a) Module layout. (b) Chip and terminal location.

values. Simulation continues on the basis of the previous state. In other words, the state at the end of the previous cycle will be used as the initial state of the next cycle, making it a iterative and successive process. In the steady-state simulation study, when the temperature difference of the power device between two contiguous iterations is inferior to the predefined threshold, it is considered that the steady-state is reached and the cycle ends.

The simulation of full operating conditions can be implemented by controlling time steps which data is exchanged, and data iteration between the two simulation software. It is worth noting that all processes are implemented through programs, enabling an fully automated co-simulation environment.

III. MULTI-PHYSICAL COUPLING SIMULATION OF SiC POWER MODULE

To check out the feasibility of co-simulation methodology, this section presents a buck converter based on a phase-leg SiC power module. Furthermore, FEA analysis are performed to emulate thermal-mechanical stress distribution in the SiC power module.

A. Phase-Leg SiC Power Module

A custom designed SiC power module is made to verify the proposed simulation methodology. The layout of the phase-leg power module is shown in Fig. 4. In this module, SiC MOSFET bare dies CPM312000075A and SiC SBD CPW41200S015B from CREE are utilized. Each switch position has three diode chips and three MOSFET chips in parallel to achieve a greater discharge capability. The packaging materials with their relevant specifications are shown in table I.

In view of the reliability and maturity of aluminum wire-bonding, aluminum wire is adopted to achieve electrical interconnection. Direct-bonded-copper (DBC) is AlN featuring much higher thermal conductivity and better coefficient of thermal expansion than Al_2O_3.

TABLE I. Packaging Materials List

Component	Material	Description
Switch	SiC MOSFET Bare die CPM312000075A	2.5 mm × 2.8 mm Thickness: 0.177 mm
Diode	SiC bare die CPW41200S015B	2.7 mm × 2.7 mm Thickness: 0.374 mm
Substrate	AlN DBC	Area: 60 mm × 40 mm AlN thickness:0.63 mm Cu thickness:0.3 mm
Bonding wire	Al	Diameter:0.254 mm
Die-attach	Sn5-Pb95	Thickness: 0.2 mm
Baseplate	Cu	Area: 104 mm × 59 mm Thickness:3 mm

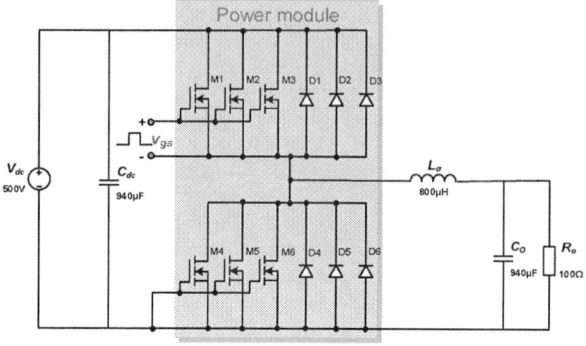

Fig. 5.   Buck circuit diagram of PSpice simulation.

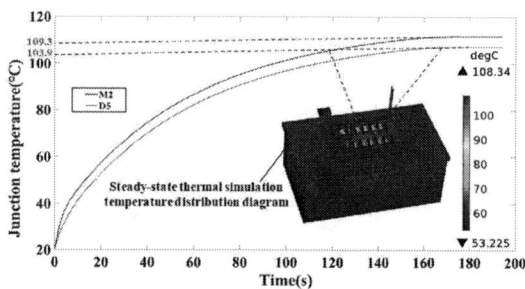

Fig. 6. Comparison of SiC device junction temperature change obtained by co-simulation and direct steady-state thermal simulation.

B. Multi-field Coupling Simulation Analysis

A buck converter is constructed in PSpice AD in the form of spice language code. The corresponding circuit schematic is demonstrated in Fig. 5. M1, M2 and M3 are used as the active switch, while M4, M5 and M6 (off state), together with D4, D5 and D6 are used as free-wheeling diodes. The buck converter works in a 500-V DC input voltage, 4-A load current,100-kHz switching frequency, 0.8 duty cycle. The MOSFET chip is driven by a 10 Ω resistor to designedly increase power loss. The initial temperature of circuit simulation is set to 20 °C. The thermal-mechanical model is built in COMSOL. The cooling method of the thermo-mechanical model is natural convection. Considering the balance between accuracy and efficiency, the calculation time steps of the circuit and the thermal-mechanical simulation are set to 1 ns and 100 ms respectively. The data iteration step between the two simulation models is set to 1 s.

Compared with traditional methods, the co-simulation method proposed in this paper is capable of simulating the coupling evolution of temperature, stress, and electrical parameters with time. A comparison of junction temperature evolution gained by co-simulation and the direct steady-state FEA thermal simulation is demonstrated in Fig. 6. In the comparison, the steady-state loss obtained by the co-simulation is used as the input parameter of the direct steady-state thermal simulation. The magenta and blue curves represent the temperature changes over time on the chip M2 and D5 during co-simulation process respectively. Steady-state temperatures of M2 and D5 gained by co-simulation are 109.3 °C and 103.9 °C, respectively. The temperature distribution diagram of the power module at the lower right of two curves is obtained by direct steady-state thermal simulation method. The junction temperatures of SiC MOSFET M2 and diode D5 are 108.3 °C and 103.1 °C respectively by direct steady-state FEA thermal simulation. The junction temperature difference of the two simulation methods is less than 1 °C, which confirms the feasibility of this proposed method.

The steady-state stress distribution is demonstrated in Fig. 7. The die attachments and bonding wires present relatively high stress in the SiC power module. As is shown in Fig. 7(b), the maximum mechanical stress occurs at the corner of the die attachment. Due to the stress at bonding wires, bonding wires are easy to lift off in a practical converter.

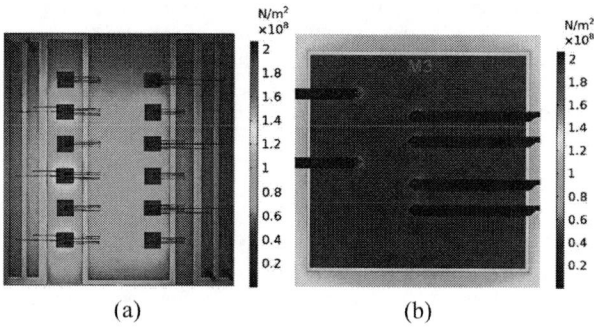

(a)   (b)

Fig. 7. The steady-state stress distributions. (a) Overall stress distribution. (b) Chip-M3 stress distribution.

IV. CONCLUSION

This paper proposes an automated electro-thermal-mechanical co-simulation methodology for SiC power module. The proposed methodology connects a device-level, electrical simulation software PSpice with a FEM simulation software COMSOL by establishing data interface program. It can acquire high accuracy on both electrical simulation and the thermal-mechanical simulation. By comparing co-simulation with the direct steady-state thermal simulation, it is found that the proposed co-simulation is only one degree different from the traditional steady-state thermal simulation, which demonstrates the high accuracy and feasibility of the proposed co-simulation methodology. Moreover, by using proposed methodology in a SiC power module, it is found that die attachments and wire bonding suffer from the highest stresses in the power module.

Future work involves the study of power module reliability

and failure mechanism analysis by taking advantage of the proposed methodology in this paper.

REFERENCES

[1] X. Ding et al., "Analytical and experimental evaluation of SiC-inverter nonlinearities for traction drives used in electric vehicles," *IEEE Trans. Veh. Technol.*, vol. 67, no. 1, pp. 146–159, Jan. 2018.

[2] R. Wu, H. Wang, and K. B. Pedersen, "A temperature-dependent thermal model of IGBT modules suitable for circuit-level simulations," *IEEE Trans. on Industry Applications*, vol. 52, no. 4, pp. 3306-3314, Sep. 2016.

[3] E. Deng, Z. Zhao, Z. Lin, R. Han, and Y. Huang, "Influence of temperature on the pressure distribution within press pack IGBTs," *IEEE Trans. Power Electron.*, vol. 33, no. 7, pp. 6048-6059, Sep. 2018.

[4] K. B. Pedersen and K. Pedersen, "Dynamic modeling method of electro-thermo-mechanical degradation in IGBT modules," *IEEE Trans. Power Electron.*, vol. 31, no. 2, pp. 975-986, Apr. 2016.

[5] Y. Jia, F. Xiao, Y. Duan, Y. Luo, B. Liu and Y. Huang, "PSpice-COMSOL-Based 3-D Electrothermal–Mechanical Modeling of IGBT Power Module, " *IEEE Journal of Emerging and Selected Topics in Power Electronics*, vol. 8, no. 4, pp. 4173-4185, Dec. 2020.

[6] B. Gao, F. Yang, and M. Chen, "A temperature gradient-based potential defects identification method for IGBT module," *IEEE Trans. Power Electron.*, vol. 32, no. 3, pp. 2227-2242, Mar. 2017.

An Accurate Analytical Model for Motor Terminal Overvoltage Prediction and Mitigation in SiC Motor Drives

Neng Wang, Cheng Qian, Zhiqiang (Jack) Wang, Yong Kang
School of Electrical and Electronic Engineering
Huazhong University of Science and Technology
Wuhan, China
zhiqiangwang@hust.edu.cn

Abstract—The motor terminal overvoltage caused by the fast switching speed of SiC devices remains a challenge for SiC based motor drive systems. To deal with this challenge, this paper presents an analytical model based on voltage frequency spectrum and distributed parameter matrix for motor terminal overvoltage prediction and mitigation. First, the analysis and derivation of this model is elaborated combining two-port network and transmission-line wave equation. Second, based on this model, the procedure to predict the motor terminal overvoltage and design *dv/dt* filter is discussed in detail. With the designed *dv/dt* filter, a Saber simulation model of a SiC motor drive system is established to verify the proposed analytical model. Finally, an experimental platform is built based on commercial SiC power modules to further validate the accuracy of the proposed model. The experimental results show that the proposed model can achieve an overvoltage prediction error less than 1.9%.

Keywords—motor drive, silicon carbide device, transmission line, two-port network, overvoltage prediction

I. INTRODUCTION

Thanks to the superior material properties of silicon carbide (SiC), SiC power semiconductor devices are expected to be widely used in high frequency and high-power-density motor drive systems, such as electric vehicle and more electric aircraft. However, due to the fast switching speed of SiC devices, the PWM voltage wave generated by SiC inverter has a high voltage slope (*dv/dt*). Through the feeder cable, the high *dv/dt* voltage wave is reflected at the motor end, leading to serious motor terminal overvoltage and insulation issues.

In order to avoid the terminal insulation failure, the reflected wave phenomenon has been studied and different *dv/dt* filters have been designed in the past decade. The existing methods for motor terminal overvoltage prediction can be roughly divided into three categories. In [1]—[3], the wave reflection coefficient is used to derive an overvoltage-peak prediction formula. This method is the most commonly used way of overvoltage estimation, whereas it can only provide a qualitative analysis for overvoltage phenomenon. Since the damping of motor cable and the frequency characteristics of the cable impedance are not considered, the error of the prediction results is quite large. In [4]—[5], the finite element simulation software is used to establish the physical model of cable and predict the overvoltage waveform. This method can build an accurate model for a particular cable, while its accuracy is heavily dependent on a myriad of cable material and structure parameters making it difficult to use in practical applications. In [6]—[7], the impedance-network models of cable and motor are derived, and by cascading multiple impedance networks the motor terminal overvoltage

This work was supported by the National Natural Science Foundation of China under Grant 51907070.

transients are obtained in circuit simulation software. This method strikes a balance between model accuracy and implementation complexity. However, it still relies on simulation software to build complex circuit models, and the number of cascade networks used in simulation can lead to significant errors.

In order to make more accurate overvoltage prediction through mathematical calculation in the design stage of SiC motor drive systems, this paper proposes a two-port analytical model combining transmission-line wave equation and cable impedance-networks. Based on the proposed analytical model, the mechanism of motor-side overvoltage is analyzed and the method of overvoltage mitigation is discussed as well.

II. DERIVATION OF ANALYTICAL MODEL

Towards the three-phase structure of a motor drive system, the overvoltage phenomena are symmetric. According to [8], the motor terminal voltage transient in one phase is independent of the other two phases. Therefore, only one of these phases is studied in this work, while the other two phases are connected to the negative busbar. The simplified system diagram of a three-phase motor drive system for motor terminal overvoltage analysis is shown in Fig. 1.

In this system, the voltage wave from the inverter side (V_{in}) and the voltage wave reflected at the motor side (V_{out}) are concerned with an interconnection cable. The system is equivalent to a two-port network shown in Fig. 2. The two-port model in Fig. 2 connects the motor phase B and C terminal together since the motor is also three-phase symmetrical, and the voltage at B and C terminal is equal. The two-port matrix of the network can be given in (1)

$$\begin{bmatrix} V_{in} \\ I_{in} \end{bmatrix} = \begin{bmatrix} A_c & B_c \\ C_c & D_c \end{bmatrix} \cdot \begin{bmatrix} V_{out} \\ I_{out} \end{bmatrix} \tag{1}$$

where V_{in} (and I_{in}) is the inverter-side input voltage (and current); V_{out} (and I_{out}) is the motor-side output voltage (and current); and the matrix coefficients A_c, B_c, C_c, D_c are determined by cable parameters. Based on this matrix, the transfer function K_u between the network output voltage and input voltage can be expressed as

$$K_u = \frac{V_{out}}{V_{in}} = \frac{Z_L}{A_c \cdot Z_L + B_c} \tag{2}$$

where Z_L is the motor impedance.

With given inverter-side input voltage and the motor load, the coefficients of this two-port matrix become the key to predict the motor terminal overvoltage V_{out}. The derivation

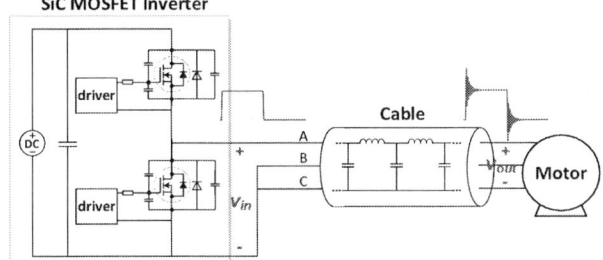

Fig. 1.   Schematic of simplified SiC motor drive system.

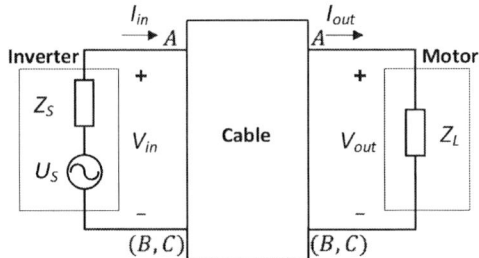

Fig. 2.   Two-port model of motor drive system.

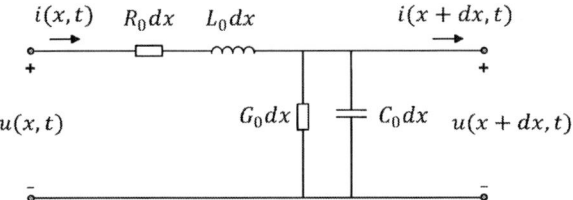

Fig. 3.   Basic distributed parameter circuit of transmission line.

of these coefficients can be divided into three steps, as discussed in the following.

A. Fundamental Expressions Based on Wave Equation

Cable is one kind of transmission line, and its basic distributed parameter circuit model is illustrated in Fig. 3. According to this circuit, the wave equation can be derived as

$$\begin{cases} \frac{\partial u(x,t)}{\partial x} = -R_0 i(x,t) - L_0 \frac{\partial i(x,t)}{\partial t} \\ \frac{\partial i(x,t)}{\partial x} = -G_0 u(x,t) - C_0 \frac{\partial u(x,t)}{\partial t} \end{cases} \quad (3)$$

where R_0, L_0, G_0, C_0 are the per-unit distributed resistance, inductance, conductance, capacitance of transmission line; x is the distance from the input of the transmission line.

Considering the inverter-side pulse voltage wave as the superposition of sinusoidal voltage waves, the complex form of (5) can be obtained by introducing harmonic variables $u(x), i(x)$ with angular frequency ω.

$$\begin{cases} u(x,t) = u(x)e^{j\omega t} \\ i(x,t) = i(x)e^{j\omega t} \end{cases} \quad (4)$$

$$\begin{cases} \frac{du(x)}{dt} = -(R_0 + j\omega L_0)i(x) \\ \frac{di(x)}{dt} = -(G_0 + j\omega C_0)u(x) \end{cases} \quad (5)$$

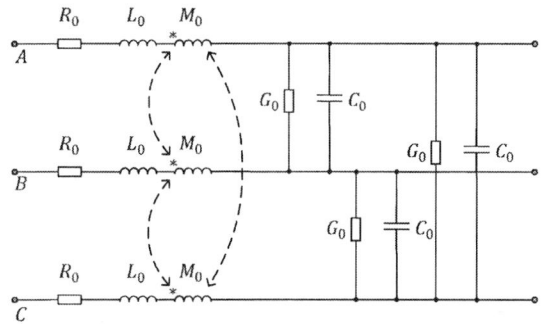

Fig. 4.   Improved distributed circuit model of three-core cable.

By defining the propagation coefficient γ and the characteristic impedance Z_0 as

$$\begin{cases} \gamma = \sqrt{(R_0 + j\omega L_0) \cdot (G_0 + j\omega C_0)} \\ Z_0 = \sqrt{(R_0 + j\omega L_0)/(G_0 + j\omega C_0)} \end{cases} \quad (6)$$

the solution of the wave equation in complex form is

$$\begin{cases} u(x) = U^+ e^{-\gamma x} + U^- e^{\gamma x} \\ i(x) = \frac{U^+}{Z_0} e^{-\gamma x} - \frac{U^-}{Z_0} e^{\gamma x} \end{cases} \quad (7)$$

By substituting $u(0) = V_{in}$, $i(0) = I_{in}$, $u(l) = V_{out}$, $i(l) = I_{out}$ (l is the length of cable) into (7), the coefficient U^+, U^- can be determined. Finally, the basic two-port matrix expressed by the transmission line distributed parameters can be given in (8).

$$\begin{bmatrix} V_{in} \\ I_{in} \end{bmatrix} = \begin{bmatrix} cosh(\gamma l) & Z_0 sinh(\gamma l) \\ \frac{sinh(\gamma l)}{Z_0} & cosh(\gamma l) \end{bmatrix} \cdot \begin{bmatrix} V_{out} \\ I_{out} \end{bmatrix} \quad (8)$$

B. Cable Circuit Modeling and Simplification

Although one phase two-port matrix can be derived based on the basic distributed parameter circuit model, the actual cable circuit is more complex than this basic circuit. For multicore cables, each core wire has symmetric impedances and there is mutual coupling between these wires. After considering these factors, the circuit model of the widely used three-core cable is established in Fig. 4.

Taking full advantage of the symmetry of system in Fig. 1, the voltage and current relationship between three phases can be given as

$$I_A = -2I_B = -2I_C, \quad V_B = V_C. \quad (9)$$

Therefore, the mutual inductance between phase lines can be decoupled and the impedance of the other two phase lines can be combined. Fig. 5 shows the simplified circuit model of cable.

By defining new cable parameters $R_c, L_c, G_c, C_c, \gamma_c, Z_c$ as

$$R_c = \frac{3R_0}{2}, L_c = \frac{3(L_0 - M_0)}{2}, G_c = 2G_0, C_c = 2C_0 \quad (10)$$

$$\begin{cases} \gamma_c = \sqrt{(R_c + j\omega L_c) \cdot (G_c + j\omega C_c)} \\ Z_c = \sqrt{(R_c + j\omega L_c)/(G_c + j\omega C_c)} \end{cases} \quad (11)$$

978-1-6654-4817-8/21 $31.00 © 2021 IEEE   505

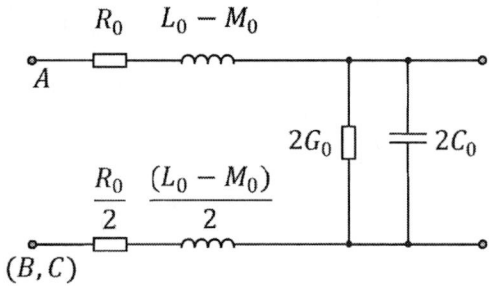

Fig. 5. Simplified circuit model of three-core cable.

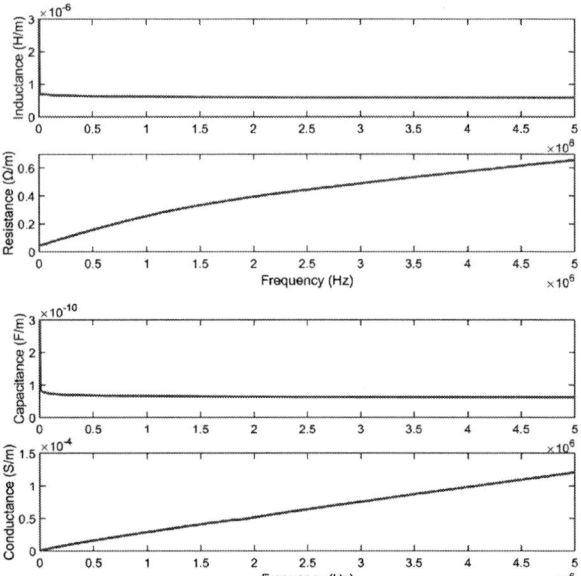

Fig. 6. Measurement results of cable per-unit impedance.

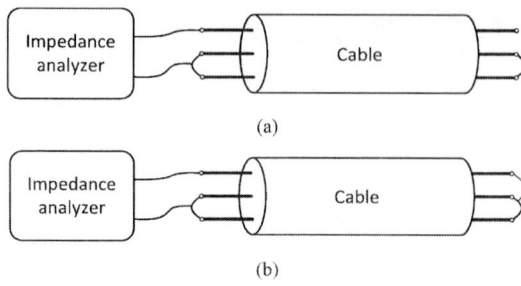

Fig. 7. Measurement schematic of three-core cable. (a) Open-circuit method. (b) Short-circuit method.

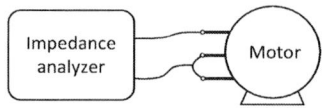

Fig. 8. Measurement schematic of motor.

final expression of the cable two-port matrix is given in (12) and (13)

$$\begin{cases} \gamma_c = \sqrt{[R_c(\omega_0)\cdot\sqrt{\omega/\omega_0}+j\omega L_c(\omega_0)]\cdot[G_c(\omega_0)\cdot(\omega/\omega_0)+j\omega C_c(\omega_0)]} \\ Z_c = \sqrt{[R_c(\omega_0)\cdot\sqrt{\omega/\omega_0}+j\omega L_c(\omega_0)]/[G_c(\omega_0)\cdot(\omega/\omega_0)+j\omega C_c(\omega_0)]} \end{cases} \quad (12)$$

$$\begin{bmatrix} V_{in} \\ I_{in} \end{bmatrix} = \begin{bmatrix} cosh(\gamma_c l) & Z_c sinh(\gamma_c l) \\ \dfrac{sinh(\gamma_c l)}{z_c} & cosh(\gamma_c l) \end{bmatrix} \cdot \begin{bmatrix} V_{out} \\ I_{out} \end{bmatrix}. \quad (13)$$

III. METHOD FOR OVERVOLTAGE PREDICTION AND MITIGATION

In order to utilize the proposed model in Section II, the impedance parameters need to be evaluated. The datasheets of commercial cables and motors usually provide a few impedance parameters with limited accuracy. This paper uses high-bandwidth impedance analyzer to comprehensively evaluate the cable and motor impedance characteristics. The measurement schematics of three-core cable and motor are shown in Fig. 7 and Fig. 8 respectively.

The cable parameters G_c and C_c can be measured directly by the open-circuit method in Fig. 7(a), and the cable parameters L_c and R_c can be measured directly by the short-circuit method in Fig. 7(b). These special measurement methods simplify the measurement procedure and reduce the measurement error. In addition, the cable impedance presents plenty of resonant peak, and the corresponding resonant frequency is inversely proportional to the length of the cable. Therefore, a shorter measuring cable should be used and measured at a relatively low frequency to avoid resonances. For example, the actual cable used in this work is 5 m, while the measuring cable is 1 m. The resonant frequency of this 1-m cable is about 40 MHz, and the measuring frequency is selected at 5 MHz.

A. Overvoltage Prediction

With the cable impedance parameters at one frequency and the motor impedance characteristic over the entire frequency range, the transfer function over the entire frequency range in (2) can be derived. According to (2), the output voltage frequency spectrum is the product of the input voltage frequency spectrum and the voltage transfer function. With a given input voltage waveform, the output voltage waveform can be predicted by Fourier transform of the input voltage and

the two-port matrix of three-core cable can be obtained just by replacing γ and Z_0 in (8) with γ_c and Z_c.

C. Frequency Correction of Impedance Parameters

In order to determine the value of the cable two-port matrix coefficient, it is necessary to extract distributed impedance parameter of the cable. Since the impedance parameters of the cable are frequency dependent due to skin effect and dielectric loss, the accurate parameter measurement and calculation become quite challenging.

Fig. 6 shows the measurement results of cable per-unit distributed impedance using an impedance analyzer. In the high frequency range, the inductance and capacitance are almost constant, while the resistance and conductance change significantly with frequency.

According to the curve fitting tool in MATLAB, the per-unit resistance is approximately proportional to the square root of frequency due to the skin effect, and the per-unit conductance is basically proportional to frequency due to the dielectric loss. Therefore, if the per-unit impedance parameters at one frequency are given, the parameters over the entire frequency range can be derived approximately.

By defining frequency ω_0 and the corresponding impedance parameter $R_c(\omega_0)$, $L_c(\omega_0)$, $G_c(\omega_0)$, $C_c(\omega_0)$, the

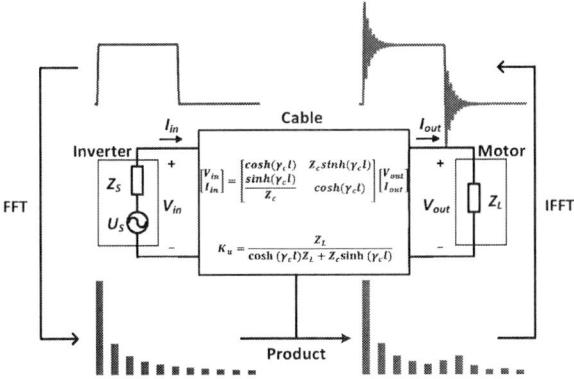

Fig. 9. Proposed analytical model and forecasting process.

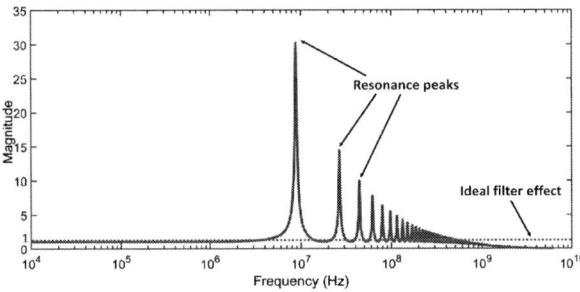

Fig. 10. Amplitude-frequency characteristic of K_u.

inverse Fourier transform of the calculated output voltage frequency spectrum. The whole process of overvoltage prediction is shown in Fig. 9.

For the acquisition of the input voltage waveform at the inverter side, measuring the actual voltage waveform with high bandwidth oscilloscope is preferred. Instead, a trapezoidal wave with constant voltage slope (dv/dt) can be used as an approximation, which is widely used in the past studies.

B. Overvoltage Mitigation

Fig. 10 illustrates the amplitude-frequency characteristic of the transfer function K_u in (2). There are many resonance peaks in this transfer function due to the series resonance of cable and motor impedance, and the motor overvoltage is caused by the amplification of the specific frequency components of input voltage. Therefore, the motor terminal overvoltage can be mitigated by suppressing these resonance effects through a dedicated filter at the inverter side. The ideal filtering effect is to keep the amplitude of the transfer function at 1 for motor terminal overvoltage suppression as well as low filtering power loss. Given that the high frequency components of the input voltage and the resonance peaks in the transfer function both decay rapidly with frequency, the first resonance frequency dominants the overall resonance effect. Consequently, in practical application the filter can be designed mainly for suppression of the first resonant frequency.

IV. SIMULATION AND EXPERIMENTAL VERIFICATION

A simulation model of the motor drive system with dv/dt filter has been built by the circuit simulation software to verify the accuracy of the overvoltage prediction and the effectiveness of the overvoltage mitigation. Moreover, an experimental platform has also been built to further verify the

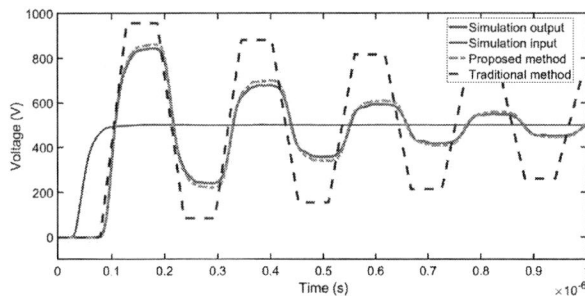

Fig. 11. Motor-side voltage waveforms predicted by different methods.

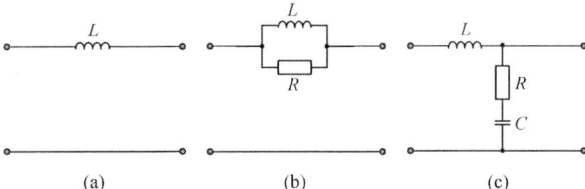

Fig. 12. dv/dt filters with different topologies. (a) L. (b) LR. (c) LRC.

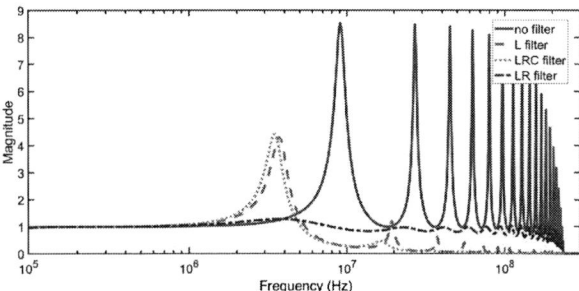

Fig. 13. Amplitude-frequency characteristics of voltage transfer functions of system with different filters.

accuracy of the overvoltage prediction based on the proposed analytical model.

A. Simulation Verification

Based on behavior model of SiC MOSFET, the distributed circuit model of cable, and the impedance model of motor, an equivalent circuit simulation model of three-phase motor drive system is established in SABER.

Combining SABER and MATLAB, the motor terminal overvoltage waveforms obtained by the simulation, predicted by the traditional method (use the wave reflection coefficient) and predicted by the proposed model are compared. As shown in Fig. 11, the traditional prediction waveform has great deviation from the simulated waveform, and the error of overvoltage peak is 13.4% (113 V). By contrast, the waveform predicted by the proposed model is basically consistent with the simulated waveform, and the error of overvoltage peak is 2.1% (18 V).

For overvoltage mitigation, three dv/dt filters with different topologies in Fig. 12 are compared. The parameters of these three filters are designed based on the first dominant resonance frequency. The corresponding voltage transfer functions of the motor drive system with different filters can be extracted in the simulation software, as is shown in Fig. 13.

Although the traditional low-pass filters (L, LRC) can suppress the original resonance peak, a new peak is generated

978-1-6654-4817-8/21 $31.00 © 2021 IEEE 507

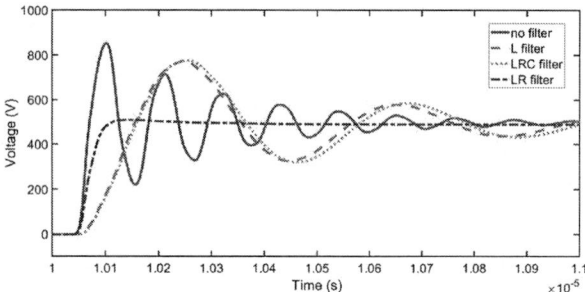

Fig. 14. Motor-side voltage waveforms using different filters.

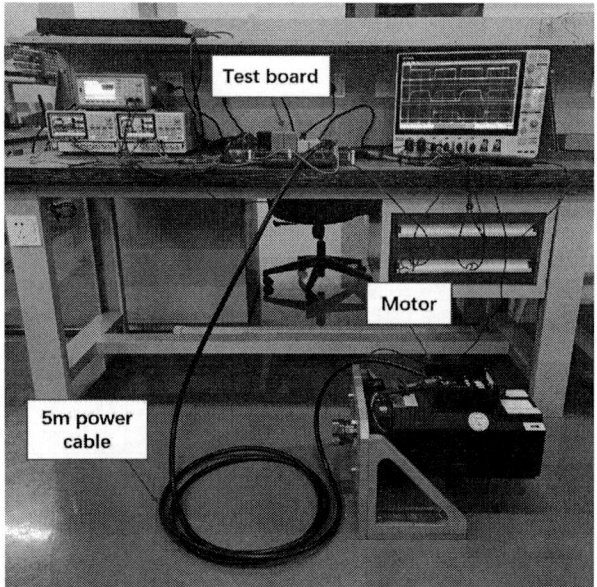

Fig. 15. Experimental test setup of motor drive system.

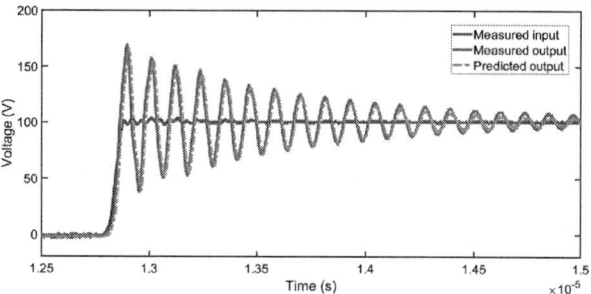

Fig. 16. Motor-side voltage waveform comparison under no load condition.

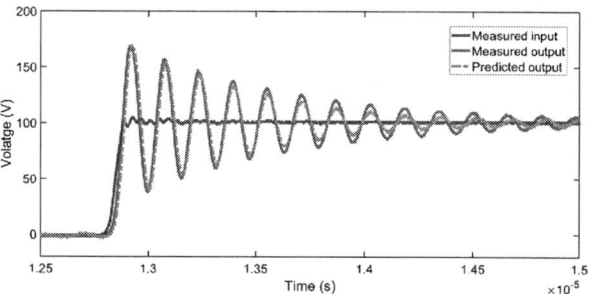

Fig. 17. Motor-side voltage waveforms comparison under inductive load.

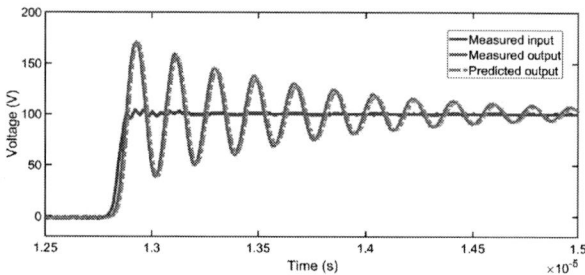

Fig. 18. Motor-side voltage waveforms comparison under motor load.

at low frequency due to the resonance effect between the filter and its load (cable and motor). Therefore, these filters present poor performance due to the amplification of low-frequency components of the input voltage. On the other hand, the LR filter designed for cable load has a filtering effect close to ideal, because it can achieve impedance matching between cable and load at high frequency.

Fig. 14 shows a comparison of motor terminal overvoltage waveforms using different filters in simulation. The results are essentially consistent with the analysis above. Specifically, the L and LRC filter both cause low-frequency overvoltage oscillation, while the LR filter almost completely suppresses the motor-side overvoltage.

B. Experimental Verification

The experimental test setup of the simplified SiC motor drive system in Fig. 1 based on Wolfspeed CAB425M12XM3 SiC power module and AlphaWire V16016 cable is built, as shown in Fig. 15. The experimental verification of terminal overvoltage prediction is carried out under three different load conditions.

a) No Load (open circuit)

Under no load condition, the comparison of overvoltage waveforms obtained by experimental measurement and the proposed model is shown in Fig. 16. In this case, the predicted

waveform and the experimental waveform are basically identical, and the error of overvoltage peak is 1.9% (3.3 V).

b) Inductive Load

In this case three 1-*mH* inductors are used as inductive load, which are often used as a substitute for motor. Under the inductive load, the comparison of overvoltage waveforms is shown in Fig. 17. In this case, the predicted waveform is also in good agreement with the experimental waveform, and the error of overvoltage peak is 0.4% (0.7 V).

c) Motor Load

Under motor load condition, the comparison of overvoltage waveforms is shown in Fig. 18. In this case, the deviation between the predicted waveform and the experimental waveform is relatively higher, and the error of overvoltage peak is 0.8% (1.4 V).

V. CONCLUSION

In this paper, an analytical model based on transmission-line wave equation and cable impedance-networks is proposed for motor terminal overvoltage prediction and mitigation in SiC motor drives. The simulation and experimental results show that this model can quickly predict the peak value or even transient waveform of overvoltage at motor side by mathematical calculation tool with an error less than 1.9%. It can be helpful for the design of SiC motor drive system, especially under the ultrafast switching speed environment. Future work involves the development of a more comprehensive model considering parasitic parameters of the system to the ground and a method to predict the inverter-side voltage waveform.

REFERENCES

[1] B. Narayanasamy, A. S. Sathyanarayanan, F. Luo, and C. Chen, "Reflected wave phenomenon in SiC motor drives: consequences, boundaries, and mitigation," *IEEE Trans. Power Electron.*, vol. 35, no. 10, pp. 10629-10642, Oct. 2020.

[2] M. Diab and X. Yuan, "A quasi-three-level PWM scheme to combat motor overvoltage in SiC-based single-phase drives," *IEEE Trans. Power Electron.*, vol. 35, no. 12, pp. 12639-12645, Dec. 2020.

[3] M. J. Scott *et al.*, "Reflected wave phenomenon in motor drive systems using wide bandgap devices," in *Proc. IEEE Workship Wide Bandgap Power Devices Appl.*, 2014, pp. 164-168.

[4] I. Tilea and C. Munteanu, "Motor cable electric parameter effects on the overvoltage phenomenon in inverter driven motors," in *Proc. 8th Int. Symp. Adv. Topics Electr. Eng.*, 2013, pp. 1-6.

[5] I. Tilea and C. Munteanu, "Overvoltage analysis in inverter driven induction motors," in *Proc. Int. Conf. Expo. Electr. Power Eng.*, 2012, pp. 509-512.

[6] H. Akagi and I. Matsumura, "Overvoltage mitigation of inverter-driven motors with long cables of different lengths," *IEEE Trans. Ind. Appl.*, vol. 47, no. 4, pp. 1741-1748, Jul./Aug. 2011.

[7] L. Wang, C. N. Ho, F. Canales, and J. Jatskevich, "High-Frequency Modeling of the Long-Cable-Fed Induction Motor Drive System Using TLM Approach for Predicting Overvoltage Transients," *IEEE Trans. Power Electron.*, vol. 25, no. 10, pp. 2653-2664, Oct. 2010.

[8] Z. Zhang, F. Wang, L. M. Tolbert, B. J. Blalock, and D. J. Costinett, "Evaluation of switching performance of SiC devices in PWM inverter-fed induction motor drives," *IEEE Trans. on Power Electron.*, vol. 30, no. 10, pp. 5701-5711, Oct. 2015.

An Improved Desaturation Protection Method with Self-Adaptive Blanking-Time for Silicon Carbide (SiC) Power MOSFETs

Jiawei Li, Cheng Qian, Zhiqiang (Jack) Wang, Yong Kang
School of Electrical and Electronic Engineering
Huazhong University of Science and Technology
Wuhan, China
zhiqiangwang@hust.edu.cn

Abstract—This paper presents a new desaturation protection method for silicon carbide (SiC) metal-oxide-semiconductor field-effect transistors (MOSFETs). The proposed method provides an extra charging loop for the fault detection part of desaturation protection circuit by utilizing the difference of drain-source voltage between fault and normal conditions. It can effectively reduce the overall response time, and the acceleration is partly self-adaptive according to the operating voltage, without compromising its noise immunity. The performance of the proposed circuit has been validated both by simulation and experimental results. The experimental results show that the blanking-time can be reduced by about 50% with the proposed method.

Keywords—Silicon Carbide, MOSFET, self-adaptive, desaturation protection

I. INTRODUCTION

High power density is one main technical goal of power electronics technology. To achieve the target, wide-band-gap (WBG) power semiconductors such as SiC MOSFETs begin to replace silicon (Si) devices gradually, thanks to their overall better performance like smaller on-state resistance, higher melting point and breakdown field, etc [1]. Based on above advantages, SiC MOSFETs have been widely used in various applications, including Photo-Voltaic (PV) systems, Electric Vehicles (EV), etc [2].

Despite of their superiorities, SiC MOSFETs have weaker short-circuit (SC) withstand capability compared to Si devices due to higher heat flux density during short circuit transients. The short-circuit withstand time of most SiC MOSFETs is less than 5 us [3] [4]. Furthermore, SiC MOSFETs often work in fast switching speed and strong electromagnetic interference (EMI) environment, leading to high risk of false-triggering of short circuit protection circuit. Thus, former protection circuits used for Si devices cannot be applied to SiC devices directly, and their response time as well as noise immunity should be further improved.

There are various protection methods for power semiconductors, including parasitic inductance method, gate voltage method [5], desaturation protection and so on. These methods have their own pros and cons. For a long time, desaturation protection has been widely used in Si devices, because it is easy to be integrated into the gate driver and has a low cost in addition to its good protection performance.

However, desaturation protection method has several defects when used in SiC MOSFETs, such as the relatively high blanking time and the large active region of SiC MOSFETs. Therefore, it is critical to optimize the desaturation protection method for SiC MOSFETs.

Currently, there are some research efforts regarding desaturation protection improvement. In [6], a desaturation protection method with self-adaptive blanking circuit is proposed. The proposed method can promote the protection response for fault under load (FUL), while it cannot be used for hard switching fault (HSF). In [7], a desaturation protection circuit with gate voltage clamped in turn-on is built. This method does not reduce the response time but decreases the SC energy instead. Besides, the circuit implementation of this method is fairly complex. Additionally, in [8]-[9], an extra resistor is connected to gate and detection circuit. Although the change is simple, it can reduce the response time in both HSF and FUL. However, the noise immunity is compromised. To overcome these issues, this paper proposes an improved desaturation protection method for fast response time without degrading the noise immunity.

II. BASIC INTRODUCTION OF DESATURATION PROTECTION

Fig. 1 shows a typical desaturation protection circuit for IGBTs. It mainly consists of a diode with high blocking voltage (D_{ss}), a RC charging network (R_{s1}, R_{s2}, R_{s3} and C_{blk}), a comparator and a discharging loop (R_{dg} and M_{dg}). The circuit protect the device by monitoring its voltage drop (V_{ce}) during turn-on transient and all on-state.

As shown in Fig. 1(a), when the device is under on-state condition, V_{ce} is low enough and the diode conducts. The voltage across the blanking capacitor (V_{blk_nor}) can be calculated according to (1)

$$V_{blk\_nor} = \left(V_{ce\_on} + V_F\right) \cdot \frac{R_{s3}}{R_{s2} + R_{s3}} \tag{1}$$

where V_F is the forward voltage of the diode D_{ss}, and V_{ce_on} is the on-state voltage of the IGBT T_1.

Under short-circuit condition, the current of device will first increase sharply and then keep nearly constant, while V_{ce} will rise continually. It means the device will pull out of saturation and enter into the detection region (i.e. active region), as shown in Fig. 2. Thus, according to Fig. 1(b), the diode will become reversed biased, and V_{blk_flt} can be calculated according to (2)

$$V_{blk\_flt} = V_{dd} \cdot \frac{R_{s3}}{R_{s1} + R_{s2} + R_{s3}} . \tag{2}$$

This work was supported by the National Natural Science Foundation of China under Grant 51907070.

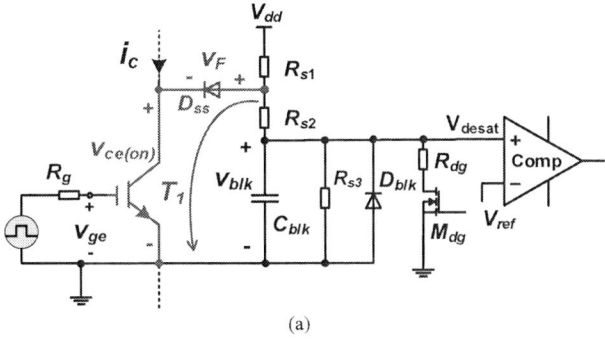

(a)

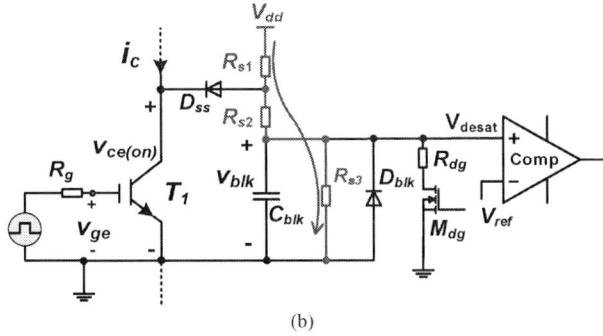

(b)

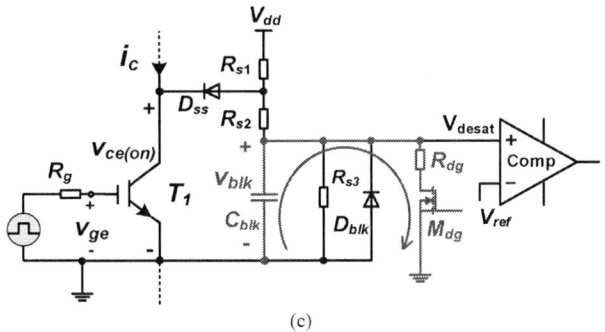

(c)

Fig. 1. Current loop of desaturation protection circuit. (a) On-state. (b) Short-circuit or overcurrent. (c) Off-state.

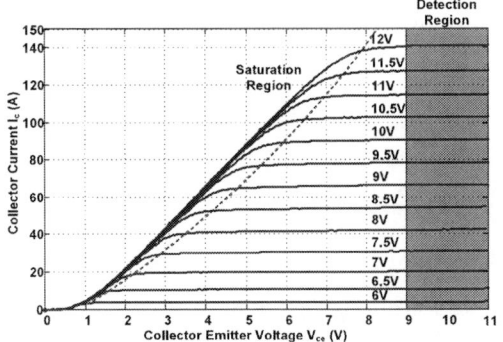

Fig. 2. Output characteristic of a Si IGBT (APT35GP120BG).

To protect the circuit correctly, the reference voltage (V_{ref}) should fall in between V_{blk_nor} and V_{blk_flt}.

Since V_{ce} remains high for a period of time during turn-on transient, a blanking capacitor (C_{blk}) is adopted to generate blanking-time delay (T_{blk}) to avoid potential false-triggering.

T_{blk} should be higher than turn-on time (T_{on}), and it can be calculated as follows

$$T_{blk} = \frac{R_{s1} + R_{s2} + R_{s3}}{(R_{s1} + R_{s2}) \cdot R_{s3}} \cdot C_{blk} \cdot \ln \frac{1}{1 - \frac{V_{ref}}{V_{dd}} \cdot \frac{R_{s3}}{R_{s1} + R_{s2} + R_{s3}}} . (3)$$

On the other hand, T_{blk} would also delay the response time inevitably when HSF occurs. Thus, its value needs to be carefully designed considering these tradeoffs.

A discharging loop is designed to release the charge of C_{blk} to zero when the device is under off-state condition, as shown in Fig. 1(c). It ensures the protection circuit is triggered at the same current level. By designing the value of RC detection network properly, the desaturation protection circuit can detect short circuit fault correctly in the active region, as shown in the shadow area of Fig. 2.

III. PROPOSED DESATURATION PROTECTION CIRCUIT

As discussed above, an excellent short circuit protection method should have the following two properties: fast response speed and high noise immunity. Generally, the two properties are contradictory for desaturation protection. Shorter T_{blk} can reduce the overall response time, whereas it increases the risk of false-triggering, i.e. worse noise immunity. Thus, it is not effective to just change the RC network for lower blanking time.

To solve this contradiction in the design of protection circuit, it is critical to identify the key differences between the fault and normal cases.

Different protection methods rely on their signature characteristics to detect short circuit faults. For instance, gate charge method generally relies on detecting the difference in the amount of charge to distinguish HSF and normal case. In addition, parasitic inductance method takes the advantage of the characteristic that the volt-seconds of the parasitic inductance is much higher during a short circuit condition. Similarly, the desaturation protection method relies on the drain-source voltage (V_{ds}) to detect a fault. If the drain-source voltage difference between fault and normal condition can be fully utilized, a desaturation protection with faster response speed and unaffected noisy immunity can be achieved.

As shown in Fig. 3, the device drain-source voltage under HSF and normal condition is quite different. Specifically, the drain-source voltage decreases from DC bus voltage (V_{bus}) to on-state voltage (V_{ds_on}) in normal case, while it almost keeps at V_{bus} during HSF. In the conventional design, V_{ds} is only used by desaturation detection circuit after it drops to an enough low level for diode conduction, which missed some available information of V_{ds} variation.

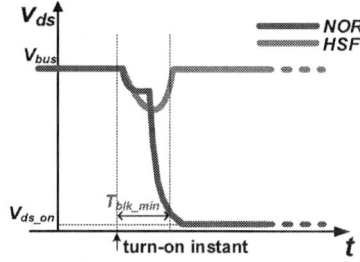

Fig. 3. Drain-source voltage under HSF and normal turn-on case.

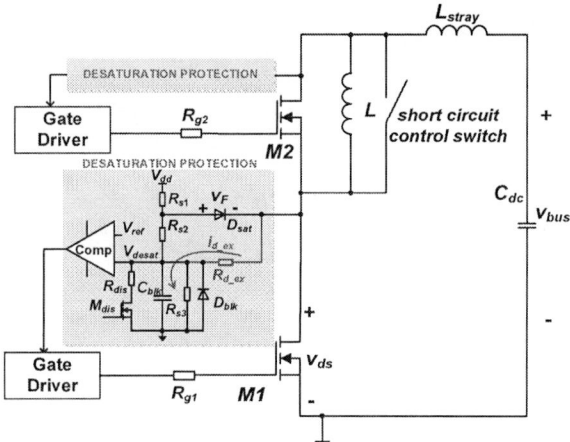

Fig. 4. The proposed desaturation protection circuit.

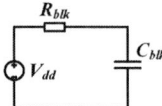

Fig. 5. Simplified circuit of traditional methods.

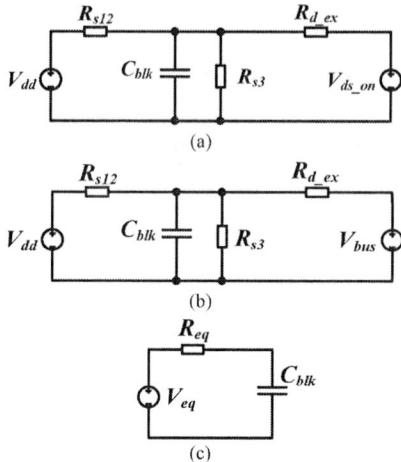

Fig. 6. Simplified circuits of the proposed method (a) Normal case. (b) HSF. (c) Simplified equivalent circuit.

To fully utilize the difference, in this paper an extra resistor R_{d_ex} is added between the drain and the blanking capacitor C_{blk}, as shown in Fig. 4. With this added resistor, extra current i_{d_ex} can be provided by V_{ds} and it can speed up the charge of C_{blk} during HSF, i.e. faster protection response.

To further analyze the impact of R_{d_ex} on protection performance improvement, the charging loops of the traditional method and the proposed method are compared.

For the traditional method, the loop can be simplified as a RC charging network for both normal turn-on and HSF, as illustrated in Fig. 5. The T_{blk} can be expressed as

$$
\begin{cases}
R_{blk} = \dfrac{R_{s12} \cdot R_{s3}}{R_{s12} + R_{s3}} \\[2ex]
V_{blk} = V_{dd} \cdot \dfrac{R_{s3}}{R_{s12} + R_{s3}} \\[2ex]
T_{blk} = R_{blk} \cdot C_{blk} \cdot \ln\left(\dfrac{V_{blk}}{V_{blk} - V_{ref}}\right)
\end{cases}
\tag{4}
$$

where R_{s12} is the sum of R_{s1} and R_{s2}.

For the proposed method, under normal turn-on case, V_{ds} would drop from V_{bus} to V_{ds_on} immediately, and thus the loop can be simplified approximately as Fig. 6(a). Similarly, under HSF condition, the loop can be simplified as Fig. 6(b). Both of the two cases can be further simplified as the equivalent circuit in Fig. 6(c). For normal case, R_{eq} and V_{eq} in Fig. 6(c) can be given as

$$
\begin{cases}
R_{eq} = \dfrac{1}{\dfrac{1}{R_{s12}} + \dfrac{1}{R_{s3}} + \dfrac{1}{R_{d\_ex}}} \\[3ex]
V_{eq\_nor} = \left(\dfrac{V_{dd}}{R_{s12}} + \dfrac{V_{ds\_on}}{R_{d\_ex}}\right) \cdot R_{eq}
\end{cases}
\tag{5}
$$

For HSF, R_{eq} is the same as (5), and V_{eq} can be expressed as

$$
V_{eq\_HSF} = \left(\frac{V_{dd}}{R_{s12}} + \frac{V_{bus}}{R_{d\_ex}}\right) \cdot R_{eq} .
\tag{6}
$$

According to Fig. 6(c), T_{blk} in both two cases can be calculated as

$$
T_{blk} = R_{eq} \cdot C_{blk} \cdot \ln\left(\frac{V_{eq}}{V_{eq} - V_{ref}}\right) .
\tag{7}
$$

Based on (5) to (7), by taking the derivative, the correlation of these parameters can be expressed as

$$
T_{blk} \propto R_{d\_ex} \propto \frac{1}{V_{eq}} \propto \frac{1}{V_{bus}} .
\tag{8}
$$

In order to accelerate the response without deteriorating the noise immunity of normal turn-on, (i.e. V_{eq_nor} barely increase), the selection of R_{d_ex} should be a compromise. A rough reference is shown in (10) and the value of R_{d_ex} is usually hundreds of kilo-ohms

$$
\frac{V_{dd}}{R_{s12}} = \frac{V_{bus}}{R_{d\_ex}} .
\tag{9}
$$

As V_{bus} is much larger than V_{ds_on} and V_{dd}, R_{d_ex} should be also much larger than R_{s12} according to (10). Thus, for normal case, V_{eq_nor} and R_{eq} is approximately equal to V_{blk} and R_{blk}, which means the T_{blk} is nearly unaffected.

In addition, resistors with high power rating and connected in series are preferable , because the voltage across R_{d_ex} is always equal to V_{bus} and its power dissipation is non-negligible.

For HSF, the acceleration of response time can be calculated by subtracting (7) from (4)

$$
\Delta T_{blk} = R_{blk} \cdot C_{blk} \cdot \ln\left(\frac{V_{blk}}{V_{eq}} \cdot \frac{V_{eq} - V_{ref}}{V_{blk} - V_{ref}}\right) .
\tag{10}
$$

978-1-6654-4817-8/21 $31.00 © 2021 IEEE   512

According to (8) and (10), it can be seen that T_{blk} is inversely proportional to V_{bus}. This is beneficial to achieve a faster protection speed under higher DC bus voltage condition, considering that a higher DC bus voltage causes higher short circuit energy and lower short circuit withstand time.

IV. SIMULATION AND EXPERIMENTAL RESULTS

A double pulse test (DPT) platform is built in Saber simulation software and experiment to verify the effectiveness of the improved desaturation protection circuit. Both the noise immunity and response time of the proposed circuit are compared with the conventional circuit.

A. Simulation Results

Firstly, the response time between the two methods is compared in simulation. In this simulation, the protection threshold is set to 2.5 V and the initial V_{bus} is 400 V.

As shown in Fig. 7, when HSF occurs, conventional circuit takes 700 ns to trigger the protection, while the proposed method takes around 300 ns under the same condition. In addition, the acceleration of proposed method is positively correlated with V_{bus}. When the V_{bus} increases from 200 V to 600 V, the response time reduces from 400 ns to 230 ns.

Fig. 8 shows the simulation comparison of the noise immunity of the two methods under normal turn-on condition. As can be seen, their V_{desat} waveforms almost overlap with each other, indicating unchanged noise immunity for the proposed method.

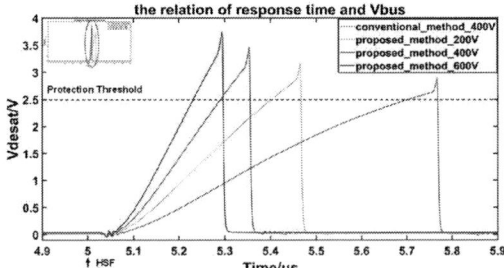

Fig. 7. The relation of response time and V_{bus} in the proposed method.

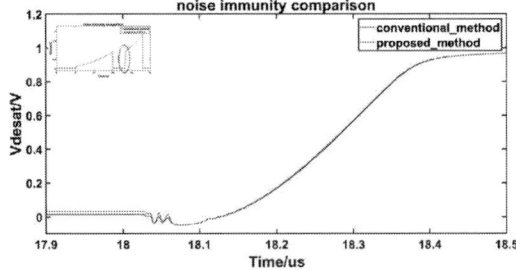

Fig. 8. Comparison of noise immunity.

B. Experimental Results

Fig. 9 shows the double pulse test setup for desaturation protection performance evaluation. In the test platform, the gate driver IC is UCC21710 and its output voltage is 16/-5 V. The device under test (DUT) is C3M0120100K, and its output characteristic is shown in Fig. 10. The parameters of desaturation protection circuit are shown in Table I. In this design, the protection threshold is about 39 A.

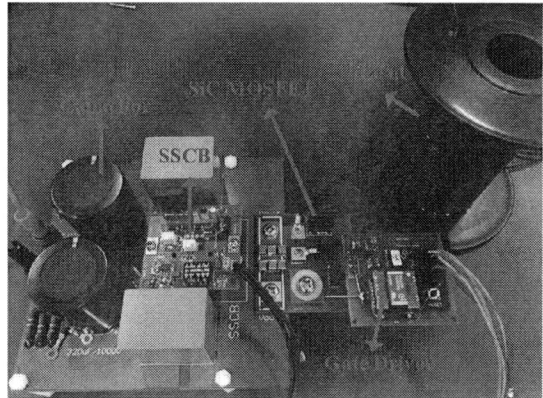

Fig. 9. Double pulse test setup for desaturation protection performance evaluation.

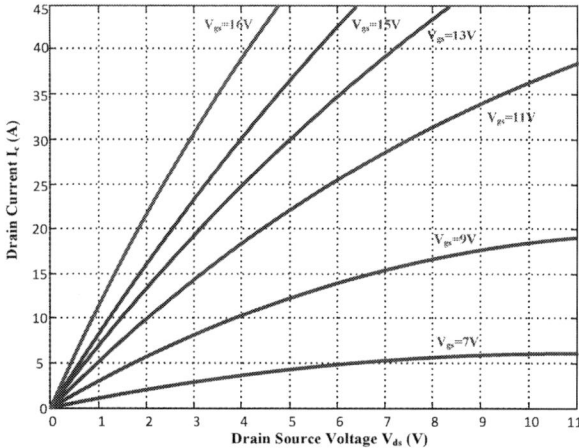

Fig. 10. Output characteristic of C3M0120100K ($T_a = 25°C$).

TABLE I. PARAMETERS OF DESATURATION PROTECTION CIRCUIT

Parameter	Value	Parameter	Value
R_{s1}	6 kΩ	V_{dd}	16 V
R_{s2}	6 kΩ	V_{ref}	0.7 V
R_{s3}	1 kΩ	V_F	1 V
R_d	450 kΩ	C_{blk}	500 pF

With these parameters, according to (4), the T_{blk} of conventional method is 388 ns, and the internal delay of the driver IC is 270 ns. Therefore, the total response time (T_{tot}) is 658 ns.

To verify the improved performance of the proposed method under high DC bus voltage , short circuit experiments are conducted under 300 V, 400 V and 500 V. According to (7), T_{tot} of these three conditions can be calculated as 490 ns, 463 ns and 441 ns respectively.

Fig. 11 shows the comparison of protection performance under different DC bus voltages. The response time of the proposed method (492 ns, 453 ns and 425 ns respectively) is shorter than the conventional method (649 ns), and the acceleration increases with DC bus voltage V_{bus}. In addition, ΔT_{tot} under different test conditions are consistent with the calculation basically, as shown in Table II.

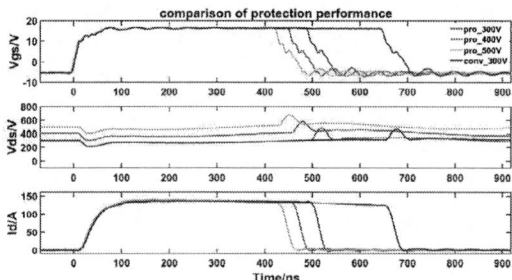

Fig. 11. Comparison of protection performance under different DC bus voltages.

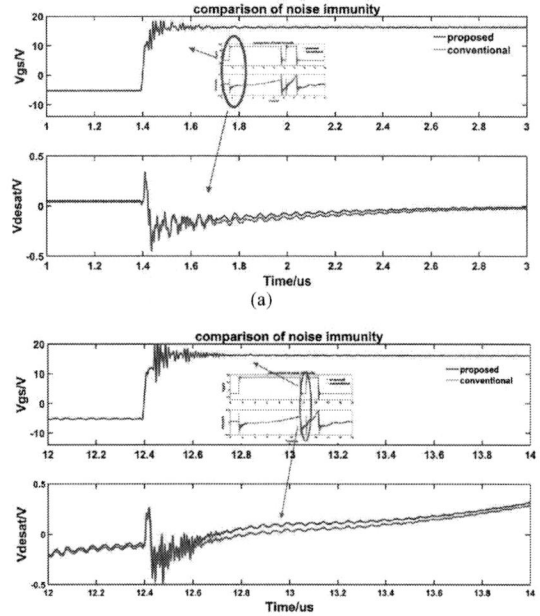

Fig. 12. Comparison of noise immunity of the conventional and proposed desaturation protection.(a) The first pulse. (b) The second pulse.

TABLE II. PREDICTIONS AND ACTUAL RESULTS OF ACCELERATION

Condition	Actual T_{tot}	Actual ΔT_{tot}	Prediction ΔT_{tot}
conv_300 V	649 ns		
pro_300 V	492 ns	157 ns	168 ns
pro_400 V	453 ns	196 ns	196 ns
pro_500 V	425 ns	224 ns	217 ns

As the inherent delay of gate driver IC is almost constant, the T_{tot} is decreased by about 25%, while T_{blk} is decreased by about 50% actually.

Fig. 12 shows the double pulse test results of the two protection circuits. Although both the two methods present severe ringing at the switching transient, V_{desat} in the proposed circuit is just slightly higher at the beginning of turn-on. The impact of the proposed circuit on noise immunity is negligible.

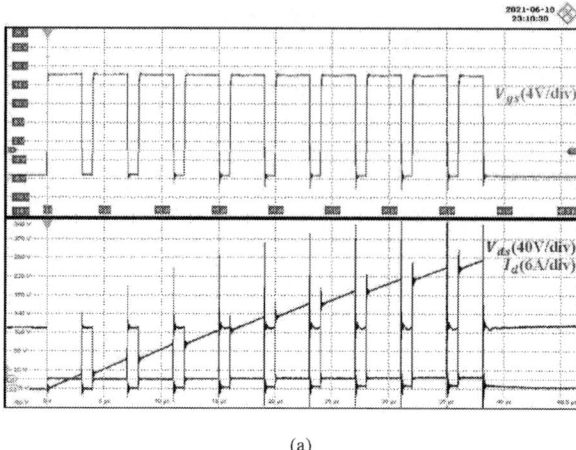

(a)

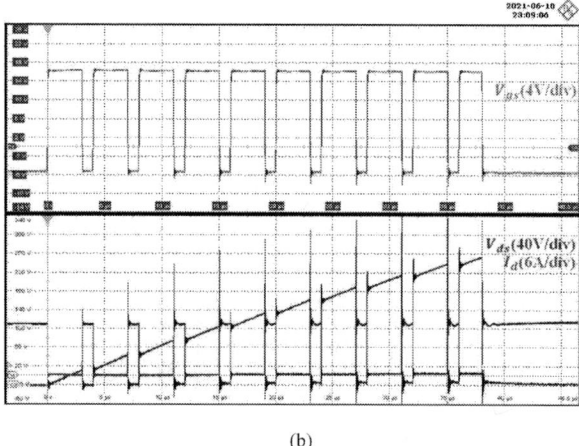

(b)

Fig. 13. The continuous pulse test (a) Conventional circuit. (b) Proposed circuit.

A continuous pulse test is also performed to further confirm the negligible change of the noise immunity. As shown in Fig. 13, both circuits trigger the protection at the same current level, which is about 41 A. Therefore, it validates that the proposed protection method would not be falsely triggered at a lower level current in normal case, i.e. the noise immunity is unaffected.

V. CONCLUSION

In this paper, an improved desaturation protection circuit has been proposed. It improves the response speed by offering an extra charging loop for the detection circuit and the improvement idea is based on the differences between fault and normal case. The theoretical analysis is given and both simulations and experiments validate its feasibility. The proposed method can reduce the overall response time and the acceleration has a positive relation with DC bus voltage, so that it is self-adaptive partly. Experimental results shows that it can shorten blanking time by 50% in this circuit without compromising noise immunity.

REFERENCES

[1] Q. Qian, J. Yu, Zhu J., W. Sun, and Y. Yi, "Isolated gate driver for SiC MOSFETs with constant negative off voltage," in *Proc. IEEE Applied Power Electronics Conference and Exposition*, 2017, pp.1990-1993.

[2] X. Wen, T. Fan, P. Ning and Q. Guo, "Technical approaches towards ultra-high power density SiC inverter in electric vehicle applications," *IEEE Trans. CES TEMS.*, vol. 1, no. 3, pp. 231-237, September 2017.

[3] A. Romero and R. Burgos, "Non-destructive and destructive shortcircuit characterization of a high-current SiC MOSFET," in *Proc IEEE Energy Conversion Congress and Exposition*, 2018, pp. 862-867.

[4] J. Sun, H. Xu, X. Wu and K. Sheng, "Comparison and analysis of short circuit capability of 1200V single-chip SiC MOSFET and Si IGBT," in *Proc. SSLChina: IFWS*, 2016, pp. 42-45.

[5] T. Horiguchi, S. Kinouchi, Y. Nakayama and H. Akagi, "A fast short-circuit protection method using gate charge characteristics of SiC MOSFETs," in *Proc. IEEE Energy Convers. Congr. Expo.*, 2015, pp. 4759-4764.

[6] X. Zhang, M. Chen, N. Zhu and D. Xu, "A self-adaptive blanking circuit for IGBT short-circuit protection based on VCE measurement," in *Proc. IEEE Energy Convers. Congr. Expo.*, 2015, pp. 4125-4131.

[7] T. Bertelshofer, A. Maerz and M. Bakran, "Design rules to adapt the desaturation detection for SiC MOSFET modules," in *Proc. Europe International Exhibition and Conference for Power Electronics, Intelligent Motion, Renewable Energy and Energy Management*, 2017, pp. 1-8.

[8] J. Kim and Y. Cho, "Overcurrent and short-circuit protection method using desaturation detection of SiC MOSFET," in *Proc. IEEE PELS Workshop on Emerging Technologies: Wireless Power Transfer*, 2020, pp. 197-200.

[9] T. Krone, C. Xu and A. Mertens, "Fast and easily implementable detection circuits for short circuits of power Semiconductors," *IEEE Trans. Ind. Appl.*, vol. 53, no. 3, pp. 2871-2879, May-June 2017.

Analysis of Dynamic Current Balancing in Multichip SiC Power Modules Based on Coupled Parasitic Network Model

Yuxin Ge, Yayong Yang, Cheng Qian, Zhiqiang (Jack) Wang, Yong Kang
School of Electrical and Electronic Engineering
Huazhong University of Science and Technology
Wuhan, China
zhiqiangwang@hust.edu.cn

Abstract—The parasitic effect induced by the interconnections in device package brings about the dynamic current mismatch among paralleled dies, thus limiting the available capacity of power modules. In this paper, a general analysis based on coupled parasitic network model is carried out to reveal the mechanism of layout-dominated dynamic current balancing in multichip silicon carbide (SiC) power modules. First, according to the interrelation of parasitic parameters in the power module, a coupled parasitic network model at switching transients is established for the analysis of dynamic current balancing. For the validation of the proposed model, the parasitic parameters of two different module layouts considering magnetic coupling are extracted from Q3D. Based on the acquired parasitic parameters, the state space frequency-dependent model is reorganized to perform the electromagnetic coupling simulation in SIMPLORER. Finally, the simulation results and switching waveforms of SiC MOSFETs are combined to verify the effectiveness of the dynamic current balancing equations derived from the proposed model.

Keywords—dynamic current balancing, multichip power modules, package layout, SiC MOSFET, coupled parasitic network model

I. INTRODUCTION

Due to defects in material characteristics, the current rating of SiC single die is inadequate to meet the needs of high-capacity applications. To increase the current rating of power modules, multichip parallel operation is in general regarded as a feasible and practical approach. However, the parasitic effect induced by the power module package, especially parasitic inductances, has a significant impact on dynamic current sharing among paralleled dies [1]. Unbalanced current sharing leads to uneven power losses and non-uniform junction temperature, and as a consequence, the reliability and lifetime of power modules are degraded.

In recent years, several research work effort on analysis of dynamic current sharing at device package level. In [2], a double-end sourced layout is proposed by incorporating an additional pair of DC- bus terminals to improve the dynamic current sharing. However, the analysis of the current balancing mechanism in the sight of an equivalent circuit is missing. In [3], a method to balance dynamic current by adjusting the connection points of bonding wires and copper traces is presented according to the response surface model and nonlinear optimization. In this method, the main power net and gate drive net are assumed to be fully decoupled to

This work was supported by the National Natural Science Foundation of China under Grant 51907070

simplify the analysis of magnetic coupling, resulting in compromised generality for various power modules. The effect of coupling inductances of commutation loop and gate loop is reported in [4], yet the self and mutual inductances of the individual gate loop are neglected for simplicity. To sum up, the magnetic coupling effect of power loop and drive loop in SiC power module has not been emphasized in existing research work, leading to limited accuracy and generality for dynamic current sharing analysis.

In this article, a coupled parasitic network model is proposed for the analysis of dynamic current balancing in multichip SiC power modules. The impact of self and mutual partial inductances among multiple sections involved in switching transients is taken into consideration. Thus, the proposed model is capable of relatively high accuracy. Moreover, dynamic coupling simulation is conducted to verify the effectiveness of the dynamic current balancing equations derived from the proposed model.

II. PROPOSED COUPLED PARASITIC NETWORK MODEL FOR DYNAMIC CURRENT BALANCING ANALYSIS

A. Studied Multichip SiC Power Modules: Configuration and DBC Layout

A multichip SiC power module with different direct bond copper (DBC) substrate layouts is studied in this work for dynamic current balancing analysis, as shown in Fig. 1. Each switch covers three SiC MOSFETs (CPM312000075A from CREE) with Kelvin source connection and three anti-parallel SiC Schottky barrier diodes (CPW41200S015B from CREE). The MOSFETs and Schottky barrier diodes (SBDs) of the high side switch are denoted as M_1-M_3 and D_1-D_3, respectively, while those of the low side is denoted as M_4-M_6 and D_4-D_6. The gate and source terminal for high and low side switches are marked as G_H, KS_H, G_L, KS_L, respectively. The DC bus and output terminal are shown as DC+, DC- (or DC1-, DC2-), AC, in the figure. For a fair comparison, the two module layouts have the same DBC substrate size. The detailed physical dimensions of the DBC substrate are listed in TABLE I.

B. Mechanism of Layout-dominated Dynamic Current Balancing Based on Coupled Parasitic Network Model

In a power module, the layout-dominated parasitic parameters mainly consist of partial inductance and equivalent series resistance (ESR). Specifically, the module partial inductance includes self partial inductances and mutual partial inductances [5]. The self partial inductance of a segment in a current loop is the ratio of the magnetic flux (between the

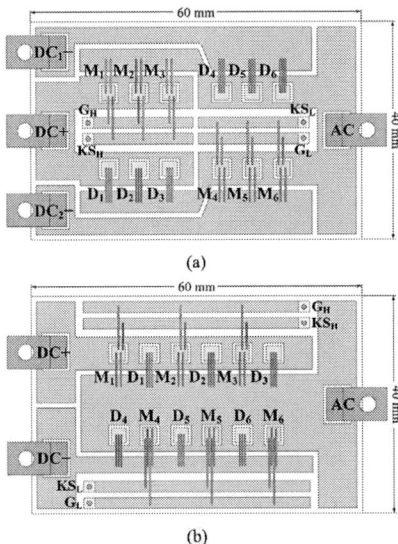

(a)

(b)

Fig. 1. Configuration and DBC layout of multichip phase-leg power modules. (a) Layout A. (b) Layout B.

TABLE I. PHYSICAL DIMENSIONS OF THE TWO MODULES

	Layout A	Layout B
DBC Size (mm)	60.0×40.0	60.0×40.0
DBC Thickness (mil)	12 (Cu) 25 (AlN)	12 (Cu) 25 (AlN)
SiC MOSFET Size (mm)	2.5×2.8	2.5×2.8
SiC SBD Size (mm)	2.7×2.7	2.7×2.7
Bonding Wires Diameter (mil)	10	10

current segment and infinity) to the current of that segment. The mutual partial inductance between two segments of the same or different current loops is the ratio of the magnetic flux (penetrating the surface between the second segment and infinity) to the current of the first segment. Intuitively, it is feasible to develop a lumped-circuit model of a closed current loop, where the segments of the loop perimeter are represented with self partial inductances and mutual partial inductances between every two segments.

Taking paralleled SiC MOSFETs of high side switch for example, a coupled parasitic network model at switching transients is proposed considering magnetic coupling among each section, as shown in Fig. 2. The proposed model consists of a power loop and gate drive loop magnetically coupled to each other. In these two loops, L_x and M_x refer to self partial inductance and mutual partial inductance of individual MOSFET. For example, L_{d1} is the self partial inductance between DC+ and the drain of MOSFET die M_1, and L_{s1} is the self partial inductance between AC and the source of the MOSFET die M_1. Similarly, L_{g1} is the self partial inductance between gate driver terminal G_H and the gate of M_1, and L_{ks1} is the self partial inductance between KS_H and the Kelvin source of M_1. M_{ds1}, M_{dg1}, M_{dk1}, M_{sg1}, M_{sk1} and M_{gk1} are the mutual partial inductances between the corresponding two segments of the loop.

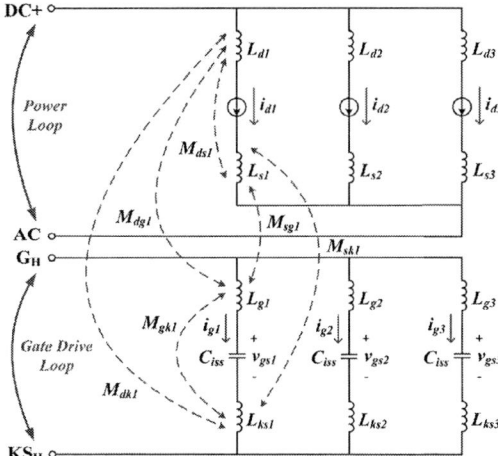

Fig. 2. Coupled parasitic network model for high side paralleled MOSFETs at switching transients.

While the mutual partial inductances between paralleled branches of different dies are not depicted in Fig. 2, an assumption is made that these inductances are equal due to the compact arrangement of the paralleled dies. As a consequence, the effect of these mutual partial inductances on the dynamic current balancing of each paralleled dies is approximately uniform and therefore is not taken into consideration. In addition, compared with partial inductances, the effect of ESR can be ignored during turn-on and turn-off periods for simplicity.

Dynamic current imbalance primarily occurs in the switching transients when MOSFETs operate in the saturation region [6]. Accordingly, the drain-source parasitic capacitance (C_{ds}) can be neglected [7]. The gate-source capacitance (C_{gs}) is dynamically in parallel with gate-drain capacitance (C_{gd}), which together forms the equivalent input capacitance. Thus, each SiC MOSFET can be modeled as a current source (i_d) controlled by its gate-source voltage (v_{gs}) in the power loop and input capacitance (C_{iss}) in the gate drive loop, as shown in Fig. 2.

The SiC MOSFET bare dies in the module are assumed to be identical so as to focus on the layout dominated current balancing issue. According to the Law of Electromagnetic Induction and the definition of inductance, $v_{d1}(t)$ and $v_{s1}(t)$ of M_1 can be expressed as

$$\begin{cases} v_{d1}(t) = L_{d1}\dfrac{di_{d1}(t)}{dt} + M_{ds1}\dfrac{di_{d1}(t)}{dt} + M_{dg1}\dfrac{di_{g1}(t)}{dt} + M_{dk1}\dfrac{di_{g1}(t)}{dt} \\ v_{s1}(t) = L_{s1}\dfrac{di_{d1}(t)}{dt} + M_{ds1}\dfrac{di_{d1}(t)}{dt} + M_{sg1}\dfrac{di_{g1}(t)}{dt} + M_{sk1}\dfrac{di_{g1}(t)}{dt} \end{cases}. \quad (1)$$

Likewise, concerning M_2 and M_3, $v_{d2}(t)$, $v_{s2}(t)$, $v_{d3}(t)$ and $v_{s3}(t)$ can be written in a similar form. Applying Kirchhoff Voltage Law (KVL) to every parallel branch of the power loop, there is

$$v_{d1}(t) + v_{s1}(t) = v_{d2}(t) + v_{s2}(t) = v_{d3}(t) + v_{s3}(t). \quad (2)$$

Based on (1) and (2), the following equation holds

978-1-6654-4817-8/21 $31.00 © 2021 IEEE

$$\left(L_{d1}+L_{s1}+2M_{ds1}\right)\frac{di_{d1}(t)}{dt}+\left(M_{dg1}+M_{dk1}+M_{sg1}+M_{sk1}\right)\frac{di_{g1}(t)}{dt}$$

$$=\left(L_{d2}+L_{s2}+2M_{ds2}\right)\frac{di_{d2}(t)}{dt}+\left(M_{dg2}+M_{dk2}+M_{sg2}+M_{sk2}\right)\frac{di_{g2}(t)}{dt}.$$

$$=\left(L_{d3}+L_{s3}+2M_{ds3}\right)\frac{di_{d3}(t)}{dt}+\left(M_{dg3}+M_{dk3}+M_{sg3}+M_{sk3}\right)\frac{di_{g3}(t)}{dt}$$

$$(3)$$

According to the characteristics of SiC MOSFETs in the saturation region, the correlation between drain current and gate-source voltage of paralleled dies is denoted as

$$\begin{cases} i_{d1}(t)=g\left[v_{gs1}(t)-V_{th}\right]^2 \\ i_{d2}(t)=g\left[v_{gs2}(t)-V_{th}\right]^2 \\ i_{d3}(t)=g\left[v_{gs3}(t)-V_{th}\right]^2 \end{cases} \qquad (4)$$

where g and V_{th} represent the transconductance and threshold voltage of paralleled SiC MOSFET dies respectively. Both parameters are related only to the device material and manufacturing process[8].

In addition, the gate current in the gate drive loop can be given as

$$\begin{cases} C_{iss}\dfrac{dv_{gs1}(t)}{dt}=i_{g1}(t) \\ C_{iss}\dfrac{dv_{gs2}(t)}{dt}=i_{g2}(t). \\ C_{iss}\dfrac{dv_{gs3}(t)}{dt}=i_{g3}(t) \end{cases} \qquad (5)$$

Based on (4) and (5), the correlation between gate current and drain current of paralleled dies can be deduced as

$$\begin{cases} i_{g1}(t)=\dfrac{C_{iss}}{2\sqrt{g}}i_{d1}(t)^{-\frac{1}{2}} \\ i_{g2}(t)=\dfrac{C_{iss}}{2\sqrt{g}}i_{d2}(t)^{-\frac{1}{2}}. \\ i_{g3}(t)=\dfrac{C_{iss}}{2\sqrt{g}}i_{d3}(t)^{-\frac{1}{2}} \end{cases} \qquad (6)$$

Substituting (6) into (3) yields

$$\left(L_{d1}+L_{s1}+2M_{ds1}\right)\frac{di_{d1}(t)}{dt}$$
$$+\left(M_{dg1}+M_{dk1}+M_{sg1}+M_{sk1}\right)\frac{d}{dt}\left[\frac{C_{iss}}{2\sqrt{g}}i_{d1}(t)^{-\frac{1}{2}}\right]$$
$$=\left(L_{d2}+L_{s2}+2M_{ds2}\right)\frac{di_{d2}(t)}{dt}$$
$$+\left(M_{dg2}+M_{dk2}+M_{sg2}+M_{sk2}\right)\frac{d}{dt}\left[\frac{C_{iss}}{2\sqrt{g}}i_{d2}(t)^{-\frac{1}{2}}\right]. \quad (7)$$
$$=\left(L_{d3}+L_{s3}+2M_{ds3}\right)\frac{di_{d3}(t)}{dt}$$
$$+\left(M_{dg3}+M_{dk3}+M_{sg3}+M_{sk3}\right)\frac{d}{dt}\left[\frac{C_{iss}}{2\sqrt{g}}i_{d3}(t)^{-\frac{1}{2}}\right]$$

Dynamic current balancing among paralleled SiC MOSFET dies occurs when

$$i_{d1}(t)=i_{d2}(t)=i_{d3}(t)=i_d(t). \qquad (8)$$

Combining (7) with (8) yields

$$\left(L_{d1}+L_{s1}+2M_{ds1}-L_{d2}-L_{s2}-2M_{ds2}\right)\frac{di_d(t)}{dt}$$
$$=\left(M_{dg2}+M_{dk2}+M_{sg2}+M_{sk2}-M_{dg1}-M_{dk1}-M_{sg1}-M_{sk1}\right)\frac{d}{dt}\left[\frac{C_{iss}}{2\sqrt{g}}i_d(t)^{-\frac{1}{2}}\right]$$

$$\left(L_{d2}+L_{s2}+2M_{ds2}-L_{d3}-L_{s3}-2M_{ds3}\right)\frac{di_d(t)}{dt}$$
$$=\left(M_{dg3}+M_{dk3}+M_{sg3}+M_{sk3}-M_{dg2}-M_{dk2}-M_{sg2}-M_{sk2}\right)\frac{d}{dt}\left[\frac{C_{iss}}{2\sqrt{g}}i_d(t)^{-\frac{1}{2}}\right]$$

$$(9)$$

According to (9), dynamic current balancing among paralleled dies can be achieved when the following two equations are met simultaneously, i.e.,

$$\begin{cases} L_{d1}+L_{s1}+2M_{ds1}=L_{d2}+L_{s2}+2M_{ds2} \\ L_{d2}+L_{s2}+2M_{ds2}=L_{d3}+L_{s3}+2M_{ds3} \end{cases} \qquad (10)$$

$$\begin{cases} M_{dg1}+M_{dk1}+M_{sg1}+M_{sk1}=M_{dg2}+M_{dk2}+M_{sg2}+M_{sk2} \\ M_{dg2}+M_{dk2}+M_{sg2}+M_{sk2}=M_{dg3}+M_{dk3}+M_{sg3}+M_{sk3} \end{cases}. \qquad (11)$$

By ignoring ESRs, which have negligible impact on dynamic current sharing, the above two equations can serve as current balancing optimization guidelines for designing the DBC layout of multichip power modules.

III. Parasitics Extraction and Electromagnetic Coupling Analysis for Studied Power Modules

The partial inductances involved in the coupled parasitic network model are extracted by ANSYS Q3D for the studied multichip SiC power modules. In accordance with the state space frequency-dependent model derived from parasitic parameters and SiC device behavior model, the electromagnetic coupling simulation is conducted in ANSYS SIMPLORER to evaluate the switching performance of paralleled SiC MOSFETs.

A. Extraction of Partial Inductance and Equivalent Series Resistance

According to the proposed coupled parasitic network model, multiple current conducting nets are assigned to divide the power loop and drive loop into several paths in Q3D. The studied power module with layout A is separated into eight nets, as depicted in Fig. 3. In the same manner, the power module with layout B can be divided into seven nets. The top surface of DC bus bars are set as Sink, while the solder layers at the bottom pads of dies and the ends of bonding wires connected with the top pads are set as Source.

In this Q3D electromagnetic field simulation, proximity and skin effects are taken into account in the solution. Consequently, the self partial inductance of each conduction path and the mutual partial inductance between every two paths can be calculated as a function of frequency. Additionally, the ESRs ignored in the coupled parasitic

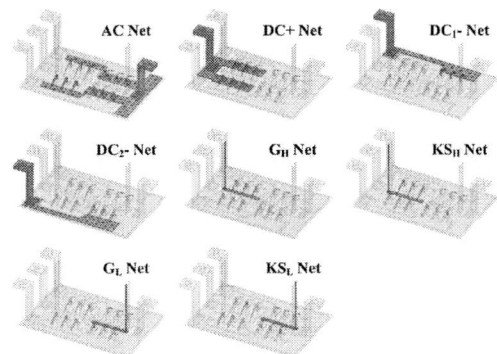

Fig. 3. Current conducting nets for layout A in ANSYS Q3D.

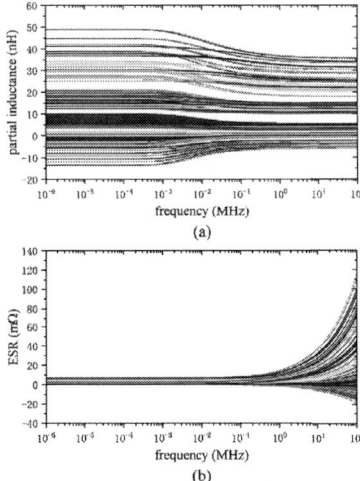

Fig. 4. Frequency dependent parasitic parameters of the power module with layout A. (a) Self and mutual partial inductances. (b) ESRs.

network model are still extracted for establishing more accurate switching current. In the studied case, the partial inductances and ESRs of the two modules are extracted through frequency sweep from 0 to 100 MHz. The frequency dependent parasitic parameters of the power module with layout A are illustrated in Fig. 4. Moreover, the reference current direction in Q3D is from Source to Sink, which is inconsistent with the actual direction of several current paths. Therefore, the positive and negative polarities of the extracted parasitic parameters are corrected in accordance with the actual current direction. The dominant partial inductances for layout A and layout B under 100 MHz are indicated in TABLE II and TABLE III, separately.

B. Electromagnetic Coupling Simulation Topology

As demonstrated in the proposed coupled parasitic network model, the effect of dynamic current sharing relies heavily on the frequency-dependent parasitic parameters in module package. In the frequency sweep mode, each frequency point links to a matrix consisting of the acquired parasitic parameters as elements. These matrixes can be reorganized into the state space frequency-dependent model as illustrated in Fig. 5. The model implements bidirectional dynamic data transmission with an application circuit through a multi-pin connection. Furthermore, the partial inductances and ESRs are fit as functions of the instantaneous switching

TABLE II. PARTIAL INDUCTANCE EXTRACTED FOR LAYOUT A

Layout A $f = 100$ MHz		High side switch			Low side switch		
		M_1	M_2	M_3	M_4	M_5	M_6
Partial inductance (nH)	L_d	13.20	13.51	13.79	14.02	13.72	13.40
	L_s	24.48	23.13	21.78	31.62	33.56	35.74
	M_{ds}	0.17	-0.01	-0.12	0.22	0.33	0.51
	M_{dg}	4.68	4.88	5.08	5.38	5.18	4.97
	M_{dk}	-4.65	-4.74	-4.86	-5.16	-5.04	-4.95
	M_{sg}	-0.98	-0.87	-0.82	-0.57	-0.68	-0.84
	M_{sk}	0.90	0.83	0.81	0.48	0.59	0.72

TABLE III. PARTIAL INDUCTANCE EXTRACTED FOR LAYOUT B

Layout B $f = 100$ MHz		High side switch			Low side switch		
		M_1	M_2	M_3	M_4	M_5	M_6
Partial inductance (nH)	L_d	13.20	13.86	14.47	14.30	13.70	13.07
	L_s	19.98	18.95	17.92	21.67	26.91	32.19
	M_{ds}	-0.23	-0.31	-0.30	0.00	0.12	0.38
	M_{dg}	0.12	0.46	0.68	0.55	0.28	-0.07
	M_{dk}	-0.28	-0.57	-0.74	-0.61	-0.38	-0.05
	M_{sg}	-3.43	-3.16	-2.85	-6.66	-8.94	-11.44
	M_{sk}	3.61	3.37	3.09	7.54	10.47	13.57

TABLE IV. CALCULATED VALUES OF THE TWO EQUATIONS BASED ON ACQUIRED PARASITIC PARAMETERS

$f = 100$ MHz		Inductance of high side switch (nH)			Inductance of low side switch (nH)		
		M_1	M_2	M_3	M_4	M_5	M_6
Layout A	(10)	38.02	36.62	35.33	46.08	47.94	50.16
	(11)	-0.05	0.10	0.21	0.13	0.05	-0.10
Layout B	(10)	32.72	32.19	31.79	35.97	40.85	46.02
	(11)	0.02	0.10	0.18	0.82	1.43	2.01

frequency. In the above model, each pin corresponds to a measurement point designated in Q3D, i.e., Source and Sink. Therefore, as a sub-circuit of the double pulse test platform, the state space frequency-dependent model can be utilized to evaluate dynamic current sharing.

The electromagnetic coupling simulation is performed using SIMPLORER, as shown in Fig. 6. In the state space frequency-dependent model, pins d1-d6 refer to the solder layers at the bottom drain pads of MOSFET dies. Pins g1-g6, ks1-ks6 and s1-s6 represent the ends of bonding wires connected with the top gate, Kelvin source and source pads of MOSFET dies. Additionally, pins a1-a6 describe the ends of bonding wires connected with the top anode pads of SBD dies. Pins c1-c6 denote the solder layers at the bottom cathode pads of SBD dies. When high side MOSFETs of layout A are as the active switch, the high side gate driver connected to the Gate$_H$ Bus pin and Kelvin Source$_H$ Bus pin is enabled, whereas the low side gate driver linked to the Gate$_L$ Bus pin and Kelvin Source$_L$ Bus pin is shorted. DC+ Bus, DC$_1$- Bus and DC$_2$- Bus are pins connected to the DC bus voltage source. DC$_1$- Bus and DC$_2$- Bus represent the paralleled double-end negative bus bars of the module with layout A.

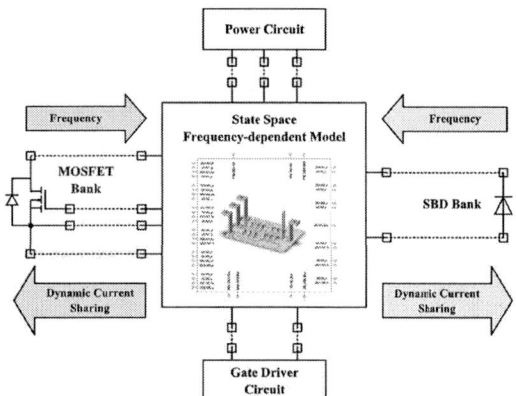

Fig. 5. Schematic view of the state space frequency-dependent model derived from the acquired parasitic parameters.

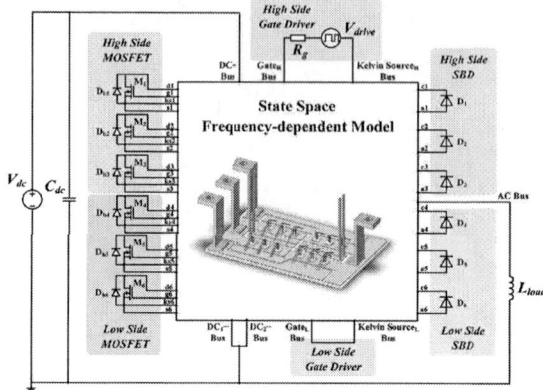

Fig. 6. Electromagnetic coupling simulation topology for high side MOSFETs of layout A as active switch.

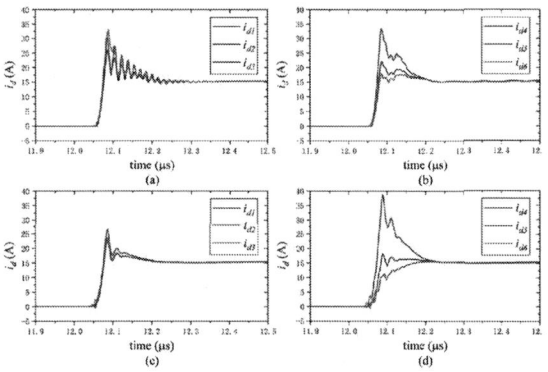

Fig. 7. Comparison of switching current at turn-on transient. (a) High side MOSFETs of layout A. (b) Low side MOSFETs of layout A. (c) High side MOSFETs of layout B. (d) Low side MOSFETs of layout B.

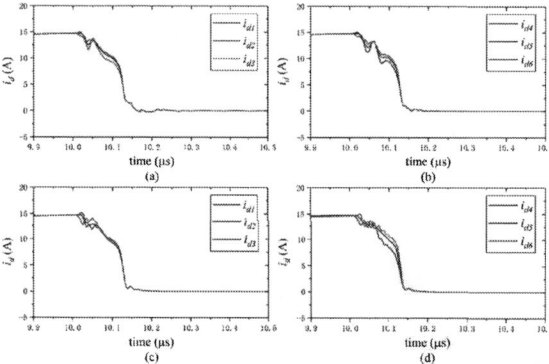

Fig. 8. Comparison of switching current at turn-off transient. (a) High side MOSFETs of layout A. (b) Low side MOSFETs of layout A. (c) High side MOSFETs of layout B. (d) Low side MOSFETs of layout B.

IV. SIMULATION VERIFICATIONS

As illustrated in the above analysis based on the coupled parasitic network model, dynamic current sharing among paralleled dies depends on whether (10) and (11) are met simultaneously in a power module. The effectiveness of this can be verified by the switching transient waveforms and the calculated value of the two equations based on parasitics extraction from Q3D.

In TABLE IV, the calculation results of (10) and (11) for the studied power module with different layouts are filled according to the partial inductances from TABLE II and TABLE III, separately. In TABLE IV, four sets of results representing the inductance of high or low side switches for layout A or B are presented. The four sets of results actually demonstrate different cases of dynamic current balancing. Among them, the high side switch of layout B presents the least difference among the three paralleled MOSFETs, indicating potentially the best dynamic current sharing performance. On the other hand, the low side switch of layout B presents the most significant discrepancy among the three paralleled MOSFETs in (10) and (11), indicating potentially the worst performance.

A double pulse test circuit is established to evaluate the switching characteristics of paralleled MOSFETs. The double pulse test circuit mainly includes the DC bus voltage source,

energy storage capacitors, decoupling capacitors, load inductor, gate driver, devices under test and the developed state space frequency-dependent model, as shown in Fig. 5. The gate driver is modeled as a programable voltage source in series with an external gate resistor. Depending on the active switch is located on the high or low side, one end of the load inductor is connected to the negative or positive pole of the DC bus voltage source. Under the condition of 600-V bus voltage and 135.6-μH load inductance, a comparison of the simulated switching waveforms at turn-on and turn-off transients are exhibited in Fig. 7 and Fig. 8, respectively. Based on the comparison, the high side switch of layout B has the best dynamic current sharing performance, which is consistent with the calculation results in TABLE IV. In contrast, the low side switch of layout B presents the worst performance, as indicated in TABLE IV.

V. CONCLUSION

This article proposes a coupled parasitic network model to analyze the mechanism of dynamic current balancing in multichip SiC power modules with Kelvin source connection. Especially, the impact of self and mutual partial inductances among each section involved in switching transients on dynamic current sharing is taken into consideration. According to the proposed model, the conditions for dynamic current balancing are identified and summarized. Furthermore, simulation results from SiC MOSFET power modules with different layouts are compared to verify the effectiveness of the proposed model and analysis. The results suggest that the dynamic current balancing equations derived from the proposed model can potentially serve as optimization guidelines for power module design.

REFERENCES

[1] H. Li, S. Munk-Nielsen, C. Pham, and S. Bęczkowski, "Circuit mismatch influence on performance of paralleling silicon carbide MOSFETs," in *Proc. European Conference on Power Electronics and Applications*, 2014, pp. 1-8.

[2] M. Wang, F. Luo, and L. Xu, "A double-end sourced wire-bonded multichip SiC MOSFET power module with improved dynamic current sharing," *IEEE Journal of Emerging and Selected Topics in Power Electronics*, vol. 5, no. 4, pp. 1828-1836, Dec. 2017.

[3] C. Zhao, L. Wang, F. Zhang, and F. Yang, "A method to balance dynamic current of paralleled SiC MOSFETs with Kelvin connection based on response surface model and nonlinear optimization," *IEEE Transactions on Power Electronics*, vol. 36, no. 2, pp. 2068-2079, Feb. 2021.

[4] M. Spang, "Experiences with coupling inductances between commutation loop and gate circuit in power modules," in *Proc. European Conference on Power Electronics and Applications*, 2018, pp. 1-10.

[5] Clayton R. Paul, *Inductance: Loop and Partial*, HOB: John Wiley & Sons, Inc., 2010, pp. 195-205.

[6] Z. Chen, Y. Yao, D. Boroyevich, K. D. T. Ngo, P. Mattavelli, and K. Rajashekara, "A 1200-V, 60-A SiC MOSFET multichip phase-leg module for high-temperature, high-frequency applications," *IEEE Transactions on Power Electronics*, vol. 29, no. 5, pp. 2307-2320, May 2014.

[7] Z. Zeng, X. Zhang, and X. Li, "Layout-dominated dynamic current imbalance in multichip power module: Mechanism modeling and comparative evaluation," *IEEE Transactions on Power Electronics*, vol. 34, no. 11, pp. 11199-11214, Nov. 2019.

[8] Z. Zeng, X. Zhang, and Z. Zhang, "Imbalance current analysis and its suppression methodology for parallel SiC MOSFETs with aid of a differential mode choke," *IEEE Transactions on Industrial Electronics*, vol. 67, no. 2, pp. 1508-1519, Feb. 2020.

A Compact 175℃ High Temperature Gate Driver with Isolated Power Supply and Advanced Protection for HybridPACK Drive SiC Module

Cheng Qian, Neng Wang, Yayong Yang, Zhiqiang (Jack) Wang, Yong Kang
State Key Laboratory of Advanced Electromagnetic Engineering and Technology
Huazhong University of Science and Technology
Wuhan, China
zhiqiangwang@hust.edu.cn

Abstract—The operation of silicon carbide (SiC) power semiconductor devices under high ambient temperature requires high temperature gate drivers adjacent to the devices. This paper presents a three-phase isolated gate driver circuit suitable for HybridPACK Drive (HP Drive) power modules with operating ambient temperature up to 175°C. First, an overall introduction of the gate driver structure and layout design method for reducing thermal coupling are presented. Second, detailed designs are presented in the form of sub-circuits such as desaturation protection circuit, isolated power supply, active Miller clamp, undervoltage lockout and so on. Finally, the proposed gate driver design is validated through double pulse test, continuous full-load operation under 145°C and continuous light-load test under 175°C. Comprehensive performance evaluation of proposed high temperature gate driver is presented.

Keywords—gate driver, high temperature, desaturation.

I. INTRODUCTION

In many industrial applications, harsh environment such as high ambient temperature is a major challenge for power converters [1]. Traditional silicon (Si) devices are limited by material characteristics to operate in high temperature (HT) environment. Leakage current and on-state resistance of Si devices increase significantly with increasing junction temperature [2], which will reduce converter performance significantly. Thanks to the excellent high temperature characteristics of silicon carbide (SiC) power devices, the operation of power converters under high ambient temperature becomes possible [3]. One key challenge for high temperature operation of HT converters is the compact gate driver as close to power devices as possible for high switching performance while capable of withstanding high ambient temperature. Traditional commercial gate driver circuits generally with maximum operation temperature below 125°C can't meet this demand. In addition, the power derating with increase of ambient temperature further prevents traditional gate driver from high ambient temperature applications.

Existing HT gate driver solutions are mainly based on silicon-on-insulator (SOI) process or HT discrete devices. In [4], a gate driver circuit based on SOI process is presented with 200°C maximum operation temperature. However, the cost of HT SOI gate driver is much higher than traditional Si gate driver. In [5], a gate driver circuit with external isolated power supply based on discrete devices is presented and tested at 180°C. Solutions using discrete devices are cheaper and more flexible, while the size, volume, and reliability are

This work was supported by the National Natural Science Foundation of China under Grant 51907070

comprised due to large number of devices, especially for gate drivers with diverse advanced functions like desaturation protection, active Miller clamp, etc. The size and volume of HT gate drivers are critical for compact SiC power modules, e.g. HybiridPACK Drive (HP Drive) power module. Currently, compact HT gate driver, especially with integrated HT isolated power supply, suitable for HP Drive module is still lacking. In this paper, the design and development of a compact 175°C three-phase isolated gate driver with on-board integrated isolated power supply is presented for HP Drive SiC power module. Design consideration of desaturation protection, cross-talk suppression and isolated power supply are discussed in this paper.

In Section II, the proposed gate driver is presented and overall circuit structure is introduced. A layout design method considering thermal coupling and parasitic inductance simultaneously is presented. In Section III, the whole circuit is divided into several sub-circuits and detailed design consideration of each sub-circuit is presented. In Section IV, proposed HT gate driver is validated through double pulse test under 145°C, continuous full-load operation under 145°C and continuous light-load operation under 175°C. Falsely turn-on phenomenon of transistors in active Miller clamp circuits is reported and analysed. No fault is observed during experiments.

II. OVERALL INTRODUCTION OF PROPOSED HIGH TEMPERATURE GATE DRIVER

A. Overall Structure of Proposed High Temperature Gate Driver

Proposed HT gate driver is designed for high temperature SiC HP Drive power module, based on SOI integrated circuits (ICs) from CISSOID. The total structure of one phase leg of proposed gate driver is shown in Fig. 1. It provides six PWM inputs, six fault signal outputs and three temperature monitoring interfaces to communicate with microcontroller. The whole gate driver incorporates three identical phase-leg drivers, and each phase-leg gate driver is mainly comprised of two isolated single-channel gate drivers, one isolated dual-channel DC/DC converter and one control section. Isolated single-channel gate driver is based on CMT-TIT0521B and it is controlled by control section. Peripheral circuits are designed to offer desaturation protection, active Miller clamp and other advanced protections. The isolated dual-channel DC/DC converter will converter 15V power supply to +15/-4V and its control is integrated in control section. The control section is based on CMT-TIT9687B and mainly used for modulation/demodulation of PWM/fault signals and voltage regulation of isolated DC/DC converters. The PWM signals are modulated with on-off-key (OOK) modulation at primary side and then transmitted to secondary side using pulse transformer. Subsequently, the modulated PWM signals are demodulated at secondary side. The fault signals of gate driver chips are transmitted in opposite direction, i.e. from secondary side to primary side. Fault will be reported to microcontroller if any of desaturation protection, over-temperature protection or undervoltage lockout protection is triggered.

B. Layout Design Method for Reducing Thermal Coupling

Different from applications under normal temperature, the components in proposed gate driver are close to upper temperature limit, especially when choosing lower temperature devices in order to reduce costs. The nominal maximum temperature of elements in their datasheet is the ambient temperature. However, there will be losses at the components, which will heat the components nearby. The thermal coupling among devices may overheat devices and put them at risk. Increasing the distance between devices can effectively reduce thermal coupling, but will increase parasitic inductance for exchange. For example, the losses mainly locate at gate resistors and the self-heating of gate resistors are most obvious in gate driver. Put all gate resistors close to each other will lead to small gate loop inductance but higher temperature. A balance between parasitic parameters and thermal coupling. A layout design method is proposed below to give a layout guidance to proposed HT gate driver.

Components are grouped according to their sensitivity to parasitic inductance. Some components are preferred to be placed as close as possible to achieve best performance and they can be regarded as a group in layout design. The thermal influence radius on PCB under different heating power is simulated in COMSOL. Based on simulation results, thermal influence radius of each group can be obtained and layout considering thermal coupling can be achieved accordingly by choosing appropriate distance between different groups. Total design method for reducing thermal coupling is presented in Fig. 2.

III. DETAILED INTRODUCTION OF EACH SUB-CICUIT OF PROPOSED HIGH TEMPERATURE GATE DRIVER

A. Isolated Single-channel Gate Driver

1) Output Stage

In order to achieve better switching performance, NMOS and bootstrap circuit are used in output stage, which integrated in gate driver chips. The output stage will be disabled if any fault occurs, and fault will be reported to control section. After a programmable period, the gate driver will try to restart automatically if not receive any command from control section. If the fault still exists, the output stage will be turned off again. The output stage is powered by a +15/-4V isolated power supply and controlled by control section.

2) Desaturation Protection

Different from traditional desaturation protection solutions [6], the threshold judgment and timing setting are separated as shown in Fig. 3. Resistors combination is used to set protection threshold and capacitor is used to adjust blanking time. V_{ref} is reference voltage and in this circuit is 2.5V. V_F is forward voltage of the diode in Fig. 3. Internal current source is used to discharge C_{blk} to set blanking time. The protection threshold voltage is shown in (1). Once the protection is triggered, the soft shut-down with 35Ω gate resistor will turn off the devices safely. Due to the wide

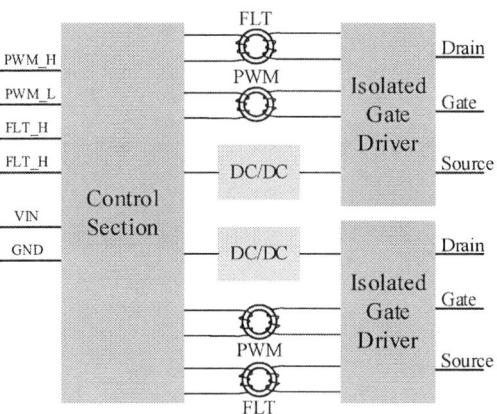

Fig. 1. Overall structure of proposed high temperature gate driver.

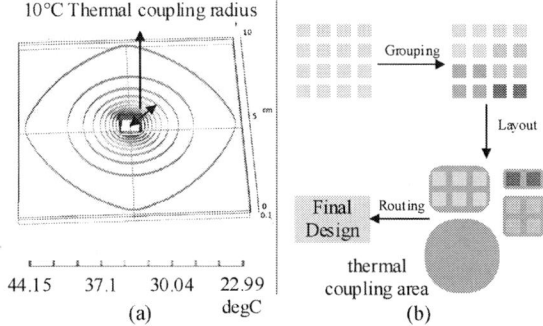

Fig. 2. Layout design method for reducing thermal coupling (a) simulation of thermal influence radius in COMSOL (b) layout design process based on simulation results.

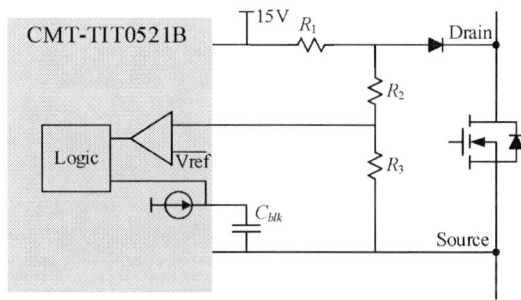

Fig. 3. Desaturation protection circuit in proposed gate driver. An integrated current source and capacitor is used to set blanking time and three resistors are used to set protection threshold.

operation temperature range of proposed gate driver, the change of V_F should be taken into consideration. V_F will decrease when temperature goes up. If the desaturation protection threshold voltage V_{dsth} is set under room temperature, the real threshold voltage at high temperature will be higher. This phenomenon will become more obvious when connecting several diodes in series.

$$V_{dsth} = V_{ref} \cdot \frac{R_2 + R_3}{R_3} - V_F \qquad (1)$$

3) Active Miller Clamp

Active Miller clamp (AMC) circuit will provide a low impedance gate path during switching transient to suppress crosstalk [7]. Typical AMC circuits used in commercial gate

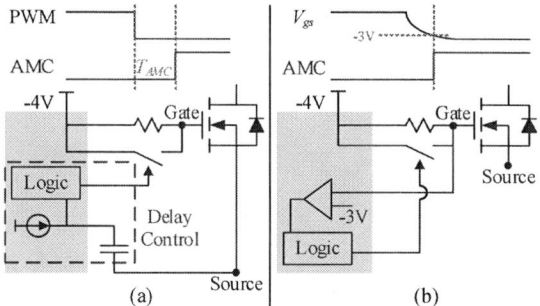

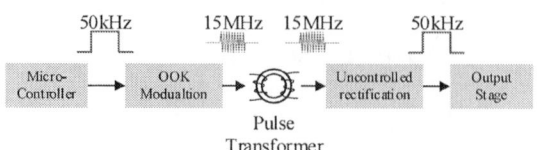

Fig. 8. Schematic of OOK modulation and demodulation. A toroidal core with 6mm outer diameter and 2mm height is selected for pulse transformers.

Fig. 7. Typical active Miller clamp circuit used in commercial gate driver chips (a) CMT-TIT0521B from CISSOID (b) UCC21710 from TI.

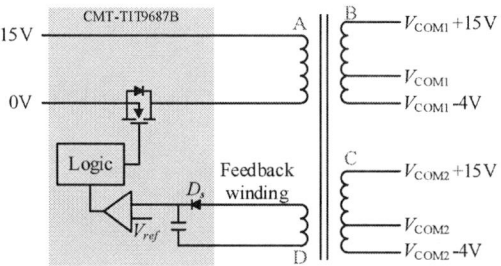

Fig. 6. Structure of isolated dual-channel power supply. Windings A, B and C are used for energy transmission, and winding D is used for voltage regulation.

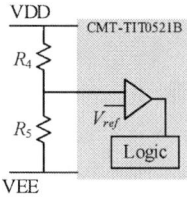

Fig. 5. Schematic of undervoltage lockout circuit.

driver chips [8]-[9] are shown in Fig. 4. AMC circuit will be triggered after PWM signal turns LOW for a programmable duration in (a), while after gate-source voltage falls below a reference voltage in (b). Plan (a) is not sensitive to parasitic ringing but can't adapt to switching speed automatically, while plan (b) is adaptive but sensitive to parasitic oscillation.

4) Undervoltage Lockout

When power supply of gate driver fails and the turn-on voltage is lower than normal value, under voltage lockout (UVLO) will be triggered to avoid abnormal operation. Low turn-on voltage will increase on-state resistance significantly and put devices at risk. External resistors are used for UVLO detection and the whole schematic is shown in Fig. 5. When V_{UVLO} is lower than reference voltage, the gate driver will be disabled. Both primary side and secondary side have UVLO protection.

B. Isolated Dual-chnnel Power Supply

Three isolated dual-channel flyback DC/DC converters are integrated in proposed HT gate driver to supply power to isolated single-channel gate driver. The schematic of this DC/DC converter is shown in Fig. 6. Winding A is connected

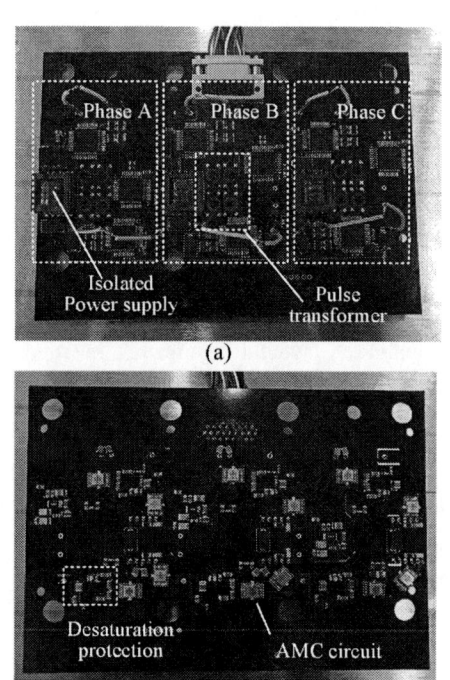

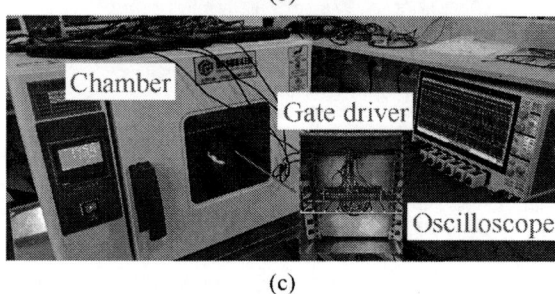

Fig. 4. Prototype of proposed HT gate driver. (a) Front side (b) Back side (c) Experimental platform.

to primary side power supply. Central tapped windings B and C are connected to secondary side and central taps are connected to source terminals of devices. Winding D is a feedback winding to regulate output voltage. The final output voltage of each isolated DC/DC converter is +15/-4V.

C. Control Section

The control section at primary side is responsible for DC/DC converter voltage regulation, modulation or demodulation of PWM/fault signals and fault logic control.

1) DC/DC Converter Voltage Regulation

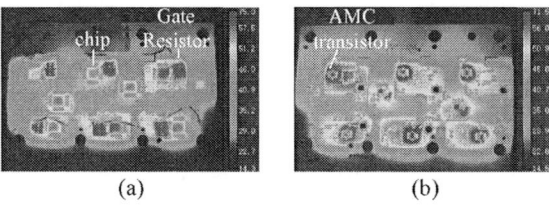

(a)   (b)

Fig. 10. Function verification of prototype at 15°C. (a) front side (b) back side.

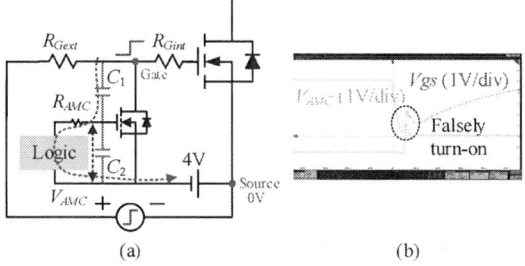

(a)   (b)

Fig. 9. Falsely turn on of AMC transistors. (a) Equivalent circuit (b) Experimental waveform.

The active switch and control circuit of DC/DC converter are integrated in this section to regulate voltage as shown in Fig. 6. Due to the wide operating temperature range of proposed HT gate driver, the output of isolated gate driver can't keep constant. The variation of output voltage is shown in (2)-(3), and should be taken into consideration to avoid too high or too low output voltage in the whole temperature range. V_{ref} is reference voltage of voltage regulation, V_s is the forward voltage drop of D_s, N_2 is the turns of winding B and N_4 is the turns of winding D.

$$V_{out} = (V_{ref} + V_s)\frac{N_2}{N_4} \qquad (2)$$

$$\frac{\partial V_{out}}{\partial T} = \frac{N_2}{N_4} \cdot \left(\frac{\partial V_{ref}}{\partial T} + \frac{\partial V_s}{\partial T} \right) \qquad (3)$$

2) Modulation/demodulation of PWM/fault Signals

HT oscillator is integrated in primary chip to modulate PWM signals, and then the modulated signals are transmitted to secondary side through pulse transformer. The carrier frequency is 15 MHz typically and thus required pulse transformer is compact. The modulated PWM signals are demodulated at secondary side through uncontrolled rectification and then used to control output stage. Fault signals are generated at secondary side and transmitted to primary side, just in an opposite direction of PWM signals. The total schematic of this part is shown in Fig. 7.

3) Fault Logic Control

For secondary side, if any of desaturation protection, UVLO protection or overtemperature protection is triggered, the output stage will be disabled for a duration and fault signal will be reported to control section. For primary side, if UVLO protection is triggered, the control section will be disabled and send turn-off signals to secondary side. All chips will try to restart after a programmable duration if microcontroller does not react to fault signals. Therefore, the higher-level fault lockout function needs to be implemented in the microcontroller.

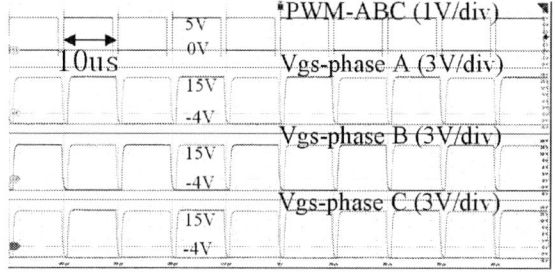

Fig. 11. Experimental waveform of continuous operation at 1W output under 145°C for 12 hours.

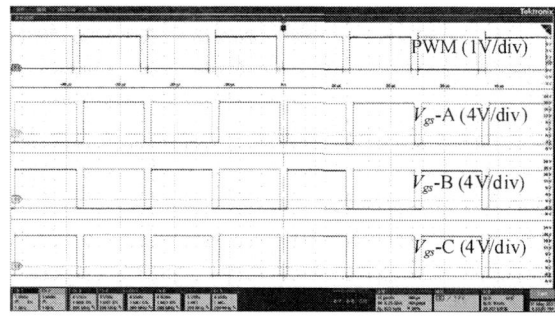

Fig. 12. Experimental waveforms of continuous operation at light-load under 175°C for 1 hour.

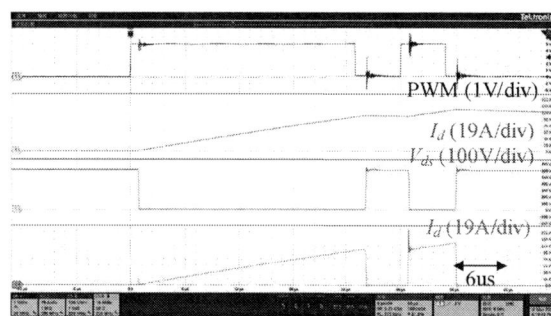

Fig. 13. Experimental waveforms of double pulse test under 145°C.

IV. EXPERIMENTS

A comprehensive evaluation of proposed gate driver is shown in TABLE. I.

TABLE. I COMPREHENSIVE EVALUATION OF PROPOSED HIGH TEMPERATURE GATE DRIVER

Total area	16.1 in²
Output voltage	+15/-4V
Maximum bus voltage	800V
Maximum output per channel	1W
Maximum operating temperature	175°C
Desaturation protection	Yes
Active Miller clamp	Yes
Undervoltage lockout	Yes

| Overtemperature protection | Yes |
| Integrated isolated power supply | Yes |

The proposed HT gate driver prototype and experimental platform are shown in Fig. 8. A 50 nF capacitor load and 50 kHz switching frequency is selected to achieve 1 W output. This prototype is tested at 15 °C first for function verification in Fig. 9. The temperature rise of gate driver ICs and gate resistors are 40 °C and 60 °C respectively, as power dissipation mainly locates at them. However, the active Miller clamp transistors have shown a temperature rise of 56 °C (nearly zero loss theoretically). A further investigation in Fig. 10 indicates that these transistors are partially turned-on due to crosstalk effect during turn on transient of gate driver. This partial turn-on of these transistors leads to considerable power loss and consequently temperature rise. Reducing R_1 or increasing C_1 will suppress this phenomenon. R_1 is integrated in driver chip and can't be changed. External capacitor could be added to change C_1, which will increase T_{AMC}. Furthermore, the gate driver has operated continuously at 1 W output power per channel for 12 hours in chamber of 145 °C and experimental waveform is shown in Fig. 11. A further light-load test in chamber of 175 °C is shown in Fig. 12. Double pulse test waveforms under 145°C are shown in Fig. 13. No fault is observed during operation.

V. CONCLUSION

A compact 175 °C high temperature gate driver with integrated isolated power supply and advanced protection is proposed in this article. A layout design method to reduce thermal coupling is presented in this paper. Detailed design considerations about dual-channel isolated power supply, desaturation protection, active Miller clamp, control section and other sub-circuits are presented. Proposed HT gate driver is validated through double pulse test and full-load test at 145 °C, and light-load test under 175 °C.

REFERENCES

[1] P. Ning et al., "High-temperature hardware: development of a 10-kW high-temperature, high-power-density three-phase ac-dc-ac SiC converter," in *IEEE Industrial Electronics Magazine*, vol. 7, no. 1, pp. 6-17, March 2013.

[2] Z. Chen, "Electrical integration of SiC power devices fo high-power-density applications," Ph. D dissertation, Virginia Polytechnic Institute and State University, Blacksburg, Virginia, USA, 2013.

[3] J. Millán, P. Godignon, X. Perpiñà, A. Pérez-Tomás and J. Rebollo, "A survey of wide bandgap power semiconductor devices," *IEEE Transactions on Power Electronics*, vol. 29, no. 5, pp. 2155-2163, May 2014.

[4] M. A. Huque, S. K. Islam, L. M. Tolbert and B. J. Blalock, "A 200 °C universal gate driver integrated circuit for extreme environment applications," *IEEE Trans. Power Electron.*, vol. 27, no. 9, pp. 4153-4162, September 2012.

[5] F. Qi and L. Xu, "Development of a high-temperature gate drive and protection circuit using discrete components," *IEEE Trans. Power Electron.*, vol. 32, no. 4, pp. 2957-2963, April 2017.

[6] Z. Wang, X. Shi, Y. Xue, L. M. Tolbert, F. Wang and B. J. Blalock, "Design and performance evaluation of overcurrent protection schemes for silicon carbide (SiC) power MOSFETs," *IEEE Trans. Ind. Electron.*, vol. 61, no. 10, pp. 5570-5581, Oct. 2014.

[7] E. Aeloiza, A. Kadavelugu and R. Rodrigues, "Novel bipolar active Miller clamp for parallel SiC MOSFET power modules," in *2018 IEEE Energy Conversion Congress and Exposition (ECCE)*, 2018, pp. 401-407.

[8] Texas Instruments, UCC21710 datasheet. 2020. [Online]. Available: https://www.ti.com/product/UCC21710.

[9] CISSOID, CMT-TIT0521B datasheet. 2020. [Online]. Available: https://www.cissoid.com/high-temperature-electronics/gate-drivers/titan-cmt/cmt-hades2s-high-temperature-175c-hadesv2-gate-driver-secondary-side-ic/.

978-1-6654-4817-8/21 $31.00 © 2021 IEEE

An Optimzed Parameter Design Method for Desaturation Protection Circuits towards Fast Response and Strong Noise Immunity

Cheng Qian, Zhiqiang (Jack) Wang, Yong Kang
State Key Laboratory of Advanced Electromagnetic Engineering and Technology
Huazhong University of Science and Technology
Wuhan, China
zhiqiangwang@hust.edu.cn

Abstract—This paper presents a theoretical analysis and an optimized parameter design method of desaturation short-circuit protection for fast response speed and strong noise immunity in silicon carbide (SiC) applications. First, the mechanism of false triggering of desaturation protection is analyzed in detail, and several key related factors are identified based on the analysis. Second, a noise suppression ability evaluation method is presented. A corresponding optimized parameter design method is also presented, which is capable of enhancing noise immunity without sacrificing response speed. Third, several influencing factors are also discussed based on evaluation method. Finally, experimental results are presented to validate the proposed parameter optimization method.

Keywords—desaturation protection, fast response, silicon carbide MOSFETs.

I. INTRODUCTION

Wide bandgap (WBG) devices such as silicon carbide (SiC) metal oxide semiconductor field effect transistors (MOSFETs) offer superior performance compared to silicon devices [1]. The lower on-state resistance, higher critical electric field and higher operating temperature enable WBG devices to build converters with higher efficiency and higher power density. However, a key obstacle of WBG devices is their relatively lower short-circuit withstand capability compared to Si devices. In [2], gallium nitride (GaN) device failure is observed after 400ns short-circuit condition under 400V dc-bus voltage. In [3], short-circuit withstand capability of SiC MOSFETs is reported as low as 8us, and degeneration happens earlier. Both of them are weaker than their Si counterparts, which could still survive after 10us short-circuit condition. Lower short-circuit withstand capability of WBG devices introduces challenges to short-circuit protection. In order to avoid failure and degeneration of devices, response speed of protection is preferred to be as faster as possible. However, fast switching speed of SiC leads to severe switching noise compared to silicon devices [4]. Faster di/dt and dv/dt not only introduce challenge to normal operation, but also introduce challenge to desaturation protection. False triggering caused by noise is the main factor limiting the improvement of response speed. Generally, fast response and strong noise immunity are contradictory with each other. Therefore, it is beneficial to achieve a short circuit protection circuit with both fast response speed and strong noise immunity.

Desaturation protection is widely used for short-circuit protection in converters based on Si devices. Traditional protection schemes need to be accelerated to meet the demand of WBG devices. Severe switching noise caused by fast switching transient and corresponding noise suppression issues are analysed in existing paper. In [5], the influence of fast dv/dt is discussed and an auxiliary circuit is proposed to suppress this noise. In [6], kelvin drain terminal is adopted to eliminate noise caused by parasitic ringing. However, for most SiC devices, kelvin drain terminals are unavailable. The suppression of parasitic ringing noise for devices without kelvin drain terminal is still lack of research.

This article presents a study on noise suppression in relation to parasitic ringing. In Section II, the mechanism of false triggering caused by parasitic ringing of desaturation protection is analysed. In Section III, a noise suppression ability evaluation method is presented. An optimized parameter design method to improve noise immunity without sacrificing response speed is presented accordingly. In Section IV, influences of reference voltage, power dissipation, circuit structure and other factors are discussed. In Section V, experimental results are presented to validate the proposed parameter optimization method.

II. MECHANISM OF FALSE TRIGGER OF DESATURATION PROTECTION

Desaturation protection is widely used for short-circuit protection in converters based on Si devices. It has two key parameters: protection threshold voltage (V_{dsth}) and blanking time (T_{blk}). If drain-source voltage (V_{ds}) is still higher than protection threshold voltage after blanking time during turn-on transient, protection will be triggered. V_{dsth} should be set higher than maximum normal V_{ds} and T_{blk} should be set larger than maximum normal turn-on duration. High V_{dsth} can enhance noise immunity, but it may put devices at risk when overcurrent happens. Large blanking time can enhance noise immunity, at the cost of increased device stresses and aging effects under short-circuit conditions. A typical desaturation protection circuit is illustrated in Fig. 1 [5]. Various combinations of R_1, R_2, R_3 and C_{blk} lead to different protection threshold voltages and blanking times as shown in (1)-(2). V_f is forward voltage of diode, V_s is power supply of protection circuit, which is 15 V typical for SiC MOSFET and k is an intermediate parameter which will be discussed in detail later. V_o is the initial value of V_b and depends on characteristic of diode D_1, which is -0.3 V is this article.

When device is turned off, reset switch S_{reset} is closed and desaturation protection circuit is disabled. When device is under normal turn-on transient, V_{ds} falls to a low value quickly and protection will not be triggered. When device is under hard-switching-fault (HSF), detection diode is blocked reversely, V_b is charged larger than reference voltage (V_{ref}) and

---

This work was supported by the National Natural Science Foundation of China under Grant 51907070

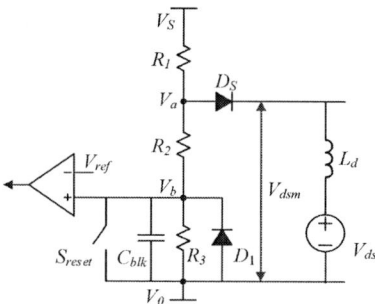

Fig. 1. Typical desaturation protection circuit.

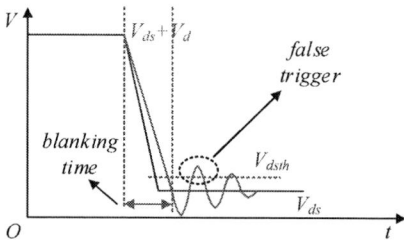

Fig. 2. Mechanism of false trigger of desaturation protection.

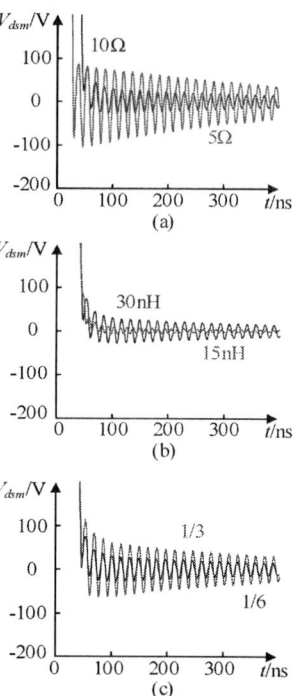

Fig. 3. Main influencing factors of V_d. CPM2-1200-0025B from Cree is selected as DUT and simulation results are obtained from LTSpice. (a) L_{loop}=30nH, L_d/L_{loop}=1/6, gate resistance changes from 10Ω to 5Ω. (b) R_{gate}=10Ω, L_d/L_{loop}=1/6, L_{loop} changes from 30nH to 15nH. (c) R_{gate}=10Ω, L_{loop}=30nH, L_d/L_{loop} changes from 1/6 to 1/3.

protection is triggered. However, for practical applications without kelvin drain terminal, the measured V_{ds} contains actual drain-source voltage and voltage drop across L_d (V_d), as shown in Fig. 2. The parasitic ringing during turn-on transient will introduce severe oscillation on parasitic inductance [7]. Under normal operation conditions, the oscillation voltage on L_d, whose amplitude is much large than actual V_{ds}, will trigger protection falsely. Extending blanking time until the oscillation decays is an intuitive method, but will slow down the protection response speed.

$$V_{dsth} = \frac{R_2 + R_3}{R_3} \cdot V_{ref} - V_f \tag{1}$$

$$T_{blk} = \frac{(R_1 + R_2) \cdot R_3}{R_1 + R_2 + R_3} \cdot C_{blk} \cdot \ln\left(\frac{k + \frac{|V_o|}{V_{ref}}}{k - 1}\right) \tag{2}$$

The main influencing factors of V_d is switching speed, L_d and the ratio of L_d to total parasitic inductance L_{loop}. From simulation in Fig. 3, fast switching speed, large L_d and large ratio of L_d to L_{loop} will lead to large oscillation amplitude of V_d. Low inductance package, kelvin drain terminal and slow switching speed will reduce V_d, however, these methods are not feasible for many applications.

III. NOISE SUPPRESSION EVALUATION METHOD AND OPTIMIZED PARAMETER DESIGN METHOD

As shown in Fig. 1, oscillation of V_d won't influence V_b directly. The R-C network between these two points can be regarded as a low pass filter, and the bandwidth of this filter (f_{bw}) is (3). The lower the f_{bw} is, the better noise suppression can be achieved.

$$f_{bw} = \frac{R_2 + R_3}{2\pi C_{blk} \cdot R_2 \cdot R_3} \tag{3}$$

Various combinations of R_1, R_2, R_3 and C_{blk} lead to different protection threshold (V_{dsth}) and blanking time (T_{blk}). However, for a certain combination of V_{dsth} and T_{blk}, R_1, R_2, R_3 and C_{blk} are not fixed. Different combinations lead to different f_{bw} and a lower f_{bw} is preferred. An intermediate parameter k is adopted to simplify the derivation as shown in (4). Relationships between V_{dsth}, T_{blk} and other parameters are shown in (4)-(6).

$$k = \frac{R_3}{R_1 + R_2 + R_3} \cdot \frac{V_s}{V_{ref}} = \frac{R_2 + R_3}{R_1 + R_2 + R_3} \cdot \frac{V_s}{V_f + V_{dsth}} \tag{4}$$

$$R_2 = \frac{k(V_{dsth} + V_f - V_{ref})}{V_s - k(V_{dsth} + V_f)} \cdot R_1 \tag{5}$$

$$R_3 = \frac{V_{ref}}{V_{dsth} + V_f - V_{ref}} \cdot R_2 \tag{6}$$

When device is turned on, the power dissipation of protection circuit is (7) and when device is turned off, the power dissipation is (8). The total power dissipation is (9) when duty cycle is 50%. Value of R_1 can be derived according to (9) for a given power dissipation limit, and R_2, R_3 and C_{blk} can be calculated according to (4)-(6) for a given combination of T_{blk}, V_{dsth} and k.

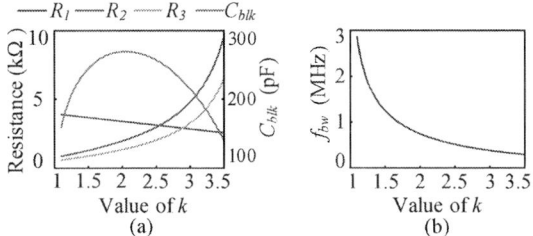

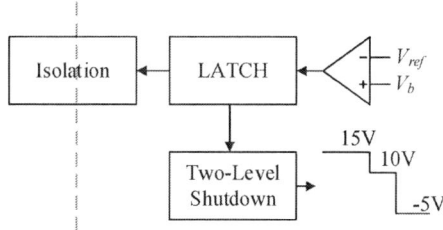

Fig. 4. Relationship between parameter values and noise immunity. V_{dsth}=4V, V_f=1.2V, T_{blk}=100ns and dissipation power is limited within 100mW.

Fig. 6. Schematic of proposed desaturation protection.

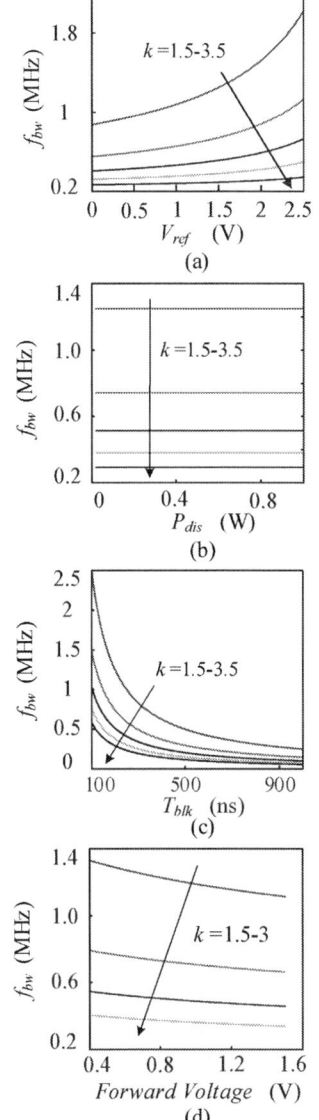

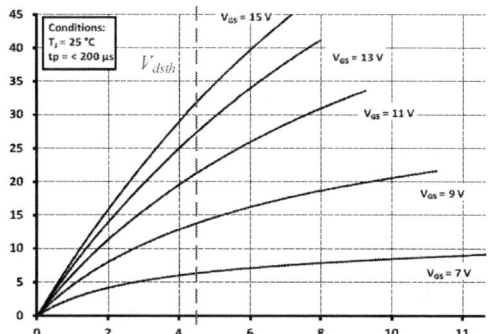

Fig. 7. Output characteristic of DUT.

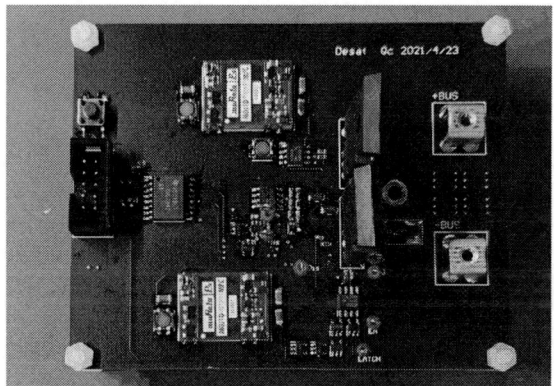

Fig. 8. Prototype of proposed protection circuit.

Fig. 5. Impact of (a) reference voltage (b) dissipation power limit (c) blanking time (d) diode forward voltage on noise immunity.

$$P_{off} = \frac{V_s^2}{R_1 + R_2} \tag{8}$$

$$P_{tot} = \frac{1}{2}\left(P_{on} + P_{off}\right)$$
$$= \frac{V_s^2}{2R_1} \cdot \frac{2V_s - k\left(V_{ref} + V_{dsth} + V_f\right)}{V_s - kV_{ref}} \tag{9}$$

For a certain combination of V_{dsth} and T_{blk}, the relationships between f_{bw} and k are shown in Fig. 4. It can be seen a larger k will lead to stronger noise immunity. Therefore, the optimized parameter design method is based on (4)-(6), (9)

$$P_{on} \approx \frac{V_s^2}{R_1} \tag{7}$$

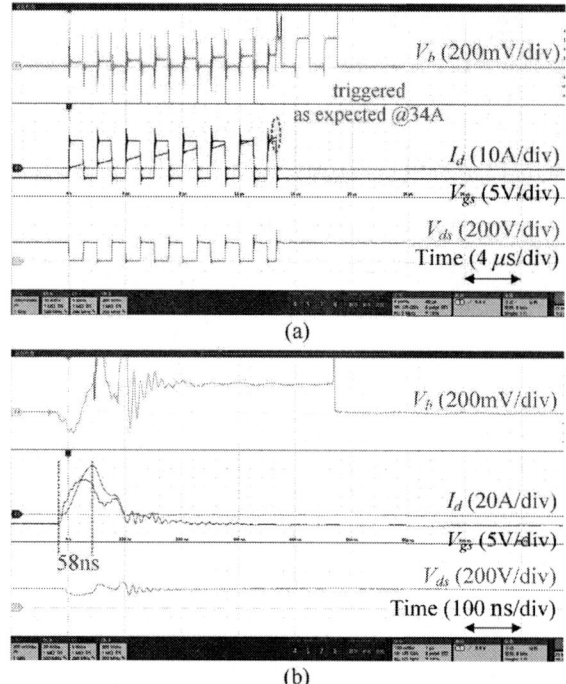

Fig. 9. (a) Normal pulse test of optimized parameter design. (b) hard switching fault test. The response time of proposed desaturation protection is 58ns, and no false trigger is observed under normal operation.

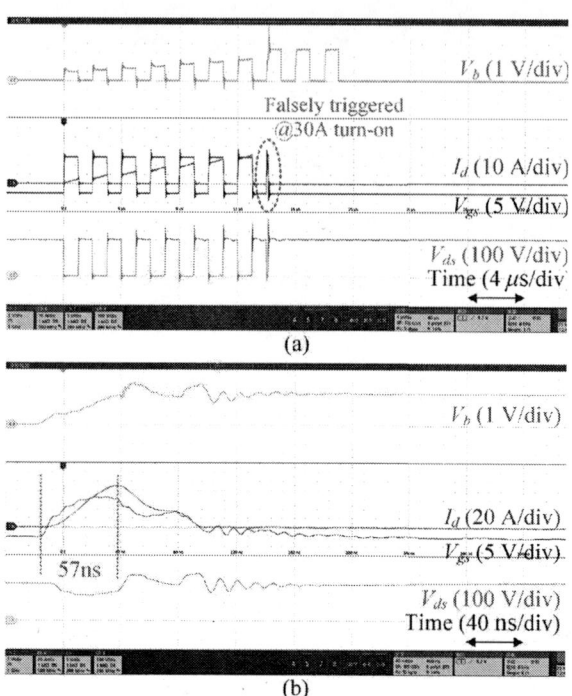

Fig. 10. (a) Normal pulse test of unoptimized parameter design. (b) Hard switching fault test. The response time of proposed desaturation protection is 57ns, but false trigger is observed under normal operation at 30A turn-on transient.

and a large value of k is recommended. The specific improvement effect varies with different applications and similar conclusion can be obtained.

IV. INFLUENCES OF OTHER PARAMETERS

In typical desaturation protection circuit shown in Fig. 1, reference voltage, power dissipation limit, forward voltage of diode and blanking time selection will also influence f_{bw}. In Fig. 5, the impacts of these factors are presented. It can be seen that for a certain V_{dsth} and T_{blk}, lower reference voltage will make benefit to noise immunity, and the influence of power dissipation limit and diode forward voltage are not obvious. In commercial gate driver, the reference voltage varies from 0.7V-2.5V [8]-[9]. According to analysis in this section, a lower reference voltage is preferred. Different from some other applications, larger dissipation power limit won't make any benefit. Forward voltage drop has limited impact, but in order to reduce junction capacitance of diode, several diodes in series are usually used in desaturation protection.

Extending blanking time will improve noise immunity, and on the other hand, oscillation of V_d will decay significantly if T_{blk} is large enough. This is consistent with experimental results that extending blanking time will avoid false trigger.

V. EXPERIMENTAL VERIFICATION

Desaturation protection based on discrete devices are proposed in this paper. The schematic of proposed desaturation protection is consistent with that in Fig. 1. A latch circuit and two-level soft shutdown circuit in Fig. 6 are used to turn-off the devices safely if fault is reported. Once the comparator output flips, the latch circuit will hold the fault signal. When device is under short-circuit condition, the

channel current is very large and in order to avoid overheating the device should be turned off immediately. But on the other hand, too fast turn off speed at such large channel current will lead to severe overvoltage of V_{ds}. A balance between overvoltage and overtemperature should be achieved. Typical turn off methods under short-circuit conditions is using large turn-off resistance or using two-level turn-off circuit. In this article, two-level turn-off is selected. Under normal conditions, the output of gate driver is 15V to achieve low on-state. When desaturation protection is triggered, the output of gate driver will fall to 10V for around 30ns to limit short-circuit current. Then the output of gate driver will fall to -5V and turn off the device safely.

TABLE I
COMPARISON BETWEEN TWO DESIGN CASES

V_{dsth}=4.2V, V_f=1V, T_{blk}=25ns, V_s=15V					
	$R_1\,(k\Omega)$	$R_2\,(k\Omega)$	$R_3\,(k\Omega)$	$C_{blk}\,(pF)$	$f_{bw}\,(MHz)$
V_{ref}=0.25V k=2.5	1.28	7.93	0.4	60	6.98
V_{ref}=2.5V k=1.5	0.4	0.22	0.2	160	9.56

Output characteristic of device under test (DUT) is shown in Fig. 7 and V_{dsth} is set to 4.2V, at which load current is around 34 A. Normal operation waveforms are obtained from double pulse test. According to the test results, the total turn on duration increases with load current and is around 25 ns at 30 A. Two different parameter designs are shown in TABLE I, it can be seen that 2.7dB noise suppression improvement can be

978-1-6654-4817-8/21 $31.00 © 2021 IEEE

achieved using proposed optimized parameter design method theoretically. Once the comparator flips, subsequent latch circuit will activate two-level turn-off circuit after around 30ns propagation delay. The total response time of proposed desaturation protection circuit is around 55ns. It should be noticed that C_{blk} is the total capacitance of protection circuit and around 60pF parasitic capacitance is inherent in proposed circuit, which is introduced by reset switch S_{reset} and D_1 in Fig. 1.

The prototype of proposed protection circuit is shown in Fig. 8, and experimental waveforms of two different parameter designs are shown in Fig. 9 and Fig. 10. V_b is measured with passive probe. Drain current is measured with Rogowski coil and V_{ds} is measured with high voltage differential probe. It can be seen that at the same response speed, the unoptimized design falsely triggers the protection at 30A load current turn-on and optimized design does not.

VI. Conclusion

Desaturation protection with fast response speed will be easily falsely triggered because of parasitic ringing across drain inductance and weak noise immunity caused by high f_{bw}. Mechanism of false trigger is analysed in this article and optimized parameter design method to improve noise immunity without sacrificing response speed is presented. The proposed method also takes dissipation power limit into consideration. Proposed design method is validated through experiments through comparing unoptimized and optimized parameter design. Under same blanking time, unoptimized design falsely triggers the protection while optimized method does not. The proposed desaturation protection will protection

DUT within 58ns and won't be falsely triggered under normal operation.

References

[1] J. Millán, P. Godignon, X. Perpiñà, A. Pérez-Tomás and J. Rebollo, "A survey of wide bandgap power semiconductor devices," *IEEE Transactions on Power Electronics*, vol. 29, no. 5, pp. 2155-2163, May 2014.

[2] M. Landel, C. Gautier, D. Labrousse, S. Levebvre, F. Zaki and Z. Khatir, "Dispersion of electrical characteristics and short-circuit robustness of 600 V Emode GaN transistors," in *Proc. PCIM Europe*, 2017, pp. 1-9.

[3] Z. Wang, X. Shi, Y. Xue, L. M. Tolbert, F. Wang and B. J. Blalock, "Design and performance evaluation of overcurrent protection schemes for silicon carbide (SiC) power MOSFETs," *IEEE Trans. Ind. Electron.*, vol. 61, no. 10, pp. 5570-5581, Oct. 2014.

[4] Z. Zhang, "Characterization and realization of high switching-speed capability of SiC power devices in voltage source converter," Ph. D. dissertation, University of Tennessee, Knoxville, USA, 2015.

[5] J. Wu, W. Meng, F. Zhang, G. Dong and J. Shu, "A short-circuit protection circuit with strong noise immunity for GaN HEMTs," *IEEE Trans. Power Electron.*, vol. 36, no. 2, pp. 2432-2442, Feb. 2021.

[6] F. Yang, Z. Wang, Z. Liang and F. Wang, "Electrical performance advancement in SiC power module package design with kelvin drain connection and low parasitic inductance," *IEEE Journal of Emerging and Selected Topics in Power Electronics*, vol. 7, no. 1, pp. 84-98, March 2019.

[7] W. Zhang, Z. Zhang, F. Wang, D. Costinett, L. M. Tolbert and B. J. Blalock, "Characterization and Modeling of a SiC MOSFET's Turn-On Overvoltage," in *2018 IEEE Energy Conversion Congress and Exposition (ECCE)*, 2018, pp. 7003-7009.

[8] Texas Instruments, UCC21710 datasheet. 2020. [Online]. Avaliable: https://www.ti.com/product/UCC21710.

[9] CISSOID, CMT-TIT0521B datasheet. 2020. [Online]. Avaliable: https://www.cissoid.com/high-temperature-electronics/gate-drivers/titan-cmt/cmt-hades2s-high-temperature-175c-hadesv2-gate-driver-secondary-side-ic/

Datasheet Driven Turn Off Overvoltage Prediction for Silicon Carbide Power MOSFETs Based on Theoretical Analysis

Cheng Qian, Yuxin Ge, Zhiqiang (Jack) Wang, Yong Kang

State Key Laboratory of Advanced Electromagnetic Engineering and Technology
Huazhong University of Science and Technology
Wuhan, China
zhiqiangwang@hust.edu.cn

Abstract—This paper presents a turn-off overvoltage prediction method for silicon carbide (SiC) MOSFETs based on theoretical analysis. First, the trajectory method is used to analyze the turn-off process and the key parameters associated with turn-off overvoltage are identified. Second, several reasonable assumptions are introduced to simplify the derivation. A quadratic equation containing power loop inductance, load current, gate resistance, transconductance and other parameters from datasheet is obtained. Third, experiments are presented to validate the proposed method. It is found that the prediction values are always larger than experimental values and the prediction error is within 15%. The prediction is always conservative, which is acceptable for safety concerns.

Keywords—SiC MOSFET, turn off overvoltage, trajectory.

I. INTRODUCTION

Wide bandgap (WBG) devices such as silicon carbide (SiC) metal oxide semiconductor field effect transistors (MOSFETs) and gallium nitride (GaN) high electron mobility transistors (HEMTs) present fast switching speed capability compared to their silicon counterparts. Faster switching speed is preferred because this will reduce switching loss and make benefit to higher efficiency and higher power density [2]. However, fast switching speed also causes severe turn-off overvoltage jeopardizing safe operation. Appropriate switching speed should be adopted to get a balance between switching loss and safe operation.

Turn-off overvoltage phenomenon of SiC MOSFETs has been widely studied in the past decade. Mechanism and solutions are discussed in many existing papers [3]-[6]. In [3], mechanism of turn-off voltage overshoot is analyzed and power loop parasitic inductance is pointed out as the main influencing factor. In [4], influences of gate resistance and parasitic capacitance are discussed. Many different solutions to suppress turn-off voltage overshoot are presented. In [5], decoupling capacitors are used to suppress voltage overshoot by reducing commutation loop inductance. In [6], low parasitic inductance package technology is used to achieve low parasitic inductance.

However, for a certain application, adjusting the gate resistors is the main method to limit overvoltage. Too large resistance will increase turn-off loss, while too small resistance could not guarantee overvoltage within the safe operation area. In other words, appropriate selection of gate resistance value is important and selection guide method is needed. In [7], simulation method is used to investigate the effect of gate resistance, however, this method would not give deep insight and the accuracy depends on simulation model significantly. In [8], an analytical method is proposed to analyze switching transient and voltage overshoot can be predicted. However, the nonlinearity of transconductance and output capacitance of complementary switch are not taken into consideration, and this method is kind of complicated and not intuitive enough. In [9], turn-off overvoltage values are obtained by experiments. The experimental method is the most accurate but with limited design value. All existing research can't provide intuitive, simple and accurate guidance in devices selection and design stage. A simple, accurate and datasheet-driven turn-off overvoltage prediction method will make benefit to gate resistance selection and is still lacking. Two turn-off overvoltage prediction equations including parasitic inductance, parasitic capacitance, load current and other parameters are proposed in this paper. Nearly all parameters of proposed methods can be extracted in devices datasheets and no complicated process is needed. Proposed methods will not only give reference to gate resistance selection in design stage, but also reveal mechanism of voltage overshoot to help further research.

In Section II, the turn-off process of a SiC MOSFET is analysed based on a turn-off transient equivalent circuit, and the key parameters associated with turn-off overvoltage are identified. The trajectory of turn-off process is decomposed into orthogonal vectors in two directions in V-I coordinate plane. In Section III, the equivalent circuit is simplified based on piecewise linearization and charge conservation. A turn-off overvoltage prediction equation is then derived accordingly. All parameters in equation are explained and their acquisition methods are discussed in detail. In Section IV, the influence of complementary switch output capacitor is taken into consideration to improve prediction accuracy. A more complicated but more accurate prediction equation is proposed. In Section V, experimental results using SiC MOSFETs from different manufacturers are presented to validate the proposed prediction methods and the prediction errors are analysed. In Section VI, conclusion and future work outlook are discussed.

II. THE MECHANISM EXPLANATION OF TURN OFF OVERVOLTAGE

A typical half-bridge circuit under inductive load shown in Fig. 1 is used for analysis and the bottom switch is selected as device under test (DUT). During the turn-off transient, the inductor load can be regarded as a current source. FWD is freewheeling diode to conduct load current during DUT is turned off. V_G is output of gate driver, which is -5V when turn off. R_{driver} is output resistance of gate driver chip, R_{Gext} is

---

This work was supported by the National Natural Science Foundation of China under Grant 51907070

978-1-6654-4817-8/21 $31.00 © 2021 IEEE

external gate resistance and R_{Gint} is internal resistance of DUT. L_{loop} is power loop parasitic inductance and L_{cs} is common source inductance. C_{gs} is gate-source capacitance, C_{gd} is gate-drain capacitance and C_{ds} is drain-source capacitance of DUT. V_{dc} is dc-bus voltage. C_{gs} will be discharged when V_G is changed from 15V to -5V and turn-off process begins.

The turn-off process can be divided into four phases: turn-off delay, voltage rise period, current fall period and V_{gs} fall period [10], as shown in Fig. 2 (a). During the voltage rise period, the instantaneous on-state resistance (R_{oni}) and drain-source voltage (V_{ds}) increase continuously. When V_{ds} reaches dc-bus voltage, FWD starts to distract load current and the current fall period starts. During the current fall period, the load current will be transferred from the bottom MOSFET channel to the upper freewheeling diode. However, due to the parasitic inductance in the power loop, this transfer process could not be completed instantly and will continue for whole period. During this period, continuously gate discharge causes V_{gs} to drop, which leads to the increase of R_{oni}. If the resistance rises much faster than the current falls, severe overvoltage will occur. When current fall period ends, decayed parasitic ringing starts.

Trajectory methods have been reported in [4], but no detailed analysis is presented. In Fig. 2 (b), the turn-off trajectory in V-I coordinate plane is decomposed into two orthogonal vectors: V_1 tends to reduce current and V_2 tends to increase R_{oni}. V_1 depends on current transfer speed, which is related to L_{loop} and instantaneous voltage overshoot ($\triangle V_{ds}$) according to (1)-(2). Because forward voltage drop of FWD is relatively small, it can be assumed that total voltage overshoot is applied upon L_{loop}. Large load current will extend the duration of V_1, which will also lead to severe overvoltage. Load current and L_{loop} are related to V_1. V_2 depends on V_{gs} drop speed, in other words, depends on gate discharge speed. Total gate discharge current I_c can be decomposed into I_{c1}, I_{c2} and I_{c3}, which is shown in Fig. 1. I_c is determined by V_G and total gate resistance R_G as shown in (3)-(4). I_{c1} is determined by transfer characteristic and instantaneous channel current. The faster the current falls, the larger V_{gs} drop and larger I_{c1} will occur, which depends on devices transfer characteristic as shown in (5). In other words, I_{c1} tends to increase R_{oni} but keep V_{ds} constant. The remaining gate current I_{c2} tends to further increase R_{oni} and I_{c3} is the charge current of C_{gd} which depends on $R_{oni} \cdot I_d$. In other words, I_{c2} and I_{c3} tend to increase R_{oni} and V_{ds} simultaneously. Transconductance (g_{fs}), C_{gs}, C_{gd} and R_G are related to V_2.

$$\Delta V_{ds} = V_{ds} - V_{dc} \tag{1}$$

$$\Delta V_{ds} = L_{loop} \frac{dI_d}{dt} \tag{2}$$

$$R_G = R_{G\,int} + R_{Gext} + R_{driver} \tag{3}$$

$$I_c = \frac{V_{gs} - V_G}{R_G} \tag{4}$$

$$dV_{gs} = \frac{dI_d}{g_{fs}} \tag{5}$$

It should be noticed that V_1 and V_2 are not independent as shown in Fig. 3. For example, larger I_{c2} and I_{c3} will lead to larger voltage overshoot, but larger voltage overshoot will accelerate current transfer, which leads to larger I_{c1} and smaller I_{c2} and I_{c3} in turn. The final trajectory depends on the balance of these two vectors. If V_2 is much larger than V_1,

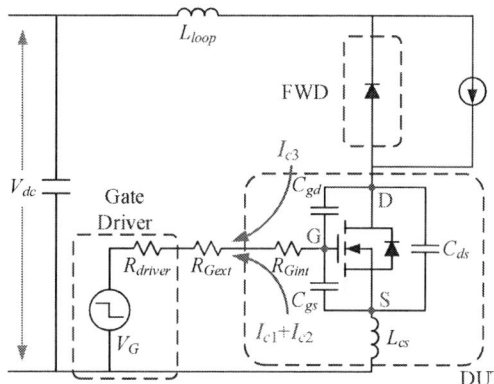

Fig. 1. Typical half-bridge circuit under inductive load during turn-off transient.

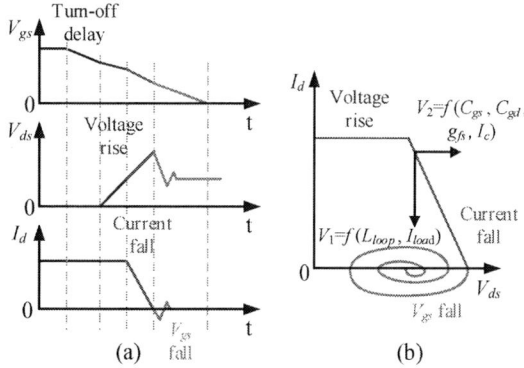

Fig. 2. Ideal switching process of turn-off transient. (a) four phases of turn-off process (b) turn-off trajectory.

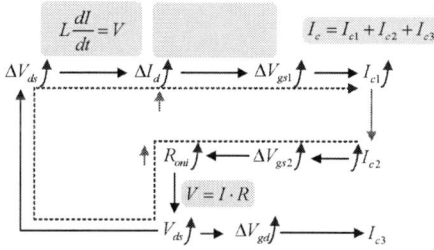

Fig. 3. Relationship between two vectors of turn-off trajectory.

severe overvoltage will happen. As analysed, fast di/dt is the result of overvoltage, not the reason.

III. TURN-OFF OVERVOLTAGE PREDICTION METHOD

Several relationships can be found from the model in Fig. 1 to derive overvoltage prediction equation.

A. Derivation of overvoltage prediction equation

1) Kirchhoff's current law of point G in Fig. 1
According to the Kirchhoff's current law, we can get:

$$I_c \cdot dt = -C_{gs} \cdot dV_{gs} + C_{gd} \cdot dV_{dg}$$
$$\approx -C_{gs} \cdot dV_{gs} + C_{gd} \cdot dV_{ds} \tag{6}$$

2) Transfer characteristic of DUT

Transconductance is not a constant as channel current changes. In this paper it is assumed to be a constant to simplify the derivation. g_{fs} in (5) will be discussed in the later part of this section. I_d is drain current of DUT.

3) Characteristic of inductance

When V_{ds} exceeds V_{dc}, the freewheeling diode conducts forward. The total voltage overshoot $\triangle V_{ds}$ is applied upon the parasitic loop inductance to force current transfer:

$$L_{loop} \cdot dI_d = -\Delta V_{ds} \cdot dt \tag{7}$$

The integral of (7) during the whole current fall period can be obtained in (8).

$$L_{loop} \cdot I_{load} = \int_{t_r} \Delta V_{ds} \cdot dt \tag{8}$$

t_r is the duration of current fall period and I_{load} is total load current.

4) Dv/dt constant assumption

According to a large number of existing experimental waveforms [12], the dv/dt of V_{ds} remains nearly unchanged and corresponding relationship can be derived:

$$\int_{t_r} \Delta V_{ds} \cdot dt = \frac{1}{2} \cdot \Delta V_{ds\,max} \cdot t_r \tag{9}$$

$\triangle V_{dsmax}$ is peak voltage overshoot, which is the object of proposed prediction method.

5) Average gate current assumption

At the beginning of current fall period, the total gate current is (10):

$$I_{c\_begin} = \frac{V_{miller} - V_G - V_{cs}}{R_G} \tag{10}$$

$$V_{miller} = V_{th} + \frac{I_{load}}{g_{fs}} \tag{11}$$

V_{miller} is Miller plateau voltage, V_G is gate driver output voltage, V_{cs} is voltage drop on common source inductance and R_G is total gate resistance. Miller plateau voltage can be expressed in (11). All of these parameters will be discussed in detail and their acquisition methods will be introduced later in this section. At the end of current fall period, the total gate current is (12):

$$I_{c\_end} = \frac{V_{th} - V_G - V_{cs}}{R_G} \tag{12}$$

V_{th} is gate threshold voltage which can be read from transfer characteristic curves. From existing experimental waveforms, di/dt is assumed to be constant during current fall period. The voltage drop on L_{cs} could be expressed as:

$$V_{cs} = L_{cs} \frac{I_{load}}{t_r} \tag{13}$$

The integral of the gate current during the whole period can be simplified as:

$$\int_{t_r} I_c \cdot dt = \frac{1}{2} \left(I_{c\_begin} + I_{c\_end} \right) \cdot t_r$$
$$= \frac{t_r}{R_G} \left[V_{th} + \frac{I_{load}}{2 \cdot g_{fs}} - V_G - L_{cs} \cdot \frac{I_{load}}{t_r} \right] \tag{14}$$

Solve (6)-(9), (14) simultaneously and eliminate t_r, the overvoltage prediction equation (15) is derived. Substitute all parameters into (15) and peak voltage overshoot $\triangle V_{dsmax}$ could be calculated in MATLAB or by hand.

$$C_{gd} \cdot \left(\Delta V_{ds\,max} \right)^2 + \left(\frac{C_{gs}}{g_{fs}} + \frac{L_{cs}}{R_G} \right) \cdot I_{load} \cdot \Delta V_{ds\,max}$$
$$= \frac{2 \cdot L_{loop} \cdot I_{load}}{R_G} \cdot \left(V_{th} + \frac{I_{load}}{2 \cdot g_{fs}} - V_G \right) \tag{15}$$

B. Acquisition method for each parameter

1) Gate-source capacitance C_{gs} and gate-drain capacitance C_{gd}

Parasitic capacitance changes with V_{ds}. However, it can be considered unchanged when V_{ds} is relatively high. C_{iss} and C_{rss} can be obtained at corresponding voltage bias from C-V curve from datasheet directly. C_{gd} and C_{gs} can be calculated from (16) and (17).

$$C_{gd} = C_{rss} \tag{16}$$

$$C_{gs} = C_{iss} - C_{gd} \tag{17}$$

2) Total gate resistance R_G

Total gate resistance consists of three parts: internal resistance of devices (R_{Gint}), external gate resistance (R_{Gext}) and output resistance of gate driver (R_{driver}). R_{Gint} can be acquired from devices datasheets and R_{driver} can be acquired from gate driver datasheet. Total gate resistance can be calculated from (3).

3) Threshold voltage V_{th} and transconductance g_{fs}

V_{th} can be acquired from transfer characteristic curve. However, nominal transconductance provided in datasheet is local transconductance and can't be used to predict overvoltage because of the nonlinearity. To simplify the derivation, transconductance is assumed to be a fixed value which depends on load current. In other words, for certain load current we assume that gfs is constant and can be acquired from transfer characteristic curve, which is called equivalent gfs in this paper. More details are illustrated in Fig. 4.

4) Common source inductance L_{cs}

For SiC MOSFETs with kelvin source terminal, common source inductance is inevitable because of magnetic field coupling between power loop and driving loop. Q3D is used to obtain common source inductance. The simulation model is shown in Fig. 5 and simulation result show that L_{cs} is 400pH approximately in this article.

5) Power loop parasitic inductance L_{loop}

Electromagnetic field simulation and direct measurement are two main methods to obtain L_{loop}. Acquisition methods of power loop parasitic inductance have been reported in existing researches. In this paper experimental method using double pulse test is chosen for convenience.

6) Gate driver output voltage V_G

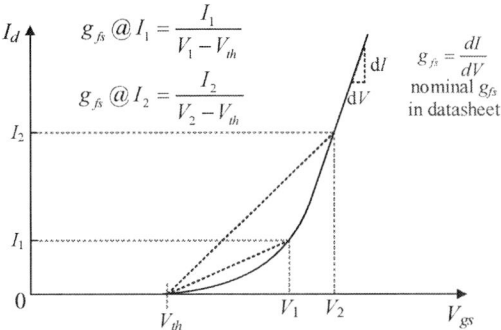

Fig. 4. Acquisition methods of gate threshold voltage and equivalent transconductance at different load current.

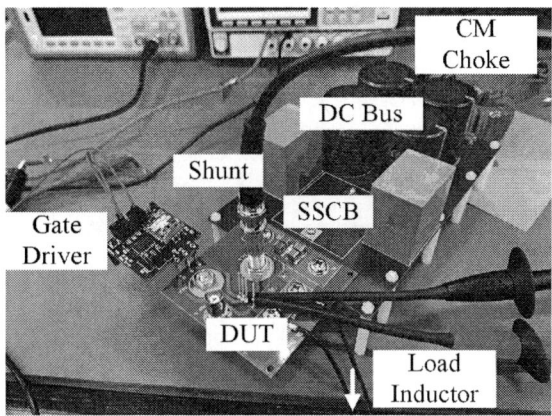

Fig. 6. Experimental platform for double pulse test.

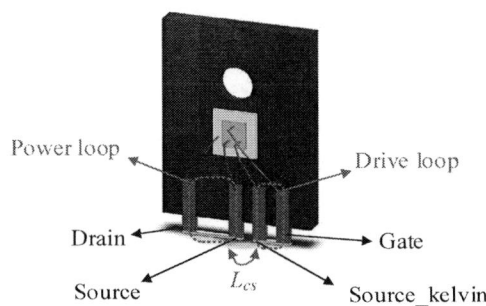

Fig. 5. Simulation model for Q3D of SiC devices with TO247-4L package. Mechanical dimensions are obtained from devices datasheet.

Gate driver output voltage varies with different applications. In this paper, V_G is -5V.

IV. EXPERIMENTAL VERIFICATION OF PREDICTION METHOD

SiC MOSFET C3M0025065K from Cree is selected as DUT to verify proposed prediction method. A brief introduction of DUT is shown in TABLE. I.

TABLE I
DUT FOR VERIFICATION OF PROPOSED METHOD

Typical Value	C3M0025065K
V_{dsmax}	650V
C_{iss}	2980pF
C_{rss}	12pF
R_{Gint}	1.3Ω
V_{th}	2.3V
g_{fs}	25S

A. Experiment platform

The double pulse test is used to verify proposed method and the experimental platform is shown in Fig. 6. The SiC Schottky diode FFSH5065B-F085 from On-semi is used as freewheeling diode. The power loop parasitic inductance is 42.4nH tested by double pulse test. The gate driver is UCC21710 from TI with R_{driver} of 0.3Ω. V_{ds} is measured with THDP0200 with bandwidth of 200MHz, V_{gs} is measured with 1GHz passive probe and I_d is measured with coaxial shunt

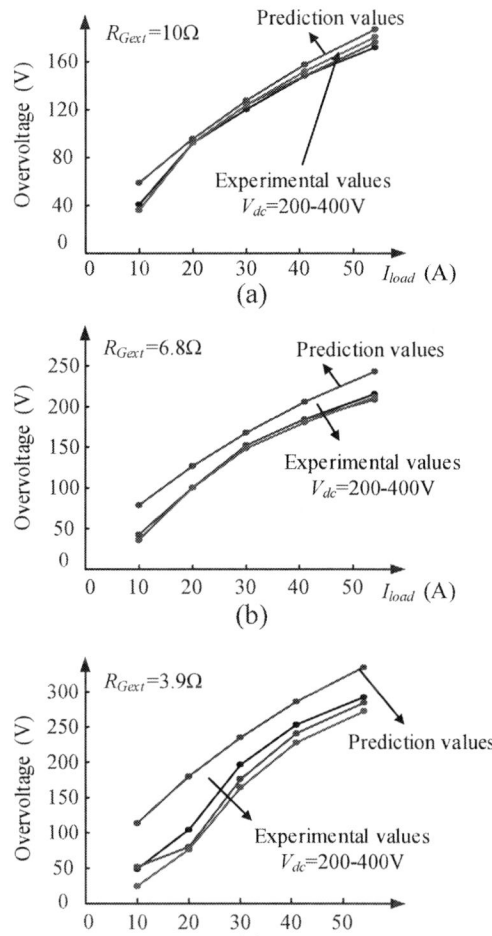

Fig. 7. Comparison between experimental results and prediction results of C3M0025065K from Cree at different dc voltage, different gate resistance and different load current (a) results of 10Ω external resistance (b) 6.8Ω external resistance (c) 3.9Ω external resistance.

SSDN-41405 with bandwidth of 2GHz. Experimental data are processed in MATLAB.

B. Experimental results

The comparisons between experimental results and prediction results are presented in Fig. 7. Conditions of different gate resistance, different DC bus voltage and different load current are tested to validate proposed prediction method. Large prediction errors occur at low load current conditions. However, this is acceptable for safety concerns, because the most severe overvoltage always occurs at large load current conditions and the prediction is relatively accurate at large current conditions. The relative errors are smaller than 20% for all test conditions. For current larger than 20A, the errors are lower than 12%. The proposed method is a conservative method and will not put devices at risk.

V. CONCLUSION

A practical datasheet driven turn-off overvoltage prediction method is proposed in this paper. One prediction equation and the acquisition methods of all parameters are discussed. Proposed method is validated through double pulse test. Experimental results show that the prediction errors are lower than 20% for all tests and the prediction results are always conservative, which is acceptable for safe concerns. The influence of parameters dispersion and junction temperature will be analysed in the future work when curve tracer is available.

REFERENCES

[1] J. Millán, P. Godignon, X. Perpiñà, A. Pérez-Tomás and J. Rebollo, "A survey of wide bandgap power semiconductor devices," *IEEE Transactions on Power Electronics*, vol. 29, no. 5, pp. 2155-2163, May 2014.

[2] F. F. Wang and Z. Zhang, "Overview of silicon carbide technology: device, converter, system, and application," in *CPSS Transactions on Power Electronics and Applications*, vol. 1, no. 1, pp. 13-32, Dec. 2016.

[3] Z. Zhang, "Characterization and realization of high switching-speed capability of SiC power devices in voltage source converter," Ph. D. dissertation, University of Tennessee, Knoxville, USA, 2015.

[4] J. Wang, H. S. Chung and R. T. Li, "Characterization and experimental assessment of the effects of parasitic elements on the MOSFET switching performance," *IEEE Transactions on Power Electronics*, vol. 28, no. 1, pp. 573-590, Jan. 2013.

[5] Z. Chen, D. Boroyevich, P. Mattavelli and K. Ngo, "A frequency-domain study on the effect of DC-link decoupling capacitors," in *2013 IEEE Energy Conversion Congress and Exposition*, 2013, pp. 1886-1893.

[6] B. Cougo, H. H. Sathler, R. Riva, V. D. Santos, N. Roux and B. Sareni, "Characterization of low-inductance SiC module with integrated capacitors for aircraft applications requiring low losses and low EMI issues," *IEEE Transactions on Power Electronics*, vol. 36, no. 7, pp. 8230-8242, July 2021.

[7] Z. Chen, "Characterization and modeling of high-switching-speed behavior of SiC active devices," master thesis, Virginia Polytechnic Institute and State University, Blacksburg, Virginia, USA, 2009.

[8] Y. Xiao, H. Shah, T. P. Chow and R. J. Gutmann, "Analytical modeling and experimental evaluation of interconnect parasitic inductance on MOSFET switching characteristics," in *Nineteenth Annual IEEE Applied Power Electronics Conference and Exposition*, 2004, pp. 516-521, Vol.1.

[9] Z. Chen, D. Boroyevich and R. Burgos, "Experimental parametric study of the parasitic inductance influence on MOSFET switching characteristics," in *2010 International Power Electronics Conference - ECCE ASIA*, 2010, pp. 164-169.

650V 4H-SiC VDMOSFET with Additional n Region: A Simulation Study

Xiuxiu Gao[1]
Research&Development
CORESING SEMICONDUCTOR
TECHNOLOGY CO., LTD
Zhuzhou, China
1507978182@qq.com

Chengzhan Li[2]
Research&Development
ZHUZHOU CRRC TIMES
SEMICONDUCTOR CO., LTD
Zhuzhou, China
licz@csrzic.com

Xiaoping Dai[1]
Research&Development
CORESING SEMICONDUCTOR
TECHNOLOGY CO., LTD
Zhuzhou, China
daixp@csrzic.com

Abstract—A novel 4H-SiC VDMOSFET with additional n region(NA-VDMOSFET) is proposed and investigated by TCAD simulation in this paper.The NA-VDMOSFET exhibits 0.96mΩ·cm² lower specific resistance ($R_{on,sp}$) and 0.2V higher threshold voltage (V_{th}) than the conventional VDMOSFET(C-VDMOSFET) with forward blocking voltage (V_{DSS}) and the maximum electric field in gate oxide ($E_{OX,Max}$) satisfy the design value.Meanwhile, the total losss (E_{total}) reduced lightly compared with C-VDMOSFET. The overall similar performance makes NA-VDMOSFET an excellent choice for manufacture of uniform channel length. It shows the doping profile and dimension of NA and thichness of SiO_2 for NA implantation.

Keywords—4H-SiC,VDMOSFET, additional n region, JFET resistance, uniform channel length

I. INTRODUCTION

Silicon carbide (SiC) metal-oxide-semiconductor field-effect transistors (MOSFETs) are very attractive devices for high temperature, high frequency, and high power applications.

Unlike the case with silicon which introduces impurities by a double diffusion process, the SiC material must form impurity diffusion layer mainly by selective doping of ion implantation. As shown in Fig.1, the p-type well(pw) and n+type source(n+) are formed by ion implantation. Therefore, if there is a displacement caused between the pw mask and the n+ mask, the uniform channel length over the entire target cannot be obtained, so that desired device properties cannot be obtained. In this case, reproducibility cannot be expected from the displacement of the masks and thus undesirable variation cannot be avoided in the device properties of the resultant vertical diffused metal-oxide-semiconductor field-effect transistors (VDMOSFET)。

In addition, in the VDMOSFET as described above, if the space between the pw is not sufficient, the current that passes there through is subject to a so-called junction field-effect transistor(JFET) resistance and causes high on-resistance. As a matter of course, this problem can be solved by arranging the pw in further spaced apart relation,however,the cell size will be enlarged. The channel resistance[1],the reduction of which has been the greatest challenge in the 4H-SiC MOSFET. As a result, it has been a difficult challenge to reduce the MOSFET on-resistance[2] (including the JFET resistance and the channel resistance)using 4H-SiC semiconductor.

Therefore, a new structure of 650V 4H-SiC VDMOSFET with addtional N-region(the abbreviation is NA-VDMOSFET)[3] is proposed in this paper. First,the static electrical parameters of the NA-VDMOSFET with varied doping profile and dimension of NA are studied. Then, the static electrical parameters compared between conventional VDMOSFET(C-VDMOSFET) and NA-VDMOSFET.

Finaly,pw and NA implantation curves and SiO_2 for NA implantation are shown.

II. DEVICE STRUCTURE

The schematic cross-sections of the proposed 4H-SiC NA-VDMOSFET and C-VDMOSFET are shown in fig.1 (a) and (b). Additional n region(NA) which in one layer is integrated with the conventional MOSFET cell. NA is employed to achieve a uniform channel length and low JFET resistance.Space of NA which are in one pw is called L_{ch_sub}.The drift region has a thickness of 6μm and a doping concentration of $1.15×10^{16}cm^{-3}$ to support a blocking voltage more than 650V.The JFET region width is 2μm.The peak concentration and junction depth of pw are $2×10^{18}cm^{-3}$ and 1μm, the thickness of gate oxide is 56nm. The space from JFET region center to the closest NA is Sn.

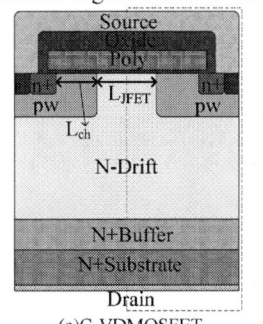 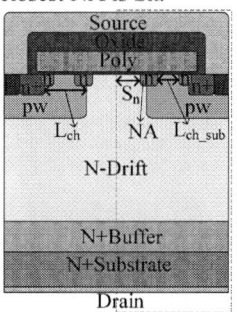

(a)C-VDMOSFET   (b)NA-VDMOSFET

Fig. 1. The schematic cross-sections of simulated 4H-SiC VDMOSFET

III. RESULTS AND DISCUSSION

A. Effect of NA Doping Profile on Blocking Characteristics

The forward blocking voltage V_{DSS}(extracted at $I_{DS}=5μA$,the same as below) and the maximum electric field in gate oxide $E_{OX,Max}$ (extracted at $V_{DS}=400V$,the same as below) of NA-VDMOSFET half cell varied with junction depth and peak concentration of NA region with Sn =0μm are shown in Fig.2(a) and (b) .The 1160(red) in Fig. 2 (a) represents the reference value of V_{DSS} that considering test data and design margin.2.1 × 10^6(red) in Fig. 2 (b) represents the $E_{OX,MAX}$ of C-VDMOSFET.

The NA-VDMOSFET cell's V_{DSS} is reduced and the $E_{OX,max}$ is increased with the junction depth and surface peak concentration of NA increase. The V_{DSS} and $E_{OX,max}$ with the surface peak concentration of $2.5×10^{17}cm^{-3}±20\%$ and junction depth of 0.2μm+10% are 1212~1227V and 3.02~3.54MV/cm, respectively. $E_{OX,max}$ with Sn=0μm for the NA-VDMOSFET are nuch higher than the C-VDMOSFET.

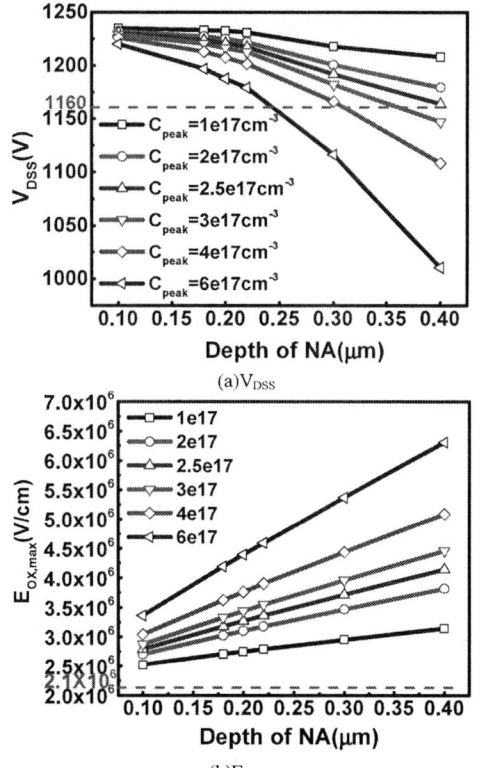

(a)V_{DSS}

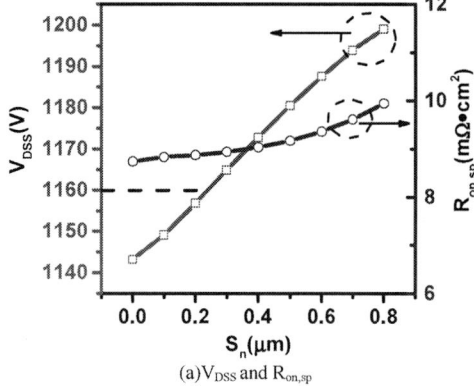

(b)$E_{OX,max}$

Fig. 2. The forward blocking voltage V_{DSS} and the maximum electric field in gate oxide $E_{OX,Max}$ of 650V 4H-SiC NA-VDMOSFET half cell varied with junction depth and peak concentration of NA region with Sn =0μm

B. Effect of Sn on Static Characteristics

The forward blocking voltage V_{DSS}, specific on-resistance $R_{on,sp}$ and the maximum electric field in gate oxide $E_{OX,Max}$ of NA-VDMOSFET half cell varied with Sn are shown in Fig.3(a) and (b).

V_{DSS} of NA-VDMOSFET is increased, while $R_{on,sp}$ and $E_{OX,max}$ is reduced with increased Sn . $E_{OX,max}\leq3MV/cm$ and V_{DSS} are 1172~1200V when 0.4μm$\leq$Sn$\leq$0.8μm.

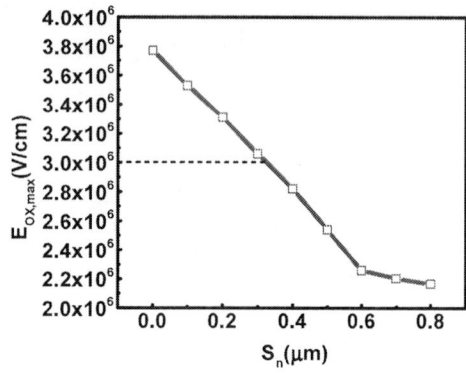

(b)$E_{OX, max}$

Fig. 3. The forward blocking voltage V_{DSS}, specific on-resistance $R_{on,sp}$ and the maximum electric field in gate oxide $E_{OX,Max}$ of 650V 4H-SiC NA-VDMOSFET half cell varied with Sn

C. Comparison of NA-VDMOSFET and C-VDMOSFET

Relationship between VDSS, Vth(extracted at on-state current density $J_{DS,on}$=1mA/mm^2), $R_{on,sp}$, $E_{OX,Max}$ and $L_{ch,min}$ for the C-VDMOSFET and NA-VDMOSFET are shown in Fig.4(a) and (b). $L_{ch,min}$ in Fig.4 refer to the L_{ch} for the C-VDMOSFET and L_{ch_Sub} for the NA-VDMOSFET. L_{ch} of the NA-VDMOSFET is equal to L_{ch_Sub}+0.4μm. The active region area of NA-VDMOSFET and C-VDMOSFET are 27.9mm^2.

The V_{DSS} of C-VDMOSFET and NA-VDMOSFET are about 1206V and 1178V,respectively. The $E_{OX,max}$ of C-VDMOSFET and NA-VDMOSFET are about 2.11MV/cm and 2.82MV/cm,respectively. V_{DSS} and $E_{OX,max}$ for C-VDMOSFET and NA-VDMOSFET are tend to stable when $L_{ch,min}$ is bigger than 0.4μm and 0.3μm,which means NA-VDMOSFET can obtain shorter channel length. The V_{th} of C-VDMOSFET and NA-VDMOSFET are 3.19V and 3.39V, $R_{on,sp}$ of C-VDMOSFET and NA-VDMOSFET are 7.69mΩ·cm^2 and 6.73mΩ·cm^2 at $L_{ch,min}$=0.5μm. This is mainly due to the higher NA concentration of the NA-VDMOSFET, which reduces the JFET resistance.

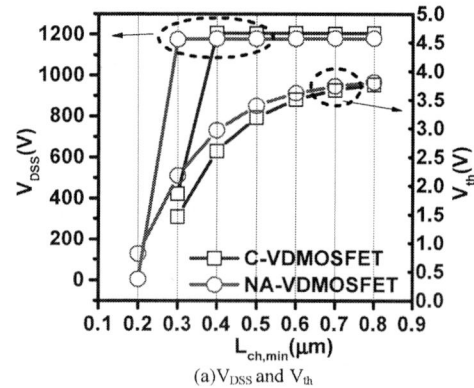

(a)V_{DSS} and V_{th}

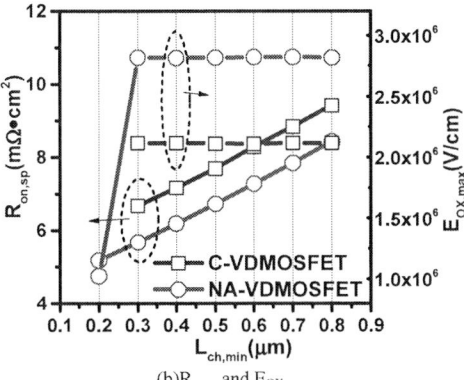

(b)$R_{on,sp}$ and $E_{OX,max}$

Fig. 4. Relationship between V_{DSS}, V_{th}, $R_{on,sp}$, $E_{OX,Max}$ and $L_{ch,min}$ for the C-VDMOSFET and NA-VDMOSFET

The total breakdown current density distribution of the C-VDMOSFET and NA-VDMOSFET are shown in Fig.5. Breakdown positions of the C-VDMOSFET and NA-VDMOSFET are the same, which at the pw's lower corner.

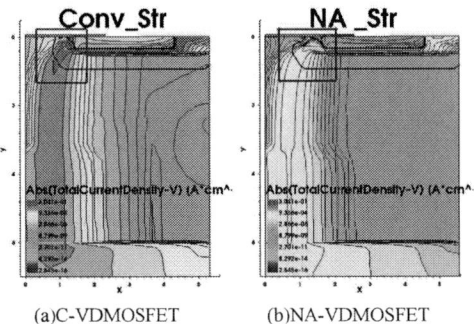

(a)C-VDMOSFET   (b)NA-VDMOSFET

Fig. 5. Total current density distribution of the 650V 4H-SiC C-VDMOSFET and NA-VDMOSFET at breakdown.

The switching characteristics of the studied VDMOSFETs are shown in Fig. 6. A double pulse test circuit is used to investigate the switching characteristics, as shown in Fig. 6(c). Stray inductance and load inductance are 10nH and 200μH, respectively. And the load current is about 200A/cm^2 . The turn-on loss (Eon) of 1.05mJ/cm2 and turn-off loss (Eoff) of 1.10mJ/cm^2 are obtained in the NA-VDMOSFET, showing a 15% increment and a 6.7% reduction compared with C-VDMOSFET.The turn-on time and turn-off time of NA-VDMOSFET, showing 3ns and 8.5ns delay compared with C-VDMOSFET.

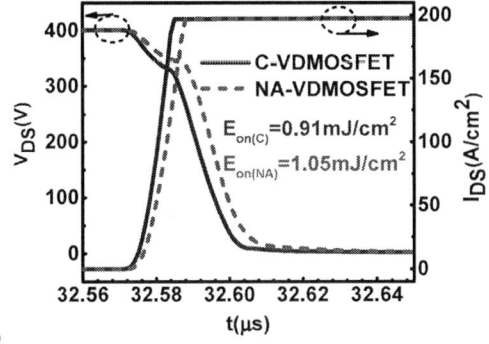

(a)

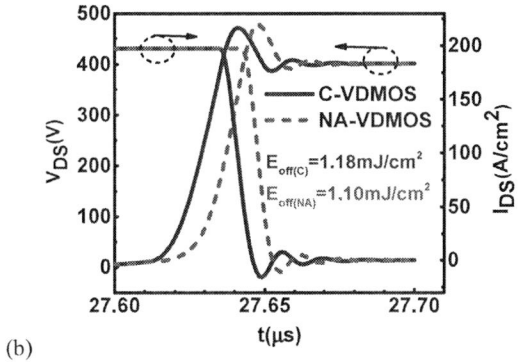

(b)

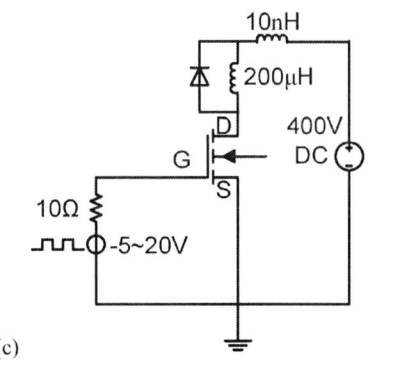

(c)

Fig. 6. (a) Turn-on characteristics and (b) turn-off characteristics for the studied devices, and (c) test circuit.

The main characteristics of the two devices are listed in Table I for comparison. The NA-MOSFET exhibits a more superior on-state performance, similar blocking characteristics and slightly reduced total loss compared with C-VDMOSFET. It makes SiC NA-VDMOSFET more suitable for the uniform manufacture.

TABLE I. PERFORMANCE COMPARISON

	C-MOSFET	NA-VDMOSFET	Unit
V_{DSS}	1206	1178	V
V_{th}	3.19	3.39	V
$R_{on,sp}$	7.69	6.73	mΩ·cm^2
$E_{ox,max}$	2.11	2.82	MV/cm
E_{on}	0.91	1.05	mJ/cm^2
E_{off}	1.18	1.10	mJ/cm^2
E_{total}	4.764	4.61	mJ/cm^2

D. Process Conditions

Ion implantation curves of the NA and pw region of NA-VDMOSFET are shown in Fig.7.Ion implantation of NA can compensate the surface concentration of pw at the distance from surface are 0.04~0.18μm.

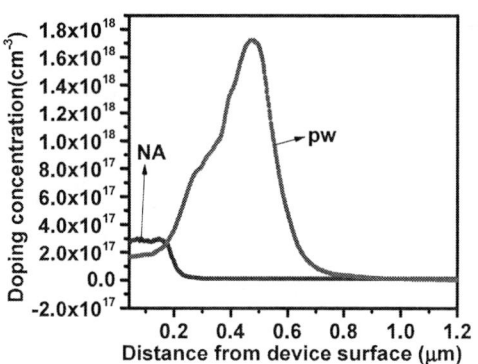

Fig. 7. Ion implantation curves of the NA and pw region of the 650V 4H-SiC NA-VDMOSFET

The masking effect of SiO_2 for NA implantation varied with SiO_2 thickness are shown in Fig.8. It shows that the thickness of SiO_2 which are $1.0\pm0.2\mu m$ can satisfy the masking effect for NA implantation.

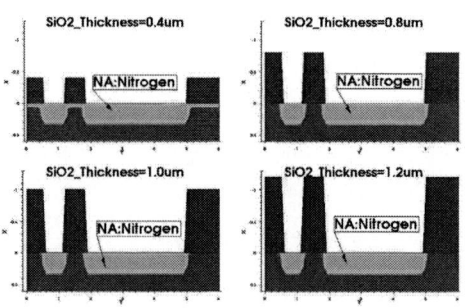

Fig. 8. Masking effect of the SiO_2 thickness on the NA ion implantation

IV. CONCLUSION

In this paper, a novel 4H-SiC VDMOSFET with additional n-region is proposed. TCAD simulations are carried out for the performance comparison with C-VDMOSFET and NA-VDMOSFET.The NA-MOSFET exhibits a more superior on-state performance, similar blocking characteristics and slightly reduced total loss compared with C-VDMOSFET. It makes SiC NA-VDMOSFET more suitable for the uniform manufacture.

ACKNOWLEDGMENT

Thanks for supported by the project,the name is Innovation Center Establish of Power Semiconductor National Manufacturing (No.2018XK2202) , which belong to Special project of Changzhutan National Independent Innovation Demonstration district.

REFERENCES

[1] Seiya Nakazawa,Takafumi Okuda, Jun Suda, Takashi Nakamura, and Tsunenobu Kimoto. Interface properties of 4H-SiC (1120) and 1100) MOS structures annealed in NO.IEEE Transactions on Electron Devices 62(2):309-315,February 2015.

[2] B.Jayant Baliga. Fundamentals of Power Semiconductor Device. Beijing,2008: pp.152-269.

[3] ROHM CO , LTD . SEMICONDUCTOR DEVICE AND METHOD FOR MANUEACTURING SAME: US, US2015069417 (A1). 2015-03-12.

Author Index

Name	Name	Name
Alan Mantooth	Chengzi Yang	Guanqun Qiu
Ao Liu	Chengzi Yang	Guiqinchang
Arman Ur Rashid	Chengzi Yang	Guoguo Yan
Baoshun Zhang	Chengzi Yang	Guohao Yu
Bin Hou	Chengzi Yang	Guolin Xu
Bin Zhou	Chenya Wang	Guosheng Sun
Binyu Wang	Chi Sun	Guoyou Liu
Bo Zhang	Chi Zhang	H. Alan Mantooth
Bo Zhang	Chi Zhang	Hai Shang
Borui Liu	Chongyu Jiang	Haihong Qin
Borui Liu	Chun Han	Haihong Qin
Boya Zhang	Chunming Tu	Haihuiluo
Boya Zhang	Chunyu Yang	Haitao Zhang
Cai Chen	Dafeng Fu	Han Peng
Cai Chen	Deliang Wu	Han Peng
Cai Chen	Dereje Woldegiorgis	Hang Kong
Cai Chen	Dianguo Xu	Hansen Chen
Cai Chen	Dianguo Xu	Hanwei Shen*
Cai Chen	Dianguo Xu	Hao Feng
Cai Chen	Dingbang Zhang	Hao Feng
Cai Chen	Dingkun Ma	Hao Fu
Chang Liu	Dingqu Ling	Hao Lu
Chang Liu	Dong Jiang	Hao Peng
Changle Xu	Du Yujie	Hao Shi
Changsong Chen	Fanghua Zhang	Hao Xu
Changwei Zheng	Fangzhou Zhao	Hao Zhang
Chao Liu	Fei Wang	Haotao Ke
Chaobiao Lin	Fei Xiao	Haotao Ke
Chaofang	Fei Yang	Haotaoke
Chaoyue Shen	Feifei Bu	Haowen Chen
Cheng Qian	Feifei Bu	Haoyuan Jin
Cheng Qian	Feng Wang	Haoyuan Jin
Cheng Qian	Feng Xie	Haoyuan Jin
Cheng Qian	Feng Zhang	Haoyuan Jin
Cheng Qian	Feng Zhang	Haoze Luo
Cheng Qian	Feng Zhang	Hong Li
Chengguo Li	Fengtao Yang	Hong Wang
Chengmin Li	Gang Cao	Hongbo Sun
Chengzhan Li	Gao Yang	Hongbo Zhao
Chengzhan Li	Guangyuanqin	Hongfei Wu

Name	Name	Name
Hongfei Wu	Jianyu Pan	Laili Wang
Hongling Lu	Jiawei Li	Laili Wang
Hongling Lu	Jiaxing Wei	Laili Wang
Hongyi Gao	Jiayue Liu	Laili Wang
Hongyi Xu	Jie Wei*	Laili Wang
Hongyi Xu	Jiejie Zhu	Laili Wang
Hongyi Xu	Jiejie Zhu	Lei Lin
Hongyu Yu	Jielong Liu	Lei Tang
Hongyue Wang	Jielong Liu	Lei Tang
Hua Mao	Jieqin Ding	Lei Wang
Hua Mao	Jing Chen	Li Liu
Hua Mao	Jing Chen	Li Liu
Hua Zhou	Jingrui Han	Li Ran
Huaizhi Zheng*	Jingshu Guo	Li Ran
Huaping Jiang*	Jingwen Zheng	Li Ran
Huaping Jiang*	Jingzhang	Li Xuan*
Huaping Jiang*	Jinhao Cai	Li Zhang
Huaqing Li	Jinhong Jiang	Lianghao Xue
Huaqing Li	Jinkunchu	Lin Li
Huaqing Li	Jiupeng Wu	Lin Liang
Huaqing Li	Jun Wang	Ling Sang
Huaqing Li	Jun Wang	Ling Yang
Huaqing Li	Junduo Wen	Liping Liu
Hui Guo*	Junjie Ye	Lixin Tian
Hui Li	Junmin Wu	Longyang Yu
Huimin Quan	Junnan Gu	Longyang Yu
Jia Zhao	Junnan Gu	Longyang Yu
Jiafa Cai	Junyu Chen	Lubin Han
Jiahuizhou	Kaixuan Li	Lufan Xi
Jiajia Guan	Kaixuan Li	Maojun Wang
Jiajia Guan	Kaiyuan Jing	Md Maksudul Hossain
Jiajia Guan	Kamalesh Hatua	Meng Lu
Jian Li	Kemeng Yang	Meng Lu
Jian Luo	Keqiu Zeng	Mengjie Jiang
Jiangtao Xu	Keqiu Zeng	Mengling Ta
Jianing Wang	Keqiu Zeng	Mengyu Zhu
Jianing Wang	Keqiu Zeng	Michael A. E. Andersen
Jianjun Shu	Kexin Gao	Michael A. E. Andersen
Jianpeng Wang	Kexin Gao	Michael A. E. Andersen
Jianwei Lv	Kuang Sheng	Min Liao
Jianwei Lv	Kuang Sheng	Min Liao*
Jianxin Ji	Kuang Sheng	Ming Hua
Jianxin Zhu	Laili Wang	Ming Yang

Name	Name	Name
Minghai Dong	Ruijun Zhang	Vamshi Krishna Miryala
Mingjie Liu	Ruitao Yan	Venkata Samhitha Machireddy
Mingkun Zhang	Ruitian Wang	Wang Liang
Minhan Mi	Ruiying Li	Wanshun Zhao
Minhan Mi	Ruiying Li	Wei Mu
Na Ren*	Ruizhou Xue	Wei Mu
Na Ren*	Runquan Meng	Wei Wang
Na Ren*	Saijun Mao	Weifeng Sun
Nan Jiang	Saijun Mao	Weining Liu
Nathan M.Ellis	Saijun Mao	Wen Lei
Neng Wang	Saijun Mao	Wenjie Chen
Neng Wang	Saravanan Dhanasekaran	Wenjie Chen
Ni Zheng	Shan Wu	Wenjie Chen
Owen Song	Shan Yin	Wenlu Wang
Peng Chen	Shanxu Duan	Wenming Chen
Peng Sun	Shaolin Yu	Wenyu Li
Pengfei Wang	Shaoxiong Wu	Wenyu Li
Penghui Yin	Shasha Jiao	Wenyu Li
Pengyuan Ren	Sheng Liu	Wu Junmin
Pengyuan Ren	Shenghe Wang	Wuhua Li
Ping Xiong	Shishan Wang	Wuji Meng
Qian Sun	Shizhao Wang	Xi Liu
Qiaozhi Yue	Shuaizhi Zheng	Xi Lu
Qihui Fu	Shuhao Yang	Xia Du
Qijun Liu	Shuhao Yang	Xiang Guo
Qinfeng Jiang	Shuhao Yang	Xiang Lin
Qing Cheng	Shuhao Yang	Xianging He
Qing Guo	Shumin Ding	Xiangtang
Qing Guo	Shuran Jia	Xiaochuan Deng
Qing Guo	Si Chen	Xiaochuan Deng*
Qing Zhu	Si Li	Xiaochuan Deng*
Qiqi Li	Sixuan Xie	Xiaofei Hu
Quan Zheng	Siyang Liu*	Xiaofeng Jiang
Rafael Garcia	Siyu Deng	Xiaohan Zhong
Rahul Iyer	Siyu Liu	Xiaohan Zhong
Robert	Siyu Liu	Xiaohan Zhong
C.N.Pilawa-Podgurski	Song Bai	Xiaohua Ma
Rongdun Hong	Stig Munk-Nielsen	Xiaohua Ma
Rongsheng Chen	Tang Xinling	Xiaohua Ma
Ru Zhang	Tao Liu	Xiaoping Dai
Rui Hu	Tao Sun	Xiaorong Luo*
Rui Li	Tianxiang Yin	Xiaowei Qi
Rui Wang	Tingting Yao	Xiaowen Yan

Name	Name	Name
Xiaoyu Ding	Yayong Yang	Yongjian Chen
Xiaping Chen	Yayong Yang	Yongzhi Liu
Xikun Chen	Yehui Luo	Yu Chen
Xikun Chen	Yi Dou	Yu Chen
Xin Li*	Yi Yu	Yu Guo
Xin Yang	Yi Zhang	Yu Ma
Xinbing Xu	Yibowu	Yu Sun
Xinbo Ruan	Yifan Zhang	Yu Zhou
Xinbo Ruan	Yifei Ding	Yuanjian Liu
Xing Wei	Yifei Luo	Yue Cao
Xing Zhang	Yifu Zhang	Yue Cao
Xingfang Liu	Yihong Huang	Yue Cao
Xingshuo Liu	Yihui Zhao	Yue Hao
Xingwei Huang	Yijian Wang	Yue Hao
Xingwei Huang	Yijie Wang	Yue Hao
Xingwen Li	Yijun Shi	Yue Xie
Xingwen Li	Yilin Chen	Yue Xie
Xinhui Gan	Yingying Ding	Yueshi Guan
Xinling Tang	Yingzhe Wu	Yueshi Guan
Xinyue Guo	Yiping Zeng	Yuhan Xie
Xiuxiu Gao	Yiqiang Chen	Yuhua Cheng
Xu Li	Yiqiang Chen*	Yujie Cheng
Xu Yang	Yiqiang Chen*	Yujie Ding
Xu Yang	Yiqun Kang	Yujie Ding
Xu Yang	Yiyang Yan	Yujie Ding
Xuan Li	Yong Kang	Yumeng Cai
Xuan Li	Yong Kang	Yumingzhou
Xudong Zou	Yong Kang	Yunfeng Liu
Xuehua Wang	Yong Kang	Yunfeng Liu
Xuehua Wang	Yong Kang	Yuning Zhang
Xuehua Wang	Yong Kang	Yunqing Pei
Xuejun Pei	Yong Kang	Yuqi Wei
Xuezheng Huang	Yong Kang	Yuqi Wei
Yameng Sun	Yong Kang	Yutao Lou
Yan Ren	Yong Kang	Yutian Wang
Yang Xiaolei	Yong Kang	Yuting Wang
Yang Fei	Yong Kang	Yuwei Wu
Yang Zhou	Yong Kang	Yuwei Zhou
Yangyang Wu	Yong Kang	Yuxin Ge
Yanjiang Jia	Yong Kang	Yuxin Ge
Yanqing Wang	Yong Kang	Yuxin Ge
Yao Luo	Yong Kang	Yuxing Dai
Yayong Yang	Yong Li	Yuxuan Chen

Name	Name	Name
Yuxuan Chen	Zhenye Wang	Zhiqiang Li
Yuxuan Chen	Zhibin Zhao	Zhiwei Fu
Zaixun Ling	Zhicheng Wang	Zhiwei Wang
Zebing Wu	Zhidong Qiu	Zhong Chen
Zehui Peng	Zhiguo Wei	Zhong Ye
Zenan Shi	Zhikun Wang	Zhuolin Jiang
Zeng Liu	Zhikun Wang	Zicheng Yu
Zhang Chen	Zhikun Wang	Zihe Peng
Zhao Zhibin	Zhikun Wang	Ziniu Wu
Zhaoliang Wen	Zhiqiang (Jack) Wang	Zipeng Ke
Zhaoxiang Wei	Zhiqiang (Jack) Wang	Zishun Peng
Zhaoyang Wei	Zhiqiang (Jack) Wang	Ziwei Ouyang
Zhe Zhang	Zhiqiang (Jack) Wang	Ziwei Ouyang
Zheng Feng	Zhiqiang (Jack) Wang	Ziyan Tang
Zhengyun Wu	Zhiqiang (Jack) Wang	Zongheng Wu
Zhengyun Zhu	Zhiqiang (Jack) Wang	Zuochen Liu
Zhengyun Zhu	Zhiqiang Li	Zuoxing Wang
Zhenxing Zhao	Zhiqiang Li	